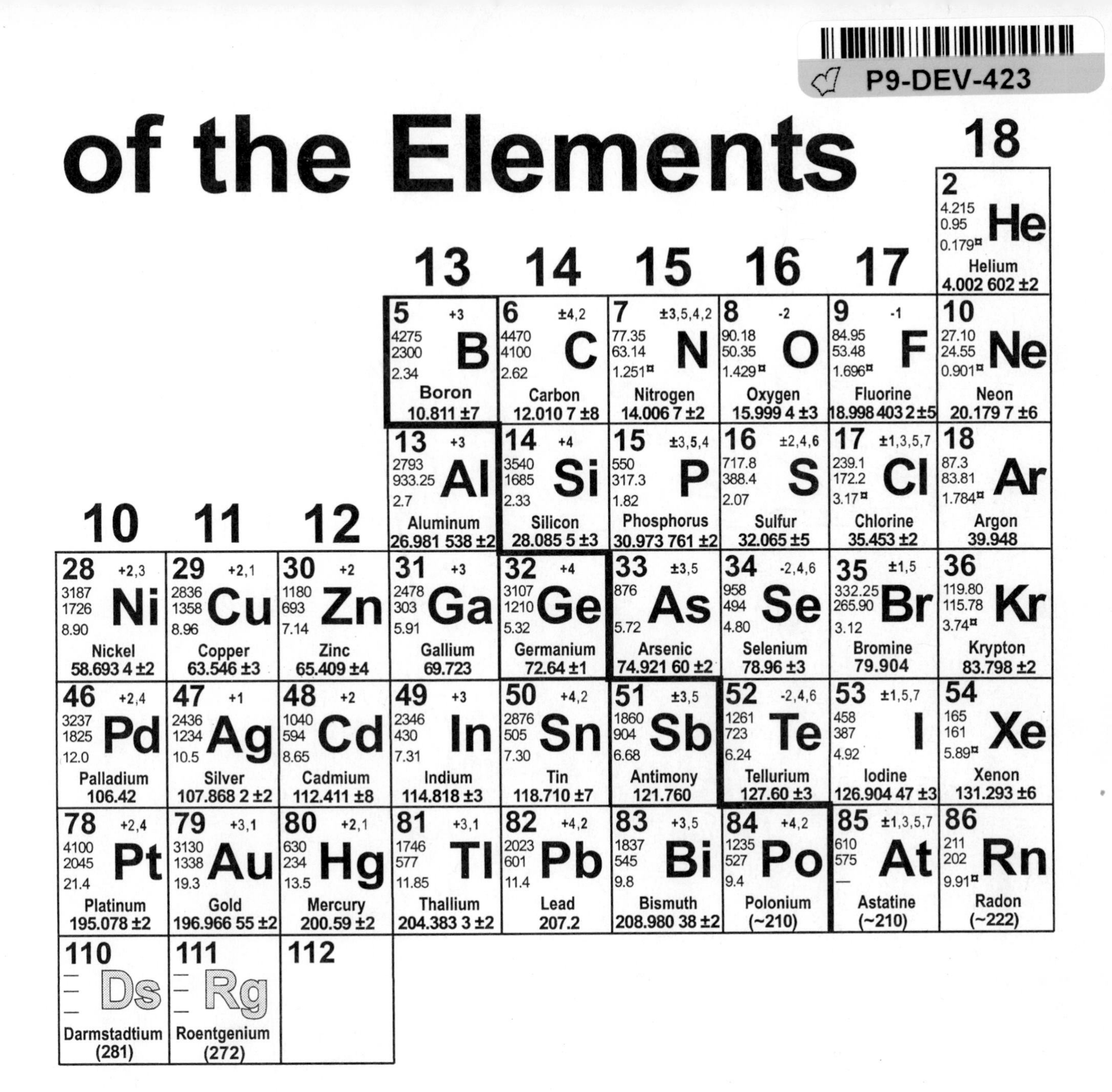
P9-DEV-423
of the Elements
18
2 4.215 0.95 0.179¤ He Helium 4.002 602 ±2
13 14 15 16 17
5 +3 4275 2300 2.34 B Boron 10.811 ±7
6 ±4,2 4470 4100 2.62 C Carbon 12.010 7 ±8
7 ±3,5,4,2 77.35 63.14 1.251¤ N Nitrogen 14.006 7 ±2
8 -2 90.18 50.35 1.429¤ O Oxygen 15.999 4 ±3
9 -1 84.95 53.48 1.696¤ F Fluorine 18.998 403 2 ±5
10 27.10 24.55 0.901¤ Ne Neon 20.179 7 ±6
13 +3 2793 933.25 2.7 Al Aluminum 26.981 538 ±2
14 +4 3540 1685 2.33 Si Silicon 28.085 5 ±3
15 ±3,5,4 550 317.3 1.82 P Phosphorus 30.973 761 ±2
16 ±2,4,6 717.8 388.4 2.07 S Sulfur 32.065 ±5
17 ±1,3,5,7 239.1 172.2 3.17¤ Cl Chlorine 35.453 ±2
18 87.3 83.81 1.784¤ Ar Argon 39.948
10 11 12
28 +2,3 3187 1726 8.90 Ni Nickel 58.693 4 ±2
29 +2,1 2836 1358 8.96 Cu Copper 63.546 ±3
30 +2 1180 693 7.14 Zn Zinc 65.409 ±4
31 +3 2478 303 5.91 Ga Gallium 69.723
32 +4 3107 1210 5.32 Ge Germanium 72.64 ±1
33 ±3,5 876 5.72 As Arsenic 74.921 60 ±2
34 -2,4,6 958 494 4.80 Se Selenium 78.96 ±3
35 ±1,5 332.25 265.90 3.12 Br Bromine 79.904
36 119.80 115.78 3.74¤ Kr Krypton 83.798 ±2
46 +2,4 3237 1825 12.0 Pd Palladium 106.42
47 +1 2436 1234 10.5 Ag Silver 107.868 2 ±2
48 +2 1040 594 8.65 Cd Cadmium 112.411 ±8
49 +3 2346 430 7.31 In Indium 114.818 ±3
50 +4,2 2876 505 7.30 Sn Tin 118.710 ±7
51 ±3,5 1860 904 6.68 Sb Antimony 121.760
52 -2,4,6 1261 723 6.24 Te Tellurium 127.60 ±3
53 ±1,5,7 458 387 4.92 I Iodine 126.904 47 ±3
54 165 161 5.89¤ Xe Xenon 131.293 ±6
78 +2,4 4100 2045 21.4 Pt Platinum 195.078 ±2
79 +3,1 3130 1338 19.3 Au Gold 196.966 55 ±2
80 +2,1 630 234 13.5 Hg Mercury 200.59 ±2
81 +3,1 1746 577 11.85 Tl Thallium 204.383 3 ±2
82 +4,2 2023 601 11.4 Pb Lead 207.2
83 +3,5 1837 545 9.8 Bi Bismuth 208.980 38 ±2
84 +4,2 1235 527 9.4 Po Polonium (~210)
85 ±1,3,5,7 610 575 — At Astatine (~210)
86 211 202 9.91¤ Rn Radon (~222)
110 — — — Ds Darmstadtium (281)
111 — — — Rg Roentgenium (272)
112
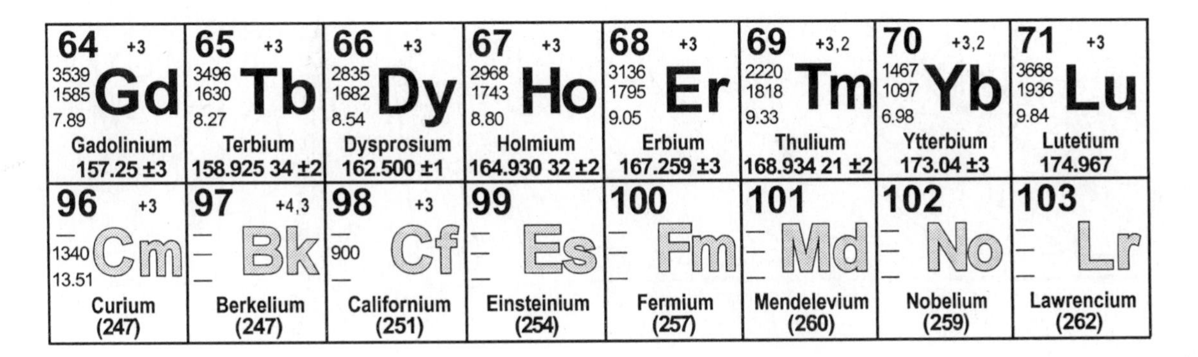
64 +3 3539 1585 7.89 Gd Gadolinium 157.25 ±3
65 +3 3496 1630 8.27 Tb Terbium 158.925 34 ±2
66 +3 2835 1682 8.54 Dy Dysprosium 162.500 ±1
67 +3 2968 1743 8.80 Ho Holmium 164.930 32 ±2
68 +3 3136 1795 9.05 Er Erbium 167.259 ±3
69 +3,2 2220 1818 9.33 Tm Thulium 168.934 21 ±2
70 +3,2 1467 1097 6.98 Yb Ytterbium 173.04 ±3
71 +3 3668 1936 9.84 Lu Lutetium 174.967
96 +3 — 1340 13.51 Cm Curium (247)
97 +4,3 — — — Bk Berkelium (247)
98 +3 — 900 — Cf Californium (251)
99 — — — Es Einsteinium (254)
100 — — — Fm Fermium (257)
101 — — — Md Mendelevium (260)
102 — — — No Nobelium (259)
103 — — — Lr Lawrencium (262)

["The Experiment" by Sempé. Copyright C. Charillon, Paris.]

Exploring Chemical Analysis

Fourth Edition

Daniel C. Harris

Michelson Laboratory
China Lake, California

W. H. Freeman and Company
New York

Publisher: Clancy Marshall
Senior Acquisitions Editor: Jessica Fiorillo
Executive Marketing Manager: Mark Santee
Media Editor: Samantha Calamari
Supplements Editor: Kathryn Treadway
Senior Project Editor: Georgia Lee Hadler
Manuscript Editor: Jodi Simpson
Design Manager: Diana Blume
Illustration Coordinator: Susan Timmins
Illustrations: Network Graphics
Photo Editor: Ted Szczepanski
Production Coordinator: Susan Wein
Composition: Aptara, Inc.
Printing and Binding: Quebecor World

Library of Congress Control Number: 2008922928

ISBN-13: 978-1-4292-0147-6
ISBN-10: 1-4292-0147-9

Printed in the United States of America
First printing

W. H. Freeman and Company
41 Madison Avenue
New York, NY 10010
Houndmills, Basingstoke RG21 6XS, England

www.whfreeman.com

Contents

Experiments

(See Web site **www.whfreeman.com/exploringchem4e**)

Applications

Environmental topics

Biological topics

Forensic topics

Green Chemistry topics

[Courtesy Ralph Keeling, Scripps Institution of Oceanography, University of California, San Diego.]

Charles David Keeling and his wife, Louise, circa 1970

This volume is dedicated to the memory of Charles David Keeling (1928–2005), who is responsible for a half-century program of precise measurements of atmospheric carbon dioxide. Keeling's work provided "the single most important environmental data set taken in the 20th century"* and awakened us to the fact that our activities have global effects. You can read about Keeling's work on pages 10–16. The half-million-year ice core record on the back cover of this book shows the abrupt change in atmospheric carbon dioxide wrought by 200 years of burning fossil fuel. Enlightened, determined, and creative leadership is required from today's generation of students to deal with consequences of the actions of past generations and to provide sustainable sources of energy for the future.

---

*C. F. Kennel, Scripps Institution of Oceanography.

Preface

This book is intended to provide a *short, interesting, elementary* introduction to analytical chemistry for students whose primary interests generally lie outside of chemistry. I selected topics that I think should be covered in a single exposure to analytical chemistry and have refrained from going into more depth than necessary.

What's New?

A new feature of this edition is a short Test Yourself question at the end of each worked example. If you understand the worked example, you should be able to answer the Test Yourself question. Compare your answer with mine to see if we agree.

You will find new content in Section 0-3 and Box 11-1 on the measurement and effects of carbon dioxide on the Earth's environment. Box 11-1 applies your knowledge of acid-base equilibria to effects of atmospheric carbon dioxide on the marine ecosystem. New applications scattered through the book include estimation of drug use from analytical measurements of river water, biochemical measurements with a nanoelectrode, systematic error in atmospheric ozone measurements, nutrition labels for *trans* fat in foods, acid-base chemistry of cocaine, nitrogen analysis of adulterated animal food, carbonate ion-selective electrode for seawater measurements, ammonium ion-selective microelectrode for river sediment, application of biological oxygen demand measurement, electrolytic treatment of wastewater, oxygen microelectrode for river sediment, enhanced explanation of blood glucose monitor, enzymatic nitrate analysis, single-molecule fluorescence in biology, isotope analysis in drug testing of athletes, superheated water as a "green" chromatography solvent, solid-phase extraction with molecularly imprinted polymer, capillary electrophoresis in the diagnosis of thalassemia, electrolytic suppression in ion chromatography, and probing brain chemistry by microdialysis attached to a lab on a chip.

New topics in this edition include "green" chemistry, Grubbs test for outliers (replacing the Q test), gravimetric titrations, acidity of metal ions, pictorial description for predicting the direction of an electrochemical reaction, and the charged aerosol detector for liquid chromatography. Applications of more of the built-in features of Excel include using LINEST for linear regression and adding error bars to graphs. A new Experiment (10), which you can download at the companion Web site www.whfreeman.com/exploringchem4e, uses Excel SOLVER® to fit an acid-base titration curve.

Problem Solving

The two most important ways to master this course are to work problems and to gain experience in the laboratory. **Worked Examples** are a principal pedagogic tool designed to teach problem solving and to illustrate how to apply what you have just read. At the end of each worked example is a similar **Test Yourself** question and answer. I recommend that you answer the question right after reading the example. The **Ask Yourself** question, at the end of most numbered sections, is broken down into elementary steps and should be tackled as you work your way through this book. **Solutions** to Ask Yourself questions are at the back of the book. **Problems** at the end of each chapter cover the entire chapter. Short **Answers** to problems are at the back of the book and complete solutions appear in a separate **Solutions Manual.** **How Would You Do It?** problems at the end of most chapters are more open-ended and might have many good answers.

Features

Chapter Openers show the relevance of analytical chemistry to the real world and to other disciplines of science. **Boxes** discuss interesting topics related to what you are currently studying or amplify points from the text. **Demonstration** boxes describe classroom demonstrations and **Color Plates** near the center of the book illustrate demonstrations or other points. **Marginal Notes** amplify what is in the text. **Spreadsheets** are introduced in Chapter 3 and applications appear throughout the book. You can study this book without ever using a spreadsheet, but your experience will be enriched by spreadsheets and they will serve you well outside of chemistry. End-of-chapter problems intended to be worked on a spreadsheet are marked by an icon: . However, you might choose to work more problems with a spreadsheet than those that are marked.

Essential vocabulary in the text is highlighted in **bold** and listed in **Important Terms** at the end of each chapter. Other unfamiliar terms are usually *italicized*. The **Glossary** at the end of the book defines all bold terms and many italicized terms. **Key Equations** are highlighted in the text and collected at the end of the chapter. **Appendixes** A–D contain tables of chemical information and a discussion of balancing redox equations. The **inside covers** contain your trusty periodic table, physical constants, and other useful information.

Media Supplements

The **Student Web Site**, www.whfreeman.com/exploringchem4e, has instructions for laboratory experiments, a list of analytical chemistry experiments from the *Journal of Chemical Education*, and chapter quizzes. All illustrations from the textbook can be found on the password-protected **Instructor's Web Site** (www.whfreeman.com/exploringchem4e).

The People

My wife Sally works on every aspect of this book. She contributes mightily to whatever clarity and accuracy we have achieved.

The guiding hand for this book at W. H. Freeman and Company is my enthusiastic editor, Jessica Fiorillo. Georgia Lee Hadler shepherded the manuscript through production. Jodi Simpson provided thorough and insightful copyediting. Diana Blume created the design. Solutions to problems were checked by Zach Sechrist at Michelson Laboratory.

I truly appreciate comments, criticism, and suggestions from students and teachers. You can reach me at the Chemistry Division, Mail Stop 6303, Michelson Laboratory, China Lake CA 93555.

Dan Harris
China Lake, 2008

Acknowledgments

I am most grateful to Ralph Keeling, Peter Guenther, David Moss, and Alane Bollenbacher of the Scripps Institution of Oceanography for their tremendous assistance in educating me about atmospheric carbon dioxide measurements and providing access to Keeling family photographs. Doug Raynie (South Dakota State University) provided "green" contributions to the instructions for experiments found at www.whfreeman.com/exploringchem4e. Edward Kremer (Kansas City, Kansas, Community College) contributed information for the cocaine acid-base chemistry box. James Gordon (Central Methodist University, Fayette, Missouri), Chongmok Lee (Ewha Womans University, Korea), Allen Vickers (Agilent Technologies), Krishnan Rajeshwar (University of Texas, Arlington), Wilbur H. Campbell (The Nitrate Elimination Company), Nebojsa Avdalovic (Dionex Corp.), D. J. Asa (ESA, Inc.), and John Birks (2B Technologies) provided helpful information and comments. Bob Kennedy (University of Michigan) kindly provided graphics and information for the new microdialysis/lab-on-a-chip section. J. M. Kelly and D. Ledwith (Trinity College, University of Dublin) provided new Color Plate 15.

Reviewers of the third edition who helped provided direction for the fourth edition were Adedoyin M. Adeyiga (Cheyney University), Jihong Cole-Dai (South Dakota State University), Nikolay G. Dimitrov (State University of New York at Binghamton), Andreas Gebauer (California State University, Bakersfield), C. Alton Hassell (Baylor University), Glen P. Jackson (Ohio University), William R. LaCourse (University of Maryland, Baltimore), Gary L. Long (Virginia Tech), David N. Rahni (Pace University), Kris Varazo (Francis Marion University), and Linda S. Zarzana (American River College).

People who reviewed parts of the manuscript for the fourth edition were Donald Land (University of California, Davis), Karl Sienerth (Elon College), Mark Anderson (Virginia Tech), Pat Castle (U.S. Air Force Academy), Tony Borgerding (University of St. Thomas), D. C. Peridan (Iowa State University), Gerald Korenowski (Rensselaer Polytechnic University), Alan Doucette (Dalhousie University), Caryn Seney (Mercer University), David Collins (Brigham Young University, Idaho), David Paul (University of Arkansas), Gary L. Long (Virginia Tech), Dan Philen (Emory University), Craig Taylor (Oakland University), Shawn White (University of Maryland East Shore), Andreas Gebauer (California State University, Bakersfield), Takashi Ito (Kansas State University), Greg Szulczewski (University of Alabama), Rosemarie Chinni (Alvernia College), Jeremy Mitchell-Koch (Emporia State University), Daryl Mincey (Youngstown State University), Heather Lord (McMaster University), Kasha Slowinska (California State University, Long Beach), Kris Slowinska (California State University, Long Beach), Larry Taylor (Virginia Tech), Marisol Vera (University of Puerto Rico Mayaguez), and Donald Stedman (University of Denver).

Cocaine Use? Ask the River

Map of Italy, showing where Po River was sampled to measure cocaine metabolite. [The reference for this cocaine study: E. Zuccato, C. Chiabrando, S. Castiglioni, D. Calamari, R. Bagnati, S. Schiarea, and R. Fanelli, *Environ. Health* **2005**, *4*, 14. The notation refers to the journal *Environmental Health* published in the year **2005**, volume *4*, page 14, available at http://www.ehjournal.net/content/4/1/14.]

Cocaine

↓

Benzoylecgonine

How honest do you expect people to be when questioned about illegal drug use? In Italy in 2001, 1.1% of people aged 15 to 34 years old acknowledged using cocaine "at least once in the preceding month." Researchers studying the occurrence of therapeutic drugs in sewage realized that they had a tool to measure illegal drug use.

After ingestion, cocaine is largely converted to benzoylecgonine before being excreted in urine. Scientists collected representative composite samples of water from the Po River and samples of waste water entering treatment plants serving four Italian cities. They concentrated minute quantities of benzoylecgonine from large volumes of water by solid-phase extraction, which is described in Chapter 22. Extracted chemicals were washed from the solid phase by a small quantity of solvent, separated by liquid chromatography, and measured by mass spectrometry. Cocaine use was estimated from the concentration of benzoylecgonine, the volume of water flowing in the river, and the fact that 5.4 million people live upstream of the collection site.

Benzoylecgonine in the Po River corresponded to 27 ± 5 100-mg doses of cocaine per 1000 people per day by the 15- to 34-year-old population. Similar results were observed in water from four treatment plants. Cocaine use is much higher than people admit in a survey.

CHAPTER 0

The Analytical Process

Chocolate has been the savior of many a student on the long night before a major assignment was due. My favorite chocolate bar, jammed with 33% fat and 47% sugar, propelled me over mountains in California's Sierra Nevada. In addition to its high energy content, chocolate packs an extra punch from the stimulant caffeine and its biochemical precursor, theobromine.

Chocolate is great to eat but not so easy to analyze. [W. H. Freeman and Company photo by K. Bendo.]

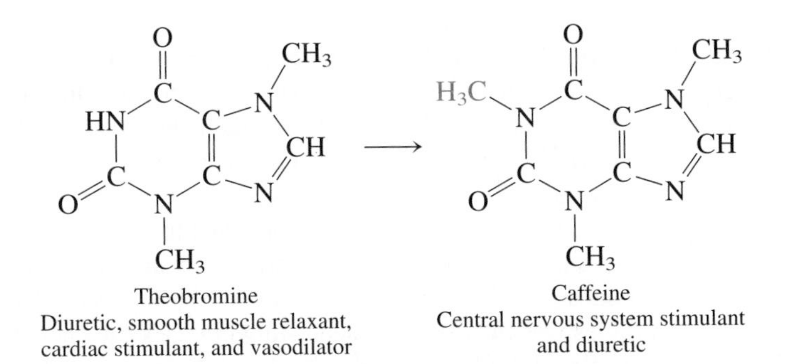

A *diuretic* makes you urinate.
A *vasodilator* enlarges blood vessels.

Too much caffeine is harmful for many people, and even small amounts cannot be tolerated by some unlucky persons. How much caffeine is in a chocolate bar? How does that amount compare with the quantity in coffee or soft drinks? At Bates College in Maine, Professor Tom Wenzel teaches his students chemical problem solving through questions such as these.[1] How *do* you measure the caffeine content of a chocolate bar?

0-1 The Analytical Chemist's Job

Two students, Denby and Scott, began their quest at the library with a computer search for analytical methods. Searching through *Chemical Abstracts* and using "caffeine" and "chocolate" as key words, they uncovered numerous articles in chemistry journals. The articles, "High Pressure Liquid Chromatographic Determination of Theobromine and Caffeine in Cocoa and Chocolate Products,"[2] described a procedure suitable for the equipment available in their laboratory.

Chemical Abstracts is the most comprehensive database of the chemical literature. It is commonly accessed on line through *Scifinder*.

Sampling

Bold terms should be learned. They are listed at the end of the chapter and in the Glossary at the back of the book. *Italicized* words are less important, but many of their definitions are also found in the Glossary.

The first step in any chemical analysis is procuring a representative, small sample to measure—a process called **sampling**. Is all chocolate the same? Of course not. Denby and Scott chose to buy chocolate in the neighborhood store and analyze pieces of it. If you wanted to make universal statements about "caffeine in chocolate," you would need to analyze a variety of chocolates from different manufacturers. You would also need to measure multiple samples of each type to determine the range of caffeine content in each kind of chocolate from the same manufacturer.

Homogeneous: same throughout

Heterogeneous: differs from region to region

A pure chocolate bar is probably fairly **homogeneous**; in other words, its composition is the same everywhere. It might be safe to assume that a piece from one end has the same caffeine content as a piece from the other end. Chocolate with a macadamia nut in the middle is an example of a **heterogeneous** material; in other words, the composition differs from place to place because the nut is different from the chocolate. If you were sampling a heterogeneous material, you would need to use a strategy different from that used to sample a homogeneous material.

Sample Preparation

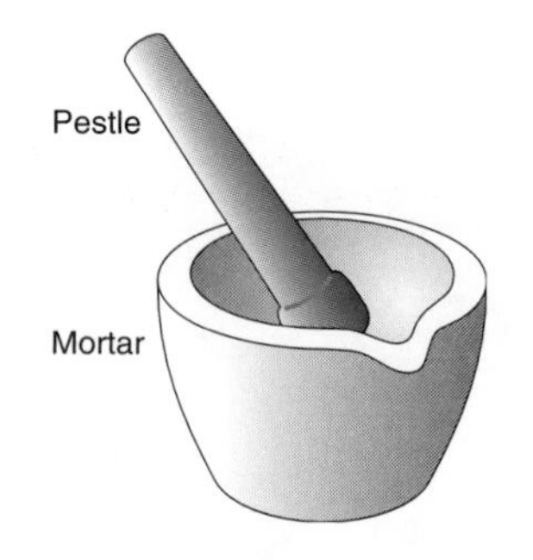

Figure 0-1 Ceramic mortar and pestle used to grind solids into fine powders.

Denby and Scott analyzed a piece of chocolate from one bar. The first step in the procedure calls for weighing a quantity of chocolate and extracting fat from it by dissolving the fat in a hydrocarbon solvent. Fat needs to be removed because it would interfere with chromatography later in the analysis. Unfortunately, shaking a chunk of chocolate with solvent does not extract much fat because the solvent has no access to the inside of the chocolate. So, our resourceful students sliced the chocolate into fine pieces and placed the pieces into a mortar and pestle (Figure 0-1), thinking they would grind the solid into small particles.

Imagine trying to grind chocolate! The solid is too soft to grind. So Denby and Scott froze the mortar and pestle with its load of sliced chocolate. Once the chocolate was cold, it was brittle enough to grind. Then small pieces were placed in a preweighed 15-milliliter (mL) centrifuge tube, and the mass of chocolate was noted.

Figure 0-2 outlines the next part of the procedure. A 10-mL portion of the organic solvent, petroleum ether, was added to the tube, and the top was capped with a stopper. The tube was shaken vigorously to dissolve fat from the solid chocolate into the solvent. Caffeine and theobromine are insoluble in this solvent. The mixture of liquid and fine particles was then spun in a centrifuge to pack the chocolate at the bottom of the tube. The clear liquid, containing dissolved fat, could now be **decanted**

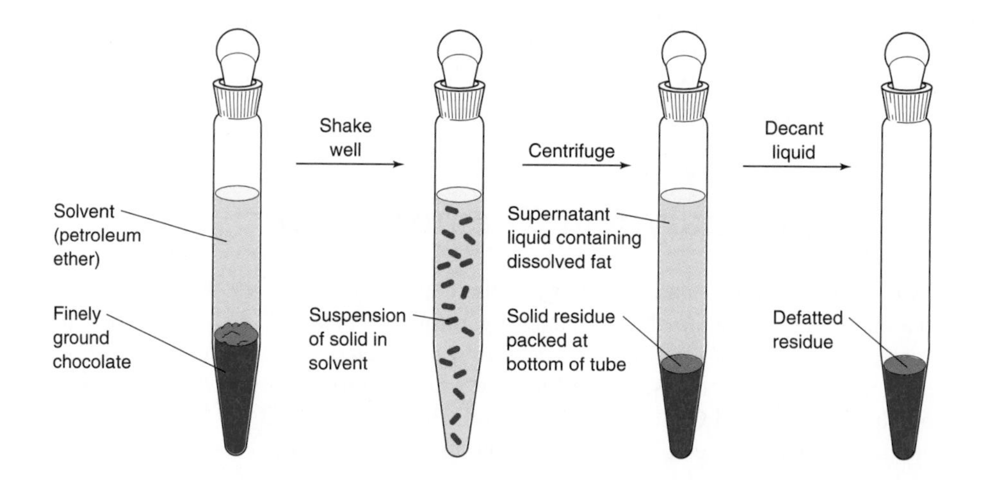

Figure 0-2 Extracting fat from chocolate to leave defatted solid residue for analysis.

(poured off) and discarded. Extraction with fresh portions of solvent was repeated twice more to ensure complete removal of fat from the chocolate. Residual solvent in the chocolate was finally removed by heating the uncapped centrifuge tube in a beaker of boiling water. By weighing the centrifuge tube plus its content of defatted chocolate residue and subtracting the known mass of the empty tube, Denby and Scott could calculate the mass of chocolate residue.

Substances being measured—caffeine and theobromine in this case—are called **analytes**. The next step in the sample preparation procedure is to make a **quantitative transfer** (a complete transfer) of the fat-free chocolate residue to an Erlenmeyer flask and to dissolve the analytes in water for the chemical analysis. If any residue were not transferred from the tube to the flask, then the final analysis would be in error because not all of the analyte would be present. To perform the quantitative transfer, Denby and Scott added a few milliliters of pure water to the centrifuge tube and used stirring and heating to dissolve or suspend as much of the chocolate as possible. The **slurry** (a suspension of solid in a liquid) was then poured from the tube into a 50-mL flask. They repeated the procedure several times with fresh portions of water to ensure that every last bit of chocolate was transferred from the centrifuge tube into the flask.

To complete the dissolution of analytes, Denby and Scott added water to bring the volume up to about 30 mL. They heated the flask in a boiling water bath to extract all the caffeine and theobromine from the chocolate into the water. To compute the quantity of analyte later, the total mass of solvent (water) must be accurately known. Denby and Scott knew the mass of chocolate residue in the centrifuge tube and they knew the mass of the empty Erlenmeyer flask. So they put the flask on a balance and added water drop by drop until there were exactly 33.3 g of water in the flask. Later, they would compare known solutions of pure analyte in water with the unknown solution containing 33.3 g of water.

A solution of anything in water is called an **aqueous** solution.

Before Denby and Scott could inject the unknown solution into a chromatograph for the chemical analysis, they had to "clean up" (purify) the unknown even further (Figure 0-3). The slurry of chocolate residue in water contained tiny solid particles that would surely clog their expensive chromatography column and ruin it. So they transferred a portion of the slurry to a centrifuge tube and centrifuged the

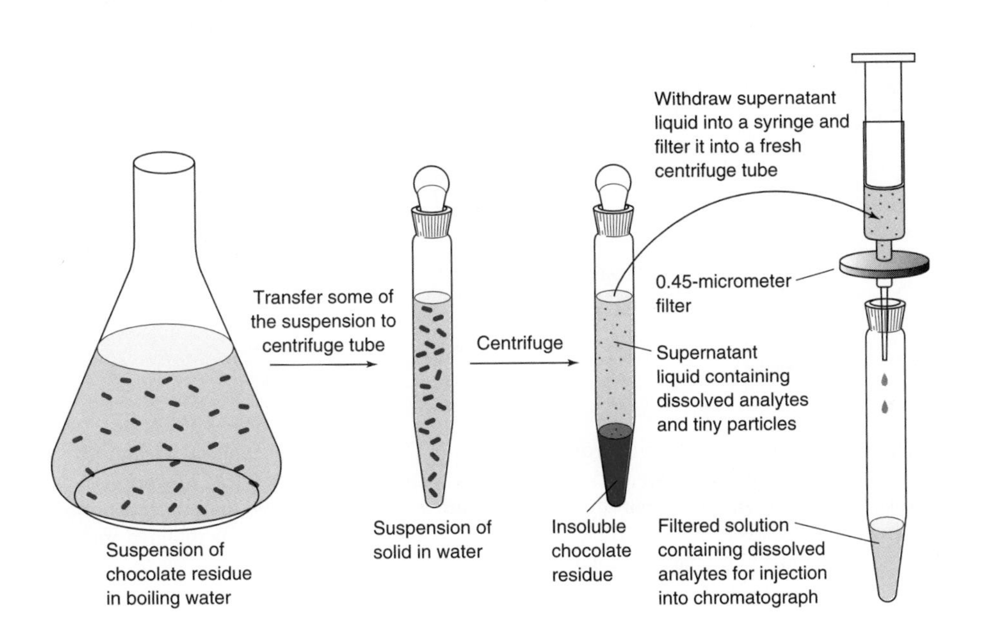

Figure 0-3 Centrifugation and filtration are used to separate undesired solid residue from the aqueous solution of analytes.

mixture to pack as much of the solid as possible at the bottom of the tube. The cloudy, tan, **supernatant liquid** (liquid above the packed solid) was then filtered in a further attempt to remove tiny particles of solid from the liquid.

Real samples rarely cooperate with you!

It is critical to avoid injecting solids into the chromatography column, but the tan liquid still looked cloudy. So Denby and Scott took turns between their classes to repeat the centrifugation and filtration steps five times. After each cycle in which supernatant liquid was filtered and centrifuged, it became a little cleaner. But the liquid was never completely clear. Given enough time, more solid always seemed to precipitate from the filtered solution.

The tedious procedure described so far is called **sample preparation**—transforming sample into a state that is suitable for analysis. In this case, fat had to be removed from the chocolate, analytes had to be extracted into water, and residual solid had to be separated from the water.

The Chemical Analysis (At Last!)

Chromatography solvent is selected by a systematic trial-and-error process. The function of the acetic acid is to react with negatively charged oxygen atoms on the silica surface that, if not neutralized, tightly bind a small fraction of caffeine and theobromine.

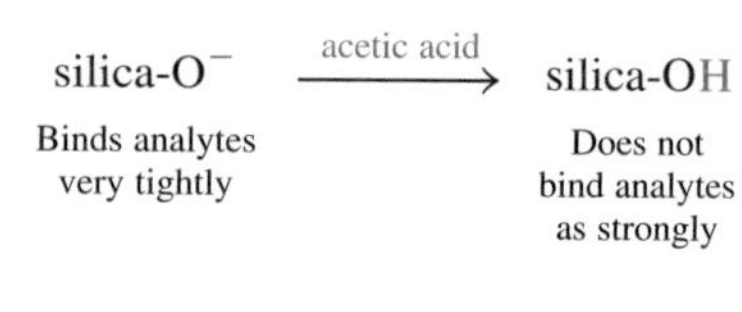

Compromising with reality, Denby and Scott decided that the solution of analytes was as clean as they could make it in the time available. The next step is to inject solution into a *chromatography* column, which separates the analytes and measures their quantity. The column in Figure 0-4a is packed with tiny particles of silica

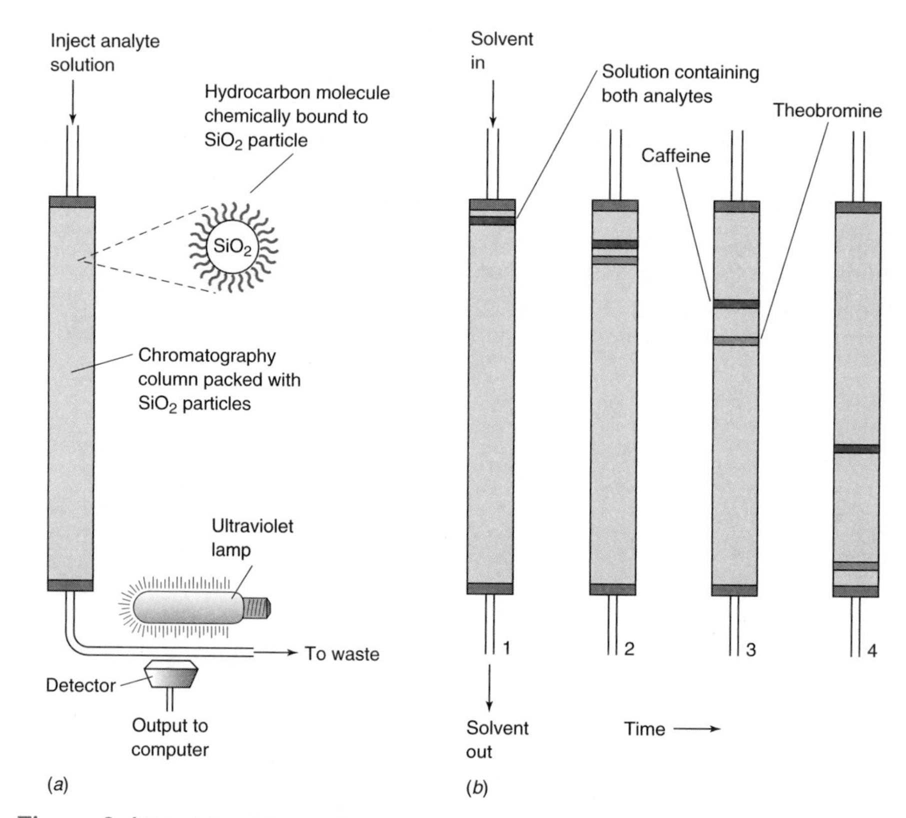

Figure 0-4 Principle of liquid chromatography. (*a*) Chromatography apparatus with an ultraviolet absorbance monitor to detect analytes at the column outlet. (*b*) Separation of caffeine and theobromine by chromatography. Caffeine is more soluble than theobromine in the hydrocarbon layer on the particles in the column. Therefore caffeine is retained more strongly and moves through the column more slowly than theobromine.

(SiO_2) on which are attached long hydrocarbon molecules. Twenty microliters $(20.0 \times 10^{-6}$ liters) of the solution of chocolate extract were injected into the column and washed through with a solvent made by mixing 79 mL of pure water, 20 mL of methanol, and 1 mL of acetic acid. Caffeine is more soluble than theobromine in the hydrocarbon on the silica surface. Therefore caffeine "sticks" to the coated silica particles in the column more strongly than theobromine does. When both analytes are flushed through the column by solvent, theobromine reaches the outlet before caffeine (Figure 0-4b).

Analytes are detected at the outlet by their ability to absorb ultraviolet radiation. As compounds emerge from the column, they absorb radiation emitted from the lamp in Figure 0-4a and less radiation reaches the detector. The graph of detector response versus time in Figure 0-5 is called a *chromatogram*. Theobromine and caffeine are the major peaks in the chromatogram. The small peaks are other components of the aqueous extract from chocolate.

Only substances that absorb ultraviolet radiation at a wavelength of 254 nanometers are observed in Figure 0-5. By far, the major components in the aqueous extract are sugars, but they are not detected in this experiment.

The chromatogram alone does not tell us what compounds are in an unknown. If you do not know beforehand what to expect, you would need to identify the peaks. One way to identify individual peaks is to measure the *mass spectrum* (Chapter 21) of each one as it emerges from the column. Another way is to add an authentic sample of either caffeine or theobromine to the unknown and see whether one of the peaks grows in magnitude.

Identifying *what* is in an unknown is called **qualitative analysis**. Identifying *how much* is present is called **quantitative analysis**. The vast majority of this book deals with quantitative analysis.

In Figure 0-5, the *area* under each peak is proportional to the quantity of that component passing through the detector. The best way to measure the area is with a computer that receives the output of the detector during the chromatography experiment. Denby and Scott did not have a computer linked to their chromatograph, so they measured the *height* of each peak instead.

Calibration Curves

In general, analytes with equal concentrations give different detector responses. Therefore, the response must be measured for known concentrations of each analyte. A graph showing detector response as a function of analyte concentration is called a **calibration curve** or a *standard curve*. To construct a calibration curve, **standard solutions** containing known concentrations of pure theobromine or caffeine were prepared and injected into the column, and the resulting peak heights were measured. Figure 0-6 is a chromatogram of one of the standard solutions, and Figure 0-7 shows calibration curves made by injecting solutions containing 10.0, 25.0, 50.0, or 100.0 micrograms of each analyte per gram of solution.

Straight lines drawn through the calibration points could then be used to find the concentrations of theobromine and caffeine in an unknown. Figure 0-7 shows that, if the observed peak height of theobromine from an unknown solution is 15.0 cm, then the concentration is 76.9 micrograms per gram of solution.

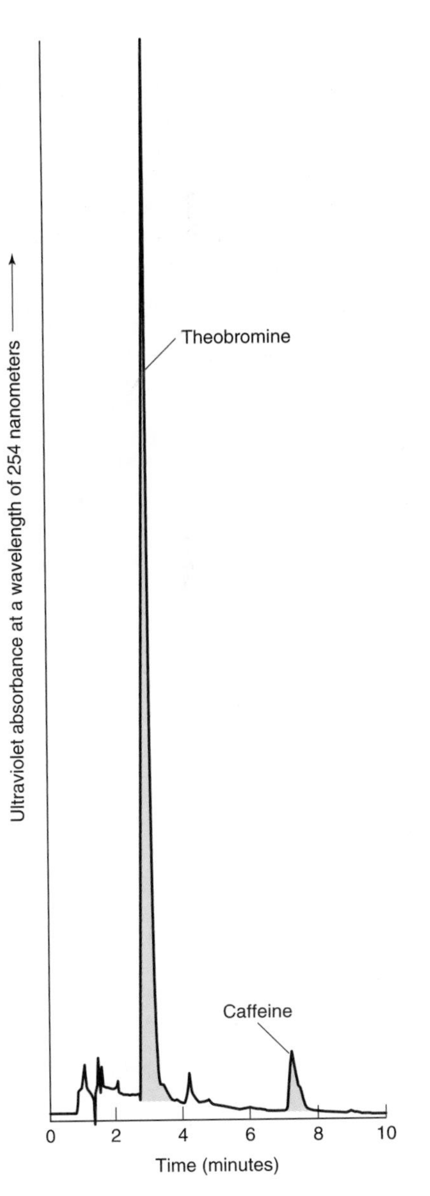

Figure 0-5 Chromatogram of 20.0 microliters of dark chocolate extract. A 4.6-mm-diameter × 150-mm-long column, packed with 5-micrometer particles of Hypersil ODS, was eluted (washed) with water:methanol:acetic acid (79:20:1 by volume) at a rate of 1.0 mL per minute.

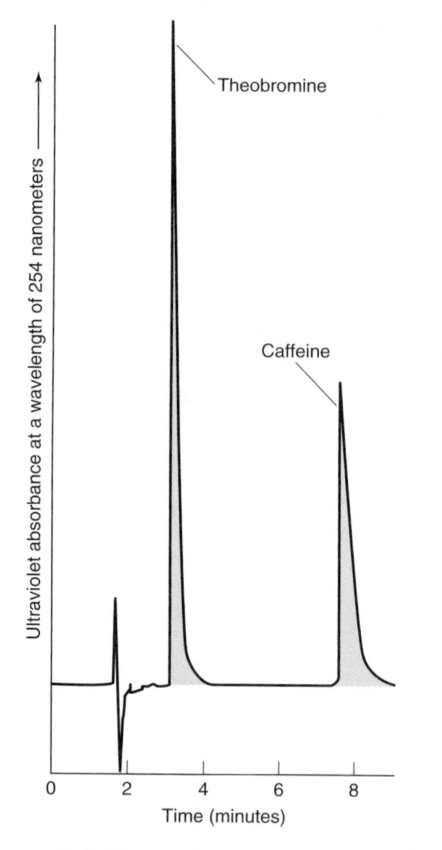

Figure 0-6 Chromatogram of 20.0 microliters of a standard solution containing 50.0 micrograms of theobromine and 50.0 micrograms of caffeine per gram of solution.

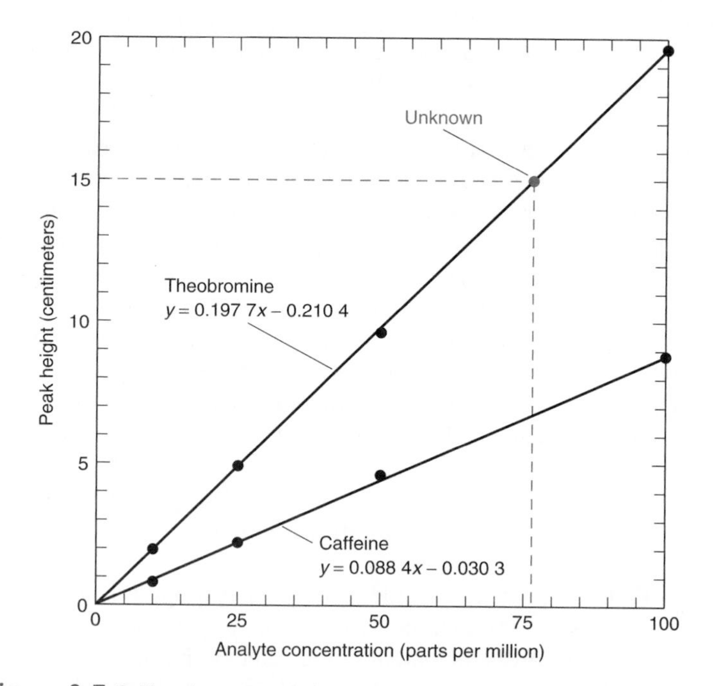

Figure 0-7 Calibration curves, showing observed peak heights for known concentrations of pure compounds. One part per million is 1 microgram of analyte per gram of solution. Equations of the straight lines drawn through the experimental data points were determined by the *method of least squares*, described in Chapter 4.

Interpreting the Results

Knowing how much analyte is in the aqueous extract of the chocolate, Denby and Scott could calculate how much theobromine and caffeine were in the original chocolate. Results for dark and white chocolates are shown in Table 0-1. The quantities found in white chocolate are only about 2% as great as the quantities in dark chocolate.

Table 0-1 Analyses of dark and white chocolate

	Grams of analyte per 100 grams of chocolate	
Analyte	Dark chocolate	White chocolate
Theobromine	0.392 ± 0.002	0.010 ± 0.007
Caffeine	0.050 ± 0.003	0.000 9 ± 0.001 4

Uncertainties are the *standard deviation* of three replicate injections of each extract.

Table 0-1 also reports the *standard deviation* of three replicate measurements for each sample. Standard deviation, which is discussed in Chapter 4, is a measure of the reproducibility of the results. If three samples were to give identical results, the standard deviation would be 0. If the standard deviation is very large, then the results are not very reproducible. For theobromine in dark chocolate, the standard deviation (0.002) is less than 1% of the average (0.392); so we say the measurement is very reproducible. For theobromine in white chocolate, the standard deviation (0.007) is nearly as great as the average (0.010); so the measurement is not very reproducible.

The arduous path to reliable analytical results is not the end of the story. The purpose of the analysis is to reach some interpretation or decision. The questions posed at the beginning of this chapter were "How much caffeine is in a chocolate bar?" and "How does it compare with the quantity in coffee or soft drinks?" After all this work, Denby and Scott discovered how much caffeine is in *one* particular chocolate bar. It would take a great deal more work to sample many chocolate bars of the same type and many different types of chocolate to gain a more universal view. Table 0-2 compares results from different kinds of analyses of different sources of caffeine. A can of soft drink or a cup of tea contains about one-quarter to one-half of the caffeine in a small cup of coffee. Chocolate contains even less caffeine, but a hungry backpacker eating enough baking chocolate can get a pretty good jolt!

Table 0-2 Caffeine content of beverages and foods

Source	Caffeine (milligrams per serving)	Serving size[a] (ounces)
Regular coffee	106–164	5
Decaffeinated coffee	2–5	5
Tea	21–50	5
Cocoa beverage	2–8	6
Baking chocolate	35	1
Sweet chocolate	20	1
Milk chocolate	6	1
Caffeinated soft drinks	36–57	12
Red Bull	80	8.2

a. 1 ounce = 28.35 grams.

SOURCES: http://www.holymtn.com/tea/caffeine_content.htm. Red Bull from http://wilstar.com/caffeine.htm.

Quality Assurance

How could Denby and Scott be sure that their analytical results are reliable? Professional analysts follow a set of practices, called **quality assurance**, intended to give themselves and their customers confidence in the quality of their results. One way that Denby and Scott could assess the reliability of their analytical method might be to melt some chocolate, add a known quantity of caffeine to the melt, mix it as well as possible, and freeze the chocolate. The added caffeine is called a *spike*. When the spiked chocolate is analyzed, they should find a quantity of caffeine equal to that in the original chocolate plus the amount in the spike. If they find the

Chapter 5 discusses quality assurance.

Box 0-1 Constructing a Representative Sample

The diagram below shows steps in transforming a complex substance into individual samples that can be analyzed. A *lot* is the total material (for example, a railroad car full of grain or a carton full of macadamia chocolates) from which samples are taken. A *bulk sample* (also called a *gross sample*) is taken from the lot for analysis and *archiving* (storing for future reference). The bulk sample must be representative of the lot or the analysis will be meaningless. From the representative bulk sample, a smaller, homogeneous *laboratory sample* is formed that must have the same composition as the bulk sample. For example, you might obtain a laboratory sample by grinding an entire solid bulk sample to a fine powder, mixing thoroughly, and keeping one bottle of powder for testing. Small test portions (called *aliquots*) of the laboratory sample are used for individual analyses.

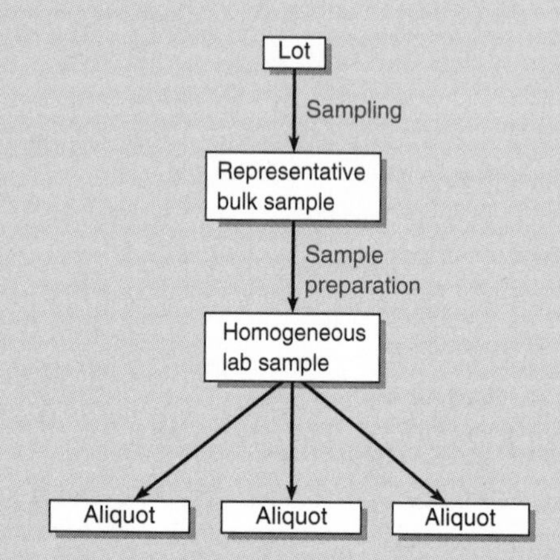

Sampling is the process of selecting a representative bulk sample from the lot. *Sample preparation* converts a bulk sample into a homogeneous laboratory sample. Sample preparation also refers to steps that eliminate interfering species or that concentrate the analyte.

In a **random heterogeneous material**, differences in composition are random and on a fine scale. When you collect a portion of the material for analysis, you obtain some of each of the different compositions. To construct a representative sample from a heterogeneous material, you can first visually divide the material into many small regions. For example, if you want to measure the magnesium content of the grass in the 10-meter × 20-meter field in panel *a*, you could divide the field into 20 000 small patches that are 10 centimeters on a side. A **random sample** is collected by taking portions from the desired number of regions chosen at random. Assign a number to each small patch and use a computer to generate 100 numbers at random from 1 to 20 000. Then harvest and combine the grass from each of the selected 100 patches to construct a representative bulk sample for analysis.

In a **segregated heterogeneous material**, large regions have obviously different compositions. To obtain a representative specimen of such a material, we construct a **composite sample**. For example, the field in panel *b* has three different types of grass in regions A, B, and C. You could draw a map of the field on graph paper and measure the area in each region. In this case, 66% of the area lies in region A, 14% lies in region B, and 20% lies in region C. To construct a representative bulk sample from this segregated material, take 66 small patches from region A, 14 from region B, and 20 from region C. You could do so by drawing random numbers from 1 to 20 000 to select patches until you have the desired number from each region.

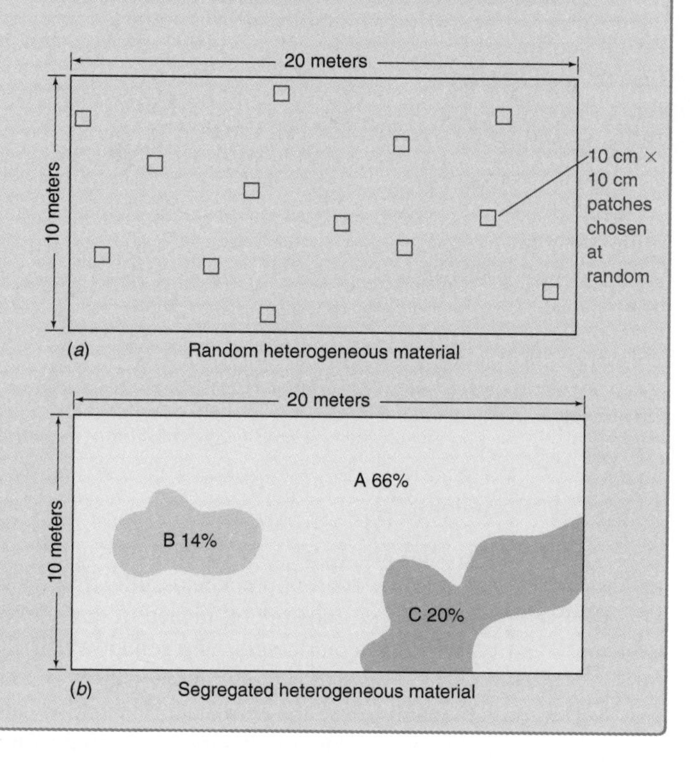

expected quantity, they can have some confidence that their method extracts all the caffeine that was present and measures it accurately.

0-2 General Steps in a Chemical Analysis

The analytical process often begins with a question such as "Is this water safe to drink?" or "Does emission testing of automobiles reduce air pollution?" A scientist translates such questions into the need for particular measurements. An analytical chemist then must choose or invent a procedure to carry out those measurements.

When the analysis is complete, the analyst must translate the results into terms that can be understood by others. A most important feature of any result is its limitations. What is the statistical uncertainty in reported results? If you took samples in a different manner, would you obtain the same results? Is a tiny amount (a *trace*) of analyte found in a sample really there or is it contamination? Once all interested parties understand the results and their limitations, then people can draw conclusions and reach decisions.

We can now summarize general steps in the analytical process:

Formulating the question — Translate general questions into specific questions to be answered through chemical measurements.

Selecting analytical procedures — Search the chemical literature to find appropriate procedures or, if necessary, devise new procedures to make the required measurements.

Sampling — Select representative material to analyze, as described in Box 0-1. If you begin with a poorly chosen sample or if the sample changes between the time it is collected and the time it is analyzed, the results are meaningless. "Garbage in—garbage out!"

Sample preparation — *Sample preparation* is the process of converting a representative sample into a form suitable for chemical analysis, which usually means dissolving the sample. Samples with a low concentration of analyte may need to be concentrated prior to analysis. It may be necessary to remove or *mask* species that interfere with the chemical analysis. For a chocolate bar, sample preparation consisted of removing fat and dissolving the desired analytes. The reason for removing fat was that it would interfere with chromatography.

Chemists use the term **species** to refer to any chemical of interest. Species is both singular and plural. **Interference** occurs when a species other than analyte increases or decreases the response of the analytical method and makes it appear that there is more or less analyte than is actually present. **Masking** is the transformation of an interfering species into a form that is not detected. For example, Ca^{2+} in lake water can be measured with a reagent called EDTA. Al^{3+} interferes with this analysis, because it also reacts with EDTA. Al^{3+} can be masked by treating the sample with excess F^- to form AlF_6^{3-}, which does not react with EDTA.

Analysis — Measure the concentration of analyte in several identical **aliquots** (portions). The purpose of *replicate measurements* (repeated measurements) is to assess the variability (uncertainty) in the analysis and to guard against a gross error in the analysis of a single aliquot. *The uncertainty of a measurement is as important as the measurement itself*, because it tells us how reliable the measurement is. If necessary, use different analytical methods on similar samples to make sure that all methods give the same result and that the choice of analytical method is not biasing the result. You may also wish to construct and analyze several different bulk samples to see what variations arise from your sampling procedure. Steps taken to demonstrate the reliability of the analysis are called *quality assurance.*

Reporting and interpretation — Deliver a clearly written, complete report of your results, highlighting any special limitations that you attach to them. Your report might be written to be read only by a specialist (such as your instructor), or it might be written for a general audience (perhaps your mother). Be sure the report is appropriate for its intended audience.

Drawing conclusions — Once a report is written, the analyst might not further participate in what is done with the information, such as modifying the raw material supply for a factory or creating new laws to regulate food additives. The more clearly a report is written, the less likely it is to be misinterpreted. The analyst should have the responsibility of ensuring that conclusions drawn from his or her data are consistent with the data.

Most of this book deals with measuring chemical concentrations in homogeneous aliquots of an unknown. The analysis is meaningless unless you have collected the sample properly, you have taken measures to ensure the reliability of the analytical method, and you communicate your results clearly and completely. The chemical analysis is only the middle part of a process that begins with a question and ends with a conclusion.

Ask Yourself

Answers to Ask Yourself questions are at the back of the book.

0-A. After reading Box 0-1, answer the following questions:
(a) What is the difference between a *heterogeneous* and a *homogeneous* material?
(b) What is the difference between a *random* heterogeneous material and a *segregated* heterogeneous material?
(c) What is the difference between a *random* sample and a *composite* sample? When would each be used?

0-3 Charles David Keeling and the Measurement of Atmospheric CO_2

There are three occasions when dedication to scientific measurements has changed all of science.

- Tycho Brahe's observations of planets laid the foundation for . . . Newton's theory of gravitation.
- Albert Michelson's measurements of the speed of light laid the foundation for . . . Einstein's theory of relativity.
- Charles David Keeling's measurements of the global accumulation of carbon dioxide in the atmosphere set the stage for today's profound concerns about climate change. They are the single most important environmental data set taken in the 20th century.

—C. F. Kennel (2005), Scripps Institution of Oceanography

Charles David Keeling (Figure 0-8) grew up near Chicago during the Great Depression.[3] His investment banker father excited an interest in astronomy in 5-year-old Keeling. His mother, who had been a graduate student in English Literature at Yale before she was married, instilled a lifelong love of music in young Keeling. Though "not predominantly interested in science," he took all the science available in high school, including a wartime course in aeronautics that exposed him to aerodynamics, meteorology, navigation, combustion engines, and radio. In 1945, he enrolled in a summer session at the University of Illinois prior to his anticipated draft into the army. When World War II ended that summer, Keeling was free to continue at Illinois where he "drifted into chemistry" and graduated in 1948.

Upon graduation, Professor Malcolm Dole of Northwestern University, who had known Keeling as a precocious child, offered him a graduate fellowship in chemistry. On Keeling's second day in the lab, Dole taught him how to make careful measurements with an analytical balance. Keeling went on to conduct research in polymer chemistry, though he never had any particular attraction to polymers or to chemistry. One of the requirements for graduate study was a minor outside of chemistry. Keeling noticed the textbook *Glacial Geology and the Pleistocene Epoch* on a friend's bookshelf and began reading it. It was so interesting that he bought a copy and read it for pleasure between experiments in the lab. He imagined himself "climbing mountains while measuring the physical properties of glaciers." Keeling completed most of the undergraduate curriculum in geology and twice interrupted his research to hike and climb in the Cascade Mountains of Washington State.

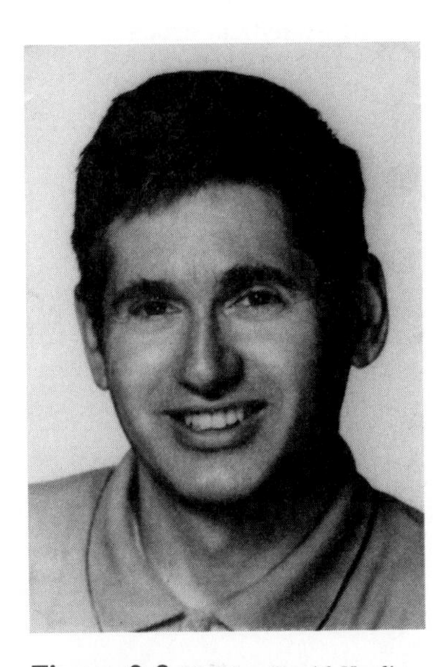

Figure 0-8 Charles David Keeling (1928–2005). [Courtesy Ralph Keeling, Scripps Institution of Oceanography, University of California, San Diego.]

Keeling graduated in 1953 when there was a shortage of Ph.D. chemists. Polymer chemists were in demand for the exciting, new plastics industry. He had many job offers from chemical manufacturers in the eastern United States. But Keeling "had trouble seeing the future this way." He had acquired a working knowledge of geology and loved the outdoors. Professor Dole considered it "foolhardy" to pass up high-paying permanent jobs for a low-paying postdoctoral position. Nonetheless, Keeling wrote letters seeking a postdoctoral position as a chemist "exclusively to geology departments west of the North American continental divide." He became the first postdoctoral fellow in the new Department of Geochemistry in Harrison Brown's laboratory at Caltech in Pasadena, California.

One day, "Brown illustrated the power of applying chemical principles to geology. He suggested that the amount of carbonate in surface water . . . might be estimated by assuming the water to be in chemical equilibrium with both limestone ($CaCO_3$) and atmospheric carbon dioxide." Keeling decided to test this idea. He "could fashion chemical apparatus to function in the real environment" and, best of all, "the work could take place outdoors."

Keeling built a vacuum extraction system to isolate CO_2 from air or acidified water. The water content of air is highly variable, so water was removed by freezing it out in a Dry Ice trap. The CO_2 in the remaining dry air was trapped as a solid in the vacuum system by using liquid nitrogen, "which had recently become available commercially." Keeling modernized the design of a gas manometer from a 1916 journal article so that he could measure gaseous CO_2 by confining it in a known volume at a known pressure and temperature above a column of mercury (Figure 0-9). The measurement was precise to 0.1%, which was as good or better than other procedures for measuring CO_2 in air or carbonate in water.

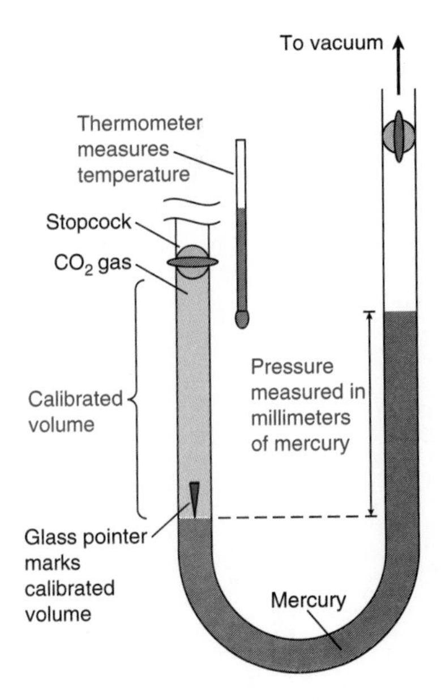

Figure 0-9 A mercury manometer made from a glass U-tube. The difference in height between the mercury on the left and the right gives the pressure of the gas in milliliters of mercury.

Keeling prepared for a field experiment at Big Sur State Park near Monterey. The area was rich in calcite ($CaCO_3$), which would, presumably, be in good contact with ground water. Before going to Big Sur, he "began to worry . . . about assuming a specified concentration for CO_2 in air." This concentration had to be known precisely for his experiments. Published values varied widely, so he decided to make his own measurements. He had a dozen 5-liter flasks built with stopcocks that would hold a vacuum. He weighed each flask empty and filled with water. From the mass of water it held, he could calculate the volume of each flask. To rehearse for field experiments, Keeling collected air samples in Pasadena and measured the CO_2 with his manometer. Concentrations varied significantly, apparently affected by urban emissions.

Not being certain that CO_2 in pristine air next to the Pacific Ocean at Big Sur would be constant, he collected air samples every few hours over a full day and night. He also collected water samples and brought everything back to the lab to measure CO_2 with his manometer. At the suggestion of Professor Sam Epstein at Caltech, Keeling provided samples of CO_2 for Epstein's group to measure carbon and oxygen isotopes with their newly built isotope ratio mass spectrometer. "I did not anticipate that the procedures established in this first experiment would be the basis for much of the research that I would pursue over the next forty-odd years," recounted Keeling. Contrary to hypothesis, Keeling found that river and ground waters contained more dissolved CO_2 than would be in equilibrium with air.

Keeling's attention was drawn to the diurnal pattern that he observed in atmospheric CO_2. Air in the afternoon had an almost constant CO_2 content of 310 parts per million (ppm) by volume of dry air. The concentration of CO_2 at night was higher and variable. Also, the higher the CO_2 content, the lower the $^{13}C/^{12}C$ ratio. It was thought that photosynthesis by plants would draw down atmospheric CO_2 near the ground during the day and respiration would restore CO_2 to the air at night. However, samples collected in daytime from many locations had nearly the same 310 ppm CO_2, regardless of whether the area was vegetated.

Keeling found an explanation in a book entitled *The Climate Near the Ground.* All of his samples were collected in fair weather when solar heating induces afternoon turbulence that mixes air near the ground with air higher in the atmosphere. At night, air cools and forms a stable layer near the ground that becomes rich in CO_2 from respiration of plants. Keeling had discovered that CO_2 is near 310 ppm in the free atmosphere over large regions of the Northern Hemisphere. By 1956, his findings were firm enough to be told to others, including Dr. Oliver Wulf of the U.S. Weather Bureau, who was working at Caltech.

Wulf passed Keeling's results to Harry Wexler, Head of Meteorological Research at the Weather Bureau. Wexler invited Keeling to Washington, DC, where he explained that the International Geophysical Year was to commence in July 1957 to collect worldwide geophysical data for a period of 18 months. The Weather Bureau had just built an observatory near the top of the Mauna Loa volcano on the big island of Hawaii at an elevation of 3 400 m and Wexler was anxious to put it to use (Figure 0-10). The Bureau was already planning to measure atmospheric CO_2 at remote locations around the world.

Keeling explained that measurements in the scientific literature might be unreliable. He proposed to measure CO_2 with an infrared spectrometer that would be precisely calibrated with gas measured by a manometer. The manometer is the most reliable way to measure CO_2, but each measurement requires about half a day of work. The spectrometer could measure several samples per hour but must be calibrated with reliable standards.

Wexler liked Keeling's proposal and declared that infrared measurements should be made on Mauna Loa and at a station in Antarctica. The next day, Wexler

Figure 0-10 Mauna Loa Observatory in 2006. [© Forrest M. Mims III, www.forrestmims.org/maunaloaobservatory.html.]

offered Keeling a job. Keeling described what happened next: "I was escorted to where I might work . . . in the dim basement of the Naval Observatory where the only activity seemed to be a cloud-seeding study being conducted by a solitary scientist."

Fortunately, Keeling's CO_2 results had also been brought to the attention of Roger Revelle, Director of the Scripps Institution of Oceanography near San Diego, California. Revelle invited Keeling for a job interview. He was given lunch outdoors "in brilliant sunshine wafted by a gentle sea breeze." "Dim basement or brilliant sunshine and sea breeze?" Keeling thought to himself. "Dim basement or brilliant sunshine and sea breeze?" Keeling chose Scripps, and Wexler graciously provided funding to support CO_2 measurements.

Keeling identified several continuous gas analyzers and tested one made by Applied Physics Corporation, "the only company in which [he] was able to get past a salesman and talk directly with an engineer." He went to great lengths to calibrate the infrared instrument with precisely measured gas standards. Keeling painstakingly constructed a manometer at Scripps that was reproducible to 1 part in 4 000, thus enabling atmospheric CO_2 measurements to be reproducible to 0.1 ppm. Contemporary experts questioned the need for such precision because existing literature indicated that CO_2 in the air varied by a factor of 2. There was general concern that measurements on Mauna Loa—an active volcano—would be confounded by CO_2 emissions from the volcano.

Roger Revelle of Scripps believed that the main value of the measurements would be to establish a "snapshot" of CO_2 around the world in 1957, which could be compared with another snapshot taken 20 years later to see if atmospheric CO_2 concentration was changing. People had considered that burning of fossil fuel could increase atmospheric CO_2 concentration, but it was thought that a good deal of this CO_2 would be absorbed by the ocean. No meaningful measurements existed to evaluate any hypothesis.

In March 1958, Ben Harlan of Scripps and Jack Pales of the Weather Bureau installed Keeling's infrared instrument on Mauna Loa. On the first day of operation, the reading was within 1 ppm of the 313-ppm value expected by Keeling from his

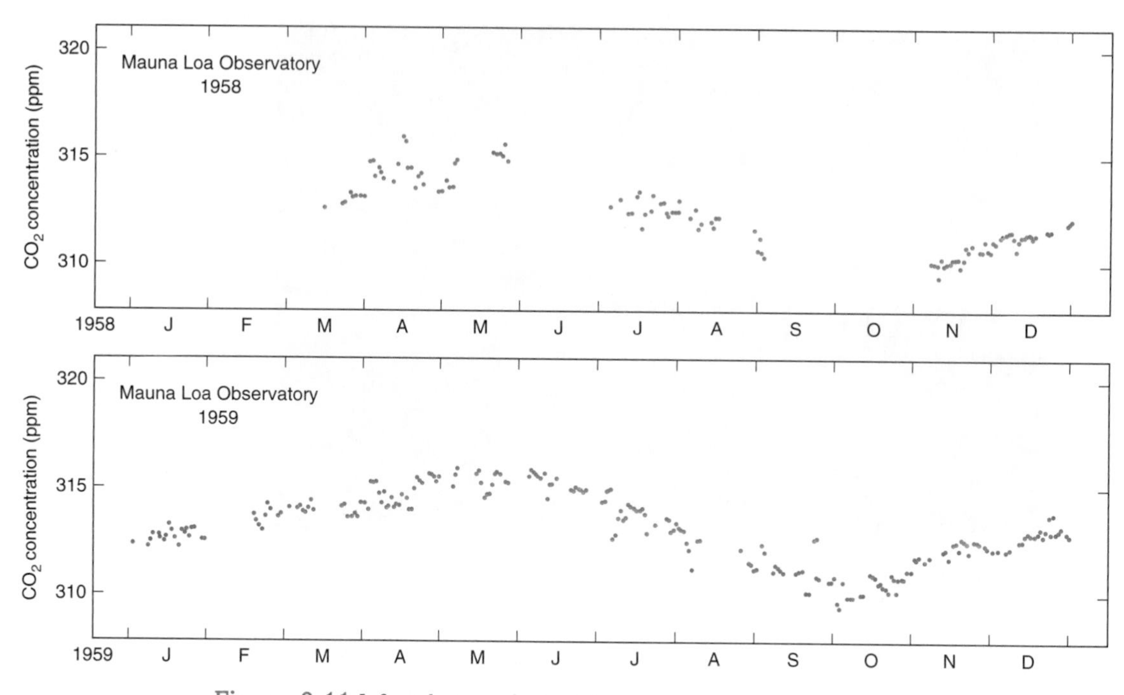

Figure 0-11 Infrared atmospheric CO_2 measurements from Mauna Loa in 1958–1959. [J. D. Pales and C. D. Keeling, *J. Geophys. Res.* **1965**, *70*, 6053.]

measurements made on the pier at Scripps. Concentrations in Figure 0-11 rose between March and May, when operation was interrupted by a power failure. Concentrations were falling in September when power failed again. Keeling was allowed to make his first trip to Mauna Loa to restart the equipment. Concentrations steadily rose from November to May 1959, before gradually falling again. Data for the full year 1959 in Figure 0-10 reproduced the pattern from 1958. These patterns could not have been detected if Keeling's measurements had not been made so carefully. Maximum CO_2 was observed just before plants in the temperate zone of the Northern Hemisphere put on new leaves in May. Minimum CO_2 was observed at the end of the growing season in October. Keeling's conclusion had global significance: "We were witnessing for the first time nature's withdrawing CO_2 from the air for plant growth during the summer and returning it each succeeding winter."

Figure 0-12, known as the *Keeling curve*, shows the results of half a century of CO_2 monitoring on Mauna Loa. Seasonal oscillations are superimposed on a steady rise of CO_2. Approximately half of the CO_2 produced by the burning of fossil fuel (principally coal, oil, and natural gas) in the last half-century resides in the atmosphere. Most of the remainder was absorbed by the ocean.

In the atmosphere, CO_2 absorbs infrared radiation from the surface of the Earth and reradiates part of that energy back to the ground. This process, called the *greenhouse effect*, warms the Earth's surface and might produce climate change. In the ocean, CO_2 forms carbonic acid, H_2CO_3, which makes the ocean more acidic. Fossil fuel burning has already lowered the pH of ocean surface waters by 0.1 unit from preindustrial values. Combustion during the twenty-first century is expected to acidify the ocean by another 0.3–0.4 pH units—more than doubling the concentration of H^+ and threatening the existence of marine life because calcium carbonate shells dissolve in acid (Box 11-1). The entire ocean food chain is jeopardized by ocean acidification.[4]

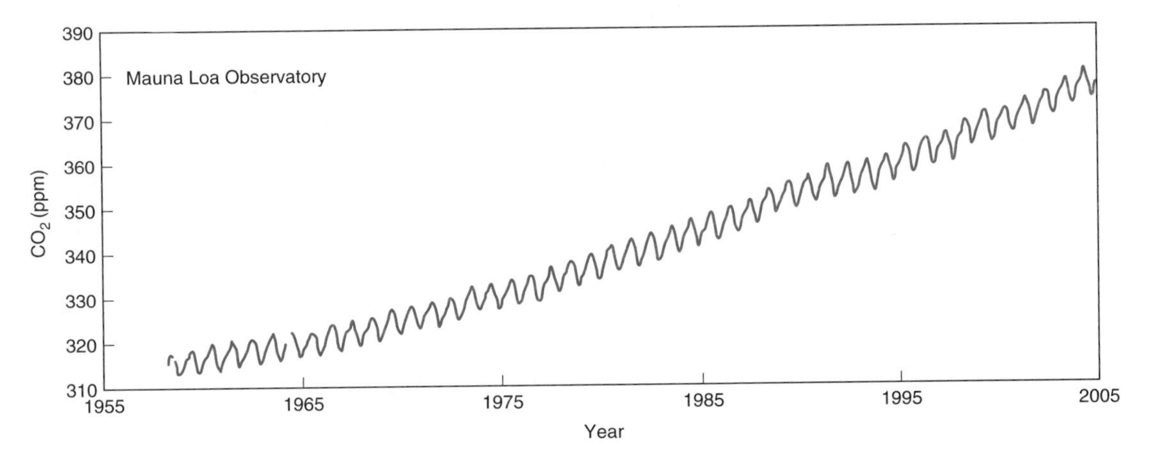

Figure 0-12 Monthly average atmospheric CO_2 measured on Mauna Loa for half a century. This graph, known as the *Keeling curve,* shows rising CO_2 and seasonal oscillations. [Data from http://scrippsco2.ucsd.edu/data/in_situ_co2/monthly_mlo.csv.]

The significance of the Keeling curve is apparent by appending Keeling's data to the record of atmospheric CO_2 preserved in Antarctic ice. Figure 0-13 shows CO_2 and temperature going back almost half a million years. During this time, temperature and CO_2 experienced four major cycles with minima around 350, 250, 150, and 25 thousand years ago.

It is thought that cyclic changes in Earth's orbit and tilt cause cyclic temperature change. Small increases in temperature drive CO_2 from the ocean into the atmosphere. Increased atmospheric CO_2 further increases warming by the greenhouse effect. Cooling brought on by orbital changes redissolves CO_2 in the ocean, thereby

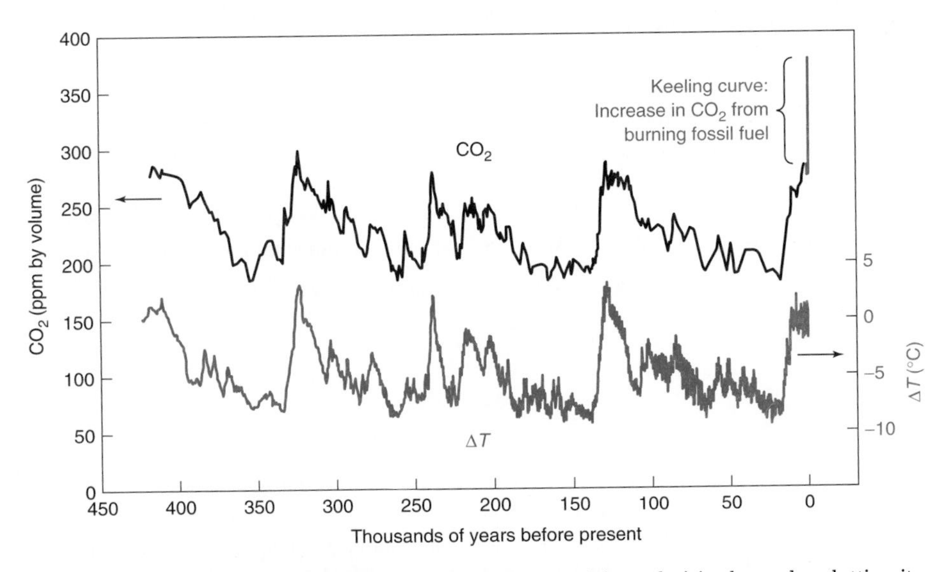

Figure 0-13 Significance of the Keeling curve (upper right, color) is shown by plotting it on the same graph with atmospheric CO_2 measured in air bubbles trapped in ice cores drilled from Antarctica. Atmospheric temperature at the level where precipitation forms is deduced from the hydrogen and oxygen isotopic composition of the ice. [Vostok ice core data from J. M. Barnola, D. Raynaud, C. Lorius, and N. I. Barkov, http://cdiac.esd.ornl.gov/ftp/trends/co2/vostok.icecore.co2.]

ΔT is the change in temperature at the level in the atmosphere where precipitation forms.

causing further cooling. Temperature and CO_2 have followed each other for more than 400 000 years.

Burning fossil fuel in the last 150 years increased CO_2 from its historic cyclic peak of 280 ppm to today's 380 ppm. There is almost no conceivable action in the present century that will prevent CO_2 from climbing to several times its historic high. Alteration of the atmosphere could lead to unprecedented effects on climate. The longer we take to reduce our use of fossil fuel, the longer this unintended global experiment will continue. Increasing population exacerbates this and many other problems.

Keeling's CO_2 measurement program at Mauna Loa survived half a century of precarious existence, during which its life was jeopardized many times by funding decisions by government agencies. Keeling's dogged persistence ensured the continuity and quality of the measurements. Manometrically measured calibration standards are extremely labor intensive and costly. More than once, funding agencies tried to lower the cost of the program by finding substitutes for manometry, but no method provided the same precision. The analytical quality of Keeling's data has enabled subtle trends, such as the effect of El Niño ocean temperature patterns on atmospheric CO_2, to be teased out of the overriding pattern of ever-increasing CO_2 and seasonal oscillations.

Important Terms†

aliquot
analyte
aqueous
calibration curve
composite sample
decant
heterogeneous
homogeneous
interference
masking
qualitative analysis
quality assurance
quantitative analysis
quantitative transfer
random heterogeneous material
random sample
sample preparation
sampling
segregated heterogeneous material
slurry
species
standard solution
supernatant liquid

Problems

0-1. What is the difference between *qualitative* and *quantitative* analysis?

0-2. List the steps in a chemical analysis.

0-3. What does it mean to *mask* an interfering species?

0-4. What is the purpose of a calibration curve?

Notes and References

1. T. J. Wenzel, *Anal. Chem.* **1995**, *67*, 470A.

2. W. R. Kreiser and R. A. Martin, Jr., *J. Assoc. Off. Anal. Chem.* **1978**, *61*, 1424; W. R. Kreiser and R. A. Martin, Jr., *J. Assoc. Off. Anal. Chem.* **1980**, *63*, 591.

3. C. D. Keeling, "Rewards and Penalties of Monitoring the Earth," *Ann. Rev. Energy Environ.* **1998**, *23*, 25–82. This spellbinding autobiographical account can be downloaded free of charge from www.arjournals.annualreviews.org.

4. J. C. Orr et al., "Anthropogenic Ocean Acidification Over the Twenty-first Century and Its Impact on Calcifying Organisms," *Nature*, **2005**, *437*, 681.

†Terms are introduced in **bold** type in the chapter and are also defined in the Glossary.

Further Reading

S. Bell, *Forensic Chemistry* (Upper Saddle River, NJ: Pearson Prentice Hall, 2006).

P. C. White, ed., *Crime Scene to Court: The Essentials of Forensic Science*, 2nd ed. (Cambridge: Royal Society of Chemistry, 2004). ISBN 0-85404-656-9.

S. M. Gerber, ed., *Chemistry and Crime: From Sherlock Holmes to Today's Courtroom* (Washington, DC: American Chemical Society, 1983). Available in paperback from Oxford University Press, ISBN 0-8412-0785-2.

S. M. Gerber and R. Saferstein, eds., *More Chemistry and Crime: From Marsh Arsenic Test to DNA Profile* (Washington, DC: American Chemical Society, 1997). Available from Oxford University Press, ISBN 0-8412-3406-X.

A. M. Pollard and C. Heron, *Archaeological Chemistry* (Cambridge: Royal Society of Chemistry, 1996). ISBN 0-85404-523-6.

Biochemical Measurements with a Nanoelectrode

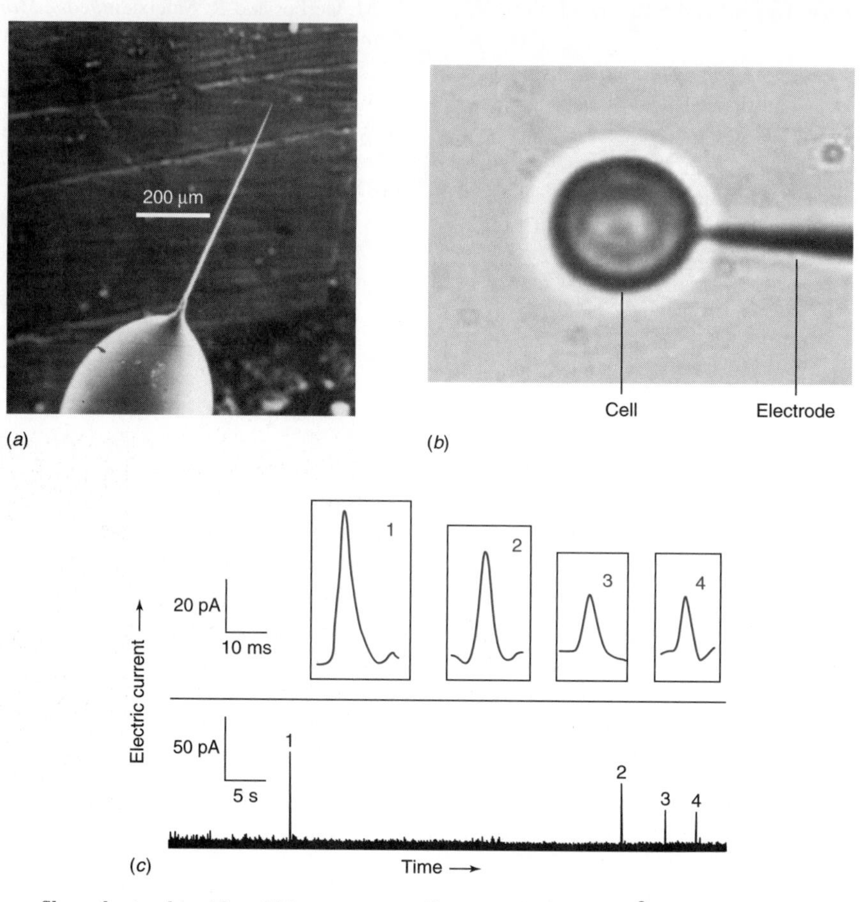

(*a*) Carbon fiber electrode with a 100-nanometer-diameter (100×10^{-9} meter) tip extending from glass capillary. The marker bar is 200 micrometers (200×10^{-6} meter). [From W.-H. Huang, D.-W. Pang, H. Tong, Z.-L. Wang, and J.-K. Cheng, *Anal. Chem.* **2001**, *73*, 1048.] (*b*) Electrode positioned adjacent to a cell detects release of the neurotransmitter dopamine from the cell. A nearby, larger counter-electrode is not shown. (*c*) Bursts of electric current detected when dopamine is released. Insets are enlargements. [From W.-Z. Wu, W.-H. Huang, W. Wang, Z.-L. Wang, J.-K. Cheng, T. Xu, R.-Y. Zhang, Y. Chen, and J. Liu, *J. Am. Chem. Soc.* **2005**, *127*, 8914.]

An electrode whose tip is smaller than a single cell allows us to measure neurotransmitter molecules released by a nerve cell in response to a chemical stimulus. We call the electrode a *nanoelectrode* because its active region has dimensions of nanometers (10^{-9} meters). Neurotransmitter molecules released from one *vesicle* (a small compartment) of a nerve cell diffuse to the electrode where they donate or accept electrons, generating an electric current measured in picoamperes (10^{-12} amperes) for a period of milliseconds (10^{-3} seconds). This chapter discusses units that describe chemical and physical measurements of objects ranging in size from atoms to galaxies.

CHAPTER 1

Chemical Measurements

Most people who practice analytical chemistry do not identify themselves as analytical chemists. For example, chemical analysis is an essential tool used by biologists to understand how organisms function and by doctors to diagnose disease and monitor the response of a patient to treatment. Environmental scientists measure chemical changes in the atmosphere, water, and soil that occur in response to the activities of both man and nature. Forensic scientists identify and sometimes measure drugs, combustion residues, and fibers from crime scenes. You are taking this course because you might make chemical measurements yourself or you will need to understand analytical results reported by others. This chapter provides basic working knowledge of measurements and equilibrium.

Figure 1-1 Of the fundamental SI units in Table 1-1, only the kilogram is defined by an artifact rather than by a reproducible physical measurement. The international kilogram in France, made from a Pt-Ir alloy in 1885, has been removed from its protective enclosure to be weighed against working copies only in 1890, 1948, and 1992. Its mass could change from reaction with the atmosphere or from wear, so there is research to define a standard for mass based on measurements that should not change over time. [Bureau International des Poids at Measures.]

1-1 SI Units and Prefixes

SI units of measurement derive their name from the French *Système International d'Unités. Fundamental units* (base units) from which all others are derived are defined in Table 1-1. Standards of length, mass, and time are the *meter* (m), *kilogram* (kg) (Figure 1-1), and *second* (s), respectively. Temperature is measured in *kelvins* (K), amount of substance in *moles* (mol), and electric current in *amperes* (A). Table 1-2 lists derived quantities that are defined in terms of the fundamental quantities. For example, force is measured in *newtons* (N), pressure is measured in *pascals* (Pa), and energy is measured in *joules* (J), each of which can be expressed in terms of the more fundamental units of length, time, and mass.

It is convenient to use the prefixes in Table 1-3 to express large or small quantities. For example, the pressure of dissolved oxygen in arterial blood is approximately 1.3×10^4 Pa. Table 1-3 tells us that 10^3 is assigned the prefix k for "kilo." We can express the pressure in multiples of 10^3 as follows:

$$1.3 \times 10^4 \cancel{\text{Pa}} \times \frac{1 \text{ kPa}}{10^3 (\cancel{\text{Pa}})} = 1.3 \times 10^1 \text{ kPa} = 13 \text{ kPa}$$

Pressure is force per unit area.
1 pascal (Pa) = 1 N/m^2.
The pressure of the atmosphere is about 100 000 Pa.

Table 1-1 Fundamental SI units

Quantity	Unit (symbol)	Definition
Length	meter (m)	One meter is the distance light travels in a vacuum during $\frac{1}{299\,792\,458}$ of a second.
Mass	kilogram (kg)	One kilogram is the mass of the prototype kilogram kept at Sèvres, France.
Time	second (s)	One second is the duration of 9 192 631 770 periods of the radiation corresponding to a certain atomic transition of ^{133}Cs.
Electric current	ampere (A)	One ampere of current produces a force of 2×10^{-7} newtons per meter of length when maintained in two straight, parallel conductors of infinite length and negligible cross section, separated by 1 meter in a vacuum.
Temperature	kelvin (K)	Temperature is defined such that the triple point of water (at which solid, liquid, and gaseous water are in equilibrium) is 273.16 K, and the temperature of absolute zero is 0 K.
Luminous intensity	candela (cd)	Candela is a measure of luminous intensity visible to the human eye. One cd is the luminous intensity in a given direction of a source that emits monochromatic radiation of frequency 540×10^{12} hertz and of which the radiant intensity in that direction is $\frac{1}{683}$ watt per steradian.
Amount of substance	mole (mol)	One mole is the number of atoms in exactly 0.012 kg of ^{12}C (approximately 6.022×10^{23}).
Plane angle	radian (rad)	There are 2π radians in a circle.
Solid angle	steradian (sr)	There are 4π steradians in a sphere.

Table 1-2 SI-derived units with special names

Quantity	Unit	Abbreviation	Expression in terms of other units	Expression in terms of SI base units
Frequency	hertz	Hz		1/s
Force	newton	N		$m \cdot kg/s^2$
Pressure	pascal	Pa	N/m^2	$kg/(m \cdot s^2)$
Energy, work, quantity of heat	joule	J	$N \cdot m$	$m^2 \cdot kg/s^2$
Power, radiant flux	watt	W	J/s	$m^2 \cdot kg/s^3$
Quantity of electricity, electric charge	coulomb	C		$s \cdot A$
Electric potential, potential difference, electromotive force	volt	V	W/A	$m^2 \cdot kg/(s^3 \cdot A)$
Electric resistance	ohm	Ω	V/A	$m^2 \cdot kg/(s^3 \cdot A^2)$

The unit kPa is read "kilopascals." Always write units beside each number in a calculation and cancel identical units in the numerator and denominator. This practice ensures that you know the units for your answer. If you intend to calculate pressure and your answer comes out with units other than pascals, something is wrong.

Table 1-3 Prefixes

Prefix	Abbreviation	Factor
yotta	Y	10^{24}
zetta	Z	10^{21}
exa	E	10^{18}
peta	P	10^{15}
tera	T	10^{12}
giga	G	10^{9}
mega	M	10^{6}
kilo	k	10^{3}
hecto	h	10^{2}
deca	da	10^{1}
deci	d	10^{-1}
centi	c	10^{-2}
milli	m	10^{-3}
micro	μ	10^{-6}
nano	n	10^{-9}
pico	p	10^{-12}
femto	f	10^{-15}
atto	a	10^{-18}
zepto	z	10^{-21}
yocto	y	10^{-24}

Example Counting Neurotransmitter Molecules with an Electrode

Box 1-1 describes the process by which neurotransmitters are released from a nerve cell in discrete bursts. The neurotransmitter measured by the electrode at the opening of this chapter is dopamine. Each dopamine molecule that diffuses to the electrode releases two electrons. The charge transferred to the electrode by burst 1 in panel (*c*) of the chapter opener is 0.27 pC (picocoulombs, 10^{-12} C). One coulomb of charge corresponds to 6.24×10^{18} electrons. How many molecules are released in burst 1?

SOLUTION Table 1-3 tells us that 1 pC equals 10^{-12} C. Therefore 0.27 pC corresponds to

$$0.27\ \cancel{\text{pC}} \times \left(\frac{10^{-12}\ \text{C}}{\cancel{\text{pC}}}\right) = 2.7 \times 10^{-13}\ \text{C}$$

The key to converting between units is to write a conversion factor such as 10^{-12} C/pC, carry out the multiplication, cancel the same units that appear in the numerator and denominator, and show that the answer has the correct units. The number of electrons in 0.27 pC is

$$(2.7 \times 10^{-13}\ \cancel{\text{C}}) \times \left(\frac{6.24 \times 10^{18}\ \text{electrons}}{\cancel{\text{C}}}\right) = 1.68 \times 10^{6}\ \text{electrons}$$

Each molecule releases two electrons, so the number of molecules in one burst is

$$(1.68 \times 10^{6}\ \cancel{\text{electrons}}) \times \left(\frac{1\ \text{molecule}}{2\ \cancel{\text{electrons}}}\right) = 8.4 \times 10^{5}\ \text{molecules}$$

Test Yourself How many molecules of dopamine are in burst 4 on page 18, panel *c*, whose total charge is 0.13 pC? (**Answer:** 4.1×10^{5} molecules)

In 1999, the $125 million *Mars Climate Orbiter* spacecraft was lost when it entered the Martian atmosphere 100 km lower than planned. *The navigation error would have been prevented if people had labeled their units of measurement.* Engineers who built the spacecraft calculated thrust in the English unit, pounds of force. Jet Propulsion Laboratory engineers thought they were receiving the information in the metric unit, newtons. Nobody caught the error.

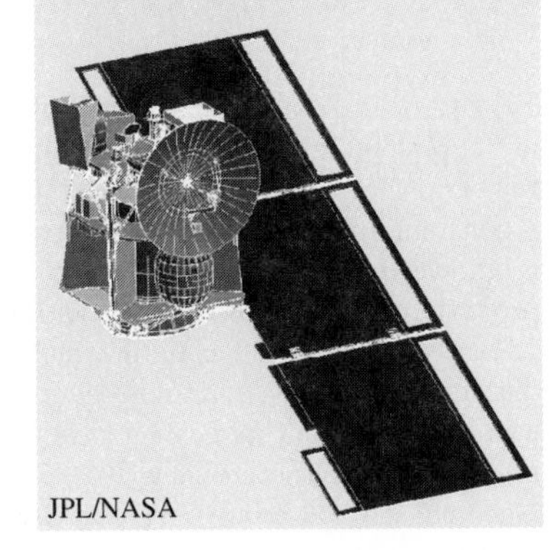

JPL/NASA

Ask Yourself

1-A. (a) What are the names and abbreviations for each of the prefixes from 10^{-24} to 10^{24}? Which abbreviations are capitalized?

(b) Figure 1-2 relates metabolic rate (W = watts = J/s = energy per unit time) to the mass of living organisms. These properties vary over many **orders of magnitude** (powers of 10). Data points on the horizontal axis (*x*-axis, also called the **abscissa**) cover a range from 10^{-19} to 10^{7} g. Data points on the vertical axis (*y*-axis, also called the **ordinate**) go from 10^{-19} to 10^{4} W. Express 10^{-19} g, 10^{7} g, and 10^{4} W with prefixes from Table 1-3.

Box 1-1 Exocytosis of Neurotransmitters

Nerve signals are transmitted from the *axon* of one nerve cell (a *neuron*) to the *dendrite* of a neighboring neuron across a junction called a *synapse*. A change in electric potential at the axon causes tiny chemical-containing packages called *vesicles* to fuse with active zones at the cell membrane. This process, called *exocytosis*, releases neurotransmitter molecules stored in the vesicles. When neurotransmitters bind to receptors on the dendrite, gates open up to allow cations to cross the dendrite membrane. Cations diffusing into the dendrite change its electric potential, thereby transmitting the nerve impulse into the second neuron.

Release of the neurotransmitter dopamine from a single active zone on a cell surface can be monitored by placing a nanoelectrode next to the cell, as shown at the opening of this chapter. When stimulated by K^+ ion injected near the cell, vesicles release dopamine by exocytosis. Each dopamine molecule that diffuses to the nanoelectrode gives up two electrons. Panel (*c*) at the beginning of the chapter shows four pulses measured over a period of 1 minute near one active zone. Each pulse lasts ~10 milliseconds and has a peak current of ~10–30 picoamperes. The number of electrons in a pulse tells us how many dopamine molecules were released from one vesicle.

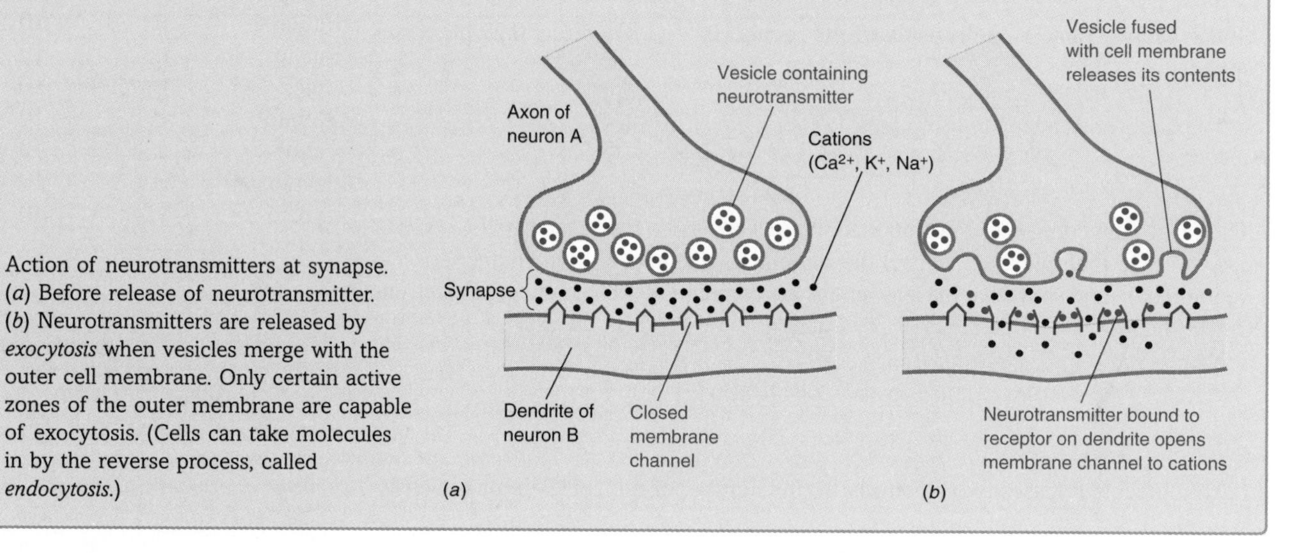

Action of neurotransmitters at synapse. (*a*) Before release of neurotransmitter. (*b*) Neurotransmitters are released by *exocytosis* when vesicles merge with the outer cell membrane. Only certain active zones of the outer membrane are capable of exocytosis. (Cells can take molecules in by the reverse process, called *endocytosis*.)

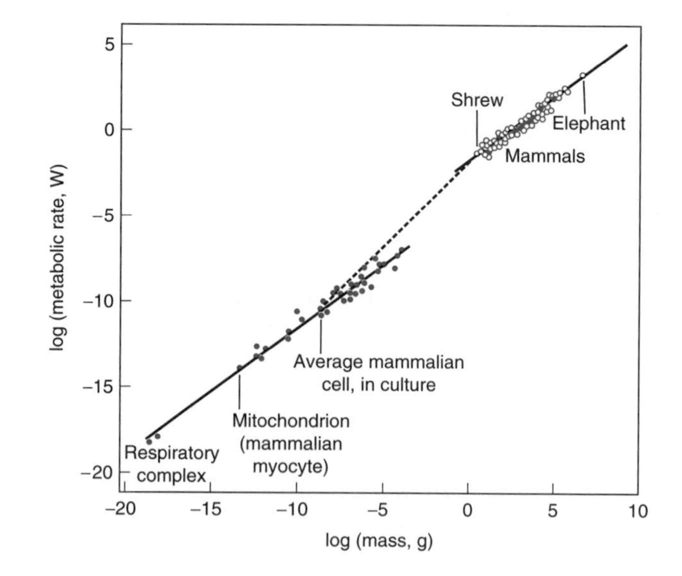

Figure 1-2 A *scaling law*. Metabolic rates of living organisms are approximately proportional to the mass (m) of the organism raised to the ¾ power: metabolic rate (watts) $\approx km^{3/4}$, where k is a constant and "$\approx$" means "is approximately equal to." This relationship is called a "scaling law" because it describes how a property of the organism scales with mass or size. The evolution of organisms toward a state in which minimum energy is used to distribute molecules and information through the organism is thought to account for scaling laws. [From G. B. West and J. H. Brown, *Physics Today* September 2004, p. 36. See also G. B. West, W. H. Woodruff, and J. H. Brown, *Proc. Natl. Acad. Sci. USA* **2002**, *99*, 2473.]

Table 1-4 Conversion factors

Quantity	Unit	Abbreviation	SI equivalent[a]
Volume	liter	L	$*10^{-3}$ m^3
	milliliter	mL	$*10^{-6}$ m^3
Length	angstrom	Å	$*10^{-10}$ m
	inch	in.	*0.025 4 m
Mass	pound	lb	*0.453 592 37 kg
	metric ton (tonne)	t	*1 000 kg
Force	dyne	dyn	$*10^{-5}$ N
Pressure	atmosphere	atm	*101 325 Pa
	atmosphere	atm	1.013 25 bar
	bar	bar	$*10^5$ Pa
	torr	mm Hg	133.322 Pa
	pound/in.^{2}	psi	6 894.76 Pa
Energy	erg	erg	$*10^{-7}$ J
	electron volt	eV	$1.602\ 176\ 487 \times 10^{-19}$ J
	calorie, thermochemical	cal	*4.184 J
	calorie (with a capital C)	Cal	*1 000 cal = 4.184 kJ
	British thermal unit	Btu	1 055.06 J
Power	horsepower		745.700 W
Temperature	centigrade (= Celsius)	°C	*K − 273.15
	Fahrenheit	°F	*1.8(K − 273.15) + 32

a. An asterisk (*) indicates that the conversion is exact (by definition).

1-2 Conversion Between Units

Although SI is the internationally accepted system of measurement in science, other units are encountered. Conversion factors are found in Table 1-4. For example, common non-SI units for energy are the *calorie* (cal) and the *Calorie* (with a capital C, which represents 1 000 calories, or 1 kcal). Table 1-4 states that 1 cal is exactly 4.184 J (joules).

1 *calorie* (cal) of energy will heat 1 gram of water from 14.5° to 15.5°C.

1 000 *joules* will raise the temperature of a cup of water by about 1°C.

1 cal = 4.184 J

1 pound (mass) ≈ 0.453 6 kg

1 mile ≈ 1.609 km

You require approximately 46 Calories per hour (h) per 100 pounds (lb) of body mass to carry out basic functions required for life, such as breathing, pumping blood, and maintaining body temperature. This minimum energy requirement for a conscious person at rest is called the *basal metabolism.* A person walking at 2 miles per hour on a level path requires approximately 45 Calories per hour per 100 pounds of body mass beyond basal metabolism. The same person swimming at 2 miles per hour consumes 360 Calories per hour per 100 pounds beyond basal metabolism.

Example **Unit Conversions**

Express the rate of energy use by a walking woman (46 + 45 = 91 Calories per hour per 100 pounds of body mass) in kilojoules per hour per kilogram of body mass.

SOLUTION We will convert each non-SI unit separately. First, note that 91 Calories equals 91 kcal. Table 1-4 states that 1 cal = 4.184 J, or 1 kcal = 4.184 kJ, so

$$91 \text{ kcal} \times \frac{4.184 \text{ kJ}}{1 \text{ kcal}} = 381 \text{ kJ}$$

Table 1-4 also says that 1 lb is 0.453 6 kg; so 100 lb = 45.36 kg. The rate of energy consumption is, therefore,

Don't panic about the number of significant digits in problems in this chapter. We will take up significant figures in Chapter 3.

$$\frac{91 \text{ kcal/h}}{100 \text{ lb}} = \frac{381 \text{ kJ/h}}{45.36 \text{ kg}} = 8.4 \frac{\text{kJ/h}}{\text{kg}}$$

You could have written this as one calculation with appropriate unit cancellations:

$$\text{rate} = \frac{91 \text{ kcal/h}}{100 \text{ lb}} \times \frac{4.184 \text{ kJ}}{1 \text{ kcal}} \times \frac{1 \text{ lb}}{0.453\ 6 \text{ kg}} = 8.4 \frac{\text{kJ/h}}{\text{kg}}$$

Test Yourself A person who is swimming at 2 miles per hour requires 360 + 46 Calories per hour per 100 pounds of body mass. Express the energy use in kJ/h per kg of body mass. (**Answer:** 37 kJ/h per kg)

Example **Watts Measure Power (Energy per Second)**

1 W = 1 J/s

One watt is 1 joule per second. The woman in the preceding example expends 8.4 kilojoules per hour per kilogram of body mass while walking. **(a)** How many watts per kilogram does she use? **(b)** If her mass is 50 kg, how many watts does she expend?

The complex unit joules per hour per kilogram (J/h/kg) is the same as the expression $\frac{\text{J}}{\text{h} \cdot \text{kg}}$.

SOLUTION **(a)** She expends 8.4×10^3 J per h per kg. We can write the units as J/h/kg, which is equivalent to writing J/(h · kg). Because an hour contains 60 s/min × 60 min/h = 3 600 s, the required power is

$$8.4 \times 10^3 \frac{\text{J}}{\text{h} \cdot \text{kg}} \times \frac{3\ 600 \text{ s}}{1 \text{ h}}$$

Oops! The units didn't cancel out. I guess we need to use the inverse conversion factor:

$$8.4 \times 10^3 \frac{\text{J}}{\text{h} \cdot \text{kg}} \times \frac{1 \text{ h}}{3\ 600 \text{ s}} = 2.33 \frac{\text{J}}{\text{s} \cdot \text{kg}} = 2.33 \frac{\text{J/s}}{\text{kg}} = 2.33 \frac{\text{W}}{\text{kg}}$$

(b) Our intrepid walker has a mass of 50 kg. Therefore her power requirement is

$$2.33 \frac{\text{W}}{\text{kg}} \times 50 \text{ kg} = 116 \text{ W}$$

Test Yourself With the conversion factor in Table 1-4, express the woman's energy use in horsepower. (**Answer:** 0.156 horsepower)

Ask Yourself

1-B. A 120-pound woman working in an office expends about 2.2×10^3 kcal/day, whereas the same woman climbing a mountain needs 3.4×10^3 kcal/day.
(a) How many joules per day does the woman expend in each activity?
(b) How many seconds are in 1 day?
(c) How many joules per second (= watts) does the woman expend in each activity?
(d) Which consumes more power (watts), the office worker or a 100-W light bulb?

1-3 Chemical Concentrations

The minor species in a solution is called the **solute** and the major species is the **solvent**. In this text, most discussions concern *aqueous* solutions, in which the solvent is water. **Concentration** refers to how much solute is contained in a given volume or mass.

Molarity and Molality

$$\text{molarity (M)} = \frac{\text{moles of solute}}{\text{liters of solution}}$$

Molarity (M) is the number of moles of a substance per liter of solution. A **mole** is Avogadro's number of atoms or molecules or ions (6.022×10^{23} mol^{-1}). A **liter** (L) is the volume of a cube that is 10 cm on each edge. Because 10 cm = 0.1 m, 1 L = $(0.1\ \text{m})^3 = (10^{-1})^3\ \text{m}^3 = 10^{-3}\ \text{m}^3$. In Figure 1-3, chemical concentrations in the ocean are expressed in micromoles per liter (10^{-6} mol/L = μM) and nanomoles per liter (10^{-9} mol/L = nM). The molarity of a species is usually designated by square brackets, as in [Cl^-].

The liter is named after the Frenchman Claude Litre (1716–1778), who named his daughter Millicent. I suppose her friends called her Millie Litre.

The **atomic mass** of an element is the number of grams containing Avogadro's number of atoms. The **molecular mass** of a compound is the sum of atomic masses

"Mole Day" is celebrated at 6:02 A.M. on October 23 at many schools.

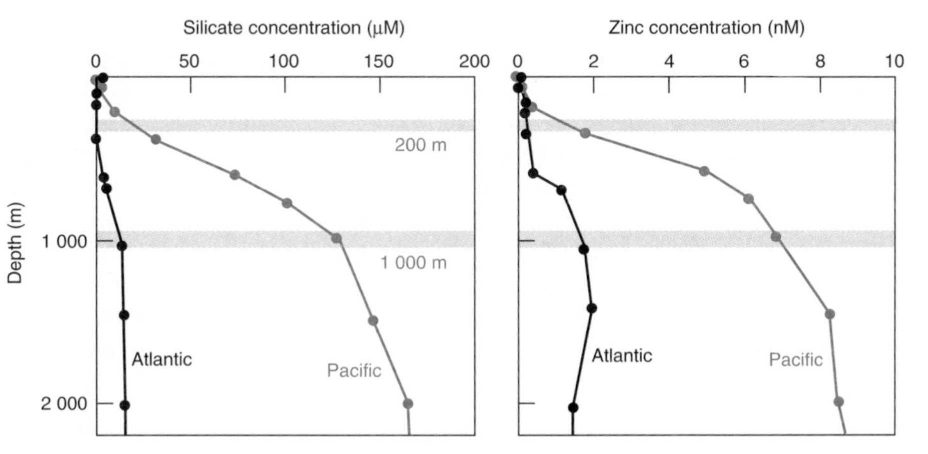

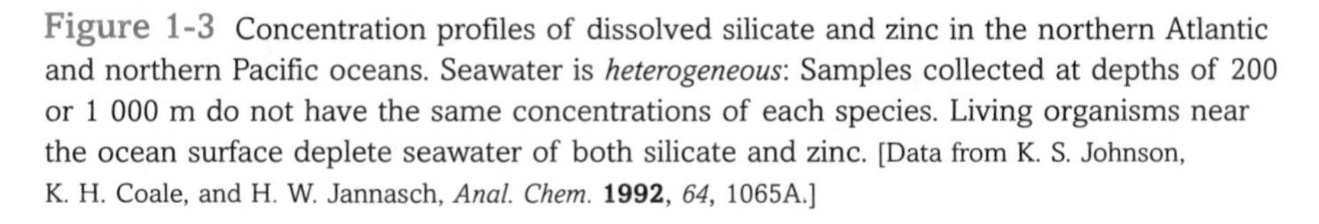
Figure 1-3 Concentration profiles of dissolved silicate and zinc in the northern Atlantic and northern Pacific oceans. Seawater is *heterogeneous*: Samples collected at depths of 200 or 1 000 m do not have the same concentrations of each species. Living organisms near the ocean surface deplete seawater of both silicate and zinc. [Data from K. S. Johnson, K. H. Coale, and H. W. Jannasch, *Anal. Chem.* **1992**, *64*, 1065A.]

of the atoms in the molecule. It is the number of grams containing Avogadro's number of molecules.

Example Molarity of Salts in the Sea

(a) Typical seawater contains 2.7 g of salt (sodium chloride, NaCl) per deciliter (= dL = 0.1 L). What is the molarity of NaCl in the ocean? **(b)** $MgCl_2$ has a typical concentration of 0.054 M in the ocean. How many grams of $MgCl_2$ are present in 25 mL of seawater?

You can find atomic masses in the periodic table inside the front cover of this book.

SOLUTION **(a)** The molecular mass of NaCl is 22.99 (Na) + 35.45 (Cl) = 58.44 g/mol. The moles of salt in 2.7 g are

$$\text{moles of NaCl} = \frac{(2.7\ \cancel{g})}{\left(58.44\ \frac{\cancel{g}}{\text{mol}}\right)} = 0.046\ \text{mol}$$

so the molarity is

$$[\text{NaCl}] = \frac{\text{mol NaCl}}{\text{L of seawater}} = \frac{0.046\ \text{mol}}{0.1\ \text{L}} = 0.46\ \text{M}$$

(b) The molecular mass of $MgCl_2$ is 24.30 (Mg) + [2 × 35.45] (Cl) = 95.20 g/mol, so the number of grams in 25 mL is

$$\text{grams of MgCl}_2 = 0.054\ \frac{\cancel{\text{mol}}}{\cancel{\text{L}}} \times 95.20\ \frac{\text{g}}{\cancel{\text{mol}}} \times (25 \times 10^{-3}\ \cancel{\text{L}}) = 0.13\ \text{g}$$

Test Yourself The sulfate ion (SO_4^{2-}) has a typical concentration of 0.038 M in seawater. Find the concentration of sulfate in grams per 100 mL. (**Answer:** 0.37 g/100 mL)

Strong electrolyte: mostly dissociated into ions in solution
Weak electrolyte: partially dissociated into ions in solution

An *electrolyte* dissociates into ions in aqueous solution. Magnesium chloride is a **strong electrolyte**, which means that it is mostly dissociated into ions in most solutions. In seawater, about 89% of the magnesium is present as Mg^{2+} and 11% is found as the *complex ion*, $MgCl^+$. The concentration of $MgCl_2$ molecules in seawater is close to 0. Sometimes the molarity of a strong electrolyte is referred to as **formal concentration** (F) to indicate that the substance is really converted to other species in solution. When we commonly, and inaccurately, say that the "concentration" of $MgCl_2$ is 0.054 M in seawater, we really mean that its formal concentration is 0.054 F. The "molecular mass" of a strong electrolyte is more properly called the **formula mass** (which we will abbreviate FM), because it is the sum of atomic masses in the formula, even though there may be few molecules with that formula.

A **weak electrolyte** such as acetic acid, CH_3CO_2H, is partially split into ions in solution:

$$\underset{\substack{\text{Acetic acid}\\ \text{(undissociated)}}}{CH_3C(=O)OH} \rightleftharpoons \underset{\substack{\text{Acetate ion}\\ \text{(dissociated)}}}{CH_3C(=O)O^-} + H^+$$

Formal concentration	Percent dissociated
0.1 F	1.3
0.01	4.1
0.001	12.4

A solution prepared by dissolving 0.010 00 mol of acetic acid in 1.000 L has a formal concentration of 0.010 00 F. The actual molarity of CH_3CO_2H is 0.009 59 M because 4.1% is dissociated into $CH_3CO_2^-$ and 95.9% remains as CH_3CO_2H. Nonetheless, we customarily say that the solution is 0.010 00 M acetic acid and understand that some of the acid is dissociated.

Molality (m) is a designation of concentration expressing the number of moles of a solute per kilogram of solvent (not total solution). The masses of solute and solvent do not change with temperature, as long as neither one is allowed to evaporate. Therefore, molality does not change when temperature changes. By contrast, molarity changes with temperature because the volume of a solution usually increases when it is heated.

Confusing abbreviations:
mol = moles

$$\text{M} = \text{molarity} = \frac{\text{mol solute}}{\text{L solution}}$$

$$m = \text{molality} = \frac{\text{mol solute}}{\text{kg solvent}}$$

Percent Composition

The percentage of a component in a mixture or solution is usually expressed as a **weight percent** (wt%):

Definition of weight percent:

$$\text{weight percent} = \frac{\text{mass of solute}}{\text{mass of total solution or mixture}} \times 100 \quad (1\text{-}1)$$

A common form of ethanol (CH_3CH_2OH) is 95 wt%; it has 95 g of ethanol per 100 g of total solution. The remainder is water. Another common expression of composition is **volume percent** (vol%):

Definition of volume percent:

$$\text{volume percent} = \frac{\text{volume of solute}}{\text{volume of total solution}} \times 100 \quad (1\text{-}2)$$

Although "wt%" or "vol%" should always be written to avoid ambiguity, wt% is usually implied when you just see "%."

Example Converting Weight Percent to Molarity

Find the molarity of HCl in a reagent labeled "37.0 wt% HCl, density = 1.188 g/mL." The **density** of a substance is the mass per unit volume.

$$\textit{density} = \frac{\text{mass}}{\text{volume}} = \frac{\text{g}}{\text{mL}}$$

SOLUTION We need to find the moles of HCl per liter of solution. To find moles of HCl, we need to find the mass of HCl. The mass of HCl in 1 L is 37.0% of the mass of 1 L of solution. The mass of 1 L of solution is (1.188 g/~~mL~~)(1 000 ~~mL~~/L) = 1 188 g/L. The mass of HCl in 1 L is

$$\text{HCl}\left(\frac{\text{g}}{\text{L}}\right) = 1\,188\,\frac{\cancel{\text{g solution}}}{\text{L}} \times 0.370\,\frac{\text{g HCl}}{\cancel{\text{g solution}}} = 439.6\,\frac{\text{g HCl}}{\text{L}}$$

↑ This is what 37.0 wt% HCl means

A closely related dimensionless quantity is

$$\textit{specific gravity} = \frac{\text{density of a substance}}{\text{density of water at 4°C}}$$

Because the density of water at 4°C is very close to 1 g/mL, specific gravity is nearly the same as density.

The molecular mass of HCl is 36.46 g/mol, so the molarity is

$$\text{molarity} = \frac{\text{mol HCl}}{\text{L solution}} = \frac{439.6\ \cancel{\text{g HCl}}/\text{L}}{36.46\ \cancel{\text{g HCl}}/\text{mol}} = 12.1\,\frac{\text{mol}}{\text{L}} = 12.1\ \text{M}$$

Test Yourself Phosphoric acid (H_3PO_4, FM = formula mass = 97.99 g/mol) is commonly sold as an 85.5 wt% aqueous solution with a density of 1.69 g solution/mL. Find the molarity of H_3PO_4. (**Answer:** See the table inside the cover for the molarity of 85.5 wt% H_3PO_4.)

Parts per Million and Parts per Billion

A familiar analogy is percent, which is *parts per hundred*:

$$\text{parts per hundred} = \frac{\text{mass of substance}}{\text{mass of sample}} \times 100$$

Concentrations of trace components of a sample can be expressed as **parts per million** (ppm) or **parts per billion** (ppb), terms that mean grams of substance per million or billion grams of total solution or mixture.

Definition of parts per million:
$$\text{ppm} = \frac{\text{mass of substance}}{\text{mass of sample}} \times 10^6 \qquad (1\text{-}3)$$

Question What would be the definition of parts per trillion?

Definition of parts per billion:
$$\text{ppb} = \frac{\text{mass of substance}}{\text{mass of sample}} \times 10^9 \qquad (1\text{-}4)$$

Masses must be expressed in the same units in the numerator and denominator.

$1 \text{ ppm} \approx 1\ \mu\text{g/mL}$
$1 \text{ ppb} \approx 1\ \text{ng/mL}$
The symbol $\approx$ is read "is approximately equal to."

The density of a dilute aqueous solution is close to 1.00 g/mL; so *we frequently equate 1 g of water with 1 mL of water*, although this equivalence is only approximate. Therefore 1 ppm corresponds to 1 μg/mL (= 1 mg/L) and 1 ppb is 1 ng/mL (= 1 μg/L).

Example **Converting Parts per Billion to Molarity**

Hydrocarbons are compounds containing only hydrogen and carbon. Plants manufacture hydrocarbons as components of the membranes of cells and vesicles. The biosynthetic pathway leads mainly to compounds with an odd number of carbon atoms. Figure 1-4 shows the concentrations of hydrocarbons washed from the air by rain in the winter and summer. The preponderance of odd-number hydrocarbons in the summer suggests that the source is mainly from plants. The more uniform

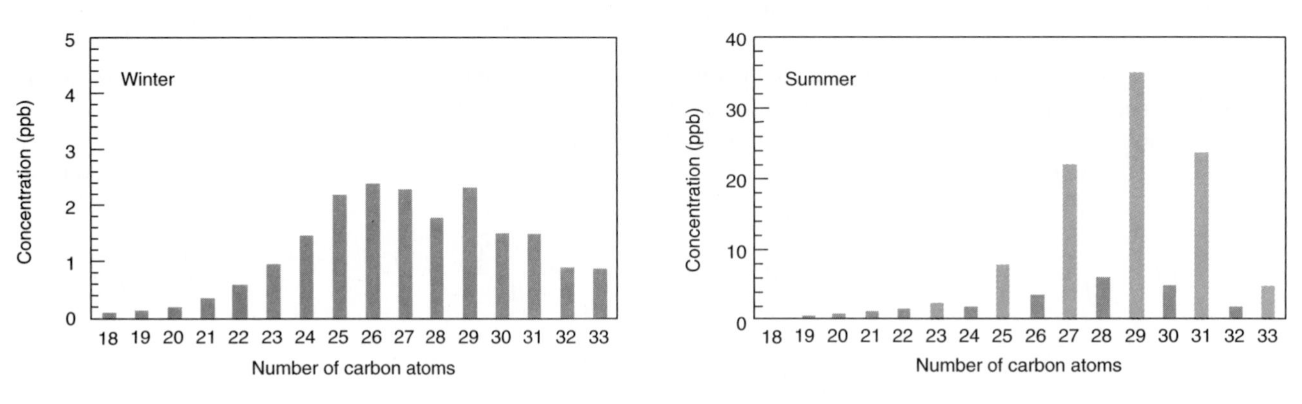

Figure 1-4 Concentrations of alkanes (hydrocarbons with the formula C_nH_{2n+2}) found in rainwater in Hannover, Germany, in the winter and summer in 1989 are measured in parts per billion (= μg hydrocarbon/L of rainwater). Summer concentrations are higher, and compounds with an odd number of carbon atoms (colored bars) predominate. Plants produce mainly hydrocarbons with an odd number of carbon atoms. [K. Levsen, S. Behnert, and H. D. Winkeler, *Fresenius J. Anal. Chem.* **1991**, *340*, 665.]

distribution of odd- and even-number hydrocarbons in the winter suggests a man-made origin. The concentration of $C_{29}H_{60}$ in summer rainwater is 34 ppb. Find the molarity of this compound in nanomoles per liter (nM).

SOLUTION A concentration of 34 ppb means 34×10^{-9} g (= 34 ng) of $C_{29}H_{60}$ per gram of rainwater, which we equate to 34 ng/mL. To find moles per liter, we first find grams per liter:

$$34 \times 10^{-9} \frac{\text{g}}{\cancel{\text{mL}}} \times \frac{1\,000\ \cancel{\text{mL}}}{\text{L}} = 34 \times 10^{-6} \frac{\text{g}}{\text{L}}$$

Because the molecular mass of $C_{29}H_{60}$ is 408.8 g/mol, the molarity is

$$\text{molarity of } C_{29}H_{60} \text{ in rainwater} = \frac{34 \times 10^{-6}\ \cancel{\text{g}}/\text{L}}{408.8\ \cancel{\text{g}}/\text{mol}} = 8.3 \times 10^{-8}\ \text{M}$$
$$= 83 \times 10^{-9}\ \text{M} = 83\ \text{nM}$$

Test Yourself The molarity of $C_{29}H_{60}$ in winter rainwater is 5.6 nM. Find the concentration in ppb. (**Answer:** 2.3 ppb)

With reference to gases, ppm usually indicates volume rather than mass. For example, 8 ppm carbon monoxide in air means 8 μL of CO per liter of air. Always write the units to avoid confusion. Figure 1-5 shows gas concentration measured in *parts per trillion* by volume (picoliters per liter).

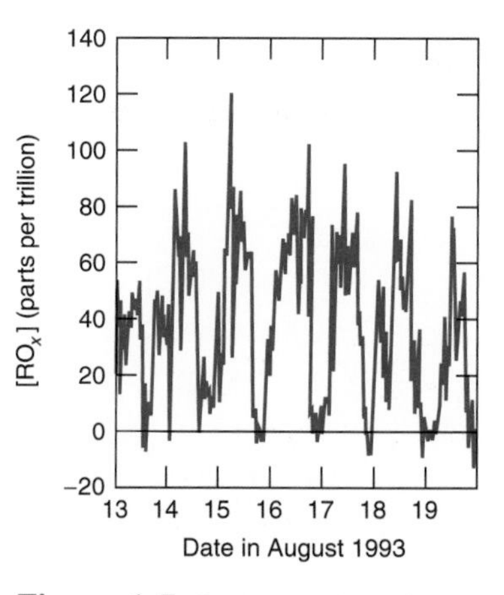

Figure 1-5 Concentration of highly reactive radicals RO_x (parts per trillion by volume = pL/L) measured outdoors at the University of Denver. RO_x refers to the combined concentrations of HO (hydroxyl radical), HO_2 (hydroperoxide radical), RO (alkoxy radical, where R is any organic group), and RO_2 (alkyl peroxide radical). These species are created mainly by photochemical reactions driven by sunlight. Concentrations peak about 2:00 P.M. each day and fall close to 0 during the night. [J. Hu and D. H. Stedman, *Anal. Chem.* **1994**, *66*, 3384.]

Ask Yourself

1-C. The density of 70.5 wt% aqueous perchloric acid is 1.67 g/mL. Note that grams refers to grams of *solution* (= g $HClO_4$ + g H_2O).
(a) How many grams of solution are in 1.00 L?
(b) How many grams of $HClO_4$ are in 1.00 L?
(c) How many moles of $HClO_4$ are in 1.00 L? This is the molarity.

1-4 Preparing Solutions

To prepare a solution with a desired molarity, weigh out the correct mass of pure reagent, dissolve it in solvent in a *volumetric flask* (Figure 1-6), dilute with more solvent to the desired final volume, and mix well by inverting the flask many times. A more complete description of the procedure is given in Section 2-5.

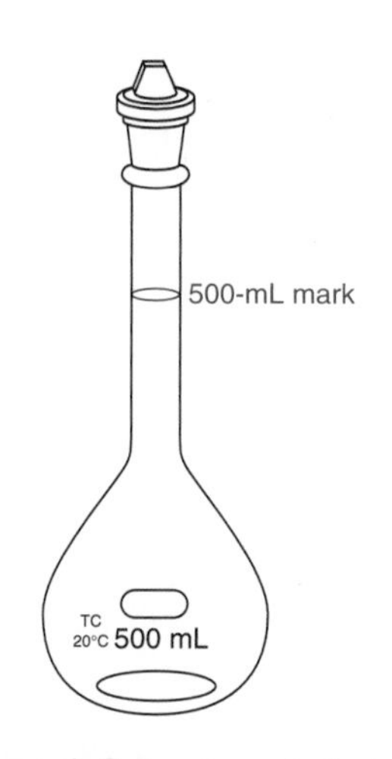

Figure 1-6 A *volumetric flask* contains a specified volume when the liquid level is adjusted to the middle of the mark in the thin neck of the flask.

Example **Preparing a Solution with Desired Molarity**

Cupric sulfate is commonly sold as the pentahydrate, $CuSO_4 \cdot 5H_2O$, which has 5 moles of H_2O for each mole of $CuSO_4$ in the solid crystal. The formula mass of $CuSO_4 \cdot 5H_2O$ (= $CuSO_9H_{10}$) is 249.69 g/mol. How many grams of $CuSO_4 \cdot 5H_2O$ should be dissolved in a 250-mL volumetric flask to make a solution containing 8.00 mM Cu^{2+}?

SOLUTION An 8.00 mM solution contains 8.00×10^{-3} mol/L. Because 250 mL is 0.250 L, we need

$$8.00 \times 10^{-3}\,\frac{\text{mol}}{\cancel{\text{L}}} \times 0.250\,\cancel{\text{L}} = 2.00 \times 10^{-3}\text{ mol CuSO}_4 \cdot 5\text{H}_2\text{O}$$

The required mass of reagent is

$$(2.00 \times 10^{-3}\,\cancel{\text{mol}})\left(249.69\,\frac{\text{g}}{\cancel{\text{mol}}}\right) = 0.499\text{ g}$$

The procedure is to weigh 0.499 g of solid $CuSO_4 \cdot 5H_2O$ into a 250-mL volumetric flask, add about 200 mL of distilled water, and swirl to dissolve the reagent. Then dilute with distilled water up to the 250-mL mark and invert the stoppered flask many times to ensure complete mixing. The solution contains 8.00 mM Cu^{2+}.

Test Yourself The anion $EDTA^{4-}$ strongly binds metal ions with a charge ≥ 2. How many grams of the reagent $Na_2H_2(EDTA) \cdot 2H_2O$ (FM 372.24 g/mol) should be dissolved in 0.500 L to give a 20.0 mM EDTA solution? What will be the molarity of Na^+ in this solution? (**Answer:** 3.72 g, 40.0 mM)

Dilute solutions can be prepared from concentrated solutions. Typically, a desired volume or mass of the concentrated solution is transferred to a volumetric flask and diluted to the intended volume with solvent. The number of moles of reagent in V liters containing M moles per liter is the product $\text{M} \cdot V = (\text{mol}/\cancel{\text{L}})(\cancel{\text{L}}) = \text{mol}$. When a solution is diluted from a high concentration to a low concentration, the number of moles of solute is unchanged. Therefore we equate the number of moles in the concentrated (conc) and dilute (dil) solutions:

Dilution formula:

$$\underbrace{\text{M}_{\text{conc}} \cdot V_{\text{conc}}}_{\text{Moles taken from concentrated solution}} = \underbrace{\text{M}_{\text{dil}} \cdot V_{\text{dil}}}_{\text{Moles placed in dilute solution}} \qquad (1\text{-}5)$$

Example Preparing 0.1 M HCl

The symbol ~ is read "approximately."

The molarity of "concentrated" HCl purchased for laboratory use is ~12.1 M. How many milliliters of this reagent should be diluted to 1.00 L to make 0.100 M HCl?

SOLUTION The required volume of concentrated solution is found with Equation 1-5:

The symbol ⇒ is read "implies that."

$$\text{M}_{\text{conc}} \cdot V_{\text{conc}} = \text{M}_{\text{dil}} \cdot V_{\text{dil}}$$
$$(12.1\text{ M}) \cdot (x\text{ mL}) = (0.100\text{ M}) \cdot (1\,000\text{ mL}) \Rightarrow x = 8.26\text{ mL}$$

It is all right to express both volumes in mL or both in L. The important point is to use the same units for volume on both sides of the equation so that the units cancel. To make 0.100 M HCl, place 8.26 mL of concentrated HCl in a 1-L volumetric flask and add ~900 mL of water. After swirling to mix, dilute to the 1-L mark with water and invert the flask many times to ensure complete mixing.

Test Yourself Concentrated nitric acid has a molarity of ~15.8 M. How many milliliters should be used to prepare 1.00 L of 1.00 M HNO_3? (**Answer:** See inside the cover of the book. Your answer will be slightly different from the printed number because of round-off errors.)

Example A More Complicated Dilution Calculation

A solution of ammonia in water is called "ammonium hydroxide" because of the equilibrium

$$\underset{\text{Ammonia}}{NH_3} + H_2O \rightleftharpoons \underset{\text{Ammonium ion}}{NH_4^+} + \underset{\text{Hydroxide ion}}{OH^-}$$

The density of concentrated ammonium hydroxide, which contains 28.0 wt% NH_3, is 0.899 g/mL. What volume of this reagent should be diluted to 500 mL to make 0.250 M NH_3?

SOLUTION To use Equation 1-5, we need to know the molarity of the concentrated reagent. The density tells us that the reagent contains 0.899 grams of solution per milliliter of solution. The weight percent tells us that the reagent contains 0.280 grams of NH_3 per gram of solution. To find the molarity of NH_3 in the concentrated reagent, we need to know the number of moles of NH_3 in 1 liter:

$$\text{grams of } NH_3 \text{ per liter} = 899\,\frac{\cancel{\text{g solution}}}{\text{L}} \times 0.280\,\frac{\text{g } NH_3}{\cancel{\text{g solution}}} = 252\,\frac{\text{g } NH_3}{\text{L}}$$

$$\text{molarity of } NH_3 = \frac{252\,\dfrac{\cancel{\text{g } NH_3}}{\text{L}}}{17.03\,\dfrac{\cancel{\text{g } NH_3}}{\text{mol } NH_3}} = 14.8\,\frac{\text{mol } NH_3}{\text{L}} = 14.8\text{ M}$$

Now we use Equation 1-5 to find the volume of 14.8 M NH_3 required to prepare 500 mL of 0.250 M NH_3:

$$M_{conc} \cdot V_{conc} = M_{dil} \cdot V_{dil}$$

$$14.8\,\frac{\text{mol}}{\text{L}} \times V_{conc} = 0.250\,\frac{\text{mol}}{\cancel{\text{L}}} \times 0.500\,\cancel{\text{L}}$$

$$\Rightarrow V_{conc} = 8.45 \times 10^{-3}\text{ L} = 8.45\text{ mL}$$

The correct procedure is to place 8.45 mL of concentrated reagent in a 500-mL volumetric flask, add about 400 mL of water, and swirl to mix. Then dilute to exactly 500 mL with water and invert the stoppered flask many times to mix well.

Test Yourself What volume of 28.0 wt% NH_3 should be diluted to 1.00 L to make 1.00 M NH_3? (**Answer:** See inside the cover of the book.)

Example **Preparing a Parts per Million Concentration**

Drinking water usually contains 1.6 ppm fluoride (F^-) to help prevent tooth decay. Consider a reservoir with a diameter of 450 m and a depth of 10 m. **(a)** How many liters of 0.10 M NaF should be added to produce 1.6 ppm F^-? **(b)** How many grams of solid NaF could be used instead?

SOLUTION **(a)** If we assume that the density of water in the reservoir is close to 1.00 g/mL, 1.6 ppm F^- corresponds to 1.6×10^{-6} g F^-/mL or

$$1.6 \times 10^{-6} \frac{\text{g F}^-}{\cancel{\text{mL}}} \times 1\,000 \frac{\cancel{\text{mL}}}{\text{L}} = 1.6 \times 10^{-3} \frac{\text{g F}^-}{\text{L}}$$

The atomic mass of fluorine is 19.00, so the desired molarity of fluoride in the reservoir is

$$\text{desired [F}^-\text{] in reservoir} = \frac{1.6 \times 10^{-3} \frac{\cancel{\text{g F}^-}}{\text{L}}}{19.00 \frac{\cancel{\text{g F}^-}}{\text{mol}}} = 8.42 \times 10^{-5}\text{ M}$$

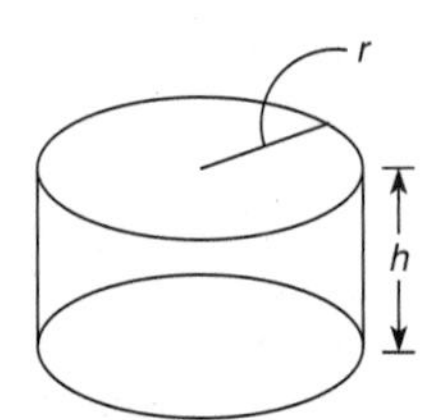

Volume of cylinder
= end area × height
= $\pi r^2 h$

The volume of the reservoir is $\pi r^2 h$, where r is the radius and h is the height.

$$\text{volume of reservoir} = \pi \times (225\text{ m})^2 \times 10\text{ m} = 1.59 \times 10^6\text{ m}^3$$

To use the dilution formula, we need to express volume in liters. Table 1-4 told us that there are 1 000 L in 1 cubic meter. Therefore the volume of the reservoir in liters is

$$\text{volume of reservoir (L)} = 1.59 \times 10^6\ \cancel{\text{m}^3} \times 1\,000 \frac{\text{L}}{\cancel{\text{m}^3}} = 1.59 \times 10^9\text{ L}$$

Finally, we are in a position to use the dilution formula 1-5:

$$\text{M}_{\text{conc}} \cdot V_{\text{conc}} = \text{M}_{\text{dil}} \cdot V_{\text{dil}}$$

$$0.10 \frac{\text{mol}}{\text{L}} \times V_{\text{conc}} = \left(8.42 \times 10^{-5} \frac{\text{mol}}{\cancel{\text{L}}}\right) \times (1.59 \times 10^9\ \cancel{\text{L}})$$

$$\Rightarrow V_{\text{conc}} = 1.3 \times 10^6\text{ L}$$

We require 1.3 million liters of 0.10 M F^-. Note that our calculation assumed that the final volume of the reservoir is 1.59×10^9 L. Even though we are adding more than 10^6 L of reagent, this amount is small relative to 10^9 L. Therefore the approximation that the reservoir volume remains 1.59×10^9 L is pretty good.

(b) The number of moles of F^- in the reservoir is $(1.59 \times 10^9\ \cancel{\text{L}}) \times (8.42 \times 10^{-5}\text{ mol}/\cancel{\text{L}}) = 1.34 \times 10^5$ mol F^-. Because 1 mole of NaF provides one mole of F^-, we need $(1.34 \times 10^5\ \cancel{\text{mol NaF}}) \times (41.99\text{ g NaF}/\cancel{\text{mol NaF}}) = 5.6 \times 10^6$ grams of NaF.

Test Yourself If the diameter of the reservoir is doubled to 900 m, how many metric tons of NaF should be added to bring the concentration of F^- to 1.6 ppm? A metric ton is 1 000 kg. (**Answer:** 22 metric tons)

Ask Yourself

1-D. A 48.0 wt% solution of HBr in water has a density of 1.50 g/mL.
(a) How many grams of solution are in 1.00 L?
(b) How many grams of HBr are in 1.00 L?
(c) What is the molarity of HBr?
(d) How much solution is required to prepare 0.250 L of 0.160 M HBr?

1-5 The Equilibrium Constant

Equilibrium describes the state that a system will reach "if you wait long enough." Most reactions of interest in analytical chemistry reach equilibrium in times ranging from fractions of a second to many minutes.

At equilibrium, the rates of the forward reaction

$$a\text{A} + b\text{B} \longrightarrow c\text{C} + d\text{D}$$

and the reverse reaction

$$c\text{C} + d\text{D} \longrightarrow a\text{A} + b\text{B}$$

are equal.

If the reactants A and B are converted to products C and D with the stoichiometry

$$a\text{A} + b\text{B} \rightleftharpoons c\text{C} + d\text{D} \quad (1\text{-}6)$$

we write the **equilibrium constant,** *K,* in the form

The equilibrium constant is more correctly expressed as a ratio of *activities* rather than of concentrations. See Section 12-2.

Equilibrium constant:

$$K = \frac{[\text{C}]^c[\text{D}]^d}{[\text{A}]^a[\text{B}]^b} \quad (1\text{-}7)$$

where the small superscript letters denote stoichiometric coefficients and each capital letter stands for a chemical species. The symbol [A] stands for the concentration of A relative to its standard state (defined below). We say that a reaction is favored whenever $K > 1$.

Equation 1-7, also called the *law of mass action,* was formulated by the Norwegians C. M. Guldenberg and P. Waage and published in 1864. Their derivation was based on the idea that the forward and reverse rates of a reaction at equilibrium must be equal.

In deriving the equilibrium constant, each quantity in Equation 1-7 is expressed as the *ratio* of the concentration of a species to its concentration in its *standard state.* For solutes, the standard state is 1 M. For gases, the standard state is 1 bar, which is very close to 1 atmosphere (Table 1-4). For solids and liquids, the standard states are the pure solid or liquid. It is understood (but rarely written) that the term [A] in Equation 1-7 really means [A]/(1 M) if A is a solute. If D is a gas, [D] really means (pressure of D in bars)/(1 bar). To emphasize that [D] means pressure of D, we usually write P_D in place of [D]. If C were a pure liquid or solid, the ratio [C]/(concentration of C in its standard state) would be unity (1) because the standard state *is* the pure liquid or solid. If [C] is a solvent, the concentration is so close to that of pure liquid C that the value of [C] is essentially 1. Each term of Equation 1-7 is dimensionless because each is a ratio in which the units cancel; therefore all equilibrium constants are dimensionless.

$1 \text{ bar} = 10^5 \text{ Pa} \approx 0.987 \text{ atm}$

The take-home lesson is this: To evaluate an equilibrium constant,

1. The concentrations of solutes should be expressed as moles per liter.
2. The concentrations of gases should be expressed in bars.
3. The concentrations of pure solids, pure liquids, and solvents are omitted because they are unity.

These conventions are arbitrary, but you must use them so that your results are consistent with tabulated values of equilibrium constants and standard reduction potentials.

Equilibrium constants are dimensionless; but, when specifying concentrations, you must use units of molarity (M) for solutes and bars for gases.

Example **Writing an Equilibrium Constant**

Write the equilibrium constant for the reaction

$$\underset{\text{Zinc}}{Zn(s)} + \underset{\text{Ammonium}}{2NH_4^+(aq)} \rightleftharpoons \underset{\text{Zinc(II)}}{Zn^{2+}(aq)} + \underset{\text{Dihydrogen}}{H_2(g)} + \underset{\text{Ammonia}}{2NH_3(aq)}$$

(In chemical equations, *s* stands for solid, *aq* stands for aqueous, *g* stands for gas, and *l* stands for liquid.)

SOLUTION Omit the concentration of pure solid and express the concentration of gas as a pressure in bars:

P_{H_2} stands for the pressure of $H_2(g)$ in bars.

$$K = \frac{[Zn^{2+}]P_{H_2}[NH_3]^2}{[NH_4^+]^2}$$

Test Yourself Write the equilibrium constant for the reaction

$$\underset{\text{Dimethyl oxalate}}{H_3C\overset{\overset{\displaystyle O}{\|}}{C}OC-\overset{\overset{\displaystyle O}{\|}}{C}OCH_3(aq)} + 2H_2O(l) \rightleftharpoons \underset{\text{Oxalate}}{C_2O_4^{2-}(aq)} + 2H^+(aq) + \underset{\text{Methanol}}{2CH_3OH(g)}$$

$$\left(\textbf{Answer: } K = \frac{[C_2O_4^{2-}][H^+]^2 P^2_{CH_3OH}}{[H_3COC(=O)-C(=O)OCH_3]}\right)$$

Manipulating Equilibrium Constants

Throughout this text, you should assume that all species in chemical equations are in aqueous solution, unless otherwise specified.

Consider the reaction of the acid HA that dissociates into H^+ and A^-:

$$HA \overset{K}{\rightleftharpoons} H^+ + A^- \qquad K = \frac{[H^+][A^-]}{[HA]}$$

If the reverse reaction is written, the new K′ is the reciprocal of the original K:

Equilibrium constant for reverse reaction:

$$H^+ + A^- \overset{K'}{\rightleftharpoons} HA \qquad \boxed{K' = 1/K} = \frac{[HA]}{[H^+][A^-]}$$

If a reaction is reversed, then $K' = 1/K$. If two reactions are added, then $K_3 = K_1K_2$.

If reactions are added, the new K is the product of the original K's. The equilibrium of H^+ between the species HA and CH^+ can be derived by adding two equations:

$$\begin{array}{llll} HA & \rightleftharpoons & \cancel{H^+} + A^- & K_1 \\ \cancel{H^+} + C & \rightleftharpoons & CH^+ & K_2 \\ \hline HA + C & \rightleftharpoons & A^- + CH^+ & K_3 \end{array}$$

Equilibrium constant for sum of reactions:

$$\boxed{K_3 = K_1K_2} = \frac{[\cancel{H^+}][A^-]}{[HA]} \cdot \frac{[CH^+]}{[\cancel{H^+}][C]} = \frac{[A^-][CH^+]}{[HA][C]}$$

If n reactions are added, the overall equilibrium constant is the product of all n individual equilibrium constants.

Example **Combining Equilibrium Constants**

From the equilibria

$$H_2O \rightleftharpoons H^+ + OH^- \qquad K_w = [H^+][OH^-] = 1.0 \times 10^{-14}$$

$$NH_3(aq) + H_2O \rightleftharpoons NH_4^+ + OH^- \qquad K_{NH_3} = \frac{[NH_4^+][OH^-]}{[NH_3(aq)]} = 1.8 \times 10^{-5}$$

H_2O is omitted from K because it is a pure liquid. Its concentration remains nearly constant.

find the equilibrium constant for the reaction

$$NH_4^+ \rightleftharpoons NH_3(aq) + H^+$$

SOLUTION The third reaction is obtained by reversing the second reaction and adding it to the first reaction:

$$\begin{array}{lcll} \cancel{H_2O} & \rightleftharpoons & H^+ + \cancel{OH^-} & K_1 = K_w \\ NH_4^+ + \cancel{OH^-} & \rightleftharpoons & NH_3(aq) + \cancel{H_2O} & K_2 = 1/K_{NH_3} \\ \hline NH_4^+ & \rightleftharpoons & H^+ + NH_3(aq) & K_3 = K_w \cdot \dfrac{1}{K_{NH_3}} = 5.6 \times 10^{-10} \end{array}$$

Test Yourself From the reactions $NH_3(aq) + H_2O \rightleftharpoons NH_4^+ + OH^-$ ($K_{NH_3} = 1.8 \times 10^{-5}$) and $CH_3NH_2(aq) + H_2O \rightleftharpoons CH_3NH_3^+ + OH^-$ ($K_{CH_3NH_2} = 4.5 \times 10^{-4}$), find the equilibrium constant for the reaction $CH_3NH_2(aq) + NH_4^+ \rightleftharpoons CH_3NH_3^+ + NH_3(aq)$. (**Answer:** 25)

Le Châtelier's Principle

Le Châtelier's principle states that if a system at equilibrium is disturbed, the direction in which the system proceeds back to equilibrium is such that the disturbance is partly offset.

Let's see what happens when we change the concentration of one species in the reaction:

$$\underset{\text{Bromate}}{BrO_3^-} + \underset{\text{Chromium(III)}}{2Cr^{3+}} + 4H_2O \rightleftharpoons \underset{\text{Bromide}}{Br^-} + \underset{\text{Dichromate}}{Cr_2O_7^{2-}} + 8H^+ \qquad (1\text{-}8)$$

for which the equilibrium constant is

$$K = \frac{[Br^-][Cr_2O_7^{2-}][H^+]^8}{[BrO_3^-][Cr^{3+}]^2} = 1 \times 10^{11} \text{ at } 25°C$$

H_2O is omitted from K because it is the solvent. Its concentration remains nearly constant.

In one particular equilibrium state of this system, the following concentrations exist:

$$[H^+] = 5.0\ M \qquad [Cr_2O_7^{2-}] = 0.10\ M \qquad [Cr^{3+}] = 0.003\,0\ M$$
$$[Br^-] = 1.0\ M \qquad [BrO_3^-] = 0.043\ M$$

Suppose that equilibrium is disturbed by increasing the concentration of dichromate from 0.10 to 0.20 M. In what direction will the reaction proceed to reach equilibrium?

According to the principle of Le Châtelier, the reaction should go in the reverse direction to partly offset the increase in dichromate, which is a product in Reaction 1-8. We can verify this algebraically by setting up the *reaction quotient*, Q, which has the same form as the equilibrium constant. The only difference is that Q is evaluated with whatever concentrations happen to exist, even though the solution is not at equilibrium. When the system reaches equilibrium, $Q = K$. For Reaction 1-8,

Q has the same form as K, but the concentrations are generally not the equilibrium concentrations.

$$Q = \frac{(1.0)(0.20)(5.0)^8}{(0.043)(0.003\,0)^2} = 2 \times 10^{11} > K$$

If $Q < K$, then the reaction must proceed to the right to reach equilibrium. If $Q > K$, then the reaction must proceed to the left to reach equilibrium.

Because $Q > K$, the reaction must go in reverse to decrease the numerator and increase the denominator, until $Q = K$.

In general,

1. If a reaction is at equilibrium and products that appear in the reaction quotient are added (or reactants that appear in the reaction quotient are removed), the reaction goes in the reverse direction (to the left).

2. If a reaction is at equilibrium and reactants that appear in the reaction quotient are added (or products that appear in the reaction quotient are removed), the reaction goes in the forward direction (to the right).

A species must appear in the reaction quotient to affect the equilibrium. If solid CaO is present in the reaction below, adding more $CaO(s)$ does not consume $CO_2(g)$.

$$\underset{\text{Calcium carbonate}}{CaCO_3(s)} \rightleftharpoons \underset{\text{Calcium oxide}}{CaO(s)} + \underset{\text{Carbon dioxide}}{CO_2(g)}$$

In equilibrium problems, we predict what must happen for a system to reach equilibrium, but not how long it will take. Some reactions are over in an instant; others do not reach equilibrium in a million years. Dynamite remains unchanged indefinitely, until a spark sets off the spontaneous, explosive decomposition. The size of an equilibrium constant tells us nothing about the rate of reaction. A large equilibrium constant does not imply that a reaction is fast.

Ask Yourself

1.E. (a) Show how the following equations can be rearranged and added to give the reaction $HOBr \rightleftharpoons H^+ + OBr^-$:

$$HOCl \rightleftharpoons H^+ + OCl^- \qquad K = 3.0 \times 10^{-8}$$

$$HOCl + OBr^- \rightleftharpoons HOBr + OCl^- \qquad K = 15$$

(b) Find the value of K for the reaction $HOBr \rightleftharpoons H^+ + OBr^-$.
(c) If the reaction $HOBr \rightleftharpoons H^+ + OBr^-$ is at equilibrium and a substance is added that consumes H^+, will the reaction proceed in the forward or reverse direction to reestablish equilibrium?

Key Equations

Molarity (M) $\qquad [A] = \dfrac{\text{moles of solute A}}{\text{liters of solution}}$

Weight percent $\qquad \text{wt\%} = \dfrac{\text{mass of solute}}{\text{mass of solution or mixture}} \times 100$

Volume percent	$\text{vol\%} = \dfrac{\text{volume of solute}}{\text{volume of solution or mixture}} \times 100$
Density	$\text{density} = \dfrac{\text{grams of substance}}{\text{milliliters of substance}}$
Parts per million	$\text{ppm} = \dfrac{\text{mass of substance}}{\text{mass of sample}} \times 10^6$
Parts per billion	$\text{ppb} = \dfrac{\text{mass of substance}}{\text{mass of sample}} \times 10^9$
Dilution formula	$M_{conc} \cdot V_{conc} = M_{dil} \cdot V_{dil}$ M_{conc} = concentration (molarity) of concentrated solution M_{dil} = concentration of dilute solution V_{conc} = volume of concentrated solution V_{dil} = volume of dilute solution
Equilibrium constant	$aA + bB \overset{K}{\rightleftharpoons} cC + dD \qquad K = \dfrac{[C]^c[D]^d}{[A]^a[B]^b}$ Concentrations of solutes are in M and gases are in bars. Omit solvents and pure solids and liquids.
Reversed reaction	$K' = 1/K$
Add two reactions	$K_3 = K_1K_2$
Le Châtelier's principle	1. Adding product (or removing reactant) drives reaction in reverse 2. Adding reactant (or removing product) drives reaction forward

Important Terms

abscissa	liter	parts per millions
atomic mass	molality	SI units
concentration	molarity	solute
density	mole	solvent
equilibrium constant	molecular mass	strong electrolyte
formal concentration	order of magnitude	volume percent
formula mass	ordinate	weak electrolyte
Le Châtelier's principle	parts per billion	weight percent

Problems

1-1. **(a)** List the SI units of length, mass, time, electric current, temperature, and amount of substance. Write the abbreviation for each.

(b) Write the units and symbols for frequency, force, pressure, energy, and power.

1-2. Write the name and number represented by each abbreviation. For example, for kW you should write kW = kilowatt = 10^3 watts. **(a)** mW **(b)** pm **(c)** kΩ **(d)** μC **(e)** TJ **(f)** ns **(g)** fg **(h)** dPa

1-3. Express the following quantities with abbreviations for units and prefixes from Tables 1-1 through 1-3: **(a)** 10^{-13} joules **(b)** $4.317\,28 \times 10^{-8}$ coulombs **(c)** $2.997\,9 \times 10^{14}$ hertz **(d)** 10^{-10} meters **(e)** 2.1×10^{13} watts **(f)** 48.3×10^{-20} moles

1-4. Table 1-4 states that 1 horsepower = 745.700 watts. Consider a 100.0-horsepower engine. Express its power output in **(a)** watts; **(b)** joules per second; **(c)** calories per second; **(d)** calories per hour.

1-5. **(a)** Refer to Table 1-4 and calculate how many meters are in 1 inch. How many inches are in 1 m?

(b) A mile is 5 280 feet and a foot is 12 inches. The speed of sound in the atmosphere at sea level is 345 m/s. Express the speed of sound in miles per second and miles per hour.

(c) There is a delay between lightning and thunder in a storm, because light reaches us almost instantaneously, but sound is slower. How many meters, kilometers, and miles away is lightning if the sound reaches you 3.00 s after the light?

1-6. Define the following measures of concentration: **(a)** molarity **(b)** molality **(c)** density **(d)** weight percent **(e)** volume percent **(f)** parts per million **(g)** parts per billion **(h)** formal concentration

1-7. What is the formal concentration (expressed as mol/L = M) of NaCl when 32.0 g are dissolved in water and diluted to 0.500 L?

1-8. If 0.250 L of aqueous solution with a density of 1.00 g/mL contains 13.7 μg of pesticide, express the concentration of pesticide in **(a)** ppm and **(b)** ppb.

1-9. The concentration of the sugar glucose ($C_6H_{12}O_6$) in human blood ranges from about 80 mg/dL before meals to 120 mg/dL after eating. Find the molarity of glucose in blood before and after eating.

1-10. Hot perchloric acid is a powerful (and potentially explosive) reagent used to decompose organic materials and dissolve them for chemical analysis.

(a) How many grams of perchloric acid, $HClO_4$, are contained in 100.0 g of 70.5 wt% aqueous perchloric acid?

(b) How many grams of water are in 100.0 g of solution?

(c) How many moles of $HClO_4$ are in 100.0 g of solution?

1-11. How many grams of boric acid [$B(OH)_3$, FM 61.83] should be used to make 2.00 L of 0.050 0 M solution?

1-12. Water is fluoridated to prevent tooth decay.

(a) How many liters of 1.0 M H_2SiF_6 should be added to a reservoir with a diameter of 100 m and a depth of 20 m to give 1.2 ppm F^-? (Remember that 1 mol H_2SiF_6 contains 6 mol F.)

(b) How many grams of solid H_2SiF_6 should be added to the same reservoir to give 1.2 ppm F^-?

1-13. How many grams of 50 wt% NaOH (FM 40.00) should be diluted to 1.00 L to make 0.10 M NaOH?

1-14. A bottle of concentrated aqueous sulfuric acid, labeled 98.0 wt% H_2SO_4, has a concentration of 18.0 M.

(a) How many milliliters of reagent should be diluted to 1.00 L to give 1.00 M H_2SO_4?

(b) Calculate the density of 98.0 wt% H_2SO_4.

1-15. How many grams of methanol (CH_3OH, FM 32.04) are contained in 0.100 L of 1.71 M aqueous methanol?

1-16. A dilute aqueous solution containing 1 ppm of solute has a density of 1.00 g/mL. Express the concentration of solute in g/L, μg/L, μg/mL, and mg/L.

1-17. The concentration of $C_{20}H_{42}$ (FM 282.55) in winter rainwater in Figure 1-4 is 0.2 ppb. Assuming that the density of rainwater is close to 1.00 g/mL, find the molar concentration of $C_{20}H_{42}$.

1-18. A 95.0 wt% solution of ethanol (CH_3CH_2OH, FM 46.07) in water has a density of 0.804 g/mL.

(a) Find the mass of 1.00 L of this solution and the grams of ethanol per liter.

(b) What is the molar concentration of ethanol in this solution?

1-19. **(a)** How many grams of nickel are contained in 10.0 g of a 10.2 wt% solution of nickel sulfate hexahydrate, $NiSO_4 \cdot 6H_2O$ (FM 262.85)?

(b) The concentration of this solution is 0.412 M. Find its density.

1-20. A 500.0-mL solution was prepared by dissolving 25.00 mL of methanol (CH_3OH, density = 0.791 4 g/mL) in chloroform. Find the molarity of the methanol.

1-21. Describe how to prepare exactly 100 mL of 1.00 M HCl from 12.1 M HCl reagent.

1-22. Cesium chloride is used to prepare dense solutions required for isolating cellular components with a centrifuge. A 40.0 wt% solution of CsCl (FM 168.36) has a density of 1.43 g/mL.

(a) Find the molarity of CsCl.

(b) How many milliliters of the concentrated solution should be diluted to 500 mL to make 0.100 M CsCl?

1-23. Protein and carbohydrates provide 4.0 Cal/g, whereas fat gives 9.0 Cal/g. (Remember that 1 Calorie, with a capital C, is really 1 kcal.) The weight percent of these components in some foods are

Food	Protein (wt%)	Carbohydrate (wt%)	Fat (wt%)
Shredded Wheat	9.9	79.9	—
Doughnut	4.6	51.4	18.6
Hamburger (cooked)	24.2	—	20.3
Apple	—	12.0	—

Calculate the number of calories per gram and calories per ounce in each of these foods. (Use Table 1-4 to convert grams into ounces, remembering that there are 16 ounces in 1 pound.)

1-24. Even though you need to express concentrations of solutes in mol/L and the concentrations of gases in bars, why do we say that equilibrium constants are dimensionless?

1-25. Write the expression for the equilibrium constant for each of the following reactions. Write the pressure of a gaseous molecule, X, as P_X.

(a) $3Ag^+(aq) + PO_4^{3-}(aq) \rightleftharpoons Ag_3PO_4(s)$

(b) $C_6H_6(l) + \frac{15}{2}O_2(g) \rightleftharpoons 3H_2O(l) + 6CO_2(g)$

1-26. For the reaction $2A(g) + B(aq) + 3C(l) \rightleftharpoons D(s) + 3E(g)$, the concentrations at equilibrium are $P_A = 2.8 \times 10^3$ Pa, $[B] = 1.2 \times 10^{-2}$ M, $[C] = 12.8$ M, $[D] = 16.5$ M, and $P_E = 3.6 \times 10^4$ torr. (There are exactly 760 torr in 1 atm.)

(a) Gas pressure should be expressed in bars when writing the equilibrium constant. Express the pressures of A and E in bars.

(b) Find the numerical value of the equilibrium constant that would appear in a table of equilibrium constants.

1-27. Suppose that the reaction $Br_2(l) + I_2(s) + 4Cl^-(aq) \rightleftharpoons 2Br^-(aq) + 2ICl_2^-(aq)$ has come to equilibrium. If more $I_2(s)$ is added, will the concentration of ICl_2^- in the aqueous phase increase, decrease, or remain unchanged?

1-28. From the reactions

$$CuN_3(s) \rightleftharpoons Cu^+ + N_3^- \qquad K = 4.9 \times 10^{-9}$$

$$HN_3 \rightleftharpoons H^+ + N_3^- \qquad K = 2.2 \times 10^{-5}$$

find the value of K for the reaction $Cu^+ + HN_3 \rightleftharpoons CuN_3(s) + H^+$.

1-29. Consider the following equilibria in aqueous solution:

(1) $Ag^+ + Cl^- \rightleftharpoons AgCl(aq)$ $\qquad K = 2.0 \times 10^3$

(2) $AgCl(aq) + Cl^- \rightleftharpoons AgCl_2^-$ $\qquad K = 93$

(3) $AgCl(s) \rightleftharpoons Ag^+ + Cl^-$ $\qquad K = 1.8 \times 10^{-10}$

(a) Find K for the reaction $AgCl(s) \rightleftharpoons AgCl(aq)$.

(b) Find $[AgCl(aq)]$ in equilibrium with excess $AgCl(s)$.

(c) Find K for $AgCl_2^- \rightleftharpoons AgCl(s) + Cl^-$.

How Would You Do It?

1-30. The ocean is a *heterogeneous* fluid with different concentrations of zinc at different depths, as shown in Figure 1-3. Suppose that you want to know how much zinc is contained in an imaginary cylinder of ocean water that is 1 m in diameter and 2 000 m deep. How would you construct a representative sample to measure the average concentration of zinc in the cylinder?

1-31. Chemical characteristics of the Naugatuck River in Connecticut were monitored by students from Sacred Heart University (J. Clark and E. Alkhatib, *Am. Environ. Lab.* February 1999, 421). The mean flow of the river is 560 cubic feet per second. The concentration of nitrate anion (NO_3^-) in the river was reported to range from 2.05 to 2.50 milligrams of nitrate nitrogen per liter during dry weather and from 0.81 to 4.01 mg nitrate nitrogen per liter during wet weather at the outlet of the river. (The unit "mg nitrate nitrogen" refers to the mass of nitrogen in the nitrate anion.) Estimate how many metric tons of nitrate anion per year flow from the river (1 metric ton = 1 000 kg). State your assumptions.

A Quartz-Crystal Sensor with an Imprinted Layer for Yeast Cells

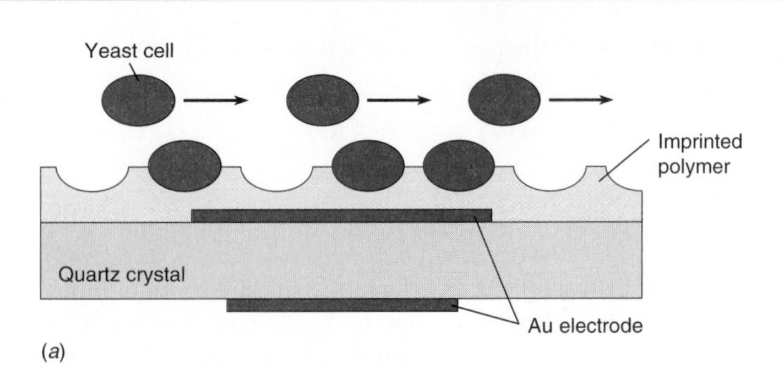

Quartz crystal with gold electrodes. One electrode is overcoated with polymer imprinted by a particular strain of yeast cells.

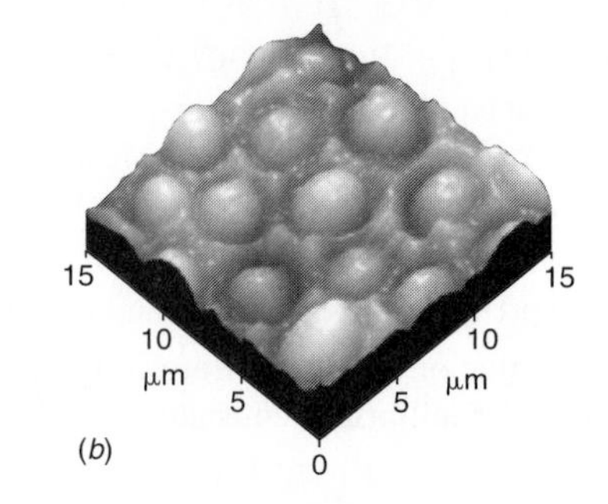

Atomic force micrograph showing yeast cells embedded in polymer layer.

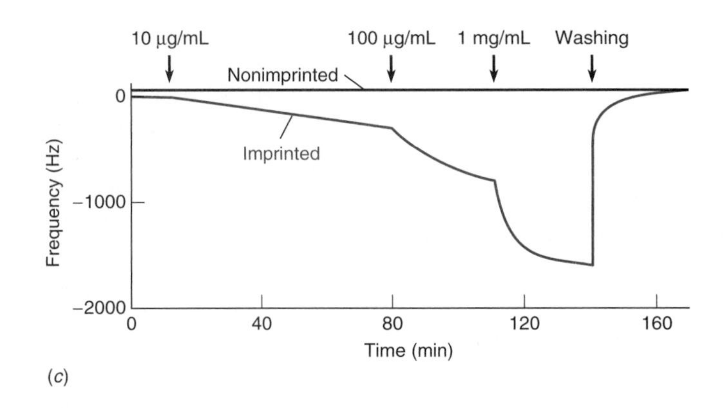

Response of crystal oscillation frequency to changes in concentration of yeast cells in liquid flowing past sensor. If the electrode is coated with unimprinted polymer, there is no response to yeast cells. [From F. L. Dickert and O. Hayden, *Anal. Chem.* **2002**, *74*, 1302.]

The laboratory balance, which we study in this chapter, descends from the earliest apparatus employed in quantitative chemical analysis. Quartz crystals, such as the one that keeps time in your wristwatch, are among the most sensitive modern means to measure very small masses.

A quartz crystal excited by an oscillating electric field has a highly reproducible vibrational frequency. When a substance is *adsorbed* (bound) on the crystal's surface, the vibrational frequency of the crystal decreases in proportion to the adsorbed mass.

Panels (*a*) and (*b*) show a polymer layer formed in the presence of yeast cells such that the polymer has the shape of the cells. The polymer is formed on top of a gold electrode used to apply a 10-MHz oscillating electric field to the quartz crystal. Cells can be removed from the polymer by hot water. Panel (*c*) shows that the oscillation frequency of the crystal decreases when the sensor is exposed to increasing concentrations of yeast cells.

Sensitive detectors for DNA or proteins can be made by attaching complementary DNA or antibodies to a gold electrode on a quartz crystal. When specific DNA or proteins in solution bind to the DNA or antibody on the electrode, the oscillation frequency decreases. From the change in frequency, we can calculate the mass that was bound.

CHAPTER 2

Tools of the Trade

Analytical chemists use a range of equipment, from simple glassware to complex instruments that measure spectroscopic or electrical properties of analytes. The object to be analyzed might be as large as a vein of ore in a mountainside or as small as a vesicle inside a living cell. This course should expose you to some of the instrumental techniques of modern analytical chemistry. Along the way, you must gain some understanding and proficiency in "wet" laboratory operations with simple glassware. *The most sophisticated instruments are useless if you cannot prepare accurate standards for calibration or accurate, representative samples of an unknown for analysis.* This chapter describes basic laboratory apparatus and manipulations associated with chemical measurements.

2-1 Safety, Waste Disposal, and Green Chemistry

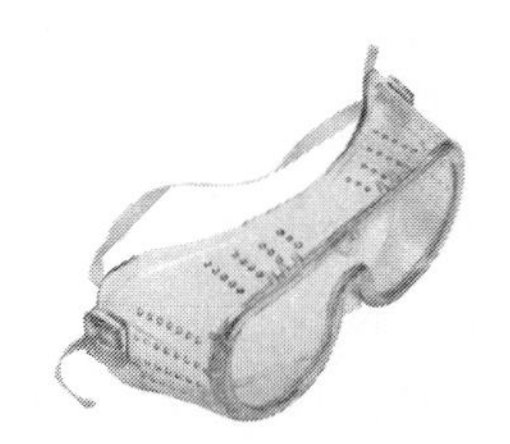

Figure 2-1 Goggles or safety glasses with side shields are required in every laboratory.

The primary safety rule is to do nothing that you (or your instructor) consider to be dangerous. If you believe that an operation is hazardous, discuss it with your instructor and do not proceed until sensible procedures and precautions are in place. If you still consider an activity to be too dangerous, don't do it.

Before beginning work, you should be familiar with safety precautions appropriate to your laboratory. Wear goggles (Figure 2-1) or safety glasses with side shields at all times in the lab to protect yourself from flying chemicals and glass. Even if you are very careful, one of your neighbors may be more accident prone. Wear a flame-resistant lab coat, long pants, and shoes that cover your feet to protect yourself from spills and flames. Rubber gloves can protect you when pouring concentrated acids, but organic solvents can penetrate rubber. Food and chemicals should not mix: Don't bring food or drink into the lab.

Limitations of gloves In 1997, popular Dartmouth College chemistry professor Karen Wetterhahn, age 48, died from a drop of dimethylmercury absorbed through the latex rubber gloves that she was wearing. Many organic compounds readily penetrate rubber. Wetterhahn was an expert in the biochemistry of metals and the first female professor of chemistry at Dartmouth. She had two children and played a major role in bringing more women into science and engineering.

Treat chemical spills on your skin *immediately* by flooding the affected area with water and then seek medical attention. Clean up spills on the bench, floor, or reagent bottles immediately to prevent accidental contact by the next person who comes along.

Solvents and concentrated acids that produce harmful fumes should be handled in a fume hood that sweeps vapors away from you and out through a vent on the roof. The hood is not meant to transfer toxic vapors from the chemistry building to the cafeteria. Never generate large quantities of toxic fumes in the hood. If you use a toxic gas in a fume hood, bubble excess gas through a chemical trap or burn it in a flame to prevent its escape from the hood.

Label every vessel to show what it contains. Without labels, you *will* forget what is in some containers. Unlabeled waste is extremely expensive to discard, because you must analyze the contents before you can legally dispose of it. Chemically incompatible wastes should never be mixed.

If we want our grandchildren to inherit a habitable planet, we need to minimize waste production and dispose of chemical waste in a responsible manner.[1] When it is economically feasible, recycling of chemicals is preferable to waste disposal. Carcinogenic dichromate ($Cr_2O_7^{2-}$) waste provides an example of an accepted disposal strategy. Cr(VI) from dichromate should be reduced to Cr(III) with sodium hydrogen sulfite ($NaHSO_3$) and precipitated with hydroxide as $Cr(OH)_3$. The solution is then evaporated to dryness and the solid is discarded in an approved landfill that is lined to prevent escape of the chemicals. Wastes such as silver and gold that can be economically recycled should be chemically treated to recover the metal.

Green chemistry is a set of principles intended to change our behavior in a manner that will help sustain the habitability of Earth.[2] Examples of unsustainable behavior are to consume a limited resource and to dispose of waste in a manner that poisons our air, water, or land. Green chemistry seeks to design chemical products and processes to reduce the use of resources and energy and the generation of hazardous waste. It is better to design a process to prevent waste than to dispose of waste. If possible, use resources that are renewable and generate waste that is not hazardous. For analytical chemistry, it is desirable to design analytical procedures to consume minimal quantities of chemicals and solvents and to substitute less toxic substances for more toxic substances. "Microscale" classroom experiments are encouraged to reduce the cost of reagents and the generation of waste.

2-2 Your Lab Notebook

The lab notebook must

1. State what was done
2. State what was observed
3. Be understandable to someone else

Without a doubt, somebody reading this book today is going to make an important discovery in the future and will seek a patent. The lab notebook is your legal record of your discovery. Therefore, each notebook page should be signed and dated. Anything of potential importance should also be signed and dated by a second person.

Do not rely on a computer for long-term storage of information. Even if a file survives, software or hardware required to read the file will become obsolete.

The critical functions of your lab notebook are to state *what you did* and *what you observed*, and it should be *understandable by a stranger* who is trained in your discipline (chemistry in our case). The greatest error is writing ambiguous notes. After a few years, you may not be able to interpret your own notebook when memories of the experiment have faded. Writing in *complete sentences* is an excellent way to reduce this problem. Box 2-1 gives an example.

The measure of scientific "truth" is the ability to reproduce an experiment. A good lab notebook will allow you or anyone else to duplicate an experiment in the exact manner in which it was conducted the first time.

Beginning students find it useful (or required!) to write a complete description of an experiment, with sections describing the purpose, methods, results, and conclusions. Arranging your notebook to accept numerical data prior to coming to the lab is an excellent way to prepare for an experiment.

It is good practice to write a balanced chemical equation for every reaction that you use. This helps you understand what you are doing and may point out what you do not understand.

Record in your notebook the names of computer files where programs and data are stored. *Printed copies* of important data collected on a computer should be pasted into your notebook. The lifetime of a printed page is 10 to 100 times greater than that of a computer file.

Box 2-1 Dan's Lab Notebook Entry

Your lab notebook should (1) state what was done, (2) state what was observed, and (3) be understandable to someone else who is trained in your discipline. The passage below was extracted in 2002 from my notebook of 1974 when, as a "postdoc" at Albert Einstein College of Medicine, I began to isolate the iron storage protein ferritin. The complete procedure, in which protein was isolated and its purity assessed, occupied 3 weeks and 17 notebook pages. Phrases in brackets were added to help you understand the passage. I do not doubt that you can improve this description.

14 Sept 1974

Isolation of Human Spleen Ferritin

Based on R. R. Crichton et al., Biochem. J. 131, 51 (1973).

Procedure: Mince and homogenize spleen in ~4 vol H_2O
Heat to 70° for 5 min and cool on ice
Centrifuge at 3300×g for 20 min
Filter through filter paper
Precipitate with 50% $(NH_4)_2SO_4$ (= 313 g solid/L solution)
Centrifuge at 3300×g for 20 min
Dissolve in H_2O and dialyze vs. 0.1 M Tris, pH 8
Chromatograph on Sepharose 6B

Today's procedure started with a frozen 41 g human half spleen thawed overnight at 4°. The spleen was healthy and taken from an autopsy about a month ago. The spleen was blended 2 min on the high setting of the Waring blender in a total volume of ~250 mL. Try a smaller volume next time. The mixture was heated to 70–73° in a preheated water bath. It took ~5 min to attain 70° and the sample was then left at that temperature with intermittent stirring for 5 min. It was then cooled in an ice bath to ~10° before centrifugation in the cold at 3300×g for 20 min (GSA head—4500 rpm). The red supernatant was filtered through Whatman #1 filter paper to give 218 mL solution, pH 6.4. The pH was raised to 7.5 with 10 M KOH and maintained between 7–8 during the addition of 68.2 g (50% saturation) $(NH_4)_2SO_4$. The solution [with precipitated protein] was left at RT [room temperature] overnight with 60 mg NaN_3 [a preservative]. Final pH = 7.6.

Later, there are tables of numerical data, graphs of results, and original, *well-labeled* instrument output pasted permanently into the notebook.

Ask Yourself

2-A. What are the three essential attributes of a lab notebook?

2-3 The Analytical Balance

Balances are delicate and expensive. Be gentle when you place objects on the pan and when you adjust the knobs. A balance should be calibrated by measuring a set of standard weights at least every year.

Figure 2-2 shows a typical analytical **electronic balance** with a capacity of 100–200 g and a sensitivity of 0.01–0.1 mg. *Sensitivity* refers to the smallest increment of mass that can be measured. A *microbalance* weighs milligram quantities with a sensitivity of 0.1 μg. An electronic balance works by generating an electromagnetic force to

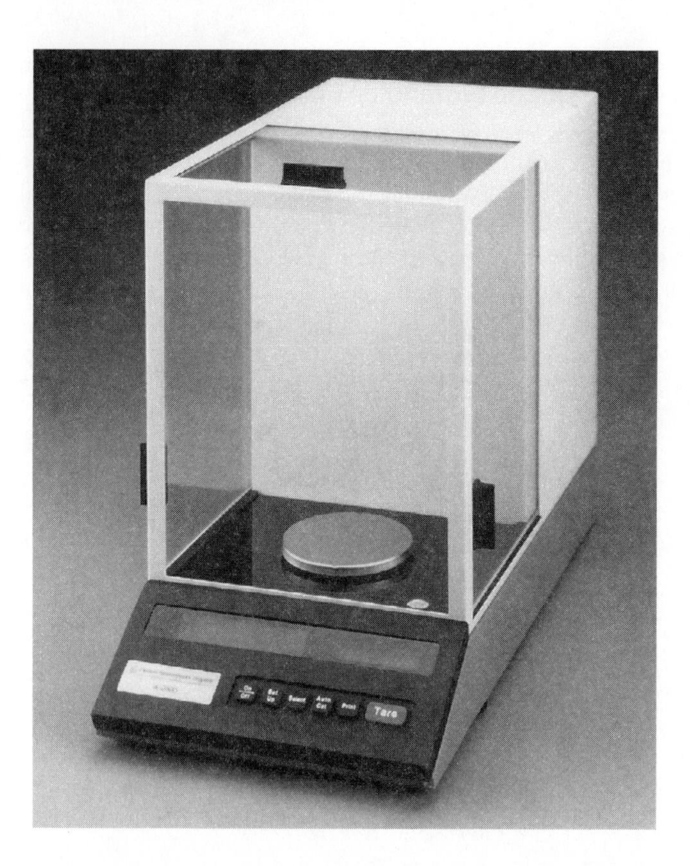

Figure 2-2 Analytical electronic balance. Good-quality balances calibrate themselves with internal weights to correct for variations in the force of gravity, which can be as great as 0.3% from place to place. [Fisher Scientific, Pittsburgh, PA.]

balance the gravitational force acting on the object being weighed. The current required by the electromagnet is proportional to the mass being weighed.

Figure 2-3 shows the principle of operation of a single-pan **mechanical balance.** The balance beam is suspended on a sharp *knife edge*. The mass of the pan hanging from the balance point (another knife edge) at the left is balanced by a counterweight at the right. You place the object to be weighed on the pan and adjust knobs that remove standard weights from the bar above the pan. The balance beam

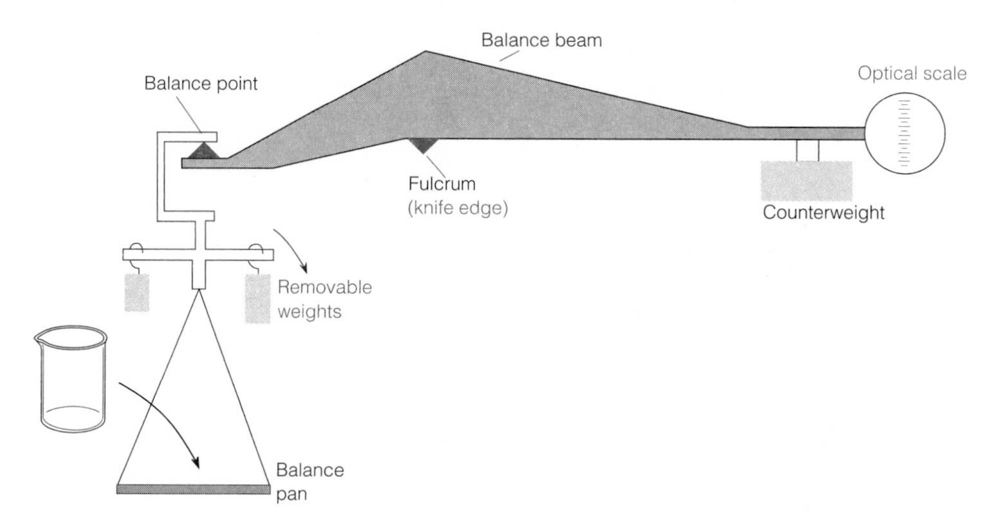

Figure 2-3 Single-pan mechanical balance. After placing an object on the pan, we detach removable weights until the balance beam is restored as close as possible to its original position. The remaining small difference is read on the optical scale.

is restored close to its original position when the weights removed from the bar are nearly equal to the mass on the pan. The slight difference from the original position is shown on an optical scale, whose reading is added to that of the knobs.

A mechanical balance should be in its arrested position when you load or unload the pan and in the half-arrested position when you are dialing weights. This practice prevents abrupt forces that would wear down the knife edges and decrease the sensitivity of the balance.

Using a Balance

To weigh a chemical, place a clean receiving vessel on the balance pan. The mass of the empty vessel is called the **tare**. On most electronic balances, the tare can be set to 0 by pressing a button. Add chemical to the vessel and read the new mass. If there is no automatic tare operation, the mass of the empty vessel should be subtracted from that of the filled vessel. Do not place chemicals directly on the weighing pan. This precaution protects the balance from corrosion and allows you to recover all the chemical being weighed.

An alternate procedure, called "weighing by difference," is necessary for **hygroscopic** reagents, which rapidly absorb moisture from the air. First weigh a capped bottle containing dry reagent. Then quickly pour some reagent from the weighing bottle into a receiver. Cap the weighing bottle and weigh it again. The difference is the mass of reagent delivered. With an electronic balance, set the initial mass of the weighing bottle to 0 with the tare button. Then deliver reagent from the bottle and reweigh the bottle. The negative reading on the balance is the mass of reagent delivered from the bottle.[3]

Clean up spills on the balance and do not allow chemicals to get into the mechanism below the pan. Use a paper towel or tissue to handle the vessel that you are weighing, because fingerprints will change its mass. Samples should be at *ambient temperature* (the temperature of the surroundings) when weighed to prevent errors due to convective air currents. The doors of the balance in Figure 2-2 must be closed during weighing so that air currents do not disturb the pan. A top-loading balance without sliding doors has a fence around the pan to deflect air currents. Sensitive balances should be placed on a heavy table, such as a marble slab, to minimize the effect of vibrations on the reading. Use the bubble meter and adjustable feet of a balance to keep it level.

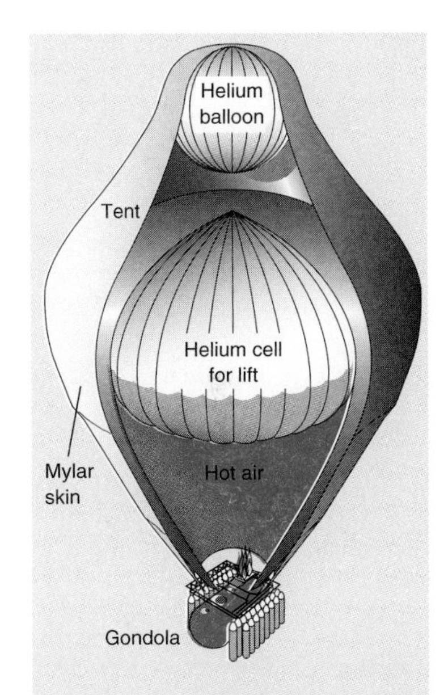

The *Breitling Orbiter 3* in 1999 was the first balloon to fly around the world. Previous balloons could not carry enough propane fuel for such a trip. The design of the *Breitling Orbiter 3* keeps the temperature of the large, inner helium cell as constant as possible, so the buoyancy variation between the warm day and the cold night is minimal. During the day, the sun heats the helium cell, which expands, thereby increasing its buoyancy and its ability to keep the vessel aloft. If the temperature rises too much, a solar-powered fan brings in cool air to keep the vessel from rising to an undesired altitude. At night, the helium cell cools and shrinks, which reduces its buoyancy. To keep the balloon aloft at night, heat from burning propane is required. The double-wall design reduces radiational cooling of the helium cell and decreases the requirement for propane.

Buoyancy

When you swim, your weight in the water is nearly zero, which is why people can float. **Buoyancy** is the upward force exerted on an object in a liquid or gaseous fluid. An object weighed in air appears lighter than its actual mass by an amount equal to the mass of air that it displaces. The true mass is the mass measured in vacuum. The standard weights in a balance also are affected by buoyancy, so they weigh less in air than they would in vacuum. A buoyancy error occurs whenever the density of the object being weighed is not equal to the density of the standard weights.

If mass m' is read from a balance, the true mass m is

Buoyancy equation:

$$m = \frac{m'\left(1 - \dfrac{d_a}{d_w}\right)}{\left(1 - \dfrac{d_a}{d}\right)} \tag{2-1}$$

where d_a is the density of air (0.001 2 g/mL near 1 bar and 25°C); d_w is the density of balance weights (8.0 g/mL); and d is the density of the object being weighed.

Example **Buoyancy Correction**

Find the true mass of water (density = 1.00 g/mL) if the apparent mass is 100.00 g.

SOLUTION Equation 2-1 gives the true mass:

$$m = \frac{100.00 \text{ g}\left(1 - \frac{0.001\,2 \text{ g/mL}}{8.0 \text{ g/mL}}\right)}{\left(1 - \frac{0.001\,2 \text{ g/mL}}{1.00 \text{ g/mL}}\right)} = 100.11 \text{ g}$$

Test Yourself Find the true mass of 28.0 wt% ammonia (density = 0.90 g/mL) when the apparent mass is 20.000 g. (**Answer:** 20.024 g)

The buoyancy error for water is 0.11%, which is significant for many purposes. For solid NaCl with a density of 2.16 g/mL, the error is 0.04%.

Ask Yourself

2-B. (a) Buoyancy corrections are most critical when you calibrate glassware such as a volumetric flask to see how much volume it actually holds. Suppose that you fill a 25-mL volumetric flask with distilled water and find that the mass of water in the flask measured in air is 24.913 g. What is the true mass of the water?
(b) You made the measurement when the lab temperature was 21°C, at which the density of water is 0.998 00 g/mL. What is the true volume of water contained in the volumetric flask?

2-4 Burets

A **buret**[4] is a precisely manufactured glass tube with graduations enabling you to measure the volume of liquid delivered through the *stopcock* (the valve) at the bottom (Figure 2-4*a*). The numbers on the buret increase from top to bottom (with 0 mL near the top). A volume measurement is made by reading the level before and after draining liquid from the buret and subtracting the first reading from the second reading. The graduations of Class A burets (the most accurate grade) are certified to meet the tolerances in Table 2-1. For example, if the reading of a 50-mL buret is 32.50 mL, the true volume can be anywhere in the range 32.45 to 32.55 mL and still be within the manufacturer's stated tolerance of ±0.05 mL.

Table 2-1 Tolerances of Class A burets

Buret volume (mL)	Smallest graduation (mL)	Tolerance (mL)
5	0.01	±0.01
10	0.05 *or* 0.02	±0.02
25	0.1	±0.03
50	0.1	±0.05
100	0.2	±0.10

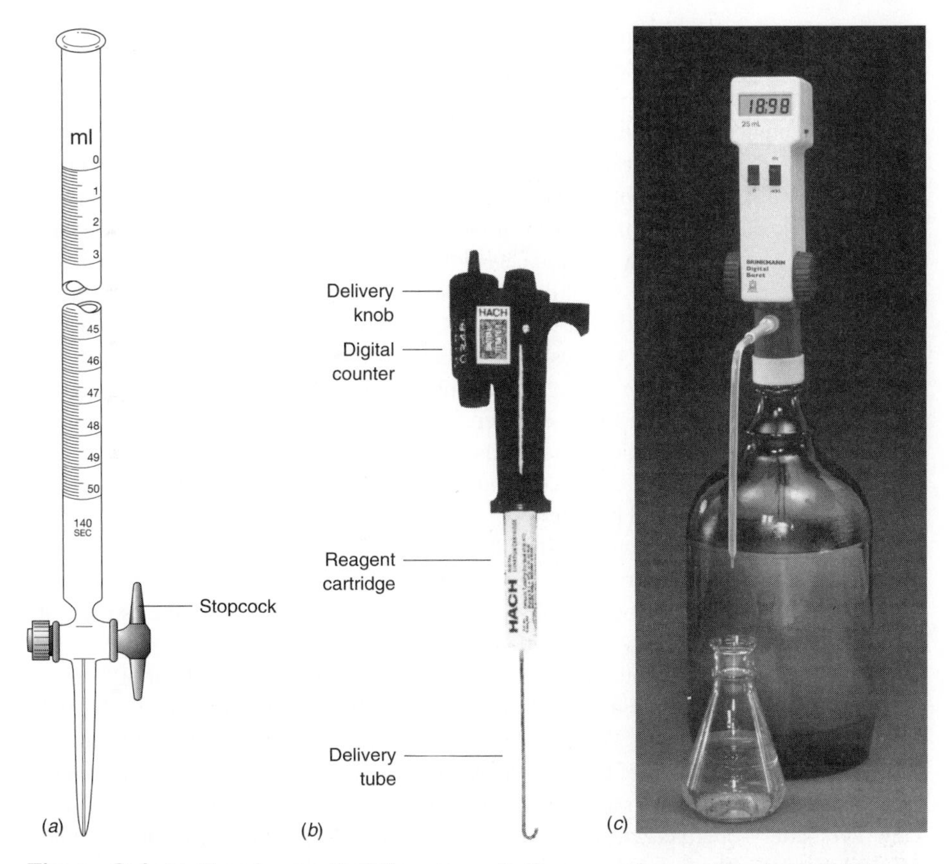

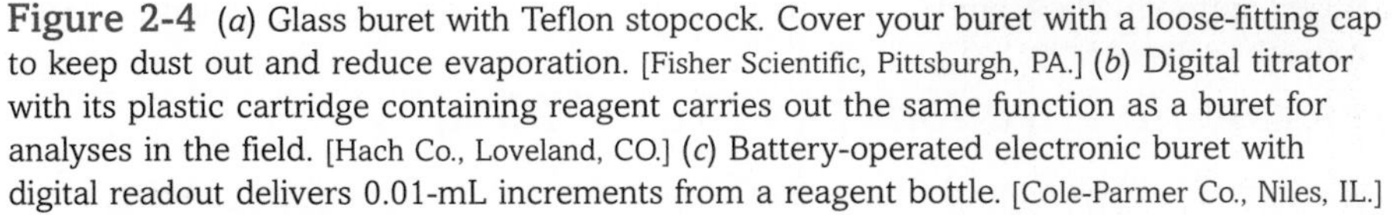

Figure 2-4 (*a*) Glass buret with Teflon stopcock. Cover your buret with a loose-fitting cap to keep dust out and reduce evaporation. [Fisher Scientific, Pittsburgh, PA.] (*b*) Digital titrator with its plastic cartridge containing reagent carries out the same function as a buret for analyses in the field. [Hach Co., Loveland, CO.] (*c*) Battery-operated electronic buret with digital readout delivers 0.01-mL increments from a reagent bottle. [Cole-Parmer Co., Niles, IL.]

When reading the liquid level in a buret, your eye should be at the same height as the top of the liquid. If your eye is too high, the liquid seems to be higher than it actually is. If your eye is too low, the liquid appears too low. The error that occurs when your eye is not at the same height as the liquid is called **parallax error**.

The **meniscus** is the curved upper surface of liquid in the glass buret in Figure 2-5. Water has a concave meniscus because water is attracted to glass and climbs slightly up the glass. It is helpful to use black tape on a white card as a background for locating the meniscus. Align the top of the tape with the bottom of the meniscus and read the position on the buret. Highly colored solutions may appear to have a double meniscus, either of which you can use. Volume is determined by subtracting one reading from another, so the important point is to read the position of the meniscus reproducibly. Always estimate the reading to the nearest tenth of the division between marks.

The thickness of a graduation line on a 50-mL buret corresponds to about 0.02 mL. To use the buret most accurately, consider the *top* of a graduation line to be 0. When the meniscus is at the bottom of the same graduation line, the reading is 0.02 mL greater.

A drop from a 50-mL buret is about 0.05 mL. Near the end point of a titration, try to deliver less than one drop at a time so that you can locate the end point more precisely than ± 0.05 mL. To deliver a fraction of a drop, carefully open the stopcock until part of a drop is hanging from the buret tip. Then touch the inside wall of the receiving flask to the buret tip to transfer the droplet to the flask. Carefully tip the flask

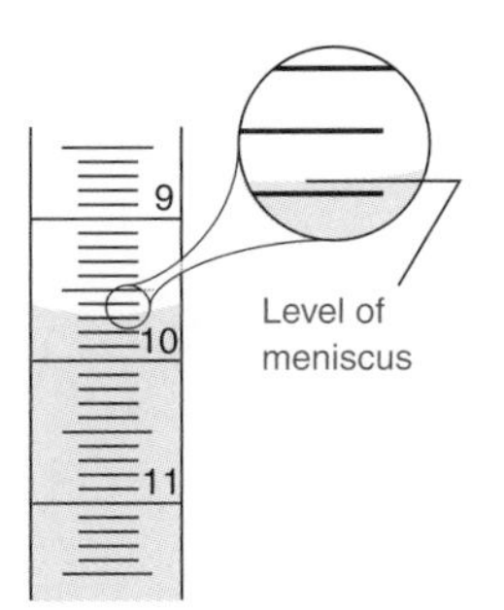

Figure 2-5 Buret with the meniscus at 9.68 mL. Estimate the reading of any scale to the nearest tenth of a division. Because this buret has 0.1-mL divisions, we estimate the reading to 0.01 mL.

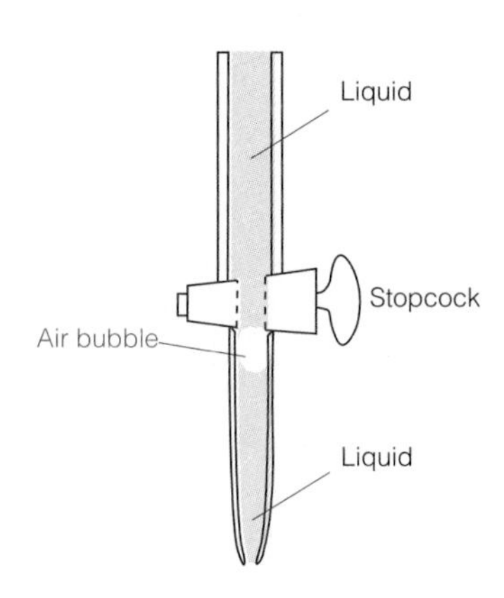

Figure 2-6 Air bubble beneath the stopcock should be expelled before you use a buret.

Operating a buret:

- read bottom of concave meniscus
- estimate reading to 1/10 of a division
- avoid parallax
- account for graduation thickness in readings
- drain liquid slowly
- wash buret with new solution
- deliver fraction of a drop near end point
- eliminate air bubble before use

so that the main body of liquid washes over the newly added droplet. Then swirl the flask to mix the contents. Near the end of a titration, tip and rotate the flask often to ensure that droplets on the wall containing unreacted analyte contact the bulk solution.

Liquid should drain evenly down the wall of a buret. The tendency of liquid to stick to the glass is reduced by draining the buret slowly (<20 mL/min). If many droplets stick to the wall, clean the buret with detergent and a buret brush. If this cleaning is insufficient, soak the buret in peroxydisulfate–sulfuric acid cleaning solution prepared by your instructor.[5] Cleaning solution eats clothing and people, as well as grease in the buret. Volumetric glassware should not be soaked in alkaline solutions, which attack glass. (A 5 wt% NaOH solution at 95°C dissolves Pyrex glass at a rate of 9 μm/h.)

A common buret error is caused by failure to expel the air bubble often found beneath the stopcock (Figure 2-6). A bubble present at the start of the titration may be filled with liquid during the titration. Therefore some volume that drained out of the graduated part of the buret did not reach the titration vessel. Usually the bubble can be dislodged by draining the buret for a second or two with the stopcock wide open. A tenacious bubble can be expelled by carefully shaking the buret while draining it into a sink.

Before you fill a buret with fresh solution, it is a wonderful idea to rinse the buret several times with small portions of the new solution, discarding each wash. It is not necessary to fill the entire buret with wash solution. Simply tilt the buret so that its whole surface contacts the wash liquid. This same technique should be used with any vessel (such as a spectrophotometer cuvet or a pipet) that is reused without drying.

The *digital titrator* in Figure 2-4*b* is useful for conducting titrations in the field where samples are collected. The counter tells how much reagent from the cartridge has been dispensed by rotation of the delivery knob. Its accuracy of 1% is 10 times poorer than the accuracy of a glass buret, but many measurements do not require higher accuracy. The battery-operated *electronic buret* in Figure 2-4*c* fits on a reagent bottle and delivers up to 99.99 mL in 0.01-mL increments displayed on a digital readout.

Microscale Titrations (A "Green" Idea)

"Microscale" experiments decrease costs, consumption of reagents, and generation of waste. A student buret can be constructed from a 2-mL pipet graduated in 0.01-mL intervals.[6] Volume can be read to 0.001 mL, and titrations can be carried out with a precision of 1%.

2-5 Volumetric Flasks

A **volumetric flask** (Figure 2-7, Table 2-2) is calibrated to contain a particular volume of solution at 20°C when the bottom of the meniscus is adjusted to the center of the mark on the neck of the flask (Figure 2-8). Most flasks bear the label "TC 20°C," which means *to contain* at 20°C. (Other types of glassware may be calibrated *to deliver,* "TD," their indicated volume.) The temperature of the container is relevant because liquid and glass expand when heated.

We use a volumetric flask to prepare a solution of known volume. Typically, reagent is weighed into the flask, dissolved, and diluted to the mark. The mass of reagent and final volume are therefore known. Dissolve the reagent in the flask in *less* than the final volume of liquid. Add more liquid and mix the solution again. Make the final volume adjustment with as much well-mixed liquid in the flask as possible. (When two different liquids are mixed, there is generally a small volume change. The total volume is *not* the sum of the two volumes that were mixed. By swirling the liquid in a nearly full volumetric flask before the liquid reaches the thin neck, you

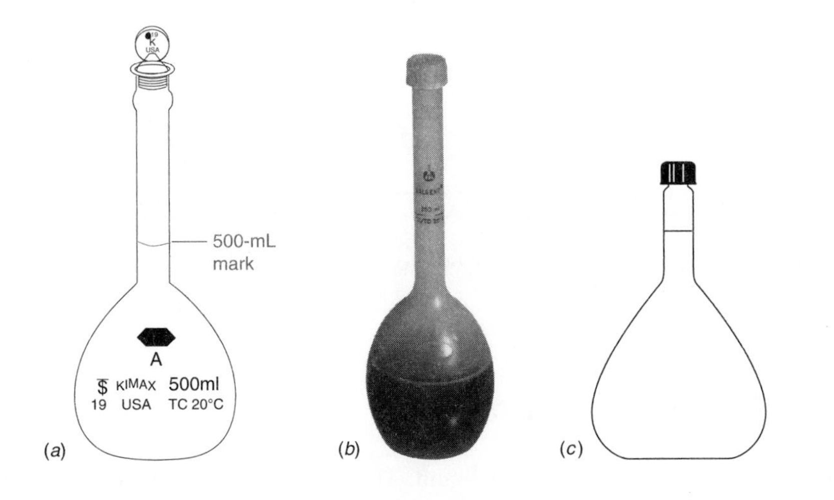

Figure 2-7 (*a*) Class A glass volumetric flask meets tolerances in Table 2-2. [A. H. Thomas Co., Philadelphia, PA.] (*b*) Class B polypropylene plastic flask for trace analysis (ppb concentrations) in which analyte might be lost by adsorption (sticking) on glass or contaminated with previously adsorbed species. A plastic flask is also required for reagents such as HF or hot, basic solutions that react with glass. Class B flasks are less accurate than Class A flasks, with twice the tolerances in Table 2-2. [Fisher Scientific, Pittsburgh, PA.] (*c*) Short-form volumetric flask with Teflon-lined screw cap fits on analytical balance. Teflon prevents solutions from attacking the inside of the cap.

minimize the change in volume when the last liquid is added.) For best control, add the final drops of liquid with a pipet, *not a squirt bottle*. After adjusting the liquid to the correct level, hold the cap firmly in place and invert the flask several times to complete mixing. Before the liquid is homogeneous, we observe streaks (called *schlieren*) arising from regions that refract light differently. After the schlieren are gone, invert the flask a few more times to ensure complete mixing.

Glass is notorious for *adsorbing* traces of chemicals—especially cations. **Adsorption** means to stick to the surface. (In contrast, **absorption** means to take inside, as a sponge takes up water.) For critical work, **acid wash** the glassware to replace low concentrations of cations on the glass surface with H^+. To do this, soak already thoroughly cleaned glassware in 3–6 M HCl or HNO_3 (in a fume hood) for >1 h, followed by several rinses with distilled water and a final soak in distilled water. The HCl can be reused many times, as long as it is only used for soaking clean glassware.

Adsorption: to bind a substance on the surface
Absorption: to bind a substance internally

As an example, high purity nitric acid was delivered from a glass pipet that had been washed normally without acid and another that had been acid washed. The level of the transition elements Ti, Cr, Mn, Fe, Co, Ni, Cu, and Zn in acid delivered from the acid-washed pipet was below the detection level of 0.01 ppb (0.01 ng/g). The concentration of each transition element in acid delivered from the pipet that had not been acid washed was in the range 0.5 to 9 ppb.[7]

Ask Yourself

2-C. How would you use a volumetric flask to prepare 250.0 mL of 0.150 0 M K_2SO_4?

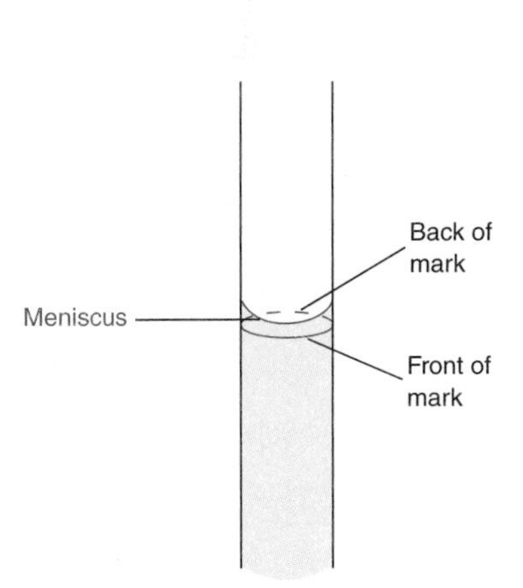

Figure 2-8 Proper position of the meniscus: at the center of the ellipse formed by the front and back of the calibration mark when viewed from above or below the level of the mark. Volumetric flasks and transfer pipets are calibrated to this position.

Table 2-2 Tolerances of Class A volumetric flasks

Flask capacity (mL)	Tolerance (mL)	Flask capacity (mL)	Tolerance (mL)
1	±0.02	100	±0.08
2	±0.02	200	±0.10
5	±0.02	250	±0.12
10	±0.02	500	±0.20
25	±0.03	1 000	±0.30
50	±0.05	2 000	±0.50

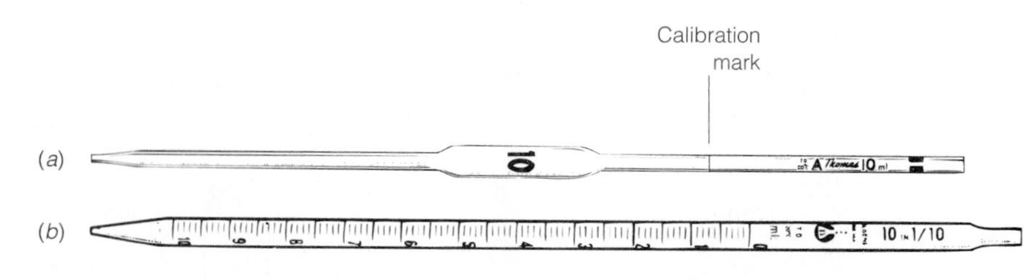

Figure 2-9 (*a*) Transfer pipet. Do not blow out the last drop. (*b*) Measuring (Mohr) pipet. [A. H. Thomas Co., Philadelphia, PA.]

2-6 Pipets and Syringes

Pipets deliver known volumes of liquid. The *transfer pipet* in Figure 2-9 is calibrated to deliver one fixed volume. The last drop of liquid does not drain out of the pipet; *it should not be blown out.* The *measuring pipet* is calibrated to deliver a variable volume, which is the difference between the initial and final volumes. A measuring pipet could be used to deliver 5.6 mL by starting delivery at the 1-mL mark and terminating at the 6.6-mL mark.

Do not blow the last drop out of a transfer pipet.

A transfer pipet is more accurate than a measuring pipet. Tolerances for Class A (the most accurate grade) transfer pipets in Table 2-3 are the allowed error in the volume that is actually delivered.

Table 2-3 Tolerances of Class A transfer pipets

Volume (mL)	Tolerance (mL)
0.5	±0.006
1	±0.006
2	±0.006
3	±0.01
4	±0.01
5	±0.01
10	±0.02
15	±0.03
20	±0.03
25	±0.03
50	±0.05
100	±0.08

Using a Transfer Pipet

Using a rubber bulb, *not your mouth,* suck liquid up past the calibration mark. It is a good idea to discard one or two pipet volumes of liquid to remove traces of previous reagents from the pipet. After taking up a third volume past the calibration mark, quickly replace the bulb with your index finger at the end of the pipet. The liquid should still be above the mark after this maneuver. Pressing the pipet against the bottom of the vessel while removing the rubber bulb helps prevent liquid from draining while you put your finger in place. Wipe the excess liquid off the outside of the pipet with a clean tissue. *Touch the tip of the pipet to the side of a beaker* and drain the liquid until the bottom of the meniscus just reaches the center of the mark, as in Figure 2-8. Touching the beaker wall draws liquid from the pipet without leaving part of a drop hanging from the pipet when the level reaches the calibration mark.

Transfer the pipet to the desired receiving vessel and drain it *while holding the tip against the wall of the vessel.* After the pipet stops draining, hold it against the wall for a few more seconds to complete draining. *Do not blow out the last drop.* The pipet should be nearly vertical at the end of delivery. When you finish with a pipet, it should be rinsed with distilled water or soaked in a pipet container until it is cleaned. Solutions should never be allowed to dry inside a pipet because dry residue is difficult to remove.

Micropipets

Accuracy: difference between delivered volume and desired volume
Precision: reproducibility of replicate deliveries

A micropipet (Figure 2-10) is used to deliver volumes of 1 to 1 000 μL (1 μL = 10^{-6} L) with accuracies given in Table 2-4. The liquid is contained in the disposable plastic tip. Micropipets may have a metal barrel on the inside that can be corroded by pipetting

Table 2-4 Accuracy (%) of micropipets

Pipet volume (μL)	At 10% of pipet volume	At 100% of pipet volume	Pipet volume (μL)	Accuracy (%)
	Adjustable volume		*Fixed volume*	
2	±8	±1.2	10	±0.8
10	±2.5	±0.8	25	±0.8
25	±4.5	±0.8	100	±0.5
100	±1.8	±0.6	500	±0.4
300	±1.2	±0.4	1 000	±0.3
1 000	±1.6	±0.3		

Precision is typically 2–3 times smaller (better) than the accuracy.

SOURCE: Hamilton Co., Reno, NV.

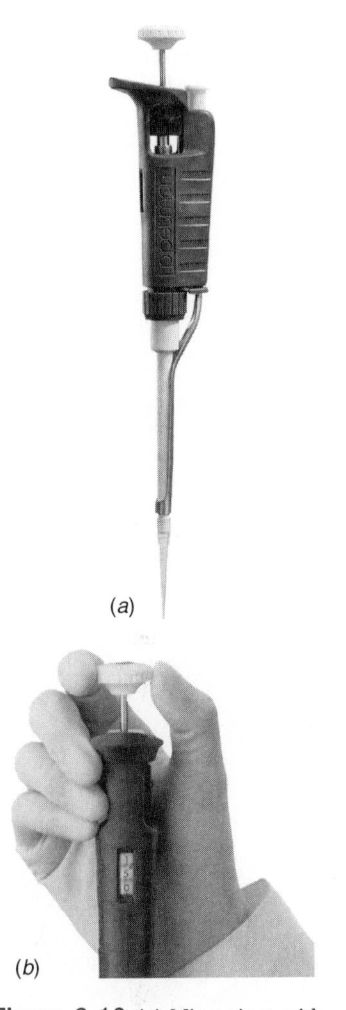

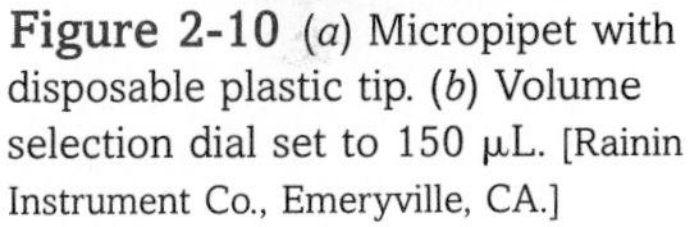

Figure 2-10 (*a*) Micropipet with disposable plastic tip. (*b*) Volume selection dial set to 150 μL. [Rainin Instrument Co., Emeryville, CA.]

volatile acids such as concentrated HCl. Corrosion slowly diminishes the accuracy of the pipet.

To use a micropipet, place a fresh tip tightly on the barrel. Tips are contained in a package or dispenser so that you do not handle (and contaminate) the points with your fingers. Set the desired volume with the knob at the top of the pipet. Depress the plunger to the first stop, which corresponds to the selected volume. Hold the pipet *vertically*, dip it 3–5 mm into the reagent solution, and *slowly* release the plunger to suck up liquid. Withdraw the tip from the liquid by sliding it along the wall of the vessel to remove liquid from the outside of the tip. To dispense liquid, touch the micropipet tip to the wall of the receiver and gently depress the plunger to the first stop. After a few seconds to allow liquid to drain down the wall of the pipet tip, depress the plunger farther to squirt out the last liquid. It is a good idea to clean and wet a fresh tip by taking up and discarding two or three squirts of reagent first. The tip can be discarded or rinsed well with a squirt bottle and reused. When you use a squirt bottle, never touch the tip of the squirt bottle to anything, to avoid contaminating the squirt bottle.

When you use a squirt bottle, *never* touch the tip to anything.

The volume of liquid taken into the tip depends on the angle at which the pipet is held and how far beneath the surface of reagent the tip is held during uptake. As internal parts wear out, both precision and accuracy can decline by an *order of magnitude* (a factor of 10). Micropipets require periodic cleaning, seal replacement, and lubrication. You can check performance by weighing the amount of water delivered from a micropipet. Monthly calibration to identify pipets in need of repair is recommended.

A microliter *syringe*, such as that in Figure 2-11, dispenses volumes in the range 1 to 500 μL with accuracy and precision near 1%. When using a syringe, take up and discard several volumes of liquid to wash the glass and remove air bubbles from the barrel. The steel needle is attacked by strong acid and will contaminate strongly acidic solutions with iron.

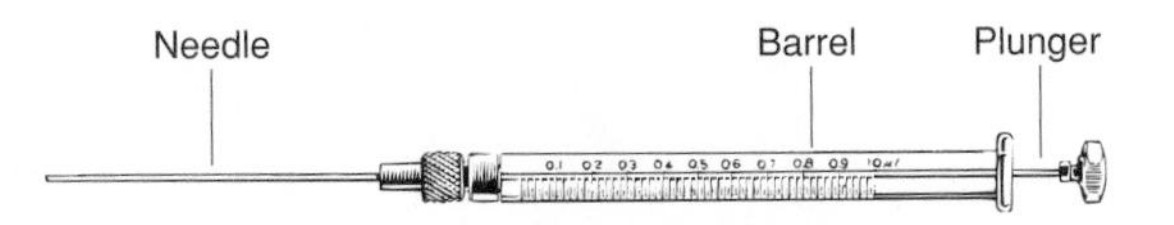

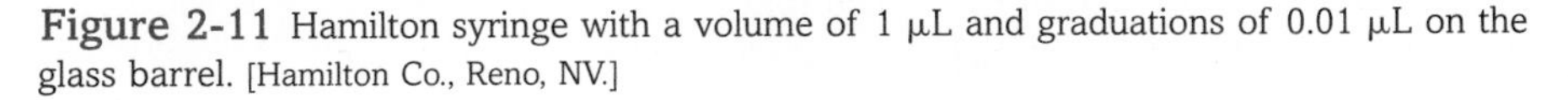

Figure 2-11 Hamilton syringe with a volume of 1 μL and graduations of 0.01 μL on the glass barrel. [Hamilton Co., Reno, NV.]

Ask Yourself

2-D. Which is more accurate, a transfer pipet or a measuring pipet? How much is the uncertainty in microliters when you deliver **(a)** 10 μL or **(b)** 100 μL from a 100-μL adjustable micropipet?

2-7 Filtration

In **gravimetric analysis**, the mass of product from a reaction is measured to determine how much unknown was present. Precipitates from gravimetric analyses are collected by filtration, washed, and then dried. Most precipitates are collected in a *fritted-glass funnel* with suction to speed filtration (Figure 2-12). The porous glass plate in the funnel allows liquid to pass but retains solids. Filters with coarse, medium, and fine pores are available to collect precipitates with large, medium, or small particle size. The finer the filter, the slower the filtration. The empty crucible is first dried at 110°C and weighed. After collecting solid and drying again, the crucible and its contents are weighed a second time to determine the mass of solid.

Liquid from which a substance precipitates or crystallizes is called the **mother liquor**. Liquid that passes through the filter is called **filtrate**.

In some gravimetric procedures, **ignition** (heating at high temperature over a burner or in a furnace) is used to convert a precipitate to a known, constant composition. For example, Fe^{3+} precipitates as hydrated $Fe(OH)_3$ with variable composition. Ignition converts it to Fe_2O_3 prior to weighing. When a gravimetric precipitate is to be ignited, it is collected in **ashless filter paper**, which leaves little residue when burned.

To use filter paper with a conical glass funnel, fold the paper into quarters, tear off one corner (to allow a firm fit into the funnel), and place the paper in the funnel (Figure 2-13). The filter paper should fit snugly and be seated with some distilled water. When liquid is poured in, an unbroken stream of liquid should fill the stem of the funnel. The weight of liquid in the stem helps speed filtration.

For filtration, pour the slurry of precipitate in the mother liquor down a glass rod to prevent splattering (Figure 2-14). (A *slurry* is a suspension of solid in liquid.) Dislodge any particles adhering to the beaker or rod with a **rubber policeman**, which is a flattened

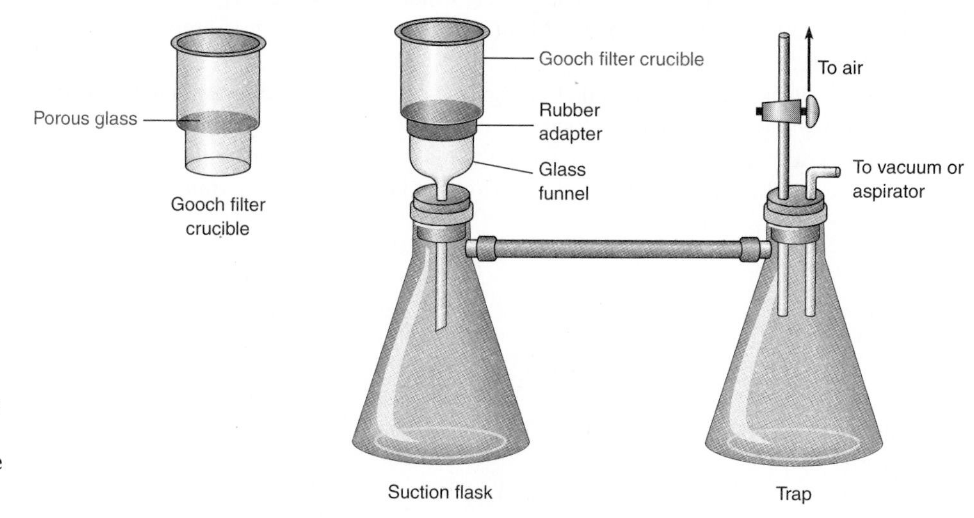

Figure 2-12 Filtration with a Gooch filter crucible that has a porous (*fritted*) glass disk through which liquid can pass. Suction is provided by a vacuum line at the lab bench or by an *aspirator* that uses flowing water from a tap to create a vacuum. The trap prevents backup of filtrate into the vacuum system or backup of water from the aspirator into the suction flask.

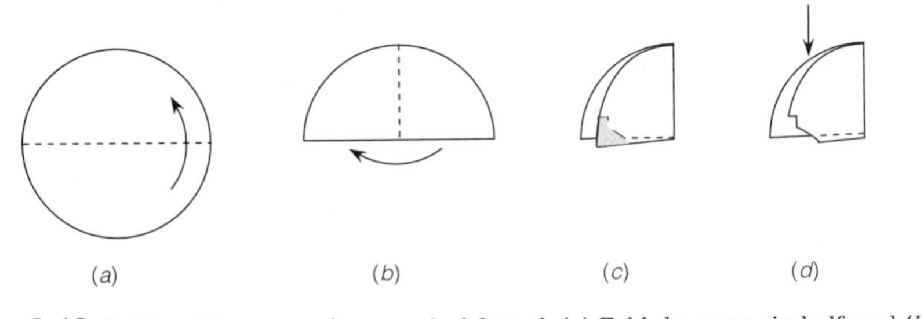

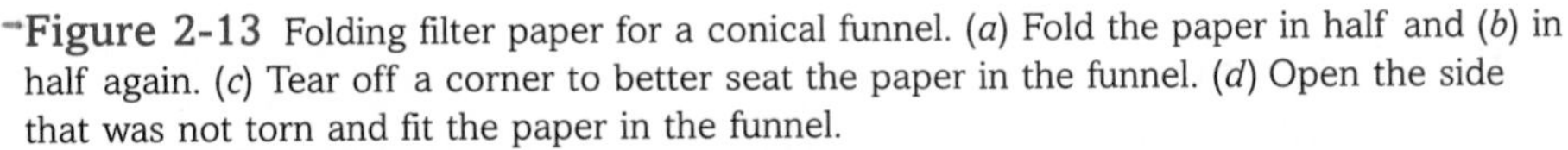

Figure 2-13 Folding filter paper for a conical funnel. (*a*) Fold the paper in half and (*b*) in half again. (*c*) Tear off a corner to better seat the paper in the funnel. (*d*) Open the side that was not torn and fit the paper in the funnel.

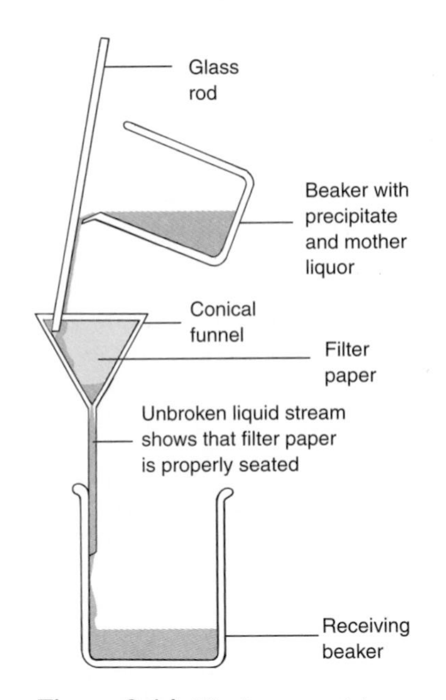

Figure 2-14 Filtering a precipitate.

piece of rubber at the end of a glass rod. Use a jet of appropriate wash liquid from a squirt bottle to transfer particles from the rubber and glassware to the filter. If the precipitate is going to be ignited, particles remaining in the beaker should be wiped onto a small piece of moist filter paper, which is then added to the filter to be ignited.

2-8 Drying

Reagents, precipitates, and glassware are usually dried in an oven at 110°C. (Some chemicals require other temperatures.) Label everything that you put in the oven. Use a beaker and watchglass (Figure 2-15) to minimize contamination by dust during drying. It is good practice to cover all vessels on the benchtop to prevent dust contamination.

We measure the mass of a gravimetric precipitate by weighing a dry, empty filter crucible before the procedure and weighing the same crucible containing dry product after the procedure. To weigh the empty crucible, first bring it to "constant mass" by drying in the oven for 1 h or longer and then cooling for 30 min in a *desiccator*. Weigh the crucible and then heat it again for about 30 min. Cool it and reweigh it. When successive weighings agree to ±0.3 mg, the filter has reached "constant mass." A microwave oven can be used instead of an electric oven for drying reagents, precipitates, and crucibles. Try an initial heating time of 4 min, with subsequent 2-min heatings.

A **desiccator** (Figure 2-16) is a closed chamber containing a drying agent called a **desiccant**. The lid is greased to make an airtight seal. Desiccant is placed in the

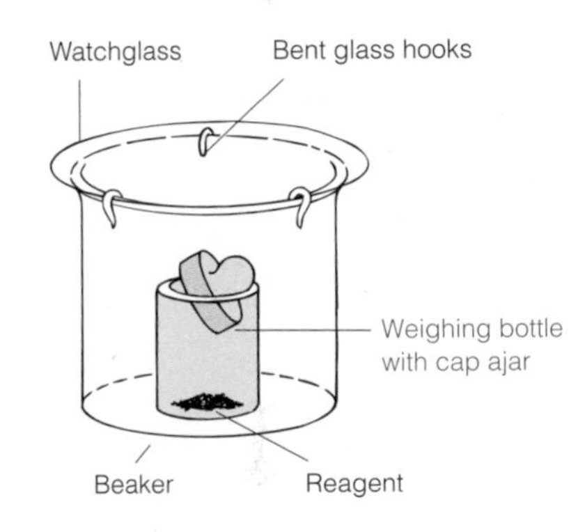

Figure 2-15 Use a watchglass as a dust cover while drying reagents or crucibles in the oven.

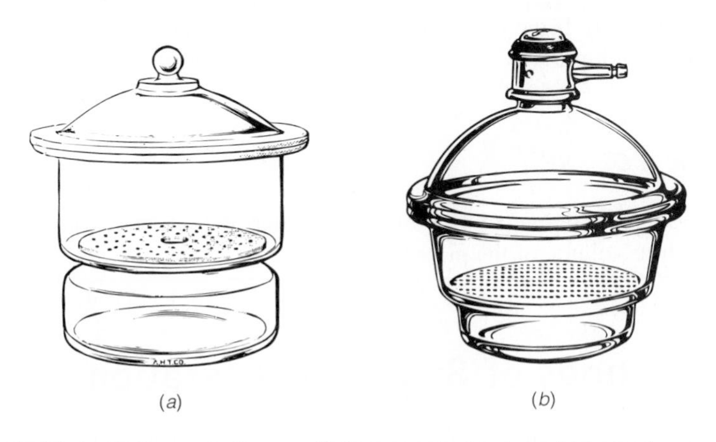

Figure 2-16 (*a*) Ordinary desiccator. (*b*) Vacuum desiccator, which can be evacuated through the side arm and then sealed by rotating the joint containing the side arm. Drying is more efficient at low pressure. Drying agents (*desiccants*) are placed at the bottom of each desiccator below the porous porcelain plate. [A. H. Thomas Co., Philadelphia, PA.]

Table 2-5 Correction factors for volumetric calibration

Temperature (°C)	Correction factor (mL/g)[a]
15	1.002 0
16	1.002 1
17	1.002 3
18	1.002 5
19	1.002 7
20	1.002 9
21	1.003 1
22	1.003 3
23	1.003 5
24	1.003 8
25	1.004 0
26	1.004 3
27	1.004 6
28	1.004 8
29	1.005 1
30	1.005 4

a. Factors are based on the density of water and are corrected for buoyancy with Equation 2-1.

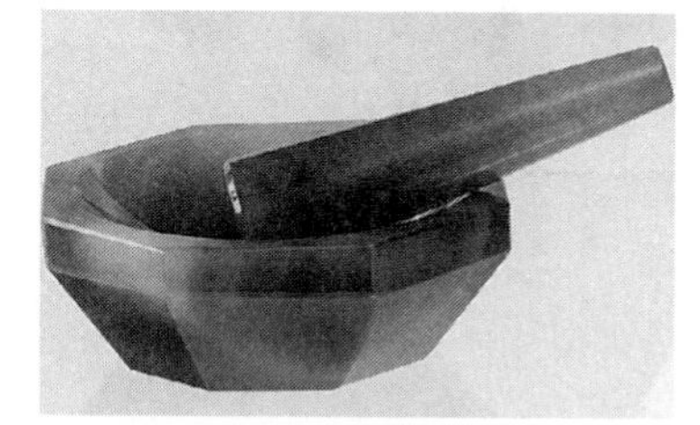

Figure 2-17 Agate mortar and pestle. The mortar is the base and the pestle is the grinding tool. Agate is very hard and expensive. Less expensive porcelain mortars are widely used, but they are somewhat porous and easily scratched. These properties can lead to contamination of the sample by porcelain particles or by traces of previous samples embedded in the porcelain. [Thomas Scientific, Swedesboro, NJ.]

bottom beneath the perforated disk. Common desiccants in approximate order of decreasing efficiency are magnesium perchlorate ($Mg(ClO_4)_2$) > barium oxide (BaO) ≈ alumina (Al_2O_3) ≈ phosphorus pentoxide (P_4O_{10}) >> calcium chloride ($CaCl_2$) ≈ calcium sulfate ($CaSO_4$, called Drierite) ≈ silica gel (SiO_2). After placing a hot object in the desiccator, leave the lid cracked open for a minute until the object has cooled slightly. This practice prevents the lid from popping open when the air inside warms up. To open a desiccator, slide the lid sideways rather than trying to pull it straight up.

2-9 Calibration of Volumetric Glassware

Volumetric glassware can be calibrated to measure the volume that is actually contained in or delivered by a particular piece of equipment. Calibration is done by measuring the mass of water contained or delivered and using Table 2-5 to convert mass to volume:

$$\text{true volume} = (\text{mass of water}) \times (\text{correction factor in Table 2-5}) \qquad (2\text{-}2)$$

To calibrate a 25-mL transfer pipet, first weigh an empty weighing bottle like the one in Figure 2-15. Then fill the pipet to the mark with distilled water, drain it into the weighing bottle, and put on the lid to prevent evaporation. Weigh the bottle again to find the mass of water delivered from the pipet. Use Equation 2-2 to convert mass to volume.

Example Calibration of a Pipet

An empty weighing bottle had a mass of 10.283 g. After water was added from a 25-mL pipet, the mass was 35.225 g. The temperature was 23°C. Find the volume of water delivered by the pipet.

SOLUTION The mass of water is 35.225 − 10.283 = 24.942 g. From Equation 2-2 and Table 2-5, the volume of water is (24.942 g)(1.003 5 mL/g) = 25.029 mL.

Test Yourself If the temperature had been 29°C, and the mass of water was 24.942 g, what volume was delivered? (**Answer:** 25.069 mL)

Ask Yourself

2-E. A 10-mL pipet delivered 10.000 0 g of water at 15°C to a weighing bottle. What is the true volume of the pipet?

2-10 Methods of Sample Preparation

In the analytical process chart in Box 0-1, a homogeneous laboratory sample must be prepared from a representative bulk sample. You can homogenize solids by grinding them to fine powder with a **mortar and pestle** (Figure 2-17) or by dissolving the entire sample.

Dissolving Inorganic Materials with Strong Acids

The acids HCl, HBr, HF, H_3PO_4, and dilute H_2SO_4 dissolve most metals (M) by the reaction

$$M(s) + nH^+(aq) \xrightarrow{\text{heat}} M^{n+}(aq) + \frac{n}{2}H_2(g)$$

HCl	hydrochloric acid
HBr	hydrobromic acid
HF	hydrofluoric acid
H_3PO_4	phosphoric acid
H_2SO_4	sulfuric acid
HNO_3	nitric acid

HF is extremely harmful to touch or breathe. Flood the affected area with water, coat the skin with calcium gluconate (or another calcium salt), and seek medical help.

Many other inorganic substances also can be dissolved. Some anions react with H^+ to form **volatile** products (species that evaporate easily), which are lost from hot solutions in open vessels. Examples include carbonate ($CO_3^{2-} + 2H^+ \rightarrow H_2CO_3 \rightarrow CO_2\uparrow + H_2O$) and sulfide ($S^{2-} + 2H^+ \rightarrow H_2S\uparrow$). Hot hydrofluoric acid dissolves silicates found in most rocks. HF also attacks glass, so it is used in Teflon, polyethylene, silver, or platinum vessels. Teflon is inert to attack by most chemicals and can be used up to 260°C.

Teflon is a *polymer* (a chain of repeating units) with the structure

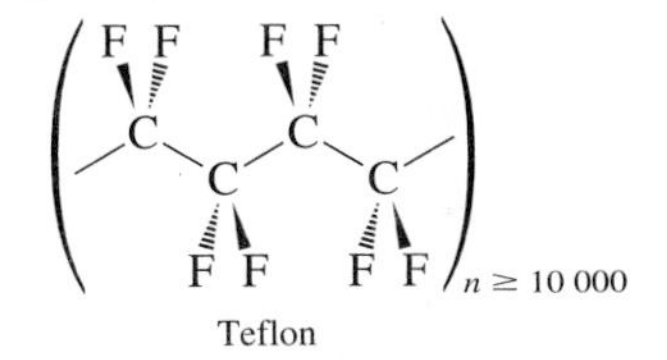

Teflon

Carbon atoms are in the plane of the page. A solid wedge is a bond coming out of the page toward you and a dashed wedge is a bond going behind the page.

Substances that do not dissolve in the acids above may dissolve as a result of oxidation by HNO_3 or concentrated H_2SO_4. Nitric acid attacks most metals, but not Au and Pt, which dissolve in the 3:1 (vol:vol) mixture of HCl:HNO_3 called *aqua regia.*

Acid dissolution is conveniently carried out with a Teflon-lined **bomb** (a sealed vessel, Figure 2-18) in a microwave oven, which heats the contents to 200°C in a minute. The bomb cannot be made of metal, which absorbs microwaves. The bomb should be cooled prior to opening to prevent loss of volatile products.

Fusion

Inorganic substances that do not dissolve in acid can usually be dissolved by a hot, molten inorganic **flux**, examples of which are lithium tetraborate ($Li_2B_4O_7$) and sodium hydroxide (NaOH). Mix the finely powdered unknown with 2 to 20 times its mass of solid flux, and **fuse** (melt) the mixture in a platinum–gold alloy crucible at

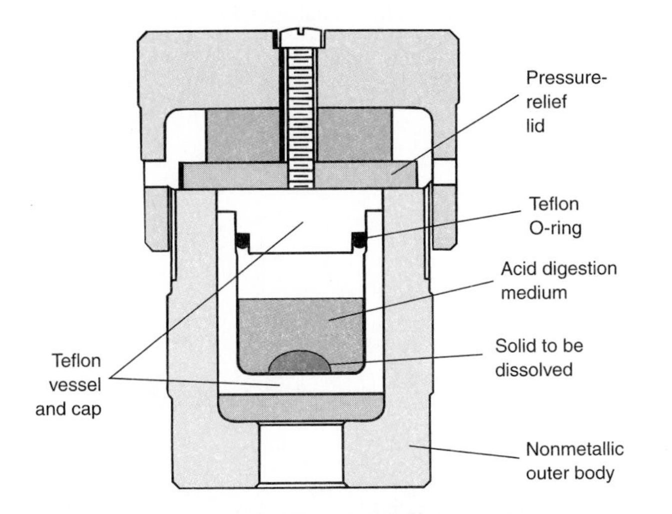

Figure 2-18 Microwave digestion bomb lined with Teflon. A typical 23-mL vessel can be used to digest as much as 1 g of inorganic material (or 0.1 g of organic material, which releases a great deal of gaseous CO_2) in as much as 15 mL of concentrated acid. The outer container maintains its strength up to 150°C, but rarely rises above 50°C. If the internal pressure exceeds 80 bar, the cap deforms and releases the excess pressure. [Parr Instrument Co., Moline, IL.]

300° to 1 200°C in a furnace or over a burner. When the sample is homogeneous, carefully pour the molten flux into a beaker containing 10 wt% aqueous HNO_3 to dissolve the product.

Digestion of Organic Substances

To analyze N, P, halogens (F, Cl, Br, I), and metals in an organic compound, first decompose the compound by combustion (described in Section 7-4) or by *digestion*. In **digestion**, a substance is decomposed and dissolved by a reactive liquid. For this purpose, add sulfuric acid or a mixture of H_2SO_4 and HNO_3 to an organic substance and gently boil (or heat it in a microwave bomb) for 10 to 20 min until all particles have dissolved and the solution has a uniform black appearance. After cooling, destroy the dark color by adding hydrogen peroxide (H_2O_2) or HNO_3, and heat again. Analyze the decomposed sample after digestion.

Extraction

In **extraction**, analyte is dissolved in a solvent that does not dissolve the entire sample and does not decompose the analyte. In a typical extraction of pesticides from soil, a mixture of soil plus the solvents acetone and hexane is placed in a Teflon-lined bomb and heated by microwaves to 150°C. This temperature is 50° to 100° higher than the boiling points of the individual solvents in an open vessel at atmospheric pressure. Soluble pesticides dissolve, but most of the soil remains behind. To complete the analysis, analyze the solution by chromatography, which is described in Chapters 21 through 23.

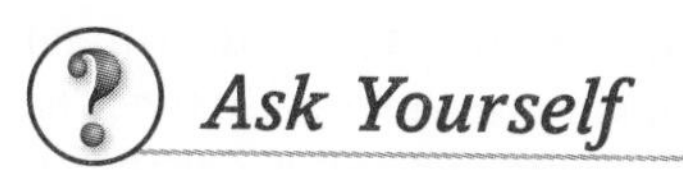

Ask Yourself

2-F. Lead sulfide (PbS) is a black solid that is sparingly soluble in water but dissolves in concentrated HCl. If such a solution is boiled to dryness, white, crystalline lead chloride ($PbCl_2$) remains. What happened to the sulfide?

Key Equation

Buoyancy $$m = m'\left(1 - \frac{d_a}{d_w}\right)\Big/\left(1 - \frac{d_a}{d}\right)$$

m = true mass; m' = mass measured in air

d_a = density of air (0.001 2 g/mL near 1 bar and 25°C)

d_w = density of balance weights (8.0 g/mL)

d = density of object being weighed

Important Terms

absorption
acid wash
adsorption
ashless filter paper
bomb
buoyancy
buret
desiccant
desiccator
digestion
electronic balance
extraction
filtrate
flux
fusion

gravimetric analysis
green chemistry
hygroscopic
ignition
mechanical balance
meniscus
mortar and pestle
mother liquor
parallax error
pipet
rubber policeman
tare
volatile
volumetric flask

Problems

2-1. What do the symbols TD and TC mean on volumetric glassware?

2-2. When would it be preferable to use a plastic volumetric flask instead of a glass flask?

2-3. What is the purpose of the trap in Figure 2-12? What does the watchglass do in Figure 2-15?

2-4. Distinguish absorption from adsorption. When you heat glassware in a drying oven, are you removing absorbed or adsorbed water?

2-5. What is the difference between digestion and extraction?

2-6. What is the true mass of water if the mass measured in air is 5.397 4 g?

2-7. Pentane (C_5H_{12}) is a liquid with a density of 0.626 g/mL. Find the true mass of pentane when the mass weighed in air is 14.82 g.

2-8. Ferric oxide (Fe_2O_3, density = 5.24 g/mL) obtained from ignition of a gravimetric precipitate weighed 0.296 1 g in the atmosphere. What is the true mass in vacuum?

2-9. Your professor has recruited you to work in her lab to help her win the Nobel Prize. It is critical that your work be as accurate as possible. Rather than using the stated volumes of glassware in the lab, you decide to calibrate each piece. An empty 10-mL volumetric flask weighed 10.263 4 g. When filled to the mark with distilled water at 20°C, it weighed 20.214 4 g. What is the true volume of the flask?

2-10. Water from a 5-mL pipet was drained into a weighing bottle whose empty mass was 9.974 g to give a new mass of 14.974 g at 26°C. Find the volume of the pipet.

2-11. Water was drained from a buret between the 0.12- and 15.78-mL marks. The apparent volume was 15.78 − 0.12 = 15.66 mL. Measured in air at 25°C, the mass of water delivered was 15.569 g. What was the true volume?

2-12. Glass is a notorious source of metal ion contamination. Three glass bottles were crushed and sieved to collect 1-mm pieces.[8] To see how much Al^{3+} could be extracted, 200 mL of a 0.05 M solution of the metal-binding compound EDTA was stirred with 0.50 g of ~1-mm glass particles in a polyethylene flask. The Al content of the solution after 2 months was 5.2 μM. The total Al content of the glass, measured after completely dissolving some glass in 48 wt% HF with microwave heating, was 0.80 wt%. What fraction of the Al was extracted from glass by EDTA?

Reference Procedure: Calibrating a 50-mL Buret

This procedure tells how to construct a graph like that in Figure 3-2 (page 66) to convert the measured volume delivered by a buret to the true volume delivered at 20°C.

0. Measure the temperature in the laboratory. Distilled water for this experiment must be at laboratory temperature.

1. Fill the buret with distilled water and force any air bubbles out the tip. See whether the buret drains without leaving drops on its walls. If drops are left, clean the buret with soap and water or soak it with cleaning solution.[5] Adjust the meniscus to be at or slightly below 0.00 mL, and touch the buret tip to a beaker to remove the suspended drop of water. Allow the buret to stand for 5 min while you weigh a 125-mL flask fitted with a rubber stopper. (Hold the flask with a paper towel to prevent fingerprints from changing its mass.) If the level of the liquid in the buret has changed, tighten the stopcock and repeat the procedure. Record the level of the liquid.

2. Drain approximately 10 mL of water at a rate of <20 mL/min into the weighed flask, and cap it tightly to prevent evaporation. Allow 30 s for the film of liquid on the walls to descend before you read the buret. Estimate all readings to the nearest 0.01 mL. Weigh the flask again to determine the mass of water delivered.

3. Drain the buret from 10 to 20 mL, and measure the mass of water delivered. Repeat the procedure for 30, 40, and 50 mL. Then do the entire procedure (10, 20, 30, 40, 50 mL) a second time.

4. Use Table 2-5 to convert the mass of water to the volume delivered. Repeat any set of duplicate buret corrections that do not agree to within 0.04 mL. Prepare a calibration graph as in Figure 3-2, showing the correction factor at each 10-mL interval.

Example Buret Calibration

When draining the buret at 24°C, you observe the following values:

Final reading	10.01	10.08 mL
Initial reading	0.03	0.04
Difference	9.98	10.04 mL
Mass	9.984	10.056 g
Actual volume delivered	10.02	10.09 mL
Correction	+0.04	+0.05 mL
Average correction	+0.045 mL	

To calculate the actual volume delivered when 9.984 g of water are delivered at 24°C, use the conversion factor 1.003 8 mL/g in Table 2-5. We find that 9.984 g occupies (9.984 g)(1.003 8 mL/g) = 10.02 mL. The average correction for both sets of data is +0.045 mL.

To obtain the correction for a volume greater than 10 mL, add successive masses of water collected in the flask. Suppose that the following masses were measured:

Volume interval (mL)	Mass delivered (g)
0.03–10.01	9.984
10.01–19.90	9.835
19.90–30.06	10.071
Sum 30.03 mL	29.890 g

The total volume of water delivered is (29.890 g)(1.003 8 mL/g) = 30.00 mL. Because the indicated volume is 30.03 mL, the buret correction at 30 mL is −0.03 mL.

What does this mean? Suppose that Figure 3-2 applies to your buret. If you begin a titration at 0.04 mL and end at 29.43 mL, you would deliver 29.39 mL if the buret were perfect. Figure 3-2 tells you that the buret delivers 0.03 mL less than the indicated amount; so only 29.36 mL were actually delivered. To use the calibration curve, either begin all titrations near 0.00 mL or correct both the initial and the final readings. Use the calibration curve whenever you use your buret.

Notes and References

1. R. J. Lewis, Sr., *Hazardous Chemicals Desk Reference*, 5th ed. (New York: Wiley, 2002); P. Patnaik, *A Comprehensive Guide to the Hazardous Properties of Chemical Substances*, 2nd ed. (New York: Wiley, 1999); G. Lunn and E. B. Sansone, *Destruction of Hazardous Chemicals in the Laboratory* (New York: Wiley, 1994); and M. A. Armour, *Hazardous Laboratory Chemical Disposal Guide*, 2nd ed. (Boca Raton, FL: CRC Press, 1996).

2. P. T. Anastas and J. C. Warner, *Green Chemistry: Theory and Practice* (New York: Oxford University Press, 1998); M. C. Cann and M. E. Connelly, *Real-World Cases in Green Chemistry* (Washington, DC: American Chemical Society, 2000); M. Lancaster, *Green Chemistry: An Introductory Text* (Cambridge: Royal Society of Chemistry, 2002); C. Baird and M. Cann, *Environmental Chemistry*, 3rd ed. (New York: W. H. Freeman and Company, 2005); J. E. Girard, *Principles of Environmental Chemistry* (Sudbury, MA: Bartlett, 2005); B. Braun, R. Charney, A. Clarens, J. Farrugia, C. Kitchens, C. Lisowski, D. Naistat, and A. O'Neil, *J. Chem. Ed.* **2006**, *83*, 1126.

3. J. M. Bonicamp, *J. Chem. Ed.* **2002**, *79*, 476.

4. Videos illustrating basic laboratory techniques are available from the *Journal of Chemical Education* at http://jchemed.chem.wisc.edu/ and also from www.academysavant.com.

5. Prepare cleaning solution by dissolving 36 g of ammonium peroxydisulfate, $(NH_4)_2S_2O_8$, in a *loosely stoppered* 2.2-L ("one gallon") bottle of 98 wt% sulfuric acid. Add ammonium

peroxydisulfate every few weeks to maintain the oxidizing strength. EOSULF is an alternative cleaning solution for removing proteins and other residues from glassware in a biochemistry lab. EOSULF contains the metal binder EDTA and a sulfonate detergent. It can be safely poured down the drain. [P. L. Manske, T. M. Stimpfel, and E. L. Gershey, *J. Chem. Ed.* **1990**, *67*, A280.]

6. M. M. Singh, C. McGowan, Z. Szafran, and R. M. Pike, *J. Chem. Ed.* **1998**, *75*, 371; *J. Chem. Ed.* **2000**, *77*, 625.

7. R. H. Obenauf and N. Kocherlakota, *Spectroscopy Applications Supplement*, March 2006, p. 12.

8. D. Bohrer, P. Cícero do Nascimento, P. Martins, and R. Binotto, *Anal. Chim. Acta* **2002**, *459*, 267.

Further Reading

H. M. Kanare, *Writing the Laboratory Notebook* (Washington, DC: American Chemical Society, 1985).

Experimental Error

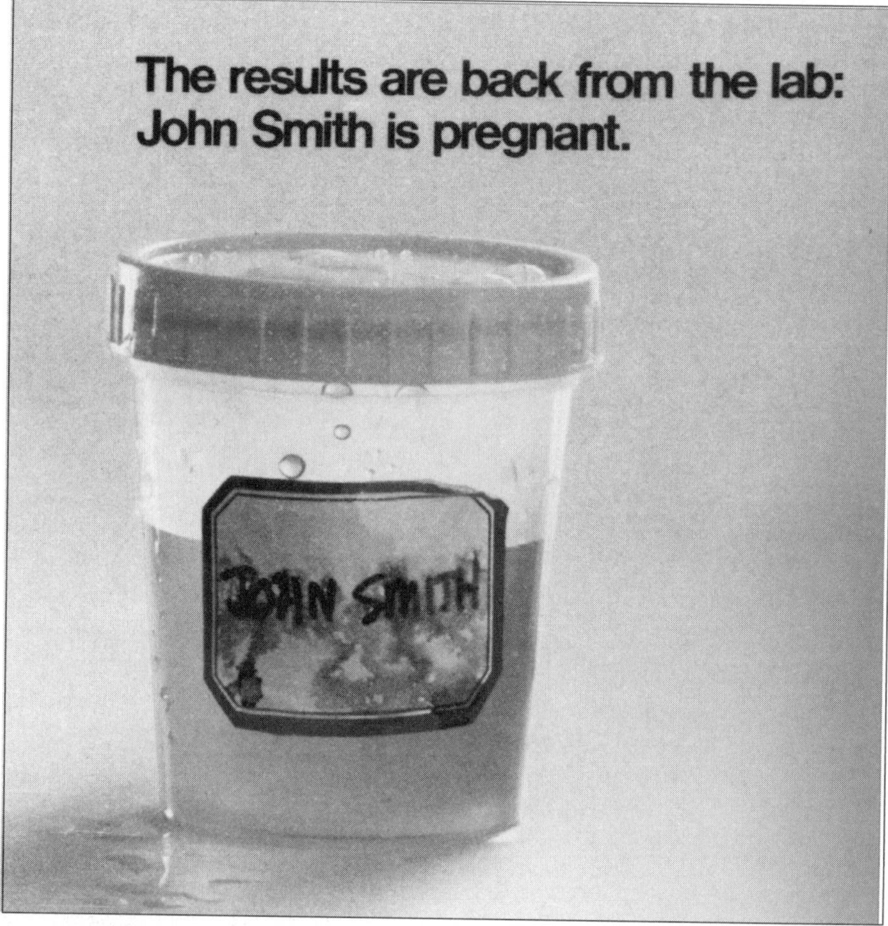

The results are back from the lab:
John Smith is pregnant.
JOHN SMITH

Courtesy 3M Company, St. Paul, MN.

Some laboratory errors are more obvious than others, but there is error associated with every measurement. There is no way to measure the "true" value of anything. The best we can do in a chemical analysis is to carefully apply a technique that experience tells us is reliable. Repetition of one type of measurement several times tells us the reproducibility (*precision*) of the measurement. Measuring the same quantity by different methods gives us confidence of nearness to the "truth" (*accuracy*), if the results agree with one another.

CHAPTER 3

Math Toolkit

Suppose that you measure the density of a mineral by finding its mass (4.635 ± 0.002 g) and its volume (1.13 ± 0.05 mL). Density is mass per unit volume: 4.635 g/1.13 mL = 4.101 8 g/mL. The uncertainties in mass and volume are ±0.002 g and ±0.05 mL, but what is the uncertainty in the computed density? And how many significant figures should be used for the density? This chapter answers these questions and introduces spreadsheets—a powerful tool that will be invaluable to you in and out of this course.

3-1 Significant Figures

The number of **significant figures** is the minimum number of digits needed to write a given value in scientific notation without loss of accuracy. The number 142.7 has four significant figures because it can be written 1.427×10^2. If you write $1.427\,0 \times 10^2$, you imply that you know the value of the digit after 7, which is not the case for the number 142.7. The number $1.427\,0 \times 10^2$ has five significant figures.

Significant figures: Minimum number of digits required to express a value in scientific notation without loss of accuracy.

The number 6.302×10^{-6} has four significant figures, because all four digits are necessary. You could write the same number as 0.000 006 302, which also has just *four* significant figures. The zeros to the left of the 6 are merely holding decimal places. The number 92 500 is ambiguous. It could mean any of the following:

9.25×10^4	3 significant figures
9.250×10^4	4 significant figures
$9.250\,0 \times 10^4$	5 significant figures

You should write one of these three numbers, instead of 92 500, to indicate how many figures are actually known.

Zeros are significant when they are (1) in the middle of a number or (2) at the end of a number on the right-hand side of a decimal point.

Significant zeros are **bold:**

1**0**6 0.01**0**6 0.106 0.106 **0**

The last (farthest to the right) significant figure in a measured quantity always has some associated uncertainty. The minimum uncertainty is ±1 in the last digit.

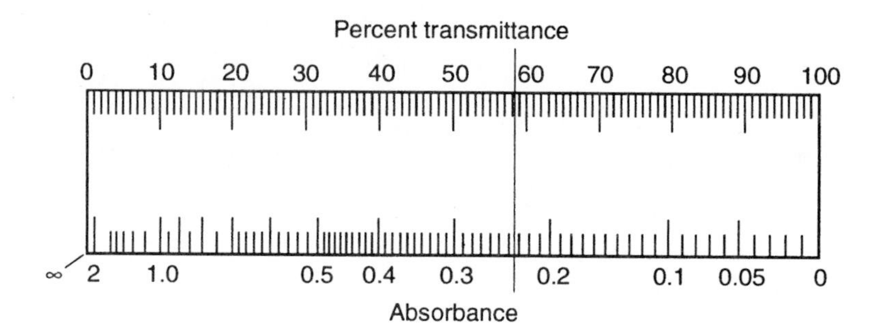

Figure 3-1 Scale of a Bausch and Lomb Spectronic 20 spectrophotometer. Percent transmittance is a linear scale and absorbance is a logarithmic scale.

The scale of a Spectronic 20 spectrophotometer is drawn in Figure 3-1. The needle in the figure appears to be at an absorbance value of 0.234. We say that this number has three significant figures because the numbers 2 and 3 are completely certain and the number 4 is an estimate. The value might be read 0.233 or 0.235 by other people. The percent transmittance is near 58.3. The transmittance scale is smaller than the absorbance scale at this point, so there is more uncertainty in the last digit of transmittance. A reasonable estimate of the uncertainty might be 58.3 ± 0.2. There are three significant figures in the number 58.3.

Interpolation: Estimate all readings to the nearest tenth of the distance between scale divisions.

When reading the scale of any apparatus, *interpolate* between the markings. Try to estimate to the nearest tenth of the distance between two marks. Thus, on a 50-mL buret, which is graduated to 0.1 mL, read the level to the nearest 0.01 mL. When using a ruler calibrated in millimeters, estimate distances to the nearest tenth of a millimeter.

Ask Yourself

3-A. How many significant figures are there in each number below?
(a) 1.903 0 **(b)** 0.039 10 **(c)** 1.40×10^4

3-2 Significant Figures in Arithmetic

We now address the question of how many digits to retain in the answer after you have performed arithmetic operations with your data. Rounding should be done only on the *final answer* (not intermediate results), to avoid accumulating round-off errors.

Addition and Subtraction

If the numbers to be added or subtracted have equal numbers of digits, the answer is given to the *same decimal place* as that in any of the individual numbers:

$$\begin{array}{r} 1.362 \times 10^{-4} \\ +\ 3.111 \times 10^{-4} \\ \hline 4.473 \times 10^{-4} \end{array}$$

The number of significant figures in the answer may exceed or be less than that in the original data.

$$\begin{array}{r} 5.345 \\ +\ 6.728 \\ \hline 12.073 \end{array} \qquad \begin{array}{r} 7.26 \times 10^{14} \\ -\ 6.69 \times 10^{14} \\ \hline 0.57 \times 10^{14} \end{array}$$

If the numbers being added do not have the same number of significant figures, we are limited by the least-certain one. For example, in a calculation of the molecular mass of KrF_2, the answer is known only to the second decimal place, because we are limited by our knowledge of the atomic mass of Kr.

$$\begin{array}{rl} 18.998\,403\,2 & \text{(F)} \\ +\ 18.998\,403\,2 & \text{(F)} \\ +\ 83.798 & \text{(Kr)} \\ \hline 121.794\,\underbrace{806\,4}_{\text{Not significant}} & \end{array}$$

The number 121.794 806 4 should be rounded to 121.795 as the final answer.

Rules for rounding off numbers.

When rounding off, look at *all* the digits *beyond* the last place desired. In the preceding example, the digits 806 4 lie beyond the last significant decimal place. Because this number is more than halfway to the next higher digit, we round the 4 up to 5 (that is, we round up to 121.795 instead of down to 121.794). If the insignificant figures were less than halfway, we would round down. For example, 121.794 3 is rounded to 121.794.

In the special case where the number is exactly halfway, round to the nearest *even* digit. Thus, 43.55 is rounded to 43.6, if we can have only three significant figures. If we are retaining only three figures, 1.425×10^{-9} becomes 1.42×10^{-9}. The number $1.425\,01 \times 10^{-9}$ would become 1.43×10^{-9}, because 501 is more than halfway to the next digit. The rationale for rounding to an even digit is to avoid systematically increasing or decreasing results through successive round-off errors. Half the round-offs will be up and half down.

Addition and subtraction: Express all numbers with the same exponent and align all numbers with respect to the decimal point. Round off the answer according to the number of decimal places in the number with the fewest decimal places.

In adding or subtracting numbers expressed in scientific notation, we should express all numbers with the same exponent:

$$\begin{array}{r} 1.632 \times 10^5 \\ +\ 4.107 \times 10^3 \\ +\ 0.984 \times 10^6 \\ \hline \end{array} \quad \Rightarrow \quad \begin{array}{rl} 1.632 & \times 10^5 \\ +\ 0.041\,07 & \times 10^5 \\ +\ 9.84 & \times 10^5 \\ \hline 11.51 & \times 10^5 \end{array}$$

The sum $11.513\,07 \times 10^5$ is rounded to 11.51×10^5 because the number 9.84×10^5 limits us to two decimal places when all numbers are expressed as multiples of 10^5.

Challenge Show that the answer has four significant figures even if all numbers are expressed as multiples of 10^4 instead of 10^5.

Multiplication and Division

In multiplication and division, we are normally limited to the number of digits contained in the number with the fewest significant figures:

$$\begin{array}{r} 3.26 \times 10^{-5} \\ \times\ 1.78\ \ \ \ \ \ \ \\ \hline 5.80 \times 10^{-5} \end{array} \qquad \begin{array}{rl} 4.317\,9 & \times 10^{12} \\ \times\ 3.6 & \times 10^{-19} \\ \hline 1.6 & \times 10^{-6} \end{array} \qquad \begin{array}{r} 34.60 \\ \div\ 2.462\,87 \\ \hline 14.05 \end{array}$$

The power of 10 has no influence on the number of figures that should be retained.

Example **Significant Figures in Molecular Mass**

Find the molecular mass of $C_{14}H_{10}$ with the correct number of significant digits.

SOLUTION Multiply the atomic mass of C by 14 and the atomic mass of H by 10 and add the products together.

We often retain extra digits beyond the significant figures to avoid introducing unnecessary round-off error into subsequent computations. In this book, we show the extra, insignificant figures as subscripts, as in 168.149_8.

$$14 \times 12.010\,7 = 168.149_8 \longleftarrow \text{6 significant figures because 12.010 7 has 6 digits}$$
$$10 \times 1.007\,94 = \underline{\;10.079\,4\;} \longleftarrow \text{6 significant figures because 1.007 94 has 6 digits}$$
$$178.229_2$$

A reasonable answer is 178.229. The molecular mass is limited to three decimal places because the last significant digit in $14 \times 12.010\,7 = 168.149_8$ is the third decimal place. We often retain an extra (subscripted) digit beyond the last significant figure to avoid introducing round-off error into subsequent calculations.

Test Yourself Find the molecular mass of $C_{14}H_{10}O_8$ with the correct number of significant digits. (**Answer:** 306.224)

Logarithms and Antilogarithms

The base 10 **logarithm** of n is the number a, whose value is such that $n = 10^a$:

Logarithm of n: $\quad n = 10^a$ means that $\log n = a \qquad$ (3-1)

$$10^{-3} = \frac{1}{10^3} = \frac{1}{1\,000} = 0.001$$

For example, 2 is the logarithm of 100 because $100 = 10^2$. The logarithm of 0.001 is -3 because $0.001 = 10^{-3}$. To find the logarithm of a number with your calculator, enter the number and press the *log* function.

In Equation 3-1, the number n is said to be the **antilogarithm** of a. That is, the antilogarithm of 2 is 100 because $10^2 = 100$, and the antilogarithm of -3 is 0.001 because $10^{-3} = 0.001$. Your calculator may have a *10^x* key or an *antilog* key or an *INV log* key. To find the antilogarithm of a number, enter it in your calculator and press *10^x* (or *antilog* or *INV log*).

A logarithm is composed of a **characteristic** and a **mantissa**. The characteristic is the integer part and the mantissa is the decimal part:

$$\log 339 = \underbrace{2}_{\text{Characteristic} = 2}.\underbrace{530}_{\text{Mantissa} = 0.530} \qquad \log 3.39 \times 10^{-5} = \underbrace{-4}_{\text{Characteristic} = -4}.\underbrace{470}_{\text{Mantissa} = 0.470}$$

Number of digits in ***mantissa*** of $\log x$ = number of significant figures in x:

$$\log(\underbrace{5.403}_{\text{4 digits}} \times 10^{-8}) = -7.\underbrace{267\,4}_{\text{4 digits}}$$

The number 339 can be written 3.39×10^2. *The number of digits in the mantissa of log 339 should equal the number of significant figures in 339.* The logarithm of 339 is properly expressed as 2.530. The *characteristic*, 2, corresponds to the exponent in 3.39×10^2.

To see that the third decimal place is the last significant place, consider the following results:

$$10^{2.531} = 340\ (339.6)$$
$$10^{2.530} = 339\ (338.8)$$
$$10^{2.529} = 338\ (338.1)$$

The numbers in parentheses are the results prior to rounding to three figures. Changing the exponent by one digit in the third decimal place changes the answer by one digit in the last (third) place of 339.

In converting a logarithm to its antilogarithm, *the number of significant figures in the antilogarithm should equal the number of digits in the mantissa.* Thus

$$\underbrace{\text{antilog}(-3.42)}_{\text{2 digits}} = 10^{-3.\underbrace{42}_{\text{2 digits}}} = \underbrace{3.8}_{\text{2 digits}} \times 10^{-4}$$

Number of digits in antilog x ($= 10^x$) = number of significant figures in ***mantissa*** of x:

$$10^{6.\underbrace{142}_{\text{3 digits}}} = \underbrace{1.39}_{\text{3 digits}} \times 10^6$$

Here are some examples showing the proper use of significant figures:

$$\log 0.001\,237 = -2.907\,6 \qquad \text{antilog } 4.37 = 2.3 \times 10^4$$
$$\log 1\,237 = 3.092\,4 \qquad 10^{4.37} = 2.3 \times 10^4$$
$$\log 3.2 = 0.51 \qquad 10^{-2.600} = 2.51 \times 10^{-3}$$

Ask Yourself

3-B. How would you express each answer with the correct number of digits?
(a) $1.021 + 2.69 = 3.711$
(b) $12.3 - 1.63 = 10.67$
(c) $4.34 \times 9.2 = 39.928$
(d) $0.060\,2 \div (2.113 \times 10^4) = 2.849\,03 \times 10^{-6}$
(e) $\log (4.218 \times 10^{12}) = ?$
(f) $\text{antilog}(-3.22) = ?$
(g) $10^{2.384} = ?$

3-3 Types of Error

Every measurement has some uncertainty, which is called *experimental error*. Scientific conclusions can be expressed with a high or low degree of confidence, but never with complete certainty. Experimental error is classified as either *systematic* or *random.*

Systematic error is a consistent error that can be detected and corrected. Standard Reference Materials described in Box 3-1 are designed to reduce systematic errors. Box 3-2 provides a case study.

Systematic Error

A **systematic error**, also called a **determinate error**, is repeatable if you make the measurement over again in the same way. In principle, a systematic error can be discovered and corrected. For example, using a pH meter that has been standardized incorrectly produces a systematic error. Suppose you think that the pH of the buffer used to standardize the meter is 7.00, but it is really 7.08. If the meter is otherwise working properly, all pH readings will be 0.08 pH unit too low. When you read a pH of 5.60, the actual pH of the sample is 5.68. This systematic error could be discovered by using another buffer of known pH to test the meter.

Another systematic error arises from an uncalibrated buret. The manufacturer's tolerance for a Class A 50-mL buret is ±0.05 mL. When you think you have delivered 29.43 mL, the real volume could be 29.40 mL and still be within tolerance. One way to correct for an error of this type is by constructing an experimental calibration

Ways to detect systematic error:

1. Analyze known sample, such as a Standard Reference Material. You should observe the known answer. (See Box 15-1 for an example.)
2. Analyze "blank" sample containing no analyte. If you observe a nonzero result, your method responds to more than you intend.
3. Use different analytical methods for the same analyte. If results do not agree, there is error in one (or more) of the methods.
4. *Round robin* experiment: Analyze identical samples in different laboratories by different people using the same or different methods. Disagreement beyond the expected random error is systematic error.

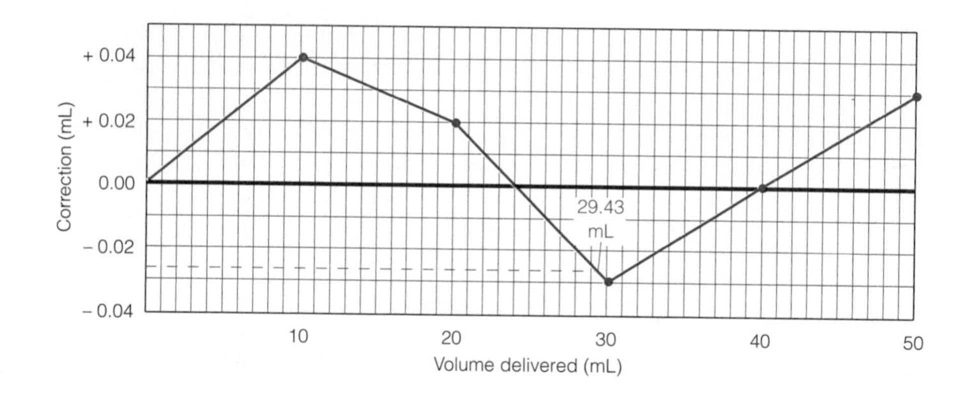

Figure 3-2 Calibration curve for a 50-mL buret.

curve (Figure 3-2). To do this, deliver distilled water from the buret into a flask and weigh it. Convert the mass of water into volume by using Table 2-5. The graph tells us to apply a correction factor of −0.03 mL to the measured value of 29.43 mL to reach the correct value of 29.40 mL.

Systematic error might be positive in some regions and negative in others. The error is repeatable and, with care and cleverness, you can detect and correct it.

Box 3-1 What Are Standard Reference Materials?

Inaccurate laboratory measurements can mean wrong medical diagnosis and treatment, lost production time, wasted energy and materials, manufacturing rejects, and product liability. To minimize errors, the U.S. National Institute of Standards and Technology and institutes of standards of other nations distribute standard reference materials, such as metals, chemicals, rubber, plastics, engineering materials, radioactive substances, and environmental and clinical standards that can be used to test the accuracy of analytical procedures.

For example, in treating patients with epilepsy, physicians depend on laboratory tests to measure blood serum concentrations of anticonvulsant drugs. Low drug levels lead to seizures, and high levels are toxic. Tests of identical serum specimens at different laboratories gave an unacceptably wide range of results. Therefore, the National Institute of Standards and Technology developed a standard reference serum containing known levels of antiepilepsy drugs. The reference serum allows different laboratories to detect and correct errors in their assay procedures.

Before introduction of this reference material, five laboratories analyzing identical samples reported a range of results with relative errors of 40% to 110% of the expected value. After distribution of the reference material, the error was reduced to 20% to 40%.

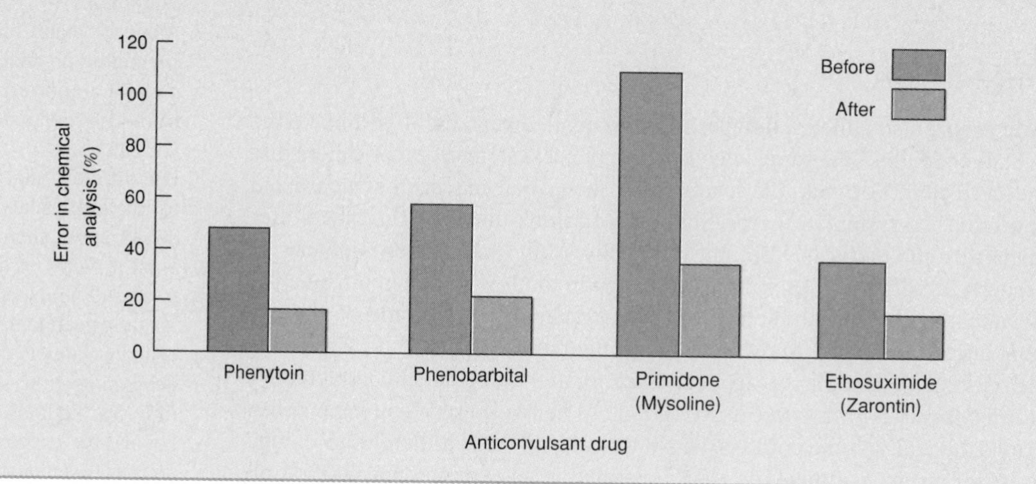

Box 3-2 Case Study: Systematic Error in Ozone Measurement

Ozone (O_3) is an oxidizing, corrosive gas that harms your lungs and all forms of life. It is formed near the surface of Earth by the action of sunlight on air pollutants largely derived from automobile exhaust. The U.S. Environmental Protection Agency sets an 8-h average O_3 limit of 80 ppb (80 nL/L by volume) in air. Regions that fail to meet this standard can be required to reduce sources of pollution that contribute to O_3 formation. Systematic error in O_3 measurement can have serious consequences for the health and economy of a region.

Prior to the work described in this Box, it was known that O_3 monitors often exhibit erratic behavior on hot, humid days. It was conjectured that half of the regions deemed to be out of compliance with the O_3 standard might actually have been under the legal limit. This error could force expensive remediation measures when none were required. Conversely, there were rumors that some unscrupulous operators of O_3 monitors were aware that zeroing their instrument at night when the humidity is higher produced lower O_3 readings the next day, thereby reducing the number of days when a region is deemed out of compliance.

The O_3-measuring instrument in the diagram pumps ambient air through a cell with a pathlength of 15 cm. Ultraviolet radiation from a mercury lamp is partially absorbed by O_3, whose absorption spectrum is shown at the opening of Chapter 18. From the measured absorbance, the instrument computes O_3 concentration. In routine use, the operator only adjusts the zero control, which sets the meter to read zero when O_3-free air is drawn through the instrument. Periodically, the instrument is recalibrated with a known source of O_3.

A study of commercial O_3 monitors found that controlled changes in humidity led to *systematic errors in the apparent O_3 concentration of tens to hundreds of ppb*—up to several times greater than the O_3 being measured. Increasing humidity produced systematic *positive* errors in some types of instruments and systematic *negative* errors in other types.

Water does not absorb ultraviolet radiation, so humidity is not interfering by absorbing radiation. A perceptive analysis of the problem led to the hypothesis that *adsorption* of moisture on the inside surface of the measurement cell changed the reflectivity of that surface. In one type of instrument, water adsorbed inside a quartz cell reflects less light than dry quartz and thus increases the amount of light lost by absorption in black paint on the outside of the cell. This instrument produces a false, *high* O_3 reading. Another instrument has a highly reflective aluminum cell coated on the inside with polyvinylidene fluoride. Adsorption of moisture on polyvinylidene fluoride reduces total internal reflection within the coating and increases the radiant energy reaching the detector, giving a false, *low* O_3 reading. These effects need not be large. A 0.03% change in light intensity reaching the detector corresponds to an O_3 change of 100 ppb in a 15-cm pathlength. Systematic error caused by variable humidity was eliminated by installing water-permeable tubing just before the absorption cell to equalize the humidity in air being measured and air used to zero the instrument.

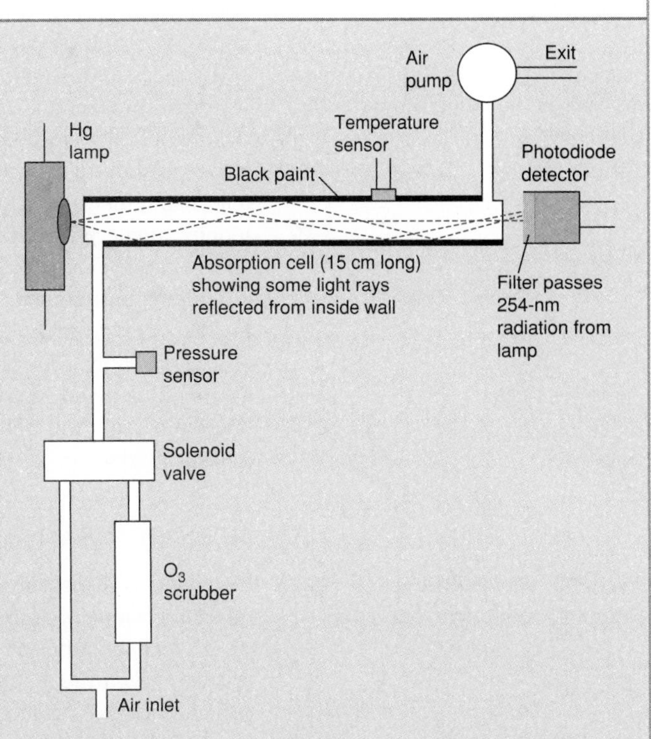

Optical path of 2B Technologies Model 202 Ozone Monitor. The solenoid alternately admits ambient air or air that is scrubbed free of O_3. Absorbance, which is proportional to O_3 concentration, is $-\log(I/I_o)$, where I is the radiant intensity reaching the photodiode with ambient air in the cell and I_o is the intensity with O_3-free air. [Adapted from www.twobtech.com/manuals/model_202_new.pdf. Case study from K. L. Wilson and J. W. Birks, *Environ. Sci. Technol.* **2006,** *40,* 6361.]

Random Error

Random error cannot be eliminated, but it might be reduced by a better experiment.

Random error, also called **indeterminate error**, arises from limitations on our ability to make physical measurements and on natural fluctuations in the quantity being measured. Random error has an equal chance of being positive or negative. It is always present and cannot be corrected. One random error is that associated with reading a scale. Different people reading the scale in Figure 3-1 report a range of values representing their subjective interpolation between the markings. One person reading an instrument several times might report several different readings. Another indeterminate error comes from random electrical noise in an instrument. Positive and negative fluctuations occur with approximately equal frequency and cannot be completely eliminated. Still another source of random error is actual variation in the quantity being measured. If you were to measure the pH of blood in your body, you would probably get different answers for blood from different parts of the body and the pH at a given location probably would vary with time. There is some random uncertainty in "the pH of your blood" even if there were no variations in the pH measuring device.

Precision and Accuracy

Precision: reproducibility
Accuracy: nearness to the "truth"

Precision is a measure of the reproducibility of a result. **Accuracy** is how close a measured value is to the "true" value.

A measurement might be reproducible, but wrong. For example, if you made a mistake preparing a solution for a titration, the solution would not have the desired concentration. You might then do a series of reproducible titrations but report an incorrect result because the concentration of the titrating solution was not what you intended. In this case, the precision is good but the accuracy is poor. Conversely, it is possible to make poorly reproducible measurements clustered around the correct value. In this case, the precision is poor but the accuracy is good. An ideal procedure is both accurate and precise.

Accuracy is defined as nearness to the "true" value. The word *true* is in quotes because somebody must *measure* the "true" value, and there is error associated with *every* measurement. The "true" value is best obtained by an experienced person using a well-tested procedure. It is desirable to test the result by using different procedures because, even though each method might be precise, systematic error could lead to poor agreement between methods. Good agreement among several methods affords us confidence, but never proof, that the results are "true."

Absolute and Relative Uncertainty

Absolute uncertainty expresses the margin of uncertainty associated with a measurement. If the estimated uncertainty in reading a calibrated buret is ± 0.02 mL, we say that ± 0.02 mL is the absolute uncertainty associated with the reading.

Relative uncertainty compares absolute uncertainty with its associated measurement. The relative uncertainty of a buret reading of 12.35 ± 0.02 mL is a dimensionless quotient:

Relative uncertainty:

$$\text{relative uncertainty} = \frac{\text{absolute uncertainty}}{\text{magnitude of measurement}} \tag{3-2}$$

$$= \frac{0.02 \text{ mL}}{12.35 \text{ mL}} = 0.002$$

Percent relative uncertainty is simply

Percent relative uncertainty:

$$\text{percent relative uncertainty} = 100 \times \text{relative uncertainty} \quad (3\text{-}3)$$

$$= 100 \times 0.002 = 0.2\%$$

If the absolute uncertainty in reading a buret is constant at ±0.02 mL, the percent relative uncertainty is ±0.2% for a volume of 10 mL and ±0.1% for a volume of 20 mL.

Ask Yourself

3-C. Cheryl, Cynthia, Carmen, and Chastity shot these targets at Girl Scout camp. Match each target with the proper description.

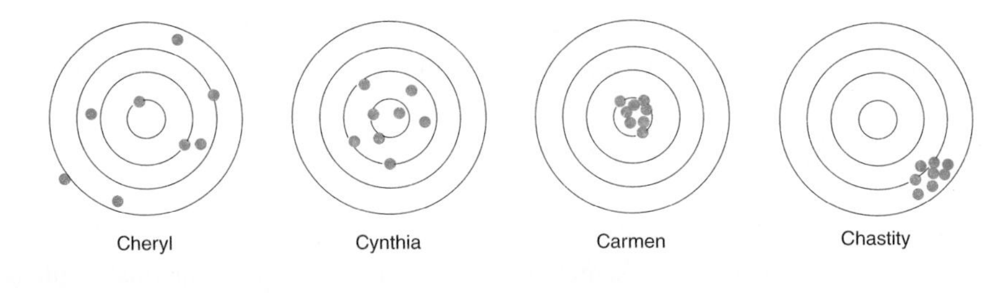

(a) accurate and precise
(b) accurate but not precise
(c) precise but not accurate
(d) neither precise nor accurate

3-4 Propagation of Uncertainty

We can usually estimate or measure the random error associated with a measurement, such as the length of an object or the temperature of a solution. The uncertainty might be based on how well we can read an instrument or on experience with a particular method. If possible, uncertainty is expressed as the *standard deviation* or as a *confidence interval*; these parameters are based on a series of replicate measurements. The following discussion applies only to random error. We assume that systematic error has been detected and corrected.

Standard deviation and confidence interval are discussed in Chapter 4.

In most experiments, it is necessary to perform arithmetic operations on several numbers, each of which has an associated random error. The most likely uncertainty in the result is not simply the sum of the individual errors, because some of them are likely to be positive and some negative. We expect some cancellation of errors.

Addition and Subtraction

Suppose you wish to perform the following arithmetic, in which experimental uncertainties, designated e_1, e_2, and e_3, are given in parentheses.

$$\begin{array}{rl} 1.76\ (\pm 0.03) & \leftarrow e_1 \\ +\ 1.89\ (\pm 0.02) & \leftarrow e_2 \\ -\ 0.59\ (\pm 0.02) & \leftarrow e_3 \\ \hline 3.06\ (\pm e_4) & \end{array} \quad (3\text{-}4)$$

The arithmetic answer is 3.06; but what is the uncertainty associated with this result?

For addition and subtraction, use *absolute* uncertainty.

For addition and subtraction, the uncertainty in the answer is obtained from the *absolute uncertainties* of the individual terms as follows:

Uncertainty in addition and subtraction:

$$e_4 = \sqrt{e_1^2 + e_2^2 + e_3^2} \tag{3-5}$$

For the sum in expression 3-4, we can write

$$e_4 = \sqrt{(0.03)^2 + (0.02)^2 + (0.02)^2} = 0.04_1$$

The absolute uncertainty e_4 is ± 0.04, and we can write the answer as 3.06 ± 0.04. Although there is only one significant figure in the uncertainty, we wrote it initially as 0.04_1, with the first insignificant figure subscripted. We retain one or more insignificant figures to avoid introducing round-off errors into later calculations through the number 0.04_1. The insignificant digit was subscripted to remind us where the last significant digit should be at the conclusion of the calculations.

To find the percent relative uncertainty in the sum of expression 3-4, we write

$$\text{percent relative uncertainty} = \frac{0.04_1}{3.06} \times 100 = 1._3\%$$

The uncertainty, 0.04_1, is $1._3\%$ of the result, 3.06. The subscript 3 in $1._3\%$ is not significant. It is sensible to drop the insignificant figures now and express the final result as

For addition and subtraction, use absolute uncertainty. Relative uncertainty can be found at the end of the calculation.

3.06 ($\pm$0.04) (absolute uncertainty)
3.06 ($\pm$1%) (relative uncertainty)

Example **Uncertainty in a Buret Reading**

The volume delivered by a buret is the difference between the final and initial readings. If the uncertainty in each reading is $\pm$0.02 mL, what is the uncertainty in the volume delivered?

SOLUTION Suppose that the initial reading is 0.05 ($\pm$0.02) mL and the final reading is 17.88 ($\pm$0.02) mL. The volume delivered is the difference:

$$\begin{array}{r} 17.88\ (\pm 0.02) \\ -\ 0.05\ (\pm 0.02) \\ \hline 17.83\ (\pm e) \end{array} \qquad e = \sqrt{0.02^2 + 0.02^2} = 0.03$$

Regardless of the initial and final readings, if the uncertainty in each one is $\pm$0.02 mL, the uncertainty in volume delivered is $\pm$0.03 mL.

Test Yourself Suppose that the uncertainty in measuring pH is $\pm$0.03 pH units. The pH of two solutions is measured to be 8.23 and 4.01. Find the difference in pH and its uncertainty. (**Answer:** 4.22 $\pm$ 0.04)

Multiplication and Division

For multiplication and division, first convert all uncertainties to percent relative uncertainties. Then calculate the error of the product or quotient as follows:

For multiplication and division, use percent relative uncertainty.

Uncertainty in multiplication and division:

$$\%e_4 = \sqrt{(\%e_1)^2 + (\%e_2)^2 + (\%e_3)^2} \tag{3-6}$$

For example, consider the following operations:

$$\frac{1.76\ (\pm 0.03) \times 1.89\ (\pm 0.02)}{0.59\ (\pm 0.02)} = 5.64 \pm e_4$$

First, convert absolute uncertainties to percent relative uncertainties:

$$\frac{1.76\ (\pm 1._7\%) \times 1.89\ (\pm 1._1\%)}{0.59\ (\pm 3._4\%)} = 5.64 \pm e_4$$

Then, find the percent relative uncertainty of the answer by using Equation 3-6.

$$\%e_4 = \sqrt{(1._7)^2 + (1._1)^2 + (3._4)^2} = 4._0\%$$

The answer is $5.6_4\ (\pm 4._0\%)$.

To convert relative uncertainty to absolute uncertainty, find $4._0\%$ of the answer:

$$4._0\% \times 5.6_4 = 0.04_0 \times 5.6_4 = 0.2_3$$

The answer is $5.6_4\ (\pm 0.2_3)$. Finally, drop the insignificant digits:

$$\begin{array}{ll} 5.6\ (\pm 0.2) & \text{(absolute uncertainty)} \\ 5.6\ (\pm 4\%) & \text{(relative uncertainty)} \end{array}$$

The denominator of the original problem, 0.59, limits the answer to two digits.

Advice: Retain one or more extra insignificant figures until you have finished your entire calculation. Then round to the correct number of digits. When storing intermediate results in a calculator, keep all digits without rounding.

For multiplication and division, use percent relative uncertainty. Absolute uncertainty can be found at the end of the calculation.

Example **Scientific Notation and Propagation of Uncertainty**

Express the absolute uncertainty in

(a) $\dfrac{3.43\ (\pm 0.08) \times 10^{-8}}{2.11\ (\pm 0.04) \times 10^{-3}}$

(b) $[3.43\ (\pm 0.08) \times 10^{-8}] + [2.11\ (\pm 0.04) \times 10^{-7}]$

SOLUTION **(a)** The uncertainty ± 0.08 applies to the number 3.43. Therefore the uncertainty in the numerator is $0.08/3.43 = 2._{332}\%$. (Remember to keep extra digits in your calculator until the end of the problem. Do not round off until the end.) The uncertainty in the denominator is $0.04/2.11 = 1._{896}\%$. The uncertainty in the answer is $\sqrt{2._{332}\%^2 + 1._{896}\%^2} = 3._{006}\%$. The quotient is $(3.43 \times 10^{-8})/(2.11 \times 10^{-3}) = 1.63 \times 10^{-5}$ and the uncertainty is $3._{006}\%$ of $1.63 = 0.05$. The answer is $1.63\ (\pm 0.05) \times 10^{-5}$.

(b) For addition and subtraction, we must express each term with the same power of 10. Let's write the second number as a multiple of 10^{-8} instead of 10^{-7}. To do this, multiply 2.11 and 0.04 by 10 and divide 10^{-7} by 10:

$$\begin{array}{r} 3.43\ (\pm 0.08) \times 10^{-8} \\ +\ 2.11\ (\pm 0.04) \times 10^{-7} \end{array} \Rightarrow \begin{array}{r} 3.43\ (\pm 0.08) \times 10^{-8} \\ +\ 21.1\ \ (\pm 0.4)\ \ \times 10^{-8} \\ \hline 24.5_3\ (\pm e)\ \ \ \times 10^{-8} \end{array}$$

$$e = \pm\sqrt{0.08^2 + 0.4^2} = 0.4_1 \Rightarrow \text{Answer: } 24.5\ (\pm 0.4) \times 10^{-8}$$

Test Yourself Find the difference $4.22\ (\pm 0.04) \times 10^{-3} - 3.8\ (\pm 0.6) \times 10^{-4}$. (**Answer:** $3.84\ (\pm 0.07) \times 10^{-3}$)

Mixed Operations

Now consider an operation containing subtraction and division:

$$\frac{[1.76\ (\pm 0.03) - 0.59\ (\pm 0.02)]}{1.89\ (\pm 0.02)} = 0.619_0 \pm ?$$

First work out the difference in the numerator, by using absolute uncertainties:

$$1.76\ (\pm 0.03) - 0.59\ (\pm 0.02) = 1.17\ (\pm 0.03_6)$$

because $\sqrt{(0.03)^2 + (0.02)^2} = 0.03_6$.

Then convert to percent relative uncertainties:

$$\frac{1.17\ (\pm 0.03_6)}{1.89\ (\pm 0.02)} = \frac{1.17\ (\pm 3._1\%)}{1.89\ (\pm 1._1\%)} = 0.619_0\ (\pm 3._3\%)$$

because $\sqrt{(3._1\%)^2 + (1._1\%)^2} = 3._3\%$.

The percent relative uncertainty is $3._3\%$, so the absolute uncertainty is $0.03_3 \times 0.619_0 = 0.02_0$. The final answer can be written

$$0.619\ (\pm 0.02_0) \quad \text{(absolute uncertainty)}$$
$$0.619\ (\pm 3._3\%) \quad \text{(relative uncertainty)}$$

The result of a calculation ought to be written in a manner consistent with the uncertainty in the result.

Because the uncertainty begins in the 0.01 decimal place, it is reasonable to round the result to the 0.01 decimal place:

$$0.62\ (\pm 0.02) \quad \text{(absolute uncertainty)}$$
$$0.62\ (\pm 3\%) \quad \text{(relative uncertainty)}$$

The Real Rule for Significant Figures

The real rule: The first uncertain figure is the last significant figure.

The first uncertain figure of the answer is the last significant figure. For example, in the quotient

$$\frac{0.002\ 364\ (\pm 0.000\ 003)}{0.025\ 00\ (\pm 0.000\ 05)} = 0.094\ 6\ (\pm 0.000\ 2)$$

the uncertainty ($\pm$0.000 2) is in the fourth decimal place. Therefore the answer is properly expressed with *three* significant figures, even though the original data have four figures. The first uncertain figure of the answer is the last significant figure. The quotient

$$\frac{0.002\ 664\ (\pm 0.000\ 003)}{0.025\ 00\ (\pm 0.000\ 05)} = 0.106\ 6\ (\pm 0.000\ 2)$$

is expressed with *four* significant figures because the uncertainty is in the fourth decimal place. The quotient

$$\frac{0.821\ (\pm 0.002)}{0.803\ (\pm 0.002)} = 1.022\ (\pm 0.004)$$

is expressed with *four* figures even though the dividend and divisor each have *three* figures.

Example Significant Figures in Laboratory Work

You prepared a 0.250 M NH_3 solution by diluting 8.45 (±0.04) mL of 28.0 (±0.5) wt% NH_3 [density = 0.899 (±0.003) g/mL] up to 500.0 (±0.2) mL. Find the uncertainty in 0.250 M. Consider the molecular mass of NH_3, 17.031 g/mol, to have negligible uncertainty.

SOLUTION To find the uncertainty in molarity, you need to find the uncertainty in moles delivered to the 500-mL flask. The concentrated reagent contains 0.899 (±0.003) g of solution per milliliter. The weight percent tells you that the reagent contains 0.280 (±0.005) g of NH_3 per gram of solution. In the following calculations, you should retain extra insignificant digits and round off only at the end.

$$\left.\begin{array}{l}\text{grams of NH}_3\\ \text{per mL in}\\ \text{concentrated}\\ \text{reagent}\end{array}\right\} = 0.899\ (\pm 0.003)\ \frac{\text{g solution}}{\text{mL}} \times 0.280\ (\pm 0.005)\ \frac{\text{g NH}_3}{\text{g solution}}$$

$$= 0.899\ (\pm 0.334\%)\ \frac{\cancel{\text{g solution}}}{\text{mL}} \times 0.280\ (\pm 1.79\%)\ \frac{\text{g NH}_3}{\cancel{\text{g solution}}}$$

$$= 0.251\ 7\ (\pm 1.82\%)\ \frac{\text{g NH}_3}{\text{mL}}$$

For multiplication and division, convert absolute uncertainty to percent relative uncertainty.

because $\sqrt{(0.334\%)^2 + (1.79\%)^2} = 1.82\%$.

Next, find the moles of ammonia in 8.45 (±0.04) mL of concentrated reagent. The relative uncertainty in volume is ±0.04/8.45 = ±0.473%.

$$\text{mol NH}_3 = \frac{0.251\,7\ (\pm 1.82\%)\ \frac{\cancel{\text{g NH}_3}}{\cancel{\text{mL}}} \times 8.45\ (\pm 0.473\%)\ \cancel{\text{mL}}}{17.031\ \frac{\cancel{\text{g NH}_3}}{\text{mol}}}$$

$$= 0.124\ 9\ (\pm 1.88\%)\ \text{mol}$$

because $\sqrt{(1.82\%)^2 + (0.473\%)^2 + (0\%)^2} = 1.88\%$.

This much ammonia was diluted to 0.500 0 (±0.000 2) L. The relative uncertainty in final volume is ±0.000 2/0.500 0 = ±0.04%. The diluted molarity is

$$\frac{\text{mol NH}_3}{\text{L}} = \frac{0.124\ 9\ (\pm 1.88\%)\ \text{mol}}{0.500\ 0\ (\pm 0.04\%)\ \text{L}}$$

$$= 0.249\ 8\ (\pm 1.88\%)\ \text{M}$$

because $\sqrt{(1.88\%)^2 + (0.04\%)^2} = 1.88\%$. The absolute uncertainty is 1.88% of 0.249 8 M = 0.018 8 × 0.249 8 M = 0.004 7 M. The result 0.004 7 tells us that the uncertainty in molarity is in the third decimal place, so your final, rounded answer is

$$[\text{NH}_3] = 0.250\ (\pm 0.005)\ \text{M}$$

By far, the largest uncertainty in the initial data is in wt%, which has an uncertainty of 0.005/0.280 = 1.79%. The only way to decrease uncertainty in the result (0.250 ± 0.005 M) is to know the wt% of NH_3 reagent more precisely. Improving the other numbers does not help.

Test Yourself The uncertainty in [NH_3] above is ±1.9%. If the uncertainty in wt% NH_3 were 1.0% instead of 1.8%, what would be the relative uncertainty in [NH_3]? (**Answer:** ±1.2%)

Ask Yourself

3-D. To help identify an unknown mineral in your geology class, you measured its mass and volume and found them to be 4.635 ± 0.002 g and 1.13 ± 0.05 mL.
(a) Find the percent relative uncertainty in the mass and in the volume.
(b) Write the density (= mass/volume) and its uncertainty with the correct number of digits.

3-5 Introducing Spreadsheets

Spreadsheets are powerful tools for manipulating quantitative information with a computer. They allow us to conduct "what if" experiments in which we investigate effects such as changing acid strength or concentration on the shape of a titration curve. Any spreadsheet is suitable for exercises in this book. Our specific instructions apply to Microsoft Excel, which is widely available. You will need directions for your particular software. Although this book can be used without spreadsheets, you will be amply rewarded far beyond this course if you invest the time to learn to use spreadsheets.

A Spreadsheet for Temperature Conversions

Let's prepare a spreadsheet to convert temperature from degrees Celsius to kelvins and degrees Fahrenheit by using formulas from Table 1-4:

$$K = °C + C_0 \tag{3-7a}$$

$$°F = \left(\frac{9}{5}\right) * °C + 32 \tag{3-7b}$$

where C_0 is the constant 273.15.

Figure 3-3a shows a blank spreadsheet as it would appear on your computer. Rows are numbered 1, 2, 3, . . . and columns are lettered A, B, C, ... Each rectangular box is called a *cell*. The fourth cell down in the second column, for example, is designated cell B4.

We adopt a standard format in this book in which constants are collected in column A. Select cell A1 and type "Constant:" as a column heading. Select cell A2 and type "C0 =" to indicate that the constant C_0 will be written in the next cell down. Now select cell A3 and type the number 273.15. Your spreadsheet should now look like Figure 3-3b.

In cell B1, type the label "°C" (or "Celsius" or whatever you like). For illustration, we enter the numbers −200, −100, 0, 100, and 200 in cells B2 through B6. This is our *input* to the spreadsheet. The *output* will be computed values of kelvins and °F in columns C and D. (If you want to enter very large or very small numbers, you can write, for example, 6.02E23 for 6.02×10^{23} and 2E-8 for 2×10^{-8}.)

The formula "=B2+$A$3" in cell C2 is equivalent to writing $K = °C + C_0$.

Label column C "kelvin" in cell C1. In cell C2, we enter our first *formula*—an entry beginning with an equals sign. Select cell C2 and type "=B2+$A$3". This expression tells the computer to calculate the contents of cell C2 by taking the contents of cell B2 and adding the contents of cell A3 (which contains the constant, 273.15). We will explain the dollar signs shortly. When this formula is entered, the computer responds by calculating the number 73.15 in cell C2. This is the kelvin equivalent of −200°C.

Columns

	A	B	C	D
1				
2				
3				
4		cell B4		
5				
6				
7				
8				
9				
10				

Rows

(a)

	A	B	C	D
1	Constant:			
2	C0 =			
3	273.15			
4				
5				
6				
7				
8				
9				
10				

(b)

	A	B	C	D
1	Constant:	°C	kelvin	°F
2	C0 =	−200	73.15	−328
3	273.15	−100	173.15	−148
4		0	273.15	32
5		100	373.15	212
6		200	473.15	392
7				
8				
9				
10				

(c)

	A	B	C	D
1	Constant:	°C	kelvin	°F
2	C0 =	−200	73.15	−328
3	273.15	−100	173.15	−148
4		0	273.15	32
5		100	373.15	212
6		200	473.15	392
7				
8	Formulas:			
9	C2 = B2+$A$3			
10	D2 = (9/5)*B2+32			

(d)

Figure 3-3 Constructing a spreadsheet for temperature conversions.

Now comes the beauty of a spreadsheet. Instead of typing many similar formulas, select cells C2, C3, C4, C5, and C6 all together and select the FILL DOWN command from the EDIT menu. This command tells the computer to do the same thing in cells C3 through C6 that was done in cell C2. The numbers 173.15, 273.15, 373.15, and 473.15 will appear in cells C3 through C6.

When computing the output in cell C3, the computer automatically uses input from cell B3 instead of cell B2. The reason for the dollar signs in $A$3 is that we do not want the computer to go down to cell A4 to find input for cell C3. $A$3 is called an *absolute reference* to cell A3. No matter what cell uses the constant C_0, we want it to come from cell A3. The reference to cell B2 is a *relative reference*. Cell C2 will use the contents of cell B2. Cell C6 will use the contents of cell B6. In general, references to constants in column A will be absolute (with dollar signs). References to numbers in the remainder of a spreadsheet will usually be relative (without dollar signs).

Absolute reference: $A$3
Relative reference: B2

In cell D1, enter the label "°F". In cell D2, type the formula "= (9/5)*B2 + 32". This is equivalent to writing °F = (9/5)*°C + 32. The slash (/) is a division sign and the asterisk (*) is a multiplication sign. Parentheses are used to make the computer do what we intend. Operations inside parentheses are carried out before operations outside the parentheses. The computer responds to this formula by writing −328 in cell D2. This is the Fahrenheit equivalent to −200°C. Select cells D2 through D6 all together and use FILL DOWN to complete the table shown in Figure 3-3c.

Order of Operations

The arithmetic operations in a spreadsheet are addition, subtraction, multiplication, division, and exponentiation (which uses the symbol ^). The order of operations in formulas is ^ first, followed by * and / (evaluated in order from left to right as they appear), finally followed by + and − (also evaluated from left to right). Make liberal use of parentheses to be sure that the computer does what you intend. The contents of parentheses are evaluated first, before carrying out operations outside the parentheses. Here are some examples:

9/5*100+32 = (9/5)*100+32 = (1.8)*100+32 = (1.8*100)+32 = (180)+32 = 212
9/5*(100+32) = 9/5*(132) = (1.8)*(132) = 237.6
9+5*100/32 = 9+(5*100)/32 = 9+(500)/32 = 9+(500/32) = 9+(15.625) = 24.625
9/5^2+32 = 9/(5^2)+32 = (9/25)+32 = (0.36)+32 = 32.36
−2^2 = 4 but −(2^2) = −4

When in doubt about how an expression will be evaluated by the computer, use parentheses to force it to do what you intend.

Documentation and Readability

If your spreadsheet cannot be read by another person without your help, it needs better documentation. (The same is true of your lab notebook!)

If you look at your spreadsheet next month, you will probably not know what formulas were used. Therefore, we *document* the spreadsheet to show how it works by adding the text in cells A8, A9, and A10 in Figure 3-3d. In cell A8, write "Formulas:". In cell A9, write "C2 = B2+$A$3" and, in cell A10, write "D2 = (9/5)*B2+32". Documentation is an excellent practice for every spreadsheet. As you learn to use your spreadsheet, you should use COPY and PASTE commands to copy the formulas used in cells C2 and D2 into the text in cells A9 and A10. This practice saves time and reduces transcription errors. Another basic form of documentation that we will add to future spreadsheets is a title in cell A1. A title such as "Temperature Conversions" tells us immediately what spreadsheet we are looking at.

For additional readability, select how many decimal places are displayed in a cell or column. The computer retains more digits for calculations. It does not throw away digits that are not displayed. You can also control whether numbers are displayed in decimal or exponential notation. To alter the format of a cell, go to the FORMAT menu and select CELLS. Select the Number tab and choose the category Number or Scientific. In either case, you will be asked how many decimal places to display. Many other cell formatting options are available in the FORMAT CELLS window.

Ask Yourself

3-E. Reproduce the spreadsheet in Figure 3-3 on your computer. The boiling point of N_2 at 1 atm pressure is −196°C. Use your spreadsheet to find the kelvin and Fahrenheit equivalents of −196°C. Check your answers with your calculator.

3-6 Graphing in Excel

Humans require a visual display to understand the relation between two columns of numbers. This section introduces the basics of creating a graph with Excel.

First, we will generate some data to plot. The spreadsheet in Figure 3-4 computes the density of water as a function of temperature (°C) with the equation

$$\text{density (g/mL)} = a_0 + a_1{*}T + a_2{*}T^2 + a_3{*}T^3 \tag{3-8}$$

Equation 3-8 is accurate to five decimal places in the range 4° to 40°C.

where $a_0 = 0.999\,89$, $a_1 = 5.332\,2 \times 10^{-5}$, $a_2 = -7.589\,9 \times 10^{-6}$, and $a_3 = 3.671\,9 \times 10^{-8}$. After writing a title in cell A1, enter the constants a_0 to a_3 in column A. Column B is labeled "Temp (°C)" and column C is labeled "Density (g/mL)". Enter values of temperature in column B. In cell C4, type the formula "=$A$5 + $A$7*B4 + $A$9*B4^2 + $A$11*B4^3", which uses the exponent symbol ^ to compute T^2 and T^3. When you enter the formula, the number 0.999 97 is computed in cell C4. The remainder of column C is completed with a FILL DOWN command from the EDIT menu. The spreadsheet is not finished until it is documented by entering text in cells A13 and A14 to show what formula was used in column C.

A spreadsheet does not know how many figures are significant. You can choose how many decimal places are displayed to be consistent with the number of significant figures associated with a cell or column.

Now we want to make a graph of density in column C versus temperature in column B. Density will appear on the y-axis (the *ordinate*) and temperature will be on the x-axis (the *abscissa*). There may be some variation from the following description in different versions of Excel.

Go to the INSERT menu and select CHART. A window appears with a variety of options. The one that you will almost always want in this book is XY (Scatter). When you highlight XY (Scatter), several new options appear. For this example, select the one that shows data points connected by a smooth curve. Click Next to move to the next window.

Now you are asked which cells contain the data to be plotted. Identify the x data by writing B4:B11 next to Data Range. Then write a comma and identify the y data by writing C4:C11. The input for Data Range now looks like B4:B11,C4:C11. Click the button to show that data are in columns not rows. Click Next.

Now a small graph of your data appears. If it does not look as expected, make sure that you selected the correct data, with x before y. The new window asks you

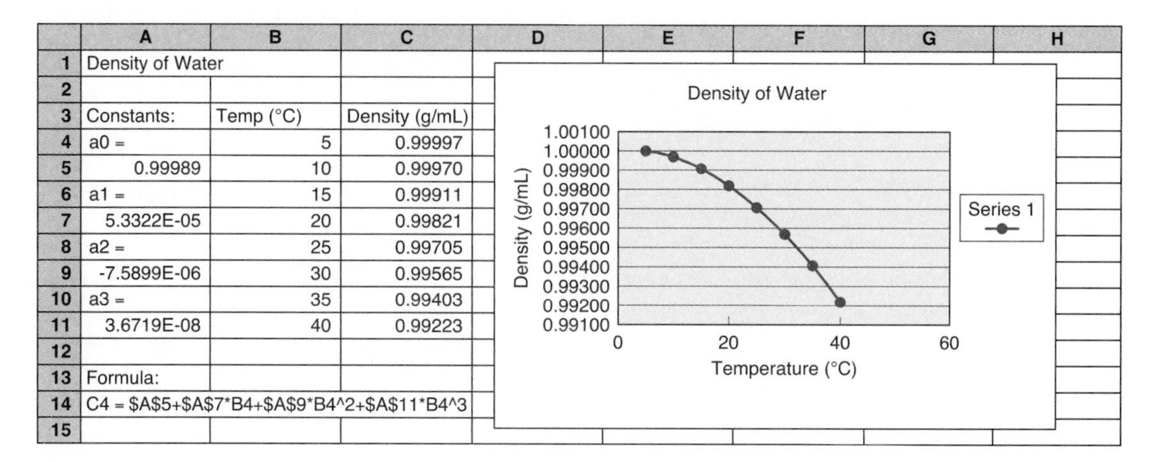

	A	B	C
1	Density of Water		
2			
3	Constants:	Temp (°C)	Density (g/mL)
4	a0 =	5	0.99997
5	0.99989	10	0.99970
6	a1 =	15	0.99911
7	5.3322E-05	20	0.99821
8	a2 =	25	0.99705
9	-7.5899E-06	30	0.99565
10	a3 =	35	0.99403
11	3.6719E-08	40	0.99223
12			
13	Formula:		
14	C4 = $A$5+$A$7*B4+$A$9*B4^2+$A$11*B4^3		
15			

Figure 3-4 Spreadsheet for computing the density of water as a function of temperature.

for axis labels and an optional title for the graph. For the title, write "Density of Water". For the x-axis, enter "Temperature (°C)" and, for the y-axis, write "Density (g/mL)". Click Next.

Now you are given the option of drawing the graph on a new sheet or on the same sheet that is already open. For this case, select "As object in Sheet 1". Click Finish and the chart will appear on your spreadsheet. Grab the chart with your mouse, size it as you wish, and move it to the right of your spreadsheet, as in Figure 3-4.

Excel provides many options for formatting the graph. Double click on the y-axis and a window appears. Select the Patterns tab. Change Minor tic mark type from None to Outside and click OK. You will see new tic marks appear on the y-axis. Double click on the y-axis again and select the Number tab. Change the decimal places to 3 and click OK. Double click on the y-axis again and select the Scale tab. Set the minimum to 0.992 and the maximum to 1.000 and click OK.

Double click on the x-axis and select Patterns tab. Change the Minor tic mark type from None to Outside. Select Scale tab and set the maximum to 40, the major unit to 10, the minor unit to 5, and click OK.

Double click on the gray area of the graph and a window called Patterns appears. Select Automatic for the Border and None for the Area. These choices remove the gray background and give a solid line around the graph. To add vertical lines at the major tic marks, select the graph with the mouse. Then go to the CHART menu and select CHART OPTIONS. In the window that appears, select Gridlines. For the Value (X) axis, check Major gridlines. Then select the tab for Legend and remove the check mark from Show Legend. The legend will disappear. Click OK. You should be getting the idea that you can format virtually any part of the chart.

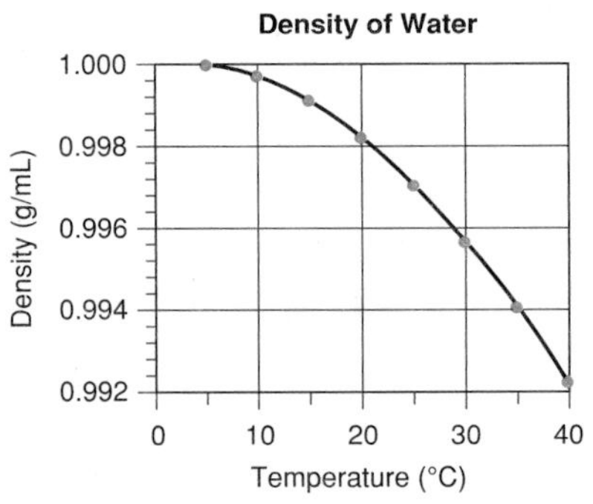

Click on the outer border of the chart and handles appear. Grab the one on the right and resize the chart so that it does not extend past column F of the spreadsheet. Grab the handle at the bottom and resize the chart so that it does not extend below row 15. When you resized the chart, letters and numbers shrank. Double click on each set of numbers and change the font to 8 points. Double click on the labels and change the letters to 9 points. Your chart should now look like the one in Figure 3-5.

Figure 3-5 Chart from Figure 3-4 after reformatting.

To write on the chart, go to the VIEW menu and select TOOLBARS and DRAWING. Select the Text Box tool from the Drawing toolbar, click inside your chart, and you can begin typing words. You can move the words around and change their format. You can draw arrows on the chart with the Arrow tool. If you double click on a data point on the chart, a box appears that allows you to change the plotting symbols.

3-F. Reproduce the spreadsheet in Figure 3-4 and the graph in Figure 3-5 on your computer.

Key Equations

Definition of logarithm	If $n = 10^a$, then a is the logarithm of n.
Definition of antilogarithm	If $n = 10^a$, then n is the antilogarithm of a.
Relative uncertainty	$\text{relative uncertainty} = \dfrac{\text{absolute uncertainty}}{\text{magnitude of measurement}}$
Percent relative uncertainty	percent relative uncertainty = 100 × relative uncertainty

Uncertainty in addition and subtraction
$e_4 = \sqrt{e_1^2 + e_2^2 + e_3^2}$ (use absolute uncertainties)
e_4 = uncertainty in final answer
e_1, e_2, e_3 = uncertainty in individual terms

Uncertainty in multiplication and division
$\%e_4 = \sqrt{\%e_1^2 + \%e_2^2 + \%e_3^2}$
(use percent relative uncertainties)

Important Terms

absolute uncertainty
accuracy
antilogarithm
characteristic
determinate error
indeterminate error
logarithm
mantissa
precision
random error
relative uncertainty
significant figure
systematic error

Problems

3-1. Round each number as indicated:
(a) 1.236 7 to 4 significant figures
(b) 1.238 4 to 4 significant figures
(c) 0.135 2 to 3 significant figures
(d) 2.051 to 2 significant figures
(e) 2.005 0 to 3 significant figures

3-2. Round each number to three significant figures:
(a) 0.216 74; **(b)** 0.216 5; **(c)** 0.216 500 3; **(d)** 0.216 49.

3-3. Indicate how many significant figures there are in
(a) 0.305 0; **(b)** 0.003 050; **(c)** 1.003×10^4.

3-4. Write each answer with the correct number of digits:
(a) $1.0 + 2.1 + 3.4 + 5.8 = 12.300\ 0$
(b) $106.9 - 31.4 = 75.500\ 0$
(c) $107.868 - (2.113 \times 10^2) + (5.623 \times 10^3) = 5\ 519.568$
(d) $(26.14/37.62) \times 4.38 = 3.043\ 413$
(e) $(26.14/37.62 \times 10^8) \times (4.38 \times 10^{-2}) = 3.043\ 413 \times 10^{-10}$
(f) $(26.14/3.38) + 4.2 = 11.933\ 7$
(g) $\log (3.98 \times 10^4) = 4.599\ 9$
(h) $10^{-6.31} = 4.897\ 79 \times 10^{-7}$

3-5. Write each answer with the correct number of digits:
(a) $3.021 + 8.99 = 12.011$
(b) $12.7 - 1.83 = 10.87$
(c) $6.345 \times 2.2 = 13.959\ 0$
(d) $0.030\ 2 \div (2.114\ 3 \times 10^{-3}) = 14.283\ 69$
(e) $\log (2.2 \times 10^{-18}) = ?$
(f) antilog $(-2.224) = ?$
(g) $10^{-4.555} = ?$

3-6. Find the formula mass of **(a)** $BaCl_2$ and **(b)** $C_{31}H_{32}O_8N_2$ with the correct number of significant figures.

3-7. Find the molecular mass of $Mn_2(CO)_{10}$ with the correct number of significant figures.

3-8. Why do we use quotation marks around the word *true* in the statement that accuracy refers to how close a measured value is to the "true" value?

3-9. **(a)** Explain the difference between systematic and random errors. State whether the errors in **(b)–(e)** are random or systematic.
(b) A 25-mL transfer pipet consistently delivers 25.031 ± 0.009 mL when drained from the mark.
(c) A 10-mL buret consistently delivers 1.98 ± 0.01 mL when drained from exactly 0 to exactly 2 mL and consistently delivers 2.03 mL ± 0.02 mL when drained from 2 to 4 mL.
(d) A 10-mL buret delivered 1.983 9 g of water when drained from exactly 0.00 to 2.00 mL. The next time I delivered water from the 0.00 to the 2.00 mL mark, the delivered mass was 1.990 0 g.
(e) Four consecutive 20.0-μL injections of a solution into a chromatograph were made (as in Figure 0-6) and the area of a particular peak was 4 383, 4 410, 4 401, and 4 390 units.
(f) A clean funnel that had been in the lab since last semester had a mass of 15.432 9 g. When filled with a solid precipitate and dried thoroughly in the oven at 110°C, the mass was 15.845 6 g. The calculated mass of precipitate was therefore 15.845 6 − 15.432 9 = 0.412 7 g. Is there systematic or random error (or both) in the mass of precipitate?

3-10. Rewrite the number 3.123 56 (±0.167 89%) in the forms **(a)** number (± absolute uncertainty) and **(b)** number (± percent relative uncertainty) with an appropriate number of digits.

3-11. Write each answer with the correct number of digits. Find the absolute uncertainty and percent relative uncertainty for each answer.

(a) 6.2 (±0.2) − 4.1 (±0.1) = ?

(b) 9.43 (±0.05) × 0.016 (±0.001) = ?

(c) [6.2 (±0.2) − 4.1 (±0.1)] ÷ 9.43 (±0.05) = ?

(d) 9.43 (±0.05) × {[6.2 (±0.2) × 10^{-3}] + [4.1 (±0.1) × 10^{-3}]} = ?

3-12. Write each answer with a reasonable number of figures. Find the absolute uncertainty and percent relative uncertainty for each answer.

(a) [12.41 (±0.09) ÷ 4.16 (±0.01)] × 7.068 2 (±0.000 4) = ?

(b) [3.26 (±0.10) × 8.47 (±0.05)] − 0.18 (±0.06) = ?

(c) 6.843 (±0.008) × 10^4 ÷ [2.09 (±0.04) − 1.63 (±0.01)] = ?

3-13. Write each answer with the correct number of digits. Find the absolute uncertainty and percent relative uncertainty for each answer.

(a) 9.23 (±0.03) + 4.21 (±0.02) − 3.26 (±0.06) = ?

(b) 91.3 (±1.0) × 40.3 (±0.2)/21.2 (±0.2) = ?

(c) [4.97 (±0.05) − 1.86 (±0.01)]/21.2 (±0.2) = ?

(d) 2.016 4 (±0.000 8) + 1.233 (±0.002) + 4.61 (±0.01) = ?

(e) 2.016 4 (±0.000 8) × 10^3 + 1.233 (±0.002) × 10^2 + 4.61 (±0.01) × 10^1 = ?

3-14. Find the absolute and percent relative uncertainty and express each answer with a reasonable number of significant figures.

(a) 3.4 (±0.2) + 2.6 (±0.1) = ?

(b) 3.4 (±0.2) ÷ 2.6 (±0.1) = ?

(c) [3.4 (±0.2) × 10^{-8}] ÷ [2.6 (±0.1) × 10^3] = ?

(d) [3.4 (±0.2) − 2.6 (±0.1)] × 3.4 (±0.2) = ?

3-15. *Uncertainty in molecular mass.* The periodic table inside the cover of this book has a note in the legend about uncertainties in atomic mass. Here is an example of how to find uncertainty in molecular mass. For the compound diborane, B_2H_6, first multiply the uncertainty in each atomic mass by the number of atoms in the formula:

$$\begin{aligned} 2\text{B}: 2 \times 10.811 \pm 0.007 &= 21.622 \quad \pm 0.014 \\ 6\text{H}: 6 \times 1.007\,94 \pm 0.000\,07 &= 6.047\,64 \pm 0.000\,42 \\ \text{sum} &= 27.669\,64 \pm ? \end{aligned}$$

Then find the uncertainty in the sum of atomic masses by the formula for addition:

$$\text{uncertainty} = \sqrt{e_1^2 + e_2^2} = \sqrt{0.014^2 + 0.000\,42^2} = 0.014$$

$$\text{molecular mass} = 27.670 \pm 0.014 \text{ (or } 27.67 \pm 0.01)$$

Express the molecular mass (± uncertainty) of benzene, C_6H_6, with the correct number of significant figures.

3-16. As in Problem 3-15, express the molecular mass of $C_6H_{13}B$ with the correct number of significant figures and find its uncertainty.

3-17. **(a)** Show that the formula mass of NaCl is 58.443 (±0.002) g/mol.

(b) You dissolve 2.634 (±0.002) g of NaCl in a volumetric flask with a volume of 100.00 (±0.08) mL. Express the molarity of NaCl and its uncertainty with an appropriate number of digits.

3-18. **(a)** For use in an iodine titration, you prepare a solution from 0.222 2 (±0.000 2) g of KIO_3 [FM 214.001 0 (±0.000 9)] in 50.00 (±0.05) mL. Find the molarity and its uncertainty with an appropriate number of significant figures.

(b) Would your answer be affected significantly if the reagent were only 99.9% pure?

3-19. A 500.0 ± 0.2-mL solution was prepared by dissolving 25.00 ± 0.03 mL of methanol (CH_3OH, density = 0.791 4 ± 0.000 2 g/mL, molecular mass = 32.041 9 ± 0.000 9 g/mol) in chloroform. Find the molarity ± uncertainty of the methanol.

3-20. Your instructor has asked you to prepare 2.00 L of 0.169 M NaOH from a stock solution of 53.4 (±0.4) wt% NaOH with a density of 1.52 (±0.01) g/mL.

(a) How many milliliters of stock solution will you need?

(b) If the uncertainty in delivering the NaOH is ±0.10 mL, calculate the absolute uncertainty in the molarity (0.169 M). Assume negligible uncertainty in the formula mass of NaOH and in the final volume, 2.00 L.

3-21. *Formula mass calculator.* Reproduce the spreadsheet shown here. Atomic masses are in column A. Numbers of atoms are in columns B to E. Write a formula in column F to compute formula mass from masses in column A and the numbers of atoms. I find this spreadsheet extremely useful. As you need to, add more atomic masses to column A and add columns between E and F for additional atoms.

	A	B	C	D	E	F	G
1	Formula Mass Calculator					Formula	
2		C	H	O	N	mass	
3	C =	1	4	1		32.0419	CH3OH
4	12.0107	5	5	1	1	95.0993	C5H5NO
5	H =						
6	1.00794						
7	O =						
8	15.9994						
9	N =						
10	14.0067						

3-22. *Graphing.* Atmospheric CO_2 has been increasing from the burning of fossil fuel since the dawn of the industrial age as shown in the graph below.

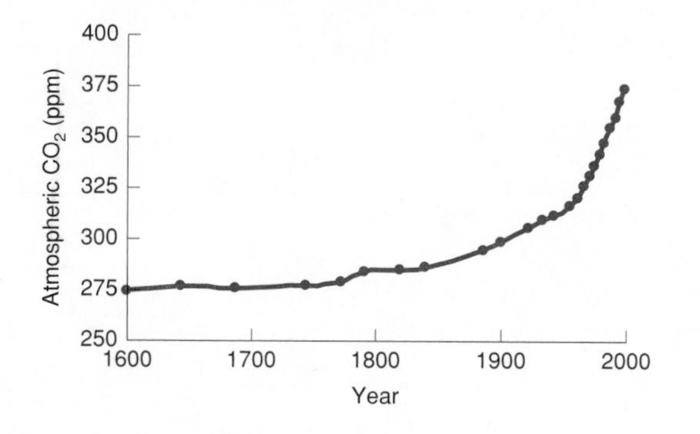

Copy the following data into two columns of a spreadsheet and use the spreadsheet to reproduce the graph. Adjust the axes so that the scales and tic marks are the same as in the graph above.

Year	CO_2 (ppm)	Year	CO_2 (ppm)	Year	CO_2 (ppm)
1603	274	1889	295	1974	330
1646	277	1903	299	1978	335
1691	276	1925	305	1982	341
1747	277	1937	309	1986	347
1776	279	1945	311	1990	354
1795	284	1959	316	1994	359
1823	285	1965	320	1998	367
1843	286	1970	326	2002	373

How Would You Do It?

3-23. Here are two methods that you might use to prepare a dilute silver nitrate solution:

Method 1: Weigh out 0.046 3 g $AgNO_3$ and dissolve it in a 100-mL volumetric flask.

Method 2: Weigh out 0.463 0 g $AgNO_3$ and dissolve it in a 100-mL volumetric flask. Then pipet 10 mL of this solution into a fresh 100-mL flask and dilute to the mark.

The uncertainty in the balance is ± 3 in the last decimal place. Which method is more accurate?

Further Reading

J. R. Taylor, An *Introduction to Error Analysis,* 2nd ed. (Sausalito, CA: University Science Books, 1997).

E. J. Billo, *Microsoft Excel for Chemists,* 2nd ed. (New York: Wiley, 2001).

R. de Levie, *How to Use Excel® in Analytical Chemistry and in General Scientific Data Analysis* (Cambridge: Cambridge University Press, 2001).

Is My Red Blood Cell Count High Today?

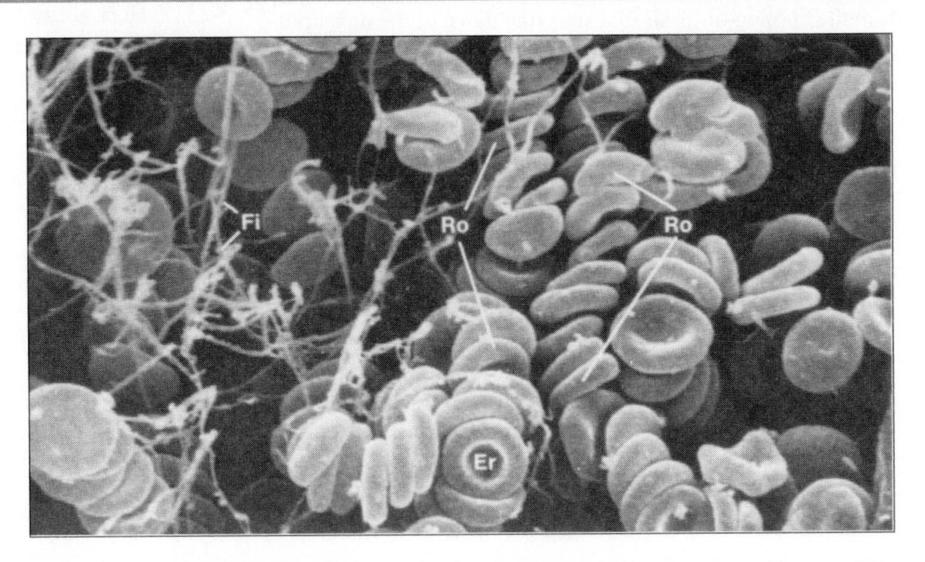

Red blood cells (erythrocytes, Er) tangled in fibrin threads (Fi) in a blood clot. Stacks of erythrocytes in a clot are called a rouleaux formation (Ro). [From R. H. Kardon, *Tissues and Organs* (San Francisco: W. H. Freeman and Company, 1978), p. 39.]

All measurements contain experimental error, so it is impossible to be completely certain of a result. Nevertheless, we seek to answer questions such as "Is my red blood cell count today higher than usual?" If today's count is twice as high as usual, it is probably truly higher than normal. But what if the "high" count is not excessively above "normal" counts?

Count on "normal" days	Today's count
5.1, 5.3, 4.8, 5.4, 5.2 $\times 10^6$ cells/μL	5.6×10^6 cells/μL

The number 5.6 is higher than the five normal values, but the random variation in normal values might lead us to expect that 5.6 will be observed on some "normal" days.

The study of statistics allows us to say that, over a long period, today's value will be observed on 1 out of 20 normal days. It is still up to you to decide what to do with this information.

CHAPTER 4

Statistics

Experimental measurements always have some random error, so no conclusion can be drawn with complete certainty. Statistics gives us tools to accept conclusions that have a high probability of being correct and to reject conclusions that do not. This chapter describes basic statistical tests and introduces the method of least squares for calibration curves.

Statistics deals only with random error—not determinate error. We must be ever vigilant and try to detect and prevent systematic errors. Two good ways to detect systematic errors are to analyze certified standards to see whether our method gives the expected results and to use different methods of analysis and see whether the results agree.

4-1 The Gaussian Distribution

Nerve cells communicate with muscle cells by releasing neurotransmitter molecules adjacent to the muscle. As shown in Figure 4-1, neurotransmitters bind to membrane proteins of the muscle cell and open up channels that permit cations to diffuse into the cell. Ions entering the cell trigger contraction of the muscle.

Bert Sakmann and Erwin Neher shared the Nobel Prize in Medicine or Physiology in 1991 for their work on signal transmission at the neuromuscular junction.

Channels are all the same size, so each should allow a similar rate of ion passage across the membrane. Because ions are charged particles, the flow of ions is equivalent to a flow of electricity across the membrane. Of the 922 ion-channel

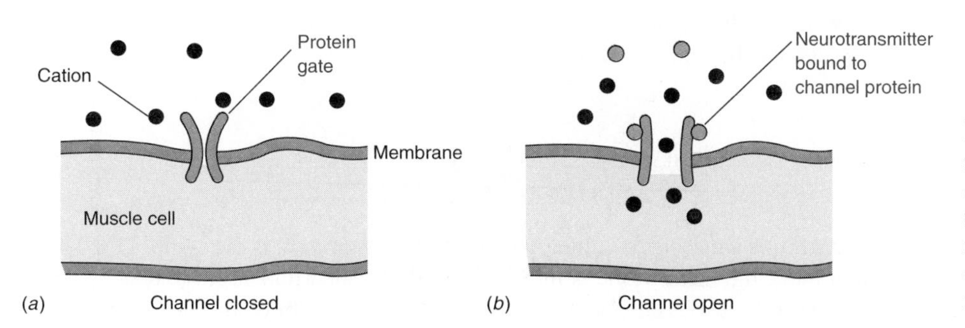

Figure 4-1 (*a*) In the absence of neurotransmitter, the ion channel is closed and cations cannot enter the muscle cell. (*b*) In the presence of neurotransmitter, the channel opens, cations enter the cell, and muscle action is initiated.

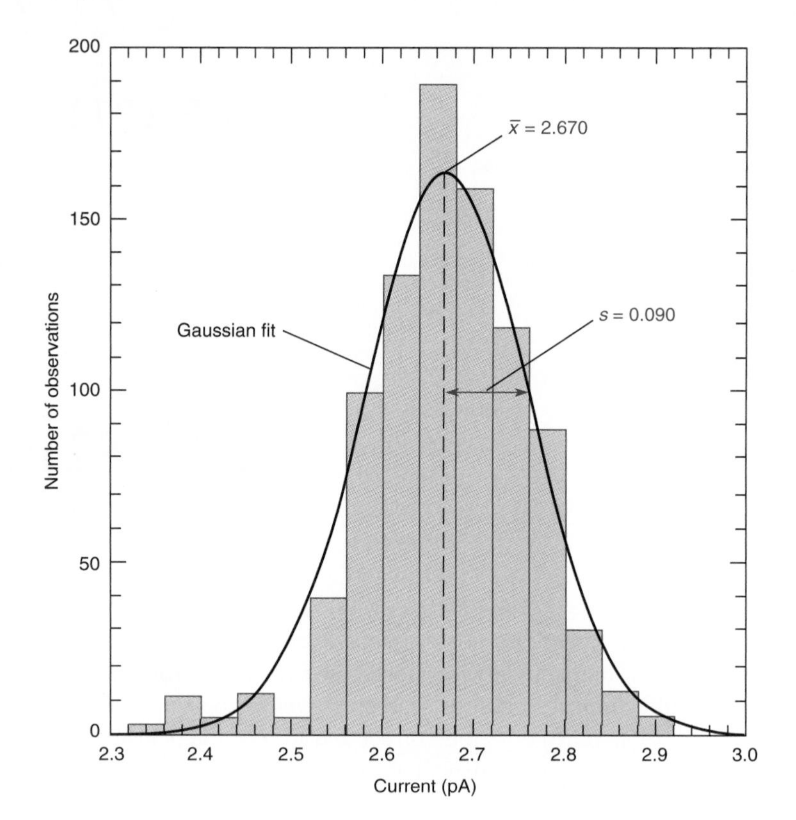

Figure 4-2 Observed cation current passing through individual channels of a frog muscle cell. The smooth line is the Gaussian curve that has the same mean and standard deviation as the measured data. The bar chart is also called a *histogram*. [Data from Nobel Lecture of B. Sakmann, *Angew. Chem. Int. Ed. Engl.* **1992**, *31*, 830.]

responses recorded for Figure 4-2, 190 are in the narrow range from 2.64 to 2.68 pA (picoamperes, 10^{-12} amperes), represented by the tallest bar at the center of the chart. The next most probable responses fall in the range just to the right of the tallest bar, and the third most probable responses fall in the range just to the left of the tallest bar.

The bar graph in Figure 4-2 is typical of many laboratory measurements: The most probable response is at the center, and the probability of observing other responses decreases as the distance from the center increases. The smooth, bell-shaped curve superimposed on the data in Figure 4-2 is called a **Gaussian distribution**. The more measurements made on any physical system, the closer the bar chart comes to the smooth curve.

Mean and Standard Deviation

Mean locates center of distribution.
Standard deviation measures width of distribution.

A Gaussian distribution is characterized by a *mean* and a *standard deviation*. The mean is the *center* of the distribution, and the standard deviation measures the *width* of the distribution.

The arithmetic **mean**, $\bar{x}$, also called the **average**, is the sum of the measured values divided by the number of measurements.

Mean:
$$\bar{x} = \frac{\sum_i x_i}{n} = \frac{1}{n}(x_1 + x_2 + x_3 + \cdots + x_n) \qquad (4\text{-}1)$$

The smaller the standard deviation, the more *precise* (reproducible) the results. Greater precision does not necessarily imply greater *accuracy*, which means nearness to the "truth."

where each x_i is a measured value. A capital Greek sigma, Σ, is the symbol for a sum. In Figure 4-2, the mean value is indicated by the dashed line at 2.670 pA.

The **standard deviation**, s, is a measure of the width of the distribution. *The smaller the standard deviation, the narrower the distribution.*

Standard deviation:

$$s = \sqrt{\frac{\sum_i (x_i - \bar{x})^2}{n - 1}} \quad (4\text{-}2)$$

In Figure 4-2, $s = 0.090$ pA. Figure 4-3 shows that if the standard deviation were doubled, the Gaussian curve for the same number of observations would be shorter and broader.

The *relative standard deviation* is the standard deviation divided by the average. It is usually expressed as a percentage. For $s = 0.090$ pA and $\bar{x} = 2.670$ pA, the relative standard deviation is $(0.090/2.670) \times 100 = 3.4\%$.

The quantity $n - 1$ in the denominator of Equation 4-2 is called the *degrees of freedom.* Initially we have n independent data points, which represent n pieces of information. After computing the average, there are only $n - 1$ independent pieces of information left, because we could calculate the nth data point if we know $n - 1$ points and the average.

The symbols $\bar{x}$ and s apply to a finite set of measurements. For an infinite set of data, the true mean (called the *population mean*) is designated μ (Greek mu) and the true standard deviation is denoted by σ (lowercase Greek sigma).

The term *variance* is found in many applications. **Variance** is the square of the standard deviation.

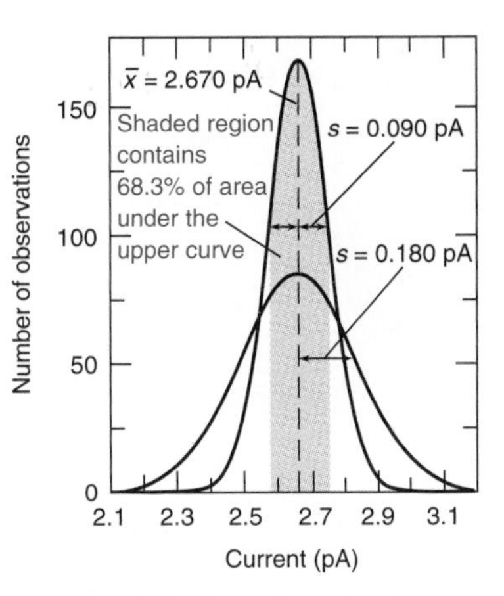

Figure 4-3 Gaussian curves showing the effect of doubling the standard deviation. The number of observations described by each curve is the same.

Example Mean and Standard Deviation

Find the mean, standard deviation, and relative standard deviation for the set of measurements (7, 18, 10, 15).

SOLUTION The mean is

$$\bar{x} = \frac{7 + 18 + 10 + 15}{4} = 12._5$$

To avoid accumulating round-off errors, retain one more digit for the mean and the standard deviation than was present in the original data. The standard deviation is

$$s = \sqrt{\frac{(7 - 12._5)^2 + (18 - 12._5)^2 + (10 - 12._5)^2 + (15 - 12._5)^2}{4 - 1}} = 4._9$$

The mean and the standard deviation should both end at the *same decimal place.* For $\bar{x} = 12._5$, we write $s = 4._9$. The relative standard deviation is $(4._9/12._5) \times 100 = 39\%$.

Test Yourself Use the mean and standard deviation functions on your calculator to show that you can reproduce the results in this example.

If your calculator gives 4.3 instead of 4.9, it is using $n = 4$ in the denominator instead of $n - 1 = 4 - 1$. The correct factor is $n - 1$.

Excel has built-in functions for average and standard deviation. Enter the numbers 7, 18, 10, and 15 in cells A1 through A4 of a spreadsheet. In cell A5, enter the formula "=Average(A1:A4)"; and in cell A6, enter the formula "=Stdev(A1:A4)".

	A	B
1	7	
2	18	
3	10	
4	15	
5	12.50	
6	4.93	
7	A5 = Average(A1:A4)	
8	A6 = Stdev(A1:A4)	

Results in the spreadsheet in the margin reproduce those in the example above. For a list of built-in functions, go to the INSERT menu and select FUNCTION. Double click on a function and a box appears to describe the function.

Other terms that you should know are the median and the range. The *median* is the middle number in a series of measurements. When (8, 17, 11, 14, 12) are ordered from lowest to highest to give (8, 11, 12, 14, 17), the middle number (12) is the median. For an even number of measurements, the median is the average of the two middle numbers. For (8, 11, 12, 14), the median is 11.5. Some people prefer to report the median instead of the average, because the median is less influenced by outlying data. The *range* is the difference between the highest and lowest values. The range of (8, 17, 11, 14, 12) is $17 - 8 = 9$.

Standard Deviation and Probability

For an ideal Gaussian distribution, 68.3% of the measurements lie within one standard deviation on either side of the mean (in the interval $\mu \pm \sigma$). That is, 68.3% of the area beneath a Gaussian curve lies in the interval $\mu \pm \sigma$, as shown in Figure 4-3. The percentage of measurements lying in the interval $\mu \pm 2\sigma$ is 95.5% and the percentage in the interval $\mu \pm 3\sigma$ is 99.7%. For real data with a standard deviation s, about 1 in 20 measurements (4.5%) will lie outside the range $\bar{x} \pm 2s$, and only 3 in 1 000 measurements (0.3%) will lie outside the range $\bar{x} \pm 3s$. Table 4-1 shows the correspondence between ideal Gaussian behavior and the observations in Figure 4-2.

Table 4-1 Percentage of observations in Gaussian distribution

Range	Gaussian distribution	Observed in Figure 4-2
$\mu \pm 1\sigma$	68.3%	71.0%
$\mu \pm 2\sigma$	95.5	95.6
$\mu \pm 3\sigma$	99.7	98.5

The Gaussian distribution is symmetric. If 4.5% of measurements lie outside the range $\mu \pm 2\sigma$, 2.25% of measurements are above $\mu + 2\sigma$ and 2.25% are below $\mu - 2\sigma$.

Question What fraction of observations in a Gaussian distribution is expected to be below $\mu - 3\sigma$?

Ask Yourself

4-A. What are the mean, standard deviation, relative standard deviation, median, and range for the numbers 821, 783, 834, and 855? All but the range should be expressed with one extra digit beyond the last significant digit.

4-2 Student's *t*

Student's *t* is the statistical tool used to express confidence intervals and to compare results from different experiments. You can use it to evaluate the probability that your red blood cell count will be found in a certain range on "normal" days.

"Student" was the pseudonym of W. S. Gosset, whose employer, the Guinness Breweries of Ireland, restricted publications for proprietary reasons. Because of the importance of his work, Gosset published it under an assumed name in 1908.

Confidence Intervals

From a limited number of measurements, it is impossible to find the true mean, μ, or the true standard deviation, σ. What we can determine are $\bar{x}$ and s, the sample mean and the sample standard deviation. The **confidence interval** is a range of values within which there is a specified probability of finding the true mean. We say that the true mean, μ, is likely to lie within a certain distance from the measured mean, $\bar{x}$. The confidence interval ranges from $-ts/\sqrt{n}$ below $\bar{x}$ to $+ts/\sqrt{n}$ above $\bar{x}$:

Using the confidence interval:

$$\mu = \bar{x} \pm \frac{ts}{\sqrt{n}} \tag{4-3}$$

Table 4-2 Values of Student's t

Degrees of freedom	Confidence level (%) 50	90	95	98	99	99.5	99.9
1	1.000	6.314	12.706	31.821	63.656	127.321	636.578
2	0.816	2.920	4.303	6.965	9.925	14.089	31.598
3	0.765	2.353	3.182	4.541	5.841	7.453	12.924
4	0.741	2.132	2.776	3.747	4.604	5.598	8.610
5	0.727	2.015	2.571	3.365	4.032	4.773	6.869
6	0.718	1.943	2.447	3.143	3.707	4.317	5.959
7	0.711	1.895	2.365	2.998	3.500	4.029	5.408
8	0.706	1.860	2.306	2.896	3.355	3.832	5.041
9	0.703	1.833	2.262	2.821	3.250	3.690	4.781
10	0.700	1.812	2.228	2.764	3.169	3.581	4.587
15	0.691	1.753	2.131	2.602	2.947	3.252	4.073
20	0.687	1.725	2.086	2.528	2.845	3.153	3.850
25	0.684	1.708	2.060	2.485	2.787	3.078	3.725
30	0.683	1.697	2.042	2.457	2.750	3.030	3.646
40	0.681	1.684	2.021	2.423	2.704	2.971	3.551
60	0.679	1.671	2.000	2.390	2.660	2.915	3.460
120	0.677	1.658	1.980	2.358	2.617	2.860	3.373
∞	0.674	1.645	1.960	2.326	2.576	2.807	3.291

In calculating confidence intervals, σ may be substituted for s in Equation 4-3 if you have a great deal of experience with a particular method and have therefore determined its "true" population standard deviation. If σ is used instead of s, the value of t to use in Equation 4-3 comes from the bottom row of this table.

where s is the measured standard deviation, n is the number of observations, and t is Student's t, taken from Table 4-2. Remember that in this table the *degrees of freedom* are equal to $n - 1$. If there are five data points, there are four degrees of freedom.

Example Calculating Confidence Intervals

In replicate analyses, the carbohydrate content of a glycoprotein (a protein with sugars attached to it) is found to be 12.6, 11.9, 13.0, 12.7, and 12.5 g of carbohydrate per 100 g of protein. Find the 50% and 90% confidence intervals for the carbohydrate content.

SOLUTION First we calculate $\bar{x} = 12.5_4$ and $s = 0.4_0$ for the five measurements. To find the 50% confidence interval, look up t in Table 4-2 under 50 and across from *four* degrees of freedom (degrees of freedom $= n - 1$). The value of t is 0.741, so the confidence interval is

$$\mu\ (50\%) = \bar{x} \pm \frac{ts}{\sqrt{n}} = 12.5_4 \pm \frac{(0.741)(0.4_0)}{\sqrt{5}} = 12.5_4 \pm 0.1_3$$

The 90% confidence interval is

$$\mu\ (90\%) = \bar{x} \pm \frac{ts}{\sqrt{n}} = 12.5_4 \pm \frac{(2.132)(0.4_0)}{\sqrt{5}} = 12.5_4 \pm 0.3_8$$

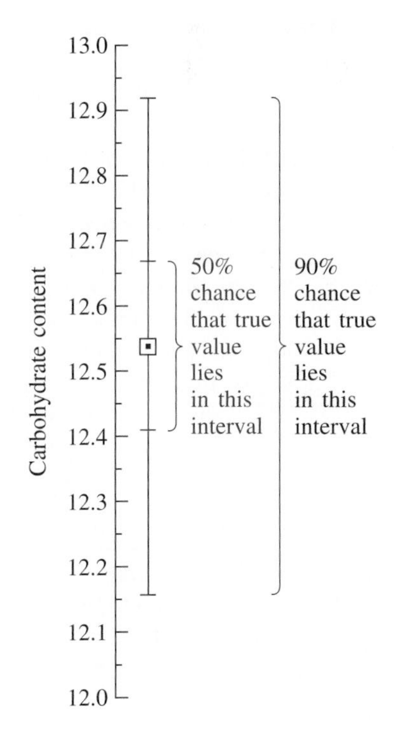

These calculations mean that there is a 50% chance that the true mean, μ, lies in the range $12.5_4 \pm 0.1_3$ (12.4_1 to 12.6_7). There is a 90% chance that μ lies in the range $12.5_4 \pm 0.3_8$ (12.1_6 to 12.9_2).

You might appreciate Box 4-1 at this time.

Test Yourself If $\bar{x}$ and s are unchanged, but there are 10 measurements instead of 5, what would be the 90% confidence interval? (**Answer:** $12.5_4 \pm 0.2_3$)

Improving the Reliability of Your Measurements

Accuracy: nearness to the "truth"
Precision: reproducibility

We wish to be as accurate and precise as possible. Systematic errors reduce the accuracy of a measurement. If a pH meter is not calibrated correctly, it will give inaccurate readings, no matter how precise (reproducible) they are. Making a measurement by two different methods is a good way to detect systematic errors. If results do not agree within the expected uncertainty, systematic error is probably present.

Replicate measurements improve reliability:

If $s = 2.0\%$, 3 measurements give a 95% confidence interval of 5.0%:

$$\frac{\pm ts}{\sqrt{n}} = \frac{(4.303)(2.0\%)}{\sqrt{3}} = \pm 5.0\%$$

Making 9 measurements reduces the 95% confidence interval to 1.5%:

$$\frac{\pm ts}{\sqrt{n}} = \frac{(2.306)(2.0\%)}{\sqrt{9}} = \pm 1.5\%$$

(Values of t came from Table 4-2.)

Better precision gives smaller confidence intervals. The confidence interval is $\pm ts/\sqrt{n}$. To reduce the size of the confidence interval, we make more measurements (increase n) or decrease the standard deviation (s). The only way to reduce s is to improve your experimental procedure. In the absence of a procedural change, the way to reduce the confidence interval is to increase the number of measurements. Doubling the number of measurements decreases the factor $1/\sqrt{n}$ by $1/\sqrt{2} = 0.71$.

Comparison of Means with Student's t

Student's t can be used to compare two sets of measurements to decide whether they are "statistically different." We will test the *null hypothesis*, which says that the two means are *not different*. We adopt the following standard: If there is less than 1 chance in 20 that the difference between two sets of measurements arises from random vari-

Box 4-1 Analytical Chemistry and the Law

Here is a quotation you should know about:[1]

> Analytical chemists must always emphasize to the public that *the single most important characteristic of any result . . . is an adequate statement of its uncertainty interval.* Lawyers usually attempt to dispense with uncertainty and try to obtain unequivocal statements; therefore, an uncertainty interval must be clearly defined in cases involving litigation and/or enforcement proceedings. Otherwise, a value of 1.001 without a specified uncertainty, for example, may be viewed as legally exceeding a permissible level of 1.

Some legal limits make no scientific sense. The Delaney Amendment to the U.S. Food, Drug, and Cosmetic Act of 1958 stated that "no additive [in processed food] shall be deemed to be safe if it is found to induce cancer when ingested by man or animal . . . ." This statement meant that no detectable level of any carcinogenic (cancer-causing) pesticide may remain in processed foods, even if the level is below that which can be shown to cause cancer. In 1958, the sensitivity of analytical procedures was relatively poor. As procedures became more sensitive, detectable levels of chemical residues decreased by 10^3 to 10^6. A concentration that may have been acceptable in 1958 was 10^6 times above the legal limit in 1995, regardless of whether there was any evidence that such a low level is harmful. In 1996, Congress finally changed the law to take a health-based approach to pesticide regulation. Pesticides are supposed to be allowed at a level where there is a "reasonable certainty of no harm." For carcinogens, the allowed level will probably be set at a concentration that produces less than one excess cancer per million persons exposed. Unfortunately, the scientific basis for predicting effects of low-level exposure on human health is slim.

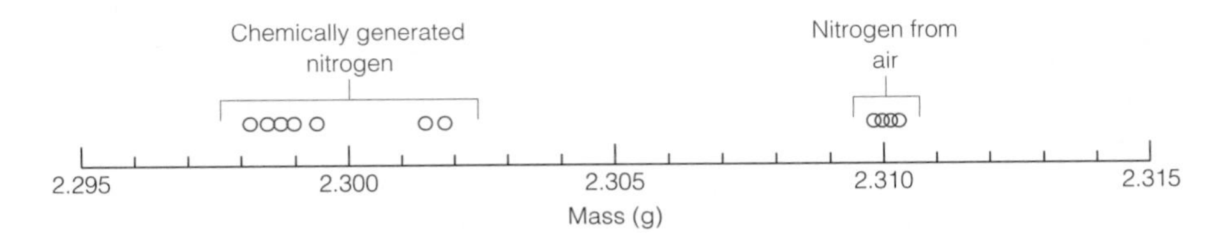

Figure 4-4 Lord Rayleigh's measurements of the mass of nitrogen isolated from air or generated by decomposition of nitrogen compounds. Rayleigh recognized that the difference between the two data sets was too great to be due to experimental error. He deduced that a heavier component, which turned out to be argon, was present in nitrogen isolated from air.

Table 4-3 Grams of nitrogen-rich gas isolated by Lord Rayleigh

From air	From chemical decomposition
2.310 17	2.301 43
2.309 86	2.298 90
2.310 10	2.298 16
2.310 01	2.301 82
2.310 24	2.298 69
2.310 10	2.299 40
2.310 28	2.298 49
—	2.298 89
Average	
2.310 10_9	2.299 47_2
Standard deviation	
0.000 14_3	0.001 37_9

SOURCE: R. D. Larsen, *J. Chem. Ed.* **1990**, *67*, 925.

ation in the data, then the difference is significant. This criterion gives us 95% confidence in concluding that two measurements are different. There is a 5% probability that our conclusion is wrong.[2]

An example comes from the work of Lord Rayleigh (John W. Strutt), who received the Nobel Prize in 1904 for discovering the inert gas argon—a discovery that came about when he noticed a discrepancy between two sets of measurements of the density of nitrogen. In Rayleigh's time, dry air was known to be ~⅕ oxygen and ~⅘ nitrogen. Rayleigh removed O_2 from air by reaction with red-hot copper [$Cu(s) + \frac{1}{2}O_2(g) \rightarrow CuO(s)$] and measured the density of the remaining gas by collecting it in a fixed volume at constant temperature and pressure. He then prepared the same volume of nitrogen by decomposition of nitrous oxide (N_2O), nitric oxide (NO), or ammonium nitrite ($NH_4^+NO_2^-$). Figure 4-4 and Table 4-3 show the mass of gas collected in each experiment. The average mass from air was 0.46% greater than the average mass of the same volume of gas from chemical sources.

If Rayleigh's measurements had not been performed with care, a 0.46% difference might have been attributed to experimental error. Instead, Rayleigh understood that the discrepancy was outside his margin of error, and he postulated that nitrogen from the air was mixed with a heavier gas, which turned out to be argon.

Challenge Rayleigh discovered a systematic error by comparing two different methods of measurement. Which method had the systematic error? Did it overestimate or underestimate the mass of nitrogen in air?

Let's see how to use the ***t* test** to decide whether nitrogen isolated from air is "significantly" heavier than nitrogen isolated from chemical sources. For two sets of data consisting of n_1 and n_2 measurements (with averages $\bar{x}_1$ and $\bar{x}_2$), we calculate a value of t from the formula

t test for comparison of means:

$$t = \frac{|\bar{x}_1 - \bar{x}_2|}{s_{pooled}}\sqrt{\frac{n_1 n_2}{n_1 + n_2}} \quad (4\text{-}4)$$

where

$$s_{pooled} = \sqrt{\frac{s_1^2(n_1 - 1) + s_2^2(n_2 - 1)}{n_1 + n_2 - 2}} \quad (4\text{-}5)$$

Here s_{pooled} is a *pooled* standard deviation making use of both sets of data. The absolute value of $\bar{x}_1 - \bar{x}_2$ is used in Equation 4-4 so that t is always positive. The value of t from Equation 4-4 is to be compared with the value of t in Table 4-2 for $(n_1 + n_2 - 2)$ degrees of freedom. *If the calculated t is greater than the tabulated t at the 95% confidence level, the two results are considered to be significantly different.*

If $t_{calculated} > t_{table}$ (95%), then the difference is significant.

Example **Is Rayleigh's N_2 from Air Denser than N_2 from Chemicals?**

The average mass of nitrogen from air in Table 4-3 is $\bar{x}_1 = 2.310\ 10_9$ g, with a standard deviation of $s_1 = 0.000\ 14_3$ (for $n_1 = 7$ measurements). The average mass from chemical sources is $\bar{x}_2 = 2.299\ 47_2$ g, with a standard deviation of $s_2 = 0.001\ 37_9$ (for $n_2 = 8$ measurements). Are the two masses significantly different?

SOLUTION To answer this question, we calculate s_{pooled} with Equation 4-5,

$$s_{\text{pooled}} = \sqrt{\frac{0.000\ 14_3^2(7-1) + 0.001\ 37_9^2(8-1)}{7+8-2}} = 0.001\ 01_7$$

and t with Equation 4-4:

$$t = \frac{|2.310\ 10_9 - 2.299\ 47_2|}{0.001\ 01_7}\sqrt{\frac{7 \cdot 8}{7+8}} = 20.2$$

For $7 + 8 - 2 = 13$ degrees of freedom in Table 4-2, t lies between 2.228 and 2.131 at the 95% confidence level. The observed value ($t = 20.2$) is greater than the tabulated t, so the difference is significant. In fact, the tabulated value of t for 99.9% confidence is about 4.3. The difference is significant beyond the 99.9% confidence level. Our eyes do not lie to us in Figure 4-4: N_2 from the air is undoubtedly denser than N_2 from chemical sources. This observation led Rayleigh to discover argon as a heavy constituent of air.

Test Yourself If the difference between the two mean values were half as great as Rayleigh found, but the pooled standard deviation were unchanged, would the difference still be significant? (**Answer:** yes)

Ask Yourself

4-B. A reliable assay of ATP (adenosine triphosphate) in a certain type of cell gives a value of $111._0$ μmol/100 mL, with a standard deviation of $2._8$ in four replicate measurements. You have developed a new assay, which gave the following values in replicate analyses: 117, 119, 111, 115, 120 μmol/100 mL.
(a) Find the mean and standard deviation for your new analysis.
(b) Can you be 95% confident that your method produces a result different from the "reliable" value?

4-3 A Spreadsheet for the *t* Test

Excel has built-in procedures for conducting tests with Student's *t*. To compare Rayleigh's two sets of results in Table 4-3, enter his data in columns B and C of the spreadsheet in Figure 4-5. In rows 13 and 14, we compute the averages and standard deviations.

In the TOOLS menu, you might find DATA ANALYSIS. If not, select ADD-INS in the TOOLS menu and find ANALYSIS TOOLPACK. Check the box beside ANALYSIS TOOLPACK and click OK. DATA ANALYSIS will then be available in the TOOLS menu.

	A	B	C	D	E	F	G
1	Analysis of Rayleigh's Data				t-Test: Two-Sample Assuming Equal Variances		
2						*Variable 1*	*Variable 2*
3		Mass of gas (g) collected from			Mean	2.310109	2.299473
4		air	chemical		Variance	2.03E-08	1.9E-06
5		2.31017	2.30143		Observations	7	8
6		2.30986	2.29890		Pooled Variance	1.03E-06	
7		2.31010	2.29816		Hypothesized Mean Diff	0	
8		2.31001	2.30182		df	13	
9		2.31024	2.29869		t Stat	20.21372	
10		2.31010	2.29940		P(T<=t) one-tail	1.66E-11	
11		2.31028	2.29849		t Critical one-tail	1.770932	
12			2.29889		P(T<=t) two-tail	3.32E-11	
13	Average	2.31011	2.29947		t Critical two-tail	2.160368	
14	Std Dev	0.00014	0.00138				
15					t-Test: Two-Sample Assuming Unequal Variances		
16	B13 = AVERAGE(B5:B12)					*Variable 1*	*Variable 2*
17	B14 = STDEV(B5:B12)				Mean	2.310109	2.299473
18					Variance	2.03E-08	1.9E-06
19					Observations	7	8
20					Hypothesized Mean Diff	0	
21					df	7	
22					t Stat	21.68022	
23					P(T<=t) one-tail	5.6E-08	
24					t Critical one-tail	1.894578	
25					P(T<=t) two-tail	1.12E-07	
26					t Critical two-tail	2.364623	

Figure 4-5 Spreadsheet for the t test.

Returning to Figure 4-5, we want to know whether the mean values of the two sets of data are statistically the same or not. In the TOOLS menu, select DATA ANALYSIS. In the window that appears, select t-Test: Two-Sample Assuming Equal Variances. Click OK. The next window asks you where the data are located. Write B5:B12 for Variable 1 and C5:C12 for Variable 2. The blank space in cell B12 is ignored. For the Hypothesized Mean Difference enter 0 and for Alpha enter 0.05. With Alpha = 0.05, we are at the 95% confidence level. For Output Range, select cell E1 and click OK.

Excel now prints results in cells E1 to G13 of Figure 4-5. Mean values are in cells F3 and G3. Cells F4 and G4 give *variance*, which is the square of the standard deviation. Cell F6 gives *pooled variance* computed from the square of Equation 4-5. Cell F8 shows degrees of freedom ($df = 13$), and $t_{\text{calculated}} = 20.2$ from Equation 4-4 appears in cell F9.

At this point in Section 4-3, we consulted Table 4-2 to find that t_{table} lies between 2.228 and 2.131 for 95% confidence and 13 degrees of freedom. Excel gives the critical value of $t = 2.160$ in cell F13 of Figure 4-5. Because $t_{\text{calculated}}$ $(= 20.2) > t_{\text{table}} (= 2.160)$, we conclude that the two means are not the same. The difference is significant. Cell F12 states that the probability of observing these two mean values and standard deviations by random chance if the mean values were really the same is 3.32×10^{-11}. The difference is *highly* significant. For any value of $P \leq 0.05$ in cell F12, we reject the *null hypothesis* and conclude that the means *are different*.

In this book, we use the 2-tail test with results in cells F12 and F13. It is beyond the scope of this book to discuss what is meant by 1-tail or 2-tail.

The standard deviations for Rayleigh's two sets of data are $s_1 = 0.000\ 14$ and $s_2 = 0.001\ 38$. s_2 is 10 times greater than s_1. We have reason to suspect that the two data sets really have different population standard deviations. So, go to the TOOLS menu and DATA ANALYSIS and select *t*-Test: Two-Sample Assuming *Unequal* Variances and fill in the blanks as before. Results are displayed in cells E15 to G26 of Figure 4-5. With the assumption that the variances are *not* equal, the spreadsheet produces $t_{calculated} = 21.7$ in cell F22 and the critical value $t = 2.36$ in cell F26. Again, $t_{calculated}$ $(= 21.7) > t_{table}$ $(= 2.36)$, so we again conclude that the two means are not the same.

Ask Yourself

4-C. Reproduce Figure 4-5 in a spreadsheet.

4-4 Grubbs Test for an Outlier

Table 4-4 Critical values of *G* for rejection of outlier

Number of observations	*G* (95% confidence)
4	1.463
5	1.672
6	1.822
7	1.938
8	2.032
9	2.110
10	2.176
11	2.234
12	2.285
15	2.409
20	2.557

$G_{calculated} = |\text{questionable value} - \text{mean}|/s$. If $G_{calculated} > G_{table}$, the value in question can be rejected with 95% confidence. Values in this table are for a one-sided test, as recommended by ASTM.

SOURCE: ASTM E 178-02 *Standard Practice for Dealing with Outlying Observations*; F. E. Grubbs and G. Beck, *Technometrics* **1972**, *14*, 847.

There was always someone in my lab section who seemed to have four thumbs. Freshmen at Phillips University perform an experiment in which they dissolve the zinc from a galvanized nail and measure the mass lost by the nail to tell how much of the nail was zinc. Several students performed the experiment in triplicate and pooled their results:

Mass loss (%):	10.2, 10.8, 11.6	9.9, 9.4, 7.8	10.0, 9.2, 11.3	9.5, 10.6, 11.6
	Sidney	Cheryl	Tien	Dick

It appears that Cheryl might be the person with four thumbs, because her value 7.8 looks out of line from the other data. A datum that is far from the other points is called an *outlier*. Should the group reject 7.8 before averaging the rest of the data or should 7.8 be retained?

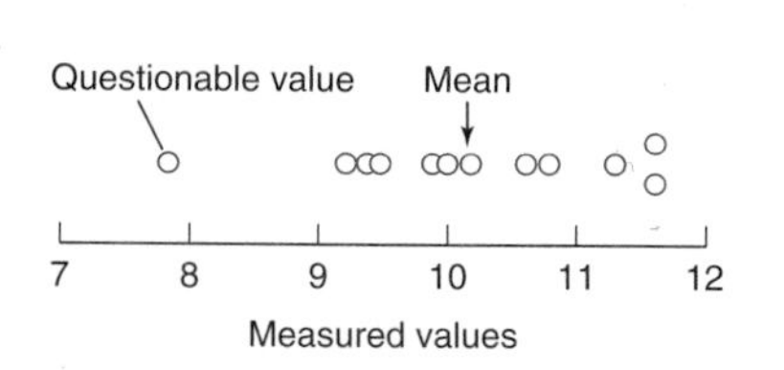

We answer this question with the **Grubbs test**. First compute the average ($\bar{x}$) and the standard deviation (s) of the complete data set (all 12 points in this example):

$$\bar{x} = 10.16 \qquad s = 1.11$$

The Grubbs test is recommended by the International Standards Organization and the American Society for Testing and Materials in place of the *Q* test, which was formerly used in this book.

Then compute the Grubbs statistic *G*, defined as

Grubbs test:
$$G = \frac{|\text{questionable value} - \bar{x}|}{s} \qquad (4\text{-}6)$$

where the numerator is the absolute value of the difference between the suspected outlier and the mean value. *If G calculated from Equation 4-6 is greater than G in Table 4-4, the questionable point should be discarded.*

If $G_{calculated} > G_{table}$, then reject the questionable point.

For the preceding numbers, $G_{calculated} = |7.8 - 10.16|/1.11 = 2.13$. The value G_{table} is 2.285 for 12 observations in Table 4-4. Because $G_{calculated}$ is smaller than G_{table}, the questionable point should be retained. There is more than a 5% chance that the value 7.8 is a member of the same population as the other measurements.

Common sense must always prevail. If Cheryl knows that her measurement was low because she spilled some of her unknown, then the probability that the result is wrong is 100% and the datum should be discarded. Any datum based on a faulty procedure should be discarded, no matter how well it fits the rest of the data.

Ask Yourself

4-D. Would you reject the value 216 from the set of results 192, 216, 202, 195, and 204?

4-5 Finding the "Best" Straight Line

The *method of least squares* finds the "best" straight line through experimental data points. We will apply this procedure to analytical chemistry calibration curves in the next section.

The equation of a straight line is

Equation of straight line:
$$y = mx + b \tag{4-7}$$

in which m is the **slope** and b is the **y-intercept** (Figure 4-6). If we measure between two points that lie on the line, the slope is $\Delta y/\Delta x$, which is constant for any pair of points on the line. The y-intercept is the point at which the line crosses the y-axis.

Figure 4-6 Parameters of a straight line:

equation $y = mx + b$

slope $m = \dfrac{\Delta y}{\Delta x} = \dfrac{y_2 - y_1}{x_2 - x_1}$

y-intercept b = crossing point on y-axis

Method of Least Squares

The **method of least squares** finds the "best" line by adjusting the line to minimize the vertical deviations between the points and the line (Figure 4-7). The reasons for

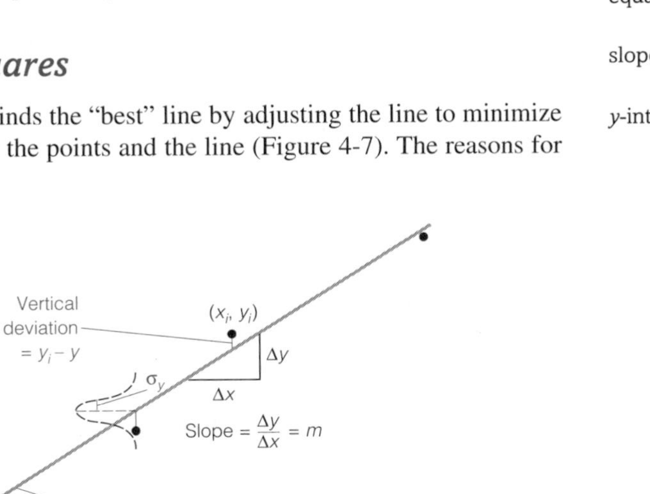

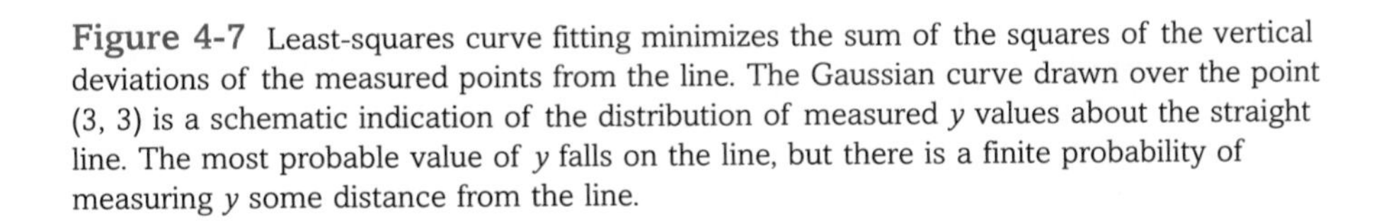

Figure 4-7 Least-squares curve fitting minimizes the sum of the squares of the vertical deviations of the measured points from the line. The Gaussian curve drawn over the point (3, 3) is a schematic indication of the distribution of measured y values about the straight line. The most probable value of y falls on the line, but there is a finite probability of measuring y some distance from the line.

The *ordinate* is the value of y that really lies on the line; y_i is the measured value that does not lie exactly on the line.

minimizing only the vertical deviations are that (1) experimental uncertainties in y values are often greater than uncertainties in x values and (2) the calculation for minimizing the vertical deviations is relatively simple.

In Figure 4-7, the vertical deviation for the point (x_i, y_i) is $y_i - y$, where y is the ordinate of the straight line when $x = x_i$.

$$\text{vertical deviation} = d_i = y_i - y = y_i - (mx_i + b) \tag{4-8}$$

Some of the deviations are positive and some are negative. To minimize the magnitude of the deviations irrespective of their signs, we square the deviations to create positive numbers:

$$d_i^2 = (y_i - y)^2 = (y_i - mx_i - b)^2$$

Because we minimize the squares of the deviations, this is called the *method of least squares.*

When we use such a procedure to minimize the sum of squares of the vertical deviations, the slope and the intercept of the "best" straight line fitted to n points are

Remember that Σ means summation: $\Sigma x_i = x_1 + x_2 + x_3 + \ldots$.

Least-squares slope:

$$m = \frac{n\,\Sigma(x_i y_i) - \Sigma x_i\,\Sigma y_i}{D} \tag{4-9}$$

Least-squares intercept:

$$b = \frac{\Sigma(x_i^2)\,\Sigma y_i - \Sigma(x_i y_i)\,\Sigma x_i}{D} \tag{4-10}$$

where the denominator, D, is given by

$$D = n\,\Sigma(x_i^2) - (\Sigma x_i)^2 \tag{4-11}$$

These equations are not as terrible as they appear. Table 4-5 sets out an example in which the four points ($n = 4$) in Figure 4-7 are treated. The first two columns list x_i and y_i for each point. The third column gives the product $x_i y_i$, and the fourth column lists the square x_i^2. At the bottom of each column is the sum for that column. That is, beneath the first column is $\Sigma\, x_i$ and beneath the third column is $\Sigma\,(x_i y_i)$. The last two columns at the right will be used later.

Table 4-5 Calculations for least-squares analysis

x_i	y_i	$x_i y_i$	x_i^2	$d_i (= y_i - mx_i - b)$	d_i^2
1	2	2	1	0.038 462	0.001 479
3	3	9	9	−0.192 308	0.036 982
4	4	16	16	0.192 308	0.036 982
6	5	30	36	−0.038 462	0.001 479
$\Sigma x_i = 14$	$\Sigma y_i = 14$	$\Sigma(x_i y_i) = 57$	$\Sigma(x_i^2) = 62$		$\Sigma(d_i^2) = 0.076\,923$

Quantities required for propagation of uncertainty with Equation 4-16:

$\bar{x} = (\Sigma x_i)/n = (1 + 3 + 4 + 6)/4 = 3.50$ $\bar{y} = (\Sigma y_i)/n = (2 + 3 + 4 + 5)/4 = 3.50$

$\Sigma(x_i - \bar{x})^2 = (1 - 3.5)^2 + (3 - 3.5)^2 + (4 - 3.5)^2 + (6 - 3.5)^2 = 13$

With the sums from Table 4-5, we compute the slope and intercept by substituting into Equations 4-11, 4-9, and 4-10:

$$D = n\Sigma(x_i^2) - (\Sigma x_i)^2 = 4 \cdot 62 - 14^2 = 52$$

$$m = \frac{n\Sigma(x_i y_i) - \Sigma x_i \Sigma y_i}{D} = \frac{4 \cdot 57 - 14 \cdot 14}{52} = 0.615\ 38$$

$$b = \frac{\Sigma(x_i^2)\Sigma y_i - \Sigma(x_i y_i)\Sigma x_i}{D} = \frac{62 \cdot 14 - 57 \cdot 14}{52} = 1.346\ 15$$

The equation of the best straight line through the points in Figure 4-7 is therefore

$$y = 0.615\ 38x + 1.346\ 15$$

Next, we will see how many figures in these numbers are significant.

How Reliable Are Least-Squares Parameters?

The uncertainties in m and b are related to the uncertainty in measuring each value of y. Therefore, we first estimate the standard deviation describing the population of y values. This standard deviation, s_y, characterizes the little Gaussian curve inscribed in Figure 4-7. The deviation of each y_i from the center of its Gaussian curve is $d_i = y_i - y = y_i - (mx_i + b)$ (Equation 4-8). The standard deviation of these vertical deviations is

$$s_y \approx \sqrt{\frac{\Sigma(d_i^2)}{n - 2}} \qquad (4\text{-}12)$$

Analysis of uncertainty for Equations 4-9 and 4-10 leads to the following results:

standard deviation of slope: $$s_m = s_y\sqrt{\frac{n}{D}} \qquad (4\text{-}13)$$

standard deviation of intercept: $$s_b = s_y\sqrt{\frac{\Sigma(x_i^2)}{D}} \qquad (4\text{-}14)$$

where s_y is given by Equation 4-12 and D is given by Equation 4-11.

At last, we can address significant figures for the slope and the intercept of the line in Figure 4-7. In Table 4-5, we see that $\Sigma(d_i^2) = 0.076\ 923$. Inserting this value into Equation 4-12 gives

$$s_y = \sqrt{\frac{0.076\ 923}{4 - 2}} = 0.196\ 12$$

Now, we can plug numbers into Equations 4-13 and 4-14 to find

$$s_m = s_y\sqrt{\frac{n}{D}} = (0.196\ 12)\sqrt{\frac{4}{52}} = 0.054\ 394$$

$$s_b = s_y\sqrt{\frac{\Sigma(x_i^2)}{D}} = (0.196\ 12)\sqrt{\frac{62}{52}} = 0.214\ 15$$

The first digit of the uncertainty is the last significant figure.

Combining the results for m, s_m, b, and s_b, we write

Slope: $\begin{matrix} 0.615\ 38 \\ \pm\ 0.054\ 39 \end{matrix} = 0.62 \pm 0.05 \quad \text{or} \quad 0.61_5 \pm 0.05_4$

Intercept: $\begin{matrix} 1.346\ 15 \\ \pm\ 0.214\ 15 \end{matrix} = 1.3 \pm 0.2 \quad \text{or} \quad 1.3_5 \pm 0.2_1$

where the uncertainties represent one standard deviation. *The first decimal place of the standard deviation is the last significant figure of the slope or intercept.*

Ask Yourself

4-E. Construct a table analogous to Table 4-5 to calculate the equation of the best straight line going through the points (1, 3), (3, 2), and (5, 0). Express your answer in the form $y\ (\pm s_y) = [m\ (\pm s_m)]x + [b\ (\pm s_b)]$, with a reasonable number of significant figures.

4-6 Constructing a Calibration Curve

Real data from a spectrophotometric analysis are given in Table 4-6. In this procedure, a color develops in proportion to the amount of protein in the sample. Color is measured by the absorbance of light recorded on a spectrophotometer. The first row in Table 4-6 shows readings obtained when no protein was present. The nonzero values arise from color in the reagents themselves. A result obtained with zero analyte is called a **blank** (or a *reagent blank*), because it measures effects due to the analytical reagents. The second row shows three readings obtained with 5 μg of protein. Successive rows give results for 10, 15, 20, and 25 μg of protein. A solution containing a known quantity of analyte (or other reagent) is called a **standard solution.**

A **calibration curve** is a graph showing how the experimentally measured property (absorbance) depends on the known concentrations of the standards. To construct the calibration curve in Figure 4-8, we first subtract the average absorbance of the blanks (0.099_3) from those of the standards to obtain *corrected absorbance.*

Table 4-6 Spectrophotometer readings for protein analysis by the Lowry method

Sample (μg)	Absorbance of three independent samples			Range	Corrected absorbance (after subtracting average blank)			
0	0.099	0.099	0.100	0.001	-0.000_3	-0.000_3	0.000_7	Data used for calibration curve
5	0.185	0.187	0.188	0.003	0.085_7	0.087_7	0.088_7	
10	0.282	0.272	0.272	0.010	0.182_7	0.172_7	0.172_7	
15	0.392	0.345	0.347	0.047	—	0.245_7	0.247_7	
20	0.425	0.425	0.430	0.005	0.325_7	0.325_7	0.330_7	
25	0.483	0.488	0.496	0.013	0.383_7	0.388_7	0.396_7	

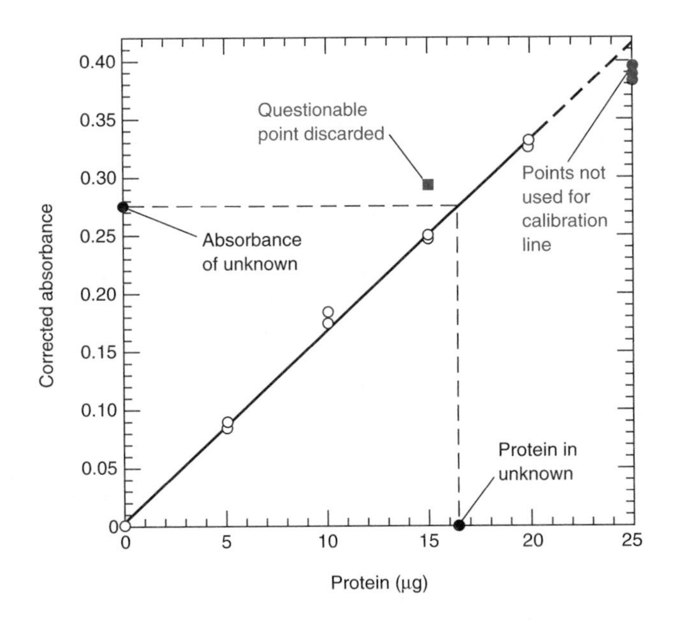

Figure 4-8 *Calibration curve* showing the average absorbance values in Table 4-6 versus micrograms of protein analyzed. The average *blank value* for 0 μg of protein has been subtracted from each point.

When all points are plotted in Figure 4-8 and a rough straight line is drawn through them, two features stand out:

1. One data point for 15 μg of protein (shown by the square in Figure 4-8) is clearly out of line. When we inspect the range of values for each set of three measurements in Table 4-6, we discover that the range for 15 μg is four times greater than the next greatest range. We discard the absorbance value 0.392 as "bad data." Perhaps the glassware was contaminated with protein from a previous experiment?

2. All three points at 25 μg lie slightly below the straight line through the remaining data. Many repetitions of this experiment show that these points are consistently below the straight line. Therefore the *linear range* for this determination extends from 0 to 20 μg, but not to 25 μg.

Inspect your data and use judgment before mindlessly asking a computer to draw a calibration curve!

In view of these observations, we discard the 0.392 absorbance value and we do not use the three points at 25 μg for the least-squares straight line. We could use a *nonlinear calibration curve* that extends to 25 μg, but we will not do so in this book.

To construct the calibration curve (the straight line) in Figure 4-8, we use the method of least squares with $n = 14$ data points from Table 4-6 (including three blank values) covering the range 0 to 20 μg of protein. The results of applying Equations 4-9 through 4-14 are

$$m = 0.016\,3_0 \qquad s_m = 0.000\,2_2$$
$$b = 0.004_7 \qquad s_b = 0.002_6$$
$$s_y = 0.005_9$$

Equation of calibration line:

$$y\,(\pm s_y) = [m\,(\pm s_m)]x + [b\,(\pm s_b)]$$
$$y\,(\pm 0.005_9) = [0.016\,3_0\,(\pm 0.000\,2_2)]x + [0.004_7\,(\pm 0.002_6)]$$

Finding the Protein in an Unknown

Suppose that the measured absorbance of an unknown sample is 0.373. How many micrograms of protein does it contain, and what uncertainty is associated with the answer?

The first question is easy. The equation of the calibration line is

$$y = mx + b = (0.016\,3_0)x + 0.004_7$$

where y is corrected absorbance (= observed absorbance − blank absorbance) and x is micrograms of protein. If the absorbance of the unknown is 0.373, its corrected absorbance is $0.373 - 0.099_3 = 0.273_7$. Plugging this value in for y in the preceding equation permits us to solve for x:

$$0.273_7 = (0.016\ 3_0)x + (0.004_7) \tag{4-15a}$$

$$x = \frac{0.273_7 - 0.004_7}{0.016\ 3_0} = 16.50\ \mu\text{g of protein} \tag{4-15b}$$

But what is the uncertainty in 16.50 μg?

The uncertainty in x in Equation 4-15 turns out to be

$$\text{uncertainty in } x\ (= s_x) = \frac{s_y}{|m|}\sqrt{\frac{1}{k} + \frac{1}{n} + \frac{(y - \bar{y})^2}{m^2 \Sigma (x_i - \bar{x})^2}} \tag{4-16}$$

Example: If 4 replicate samples of an unknown have an average absorbance of 0.373, use the corrected absorbance $y = 0.373 - 0.099_3 = 0.273_7$. In Equation 4-16, $k = 4$ for 4 replicate measurements and $n = 14$ because there are 14 points on the calibration curve (Table 4-6). Other values in Equation 4-16 are

x_i = μg of protein in standards in Table 4-6
= (0, 0, 0, 5.0, 5.0, 5.0, 10.0, 10.0, 10.0, 15.0, 15.0, 20.0, 20.0, 20.0)

$\bar{x}$ = average of 14 x values = 9.64_3 μg

$\bar{y}$ = average of 14 corrected y values = 0.161_8

From these values we calculate s_x = 0.22 μg.

where s_y is the standard deviation of y (Equation 4-12), $|m|$ is the absolute value of the slope, k is the number of replicate measurements of the unknown, n is the number of data points for the calibration line (14 in Table 4-6), $\bar{y}$ is the mean value of y for the points on the calibration line, x_i are individual values of x for the points on the calibration line, and $\bar{x}$ is the mean value of x for the points on the calibration line. For a single measurement of the unknown, $k = 1$ and Equation 4-16 gives $s_x = \pm 0.3_8$ μg. The result of the analysis can therefore be expressed with a reasonable number of significant digits as

$$x = 16.5\ (\pm 0.4)\ \mu\text{g of protein}$$

If you measured four replicate unknowns ($k = 4$) and the average corrected absorbance was still 0.273_7, the uncertainty is reduced from $\pm 0.3_8$ to $\pm 0.2_2$ μg.

Ask Yourself

4-F. Using results from Ask Yourself 4-E, find the value of x (and its uncertainty) corresponding to a mean value of $y = 1.00$ for $k = 5$ replicate measurements.

4-7 A Spreadsheet for Least Squares

Figure 4-9 uses built-in power of Excel for least-squares calculations of straight lines. As an example, enter the x- and y-coordinates from Table 4-5 into cells B4 through B7 and C4 through C7. This range is abbreviated B4:C7. The key portion of the spreadsheet uses the Excel function LINEST to compute the least-squares parameters in cells B10:C12.

Highlight the 3-row × 2-column region B10:C12 with your mouse. Under the INSERT menu, select FUNCTION. In the window that appears, go to Statistical and double click on LINEST. A new window asks for four inputs to the function. For y values, enter C4:C7. Then enter B4:B7 for x values. The next two entries are both

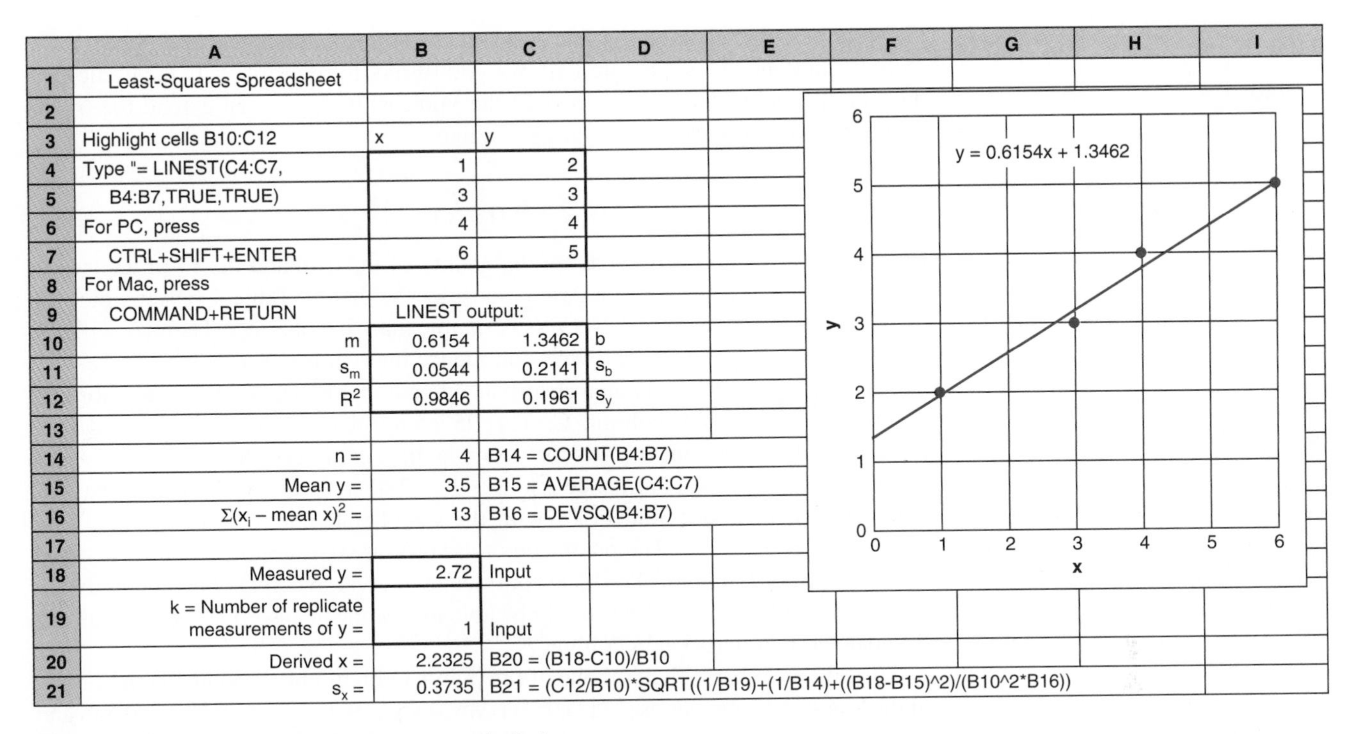

	A	B	C	D
1	Least-Squares Spreadsheet			
2				
3	Highlight cells B10:C12	x	y	
4	Type "= LINEST(C4:C7,	1	2	
5	B4:B7,TRUE,TRUE)	3	3	
6	For PC, press	4	4	
7	CTRL+SHIFT+ENTER	6	5	
8	For Mac, press			
9	COMMAND+RETURN	LINEST output:		
10	m	0.6154	1.3462	b
11	s_m	0.0544	0.2141	s_b
12	R^2	0.9846	0.1961	s_y
13				
14	n =	4	B14 = COUNT(B4:B7)	
15	Mean y =	3.5	B15 = AVERAGE(C4:C7)	
16	$\Sigma(x_i - \text{mean } x)^2$ =	13	B16 = DEVSQ(B4:B7)	
17				
18	Measured y =	2.72	Input	
19	k = Number of replicate measurements of y =	1	Input	
20	Derived x =	2.2325	B20 = (B18-C10)/B10	
21	s_x =	0.3735	B21 = (C12/B10)*SQRT((1/B19)+(1/B14)+((B18-B15)^2)/(B10^2*B16))	

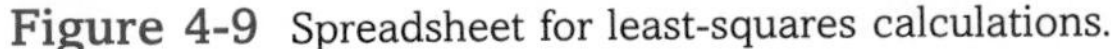

Figure 4-9 Spreadsheet for least-squares calculations.

"TRUE". The first TRUE tells Excel that we want to compute the y-intercept of the least-squares line and not force the intercept to be 0. The second TRUE tells Excel to compute standard deviations of the slope and intercept. The formula you just entered is "=LINEST(C4:C7,B4:B7,TRUE,TRUE)". Now press CONTROL+SHIFT+ENTER on a PC or COMMAND(⌘)+RETURN on a Mac. Excel dutifully prints out a matrix in cells B10:C12. Write labels around the block to indicate what is in each cell. The slope (m) and intercept (b) are on the top line. The second line contains the standard deviations of slope and intercept, s_m and s_b. Cell C12 contains s_y and cell B12 contains a quantity called R^2, which measures the goodness of fit of the data to the line. The closer R^2 is to unity, the better the fit.

Cell B14 gives the number of data points with the formula = COUNT(B4:B7). Cell B15 computes the mean value of y. Cell B16 computes the sum $\Sigma (x_i - \bar{x})^2$ that we need for Equation 4-16. This sum is common enough that Excel has a built-in function called DEVSQ that you can find in the Statistics menu of the INSERT FUNCTION menu. Formulas in Figure 4-9 are documented beside the cell where they are used.

Enter the measured value $y = 2.72$ for replicate measurements of the unknown in cell B18. In cell B19, enter the number of replicate measurements ($k = 1$) of the unknown. In this example, we compute $x = 2.2_3$ in cell B20 with an uncertainty of 0.3_7 in cell B21.

We always want a graph to see if the calibration points lie on a straight line. Follow instructions in Section 3-6 to plot the calibration data. To add a straight line, click on one data point and they will all be highlighted. Go to the CHART menu and select ADD TRENDLINE. In some versions of Excel, there is no CHART menu. In this case, go to the INSERT menu and select TRENDLINE. In the window that appears, select Linear. Go to Options in the TRENDLINE box and select Display Equation on Chart. When you click OK, the least-squares straight line and its equation appear on the graph. Double click on the line and you can adjust its thickness and appearance. Double clicking on the equation allows you to modify its format. Double click on the straight line and select Options. In the Forecast box, you can extend the trendline Forward and Backward as far as you like.

Adding Error Bars to a Graph

Error bars on a graph help us judge the quality of the data and the fit of a curve to the data. To add error bars to a graph, consider the data in Table 4-6. Let's plot the mean absorbance of columns 2–4 versus sample mass in column 1. Then we will add error bars corresponding to the 95% confidence interval for each point. Figure 4-10 lists mass in column A and mean absorbance in column B. The standard deviation of absorbance is given in column C. The 95% confidence interval for absorbance is computed in column D with the formula in the margin. Student's $t = 4.303$ can be found for 95% confidence and $3 - 1 = 2$ degrees of freedom in Table 4-2. Alternatively, compute Student's t with the Excel statement "=TINV(0.05,2)" in cell B11 of Figure 4-10. The parameters for TINV are 0.05 for 95% confidence and 2 for degrees of freedom. The 95% confidence interval in cell D4 is computed with "=$B$11*C4/SQRT(3)". You should be able to plot mean absorbance (y) in column B versus protein mass (x) in column A.

Confidence interval $= \pm ts/\sqrt{n}$
t = Student's t for 95% confidence and $n - 1 = 2$ degrees of freedom
s = standard deviation
n = number of values in average = 3

To add error bars, click on one of the points to highlight all points on the graph. In the FORMAT menu, choose SELECTED DATA SERIES. Select the Y Error Bars tab and the window in Figure 4-11 appears. Click Custom. Click in the positive error bar box and select cells D4:D9. Click the negative error bar box and select cells D4:D9 again. You just told Excel to use values in cells D4:D9 for the lengths of the error

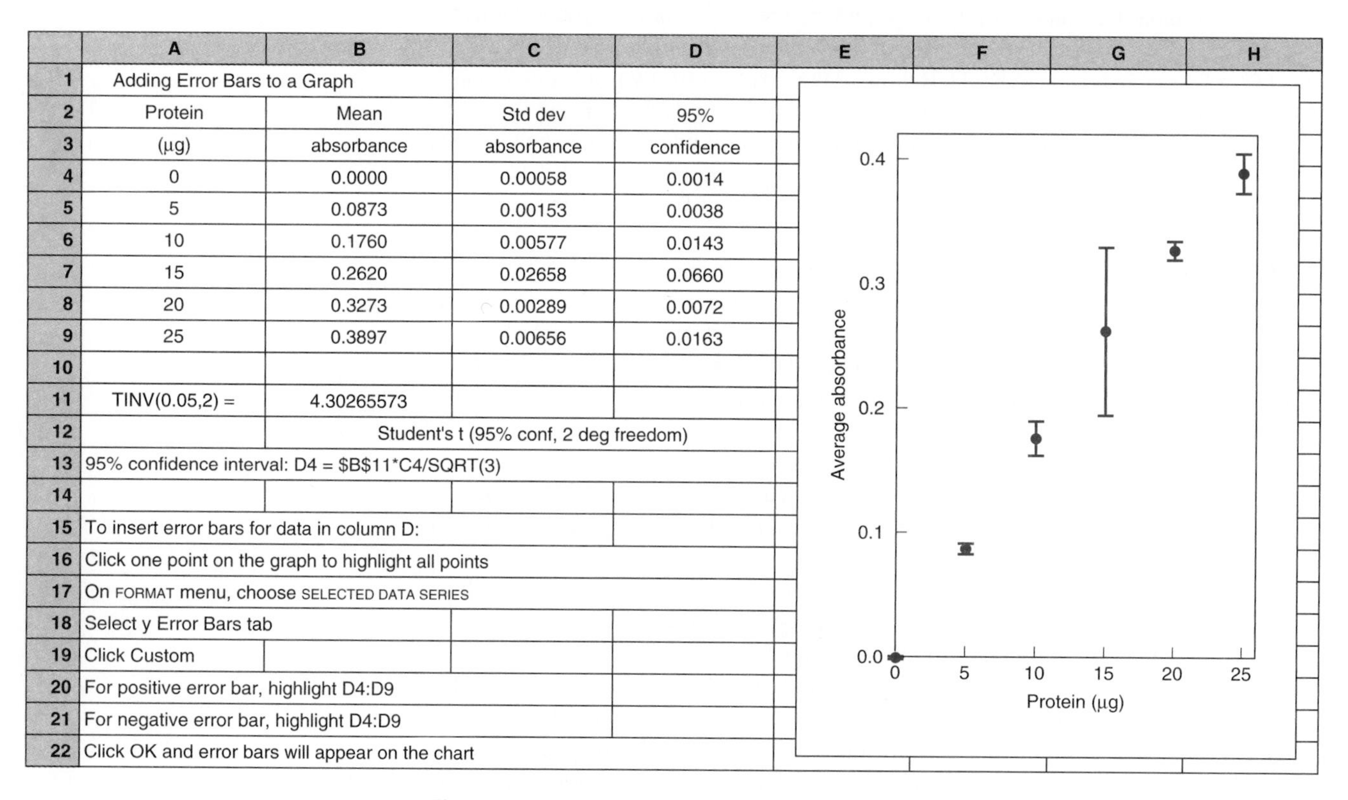

	A	B	C	D
1	Adding Error Bars to a Graph			
2	Protein	Mean	Std dev	95%
3	(μg)	absorbance	absorbance	confidence
4	0	0.0000	0.00058	0.0014
5	5	0.0873	0.00153	0.0038
6	10	0.1760	0.00577	0.0143
7	15	0.2620	0.02658	0.0660
8	20	0.3273	0.00289	0.0072
9	25	0.3897	0.00656	0.0163
10				
11	TINV(0.05,2) =	4.30265573		
12		Student's t (95% conf, 2 deg freedom)		
13	95% confidence interval: D4 = $B$11*C4/SQRT(3)			
14				
15	To insert error bars for data in column D:			
16	Click one point on the graph to highlight all points			
17	On FORMAT menu, choose SELECTED DATA SERIES			
18	Select y Error Bars tab			
19	Click Custom			
20	For positive error bar, highlight D4:D9			
21	For negative error bar, highlight D4:D9			
22	Click OK and error bars will appear on the chart			

Figure 4-10 Adding 95% confidence error bars to a graph.

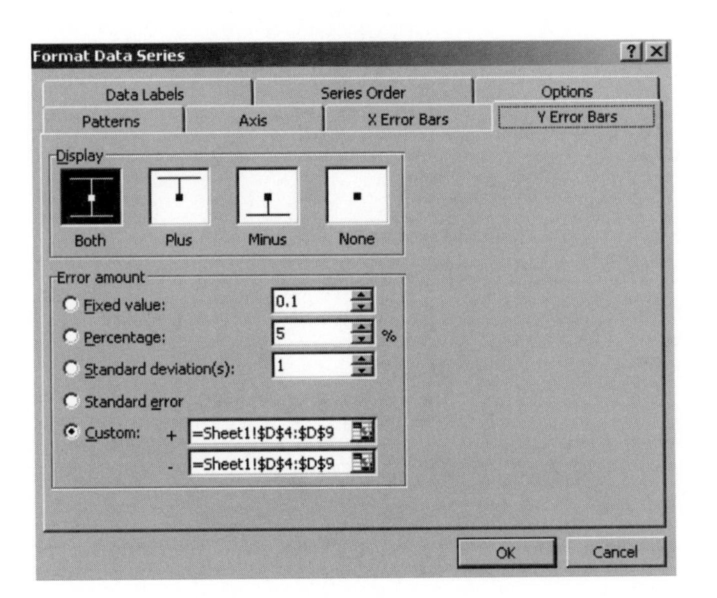

Figure 4-11 Format Data Series window for adding error bars to a graph.

bars. Click OK and error bars appear in the graph. Figure 4-11 gives other options for error bars, such as a percentage of y.

Ask Yourself

4-G. **(a)** Reproduce the spreadsheet in Figure 4-9 to solve linear least-squares problems.
(b) Use your spreadsheet to plot the data points and its least-squares line, as in Figure 4-9.
(c) Reproduce the error bar chart in Figure 4-10.

Key Equations

Mean
$$\bar{x} = \frac{1}{n}\sum_i x_i = \frac{1}{n}(x_1 + x_2 + x_3 + \cdots + x_n)$$
x_i = individual observation, n = number of observations

Standard deviation
$$s = \sqrt{\frac{\sum_i (x_i - \bar{x})^2}{(n-1)}} \quad (\bar{x} = \text{average})$$

Confidence interval
$$\mu = \bar{x} \pm \frac{ts}{\sqrt{n}} \quad (\mu = \text{true mean})$$
Student's t comes from Table 4-2 for $n - 1$ degrees of freedom at the selected confidence level.

t test
$$t = \frac{\bar{x}_1 - \bar{x}_2}{s_{\text{pooled}}}\sqrt{\frac{n_1 n_2}{n_1 + n_2}}$$
$$s_{\text{pooled}} = \sqrt{\frac{s_1^2(n_1 - 1) + s_2^2(n_2 - 1)}{n_1 + n_2 - 2}}$$
If $t_{\text{calculated}} > t_{\text{table}}$ (for 95% confidence and $n_1 + n_2 - 2$ degrees of freedom), the difference is significant.

Grubbs test	$G = \dfrac{\lvert \text{questionable value} - \bar{x} \rvert}{s}$
	If $G_{\text{calculated}} > G_{\text{table}}$, reject questionable point.
Straight line	$y = mx + b$ $m = \text{slope} = \Delta y/\Delta x$
	$b = y\text{-intercept}$
Least-squares equations	You should know how to use Equations 4-9 through 4-14 to derive least-squares slope and intercept and uncertainties.
Calibration curve	You should be able to use Equation 4-16 to find the uncertainty in a result derived from a calibration curve.

Important Terms

average	Grubbs test	standard solution
blank	mean	Student's *t*
calibration curve	method of least squares	*t* test
confidence interval	slope	variance
Gaussian distribution	standard deviation	*y*-intercept

Problems

4-1. What is the relation between the standard deviation and the precision of a procedure? What is the relation between standard deviation and accuracy?

4-2. What fraction of observations in an ideal Gaussian distribution lies within $\mu \pm \sigma$? Within $\mu \pm 2\sigma$? Within $\mu \pm 3\sigma$?

4-3. The ratio of the number of atoms of the isotopes ^{69}Ga and ^{71}Ga in samples from different sources is listed here:

Sample	^{69}Ga/^{71}Ga	Sample	^{69}Ga/^{71}Ga
1	1.526 60	5	1.528 94
2	1.529 74	6	1.528 04
3	1.525 92	7	1.526 85
4	1.527 31	8	1.527 93

(a) Find the mean value of ^{69}Ga/^{71}Ga.

(b) Find the standard deviation and relative standard deviation.

(c) Sample 8 was analyzed seven times, with $\bar{x} = 1.527\ 93$ and $s = 0.000\ 07$. Find the 99% confidence interval for sample 8.

4-4. **(a)** What is the meaning of a confidence interval? **(b)** For a given set of measurements, will the 95% confidence interval be larger or smaller than the 90% confidence interval? Why?

4-5. For the numbers 116.0, 97.9, 114.2, 106.8, and 108.3, find the mean, standard deviation, and 90% confidence interval for the mean.

4-6. The calcium content of a mineral was analyzed five times by each of two methods. Are the mean values for the two methods significantly different at the 95% confidence level?

Method	Ca (wt%, five replications)				
1	0.027 1	0.028 2	0.027 9	0.027 1	0.027 5
2	0.027 1	0.026 8	0.026 3	0.027 4	0.026 9

4-7. Find the 95% and 99% confidence intervals for the mean mass of nitrogen from chemical sources given in Table 4-3.

4-8. Two methods were used to measure the specific activity (units of enzyme activity per milligram of protein) of an enzyme. One unit of enzyme activity is defined as the amount of enzyme that catalyzes the formation of 1 micromole of product per minute under specified conditions.

Method	Enzyme activity (five replications)				
1	139	147	160	158	135
2	148	159	156	164	159

Is the mean value of method 1 significantly different from the mean value of method 2 at the 95% confidence level? Answer with a calculator or spreadsheet.

4-9. Students measured the concentration of HCl in a solution by various titrations in which different indicators were used to find the end point.

Indicator	Mean HCl concentration (M) (± standard deviation)	Number of measurements
Bromothymol blue	0.095 65 ± 0.002 25	28
Methyl red	0.086 86 ± 0.000 98	18
Bromocresol green	0.086 41 ± 0.001 13	29

SOURCE: D. T. Harvey, *J. Chem. Ed.* **1991**, *68*, 329.

Is the difference between indicators 1 and 2 significant at the 95% confidence level? Answer the same question for indicators 2 and 3.

4-10. The calcium content of a person's urine was determined on two different days.

Day	[Ca] (mg/L) Average ± standard deviation	Number of measurements
1	238 ± 8	4
2	255 ± 10	5

Are the average values significantly different at the 95% confidence level?

4-11. Lithium isotope ratios are important to medicine, geology, astrophysics, and nuclear chemistry. The $^6Li/^7Li$ ratio in a Standard Reference Material was measured by two methods.

Method 1:
0.082 601, 0.082 621, 0.082 589, 0.082 617, 0.082 598

Method 2:
0.082 604, 0.082 542, 0.082 599, 0.082 550, 0.082 583, 0.082 561

Do the two methods give statistically equivalent results? Answer with a calculator or spreadsheet.

4-12. Students at Butler University compared the accuracy and precision of delivering 10 mL from a 50-mL buret, a 10-mL volumetric pipet, and a 10-mL volumetric flask. The table shows results for 6 replicate measurements by each of two students.

Apparatus	Student 1 $\bar{x} \pm s$ (mL)	Student 2 $\bar{x} \pm s$ (mL)
Buret	10.01 ± 0.09	9.98 ± 0.2
Pipet	9.98 ± 0.02	10.004 ± 0.009
Flask	9.80 ± 0.03	9.84 ± 0.02

SOURCE: M. J. Samide, *J. Chem. Ed.* **2004**, *81*, 1641.

(a) Do the volumes delivered by student 1 from the buret and pipet differ at the 95% confidence level?

(b) Do the volumes delivered by student 1 from the buret and flask differ at the 95% confidence level?

(c) Do the volumes delivered from the pipet by students 1 and 2 differ at the 95% confidence level?

(d) What can you conclude about the accuracy of the three methods of delivery? Is the observed accuracy within the manufacturer's tolerance for Class A glassware?

4-13. Using the Grubbs test, decide whether the value 0.195 should be rejected from the set of results 0.217, 0.224, 0.195, 0.221, 0.221, 0.223.

4-14. Students at the University of North Dakota measured visible light absorbance of food colorings. Replicate measurements of a solution of the drink Kool-Aid at a wavelength of 502 nm gave the following values: 0.189, 0.169, 0.187, 0.183, 0.186, 0.182, 0.181, 0.184, 0.181, and 0.177. Identify the outlier and decide whether to exclude it from the data set.

4-15. Find the values of m and b in the equation $y = mx + b$ for the straight line going through the points $(x_1, y_1) = (6, 3)$ and $(x_2, y_2) = (8, -1)$. You can do so by writing

$$m = \frac{\Delta y}{\Delta x} = \frac{(y_2 - y_1)}{(x_2 - x_1)} = \frac{(y - y_1)}{(x - x_1)}$$

and rearranging to the form $y = mx + b$. Sketch the curve and satisfy yourself that the value of b is sensible.

4-16. A straight line is drawn through the points $(3.0, -3.87 \times 10^4)$, $(10.0, -12.99 \times 10^4)$, $(20.0, -25.93 \times 10^4)$, $(30.0, -38.89 \times 10^4)$, and $(40.0, -51.96 \times 10^4)$ by using the method of least squares. The results are $m = -1.298\ 72 \times 10^4$, $b = 256.695$, $s_m = 13.190$, $s_b = 323.57$, and $s_y = 392.9$. Express the slope and intercept and their uncertainties with the correct significant figures.

4-17. Consider the least-squares problem illustrated in Figure 4-7. Suppose that a single new measurement produces a y value of 2.58.

(a) Calculate the corresponding x value and its uncertainty.

(b) Suppose you measure y four times and the average is 2.58. Calculate the uncertainty in x on the basis of four measurements, not one.

4-18. In a common protein analysis, a dye binds to the protein and the color of the dye changes from brown to blue. The intensity of blue color is proportional to the amount of protein present.

Protein (μg):	0.00	9.36	18.72	28.08	37.44
Absorbance:	0.466	0.676	0.883	1.086	1.280

(a) After subtracting the blank absorbance (0.466) from the remaining absorbances, use the method of least squares to determine the equation of the best straight line through these five points ($n = 5$). Use the standard deviation of the slope and intercept to express the equation in the form $y\ (\pm s_y) = [m\ (\pm s_m)]x + [b\ (\pm s_b)]$ with a reasonable number of significant figures.

(b) Make a graph showing the experimental data and the calculated straight line.

(c) An unknown gave an observed absorbance of 0.973. Calculate the number of micrograms of protein in the unknown, and estimate its uncertainty.

4-19. The equation for a Gaussian curve is

$$y = \frac{1}{\sigma\sqrt{2\pi}} e^{-(x-\mu)^2/2\sigma^2}$$

In Excel, the square root function is Sqrt and the exponential function is Exp. To find $e^{-3.4}$, write Exp(−3.4). The preceding exponential function is written

$$e^{-(x-\mu)^2/2\sigma^2} = \text{Exp}(-((x - \mu)\text{^}2)/(2*\sigma\text{^}2))$$

(a) For $\mu = 10$ and $\sigma = 1$, compute values of y for the range $4 \le x \le 16$.

(b) Repeat the calculation for $\sigma = 2$.

(c) Plot the results of **(a)** and **(b)** on one graph. For $\sigma = 2$, label the regions that contain 68.3% and 95.5% of all observations.

How Would You Do It?

4-20. Students at Eastern Illinois University[3] intended to prepare copper(II) carbonate by adding a solution of $CuSO_4 \cdot 5H_2O$ to a solution of Na_2CO_3.

$$CuSO_4 \cdot 5H_2O(aq) + Na_2CO_3(aq) \longrightarrow \underset{\text{Copper(II) carbonate}}{CuCO_3(s)} + Na_2SO_4(aq) + 5H_2O(l)$$

After warming the mixture to 60°C, the gelatinous blue precipitate coagulated into an easily filterable pale green solid. The product was filtered, washed, and dried at 70°C. Copper in the product was measured by heating 0.4 g of solid in a stream of methane at high temperature to reduce the solid to pure Cu, which was weighed.

$$4CuCO_3(s) + CH_4(g) \xrightarrow{\text{heat}} 4Cu(s) + 5CO_2(g) + 2H_2O(g)$$

In 1995, 43 students found a mean value of 55.6 wt% Cu with a standard deviation of 2.7 wt%. In 1996, 39 students found 55.9 wt% with a standard deviation of 3.8 wt%. The instructor tried the experiment nine times and measured 55.8 wt% with a standard deviation of 0.5 wt%. Was the product of the synthesis $CuCO_3$? Could it have been a hydrate, $CuCO_3 \cdot xH_2O$?

4-21. Strontium isotopes vary in different rocks, depending on the original content of radioactive elements in those rocks. The isotope ratio $^{87}Sr/^{86}Sr$ is used in environmental studies to determine sources of particles and solutes in water and ice. Materials originating from one source should have the same $^{87}Sr/^{86}Sr$ ratio. Materials from different sources could have different isotope ratios. Observations for Sr found in microscopic dust particles in ice drilled from Antarctica are shown below. Age is determined from depth in the ice. Uncertainties are expressed as 95% confidence intervals. EH stands for early Holocene and LGM stands for last glacial maximum; both are measures of geologic time.

Location	Age (years before present)	Sr(pg/g)	$^{87}Sr/^{86}Sr$
Dome C	7 500 [EH]	30.8 ± 0.4	0.706 8 ± 0.000 6
Dome C	23 000 [LGM]	324 ± 4	0.708 2 ± 0.000 5
Law Dome	6 500 [EH]	45.6 ± 0.6	0.709 7 ± 0.000 4
Law Dome	34 000 [LGM]	96 ± 2	0.709 3 ± 0.001 1

SOURCE: G. R. Burton, V. I. Morgan, C. F. Boutron, and K. J. R. Rosman, *Anal. Chim. Acta* **2002**, *469*, 225.

Does it appear that the EH dust at Dome C comes from the same source as EH dust at Law Dome? Does LGM dust at Dome C come from the same source as LGM dust at Law Dome? What does the unit pg/g mean? Express pg/g with a term such as "parts per million."

Notes and References

1. L. H. Keith, W. Crummett, J. Deegan, Jr., R. A. Libby, J. K. Taylor, and G. Wentler, *Anal. Chem.* **1983**, *55*, 2210.

2. When t calculated with Equation 4-4 is greater than the tabulated t, we conclude that the two means are different with a chosen confidence level. This test does not provide the same confidence level that the two means *are equal*. For discussions of how to show that two means are equal at a certain confidence level, see S. E. Lewis and J. E. Lewis, *J. Chem. Ed.* **2005**, *82*, 1408 and G. B. Limentani, M. C. Ringo, F. Ye, M. L. Bergquist, and E. O. McSorley, *Anal. Chem.* **2005**, *77*, 221A.

3. D. Sheeran, *J. Chem. Ed.* **1998**, *75*, 453. See also H. Gamsjäger and W. Preis, *J. Chem. Ed.* **1999**, *76*, 1339.

Further Reading

D. B. Hibbert and J. J. Gooding, *Data Analysis for Chemistry* (Oxford: Oxford University Press, 2006).

J. C. Miller and J. N. Miller, *Statistics and Chemometrics for Analytical Chemistry,* 4th ed. (Harlow, UK: Prentice Hall, 2000).

P. C. Meier and R. E. Zünd, *Statistical Methods in Analytical Chemistry,* 2nd ed. (New York: Wiley, 2000).

The Need for Quality Assurance

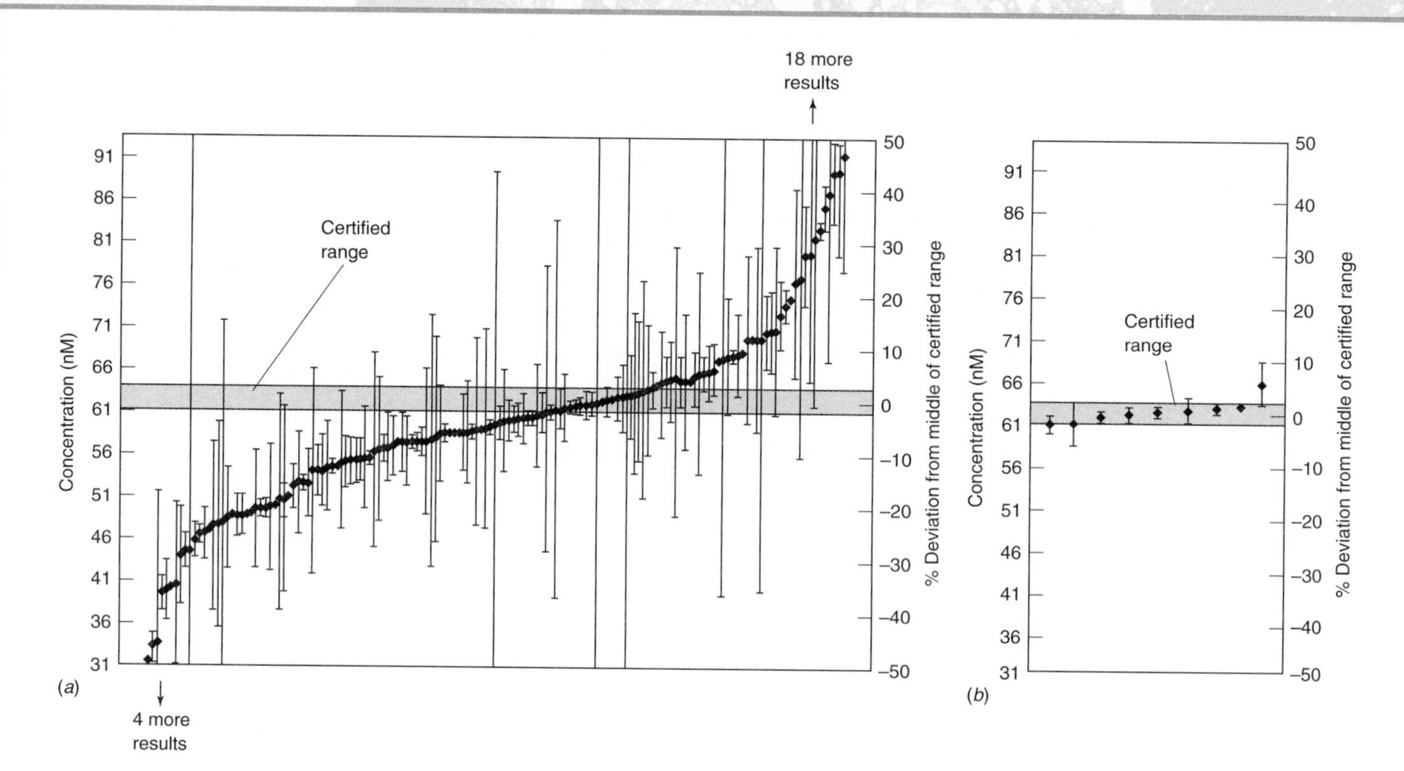

(*a*) Scattered measurements of Pb in river water by different laboratories, each of which employed a recognized quality management system. (*b*) Reproducible results from national measurement institutes. [From P. De Bièvre and P. D. P. Taylor, *Fresenius J. Anal. Chem.* **2000**, *368*, 567.]

The Institute for Reference Materials and Measurements in Belgium conducts an International Measurement Evaluation Program to allow laboratories to assess the reliability of their analyses. Panel *a* shows results for lead in river water. Of 181 labs, 18 reported results more than 50% above and 4 reported results more than 50% below the certified level of 62.3 ± 1.3 nM. Even though most labs in the study employed recognized quality management procedures, a large fraction of results did not include the certified range. Panel *b* shows that when this same river water was analyzed by nine different national measurement institutes, where the most care is taken, all results were close to the certified range.

This example illustrates that there is no guarantee that results are reliable, even if they are obtained by "accredited" laboratories using accepted procedures. A good way to assess the reliability of a lab working for you is to provide the lab with "blind" samples—similar to your unknowns—for which you know the "right" answer, but the analyst does not. If the lab does not find the known result, there is a problem. Periodic checks with blind samples are required to demonstrate continuing reliability.

CHAPTER 5

Quality Assurance and Calibration Methods

Quality assurance is what we do to get the right answer for our purpose. The answer should have sufficient accuracy and precision to support subsequent decisions. There is no point spending extra money to obtain a more accurate or more precise answer if it is not necessary. This chapter describes basic issues and procedures in quality assurance and introduces two more calibration methods. In Chapter 4, we discussed how to make a calibration curve. In this chapter, we consider the methods of *standard addition* and *internal standards.*

5-1 Basics of Quality Assurance

"Suppose you are cooking for some friends. While making spaghetti sauce, you taste it, season it, taste it some more. Each tasting is a sampling event with a quality control test. You can taste the whole batch because there is only one batch. Now suppose you run a spaghetti sauce plant that makes 1 000 jars a day. You can't taste each one, so you decide to taste three a day, one each at 11 A.M., 2 P.M., and 5 P.M. If the three jars all taste OK, you conclude all 1 000 are OK. Unfortunately, that may not be true, but the relative risk—that a jar has too much or too little seasoning—is not very important because you agree to refund the money of any customer who is dissatisfied. If the number of refunds is small, say, 100 a year, there is no apparent benefit in tasting 4 jars a day." There would be 365 additional tests to avoid refunds on 100 jars, giving a net loss of 265 jars worth of profit.

Quotation from Ed Urbansky, U.S. Environmental Protection Agency, Cincinnati, OH. Section 5-1 is adapted from a description written by Ed Urbansky.

In analytical chemistry, the product is not spaghetti sauce, but, rather, raw data, treated data, and results. *Raw data* are individual values of a measured quantity, such as peak areas from a chromatogram or volumes from a buret. *Treated data* are concentrations or amounts found by applying a calibration procedure to the raw data. *Results* are what we ultimately report, such as the mean, standard deviation, and confidence interval, after applying statistics to treated data.

Raw data: individual measurements
Treated data: concentrations derived from raw data by use of calibration method
Results: quantities reported after statistical analysis of treated data

Use Objectives

If you manufacture a drug whose therapeutic dose is just a little less than the lethal dose, you should be more careful than if you make spaghetti sauce. The kind of data that you collect and the way in which you collect them depend on how you plan to use those data. An important goal of quality assurance is making sure that results meet the customer's needs. A bathroom scale does not have to measure mass to the nearest milligram, but a drug tablet required to contain 2 mg of active ingredient probably cannot contain 2 ± 1 mg. Writing clear, concise **use objectives** for data and results is a critical step in quality assurance and helps prevent misuse of data and results.

Use objective: states purpose for which results will be used

Here is an example of a use objective. Drinking water is usually disinfected with chlorine, which kills microorganisms. Unfortunately, chlorine also reacts with organic matter in water to produce "disinfection by-products"—compounds that might harm humans. A disinfection plant was planning to introduce a new chlorination process and wrote the following analytical use objective:

> Analytical data and results shall be used to determine whether the modified chlorination process results in at least a 10% reduction of formation of selected disinfection by-products.

The new process was expected to produce fewer disinfection by-products. The use objective says that uncertainty in the analysis must be small enough so that a 10% decrease in selected by-products is clearly distinguishable from experimental error. In other words, is an observed decrease of 10% real?

Specifications

Specifications might include
- sampling requirements
- accuracy and precision
- rate of false results
- selectivity
- sensitivity
- acceptable blank values
- recovery of fortification
- calibration checks
- quality control samples

Once you have use objectives, you are ready to write **specifications** stating how good the numbers need to be and what precautions are required in the analytical procedure. How shall samples be taken and how many are needed? Are special precautions required to protect samples and ensure that they are not degraded? Within practical restraints, such as cost, time, and limited amounts of material available for analysis, what level of accuracy and precision will satisfy the use objectives? What rate of false positives or false negatives is acceptable? These questions need to be answered in detailed specifications.

Quality assurance begins with sampling. We must collect representative samples, and analyte must be preserved after sample is collected. If our sample is not representative or if analyte is lost after collection, then even the most accurate analysis is meaningless.

What do we mean by false positives and false negatives? Suppose you must certify that a contaminant in drinking water is below a legal limit. A *false positive* says that the concentration exceeds the legal limit when, in fact, the concentration is below the limit. A *false negative* says the concentration is below the limit when it is actually above the limit. Even a well-executed procedure produces some false conclusions because of the statistical nature of sampling and measurement. More stringent procedures are required to obtain lower rates of false conclusions. For drinking water, it is more important to have a low rate of false negatives than a low rate of false positives. It would be worse to certify that contaminated water is safe than to certify that safe water is contaminated.

Drug testing of athletes is designed to minimize false positives so that an innocent athlete is not falsely accused of doping. When there is any doubt about a drug test result, it is considered to be negative. In drug testing, the person who collects a sample is not the person who analyzes the sample. The identity of the athlete is not known to the analyst to prevent deliberate falsification of a result by an analyst.

In choosing a method, we also consider selectivity and sensitivity. **Selectivity** (also called *specificity*) means being able to distinguish analyte from other species in the sample (avoiding interference). **Sensitivity** is the capability of responding reliably and measurably to changes in analyte concentration. A method must have a *detection limit* (Section 5-2) lower than the concentrations to be measured.

Sensitivity:
$= \text{slope of calibration curve}$
$= \dfrac{\text{change in signal}}{\text{change in analyte concentration}}$

Specifications could include required accuracy and precision, reagent purity, tolerances for apparatus, the use of standard reference materials, and acceptable values for blanks. *Standard reference materials* (see Box 3-1) contain certified levels of analyte in realistic materials that you might be analyzing, such as blood or coal or metal alloys. Your analytical method should produce an answer acceptably close to the certified level or there is something wrong with the accuracy of your method. Blanks account for interference by other species in the sample and for traces of analyte found in reagents used for sample preservation, preparation, and analysis. Frequent measurements of blanks detect whether analyte from previous samples is carried into subsequent analyses by adhering to vessels or instruments.

A **method blank** is a sample containing all components except analyte, and it is taken through all steps of the analytical procedure. We subtract the response of the method blank from the response of a real sample prior to calculating the quantity of analyte in the sample. A **reagent blank** is similar to a method blank, but it has not been subjected to all sample preparation procedures. The method blank is a more complete estimate of the blank contribution to the analytical response.

A **field blank** is similar to a method blank, but it has been exposed to the site of sampling. For example, to analyze particulates in air, a certain volume of air could be sucked through a filter, which is then dissolved and analyzed. A field blank would be a filter carried to the collection site in the same package with the collection filters. The filter for the blank would be taken out of its package in the field and placed in the same kind of sealed container used for collection filters. The difference between the blank and the collection filters is that air was not sucked through the blank filter. Volatile organic compounds encountered during transportation or in the field are conceivable contaminants of a field blank.

Another performance requirement often specified is *spike recovery.* Sometimes, response to analyte is affected by something else in the sample. We use the word **matrix** to refer to everything else in the sample other than analyte. A **spike**, also called a *fortification*, is a known quantity of analyte added to a sample to test whether the response to a sample is the same as that expected from a calibration curve. Spiked samples are then analyzed in the same manner as unknowns. For example, if drinking water is found to contain 10.0 μg/L of nitrate upon analysis, a spike of 5.0 μg/L could be added. Ideally, the concentration in the spiked portion found by analysis will be 15.0 μg/L. If a number other than 15.0 μg/L is found, then the matrix could be interfering with the analysis.

Add a small volume of concentrated standard to avoid changing the volume of the sample significantly. For example, add 50.5 μL of 500 μg/L standard to 5.00 mL of sample to increase analyte by 5.00 μg/L.

Example Spike Recovery

Let C stand for concentration. One definition of spike recovery is

$$\%\ \text{recovery} = \frac{C_{\text{spiked sample}} - C_{\text{unspiked sample}}}{C_{\text{added}}} \times 100 \qquad (5\text{-}1)$$

An unknown was found to contain 10.0 μg of analyte per liter. A spike of 5.0 μg/L was added to a replicate portion of unknown. Analysis of the spiked sample gave a concentration of 14.6 μg/L. Find the percent recovery of the spike.

SOLUTION The percent of the spike found by analysis is

$$\% \text{ recovery} = \frac{14.6\ \mu\text{g/L} - 10.0\ \mu\text{g/L}}{5.0\ \mu\text{g/L}} \times 100 = 92\%$$

If the acceptable recovery is specified to be in the range from 96% to 104%, then 92% is unacceptable. Something in your method or techniques needs improvement.

Test Yourself An unknown containing 93.2 μg of analyte per liter was spiked with an additional 80.0 μg/L. Analysis of the spiked sample gave a concentration of 179.4 μg/L. Find the percent recovery of the spike. (**Answer:** 103.6%)

When dealing with large numbers of samples and replicates, we perform periodic calibration checks to make sure that our instrument continues to work properly and the calibration remains valid. In a **calibration check**, we analyze solutions formulated to contain known concentrations of analyte. A specification might, for example, call for one calibration check for every 10 samples. Solutions for calibration checks should be different from the ones used to prepare the original calibration curve. This practice helps to verify that the initial calibration standards were made properly.

To gauge accuracy:
- calibration checks
- fortification recoveries
- quality control samples
- blanks

To gauge precision:
- replicate samples
- replicate portions of same sample

Performance test samples (also called *quality control samples* or *blind samples*) are a quality control measure to help eliminate bias introduced by the analyst who knows the concentration of the calibration check sample. These samples of known composition are provided to the analyst as unknowns. Results are then compared with the known values, usually by a quality assurance manager. For example, the U.S. Department of Agriculture maintains a bank of quality control homogenized food samples for distribution as blind samples to laboratories that measure nutrients in foods.

Together, raw data and results from calibration checks, spike recoveries, quality control samples, and blanks are used to gauge accuracy. Analytical performance on replicate samples and replicate portions of the same sample measures precision. Fortification also helps ensure that qualitative identification of analyte is correct. If you spike the unknown in Figure 0-5 with extra caffeine and the area of a chromatographic peak not thought to be caffeine increases, then you have misidentified the caffeine peak.

Many labs have their own standard practices, such as recording temperatures in refrigerators, calibrating balances, conducting routine instrument maintenance, or replacing reagents. These are part of the overall quality management plan. The rationale behind standard practices is that some equipment is used by multiple people for different analyses. We save money by having one program to ensure that the most rigorous needs are met.

Assessment

Assessment is the process of (1) collecting data to show that analytical procedures are operating within specified limits and (2) verifying that final results meet use objectives.

Documentation is critical for assessment. Standard *protocols* provide directions for what must be documented and how the documentation is to be done, including how to record information in notebooks. For labs that rely on manuals of standard practices, it is imperative that tasks done to comply with the manuals be monitored and recorded.

Box 5-1 Control Charts

A **control chart** is a visual representation of confidence intervals for a Gaussian distribution. A control chart warns us when a property being monitored strays dangerously far from an intended *target value*.

Consider a manufacturer making vitamin C tablets intended to have μ milligrams of vitamin C per tablet. μ is the target value. Many analyses over a long time tell us the population standard deviation, σ, associated with the manufacturing process.

For quality control, 25 tablets are removed at random from the manufacturing line each hour and analyzed. The mean value of vitamin C in the 25 tablets is shown by a data point on the following control chart.

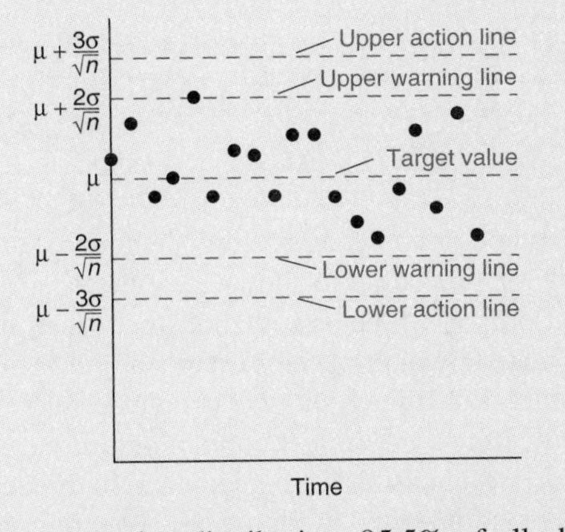

For a Gaussian distribution, 95.5% of all observations are within $\pm 2\sigma/\sqrt{n}$ from the mean and 99.7% are within $\pm 3\sigma/\sqrt{n}$. In these expressions, n is the number of tablets (= 25) that are averaged each hour. The $\pm 2\sigma/\sqrt{n}$ limits are designated *warning lines* and the $\pm 3\sigma/\sqrt{n}$ limits are designated *action lines*. We expect ~4.5% of measurements to be outside the warning lines and ~0.3% to be outside the action lines. It is unlikely that we would observe two consecutive measurements at the warning line (probability = $0.045 \times 0.045 = 0.002\,0$).

The following conditions are considered to be so unlikely that, if they occur, the process should be shut down for troubleshooting:

- 1 observation outside the action lines
- 2 out of 3 consecutive measurements between the warning and the action lines
- 7 consecutive measurements all above or all below the center line
- 6 consecutive measurements all steadily increasing or all steadily decreasing, wherever they are located
- 14 consecutive points alternating up and down, regardless of where they are located
- an obvious nonrandom pattern

For quality assessment of an analytical process, a control chart could show the relative deviation of measured values of calibration check samples or quality control samples from their known values. Another control chart could display the precision of replicate analyses of unknowns or standards as a function of time.

Control charts (Box 5-1) can be used to monitor performance on blanks, calibration checks, and spiked samples to see whether results are stable over time or to compare the work of different employees. Control charts can also monitor sensitivity or selectivity, especially if a laboratory encounters a wide variety of matrixes.

Government agencies such as the U.S. Environmental Protection Agency set requirements for quality assurance for their own labs and for certification of other labs. Published standard methods specify precision, accuracy, numbers of blanks, replicates, and calibration checks. To monitor drinking water, regulations state how often and how many samples are to be taken. Documentation is necessary to demonstrate that all requirements have been met. Table 5-1 summarizes the quality assurance process.

Ask Yourself

5-A. What are the three parts of quality assurance? What questions are asked in each part and what actions are taken in each part?

Table 5-1 Quality assurance process

Question	Actions
Use Objectives Why do you want the data and results and how will you use the results?	• Write use objectives
Specifications How good do the numbers have to be?	• Write specifications • Pick methods to meet specifications • Consider sampling, precision, accuracy, selectivity, sensitivity, detection limit, robustness, rate of false results • Employ blanks, fortification, calibration checks, quality control samples, and control charts to monitor performance
Assessment Were the specifications achieved?	• Compare data and results with specifications • Document procedures and keep records suitable to meet use objectives • Verify that use objectives were met.

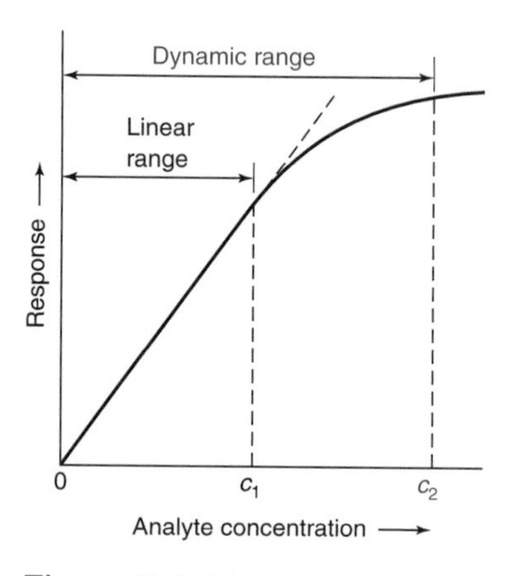

Figure 5-1 Schematic calibration curve with linear and nonlinear regions.

Linear range: concentration range over which calibration curve is linear

Dynamic range: concentration range over which there is measurable response

Range: concentration range over which linearity, accuracy, and precision meet specifications for analytical method

5-2 Validation of an Analytical Procedure

If you develop a new analytical method or apply an existing method to a new kind of sample, it is necessary to *validate* the procedure to show that it meets specifications and is acceptable for its intended purpose. Standard methods published by government and private agencies are validated by multiple labs before they are published. In pharmaceutical chemistry, **method validation** requirements include studies of *selectivity, accuracy, precision, linearity, range, robustness, limit of detection,* and *limit of quantitation.* Selectivity, accuracy, and precision were mentioned in the Section 5-1.

Linearity of a calibration curve is illustrated in Figure 5-1. We prefer a response in which the corrected analytical signal (= signal from sample − signal from blank) is proportional to the quantity of analyte, as it is between 0 and c_1 in Figure 5-1. However, you can obtain results beyond the linear region between c_1 and c_2 by fitting the data to a curve such as a polynomial. **Linear range** is the analyte concentration range over which response is proportional to concentration. **Dynamic range** is the concentration range over which there is a measurable response to analyte, even if the response is not linear. For an analytical method, the word **range** means the concentration interval over which linearity, accuracy, and precision are all acceptable.

Another goal of validation is to show that a method is **robust**, which means that it is not affected by small changes in conditions. For example, a robust chromatographic procedure gives reliable results despite small changes in solvent composition, pH, buffer concentration, temperature, injection volume, and detector wavelength.

Limits of Detection and Quantitation

The **detection limit** (also called the *lower limit of detection*) is the smallest quantity of analyte that is "significantly different" from the blank. Here is a procedure that produces a detection limit that has approximately a 99% chance of being greater than the blank. That is, only ~1% of samples containing no analyte will give a signal greater than the detection limit (Figure 5-2). We assume that the standard deviation of the signal from samples near the detection limit is similar to the standard deviation from blanks.

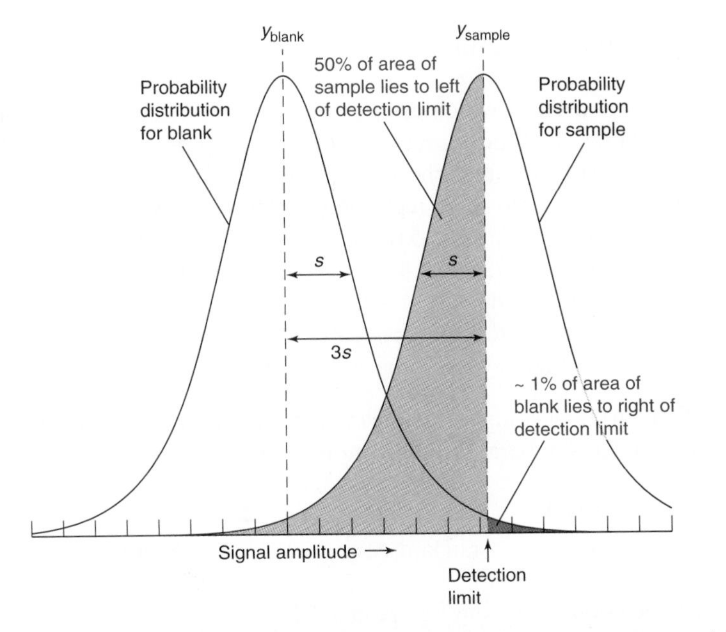

Figure 5-2 Detection limit. Curves show distribution of measurements expected for a blank and a sample whose concentration is at the detection limit. The area of any region is proportional to the number of measurements in that region. Only ~1% of measurements for a blank are expected to exceed the detection limit. However, 50% of measurements for a sample containing analyte at the detection limit will be below the detection limit. There is a 1% chance of concluding that a blank has analyte above the detection limit. If a sample contains analyte at the detection limit, there is a 50% chance of concluding that analyte is *absent* because its signal is below the detection limit. Curves in this figure are Student's *t* distributions, which are broader than the Gaussian distribution.

1. After estimating the detection limit from previous experience with the method, prepare a sample whose concentration is ~1 to 5 times the detection limit.
2. Measure the signal from n replicate samples ($n \geq 7$).
3. Compute the standard deviation (s) of the n measurements.
4. Measure the signal from n blanks (containing no analyte) and find the mean value, y_{blank}.
5. The minimum detectable signal, y_{dl}, is defined as

Signal detection limit: $$y_{dl} = y_{blank} + 3s \qquad (5\text{-}2)$$

6. The corrected signal, $y_{sample} - y_{blank}$, is proportional to sample concentration:

Calibration line: $$y_{sample} - y_{blank} = m \times \text{sample concentration} \qquad (5\text{-}3)$$

where y_{sample} is the signal observed for the sample and m is the slope of the linear calibration curve. The *minimum detectable concentration*, also called the *detection limit*, is obtained by substituting y_{dl} from Equation 5-2 for y_{sample} in Equation 5-3:

Detection limit: $$\text{minimum detectable concentration} = \frac{3s}{m} \qquad (5\text{-}4)$$

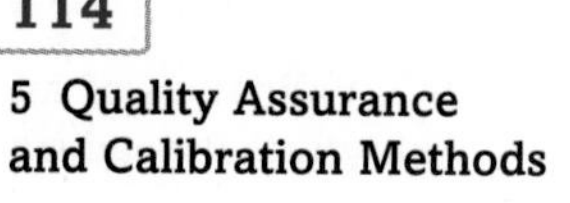

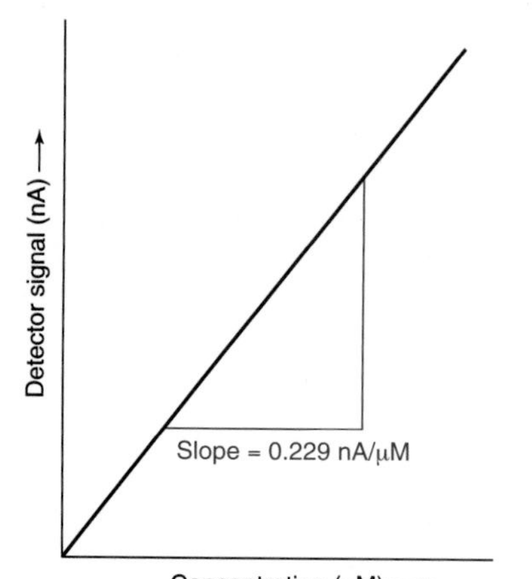

Example Detection Limit

A procedure was used in which electric current in a detector is proportional to analyte concentration. From previous measurements of a low concentration of analyte, it was estimated that the signal detection limit was in the low nanoampere range. Signals from seven replicate samples with a concentration about three times the detection limit were 5.0, 5.0, 5.2, 4.2, 4.6, 6.0, and 4.9 nA. Reagent blanks gave values of 1.4, 2.2, 1.7, 0.9, 0.4, 1.5, and 0.7 nA. The slope of the calibration curve for higher concentrations is $m = 0.229$ nA/μM. **(a)** Find the signal detection limit and the minimum detectable concentration. **(b)** What is the concentration of analyte in a sample that gave a signal of 7.0 nA?

SOLUTION (a) First compute the mean for the blanks and the standard deviation of the samples. Retain extra, insignificant digits to reduce round-off errors.

$$\text{Blank:} \quad \text{average} = y_{\text{blank}} = 1.2_6 \text{ nA}$$
$$\text{Sample:} \quad \text{standard deviation} = s = 0.5_6 \text{ nA}$$

The signal detection limit from Equation 5-2 is

$$y_{\text{dl}} = y_{\text{blank}} + 3s = 1.2_6 \text{ nA} + (3)(0.5_6 \text{ nA}) = 2.9_4 \text{ nA}$$

The minimum detectable concentration is obtained from Equation 5-4:

$$\text{detection limit} = \frac{3s}{m} = \frac{(3)(0.5_6 \text{ nA})}{0.229 \text{ nA}/\mu\text{M}} = 7._3 \ \mu\text{M}$$

(b) To find the concentration of a sample whose signal is 7.0 nA, use Equation 5-3:

$$y_{\text{sample}} - y_{\text{blank}} = m \times \text{concentration}$$
$$\Rightarrow \text{concentration} = \frac{y_{\text{sample}} - y_{\text{blank}}}{m} = \frac{7.0 \text{ nA} - 1.2_6 \text{ nA}}{0.229 \text{ nA}/\mu\text{M}} = 25._1 \ \mu\text{M}$$

Test Yourself Suppose that $s = 0.2_8$ nA instead of 0.5_6 nA, but y_{blank} is still 1.2_6 nA. Find the signal detection limit, the minimum detectable concentration, and the concentration of analyte in a sample with a signal of 7.0 nA (**Answer:** 2.1_0 nA, 3.7 μM, $25._1$ μM)

The lower limit of detection given in Equation 5-4 is $3s/m$, where s is the standard deviation of a low-concentration sample and m is the slope of the calibration curve. The standard deviation is a measure of the *noise* (random variation) in a blank or a small signal. When the signal is 3 times greater than the noise, it is readily detectable, but still too small for accurate measurement. A signal that is 10 times greater than the noise is defined as the **lower limit of quantitation**, or the smallest amount that can be measured with reasonable accuracy.

Quantitation limit = (10/3) × detection limit

The symbol ≡ means "is defined as."

Factors of 3 in the detection limit and 10 in the quantitation limit are arbitrary conventions.

$$\text{lower limit of quantitation} \equiv \frac{10s}{m} \qquad (5\text{-}5)$$

The **reporting limit** is the concentration below which regulations dictate that a given analyte is reported as "not detected." "Not detected" does not mean that analyte

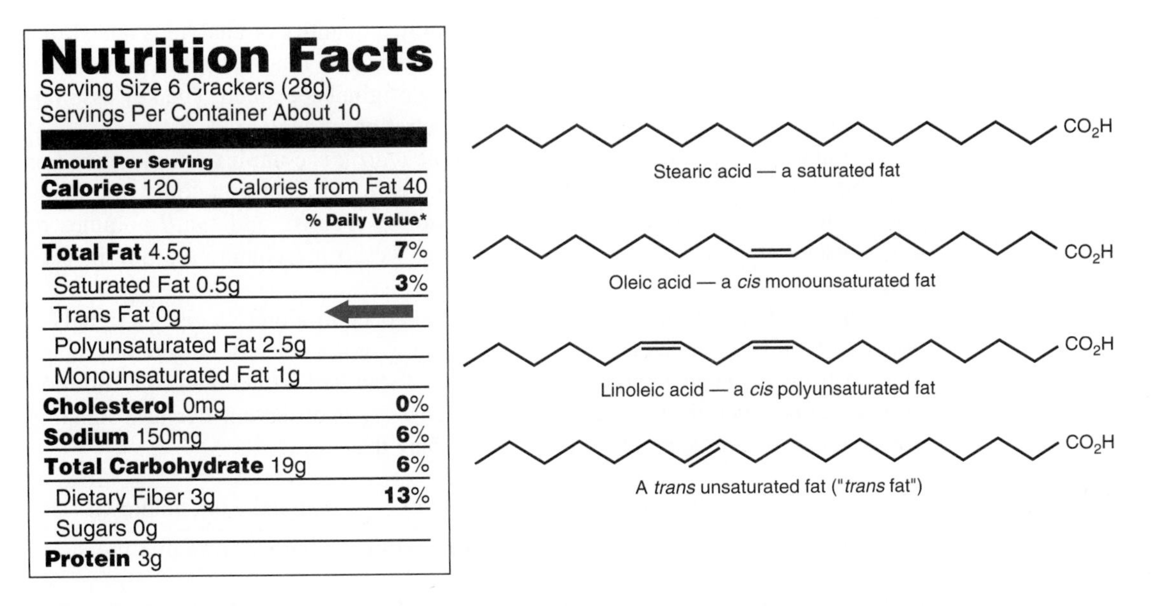

Nutrition Facts
Serving Size 6 Crackers (28g)
Servings Per Container About 10
Amount Per Serving
Calories 120 Calories from Fat 40
% Daily Value*
Total Fat 4.5g 7%
Saturated Fat 0.5g 3%
Trans Fat 0g
Polyunsaturated Fat 2.5g
Monounsaturated Fat 1g
Cholesterol 0mg 0%
Sodium 150mg 6%
Total Carbohydrate 19g 6%
Dietary Fiber 3g 13%
Sugars 0g
Protein 3g
CO_2H
Stearic acid — a saturated fat
CO_2H
Oleic acid — a *cis* monounsaturated fat
CO_2H
Linoleic acid — a *cis* polyunsaturated fat
CO_2H
A *trans* unsaturated fat ("*trans* fat")

Figure 5-3 Nutritional label from a package of crackers. The reporting limit for *trans* fat is 0.5 g/serving. Any amount less than this is reported as 0. Structures of representative saturated, monounsaturated, polyunsaturated, and *trans* fats are shown. Box 7-1 explains the shorthand used to draw these 18-carbon compounds.

is not observed, but that it is below a prescribed level. Reporting limits are set at least 5 to 10 times higher than the detection limit so that detecting analyte at the reporting limit is not ambiguous.

Beginning in 2006, labels on U.S. packaged foods must state how much *trans* fat is present. This type of fat is derived mainly from partial hydrogenation of vegetable oil and is a major component of margarine and shortening. Consumption of *trans* fat increases risk of heart disease, stroke, and some types of cancer. However, the *reporting limit* for *trans* fat is 0.5 g per serving. If the concentration is <0.5 g/serving, it is reported as 0, as in Figure 5-3. By reducing the serving size, a manufacturer can state that the *trans* fat content is 0. The reason given by the government for the high reporting limit is that many labs use infrared analysis, whose detection limit is poor. Gas chromatography provides a lower detection limit (Figure 22-4). If your favorite snack food is made with partially hydrogenated oil, it contains *trans* fat even if the label says otherwise.

Ask Yourself

5-B. Method validation includes studies of precision and accuracy of a proposed method. How would you validate precision and accuracy? (*Hint:* Review Section 5-1.)

5-3 Standard Addition

Calibration curves are usually used to determine the relation between signal and concentration in a chemical analysis. In cases where a calibration curve would be inappropriate and unreliable, we can use *standard addition* or *internal standards*.

Sections 5-3 and 5-4 could be postponed until you need to use standard addition or internal standards in the lab.

In the method of **standard addition**, a known quantity of analyte is added to a specimen and the increase in signal is measured. The relative increase in signal allows us to infer how much analyte was in the original specimen. The key assumption is that signal is proportional to the concentration of analyte.

Standard additions are most appropriate when the sample matrix is complex and difficult to reproduce in standard solutions.

Standard addition is used when the sample *matrix* is complex or unknown. For example, a matrix such as blood has many constituents that you could not incorporate into standard solutions for a calibration curve. We add *small* volumes of concentrated standard to the unknown so that we do not change the matrix very much.

Suppose that a sample with unknown initial concentration $[X]_i$ gives a signal I_X, where I might be the peak area in chromatography or the detector current or voltage of an instrument. Then a known concentration of standard S (a known concentration of analyte) is added to the sample and a signal I_{S+X} is observed. Because signal is proportional to analyte concentration, we can say that

Bear in mind that the species X and S are the same.

$$\frac{\text{concentration of analyte in unknown}}{\text{concentration of analyte + standard in mixture}} = \frac{\text{signal from unknown}}{\text{signal from mixture}}$$

Standard addition equation:

$$\frac{[X]_i}{[X]_f + [S]_f} = \frac{I_X}{I_{S+X}} \qquad (5\text{-}6)$$

where $[X]_f$ is the final concentration of unknown analyte after adding the standard and $[S]_f$ is the final concentration of standard after addition to the unknown. If we began with an initial volume V_0 of unknown and added the volume V_s of standard with initial concentration $[S]_i$, the total volume is $V = V_0 + V_s$ and the concentrations in Equation 5-6 are

$$[X]_f = [X]_i\underbrace{\left(\frac{V_0}{V}\right)}_{\text{Dilution factor}} \qquad [S]_f = [S]_i\underbrace{\left(\frac{V_s}{V}\right)}_{\text{Dilution factor}} \qquad (5\text{-}7)$$

Equations 5-7 follow from the dilution formula 1-5:

$$[X]_f V_f = [X]_i V_i$$

where f stands for "final" and i stands for "initial."

In Equations 5-7, $V = V_f$ and $V_0 = V_i$.

The factors V_0/V and V_s/V relating concentrations before and after dilution are called *dilution factors.*

Example Standard Addition

Ascorbic acid (vitamin C) in a 50.0-mL sample of orange juice was analyzed by an electrochemical method that gave a detector current of 1.78 μA. A standard addition of 0.400 mL of 0.279 M ascorbic acid increased the current to 3.35 μA. Find the concentration of ascorbic acid in the orange juice.

SOLUTION If the initial concentration of ascorbic acid in the juice is $[X]_i$, the concentration after dilution of 50.0 mL of juice with 0.400 mL of standard is

$$\text{final concentration of analyte} = [X]_f = [X]_i\left(\frac{V_0}{V}\right) = [X]_i\left(\frac{50.0}{50.4}\right)$$

The final concentration of the added standard after addition to the orange juice is

$$[S]_f = [S]_i\left(\frac{V_s}{V}\right) = [0.279\ \text{M}]\left(\frac{0.400}{50.4}\right) = 2.21_4\ \text{mM}$$

The standard addition equation 5-6 therefore becomes

$$\frac{[X]_i}{[X]_f + [S]_f} = \frac{[X]_i}{\left(\frac{50.0}{50.4}\right)[X]_i + 2.21_4 \text{ mM}} = \frac{1.78\ \mu A}{3.35\ \mu A} \Rightarrow [X]_i = 2.49 \text{ mM}$$

Test Yourself Find the concentration of ascorbic acid in the juice if the standard addition gave a current of 2.50 μA instead of 3.35 μA. (**Answer:** 5.37 mM)

Graphical Procedure for Standard Addition

A more accurate procedure is to make a series of standard additions that increase the original signal by a factor of 1.5 to 3 and use all the results together. If you take the expressions for $[X]_f$ and $[S]_f$ from Equations 5-7, plug them into the standard addition equation 5-6, and do some rearranging, you would find

Graphing equation for standard addition:

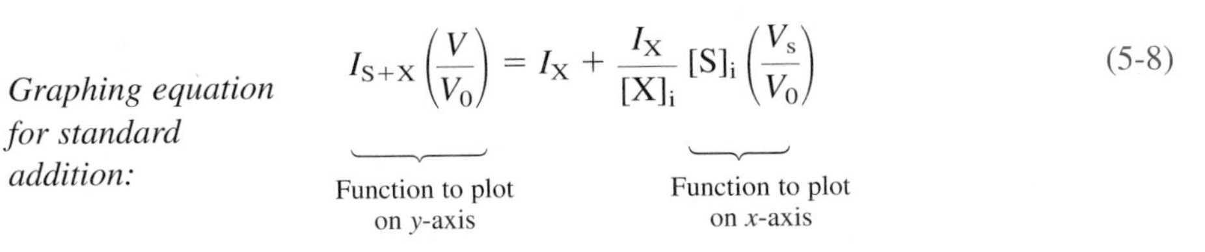

$$I_{S+X}\left(\frac{V}{V_0}\right) = I_X + \frac{I_X}{[X]_i}[S]_i\left(\frac{V_s}{V_0}\right) \quad (5\text{-}8)$$

I_{S+X} is the signal measured for a sample containing unknown plus standard. The factor V/V_0 is the final volume divided by the initial volume of the sample. The product $I_{S+X}(V/V_0)$ is the "corrected signal" because it is the signal that would be measured if the sample were not diluted by addition of the standard. On the right-hand side of Equation 5-8, the function of interest is the product $[S]_i(V_s/V_0)$, where $[S]_i$ is the concentration of standard prior to adding it to the sample, V_s is the volume of standard added, and V_0 is the initial volume of sample. A graph of $I_{S+X}(V/V_0)$ on the y-axis versus $[S]_i(V_s/V_0)$ on the x-axis should be a straight line. The intercept on the x-axis is the initial concentration of unknown, $[X]_i$.

Consider the data in Figure 5-4. The current in column D is the observed detector response, I_{S+X}. Equation 5-8 is plotted in Figure 5-5. From the x-intercept,

	A	B	C	D	E
1	Vitamin C Standard Addition Experiment				
2	Add 0.279 M ascorbic acid to 50.0 mL of orange juice				
3					
4		Vs =			
5	Vo (mL) =	mL ascorbic	x-axis function	I(s+x) =	y-axis function
6	50	acid added	Si*Vs/Vo	signal (μA)	I(s+x)*V/Vo
7	[S]i (mM) =	0.000	0.000	1.78	1.780
8	279	0.050	0.279	2.00	2.002
9		0.250	1.395	2.81	2.824
10		0.400	2.232	3.35	3.377
11		0.550	3.069	3.88	3.923
12		0.700	3.906	4.37	4.431
13		0.850	4.743	4.86	4.943
14		1.000	5.580	5.33	5.437
15		1.150	6.417	5.82	5.954
16					
17	C7 = $A$8*B7/$A$6		E7 = D7*($A$6+B7)/$A$6		

Figure 5-4 Spreadsheet with standard addition data for graphing Equation 5-8.

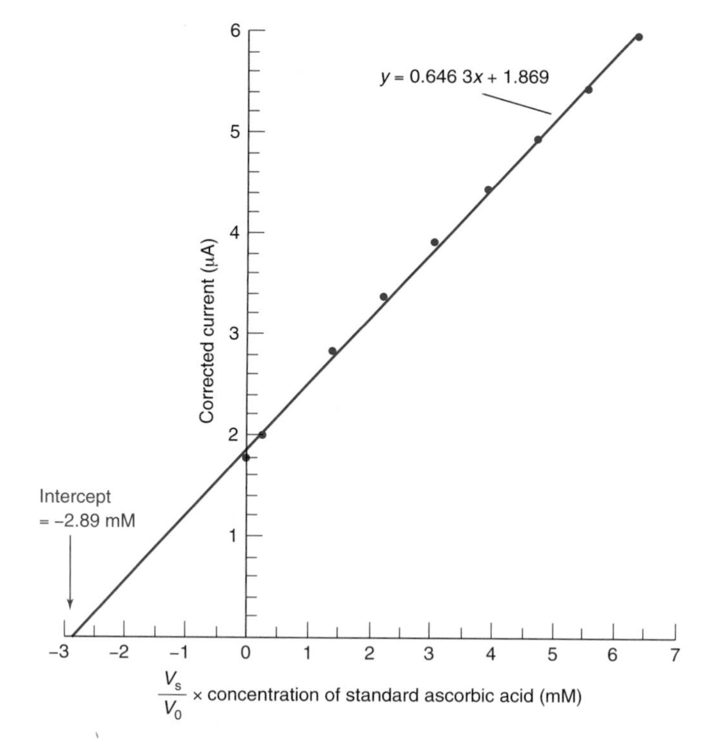

Figure 5-5 Graphical treatment of the method of standard addition, using Equation 5-8. The range of standard additions should increase the original signal by a factor of 1.5 to 3.

we conclude that the concentration of ascorbic acid in the original orange juice was $[X]_i = 2.89$ mM. In the preceding example, we found $[X]_i = 2.49$ mM from a single standard addition. The 14% difference between the two results is experimental error attributable to using one point instead of all the points. Uncertainty in the results from a standard addition graph is discussed in Problem 5-19.

If all solutions have *same final volume*:

plot I_{S+X} versus $[S]_f$

x-intercept is $[X]_f$

Often, all solutions in a standard addition experiment are made up to the same total volume by the addition of solvent. In this case, V is a constant multiple of V_0. *When all solutions are made up to the same final volume, we plot signal versus* $[S]_f$. The x-intercept of the graph is the *diluted* concentration of the unknown, $[X]_f$. For the graph in Figure 5-5, the x-intercept is the *initial* concentration of unknown, $[X]_i$.

Ask Yourself

5-C. Successive standard additions of 1.00 mL of 25.0 mM ascorbic acid were made to 50.0 mL of orange juice. Prepare a graph similar to Figure 5-5 to find the concentration of ascorbic acid in the orange juice.

Total volume of added standard (mL)	Peak current (μA)
0	1.66
1.00	2.03
2.00	2.39
3.00	2.79
4.00	3.16
5.00	3.51

5-4 Internal Standards

Section 5-4 can be postponed until you need to use an internal standard in the lab or for a problem.

An **internal standard** is a known amount of a compound, different from analyte, that is added to an unknown. Signal from analyte is compared with signal from the internal standard to find out how much analyte is present.

An *internal standard* is different from the analyte. In *standard addition* (Section 5-3), the standard is the same substance as the analyte.

Internal standards are especially useful for analyses in which the quantity of sample analyzed or the instrument response varies slightly from run to run for reasons that are difficult to control. For example, gas or liquid flow rates that vary by a few percent in chromatography could change the detector response. A calibration curve is only accurate for the one set of conditions under which it was obtained. However, the *relative* response of the detector to the analyte and standard is usually constant over a wide range of conditions. If signal from the standard increases by 8.4% because of a change in flow rate, signal from the analyte usually increases by 8.4% also. As long as the concentration of standard is known, the correct concentration of analyte can be derived. Internal standards are used in chromatography because the microliter volume of sample injected into the chromatograph is not very reproducible, so a calibration curve would not be accurate.

Internal standards are also desirable when sample loss can occur in sample preparation steps prior to analysis. If a known quantity of standard is added to the unknown prior to any manipulations, the ratio of standard to analyte remains constant because the same fraction of each is lost in any operation.

To use an internal standard, we prepare a known mixture of standard and analyte and measure the relative response of the detector to the two species. In Figure 5-6, the area under each peak is proportional to the concentration of each compound injected into the column. However, the detector generally has a different response to each component. For example, if both the analyte (X) and the internal standard (S) have concentrations of 10.0 mM, the area under the analyte peak might be 2.30 times greater than the area under the standard peak. We say that the **response factor**, F, is 2.30 times greater for X than for S.

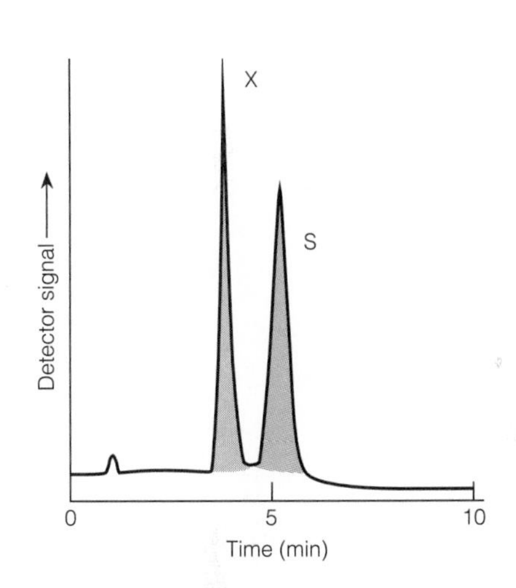

Figure 5-6 Chromatogram illustrating the use of an internal standard. A known amount of standard S is added to unknown X. From the areas of the peaks, we can tell how much X is in the unknown. To do so, we needed to measure the relative response to known amounts of each compound in a separate experiment.

Internal standard:

$$\frac{\text{area of analyte signal}}{\text{concentration of analyte}} = F\left(\frac{\text{area of standard signal}}{\text{concentration of standard}}\right) \tag{5-9}$$

$$\frac{A_X}{[X]} = F\left(\frac{A_S}{[S]}\right)$$

[X] and [S] are the concentrations of analyte and standard *after they have been mixed together.* Equation 5-9 is predicated on linear response to both the analyte and the standard.

Example Using an Internal Standard

In a chromatography experiment, a solution containing 0.083 7 M X and 0.066 6 M S gave peak areas of $A_X = 423$ and $A_S = 347$. (Areas are measured in arbitrary units by the instrument's computer.) To analyze the unknown, 10.0 mL of 0.146 M S were added to 10.0 mL of unknown, and the mixture was diluted to 25.0 mL in a volumetric flask. This mixture gave the chromatogram in Figure 5-6, with peak areas $A_X = 553$ and $A_S = 582$. Find the concentration of X in the unknown.

SOLUTION First use the standard mixture to find the response factor in Equation 5-9:

Standard mixture:

$$\frac{A_X}{[X]} = F\left(\frac{A_S}{[S]}\right)$$

$$\frac{423}{0.083\,7\text{ M}} = F\left(\frac{347}{0.066\,6\text{ M}}\right) \Rightarrow F = 0.970_0$$

In the mixture of unknown plus standard, the standard has been diluted from 10.0 mL up to 25.0 mL. The concentration of S is

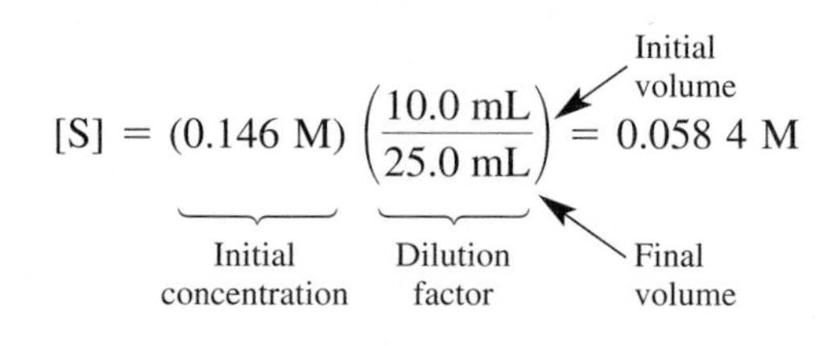

[S] = (0.146 M) (10.0 mL / 25.0 mL) = 0.058 4 M
Initial volume
Final volume
Initial concentration
Dilution factor

Using the known response factor, we substitute back into Equation 5-9 to find the concentration of unknown in the mixture:

Unknown mixture:

$$\frac{A_X}{[X]} = F\left(\frac{A_S}{[S]}\right)$$

$$\frac{553}{[X]} = 0.970_0\left(\frac{582}{0.058\,4\text{ M}}\right) \Rightarrow [X] = 0.057\,2_1\text{ M}$$

Because X was diluted from 10.0 to 25.0 mL when the mixture with S was prepared, the original concentration of X in the unknown was (25.0/10.0)($0.057\,2_1$ M) = 0.143 M.

Test Yourself A solution with 0.083 7 M X and 0.050 0 M S′ gave areas $A_X = 423$ and $A_{S'} = 372$. Then 10.0 mL of 0.05 0 M S′ plus 10.0 mL of unknown were diluted to 25.0 mL. The chromatogram gave $A_X = 553$ and $A_{S'} = 286$. Find [X] in the unknown. (**Answer:** 0.142 M)

Ask Yourself

5-D. A mixture of 52.4 nM analyte (X) and 38.9 nM standard (S) gave the relative response (area of X)/(area of S) = 0.644/1.000. A second solution containing an unknown quantity of X plus 742 nM S had (area of X)/(area of S) = 1.093/1.000. Find [X] in the second solution.

Key Equations

Detection and quantitation limits

$$\text{Minimum detectable concentration} = \frac{3s}{m}$$

$$\text{Lower limit of quantitation} \equiv \frac{10s}{m}$$

s = standard deviation of sample at 1–5 times detection limit

m = slope of calibration curve

Standard addition $\frac{[X]_i}{[X]_f + [S]_f} = \frac{I_X}{I_{S+X}}$

$[X]_i$ = concentration of analyte in initial unknown

$[X]_f$ = concentration of analyte after standard addition

$[S]_f$ = concentration of standard after addition to unknown

Standard addition graph: You should be able to graph Equation 5-8 to interpret a standard addition experiment.

Internal standard $\frac{\text{area of analyte signal}}{\text{concentration of analyte}} = F\left(\frac{\text{area of standard signal}}{\text{concentration of standard}}\right)$

F = response factor measured in a separate experiment with known concentrations of analyte and standard

Important Terms

assessment	matrix	robustness
calibration check	method blank	selectivity
control chart	method validation	sensitivity
detection limit	performance test sample	specifications
dynamic range	quality assurance	spike
field blank	range	standard addition
internal standard	reagent blank	use objectives
linear range	reporting limit	
lower limit of quantitation	response factor	

Problems

5-1. Distinguish *raw data*, *treated data*, and *results*.

5-2. What is the difference between a *calibration check* and a *performance test sample*?

5-3. What is the purpose of a blank? Distinguish *method blank*, *reagent blank*, and *field blank*.

5-4. Distinguish *linear range*, *dynamic range*, and *range*.

5-5. What is the difference between a *false positive* and a *false negative*?

5-6. Consider a sample that contains analyte at the detection limit defined by Equation 5-4. Refer to Figure 5-2 to explain the following statements: There is approximately a 1% chance of falsely concluding that a sample containing no analyte contains analyte above the detection limit. There is a 50% chance of concluding that a sample that really contains analyte at the detection limit does not contain analyte above the detection limit.

5-7. How is a control chart used? State six indications that a process is going out of control.

5-8. Here is a use objective for a chemical analysis to be performed at a drinking water purification plant: "Data and results collected quarterly shall be used to determine whether the concentrations of haloacetates in the treated water demonstrate compliance with the levels set by the Stage 1 Disinfection By-products Rule using Method 552.2" (a specification that sets precision, accuracy, and other requirements). Which of the following questions best summarizes the meaning of the use objective?

(a) Are haloacetate concentrations known within specified precision and accuracy?

(b) Are any haloacetates detectable in the water?

(c) Do any haloacetate concentrations exceed the regulatory limit?

5-9. *False positives and negatives.* Drinking water from wells in Bangladesh and much of Southeast Asia has unsafe levels of naturally occurring arsenic. Colorimetric test kits are used to combat this severe public health problem. If the color response indicates As > 50 μg/L, the well is painted red and not used for drinking. If As < 50 μg/L, the well is painted green and used for drinking. (By comparision, the allowed level of As in Europe and North America is 10 μg/L). We say that a positive result in the colorimetric test means that As > 50 μg/L. A study in 2002 found 50% false positives and 8% false negatives. What percentage of green wells should be red and what percentage of red wells should be green? Would it be better or worse for public health to have a 50% false negative rate?

5-10. *Blind samples* of homogenized beef baby food were provided to three laboratories for analysis. Results from the three labs agreed well for protein, fat, zinc, riboflavin, and palmitic acid. Results for iron were questionable: Lab A: 1.59 ± 0.14 (13); Lab B: 1.65 ± 0.56 (8); Lab C: 2.68 ± 0.78 (3) mg Fe/100 g, with numbers of replicate analyses in parentheses. Lab C produced a higher result than those from Labs A and B. Is the result from Lab C different at the 95% confidence level from that of Lab B?

5-11. *Detection limit.* In spectrophotometry, we measure the concentration of analyte by its absorbance of light. A low-concentration sample was prepared, and nine replicate measurements gave absorbances of 0.004 7, 0.005 4, 0.006 2, 0.006 0, 0.004 6, 0.005 6, 0.005 2, 0.004 4, and 0.005 8 in a 1.000-cm cell. Nine reagent blanks gave values of 0.000 6, 0.001 2, 0.002 2, 0.000 5, 0.001 6, 0.000 8, 0.001 7, 0.001 0, and 0.001 1.

(a) Find the absorbance detection limit with Equation 5-2.

(b) The calibration curve is a graph of absorbance versus concentration in a cell with a pathlength of 1.000 cm. Absorbance is a dimensionless quantity. The slope of the calibration curve is $m = 2.24 \times 10^4\ M^{-1}$. Find the concentration detection limit with Equation 5-4.

(c) Find the lower limit of quantitation with Equation 5-5.

5-12. *Control chart.* Volatile compounds in human blood serum were measured by purge and trap gas chromatography–mass spectrometry. For quality control, serum was periodically *spiked* (treated) with a constant amount of 1,2-dichlorobenzene, and the concentration (ng/g = ppb) was measured. Find the mean and standard deviation for the following spike data and prepare a control chart. State whether or not the observations meet each of the criteria in Box 5-1 for stability in a control chart.

Observed		Observed		Observed		Observed	
Day	ppb	Day	ppb	Day	ppb	Day	ppb
0	1.05	3	0.42	7	0.55	70	0.83
1	0.70	6	0.95	30	0.68	72	0.97
76	0.60	128	0.81	199	0.85	290	1.04
80	0.87	134	0.84	212	1.03	294	0.85
84	1.03	147	0.83	218	0.90	296	0.59
91	1.13	149	0.88	220	0.86	300	0.83
101	1.64	154	0.89	237	1.05	302	0.67
104	0.79	156	0.72	251	0.79	304	0.66
106	0.66	161	1.18	259	0.94	308	1.04
112	0.88	167	0.75	262	0.77	311	0.86
113	0.79	175	0.76	277	0.85	317	0.88
115	1.07	182	0.93	282	0.72	321	0.67
119	0.60	185	0.72	286	0.68	323	0.68
125	0.80	189	0.87	288	0.86		

SOURCE: D. L. Ashley, M. A. Bonin, F. L. Cardinali, J. M. McCraw, J. S. Holler, L. L. Needham, and D. G. Patterson, Jr., *Anal. Chem.* **1992**, *64*, 1021.

5-13. *Control chart.* A laboratory monitoring perchlorate (ClO_4^-) in human urine measured quality control samples made from synthetic urine spiked with perchlorate. The graph shows consecutive quality control measurements. The standard operating procedure called for stopping work to identify the source of error if any quality control sample deviated outside the action lines ($\pm 3\sigma/\sqrt{n}$). This condition does not occur in the data shown here. Are any other rejection conditions from Box 5-1 observed in this data?

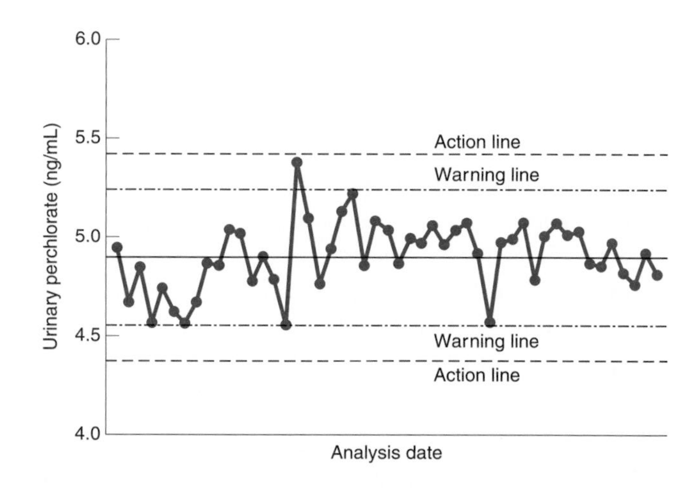

Control chart for ClO_4^- in urine, [Data from L. Valentin-Blasini, J. P. Mauldin, D. Maple, and B. C. Blount, *Anal. Chem.* **2005**, *77*, 2475.]

5-14. In a murder trial in the 1990s, the defendant's blood was found at the crime scene. The prosecutor argued that blood was left by the defendant during the crime. The defense argued that police "planted" the defendant's blood from a sample collected later. Blood is normally collected in a vial containing the metal-binding compound EDTA as an anticoagulant with a concentration of ~4.5 mM after being filled with blood. At the time of the trial, procedures to measure EDTA in blood were not well established. Even though the amount of EDTA found in the crime-scene blood was orders of magnitude below 4.5 mM, the jury acquitted the defendant. This trial motivated the development of a new method to measure EDTA in blood.

(a) *Precision and accuracy.* To measure accuracy and precision of the method, blood was fortified with EDTA to known levels.

$$\text{accuracy} = 100 \times \frac{\text{mean value found} - \text{known value}}{\text{known value}}$$

$$\text{precision} = 100 \times \frac{\text{standard deviation}}{\text{mean}} \equiv \textit{coefficient of variation}$$

For each of the three spike levels in the table, find the precision and accuracy of the quality control samples.

EDTA measurements (ng/mL) at three fortification levels

Spike:	22.2 ng/mL	88.2 ng/mL	314 ng/mL
Found:	33.3	83.6	322
	19.5	69.0	305
	23.9	83.4	282
	20.8	100	329
	20.8	76.4	276

SOURCE: R. L. Sheppard and J. Henion, *Anal. Chem.* **1997**, *69*, 477A, 2901.

(b) *Detection and quantitiation limits.* Low concentrations of EDTA near the detection limit gave the following dimensionless instrument readings: 175, 104, 164, 193, 131, 189, 155, 133, 151, and 176. Ten blanks had a mean reading of $45._0$. The slope of the calibration curve is $1.75 \times 10^9\ M^{-1}$. Estimate the signal and concentration detection limits and the lower limit of quantitation for EDTA.

5-15. *Spike recovery and detection limit.* Species of arsenic found in drinking water include AsO_3^{3-} (arsenite), AsO_4^{3-} (arsenate), $(CH_3)_2AsO_2^-$ (dimethylarsinate), and $(CH_3)AsO_3^{2-}$ (methylarsonate). Pure water containing no arsenic was spiked with 0.40 μg arsenate/L. Seven replicate determinations gave 0.39, 0.40, 0.38, 0.41, 0.36, 0.35, and 0.39 μg/L (J. A. Day, M. Montes-Bayón, A. P. Vonderheide, and J. A. Caruso, *Anal. Bioanal. Chem.* **2002**, *373*, 664). Find the mean percent recovery of the spike and the concentration detection limit.

5-16. *Standard addition.* Vitamin C was measured by an electrochemical method in a 50.0-mL sample of lemon juice. A detector signal of 2.02 μA was observed. A standard addition of 1.00 mL of 29.4 mM vitamin C increased the signal to 3.79 μA. Find the concentration of vitamin C in the juice.

5-17. *Standard addition.* Analyte in an unknown gave a signal of 10.0 mV. When 1.00 mL of 0.050 0 M standard was added to 100.0 mL of unknown, the signal increased to 14.0 mV. Find the concentration of the original unknown.

5-18. *Standard addition graph.* Tooth enamel consists mainly of the mineral calcium hydroxyapatite, $Ca_{10}(PO_4)_6(OH)_2$. Trace elements in teeth of archeological specimens provide anthropologists with clues about diet and diseases of ancient people. Students at Hamline University measured the trace element strontium in enamel from extracted wisdom teeth by atomic absorption spectroscopy. Solutions were prepared with a constant total volume of 10.0 mL containing 0.750 mg dissolved tooth enamel plus variable concentrations of added Sr.

Added Sr (ng/mL = ppb)	Signal (arbitrary units)
0	28.0
2.50	34.3
5.00	42.8
7.50	51.5
10.00	58.6

SOURCE: V. J. Porter, P. M. Sanft, J. C. Dempich, D. D. Dettmer, A. E. Erickson, N. A. Dubauskie, S. T. Myster, E. H. Matts, and E. T. Smith, *J. Chem. Ed.* **2002**, *79*, 1114.

(a) Prepare a graph to find the concentration of Sr in the 10-mL sample solution in parts per billion = ng/mL.

(b) Find the concentration of Sr in tooth enamel in parts per million = μg/g.

5-19. *Uncertainty in standard addition.* We now find the uncertainty in the x-intercept of the standard addition graph of Problem 5-18 with the formula

$$\begin{array}{l}\text{standard deviation}\\ \text{of } x\text{-intercept}\end{array} = \frac{s_y}{|m|}\sqrt{\frac{1}{n} + \frac{\bar{y}^2}{m^2 \sum (x_i - \bar{x})^2}}$$

where s_y is the standard deviation of y (Equation 4-12), $|m|$ is the absolute value of the slope of the least-squares line (Equation 4-9), n is the number of data points ($n = 5$ for Problem 5-18), $\bar{y}$ is the mean value of y for the five points,

x_i are the individual values of x for the five points, and $\bar{x}$ is the mean value of x for the five points.

(a) Create a spreadsheet modeled after the one in Figure 4-9 to find the straight line for standard additions and include the formula for uncertainty in the x-intercept. Find the uncertainty in the Sr concentration found in Problem 5-18.

(b) If the standard addition intercept is the major source of uncertainty, find the uncertainty in the concentration of Sr in tooth enamel in parts per million.

5-20. *Standard addition graph.* The figure shows standard additions of Cu^{2+} to acidified tap water measured by an electrochemical method. The standard additions had negligible volume compared with that of the tap water sample, so you can consider all solutions to have the same volume. The signal is the peak height (in μA), which you will need to measure relative to the baseline in the figure. Find the concentration of Cu^{2+} in the tap water.

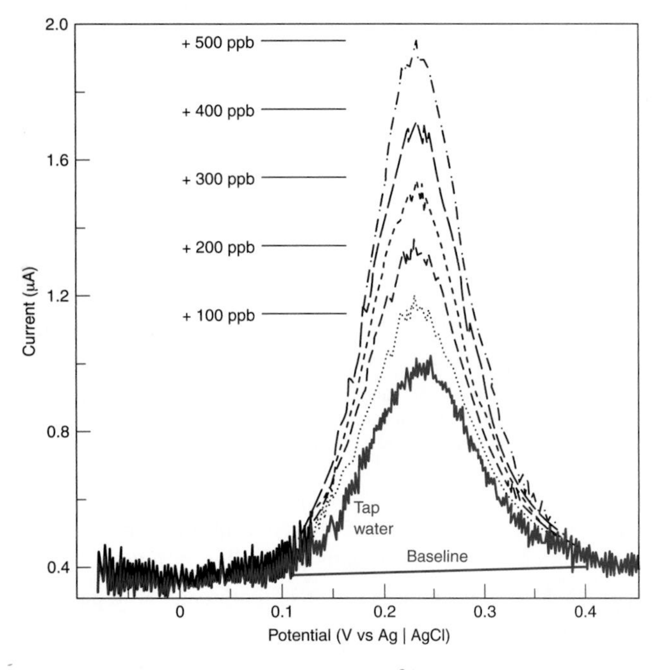

Five standard additions of 100 ppb Cu^{2+} to tap water. [From M. A. Nolan and S. P. Kounaves, *Anal. Chem.* **1999**, *71*, 3567.]

5-21. *Internal standard.* A mixture containing 12.8 μM analyte (X) and 44.4 μM standard (S) gave chromatographic peak areas of 306 for X and 511 for S. A second solution containing an unknown quantity of X plus 55.5 μM S had peak areas of 251 for X and 563 for S. Find [X] in the second solution.

5-22. *Internal standard.* A solution was prepared by mixing 10.00 mL of unknown (X) with 5.00 mL of standard (S) containing 8.24 μg S/mL and diluting to 50.0 mL. The measured signal quotient (signal due to X/signal due to S) was 1.69. In a separate experiment, it was found that, for equal concentrations of X and S, the signal due to X was 0.930 times as intense as the signal due to S. Find the concentration of X in the unknown.

5-23. *Internal standard.* When 1.06 mmol of 1-pentanol and 1.53 mmol of 1-hexanol were separated by gas chromatography, they gave relative peak areas of 922 and 1 570 units, respectively. When 0.57 mmol of pentanol was added to an unknown containing hexanol, the relative chromatographic peak areas were 843:816 (pentanol:hexanol). How much hexanol did the unknown contain?

How Would You Do It?

5-24. Olympic athletes are tested to see whether or not they are using illegal performance-enhancing drugs. Suppose that urine samples are taken and analyzed, and the rate of false positive results is 1%. Suppose also that it is too expensive to refine the method to reduce the rate of false positive results. We certainly do not want to accuse innocent people of using illegal drugs. What can you do to reduce the rate of false accusations even though the test always has a false positive rate of 1%?

Further Reading

B. W. Wenclawiak, M. Koch, and E. Hadjiscostas, eds., *Quality Assurance in Analytical Chemistry* (Heidelberg: Springer-Verlag, 2004).

E. Mullins, *Statistics for the Quality Control Chemistry Laboratory* (Cambridge: Royal Society of Chemistry, 2003).

P. Quevauviller, *Quality Assurance for Water Analysis* (Chichester: Wiley, 2002).

J. Kenkel, *A Primer on Quality in the Analytical Laboratory* (Boca Raton, FL: Lewis Press, 1999).

F. E. Prichard, *Quality in the Analytical Chemistry Laboratory* (New York: Wiley, 1995).

J. K. Taylor, *Quality Assurance of Chemical Measurements* (Chelsea, MI: Lewis Publishers, 1987).

E. P. Popek, *Sampling and Analysis of Environmental Chemical Pollutants* (Amsterdam: Academic Press, 2003).

The Earliest Known Buret

Good Titrations—Ode to a Lab Partner

Kurt Wood and Jeff Lederman

(University of California, Davis, 1977)

(Sung to the tune of *Good Vibrations* by the Beach Boys)

Ah! I love the color of pink you get,
And the way the acid drips from your buret.
All the painful things in life seem alien
As I mix in several drops of phenolphthalein.
I'm pickin' up good titrations
She's givin' me neutralizations
Dew drop drop, good titrations . . .

I look at you and drift away;
The ruby red turns slowly to rosé.
You gaze at me and light my fire
As drop on drop falls to the Erlenmeyer . . .
I'm pickin' up good titrations
She's givin' me neutralizations
Dew drop drop, good titrations . . .

I look longingly to your eyes,
But you stare down at the lab bench in surprise.
Gone our love before it grew much sweeter,
'Cause we passed the end point by a milliliter.
I'm pickin' up back titrations
She's prayin' for neutralizations
Dew drop drop, good titrations . . .

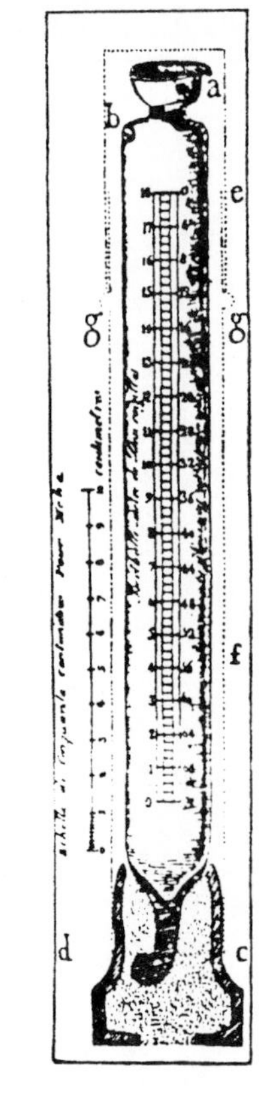

The buret, invented by F. Descroizilles in the early 1800s, was used in the same manner as a graduated cylinder is used today. The stopcock was introduced in 1846. This buret and its progeny have terrorized generations of analytical chemistry students. [From E. R. Madsen, *The Development of Titrimetric Analysis 'till 1806* (Copenhagen: E. E. C. Gad Publishers, 1958).]

CHAPTER 6

Good Titrations

In **volumetric analysis**, the volume of a known reagent required for complete reaction with analyte by a known reaction is measured. From this volume and the stoichiometry of the reaction, we calculate how much analyte is in the unknown. In this chapter we discuss general principles that apply to any volumetric procedure, and then we illustrate some analyses based on precipitation reactions. Along the way, we introduce the solubility product as a means of understanding precipitation reactions.

6-1 Principles of Volumetric Analysis

A **titration** is a procedure in which increments of the known reagent solution—the **titrant**—are added to *analyte* until the reaction is complete. Titrant is usually delivered from a buret, as shown in Figure 6-1. Each increment of titrant should be completely and quickly consumed by reaction with analyte until the analyte is used up. Common titrations are based on acid-base, oxidation-reduction, complex formation, or precipitation reactions.

Methods of determining when analyte has been consumed include (1) detecting a sudden change in voltage or current between a pair of electrodes, (2) observing an *indicator* color change (Color Plate 1), and (3) monitoring the absorbance of light by species in the reaction. An **indicator** is a compound with a physical property (usually color) that changes abruptly when the titration is complete. The change is caused by the disappearance of analyte or the appearance of excess titrant.

We will study end point detection methods later:
electrodes: Section 10-4 and Chapter 15
indicators: Sections 6-6, 9-6, 10-4, 13-3, and 16-2
absorbance: Section 19-3

The **equivalence point** is reached when the quantity of titrant added is the exact amount necessary for stoichiometric reaction with the analyte. For example, 5 mol of oxalic acid react with 2 mol of permanganate in hot acidic solution:

$$\underset{\substack{\textbf{Analyte}\\ \text{Oxalic acid}\\ \text{(colorless)}}}{5HO{-}\overset{\overset{\displaystyle O}{\|}}{C}{-}\overset{\overset{\displaystyle O}{\|}}{C}{-}OH} + \underset{\substack{\textbf{Titrant}\\ \text{Permanganate}\\ \text{(purple)}}}{2MnO_4^-} + 6H^+ \longrightarrow 10CO_2 + \underset{\text{(colorless)}}{2Mn^{2+}} + \underset{\text{(colorless)}}{8H_2O} \qquad (6\text{-}1)$$

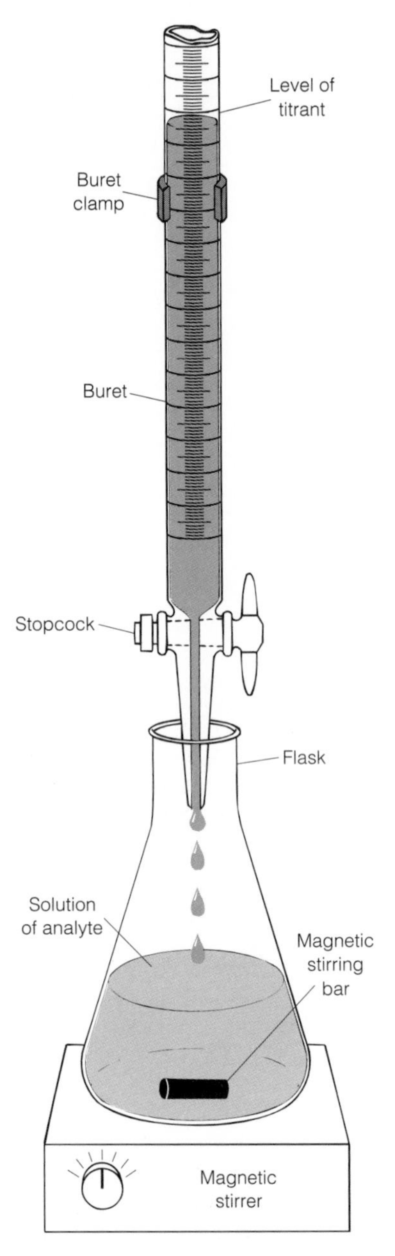

Figure 6-1 Typical setup for a titration. Analyte is contained in the flask, and titrant in the buret. The stirring bar is a magnet coated with Teflon, which is inert to most solutions. The bar is spun by a rotating magnet inside the stirrer.

If the unknown contains 5.00 mmol of oxalic acid, the equivalence point is reached when 2.00 mmol of MnO_4^- have been added.

The equivalence point is the ideal result that we seek in a titration. What we actually measure is the **end point**, which is marked by a sudden change in a physical property of the solution. For Reaction 6-1, a convenient end point is the abrupt appearance of the purple color of permanganate in the flask. Up to the equivalence point, all the added permanganate is consumed by oxalic acid, and the titration solution remains colorless. After the equivalence point, unreacted MnO_4^- ion builds up until there is enough to see. The *first trace* of purple color marks the end point. The better your eyes, the closer the measured end point will be to the true equivalence point. The end point cannot exactly equal the equivalence point because extra MnO_4^-, more than that needed to react with oxalic acid, is required to create perceptible color.

The difference between the end point and the equivalence point is an inescapable **titration error**. By choosing an appropriate physical property, in which a change is easily observed (such as indicator color, optical absorbance of a reactant or product, or pH), we can have an end point that is very close to the equivalence point. We can also estimate the titration error with a **blank titration**, in which the same procedure is carried out without analyte. For example, a solution containing no oxalic acid could be titrated with MnO_4^- to see how much titrant is needed to create observable purple color. We subtract this volume of MnO_4^- from the volume observed in the titration of unknown.

The validity of an analytical result depends on knowing the amount of one of the reactants used. A **primary standard** is a reagent that is pure enough to weigh out and use directly to provide a known number of moles. For example, if you want to titrate unknown hydrochloric acid with base, you could weigh primary standard grade sodium carbonate and dissolve it in water to make titrant:

$$\underset{\text{(unknown)}}{2HCl} + \underset{\substack{\text{Sodium carbonate} \\ \text{FM 105.99} \\ \text{(primary standard)}}}{Na_2CO_3} \longrightarrow H_2CO_3 + 2NaCl \qquad (6\text{-}2)$$

Two moles of HCl react with one mole of Na_2CO_3, which has a mass of 105.99 g. You could not carry out the same procedure with solid NaOH because the solid is not pure.

$$HCl + \underset{\substack{\text{Sodium hydroxide} \\ \text{FM 40.00}}}{NaOH} \longrightarrow H_2O + NaCl$$

NaOH is normally contaminated with some Na_2CO_3 (from reaction with CO_2 in the air) and H_2O (also from the air). If you weighed out 40.00 g of sodium hydroxide, it would not contain exactly 1 mol.

A primary standard should be 99.9% pure or better. It should not decompose under ordinary storage, and it should be stable when dried by heating or vacuum, because drying is required to remove traces of water adsorbed from the atmosphere.

Box 3-1 describes Standard Reference Materials that allow different laboratories to test the accuracy of their procedures.

In most cases, titrant is not available as a primary standard. Instead, a solution having approximately the desired concentration is used to titrate a weighed, primary standard. From the volume of titrant required to react with the primary standard, we calculate the concentration of titrant. The process of titrating a standard to determine the concentration of titrant is called **standardization**. We say that a solution whose concentration is known is a **standard solution**. The validity of the analytical result ultimately depends on knowing the composition of some primary standard.

In a **direct titration**, titrant is added to analyte until the end point is observed.

$$\text{Direct titration:} \quad \underset{\text{Unknown}}{\text{analyte}} + \underset{\text{Known}}{\text{titrant}} \longrightarrow \text{product}$$

The addition of permanganate titrant to oxalic acid analyte in Reaction 6-1 is a direct titration.

In a **back titration**, a known *excess* of a standard reagent is added to the analyte. Then a second standard reagent is used to titrate the excess of the first reagent.

$$\underset{\text{Unknown}}{\text{analyte}} + \underset{\text{Known}}{\text{reagent 1}} \longrightarrow \text{product} + \underset{\text{Unknown quantity}}{\text{excess reagent 1}} \quad (6\text{-}3a)$$

$$\text{Back titration:} \quad \underset{\text{Unknown}}{\text{excess reagent 1}} + \underset{\text{Known}}{\text{reagent 2}} \longrightarrow \text{product} \quad (6\text{-}3b)$$

Back titrations are useful when the end point of the back titration is clearer than the end point of the direct titration or when an excess of the first reagent is required for complete reaction with analyte.

For example, a back titration is used in the determination of peroxydisulfate ($S_2O_8^{2-}$). An unknown such as impure $K_2S_2O_8$ is treated with excess standard $Na_2C_2O_4$ in H_2SO_4 containing Ag_2SO_4 catalyst.

Peroxydisulfate is a powerful oxidizing agent used to destroy organic matter in environmental analysis. It is also the active ingredient of "cleaning solution" used to destroy grease and organic matter in glassware.

$$\underset{\text{Unknown}}{H_2S_2O_8} + \underset{\substack{\text{Excess standard}\\ \text{reagent}}}{H_2C_2O_4} \xrightarrow{Ag^+} 2H_2SO_4 + 2CO_2 \quad (6\text{-}4)$$

Excess standard reagent ensures complete reaction of the unknown. The mixture is heated until $CO_2(g)$ evolution is complete. Then the solution is cooled to 40°C and the excess $H_2C_2O_4$ is back titrated with standard $KMnO_4$ by Reaction 6-1. The quantity of $KMnO_4$ required for back titration tells us how much unreacted $H_2C_2O_4$ was left over from Reaction 6-4.

In a **gravimetric titration**, titrant is measured by mass, not volume. Titrant concentration is expressed as moles of reagent per kg of solution. Precision is improved from 0.3% attainable with a buret to 0.1% with a balance. Experiments by Guenther[1] and by Butler and Swift[2] provide examples. In a gravimetric titration, there is no need for a buret. Titrant can be delivered from a pipet. "Gravimetric titrations should become the gold standard, and volumetric glassware should be seen in museums only."[3]

The Web site www.whfreeman.com/eca includes lists of experiments from the *Journal of Chemical Education* keyed to the chapters of this book.

Ask Yourself

6-A. **(a)** Why does the validity of an analytical result ultimately depend on knowing the composition of a primary standard?
(b) How does a blank titration reduce titration error?
(c) What is the difference between a direct titration and a back titration?
(d) Suppose that the uncertainty in locating the equivalence point in a titration is ±0.04 mL. Why is it more accurate to use enough of an unknown to require ~40 mL in a titration with a 50-mL buret instead of ~20 mL?

6-2 Titration Calculations

To interpret the results of a direct titration, the key steps are

1. From the volume of titrant, calculate the number of moles of titrant consumed.
2. From the stoichiometry of the titration reaction, relate the unknown moles of analyte to the known moles of titrant.

1:1 Stoichiometry

Consider the titration of an unknown chloride solution with standard Ag^+:

When K is large, we sometimes write $\rightarrow$ instead of $\rightleftharpoons$:

$$Ag^+ + Cl^- \longrightarrow AgCl(s)$$

$$Ag^+ + Cl^- \overset{K}{\rightleftharpoons} AgCl(s)$$

The reaction is rapid and the equilibrium constant is large ($K = 5.6 \times 10^9$), so the reaction essentially goes to completion with each addition of titrant. White AgCl precipitate forms as soon as the two reagents are mixed.

Suppose that 10.00 mL of unknown chloride solution (measured with a transfer pipet) requires 22.97 mL of 0.052 74 M $AgNO_3$ solution (delivered from a buret) for complete reaction. What is the concentration of Cl^- in the unknown?

Following the two-step recipe, we first find the moles of Ag^+:

Retain an extra, insignificant digit until the end of the calculations to prevent round-off error.

$$\text{mol } Ag^+ = \text{volume} \times \text{molarity} = (0.022\ 97\ \text{L})\left(0.052\ 74\ \frac{\text{mol}}{\text{L}}\right) = 0.001\ 211_4\ \text{mol}$$

Next, relate the unknown moles of Cl^- to the known moles of Ag^+. We know that 1 mol of Cl^- reacts with 1 mol of Ag^+. If $0.001\ 211_4$ mol of Ag^+ is required, then $0.001\ 211_4$ mol of Cl^- must have been in 10.00 mL of unknown. Therefore

$$[Cl^-] \text{ in unknown} = \frac{\text{mol } Cl^-}{\text{L of unknown}} = \frac{0.001\ 211_4\ \text{mol}}{0.010\ 00\ \text{L}} = 0.121\ 1_4\ \text{M}$$

The solution that we titrated was made by dissolving 1.004 g of unknown solid in a total volume of 100.0 mL. What is the weight percent of chloride in the solid? We know that 10.00 mL of unknown solution contain $0.001\ 211_4$ mol of Cl^-. Therefore 100.0 mL must contain 10 times as much, or $0.012\ 11_4$ mol of Cl^-. This much Cl^- weighs ($0.012\ 11_4$ mol Cl^-)(35.453 g/mol Cl^-) = $0.429\ 4_8$ g Cl^-. The weight percent of Cl^- in the unknown is

$$\text{wt\% } Cl^- = \frac{\text{g } Cl^-}{\text{g unknown}} \times 100 = \frac{0.429\ 4_8\ \text{g } Cl^-}{1.004\ \text{g unknown}} \times 100 = 42.78\ \text{wt\%}$$

Example **A Case Involving a Dilution**

(a) Standard Ag^+ solution was prepared by dissolving 1.224 3 g of dry $AgNO_3$ (FM 169.87) in water in a 500.0-mL volumetric flask. A dilution was made by delivering 25.00 mL of solution with a pipet to a second 500.0-mL volumetric flask and diluting to the mark. Find the concentration of Ag^+ in the dilute solution. **(b)** A 25.00-mL aliquot of unknown containing Cl^- was titrated with the dilute Ag^+ solution, and the equivalence point was reached when 37.38 mL of Ag^+ solution had been delivered. Find the concentration of Cl^- in the unknown.

SOLUTION **(a)** The concentration of the initial $AgNO_3$ solution is

$$[Ag^+] = \frac{(1.224\,3\text{ g})/(169.87\text{ g/mol})}{0.500\,0\text{ L}} = 0.014\,41_5\text{ M}$$

To find the concentration of the dilute solution, use the dilution formula 1-5:

$$[Ag^+]_{conc} \cdot V_{conc} = [Ag^+]_{dil} \cdot V_{dil} \quad (1\text{-}5)$$

$$(0.014\,41_5\text{ M}) \cdot (25.00\text{ mL}) = [Ag^+]_{dil} \cdot (500.0\text{ mL})$$

$$[Ag^+]_{dil} = \left(\frac{25.00\text{ mL}}{500.0\text{ mL}}\right)(0.014\,41_5\text{ M}) = 7.207_3 \times 10^{-4}\text{ M}$$

Notice that the general form of all dilution problems is

$$[X]_{final} = \underbrace{\frac{V_{initial}}{V_{final}}}_{\text{Dilution factor}} \cdot [X]_{initial}$$

(b) One mole of Cl^- requires one mole of Ag^+. The number of moles of Ag^+ required to reach the equivalence point is

$$\text{mol } Ag^+ = (7.207_3 \times 10^{-4}\text{ M})(0.037\,38\text{ L}) = 2.694_1 \times 10^{-5}\text{ mol}$$

The concentration of Cl^- in 25.00 mL of unknown is therefore

$$[Cl^-] = \frac{2.694_1 \times 10^{-5}\text{ mol}}{0.025\,00\text{ L}} = 1.078 \times 10^{-3}\text{ M} = 1.078\text{ mM}$$

mM stands for "millimolar" = 10^{-3} M

Test Yourself Suppose that 25.00 mL of standard $AgNO_3$ were diluted to 250.0 mL (instead of 500.0 mL). A 10.00-mL aliquot of unknown containing Cl^- required 15.77 mL of Ag^+ solution for titration. Find $[Cl^-]$ in the unknown. (**Answer:** 2.273 mM)

Silver ion titrations are nice because $AgNO_3$ is a primary standard. After drying at 110°C for 1 h to remove moisture, the solid has the exact composition $AgNO_3$. Methods for finding the end point in silver titrations are described in Section 6-6. Silver compounds and solutions should be stored in the dark to protect against photodecomposition and should never be exposed to direct sunlight.

$AgNO_3$ solution is an antiseptic. If you spill $AgNO_3$ solution on yourself, your skin will turn black for a few days until the affected skin is shed.

Photodecomposition of AgCl(*s*):

$$AgCl(s) \xrightarrow{\text{light}} Ag(s) + \tfrac{1}{2}Cl_2(g)$$

Finely divided Ag(*s*) imparts a faint violet color to the white solid.

x:*y* Stoichiometry

In Reaction 6-1, 5 mol of oxalic acid ($H_2C_2O_4$) react with 2 mol of permanganate (MnO_4^-). If an unknown quantity of $H_2C_2O_4$ consumed 2.345×10^{-4} mol of MnO_4^-, there must have been

$$\text{mol } H_2C_2O_4 = (\text{mol } MnO_4^-)\left(\frac{5\text{ mol } H_2C_2O_4}{2\text{ mol } MnO_4^-}\right)$$

$$= (2.345 \times 10^{-4}\text{ mol } MnO_4^-)\left(\frac{5\text{ mol } H_2C_2O_4}{2\text{ mol } MnO_4^-}\right) = 5.862 \times 10^{-4}\text{ mol}$$

Ask Yourself

6-B. Vitamin C (ascorbic acid) from foods can be measured by titration with I_3^-:

$$\underset{\substack{\text{Ascorbic acid}\\ \text{FM 176.126}}}{C_6H_8O_6} + \underset{\text{Triiodide}}{I_3^-} + H_2O \longrightarrow C_6H_8O_7 + 3I^- + 2H^+$$

Linda A. Hughes

Starch is used as an indicator in the reaction. The end point is marked by the appearance of a deep blue starch-iodine complex when unreacted I_3^- is present.

(a) If 29.41 mL of I_3^- solution are required to react with 0.197 0 g of pure ascorbic acid, what is the molarity of the I_3^- solution?

(b) A vitamin C tablet containing ascorbic acid plus inert binder was ground to a powder, and 0.424 2 g was titrated by 31.63 mL of I_3^-. How many moles of ascorbic acid are present in the 0.424 2-g sample?

(c) Find the weight percent of ascorbic acid in the tablet.

6-3 Chemistry in a Fishtank

Students of Professor Kenneth Hughes at Georgia Institute of Technology study analytical chemistry by measuring chemical changes in a saltwater aquarium in their laboratory. One of the chemicals measured is nitrite, NO_2^-, which is a key species in the natural cycle of nitrogen (Figure 6-2). Box 6-1 shows concentrations of ammonia (NH_3), nitrite, and nitrate (NO_3^-) measured in the aquarium. Concentrations are expressed in parts per million of nitrogen, which means micrograms (μg) of nitrogen per gram of seawater. Because 1 g of water ≈ 1 mL, we will consider 1 ppm to be 1 μg/mL.

NH_3 (FM 17.031) contains 82.24 wt% N. Therefore, a solution with 1.216 mg NH_3/L contains

$$\left(1.216\ \frac{\text{mg NH}_3}{\text{L}}\right)\left(0.822\ 4\ \frac{\text{mg N}}{\text{mg NH}_3}\right)$$
$$= 1.000 \text{ mg N/L} = 1.000\ \mu\text{g N/mL}$$
$$= 1 \text{ ppm N}$$

The nitrite measurement was done by a spectrophotometric procedure described in Section 18-4. Because there is no convenient primary standard for nitrite, a titration is used to standardize a $NaNO_2$ solution that serves as the standard for the spectrophotometric procedure. As always, the validity of any analytical procedure ultimately depends on knowing the composition of a primary standard, which is sodium oxalate in this case.

Three solutions are required for the measurement of nitrite:

The concentrations of $NaNO_2$ and $KMnO_4$ are only approximate. The solutions will be standardized in subsequent titrations.

a. Prepare ~0.018 M $NaNO_2$ (FM 68.995) by dissolving 1.25 g of $NaNO_2$ in 1.00 L of distilled water. Standardization of this solution is described later.

b. Prepare ~0.025 00 M $Na_2C_2O_4$ (FM 134.00) by dissolving ~3.350 g of primary standard grade $Na_2C_2O_4$ in 1.0 M H_2SO_4 and diluting to 1.000 L with 1.0 M H_2SO_4. It is not necessary to weigh out exactly 3.350 g. What is important is that you know the exact mass so that you can calculate the molarity of the reagent.

c. Prepare ~0.010 M $KMnO_4$ (FM 158.03) by dissolving 1.6 g of $KMnO_4$ in 1.00 L of distilled water. $KMnO_4$ is not pure enough to be a primary standard. Also, traces of organic impurities in distilled water consume some of the freshly dissolved MnO_4^- to produce $MnO_2(s)$. Therefore, dissolve $KMnO_4$ in distilled water to give the approximately desired concentration and boil the solution for 1 h to complete the reaction between MnO_4^- and organic impurities. Filter the mixture through a sintered glass filter (not a paper filter, which is organic) to remove $MnO_2(s)$, cool the solution, and standardize it against primary standard $Na_2C_2O_4$:

$$\underset{\text{Oxalate}}{5C_2O_4^{2-}} + \underset{\text{Permanganate}}{2MnO_4^-} + 16H^+ \longrightarrow 10CO_2 + 2Mn^{2+} + 8H_2O \qquad (6\text{-}5)$$

For best results, treat the oxalic acid solution at 25°C with 90% to 95% of the expected volume of $KMnO_4$. Then heat the solution to 60°C and complete the titration.

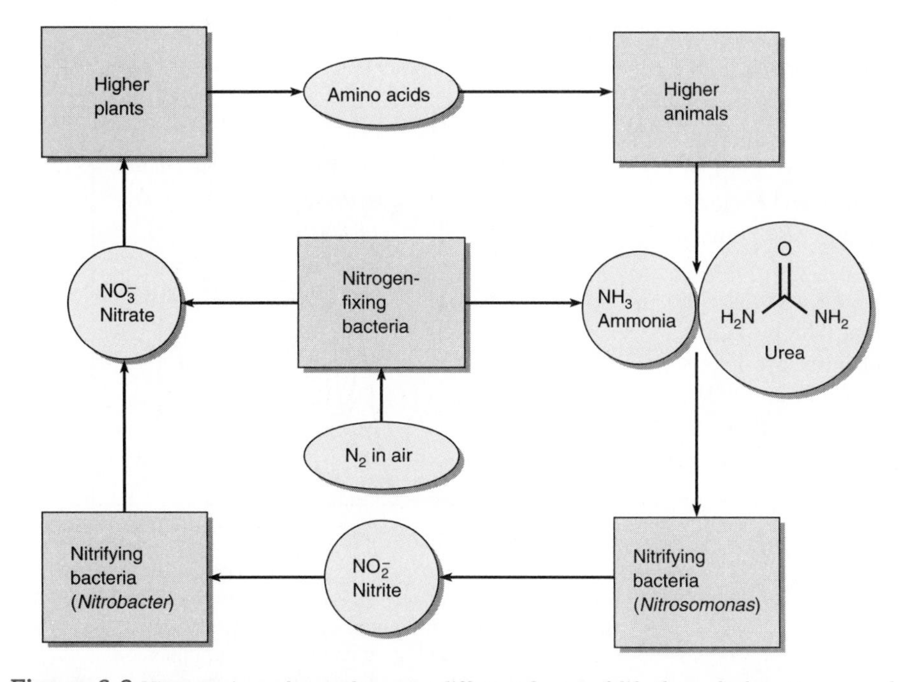

Figure 6-2 Nitrogen is exchanged among different forms of life through the *nitrogen cycle*. Only a few organisms, such as blue-green algae, are able to use N_2 directly from the air. Our existence depends on the health of all organisms in the nitrogen cycle.

Example Standardizing $KMnO_4$ by a Direct Titration of Oxalate

Standard oxalate solution was made by dissolving 3.299 g of $Na_2C_2O_4$ in 1.000 L of 1 M H_2SO_4. A 25.00-mL aliquot required 28.39 mL of $KMnO_4$ for titration, and a blank titration of 25 mL of 1 M H_2SO_4 required 0.03 mL of $KMnO_4$. Find the molarity of $KMnO_4$.

SOLUTION The number of moles of $Na_2C_2O_4$ dissolved in 1 L is (3.299 g)/(134.00 g/mol) = 0.024 61_9 mol. The $C_2O_4^{2-}$ in 25.00 mL is (0.024 61_9 M) × (0.025 00 L) = $6.154_9 \times 10^{-4}$ mol. Reaction 6-5 requires 2 mol of permanganate for 5 mol of oxalate. Therefore

$$\text{mol } MnO_4^- = (\text{mol } C_2O_4^{2-})\left(\frac{2 \text{ mol } MnO_4^-}{5 \text{ mol } C_2O_4^{2-}}\right)$$

$$= (6.154_9 \times 10^{-4} \text{ mol})\left(\frac{2}{5}\right) = 2.461_9 \times 10^{-4} \text{ mol}$$

You can use either ratio

$$\frac{5 \text{ mol } C_2O_4^{2-}}{2 \text{ mol } MnO_4^-} \text{ or } \frac{2 \text{ mol } MnO_4^-}{5 \text{ mol } C_2O_4^{2-}}$$

whenever you please, as long as the units work out.

The equivalence volume of $KMnO_4$ is 28.39 − 0.03 = 28.36 mL. The concentration of MnO_4^- titrant is

$$[MnO_4^-] = \frac{2.461_9 \times 10^{-4} \text{ mol}}{0.028\ 36 \text{ L}} = 8.681_0 \times 10^{-3} \text{ M}$$

Test Yourself A 25.00-mL aliquot of the same standard oxalate solution required 22.05 mL of $KMnO_4$ for titration, and a blank titration required 0.05 mL of $KMnO_4$. Find the molarity of $KMnO_4$. (**Answer:** 0.011 19 M)

Box 6-1 Studying a Marine Ecosystem

A saltwater aquarium was set up at Georgia Tech to study the chemistry of a marine ecosystem. When fish and food were introduced into the aquarium, bacteria began to grow and to metabolize organic compounds into ammonia (NH_3). Ammonia is toxic to marine animals when the level exceeds 1 ppm; but, fortunately, it is removed by *Nitrosomonas* bacteria, which oxidize NH_3 to nitrite (NO_2^-). Alas, NO_2^- is also toxic at levels above 1 ppm, but it is further oxidized to nitrate (NO_3^-) by *Nitrobacter* bacteria. The natural process of oxidation of NH_3 to NO_2^- and NO_3^- is called *nitrification*.

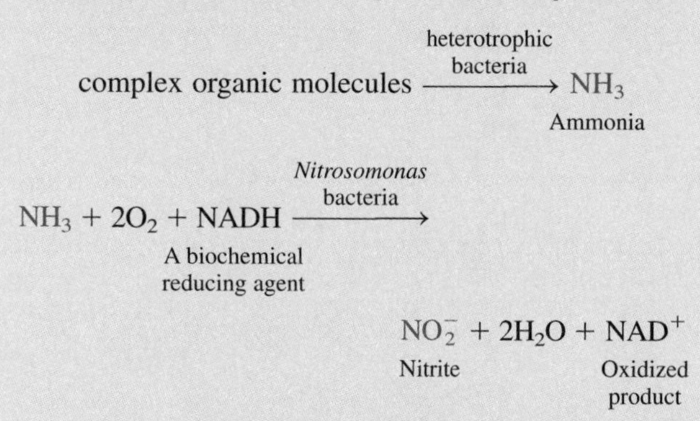

Some actions of nitrogen-metabolizing bacteria. *Heterotrophic* bacteria require complex organic molecules from the breakdown of other organisms for nourishment. By contrast, *autotrophic* bacteria can utilize CO_2 as their carbon source for biosynthesis.

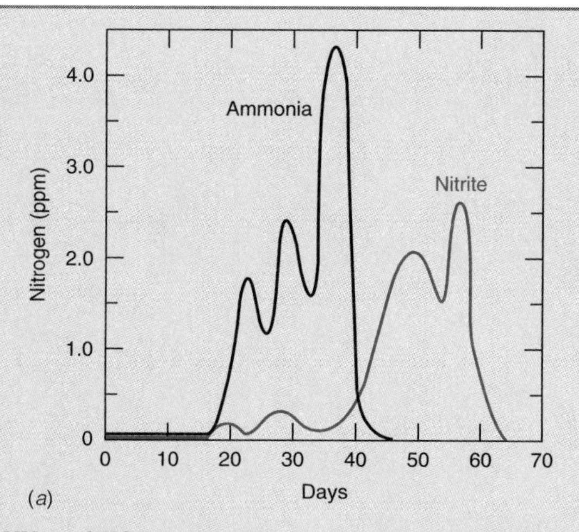

(*a*) NH_3 and NO_2^- concentrations in a saltwater aquarium at Georgia Tech after fish and food were introduced into the "sterile" tank, beginning on day 0. Concentrations are expressed in parts per million of nitrogen (that is, μg of N per mL of solution). [From K. D. Hughes, *Anal. Chem.* **1993**, *65*, 883A.]

Panel *a* shows NH_3 and NO_2^- concentrations observed by students at Georgia Tech. About 18 days after the introduction of fish, significant levels of NH_3 were observed. The first dip in NH_3 concentration occurred in a 48-h period when no food was added to the tank. The third peak in NH_3 concentration arose when changing flow patterns through the aquarium filter exposed fresh surfaces devoid of the bacteria that remove ammonia. When the population of *Nitrosomonas* bacteria was sufficient, NH_3 levels decreased but NO_2^-

Here is the procedure for standardizing the solution of $NaNO_2$ prepared in step (a):

1. Pipet 25.00 mL of standard $KMnO_4$ into a 500-mL flask, add 300 mL of 0.4 M H_2SO_4, and warm to 40°C on a hot plate.
2. Titrate the $KMnO_4$ with the $NaNO_2$ solution whose concentration is to be determined. Add titrant slowly until the permanganate is just decolorized. Add titrant very slowly near the end point because the reaction is slow. Best results are obtained with the tip of the buret immersed beneath the surface of the $KMnO_4$ solution.

$$5NO_2^- + 2MnO_4^- + 6H^+ \longrightarrow 5NO_3^- + 2Mn^{2+} + 3H_2O \qquad (6\text{-}6)$$

Example Standardizing $NaNO_2$ with $KMnO_4$

Step 2 above required 34.76 mL of $NaNO_2$ solution **(a)**. Find the concentration of $NaNO_2$.

became perilously high. After 60 days, the population of *Nitrobacter* bacteria was great enough to convert most NO_2^- into NO_3^-.

Nitrification requires an oxidizing agent to convert NH_3 to NO_2^-. Panels *b* and *c* show the relationship between photosynthesis and nitrification in sediment from the Niida River in Japan. In the presence of light, photosynthetic bacteria generate O_2, which serves as the oxidizing agent. Ammonia consumption in the top 1 mm of sediment approximately doubles in the presence of light. Ammonia and oxygen were measured with microelectrodes described in Box 15-2 and Section 17-2.

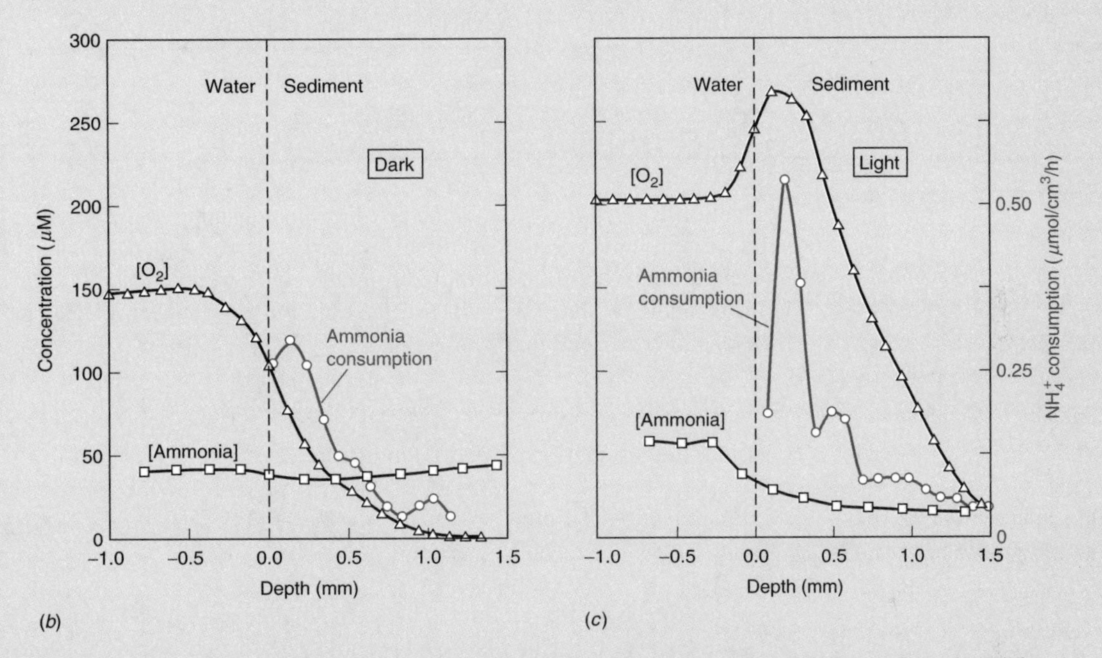

Concentrations of O_2 and ammonia and the rate of ammonia consumption in sediment from the Niida River in Japan in the dark (*b*) or exposed to light (*c*). Ammonia was measured as ammonium ion (NH_4^+), which is the predominant form at pH 7 to 8 in the river. [Data from Y. Nakamura, H. Satoh, T. Kindaichi, and S. Okabe, *Environ. Sci. Technol.* **2006**, *40*, 1532.]

SOLUTION We required 34.76 mL of $NaNO_2$ to titrate 25.00 mL of $8.681_0 \times 10^{-3}$ M $KMnO_4$. The number of moles of $KMnO_4$ consumed is (0.025 00 L) $\times$ $(8.681_0 \times 10^{-3}\text{ M}) = 2.170_3 \times 10^{-4}$ mol. In Reaction 6-6, 2 mol MnO_4^- require 5 mol NO_2^-, so the amount of $NaNO_2$ that reacted is

mol $NaNO_2$ reacting with $KMnO_4$ =

$$(2.170_3 \times 10^{-4}\text{ mol KMnO}_4)\left(\frac{5\text{ mol NaNO}_2}{2\text{ mol KMnO}_4}\right) = 5.425_6 \times 10^{-4}\text{ mol}$$

The concentration of $NaNO_2$ reagent is

$$[\text{NaNO}_2] = \frac{5.425_6 \times 10^{-4}\text{ mol}}{0.034\ 76\text{ L}} = 0.015\ 61\text{ M}$$

Test Yourself If $[KMnO_4] = 6.666 \times 10^{-3}$ M instead of $8.681_0 \times 10^{-3}$ M, what is the concentration of $NaNO_2$? (**Answer:** 0.011 99 M)

Ask Yourself

6-C. **(a)** Standard oxalate solution was made by dissolving 3.514 g of $Na_2C_2O_4$ in 1.000 L of 1 M H_2SO_4. A 25.00-mL aliquot required 24.44 mL of $KMnO_4$ for titration, and a blank titration required 0.03 mL of $KMnO_4$. Find the molarity of $KMnO_4$. **(b)** To standardize $NaNO_2$, 25.00 mL of $KMnO_4$ solution from part (a) of this problem required 38.11 mL of $NaNO_2$. Find the molarity of the $NaNO_2$.

6-4 Solubility Product

The **solubility product**, K_{sp}, is the equilibrium constant for the reaction in which a solid *salt* (an ionic compound) dissolves to give its constituent ions in solution. The concentration of the solid is omitted from the equilibrium constant because the solid is in its standard state. We will use the solubility product to discuss precipitation titrations later in Section 6-5.

Calculating the Solubility of an Ionic Compound

Consider the dissolution of lead(II) iodide in water:

Pure solid is omitted from the equilibrium constant because $PbI_2(s)$ is in its standard state.

$$\underset{\text{Lead(II) iodide}}{PbI_2(s)} \rightleftharpoons \underset{\text{Lead(II)}}{Pb^{2+}} + \underset{\text{Iodide}}{2I^-} \qquad K_{sp} = [Pb^{2+}][I^-]^2 = 7.9 \times 10^{-9} \quad (6\text{-}7)$$

for which the solubility product is listed in Appendix A. A solution that contains all the solid capable of being dissolved is said to be **saturated**. What is the concentration of Pb^{2+} in a solution saturated with PbI_2?

Reaction 6-7 produces two I^- ions for each Pb^{2+} ion. If the concentration of dissolved Pb^{2+} is x M, the concentration of dissolved I^- must be $2x$ M. We can display this relation neatly in a little concentration table:

	$PbI_2(s)$	$\rightleftharpoons$	Pb^{2+}	+	$2I^-$
Initial concentration	solid		0		0
Final concentration	solid		x		$2x$

Putting these concentrations into the solubility product gives

To find the cube root of a number with a calculator, raise the number to the 0.333 333 33 . . . power with the y^x key.

$$[Pb^{2+}][I^-]^2 = (x)(2x)^2 = 7.9 \times 10^{-9}$$
$$4x^3 = 7.9 \times 10^{-9}$$
$$x = \left(\frac{7.9 \times 10^{-9}}{4}\right)^{1/3} = 0.001\ 2_5 \text{ M}$$

The concentration of Pb^{2+} is $0.001\ 2_5$ M and the concentration of I^- is $2x = (2)(0.001\ 2_5) = 0.002\ 5$ M.

The physical meaning of the solubility product is: If an aqueous solution is left in contact with excess solid PbI_2, solid dissolves until the condition $[Pb^{2+}][I^-]^2 = K_{sp}$

is satisfied. Thereafter, no more solid dissolves. Unless excess solid remains, there is no guarantee that $[Pb^{2+}][I^-]^2 = K_{sp}$. If Pb^{2+} and I^- are mixed together (with appropriate counterions) such that the product $[Pb^{2+}][I^-]^2$ exceeds K_{sp}, then $PbI_2(s)$ will precipitate (Figure 6-3).

The solubility product does not tell the entire story of the solubility of ionic compounds. The concentration of *undissociated* species may be significant. In a solution of calcium sulfate, for example, about half of the dissolved material dissociates to Ca^{2+} and SO_4^{2-} and half dissolves as undissociated $CaSO_4(aq)$ (a tightly bound *ion pair*). In the lead(II) iodide case, species such as PbI^+, $PbI_2(aq)$, and PbI_3^- also contribute to the total solubility.

Figure 6-3 The yellow solid, lead(II) iodide (PbI_2), precipitates when a colorless solution of lead nitrate ($Pb(NO_3)_2$) is added to a colorless solution of potassium iodide (KI). [Photo by Chip Clark.]

The Common Ion Effect

Consider what happens when we add a second source of I^- to a solution saturated with $PbI_2(s)$. Let's add 0.030 M NaI, which dissociates completely to Na^+ and I^-. What is the concentration of Pb^{2+} in this solution?

	$PbI_2(s)$	$\rightleftharpoons$	Pb^{2+}	+	$2I^-$	(6-8)
Initial concentration	solid		0		0.030	
Final concentration	solid		x		$2x + 0.030$	

The initial concentration of I^- is from dissolved NaI. The final concentration of I^- has contributions from NaI and PbI_2.

The solubility product is

$$[Pb^{2+}][I^-]^2 = (x)(2x + 0.030)^2 = K_{sp} = 7.9 \times 10^{-9} \qquad (6\text{-}9)$$

But think about the size of x. With no added I^-, we found $x = 0.001\ 2_5$ M. In the present case, we anticipate that x will be smaller than $0.001\ 2_5$ M, because of Le Châtelier's principle. Addition of I^- to Reaction 6-8 displaces the reaction in the reverse direction. In the presence of extra I^-, there will be less dissolved Pb^{2+}. This application of Le Châtelier's principle is called the **common ion effect**. *A salt will be less soluble if one of its constituent ions is already present in the solution.*

Common ion effect: A salt is less soluble if one of its ions is already present in the solution.

In Equation 6-9, we suspect that $2x$ may be much smaller than 0.030. As an approximation, we ignore $2x$ in comparison with 0.030. The equation simplifies to

$$(x)(0.030)^2 = K_{sp} = 7.9 \times 10^{-9}$$
$$x = 7.9 \times 10^{-9}/(0.030)^2 = 8.8 \times 10^{-6}$$

It is important to confirm at the end of the calculation that the approximation $2x \ll 0.030$ is valid. Box 6-2 discusses approximations.

Because $2x = 1.8 \times 10^{-5} << 0.030$, the decision to ignore $2x$ to solve the problem was justified. The answer also illustrates the common ion effect. In the absence of added I^-, the solubility of Pb^{2+} was 0.001 3 M. In the presence of 0.030 M I^-, $[Pb^{2+}]$ is reduced to 8.8×10^{-6} M.

What is the maximum I^- concentration at equilibrium in a solution in which $[Pb^{2+}]$ is somehow *fixed* at 1.0×10^{-4} M? Our concentration table looks like this now:

	$PbI_2(s)$	$\rightleftharpoons$	Pb^{2+}	+	$2I^-$
Initial concentration	solid		1.0×10^{-4}		0
Final concentration	solid		1.0×10^{-4}		x

Box 6-2 The Logic of Approximations

Many problems are difficult to solve without judicious approximations. For example, rather than solving the equation

$$(x)(2x + 0.030)^2 = 7.9 \times 10^{-9}$$

we hope and pray that $2x << 0.030$ and therefore solve the much simpler equation

$$(x)(0.030)^2 = 7.9 \times 10^{-9}$$

But how can we be sure that our solution fits the original problem?

When we use an approximation, we assume it is true. *If the assumption is true, it does not create a contradiction. If the assumption is false, it leads to a contradiction.* You can test an assumption by using it and seeing whether you are right or wrong afterward.

You may object to this reasoning, feeling "How can the truth of an assumption be tested by using the assumption?" Suppose you wish to test the statement "Gail can swim 100 meters." To see whether or not the statement is true, you can assume it is true. If Gail can swim 100 m, then you could dump her in the middle of a lake with a radius of 100 m and expect her to swim to shore. If she comes ashore alive, then your assumption was correct and no contradiction is created. If she does not make it to shore, then there is a contradiction. Your assumption must have been wrong. There are only two possibilities: Either the assumption is correct and using it is correct or the assumption is wrong and using it is wrong. (A third possibility in this case is that there are freshwater sharks in the lake.)

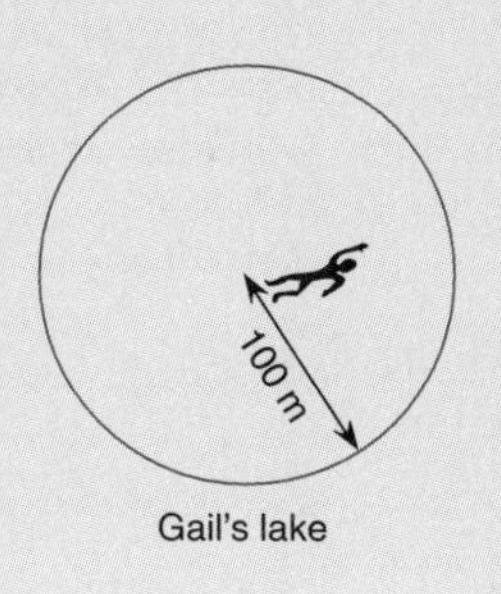

Gail's lake

Example 1. $(x)(2x + 0.030)^2 = 7.9 \times 10^{-9}$
$(x)(0.030)^2 = 7.9 \times 10^{-9}$
(assuming $2x << 0.030$)
$x = (7.9 \times 10^{-9})/(0.030)^2 = 8.8 \times 10^{-6}$

No contradiction:
$2x = 1.76 \times 10^{-5} << 0.030$
The assumption is true.

Example 2. $(x)(2x + 0.030)^2 = 7.9 \times 10^{-5}$
$(x)(0.030)^2 = 7.9 \times 10^{-5}$
(assuming $2x << 0.030$)
$x = (7.9 \times 10^{-5})/(0.030)^2 = 0.088$

A contradiction:
$2x = 0.176 > 0.030$
The assumption is false.

In Example 2, the assumption leads to a contradiction, so the assumption cannot be correct. When this happens, you must solve the cubic equation $(x)(2x + 0.030)^2 = 7.9 \times 10^{-5}$.

A reasonable way to solve this equation is by trial and error, as shown in the following table. You can create this table by hand or, even more easily, with a spreadsheet. In cell A1, enter a guess for x. In cell A2, enter the formula "=A1*(2*A1 + 0.030)^2". When you guess x correctly in cell A1, cell A2 will have the value 7.9×10^{-5}. Problem 6-24 gives an even better way to solve this problem with Excel GOAL SEEK.

Guess	$x(2x + 0.030)^2$	Result is
$x = 0.01$	2.5×10^{-5}	too low
$x = 0.02$	9.8×10^{-5}	too high
$x = 0.015$	5.4×10^{-5}	too low
$x = 0.018$	7.84×10^{-5}	too low
$x = 0.019$	8.79×10^{-5}	too high
$x = 0.018\ 1$	7.93×10^{-5}	too high
$x = 0.018\ 05$	7.89×10^{-5}	too low
$x = 0.018\ 06$	7.90×10^{-5}	not bad!

$[Pb^{2+}]$ is not x in this example, so there is no reason to set $[I^-] = 2x$. The problem is solved by plugging each concentration into the solubility product:

$$[Pb^{2+}][I^-]^2 = K_{sp}$$
$$(1.0 \times 10^{-4})(x)^2 = 7.9 \times 10^{-9}$$
$$x = [I^-] = 8.9 \times 10^{-3}\ M$$

If I^- is added above a concentration of 8.9×10^{-3} M, then $PbI_2(s)$ will precipitate.

Example Using the Solubility Product

The compound silver ferrocyanide dissociates into Ag^+ and $Fe(CN)_6^{4-}$ in solution. Find the solubility of $Ag_4Fe(CN)_6$ in water expressed as **(a)** moles of $Ag_4Fe(CN)_6$ per liter and **(b)** ppb Ag^+ ($\approx$ ng Ag^+/mL).

SOLUTION **(a)** We start off with a concentration table:

	$Ag_4Fe(CN)_6(s)$ Silver ferrocyanide FM 643.42	$\rightleftharpoons$ $4Ag^+$	+ $Fe(CN)_6^{4-}$ Ferrocyanide
Initial concentration	solid	0	0
Final concentration	solid	$4x$	x

The solubility product of $Ag_4Fe(CN)_6$ in Appendix A is 8.5×10^{-45}. Therefore

In calculating the solubility of $Ag_4Fe(CN)_6$, we have neglected species such as $AgFe(CN)_6^{3-}$, which may be significant.

$$[Ag^+]^4[Fe(CN)_6^{4-}] = (4x)^4(x) = 8.5 \times 10^{-45}$$
$$256x^5 = 8.5 \times 10^{-45}$$
$$x^5 = \frac{8.5 \times 10^{-45}}{256}$$
$$x = \left(\frac{8.5 \times 10^{-45}}{256}\right)^{1/5} = 5.0_6 \times 10^{-10}\ M$$

There are $5.0_6 \times 10^{-10}$ mol of $Ag_4Fe(CN)_6$ dissolved per liter. The concentration of Ag^+ is $4x$ and the concentration of $Fe(CN)_6^{4-}$ is x.

(b) The concentration of Ag^+ is $4x = 4(5.0_6 \times 10^{-10}\ M) = 2.0_2 \times 10^{-9}$ M. Use the atomic mass of Ag to convert mol/L into g/L:

$$\left(2.0_2 \times 10^{-9}\ \frac{\text{mol}}{\text{L}}\right)\left(107.868\ 2\ \frac{\text{g}}{\text{mol}}\right) = 2.1_8 \times 10^{-7}\text{g/L}$$

Atomic mass of Ag

To find ppb, we need units of ng/mL:

$$\left(2.1_8 \times 10^{-7}\ \frac{\text{g}}{\text{L}}\right)\left(\frac{1}{1\,000}\ \frac{\text{L}}{\text{mL}}\right)\left(10^9\ \frac{\text{ng}}{\text{g}}\right) = 0.21_8\ \frac{\text{ng}}{\text{mL}} = 0.22\ \text{ppb}$$

Test Yourself At a different temperature, K_{sp} for $Ag_4Fe(CN)_6$ is doubled from 8.5×10^{-45} to 17×10^{-45}. Find ppb Ag^+ in a saturated aqueous solution of $Ag_4Fe(CN)_6$ at this second temperature. Is the value doubled from that we just found? (**Answer:** 0.25 ppb, not doubled)

Ask Yourself

6-D. What is the concentration of Pb^{2+} in **(a)** a saturated solution of $PbBr_2$ in water or **(b)** a saturated solution of $PbBr_2$ in which $[Br^-]$ is somehow fixed at 0.10 M?

6-5 Titration of a Mixture

When Ag^+ is added to a solution containing Cl^- and I^-, the less soluble AgI(*s*) precipitates first:

$$\left.\begin{array}{l} Ag^+ + I^- \longrightarrow AgI(s) \\ Ag^+ + Cl^- \longrightarrow AgCl(s) \end{array}\right\} \begin{array}{l} \text{AgI } (K_{sp} = 8.3 \times 10^{-17}) \text{ precipitates} \\ \text{before AgCl } (K_{sp} = 1.8 \times 10^{-10}) \end{array} \quad \begin{array}{r} (6\text{-}10a) \\ (6\text{-}10b) \end{array}$$

Because the two solubility products are sufficiently different, the first precipitation is nearly complete before the second commences.

Figure 6-4 shows how the reaction is monitored with a silver electrode to find both end points. We will learn how this electrode responds to silver ion concentration in Section 15-1. Figure 6-5 shows experimental curves for the titration of I^- or a mixture of I^- plus Cl^- by Ag^+.

In the titration of I^- (curve *b* in Figure 6-5), the voltage remains almost constant near 650 mV until the equivalence point (23.76 mL) when I^- is used up. At this point, there is an abrupt decrease in the electrode potential. The reason for the abrupt

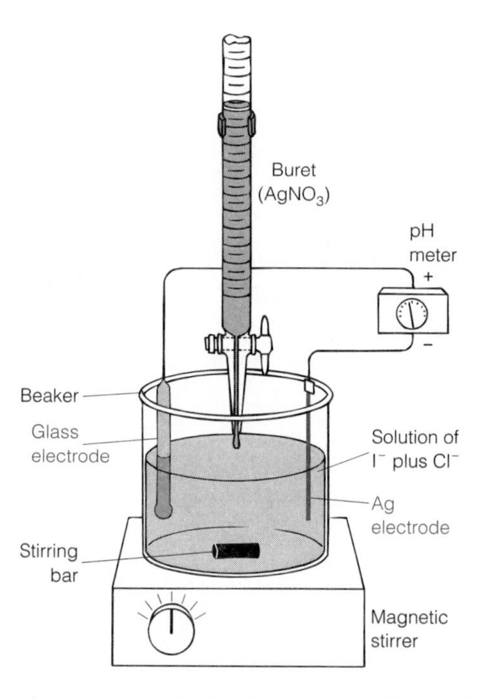

Figure 6-4 Apparatus for measuring the titration curves in Figure 6-5. The silver electrode responds to changes in Ag^+ concentration, and the glass electrode provides a constant reference potential in this experiment. The voltage changes by approximately 59 mV for each factor-of-10 change in $[Ag^+]$. All solutions, including $AgNO_3$, were maintained at pH 2.0 by using 0.010 M sulfate buffer prepared from H_2SO_4 and KOH.

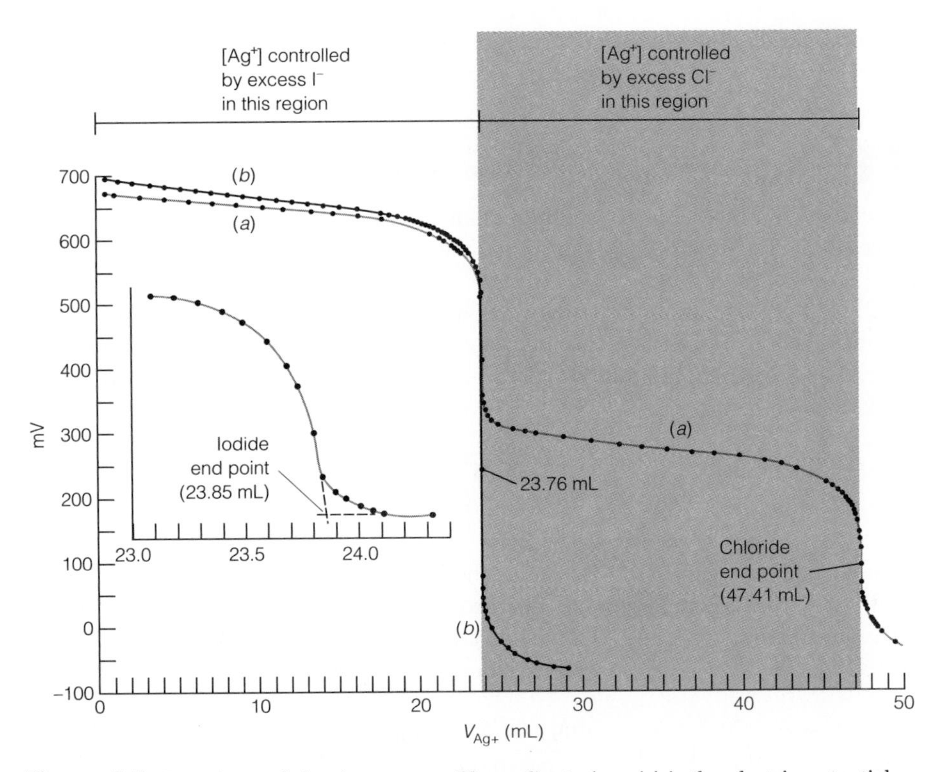

Figure 6-5 Experimental titration curves. The ordinate (y-axis) is the electric potential difference (in millivolts) between the two electrodes in Figure 6-4. (a) Titration of KI plus KCl with 0.084 5 M $AgNO_3$. The inset is an expanded view of the region near the first equivalence point. (b) Titration of I^- with 0.084 5 M Ag^+.

change is that the electrode is responding to the concentration of Ag^+ in the solution. Prior to the equivalence point, virtually all the added Ag^+ reacts with I^- to precipitate AgI(s). The concentration of Ag^+ in solution is very low and nearly constant. When I^- has been consumed, the concentration of Ag^+ suddenly increases because Ag^+ is being added from the buret and is no longer consumed by I^-. This change gives rise to the abrupt decrease in electrode potential.

In the titration of $I^- + Cl^-$ (curve a in Figure 6-5), there are two abrupt changes in electrode potential. The first occurs when I^- is used up, and the second comes when Cl^- is used up. Prior to the first equivalence point, the very low concentration of Ag^+ is governed by the solubility of AgI. Between the first and the second equivalence points, essentially all I^- has precipitated and Cl^- is in the process of being consumed. The concentration of Ag^+ is still small but governed by the solubility of AgCl, which is greater than that of AgI. After the second equivalence point, the concentration of Ag^+ shoots upward as Ag^+ is added from the buret. Therefore we observe two abrupt changes of electric potential in this experiment.

The I^- end point is taken as the intersection of the steep and nearly horizontal curves shown at 23.85 mL in the inset of Figure 6-5. The reason for using the intersection is that the precipitation of I^- is not quite complete when Cl^- begins to precipitate. Therefore the end of the steep part (the intersection) is a better approximation of the equivalence point than is the middle of the steep section. The Cl^- end point is taken as the midpoint of the second steep section, at 47.41 mL. The moles of Cl^- in the sample correspond to the moles of Ag^+ delivered between the first and the second end points. That is, it requires 23.85 mL of Ag^+ to precipitate I^-, and (47.41 − 23.85) = 23.56 mL of Ag^+ to precipitate Cl^-.

The elements F, Cl, Br, I, and At are called *halogens*. Their anions are called *halides*.

Example **Extracting Results from Figure 6-5**

In curve *a* of Figure 6-5, 40.00 mL of unknown solution containing both I^- and Cl^- were titrated with 0.084 5 M Ag^+. Find the concentrations of each halide ion.

SOLUTION The inset shows the first end point at 23.85 mL. Reaction 6-10a tells us that 1 mol of I^- consumes 1 mol of Ag^+. The moles of Ag^+ delivered at this point are (0.084 5 M)(0.023 85 L) = 2.015×10^{-3} mol. The molarity of iodide in the unknown is therefore

$$[I^-] = \frac{2.015 \times 10^{-3}\text{ mol}}{0.040\,00\text{ L}} = 0.050\,3_8\text{ M}$$

The second end point is at 47.41 mL. The quantity of Ag^+ titrant required to react with Cl^- is the difference between the two end points: (47.41 − 23.85) = 23.56 mL. The number of moles of Ag^+ required to react with Cl^- is (0.084 5 M)(0.023 56 L) = 1.991×10^{-3} mol. The molarity of chloride in the unknown is

$$[Cl^-] = \frac{1.991 \times 10^{-3}\text{ mol}}{0.040\,00\text{ L}} = 0.049\,7_7\text{ M}$$

Test Yourself Suppose that the two end points were at 24.85 and 47.41 mL. Find $[I^-]$ and $[Cl^-]$ in the unknown. (**Answer:** 0.052 5_0 M, 0.047 6_6 M)

Ask Yourself

6-E. A 25.00-mL solution containing Br^- and Cl^- was titrated with 0.033 33 M $AgNO_3$.
(a) Write the two titration reactions and use solubility products to find which takes place first.
(b) In an experiment analogous to that in Figures 6-4 and 6-5, the first end point was observed at 15.55 mL. Find the concentration of the first halide that precipitated. Is it Br^- or Cl^-?
(c) The second end point was observed at 42.23 mL. Find the concentration of the other halide.

6-6 Titrations Involving Silver Ion

The Latin word for silver is *argentum*, from which the symbol Ag is derived.

We now introduce two widely used indicator methods for titrations involving Ag^+, which are called *argentometric titrations*. The methods are

1. **Volhard titration:** formation of a soluble, colored complex at the end point
2. **Fajans titration:** adsorption of a colored indicator on the precipitate at the end point

Volhard Titration

The Volhard titration is actually a titration of Ag^+ in 0.5–1.5 M HNO_3. To determine Cl^-, a back titration is necessary. First, Cl^- is precipitated by a known, excess quantity of standard $AgNO_3$:

$$Ag^+ + Cl^- \longrightarrow AgCl(s)$$

The AgCl is isolated, and excess Ag^+ is titrated with standard KSCN in the presence of Fe^{3+}:

$$Ag^+ + SCN^- \longrightarrow AgSCN(s)$$

When all the Ag^+ has been consumed, the SCN^- reacts with Fe^{3+} to form a red complex:

$$Fe^{3+} + SCN^- \longrightarrow \underset{\text{Red}}{FeSCN^{2+}}$$

The appearance of red color signals the end point. Knowing how much SCN^- was required for the back titration tells us how much Ag^+ was left over from the reaction with Cl^-. Because the total amount of Ag^+ is known, the amount consumed by Cl^- can then be calculated.

In the analysis of Cl^- by the Volhard method, the end point slowly fades because AgCl is more soluble than AgSCN. The AgCl slowly dissolves and is replaced by AgSCN. To prevent this secondary reaction from happening, we filter off the AgCl and titrate Ag^+ in the filtrate. Br^- and I^-, whose silver salts are *less* soluble than AgSCN, can be titrated by the Volhard method without isolating the silver halide precipitate.

Because the Volhard method is a titration of Ag^+, it can be adapted for the determination of any anion that forms an insoluble silver salt.

Fajans Titration

The Fajans titration uses an **adsorption indicator**. To see how this works, consider the electric charge of a precipitate. When Ag^+ is added to Cl^-, there is excess Cl^- in solution prior to the equivalence point. Some Cl^- is selectively adsorbed on the AgCl surface, imparting a negative charge to the crystal surface (Figure 6-6*a*). After the equivalence point, there is excess Ag^+ in solution. Adsorption of Ag^+ cations on the crystal creates a positive charge on the particles of precipitate (Figure 6-6*b*). The abrupt change from negative charge to positive charge occurs at the equivalence point.

Common adsorption indicators are anionic (negatively charged) dyes, which are attracted to the positively charged precipitate produced immediately after the equivalence point. Adsorption of the dye on the surface of the solid precipitate changes the color of the dye by interactions that are not well understood. The color change signals the end point in the titration. Because the indicator reacts with the precipitate surface, it is desirable to have as much surface area as possible. The titration is performed under conditions that tend to keep the particles as small as possible, because small particles have more surface area than an equal volume of large particles. Low electrolyte concentration helps prevent coagulation of the precipitate. As in all silver titrations, strong light should be avoided.

The indicator most commonly used for AgCl is dichlorofluorescein, which has a greenish yellow color in solution but turns pink when adsorbed on AgCl

Dichlorofluorescein

Notation for drawing organic compounds is discussed in Box 7-1.

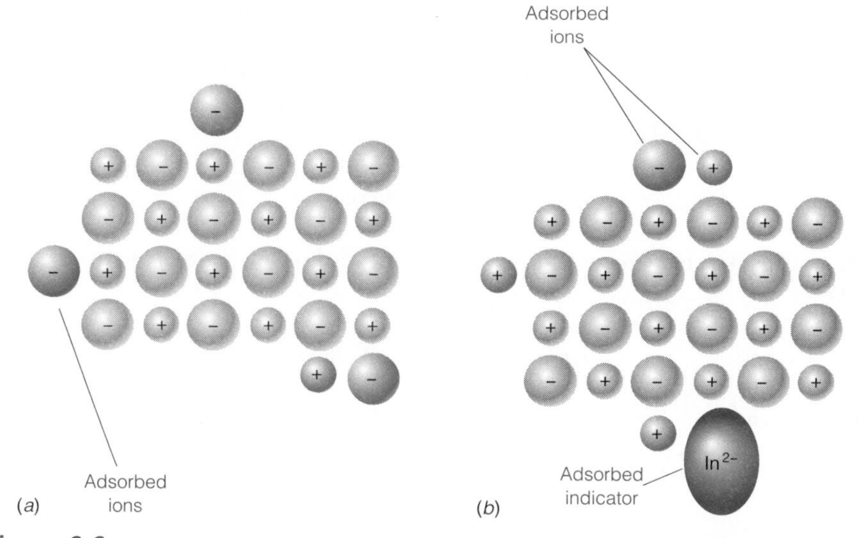

Figure 6-6 Ions from a solution are adsorbed on the surface of a growing crystallite. (*a*) A crystal growing in the presence of excess lattice anions (anions that belong in the crystal) has a negative charge because anions are predominantly adsorbed. (*b*) A crystal growing in the presence of excess lattice cations has a positive charge and can therefore adsorb a negative indicator ion. Anions and cations in the solution that do not belong in the crystal lattice are less likely to be adsorbed than are ions belonging to the lattice.

(Demonstration 6-1). To maintain the indicator in its anionic form, there must not be too much H^+ in the solution. The dye eosin is useful in the titration of Br^-, I^-, and SCN^-. It gives a sharper end point than dichlorofluorescein and is more sensitive (that is, less halide can be titrated). It cannot be used for AgCl, because the eosin anion is more strongly bound to AgCl than is Cl^- ion. Eosin will bind to the AgCl crystallites even before the particles become positively charged.

Applications of precipitation titrations are listed in Table 6-1. Whereas the Volhard method is specifically for argentometric titrations, the Fajans method has wider application. Because the Volhard titration is carried out in acidic solution (typically 0.2 M HNO_3), it avoids certain interferences that would affect other titrations. Silver salts of anions such as CO_3^{2-}, $C_2O_4^{2-}$, and AsO_4^{3-} are soluble in acidic solution, so these anions do not interfere with the analysis.

Challenge Consider the equilibrium

$$\underset{\text{Carbonate}}{CO_3^{2-}} + H^+ \rightleftharpoons \underset{\substack{\text{Hydrogen carbonate} \\ \text{also called bicarbonate}}}{HCO_3^-}$$

Use Le Châtelier's principle to explain why carbonate salts are soluble in acidic solution (which contains a high concentration of H^+).

Demonstration 6-1 Fajans Titration

The Fajans titration of Cl^- with Ag^+ demonstrates indicator end points in precipitation titrations. Dissolve 0.5 g of NaCl plus 0.15 g of dextrin in 400 mL of water. The purpose of the dextrin is to retard coagulation of the AgCl precipitate. Add 1 mL of dichlorofluorescein indicator solution containing 1 mg/mL of dichlorofluorescein in 95 wt% aqueous ethanol or 1 mg/mL of the sodium salt in water. Titrate the NaCl solution with a solution containing 2 g of $AgNO_3$ in 30 mL of water. About 20 mL are required to reach the end point.

Color Plate 1*a* shows the yellow color of the indicator in the NaCl solution prior to the titration. Color Plate 1*b* shows the milky-white appearance of the AgCl suspension during titration, before the end point is reached. The pink suspension in Color Plate 1*c* appears at the end point, when the anionic indicator becomes adsorbed to the cationic particles of precipitate.

Table 6-1 Applications of precipitation titrations

Species analyzed	Notes
	VOLHARD METHOD
Br^-, I^-, SCN^-, CNO^-, AsO_4^{3-}	Precipitate removal is unnecessary.
Cl^-, PO_4^{3-}, CN^-, $C_2O_4^{2-}$, CO_3^{2-}, S^{2-}, CrO_4^{2-}	Precipitate removal required.
	FAJANS METHOD
Cl^-, Br^-, I^-, SCN^-, $Fe(CN)_6^{4-}$	Titrate with Ag^+. Detection with dyes such as fluorescein, dichlorofluorescein, eosin, bromophenol blue.
F^-	Titrate with $Th(NO_3)_4$ to produce ThF_4. End point detected with alizarin red S.
Zn^{2+}	Titrate with $K_4Fe(CN)_6$ to produce $K_2Zn_3[Fe(CN)_6]_2$. End point detected with diphenylamine.
SO_4^{2-}	Titrate with $Ba(OH)_2$ in 50 vol% aqueous methanol; use alizarin red S as indicator.
Hg_2^{2+}	Titrate with NaCl to produce Hg_2Cl_2. End point detected with bromophenol blue.
PO_4^{3-}, $C_2O_4^{2-}$	Titrate with $Pb(CH_3CO_2)_2$ to give $Pb_3(PO_4)_2$ or PbC_2O_4. End point detected with dibromofluorescein (PO_4^{3-}) or fluorescein ($C_2O_4^{2-}$).

Ask Yourself

6-F. **(a)** Why is precipitated AgCl filtered off in the Volhard titration of chloride?
(b) Why does the surface charge of a precipitate change sign at the equivalence point?
(c) In the Fajans titration of Zn^{2+} in Table 6-1, do you expect the charge on the precipitate to be positive or negative after the equivalence point?

Key Equations

Stoichiometry — For the reaction $aA + bB \longrightarrow$ products, use the ratio $\left(\frac{a \text{ mol A}}{b \text{ mol B}}\right)$ for stoichiometry calculations.

Solubility product — $PbI_2(s) \overset{K_{sp}}{\rightleftharpoons} Pb^{2+} + 2I^-$ $\qquad K_{sp} = [Pb^{2+}][I^-]^2$

Common ion effect: A salt is less soluble in the presence of one of its constituent ions.

Important Terms

adsorption indicator	direct titration	gravimetric titration
back titration	end point	indicator
blank titration	equivalence point	primary standard
common ion effect	Fajans titration	saturated solution

solubility product
standardization
standard solution
titrant
titration
titration error
Volhard titration
volumetric analysis

Problems

6-1. Distinguish the terms *end point* and *equivalence point.*

6-2. For Reaction 6-1, how many milliliters of 0.165 0 M $KMnO_4$ are needed to react with 108.0 mL of 0.165 0 M oxalic acid? How many milliliters of 0.165 0 M oxalic acid are required to react with 108.0 mL of 0.165 0 M $KMnO_4$?

6-3. A 10.00-mL aliquot of unknown oxalic acid solution required 15.44 mL of 0.011 17 M $KMnO_4$ solution to reach the purple end point. A blank titration of 10 mL of similar solution containing no oxalic acid required 0.04 mL to exhibit detectable color. Find the concentration of oxalic acid in the unknown.

6-4. Ammonia reacts with hypobromite, OBr^- as follows:

$$2NH_3 + 3OBr^- \longrightarrow N_2 + 3Br^- + 3H_2O$$

Find the molarity of OBr^- if 1.00 mL of OBr^- solution reacts with 1.69 mg of NH_3 (FM 17.03)?

6-5. How many milliliters of 0.100 M KI are needed to react with 40.0 mL of 0.040 0 M $Hg_2(NO_3)_2$ if the reaction is $Hg_2^{2+} + 2I^- \longrightarrow Hg_2I_2(s)$?

6-6. Cl^- in blood serum, cerebrospinal fluid, or urine can be measured by titration with mercuric ion: $Hg^{2+} + 2Cl^- \longrightarrow HgCl_2(aq)$. When the reaction is complete, excess Hg^{2+} reacts with the indicator diphenylcarbazone, which forms a violet-blue color.

(a) Mercuric nitrate was standardized by titrating a solution containing 147.6 mg of NaCl (FM 58.44), which required 28.06 mL of $Hg(NO_3)_2$ solution. Find the molarity of the $Hg(NO_3)_2$.

(b) When this same $Hg(NO_3)_2$ solution was used to titrate 2.000 mL of urine, 22.83 mL were required. Find the concentration of Cl^-(mg/mL) in the urine.

6-7. *Volhard titration.* A 30.00-mL solution of unknown I^- was treated with 50.00 mL of 0.365 0 M $AgNO_3$. The precipitated AgI was filtered off, and the filtrate (plus Fe^{3+}) was titrated with 0.287 0 M KSCN. When 37.60 mL had been added, the solution turned red. How many milligrams of I^- were present in the original solution?

6-8. How many milligrams of oxalic acid dihydrate, $H_2C_2O_4 \cdot 2H_2O$ (FM 126.07), will react with 1.00 mL of 0.027 3 M ceric sulfate, $Ce(SO_4)_2$, if the reaction is $H_2C_2O_4 + 2Ce^{4+} \longrightarrow 2CO_2 + 2Ce^{3+} + 2H^+$?

6-9. Arsenic(III) oxide (As_2O_3) is available in pure form and is a useful (and poisonous) primary standard for many oxidizing agents, such as MnO_4^-. As_2O_3 is first dissolved in base and titrated with MnO_4^- in acidic solution. A small amount of iodide (I^-) or iodate (IO_3^-) is used to catalyze the reaction between H_3AsO_3 and MnO_4^-. The reactions are

$$As_2O_3 + 4OH^- \rightleftharpoons 2HAsO_3^{2-} + H_2O$$

$$HAsO_3^{2-} + 2H^+ \rightleftharpoons H_3AsO_3$$

$$5H_3AsO_3 + 2MnO_4^- + 6H^+ \rightarrow 5H_3AsO_4 + 2Mn^{2+} + 3H_2O$$

(a) A 3.214-g aliquot of $KMnO_4$ (FM 158.03) was dissolved in 1.000 L of water, heated to cause any reactions with impurities to occur, cooled, and filtered. What is the theoretical molarity of this solution if no MnO_4^- was consumed by impurities?

(b) What mass of As_2O_3 (FM 197.84) would be sufficient to react with 25.00 mL of the $KMnO_4$ solution in part **(a)**?

(c) It was found that 0.146 8 g of As_2O_3 required 29.98 mL of $KMnO_4$ solution for the faint color of unreacted MnO_4^- to appear. A blank titration required 0.03 mL of MnO_4^- for color to appear. Find the molarity of the $KMnO_4$ solution.

6-10. *Gravimetric titration.* A solution of NaOH was standardized by titration of a known quantity of the primary standard, potassium hydrogen phthalate:

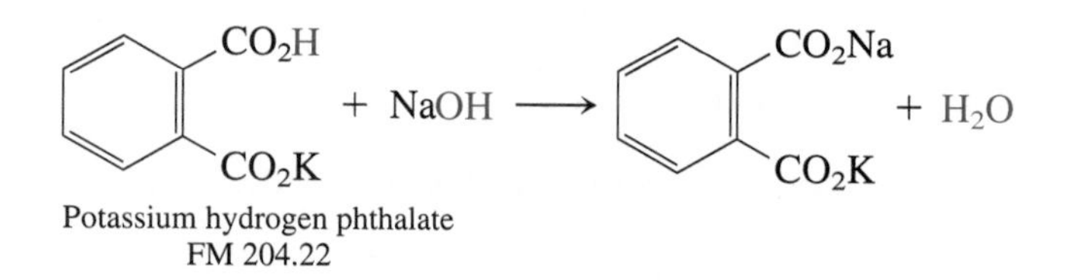

The NaOH was then used to standardize H_2SO_4:

$$H_2SO_4 + 2NaOH \longrightarrow Na_2SO_4 + H_2O$$

(a) Titration of 0.824 g of potassium hydrogen phthalate required 38.314 g of NaOH solution to reach the end point detected by phenolphthalein indicator. Find the concentration of NaOH expressed as mol NaOH/kg solution.

(b) A 10.063-gram aliquot of H_2SO_4 solution required 57.911 g of NaOH solution to reach the phenolphthalein end point. Find the concentration of H_2SO_4 in mol/kg of solution.

6-11. *Uncertainty in volumetric and gravimetric procedures.* **(a)** A silver nitrate primary standard for a gravimetric titration of chloride was prepared by dissolving 4.872 ± 0.003 g $AgNO_3$ (FM 169.873 1) in 498.633 ± 0.003 g H_2O. Then 26.207 ± 0.003 g of the solution were weighed out for titration. Find the moles of Ag^+ (and its relative uncertainty) delivered for titration. The relative uncertainty in formula mass is negligible.

(b) In a volumetric titration, 4.872 ± 0.003 g $AgNO_3$ were dissolved in a 500.00 ± 0.20 mL volumetric flask. Then 25.00 ± 0.03 mL were withdrawn for titration. Find the moles of Ag^+ (and its relative uncertainty) delivered for titration.

(c) How much greater is the relative uncertainty in the volumetric delivery than the gravimetric delivery? What is the largest source of uncertainty in each method?

6-12. *Back titration.* Impure $K_2S_2O_8$ (FM 270.32, 0.507 3 g) was analyzed by treatment with excess standard $Na_2C_2O_4$ by Reaction 6-4. After reaction with 50.00 mL of 0.050 06 M $Na_2C_2O_4$, the excess oxalate required 16.52 mL of 0.020 13 M $KMnO_4$ in Reaction 6–1. Find the weight percent of $K_2S_2O_8$ in the impure reagent.

6-13. An unknown molybdate (MoO_4^{2-}) solution (50.00 mL) was passed through a column containing Zn(*s*) to convert molybdate into Mo^{3+}. One mole of MoO_4^{2-} gives one mole of Mo^{3+}. The resulting sample required 22.11 mL of 0.012 34 M $KMnO_4$ to reach a purple end point from the reaction

$$3MnO_4^- + 5Mo^{3+} + 4H^+ \longrightarrow 3Mn^{2+} + 5MoO_2^{2+} + 2H_2O$$

A blank required 0.07 mL. Find the molarity of molybdate in the unknown.

6-14. A 25.00-mL sample of La^{3+} was treated with excess $Na_2C_2O_4$ to precipitate $La_2(C_2O_4)_3$, which was washed to remove excess $C_2O_4^{2-}$ and then dissolved in acid. The oxalate from $La_2(C_2O_4)_3$ required 12.34 mL of 0.004 321 M $KMnO_4$ to reach the purple end point of Reaction 6-1. Find the molarity of La^{3+} in the unknown.

6-15. *Back titration.* A glycerol solution weighing 153.2 mg was treated with 50.0 mL of 0.089 9 M Ce^{4+} in 4 M $HClO_4$ at 60°C for 15 min to convert glycerol into formic acid:

$$\underset{\substack{\text{Glycerol}\\ \text{FM 92.09}}}{C_3H_8O_3} + 8Ce^{4+} + 3H_2O \longrightarrow \underset{\text{Formic acid}}{3HCO_2H} + 8Ce^{3+} + 8H^+$$

The excess Ce^{4+} required 10.05 mL of 0.043 7 M Fe^{2+} for a back titration by the reaction $Ce^{4+} + Fe^{2+} \longrightarrow Ce^{3+} + Fe^{3+}$. Find wt% glycerol in the unknown.

6-16. *Propagation of uncertainty.* Consider the titration of 50.00 (±0.05) mL of a mixture of I^- and SCN^- with 0.068 3 (±0.000 1) M Ag^+. From the solubility products of AgI and AgSCN, decide which precipitate is formed first. The first equivalence point is observed at 12.6 (±0.4) mL, and the second occurs at 27.7 (±0.3) mL. Find the molarity and the uncertainty in molarity of thiocyanate in the original mixture.

6-17. Calculate the solubility of CuBr (FM 143.45) in water expressed as **(a)** moles per liter and **(b)** grams per 100 mL. (This question presumes that Cu^+ and Br^- are the only significant soluble species. $CuBr_2^-$ is negligible, and we suppose that the ion pair CuBr(*aq*) also is negligible.)

6-18. Find the solubility of silver chromate (FM 331.73) in water. Express your answer as **(a)** moles of chromate per liter and **(b)** ppm Ag^+(≈ μg Ag^+/mL).

6-19. Ag^+ at 10–100 ppb (ng/mL) disinfects swimming pools. One way to maintain an appropriate concentration of Ag^+ is to add a slightly soluble silver salt to the pool. Calculate the ppb of Ag^+ in saturated solutions of AgCl, AgBr, and AgI.

6-20. Mercury(I) is a diatomic ion (Hg_2^{2+}, also called mercurous ion) with a charge of +2. Mercury(I) iodate dissociates into three ions:

$$\underset{\text{FM 750.99}}{Hg_2(IO_3)_2(s)} \rightleftharpoons Hg_2^{2+} + 2IO_3^- \quad K_{sp} = [Hg_2^{2+}][IO_3^-]^2$$

Find the concentrations of Hg_2^{2+} and IO_3^- in **(a)** a saturated solution of $Hg_2(IO_3)_2(s)$ and **(b)** a 0.010 M solution of KIO_3 saturated with $Hg_2(IO_3)_2(s)$.

6-21. If a solution containing 0.10 M Cl^-, Br^-, I^-, and CrO_4^{2-} is treated with Ag^+, in what order will the anions precipitate?

6-22. *Volhard titration.* The concentration of Cl^- in "concentrated HCl," a ~12 M solution commonly purchased for laboratories, was determined as follows:

1. Standard Ag^+ was prepared by mixing 25.00 mL of 0.102 6 M $AgNO_3$, 5 mL of 6 M HNO_3, and 1 mL of Fe^{3+} indicator solution (40 wt% aqueous $(NH_4)Fe(SO_4)_2$ with a few drops of 6 M HNO_3). To standardize a solution of potassium thiocyanate, KSCN is delivered to the standard Ag^+ solution from a buret. The initial precipitate is white and later it is reddish brown. The color disappears on shaking. At the end point, one drop of KSCN solution produces a faint brown color that does not disappear on shaking. A volume of 24.22 mL was required to reach the end point and the blank correction was 0.02 mL.

2. 10.00 mL of concentrated HCl was diluted with water to 1.000 L in a volumetric flask. Then 20.00 mL of dilute HCl were mixed with 5 mL of 6 M HNO_3 and 25.00 mL of standard 0.102 6 M $AgNO_3$. AgCl precipitate was filtered, washed with 0.16 M HNO_3, and the washings were combined with the filtrate. Filtrate was treated with 1 mL of Fe^{3+} indicator and titrated with 2.43 mL of KSCN to reach the faint brown end point. The blank correction was 0.02 mL.

Find the concentration of KSCN in step 1 and the concentration of "concentrated HCl" reagent from step 2.

6-23. *Managing a saltwater aquarium.* The New Jersey State Aquarium, Ocean Tank, has a volume of 2.9 million liters. Bacteria remove nitrate that would otherwise build up to toxic levels. Aquarium water is first pumped into a 2 700-L deaeration tank containing bacteria that consume O_2 in the presence of added methanol:

$$\underset{\text{Methanol}}{2CH_3OH} + 3O_2 \xrightarrow{\text{bacteria}} 2CO_2 + 4H_2O \quad (1)$$

Anoxic (deoxygenated) water from the deaeration tank flows into a 1 500-L denitrification tank containing colonies of *Pseudomonas* bacteria in a porous medium. Methanol is injected continuously and nitrate is converted into nitrite and then into nitrogen:

$$\underset{\text{Nitrate}}{3NO_3^-} + CH_3OH \xrightarrow{\text{bacteria}} \underset{\text{Nitrite}}{3NO_2^-} + CO_2 + 2H_2O \quad (2)$$

$$2NO_2^- + CH_3OH \xrightarrow{\text{bacteria}} N_2 + CO_2 + H_2O + 2OH^- \quad (3)$$

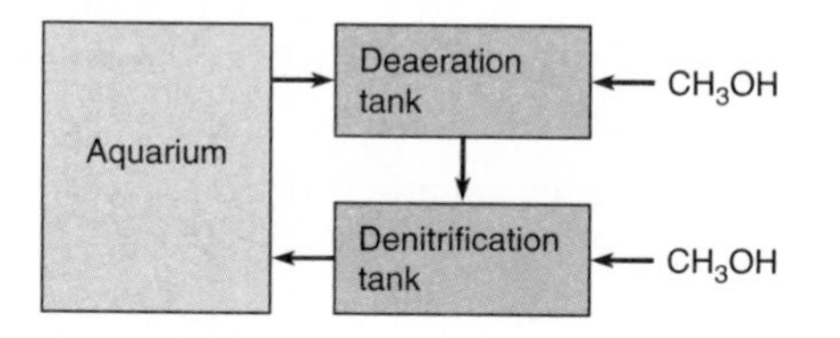

[The aquarium problem comes from G. Grguric, *J. Chem Ed.* **2002**, *79*, 179.]

(a) Deaeration can be thought of as a slow, bacteria-mediated titration of O_2 by CH_3OH. The concentration of O_2 in seawater at 24°C is 220 μM. How many liters of CH_3OH (FM 32.04, density = 0.791 g/mL) are required by Reaction 1 for 2.9 million liters of aquarium water?

(b) Multiply Reaction 2 by 2 and multiply Reaction 3 by 3 so that they can be added and NO_2^- will cancel out. Write the net reaction showing nitrate plus methanol going to nitrogen. How many liters of CH_3OH are required by the net reaction for 2.9 million liters of aquarium water with a nitrate concentration of 8 100 μM?

(c) In addition to requiring methanol for Reactions 1 through 3, the bacteria require 30% more methanol for their own growth. What is the total volume of methanol required to denitrify 2.9 million liters of aquarium water, including Reactions 1–3 plus 30% more methanol for the bacteria?

6-24. *Solving equations with Excel* GOAL SEEK. Suppose we saturate a solution of 0.001 0 M NaI with PbI_2. From the solubility product of PbI_2, we find $[Pb^{2+}]$ as follows:

$$[Pb^{2+}][I^-]^2 = (x)(2x + 0.001\,0)^2 = K_{sp} = 7.9 \times 10^{-9}$$

(a) *Solution by trial and error.* Following the procedure at the end of Box 6-2, solve for x by guessing the value that makes the expression $(x)(2x + 0.001\,0)^2$ equal to 7.9×10^{-9}. Write any value of x in cell A4. In cell B4, enter the formula "=A4*(2*A4+0.001 0)^2". Now vary the value of x in cell A4 until you get 7.9×10^{-9} in cell B4.

	A	B
1	Guessing the Answer	
2		
3	x	x(2x+0.0010)^2
4	0.01	4.41E-06

(b) *Using Excel* GOAL SEEK. Set up the same spreadsheet used in part **(a)** and guess a value of 0.01 in cell A4. We will use a built-in procedure to vary cell A4 until cell B4 has the desired value. We want cell B4 to be a small number (7.9×10^{-9}), so we need to set a small tolerance for it. In the TOOLS menu select OPTIONS on a PC or PREFERENCES on a Mac. Click the Calculation tab and enter 1E-12 for Maximum change. Click OK. We just told the computer that the value we will be finding in cell B4 needs to be precise to 10^{-12}. In the TOOLS menu select GOAL SEEK. For Set cell, enter B4. For To value, enter 7.9E-9. For By changing cell, enter A4. When you click OK, the computer finds x = 0.000 945 4 in cell A4. If you do not see enough digits in cell A4, drag the separator between columns A and B to the right to expand the cell. You can control the number of digits displayed by highlighting cell A4 and selecting CELLS in the FORMAT menu. Highlight Number and enter the number of decimal places you desire.

	A	B
1	Using Excel GOAL SEEK	
2		
3	x	x(2x+0.0010)^2
4	0.01	4.41E-06

Before executing GOAL SEEK

	A	B
1	Using Excel GOAL SEEK	
2		
3	x	x(2x+0.0010)^2
4	0.0009454	7.90E-09

After executing GOAL SEEK

How Would You Do It?

6-25. *Thermometric titration.* Chemical reactions liberate or absorb heat. One physical property that can be measured to determine the equivalence point in a titration is temperature rise. The figure shows temperature changes measured by an electrical resistor during the titration of a mixture of the bases "tris" and pyridine by HCl:

$$\text{tris} + H^+ \longrightarrow \text{trisH}^+ \qquad (1)$$

$$\text{pyridine} + H^+ \longrightarrow \text{pyridineH}^+ \qquad (2)$$

Prior to the addition of H^+, the temperature drifted down from point A to point B. Then 1.53 M HCl was added at a rate of 0.198 mL/min to the 40-mL solution between points B and E. The temperature rose rapidly during Reaction 1 between points B and C and less rapidly during Reaction 2 between points C and D. At point D, the pyridine was used up, and the temperature of the solution drifted down steadily as more HCl continued to be added between points D and E. After point E, the flow of HCl was stopped, but the temperature continued to drift down.

(a) How many mmol of tris and how many mmol of pyridine were present?

(b) Reaction 1 liberates 47.5 kJ/mol. Does Reaction 2 absorb or liberate heat?

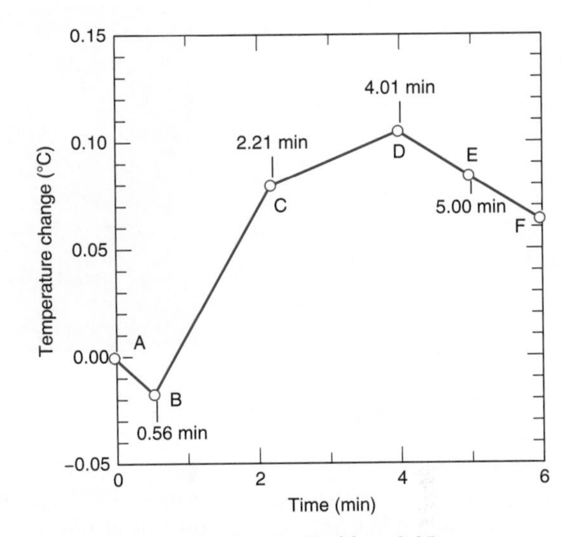

Thermometric titration for Problem 6-25.
[Data from L. D. Hansen, D. Kenney, W. M. Litchman and E. A. Lewis, *J. Chem. Ed.* **1971**, *48*, 851.]

Notes and References

1. W. B. Guenther, "Supertitrations: High-Precision Methods," *J. Chem. Ed.* **1988**, *65*, 1097.

2. E. A. Butler and E. H. Swift, "Gravimetric Titrimetry: A Neglected Technique," *J. Chem. Ed.* **1972**, *49*, 425.

3. R. W. Ramette, "In Support of Weight Titrations," *J. Chem. Ed.* **2004**, *81*, 1715.

The Geologic Time Scale and Gravimetric Analysis

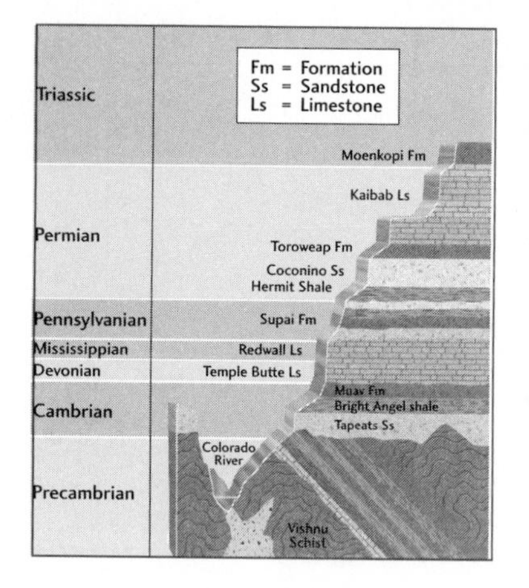

Layers of rock exposed in the Grand Canyon by the erosive action of the Colorado River provide a window on a billion years of Earth's history. [Left: Adapted from F. Press, R. Siever, J. Grotzinger, and T. H. Jordan, *Understanding Earth*, 4th ed. (New York: W. H. Freeman and Company, 2004). Right: Carol Polich/Lonely Planet Images.]

In the 1800s, geologists understood that new layers *(strata)* of rock are deposited on top of older layers. Characteristic fossils in each layer helped geologists to identify strata from the same geologic era all around the world. However, the actual age of each layer was unknown.

In 1910, Arthur Holmes, a 20-year-old geology student at Imperial College in London, used radioactive decay to measure the age of minerals. Physicists had discovered that U decays with a half-life of 4.5 billion years and suspected that the final product was Pb. Holmes conjectured that when a U-containing mineral crystallized, it should be relatively free of impurities. Once the mineral solidified, Pb would begin to accumulate. The ratio Pb/U is a "clock" giving the age of the mineral.

Holmes measured U by its rate of production of radioactive Rn gas. To measure Pb, he *fused* each mineral in borax (dissolving it in molten borax, Section 2-10), dissolved the fused mass in acid, and quantitatively precipitated milligram quantities of $PbSO_4$, which he weighed. The nearly constant ratio Pb/U = 0.045 g/g in 15 minerals from one stratum was consistent with the hypotheses that Pb is the end product of radioactive decay and that little Pb had been present when the minerals crystallized.

Prior to Holmes, the most widely accepted estimate of the age of Earth—from Lord Kelvin—was 100 million years. Holmes found much greater ages, which are shown in the table.

Geologic ages deduced by Holmes in 1911

Geologic period	Pb/U (g/g)	Millions of years	Today's accepted value
Carboniferous	0.041	340	362–330
Devonian	0.045	370	380–362
Silurian	0.053	430	443–418
Precambrian	0.125–0.20	1 025–1 640	900–2 500

SOURCE: C. Lewis, *The Dating Game* (Cambridge: Cambridge University Press, 2000); A. Holmes, *Proc. Royal Soc. London A* **1911**, *85*, 248.

CHAPTER 7

Gravimetric and Combustion Analysis

In **gravimetric analysis**, the mass of a product is used to calculate the quantity of original analyte. In the early 1900s, Nobel Prize–winning gravimetric analysis by T. W. Richards and his colleagues measured the atomic mass of Ag, Cl, and N to six-figure accuracy and formed the basis for accurate atomic mass determinations. In **combustion analysis**, a sample is burned in excess oxygen and the products are measured. Combustion is typically used to measure C, H, N, S, and halogens in organic compounds.

Gravimetric analysis has been largely replaced by instrumental methods of analysis, which are faster and less labor intensive. However, gravimetric determinations executed by a skilled analyst remain among the most accurate methods available to produce standards for instrumental analysis. Students are still exposed to some gravimetric analysis early in their laboratory career because gravimetric procedures demand excellent laboratory technique to produce accurate and precise results.

Gravimetric procedures were the mainstay of chemical analyses of ores and industrial materials in the eighteenth and nineteenth centuries, long before the chemical basis for the procedures was understood.

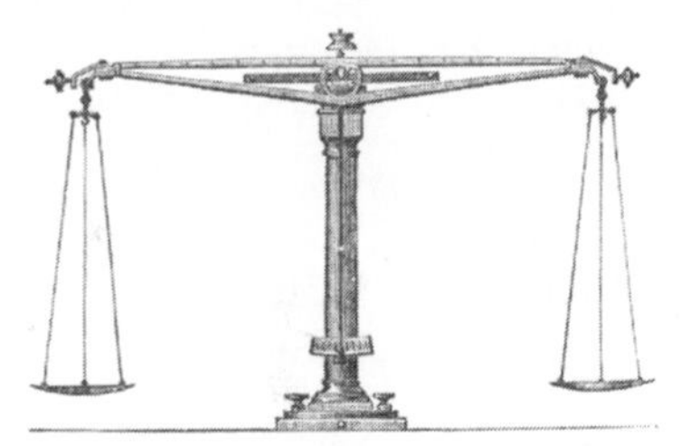

Nineteenth-century balance reproduced from *Fresenius' Quantitative Chemical Analysis,* 2nd American ed., 1881.

7-1 Examples of Gravimetric Analysis

An important industrial gravimetric analysis is the Rose Gottlieb method for measuring fat in food. First, a weighed sample is dissolved in an appropriate manner to solubilize the protein. Ammonia and ethanol are then added to break up microscopic droplets of fat, which are extracted into an organic solvent. Proteins and carbohydrates remain in the aqueous phase. After the organic phase has been separated from the aqueous phase, the organic phase is evaporated to dryness at 102°C and the dried residue is weighed. The residue consists of the fats from the food.

A simple gravimetric analysis that you might encounter is the determination of Cl^- by precipitation with Ag^+:

$$Ag^+ + Cl^- \longrightarrow AgCl(s)$$

The mass of AgCl product tells us how many moles of AgCl were produced. For every mole of AgCl produced, there must have been 1 mol of Cl^- in the unknown solution.

Example **A Simple Gravimetric Calculation**

A 10.00-mL solution containing Cl^- was treated with excess $AgNO_3$ to precipitate 0.436 8 g of AgCl (FM 143.321). What was the molarity of Cl^- in the unknown?

Table 7-1 Representative gravimetric analyses

Species analyzed	Precipitated form	Form weighed	Some interfering species
K^+	$KB(C_6H_5)_4$	$KB(C_6H_5)_4$	NH_4^+, Ag^+, Hg^{2+}, Tl^+, Rb^+, Cs^+
Mg^{2+}	$Mg(NH_4)PO_4 \cdot 6H_2O$	$Mg_2P_2O_7$	Many metals except Na^+ and K^+
Ca^{2+}	$CaC_2O_4 \cdot H_2O$	$CaCO_3$ or CaO	Many metals except Mg^{2+}, Na^+, K^+
Ba^{2+}	$BaSO_4$	$BaSO_4$	Na^+, K^+, Li^+, Ca^{2+}, Al^{3+}, Cr^{3+}, Fe^{3+}, Sr^{2+}, Pb^{2+}, NO_3^-
Cr^{3+}	$PbCrO_4$	$PbCrO_4$	Ag^+, NH_4^+
Mn^{2+}	$Mn(NH_4)PO_4 \cdot H_2O$	$Mn_2P_2O_7$	Many metals
Fe^{3+}	$Fe(HCO_2)_3$	Fe_2O_3	Many metals
Co^{2+}	Co(1-nitroso-2-naphtholate)$_2$	$CoSO_4$ (by reaction with H_2SO_4)	Fe^{3+}, Pd^{2+}, Zr^{4+}
Ni^{2+}	Ni(dimethylglyoximate)$_2$	Same	Pd^{2+}, Pt^{2+}, Bi^{3+}, Au^{3+}
Cu^{2+}	CuSCN (after reduction to Cu^+)	CuSCN	NH_4^+, Pb^{2+}, Hg^{2+}, Ag^+
Zn^{2+}	$Zn(NH_4)PO_4 \cdot H_2O$	$Zn_2P_2O_7$	Many metals
Al^{3+}	Al(8-hydroxyquinolate)$_3$	Same	Many metals
Sn^{4+}	Sn(cupferron)$_4$	SnO_2	Cu^{2+}, Pb^{2+}, As(III)
Pb^{2+}	$PbSO_4$	$PbSO_4$	Ca^{2+}, Sr^{2+}, Ba^{2+}, Hg^{2+}, Ag^+, HCl, HNO_3
NH_4^+	$NH_4B(C_6H_5)_4$	$NH_4B(C_6H_5)_4$	K^+, Rb^+, Cs^+
Cl^-	AgCl	AgCl	Br^-, I^-, SCN^-, S^{2-}, $S_2O_3^{2-}$, CN^-
Br^-	AgBr	AgBr	Cl^-, I^-, SCN^-, S^{2-}, $S_2O_3^{2-}$, CN^-
I^-	AgI	AgI	Cl^-, Br^-, SCN^-, S^{2-}, $S_2O_3^{2-}$, CN^-
SCN^-	CuSCN	CuSCN	NH_4^+, Pb^{2+}, Hg^{2+}, Ag^+
CN^-	AgCN	AgCN	Cl^-, Br^-, I^-, SCN^-, S^{2-}, $S_2O_3^{2-}$.
F^-	$(C_6H_5)_3SnF$	$(C_6H_5)_3SnF$	Many metals (except alkali metals), SiO_4^{4-}, CO_3^{2-}
ClO_4^-	$KClO_4$	$KClO_4$	
SO_4^{2-}	$BaSO_4$	$BaSO_4$	Na^+, K^+, Li^+, Ca^{2+}, Al^{3+}, Cr^{3+}, Fe^{3+}, Sr^{2+}, Pb^{2+}, NO_3^-
PO_4^{3-}	$Mg(NH_4)PO_4 \cdot 6H_2O$	$Mg_2P_2O_7$	Many metals except Na^+, K^+
NO_3^-	Nitron nitrate	Nitron nitrate	ClO_4^-, I^-, SCN^-, CrO_4^{2-}, ClO_3^-, NO_2^-, Br^-, $C_2O_4^{2-}$

SOLUTION A precipitate weighing 0.463 8 g contains

$$\frac{0.436\,8\ \cancel{\text{g AgCl}}}{143.321\ \cancel{\text{g AgCl}}/\text{mol AgCl}} = 3.048 \times 10^{-3}\ \text{mol AgCl}$$

Because 1 mol of AgCl contains 1 mol of Cl^-, there must have been 3.048×10^{-3} mol of Cl^- in the unknown. The molarity of Cl^- in the unknown is therefore

$$[Cl^-] = \frac{3.048 \times 10^{-3}\ \text{mol}}{0.010\,00\ \text{L}} = 0.304\,8\ \text{M}$$

Test Yourself A 25.00-mL solution containing NaCl plus KCl was treated with excess $AgNO_3$ to precipitate 0.436 8 g AgCl. What was the molarity of Cl^-? (**Answer:** 0.121 9 M)

Table 7-2 Common organic precipitating agents

Name	Structure	Some ions precipitated
Dimethylglyoxime	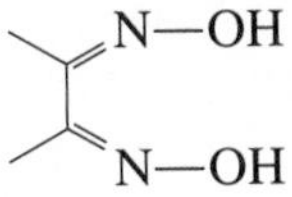 (structure: N—OH, N—OH)	Ni^{2+}, Pd^{2+}, Pt^{2+}
Cupferron	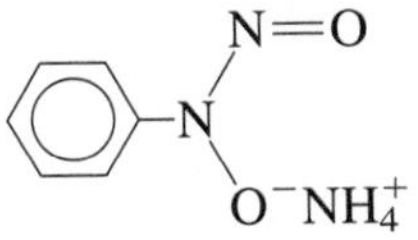 (structure: N=O, N, $O^-NH_4^+$)	Fe^{3+}, VO_2^+, Ti^{4+}, Zr^{4+}, Ce^{4+}, Ga^{3+}, Sn^{4+}
8-Hydroxyquinoline (oxine)	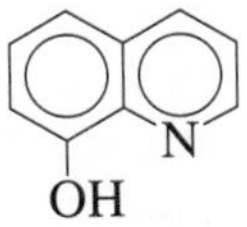 (structure: N, OH)	Mg^{2+}, Zn^{2+}, Cu^{2+}, Cd^{2+}, Pb^{2+}, Al^{3+}, Fe^{3+}, Bi^{3+}, Ga^{3+}, Th^{4+}, Zr^{4+}, UO_2^{2+}, TiO^{2+}
1-Nitroso-2-naphthol	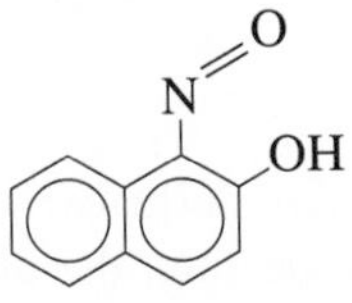 (structure: N=O, OH)	Co^{2+}, Fe^{3+}, Pd^{2+}, Zr^{4+}
Nitron	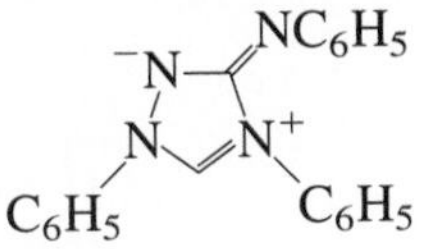 (structure: NC_6H_5, ^-N, N, N^+, C_6H_5, C_6H_5)	NO_3^-, ClO_4^-, BF_4^-, WO_4^{2-}
Sodium tetraphenylborate	$Na^+B(C_6H_5)_4^-$	K^+, Rb^+, Cs^+, NH_4^+, Ag^+, organic ammonium ions
Tetraphenylarsonium chloride	$(C_6H_5)_4As^+Cl^-$	$Cr_2O_7^{2-}$, MnO_4^-, ReO_4^-, MoO_4^{2-}, WO_4^{2-}, ClO_4^-, I_3^-

Representative analytical precipitations are listed in Table 7-1. Potentially interfering substances listed in the table may need to be removed prior to analysis. A few common organic **precipitants** (agents that cause precipitation) are listed in Table 7-2.

For those who are not familiar with drawing structures of organic compounds, Box 7-1 provides a primer.

Ask Yourself

7-A. A 50.00-mL solution containing NaBr was treated with excess $AgNO_3$ to precipitate 0.214 6 g of AgBr (FM 187.772).
(a) How many moles of AgBr product were isolated?
(b) What was the molarity of NaBr in the solution?

7-2 Precipitation

The ideal gravimetric precipitate should be insoluble, be easily filtered, and possess a known, constant composition. The precipitate should be stable when you heat it to remove the last traces of solvent. Although few substances meet these requirements, techniques described in this section help optimize the properties of precipitates.

Box 7-1 Shorthand for Organic Structures

Chemists and biochemists use simple conventions for drawing structures of carbon-containing compounds to avoid drawing every atom. Each vertex of a structure is understood to be a carbon atom, unless otherwise labeled. In the shorthand, we usually omit bonds from carbon to hydrogen. Carbon forms four chemical bonds. If you see carbon forming fewer than four bonds, the remaining bonds are assumed to go to hydrogen atoms that are not written. Here are some examples:

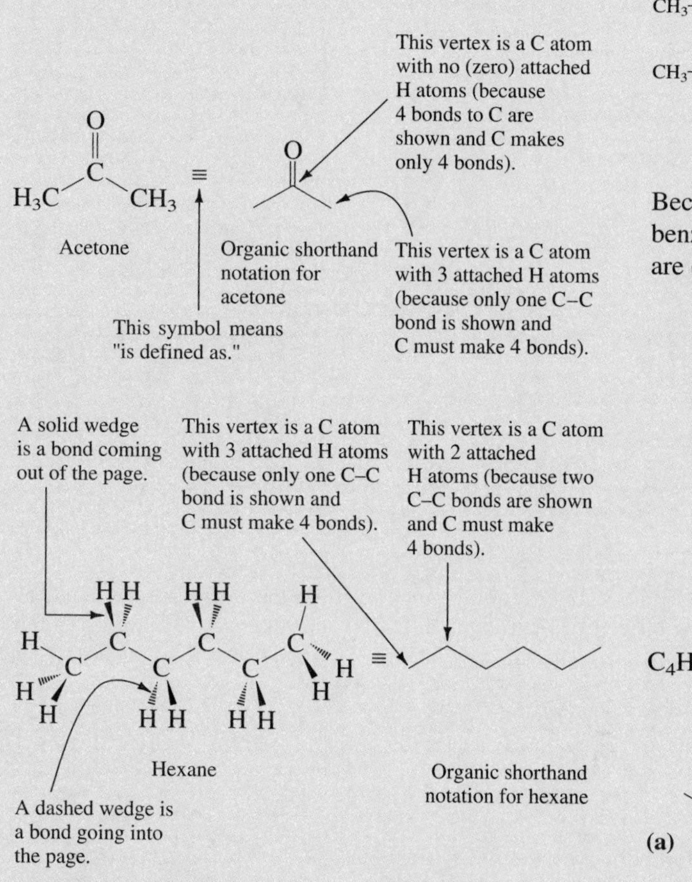

This vertex is a C atom with no (zero) attached H atoms (because 4 bonds to C are shown and C makes only 4 bonds).
Acetone
Organic shorthand notation for acetone
This vertex is a C atom with 3 attached H atoms (because only one C–C bond is shown and C must make 4 bonds).
This symbol means "is defined as."
A solid wedge is a bond coming out of the page.
This vertex is a C atom with 3 attached H atoms (because only one C–C bond is shown and C must make 4 bonds).
This vertex is a C atom with 2 attached H atoms (because two C–C bonds are shown and C must make 4 bonds).
Hexane
Organic shorthand notation for hexane
A dashed wedge is a bond going into the page.

Atoms other than carbon and hydrogen are always shown. Hydrogen atoms attached to atoms other than carbon are always shown. Oxygen and sulfur normally make two bonds. Nitrogen makes three bonds if it is neutral and four bonds if it is a cation. Here are some examples:

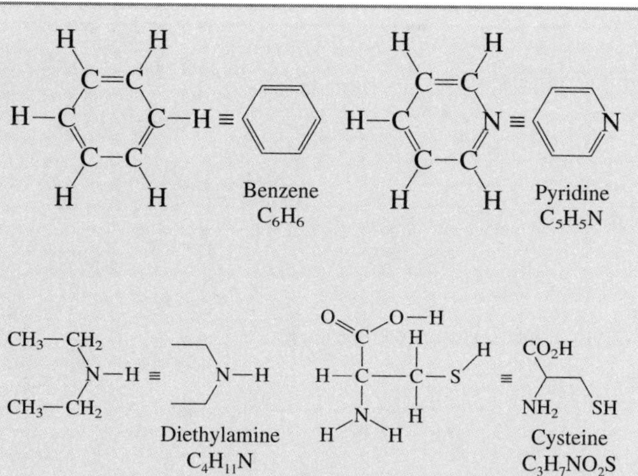

Benzene
C_6H_6
Pyridine
C_5H_5N
Diethylamine
$C_4H_{11}N$
Cysteine
$C_3H_7NO_2S$

Because of the two equivalent resonance structures of a benzene ring, the alternating single and double bonds are often replaced by a circle:

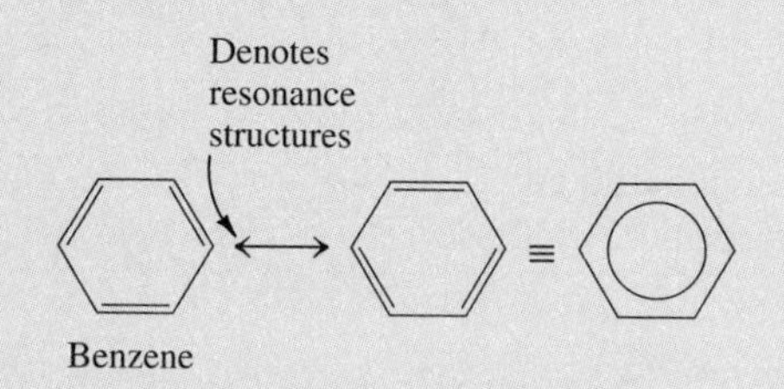

Denotes resonance structures
Benzene

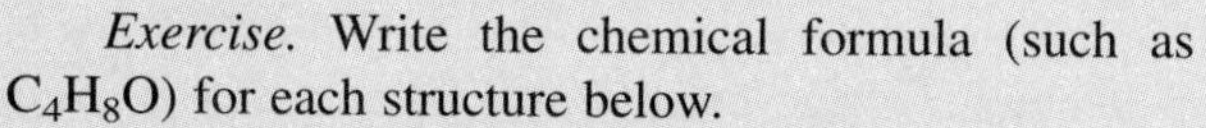

Exercise. Write the chemical formula (such as C_4H_8O) for each structure below.

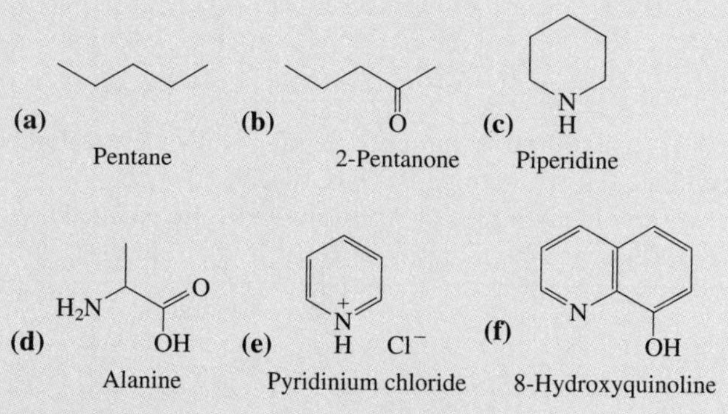

(a) Pentane
(b) 2-Pentanone
(c) Piperidine
(d) Alanine
(e) Pyridinium chloride
(f) 8-Hydroxyquinoline

Answers: **(a)** C_5H_{12}; **(b)** $C_5H_{10}O$; **(c)** $C_5H_{11}N$; **(d)** $C_3H_7NO_2$; **(e)** C_5H_6NCl; **(f)** C_9H_7NO

Precipitate particles should be large enough to be collected by filtration; they should not be so small that they clog or pass through the filter. Large crystals also have less surface area to which foreign species may become attached. At the other extreme is a *colloid*, whose particles are so small (~1–500 nm) that they pass through most filters (Figure 7-1 and Demonstration 7-1).

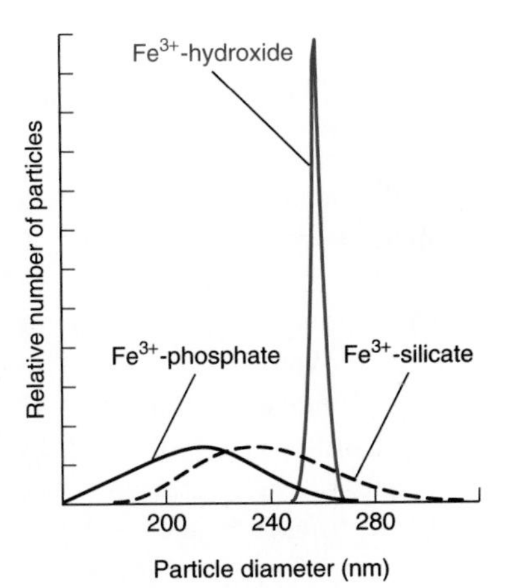

Figure 7-1 Particle size distributions of colloids formed when $FeSO_4$ was oxidized to Fe^{3+} in 10^{-4} M OH^- in the presence of phosphate (PO_4^{3-}), silicate (SiO_4^{4-}), or no added anions. [From M. L. Magnuson, D. A. Lytle, C. M. Frietch, and C. A. Kelty, *Anal. Chem.* **2001**, *73*, 4815.]

Crystal Growth

Crystallization occurs in two phases: *nucleation* and *particle growth.* In **nucleation**, dissolved molecules or ions form small crystalline aggregates capable of growing into larger particles. Nucleation tends to occur on preexisting surfaces that attract and hold solutes. Insoluble impurity particles in a liquid or scratches on a glass surface are potentially capable of initiating nucleation. In **particle growth**, solute molecules or ions add to an existing aggregate to form a crystal.

A solution containing more dissolved solute than should be present at equilibrium is said to be **supersaturated**. In a highly supersaturated solution, nucleation proceeds faster than particle growth and therefore creates tiny particles. In less concentrated solution, nucleation is slower, so the nuclei have a chance to grow into larger, more tractable particles. Techniques that promote particle growth include

1. Raising the temperature to increase solubility and thereby decrease supersaturation
2. Adding precipitant slowly with vigorous mixing to avoid high local supersaturation where the stream of precipitant first enters the analyte
3. Keeping the volume of solution large so that the concentrations of analyte and precipitant are low

Homogeneous Precipitation

In **homogeneous precipitation**, precipitant is generated slowly by a chemical reaction. This is beneficial because, when precipitation is slow, particle growth dominates over nucleation to give larger, purer particles that are easier to filter. When precipitation is rapid, nucleation tends to dominate over crystallization and the resulting particles are small and hard to filter. An example of homogeneous precipitation is the slow formation of Fe(III) formate by first decomposing urea in boiling water to slowly produce OH^-:

$$\underset{\text{Urea}}{H_2N{-}C({=}O){-}NH_2} + 3H_2O \xrightarrow{\text{heat}} CO_2 + 2NH_4^+ + 2OH^-$$

The OH^- reacts with formic acid to produce formate, which precipitates Fe(III):

$$\underset{\text{Formic acid}}{H{-}C({=}O){-}OH} + OH^- \longrightarrow \underset{\text{Formate}}{HCO_2^-} + H_2O$$

$$3HCO_2^- + Fe^{3+} \longrightarrow \underset{\text{Iron(III) formate (ferric formate)}}{Fe(HCO_2)_3 \cdot nH_2O(s)\downarrow}$$

Demonstration 7-1 Colloids and Dialysis

Colloids are particles with diameters in the approximate range 1–500 nm. They are larger than molecules, but too small to precipitate. Colloids remain in solution indefinitely, suspended by the Brownian motion (random movement) of solvent molecules.

To make a colloid, heat a beaker containing 200 mL of distilled water to 70°–90°C and leave an identical beaker of water at room temperature. Add 1 mL of 1 M $FeCl_3$ to each beaker and stir. The warm solution turns brown-red in a few seconds, whereas the cold solution remains yellow (Color Plate 2*a*). The yellow color is characteristic of low-molecular-mass Fe^{3+}compounds. The red color results from colloidal aggregates of Fe^{3+} ions held together by hydroxide, oxide, and some chloride ions. These particles have a molecular mass of $\sim 10^5$ and a diameter of $\sim$10 nm, and contain $\sim 10^3$ atoms of Fe.

To demonstrate the size of colloidal particles, we perform a **dialysis** experiment in which two solutions are separated by a *semipermeable membrane.* The membrane has pores through which small molecules, but not large molecules and colloids, can diffuse. Cellulose dialysis tubing (such as catalog number 3787 from A. H. Thomas Co.) has 1–5 nm pores.

Pour some of the brown-red colloidal Fe solution into a dialysis tube knotted at one end; then tie off the other end. Drop the tube into a flask of distilled water to show that the color remains entirely within the bag even after several days (Color Plates 2*b* and 2*c*). For comparison, leave an identical bag containing a dark blue solution of 1 M $CuSO_4 \cdot 5H_2O$ in another flask. Cu^{2+} diffuses out of the bag, and the solution in the flask becomes light blue in 24 h. Alternatively, the yellow food coloring, tartrazine, can be used in place of Cu^{2+}. If dialysis is conducted in hot water, it is completed in one class period.

Dialysis is used to treat patients suffering from kidney failure. Blood is run over a dialysis membrane having a very large surface area. Small metabolic waste products in the blood diffuse across the membrane and are diluted into a large volume of liquid going out as waste. Large proteins, which are a necessary part of the blood plasma, cannot cross the membrane and are retained in the blood.

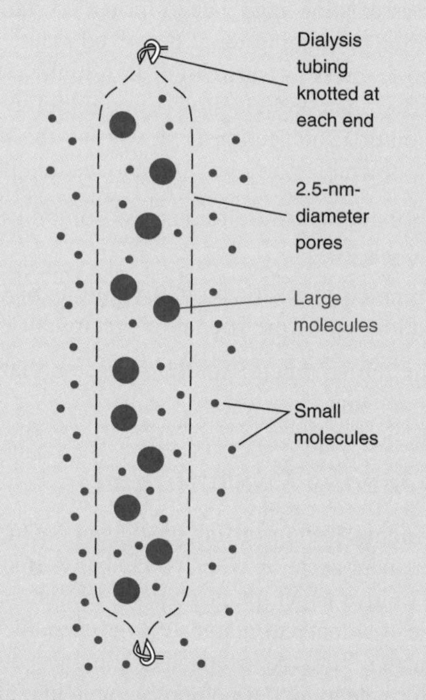

Large molecules remain trapped inside a dialysis bag, whereas small molecules diffuse through the membrane in both directions.

Table 7-3 lists several common reagents for homogeneous precipitation.

Precipitation in the Presence of Electrolyte

An *electrolyte* is a compound that dissociates into ions when it dissolves. We say that the electrolyte ionizes when it dissolves.

Ionic compounds are usually precipitated in the presence of added *electrolyte.* To understand why, consider how tiny crystallites *coagulate* (come together) into larger crystals. We will illustrate the case of AgCl, which is commonly formed in the presence of 0.1 M HNO_3.

Figure 7-2 shows a colloidal particle of AgCl growing in a solution containing excess Ag^+, H^+, and NO_3^-. The particle has an excess positive charge due to

Table 7-3 Common reagents for homogeneous precipitation

Precipitant	Reagent	Reaction	Some elements precipitated
OH^-	Urea	$(H_2N)_2CO + 3H_2O \longrightarrow CO_2 + 2NH_4^+ + 2OH^-$	Al, Ga, Th, Bi, Fe, Sn
S^{2-}	Thioacetamide[a]	$CH_3C(=S)NH_2 + H_2O \longrightarrow CH_3C(=O)NH_2 + H_2S$	Sb, Mo, Cu, Cd
SO_4^{2-}	Sulfamic acid	$H_3\overset{+}{N}SO_3^- + H_2O \longrightarrow NH_4^+ + SO_4^{2-} + H^+$	Ba, Ca, Sr, Pb
$C_2O_4^{2-}$	Dimethyl oxalate	$CH_3OC(=O)C(=O)OCH_3 + 2H_2O \longrightarrow 2CH_3OH + C_2O_4^{2-} + 2H^+$	Ca, Mg, Zn
PO_4^{3-}	Trimethyl phosphate	$(CH_3O)_3P{=}O + 3H_2O \longrightarrow 3CH_3OH + PO_4^{3-} + 3H^+$	Zr, Hf

a. Hydrogen sulfide is volatile and toxic; it should be handled only in a well-vented hood. Thioacetamide is a carcinogen that should be handled with gloves. If thioacetamide contacts your skin, wash yourself thoroughly immediately. Leftover reagent is destroyed by heating at 50°C with 5 mol of NaOCl per mole of thioacetamide and then washing the products down the drain.

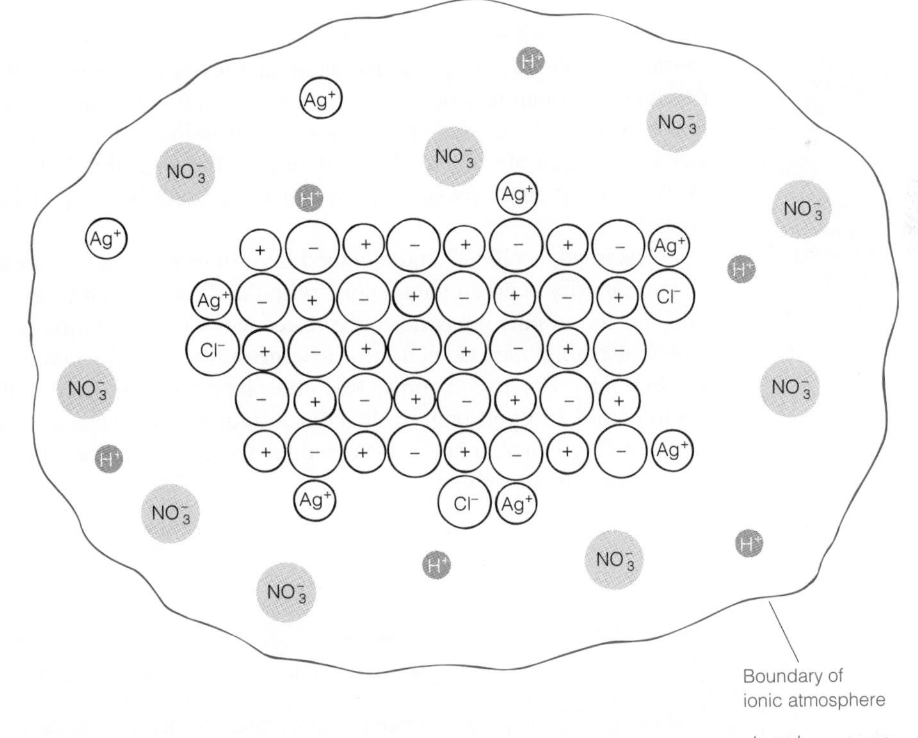

Figure 7-2 Colloidal particle of AgCl in a solution containing excess Ag^+, H^+, and NO_3^-. The particle has a net positive charge because of adsorbed Ag^+ ions. The region of solution surrounding the particle is called the *ionic atmosphere*. It has a net negative charge, because the particle attracts anions and repels cations.

Although it is common to find the excess common ion adsorbed on the crystal surface, it is also possible to find other ions selectively adsorbed. In the presence of citrate and sulfate, there is more citrate than sulfate adsorbed on $BaSO_4(s)$.

adsorption of extra silver ions on exposed chloride ions. (To be adsorbed means to be attached to the surface. In contrast, **absorption** entails penetration beyond the surface, to the inside.) The positively charged surface of the solid attracts anions and repels cations from the *ionic atmosphere* in the liquid surrounding the particle.

Colloidal particles must collide with one another to coalesce. However, the negatively charged ionic atmospheres of the particles repel one another. The particles must have enough kinetic energy to overcome electrostatic repulsion before they can coalesce. Heating promotes coalescence by increasing the particles' kinetic energy.

Increasing electrolyte concentration (HNO_3 for AgCl) decreases the volume of the ionic atmosphere and allows particles to approach closer together before repulsion becomes significant. Therefore, most gravimetric precipitations are done in the presence of electrolyte.

Digestion

Mother liquor is the solution from which a substance crystallized.

Digestion is the process of allowing a precipitate to stand in contact with the *mother liquor* for some period of time, usually with heating. Digestion promotes slow recrystallization of the precipitate. Particle size increases and impurities tend to be expelled from the crystal.

Purity

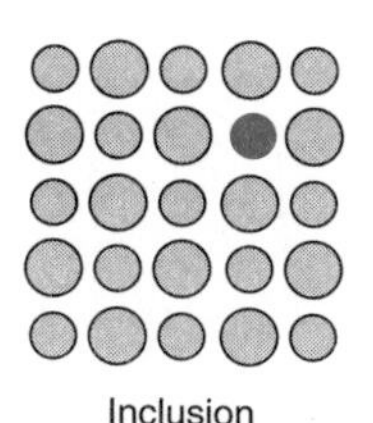

Inclusion

Occlusion (possibly containing solvent)

Adsorbed impurities are bound to the surface of a crystal. *Absorbed* impurities (within the crystal) are classified as *inclusions* or *occlusions*. Inclusions are impurity ions that randomly occupy sites in the crystal lattice normally occupied by ions that belong in the crystal. Inclusions are more likely when the impurity ion has a size and charge similar to those of one of the ions that belongs to the product. Occlusions are pockets of impurity that are literally trapped inside the growing crystal.

Adsorbed, occluded, and included impurities are said to be **coprecipitated**. That is, the impurity is precipitated along with the desired product, even though the solubility of the impurity has not been exceeded. Coprecipitation tends to be worst in colloidal precipitates (which have a large surface area), such as $BaSO_4$, $Al(OH)_3$, and $Fe(OH)_3$. Figure 7-3 shows that phosphate coprecipitated with calcium carbonate in coral is proportional to the concentration of phosphate in seawater. By measuring P/Ca in ancient coral, we can infer the concentration of phosphate in the sea at the time the coral lived.

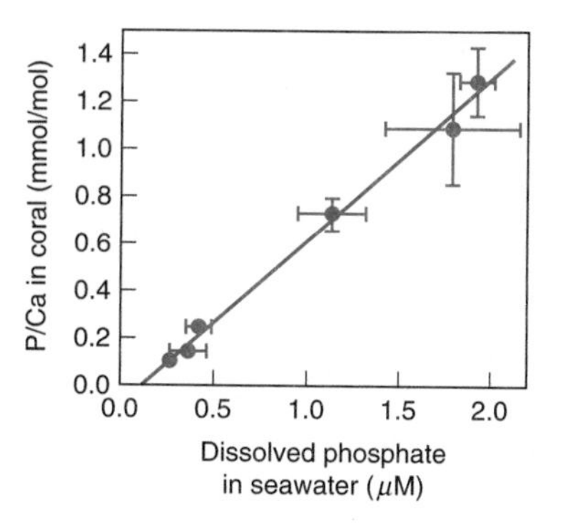

Figure 7-3 Coprecipitation of phosphate with calcium carbonate in the skeleton of coral. The skeleton is composed of the mineral aragonite ($CaCO_3$). The graph shows that the P/Ca ratio in modern coral is proportional to the concentration of phosphate in the water in which the coral is growing. From this graph and the measured P/Ca in 1.12×10^4-year-old fossil corals from the western Mediterranean Sea (not shown), we conclude that the phosphate concentration in the water was then more than twice as high as current values. [Data from P. Montagna, M. McCulloch, M. Taviani, C. Mazzoli, and B. Vendrell, *Science* **2006**, *312*, 1788.]

Table 7-4 Removal of occluded NO_3^- from $BaSO_4$ by reprecipitation

	$[NO_3^-]/[SO_4^-]$ in precipitate
Initial precipitate	0.279
First reprecipitation	0.028
Second reprecipitation	0.001

SOURCE: Data from H. Bao, *Anal. Chem.* **2006**, *78*, 304.

Some procedures call for washing away the mother liquor, redissolving the precipitate, and *reprecipitating* the product. In the second precipitation, the concentration of impurities is lower than in the first precipitation, and the degree of coprecipitation therefore tends to be lower (Table 7-4).

Occasionally, a trace component that is too dilute to be measured is intentionally concentrated by coprecipitation with a major component of the solution. The procedure is called **gathering**, and the precipitate used to collect the trace component is said to be a *gathering agent.* When the precipitate is dissolved in a small volume of solvent, the concentration of the trace component is high enough for accurate analysis.

Example of *gathering*: Se(IV) at concentrations as low as 25 ng/L is gathered by coprecipitation with $Fe(OH)_3$. Precipitate is then dissolved in a small volume of concentrated acid to obtain a more concentrated Se(IV) for analysis.

Question How many ppb is 25 ng/L?

Some impurities can be treated with a **masking agent**, which prevents them from reacting with the precipitant. In the gravimetric analysis of Be^{2+}, Mg^{2+}, Ca^{2+}, or Ba^{2+} with the reagent *N-p*-chlorophenylcinnamohydroxamic acid (designated RH), impurities such as Ag^+, Mn^{2+}, Zn^{2+}, Cd^{2+}, Hg^{2+}, Fe^{2+}, and Ga^{3+} are kept in solution by excess KCN.

$$\underset{\text{Analyte}}{Ca^{2+}} + 2RH \longrightarrow \underset{\text{Precipitate}}{CaR_2(s)\downarrow} + 2H^+$$

$$\underset{\text{Impurity}}{Mn^{2+}} + \underset{\text{Masking agent}}{6CN^-} \longrightarrow \underset{\text{Stays in solution}}{Mn(CN)_6^{4-}}$$

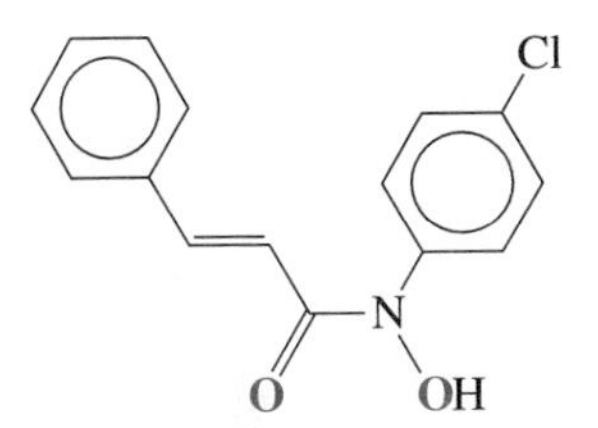

N-p-Chlorophenylcinnamohydroxamic acid (RH)
(The two oxygen atoms that bind to metal ions are **bold.**)

Impurities might collect on the product while it is standing in the mother liquor. This process is called *postprecipitation* and usually entails a supersaturated impurity that does not readily crystallize. An example is the crystallization of magnesium oxalate (MgC_2O_4) on calcium oxalate (CaC_2O_4).

Washing precipitate on a filter removes droplets of liquid containing excess solute. Some precipitates can be washed with water, but many require electrolyte to maintain coherence. For these precipitates, the ions in solution are required to neutralize the surface charge of the particles. If electrolyte is washed away with water, charged solid particles repel one another and the product breaks up. This breaking up, called **peptization**, results in loss of product through the filter. Silver chloride peptizes if washed with water, so it is washed with dilute HNO_3 instead. Volatile electrolytes including HNO_3, HCl, NH_4NO_3, NH_4Cl, and $(NH_4)_2CO_3$ are used for washing because they evaporate during drying.

Ammonium chloride, for example, decomposes as follows when it is heated:

$$NH_4Cl(s) \xrightarrow{\text{heat}} NH_3(g) + HCl(g)$$

Product Composition

The final product must have a known, stable composition. A **hygroscopic substance** is one that picks up water from the air and is therefore difficult to weigh accurately.

Many precipitates contain a variable quantity of water and must be dried under conditions that give a known (possibly zero) stoichiometry of H_2O.

Ignition (strong heating) is used to change the chemical form of some precipitates that do not have a constant composition after drying at moderate temperatures. For example, $Fe(HCO_2)_3 \cdot nH_2O$ is ignited at 850°C for 1 h to give Fe_2O_3, and $Mg(NH_4)PO_4 \cdot 6H_2O$ is ignited at 1 100°C to give $Mg_2P_2O_7$.

In **thermogravimetric analysis**, a sample is heated, and its mass is measured as a function of temperature. Figure 7-4 shows how the composition of calcium salicylate changes in four stages:

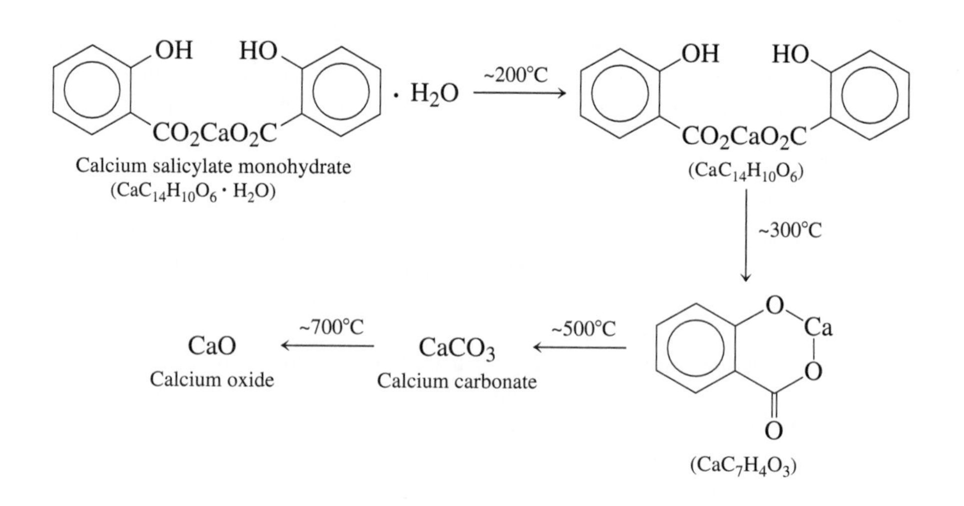

The composition of the product depends on the temperature and duration of heating.

Figure 7-4 Thermogravimetric curve for calcium salicylate. [From G. Liptay, ed., *Atlas of Thermoanalytical Curves* (London: Heyden and Son, 1976).]

Ask Yourself

7-B. See if you have digested this section by answering the following questions.
(a) What is the difference between absorption and adsorption?
(b) How is an inclusion different from an occlusion?
(c) What are desirable properties of a gravimetric precipitate?
(d) Why is high supersaturation undesirable in a gravimetric precipitation?
(e) How can you decrease supersaturation during a precipitation?
(f) Why are many ionic precipitates washed with electrolyte solution instead of pure water?
(g) Why is it less desirable to wash AgCl precipitate with aqueous $NaNO_3$ than with HNO_3 solution?
(h) Why would a reprecipitation be employed in a gravimetric analysis?
(i) What is done in thermogravimetric analysis?

7-3 Examples of Gravimetric Calculations

We now illustrate how to relate the mass of a gravimetric precipitate to the quantity of original analyte. *The general approach is to relate the moles of product to the moles of reactant.*

Example Relating Mass of Product to Mass of Reactant

The piperazine content of an impure commercial material can be determined by precipitating and weighing piperazine diacetate:

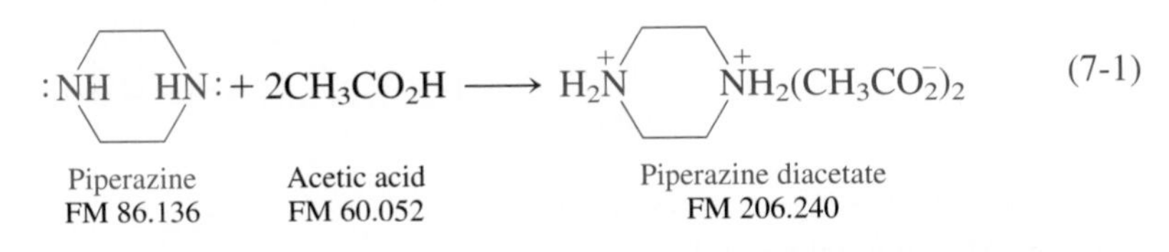

(7-1)

If you were performing this analysis, it would be important to determine that impurities in the piperazine are not also precipitated.

In one experiment, 0.312 6 g of sample was dissolved in 25 mL of acetone, and 1 mL of acetic acid was added. After 5 min, the precipitate was filtered, washed with acetone, dried at 110°C, and found to be 0.712 1 g. Find the wt% piperazine in the commercial material.

Reminder:

$$\text{wt\%} = \frac{\text{mass of analyte}}{\text{mass of unknown}} \times 100$$

SOLUTION We cannot convert grams of starting sample to moles of piperazine because the sample is not pure material. However, for each mole of piperazine in the impure material, 1 mol of product is formed.

$$\text{moles of piperazine} = \text{moles of product} = \frac{0.712\ 1\ \cancel{\text{g product}}}{206.240\ \frac{\cancel{\text{g product}}}{\text{mol product}}}$$
$$= 3.453 \times 10^{-3}\ \text{mol}$$

This many moles of piperazine corresponds to

grams of piperazine =

$$(3.453 \times 10^{-3}\ \cancel{\text{mol piperazine}})\left(86.136\ \frac{\text{g piperazine}}{\cancel{\text{mol piperazine}}}\right) = 0.297\ 4\ \text{g}$$

So, of the 0.312 6 g sample, 0.297 4 g is piperazine. Therefore

$$\text{wt\% piperazine in analyte} = \frac{0.297\ 4\ \text{g piperazine}}{0.312\ 6\ \text{g unknown}} \times 100 = 95.14\%$$

Test Yourself What is the wt% of piperazine if 0.288 g of commercial product gave 0.555 g of precipitate? (**Answer:** 80.5%)

Example When the Stoichiometry Is Not 1:1

Solid residue weighing 8.444 8 g from an aluminum refining process was dissolved in acid to give Al(III) in solution. The solution was treated with 8-hydroxyquinoline to precipitate (8-hydroxyquinoline)$_3$Al, which was ignited to give Al_2O_3 weighing 0.855 4 g. Find the weight percent of Al in the original mixture.

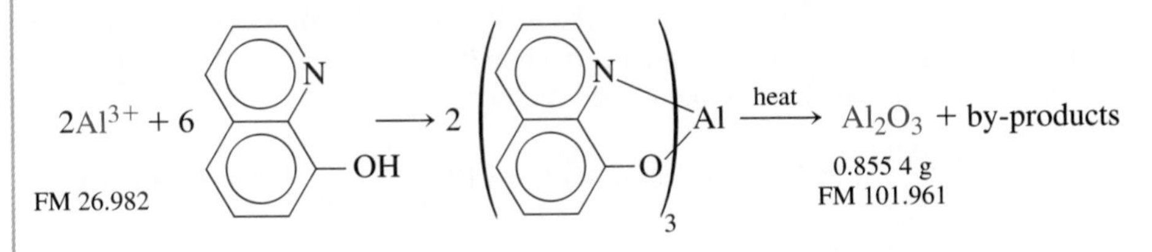

A note from Dan: Whether or not I show it, I keep at least one extra, insignificant figure in my calculations and do not round off until the final answer. Usually, I keep all the digits in my calculator.

SOLUTION Each mole of product (Al_2O_3) contains two moles of Al. The mass of product tells us the moles of product, and from this we can find the moles of Al. The moles of product are (0.855 4 g)/(101.961 g/mol) = 0.008 389_5 mol Al_2O_3. Because each mole of product contains two moles of Al, there must have been

$$\text{moles of Al in unknown} = \frac{2 \text{ mol Al}}{\cancel{\text{mol Al}_2\text{O}_3}} \times 0.008\ 389_5\ \cancel{\text{mol Al}_2\text{O}_3} = 0.016\ 77_9 \text{ mol Al}$$

The mass of Al is (0.016 77_9 mol)(26.982 g/mol) = 0.452 7_3 g Al. The weight percent of Al in the unknown is

$$\text{wt\% Al} = \frac{0.452\ 7_3 \text{ g Al}}{8.444\ 8 \text{ g unknown}} \times 100 = 5.361\%$$

Test Yourself Residue weighing 10.232 g gave 1.023 g Al_2O_3 after ignition. Find wt% Al in the original residue. (**Answer:** 5.292%)

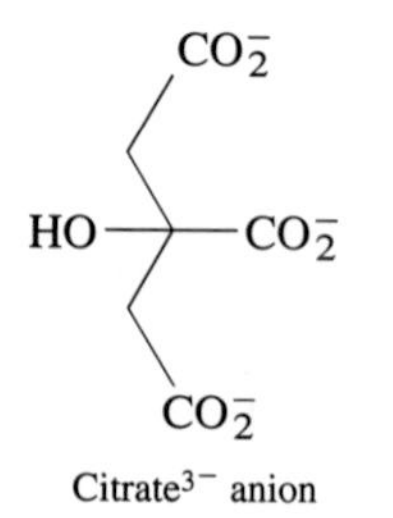

Citrate^{3-} anion

Example Calculating How Much Precipitant to Use

(a) To measure the nickel content in steel, the steel is dissolved in 12 M HCl and neutralized in the presence of citrate ion, which binds iron and keeps it in solution. The slightly basic solution is warmed, and dimethylglyoxime (DMG) is added to precipitate the red DMG-nickel complex. The product is filtered, washed with cold water, and dried at 110°C.

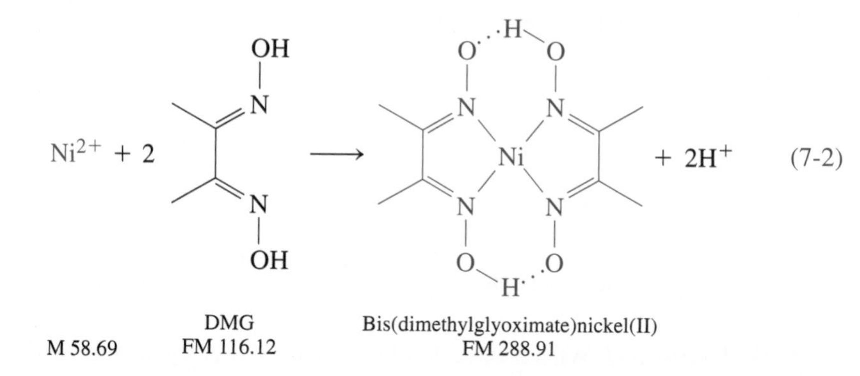

If the nickel content is known to be near 3 wt% and you wish to analyze 1.0 g of the steel, what volume of 1.0 wt% DMG in alcohol solution should be used to give a 50% excess of DMG for the analysis? Assume that the density of the alcohol solution is 0.79 g/mL.

1.0 wt% DMG means

$$\frac{1.0 \text{ g DMG}}{100 \text{ g solution}}$$

Density means

$$\frac{\text{grams of solution}}{\text{milliliters of solution}}$$

SOLUTION Our strategy is to estimate the moles of Ni in 1.0 g of steel. Equation 7-2 tells us that 2 mol of DMG are required for each mole of Ni. After finding the required number of moles of DMG, we will multiply it by 1.5 to get a 50% excess to be sure we have enough.

The Ni content of the steel is about 3%, so 1.0 g of steel contains about (0.03)(1.0 g) = 0.03 g of Ni, which corresponds to (0.03 g Ni)/(58.69 g/mol Ni) = 5.1×10^{-4} mol Ni. This amount of Ni requires

$$2\left(\frac{\cancel{\text{mol DMG}}}{\cancel{\text{mol Ni}}}\right)(5.1 \times 10^{-4}\ \cancel{\text{mol Ni}})\left(116.12\frac{\text{g DMG}}{\cancel{\text{mol DMG}}}\right) = 0.12 \text{ g DMG}$$

A 50% excess of DMG would be (1.5)(0.12 g) = 0.18 g.

The DMG solution is 1.0 wt%, which means that there are 0.010 g of DMG per gram of solution. The required mass of solution is

$$\left(\frac{0.18 \text{ g DMG}}{0.010 \text{ g DMG/g solution}}\right) = 18 \text{ g solution}$$

The volume of solution is found from the mass of solution and the density:

$$\text{volume} = \frac{\text{mass}}{\text{density}} = \frac{18 \text{ g solution}}{0.79 \text{ g solution/mL}} = 23 \text{ mL}$$

$$\text{density} = \frac{\text{mass}}{\text{volume}}$$

(b) If 1.163 4 g of steel gave 0.179 5 g of $Ni(DMG)_2$ precipitate, what is the weight percent of Ni in the steel?

SOLUTION Here is the strategy: From the mass of precipitate, we find the moles of precipitate. We know that 1 mol of precipitate comes from 1 mol of Ni in Equation 7-2. From the moles of Ni, we compute the mass of Ni and its weight percent in the steel:

Strategy:
1. Write balanced reaction
2. Calculate moles of product from mass of pure product
3. From balanced reaction, relate moles of unknown (reactant) to moles of product
4. From moles of unknown, compute mass or wt% or whatever is asked about the unknown

First, find the moles of precipitate in 0.179 5 g of precipitate:

$$\frac{0.179\,5 \text{ g Ni(DMG)}_2}{288.91 \text{ g Ni(DMG)}_2\text{/mol Ni(DMG)}_2} = 6.213 \times 10^{-4} \text{ mol Ni(DMG)}_2$$

There must have been 6.213×10^{-4} mol of Ni in the steel. The mass of Ni in the steel is $(6.213 \times 10^{-4} \text{ mol Ni})(58.69 \text{ g/mol Ni}) = 0.036\,46$ g and the weight percent of Ni in steel is

$$\text{wt\% Ni} = \frac{0.036\,46 \text{ g Ni}}{1.163\,4 \text{ g steel}} \times 100 = 3.134\%$$

Figure 7-5 reviews the problem-solving procedure, which should make sense to you now. The point is not to memorize an algorithm, but to assimilate the general approach of working backward from what is known to what is unknown.

Test Yourself If 2.376 g of steel gave 0.402 g of $Ni(DMG)_2$, what is the wt% Ni in the steel? (**Answer:** 3.44%)

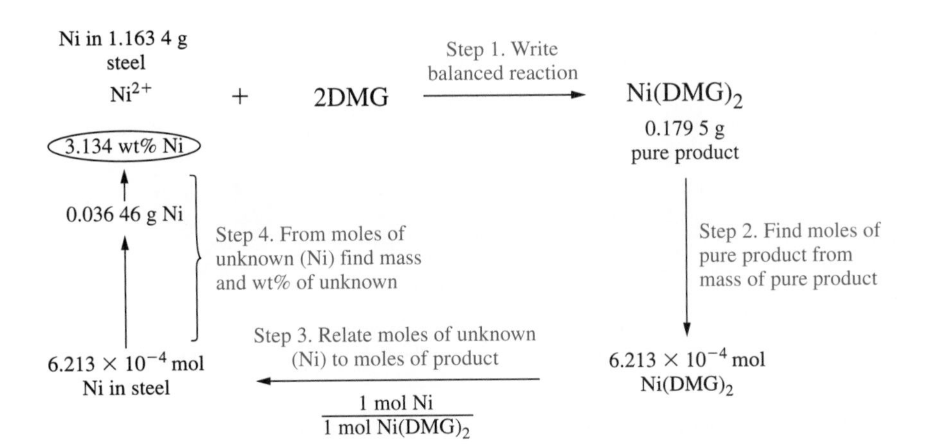

Figure 7-5 Steps taken to find wt% Ni in steel from mass of pure $Ni(DMG)_2$ product.

Ask Yourself

7-C. The element cerium, discovered in 1839 and named for the asteroid Ceres, is a major component of flint lighters. To find the Ce^{4+} content of a solid, an analyst dissolved 4.37 g of the solid and treated it with excess iodate to precipitate $Ce(IO_3)_4$. The precipitate was collected, washed, dried, and ignited to produce 0.104 g of CeO_2.

$$Ce^{4+} + 4IO_3^- \longrightarrow Ce(IO_3)_4(s) \xrightarrow{\text{heat}} \underset{\text{FM 172.115}}{CeO_2(s)}$$

(a) How much cerium is contained in 0.104 g of CeO_2?
(b) What was the weight percent of Ce in the original solid?

7-4 Combustion Analysis

A historically important form of gravimetric analysis is *combustion analysis*, used to determine the carbon and hydrogen content of organic compounds burned in excess O_2. Modern combustion analyzers use thermal conductivity, infrared absorption, or electrochemical methods to measure the products.

Gravimetric Combustion Analysis

In gravimetric combustion analysis (Figure 7-6), partly combusted product is passed through catalysts such as Pt gauze, CuO, PbO_2, or MnO_2 at elevated temperature to complete the oxidation to CO_2 and H_2O. The products are flushed through a chamber containing P_4O_{10} ("phosphorus pentoxide"), which absorbs water, and then through a chamber of Ascarite (NaOH on asbestos), which absorbs CO_2. The increase in mass of each chamber tells how much hydrogen and carbon, respectively, were initially present. A guard tube prevents atmospheric H_2O or CO_2 from entering the chambers from the reverse direction.

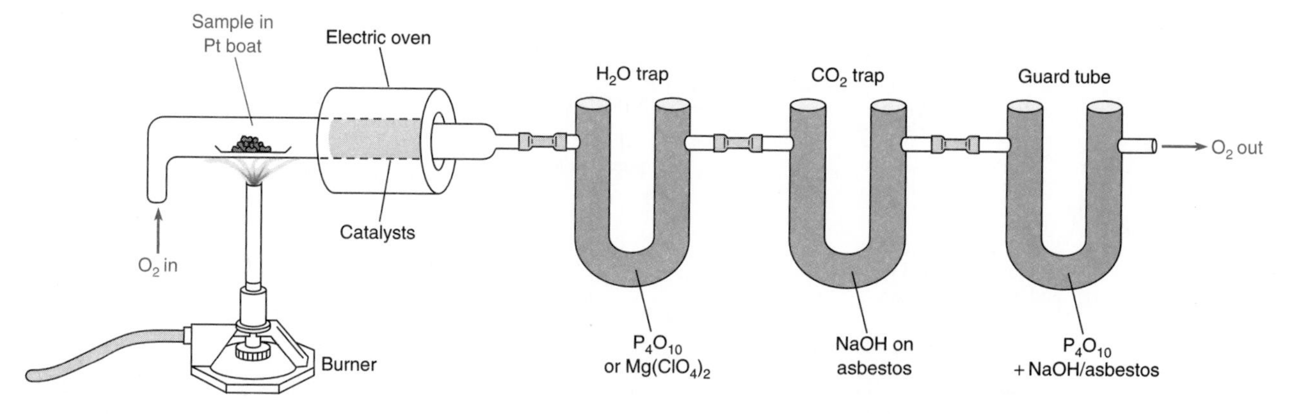

Figure 7-6 Gravimetric combustion analysis for carbon and hydrogen.

Example **Combustion Analysis Calculations**

A compound weighing 5.714 mg produced 14.414 mg of CO_2 and 2.529 mg of H_2O upon combustion. Find the weight percent of C and H in the sample.

SOLUTION One mole of CO_2 contains one mole of carbon. Therefore

$$\text{moles of C in sample} = \text{moles of } CO_2 \text{ produced}$$
$$= \frac{14.414 \times 10^{-3} \cancel{\text{g } CO_2}}{44.010 \cancel{\text{g } CO_2}/\text{mol}} = 3.275 \times 10^{-4} \text{ mol}$$

$$\text{mass of C in sample} = (3.275 \times 10^{-4} \cancel{\text{mol C}})\left(12.010\,7 \frac{\text{g}}{\cancel{\text{mol C}}}\right) = 3.934 \text{ mg}$$

$$\text{wt\% C} = \frac{3.934 \text{ mg C}}{5.714 \text{ mg sample}} \times 100 = 68.84\%$$

One mole of H_2O contains two moles of H. Therefore

$$\text{moles of H in sample} = 2(\text{moles of } H_2O \text{ produced})$$
$$= 2\left(\frac{2.529 \times 10^{-3} \cancel{\text{g } H_2O}}{18.015 \cancel{\text{g } H_2O}/\text{mol}}\right) = 2.808 \times 10^{-4} \text{ mol}$$

$$\text{mass of H in sample} = (2.808 \times 10^{-4} \cancel{\text{mol H}})\left(1.007\,94 \frac{\text{g}}{\cancel{\text{mol H}}}\right) = 2.830 \times 10^{-4} \text{ g}$$

$$\text{wt\% H} = \frac{0.283\,0 \text{ mg H}}{5.714 \text{ mg sample}} \times 100 = 4.952\%$$

Test Yourself A sample weighing 6.603 mg produced 2.603 mg of H_2O by combustion. Find wt% H in the sample. (**Answer:** 4.411%)

Combustion Analysis Today

Figure 7-7 shows how C, H, N, and S are measured in a single operation. An accurately weighed 2-mg sample is sealed in a tin or silver capsule. The analyzer is swept

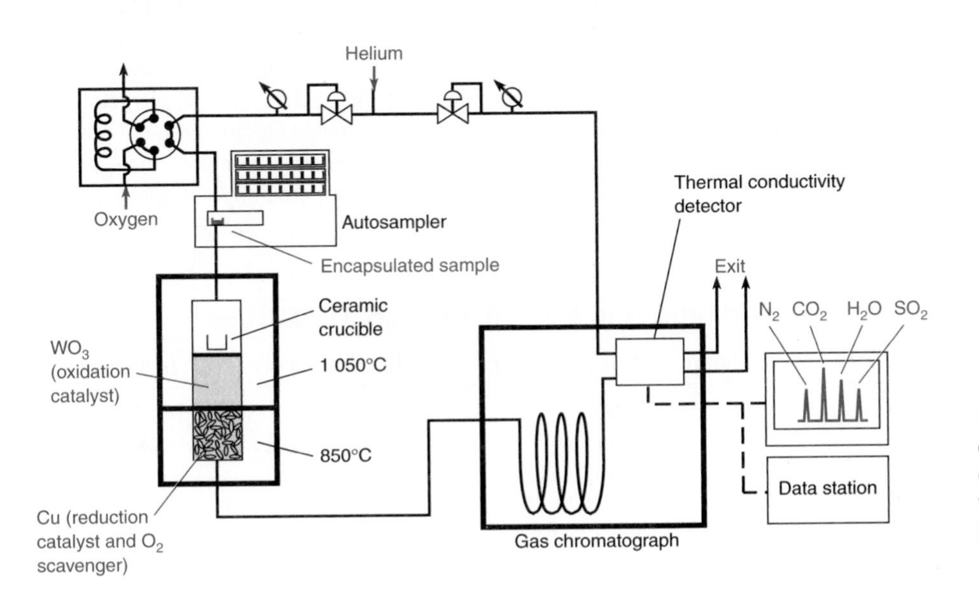

Figure 7-7 Diagram of C, H, N, S elemental analyzer that uses gas chromatographic separation and thermal conductivity detection. [From E. Pella, *Am. Lab.* August 1990, 28.]

with He gas that has been treated to remove traces of O_2, H_2O, and CO_2. At the start of a run, a measured excess volume of O_2 is added to the He stream. Then the sample capsule is dropped into a preheated ceramic crucible, where the capsule melts and the sample is rapidly oxidized.

$$\text{C, H, N, S} \xrightarrow[O_2]{1\,050°C} CO_2(g) + H_2O(g) + N_2(g) + \underbrace{SO_2(g) + SO_3(g)}_{95\%\ SO_2}$$

Elemental analyzers use an *oxidation catalyst* to complete the oxidation of sample and a *reduction catalyst* to carry out any required reduction and to remove excess O_2.

Products pass through a hot WO_3 catalyst to complete the combustion of carbon to CO_2. In the next zone, metallic Cu at 850°C reduces SO_3 to SO_2 and removes excess O_2:

$$Cu + SO_3 \xrightarrow{850°C} SO_2 + CuO(s)$$

$$Cu + \tfrac{1}{2}O_2 \xrightarrow{850°C} CuO(s)$$

The mixture of CO_2, H_2O, N_2, and SO_2 is separated by gas chromatography, and each component is measured with a thermal conductivity detector described in Section 22-1. Another common instrument uses infrared absorbance to measure CO_2, H_2O, and SO_2 and thermal conductivity to measure N_2.

A key to successful analysis in Figure 7-7 is *dynamic flash combustion*, which creates a short burst of gaseous products, instead of slowly bleeding products out over several minutes. Chromatographic separation requires that the whole sample be injected at once. Otherwise, the injection zone is so broad that products cannot be separated.

The Sn capsule is oxidized to SnO_2, which

1. Liberates heat to vaporize and crack (decompose) sample
2. Uses available oxygen immediately
3. Ensures that sample oxidation occurs in gas phase
4. Acts as an oxidation catalyst

In dynamic flash combustion, the sample is encapsulated in tin and dropped into the preheated furnace shortly after the flow of a 50 vol% O_2:50 vol% He mixture is started. The Sn capsule melts at 235°C and is instantly oxidized to SnO_2, liberating 594 kJ/mol and heating the sample to 1 700°–1 800°C. Because the sample is dropped in before very much O_2 is present, the sample decomposes prior to oxidation, which minimizes the formation of nitrogen oxides.

Oxygen analysis requires a different strategy. The sample is thermally decomposed (a process called **pyrolysis**) without adding oxygen. The gaseous products are passed through nickel-coated carbon at 1 075°C to convert oxygen from the analyte into CO (not CO_2). Other products include N_2, H_2, CH_4, and hydrogen halides. Acidic products are absorbed by NaOH-coated asbestos, and the remaining gases are separated and measured by gas chromatography with a thermal conductivity detector.

For halogen analysis, the combustion product contains HX (X = Cl, Br, I). HX is trapped in water and titrated with Ag^+ ions by an automated electrochemical process.

Table 7-5 shows analytical results for pure acetanilide from two instruments. Chemists consider a result within ±0.3 of the theoretical percentage of an element to

Table 7-5 C, H, and N in acetanilide: $C_6H_5NHC(=O)CH_3$

Element	Theoretical value (wt%)	Instrument 1	Instrument 2
C	71.09	71.17 ± 0.41	71.22 ± 1.1
H	6.71	6.76 ± 0.12	6.84 ± 0.10
N	10.36	10.34 ± 0.08	10.33 ± 0.13

Uncertainties are standard deviations from five replicate determinations.

SOURCE: E. M. Hodge, H. P. Patterson, M. C. Williams, and E. S. Gladney, *Anal. Chem.* **1994**, *66*, 1119.

be good evidence that the compound has the expected formula. For N in acetanilide, the theoretical percentage is 10.36%, so the range 10.06% to 10.66% is considered acceptable. An uncertainty of ±0.3 is a relative error of 0.3/10.36 = 3%, which is not hard to achieve. For C, ±0.3 corresponds to a relative error of 0.3/71.09 = 0.4%, which is not so easy. The standard deviation for C in Instrument 1 is 0.41/71.17 = 0.6%; for Instrument 2, it is 1.1/71.22 = 1.5%.

7-D. **(a)** What is the difference between combustion and pyrolysis?
(b) What is the purpose of the WO_3 and Cu in Figure 7-7?
(c) Why is tin used to encapsulate a sample for combustion analysis?
(d) Why is sample dropped into the preheated furnace before the oxygen concentration reaches its peak in dynamic flash combustion?
(e) What is the balanced equation for the combustion of $C_8H_7NO_2SBrCl$ in a C,H,N,S elemental analyzer?

Important Terms

absorption	gathering	particle growth
adsorption	gravimetric analysis	peptization
colloid	homogeneous precipitation	precipitant
combustion analysis	hygroscopic substance	pyrolysis
coprecipitation	ignition	supersaturated solution
dialysis	masking agent	thermogravimetric analysis
digestion	nucleation	

Problems

7-1. $BaSO_4$ precipitate in Table 7-4 contains occluded nitrate impurity.

(a) What is the difference between occluded, included, and adsorbed impurity?

(b) Why does the ratio $[NO_3^-]/[SO_4^-]$ in the precipitate decrease with each reprecipitation?

7-2. An organic compound with a molecular mass of 417 was analyzed for ethoxyl (CH_3CH_2O—) groups by the reactions

$$ROCH_2CH_3 + HI \longrightarrow ROH + CH_3CH_2I$$

(R = remainder of molecule)

$$CH_3CH_2I + Ag^+ + OH^- \longrightarrow AgI(s) + CH_3CH_2OH$$

A 25.42-mg sample of compound produced 29.03 mg of AgI (FM 234.77). How many ethoxyl groups are there in each molecule?

7-3. A 0.050 02-g sample of impure piperazine contained 71.29 wt% piperazine. How many grams of product will be formed if this sample is analyzed by Reaction 7-1?

7-4. A 1.000-g sample of unknown analyzed by Reaction 7-2 gave 2.500 g of bis(dimethylglyoximate)nickel(II). Find the wt% of Ni in the unknown.

7-5. How many milliliters of 2.15 wt% dimethylglyoxime solution should be used to provide a 50.0% excess for Reaction 7-2 with 0.998 4 g of steel containing 2.07 wt% Ni? The density of the dimethylglyoxime solution is 0.790 g/mL.

7-6. A solution containing 1.263 g of unknown potassium compound was dissolved in water and treated with excess sodium tetraphenylborate, $Na^+B(C_6H_5)_4^-$ solution to precipitate 1.003 g of insoluble $K^+B(C_6H_5)_4^-$ (FM 358.33). Find the wt% of K in the unknown.

7-7. Twenty dietary iron tablets with a total mass of 22.131 g were ground and mixed thoroughly. Then 2.998 g of the powder were dissolved in HNO_3 and heated to convert all the iron to Fe^{3+}. Addition of NH_3 caused quantitative precipitation of $Fe_2O_3 \cdot x\,H_2O$, which was ignited to give 0.264 g of Fe_2O_3 (FM 159.69). What is the average mass of $FeSO_4 \cdot 7H_2O$ (FM 278.01) in each tablet?

7-8. *The man in the vat.* Once upon a time, a workman at a dye factory fell into a vat containing a hot concentrated mixture of sulfuric and nitric acids, and he dissolved! Because nobody witnessed the accident, it was necessary to prove that he fell in so that the man's wife could collect his insurance money. The man weighed 70 kg, and a human body contains about 6.3 parts per thousand phosphorus. The acid in the vat was analyzed for phosphorus to see if it contained a dissolved human.

(a) The vat had 8.00×10^3 L of liquid, and 100.0 mL were analyzed. If the man did fall into the vat, what is the expected quantity of phosphorus in 100.0 mL?

(b) The 100.0-mL sample was treated with a molybdate reagent that precipitates ammonium phosphomolybdate, $(NH_4)_3[P(Mo_{12}O_{40})] \cdot 12H_2O$. This substance was dried at 110°C to remove waters of hydration and heated to 400°C until it reached a constant composition corresponding to the formula $P_2O_5 \cdot 24MoO_3$, which weighed 0.371 8 g. When a fresh mixture of the same acids (not from the vat) was treated in the same manner, 0.033 1 g of $P_2O_5 \cdot 24MoO_3$ (FM 3 596.46) was produced. This *blank determination* gives the amount of phosphorus in the starting reagents. The $P_2O_5 \cdot 24MoO_3$ that could have come from the dissolved man is therefore $0.371\ 8 - 0.033\ 1 = 0.338\ 7$ g. How much phosphorus was present in the 100.0-mL sample? Is this quantity consistent with a dissolved man?

7-9. Consider a mixture of the two solids $BaCl_2 \cdot 2H_2O$ (FM 244.26) and KCl (FM 74.551). When the mixture is heated to 160°C for 1 h, the water of crystallization is driven off:

$$BaCl_2 \cdot 2H_2O(s) \xrightarrow{160^\circ C} BaCl_2(s) + 2H_2O(g)$$

A sample originally weighing 1.783 9 g weighed 1.562 3 g after heating. Calculate the weight percent of Ba, K, and Cl in the original sample. (*Hint:* The mass loss tells how much water was lost, which tells how much $BaCl_2 \cdot 2H_2O$ was present. The remainder of the sample is KCl.)

7-10. Finely ground mineral (0.632 4 g) was dissolved in 25 mL of boiling 4 M HCl and diluted with 175 mL H_2O containing two drops of methyl red indicator. The solution was heated to 100°C, and 50 mL of warm solution containing 2.0 g $(NH_4)_2C_2O_4$ were slowly added to precipitate CaC_2O_4. Then 6 M NH_3 was added until the indicator changed from red to yellow, showing that the liquid was neutral or slightly basic. After slow cooling for 1 h, the liquid was decanted and the solid transferred to a filter crucible and washed with cold 0.1 wt% $(NH_4)_2C_2O_4$ solution five times until no Cl^- was detected in the filtrate on addition of $AgNO_3$ solution. The crucible was dried at 105°C for 1 h and then at 500° ± 25°C in a furnace for 2 h.

$$\underset{\text{FM } 40.078}{Ca^{2+}} + C_2O_4^{2-} \xrightarrow{105^\circ C} CaC_2O_4 \cdot H_2O(s) \xrightarrow{500^\circ C} \underset{\text{FM } 100.087}{CaCO_3(s)}$$

The mass of the empty crucible was 18.231 1 g and the mass of the crucible with $CaCO_3(s)$ was 18.546 7 g.

(a) Find the wt% Ca in the mineral.

(b) Why is the unknown solution heated to boiling and the precipitant solution, $(NH_4)_2C_2O_4$, also heated before slowly mixing the two?

(c) What is the purpose of washing the precipitate with 0.1 wt% $(NH_4)_2C_2O_4$?

(d) What is the purpose of testing the filtrate with $AgNO_3$ solution?

7-11. Write a balanced equation for the combustion of benzoic acid, $C_6H_5CO_2H$, to give CO_2 and H_2O. How many milligrams of CO_2 and of H_2O will be produced by the combustion of 4.635 mg of benzoic acid?

7-12. Combustion of 8.732 mg of an unknown organic compound gave 16.432 mg of CO_2 and 2.840 mg of H_2O.

(a) Find the wt% of C and H in the substance.

(b) Find the smallest reasonable integer mole ratio of C:H in the compound.

7-13. Combustion analysis of a compound known to contain just C, H, N, and O demonstrated that it contains 46.21 wt% C, 9.02 wt% H, 13.74 wt% N, and, by difference, $100 - 46.21 - 9.02 - 13.74 = 31.03$ wt% O. This means that 100 g of unknown would contain 46.21 g of C, 9.02 g of H, and so on. Find the atomic ratio C:H:N:O. Then divide each stoichiometry coefficient by the smallest one and express the atomic composition in the lowest reasonable integer ratio ($C_xH_yN_zO_w$, where x, y, z, and w are integers and one of them is 1).

7-14. A method for measuring organic carbon in seawater consists of oxidation of the organic materials to CO_2 with $K_2S_2O_8$, followed by gravimetric determination of the CO_2 trapped by a column of NaOH-coated asbestos. A water sample weighing 6.234 g produced 2.378 mg of CO_2 (FM 44.010). Calculate the ppm carbon in the seawater.

7-15. The reagent nitron forms a fairly insoluble salt with nitrate, the product having a solubility of 0.99 g/L near 20°C. Sulfate and acetate do not precipitate with nitron, but many other anions, including ClO_4^-, ClO_3^-, I^-, SCN^-, and $C_2O_4^{2-}$ do precipitate and interfere with the analysis. A 50.00-mL unknown solution containing KNO_3 and $NaNO_3$ was treated with 1 mL of acetic acid and heated to near boiling; 10 mL of solution containing excess nitron was added with stirring. After cooling to 0°C for 2 h, the crystalline product was filtered, washed with three 5-mL portions of ice-cold, saturated nitron nitrate solution, and finally washed with two 3-mL portions of ice-cold water. The product weighed 0.513 6 g after drying at 105°C for 1 h.

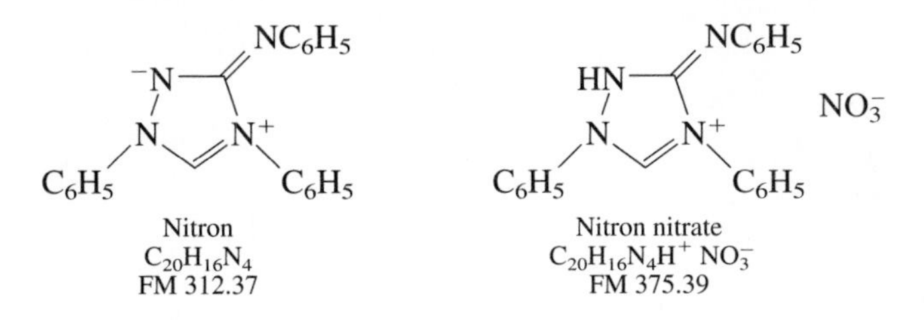

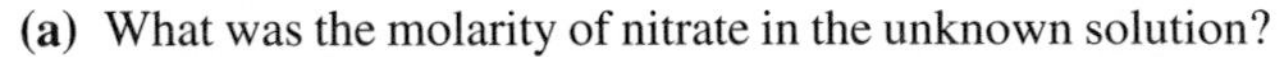

(a) What was the molarity of nitrate in the unknown solution?

(b) Some of the nitron nitrate dissolves in the final ice-cold wash water. Does this dissolution lead to a random or a systematic error in the analysis?

7-16. A mixture of $Al_2O_3(s)$ and $CuO(s)$ weighing 18.371 mg was heated under $H_2(g)$ in a thermogravimetric experiment. On reaching a temperature of 1 000°C, the mass was 17.462 mg and the final products were $Al_2O_3(s)$, $Cu(s)$, and $H_2O(g)$. Find the weight percent of Al_2O_3 in the original solid mixture.

7-17. Use the uncertainties from Instrument 1 in Table 7-5 to estimate the uncertainties in the stoichiometry coefficients in the formula $C_8H_{h \pm x}N_{n \pm y}$.

7-18. One way to determine sulfur is by combustion, which produces a mixture of SO_2 and SO_3 that can be passed through H_2O_2 to convert both into H_2SO_4. When 6.123 mg of a substance were burned, the H_2SO_4 required 3.01 mL of 0.015 76 M NaOH for titration by the reaction $H_2SO_4 + 2NaOH \rightarrow Na_2SO_4 + 2H_2O$. What is the weight percent of sulfur in the sample?

7-19. Some analyte ions can be *gathered* by $LaPO_4$ at pH 3 ($[H^+] = 10^{-3}$ M) (S. Kagaya, M. Saiki, Z. A. Malek, Y. Araki, and K. Hasegawa, *Fresenius J. Anal. Chem.* **2001**, *371*, 391.) To 100.0 mL of aqueous test sample were added 2 mL of La^{3+} solution (containing 5 mg La^{3+}/mL in 0.6 M HCl) and 0.3 mL of 0.5 M H_3PO_4. The pH was adjusted to 3.0 by adding NH_3. The precipitate was allowed to settle and collected on a filter with 0.2-μm pore size. The filter, which has negligible volume (<0.01 mL) was placed in a 10-mL volumetric flask and treated with 1 mL of 16 M HNO_3 to dissolve the precipitate. The flask was made up to 10 mL with H_2O.

(a) A 100.0-mL distilled water sample was spiked with 10.0 μg of each of the following elements: Fe^{3+}, Pb^{2+}, Cd^{2+}, In^{3+}, Cr^{3+}, Mn^{2+}, Co^{2+}, Ni^{2+}, and Cu^{2+}. Analysis of the final solution in the 10-mL volumetric flask by atomic spectroscopy gave the following concentrations: $[Fe^{3+}] = 17.6$ μM, $[Pb^{2+}] = 5.02$ μM, $[Cd^{2+}] = 8.77$ μM, $[In^{3+}] = 8.50$ μM, $[Cr^{3+}] < 0.05$ μM, $[Mn^{2+}] = 6.64$ μM, $[Co^{2+}] = 1.09$ μM, $[Ni^{2+}] < 0.05$ μM, $[Cu^{2+}] = 6.96$ μM. Find the percent recovery of each element, defined as

$$\% \text{ recovery} = \frac{\mu\text{g found}}{\mu\text{g added}} \times 100$$

(b) Which of the tested cations are quantitatively gathered?

(c) In the process of gathering, analyte is *preconcentrated* from a dilute sample into a more concentrated solution that is analyzed. By what factor are the elements preconcentrated in this procedure?

How Would You Do It?

7-20. The fat content of homogenized whole milk was measured by the Rose Gottlieb method described at the beginning of Section 7-1. Replicate measurements were made by a manual method and by an automated method. Do the two methods give the same or statistically different results?

Weight percent fat in milk

Manual method		Automated method		
2.934	2.925	2.967	2.958	3.022
2.981	2.948	3.034	3.052	2.974
2.906	2.981	3.022	2.983	2.946
2.976	2.913	2.982	2.966	2.997
2.958	2.881	2.992	3.006	3.027
2.945	2.847	2.950	2.982	2.979
2.893	2.880	2.965	2.951	3.047

7-21. *Excel's built-in t test.* Enter data from Problem 7-20 in columns B and C of a spreadsheet like Figure 4-5. Compute the mean and standard deviation for each data set. Follow directions in Section 4-3 to conduct the *t* test with the Excel procedure t-Test: Two-Sample Assuming Equal Variances. Check that the computed pooled variance is the square of the pooled standard deviation you found in Problem 7-20. Check that the computed value of *t* (called t Stat in the spreadsheet) is the same as that you calculated in Problem 7-20. Check that the value of t Critical two-tail agrees with what you estimate from Table 4-2.

Acid Rain

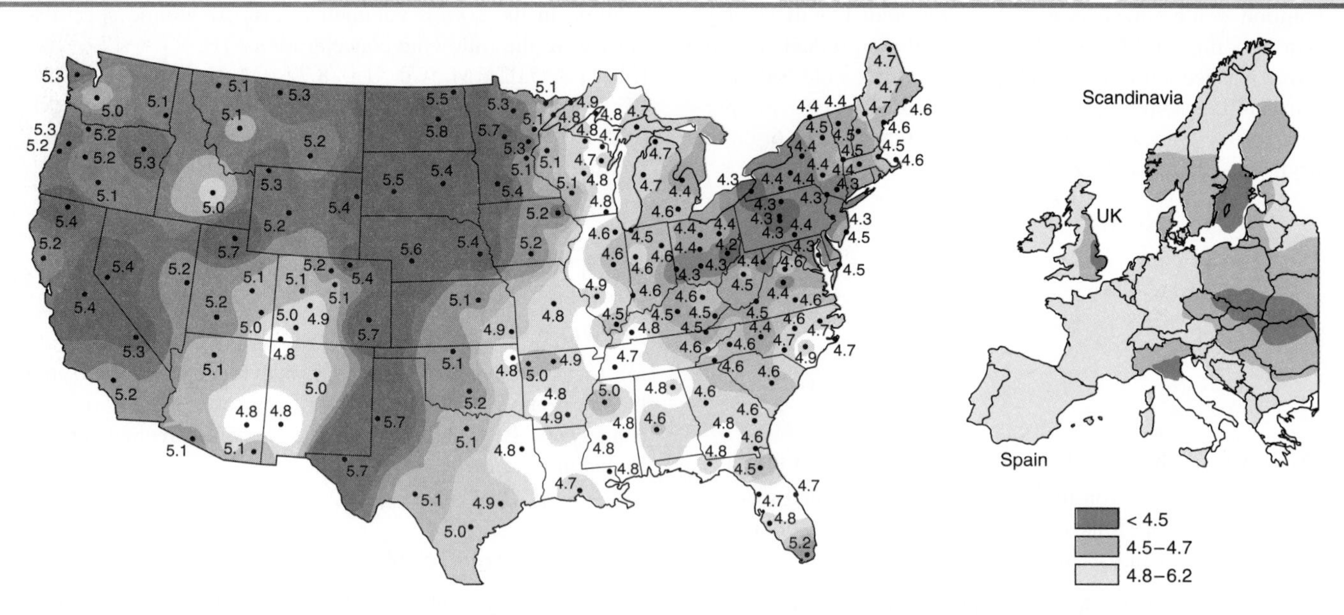

pH of precipitation in the United States in 2001. The lower the pH, the more acidic the water. [Redrawn from National Atmospheric Deposition Program (NRSP-3)/National Trends Network (2002). Illinois State Water Survey, 2204 Griffith Dr., Champaign, IL 61820. See also http://nadp.sws.uiuc.edu and www.epa.gov/acidrain.]

pH of rain in Europe. [From H. Rodhe, F. Dentener, and M. Schulz, *Environ. Sci. Technol.* **2002**, 36, 4382.]

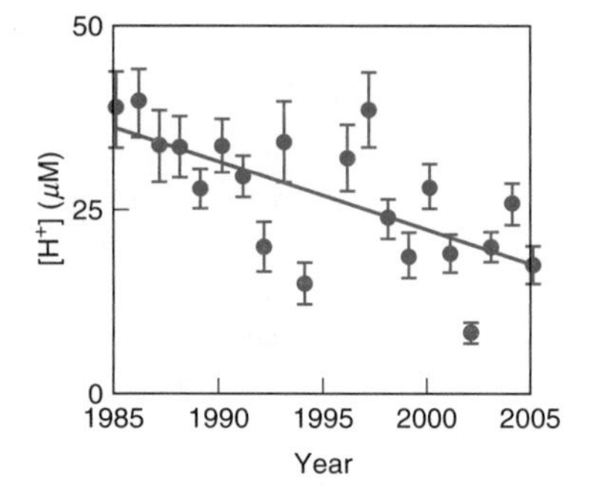

Volume-weighted H^+ concentration in precipitation in Wilmington, NC, decreased by a factor of 2 between 1985 and 2005, typical of the nationwide decrease of rainfall acidity. Error bars are ±1 standard deviation. [From J. D. Willey, R. J. Kieber, and G. B. Avery, Jr., *Environ. Sci. Technol.* **2006**, *40*, 5675.] In volume weighting, $[H^+]$ in each rainfall is multiplied by the volume of that rainfall to get moles. Total moles for the year are divided by total volume for the year.

Combustion products from automobiles and power plants include nitrogen oxides and sulfur dioxide, which react with oxidizing agents in the atmosphere to produce acids.

$$\underbrace{NO + NO_2}_{\text{Nitrogen oxides designated } NO_x} \xrightarrow[H_2O]{\text{oxidation}} \underset{\text{Nitric acid}}{HNO_3} \qquad \underset{\text{Sulfur dioxide}}{SO_2} \xrightarrow[H_2O]{\text{oxidation}} \underset{\text{Sulfuric acid}}{H_2SO_4}$$

Acid rain **in the United States is worst in the Northeast, downwind from coal-burning power plants and factories.[1] Rain in portions of Europe is similarly acidic. Acid rain kills fish in lakes and rivers and threatens forests. Half of the essential nutrients Ca^{2+} and Mg^{2+} have been leached from the soil in Sweden since 1950. Acid increases the solubility of toxic Al^{3+} and other metals in groundwater. U.S. legislation in 1990 to reduce sulfur and nitrogen emissions has decreased the acidity of precipitation, as shown in the graph. However, much damage has been done, and 16 of 50 states *increased* their SO_2 emissions between 1990 and 2000.**

CHAPTER 8

Introducing Acids and Bases

The chemistry of acids and bases is probably the most important topic you will study in chemical equilibrium. It is difficult to have a meaningful discussion of subjects ranging from protein folding to the weathering of rocks without understanding acids and bases. It will take us several chapters to provide meaningful detail to the study of acid-base chemistry.

8-1 What Are Acids and Bases?

In aqueous chemistry, an **acid** is a substance that increases the concentration of H_3O^+ (**hydronium ion**). Conversely, a **base** decreases the concentration of H_3O^+ in aqueous solution. As we shall see shortly, a decrease in H_3O^+ concentration necessarily requires an increase in OH^- concentration. Therefore, a base is also a substance that increases the concentration of OH^- in aqueous solution.

The species H^+ is called a *proton* because a proton is all that remains when a hydrogen atom loses its electron. Hydronium ion, H_3O^+, is a combination of H^+ with H_2O (Figure 8-1). Although H_3O^+ is a more accurate representation than H^+ for the hydrogen ion in aqueous solution, we will use H_3O^+ and H^+ interchangeably in this book.

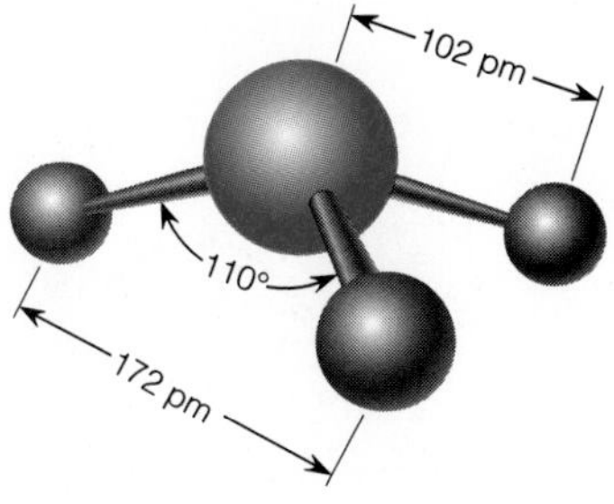

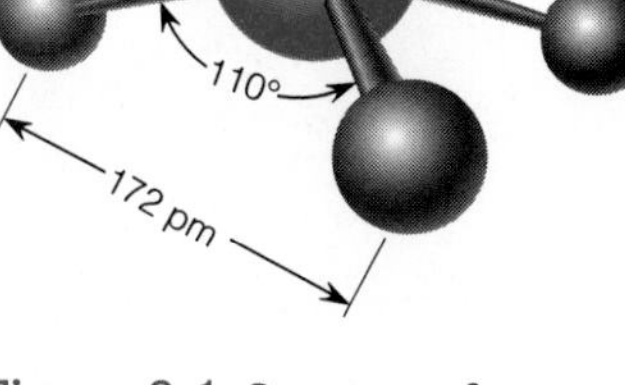

Figure 8-1 Structure of hydronium ion, H_3O^+.

A more general definition of acids and bases given by Brønsted and Lowry is that an *acid* is a *proton donor* and a *base* is a *proton acceptor.* This definition includes the one already stated. For example, HCl is an acid because it donates a proton to H_2O to form H_3O^+:

$$HCl + H_2O \rightleftharpoons H_3O^+ + Cl^-$$

The Brønsted-Lowry definition can be extended to nonaqueous solvents and to the gas phase:

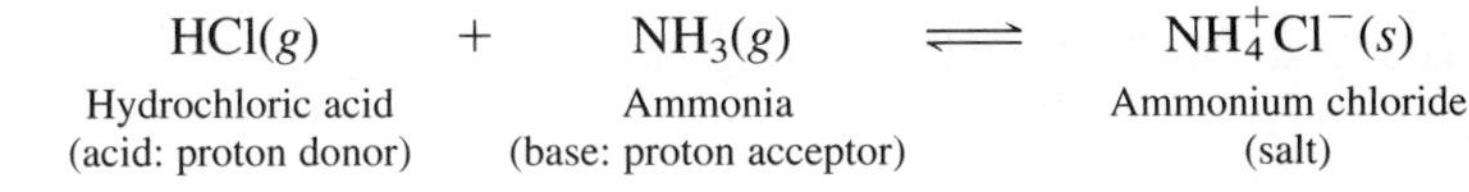

$$\underset{\substack{\text{Hydrochloric acid} \\ \text{(acid: proton donor)}}}{HCl(g)} + \underset{\substack{\text{Ammonia} \\ \text{(base: proton acceptor)}}}{NH_3(g)} \rightleftharpoons \underset{\substack{\text{Ammonium chloride} \\ \text{(salt)}}}{NH_4^+Cl^-(s)}$$

Brønsted-Lowry acid: proton donor
Brønsted-Lowry base: proton acceptor

Salts

Any ionic solid, such as ammonium chloride, is called a **salt**. In a formal sense, a salt can be thought of as the product of an acid-base reaction. When an acid and a base react stoichiometrically, they are said to **neutralize** each other. Most salts are *strong electrolytes*, meaning that they dissociate almost completely into their component ions when dissolved in water. Thus, ammonium chloride gives NH_4^+ and Cl^- in aqueous solution:

$$NH_4^+Cl^-(s) \longrightarrow NH_4^+(aq) + Cl^-(aq)$$

Conjugate Acids and Bases

The products of a reaction between an acid and a base also are acids and bases:

A solid wedge is a bond coming out of the page toward you. A dashed wedge is a bond going behind the page.

Conjugate acids and bases are related by the gain or loss of one proton.

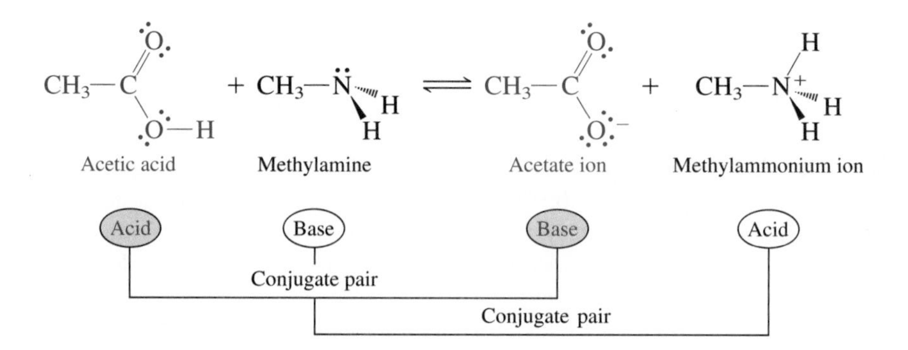

Acetate is a base because it can accept a proton to make acetic acid. The methylammonium ion is an acid because it can donate a proton and become methylamine. Acetic acid and the acetate ion are said to be a **conjugate acid-base pair**. Methylamine and the methylammonium ion are likewise conjugate. *Conjugate acids and bases are related to each other by the gain or loss of one* H^+.

Ask Yourself

8-A. When an acid and base react, they are said to __________ each other. Acids and bases related by the gain or loss of one proton are said to be __________.

8-2 Relation Between [H$^+$], [OH$^-$], and pH

In **autoprotolysis**, one substance acts as both an acid and a base:

Autoprotolysis of water:

$$H_2O + H_2O \overset{K_w}{\rightleftharpoons} \underset{\text{Hydronium ion}}{H_3O^+} + \underset{\text{Hydroxide ion}}{OH^-} \quad (8\text{-}1a)$$

We abbreviate Reaction 8-1a in the following manner:

$$H_2O \overset{K_w}{\rightleftharpoons} H^+ + OH^- \quad (8\text{-}1b)$$

and we designate its equilibrium constant as K_w.

Autoprotolysis constant for water:

$$K_w = [H^+][OH^-] = 1.0 \times 10^{-14} \text{ at } 25°C \quad (8\text{-}2)$$

Equation 8-2 provides a tool with which we can find the concentration of H^+ and OH^- in pure water. Also, given that the product $[H^+][OH^-]$ is constant, we can always find the concentration of either species if the concentration of the other is known. Because the product is a constant, *as the concentration of H^+ increases, the concentration of OH^- necessarily decreases, and vice versa.*

Example Concentration of H^+ and OH^- in Pure Water at 25°C

Calculate the concentrations of H^+ and OH^- in pure water at 25°C.

SOLUTION H^+ and OH^- are produced in a 1:1 mole ratio in Reaction 8-1b. Calling each concentration x, we write

$$K_w = 1.0 \times 10^{-14} = [H^+][OH^-] = [x][x] \Rightarrow x = \sqrt{1.0 \times 10^{-14}} = 1.0 \times 10^{-7} \text{ M}$$

The concentrations of H^+ and OH^- are both 1.0×10^{-7} M.

Test Yourself At 0°C, the equilibrium constant K_w has the value 1.2×10^{-15}. Find $[H^+]$ and $[OH^-]$ in pure water at 0°C. (**Answer:** both 3.5×10^{-8} M)

Example Finding $[OH^-]$ When $[H^+]$ Is Known

What is the concentration of OH^- if $[H^+] = 1.0 \times 10^{-3}$ M at 25°C?

SOLUTION Setting $[H^+] = 1.0 \times 10^{-3}$ M gives

$$K_w = [H^+][OH^-] \Rightarrow [OH^-] = \frac{K_w}{[H^+]} = \frac{1.0 \times 10^{-14}}{1.0 \times 10^{-3}} = 1.0 \times 10^{-11} \text{ M}$$

When $[H^+] = 1.0 \times 10^{-3}$ M, $[OH^-] = 1.0 \times 10^{-11}$ M. If $[OH^-] = 1.0 \times 10^{-3}$ M, then $[H^+] = 1.0 \times 10^{-11}$ M. As one concentration increases, the other decreases.

Test Yourself Find $[H^+]$ in water when $[OH^-] = 1.0 \times 10^{-4}$ M. (**Answer:** 1.0×10^{-10} M)

To simplify the writing of H^+ concentration, we define **pH** as

Approximate definition of pH:

$$pH = -\log[H^+] \quad (8\text{-}3)$$

pH is really defined in terms of the *activity* of H^+, which is related to concentration. Section 12-2 discusses activity.

Here are some examples:

$$[H^+] = 10^{-3} \text{ M} \Rightarrow pH = -\log(10^{-3}) = 3$$
$$[H^+] = 10^{-10} \text{ M} \Rightarrow pH = -\log(10^{-10}) = 10$$
$$[H^+] = 3.8 \times 10^{-8} \text{ M} \Rightarrow pH = -\log(3.8 \times 10^{-8}) = 7.42$$

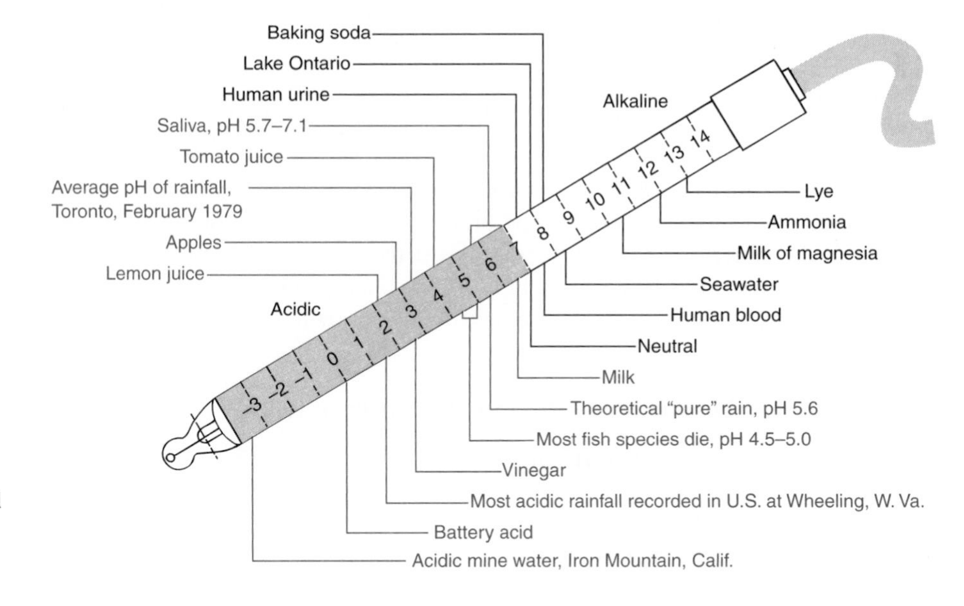

Figure 8-2 pH values of various substances. The most acidic rainfall in the United States is more acidic than lemon juice.

Example: An acidic solution has pH = 4, which means $[H^+] = 10^{-4}$ M and $[OH^-] = K_w/[H^+] = 10^{-10}$ M. Therefore, $[H^+] > [OH^-]$.

Changing the pH by 1 unit changes $[H^+]$ by a factor of 10. When the pH changes from 3 to 4, $[H^+]$ changes from 10^{-3} to 10^{-4} M.

A solution is **acidic** if $[H^+] > [OH^-]$. A solution is **basic** if $[H^+] < [OH^-]$. An earlier example demonstrated that in pure water (which is neither acidic nor basic and is said to be *neutral*), $[H^+] = [OH^-] = 10^{-7}$ M, so the pH is $-\log(10^{-7}) = 7$. At 25°C, *an acidic solution has a pH below 7, and a basic solution has a pH above 7* (Figure 8-2).

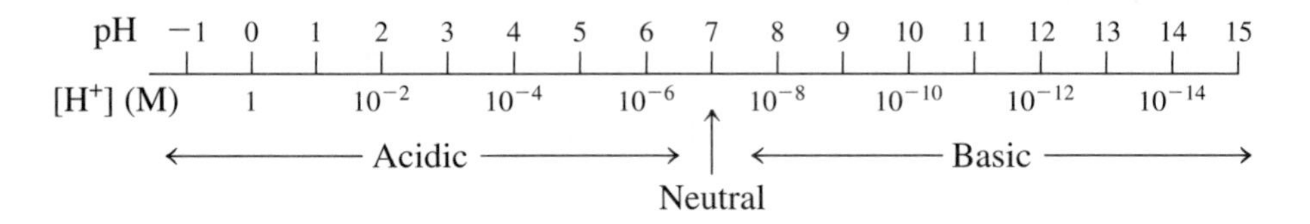

Although pH generally falls in the range 0 to 14, these are not limits. A pH of −1, for example, means $-\log[H^+] = -1$, or $[H^+] = 10^{+1} = 10$ M. This pH is attained in a concentrated solution of a strong acid such as HCl.

Ask Yourself

8-B. A solution of 0.050 M Mg^{2+} is treated with NaOH until $Mg(OH)_2$ precipitates. **(a)** At what concentration of OH^- does this occur? (Remember the solubility product in Section 6-4. Use K_{sp} for brucite $Mg(OH)_2$ in Appendix A.) **(b)** At what pH does this occur?

8-3 Strengths of Acids and Bases

Acids and bases are classified as strong or weak, depending on whether they react "completely" or only "partly" to produce H^+ or OH^-. Because there is a continuous range for a "partial" reaction, there is no sharp distinction between weak and strong.

However, some compounds react so completely that they are unquestionably strong acids or bases—and everything else is defined as weak.

Strong Acids and Bases

Common strong acids and bases are listed in Table 8-1, which you must memorize. Note that even though HCl, HBr, and HI are strong acids, HF is *not.* A **strong acid** or **strong base** is completely dissociated in aqueous solution. That is, the equilibrium constants for the following reactions are very large:

$$HCl(aq) \longrightarrow H^+ + Cl^-$$
$$KOH(aq) \longrightarrow K^+ + OH^-$$

Virtually no undissociated HCl or KOH exists in aqueous solution. Demonstration 8-1 shows one consequence of the strong-acid behavior of HCl.

Table 8-1 Common strong acids and bases

Formula	Name
ACIDS	
HCl	Hydrochloric acid (hydrogen chloride)
HBr	Hydrogen bromide
HI	Hydrogen iodide
H_2SO_4[a]	Sulfuric acid
HNO_3	Nitric acid
$HClO_4$	Perchloric acid
BASES	
LiOH	Lithium hydroxide
NaOH	Sodium hydroxide
KOH	Potassium hydroxide
RbOH	Rubidium hydroxide
CsOH	Cesium hydroxide
R_4NOH[b]	Quaternary ammonium hydroxide

a. For H_2SO_4, only the first proton ionization is complete. Dissociation of the second proton has an equilibrium constant of 1.0×10^{-2}.
b. This is a general formula for any hydroxide salt of an ammonium cation containing four organic groups. An example is tetrabutylammonium hydroxide: $(CH_3CH_2CH_2CH_2)_4N^+OH^-$.

Weak Acids and Bases

All **weak acids**, HA, react with water by donating a proton to H_2O:

$$HA + H_2O \overset{K_a}{\rightleftharpoons} H_3O^+ + A^-$$

which means exactly the same as

Dissociation of weak acid:
$$HA \overset{K_a}{\rightleftharpoons} H^+ + A^- \qquad K_a = \frac{[H^+][A^-]}{[HA]} \tag{8-4}$$

The equilibrium constant, K_a, is called the **acid dissociation constant**. A weak acid is only partly dissociated in water, which means that some undissociated HA remains.

Weak bases, B, react with water by abstracting (grabbing) a proton from H_2O:

Base hydrolysis:
$$B + H_2O \overset{K_b}{\rightleftharpoons} BH^+ + OH^- \qquad K_b = \frac{[BH^+][OH^-]}{[B]} \tag{8-5}$$

The equilibrium constant, K_b, is called the **base hydrolysis constant**. A weak base is one for which some unreacted B remains.

Carboxylic Acids Are Weak Acids and Amines Are Weak Bases

Roughly speaking, an acid is weak if $K_a < 1$ and a base is weak if $K_b < 1$.

Acetic acid is a typical weak acid:

$$CH_3{-}C({=}O){-}O{-}H \rightleftharpoons CH_3{-}C({=}O){-}O^- + H^+ \qquad K_a = 1.75 \times 10^{-5} \tag{8-6}$$

Acetic acid HA   Acetate A^-

Demonstration 8-1 HCl Fountain

The complete dissociation of HCl into H^+ and Cl^- makes HCl(g) extremely soluble in water.

$$HCl(g) \rightleftharpoons HCl(aq) \quad \text{(A)}$$

$$HCl(aq) \longrightarrow H^+(aq) + Cl^-(aq) \quad \text{(B)}$$

Reaction B consumes the product of reaction A, thereby pulling reaction A to the right.

An HCl fountain is assembled as shown below.[2] In panel a, an inverted 250-mL round-bottom flask containing air is set up with its inlet tube leading to a source of HCl(g) and its outlet tube directed into an inverted bottle of water. As HCl is admitted to the flask, air is displaced into the bottle of water. When the bottle is filled with air, the flask contains mostly HCl(g).

The hoses are disconnected and replaced with a beaker of indicator and a rubber bulb (panel b). For an indicator, we use slightly alkaline, commercial methyl purple solution, which is green above pH 5.4 and purple below pH 4.8. When 1 mL of water is squirted from the rubber bulb into the flask, a vacuum is created and indicator solution is drawn up into the flask, creating a colorful fountain (Color Plate 3).

Questions Why is vacuum created when water is squirted into the flask? Why does the indicator change color when it enters the flask?

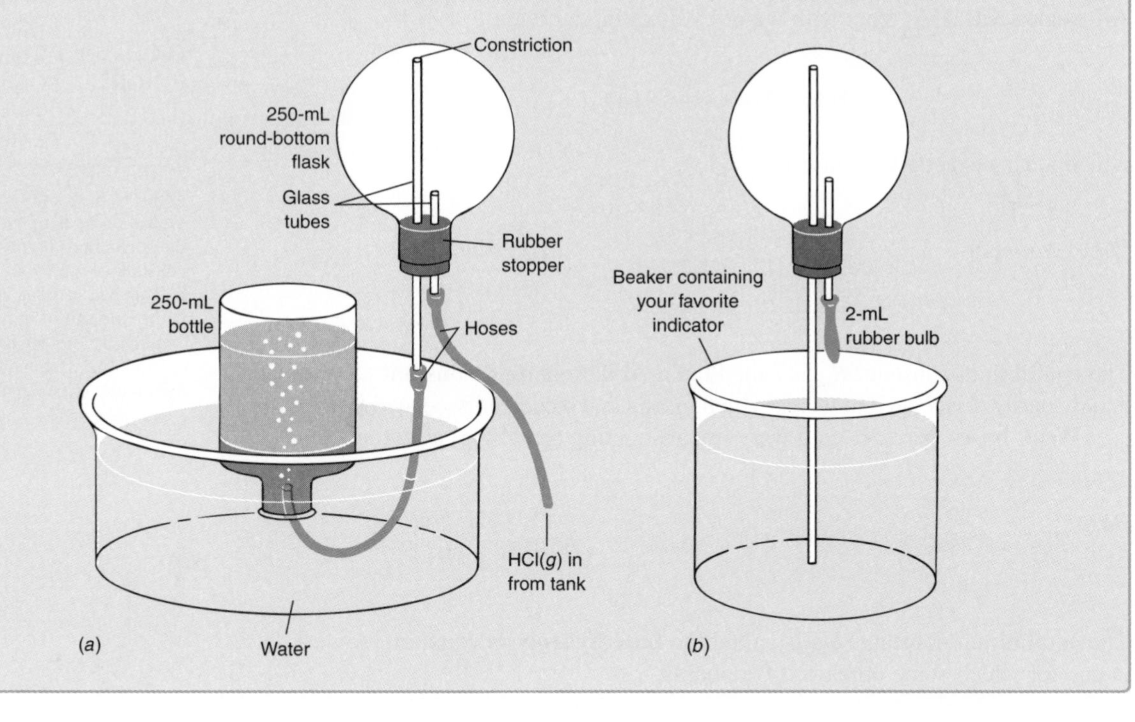

Acetic acid is representative of carboxylic acids, which have the general structure shown below, where R is an organic substituent. *Most* **carboxylic acids** *are weak acids, and most* **carboxylate anions** *are weak bases.*

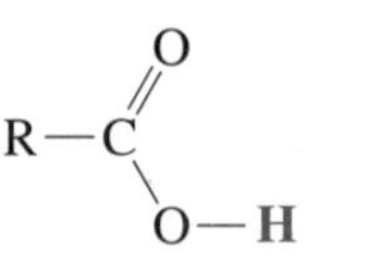

A carboxylic acid (weak acid, HA) A carboxylate anion (weak base, A^-)

Methylamine is a typical weak base. It forms a bond to H^+ by sharing the lone pair of electrons from the nitrogen atom of the *amine*:

$$\underset{\substack{\text{Methylamine}\\ \text{B}}}{CH_3\ddot{N}H_2} + H_2O \rightleftharpoons \underset{\substack{\text{Methylammonium ion}\\ BH^+}}{CH_3\overset{+}{N}H_3} + OH^- \quad K_b = 4.42 \times 10^{-4} \qquad (8\text{-}7)$$

Carboxylic acids (RCO_2H) and ammonium ions (R_3NH^+) are weak acids. Carboxylate anions (RCO_2^-) and amines (R_3N) are weak bases.

Methylamine is a representative **amine**, a nitrogen-containing compound:

$R\ddot{N}H_2$	a primary amine	RNH_3^+	**ammonium ions**
$R_2\ddot{N}H$	a secondary amine	$R_2NH_2^+$	
$R_3\ddot{N}$	a tertiary amine	R_3NH^+	

Amines are weak bases, and ammonium ions are weak acids. The "parent" of all amines is ammonia, NH_3. When methylamine reacts with water, the product is the conjugate acid. That is, the methylammonium ion produced in Reaction 8-7 is a weak acid:

$$\underset{BH^+}{CH_3\overset{+}{N}H_3} \overset{K_a}{\rightleftharpoons} \underset{B}{CH_3\ddot{N}H_2} + H^+ \qquad K_a = 2.26 \times 10^{-11} \qquad (8\text{-}8)$$

Weak acids: **HA** and $\mathbf{BH^+}$
Weak bases: $\mathbf{A^-}$ and **B**

The methylammonium ion (BH^+) is the conjugate acid of methylamine (B).

You should learn to recognize whether a compound is acidic or basic. For example, the salt methylammonium chloride dissociates completely in water to give methylammonium cation and chloride anion:

$$\underset{\substack{\text{Methylammonium}\\ \text{chloride}}}{CH_3\overset{+}{N}H_3Cl^-(s)} \longrightarrow CH_3\overset{+}{N}H_3(aq) + Cl^-(aq)$$

Methylammonium chloride is a weak acid because

1. It dissociates into $CH_3NH_3^+$ and Cl^-.
2. $CH_3NH_3^+$ is a weak acid, being conjugate to CH_3NH_2, a weak base.
3. Cl^- has no basic properties. It is conjugate to HCl, a strong acid. That is, HCl dissociates completely.

The methylammonium ion, being the conjugate acid of methylamine, is a weak acid (Reaction 8-8). The chloride ion is neither an acid nor a base. It is the conjugate base of HCl, a strong acid. In other words, Cl^- *has virtually no tendency to associate with* H^+; otherwise, HCl would not be classified as a strong acid. We predict that methylammonium chloride solution is acidic, because the methylammonium ion is an acid and Cl^- is not a base.

Challenge Phenol (C_6H_5OH) is a weak acid. Explain why a solution of the ionic compound potassium phenolate ($C_6H_5O^-K^+$) is basic.

Metal Ions with Charge ≥2 Are Weak Acids

Metal ions with a charge of +2 or higher are acidic. In aqueous solution, metal ions bind several water molecules to form $M(H_2O)_w^{n+}$ in which electrons from oxygen are shared with the metal ion. Many metal ions bind $w = 6$ water molecules, but large metal ions can bind more water. A proton can dissociate from $M(H_2O)_w^{n+}$ to reduce the positive charge on the metal complex.

$$M(H_2O)_w^{n+} \overset{K_a}{\rightleftharpoons} M(H_2O)_{w-1}(OH)^{(n-1)+} + H^+ \qquad (8\text{-}9)$$

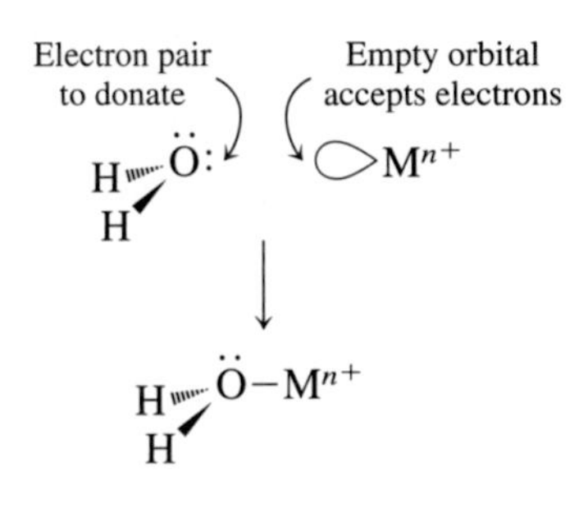

The higher the charge on the metal, the more acidic it tends to be. For example, K_a for Fe^{2+} is 4×10^{-10}, but K_a for Fe^{3+} is 6.5×10^{-3}. Cations with a charge of +1 have negligible acidity. Now you should understand why solutions of metal salts such as $Fe(NO_3)_3$ are acidic.

Relation Between K_a and K_b

An important relation exists between K_a and K_b of a conjugate acid-base pair in aqueous solution. We can derive this result with the acid HA and its conjugate base A^-.

$$\cancel{HA} \rightleftharpoons H^+ + \cancel{A^-} \qquad K_a = \frac{[H^+][A^-]}{[HA]}$$

$$\cancel{A^-} + H_2O \rightleftharpoons \cancel{HA} + OH^- \qquad K_b = \frac{[HA][OH^-]}{[A^-]}$$

$$H_2O \rightleftharpoons H^+ + OH^- \qquad K_a \cdot K_b = \frac{[H^+][\cancel{A^-}]}{[\cancel{HA}]}\,\frac{[\cancel{HA}][OH^-]}{[\cancel{A^-}]} = K_w$$

When reactions are added, their equilibrium constants must be multiplied, thereby giving a most useful result:

$K_a \cdot K_b = K_w$ for a conjugate acid-base pair in aqueous solution.

Relation between K_a and K_b for a conjugate pair:

$$K_a \cdot K_b = K_w \qquad (8\text{-}10)$$

Equation 8-10 applies to any acid and its conjugate base in aqueous solution.

Example **Finding K_b for the Conjugate Base**

The value of K_a for acetic acid is 1.75×10^{-5} (Reaction 8-6). Find K_b for acetate ion.

SOLUTION

$$K_b = \frac{K_w}{K_a} = \frac{1.0 \times 10^{-14}}{1.75 \times 10^{-5}} = 5.7 \times 10^{-10}$$

Test Yourself K_a for ammonium ion (NH_4^+) is 5.7×10^{-10}. Find K_b for ammonia (NH_3). (**Answer:** 1.8×10^{-5} M)

Example **Finding K_a for the Conjugate Acid**

K_b for methylamine is 4.42×10^{-4} (Reaction 8-7). Find K_a for methylammonium ion.

SOLUTION

$$K_a = \frac{K_w}{K_b} = \frac{1.0 \times 10^{-14}}{4.42 \times 10^{-4}} = 2.3 \times 10^{-11}$$

Test Yourself K_b for formate (HCO_2^-) is 5.6×10^{-11}. Find K_a for formic acid (HCO_2H). (**Answer:** 1.8×10^{-4} M)

Ask Yourself

8-C. Which is a stronger acid, **A** or **B**? Write the K_a reaction for each.

A $Cl_2HC\overset{O}{\overset{\|}{C}}OH$ Dichloroacetic acid $K_a = 8 \times 10^{-2}$

B $ClH_2C\overset{O}{\overset{\|}{C}}OH$ Chloroacetic acid $K_a = 1.36 \times 10^{-3}$

Which is a stronger base, **C** or **D**? Write the K_b reaction for each.

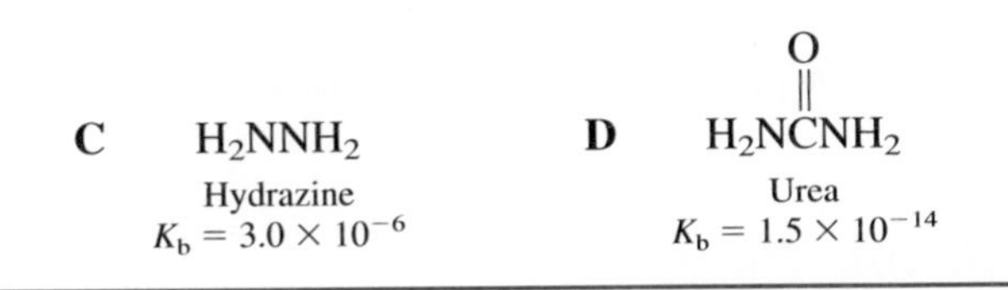

8-4 pH of Strong Acids and Bases

The principal components that make rainfall acidic are nitric and sulfuric acids, which are strong acids. Each molecule of *strong acid* or *strong base* in aqueous solution dissociates completely to provide one molecule of H^+ or OH^-. Nitric acid is a strong acid, so the reaction

$$\underset{\text{Nitric acid}}{HNO_3} \longrightarrow H^+ + \underset{\text{Nitrate}}{NO_3^-}$$

goes to completion. In the case of sulfuric acid, one proton is completely dissociated, but the second is only partly dissociated (depending on conditions):

$$\underset{\text{Sulfuric acid}}{H_2SO_4} \longrightarrow H^+ + \underset{\substack{\text{Hydrogen sulfate} \\ \text{(also called bisulfate)}}}{HSO_4^-} \overset{K_a = 0.010}{\rightleftharpoons} H^+ + \underset{\text{Sulfate}}{SO_4^{2-}}$$

pH of a Strong Acid

Because HBr is completely dissociated, the pH of 0.010 M HBr is

$$pH = -\log[H^+] = -\log(0.010) = 2.00$$

Is pH 2 sensible? (Always ask yourself that question at the end of a calculation.) Yes—because pH values below 7 are acidic and pH values above 7 are basic.

Acid: $pH < 7$
Base: $pH > 7$

Example **pH of a Strong Acid**

Find the pH of 4.2×10^{-3} M $HClO_4$.

SOLUTION $HClO_4$ is completely dissociated, so $[H^+] = 4.2 \times 10^{-3}$ M.

$$pH = -\log[H^+] = -\log(\underbrace{4.2}_{\text{2 significant figures}} \times 10^{-3}) = 2.\underbrace{38}_{\text{2 digits in mantissa}}$$

How about significant figures? The two significant figures in the mantissa of the logarithm correspond to the two significant figures in the number 4.2×10^{-3}.

Test Yourself What is the pH of 0.055 M HBr? (**Answer:** 1.26)

For consistency in working problems in this book, *we are generally going to express pH values to the 0.01 decimal place regardless of what is justified by significant figures.* Real pH measurements are rarely more accurate than ±0.02, although differences in pH between two solutions can be accurate to ±0.002 pH units.

pH of a Strong Base

Now we ask, "What is the pH of 4.2×10^{-3} M KOH?" The concentration of OH^- is 4.2×10^{-3} M, and we can calculate $[H^+]$ from the K_w equation, 8-2:

Keep at least one extra insignificant figure (or all the digits in your calculator) in the middle of a calculation to avoid round-off errors in the final answer.

$$[H^+] = \frac{K_w}{[OH^-]} = \frac{1.0 \times 10^{-14}}{4.2 \times 10^{-3}} = 2.3_8 \times 10^{-12}\ M$$

$$pH = -\log[H^+] = -\log(2.3_8 \times 10^{-12}) = 11.62$$

Here is a trick question: What is the pH of 4.2×10^{-9} M KOH? By our previous reasoning, we might first say

$$[H^+] = \frac{K_w}{[OH^-]} = \frac{1.0 \times 10^{-14}}{4.2 \times 10^{-9}} = 2.3_8 \times 10^{-6}\ M \Rightarrow pH = 5.62$$

Is this reasonable? Can we dissolve base in water and obtain an acidic pH (<7)? No way!

The fallacy is that we neglected the contribution of the reaction $H_2O \rightleftharpoons H^+ + OH^-$ to the concentration of OH^-. Pure water creates 10^{-7} M OH^-, which is more OH^- than the KOH that we added. The pH of water plus added KOH cannot fall below 7. The pH of 4.2×10^{-9} M KOH is very close to 7. Similarly, the pH of 10^{-10} M HNO_3 is very close to 7, not 10. Figure 8-3 shows how pH depends on concentration for a strong acid and a strong base. In a very dilute solution exposed to air, the acid-base chemistry of dissolved carbon dioxide ($CO_2 + H_2O \rightleftharpoons HCO_3^- + H^+$) would overwhelm the effect of the added acid or base.

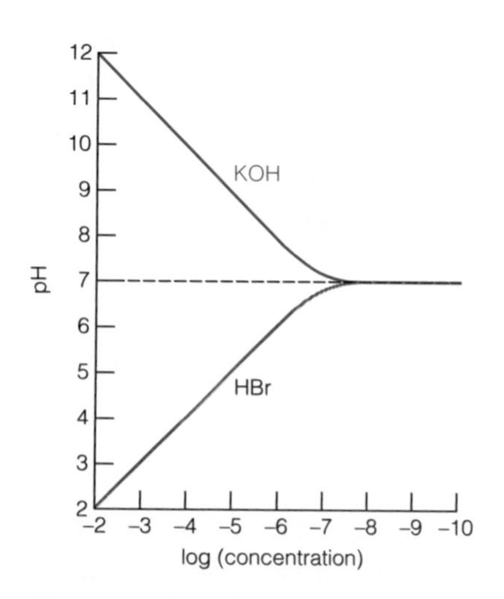

Figure 8-3 Calculated pH as a function of the concentration of a strong acid or strong base dissolved in water.

Water Almost Never Produces 10^{-7} M H^+ and 10^{-7} M OH^-

The ion concentrations 10^{-7} M H^+ and 10^{-7} M OH^- occur *only* in extremely pure water with no added acid or base. In a 10^{-4} M solution of HBr, for example, the pH is 4. The concentration of OH^- is $K_w/[H^+] = 10^{-10}$ M. But the source of $[OH^-]$ is dissociation of water. If water produces only 10^{-10} M OH^-, it must also produce only 10^{-10} M H^+, because it makes one H^+ for every OH^-. In a 10^{-4} M HBr solution, water dissociation produces only 10^{-10} M OH^- and 10^{-10} M H^+.

Acids and bases suppress water ionization, as predicted by Le Châtelier's principle.

Question What concentrations of H^+ and OH^- are produced by H_2O dissociation in 10^{-2} M NaOH?

Ask Yourself

8-D. **(a)** What is the pH of **(i)** 1.0×10^{-3} M HBr and **(ii)** 1.0×10^{-2} M KOH?
(b) Calculate the pH of **(i)** 3.2×10^{-5} M HI and **(ii)** 7.7 mM LiOH.
(c) Find the concentration of H^+ in a solution whose pH is 4.44.
(d) Find $[H^+]$ in 7.7 mM LiOH solution. What is the source of this H^+?
(e) Find the pH of 3.2×10^{-9} M tetramethylammonium hydroxide, $(CH_3)_4N^+OH^-$.

8-5 Tools for Dealing with Weak Acids and Bases

By analogy to the definition of pH, we define **p*K*** as the negative logarithm of an equilibrium constant. For the acid dissociation constant in Equation 8-4 and the base hydrolysis constant in Equation 8-5, we can write

$$pK_a = -\log K_a \qquad pK_b = -\log K_b \tag{8-11}$$

The *stronger* an acid, the *smaller* its pK_a.

Stronger acid	*Weaker acid*
$K_a = 10^{-4}$	$K_a = 10^{-8}$
$pK_a = 4$	$pK_a = 8$

However, both $K_a = 10^{-4}$ and $K_a = 10^{-8}$ are classified as weak acids.

Remember from Equation 8-10 the very important relation between K_a and K_b for a conjugate acid-base pair, which are related by gain or loss of a single proton: $K_a \cdot K_b = K_w$.

Weak Is Conjugate to Weak

The conjugate base of a weak acid is a weak base. The conjugate acid of a weak base is a weak acid. Consider a weak acid, HA, with $K_a = 10^{-4}$. The conjugate base, A^-, has $K_b = K_w/K_a = 10^{-10}$. That is, if HA is a weak acid, A^- is a weak base. If the K_a value were 10^{-5}, then the K_b value would be 10^{-9}. As HA becomes a weaker acid, A^- becomes a stronger base (but never a strong base). Conversely, the greater the acid strength of HA, the less the base strength of A^-. However, if either A^- or HA is weak, so is its conjugate. If HA is strong (such as HCl), its conjugate base (Cl^-) is *so* weak that it is not a base at all in water.

The conjugate base of a weak acid is a weak base. The conjugate acid of a weak base is a weak acid.
Weak is conjugate to weak.

Using Appendix B

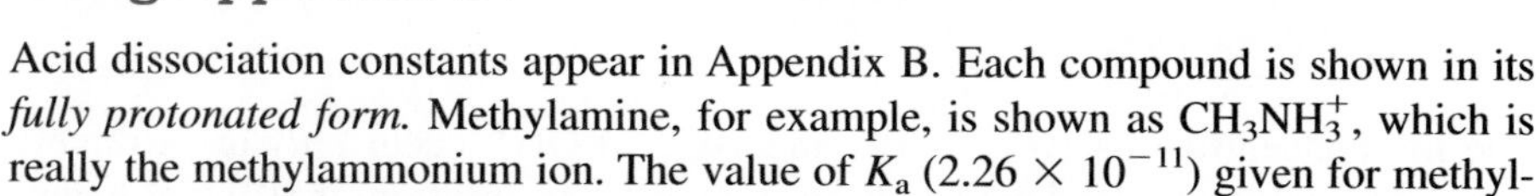

Acid dissociation constants appear in Appendix B. Each compound is shown in its *fully protonated form.* Methylamine, for example, is shown as $CH_3NH_3^+$, which is really the methylammonium ion. The value of K_a (2.26×10^{-11}) given for methylamine is actually K_a for the methylammonium ion. To find K_b for methylamine, we write $K_b = K_w/K_a = (1.0 \times 10^{-14})/(2.26 \times 10^{-11}) = 4.4_2 \times 10^{-4}$.

CH_3NH_2	$CH_3NH_3^+$
Methylamine	Methylammonium ion

For polyprotic acids and bases, several K_a values are given, beginning with the most acidic group. Pyridoxal phosphate is given in its fully protonated form as follows:

Pyridoxal phosphate is derived from vitamin B_6, which is essential in amino acid metabolism in your body. Box 7-1 discussed the drawing of organic structures.

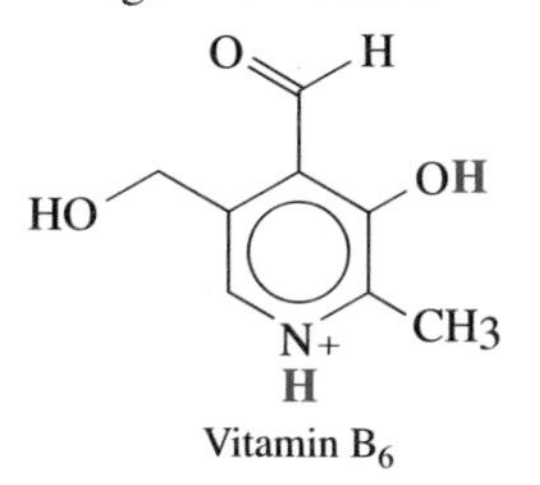

Vitamin B_6

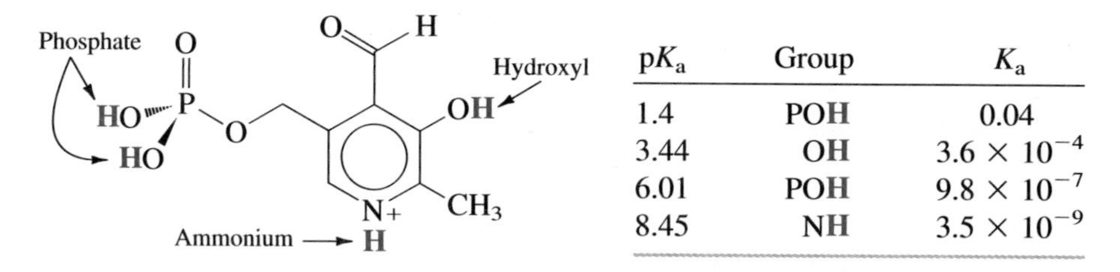

pK_a	Group	K_a
1.4	POH	0.04
3.44	OH	3.6×10^{-4}
6.01	POH	9.8×10^{-7}
8.45	NH	3.5×10^{-9}

pK_1 (1.4) is for dissociation of one of the phosphate protons, and pK_2 (3.44) is for the hydroxyl proton. The third most acidic proton is the other phosphate proton, for which $pK_3 = 6.01$, and the NH^+ group is the least acidic ($pK_4 = 8.45$).

Ask Yourself

8-E. (a) Which acid is stronger, $pK_a = 3$ or $pK_a = 4$?
(b) Which base is stronger, $pK_b = 3$ or $pK_b = 4$?
(c) Write the acid dissociation reaction for formic acid, HCO_2H.
(d) What is the conjugate base of formic acid?
(e) Write the K_a equilibrium expression for formic acid and look up its value.
(f) Write the K_b equilibrium expression for formate ion, HCO_2^-.
(g) Find the base hydrolysis constant for formate.

8-6 Weak-Acid Equilibrium

Let's find the pH and composition of a solution containing 0.020 0 mol benzoic acid in 1.00 L of water.

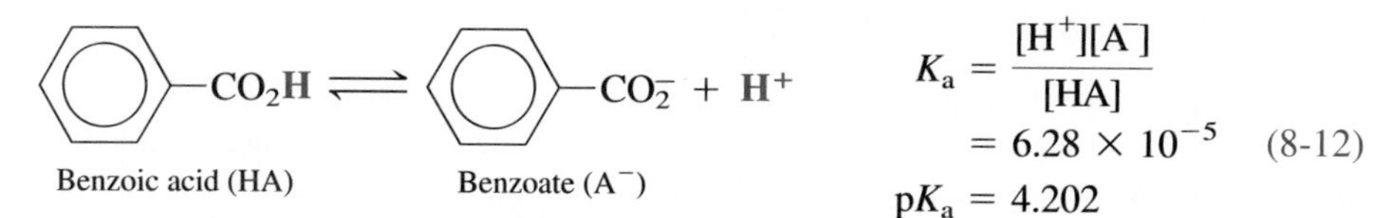

$$K_a = \frac{[H^+][A^-]}{[HA]} = 6.28 \times 10^{-5} \quad (8\text{-}12)$$

$$pK_a = 4.202$$

For every mole of A^- formed by dissociation of HA, one mole of H^+ is created. That is, $[A^-] = [H^+]$. (For reasonable concentrations of weak acids of reasonable strength, the contribution of H^+ from the acid is much greater than the contribution from $H_2O \rightleftharpoons H^+ + OH^-$.) Abbreviating the formal concentration of HA as F and using x for the concentration of H^+, we can make a table showing concentrations before and after dissociation of the weak acid:

	HA	$\rightleftharpoons$	A^-	$+$	H^+
Initial concentration	F		0		0
Final concentration	$F - x$		x		x

Putting these values into the K_a expression in Equation 8-12 gives

$$K_a = \frac{[H^+][A^-]}{[HA]} = \frac{(x)(x)}{F - x} \tag{8-13}$$

Setting F = 0.020 0 M and $K_a = 6.28 \times 10^{-5}$ gives

$$\frac{x^2}{0.020\,0 - x} = 6.28 \times 10^{-5} \tag{8-14}$$

The first step in solving Equation 8-14 for x is to multiply both sides by $(0.020\,0 - x)$:

$$\frac{x^2}{\cancel{0.020\,0 - x}}(\cancel{0.020\,0 - x}) = (6.28 \times 10^{-5})(0.020\,0 - x)$$
$$= (1.25_6 \times 10^{-6}) - (6.28 \times 10^{-5})x$$

Collecting terms gives a quadratic equation (in which the highest power is x^2):

$$x^2 + (6.28 \times 10^{-5})x - (1.25_6 \times 10^{-6}) = 0 \tag{8-15}$$

Equation 8-15 has two solutions (called *roots*) described in Box 8-1. One root is positive and the other is negative. Because a concentration cannot be negative, we reject the negative solution:

$$x = 1.09 \times 10^{-3} \text{ M (negative root rejected)}$$

You should **check your answer** by plugging it back into Equation 8-13 and seeing whether the equation is satisfied.

From the value of x, we can find concentrations and pH:

$$[H^+] = [A^-] = x = 1.09 \times 10^{-3} \text{ M}$$
$$[HA] = F - x = 0.020\,0 - (1.09 \times 10^{-3}) = 0.018\,9 \text{ M}$$
$$pH = -\log x = 2.96$$

For uniformity, we are going to express pH to the 0.01 decimal place, even though significant figures in this problem justify another digit.

Was the approximation $[H^+] \approx [A^-]$ justified? The concentration of $[H^+]$ is 1.09×10^{-3} M, which means $[OH^-] = K_w/[H^+] = 9.20 \times 10^{-12}$ M.

$$[H^+] \text{ from HA dissociation} = [A^-] \text{ from HA dissociation} = 1.09 \times 10^{-3} \text{ M}$$
$$[H^+] \text{ from } H_2O \text{ dissociation} = [OH^-] \text{ from } H_2O \text{ dissociation} = 9.20 \times 10^{-12} \text{ M}$$

In a weak-acid solution, H^+ is derived almost entirely from HA, not from H_2O.

The assumption that H^+ is derived mainly from HA is excellent because 1.09×10^{-3} M >> 9.20×10^{-12} M.

Box 8-1 Quadratic Equations

A quadratic equation of the general form $ax^2 + bx + c = 0$ has two solutions:

$$x = \frac{-b + \sqrt{b^2 - 4ac}}{2a} \qquad x = \frac{-b - \sqrt{b^2 - 4ac}}{2a}$$

The solutions to Equation 8-15

$$\underbrace{(1)}_{a=1}[\mathrm{H}^+]^2 + \underbrace{(6.28 \times 10^{-5})}_{b = 6.28 \times 10^{-5}}[\mathrm{H}^+] - \underbrace{(1.25_6 \times 10^{-6})}_{c = -1.25_6 \times 10^{-6}} = 0$$

are

$$[\mathrm{H}^+] = \frac{-(6.28 \times 10^{-5}) + \sqrt{(6.28 \times 10^{-5})^2 - 4(1)(-1.25_6 \times 10^{-6})}}{2(1)} = 1.09 \times 10^{-3}\ \mathrm{M}$$

and

$$[\mathrm{H}^+] = \frac{-(6.28 \times 10^{-5}) - \sqrt{(6.28 \times 10^{-5})^2 - 4(1)(-1.25_6 \times 10^{-6})}}{2(1)} = -1.09 \times 10^{-3}\ \mathrm{M}$$

Because the concentration of $[\mathrm{H}^+]$ cannot be negative, we reject the negative solution and choose 1.09×10^{-3} M as the correct answer.

When you solve a quadratic equation, retain all the digits in your calculator during the computation, or serious round-off errors can occur in some cases. Alternatively, create a spreadsheet to solve quadratic equations (Problem 8-36) and use it often.

Sometimes, the equation $x^2/(\mathrm{F} - x) = K$ has an easy solution. If $x \ll \mathrm{F}$, then x can be neglected in comparison with F and the denominator can be simplified to just F. In this case, the solution is $x \approx \sqrt{K\mathrm{F}}$. If $K < 10^{-4}$ F, then the error in the approximation $x \approx \sqrt{K\mathrm{F}}$ is <0.5%.

Fraction of Dissociation

What fraction of HA is dissociated? If the total concentration of acid (= [HA] + $[\mathrm{A}^-]$) is 0.020 0 M and the concentration of A^- is 1.09×10^{-3} M, then the *fraction of dissociation* is

Fraction of dissociation of an acid:

$$\frac{[\mathrm{A}^-]}{[\mathrm{A}^-] + [\mathrm{HA}]} = \frac{1.09 \times 10^{-3}}{0.020\,0} = 0.054 \qquad (8\text{-}16)$$

The acid is indeed weak. It is only 5.4% dissociated.

Figure 8-4 compares the fraction of dissociation of two weak acids as a function of formal concentration. As the solution becomes more dilute, the fraction of dissociation increases. The stronger acid has a greater fraction of dissociation at any formal concentration. Demonstration 8-2 compares properties of strong and weak acids.

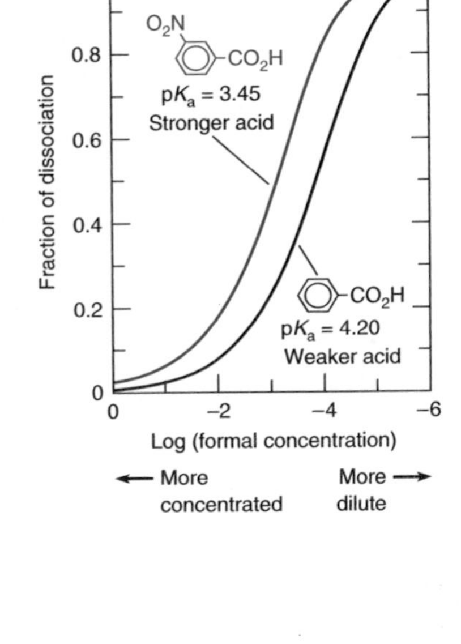

Figure 8-4 The fraction of dissociation of a weak acid increases as it is diluted. The stronger of the two acids is more dissociated at any given concentration. (pK_a = 3.45 is stronger than pK_a = 4.20.)

The Essence of a Weak-Acid Problem

When faced with a weak acid, you should immediately realize that $[H^+] \approx [A^-] = x$ and proceed to set up and solve the equation

Equation for weak acids:

$$\frac{[H^+][A^-]}{[HA]} = \frac{x^2}{F - x} = K_a \tag{8-17}$$

where F is the formal concentration of HA. The approximation $[H^+] \approx [A^-]$ would be poor only if the acid were very dilute or very weak, neither of which constitutes a practical problem.

Finding the pH of a Weak Acid

Find the pH of 0.100 M trimethylammonium chloride.

$$\left[(H_3C)_3N\text{–}H\right]^+ \quad Cl^- \qquad \text{Trimethylammonium chloride}$$

SOLUTION We must first realize that salts of this type are *completely dissociated* to give $(CH_3)_3NH^+$ and Cl^-. We then recognize that trimethylammonium ion is a weak acid, being conjugate to trimethylamine, $(CH_3)_3N$, a weak base. The ion Cl^- has no basic or acidic properties and should be ignored. In Appendix B, we find

Demonstration 8-2 Conductivity of Weak Electrolytes

The relative conductivity of strong and weak acids is directly related to their different degrees of dissociation in aqueous solution. To demonstrate conductivity, we use a Radio Shack piezo alerting buzzer, but any kind of buzzer or light bulb could be substituted for the electric horn. The voltage required will depend on the buzzer or light chosen.

When a conducting solution is placed in the beaker, the horn sounds. First show that distilled water and sucrose solution are nonconductive. Solutions of the strong electrolytes NaCl or HCl are conductive. Compare strong and weak electrolytes by demonstrating that 1 mM HCl gives a loud sound, whereas 1 mM acetic acid gives little or no sound. With 10 mM acetic acid, the strength of the sound varies noticeably as the electrodes are moved away from each other in the beaker.

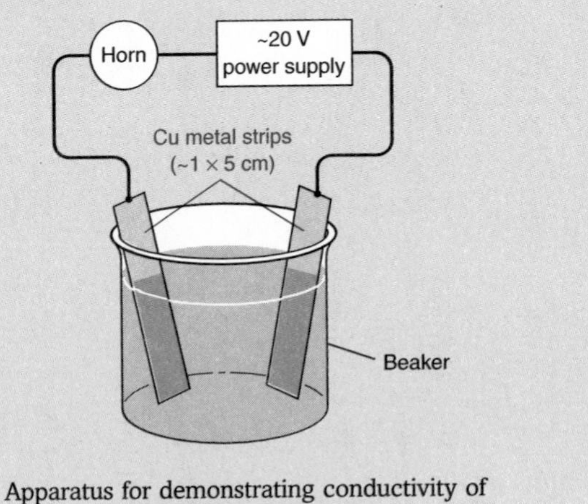

Apparatus for demonstrating conductivity of electrolyte solutions.

trimethylammonium ion listed under the name trimethylamine but drawn as the trimethylammonium ion. The value of pK_a is 9.799, so

$$K_a = 10^{-pK_a} = 10^{-9.799} = 1.59 \times 10^{-10}$$

It's all downhill from here:

$$\underset{F - x}{(CH_3)_3NH^+} \overset{K_a}{\rightleftharpoons} \underset{x}{(CH_3)_3N} + \underset{x}{H^+}$$

$$\frac{x^2}{0.100 - x} = 1.59 \times 10^{-10}$$

$$x = 3.99 \times 10^{-6}\ M \Rightarrow pH = -\log(3.99 \times 10^{-6}) = 5.40$$

Test Yourself What is the pH of 0.010 M dimethylammonium nitrate? (**Answer:** 6.39)

Example **Finding pK_a of a Weak Acid**

A 0.100 M solution of hydrazoic acid has pH = 2.83. Find pK_a for this acid.

$$\underset{\substack{\text{Hydrazoic acid} \\ 0.100 - x}}{:\overset{-\cdot\cdot}{N}=\overset{+}{N}=\overset{\cdot\cdot}{N}-H} \overset{K_a}{\rightleftharpoons} \underset{\substack{\text{Azide } (N_3^-) \\ x}}{:\overset{-\cdot\cdot}{N}=\overset{+}{N}=\overset{\cdot\cdot -}{N}:} + \underset{10^{-pH}}{H^+}$$

SOLUTION We know that $[H^+] = 10^{-pH} = 10^{-2.83} = 1.4_8 \times 10^{-3}$ M. Because $[N_3^-] = [H^+]$ in this solution, $[N_3^-] = 1.4_8 \times 10^{-3}$ M and $[HN_3] = 0.100 - 1.4_8 \times 10^{-3} = 0.098_5$ M. From these concentrations, we calculate K_a and pK_a:

$$K_a = \frac{[N_3^-][H^+]}{[HN_3]} = \frac{(1.4_8 \times 10^{-3})^2}{0.098_5} = 2.2_2 \times 10^{-5}$$

$$\Rightarrow pK_a = -\log(2.2_2 \times 10^{-5}) = 4.65$$

Test Yourself A 0.063 M solution of hydroxybenzene has pH = 5.60. Find pK_a for this acid. (**Answer:** 10.00)

? *Ask Yourself*

8-F. **(a)** What is the pH and fraction of dissociation of a 0.100 M solution of the weak acid HA with $K_a = 1.00 \times 10^{-5}$?
(b) A 0.045 0 M solution of HA has a pH of 2.78. Find pK_a for HA.
(c) A 0.045 0 M solution of HA is 0.60% dissociated. Find pK_a for HA.

8-7 Weak-Base Equilibrium

The treatment of weak bases is almost the same as that of weak acids.

$$B + H_2O \overset{K_b}{\rightleftharpoons} BH^+ + OH^- \qquad K_b = \frac{[BH^+][OH^-]}{[B]}$$

Nearly all OH^- comes from the reaction of $B + H_2O$, and little comes from dissociation of H_2O. Setting $[OH^-] = x$, we must also set $[BH^+] = x$, because one BH^+ is produced for each OH^-. Setting $F = [B] + [BH^+]$, we can write

$$[B] = F - [BH^+] = F - x$$

Plugging these values into the K_b equilibrium expression, we get

Equation for weak base:
$$\frac{[BH^+][OH^-]}{[B]} = \frac{x^2}{F - x} = K_b \qquad (8\text{-}18)$$

A weak-base problem has the same algebra as a weak-acid problem, except $K = K_b$ and $x = [OH^-]$.

which looks a lot like a weak-acid problem, except that now $x = [OH^-]$.

Example Finding the pH of a Weak Base

Find the pH of a 0.037 2 M solution of the commonly encountered weak base, cocaine:

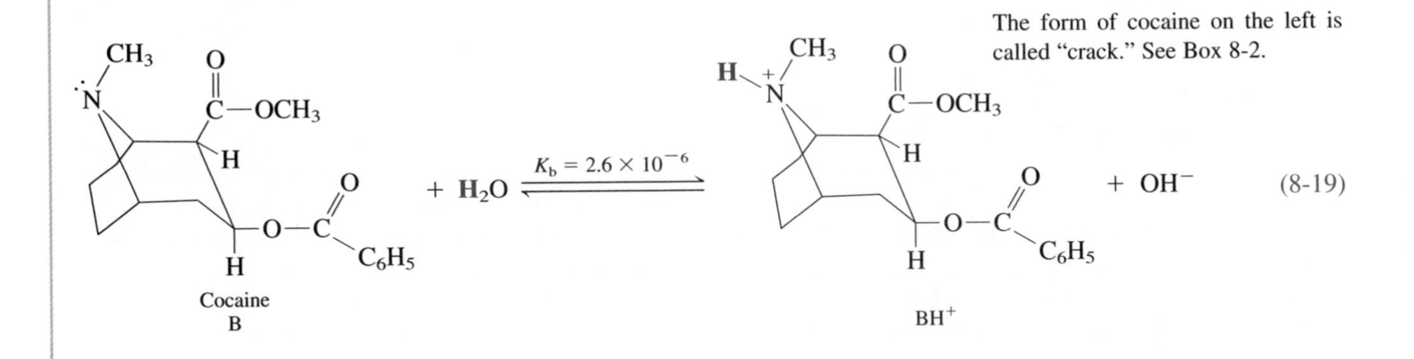

The form of cocaine on the left is called "crack." See Box 8-2.

SOLUTION Designating cocaine as B, we formulate the problem as follows:

$$\underset{0.037\,2 - x}{B} + H_2O \rightleftharpoons \underset{x}{BH^+} + \underset{x}{OH^-}$$

$$\frac{x^2}{0.037\,2 - x} = 2.6 \times 10^{-6} \Rightarrow x = 3.1_0 \times 10^{-4}\ \text{M}$$

Because $x = [OH^-]$, we can write

$$[H^+] = K_w/[OH^-] = (1.0 \times 10^{-14})/(3.1_0 \times 10^{-4}) = 3.2_2 \times 10^{-11}\ \text{M}$$
$$\text{pH} = -\log[H^+] = 10.49$$

This is a reasonable pH for a weak base.

Question What concentration of OH^- is produced by H_2O dissociation in this solution? Were we justified in neglecting water dissociation as a source of OH^-?

Test Yourself Find the pH of a 0.010 M solution of morphine, another weak base with $K_b = 1.6 \times 10^{-6}$. (**Answer:** 10.10)

Box 8-2 Five Will Get You Ten: Crack Cocaine

Cocaine is a white crystalline powder obtained as the hydrochloride salt, BH^+Cl^- in Reaction 8-19, by extraction from coca leaves. In the 1980s, drug traffickers discovered the enhanced value of the free base or "crack" form of cocaine (B in Reaction 8-19). "Crack" is soluble in organic solvents and cell membranes, melts at 98°C, and has a moderate vapor pressure. The ionic hydrochloride is soluble in water (but not in cell membranes), melts at 195°C, and has lower vapor pressure. Its physiologic action is slower than that of "crack."

The free base is obtained from the hydrochloride by reaction with baking soda:

Dried leaves from coca, a shrub native to the Andes, yield cocaine and other alkaloids. [© blickwinkel/Alamy.]

$$\underset{\substack{\text{Bicarbonate} \\ \text{from baking} \\ \text{soda (NaHCO}_3\text{)}}}{HCO_3^-} + \underset{\substack{\text{Cocaine} \\ \text{hydrochloride}}}{BH^+} \longrightarrow H_2O + \underset{\substack{\text{Carbon dioxide} \\ \text{escapes, driving} \\ \text{reaction to completion}}}{CO_2(g)\uparrow} + \underset{\substack{\text{Free base} \\ \text{"crack" cocaine}}}{B}$$

Residue from the reaction is dried under a heat lamp and then broken into small "rocks." These rocks, when smoked in a pipe, produce a cracking sound from which the name "crack" is derived.

The perceived difference between the acid and base forms of cocaine is evident from different penalties attached to their possession. In the United States, 0.5 to 5 *kilograms* of the acid earns a jail sentence of 5 to 40 years, whereas 5 to 50 *grams* of the base gets the same sentence.[3] The person who drafted the bill in 1986 that created these disparate sentences stated in 2006 that it was "one of the most unjust laws passed in recent memory."[4] He now believes that the penalties for possession of "crack" or the hydrochloride should be equal and that the federal government should focus on the prosecution of high-level distributors.

What fraction of cocaine reacted with water?

$$\textit{Fraction of association of a base:} \quad \frac{[BH^+]}{[BH^+] + [B]} = \frac{3.1_0 \times 10^{-4}}{0.037\,2} = 0.008\,3 \qquad (8\text{-}20)$$

because $[BH^+] = [OH^-] = 3.1 \times 10^{-4}$ M. Only 0.83% of the base has reacted.

Conjugate Acids and Bases—Revisited

HA and A^- are a conjugate acid-base pair. So are BH^+ and B.

The conjugate base of a weak acid is a weak base, and **the conjugate acid of a weak base is a weak acid**. The exceedingly important relation between the equilibrium constants for a conjugate acid-base pair is

$$K_a \cdot K_b = K_w$$

Reaction 8-12 featured benzoic acid, designated HA. Now consider 0.050 M sodium benzoate, Na^+A^-, which contains the conjugate base of benzoic acid. When this salt dissolves in water, it dissociates to Na^+ and A^-. Na^+ does not react with water, but A^- is a weak base:

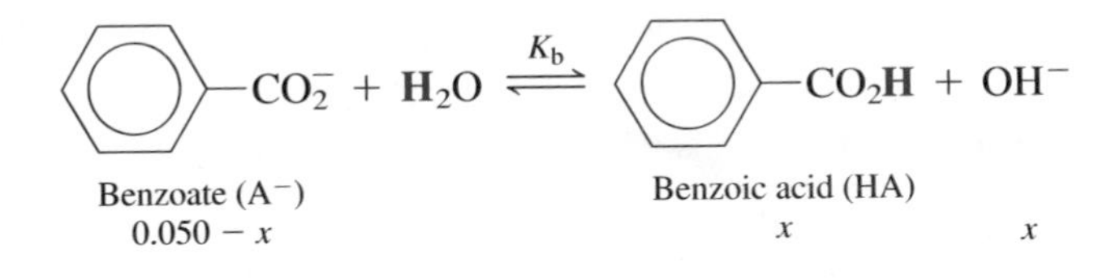

$$K_b = \frac{K_w}{K_a(\text{for benzoic acid})} = \frac{1.0 \times 10^{-14}}{6.28 \times 10^{-5}} = 1.5_9 \times 10^{-10}$$

To find the pH of this solution, we write

$$\frac{[HA][OH^-]}{[A^-]} = \frac{x^2}{0.050 - x} = 1.5_9 \times 10^{-10} \Rightarrow x = [OH^-] = 2.8 \times 10^{-6}\ M$$

$$[H^+] = K_w/[OH^-] = 3.5 \times 10^{-9}\ M \Rightarrow pH = 8.45$$

This is a reasonable pH for a solution of a weak base.

Example A Weak-Base Problem

Find the pH of 0.10 M ammonia.

SOLUTION When ammonia dissolves in water, its reaction is

$$\underset{\substack{\text{Ammonia} \\ F - x}}{NH_3} + H_2O \overset{K_b}{\rightleftharpoons} \underset{\substack{\text{Ammonium ion} \\ x}}{NH_4^+} + \underset{x}{OH^-}$$

In Appendix B, we find the ammonium ion, NH_4^+, listed next to ammonia. K_a for ammonium ion is 5.69×10^{-10}. Therefore K_b for NH_3 is

$$K_b = \frac{K_w}{K_a} = \frac{1.0 \times 10^{-14}}{5.69 \times 10^{-10}} = 1.7_6 \times 10^{-5}$$

To find the pH of 0.10 M NH_3, we set up and solve the equation

$$\frac{[NH_4^+][OH^-]}{[NH_3]} = \frac{x^2}{0.10 - x} = K_b = 1.7_6 \times 10^{-5}$$

$$x = [OH^-] = 1.3_2 \times 10^{-3}\ M$$

$$[H^+] = \frac{K_w}{[OH^-]} = 7.5_9 \times 10^{-12}\ M \Rightarrow pH = -\log[H^+] = 11.12$$

Test Yourself Find the pH of 5.0 mM diethylamine. (**Answer:** 11.25)

Ask Yourself

8-G. (a) What is the pH and fraction of association of a 0.100 M solution of a weak base with $K_b = 1.00 \times 10^{-5}$?
(b) A 0.10 M solution of a base has pH = 9.28. Find K_b.
(c) A 0.10 M solution of a base is 2.0% associated. Find K_b.

Key Equations

Conjugate acids and bases

$$\underset{\text{Acid}}{HA} + \underset{\text{Base}}{B} \rightleftharpoons \underset{\text{Base}}{A^-} + \underset{\text{Acid}}{BH^+}$$

Autoprotolysis of water

$$H_2O \overset{K_w}{\rightleftharpoons} H^+ + OH^- \qquad K_w = [H^+][OH^-] = 1.0 \times 10^{-14}\ (25°C)$$

Finding $[OH^-]$ from $[H^+]$

$$[OH^-] = K_w/[H^+]$$

Definition of pH

$pH = -\log[H^+]$ (This relation is only approximate—but it is the one we use in this book.)

Acid dissociation constant

$$HA \overset{K_a}{\rightleftharpoons} H^+ + A^- \qquad K_a = \frac{[H^+][A^-]}{[HA]}$$

Base hydrolysis constant

$$B + H_2O \overset{K_b}{\rightleftharpoons} BH^+ + OH^- \qquad K_b = \frac{[BH^+][OH^-]}{[B]}$$

Relation between K_a and K_b for conjugate acid-base pair

$$K_a \cdot K_b = K_w$$

Common weak acids

$$\underset{\text{Carboxylic acid}}{RCO_2H} \qquad \underset{\text{Ammonium ion}}{R_3NH^+}$$

Common weak bases

$$\underset{\text{Carboxylate anion}}{RCO_2^-} \qquad \underset{\text{Amine}}{R_3N}$$

Definitions of pK

$$pK_a = -\log K_a \qquad pK_b = -\log K_b$$

Weak-acid equilibrium

$$\underset{F-x}{HA} \overset{K_a}{\rightleftharpoons} \underset{x}{H^+} + \underset{x}{A^-} \qquad K_a = \frac{[H^+][A^-]}{[HA]} = \frac{x^2}{F-x}$$

(F = formal concentration of weak acid or weak base)

Weak-base equilibrium

$$\underset{F-x}{B} + H_2O \overset{K_b}{\rightleftharpoons} \underset{x}{BH^+} + \underset{x}{OH^-} \qquad K_b = \frac{[BH^+][OH^-]}{[B]} = \frac{x^2}{F-x}$$

Fraction of dissociation of HA

$$\text{fraction of dissociation} = \frac{[A^-]}{[A^-] + [HA]}$$

Fraction of association of B

$$\text{fraction of association} = \frac{[BH^+]}{[BH^+] + [B]}$$

Important Terms

acid	base hydrolysis constant	pH
acid dissociation constant	basic solution	pK
acidic solution	carboxylate anion	salt
amine	carboxylic acid	strong acid
ammonium ion	conjugate acid-base pair	strong base
autoprotolysis	hydronium ion	weak acid
base	neutralization	weak base

Problems

8-1. Identify the conjugate acid-base pairs in the following reactions:

(a) $CN^- + HCO_2H \rightleftharpoons HCN + HCO_2^-$

(b) $PO_4^{3-} + H_2O \rightleftharpoons HPO_4^{2-} + OH^-$

(c) $HSO_3^- + OH^- \rightleftharpoons SO_3^{2-} + H_2O$

8-2. A solution is *acidic* if __________. A solution is *basic* if __________.

8-3. Find the pH of a solution containing

(a) 10^{-4} M H^+

(b) 10^{-5} M OH^-

(c) 5.8×10^{-4} M H^+

(d) 5.8×10^{-5} M OH^-

8-4. The concentration of H^+ in your blood is 3.5×10^{-8} M.

(a) What is the pH of blood?

(b) Find the concentration of OH^- in blood.

8-5. Sulfuric acid is the principal acidic component of acid rain. The mean pH of rainfall in southern Norway in 2000 was 4.6. What concentration of H_2SO_4 will produce this pH by the reaction $H_2SO_4 \rightleftharpoons H^+ + HSO_4^-$?

8-6. An acidic solution containing 0.010 M La^{3+} is treated with NaOH until $La(OH)_3$ precipitates. Use the solubility product for $La(OH)_3$ to find the concentration of OH^- when La^{3+} first precipitates. At what pH does this occur?

8-7. Make a list of the common strong acids and strong bases. Memorize this list.

8-8. Write the structures and names for two classes of weak acids and two classes of weak bases.

8-9. Calculate $[H^+]$ and pH for the following solutions:

(a) 0.010 M HNO_3

(b) 0.035 M KOH

(c) 0.030 M HCl

(d) 3.0 M $HClO_4$

(e) 0.010 M $[(CH_3)_4N^+]OH^-$
Tetramethylammonium hydroxide

8-10. **(a)** Write the K_a reaction for trichloroacetic acid, Cl_3CCO_2H ($K_a = 0.3$), for anilinium ion, and for Cu^{2+} ($K_a = 3 \times 10^{-8}$).

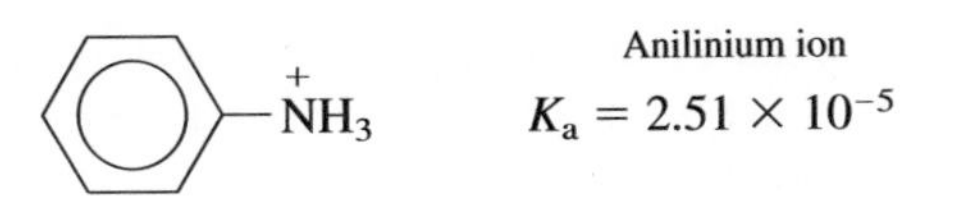

(b) Which of the three is the strongest acid?

8-11. **(a)** Write the K_b reactions for pyridine, sodium 2-mercaptoethanol, and cyanide.

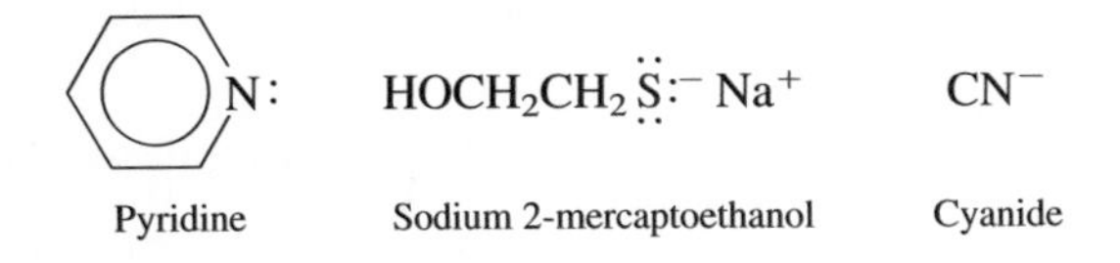

H^+ binds to S in sodium 2-mercaptoethanol and to C in cyanide.

(b) Values of K_a for the conjugate acids are shown here. Which base in part **(a)** is strongest?

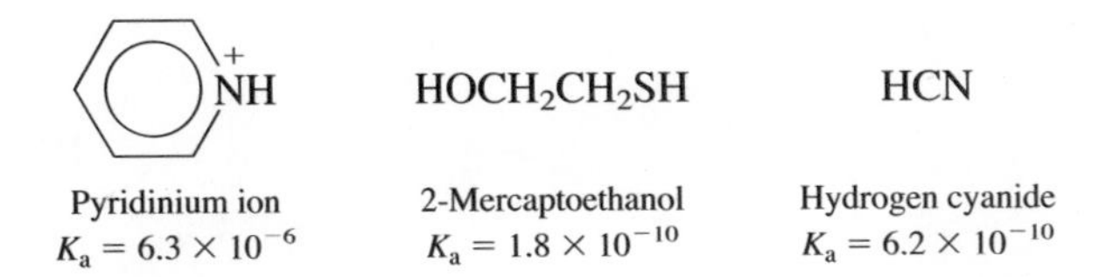

8-12. Write the autoprotolysis reaction of H_2SO_4, whose structure is

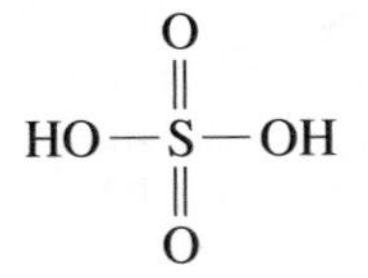

8-13. Write the K_b reactions for piperidine and benzoate.

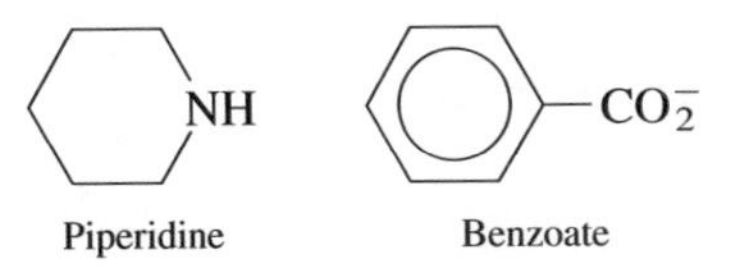

8-14. Hypochlorous acid has the structure H—O—Cl. Write the base hydrolysis reaction of hypochlorite, OCl^-. Given that K_a for HOCl is 3.0×10^{-8}, find K_b for hypochlorite.

8-15. Calculate the pH of 3.0×10^{-5} M $Mg(OH)_2$, which completely dissociates to Mg^{2+} and OH^-.

8-16. Find the concentration of H^+ in a solution whose pH is 11.65.

8-17. Find the pH and fraction of dissociation of a 0.010 0 M solution of the weak acid HA with $K_a = 1.00 \times 10^{-4}$.

8-18. Find the pH and fraction of dissociation in a 0.150 M solution of hydroxybenzene (also called phenol).

8-19. Calculate the pH of 0.085 0 M pyridinium bromide, $C_5H_5NH^+Br^-$. Find the concentrations of pyridine (C_5H_5N), pyridinium ion ($C_5H_5NH^+$), and Br^- in the solution.

8-20. Calculate the pH of 0.10 M $Zn(NO_3)_2$, which dissolves to give 0.10 M Zn^{2+} ($pK_a = 9.0$) and 0.20 M NO_3^-.

8-21. A 0.100 M solution of the weak acid HA has a pH of 2.36. Calculate pK_a for HA.

8-22. A 0.022 2 M solution of HA is 0.15% dissociated. Calculate pK_a for this acid.

8-23. Find the pH and concentrations of $(CH_3)_3N$ and $(CH_3)_3NH^+$ in a 0.060 M solution of trimethylammonium chloride.

8-24. Calculate the pH and fraction of dissociation of **(a)** $10^{-2.00}$ M and **(b)** $10^{-10.00}$ M barbituric acid.

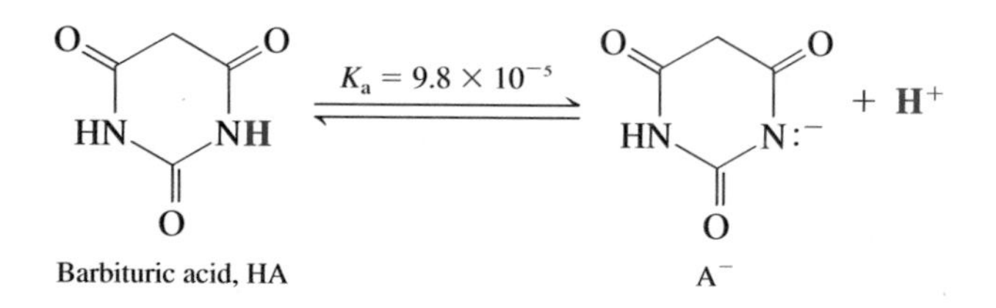

8-25. $BH^+ClO_4^-$ is a salt formed from the base B ($K_b = 1.00 \times 10^{-4}$) and perchloric acid. It dissociates into BH^+, a weak acid, and ClO_4^-, which is neither an acid nor a base. Find the pH of 0.100 M $BH^+ClO_4^-$.

8-26. Find K_a for cyclohexylammonium ion and K_b for cyclohexylamine.

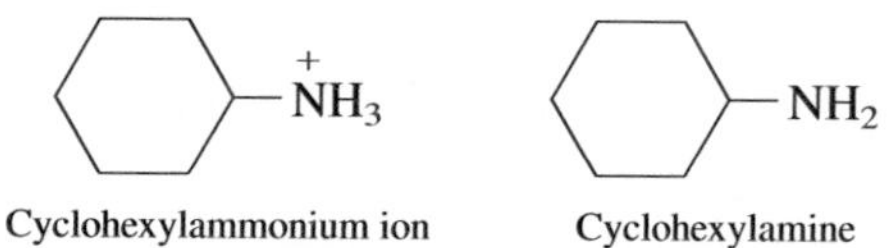

8-27. Write the chemical reaction whose equilibrium constant is **(a)** K_b for 2-aminoethanol and **(b)** K_a for 2-aminoethanol hydrobromide.

$HOCH_2CH_2NH_2$ — 2-Aminoethanol

$HOCH_2CH_2NH_3^+Br^-$ — 2-Aminoethanol hydrobromide

8-28. Find the pH and concentrations of $(CH_3)_3N$ and $(CH_3)_3NH^+$ in a 0.060 M solution of trimethylamine.

8-29. Calculate the pH and fraction of association of 1.00×10^{-1}, 1.00×10^{-2}, and 1.00×10^{-12} M sodium acetate.

8-30. Find the pH of 0.050 M NaCN.

8-31. Find the pH and fraction of association of 0.026 M NaOCl.

8-32. If a 0.030 M solution of a base has pH = 10.50, find K_b for the base.

8-33. In a 0.030 M solution of a base, 0.27% of B underwent hydrolysis to make BH^+. Find K_b for the base.

8-34. The smell (and taste) of fish arises from amine compounds. Suggest a reason why adding lemon juice (which is acidic) reduces the "fishy" smell.

8-35. Cr^{3+} has $pK_a = 3.80$. Find the pH of 0.010 M $Cr(ClO_4)_3$. What fraction of chromium is in the form $Cr(H_2O)_{w-1}(OH)^{2+}$?

8-36. Create a spreadsheet to solve the equation $x^2/(F - x) = K$ by using the formula for the roots of a quadratic equation. The input will be F and K. The output is the positive value of x. Use your spreadsheet to check your answer to **(a)** of Ask Yourself Problem 8-F.

8-37. *Excel* GOAL SEEK. Solve the equation $x^2/(F - x) = K$ by using GOAL SEEK described in Problem 6-24. Guess a value of x in cell A4 and evaluate $x^2/(F - x)$ in cell B4. Use GOAL SEEK to vary the value of x until $x^2/(F - x)$ is equal to K. Use your spreadsheet to check your answer to **(a)** of Ask Yourself Problem 8-F.

	A	B
1	Using Excel GOAL SEEK	
2		
3	x	x^2/(F-x)
4	0.01	1.11E-03
5	F =	
6	0.1	

How Would You Do It?

8-38. *Weighted average pH in precipitation.* The following data were reported for rainfall at the Philadelphia airport in 1990:

Season	Precipitation (cm)	Weighted average pH
Winter	17.3	4.40
Spring	30.5	4.68
Summer	17.8	4.68
Fall	14.7	5.10

Suggest a procedure to calculate the average pH for the entire year, which would be the pH observed if all the rain for the whole year were pooled in one container.

Notes and References

1. For "A Demonstration of Acid Rain and Lake Acidification: Wet Deposition of Sulfur Dioxide," see L. M. Goss, *J. Chem. Ed.* **2003**, *80*, 39.

2. For related demonstrations, see S.-J. Kang and E.-H. Ryu, "Carbon Dioxide Fountain," *J. Chem. Ed.* **2007**, *84*, 1671; M. D. Alexander, "The Ammonia Smoke Fountain," *J. Chem. Ed.* **1999**, *76*, 210.

3. http:/www.dea.gov/pubs/abuse/1-csa.htm#penalties, U.S. Department of Justice, Drug Enforcement Administration.

4. E. E. Sterling, "Take Another Crack at That Cocaine Law," *Los Angeles Times*, 13 November 2006, p. A17.

Measuring pH Inside Single Cells

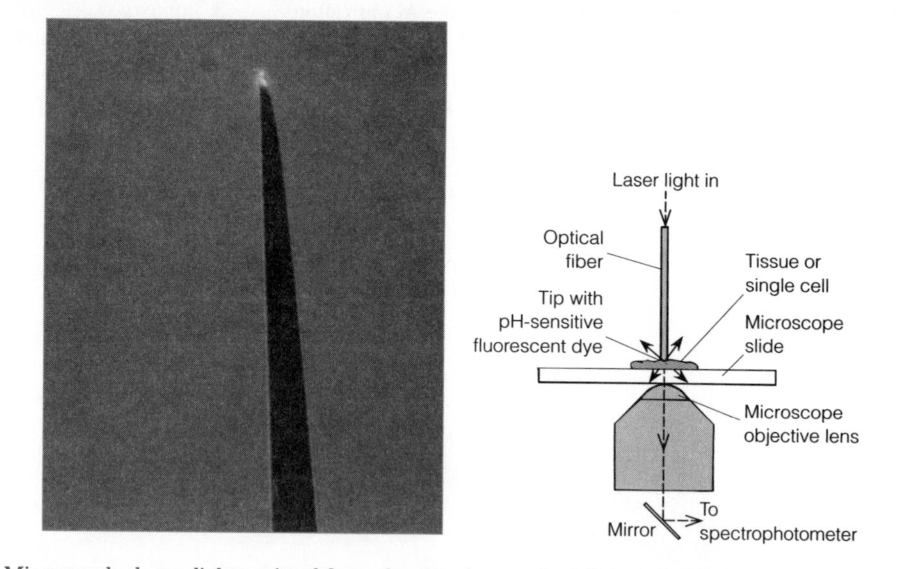

Micrograph shows light emitted from the tip of an optical fiber inserted into a rat embryo. The emission spectrum depends on the pH of fluid surrounding the fiber. Prior to the development of this microscopic method, 1 000 embryos had to be homogenized for a single measurement. [Courtesy R. Kopelman and W. Tan, University of Michigan.]

An exquisitely wrought optical fiber with a pH-sensitive, fluorescent dye bound to its tip can be used to measure the pH inside embryos and even single cells.[1] The fiber is inserted into the specimen and laser light is directed down the fiber. Dye molecules at the tip of the fiber absorb the laser light and then emit fluorescence whose spectrum depends on the pH of the surrounding medium. pH is one of the most important controlling factors in the rates and thermodynamics of every biological process.

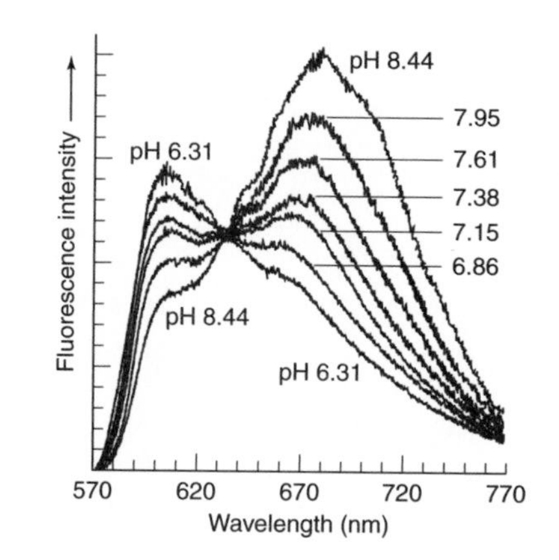

Spectra showing how fluorescence depends on pH. The ratio of the intensity of the peak near 680 nanometers to the intensity of the peak near 600 nanometers is sensitive to changes in pH in the physiologic range about pH 7. [From A. Song, S. Parus, and R. Kopelman, *Anal. Chem.* **1997**, *69*, 863.]

Embryo age (days)	pH[a]
10	**7.51 ± 0.035**
11	**7.40 ± 0.026**
12	**7.31 ± 0.021**

a. Measured on seven embryos.

CHAPTER 9

Buffers

A buffered solution resists changes in pH when small amounts of acids or bases are added or when it is diluted. The **buffer** consists of a mixture of a weak acid and its conjugate base.

Biochemists are particularly interested in buffers because the functioning of biological systems depends critically on pH. For example, Figure 9-1 shows how the rate of a particular enzyme-catalyzed reaction varies with pH. *Enzymes* are proteins that *catalyze* (increase the rate of) selected chemical reactions. For any organism to survive, it must control the pH of each subcellular compartment so that each of its enzyme-catalyzed reactions can proceed at the proper rate.

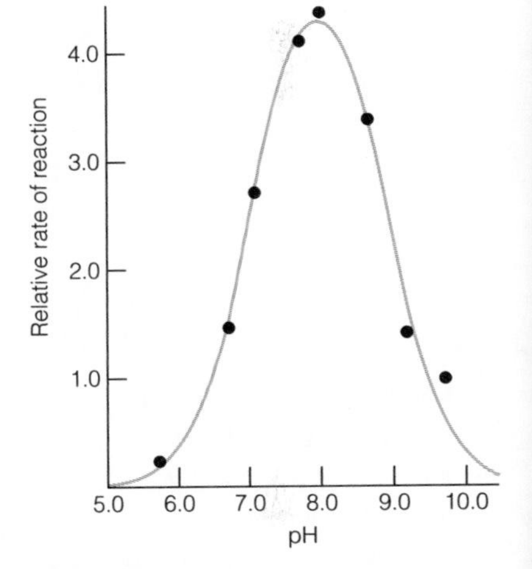

Figure 9-1 pH dependence of the rate of cleavage of an amide bond by the enzyme chymotrypsin. The rate near pH 8 is twice as great as the rate near pH 7 or pH 9. Chymotrypsin helps digest proteins in your intestine. [M. L. Bender, G. E. Clement, F. J. Kézdy, and H. A. Heck, *J. Am. Chem. Soc.* **1964**, *86*, 3680.]

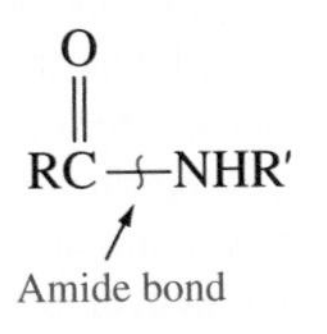

9-1 What You Mix Is What You Get

If you mix A moles of a weak acid with B moles of its conjugate base, the moles of acid remain close to A and the moles of base remain close to B. Little reaction occurs to change either concentration.

To understand why this should be so, look at the K_a and K_b reactions in terms of Le Châtelier's principle. Consider an acid with $pK_a = 4.00$ and its conjugate base with $pK_b = 10.00$. We will calculate the fraction of acid that dissociates in a 0.100 M solution of HA.

$$\underset{0.100-x}{\text{HA}} \underset{}{\overset{K_a}{\rightleftharpoons}} \underset{x}{\text{H}^+} + \underset{x}{\text{A}^-} \qquad pK_a = 4.00$$

$$\frac{x^2}{\text{F}-x} = K_a = 1.0 \times 10^{-4} \Rightarrow x = 3.1 \times 10^{-3}\ \text{M}$$

$$\text{fraction of dissociation} = \frac{[\text{A}^-]}{[\text{A}^-]+[\text{HA}]} = \frac{x}{\text{F}} = 0.031$$

F is the formal concentration of HA, which is 0.100 M in this example.

The acid is only 3.1% dissociated under these conditions.

In a solution containing 0.100 mol of A^- dissolved in 1.00 L of water, the extent of reaction of A^- with water is even smaller:

$$\underset{0.100-x}{A^-} + H_2O \underset{}{\overset{K_b}{\rightleftharpoons}} \underset{x}{HA} + \underset{x}{OH^-} \qquad pK_b = 10.00$$

$$\frac{x^2}{F-x} = K_b = 1.0 \times 10^{-10} \Rightarrow x = 3.2 \times 10^{-6}\ M$$

$$\text{fraction of association} = \frac{[HA]}{[A^-]+[HA]} = \frac{x}{F} = 3.2 \times 10^{-5}$$

When you mix a weak acid and its conjugate base, what you mix is what you get. This approximation breaks down if the solution is too dilute or the acid or base is too weak. We will not consider these cases.

HA dissociates very little, and Le Châtelier's principle tells us that adding extra A^- to the solution will make the HA dissociate even less. Similarly, A^- does not react very much with water, and adding extra HA makes A^- react even less. If 0.050 mol of A^- plus 0.036 mol of HA are added to water, there will be close to 0.050 mol of A^- and close to 0.036 mol of HA in the solution at equilibrium.

9-2 The Henderson-Hasselbalch Equation

The central equation for buffers is the **Henderson-Hasselbalch equation**, which is merely a rearranged form of the K_a equilibrium expression:

$$K_a = \frac{[H^+][A^-]}{[HA]}$$

$$\log K_a = \log\left(\frac{[H^+][A^-]}{[HA]}\right) = \log[H^+] + \log\left(\frac{[A^-]}{[HA]}\right)$$

Useful logarithm rules:
$\log xy = \log x + \log y$
$\log x/y = \log x - \log y$
$\log x^y = y \log x$

Now multiply both sides by -1 and rearrange to isolate $-\log[H^+]$:

$$\underbrace{-\log[H^+]}_{pH} = \underbrace{-\log K_a}_{pK_a} + \log\left(\frac{[A^-]}{[HA]}\right)$$

Henderson-Hasselbalch equation for an acid:

$$HA \overset{K_a}{\rightleftharpoons} H^+ + A^-$$

$$pH = pK_a + \log\left(\frac{[A^-]}{[HA]}\right) \qquad (9\text{-}1)$$

Base in numerator
Acid in denominator

L. J. Henderson was a physician who wrote $[H^+] = K_a[\text{acid}]/[\text{salt}]$ in a physiology article in 1908, a year before the word "buffer" and the concept of pH were invented by the biochemist Sørensen. Henderson's contribution was the approximation of setting [acid] equal to the concentration of HA placed in solution and [salt] equal to the concentration of A^- placed in solution. In 1916, K. A. Hasselbalch wrote what we call the Henderson-Hasselbalch equation in a biochemical journal.[2]

The Henderson-Hasselbalch equation tells us the pH of a solution, provided we know the ratio of concentrations of conjugate acid and base, as well as pK_a for the acid.

If a solution is prepared from the weak base B and its conjugate acid, the analogous equation is

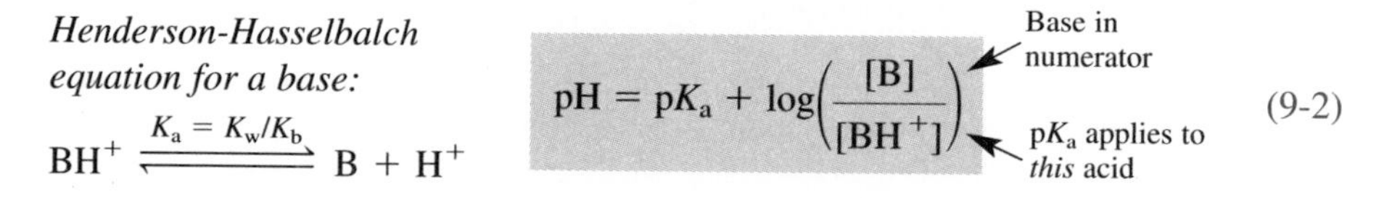

$$BH^+ \overset{K_a = K_w/K_b}{\rightleftharpoons} B + H^+$$

$$pH = pK_a + \log\left(\frac{[B]}{[BH^+]}\right) \qquad (9\text{-}2)$$

where pK_a is the acid dissociation constant of the weak acid BH^+. The important features of Equations 9-1 and 9-2 are that (1) the base (A^- or B) appears in the numerator and (2) pK_a applies to the acid in the denominator.

When $[A^-] = [HA]$, $pH = pK_a$

When the concentrations of A^- and HA are equal in Equation 9-1, the log term is 0 because $\log(1) = 0$. Therefore, when $[A^-] = [HA]$, $pH = pK_a$.

$\log(1) = 0$ because $10^0 = 1$

$$pH = pK_a + \log\left(\frac{[A^-]}{[HA]}\right) = pK_a + \log(1) = pK_a$$

Regardless of how complex a solution may be, whenever $pH = pK_a$, $[A^-]$ must equal [HA]. This relation is true because *all equilibria must be satisfied simultaneously in any solution at equilibrium.* If there are 10 different acids and bases in the solution, the 10 forms of Equation 9-1 must all give the same pH, because **there can be only one concentration of H^+ in a solution**.

Table 9-1 Change of pH with change of $[A^-]/[HA]$

$[A^-]/[HA]$	pH
100:1	$pK_a + 2$
10:1	$pK_a + 1$
1:1	pK_a
1:10	$pK_a - 1$
1:100	$pK_a - 2$

When $[A^-]/[HA]$ Changes by a Factor of 10, the pH Changes by One Unit

Another feature of the Henderson-Hasselbalch equation is that, for every power-of-10 change in the ratio $[A^-]/[HA]$, the pH changes by one unit (Table 9-1). As the concentration of base (A^-) increases, the pH goes up. As the concentration of acid (HA) increases, the pH goes down. For any conjugate acid-base pair, you can say, for example, that, if $pH = pK_a - 1$, there must be 10 times as much HA as A^-. HA is ten-elevenths of the mixture and A^- is one-eleventh.

If $pH = pK_a$, $[HA] = [A^-]$
If $pH < pK_a$, $[HA] > [A^-]$
If $pH > pK_a$, $[HA] < [A^-]$

Example **Using the Henderson-Hasselbalch Equation**

Sodium hypochlorite (NaOCl, the active ingredient of bleach) was dissolved in a solution buffered to pH 6.20. Find the ratio $[OCl^-]/[HOCl]$ in this solution.

When we say that the solution is buffered to pH 6.20, we mean that unspecified acids and bases were used to set the pH to 6.20. If buffering is sufficient, then adding a little more acid or base does not change the pH appreciably.

SOLUTION OCl^- is the conjugate base of hypochlorous acid, HOCl. In Appendix B, we find that $pK_a = 7.53$ for HOCl. Knowing the pH, we calculate the ratio $[OCl^-]/[HOCl]$ from the Henderson-Hasselbalch equation:

$$HOCl \rightleftharpoons H^+ + OCl^- \qquad pH = pK_a + \log\left(\frac{[OCl^-]}{[HOCl]}\right)$$

$$6.20 = 7.53 + \log\left(\frac{[OCl^-]}{[HOCl]}\right)$$

$$-1.33 = \log\left(\frac{[OCl^-]}{[HOCl]}\right)$$

It doesn't matter how the pH got to 6.20. If we know that pH = 6.20, the Henderson-Hasselbalch equation tells us the ratio $[OCl^-]/[HOCl]$.

To solve this equation, raise 10 to the power shown on each side:

If $a = b$, then $10^a = 10^b$.

$$10^{-1.33} = 10^{\log([OCl^-]/[HOCl])} = \frac{[OCl^-]}{[HOCl]}$$

$$0.047 = \frac{[OCl^-]}{[HOCl]}$$

$10^{\log a} = a$.

To find $10^{-1.33}$ with my calculator, I use the *10^x* function, with $x = -1.33$. If you have the *antilog* function instead of *10^x* on your calculator, you should compute antilog(−1.33).

Finding the ratio $[OCl^-]/[HOCl]$ requires only pH and pK_a. We do not care what else is in the solution, how much NaOCl was added, or what the volume of the solution is.

Test Yourself Find the ratio $[OCl^-]/[HOCl]$ if pH = 7.20. (**Answer:** 0.47, which makes sense: When pH changes by 1 unit, the ratio changes by one power of 10.)

Ask Yourself

9-A. (a) What is the pH of a buffer prepared by dissolving 0.100 mol of the weak acid HA ($K_a = 1.0 \times 10^{-5}$) plus 0.050 mol of its conjugate base Na^+A^- in 1.00 L? **(b)** Write the Henderson-Hasselbalch equation for a solution of formic acid, HCO_2H. What is the quotient $[HCO_2^-]/[HCO_2H]$ at pH 3.00, 3.744, and 4.00?

9-3 A Buffer in Action

For illustration, we choose a widely used buffer called "tris."

$$\underset{\substack{BH^+ \\ pK_a = 8.07}}{(HOCH_2)_3C\text{-}\overset{+}{N}H_3} \rightleftharpoons \underset{\substack{B = \text{"tris"} \\ \text{Tris(hydroxymethyl)aminomethane}}}{(HOCH_2)_3C\text{-}NH_2} + H^+ \tag{9-3}$$

In Appendix B, we find pK_a for BH^+, the conjugate acid of tris, to be 8.07. An example of a salt containing BH^+ is tris hydrochloride, which is BH^+Cl^-. When BH^+Cl^- is dissolved in water, it dissociates completely to BH^+ and Cl^-. To find the pH of a known mixture of B and BH^+, simply plug their concentrations into the Henderson-Hasselbalch equation.

Example **A Buffer Solution**

Find the pH of a 1.00-L aqueous solution containing 12.43 g of tris (FM 121.14) plus 4.67 g of tris hydrochloride (FM 157.60).

SOLUTION The concentrations of B and BH^+ added to the solution are

$$[B] = \frac{12.43\ \text{g/L}}{121.14\ \text{g/mol}} = 0.102\ 6\ \text{M} \qquad [BH^+] = \frac{4.67\ \text{g/L}}{157.60\ \text{g/mol}} = 0.029\ 6\ \text{M}$$

Assuming that what we mixed stays in the same form, we insert the concentrations into the Henderson-Hasselbalch equation to find the pH:

$$pH = pK_a + \log\left(\frac{[B]}{[BH^+]}\right) = 8.07 + \log\left(\frac{0.102\ 6}{0.029\ 6}\right) = 8.61$$

Don't panic over significant figures in pH. For consistency, we are almost always going to express pH to the 0.01 place.

Test Yourself Find the pH if we had mixed 4.67 g of tris with 12.43 g of tris hydrochloride. (**Answer:** 7.76; makes sense—more acid and less base give a lower pH)

Notice that *the volume of solution is irrelevant to finding the pH*, because volume cancels in the numerator and denominator of the log term:

The pH of a buffer is nearly independent of dilution.

$$\mathrm{pH} = \mathrm{p}K_a + \log\left(\frac{\text{moles of B}/\cancel{\text{L of solution}}}{\text{moles of BH}^+/\cancel{\text{L of solution}}}\right)$$

$$= \mathrm{p}K_a + \log\left(\frac{\text{moles of B}}{\text{moles of BH}^+}\right)$$

If strong acid is added to a buffered solution, some of the buffer base will be converted into the conjugate acid and the quotient $[B]/[BH^+]$ changes. If a strong base is added to a buffered solution, some BH^+ is converted into B. Knowing how much strong acid or base is added, we can compute the new quotient $[B]/[BH^+]$ and the new pH.

Example **Effect of Adding Acid to a Buffer**

If we add 12.0 mL of 1.00 M HCl to the solution in the preceding example, what will the new pH be?

SOLUTION The key to this problem is to realize that, **when a strong acid is added to a weak base, they react completely to give BH^+** (Box 9-1). We add

Box 9-1 Strong Plus Weak Reacts Completely

A strong acid reacts with a weak base "completely" because the equilibrium constant is large:

$$\underset{\text{Weak base}}{B} + \underset{\text{Strong acid}}{H^+} \rightleftharpoons BH^+ \qquad K = \frac{1}{K_a(\text{for BH}^+)}$$

If B is tris, then the equilibrium constant for reaction with HCl is

$$K = \frac{1}{K_a} = \frac{1}{10^{-8.07}} = 1.2 \times 10^8 \leftarrow \text{A big number}$$

A strong base reacts "completely" with a weak acid because the equilibrium constant is, again, very large:

$$\underset{\text{Strong base}}{OH^-} + \underset{\text{Weak acid}}{HA} \rightleftharpoons A^- + H_2O \qquad K = \frac{1}{K_b(\text{for A}^-)}$$

If HA is acetic acid, then the equilibrium constant for reaction with NaOH is

$$K = \frac{1}{K_b} = \frac{K_a(\text{for HA})}{K_w} = 1.7 \times 10^9 \leftarrow \text{Another big number}$$

The reaction of a strong acid with a strong base is even more complete than a strong plus weak reaction:

$$\underset{\text{Strong acid}}{H^+} + \underset{\text{Strong base}}{OH^-} \rightleftharpoons H_2O \qquad K = \frac{1}{K_w} = 10^{14}$$
$$\uparrow \text{Ridiculously big!}$$

If you mix a strong acid, a strong base, a weak acid, and a weak base, the strong acid and base will neutralize each other until one is used up. The remaining strong acid or base will then react with the weak base or weak acid.

Demonstration 9-1 How Buffers Work

A buffer resists changes in pH because the added acid or base is consumed by the buffer. As the buffer is used up, it becomes less resistant to changes in pH.

In this demonstration,[3] a mixture containing a 10:1 mole ratio of $HSO_3^-:SO_3^{2-}$ is prepared. Because pK_a for HSO_3^- is 7.2, the pH should be approximately

$$pH = pK_a + \log\left(\frac{[SO_3^{2-}]}{[HSO_3^-]}\right) = 7.2 + \log\left(\frac{1}{10}\right) = 6.2$$

When formaldehyde is added, the net reaction is the consumption of HSO_3^-, but not of SO_3^{2-}.

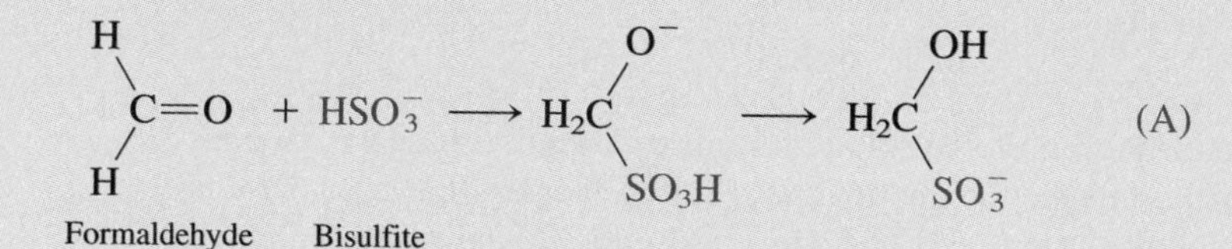

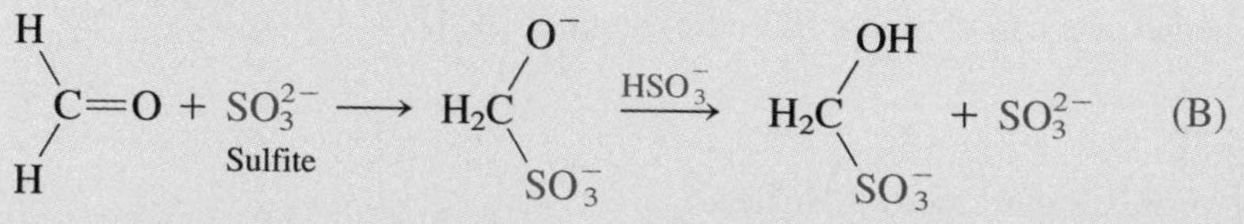

(Sequence A consumes bisulfite. In sequence B, the net reaction is destruction of HSO_3^-, with no change in the SO_3^{2-} concentration.)

We can prepare a table showing how the pH should change as the HSO_3^- reacts:

Percentage of reaction completed	$[SO_3^{2-}]:[HSO_3^-]$	Calculated pH
0	1:10	6.2
90	1:1	7.2
99	1:0.1	8.2
99.9	1:0.01	9.2
99.99	1:0.001	10.2

You can see that through 90% completion the pH rises by just one unit. In the next 9% of the reaction, the pH rises by another unit. At the end of the reaction, the change in pH is very abrupt.

In the formaldehyde clock reaction, formaldehyde is added to a solution containing HSO_3^-, SO_3^{2-}, and phenolphthalein indicator. Phenolphthalein is colorless below pH 8.5 and red above this pH. We observe that the solution remains colorless for more than a minute after the addition of formaldehyde. Suddenly the pH shoots up and the liquid turns pink. Monitoring the pH with a glass electrode gave the results shown in the graph.

The hydrogen sulfite ion, HSO_3^-, is obtained for this demonstration by dissolving sodium metabisulfite, $Na_2S_2O_5$, in water: $S_2O_5^{2-} + H_2O \rightleftharpoons 2HSO_3^-$.

Procedure: Prepare phenolphthalein solution by dissolving 50 mg of solid indicator in 50 mL of ethanol and diluting with 50 mL of water. The following solutions should be fresh: Dilute 9 mL of 37 wt% formaldehyde to 100 mL. Dissolve 1.4 g of $Na_2S_2O_5$ and 0.18 g of Na_2SO_3 in 400 mL of water and add 1 mL of phenolphthalein indicator solution. To initiate the clock reaction, add 23 mL of formaldehyde solution to the well-stirred buffer solution. Reaction time is affected by the temperature, concentrations, and volume.

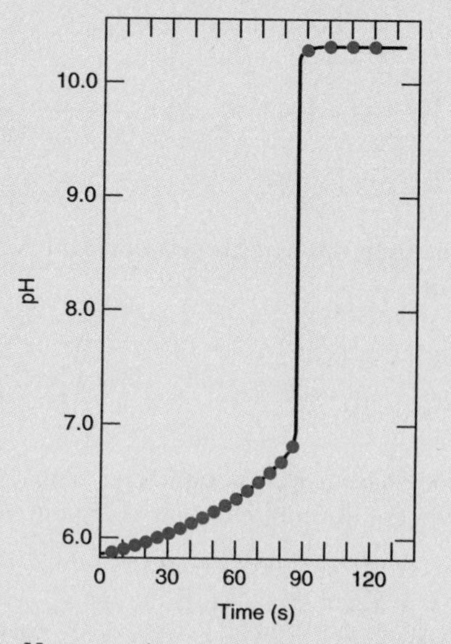

Graph of pH versus time in the formaldehyde clock reaction.

12.0 mL of 1.00 M HCl, which contain (0.012 0 L)(1.00 mol/L) = 0.012 0 mol of H^+. The H^+ consumes 0.012 0 mol of B to create 0.012 0 mol of BH^+:

	B Tris	+	H^+ From HCl	$\longrightarrow$	BH^+
Initial moles	0.102 6		0.012 0		0.029 6
Final moles	0.1026 − 0.0120		—		0.029 6 + 0.012 0
	0.090 6				0.041 6

$$\mathrm{pH} = \mathrm{p}K_a + \log\left(\frac{\text{moles of B}}{\text{moles of BH}^+}\right)$$

$$= 8.07 + \log\left(\frac{0.090\ 6}{0.041\ 6}\right) = 8.41$$

Again, the volume of solution is irrelevant.

Question Does the pH change in the right direction when HCl is added?

Test Yourself What would be the pH if we had added 12.0 mL of 1.00 M NaOH instead of HCl? (**Answer:** 8.88, which makes sense—pH goes up when we add base)

The example shows that *the pH of a buffer does not change very much when a limited amount of a strong acid or base is added.* Addition of 12.0 mL of 1.00 M HCl changed the pH from 8.61 to 8.41. Addition of 12.0 mL of 1.00 M HCl to 1.00 L of pure water would have lowered the pH to 1.93.

A buffer resists changes in pH . . .

But *why* does a buffer resist changes in pH? **It does so because the strong acid or base is consumed by B or BH^+.** If you add HCl to tris, B is converted into BH^+. If you add NaOH, BH^+ is converted into B. As long as you don't use up B or BH^+ by adding too much HCl or NaOH, the log term of the Henderson-Hasselbalch equation does not change very much and the pH does not change very much. Demonstration 9-1 illustrates what happens when the buffer does get used up. The buffer has its maximum capacity to resist changes of pH when pH = pK_a. We will return to this point later.

. . . because the buffer "consumes" the added acid or base.

Ask Yourself

9-B. **(a)** What is the pH of a solution prepared by dissolving 10.0 g of tris plus 10.0 g of tris hydrochloride in 0.250 L water?
(b) What will the pH be if 10.5 mL of 0.500 M $HClO_4$ are added to **(a)**?
(c) What will the pH be if 10.5 mL of 0.500 M NaOH are added to **(a)**?

9-4 Preparing Buffers

Buffers are usually prepared by starting with a measured amount of either a weak acid (HA) or a weak base (B). Then OH^- is added to HA to make a mixture of HA and A^- (a buffer) or H^+ is added to B to make a mixture of B and BH^+ (a buffer).

Example **Calculating How to Prepare a Buffer Solution**

How many milliliters of 0.500 M NaOH should be added to 10.0 g of tris hydrochloride (BH^+, Equation 9-3) to give a pH of 7.60 in a final volume of 250 mL?

SOLUTION The number of moles of tris hydrochloride in 10.0 g is (10.0 g)/(157.60 g/mol) = 0.063 5. We can make a table to help solve the problem:

Reaction with OH^-:	BH^+	+	OH^-	$\longrightarrow$	B
Initial moles	0.063 5		x		—
Final moles	$0.063\,5 - x$		—		x

The Henderson-Hasselbalch equation allows us to find x, because we know pH and pK_a:

$$\mathrm{pH} = \mathrm{p}K_a + \log\left(\frac{\text{mol B}}{\text{mol BH}^+}\right)$$

$$7.60 = 8.07 + \log\left(\frac{x}{0.063\,5 - x}\right)$$

$$-0.47 = \log\left(\frac{x}{0.063\,5 - x}\right)$$

To solve for x, raise 10 to the power of the terms on both sides, remembering that $10^{\log z} = z$:

If your calculator has the *antilog* function, solve the equation $-0.47 = \log z$ by calculating $z = \text{antilog}(-0.47)$.

$$10^{-0.47} = 10^{\log[x/(0.063\,5 - x)]}$$

$$0.339 = \frac{x}{0.063\,5 - x} \Rightarrow x = 0.016\,1 \text{ mol}$$

This many moles of NaOH is contained in

$$\frac{0.016\,1 \text{ mol}}{0.500 \text{ mol/L}} = 0.032\,2 \text{ L} = 32.2 \text{ mL}$$

Our calculation tells us to mix 32.2 mL of 0.500 M NaOH with 10.0 g of tris hydrochloride to give a pH of 7.60.

Test Yourself How much 0.500 M NaOH should be added to 12.0 g of tris hydrochloride to give a pH of 7.77 in a final volume of 317 mL? (**Answer:** 50.8 mL; volume is irrelevant)

Preparing a Buffer in Real Life

When you mix calculated quantities of acid and base to make a buffer, the pH *does not* come out exactly as expected. The main reason for the discrepancy is that pH is governed by the *activities* of the conjugate acid-base pair, not by the concentrations. (Activity is discussed in Section 12-2.) If you really want to prepare tris buffer at pH 7.60, you should use a pH electrode to get exactly what you need.

Suppose you wish to prepare 1.00 L of buffer containing 0.100 M tris at pH 7.60. *When we say 0.100 M tris, we mean that the total concentration of tris plus* $trisH^+$ *will be 0.100 M.* You have available solid tris hydrochloride and ~1 M NaOH. Here's how to do it:

1. Weigh out 0.100 mol tris hydrochloride and dissolve it in a beaker containing about 800 mL water and a stirring bar.
2. Place a pH electrode in the solution and monitor the pH.
3. Add NaOH solution until the pH is exactly 7.60. The electrode does not respond instantly. Be sure to allow the pH reading to stabilize after each addition of reagent.
4. Transfer the solution to a volumetric flask and wash the beaker and stirring bar a few times. Add the washings to the volumetric flask.
5. Dilute to the mark and mix.

The reason for using 800 mL of water in the first step is so that the volume will be close to the final volume during pH adjustment. Otherwise, the pH will change slightly when the sample is diluted to its final volume and the *ionic strength* changes.

Before making the buffer, it is useful to calculate how much strong acid or base will be required. It helps to know if you are going to need 10 mL or 10 drops of reagent before you dispense it.

Ionic strength is a measure of the total concentration of ions in a solution. Changing ionic strength changes activities of ionic species such as H^+ and A^-, even though their concentrations are constant. Diluting a buffer with water changes the ionic strength and therefore changes the pH slightly.

Ask Yourself

9-C. **(a)** How many milliliters of 1.20 M HCl should be added to 10.0 g of tris (B, Equation 9-3) to give a pH of 7.60 in a final volume of 250 mL?
(b) What operations would you perform to prepare exactly 100 mL of 0.200 M acetate buffer, pH 5.00, starting with pure liquid acetic acid and solutions containing ~3 M HCl and ~3 M NaOH?

9-5 Buffer Capacity

Buffer capacity measures how well a solution resists changes in pH when acid or base is added. The greater the buffer capacity, the less the pH changes. We will find that *buffer capacity is maximum when* $pH = pK_a$ *for the buffer.*

The greater the buffer capacity, the less the pH changes when H^+ or OH^- is added. Buffer capacity is maximum when $pH = pK_a$.

Figure 9-2 shows the calculated response of a buffer to small additions of H^+ or OH^-. The buffer is a mixture of HA (with acid dissociation constant $K_a = 10^{-5}$) plus A^-. The total moles of HA + A^- were set equal to 1. The relative quantities of HA and A^- were varied to give initial pH values from 3.4 to 6.6. Then 0.01 mol of either H^+ or OH^- was added to the solution and the new pH was calculated. Figure 9-2 shows the change in pH, that is, Δ(pH), as a function of the initial pH of the buffer.

For example, mixing 0.038 3 mol of A^- and 0.961 7 mol of HA gives an initial pH of 3.600:

$$\text{pH} = \text{p}K_a + \log\left(\frac{\text{mol A}^-}{\text{mol HA}}\right) = 5.000 + \log\left(\frac{0.038\ 3}{0.961\ 7}\right) = 3.600$$

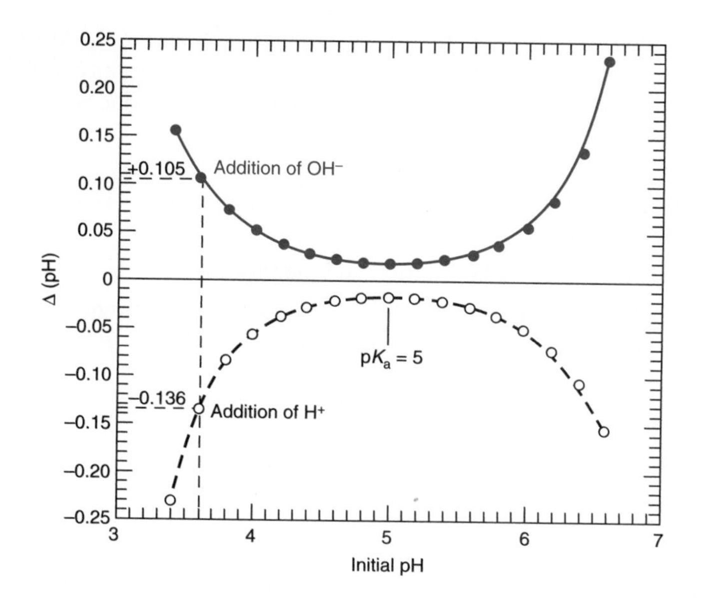

Figure 9-2 *Buffer capacity:* Effect of adding 0.01 mol of H^+ or OH^- to a buffer containing HA and A^- (total quantity of HA + A^- = 1 mol). The minimum change in pH occurs when the initial pH of the buffer equals pK_a for HA. That is, the buffer capacity is maximum when pH = pK_a.

When 0.010 0 mol of OH^- is added to this mixture, the concentrations and pH change in a manner that you are now smart enough to calculate:

Reaction with OH^-:	HA	+	OH^-	$\longrightarrow$	A^-	+	H_2O
Initial moles	0.961 7		0.010 0		0.038 3		
Final moles	0.961 7 − 0.010 0		—		0.038 3 + 0.010 0		
	0.951 7				0.048 3		

$$\text{pH} = \text{p}K_a + \log\left(\frac{\text{mol A}^-}{\text{mol HA}}\right) = 5.000 + \log\left(\frac{0.048\ 3}{0.951\ 7}\right) = 3.705$$

The change in pH is Δ(pH) = 3.705 − 3.600 = +0.105. This is the value plotted in Figure 9-2 on the upper curve for an initial pH of 3.600. If the same starting mixture had been treated with 0.010 0 mol of H^+, the pH would change by −0.136, which is shown on the lower curve.

We see in Figure 9-2 that the magnitude of the pH change is smallest when the initial pH is equal to pK_a for the buffer. That is, *the buffer capacity is greatest when pH =* pK_a.

Choose a buffer whose pK_a is close to the desired pH.

In choosing a buffer for an experiment, you should *seek one whose* pK_a *is as close as possible to the desired pH. The useful pH range of a buffer is usually considered to be* $pK_a \pm 1$ *pH unit.* Outside this range, there is not enough of either the weak acid or the weak base to react with added base or acid. Buffer capacity increases with increasing concentration of the buffer. To maintain a stable pH, you must use enough buffer to react with the quantity of acid or base expected to be encountered.

Table 9-2 lists pK_a values for common buffers. Some of the buffers have more than one acidic proton, so more than one pK is listed. We will consider polyprotic acids and bases in Chapter 11.

Table 9-2 Structures and pK_a values for common buffers

Name	Structure[a]	pK_a [b,c]	$\Delta(pK_a)/\Delta T^c$ (K^{-1})	Formula mass
Phosphoric acid	H_3PO_4	2.15 (pK_1)	0.005	98.00
Citric acid	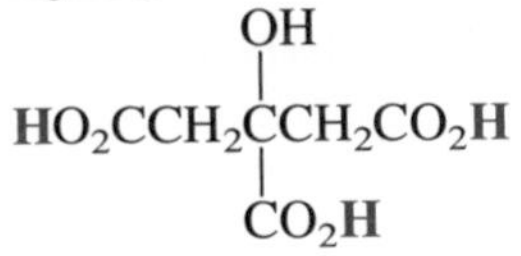 $HO_2CCH_2C(OH)(CO_2H)CH_2CO_2H$	3.13 (pK_1)	−0.002	192.12
Citric acid	$H_2(citrate)^-$	4.76 (pK_2)	−0.001	192.12
Acetic acid	CH_3CO_2H	4.76	0.000	60.05
2-(*N*-Morpholino)ethanesulfonic acid (MES)	 $O(CH_2CH_2)_2\overset{+}{N}HCH_2CH_2SO_3^-$	6.27	−0.009	195.24
Citric acid	$H(citrate)^{2-}$	6.40 (pK_3)	0.002	192.12
3-(*N*-Morpholino)-2-hydroxypropanesulfonic acid (MOPSO)	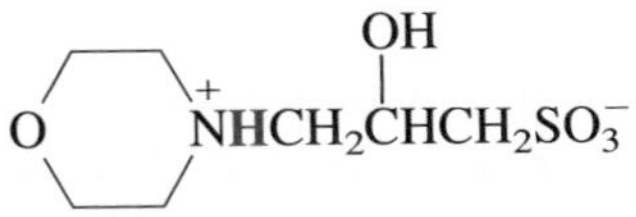 $O(CH_2CH_2)_2\overset{+}{N}HCH_2CH(OH)CH_2SO_3^-$	6.90	−0.015	225.26
Imidazole hydrochloride	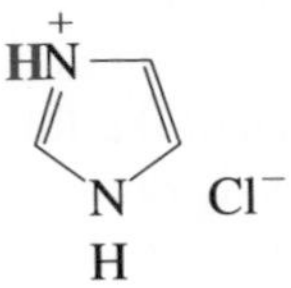 $\overset{+}{H}N$(imidazole ring)NH Cl^-	6.99	−0.022	104.54
Piperazine-*N*,*N*′-bis(2-ethanesulfonic acid) (PIPES)	 $^-O_3SCH_2CH_2\overset{+}{N}H(CH_2CH_2)_2\overset{+}{N}HCH_2CH_2SO_3^-$	7.14	−0.007	302.37
Phosphoric acid	$H_2PO_4^-$	7.20 (pK_2)	−0.002	98.00
N-2-Hydroxyethylpiperazine-*N*′-2-ethanesulfonic acid (HEPES)	 $HOCH_2CH_2N(CH_2CH_2)_2\overset{+}{N}HCH_2CH_2SO_3^-$	7.56	−0.012	238.30
Tris(hydroxymethyl)aminomethane hydrochloride (tris hydrochloride)	$(HOCH_2)_3C\overset{+}{N}H_3$ Cl^-	8.07	−0.028	157.60
Glycylglycine	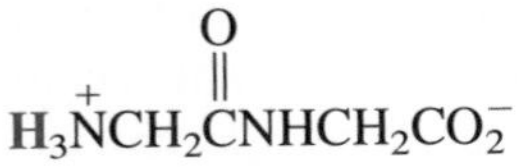 $H_3\overset{+}{N}CH_2C(=O)NHCH_2CO_2^-$	8.26	−0.026	132.12
Ammonia	NH_3	9.24	−0.031	17.03
Boric acid	$B(OH)_3$	9.24 (pK_1)	−0.008	61.83
Cyclohexylaminoethanesulfonic acid (CHES)	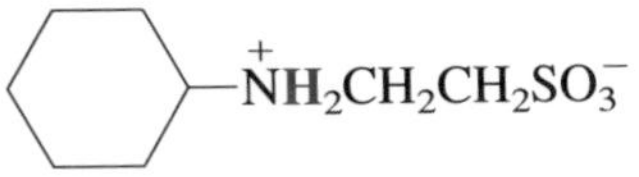 $C_6H_{11}-\overset{+}{N}H_2CH_2CH_2SO_3^-$	9.39	−0.023	207.29
3-(Cyclohexylamino)propanesulfonic acid (CAPS)	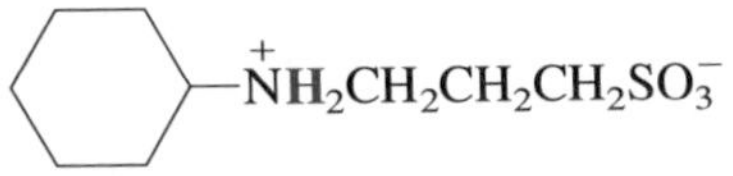 $C_6H_{11}-\overset{+}{N}H_2CH_2CH_2CH_2SO_3^-$	10.50	−0.028	221.32
Phosphoric acid	HPO_4^{2-}	12.35 (pK_3)	−0.009	98.00
Boric acid	$OB(OH)_2^-$	12.74 (pK_2)		61.83

a. The protonated form of each molecule is shown. Acidic hydrogen atoms are shown in **bold** type.
b. pK_a is generally for 25°C and zero ionic strength.
c. SOURCES: R. N. Goldberg, N. Kishore, and R. M. Lennen, *J. Phys. Chem. Ref. Data* **2002**, *31*, 231; A. E. Martell and R. J. Motekaitis, *NIST Database 46* (Gaithersburg, MD: National Institute of Standards and Technology, 2001).

Buffer pH Depends on Temperature and Ionic Strength

For tris, Table 9-2 shows $\Delta(pK_a)/\Delta T = -0.028$ per degree near 25°C.

pK_a (at 37°C)
$= pK_a$ (at 25°C) $+ [\Delta(pK_a)/\Delta T](\Delta T)$
$= 8.07 + (-0.028°C^{-1})(12°C)$
$= 7.73$

Most buffers exhibit a dependence of pK_a on temperature. Table 9-2 shows that the change of pK_a for tris is -0.028 pK_a units per degree, near 25°C. A solution of tris made up to pH 8.07 at 25°C will have pH ≈ 8.7 at 4°C and pH ≈ 7.7 at 37°C. When a 0.5 M stock solution of phosphate buffer at pH 6.6 is diluted to 0.05 M, the pH rises to 6.9, because the ionic strength and activities of the buffering species, $H_2PO_4^-$ and HPO_4^{2-}, change.

Summary

A buffer is a mixture of a weak acid and its conjugate base. The buffer is most useful when pH ≈ pK_a. Over a reasonable range of concentrations, the pH of a buffer is nearly independent of concentration. A buffer resists changes in pH because it reacts with added acids or bases. If too much acid or base is added, the buffer will be consumed and no longer resists changes in pH.

Ask Yourself

9-D. **(a)** Look up pK_a for each of the following acids and decide which one would be best for preparing a buffer of pH 3.10: **(i)** hydroxybenzene; **(ii)** propanoic acid; **(iii)** cyanoacetic acid; **(iv)** sulfuric acid.
(b) Why does buffer capacity increase as the concentration of buffer increases?
(c) From their K_b values, which of the following bases would be best for preparing a buffer of pH 9.00? **(i)** NH_3 (ammonia, $K_b = 1.8 \times 10^{-5}$); **(ii)** $C_6H_5NH_2$ (aniline, $K_b = 4.0 \times 10^{-10}$); **(iii)** H_2NNH_2 (hydrazine, $K_b = 3.0 \times 10^{-6}$); **(iv)** C_5H_5N (pyridine, $K_b = 1.6 \times 10^{-9}$).

9-6 How Acid-Base Indicators Work

An *indicator* is an acid or a base whose various protonated forms have different colors.

An acid-base **indicator** is itself an acid or base whose various protonated species have different colors. The indicator is added at such a low concentration that it has negligible effect on any acid-base equilibria of the major components of the solution. In the next chapter, we will learn how to choose an indicator to find the end point for a titration. For now, let's try to use the Henderson-Hasselbalch equation to understand the pH range over which color changes are observed.

Consider bromocresol green as an example. We will call pK_a of the indicator pK_{HIn}, to allow us to distinguish it from pK_a of an acid being titrated.

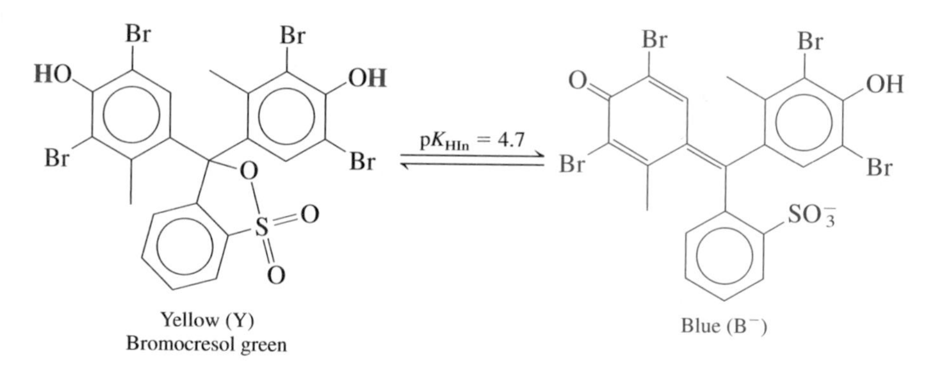

pK_{HIn} for bromocresol green is 4.7. Below pH 4.7, the predominant species is yellow (Y); above pH 4.7, the predominant species is blue (B^-).

The equilibrium between Y and B^- can be written

$$Y \rightleftharpoons B^- + H^+ \qquad K_{HIn} = \frac{[B^-][H^+]}{[Y]}$$

for which the Henderson-Hasselbalch equation is

$$pH = pK_{HIn} + \log\left(\frac{[B^-]}{[Y]}\right) \tag{9-4}$$

At pH = pK_a = 4.7, there will be a 1:1 mixture of the yellow and blue species, which will appear green. As a crude rule of thumb, we can say that the solution will appear yellow when $[Y]/[B^-] \gtrsim 10/1$ and blue when $[B^-]/[Y] \gtrsim 10/1$. (The symbol $\gtrsim$ means "is approximately equal to or greater than.") From Equation 9-4, we can see that the solution will be yellow when pH $\lesssim pK_{HIn} - 1$ (= 3.7) and blue when pH $\gtrsim pK_{HIn} + 1$ (= 5.7). By comparison, Table 9-3 lists bromocresol green as yellow below pH 3.8 and blue above pH 5.4. Between pH 3.8 and 5.4, various shades of green are seen. Demonstration 9-2 illustrates indicator color changes, and Box 9-2 describes an everyday application of indicators.

pH	$[B^-]:[Y]$	Color
3.7	1:10	Yellow
4.7	1:1	Green
5.7	10:1	Blue

Several indicators are listed twice in Table 9-3, with two different sets of colors. An example is thymol blue, which loses one proton with a p*K* of 1.7 and a second proton with a p*K* of 8.9:

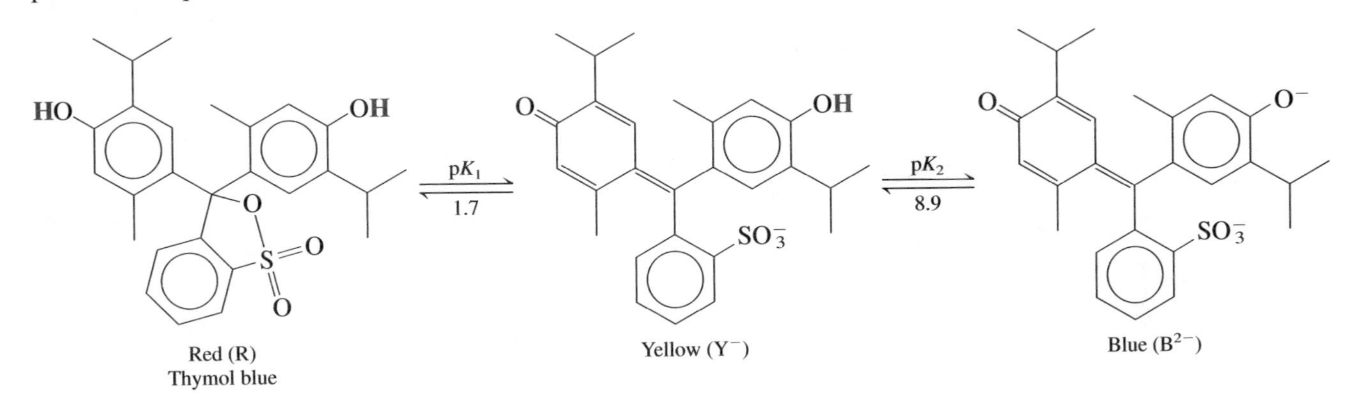

Table 9-3 Common indicators

Indicator	Transition range (pH)	Acid color	Base color
Methyl violet	0.0–1.6	Yellow	Violet
Cresol red	0.2–1.8	Red	Yellow
Thymol blue	1.2–2.8	Red	Yellow
Cresol purple	1.2–2.8	Red	Yellow
Erythrosine, disodium	2.2–3.6	Orange	Red
Methyl orange	3.1–4.4	Red	Yellow
Congo red	3.0–5.0	Violet	Red
Ethyl orange	3.4–4.8	Red	Yellow
Bromocresol green	3.8–5.4	Yellow	Blue
Methyl red	4.8–6.0	Red	Yellow
Chlorophenol red	4.8–6.4	Yellow	Red
Bromocresol purple	5.2–6.8	Yellow	Purple
p-Nitrophenol	5.6–7.6	Colorless	Yellow
Litmus	5.0–8.0	Red	Blue
Bromothymol blue	6.0–7.6	Yellow	Blue
Phenol red	6.4–8.0	Yellow	Red
Neutral red	6.8–8.0	Red	Yellow
Cresol red	7.2–8.8	Yellow	Red
α-Naphtholphthalein	7.3–8.7	Pink	Green
Cresol purple	7.6–9.2	Yellow	Purple
Thymol blue	8.0–9.6	Yellow	Blue
Phenolphthalein	8.0–9.6	Colorless	Pink
Thymolphthalein	8.3–10.5	Colorless	Blue
Alizarin yellow	10.1–12.0	Yellow	Orange-red
Nitramine	10.8–13.0	Colorless	Orange-brown
Tropaeolin O	11.1–12.7	Yellow	Orange

Demonstration 9-2 Indicators and Carbonic Acid[4]

This one is just plain fun. Fill two 1-L graduated cylinders with 900 mL of water and place a magnetic stirring bar in each. Add 10 mL of 1 M NH_3 to each. Then put 2 mL of phenolphthalein indicator solution in one and 2 mL of bromothymol blue indicator solution in the other. Both indicators will have the color of their basic species.

Drop a few chunks of Dry Ice (solid CO_2) into each cylinder. As the CO_2 bubbles through each cylinder, the solutions become more acidic. First the pink phenolphthalein color disappears. After some time, the pH drops just low enough for bromothymol blue to change from blue to its pale green intermediate color. The pH does not go low enough to turn bromothymol blue into its yellow color.

Add about 20 mL of 6 M HCl to *the bottom* of each cylinder, using a length of Tygon tubing attached to a funnel. Then stir each solution for a few seconds on a magnetic stirrer. Explain what happens. The sequence of events is shown in Color Plate 4.

When CO_2 dissolves in water, it makes carbonic acid, which has two acidic protons:

$$CO_2(g) \rightleftharpoons CO_2(aq) \qquad K = \frac{[CO_2(aq)]}{P_{CO_2}} = 0.034\,4$$

$$CO_2(aq) + H_2O \rightleftharpoons \underset{\text{Carbonic acid}}{HO{-}C({=}O){-}OH} \qquad K = \frac{[H_2CO_3]}{[CO_2(aq)]} \approx 0.002$$

$$H_2CO_3 \rightleftharpoons \underset{\text{Bicarbonate}}{HCO_3^-} + H^+ \qquad K_{a1} = 4.46 \times 10^{-7}$$

$$HCO_3^- \rightleftharpoons \underset{\text{Carbonate}}{CO_3^{2-}} + H^+ \qquad K_{a2} = 4.69 \times 10^{-11}$$

The value $K_{a1} = 4.46 \times 10^{-7}$ applies to the equation

$$\underset{CO_2(aq)\,+\,H_2CO_3}{\text{all dissolved } CO_2} \rightleftharpoons HCO_3^- + H^+$$

$$K_{a1} = \frac{[HCO_3^-][H^+]}{[CO_2(aq) + H_2CO_3]} = 4.46 \times 10^{-7}$$

Only about 0.2% of dissolved CO_2 is in the form H_2CO_3. If the true value of $[H_2CO_3]$ were used instead of $[H_2CO_3 + CO_2(aq)]$, the equilibrium constant would be $\sim 2 \times 10^{-4}$.

Below pH 1.7, the predominant species is red (R); between pH 1.7 and pH 8.9, the predominant species is yellow (Y^-); and above pH 8.9, the predominant species is blue (B^{2-}). The sequence of color changes for thymol blue is shown in Color Plate 5.

The equilibrium between R and Y^- can be written

$$R \rightleftharpoons Y^- + H^+ \qquad K_1 = \frac{[Y^-][H^+]}{[R]}$$

$$pH = pK_1 + \log\left(\frac{[Y^-]}{[R]}\right) \tag{9-5}$$

pH	$[Y^-]:[R]$	Color
0.7	1:10	Red
1.7	1:1	Orange
2.7	10:1	Yellow

At pH $= 1.7$ ($= pK_1$), there will be a 1:1 mixture of the yellow and red species, which appears orange. We expect that the solution will appear red when $[R]/[Y^-] \gtrsim 10/1$ and yellow when $[Y^-]/[R] \gtrsim 10/1$. From the Henderson-Hasselbalch equation 9-5, we can see that the solution will be red when pH $\lesssim pK_1 - 1$ ($= 0.7$) and yellow when pH $\gtrsim pK_1 + 1$ ($= 2.7$). In Table 9-3, thymol blue is listed as red below pH 1.2 and yellow above pH 2.8. Thymol blue has another transition, from yellow to blue, between pH 8.0 and pH 9.6. In this range, various shades of green are seen.

Box 9-2 The Secret of Carbonless Copy Paper[5]

Back in the days when dinosaurs roamed the Earth and I was a child, people would insert a messy piece of carbon paper between two pages so that a copy of what was written on the upper page would appear on the lower page. Then carbonless copy paper was invented to perform the same task without the extra sheet of carbon paper.

The secret of carbonless copy paper is an acid-base indicator contained inside polymeric microcapsules stuck to the underside of the upper sheet of paper. When you write on the upper sheet, the pressure of your pen breaks open the microcapsules on the bottom side, which then release their indicator.

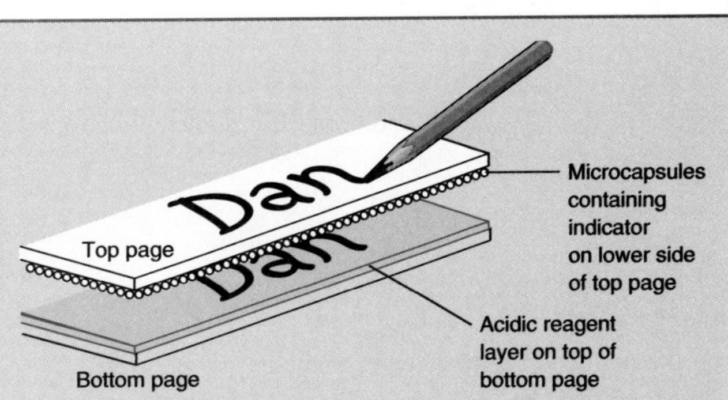

The indicator becomes adsorbed on the upper surface of the lower page, which is coated with microscopic particles of an acidic material such as the clay known as bentonite. This clay contains negatively charged aluminosilicate layers with hydronium ion (H_3O^+) between the layers to balance the negative charge. The adsorbed indicator reacts with the hydronium ion to give a colored product that appears as a copy of your writing on the lower page.

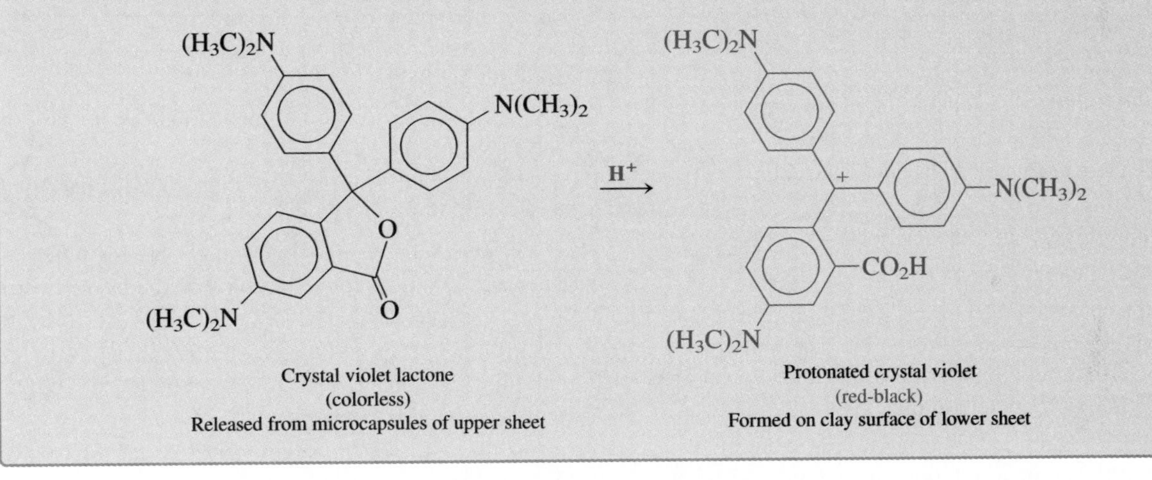

Crystal violet lactone
(colorless)
Released from microcapsules of upper sheet

Protonated crystal violet
(red-black)
Formed on clay surface of lower sheet

Ask Yourself

9-E. **(a)** What is the reasoning behind the rule of thumb that indicator color changes occur at $pK_{HIn} \pm 1$?
(b) What color do you expect to observe for cresol purple indicator (Table 9-3) at the following pH values: 1.0, 2.0, 3.0?

Key Equations

Henderson-Hasselbalch

$$pH = pK_a + \log\left(\frac{[A^-]}{[HA]}\right)$$

$$pH = pK_a + \log\left(\frac{[B]}{[BH^+]}\right)$$

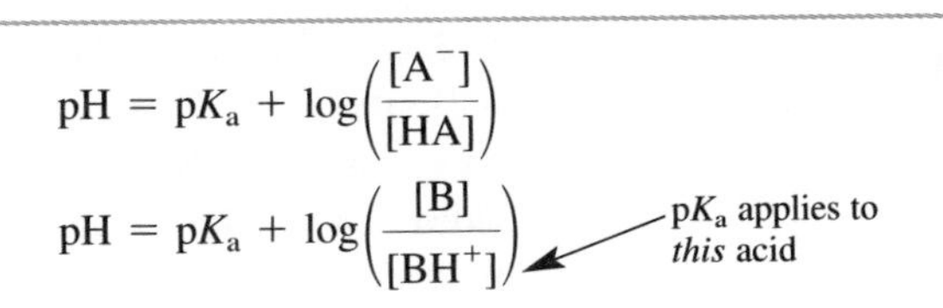

Important Terms

buffer | Henderson-Hasselbalch equation | indicator

Problems

9-1. Explain what happens when acid is added to a buffer and the pH does not change very much.

9-2. Why is the pH of a buffer nearly independent of concentration?

9-3. A solution contains 63 different conjugate acid-base pairs. Among them is acrylic acid and acrylate ion, with the ratio [acrylate]/[acrylic acid] = 0.75. What is the pH of the solution?

$$\underset{\text{Acrylic acid}}{H_2C{=}CHCO_2H} \rightleftharpoons \underset{\text{Acrylate}}{H_2C{=}CHCO_2^-} + H^+ \qquad pK_a = 4.25$$

9-4. Table 9-1 shows the relation of pH to the quotient $[A^-]/[HA]$.

(a) Use the Henderson-Hasselbalch equation to show that pH $= pK_a + 2$ when $[A^-]/[HA] = 100$.

(b) Find the quotient $[A^-]/[HA]$ when pH $= pK_a - 3$.

(c) Find the pH when $[A^-]/[HA] = 10^{-4}$.

9-5. Explain why the indicator cresol red changes color when the pH is lowered from 10 to 6. What colors will be observed at pH 10, 8, and 6? Why does the color transition require ~2 pH units for completion?

9-6. Given that pK_b for iodate ion (IO_3^-) is 13.83, find the quotient $[HIO_3]/[IO_3^-]$ in a solution of sodium iodate at **(a)** pH 7.00; **(b)** pH 1.00.

9-7. Given that pK_b for nitrite ion (NO_2^-) is 10.85, find the quotient $[HNO_2]/[NO_2^-]$ in a solution of sodium nitrite at **(a)** pH 2.00; **(b)** pH 10.00.

9-8. Write the Henderson-Hasselbalch equation for a solution of methylamine. Calculate the quotient $[CH_3NH_2]/[CH_3NH_3^+]$ at **(a)** pH 4.00; **(b)** pH 10.645; **(c)** pH 12.00.

9-9. Find the pH of a solution prepared from 2.53 g of oxoacetic acid, 5.13 g of potassium oxoacetate, and 103 g of water.

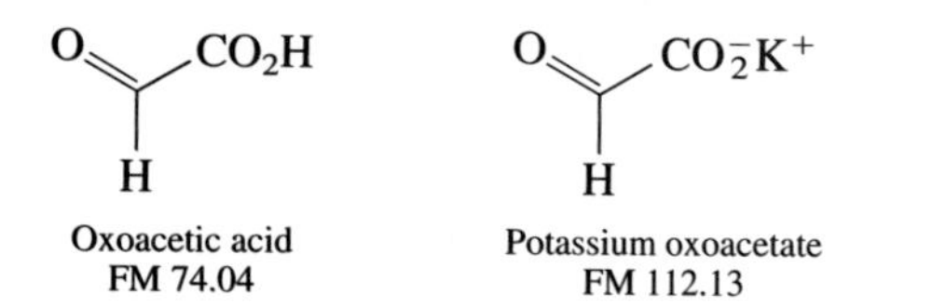

Oxoacetic acid FM 74.04

Potassium oxoacetate FM 112.13

9-10. **(a)** Find the pH of a solution prepared by dissolving 1.00 g of glycine amide hydrochloride plus 1.00 g of glycine amide in 0.100 L.

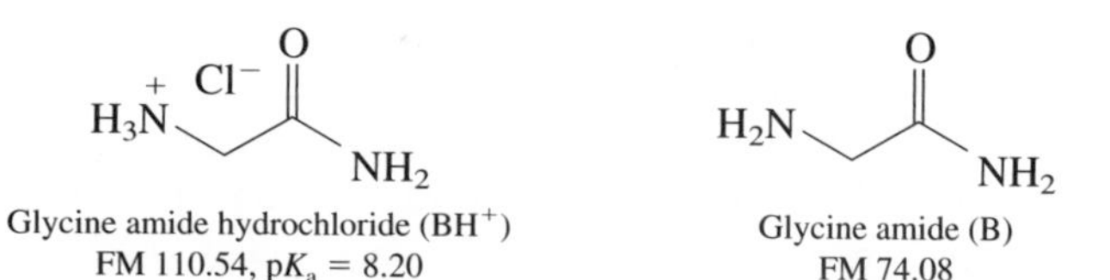

Glycine amide hydrochloride (BH^+) FM 110.54, $pK_a = 8.20$

Glycine amide (B) FM 74.08

(b) How many grams of glycine amide should be added to 1.00 g of glycine amide hydrochloride to give 100 mL of solution with pH 8.00?

(c) What would be the pH if the solution in **(a)** is mixed with 5.00 mL of 0.100 M HCl?

(d) What would be the pH if the solution in **(c)** is mixed with 10.00 mL of 0.100 M NaOH?

9-11. **(a)** Write the chemical reactions whose equilibrium constants are K_b and K_a for imidazole and imidazole hydrochloride, respectively.

(b) Calculate the pH of a 100-mL solution containing 1.00 g of imidazole (FM 68.08) and 1.00 g of imidazole hydrochloride (FM 104.54).

(c) Calculate the pH of the solution if 2.30 mL of 1.07 M $HClO_4$ are added to the solution.

(d) How many milliliters of 1.07 M $HClO_4$ should be added to 1.00 g of imidazole to give a pH of 6.993?

9-12. **(a)** Calculate the pH of a solution prepared by mixing 0.080 0 mol of chloroacetic acid plus 0.040 0 mol of sodium chloroacetate in 1.00 L of water.

(b) Using first your head and then the Henderson-Hasselbalch equation, find the pH of a solution prepared by dissolving all of the following compounds in a total volume of 1.00 L: 0.180 mol $ClCH_2CO_2H$, 0.020 mol $ClCH_2CO_2Na$, 0.080 mol HNO_3, and 0.080 mol $Ca(OH)_2$. Assume that $Ca(OH)_2$ dissociates completely.

9-13. How many milliliters of 0.246 M HNO_3 should be added to 213 mL of 0.006 66 M 2,2′-bipyridine to give a pH of 4.19?

9-14. How many milliliters of 0.626 M KOH should be added to a solution containing 5.00 g of HEPES (Table 9-2) to give a pH of 7.40?

9-15. How many milliliters of 0.113 M HBr should be added to 52.2 mL of 0.013 4 M morpholine to give a pH of 8.00?

9-16. For a fixed buffer concentration, such as 0.05 M, which buffer from Table 9-2 will provide the highest buffer capacity at pH **(a)** 4.00; **(b)** 7.00; **(c)** 10.00? **(d)** What other buffers in Table 9-2 would be useful at pH 10.00?

9-17. Which buffer system will have the greatest buffer capacity at pH 9.0? **(i)** dimethylamine/dimethylammonium ion; **(ii)** ammonia/ammonium ion; **(iii)** hydroxylamine/hydroxylammonium ion; **(iv)** 4-nitrophenol/4-nitrophenolate ion.

9-18. **(a)** Would you need NaOH or HCl to bring the pH of 0.050 0 M HEPES (Table 9-2) to 7.45?

(b) Describe how to prepare 0.250 L of 0.050 0 M HEPES, pH 7.45.

9-19. **(a)** Describe how to prepare 0.500 L of 0.100 M imidazole buffer, pH 7.50, starting with imidazole hydrochloride. Would you use NaOH or HCl to bring the pH to 7.50?

(b) Starting with imidazole, would you need NaOH or HCl to bring the pH to 7.50?

9-20. **(a)** Calculate how many milliliters of 0.100 M HCl should be added to how many grams of sodium acetate dihydrate ($NaOAc \cdot 2H_2O$, FM 118.06) to prepare 250.0 mL of 0.100 M buffer, pH 5.00.

(b) If you mixed what you calculated, the pH would not be 5.00. Describe how you would actually prepare this buffer in the lab.

9-21. Cresol red has *two* transition ranges listed in Table 9-3. What color would you expect it to be at the following pH values?

(a) 0 **(b)** 1 **(c)** 6 **(d)** 9

9-22. Consider the buffer in Figure 9-2, which has an initial pH of 6.200 and pK_a of 5.000. It is made from p mol of A^- plus q mol of HA such that $p + q = 1$ mol.

(a) Setting mol HA $= q$ and mol $A^- = 1 - q$, use the Henderson-Hasselbalch equation to find the values of p and q.

(b) Calculate the pH change when 0.010 0 mol of H^+ is added to the buffer. Did you get the same value shown in Figure 9-2? For the sake of this problem, use three decimal places for pH.

9-23. Ammonia buffer is made up to pH 9.50 at 25°C and then warmed to 37°C. The pH of the solution changes because pK_a changes. Using $\Delta(pK_a)/\Delta T$ from Table 9-2, predict the pH that would be observed at 37°C.

9-24. *Henderson-Hasselbalch spreadsheet.* Prepare a spreadsheet with the constant $pK_a = 4$ in column A. Type the heading $[A^-]/[HA]$ for column B and enter values ranging from 0.001 to 1 000, with a selection of values in between. In column C, compute pH from the Henderson-Hasselbalch equation, by using pK_a from column A and $[A^-]/[HA]$ from column B. In column D, compute $\log([A^-]/[HA])$. Use the values from columns C and D to prepare a graph of pH versus $\log([A^-]/[HA])$. Explain the shape of the resulting graph.

How Would You Do It?

9-25. We can measure the concentrations of the two forms of an indicator such as bromocresol green in a solution by measuring the absorption of visible light at two appropriate wavelengths. One wavelength would be where the yellow species has maximum absorption and the other wavelength would be where the blue species has maximum absorption. If we know how much each species absorbs at each wavelength, we can deduce the concentrations of both species from the two measurements. Suggest a procedure for using an optical absorption measurement to find the pH of a solution.

Notes and References

1. W. Tan, S.-Y. Shi, S. Smith, D. Birnbaum, and R. Kopelman, *Science* **1992**, *258*, 778; W. Tan, S.-Y. Shi, and R. Kopelman, *Anal. Chem.* **1992**, *64*, 2985; J. Ji, N. Rosenzweig, C. Griffin, and Z. Rosenzweig, *Anal. Chem.* **2000**, *72*, 3497; K. P. McNamara, T. Nguyen, G. Dumitrascu, J. Ji, N. Rosenzweig, and Z. Rosenzweig, *Anal. Chem.* **2001**, *73*, 3240.
2. H. N. Po and N. M. Senozan, *J. Chem. Ed.* **2001**, *78*, 1499; R. de Levie, *J. Chem. Ed.* **2003**, *80*, 146.
3. R. L. Barrett, *J. Chem. Ed.* **1955**, *32*, 78.
4. You can find more indicator demonstrations in J. T. Riley, *J. Chem. Ed.* **1977**, *54*, 29. The chemistry of carbonic acid is discussed by M. Kern, *J. Chem. Ed.* **1960**, *37*, 14.
5. M. A. White, *J. Chem. Ed.* **1998**, *75*, 1119.

Kjeldahl Nitrogen Analysis: Chemistry Behind the Headline

In 2007, pet dogs and cats in North America suddenly began to die, apparently from kidney failure. Within a few weeks, the mysterious illness was traced to animal food containing ingredients imported from China. It appears that melamine, used to make plastics, had been deliberately added to food ingredients "in a bid to meet the contractual demand for the amount of protein in the products."[1] Cyanuric acid, used to disinfect swimming pools, was also found in the food. Melamine alone does not cause kidney failure, but the combination of melamine and cyanuric acid does.

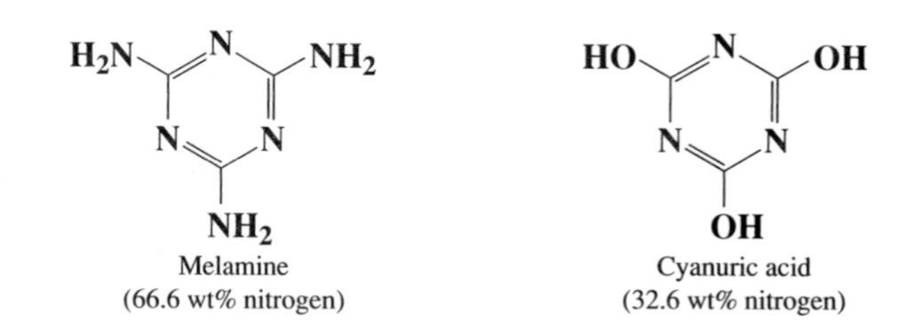

Melamine (66.6 wt% nitrogen)

Cyanuric acid (32.6 wt% nitrogen)

Protein source	Weight % nitrogen
Meat	16.0
Blood plasma	15.3
Milk	15.6
Flour	17.5
Egg	14.9

D. J. Holme and H. Peck, *Analytical Biochemistry*, 3rd ed. (New York: Addison Wesley Longman, 1998), p. 388.

What do these compounds have to do with protein? Nothing—except that they are high in nitrogen. Protein, which contains ~16 wt% nitrogen, is the main source of nitrogen in food. The Kjeldahl analysis for nitrogen described in Section 10-6 is used as a surrogate measurement for protein in food. (Combustion analysis described in Section 7-4 is another common method for measuring nitrogen in food.) For example, if food contains 10 wt% protein, it will contain ~16% of 10% = 1.6 wt% N. If you measure 1.6 wt% N in food, you could conclude that the food contains ~10 wt% protein. Melamine contains 66.6 wt% N, which is 4 times more than protein. Adding 1 wt% melamine to food makes it appear that the food contains an additional 4 wt% protein according to nitrogen analysis.

CHAPTER 10

Acid-Base Titrations

In a *titration*, we measure the quantity of a known reagent required to react with an unknown sample. From this quantity, we deduce the concentration of analyte in the unknown. Titrations of acids and bases are among the most widespread procedures in chemical analysis. For example, at the end of this chapter, we will see how acid-base titrations are used to measure nitrogen in foods, which is a measure of protein content.

10-1 Titration of Strong Base with Strong Acid

For each type of titration in this chapter, *our goal is to construct a graph showing how the pH changes as titrant is added.* If you can do this, then you understand what is happening during the titration, and you will be able to interpret an experimental titration curve.

The first step is to write the balanced chemical reaction between titrant and analyte. Then use that reaction to calculate the composition and pH after each addition of titrant. Let's consider the titration of 50.00 mL of 0.020 00 M KOH with 0.100 0 M HBr. The chemical reaction between titrant and analyte is merely

First write the reaction between *titrant* and *analyte*.

$$H^+ + OH^- \longrightarrow H_2O \qquad K = \frac{1}{K_w} = \frac{1}{10^{-14}} = 10^{14}$$

The titration reaction.

Because the equilibrium constant for this reaction is 10^{14}, it is fair to say that it "goes to completion." Prior to the equivalence point, *any amount of H^+ added will consume a stoichiometric amount of OH^-.*

Equivalence point: when moles of added titrant are exactly sufficient for stoichiometric reaction with analyte.

A useful starting point is to calculate the volume of HBr (V_e) needed to reach the equivalence point:

$$\underbrace{(V_e\,(\text{mL}))(0.100\;0\;\text{M})}_{\text{mmol of HBr at equivalence point}} = \underbrace{(50.00\;\text{mL})(0.020\;00\;\text{M})}_{\text{mmol of OH}^- \text{ being titrated}} \Rightarrow V_e = 10.00\;\text{mL}$$

$$\text{mL} \times \frac{\text{mol}}{\text{L}} = \text{mmol}$$

You can do all calculations with mol and L instead of mmol and mL, if you wish. I find mmol and mL to be more convenient.

When 10.00 mL of HBr have been added, the titration is complete. Prior to V_e, excess, unreacted OH^- is present. After V_e, there is excess H^+ in the solution.

In the titration of a strong base with a strong acid, three regions of the titration curve require different kinds of calculations:

1. Before the equivalence point, the pH is determined by excess OH^- in the solution.
2. At the equivalence point, added H^+ is just sufficient to react with all the OH^- to make H_2O. The pH is determined by the dissociation of water.
3. After the equivalence point, pH is determined by excess H^+ in the solution.

We will do one sample calculation for each region.

Region 1: Before the Equivalence Point

Before the equivalence point, there is excess OH^-.

Before we add any HBr titrant from the buret, the flask of analyte contains 50.00 mL of 0.020 00 M KOH, which amounts to (50.00 mL)(0.020 00 M) = 1.000 mmol of OH^-. Remember that mL × (mol/L) = mmol.

If we add 3.00 mL of HBr, we add (3.00 mL)(0.100 0 M) = 0.300 mmol of H^+, which consumes 0.300 mmol of OH^-.

$$\text{OH}^- \text{ remaining} = \underbrace{1.000 \text{ mmol}}_{\text{Initial OH}^-} - \underbrace{0.300 \text{ mmol}}_{\substack{\text{OH}^- \text{ consumed} \\ \text{by HBr}}} = 0.700 \text{ mmol}$$

The total volume in the flask is now 50.00 mL + 3.00 mL = 53.00 mL. Therefore the concentration of OH^- in the flask is

$$\frac{\text{mmol}}{\text{mL}} = \frac{\text{mol}}{\text{L}} = \text{M}$$

$$[\text{OH}^-] = \frac{0.700 \text{ mmol}}{53.00 \text{ mL}} = 0.013\ 2 \text{ M}$$

From the concentration of OH^-, it is easy to find the pH:

$$[\text{H}^+] = \frac{K_w}{[\text{OH}^-]} = \frac{1.0 \times 10^{-14}}{0.013\ 2} = 7.5_8 \times 10^{-13} \text{ M}$$
$$\Rightarrow \text{pH} = -\log(7.5_8 \times 10^{-13}) = 12.12$$

Challenge Calculate $[OH^-]$ and pH when 6.00 mL of HBr have been added. Check your answers against Table 10-1.

If you had to, you could reproduce all the calculations prior to the equivalence point in Table 10-1 in the same way as the 3.00-mL point. (A spreadsheet would be helpful.) The volume of acid added is designated V_a and pH is expressed to the 0.01 decimal place, regardless of what is justified by significant figures. We do this for the sake of consistency and also because 0.01 is near the limit of accuracy in pH measurements.

Region 2: At the Equivalence Point

At the equivalence point, enough H^+ has been added to react with all the OH^-. We could prepare the same solution by dissolving KBr in water. The equivalence

Table 10-1 Calculation of the titration curve for 50.00 mL of 0.020 00 M KOH treated with 0.100 0 M HBr

	mL HBr added (V_a)	Concentration of unreacted OH^- (M)	Concentration of excess H^+ (M)	pH
Region 1 (excess OH^-)	0.00	0.020 0		12.30
	1.00	0.017 6		12.24
	2.00	0.015 4		12.18
	3.00	0.013 2		12.12
	4.00	0.011 1		12.04
	5.00	0.009 09		11.95
	6.00	0.007 14		11.85
	7.00	0.005 26		11.72
	8.00	0.003 45		11.53
	9.00	0.001 69		11.22
	9.50	0.000 840		10.92
	9.90	0.000 167		10.22
	9.99	0.000 016 6		9.22
Region 2	10.00	—	—	7.00
Region 3 (excess H^+)	10.01		0.000 016 7	4.78
	10.10		0.000 166	3.78
	10.50		0.000 826	3.08
	11.00		0.001 64	2.79
	12.00		0.003 23	2.49
	13.00		0.004 76	2.32
	14.00		0.006 25	2.20
	15.00		0.007 69	2.11
	16.00		0.009 09	2.04

point pH of a strong acid-strong base titration is determined by the dissociation of water:

$$\underset{}{H_2O} \rightleftharpoons \underset{x}{H^+} + \underset{x}{OH^-}$$

$$K_w = x^2 = 1.0 \times 10^{-14} \Rightarrow x = 1.0 \times 10^{-7} \text{ M} \Rightarrow \text{pH} = 7.00$$

The pH at the equivalence point in the titration of any strong base (or acid) with strong acid (or base) will be 7.00 at 25°C.

As we will soon discover, *the pH is **not** 7.00 at the equivalence point in the titration of weak acids or bases.* The pH is 7.00 only if the titrant and analyte are both strong.

At the equivalence point, pH = 7.00, but *only* in a strong-acid–strong-base reaction.

Region 3: After the Equivalence Point

Beyond the equivalence point, excess HBr is present. For example, at the point where 10.50 mL of HBr have been added, there is an excess of 10.50 − 10.00 = 0.50 mL. The excess H^+ amounts to

$$\text{excess } H^+ = (0.50 \text{ mL})(0.100\ 0 \text{ M}) = 0.050 \text{ mmol}$$

After the equivalence point, there is excess H^+.

Using the total volume of solution (50.00 + 10.50 = 60.50 mL), we find the pH:

$$[H^+] = \frac{0.050\text{ mmol}}{60.50\text{ mL}} = 8.2_6 \times 10^{-4}\text{ M} \Rightarrow \text{pH} = -\log(8.2_6 \times 10^{-4}) = 3.08$$

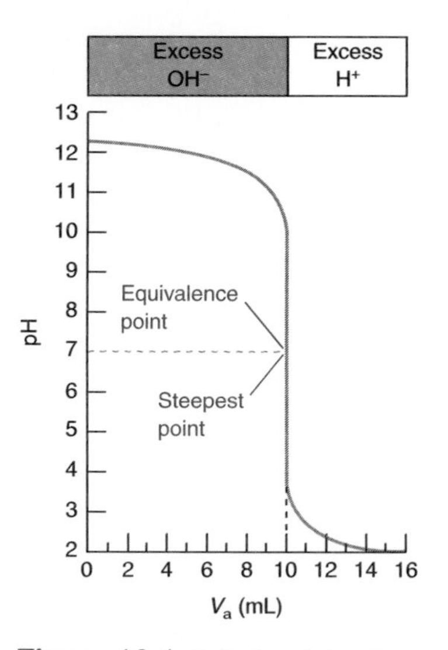

Figure 10-1 Calculated titration curve showing how pH changes as 0.100 0 M HBr is added to 50.00 mL of 0.020 00 M KOH. At the equivalence point, the curve is steepest. The first derivative reaches a maximum and the second derivative is 0. We discuss derivatives later in this chapter.

The Titration Curve

The titration curve in Figure 10-1 is a graph of pH versus V_a, the volume of acid added. The sudden change in pH near the equivalence point is characteristic of all analytically useful titrations. The curve is steepest at the equivalence point, which means that the slope is greatest. The pH at the equivalence point is 7.00 *only* in a strong-acid–strong-base titration. If one or both of the reactants are weak, the equivalence point pH is *not* 7.00.

Example **Titration of Strong Acid with Strong Base**

Find the pH when 12.74 mL of 0.087 42 M NaOH have been added to 25.00 mL of 0.066 66 M $HClO_4$.

SOLUTION The titration reaction is $H^+ + OH^- \longrightarrow H_2O$. The equivalence point is

$$\underbrace{(V_e\text{ (mL)})(0.087\ 42\text{ M})}_{\substack{\text{mmol of NaOH}\\ \text{at equivalence point}}} = \underbrace{(25.00\text{ mL})(0.066\ 66\text{ M})}_{\substack{\text{mmol of HClO}_4\\ \text{being titrated}}} \Rightarrow V_e = 19.06\text{ mL}$$

At V_b (volume of base) = 12.74 mL, there is excess acid in the solution:

$$H^+\text{ remaining} = \underbrace{(25.00\text{ mL})(0.066\ 66\text{ M})}_{\text{Initial mmol of HClO}_4} - \underbrace{(12.74\text{ mL})(0.087\ 42\text{ M})}_{\text{Added mmol of NaOH}} = 0.553\text{ mmol}$$

$$[H^+] = \frac{0.553\text{ mmol}}{(25.00 + 12.74)\text{ mL}} = 0.014\ 7\text{ M}$$

$$\text{pH} = -\log(0.014\ 7) = 1.83$$

Test Yourself Find the pH when 20.00 mL of 0.087 42 M NaOH have been added to 25.00 mL of 0.066 66 M $HClO_4$. (**Answer:** 11.26)

Ask Yourself

10-A. What is the equivalence volume in the titration of 50.00 mL of 0.010 0 M NaOH with 0.100 M HCl? Calculate the pH at the following points: V_a = 0.00, 1.00, 2.00, 3.00, 4.00, 4.50, 4.90, 4.99, 5.00, 5.01, 5.10, 5.50, 6.00, 8.00, and 10.00 mL. Make a graph of pH versus V_a.

10-2 Titration of Weak Acid with Strong Base

The titration of a weak acid with a strong base puts all our knowledge of acid-base chemistry to work. The example we treat is the titration of 50.00 mL of 0.020 00 M MES with 0.100 0 M NaOH. MES is an abbreviation for 2-(*N*-morpholino)ethanesulfonic acid, a weak acid with $pK_a = 6.27$. It is widely used in biochemistry as a buffer for the pH 6 region.

The *titration reaction* is

$$\underset{\substack{\text{HA} \\ \text{MES, } pK_a = 6.27}}{\text{O}\langle\rangle\overset{+}{\text{N}}\text{HCH}_2\text{CH}_2\text{SO}_3^-} + \text{OH}^- \longrightarrow \underset{\text{A}^-}{\text{O}\langle\rangle\text{NCH}_2\text{CH}_2\text{SO}_3^-} + \text{H}_2\text{O} \quad (10\text{-}1)$$

First write the titration reaction.

Reaction 10-1 is the reverse of the K_b reaction for the base A^-. The equilibrium constant is $1/K_b = 1/(K_w/K_{HA}) = 5.4 \times 10^7$. The equilibrium constant is so large that we can say that the reaction goes "to completion" after each addition of OH^-. As we saw in Box 9-1, *strong + weak react completely.*

strong + weak $\longrightarrow$ complete reaction

It is helpful first to calculate the volume of base needed to reach the equivalence point. Because 1 mol of OH^- reacts with 1 mol of MES, we can say

$$\underbrace{(V_e\,(\text{mL}))(0.100\,0\text{ M})}_{\text{mmol of base}} = \underbrace{(50.00\text{ mL})(0.020\,00\text{ M})}_{\text{mmol of HA}} \Rightarrow V_e = 10.00\text{ mL}$$

The titration calculations for this problem are of four types:

The four regions of the titration curve are important enough to be shown inside the back cover of this book.

1. Before any base is added, the solution contains just HA in water. This is a weak-acid problem in which the pH is determined by the equilibrium

$$\text{HA} \overset{K_a}{\rightleftharpoons} \text{H}^+ + \text{A}^-$$

2. From the first addition of NaOH until immediately before the equivalence point, there is a mixture of unreacted HA plus the A^- produced by Reaction 10-1. *Aha! A buffer!* We can use the Henderson-Hasselbalch equation to find the pH.

3. At the equivalence point, "all" HA has been converted into A^-, the conjugate base. Therefore, the pH will be higher than 7. The problem is the same as if the solution had been made by dissolving A^- in water. We have a weak-base problem in which pH is determined by the reaction

$$\text{A}^- + \text{H}_2\text{O} \overset{K_b}{\rightleftharpoons} \text{HA} + \text{OH}^- \qquad K_b = K_w/K_a$$

4. Beyond the equivalence point, excess NaOH is being added to a solution of A^-, creating a mixture of strong and weak base. We calculate the pH as if we had simply added excess NaOH to water. We ignore the small effect from A^-.

Region 1: Before Base Is Added

Before adding any base, we have a solution of 0.020 00 M HA with $pK_a = 6.27$. This is simply a weak-acid problem.

The initial solution contains just the *weak acid* HA.

$$\underset{\text{F}-x}{\text{HA}} \rightleftharpoons \underset{x}{\text{H}^+} + \underset{x}{\text{A}^-} \qquad K_a = 10^{-6.27}$$

F is the formal concentration of HA, which is 0.020 00 M.

$$\frac{x^2}{0.020\,00 - x} = K_a \Rightarrow x = 1.0_3 \times 10^{-4} \Rightarrow \text{pH} = 3.99$$

Before the equivalence point, there is a mixture of HA and A^-, which is a *buffer. Aha! A buffer!*

Region 2: Before the Equivalence Point

Once we begin to add OH^-, a mixture of HA and A^- is created by the titration reaction 10-1. This mixture is a buffer whose pH can be calculated with the Henderson-Hasselbalch equation (9-1) once we know the quotient $[A^-]/[HA]$.

Henderson-Hasselbalch equation:

$$\mathrm{pH} = \mathrm{p}K_a + \log\left(\frac{[A^-]}{[HA]}\right) \tag{9-1}$$

Consider the point where 3.00 mL of OH^- have been added:

Titration reaction:	HA	+	OH^-	$\longrightarrow$	A^-	+	H_2O
Initial mmol	1.000		0.300		—		
Final mmol	0.700		—		0.300		

Aha! A buffer! ($HA + A^-$)

The Henderson-Hasselbalch equation needs only mmol because volumes cancel in the quotient $[A^-]/[HA]$.

Once we know the *quotient* $[A^-]/[HA]$ in any solution, we know its pH:

$$\mathrm{pH} = \mathrm{p}K_a + \log\left(\frac{[A^-]}{[HA]}\right) = 6.27 + \log\left(\frac{0.300}{0.700}\right) = 5.90$$

The point at which the volume of titrant is $\frac{1}{2}V_e$ is a special one in any titration.

Titration reaction:	HA	+	OH^-	$\longrightarrow$	A^-	+	H_2O
Initial mmol	1.000		0.500		—		
Final mmol	0.500		—		0.500		

Landmark point:

When $V_b = \frac{1}{2}V_e$, $\mathrm{pH} = \mathrm{p}K_a$.

(When activities are taken into account [Section 12-2], this statement is not exactly true, but it is a good approximation.)

$$\mathrm{pH} = \mathrm{p}K_a + \log\left(\frac{0.500}{0.500}\right) = \mathrm{p}K_a$$

When the volume of titrant is $\frac{1}{2}V_e$, $\mathrm{pH} = \mathrm{p}K_a$ for the acid HA. From the experimental titration curve, you can find $\mathrm{p}K_a$ by reading the pH when $V_b = \frac{1}{2}V_e$, where V_b is the volume of added base.

Advice. As soon as you recognize a mixture of HA and A^- in any solution, *you have a buffer!* Stop right there. You can find the pH from the quotient $[A^-]/[HA]$ with the Henderson-Hasselbalch equation.

At the equivalence point, HA has been converted into A^-, a *weak base.*

Region 3: At the Equivalence Point

At the equivalence point (V_b = 10.00 mL), the quantity of NaOH is exactly enough to consume the HA.

Titration reaction:	HA	+	OH^-	$\longrightarrow$	A^-	+	H_2O
Initial mmol	1.000		1.000		—		
Final mmol	—		—		1.000		

The resulting solution contains "just" A^-. We could have prepared the same solution by dissolving the salt Na^+A^- in water. *Na^+A^- is a weak base*, so the pH must be >7.

To compute the pH of a weak base, write the reaction of the base with water:

$$\underset{F'-x}{A^-} + H_2O \rightleftharpoons \underset{x}{HA} + \underset{x}{OH^-} \qquad K_b = \frac{K_w}{K_a} \qquad (10\text{-}2)$$

The only tricky point is that the formal concentration of A^- is no longer 0.020 00 M, which was the initial concentration of HA. The initial 1.000 mmol of HA in 50.00 mL has been diluted with 10.00 mL of titrant:

$$[A^-] = \frac{1.000 \text{ mmol}}{(50.00 + 10.00) \text{ mL}} = 0.016\ 67 \text{ M} \equiv F'$$

Designating the formal concentration of A^- as F′, we can find the pH from Reaction 10-2:

$$\frac{x^2}{F' - x} = K_b = \frac{K_w}{K_a} = 1.8_6 \times 10^{-8} \Rightarrow x = 1.7_6 \times 10^{-5} \text{ M}$$

$$\text{pH} = -\log[H^+] = -\log\left(\frac{K_w}{x}\right) = 9.25$$

The pH at the equivalence point in this titration is 9.25. **It is not 7.00.** The equivalence-point pH will *always* be above 7 for the titration of a weak acid with a strong base, because the acid is converted into its conjugate base at the equivalence point.

The pH is higher than 7 at the equivalence point in the titration of a weak acid with a strong base.

Region 4: After the Equivalence Point

Now we are adding NaOH to a solution of A^-. The base NaOH is so much stronger than the base A^- that it is a fair approximation to say that the pH is determined by the concentration of excess OH^- in the solution.

Here we assume that the pH is governed by the excess OH^-.

Let's calculate the pH when V_b = 10.10 mL, which is just 0.10 mL past V_e. The quantity of excess OH^- is (0.10 mL)(0.100 0 M) = 0.010 mmol, and the total volume of solution is 50.00 + 10.10 mL = 60.10 mL.

$$[OH^-] = \frac{0.010 \text{ mmol}}{50.00 + 10.10 \text{ mL}} = 1.66 \times 10^{-4} \text{ M}$$

$$\text{pH} = -\log\left(\frac{K_w}{[OH^-]}\right) = 10.22$$

Challenge At V_b = 10.10 mL, show that excess NaOH = 0.17 mM. Show that F_{A^-} = 17 mM. Show that, in the presence of 0.17 mM NaOH, 17 mM A^- produces only 1.9 μM OH^-. That is, ignoring A^- in comparison to excess NaOH is justified.

The Titration Curve

A summary of the calculations for the titration of MES with NaOH is shown in Table 10-2. The titration curve in Figure 10-2 has two easily identified points. One is the equivalence point, which is the steepest part of the curve. The other landmark is the point where $V_b = \frac{1}{2}V_e$ and pH = pK_a. This latter point has the minimum slope, which means that the pH changes least for a given addition of NaOH. This is another way of saying that *buffer capacity* is maximum when pH = pK_a and [HA] = [A^-].

Landmarks in a titration:

At $V_b = V_e$, curve is steepest.

At $V_b = \frac{1}{2}V_e$, pH = pK_a and the slope is minimal.

The *buffer capacity* measures the ability of the solution to resist changes in pH.

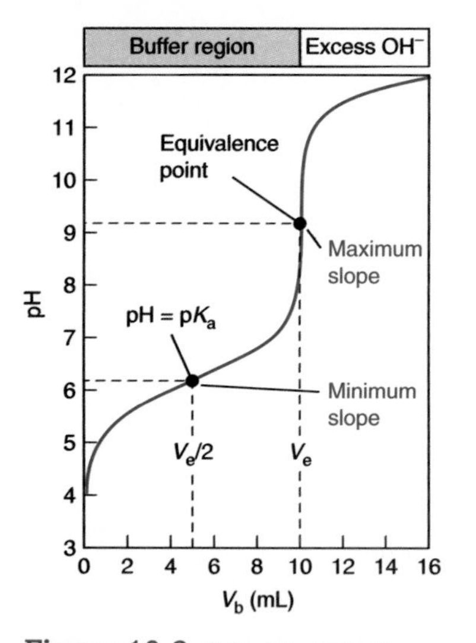

Figure 10-2 Calculated titration curve for the reaction of 50.00 mL of 0.020 00 M MES with 0.100 0 M NaOH. Landmarks occur at half of the equivalence volume (pH = pK_a) and at the equivalence point, which is the steepest part of the curve.

Table 10-2 Calculation of the titration curve for 50.00 mL of 0.020 00 M MES treated with 0.100 0 M NaOH

	mL base added (V_b)	pH
Region 1 (weak acid)	0.00	3.99
Region 2 (buffer)	0.50	4.99
	1.00	5.32
	2.00	5.67
	3.00	5.90
	4.00	6.09
	5.00	6.27
	6.00	6.45
	7.00	6.64
	8.00	6.87
	9.00	7.22
	9.50	7.55
	9.90	8.27
Region 3 (weak base)	10.00	9.25
Region 4 (excess OH^-)	10.10	10.22
	10.50	10.91
	11.00	11.21
	12.00	11.50
	13.00	11.67
	14.00	11.79
	15.00	11.88
	16.00	11.95

Ask Yourself

10-B. Write the reaction between formic acid (Appendix B) and KOH. What is the equivalence volume (V_e) in the titration of 50.0 mL of 0.050 0 M formic acid with 0.050 0 M KOH? Calculate the pH at the points V_b = 0.0, 10.0, 20.0, 25.0, 30.0, 40.0, 45.0, 48.0, 49.0, 49.5, 50.0, 50.5, 51.0, 52.0, 55.0, and 60.0 mL. Draw a graph of pH versus V_b. Without doing any calculations, what should the pH be at $V_b = \frac{1}{2}V_e$? Does your calculated result agree with the prediction?

10-3 Titration of Weak Base with Strong Acid

The titration of a weak base with a strong acid is just the reverse of the titration of a weak acid with a strong base. The *titration reaction* is

$$B + H^+ \longrightarrow BH^+$$

Because the reactants are weak + strong, the reaction goes essentially to completion after each addition of acid. There are four distinct regions of the titration curve:

1. Before acid is added, the solution contains just the weak base, B, in water. The pH is determined by the K_b reaction:

When $V_a = 0$, we have a *weak-base* problem.

$$\underset{F-x}{B} + H_2O \overset{K_b}{\rightleftharpoons} \underset{x}{BH^+} + \underset{x}{OH^-}$$

2. Between the initial point and the equivalence point, there is a mixture of B and BH^+—*Aha! A buffer!* The pH is computed by using

$$\text{pH} = \text{p}K_a \text{ (for BH}^+\text{)} + \log\left(\frac{[\text{B}]}{[\text{BH}^+]}\right)$$

When $0 < V_a < V_e$, we have a *buffer.*

At the special point where $V_a = \frac{1}{2}V_e$, pH = pK_a (for BH^+).

3. At the equivalence point, B has been converted into BH^+, a weak acid. The pH is calculated by considering the acid dissociation reaction of BH^+:

$$\underset{F' - x}{\text{BH}^+} \rightleftharpoons \underset{x}{\text{B}} + \underset{x}{\text{H}^+} \qquad K_a = \frac{K_w}{K_b}$$

When $V_a = V_e$, the solution contains the *weak acid* BH^+.

The formal concentration of BH^+, F′, is not the same as the original formal concentration of B, because there has been some dilution. Because the solution contains BH^+ at the equivalence point, it is acidic. *The pH at the equivalence point must be below 7.*

4. After the equivalence point, there is excess strong acid in the solution. We treat this problem by considering only the concentration of excess H^+ and ignoring the contribution of weak acid, BH^+.

For $V_a > V_e$, there is excess *strong acid.*

Example Titration of Pyridine with HCl

Consider the titration of 25.00 mL of 0.083 64 M pyridine with 0.106 7 M HCl, for which the equivalence point is V_e = 19.60 mL.

Titration reaction: Pyridine (B) + H^+ ⟶ BH^+

Pyridine (B)
$K_b = 1.6 \times 10^{-9}$

$$\underbrace{(V_e \text{ (mL)})(0.106\ 7 \text{ M})}_{\text{mmol of HCl}} = \underbrace{(25.00 \text{ mL})(0.083\ 64 \text{ M})}_{\text{mmol of pyridine}} \Rightarrow V_e = 19.60 \text{ mL}$$

(a) Find the pH when V_a = 4.63 mL, which is before V_e, and **(b)** find the pH at V_e.

SOLUTION (a) At 4.63 mL, part of the pyridine has been neutralized, so there is a mixture of pyridine and pyridinium ion—*Aha! A buffer!* The initial millimoles of pyridine are (25.00 mL)(0.083 64 M) = 2.091 mmol. The added H^+ is (4.63 mL) × (0.106 7 M) = 0.494 mmol. Therefore we can write

Titration reaction:	B	+	H^+	⟶	BH^+
Initial mmol	2.091		0.494		—
Final mmol	1.597		—		0.494

Aha! A buffer! (B + BH^+)

$$\text{pH} = \underset{-\log(K_w/K_b)}{\text{p}K_{BH^+}} + \log\left(\frac{[\text{B}]}{[\text{BH}^+]}\right) = 5.20 + \log\left(\frac{1.597}{0.494}\right) = 5.71$$

pK_{BH^+} is the *acid* pK_a for the *acid* BH^+.

(b) At the equivalence point (19.60 mL), enough acid has been added to convert all of the pyridine (B) into BH^+. The pH is governed by dissociation of the weak acid, BH^+, whose acid dissociation constant is $K_a = K_w/K_b = 6.3 \times 10^{-6}$. The formal concentration of BH^+ is equal to the initial mmol of pyridine divided by the mL of solution at the equivalence point: F′ = (2.091 mmol)/(25.00 + 19.60 mL) = 0.046 88 M.

$$\underset{F'-x}{BH^+} \rightleftharpoons \underset{x}{B} + \underset{x}{H^+} \qquad K_a = 6.3 \times 10^{-6}$$

$$\frac{x^2}{F'-x} = \frac{x^2}{0.046\,88 - x} = K_a = 6.3 \times 10^{-6} \Rightarrow x = [H^+] = 5.4_0 \times 10^{-4}\ M$$

$$pH = -\log[H^+] = 3.27$$

The pH at the equivalence point is acidic because the weak base has been converted into a weak acid.

Test Yourself Find the pH when 19.00 mL of HCl have been added. (**Answer:** 3.70)

Ask Yourself

10-C. (a) Why is the equivalence point pH necessarily below 7 when a weak base is titrated with strong acid?
(b) What is the equivalence volume in the titration of 100.0 mL of 0.100 M cocaine (Reaction 8-19, $K_b = 2.6 \times 10^{-6}$) with 0.200 M HNO_3? Calculate the pH at V_a = 0.0, 10.0, 20.0, 25.0, 30.0, 40.0, 49.0, 49.9, 50.0, 50.1, 51.0, and 60.0 mL. Draw a graph of pH versus V_a.

10-4 Finding the End Point

The *equivalence point* in a titration is defined by the stoichiometry of the reaction. The *end point* is the abrupt change in a physical property (such as pH) that we measure to locate the equivalence point. Indicators and pH measurements are commonly used to find the end point in an acid-base titration.

Using Indicators to Find the End Point

In Section 9-6, we learned that an indicator is an acid or base whose various protonated species have different colors. For the weak-acid indicator, HIn, the solution takes on the color of HIn when $pH \lesssim pK_{HIn} - 1$ and has the color of In^- when $pH \gtrsim pK_{HIn} + 1$. In the interval $pK_{HIn} - 1 \lesssim pH \lesssim pK_{HIn} + 1$, a mixture of both colors is observed.

Choose an indicator whose color change comes as close as possible to the theoretical pH of the equivalence point.

A titration curve for which pH = 5.54 at the equivalence point is shown in Figure 10-3. The pH drops steeply (from 7 to 4) over a small volume interval. An indicator with a color change in this pH interval would provide a fair approximation to the equivalence point. The closer the color change is to pH 5.54, the more accurate will be the end point. The difference between the observed end point (color change) and the true equivalence point is called the **indicator error.**

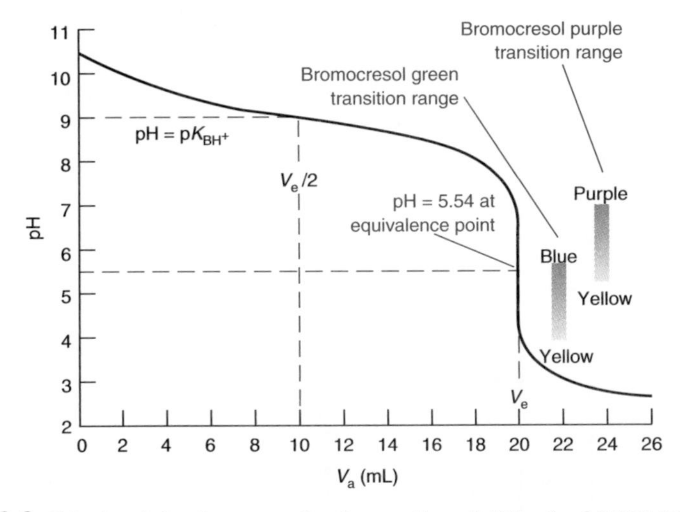

Figure 10-3 Calculated titration curve for the reaction of 100 mL of 0.010 0 M base (pK_b = 5.00) with 0.050 0 M HCl. As in the titration of HA with OH^-, pH = pK_{BH^+} when $V_a = \frac{1}{2}V_e$.

If you dump half a bottle of indicator into your reaction, you will introduce a different indicator error. Because indicators are acids or bases, they consume analyte or titrant. We use indicators under the assumption that the moles of indicator are negligible relative to the moles of analyte. Never use more than a few drops of dilute indicator.

Many indicators in Table 9-3 would be useful for the titration in Figure 10-3. For example, if bromocresol purple were used, we would use the purple-to-yellow color change as the end point. The last trace of purple should disappear near pH 5.2, which is quite close to the true equivalence point in Figure 10-3. If bromocresol green were used as the indicator, a color change from blue to green (= yellow + blue) would mark the end point.

In general, *we seek an indicator whose transition range overlaps the steepest part of the titration curve as closely as possible.* The steepness of the titration curve near the equivalence point in Figure 10-3 ensures that the indicator error caused by the noncoincidence of the color change and equivalence point will not be large. For example, if the indicator color change were at pH 6.4 (instead of 5.54), the error in V_e would be only 0.25% in this particular case.

One of the most common indicators is phenolphthalein, which changes from colorless in acid to pink in base:

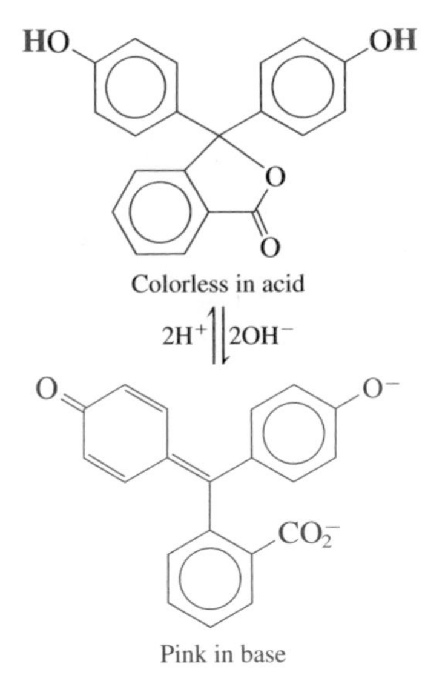

Using a pH Electrode to Find the End Point

Figure 10-4 shows experimental results for the titration of the weak acid, H_6A, with NaOH. Because the compound is difficult to purify, only a tiny amount was available for titration. Just 1.430 mg was dissolved in 1.00 mL of aqueous solution and titrated with microliter (μL) quantities of 0.065 92 M NaOH delivered with a Hamilton syringe.

When H_6A is titrated, we might expect to see an abrupt change in pH at all six equivalence points. The curve in Figure 10-4 shows two clear breaks, near 90 and 120 μL, which correspond to titration of the *third* and *fourth* protons of H_6A.

$$H_4A^{2-} + OH^- \longrightarrow H_3A^{3-} + H_2O \quad (\sim 90\ \mu L \text{ equivalence point})$$
$$H_3A^{3-} + OH^- \longrightarrow H_2A^{4-} + H_2O \quad (\sim 120\ \mu L \text{ equivalence point})$$

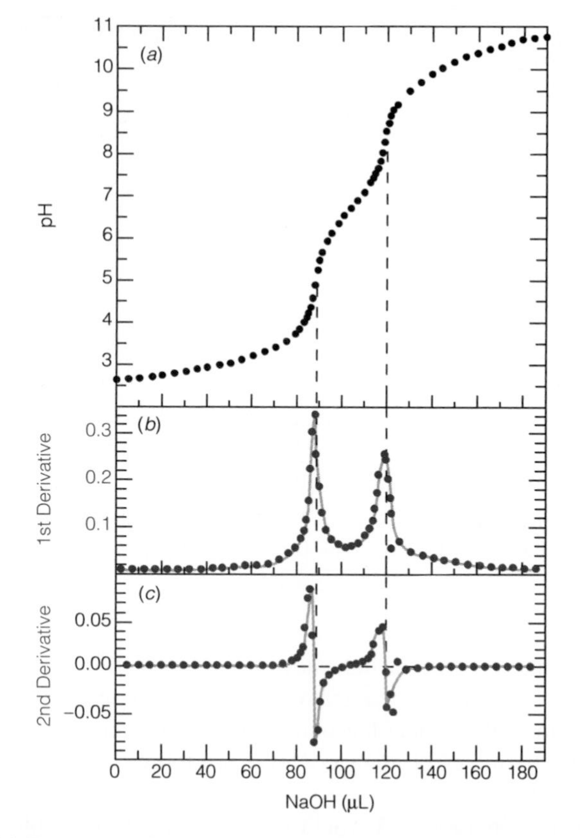

Figure 10-4 (*a*) Experimental points in the titration of 1.430 mg of xylenol orange, a hexaprotic acid, dissolved in 1.000 mL of 0.10 M $NaNO_3$. The titrant was 0.065 92 M NaOH. (*b*) The first derivative, $\Delta pH/\Delta V$, of the titration curve. (*c*) The second derivative, $\Delta(\Delta pH/\Delta V)/\Delta V$, which is the derivative of the curve in panel *b*. Derivatives for the first end point are calculated in Figure 10-5. End points are taken as maxima in the derivative curve and zero crossings of the second derivative.

The first two and last two equivalence points give unrecognizable end points, because they occur at pH values that are too low or too high.

The end point has maximum slope.

The end point is where the slope of the titration curve is greatest. The slope is the change in pH (ΔpH) divided by the change in volume (ΔV) between the points:

Slope of titration curve:

$$\text{slope} = \frac{\Delta pH}{\Delta V} \tag{10-3}$$

The slope (which is also called the *first derivative*) displayed in the middle of Figure 10-4 is calculated in Figure 10-5. The first two columns of this spreadsheet give experimental volumes and pH measurements. (The pH meter was precise to three digits, even though accuracy ends in the second decimal place.) To compute the first derivative, each pair of volumes is averaged and the quantity $\Delta pH/\Delta V$ is calculated.

The last two columns of Figure 10-5 and the graph in Figure 10-4*c* give the slope of the slope (called the *second derivative*), computed as follows:

The slope of the slope (the second derivative) is 0 at the end point.

Second derivative:

$$\frac{\Delta(\text{slope})}{\Delta V} = \frac{\Delta(\Delta pH/\Delta V)}{\Delta V} \tag{10-4}$$

	A	B	C	D	E	F
1	Derivatives of a Titration Curve					
2	Data		1st derivative		2nd derivative	
3	µL NaOH	pH	µL	ΔpH/ΔµL		Δ(ΔpH/ΔµL)
4	85.0	4.245			µL	ΔµL
5			85.5	0.155		
6	86.0	4.400			86.0	0.0710
7			86.5	0.226		
8	87.0	4.626			87.0	0.0810
9			87.5	0.307		
10	88.0	4.933			88.0	0.0330
11			88.5	0.340		
12	89.0	5.273			89.0	−0.0830
13			89.0	0.257		
14	90.0	5.530			90.0	−0.0680
15			90.5	0.189		
16	91.0	5.719			91.25	−0.0390
17			92.0	0.131		
18	93.0	5.980				
19	Representative formulas:					
20	C5 = (A6+A4)/2			E6 = (C7+C5)/2		
21	D5 = (B6−B4)/(A6−A4)			F6 = (D7−D5)/(C7−C5)		

Figure 10-5 Spreadsheet for computing first and second derivatives near 90 µL in Figure 10-4.

The end point is the volume at which the second derivative is 0. A graph on the scale of Figure 10-6 allows us to make a good estimate of the end-point volume.

Example Computing Derivatives of a Titration Curve

Let's see how the first and second derivatives in Figure 10-5 are calculated.

SOLUTION The volume in cell C5, 85.5, is the average of the first two volumes (85.0 and 86.0) in column A. The slope (first derivative) $\Delta pH/\Delta V$ in cell D5 is calculated from the first two pH values and the first two volumes:

$$\frac{\Delta pH}{\Delta V} = \frac{4.400 - 4.245}{86.0 - 85.0} = 0.15_5$$

Retain extra, insignificant digits for these calculations.

The coordinates ($x = 85.5$, $y = 0.15_5$) are one point in the graph of the first derivative in Figure 10-4*b*.

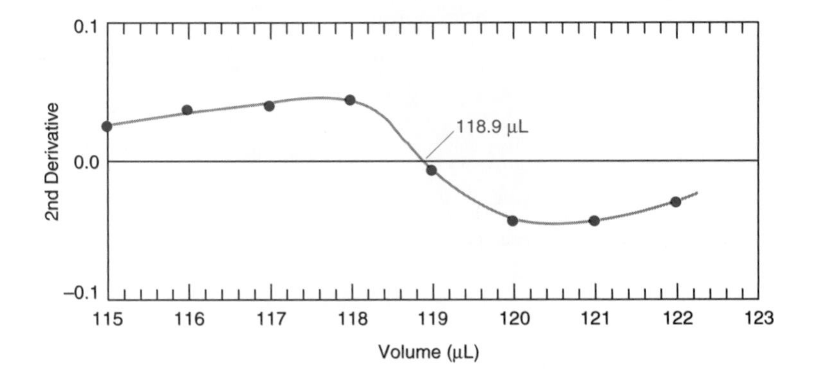

Figure 10-6 Enlargement of the second end point in the second derivative curve of Figure 10-4*c*.

The second derivative is computed from the first derivative. The volume in cell E6 is 86.0, which is the average of 85.5 and 86.5. The second derivative in cell F6 is

$$\frac{\Delta(\Delta \text{pH}/\Delta V)}{\Delta V} = \frac{0.22_6 - 0.15_5}{86.5 - 85.5} = 0.071_0$$

The coordinates ($x = 86.0$, $y = 0.071_0$) are plotted in the second derivative graph in Figure 10-4*c*. These calculations are tedious by hand, but not bad in a spreadsheet.

Test Yourself Verify the first and second derivatives in cells D17 and F16 of Figure 10-5. (**Answer:** 0.130 5 and −0.038 67. Round-off errors arise because the spreadsheet uses more digits than are displayed in Figure 10-5.)

Figure 10-7 shows an *autotitrator*, which performs titrations automatically and sends results directly to a spreadsheet. Titrant from the bottle is dispensed in small increments by a syringe pump, and pH is measured by the electrode in the beaker. The instrument waits for the pH to stabilize after each addition, before adding the next increment.

Ask Yourself

10-D. (a) Select indicators from Table 9-3 that would be useful for the titrations in Figures 10-1 and 10-2 and for the $pK_a = 8$ curve in Figure 10-11. Select a different indicator for each titration and state what color change you would use as the end point.

(b) Data near the second end point in Figure 10-4 are given in the table. Prepare a spreadsheet like Figure 10-5, showing the first and second derivatives. Plot both derivatives versus V_b and locate the end point from each plot.

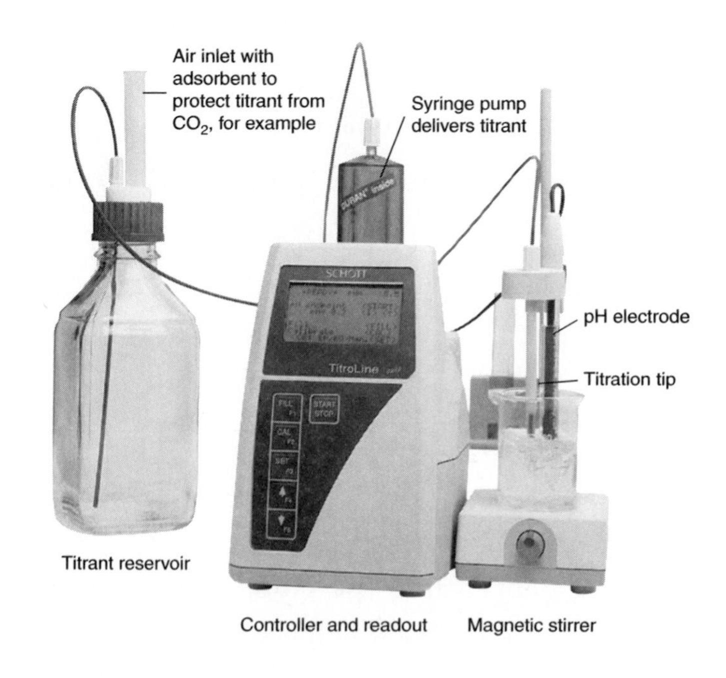

Figure 10-7 Autotitrator delivers titrant from the bottle at the left to the beaker of analyte on the stirring motor at the right. The electrode immersed in the beaker monitors pH or the concentrations of specific ions. Volume and pH readings can go directly to a spreadsheet. [Courtesy of Schott Instruments, Mainz, Germany, and Cole-Parmer Instruments, Vernon Hills, IL.]

V_b (μL)	pH	V_b (μL)	pH	V_b (μL)	pH	V_b (μL)	pH
107.0	6.921	114.0	7.457	117.0	7.878	120.0	8.591
110.0	7.117	115.0	7.569	118.0	8.090	121.0	8.794
113.0	7.359	116.0	7.705	119.0	8.343	122.0	8.952

10-5 Practical Notes

Acids and bases listed in Table 10-3 can be purchased in forms pure enough to be *primary standards*. NaOH and KOH are not primary standards because the reagent-grade materials contain carbonate (from reaction with atmospheric CO_2) and adsorbed water. Solutions of NaOH and KOH must be standardized against a primary standard. Potassium hydrogen phthalate is convenient for this purpose. Solutions of NaOH for titrations are prepared by diluting a stock solution of 50 wt% aqueous NaOH. Sodium carbonate is relatively insoluble in this stock solution and settles to the bottom.

Primary standards must be pure, stable, easily dried, and not hygroscopic. *Hygroscopic* compounds adsorb water while you are weighing them. NaOH and KOH are not primary standards.

Table 10-3 Primary standards

Compound	Formula mass	Notes
ACIDS		
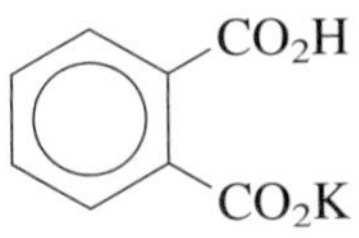 $C_6H_4(CO_2H)(CO_2K)$ (ortho) Potassium hydrogen phthalate	204.22	The pure solid is dried at 105°C and used to standardize base. A phenolphthalein end point is satisfactory. 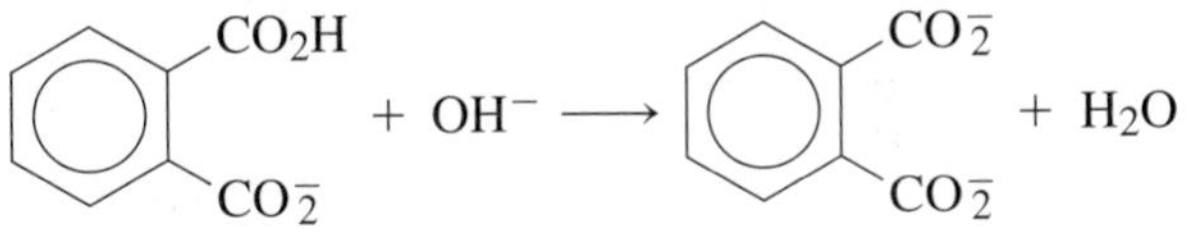 $C_6H_4(CO_2H)(CO_2^-) + OH^- \longrightarrow C_6H_4(CO_2^-)_2 + H_2O$
$KH(IO_3)_2$ Potassium hydrogen iodate	389.91	This is a strong acid, so any indicator with an end point between ~5 and ~9 is adequate.
BASES		
$H_2NC(CH_2OH)_3$ Tris(hydroxymethyl)aminomethane (also called tris or tham)	121.14	The pure solid is dried at 100°–103°C and titrated with strong acid. The end point is in the range pH 4.5–5. $H_2NC(CH_2OH)_3 + H^+ \longrightarrow H_3\overset{+}{N}C(CH_2OH)_3$
Na_2CO_3 Sodium carbonate	105.99	Primary standard grade Na_2CO_3 is titrated with acid to an end point of pH 4–5. Just before the end point, the solution is boiled to expel CO_2.
$Na_2B_4O_7 \cdot 10H_2O$ Borax	381.37	The recrystallized material is dried in a chamber containing an aqueous solution saturated with NaCl and sucrose. This procedure gives the decahydrate in pure form. The standard is titrated with acid to a methyl red end point. $\text{“}B_4O_7^{2-} \cdot 10H_2O\text{”} + 2H^+ \longrightarrow 4B(OH)_3 + 5H_2O$

Alkaline (basic) solutions must be protected from the atmosphere because they absorb CO_2:

$$OH^- + CO_2 \longrightarrow HCO_3^-$$

CO_2 changes the concentration of base over a period of time and reduces the sharpness of the end point in the titration of weak acids. If base is kept in a tightly capped polyethylene bottle, it can be used for weeks with little change. Strong base attacks glass and should not be kept in a buret longer than necessary.

Ask Yourself

10-E. **(a)** Give the name and formula of a primary standard used to standardize **(i)** HCl and **(ii)** NaOH.
(b) Referring to Table 10-3, determine how many grams of potassium hydrogen phthalate should be used to standardize ~0.05 M NaOH if you wish to use ~30 mL of base for the titration.

10-6 Kjeldahl Nitrogen Analysis

Developed in 1883, the **Kjeldahl nitrogen analysis** remains one of the most widely used methods for determining nitrogen in organic substances such as protein, cereal, and flour. The solid is *digested* (decomposed and dissolved) in boiling sulfuric acid to convert nitrogen into ammonium ion, NH_4^+:

Each atom of nitrogen in the unknown is converted into one NH_4^+ ion.

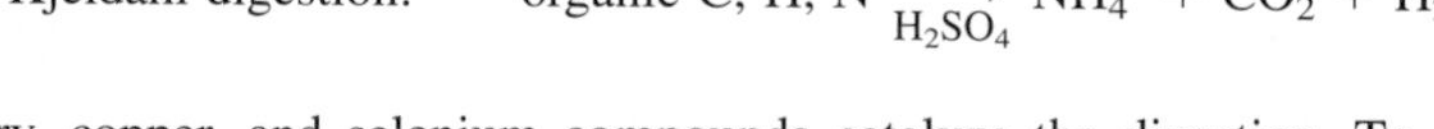

$$\text{Kjeldahl digestion:} \qquad \text{organic C, H, N} \xrightarrow[H_2SO_4]{\text{boiling}} NH_4^+ + CO_2 + H_2O$$

Mercury, copper, and selenium compounds catalyze the digestion. To speed the reaction, the boiling point of concentrated (98 wt%) sulfuric acid (338°C) is raised by adding K_2SO_4. Digestion is carried out in a long-neck *Kjeldahl flask* (Figure 10-8) that prevents loss of sample by spattering. (An alternative to the Kjeldahl flask is a microwave bomb containing H_2SO_4 and H_2O_2, like that in Figure 2-18.)

After digestion is complete, the solution containing NH_4^+ is made basic, and the liberated NH_3 is distilled (with a large excess of steam) into a receiver containing a

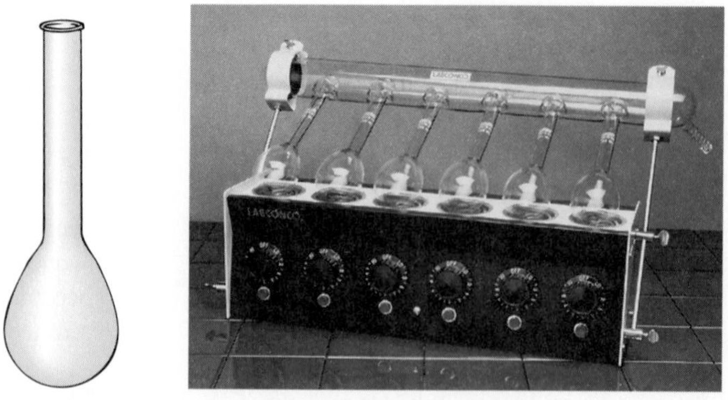

(a) (b)

Figure 10-8 Kjeldahl digestion flask has a long neck to minimize loss by spattering. (*b*) Six-port manifold for multiple samples provides for exhaust of fumes. [Fisher Scientific, Pittsburgh, PA.]

known amount of HCl (Figure 10-9). Excess, unreacted HCl is titrated with standard NaOH to determine how much HCl was consumed by NH_3.

Neutralization of NH_4^+:	$NH_4^+ + OH^- \longrightarrow NH_3(g) + H_2O$	(10-5)
Distillation of NH_3 *into standard HCl*:	$NH_3 + H^+ \longrightarrow NH_4^+$	(10-6)
Titration of unreacted HCl with NaOH:	$H^+ + OH^- \longrightarrow H_2O$	(10-7)

An alternative to the acid-base titration is to neutralize the acid and raise the pH with a buffer, followed by addition of reagents that form a colored product with NH_3.[2] The absorbance of the colored product gives the concentration of NH_3 from the digestion.

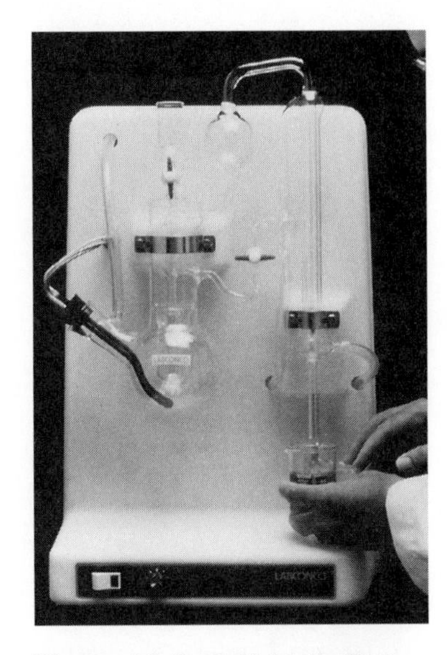

Figure 10-9 Kjeldahl distillation unit employs electric immersion heater in flask at left to carry out distillation in 5 min. Beaker at right collects liberated NH_3 in standard HCl. [Fisher Scientific, Pittsburgh, PA.]

Example Kjeldahl Analysis

A typical protein contains 16.2 wt% nitrogen. A 0.500-mL aliquot of protein solution was digested, and the liberated NH_3 was distilled into 10.00 mL of 0.021 40 M HCl. The unreacted HCl required 3.26 mL of 0.019 8 M NaOH for complete titration. Find the concentration of protein (mg protein/mL) in the original sample.

SOLUTION The original amount of HCl in the receiver was (10.00 mL) (0.021 40 M) = 0.214 0 mmol. The NaOH required for titration of unreacted HCl in Reaction 10-7 was (3.26 mL)(0.019 8 M) = 0.064 5 mmol. The difference, 0.214 0 − 0.064 5 = 0.149 5 mmol, must equal the quantity of NH_3 produced in Reaction 10-5 and distilled into the HCl.

Because 1 mmol of nitrogen in the protein gives rise to 1 mmol of NH_3, there must have been 0.149 5 mmol of nitrogen in the protein, corresponding to

$$(0.149\ 5\ \text{mmol})\left(14.006\ 74\ \frac{\text{mg N}}{\text{mmol}}\right) = 2.093\ \text{mg N}$$

If the protein contains 16.2 wt% N, there must be

$$\frac{2.093\ \text{mg N}}{0.162\ \text{mg N/mg protein}} = 12.9\ \text{mg protein}$$

$$\frac{12.9\ \text{mg protein}}{0.500\ \text{mL}} = 25.8\ \frac{\text{mg protein}}{\text{mL}}$$

 ***Test Yourself*** Find the protein concentration if 4.00 mL, instead of 3.26 mL, NaOH were required. (**Answer:** 23.3 mg/mL)

Ask Yourself

10-F. The Kjeldahl procedure was used to analyze 256 μL of a solution containing 37.9 mg protein/mL. The liberated NH_3 was collected in 5.00 mL of 0.033 6 M HCl, and the remaining acid required 6.34 mL of 0.010 M NaOH for complete titration.
(a) How many moles of NH_3 were liberated?
(b) How many grams of nitrogen are contained in the NH_3 in **(a)**?
(c) How many grams of protein were analyzed?
(d) What is the weight percent of nitrogen in the protein?

10-7 Putting Your Spreadsheet to Work

In Sections 10-1 to 10-3, we calculated titration curves because they helped us understand the chemistry behind a titration curve. Now we will see how a spreadsheet and graphics program decrease the agony and mistakes of titration calculations. First we must derive equations relating pH to volume of titrant for use in the spreadsheet.

Charge Balance

The **charge balance** states that, in any solution, the sum of positive charges must equal the sum of negative charges, because the solution must have zero net charge. For a solution of the weak acid HA plus NaOH, the charge balance is

If the solution contained HA and $Ca(OH)_2$, the charge balance would be

$$[H^+] + 2[Ca^{2+}] = [A^-] + [OH^-]$$

because one mole of Ca^{2+} provides two moles of charge. If $[Ca^{2+}] = 0.1$ M, the positive charge it contributes is 0.2 M.

Charge balance:
$$[H^+] + [Na^+] = [A^-] + [OH^-] \qquad (10\text{-}8)$$

The sum of the positive charges of H^+ and Na^+ equals the sum of the negative charges of A^- and OH^-.

Titrating a Weak Acid with a Strong Base

Consider the titration of a volume V_a of acid HA (initial concentration C_a) with a volume V_b of NaOH of concentration C_b. The concentration of Na^+ is just the moles of NaOH (C_bV_b) divided by the total volume of solution ($V_a + V_b$):

$$[Na^+] = \frac{C_bV_b}{V_a + V_b} \qquad (10\text{-}9)$$

Similarly, the formal concentration of the weak acid is

$$F = [HA] + [A^-] = \frac{C_aV_a}{V_a + V_b} \qquad (10\text{-}10)$$

because we have diluted C_aV_a moles of HA to a total volume of $V_a + V_b$.

Now we introduce two equations that are derived in Section 12-5:

Fraction of weak acid in the form HA:
$$\alpha_{HA} = \frac{[HA]}{F} = \frac{[H^+]}{[H^+] + K_a} \qquad (10\text{-}11)$$

Fraction of weak acid in the form A^-:
$$\alpha_{A^-} = \frac{[A^-]}{F} = \frac{K_a}{[H^+] + K_a} \qquad (10\text{-}12)$$

$\alpha_{HA} + \alpha_{A^-} = 1$

Equations 10-11 and 10-12 say that, if a weak acid has a formal concentration F, the concentration of HA is $\alpha_{HA} \cdot F$ and the concentration of A^- is $\alpha_{A^-} \cdot F$. These fractions must add up to 1.

Getting back to our titration, we can write an expression for the concentration of A^- by combining Equation 10-12 with Equation 10-10:

$$[A^-] = \alpha_{A^-} \cdot F = \frac{\alpha_{A^-} \cdot C_aV_a}{V_a + V_b} \qquad (10\text{-}13)$$

Substituting for $[Na^+]$ (Equation 10-9) and $[A^-]$ (Equation 10-13) in the charge balance (Equation 10-8) gives

$$[H^+] + \frac{C_b V_b}{V_a + V_b} = \frac{\alpha_{A^-} \cdot C_a V_a}{V_a + V_b} + [OH^-]$$

which can be rearranged to

Fraction of titration for weak acid by strong base:

$$\phi = \frac{C_b V_b}{C_a V_a} = \frac{\alpha_{A^-} - \dfrac{[H^+] - [OH^-]}{C_a}}{1 + \dfrac{[H^+] - [OH^-]}{C_b}} \qquad (10\text{-}14)$$

$\phi = C_b V_b / C_a V_a$ is the fraction of the way to the equivalence point:

ϕ	Volume of base
0.5	$V_b = \frac{1}{2} V_e$
1	$V_b = V_e$
2	$V_b = 2V_e$

At last! Equation 10-14 is really useful. It relates the volume of titrant (V_b) to the pH. The quantity ϕ ($= C_b V_b / C_a V_a$) is the fraction of the way to the equivalence point, V_e. When $\phi = 1$, the volume of base added, V_b, is equal to V_e. Equation 10-14 works backward from the way that you are accustomed to thinking, because you need to put in pH (on the right) to get out volume (on the left).

Let's set up a spreadsheet for Equation 10-14 to calculate the titration curve for 50.00 mL of the weak acid 0.020 00 M MES with 0.100 0 M NaOH, which was shown in Figure 10-2 and Table 10-2. The equivalence volume is $V_e = 10.00$ mL. The quantities in Equation 10-14 are

2-(*N*-Morpholino)ethanesulfonic acid
MES, $pK_a = 6.27$

$$C_b = 0.1 \text{ M} \qquad [H^+] = 10^{-pH}$$
$$C_a = 0.02 \text{ M} \qquad [OH^-] = K_w/[H^+]$$
$$V_a = 50 \text{ mL}$$
$$K_a = 5.3_7 \times 10^{-7} \qquad \alpha_{A^-} = \frac{K_a}{[H^+] + K_a}$$
$$K_w = 10^{-14}$$

pH is the input $\qquad V_b = \dfrac{\phi C_a V_a}{C_b}$ is the output

The input to the spreadsheet in Figure 10-10 is pH in column B and the output is V_b in column G. From the pH, the values of $[H^+]$, $[OH^-]$, and α_{A^-} are computed

	A	B	C	D	E	F	G
1	Titration of Weak Acid with Strong Base						
2							
3	Cb =	pH	[H+]	[OH-]	Alpha(A-)	Phi	Vb (mL)
4	0.1	3.00	1.00E-03	1.00E-11	0.001	-0.049	-0.490
5	Ca =	3.99	1.02E-04	9.77E-11	0.005	0.000	0.001
6	0.02	4.00	1.00E-04	1.00E-10	0.005	0.000	0.003
7	Va =	5.00	1.00E-05	1.00E-09	0.051	0.050	0.505
8	50	6.27	5.37E-07	1.86E-08	0.500	0.500	5.000
9	Ka =	7.00	1.00E-07	1.00E-07	0.843	0.843	8.430
10	5.37E-07	8.00	1.00E-08	1.00E-06	0.982	0.982	9.818
11	Kw =	9.25	5.62E-10	1.78E-05	0.999	1.000	10.000
12	1.00E-14	10.00	1.00E-10	1.00E-04	1.000	1.006	10.058
13		11.00	1.00E-11	1.00E-03	1.000	1.061	10.606
14		12.00	1.00E-12	1.00E-02	1.000	1.667	16.667
15							
16	C4 = 10^-B4						
17	D4 = $A$12/C4						
18	E4 = $A$10/(C4+$A$10)						
19	F4 = (E4-(C4-D4)/$A$6)/(1+(C4-D4)/$A$4) [Equation 10-14]						
20	G4 = F4*$A$6*$A$8/$A$4						

Figure 10-10 Spreadsheet uses Equation 10-14 to calculate the titration curve for 50 mL of the weak acid 0.02 M MES ($pK_a = 6.27$), treated with 0.1 M NaOH. You provide pH as input in column B, and the spreadsheet tells what volume of base in column G is required to generate that pH.

The spreadsheet in Figure 10-10 can be used to find the pH of a weak acid. Just search for the pH at which $V_b = 0$.

To home in on an exact volume (such as V_e), set the spreadsheet to show extra digits in the cell of interest. Tables in this book have been formatted to reduce the number of digits.

in columns C, D, and E. Equation 10-14 is used in column F to find the fraction of titration, ϕ. From this value, we calculate the volume of titrant, V_b, in column G.

How do we know what pH values to put in? Trial and error allows us to find the starting pH, by putting in a pH and seeing whether V_b is positive or negative. In a few tries, it is easy to home in on the pH at which $V_b = 0$. In Figure 10-10, we see that a pH of 3.00 is too low, because ϕ and V are both negative. Input values of pH are spaced as closely as needed to allow you to generate a smooth titration curve. To save space, we show only a few points in Figure 10-10, including the midpoint (pH 6.27 $\Rightarrow V_b = 5.00$ mL) and the end point (pH 9.25 $\Rightarrow V_b = 10.00$ mL). The spreadsheet agrees with Table 10-2 without dividing the titration into different regions that use different approximations.

The Power of a Spreadsheet

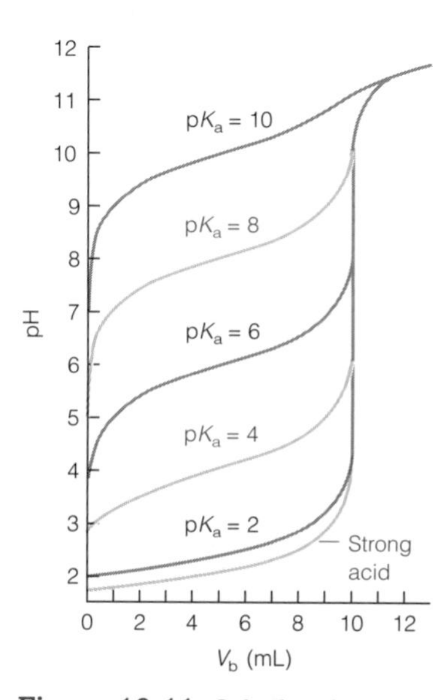

Figure 10-11 Calculated curves showing the titration of 50.0 mL of 0.020 0 M HA with 0.100 M NaOH. As the acid becomes weaker, the equivalence point becomes less distinct.

By changing K_a in cell A10 of Figure 10-10, we can calculate a family of curves for different acids. Figure 10-11 shows how the titration curve depends on the acid dissociation constant of HA. The strong-acid curve at the bottom of Figure 10-11 was computed with a large value of K_a ($K_a = 10^3$) in cell A10. Figure 10-11 shows that, as K_a decreases (pK_a increases), the pH change near the equivalence point decreases, until it becomes too shallow to detect. Similar behavior occurs as the concentrations of analyte and titrant decrease. *It is not practical to titrate an acid or a base when its strength is too weak or its concentration too dilute.*

Titrating a Weak Base with a Strong Acid

By logic similiar to that used to derive Equation 10-14, we can derive an equation for the titration of weak base B with strong acid:

Fraction of titration for weak base by strong acid:

$$\phi = \frac{C_aV_a}{C_bV_b} = \frac{\alpha_{BH^+} + \dfrac{[H^+] - [OH^-]}{C_b}}{1 - \dfrac{[H^+] - [OH^-]}{C_a}} \qquad (10\text{-}15)$$

where C_a is the concentration of strong acid in the buret, V_a is the volume of acid added, C_b is the initial concentration of weak base being titrated, V_b is the initial volume of weak base, and α_{BH^+} is the fraction of base in the form BH^+:

Fraction of weak base in the form BH^+:

$$\alpha_{BH^+} = \frac{[BH^+]}{F} = \frac{[H^+]}{[H^+] + K_{BH^+}} \qquad (10\text{-}16)$$

where K_{BH^+} is the acid dissociation constant of BH^+.

Experiment 10 at the Web site www.whfreeman.com/exploringchem4e teaches you how to fit the theoretical expressions 10-14 or 10-15 to experimental titration data. Excel SOLVER is used to find the best values of analyte concentration and pK to fit the measured data.

Ask Yourself

10-G. **(a)** *Effect of pK_a in the titration of weak acid with strong base.* Use the spreadsheet in Figure 10-10 to compute and plot the family of curves in Figure 10-11. For strong acid, use $K_a = 10^3$.

(b) *Effect of concentration in the titration of weak acid with strong base.* Use your spreadsheet to prepare a family of titration curves for $pK_a = 6$, with the following combinations of concentrations: **(i)** $C_a = 20$ mM, $C_b = 100$ mM; **(ii)** $C_a = 2$ mM, $C_b = 10$ mM; and **(iii)** $C_a = 0.2$ mM, $C_b = 1$ mM.

Key Equations

Useful shortcut

$$\text{mL} \times \frac{\text{mol}}{\text{L}} = \text{mmol}$$

Equivalence volume (V_e)

$$\underbrace{C_aV_a = C_bV_e}_{\text{Titrating acid with base}} \qquad \text{or} \qquad \underbrace{C_aV_e = C_bV_b}_{\text{Titrating base with acid}}$$

C_a = acid concentration C_b = base concentration
V_a = acid volume V_b = base volume
V_e = equivalence volume

Titration of weak acid (see inside cover)

1. Initial solution—weak acid

$$\underset{F-x}{HA} \overset{K_a}{\rightleftharpoons} \underset{x}{H^+} + \underset{x}{A^-} \qquad \frac{x^2}{F-x} = K_a$$

2. Before equivalence point—buffer

   Titration reaction tells how much HA and A^- are present

$$pH = pK_a + \log\left(\frac{[A^-]}{[HA]}\right)$$

3. Equivalence point—weak base—pH > 7

$$\underset{F'-x}{A^-} + H_2O \overset{K_b}{\rightleftharpoons} \underset{x}{HA} + \underset{x}{OH^-} \qquad K_b = \frac{K_w}{K_a}$$

   F′ is diluted concentration

4. After equivalence point—excess strong base

$$pH = -\log(K_w/[OH^-]_{excess})$$

Titration of weak base (see inside cover)

1. Initial solution—weak base

$$\underset{F-x}{B} + H_2O \overset{K_b}{\rightleftharpoons} \underset{x}{BH^+} + \underset{x}{OH^-} \qquad \frac{x^2}{F-x} = K_b$$

2. Before equivalence point—buffer

   Titration reaction tells how much B and BH^+ are present

$$pH = pK_{BH^+} + \log\left(\frac{[B]}{[BH^+]}\right)$$

3. Equivalence point—weak acid—pH < 7

$$\underset{F'-x}{BH^+} \overset{K_{BH^+}}{\rightleftharpoons} \underset{x}{B} + \underset{x}{H^+}$$

4. After equivalence point—excess strong acid

   $pH = -\log([H^+]_{excess})$

Spreadsheet titration equations	Use Equations 10-14 and 10-15 Input is pH and output is volume
Choosing indicator	Use indicator with color change close to theoretical pH at equivalence point of titration
Using electrodes for end point	End point has greatest slope: $\Delta pH/\Delta V$ is maximum End point has zero second derivative: $\frac{\Delta(\Delta pH/\Delta V)}{\Delta V} = 0$

Important Terms

charge balance
indicator error
Kjeldahl nitrogen analysis

Problems

10-1. Explain what chemistry occurs in each region of the titration of OH^- with H^+. State how you would calculate the pH in each region.

10-2. Explain what chemistry occurs in each region of the titration of the weak acid, HA, with OH^-. State how you would calculate the pH in each region.

10-3. Explain what chemistry occurs in each region of the titration of the weak base, A^-, with strong acid, H^+. State how you would calculate the pH in each region.

10-4. Why is the titration curve in Figure 10-3 steepest at the equivalence point?

10-5. Why do we use the maximum of the first derivative curve or the zero crossing of the second derivative curve to locate the end points in Figure 10-4?

10-6. Consider the titration of 100.0 mL of 0.100 M NaOH with 1.00 M HBr. What is the equivalence volume? Find the pH at the following volumes of HBr and make a graph of pH versus V_a: V_a = 0, 1.00, 5.00, 9.00, 9.90, 10.00, 10.10, and 12.00 mL.

10-7. Consider the titration of 25.0 mL of 0.050 0 M $HClO_4$ with 0.100 M KOH. Find the equivalence volume. Find the pH at the following volumes of KOH and plot pH versus V_b: V_b = 0, 1.00, 5.00, 10.00, 12.40, 12.50, 12.60, and 13.00 mL.

10-8. A 50.0-mL volume of 0.050 0 M weak acid HA (pK_a = 4.00) was titrated with 0.500 M NaOH. Write the titration reaction and find V_e. Find the pH at V_b = 0, 1.00, 2.50, 4.00, 4.90, 5.00, 5.10, and 6.00 mL and plot pH versus V_b.

10-9. When methylammonium chloride is titrated with tetramethylammonium hydroxide, the titration reaction is

$$\underset{\substack{BH^+ \\ \text{Weak acid}}}{CH_3NH_3^+} + \underset{\substack{\text{From} \\ (CH_3)_4N^+OH^-}}{OH^-} \longrightarrow \underset{\substack{B \\ \text{Weak base}}}{CH_3NH_2} + H_2O$$

Find the equivalence volume in the titration of 25.0 mL of 0.010 0 M methylammonium chloride with 0.050 0 M tetramethylammonium hydroxide. Calculate the pH at V_b = 0, 2.50, 5.00, and 10.00 mL. Sketch the titration curve.

10-10. Write the reaction for the titration of 100 mL of 0.100 M anilinium bromide ("aminobenzene · HBr") with 0.100 M NaOH. Sketch the titration curve for the points V_b = 0, $0.100V_e$, $0.500V_e$, $0.900V_e$, V_e, and $1.200V_e$.

10-11. What is the pH at the equivalence point when 0.100 M hydroxyacetic acid is titrated with 0.050 0 M KOH?

10-12. When 16.24 mL of 0.064 3 M KOH were added to 25.00 mL of 0.093 8 M weak acid, HA, the observed pH was 3.62. Find pK_a for the acid.

10-13. When 22.63 mL of aqueous NaOH were added to 1.214 g of CHES (FM 207.29, structure in Table 9-2) dissolved in 41.37 mL of water, the pH was 9.13. Calculate the molarity of the NaOH.

10-14. **(a)** When 100.0 mL of weak acid HA were titrated with 0.093 81 M NaOH, 27.63 mL were required to reach the equivalence point. Find the molarity of HA.

(b) What is the formal concentration of A^- at the equivalence point?

(c) The pH at the equivalence point was 10.99. Find pK_a for HA.

(d) What was the pH when only 19.47 mL of NaOH had been added?

10-15. A 100.0-mL aliquot of 0.100 M weak base B (pK_b = 5.00) was titrated with 1.00 M $HClO_4$. Find V_e and calculate the pH at V_a = 0, 1.00, 5.00, 9.00, 9.90, 10.00, 10.10, and 12.00 mL and make a graph of pH versus V_a.

10-16. A solution of 100.0 mL of 0.040 0 M sodium propanoate (the sodium salt of propanoic acid) was titrated with

0.083 7 M HCl. Find V_e and calculate the pH at $V_a = 0, \frac{1}{4}V_e$, $\frac{1}{2}V_e$, $\frac{3}{4}V_e$, V_e, and $1.1V_e$. Sketch the titration curve.

10-17. A 50.0-mL solution of 0.031 9 M benzylamine was titrated with 0.050 0 M HCl.

(a) What is the equilibrium constant for the titration reaction?

(b) Find V_e and calculate the pH at V_a = 0, 12.0, $\frac{1}{2}V_e$, 30.0, V_e, and 35.0 mL.

10-18. Don't ever mix acid with cyanide (CN^-) because it liberates poisonous HCN(*g*). But, just for fun, calculate the pH of a solution made by mixing 50.00 mL of 0.100 M NaCN with

(a) 4.20 mL of 0.438 M $HClO_4$.

(b) 11.82 mL of 0.438 M $HClO_4$.

(c) What is the pH at the equivalence point with 0.438 M $HClO_4$?

10-19. A 25.00-mL volume of 0.050 00 M imidazole was titrated with 0.125 0 M HNO_3. Find V_e and calculate the pH at V_a = 0, 1.00, 5.00, 9.00, 9.90, 10.00, 10.10, and 12.00 mL and make a graph of pH versus V_a.

10-20. Would the indicator bromocresol green, with a transition range of pH 3.8–5.4, ever be useful in the titration of a weak acid with a strong base? Why?

10-21. Consider the titration in Figure 10-2, for which the pH at the equivalence point is calculated to be 9.25. If thymol blue is used as an indicator, what color will be observed through most of the titration prior to the equivalence point? At the equivalence point? After the equivalence point?

10-22. Why would an indicator end point not be very useful in the titration curve for pK_a = 10.00 in Figure 10-11?

10-23. Phenolphthalein is used as an indicator for the titration of HCl with NaOH.

(a) What color change is observed at the end point?

(b) The basic solution just after the end point slowly absorbs CO_2 from the air and becomes more acidic by virtue of the reaction $CO_2 + OH^- \rightleftharpoons HCO_3^-$, causing the color to fade from pink to colorless. If you carry out the titration too slowly, does this reaction lead to a systematic or a random error in finding the end point?

10-24. In the titration of 0.10 M pyridinium bromide (the salt of pyridine plus HBr) by 0.10 M NaOH, the pH at $0.99V_e$ is 7.20. At V_e, pH = 8.95, and, at $1.01V_e$, pH = 10.70. Select an indicator from Table 9-3 that would be suitable for this titration and state what color change will be used.

10-25. Prepare a graph of the second derivative to find the end point from the following titration data:

mL NaOH	pH	mL NaOH	pH
10.679	7.643	10.729	5.402
10.696	7.447	10.733	4.993
10.713	7.091	10.738	4.761
10.721	6.700	10.750	4.444
10.725	6.222	10.765	4.227

10-26. Borax (Table 10-3) was used to standardize a solution of HNO_3. Titration of 0.261 9 g of borax required 21.61 mL. What is the molarity of the HNO_3?

10-27. A 10.231-g sample of window cleaner containing ammonia was diluted with 39.466 g of water. Then 4.373 g of solution were titrated with 14.22 mL of 0.106 3 M HCl to reach a bromocresol green end point.

(a) What fraction of the 10.231-g sample of window cleaner is contained in the 4.373 g that were analyzed?

(b) How many grams of NH_3 (FM 17.031) were in the 4.373-g sample?

(c) Find the weight percent of NH_3 in the cleaner.

10-28. In the Kjeldahl nitrogen analysis, the final product is NH_4^+ in HCl solution. It is necessary to titrate the HCl without titrating the NH_4^+ ion.

(a) Calculate the pH of pure 0.010 M NH_4Cl.

(b) The steep part of the titration curve when HCl is titrated with NaOH runs from pH ≈ 4 to pH ≈ 10. Select an indicator that allows you to titrate HCl but not NH_4^+.

10-29. Prepare a spreadsheet like the one in Figure 10-10, but, using Equation 10-15, reproduce the titration curve in Figure 10-3.

10-30. *Effect of pK_b in the titration of weak base with strong acid.* Use the spreadsheet from Problem 10-29 to compute and plot a family of curves analogous to Figure 10-11 for the titration of 50.0 mL of 0.020 0 M B (pK_b = −2.00, 2.00, 4.00, 6.00, 8.00, and 10.00) with 0.100 M HCl. (pK_b = −2.00 corresponds to $K_b = 10^{+2.00}$, which represents strong base.)

How Would You Do It?

10-31. The table at the top of page 236 gives data for 100.0 mL of solution of a single unknown base titrated with 0.111 4 M HCl. Provide an argument for how many protons the base is able to accept (Is it monoprotic, diprotic, etc.?) and find the molarity of the base. A high-quality pH meter provides pH to the 0.001 place, even though accuracy is limited to the 0.01 place. How could you deliver volumes up to 50 mL with a precision of 0.001 mL?

Coarse titration				Fine data near first end point				Fine data near second end point			
mL	pH	mL	pH	mL	pH	mL	pH	mL	pH	mL	pH
0.595	12.148	29.157	5.785	26.939	8.217	27.481	7.228	40.168	3.877	41.542	2.999
1.711	12.006	31.512	5.316	27.013	8.149	27.501	7.158	40.403	3.767	41.620	2.949
3.540	11.793	33.609	5.032	27.067	8.096	27.517	7.103	40.498	3.728	41.717	2.887
5.250	11.600	36.496	4.652	27.114	8.050	27.537	7.049	40.604	3.669	41.791	2.845
7.258	11.390	38.222	4.381	27.165	7.987	27.558	6.982	40.680	3.618	41.905	2.795
9.107	11.179	39.898	3.977	27.213	7.916	27.579	6.920	40.774	3.559	42.033	2.735
11.557	10.859	40.774	3.559	27.248	7.856	27.600	6.871	40.854	3.510	42.351	2.617
13.967	10.486	41.791	2.845	27.280	7.791	27.622	6.825	40.925	3.457	42.709	2.506
16.042	10.174	42.709	2.506	27.309	7.734	27.649	6.769	40.994	3.407	43.192	2.401
18.474	9.850	45.049	2.130	27.338	7.666	27.675	6.717	41.057	3.363	43.630	2.312
20.338	9.627	47.431	1.937	27.362	7.603	27.714	6.646	41.114	3.317		
22.136	9.402	49.292	1.835	27.386	7.538	27.747	6.594	41.184	3.263		
24.836	8.980			27.406	7.485	27.793	6.535	41.254	3.210		
26.216	8.608			27.427	7.418	27.846	6.470	41.329	3.150		
27.013	8.149			27.444	7.358	27.902	6.411	41.406	3.093		
27.969	6.347			27.463	7.287	27.969	6.347	41.466	3.047		

Reference Procedure: Preparing standard acid and base

Hydrochloric acid and sodium hydroxide are the most common strong acids and bases used in the laboratory. Both reagents need to be standardized to learn their exact concentrations. Section 10-5 provides background information for the procedures described below.

Reagents

50 wt% NaOH: (3 mL/student) Dissolve 50 g of reagent-grade NaOH in 50 mL of distilled water and allow the suspension to settle overnight. Na_2CO_3 is insoluble in the solution and precipitates. Store the solution in a tightly sealed polyethylene bottle and handle it gently to avoid stirring the precipitate when liquid is withdrawn.

Phenolphthalein indicator: Dissolve 50 mg in 50 mL of ethanol and add 50 mL of distilled water.

Bromocresol green indicator: Dissolve 100 mg in 14.3 mL of 0.01 M NaOH and dilute to 250 mL with distilled water.

Concentrated (37 wt%) HCl: 10 mL/student.

Primary standards: Potassium hydrogen phthalate (~2.5 g/student) and sodium carbonate (~1.0 g/student).

0.05 M NaCl: 50 mL/student.

Standardizing NaOH

1. Dry primary standard grade potassium hydrogen phthalate for 1 h at 105°C and store it in a capped bottle in a desiccator.

$CO_2^-K^+$, CO_2H (benzene ring) + NaOH ⟶

Potassium hydrogen phthalate
FM 204.22

$CO_2^-K^+$, $CO_2^-Na^+$ (benzene ring) + H_2O

2. Boil 1 L of distilled water for 5 min to expel CO_2. Pour the water into a polyethylene bottle, which should be tightly capped whenever possible. Calculate the volume of 50 wt% NaOH needed to prepare 1 L of 0.1 M NaOH. (The density of 50 wt% NaOH is 1.50 g per milliliter of solution.) Use a graduated cylinder to transfer this much NaOH to the bottle of water. (CAUTION: 50 wt% NaOH eats people. Flood any spills on your skin with water.) Mix well and cool the solution to room temperature (preferably overnight).

3. Weigh four samples of solid potassium hydrogen phthalate and dissolve each in ~25 mL of distilled water in a 125-mL flask. Each sample should contain enough solid to react with ~25 mL of 0.1 M NaOH. Add 3 drops of phenolphthalein to each flask and titrate one rapidly to find the end point. The buret should have a loosely fitted cap to minimize entry of CO_2 from the air.

4. Calculate the volume of NaOH required for each of the other three samples and titrate them carefully. During each titration, periodically tilt and rotate the flask to wash all liquid from the walls into the bulk solution. Near the end, deliver less than 1 drop of titrant at a time. To do so, carefully suspend a fraction of a drop from the buret tip, touch it to the inside wall of the flask, wash it into the bulk solution by careful tilting, and swirl the solution. The end point is the first appearance of faint pink color that persists for 15 s. (The color will slowly fade as CO_2 from the air dissolves in the solution.)

5. Calculate the average molarity ($\bar{x}$), the standard deviation (s), and the percent relative standard deviation ($=100 \times s/\bar{x}$). If you were careful, the relative standard deviation should be $<0.2\%$.

Standardizing HCl

1. Use the table inside the cover of this book to calculate the volume of ~37 wt% HCl that should be added to 1 L of distilled water to produce 0.1 M HCl and prepare this solution.

2. Dry primary standard grade sodium carbonate for 1 h at 105°C and cool it in a desiccator.

3. Weigh four samples, each containing enough Na_2CO_3 to react with ~25 mL of 0.1 M HCl and place each in a 125-mL flask. When you are ready to titrate each one, dissolve it in ~25 mL of distilled water. Add 3 drops of bromocresol green indicator and titrate one rapidly to a green color to find the approximate end point.

$$2HCl + \underset{\text{FM } 105.99}{Na_2CO_3} \longrightarrow CO_2 + 2NaCl + H_2O$$

4. Carefully titrate each sample until it turns from blue to green. Then boil the solution to expel CO_2. The color should return to blue. Carefully add HCl from the buret until the solution turns green again and report the volume of acid at this point.

5. Perform one blank titration of 50 mL of 0.05 M NaCl containing 3 drops of indicator. Subtract the volume of HCl needed for the blank from that required to titrate Na_2CO_3.

6. Calculate the mean HCl molarity, standard deviation, and percent relative standard deviation.

Notes and References

1. D. Lee and A. Goldman, "Plant Linked to Pet Death Had History of Polluting," *Los Angeles Times*, 9 May 2007, p. C1. B. Puschner, R. H. Poppenga, L. J. Lowenstine, M. S. Filigenzi, and P. A. Pesavento, *J. Vet. Diagn. Invest.* **2007,** *19,* 616.

2. Colorimetric Kjeldahl NH_3 measurement: http://www.epa.gov/grtlakes/lmmb/methods/tknalr2.pdf.

Acid Dissolves Buildings and Teeth

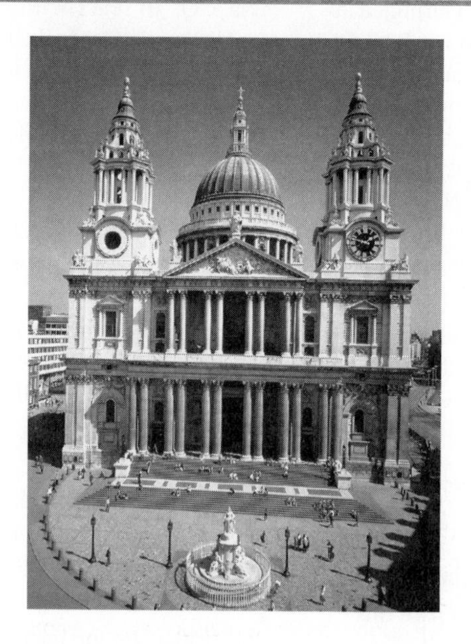

Erosion of carbonate stone. Between 1980 and 1990, acid rain dissolved ½ mm from the thickness of the external stone walls of St. Paul's Cathedral in London. Reduction of heavy industry decreased atmospheric SO_2 (a major source of acid rain) from as high as 100 ppb in the 1970s to 10 ppb in 2000. Correspondingly, only ¼ mm of St. Paul's external stone disappeared between 1990 and 2000. A corner of the building facing a power station dissolved at 10 times the rate of the rest of the building until the station was closed.[1] The opening of Chapter 8 showed rainfall pH in Europe. [Pictor International/Picture Quest.]

The main constituent of limestone and marble, which are used in many buildings, is calcite, a crystalline form of calcium carbonate. This mineral is insoluble in neutral or basic solution but dissolves in acid by virtue of two *coupled equilibria* in which the product of one reaction is consumed in the next reaction:

$$\underset{\text{Calcite}}{CaCO_3(s)} \rightleftharpoons Ca^{2+} + \underset{\text{Carbonate}}{CO_3^{2-}}$$

$$CO_3^{2-} + H^+ \rightleftharpoons \underset{\text{Bicarbonate}}{HCO_3^-}$$

Le Châtelier's principle tells us that if we remove a product of the first reaction, we will draw the reaction to the right, making calcite more soluble.

Tooth enamel contains the mineral hydroxyapatite, a calcium hydroxyphosphate. This mineral also dissolves in acid, because both PO_4^{3-} and OH^- react with H^+:

$$\underset{\text{Hydroxyapatite}}{Ca_{10}(PO_4)_6(OH)_2} + 14H^+ \rightleftharpoons 10Ca^{2+} + 6H_2PO_4^- + 2H_2O$$

Bacteria residing on your teeth metabolize sugar into lactic acid, which lowers the pH below 5 at the surface of a tooth. Acid dissolves the hydroxyapatite, thereby creating tooth decay.

$$CH_3\underset{}{\overset{\overset{\Large OH}{|}}{C}}HCO_2H \quad \text{Lactic acid}$$

CHAPTER 11

Polyprotic Acids and Bases

Carbonic acid from limestone, phosphoric acid from teeth, and amino acids from proteins are all **polyprotic acids**—those having more than one acidic proton. This chapter extends our discussion of acids, bases, and buffers to polyprotic systems encountered throughout nature.

11-1 Amino Acids Are Polyprotic

Amino acids from which proteins are built have an acidic carboxylic acid group, a basic amino group, and a variable substituent designated R:

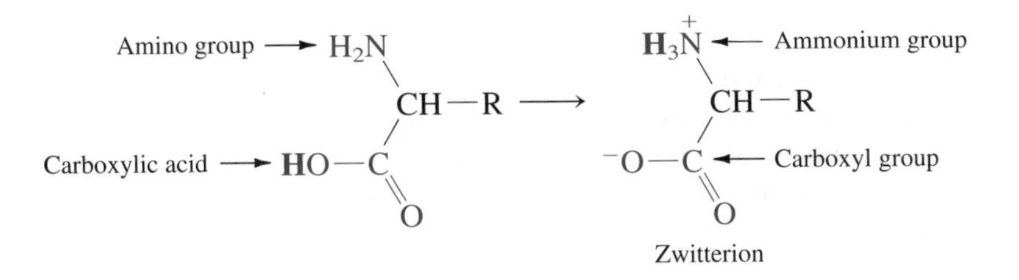

A *zwitterion* is a molecule with positive and negative charges.

Because the amino group is more basic than the carboxyl group, the acidic proton resides on nitrogen of the amino group instead of on oxygen of the carboxyl group. The resulting structure, with positive and negative sites, is called a **zwitterion**.

At low pH, both the ammonium group and the carboxyl group are protonated. At high pH, neither is protonated. The substituent may also have acidic or basic properties. Acid dissociation constants of the 20 common amino acids are given in

Table 11-1 Acid dissociation constants of amino acids[a]

Amino acid[b]	Substituent	Carboxylic acid pK_a	Ammonium pK_a	Substituent pK_a	Molecular mass
Alanine (A)	$—CH_3$	2.344	9.868		89.09
Arginine (R)	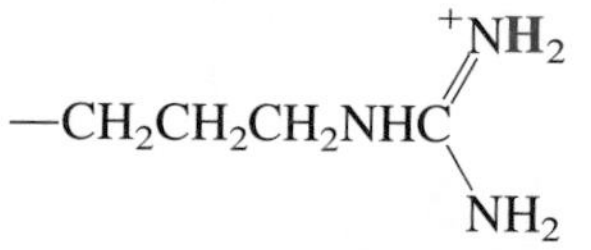 $—CH_2CH_2CH_2NHC(=\overset{+}{N}H_2)NH_2$	1.823	8.991	12.1	174.20
Asparagine (N)	$—CH_2C(=O)NH_2$	2.16	8.73		132.12
Aspartic acid (D)	$—CH_2CO_2\mathbf{H}$	1.990	10.002	3.900	133.10
Cysteine (C)	$—CH_2S\mathbf{H}$	(1.7)	10.74	8.36	121.16
Glutamic acid (E)	$—CH_2CH_2CO_2\mathbf{H}$	2.16	9.96	4.30	147.13
Glutamine (Q)	$—CH_2CH_2C(=O)NH_2$	2.19	9.00		146.15
Glycine (G)	—H	2.350	9.778		75.07
Histidine (H)	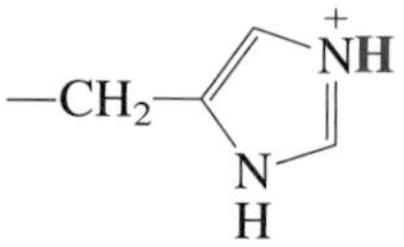 —CH₂–(imidazolium ring, $\overset{+}{N}\mathbf{H}$, NH)	(1.6)	9.28	5.97	155.16
Isoleucine (I)	$—CH(CH_3)(CH_2CH_3)$	2.318	9.758		131.17
Leucine (L)	$—CH_2CH(CH_3)_2$	2.328	9.744		131.17
Lysine (K)	$—CH_2CH_2CH_2CH_2N\mathbf{H}_3^+$	(1.77)	9.07	10.82	146.19
Methionine (M)	$—CH_2CH_2SCH_3$	2.18	9.08		149.21
Phenylalanine (F)	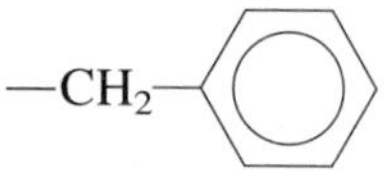 $—CH_2$–(benzene ring)	2.20	9.31		165.19
Proline (P)	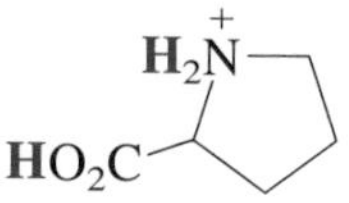 $H_2\overset{+}{N}$ (pyrrolidine ring) with HO_2C — Structure of entire amino acid	1.952	10.640		115.13
Serine (S)	$—CH_2OH$	2.187	9.209		105.09
Threonine (T)	$—CH(CH_3)(OH)$	2.088	9.100		119.12
Tryptophan (W)	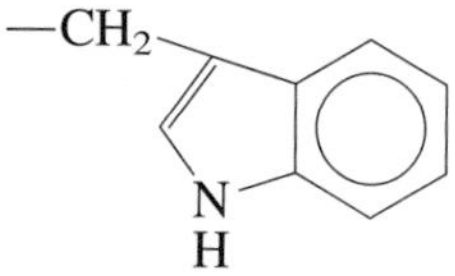 $—CH_2$–(indole ring, NH)	2.37	9.33		204.23
Tyrosine (Y)	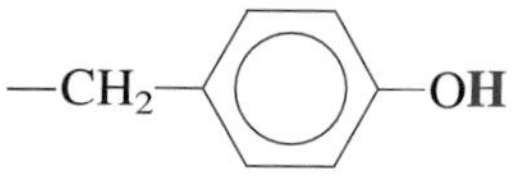 $—CH_2$–(benzene ring)–O**H**	2.41	8.67	11.01	181.19
Valine (V)	$—CH(CH_3)_2$	2.286	9.719		117.15

[a] pK_a at 25°C.

[b] Standard abbreviations are shown in parentheses. Acidic protons are **bold**. Each substituent is written in its fully protonated form.

SOURCE: A. E. Martell, R. M. Smith, and R. J. Motekaitis, *NIST Critically Selected Stability Constants of Metal Complexes,* NIST Standard Reference Database 46, Gaithersburg, MD, 2001.

Table 11-1, in which each substituent (R) is shown in its fully protonated form. For example, the amino acid cysteine has three acidic protons:

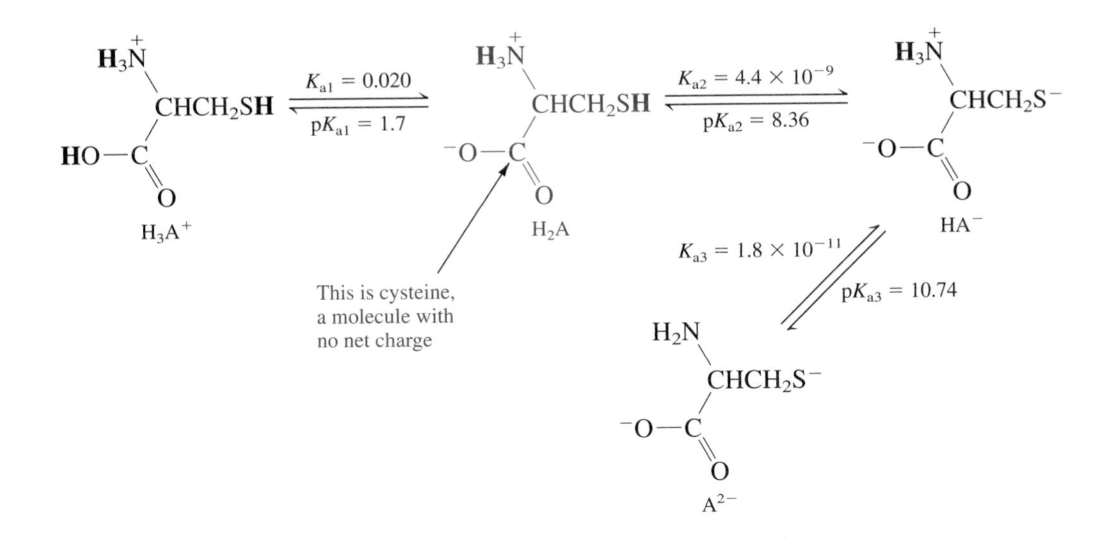

In general, a *diprotic* acid has two acid dissociation constants, designated K_{a1} and K_{a2} (where $K_{a1} > K_{a2}$):

$$H_2A \overset{K_{a1}}{\rightleftharpoons} HA^- + H^+ \qquad HA^- \overset{K_{a2}}{\rightleftharpoons} A^{2-} + H^+$$

K_{a1} applies to the *most acidic* proton. The subscript a in K_{a1} and K_{a2} is customarily omitted, and we will write K_1 and K_2 throughout most of this chapter.

The two base association constants are designated K_{b1} and K_{b2} ($K_{b1} > K_{b2}$):

$$A^{2-} + H_2O \overset{K_{b1}}{\rightleftharpoons} HA^- + OH^- \qquad HA^- + H_2O \overset{K_{b2}}{\rightleftharpoons} H_2A + OH^-$$

Relation Between K_a and K_b

If you add the K_{a1} reaction to the K_{b2} reaction, the sum is $H_2O \rightleftharpoons H^+ + OH^-$—the K_w reaction. By this means, you can derive a most important set of relations between the acid and the base equilibrium constants:

Relation between K_a and K_b for diprotic system:

$$K_{a1} \cdot K_{b2} = K_w \qquad (11\text{-}1)$$
$$K_{a2} \cdot K_{b1} = K_w$$

Challenge Add the K_{a1} and K_{b2} reactions to prove that $K_{a1} \cdot K_{b2} = K_w$.

For a *triprotic* system, with three acidic protons, the corresponding relations are

Relation between K_a and K_b for triprotic system:

$$K_{a1} \cdot K_{b3} = K_w$$
$$K_{a2} \cdot K_{b2} = K_w \qquad (11\text{-}2)$$
$$K_{a3} \cdot K_{b1} = K_w$$

The standard notation for successive acid dissociation constants of a polyprotic acid is K_1, K_2, K_3, and so on, with the subscript a usually omitted. We retain or omit the subscript a as dictated by clarity. For successive base hydrolysis constants, we retain the subscript b. *K_{a1} (or K_1) refers to the acidic species with the most protons, and K_{b1} refers to the basic species with no acidic protons.*

Ask Yourself

11-A. **(a)** Each of the following ions undergoes two consecutive acid-base reactions (called the stepwise acid-base reactions) when placed in water. Write the reactions and the correct symbol (for example, K_2 or K_{b1}) for the equilibrium constant for each. Use Appendix B to find the numerical value of each acid or base equilibrium constant.

(i) $H_3\overset{+}{N}CH_2CH_2\overset{+}{N}H_3$
Ethylenediammonium ion

(ii) $^{-}O\overset{O}{\overset{\|}{C}}CH_2\overset{O}{\overset{\|}{C}}O^{-}$
Malonate ion

(b) Starting with the following fully protonated species, write the stepwise acid dissociation reactions of the amino acids aspartic acid and arginine. Be sure to remove the protons in the correct order on the basis of the pK_a values in Table 11-1. Remember that the proton with the lowest pK_a (that is, the greatest K_a) comes off first. Label the neutral molecules that we call aspartic acid and arginine.

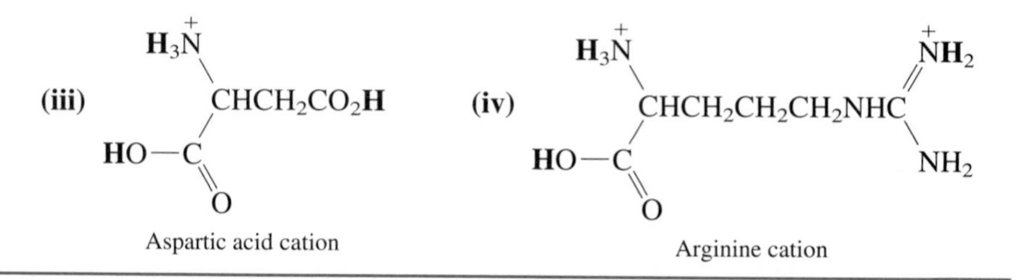

Aspartic acid cation
Arginine cation

11-2 Finding the pH in Diprotic Systems

Consider the amino acid leucine, designated HL:

$$\underset{H_2L^+}{H_3\overset{+}{N}\underset{|}{\overset{R}{C}}HCO_2H} \underset{}{\overset{pK_{a1} = 2.328}{\rightleftharpoons}} \underset{\substack{HL\\ \text{Leucine}}}{H_3\overset{+}{N}\overset{R}{C}HCO_2^-} \overset{pK_{a2} = 9.744}{\rightleftharpoons} \underset{L^-}{H_2N\overset{R}{C}HCO_2^-}$$

The side chain in leucine is an isobutyl group: R = —$CH_2CH(CH_3)_2$.

The equilibrium constants apply to the following reactions:

Diprotic acid:

$$H_2L^+ \rightleftharpoons HL + H^+ \qquad K_{a1} \equiv K_1 \qquad (11\text{-}3)$$

$$HL \rightleftharpoons L^- + H^+ \qquad K_{a2} \equiv K_2 \qquad (11\text{-}4)$$

Diprotic base:

$$L^- + H_2O \rightleftharpoons HL + OH^- \qquad K_{b1} \qquad (11\text{-}5)$$

$$HL + H_2O \rightleftharpoons H_2L^+ + OH^- \qquad K_{b2} \qquad (11\text{-}6)$$

We now set out to calculate the pH and composition of individual solutions of 0.050 0 M H_2L^+, 0.050 0 M HL, and 0.050 0 M L^-. Our methods do not depend on the charge of the acids and bases. We use the same procedure to find the pH of the diprotic H_2A, where A is anything, or H_2L^+, where HL is leucine.

The Acidic Form, H_2L^+

A salt such as leucine hydrochloride contains the protonated species H_2L^+, which can dissociate twice, as indicated in Reactions 11-3 and 11-4. Because $K_1 = 4.70 \times 10^{-3}$, H_2L^+ is a weak acid. HL is an even weaker acid, with $K_2 = 1.80 \times 10^{-10}$. It appears that H_2L^+ will dissociate only partly, and the resulting HL will hardly dissociate at all. For this reason, we make the (superb) approximation that a solution of H_2L^+ behaves as a *monoprotic* acid, with $K_a = K_1$.

Easy stuff.

With this approximation, the calculation of the pH of 0.050 0 M H_2L^+ is simple:

$$\underset{\substack{H_2L^+ \\ 0.050\,0 - x}}{H_3\overset{+}{N}\overset{\overset{\displaystyle R}{|}}{C}HCO_2H} \xrightleftharpoons{K_1 = 4.70 \times 10^{-3}} \underset{\substack{HL \\ x}}{H_3\overset{+}{N}\overset{\overset{\displaystyle R}{|}}{C}HCO_2^-} + \underset{x}{H^+}$$

H_2L^+ can be treated as monoprotic, with $K_a = K_1$.

$$\frac{x^2}{F - x} = K_1 \Longrightarrow x = 1.32 \times 10^{-2}\ M$$

$$[HL] = x = 1.32 \times 10^{-2}\ M$$

$$[H^+] = x = 1.32 \times 10^{-2}\ M \Longrightarrow pH = 1.88$$

$$[H_2L^+] = F - x = 3.68 \times 10^{-2}\ M$$

F is the formal concentration of H_2L^+ (= 0.050 0 M in this example).

What is the concentration of L^- in the solution? We can find $[L^-]$ with K_2:

$$HL \overset{K_2}{\rightleftharpoons} L^- + H^+ \qquad K_2 = \frac{[H^+][L^-]}{[HL]} \Longrightarrow [L^-] = \frac{K_2[HL]}{[H^+]}$$

$$[L^-] = \frac{(1.80 \times 10^{-10})(1.32 \times 10^{-2})}{(1.32 \times 10^{-2})} = 1.80 \times 10^{-10}\ M\ (= K_2)$$

Our approximation that the second dissociation of a diprotic acid is much less than the first dissociation is confirmed by this last result. The concentration of L^- is about eight orders of magnitude smaller than that of HL. As a source of protons, the dissociation of HL is negligible relative to the dissociation of H_2L^+. For most diprotic acids, K_1 is sufficiently larger than K_2 for this approximation to be valid. Even if K_2 were just 10 times less than K_1, the value of $[H^+]$ calculated by ignoring the second ionization would be in error by only 4%. The error in pH would be only 0.01 pH unit. In summary, *a solution of a diprotic acid behaves like a solution of a monoprotic acid, with $K_a = K_1$.*

Dissolved carbon dioxide is one of the most important diprotic acids in Earth's ecosystem. Box 11-1 describes imminent danger to the entire ocean food chain as a result of increasing atmospheric CO_2 dissolving in the oceans. Reaction A in Box 11-1 lowers the concentration of CO_3^{2-} in the oceans. As a result, $CaCO_3$ shells and skeletons of creatures at the bottom of the food chain will dissolve by Reaction B in Box 11-1. This effect is far more certain than the effects of atmospheric CO_2 on Earth's climate.

The Basic Form, L^-

The fully basic species, L^-, would be found in a salt such as sodium leucinate, which could be prepared by treating leucine with an equimolar quantity of NaOH.

More easy stuff.

Box 11-1 Carbon Dioxide in the Air and Ocean

Carbon dioxide behaves as a *greenhouse gas* in the atmosphere, with a significant role in regulating the temperature of Earth's surface and, therefore, Earth's climate.[2] Earth absorbs sunlight and then emits infrared radiation to space. The balance between sunlight absorbed and radiation to space determines the surface temperature. A greenhouse gas is so named because it absorbs infrared radiation emitted from the ground and reradiates it back to the ground. By intercepting Earth's radiation, atmospheric carbon dioxide keeps our planet warmer than it would otherwise be in the absence of this gas. Atmospheric carbon dioxide was constant at ~285 ppm (285 μL/L) for a millennium (or more) prior to the industrial revolution. Burning of fossil fuel and the destruction of Earth's forests since 1800 caused an exponential increase in CO_2 that threatens to alter Earth's climate in your lifetime.

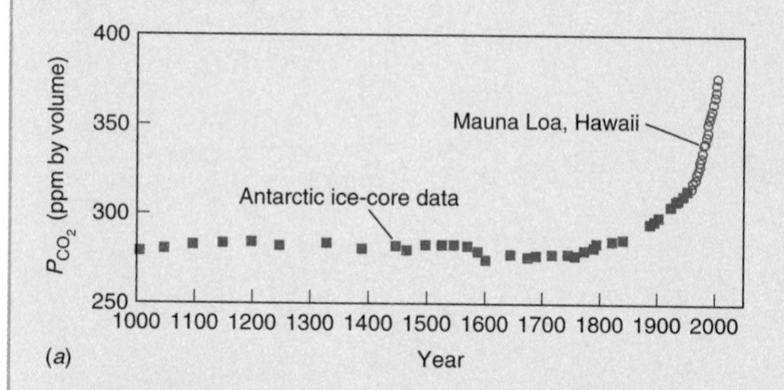

Thousand-year record of atmospheric CO_2 is derived from measurements of CO_2 trapped in ice cores in Antarctica and from direct atmospheric measurements. [Antarctica data from D. M. Etheridge, L. P. Steele, R. L. Langenfelds, R. J. Francey, J.-M. Barnola, and V. I. Morgan, in *Trends: A Compendium of Data on Global Change*, Carbon Dioxide Information Analysis Center, Oak Ridge National Laboratory, Oak Ridge, TN, 1998. Hawaii data from C. D. Keeling and T. P. Whorf, Scripps Institution of Oceanography, http://cdiac.ornl.gov/ftp/trends/co2/maunaloa.co2.]

Increasing atmospheric CO_2 increases the concentration of dissolved CO_2 in the ocean, which consumes carbonate and lowers the pH:

$$\underset{}{CO_2(aq)} + H_2O + \underset{\text{Carbonate}}{CO_3^{2-}} \longrightarrow \underset{\text{Bicarbonate}}{2HCO_3^-} \qquad \text{(A)}$$

The pH of the ocean has already decreased from its preindustrial value of 8.16 to 8.04 today.[3] Without changes in our activities, the pH could be 7.7 by 2100.

Low carbonate concentration promotes dissolution of solid calcium carbonate:

$$\underset{\text{Calcium carbonate}}{CaCO_3(s)} \rightleftharpoons Ca^{2+} + CO_3^{2-} \qquad \text{(B)}$$

Le Châtelier's principle tells us that decreasing $[CO_3^{2-}]$ draws the reaction to the right

If $[CO_3^{2-}]$ in the ocean decreases enough, organisms such as plankton and coral with $CaCO_3$ shells or skeletons will not survive.[4] Calcium carbonate has two crystalline forms called calcite and aragonite. Aragonite is more soluble than calcite. Different organisms have either calcite or aragonite in their shells and skeletons.

Pteropods are a type of plankton also known as winged snails (Panel *b*). When pteropods collected from the subarctic Pacific Ocean are kept in water that is less than saturated with aragonite, their shells begin to dissolve within 48 h. Animals such as the pteropod lie at the base of the food chain. Their destruction would reverberate through the entire ocean ecosystem.

Today, ocean surface waters contain more than enough CO_3^{2-} to sustain aragonite and calcite. As atmospheric CO_2 inexorably increases during the 21st

Dissolving sodium leucinate in water gives a solution of L^-, the fully basic species. The K_b values for this dibasic anion are

$$L^- + H_2O \rightleftharpoons HL + OH^- \qquad K_{b1} = K_w/K_{a2} = 5.55 \times 10^{-5}$$
$$HL + H_2O \rightleftharpoons H_2L^+ + OH^- \qquad K_{b2} = K_w/K_{a1} = 2.13 \times 10^{-12}$$

Hydrolysis is the reaction of anything with water. Specifically, the reaction $L^- + H_2O \rightleftharpoons HL + OH^-$ is called hydrolysis.

K_{b1} tells us that L^- will not **hydrolyze** (react with water) very much to give HL. Furthermore, K_{b2} tells us that the resulting HL is such a weak base that hardly any further reaction to make H_2L^+ will take place.

century, ocean surface waters will become undersaturated with respect to aragonite—killing off organisms that depend on this mineral for their structure. Polar regions will suffer this fate first because the acid dissociation constants K_{a1} and K_{a2} at low temperature in seawater favor HCO_3^- and $CO_2(aq)$ relative to CO_3^{2-} (Problem 11-33).

Panel *c* shows the predicted concentration of CO_3^{2-} in polar ocean surface water as a function of atmospheric CO_2. The upper horizontal line is the concentration of CO_3^{2-} below which aragonite dissolves. Atmospheric CO_2 is presently near 400 ppm and $[CO_3^{2-}]$ is near 100 μmol/kg of seawater—more than enough to precipitate aragonite or calcite. When atmospheric CO_2 reaches 600 ppm near the middle of the present century, $[CO_3^{2-}]$ will decrease to 60 μmol/kg and creatures with aragonite structures will begin to disapper from polar waters. At still higher atmospheric CO_2 concentration, extinctions will move to lower latitudes and will overtake organisms with calcite structures as well as aragonite structures. *How long will we put CO_2 into the atmosphere to see if these predictions are borne out?*

(*b*)

Pteropods. The shell of a live pteropod begins to dissolve after 48 h in water that is undersaturated with aragonite [David Wrobel/Visuals Unlimited.]

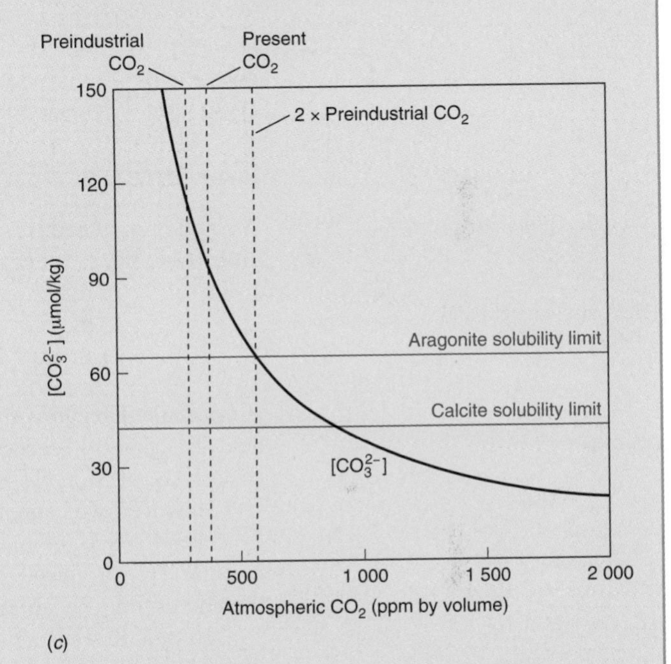

(*c*)

Calculated $[CO_3^{2-}]$ in polar ocean surface waters as a function of atmospheric CO_2. When $[CO_3^{2-}]$ drops below the upper horizontal line, aragonite dissolves. [Adapted from J. C. Orr et al., *Nature* **2005**, *437*, 681.]

We therefore treat L^- as a monobasic species, with $K_b = K_{b1}$. The results of this (fantastic) approximation can be outlined as follows:

$$\underset{\substack{L^- \\ 0.050\,0 - x}}{\overset{\displaystyle R \atop |}{H_2NCHCO_2^-}} + H_2O \xrightleftharpoons{K_{b1} = 5.55 \times 10^{-5}} \underset{\substack{HL \\ x}}{\overset{\displaystyle R \atop +\,|}{H_3NCHCO_2^-}} + \underset{x}{OH^-}$$

L^- can be treated as monobasic with $K_b = K_{b1}$.

$$\frac{x^2}{F - x} = 5.55 \times 10^{-5} \Longrightarrow x = [OH^-] = 1.64 \times 10^{-3}\ M$$

$$[HL] = x = 1.64 \times 10^{-3}\ M$$

$$[H^+] = K_w/[OH^-] = K_w/x = 6.10 \times 10^{-12}\ M \Longrightarrow pH = 11.21$$

$$[L^-] = F - x = 4.84 \times 10^{-2}\ M$$

The concentration of H_2L^+ can be found from the K_{b2} equilibrium:

$$HL + H_2O \overset{K_{b2}}{\rightleftharpoons} H_2L^+ + OH^-$$

$$K_{b2} = \frac{[H_2L^+][OH^-]}{[HL]} = \frac{[H_2L^+]x}{x} = [H_2L^+]$$

We find that $[H_2L^+] = K_{b2} = 2.13 \times 10^{-12}$ M, and the approximation that $[H_2L^+]$ is insignificant relative to [HL] is well justified. In summary, if there is any reasonable separation between K_1 and K_2 (and, therefore, between K_{b1} and K_{b2}), *the fully basic form of a diprotic acid can be treated as monobasic, with $K_b = K_{b1}$.*

The Intermediate Form, HL

A tougher problem.

A solution prepared from leucine, HL, is more complicated than one prepared from either H_2L^+ or L^-, because HL is both an acid and a base.

HL is both an acid and a base.

$$HL \rightleftharpoons H^+ + L^- \qquad K_a = K_2 = 1.80 \times 10^{-10} \qquad (11\text{-}7)$$

$$HL + H_2O \rightleftharpoons H_2L^+ + OH^- \qquad K_b = K_{b2} = 2.13 \times 10^{-12} \qquad (11\text{-}8)$$

A molecule that can both donate and accept a proton is said to be **amphiprotic**. The acid dissociation reaction (11-7) has a larger equilibrium constant than the base hydrolysis reaction (11-8), so we expect the solution of leucine to be acidic.

However, we cannot simply ignore Reaction 11-8, even when K_a and K_b differ by several orders of magnitude. Both reactions proceed to a nearly equal extent, because H^+ produced in Reaction 11-7 reacts with OH^- from Reaction 11-8, thereby driving Reaction 11-8 to the right.

The charge balance is discussed further in Section 12-3.

To treat this case, we write a *charge balance*, which says that the sum of positive charges in solution must equal the sum of negative charges. The procedure is applied to leucine, whose intermediate form (HL) has no net charge. However, results apply to the intermediate form of *any* diprotic acid, regardless of its charge.

Our problem concerns 0.050 0 M leucine, in which both Reactions 11-7 and 11-8 can happen. The charge balance is

Charge balance for HL:
sum of positive charges = sum of negative charges

$$\underbrace{[H_2L^+] + [H^+]}_{\text{Sum of positive charges}} = \underbrace{[L^-] + [OH^-]}_{\text{Sum of negative charges}} \qquad (11\text{-}9)$$

From the acid dissociation equilibria (Equations 11-3 and 11-4), we replace $[H_2L^+]$ with $[HL][H^+]/K_1$ and $[L^-]$ with $K_2[HL]/[H^+]$. Also, we can always write $[OH^-] = K_w/[H^+]$. Putting these expressions into Equation 11-9 gives

$$\frac{[HL][H^+]}{K_1} + [H^+] = \frac{K_2[HL]}{[H^+]} + \frac{K_w}{[H^+]}$$

which can be solved for $[H^+]$. First, multiply all terms by $[H^+]$:

$$\frac{[HL][H^+]^2}{K_1} + [H^+]^2 = K_2[HL] + K_w$$

Then factor out $[H^+]^2$ and rearrange:

$$[H^+]^2 \left(\frac{[HL]}{K_1} + 1\right) = K_2[HL] + K_w$$

$$[H^+]^2 = \frac{K_2[HL] + K_w}{\frac{[HL]}{K_1} + 1}$$

Multiplying the numerator and denominator by K_1 and taking the square root of both sides gives

$$[H^+] = \sqrt{\frac{K_1K_2[HL] + K_1K_w}{K_1 + [HL]}} \quad (11\text{-}10)$$

We solved for $[H^+]$ in terms of known constants plus the single unknown, [HL]. Where do we proceed from here?

In our moment of despair, a chemist gallops down from the mountain mists on her snow-white unicorn to provide the missing insight: "The major species will be HL, because it is both a weak acid and a weak base. Neither Reaction 11-7 nor Reaction 11-8 goes very far. For the concentration of HL in Equation 11-10, you can substitute the value 0.050 0 M."

The missing insight!

Taking the chemist's advice, we rewrite Equation 11-10 as follows:

$$[H^+] \approx \sqrt{\frac{K_1K_2F + K_1K_w}{K_1 + F}} \quad (11\text{-}11)$$

where F is the formal concentration of HL (= 0.050 0 M). Equation 11-11 can be further simplified under most conditions. The first term in the numerator is almost always much greater than the second term, so the second term can be dropped:

$$[H^+] \approx \sqrt{\frac{K_1K_2F + \cancel{K_1K_w}}{K_1 + F}}$$

Then, if $K_1 \ll F$, the first term in the denominator can also be neglected.

$$[H^+] \approx \sqrt{\frac{K_1K_2F}{\cancel{K_1} + F}}$$

Canceling F in the numerator and denominator gives

$$[H^+] \approx \sqrt{K_1K_2} = (K_1K_2)^{1/2} \quad (11\text{-}12)$$

Making use of the identity $\log(x^{1/2}) = \frac{1}{2}\log x$, we rewrite Equation 11-12 in the form

$$\log[H^+] \approx \log(K_1K_2)^{1/2} = \frac{1}{2}\log(K_1K_2)$$

Noting that log xy = log x + log y, we can rewrite the equation once more:

$$\log[H^+] \approx \tfrac{1}{2}(\log K_1 + \log K_2)$$

Multiplying both sides by -1 converts the terms into pH and pK:

$$\underbrace{-\log[H^+]}_{pH} \approx \tfrac{1}{2}(\underbrace{-\log K_1}_{pK_1} \underbrace{- \log K_2}_{pK_2})$$

The pH of the intermediate form of a diprotic acid is close to midway between the two pK_a values and is almost independent of concentration.

Intermediate form of diprotic acid:

$$pH \approx \tfrac{1}{2}(pK_1 + pK_2) \quad (11\text{-}13)$$

where K_1 and K_2 are the acid dissociation constants (= K_{a1} and K_{a2}) of the diprotic acid.

Equation 11-13 is a good one to keep in your head. It says that *the pH of a solution of the intermediate form of a diprotic acid is approximately midway between* pK_1 *and* pK_2, *regardless of the formal concentration.*

For leucine, Equation 11-13 gives a pH of $\frac{1}{2}(2.328 + 9.744) = 6.036$, or $[H^+] = 10^{-pH} = 9.20 \times 10^{-7}$ M. The concentrations of H_2L^+ and L^- can be found from the K_1 and K_2 equilibria, using [HL] = 0.050 0 M.

$$[H_2L^+] = \frac{[H^+][HL]}{K_1} = \frac{(9.20 \times 10^{-7})(0.050\,0)}{4.70 \times 10^{-3}} = 9.79 \times 10^{-6}\text{ M}$$

$$[L^-] = \frac{K_2[HL]}{[H^+]} = \frac{(1.80 \times 10^{-10})(0.050\,0)}{9.20 \times 10^{-7}} = 9.78 \times 10^{-6}\text{ M}$$

Was the approximation [HL] ≈ 0.050 0 M a good one? It certainly was, because $[H_2L^+]$ (= 9.79×10^{-6} M) and $[L^-]$ (= 9.78×10^{-6} M) are small in comparison with [HL] (≈ 0.050 0 M). Nearly all the leucine remained in the form HL.

Example pH of the Intermediate Form of a Diprotic Acid

Potassium hydrogen phthalate, KHP, is a salt of the intermediate form of phthalic acid. Calculate the pH of 0.10 M KHP and of 0.010 M KHP.

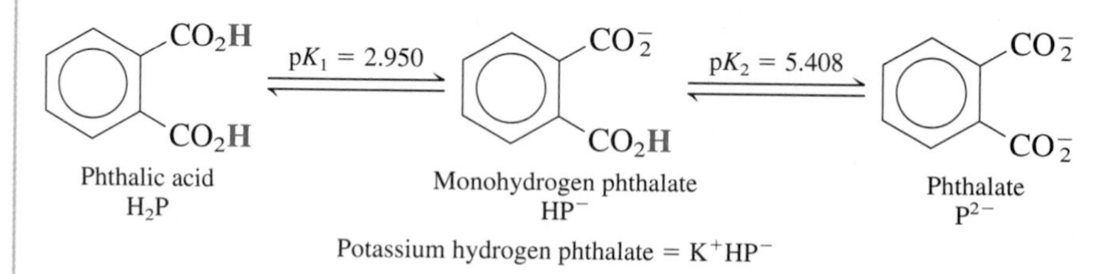

SOLUTION From Equation 11-13, the pH of potassium hydrogen phthalate is estimated as $\frac{1}{2}(pK_1 + pK_2) = 4.18$, regardless of concentration.

Test Yourself Predict the pH of 29 mM serine. (**Answer:** 5.70)

Diprotic systems:

- treat H_2A and BH_2^{2+} as monoprotic weak acids
- treat A^{2-} and B as monoprotic weak bases
- treat HA^- and BH^+ as intermediates: pH ≈ $\frac{1}{2}(pK_1 + pK_2)$

Summary of Diprotic Acid Calculations

The fully protonated species H_2A is treated as a monoprotic acid with acid dissociation constant K_1. The fully basic species A^{2-} is treated as a monoprotic base, with base association constant $K_{b1} = K_w/K_{a2}$. For the intermediate form HA^-, use the equation $pH \approx \frac{1}{2}(pK_1 + pK_2)$, where K_1 and K_2 are the acid dissociation constants for H_2A. The same considerations apply to a diprotic base ($B \longrightarrow BH^+ \longrightarrow BH_2^{2+}$): B is treated as monobasic; BH_2^{2+} is treated as monoprotic; and BH^+ is treated as an intermediate with $pH \approx \frac{1}{2}(pK_1 + pK_2)$, where K_1 and K_2 are the *acid* dissociation constants of BH_2^{2+}.

Ask Yourself

11-B. Find the pH and the concentrations of H_2SO_3, HSO_3^-, and SO_3^{2-} in each of the following solutions: **(a)** 0.050 M H_2SO_3; **(b)** 0.050 M $NaHSO_3$; **(c)** 0.050 M Na_2SO_3.

11-3 Which Is the Principal Species?

We sometimes need to identify which species of acid, base, or intermediate is predominant under given conditions. For example, what is the principal form of benzoic acid at pH 8? The pH of 8 is the net result of all reagents in the solution. The pH might be 8 because a phosphate buffer was added or because NaOH was added to benzoic acid. It does not matter how the pH came to be 8. That is just where it happens to be.

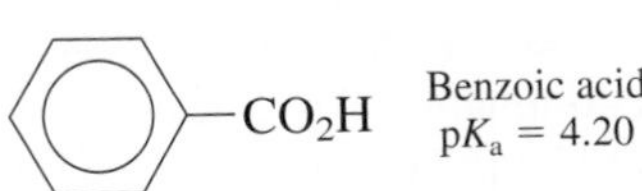

Benzoic acid
$pK_a = 4.20$

The pK_a for benzoic acid is 4.20. Therefore, at pH 4.20 there is a 1:1 mixture of benzoic acid (HA) and benzoate ion (A^-). At $pH = pK_a + 1$ (= 5.20), the quotient $[A^-]/[HA]$ is 10:1, which we deduce from the Henderson-Hasselbalch equation:

At $pH = pK_a$, $[A^-] = [HA]$ because
$$pH = pK_a + \log\left(\frac{[A^-]}{[HA]}\right) = pK_a + \log 1 = pK_a.$$

$$pH = pK_a + \log\left(\frac{[A^-]}{[HA]}\right)$$

Setting pH equal to $pK_a + 1$ gives

$$\cancel{pK_a} + 1 = \cancel{pK_a} + \log\left(\frac{[A^-]}{[HA]}\right) \Longrightarrow 1 = \log\left(\frac{[A^-]}{[HA]}\right)$$

To solve for $[A^-]/[HA]$, raise 10 to the power on each side of the equation:

$$10^1 = 10^{\log([A^-]/[HA])} \Longrightarrow \frac{[A^-]}{[HA]} = 10$$

At $pH = pK_a + 2$ (= 6.20), the quotient $[A^-]/[HA]$ would be 100:1. As the pH increases, the quotient $[A^-]/[HA]$ increases still further.

For a monoprotic system, the basic species, A^-, is the predominant form when $pH > pK_a$. The acidic species, HA, is the predominant form when $pH < pK_a$. The predominant form of benzoic acid at pH 8 is the benzoate anion, $C_6H_5CO_2^-$.

pH	Major species
$<pK_a$	HA
$>pK_a$	A^-

← More acidic pH More basic →

Predominant form	HA	A^-

↑
pK_a
$[HA] = [A^-]$

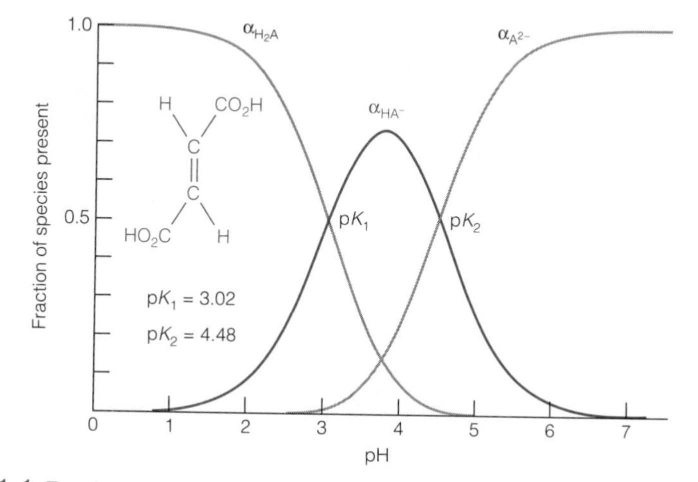

Figure 11-1 Fractional composition diagram for fumaric acid (*trans*-butenedioic acid). α_i is the fraction of species i at each pH. At low pH, H_2A is the dominant form. Between pH = pK_1 and pH = pK_2, HA^- is dominant. Above pH = pK_2, A^{2-} dominates. Because pK_1 and pK_2 are not very different, the fraction of HA^- never gets very close to unity.

Example Principal Species—Which One and How Much?

What is the predominant form of ammonia in a solution at pH 7.0? Approximately what fraction is in this form?

SOLUTION In Appendix B, we find $pK_a = 9.24$ for the ammonium ion (NH_4^+, the conjugate acid of ammonia, NH_3). At pH = 9.24, $[NH_4^+] = [NH_3]$. Below pH 9.24, NH_4^+ will be the predominant form. Because pH = 7.0 is about 2 pH units below pK_a, the quotient $[NH_3]/[NH_4^+]$ will be about 1:100. Approximately 99% is in the form NH_4^+.

Test Yourself What is the predominant form of cyclohexylamine at pH 9.5? Approximately what fraction is in this form? (**Answer:** ~10:1 $RNH_3^+ : RNH_2$)

pH	Major species
$pH < pK_1$	H_2A
$pK_1 < pH < pK_2$	HA^-
$pH > pK_2$	A^{2-}

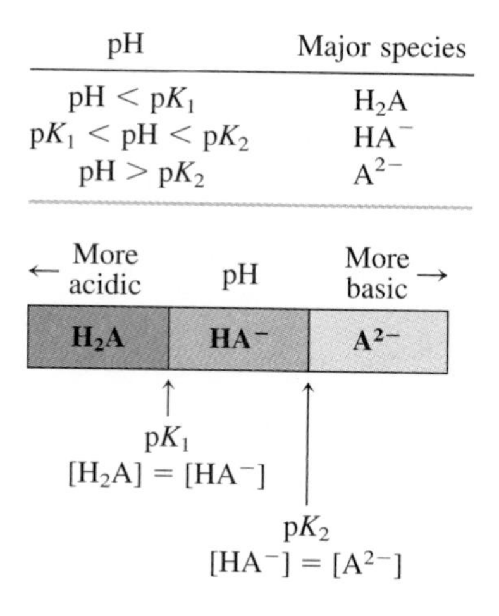

For diprotic systems, the reasoning is similar, but there are two pK_a values. Consider fumaric acid (*trans*-butenedioic acid), H_2A, with $pK_1 = 3.02$ and $pK_2 = 4.48$. At pH = pK_1, $[H_2A] = [HA^-]$. At pH = pK_2, $[HA^-] = [A^{2-}]$. The chart in the margin shows the major species in each pH region. At pH values below pK_1, H_2A is dominant. At pH values above pK_2, A^{2-} is dominant. At pH values between pK_1 and pK_2, HA^- is dominant. Figure 11-1 shows the fraction of each species as a function of pH.

The diagram below shows the major species for a triprotic system and introduces an important extension of what we learned in the previous section.

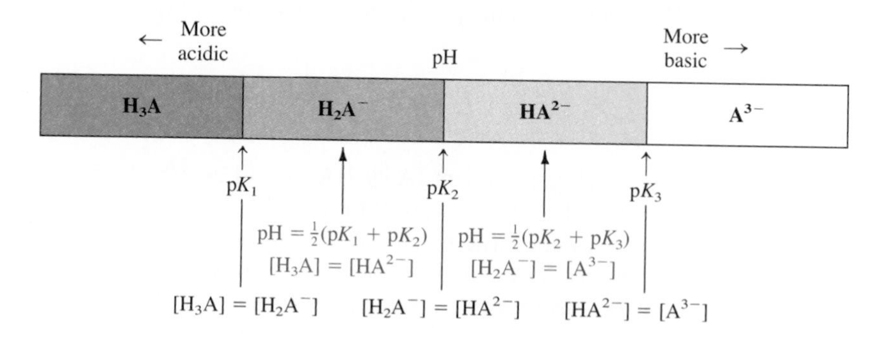

H_3A is the dominant species of a triprotic system in the most acidic solution at $pH < pK_1$. H_2A^- is dominant between pK_1 and pK_2. HA^{2-} is the major species between pK_2 and pK_3, and A^{3-} dominates in the most basic solution at $pH > pK_3$.

The preceding diagram shows that the pH of the first intermediate species, H_2A^-, is $\frac{1}{2}(pK_1 + pK_2)$. At this pH, the concentrations of H_3A and HA^{2-} are small and equal to each other. *The new feature of the diagram is that the pH of the second intermediate species, HA^{2-}, is $\frac{1}{2}(pK_2 + pK_3)$.* At this pH, the concentrations of H_2A^- and A^{3-} are small and equal.

Triprotic systems:

- treat H_3A as monoprotic weak acid
- treat A^{3-} as monoprotic weak base
- treat H_2A^- as intermediate: $pH \approx \frac{1}{2}(pK_1 + pK_2)$
- treat HA^{2-} as intermediate: $pH \approx \frac{1}{2}(pK_2 + pK_3)$

Example Principal Species in a Polyprotic System

The amino acid arginine has the following forms:

H_3N^+, H, N, NH_2, HO_2C, $^+NH_2$ — H_3A^{2+} $\underset{}{\overset{pK_1 = 1.82}{\rightleftharpoons}}$

H_3N^+, H, N, NH_2, ^-O_2C, $^+NH_2$ — H_2A^+ $\overset{pK_2 = 8.99}{\rightleftharpoons}$

H_2N, H, N, NH_2, ^-O_2C, $^+NH_2$ — HA $\overset{pK_3 = 12.1}{\rightleftharpoons}$ H_2N, H, N, NH_2, ^-O_2C, NH — A^-

This is the neutral molecule called arginine

The ammonium group next to the carboxyl group is more acidic than the substituent ammonium group at the right. What is the principal form of arginine at pH 10.0? Approximately what fraction is in this form? What is the second most abundant form at this pH?

SOLUTION It helps to draw a diagram showing which species predominates at each pH:

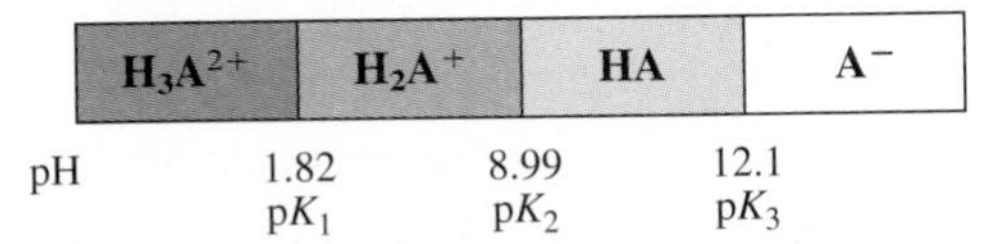

The predominant species between $pK_2 = 8.99$ and $pK_3 = 12.1$ is HA. At pK_2, $[H_2A^+] = [HA]$. At pK_3, $[HA] = [A^-]$. Because pH 10.0 is about one pH unit higher than pK_2, we can say that $[HA]/[H_2A^+] \approx 10:1$. About 90% of arginine is in the form HA. The second most abundant species is H_2A^+ at ~10%.

Test Yourself What is the principal form of arginine at pH 8.0? Approximately what fraction is in this form? (**Answer:** H_2A^+, ~90%)

Example **More on Polyprotic Systems**

In the pH range 1.82 to 8.99, H_2A^+ is the principal form of arginine. Which is the second most prominent species at pH 6.0? At pH 5.0?

SOLUTION We know that the pH of the pure intermediate (amphiprotic) species, H_2A^+, is

$$\text{pH of } H_2A^+ \approx \tfrac{1}{2}(pK_1 + pK_2) = 5.40$$

Above pH 5.40 (and below pH = pK_2), HA is the second most important species. Below pH 5.40 (and above pH = pK_1), H_3A^{2+} is the second most important species.

Test Yourself What is the second major form of arginine at pH 8.0 and what fraction is in this form? (**Answer:** HA, ~10%)

Ask Yourself

11-C. **(a)** Draw the structure of the predominant form (principal species) of 1,3-dihydroxybenzene at pH 9.00 and at pH 11.00. What is the second most prominent species at each pH?
(b) Cysteine is a triprotic system whose fully protonated form could be designated H_3C^+. Which form of cysteine is drawn here: H_3C^+, H_2C, HC^-, or C^{2-}? What would be the pH of a 0.10 M solution of this form of cysteine?

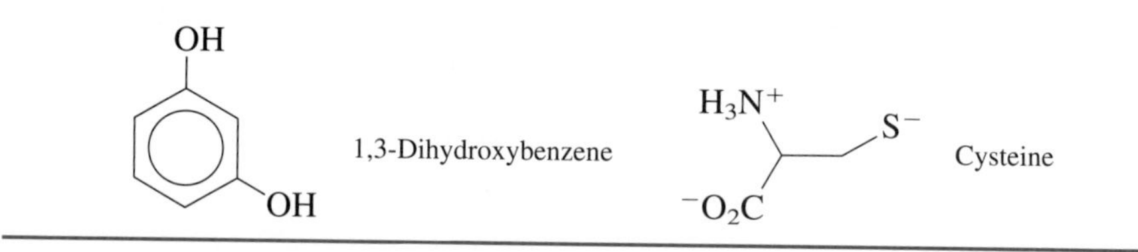

11-4 Titrations in Polyprotic Systems

Figure 10-2 showed the titration curve for the monoprotic acid HA treated with OH^-. As a brief reminder, the pH at several critical points was computed as follows:

Initial solution:	Has the pH of the weak acid HA
$V_e/2$:	pH = pK_a because [HA] = [A^-]
V_e:	Has the pH of the conjugate base, A^-
Past V_e:	pH is governed by concentration of excess OH^-

The slope of the titration curve is minimum at $V_e/2$ when pH = pK_a. The slope is maximum at the equivalence point.

Before delving into titration curves for diprotic systems, you should realize that there are two buffer pairs derived from the acid H_2A. H_2A and HA^- constitute one buffer pair and HA^- and A^{2-} constitute a second pair. For the acid H_2A, there are *two* Henderson-Hasselbalch equations, both of which are *always* true. If you happen to know the concentrations $[H_2A]$ and $[HA^-]$, then use the pK_1 equation. If you know $[HA^-]$ and $[A^{2-}]$, use the pK_2 equation.

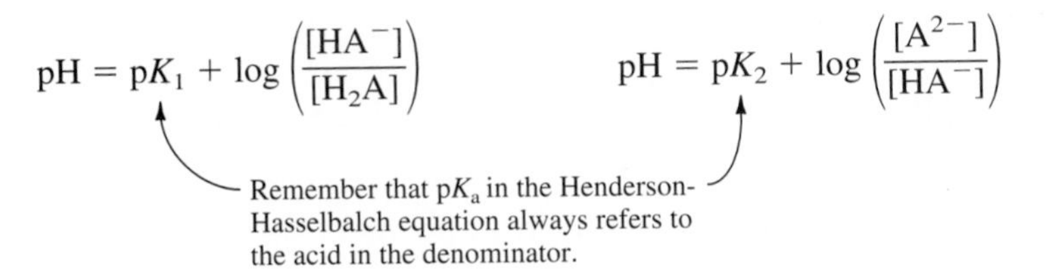

$pH = pK_1 + \log\left(\frac{[HA^-]}{[H_2A]}\right)$
$pH = pK_2 + \log\left(\frac{[A^{2-}]}{[HA^-]}\right)$
Remember that pK_a in the Henderson-Hasselbalch equation always refers to the acid in the denominator.

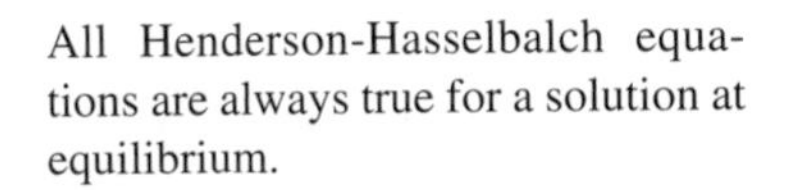

All Henderson-Hasselbalch equations are always true for a solution at equilibrium.

So now let's turn our attention to the titration of diprotic acids. Figure 11-2 shows calculated curves for 50.0 mL each of three different 0.020 0 M diprotic acids, H_2A, titrated with 0.100 M OH^-. For all three curves, H_2A has $pK_1 = 4.00$. In the lowest curve, pK_2 is 6.00. In the middle curve, pK_2 is 8.00; and, in the upper curve, pK_2 is 10.00. The first equivalence volume (V_{e1}) occurs when the moles of added OH^- equal the moles of H_2A. The second equivalence volume (V_{e2}) is always exactly twice as great as the first equivalence volume, because we must add the same amount of OH^- to convert HA^- to A^{2-}. Let's consider why pH varies as it does during these titrations.

$V_{e2} = 2V_{e1}$ (always!)

Point A has the same pH in all three cases. It is the pH of the acid H_2A, which is treated as a monoprotic acid with $pK_a = pK_1 = 4.00$ and formal concentration F.

Point A: weak acid H_2A

$$\underset{F-x}{HA} \rightleftharpoons \underset{x}{A^-} + \underset{x}{H^+} \qquad \frac{x^2}{F-x} = K_1 \tag{11-14}$$

$$\frac{x^2}{0.020\,0 - x} = 10^{-4.00} \Longrightarrow x = 1.37 \times 10^{-3}\ M \Longrightarrow pH = -\log(x) = 2.86$$

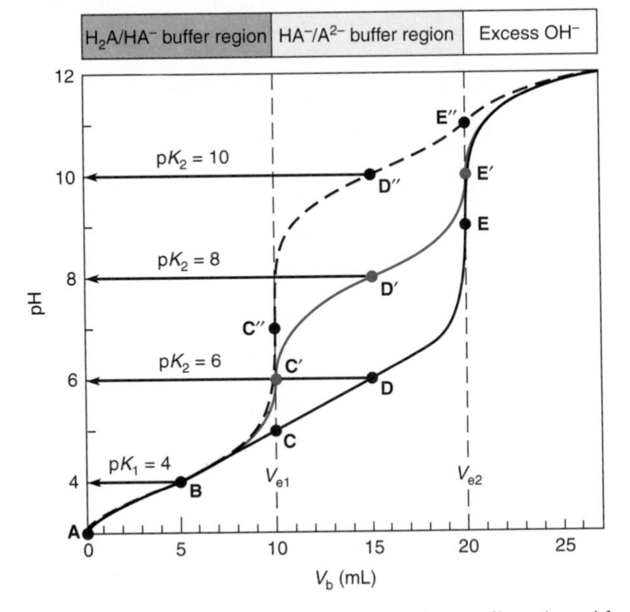

Figure 11-2 Calculated titration curves for three different diprotic acids, H_2A. For each curve, 50.0 mL of 0.020 0 M H_2A are titrated with 0.100 M NaOH. *Lowest curve:* $pK_1 = 4.00$ and $pK_2 = 6.00$. *Middle curve:* $pK_1 = 4.00$ and $pK_2 = 8.00$. *Upper curve:* $pK_1 = 4.00$ and $pK_2 = 10.00$.

Point B, which is halfway to the first equivalence point, has the same pH in all three cases. It is the pH of a 1:1 mixture of H_2A:HA^-, which is treated as a monoprotic acid with $pK_a = pK_1 = 4.00$.

Point B: buffer containing H_2A + HA^-

$$pH = pK_1 + \log\left(\frac{[HA^-]}{[H_2A]}\right) = pK_1 + \log 1 = pK_1 = 4.00 \qquad (11\text{-}15)$$

Because all three acids have the same pK_1, the pH at this point is the same in all three cases.

Point C (and C′ and C″) is the first equivalence point. H_2A has been converted into HA^-, the intermediate form of a diprotic acid. The pH is calculated from Equation 11-13:

Point C: intermediate form HA^-

$$pH \approx \tfrac{1}{2}(pK_1 + pK_2) = \begin{cases} 5.00 \text{ at C} \\ 6.00 \text{ at C}' \\ 7.00 \text{ at C}'' \end{cases} \qquad (11\text{-}13)$$

The three acids have the same pK_1 but different values of pK_2. Therefore the pH at the first equivalence point is different in all three cases.

Point D is halfway from the first equivalence point to the second equivalence point. Half of the HA^- has been converted into A^{2-}. The pH is

Point D: buffer containing HA^- + A^{2-}

$$pH = pK_2 + \log\left(\frac{[A^{2-}]}{[HA^-]}\right) = pK_2 + \log 1 = pK_2 = \begin{cases} 6.00 \text{ at D} \\ 8.00 \text{ at D}' \\ 10.00 \text{ at D}'' \end{cases} \qquad (11\text{-}16)$$

Look at Figure 11-2 and you will see that points D, D′, and D″ come at pK_2 for each of the acids.

Point E is the second equivalence point. All of the acid has been converted into A^{2-}, a weak base with concentration F′. We find the pH by treating A^{2-} as a monoprotic base with $K_{b1} = K_w/K_{a2}$:

Point E: weak base A^{2-}

$$\underset{F'-x}{A^{2-}} + H_2O \rightleftharpoons \underset{x}{HA^-} + \underset{x}{OH^-} \qquad \frac{x^2}{F'-x} = K_{b1} \qquad (11\text{-}17)$$

The pH is different for the three acids, because K_{b1} is different for all three. Here is how we find the pH at the second equivalence point:

$$F' = [A^{2-}] = \frac{\text{mmol } A^{2-}}{\text{total mL}} = \frac{(0.020\,0 \text{ M})(50.0 \text{ mL})}{70.0 \text{ mL}} = 0.014\,3 \text{ M}$$

$$\frac{x^2}{0.014\,3 - x} = K_{b1} = \begin{cases} 10^{-8.00} \text{ at E} \\ 10^{-6.00} \text{ at E}' \\ 10^{-4.00} \text{ at E}'' \end{cases} \Longrightarrow x = \begin{cases} 1.20 \times 10^{-5} \text{ at E} \\ 1.19 \times 10^{-4} \text{ at E}' \\ 1.15 \times 10^{-3} \text{ at E}'' \end{cases} = [OH^-]$$

$$\Longrightarrow pH = -\log(K_w/x) = \begin{cases} 9.08 \text{ at E} \\ 10.08 \text{ at E}' \\ 11.06 \text{ at E}'' \end{cases}$$

Beyond V_{e2}, the pH is governed mainly by the concentration of excess OH^-. pH rapidly converges to the same value for all three titrations.

You can see in Figure 11-2 that, in a favorable case (the middle curve), we observe two obvious, steep equivalence points in the titration curve. When the pK

values are too close together or when the pK values are too low or too high, there may not be a distinct break at each equivalence point.

Example Titration of Sodium Carbonate

Let's reverse the process of Figure 11-2 and calculate the pH at points A–E in the titration of a diprotic base. Figure 11-3 shows the calculated titration curve for 50.0 mL of 0.020 0 M Na_2CO_3 treated with 0.100 M HCl. The first equivalence point is at 10.0 mL, and the second is at 20.0 mL. Find the pH at points A–E.

$$\begin{array}{llll} Na_2CO_3 & pK_1 = 6.351 & K_{a1} = 4.46 \times 10^{-7} & K_{b1} = K_w/K_{a2} = 2.13 \times 10^{-4} \\ & pK_2 = 10.329 & K_{a2} = 4.69 \times 10^{-11} & K_{b2} = K_w/K_{a1} = 2.24 \times 10^{-8} \end{array}$$

$$\underset{\text{Carbonic acid}}{H_2CO_3} \overset{pK_1}{\rightleftharpoons} \underset{\text{Bicarbonate}}{HCO_3^-} \overset{pK_2}{\rightleftharpoons} \underset{\text{Carbonate}}{CO_3^{2-}}$$

SOLUTION

Point A: The initial point in the titration is just a solution of 0.020 0 M Na_2CO_3, which can be treated as a monoprotic base:

$$\underset{0.020\,0 - x}{CO_3^{2-}} + H_2O \overset{K_{b1}}{\rightleftharpoons} \underset{x}{HCO_3^-} + \underset{x}{OH^-} \qquad \frac{x^2}{0.020\,0 - x} = K_{b1}$$

$$\Longrightarrow x = 1.96 \times 10^{-3} = [OH^-] \Longrightarrow pH = -\log(K_w/x) = 11.29$$

Point B: Now we are halfway to the first equivalence point. Half of the carbonate has been converted into bicarbonate, so there is a 1:1 mixture of CO_3^{2-} and HCO_3^-—*Aha! A buffer!*

$$pH = pK_2 + \log\left(\frac{[CO_3^{2-}]}{[HCO_3^-]}\right) = pK_2 + \log 1 = pK_2 = 10.33$$

↑

Use pK_2 (= pK_{a2}) because the equilibrium involves CO_3^{2-} and HCO_3^-

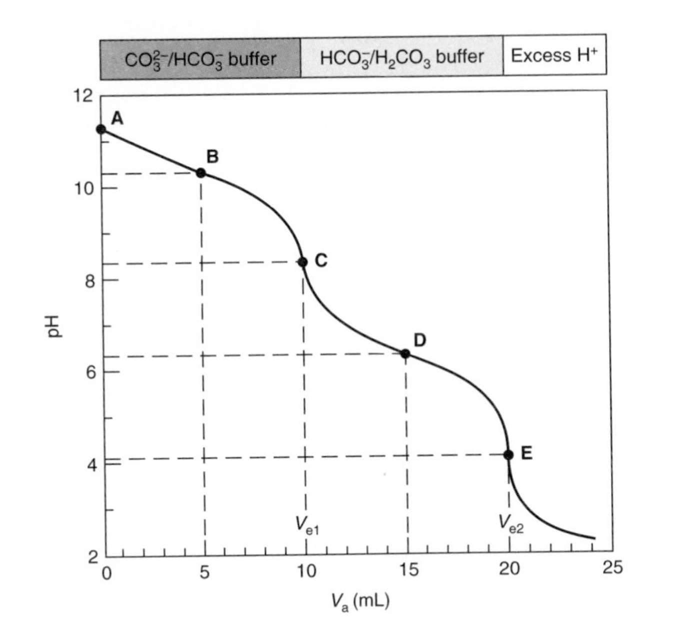

Figure 11-3 Calculated titration curve for 50.0 mL of 0.020 0 M Na_2CO_3 titrated with 0.100 M HCl.

Point C: At the first equivalence point, we have a solution of HCO_3^-, the intermediate form of a diprotic acid. To a good approximation, the pH is independent of concentration and is given by

$$\text{pH} \approx \tfrac{1}{2}(\text{p}K_1 + \text{p}K_2) = \tfrac{1}{2}(6.351 + 10.329) = 8.34$$

Point D: We are halfway to the second equivalence point. Half of the bicarbonate has been converted into carbonic acid, so there is a 1:1 mixture of HCO_3^- and H_2CO_3—*Aha! Another buffer!*

$$\text{pH} = \underset{\substack{\uparrow \\ \text{Use p}K_1\text{ because the equilibrium} \\ \text{involves HCO}_3^-\text{ and H}_2\text{CO}_3}}{\text{p}K_1} + \log\left(\frac{[\text{HCO}_3^-]}{[\text{H}_2\text{CO}_3]}\right) = \text{p}K_1 + \log 1 = \text{p}K_1 = 6.35$$

Point E: At the second equivalence point, all carbonate has been converted into carbonic acid, which has been diluted from its initial volume of 50.0 mL to a volume of 70.0 mL.

$$\text{F}' = [\text{H}_2\text{CO}_3] = \frac{\text{mmol H}_2\text{CO}_3}{\text{total mL}} = \frac{(0.020\,0\text{ M})(50.0\text{ mL})}{70.0\text{ mL}} = 0.014\,3\text{ M}$$

$$\underset{\text{F}'-x}{\text{H}_2\text{CO}_3} \overset{K_1}{\rightleftharpoons} \underset{x}{\text{HCO}_3^-} + \underset{x}{\text{H}^+} \qquad \frac{x^2}{0.014\,3 - x} = K_1$$

$$\Longrightarrow x = 7.96 \times 10^{-5} = [\text{H}^+] \Longrightarrow \text{pH} = -\log x = 4.10$$

Test Yourself What would be the pH at $\frac{1}{2}V_{e1}$, V_{e1}, and $\frac{3}{2}V_{e1}$ for the titration of mercaptoacetic acid with NaOH? (**Answer:** 3.64, 7.12, 10.61)

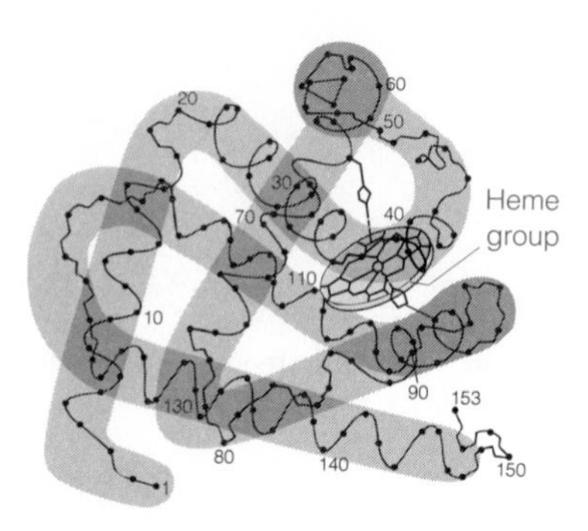

(a)

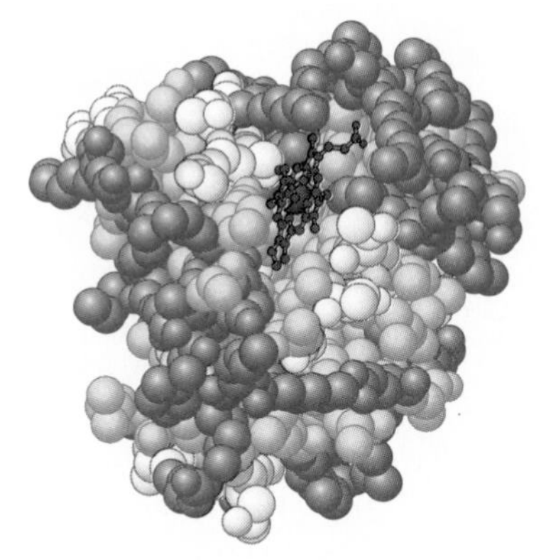

(b)

Figure 11-4 (*a*) Amino acid backbone of the protein myoglobin, which stores oxygen in muscle tissue. Substituents (R groups from Table 11-1) are omitted for clarity. The flat *heme* group at the right side of the protein contains an iron atom that can bind O_2, CO, and other small molecules. [From M. F. Perutz, "The Hemoglobin Molecule." Copyright © 1964 by Scientific American, Inc.] (*b*) Space-filling model of myoglobin with charged acidic and basic amino acids in dark color and *hydrophobic* (nonpolar, water-repelling) amino acids in light color. White amino acids are *hydrophilic* (polar, water-loving), but not charged. The surface of this water-soluble protein is dominated by charged and hydrophilic groups. [From J. M. Berg, J. L. Tymoczko, and L. Stryer, *Biochemistry*, 5th ed. (New York: W. H. Freeman and Company, 2002).]

Proteins Are Polyprotic Acids and Bases

Proteins are polymers made of amino acids:

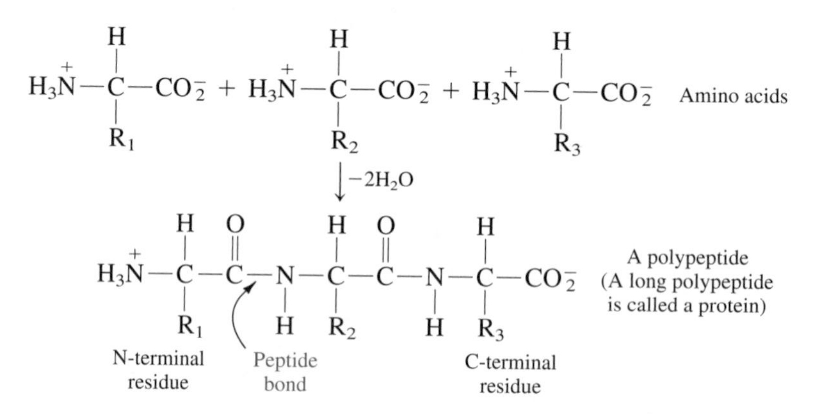

Proteins have biological functions such as structural support, catalysis of chemical reactions, immune response to foreign substances, transport of molecules across membranes, and control of genetic expression. The three-dimensional structure and function of a protein are determined by the sequence of amino acids from which the protein is made. Figure 11-4 shows the protein myoglobin, whose function is to store O_2 in muscle cells. Of the 153 amino acids in sperm-whale myoglobin, 35 have basic side groups and 23 are acidic.

At high pH, most proteins have lost so many protons that they have a negative charge. At low pH, most proteins have gained so many protons that they have a positive charge. At some intermediate pH, called the *isoelectric pH* (or isoelectric point), each protein has exactly zero net charge. Box 11-2 explains how proteins can be separated from one another because of their different isoelectric points.

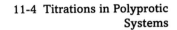

Box 11-2 What Is Isoelectric Focusing?

At its *isoelectric pH,* a protein has zero net charge and will therefore not migrate in an electric field. This effect is the basis of a very sensitive technique of protein separation called **isoelectric focusing**. A mixture of proteins is subjected to a strong electric field in a medium designed to have a pH gradient. Positively charged molecules move toward the negative pole and negatively charged molecules move toward the positive pole. Each protein migrates until it reaches the point where the pH is the same as its isoelectric pH. At this point, the protein has no net charge and no longer moves. Each protein in the mixture is focused in one small region at its isoelectric pH.

An example of isoelectric focusing is shown in the figure. A mixture of proteins was applied to a polyacrylamide gel containing a mixture of polyprotic compounds called *ampholytes.* Several hundred volts were applied across the length of the gel. The ampholytes migrated until they formed a stable pH gradient ranging from pH 3 at one end of the gel to pH 10 at the other. Each protein migrated until it reached the zone with its isoelectric pH, at which point the protein had no net charge and ceased migrating. If a molecule diffuses out of its isoelectric region, it becomes charged and migrates back to its isoelectric zone. When the proteins finished migrating, the electric field was removed and proteins were precipitated in place on the gel and stained with a dye to make them visible.

The stained gel is shown at the bottom of the figure. A spectrophotometric scan of the dye peaks is shown on the graph, and a profile of measured pH is also plotted. Each dark band of stained protein gives an absorbance peak.

Living cells also have isoelectric points and can be separated from one another by isoelectric focusing.

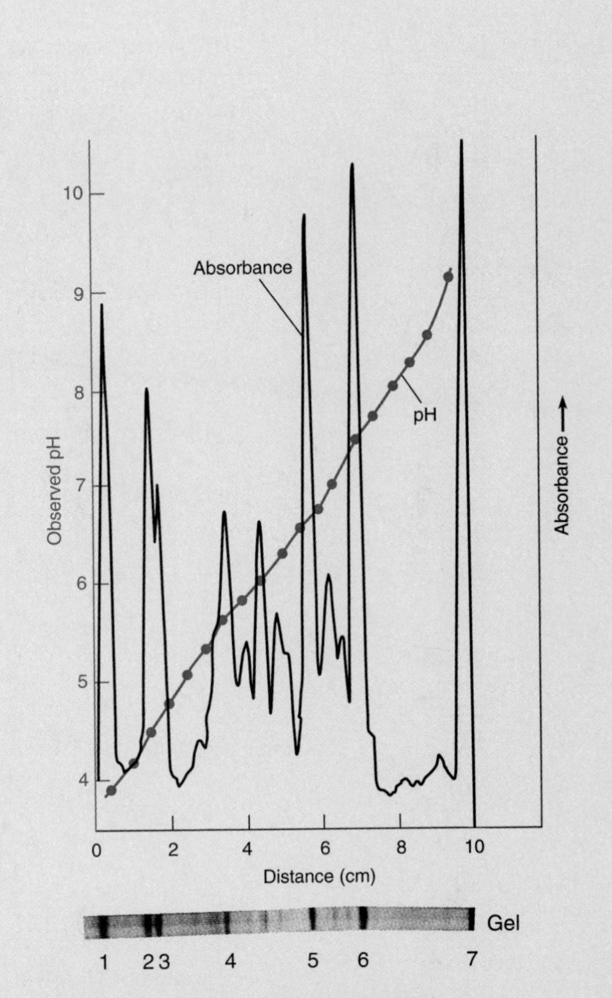

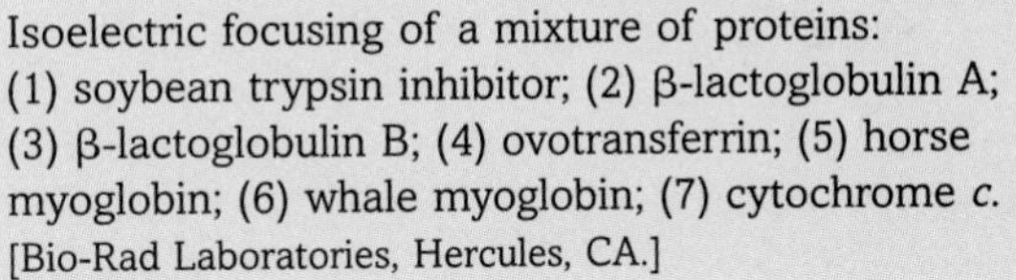

Isoelectric focusing of a mixture of proteins: (1) soybean trypsin inhibitor; (2) β-lactoglobulin A; (3) β-lactoglobulin B; (4) ovotransferrin; (5) horse myoglobin; (6) whale myoglobin; (7) cytochrome *c.* [Bio-Rad Laboratories, Hercules, CA.]

Ask Yourself

11-D. Consider the titration of 50.0 mL of 0.050 0 M malonic acid with 0.100 M NaOH.
(a) How many milliliters of titrant are required to reach each equivalence point?
(b) Calculate the pH at V_b = 0.0, 12.5, 25.0, 37.5, 50.0, and 55.0 mL.
(c) Put the points from **(b)** on a graph and sketch the titration curve.

Key Equations

Diprotic acid equilibria

$$H_2A \rightleftharpoons HA^- + H^+ \qquad K_{a1} \equiv K_1$$
$$HA^- \rightleftharpoons A^{2-} + H^+ \qquad K_{a2} \equiv K_2$$

Diprotic base equilibria

$$A^{2-} + H_2O \rightleftharpoons HA^- + OH^- \qquad K_{b1}$$
$$HA^- + H_2O \rightleftharpoons H_2A + OH^- \qquad K_{b2}$$

Relation between K_a and K_b

Monoprotic system $\quad K_aK_b = K_w$

Diprotic system $\quad K_{a1}K_{b2} = K_w$

$\quad K_{a2}K_{b1} = K_w$

Triprotic system $\quad K_{a1}K_{b3} = K_w$

$\quad K_{a2}K_{b2} = K_w$

$\quad K_{a3}K_{b1} = K_w$

pH of H_2A (or BH_2^{2+})

$$\underset{F-x}{H_2A} \overset{K_{a1}}{\rightleftharpoons} \underset{x}{H^+} + \underset{x}{HA^-} \qquad \frac{x^2}{F-x} = K_{a1}$$

(This calculation gives $[H^+]$, $[HA^-]$, and $[H_2A]$. You can solve for $[A^{2-}]$ from the K_{a2} equilibrium.)

pH of HA^- (or diprotic BH^+) $\quad pH \approx \frac{1}{2}(pK_1 + pK_2)$

pH of A^{2-} (or diprotic B)

$$\underset{F-x}{A^{2-}} + H_2O \overset{K_{b1}}{\rightleftharpoons} \underset{x}{HA^-} + \underset{x}{OH^-}$$

$$\frac{x^2}{F-x} = K_{b1} = \frac{K_w}{K_{a2}}$$

(This calculation gives $[OH^-]$, $[HA^-]$, and $[A^{2-}]$. You can find $[H^+]$ from the K_w equilibrium and $[H_2A]$ from the K_{a1} equilibrium.)

Diprotic buffer

$$pH = pK_1 + \log\left(\frac{[HA^-]}{[H_2A]}\right)$$

$$pH = pK_2 + \log\left(\frac{[A^{2-}]}{[HA^-]}\right)$$

Both equations are always true and either can be used, depending on which set of concentrations you happen to know.

Titration of H_2A with OH^-

$V_b = 0$	Find pH of H_2A
$V_b = \frac{1}{2}V_{e1}$	$pH = pK_1$
$V_b = V_{e1}$	$pH \approx \frac{1}{2}(pK_1 + pK_2)$
$V_b = \frac{3}{2}V_{e1}$	$pH = pK_2$
$V_b = V_{e2}$	Find pH of A^{2-}
$V_b > V_{e2}$	Find concentration of excess OH^-

Titration of diprotic B with H^+	$V_a = 0$	Find pH of B
	$V_a = \frac{1}{2}V_{e1}$	$pH = pK_{a2}$ (for BH_2^{2+})
	$V_a = V_{e1}$	$pH \approx \frac{1}{2}(pK_{a1} + pK_{a2})$
	$V_a = \frac{3}{2}V_{e1}$	$pH = pK_{a1}$ (for BH_2^{2+})
	$V_a = V_{e2}$	Find pH of BH_2^{2+}
	$V_a > V_{e2}$	Find concentration of excess H^+

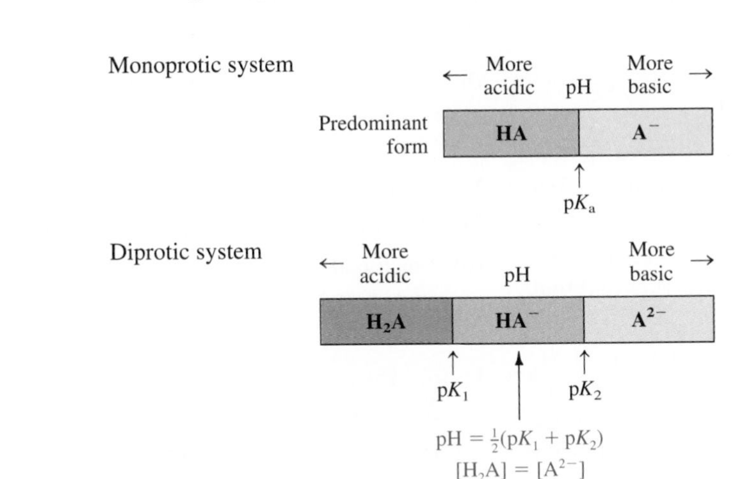

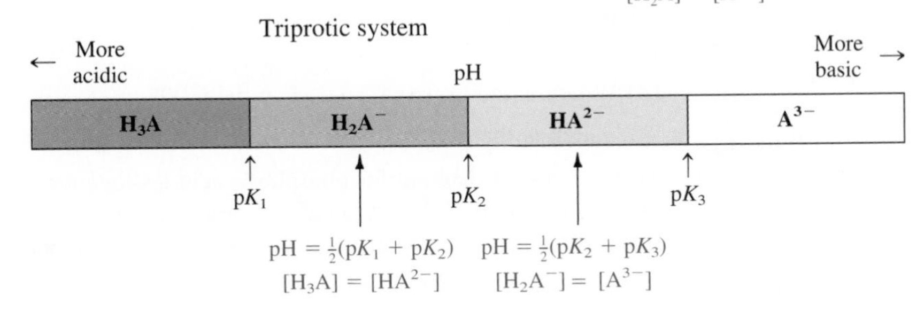

Important Terms

amino acid	hydrolysis	polyprotic acid
amphiprotic	isoelectric focusing	zwitterion

Problems

11-1. State what chemistry governs the pH at each point A through E in Figure 11-2.

11-2. Write the K_{a2} reaction of sulfuric acid (H_2SO_4) and the K_{b2} reaction of disodium oxalate ($Na_2C_2O_4$) and find their numerical values.

11-3. The base association constants of phosphate are $K_{b1} = 0.024$, $K_{b2} = 1.58 \times 10^{-7}$, and $K_{b3} = 1.41 \times 10^{-12}$. From the K_b values, calculate K_{a1}, K_{a2}, and K_{a3} for H_3PO_4.

11-4. Write the general structure of an amino acid. Why do some amino acids in Table 11-1 have two pK values and others three?

11-5. Write the stepwise acid-base reactions for the following species in water. Write the correct symbol (for example, K_{b1}) for the equilibrium constant for each reaction and find its numerical value.

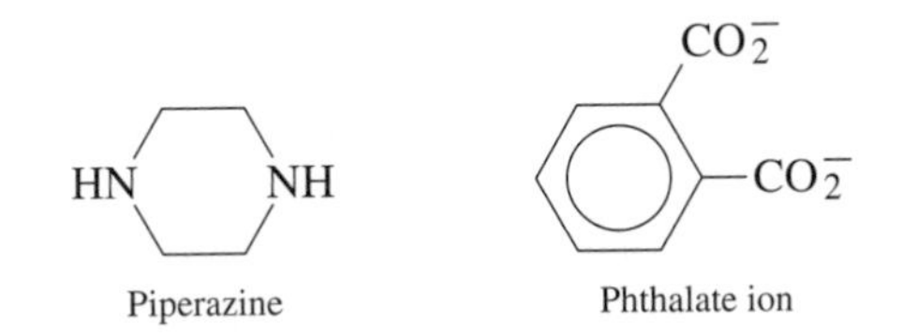

11-6. Write the K_{a2} reaction of proline and the K_{b2} reaction of the following trisodium salt.

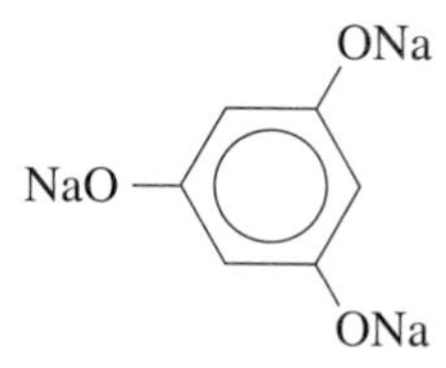

11-7. From the K_a values for citric acid in Appendix B, find K_{b1}, K_{b2}, and K_{b3} for trisodium citrate.

11-8. Write the chemical reactions whose equilibrium constants are K_{b1} and K_{b2} for the amino acid serine, and find their numerical values.

11-9. Abbreviating malonic acid, $CH_2(CO_2H)_2$, as H_2M, find the pH and concentrations of H_2M, HM^-, and M^{2-} in each of the following solutions: **(a)** 0.100 M H_2M; **(b)** 0.100 M NaHM; **(c)** 0.100 M Na_2M. For **(b),** use the approximation $[HM^-] \approx 0.100$ M.

11-10. The dibasic compound B forms BH^+ and BH_2^{2+} with $K_{b1} = 1.00 \times 10^{-5}$ and $K_{b2} = 1.00 \times 10^{-9}$. Find the pH and concentrations of B, BH^+, and BH_2^{2+} in each of the following solutions: **(a)** 0.100 M B; **(b)** 0.100 M BH^+Br^-; **(c)** 0.100 M $BH_2^{2+}(Br^-)_2$. For **(b)**, use the approximation $[BH^+] \approx 0.100$ M.

11-11. Calculate the pH of a 0.300 M solution of the dibasic compound piperazine, which we will designate B. Calculate the concentration of each form of piperazine (B, BH^+, BH_2^{2+}).

11-12. Piperazine monohydrochloride is formed when 1 mol of HCl is added to the dibasic compound piperazine:

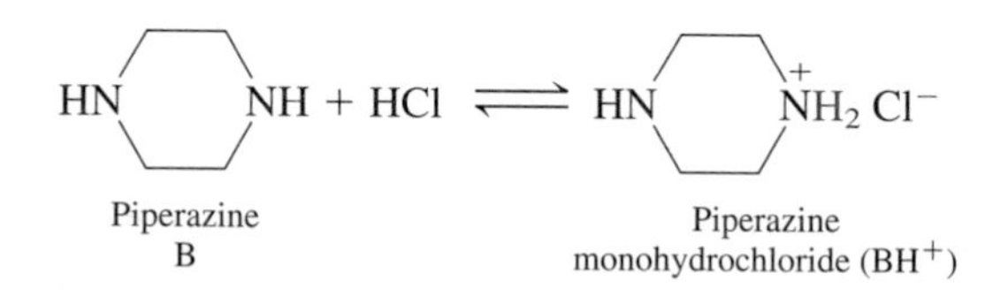

Find the pH of 0.150 M piperazine monohydrochloride and calculate the concentration of each form of piperazine. Assume $[BH^+] \approx 0.150$ M.

11-13. Draw the structure of the amino acid glutamine and satisfy yourself that it is the intermediate form of a diprotic system. Find the pH of 0.050 M glutamine.

11-14. **(a)** The diagram in the next column shows the pH range in which each species of a diprotic acid is predominant. For each of the three pH values indicated (pH = pK_1, $\frac{1}{2}(pK_1 + pK_2)$, pK_2), state which species are present in equal concentrations.

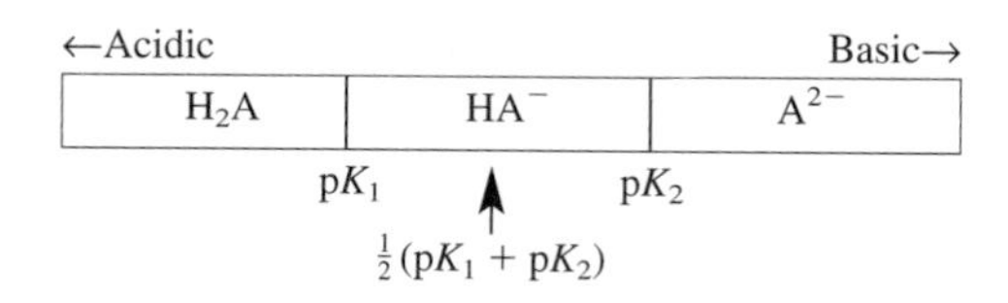

(b) Draw analogous diagrams for monoprotic and triprotic systems. Label the key pH values and state which species are present in equal concentrations at each pH.

11-15. The acid HA has $pK_a = 7.00$.

(a) Which is the principal species, HA or A^-, at pH 6.00?

(b) Which is the principal species at pH 8.00?

(c) What is the quotient $[A^-]/[HA]$ at **(i)** pH 7.00; **(ii)** at pH 6.00?

11-16. The acid H_2A has $pK_1 = 4.00$ and $pK_2 = 8.00$.

(a) At what pH is $[H_2A] = [HA^-]$?

(b) At what pH is $[HA^-] = [A^{2-}]$?

(c) Which is the principal species, H_2A, HA^-, or A^{2-} at pH 2.00?

(d) Which is the principal species at pH 6.00?

(e) Which is the principal species at pH 10.00?

11-17. Draw a diagram for phosphoric acid analogous to the one in Problem 11-14 showing the principal species as a function of pH. Label numerical values of pH at key points. State the principal species of phosphoric acid and the second most abundant species at pH 2, 3, 4, 5, 6, 7, 8, 9, 10, 11, 12, and 13.

11-18. The base B has $pK_b = 5.00$.

(a) What is the value of pK_a for the acid BH^+?

(b) At what pH is $[BH^+] = [B]$?

(c) Which is the principal species, B or BH^+, at pH 7.00?

(d) What is the quotient $[B]/[BH^+]$ at pH 12.00?

11-19. Ethylenediamine (B) is dibasic with $pK_{b1} = 4.07$ and $pK_{b2} = 7.15$.

(a) Find the two pK_a values for BH_2^{2+} and draw a diagram like the one in Problem 11-14 showing the major species in each pH region.

(b) At what pH is $[BH^+] = [B]$?

(c) At what pH is $[BH^+] = [BH_2^{2+}]$?

(d) Which is the principal species and which is the second most abundant species at pH 4, 5, 6, 7, 8, 9, 10, and 11?

(e) What is the quotient $[B]/[BH^+]$ at pH 12.00?

(f) What is the quotient $[BH_2^{2+}]/[BH^+]$ at pH 2.00?

11-20. Draw the structures of the predominant forms of glutamic acid and tyrosine at pH 9.0 and pH 10.0. What is the second most abundant species at each pH?

11-21. Calculate the pH of a 0.10 M solution of each of the following amino acids in the form drawn:

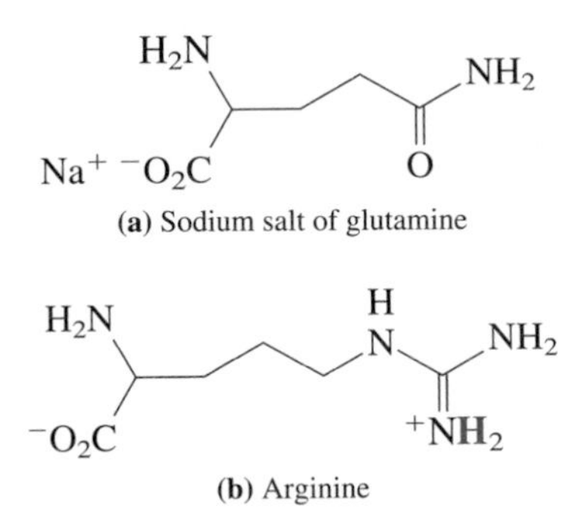

(a) Sodium salt of glutamine

(b) Arginine

11-22. Draw the structure of the predominant form of pyridoxal-5-phosphate at pH 7.00.

11-23. Find the pH and concentration of each species of arginine in 0.050 M arginine · HCl solution.

11-24. What is the charge of the predominant form of citric acid at pH 5.00?

11-25. A 100.0-mL aliquot of 0.100 M diprotic acid H_2A ($pK_1 = 4.00$, $pK_2 = 8.00$) was titrated with 1.00 M NaOH. At what volumes are the two equivalence points? Find the pH at the following volumes of base added (V_b) and sketch a graph of pH versus V_b: 0, 5.0, 10.0, 15.0, 20.0, and 22.0 mL.

11-26. The dibasic compound B ($pK_{b1} = 4.00$, $pK_{b2} = 8.00$) was titrated with 1.00 M HCl. The initial solution of B was 0.100 M and had a volume of 100.0 mL. At what volumes are the two equivalence points? Find the pH at the following volumes of acid added (V_a) and sketch a graph of pH versus V_a: 0, 5.0, 10.0, 15.0, 20.0, and 22.0 mL.

11-27. Select an indicator from Table 9-3 that would be useful for detecting each equivalence point shown in Figure 11-2 and listed below. State the color change you would observe in each case.

(a) Second equivalence point of the lowest curve

(b) First equivalence point of the upper curve

(c) First equivalence point of the middle curve

(d) Second equivalence point of the middle curve

11-28. Write two consecutive reactions that take place when 40.0 mL of 0.100 M piperazine are titrated with 0.100 M HCl and find the equivalence volumes. Find the pH at $V_a = 0$, 20.0, 40.0, 60.0, 80.0, and 100.0 mL and sketch the titration curve.

11-29. A 25.0-mL volume of 0.040 0 M phosphoric acid was titrated with 0.050 0 M tetramethylammonium hydroxide. Write the series of titration reactions and find the pH at the following volumes of added base: $V_b = 0$, 10.0, 20.0, 30.0, 40.0, and 42.0 mL. Sketch the titration curve and predict what the shape will be beyond 42.0 mL.

11-30. Write the chemical reactions (including structures of reactants and products) that take place when the amino acid histidine is titrated with perchloric acid. (Histidine has no net charge.) A solution containing 25.0 mL of 0.050 0 M histidine was titrated with 0.050 0 M $HClO_4$. List the equivalence volumes and calculate the pH at $V_a = 0$, 12.5, 25.0, and 50.0 mL.

11-31. An aqueous solution containing ~1 g of oxobutanedioic acid (FM 132.07) per 100 mL was titrated with 0.094 32 M NaOH to measure the acid molarity.

(a) Calculate the pH at the following volumes of added base: $\frac{1}{2}V_{e1}$, V_{e1}, $\frac{3}{2}V_{e1}$, V_{e2}, $1.05V_{e2}$. Sketch the titration curve.

(b) Which equivalence point would be best to use in this titration?

(c) You have the indicators erythrosine, ethyl orange, bromocresol green, bromothymol blue, thymolphthalein, and alizarin yellow. Which indicator will you use and what color change will you look for?

How Would You Do It?

11-32. *Finding the mean molecular mass of a polymer.*[5] Poly(ethylene glycol), abbreviated PEG in the scheme at the top of page 262, is a polymer with $—OCH_2CH_2—$ repeat groups capped at each end by $—CH_2CH_2OH$ (alcohol) groups. The alcohols react with the reagent PMDA shown in the scheme to make Product 1. After the reaction is complete, water converts Product 1 into Product 2, which has six carboxylic acid groups.

Standardization of PMDA: A 10.00-mL solution of PMDA in a dry organic solvent was treated with water to hydrolyze it to Product 3. Titration of the carboxylic acid groups in Product 3 required 28.98 mL of 0.294 4 M NaOH to reach a phenolphthalein end point.

Analysis of PEG: A 220.0-mg quantity of PEG was treated with 10.00 mL of standard PMDA solution. After 30 min to complete the reaction, 30 mL of water were added to convert Product 1 into Product 2 and to convert unused PMDA into Product 3. The mixture of Products 2 and 3 required 24.41 mL of 0.294 4 M NaOH to reach a phenolphthalein end point. Find the mean molecular mass of PEG and the mean number of repeat groups, n, in the formula for PEG.

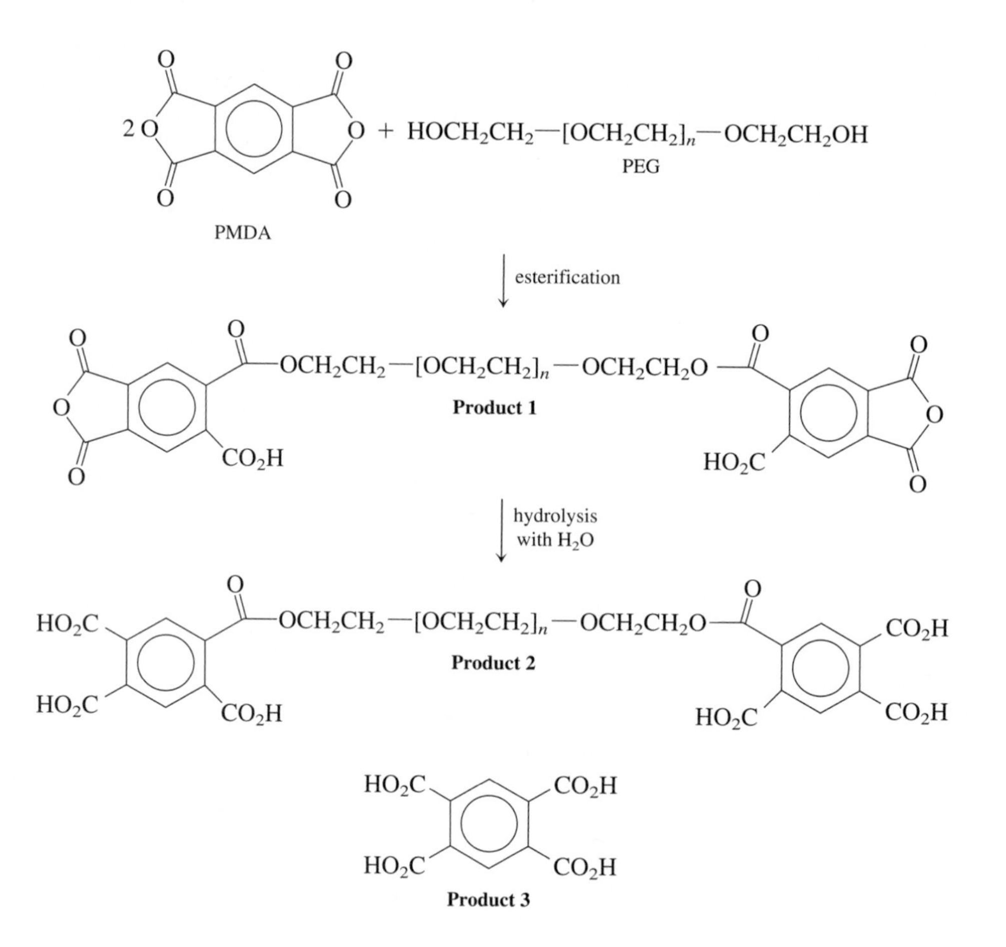

11-33. *Effect of temperature on carbonic acid acidity and the solubility of $CaCO_3$.*[6] Box 11-1 states that marine life with $CaCO_3$ shells and skeletons will be threatened with extinction in cold polar waters before that will happen in warm tropical waters. The following equilibrium constants apply to seawater at 0° and 30°C, when concentrations are measured in moles per kg of seawater and pressure is in bars:

$$CO_2(g) \rightleftharpoons CO_2(aq) \qquad \text{(A)}$$

$$K_H = \frac{[CO_2(aq)]}{P_{CO_2}} = 10^{-1.2073} \text{ mol kg}^{-1} \text{ bar}^{-1} \text{ at } 0°C$$
$$= 10^{-1.6048} \text{ mol kg}^{-1} \text{ bar}^{-1} \text{ at } 30°C$$

$$CO_2(aq) + H_2O \rightleftharpoons HCO_3^- + H^+ \qquad \text{(B)}$$

$$K_{a1} = \frac{[HCO_3^-][H^+]}{[CO_2(aq)]} = 10^{-6.1004} \text{ mol kg}^{-1} \text{ at } 0°C$$
$$= 10^{-5.8008} \text{ mol kg}^{-1} \text{ at } 30°C$$

$$HCO_3^- \rightleftharpoons CO_3^{2-} + H^+ \qquad \text{(C)}$$

$$K_{a2} = \frac{[CO_3^{2-}][H^+]}{[HCO_3^-]} = 10^{-9.3762} \text{ mol kg}^{-1} \text{ at } 0°C$$
$$= 10^{-8.8324} \text{ mol kg}^{-1} \text{ at } 30°C$$

$$CaCO_3(s, \textit{aragonite}) \rightleftharpoons Ca^{2+} + CO_3^{2-} \qquad \text{(D)}$$

$$K_{sp}^{arg} = [Ca^{2+}][CO_3^{2-}] = 10^{-6.1113} \text{ mol}^2 \text{ kg}^{-2} \text{ at } 0°C$$
$$= 10^{-6.1391} \text{ mol}^2 \text{ kg}^{-2} \text{ at } 30°C$$

$$CaCO_3(s, \textit{calcite}) \rightleftharpoons Ca^{2+} + CO_3^{2-} \qquad \text{(E)}$$

$$K_{sp}^{cal} = [Ca^{2+}][CO_3^{2-}] = 10^{-6.3652} \text{ mol}^2 \text{ kg}^{-2} \text{ at } 0°C$$
$$= 10^{-6.3713} \text{ mol}^2 \text{ kg}^{-2} \text{ at } 30°C$$

The first equilibrium constant is called K_H for Henry's law, which states that the solubility of a gas in a liquid is proportional to the pressure of the gas. Units are given to remind you what units you must use.

(a) Combine the expressions for K_H, K_{a1}, and K_{a2} to find an expression for $[CO_3^{2-}]$ in terms of P_{CO_2} and $[H^+]$.

(b) From the result of **(a)**, calculate $[CO_3^{2-}]$ (mol kg^{-1}) at P_{CO_2} = 800 μbar and pH = 7.8 at temperatures of 0° (polar ocean) and 30°C (tropical ocean). These are conditions that could be reached around the year 2100 if we continue to release CO_2 at the present rate.

(c) The concentration of Ca^{2+} in the ocean is 0.010 M. Predict whether aragonite and calcite will dissolve under the conditions in **(b)**.

Notes and References

1. Story from J. Gorman in *Science News*, 9 September 2000, p. 165.

2. Global climate change research explorer: http://www/exploratorium.com/climate/index.html.

3. P. D. Thacker, "Global Warming's Other Effects on the Oceans," *Environ. Sci. Technol.* **2005**, *39*, 10A.

4. R. E. Weston, Jr. "Climate Change and Its Effect on Coral Reefs," *J. Chem. Ed.* **2000**, *77*, 1574.

5. The actual procedure is slightly more complicated than described in this problem. Instructions for a student experiment based on the reaction scheme are given by K. R. Williams and U. R. Bernier, *J. Chem. Ed.* **1994**, *71*, 265.

6. W. Stumm and J. J. Morgan, *Aquatic Chemistry*, 3rd ed. (New York: Wiley, 1996), pp. 343–348; F. J. Millero, "Thermodynamics of the Carbon Dioxide System in the Oceans," *Geochim. Cosmochim. Acta* **1995**, *59*, 661.

Chemical Equilibrium in the Environment

Paper mill on the Potomac River near Westernport, Maryland, neutralizes acid mine drainage in the water. Upstream of the mill, the river is acidic and lifeless; below the mill, the river teems with life [C. Dalpra, Potomac River Basin Commission.]

Part of the North Branch of the Potomac River runs crystal clear through the scenic Appalachian Mountains; but it is lifeless—a victim of acid drainage from abandoned coal mines. As the river passes a paper mill and wastewater treatment plant near Westernport, Maryland, the pH rises from a lethal value of 4.5 to a neutral value of 7.2, at which fish and plants thrive. This fortunate accident comes about when the calcium carbonate by-product from papermaking exits the paper mill and reacts with massive quantities of carbon dioxide from bacterial respiration at the sewage treatment plant. The resulting soluble bicarbonate neutralizes the acidic river and restores life downstream of the plant. In the absence of CO_2, solid $CaCO_3$ would be trapped at the treatment plant and would never enter the river.

$$\underset{\text{trapped at treatment plant}}{\underset{\text{Calcium carbonate}}{CaCO_3(s)}} + CO_2(aq) + H_2O(l) \rightleftharpoons \underset{\text{enters river and neutralizes acid}}{\underset{\text{Dissolved calcium bicarbonate}}{Ca^{2+}(aq) + 2HCO_3^-(aq)}}$$

$$HCO_3^-(aq) + \underset{\text{Acid from river}}{H^+(aq)} \xrightarrow{\text{neutralization}} CO_2(g)\uparrow + H_2O(l)$$

We call these two reactions *coupled equilibria.* Consumption of bicarbonate in the second reaction drives the first reaction to make more product.

CHAPTER 12

A Deeper Look at Chemical Equilibrium

We now pause to look more carefully at chemical equilibrium. This chapter is optional in that later chapters do not depend on it in any critical way. However, many instructors consider the treatment of equilibrium in this chapter to be a fundamental component of an education in chemistry.

12-1 The Effect of Ionic Strength on Solubility of Salts

When slightly soluble lead(II) iodide dissolves in pure water, many species are formed:

$$PbI_3^- \ (0.01\%)$$
$$\Updownarrow I^-$$
$$PbI_2(s) \rightleftharpoons PbI_2(aq) \ (0.8\%) \rightleftharpoons Pb^{2+} + 2I^- \ (81\%)$$
$$PbI_2(aq) \rightleftharpoons PbI^+ + I^- \ (18\%) \qquad Pb^{2+} \overset{H_2O}{\rightleftharpoons} PbOH^+ + H^+ \ (0.3\%)$$

Approximately 81% of the lead is found as Pb^{2+}, 18% is PbI^+, 0.8% is $PbI_2(aq)$, 0.3% is $PbOH^+$, and 0.01% is PbI_3^-. The *solubility product* is the equilibrium constant for the reaction $PbI_2(s) \rightleftharpoons Pb^{2+} + 2I^-$, which tells only part of the story.

Now a funny thing happens when the "inert" salt KNO_3 is added to the saturated PbI_2 solution. (By "inert" we mean that there is no chemical reaction between K^+ or NO_3^- with any of the lead iodide species.) As more KNO_3 is added, the total concentration of dissolved iodine increases, as shown in Figure 12-1. (Dissolved iodine includes free iodide and iodide attached to lead.) It turns out that adding any inert salt,

Composition was computed from the following equilibria with activity coefficients:[1]

$$PbI_2(s) \rightleftharpoons Pb^{2+} + 2I^- \qquad K_{sp} = 7.9 \times 10^{-9}$$
$$Pb^{2+} + I^- \rightleftharpoons PbI^+ \qquad K \equiv \beta_1 = 1.0 \times 10^2$$
$$Pb^{2+} + 2I^- \rightleftharpoons PbI_2(aq) \qquad K \equiv \beta_2 = 1.6 \times 10^3$$
$$Pb^{2+} + 3I^- \rightleftharpoons PbI_3^- \qquad K \equiv \beta_3 = 7.9 \times 10^3$$
$$Pb^{2+} + H_2O \rightleftharpoons PbOH^+ + H^+ \qquad K_a = 2.5 \times 10^{-8}$$

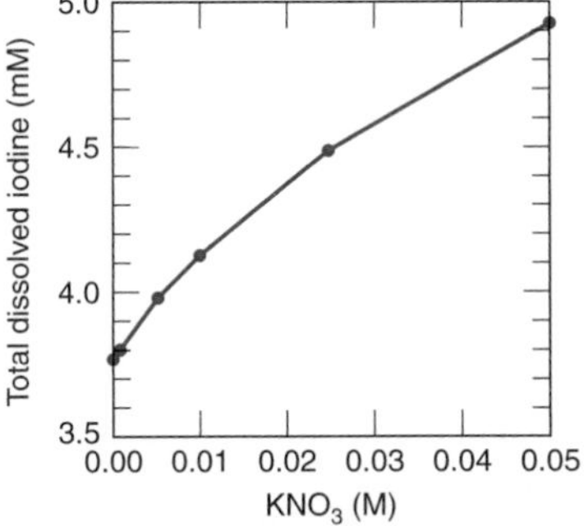

Figure 12-1 Observed effect of KNO_3 on the solubility of PbI_2. [Data from D. B. Green, G. Rechtsteiner, and A. Honodel, *J. Chem. Ed.* **1996**, *73*, 789.]

such as KNO_3, to a sparingly soluble salt, such as PbI_2, increases the solubility of the sparingly soluble substance. Why does solubility increase when salts are added?

The Explanation

An anion is surrounded by more cations than anions. A cation is surrounded by more anions than cations.

Consider one particular Pb^{2+} ion and one particular I^- ion in the solution. The I^- ion is surrounded by cations (K^+, Pb^{2+}) and anions (NO_3^-, I^-) in the solution. However, the typical anion has more cations than anions near it because cations are attracted but anions are repelled. These interactions create a region of net positive charge around any particular anion. We call this region the **ionic atmosphere** (Figure 12-2). Ions continually diffuse into and out of the ionic atmosphere. The net charge in the atmosphere, averaged over time, is less than the charge of the anion at the center. Similarly, an atmosphere of negative charge surrounds any cation in solution.

The ionic atmosphere *attenuates* (decreases) the attraction between ions in solution. The cation plus its negative atmosphere has less positive charge than the cation alone. The anion plus its ionic atmosphere has less negative charge than the anion alone. The net attraction between the cation with its ionic atmosphere and the anion with its ionic atmosphere is smaller than it would be between pure cation and anion in the absence of ionic atmospheres. *The higher the concentration of ions in a solution, the higher the charge in the ionic atmosphere. Each ion-plus-atmosphere contains less net charge and there is less attraction between any particular cation and anion.*

Increasing the concentration of ions in a solution therefore reduces the attraction between any particular Pb^{2+} ion and any I^- ion, relative to their attraction to each other in pure water. The effect is to reduce their tendency to come together, thereby increasing the solubility of PbI_2.

Ion dissociation is increased by increasing the ionic strength of the solution.

Increasing the concentration of ions in a solution promotes dissociation of ions. Thus, each of the following reactions is driven to the right if KNO_3 is added (Demonstration 12-1):

$$Fe(SCN)^{2+} \rightleftharpoons Fe^{3+} + \underset{\text{Thiocyanate}}{SCN^-}$$

$$\underset{\text{Phenol}}{C_6H_5\text{—}OH} \rightleftharpoons \underset{\text{Phenolate}}{C_6H_5\text{—}O^-} + H^+$$

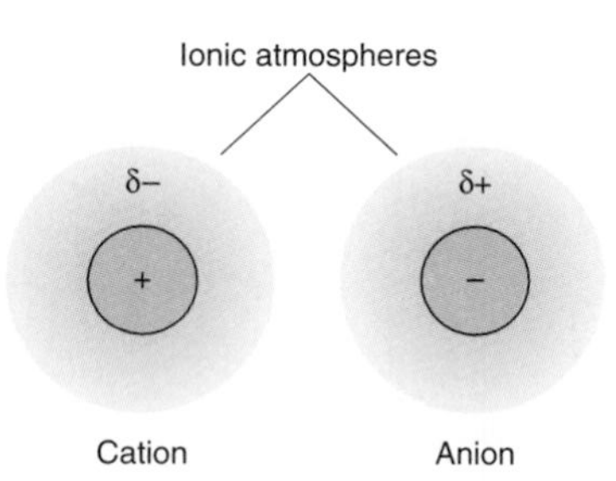

Figure 12-2 An ionic atmosphere, shown as a spherical cloud of charge δ+ or δ−, surrounds each ion in solution. The charge of the atmosphere is less than the charge of the central ion. The greater the ionic strength of the solution, the greater the charge in each ionic atmosphere.

What Do We Mean by "Ionic Strength"?

Ionic strength, μ, is a measure of the total concentration of ions in solution. The more highly charged an ion, the more it is counted.

Ionic strength:

$$\mu = \tfrac{1}{2}(c_1 z_1^2 + c_2 z_2^2 + \ldots) = \tfrac{1}{2}\sum_i c_i z_i^2 \qquad (12\text{-}1)$$

where c_i is the concentration of the ith species and z_i is its charge. The sum extends over *all* ions in solution.

Example Calculation of Ionic Strength

Find the ionic strength of **(a)** 0.10 M $NaNO_3$; **(b)** 0.010 M Na_2SO_4; and **(c)** 0.020 M KBr plus 0.010 M Na_2SO_4.

Demonstration 12-1 Effect of Ionic Strength on Ion Dissociation[2]

This experiment demonstrates the effect of ionic strength on the dissociation of the red iron(III) thiocyanate complex:

$$\underset{\text{Red}}{Fe(SCN)^{2+}} \rightleftharpoons \underset{\text{Pale yellow}}{Fe^{3+}} + \underset{\text{Colorless}}{SCN^-}$$

Prepare a solution of 1 mM $FeCl_3$ by dissolving 0.27 g of $FeCl_3 \cdot 6H_2O$ in 1 L of water containing 3 drops of 15 M (concentrated) HNO_3. The acid slows the precipitation of $Fe(OH)_3$, which occurs in a few days and necessitates the preparation of fresh solution for this demonstration.

To demonstrate the effect of ionic strength on the dissociation reaction, mix 300 mL of the 1 mM $FeCl_3$ solution with 300 mL of 1.5 mM NH_4SCN or KSCN. Divide the pale red solution into two equal portions and add 12 g of KNO_3 to one of them to increase the ionic strength to 0.4 M. As the KNO_3 dissolves, the red $Fe(SCN)^{2+}$ complex dissociates and the color fades noticeably (Color Plate 6).

Add a few crystals of NH_4SCN or KSCN to either solution to drive the reaction toward formation of $Fe(SCN)^{2+}$, thereby intensifying the red color. This reaction demonstrates Le Châtelier's principle—adding a product creates more reactant.

SOLUTION

(a) $\mu = \frac{1}{2}\{[Na^+]\cdot(+1)^2 + [NO_3^-]\cdot(-1)^2\}$
$= \frac{1}{2}\{(0.10\cdot 1) + (0.10\cdot 1)\} = 0.10\text{ M}$

(b) $\mu = \frac{1}{2}\{[Na^+]\cdot(+1)^2 + [SO_4^{2-}]\cdot(-2)^2\}$
$= \frac{1}{2}\{(0.020\cdot 1) + (0.010\cdot 4)\} = 0.030\text{ M}$

Note that $[Na^+] = 0.020$ M because there are two moles of Na^+ per mole of Na_2SO_4.

(c) $\mu = \frac{1}{2}\{[K^+]\cdot(+1)^2 + [Br^-]\cdot(-1)^2 + [Na^+]\cdot(+1)^2 + [SO_4^{2-}]\cdot(-2)^2\}$
$= \frac{1}{2}\{(0.020\cdot 1) + (0.020\cdot 1) + (0.020\cdot 1) + (0.010\cdot 4)\} = 0.050\text{ M}$

Test Yourself Find the ionic strength of 1.0 mM $Ca(ClO_4)_2$. (**Answer**: 3.0 mM)

Electrolyte	Molarity	Ionic strength
1:1	M	M
2:1	M	3M
3:1	M	6M
2:2	M	4M

$NaNO_3$ is called a 1:1 electrolyte because the cation and the anion both have a charge of 1. For 1:1 electrolytes, the ionic strength equals the molarity. For any other stoichiometry (such as the 2:1 electrolyte Na_2SO_4), the ionic strength is greater than the molarity.

Ask Yourself

12-A. (a) From the solubility product of PbI_2, calculate the expected concentration of dissolved iodine in a saturated solution of PbI_2. Why is your result different from the experimental observation in Figure 12-1, and why does the concentration of dissolved iodine increase with increasing KNO_3 concentration?
(b) If PbI_2 dissolved to give 1.0 mM Pb^{2+} plus 2.0 mM I^- and if there were no other ionic species in the solution, what would be the ionic strength?

12-2 Activity Coefficients

Until now, we have written the equilibrium for the reaction $aA + bB \rightleftharpoons cC + dD$ in the form $K = [C]^c[D]^d/[A]^a[B]^b$. This equilibrium constant does not account for any effect of ionic strength on the chemical reaction. To account for ionic strength, concentrations are replaced by **activities**:

Activity of C: $$\mathcal{A}_C = [C]\gamma_C \qquad (12\text{-}2)$$

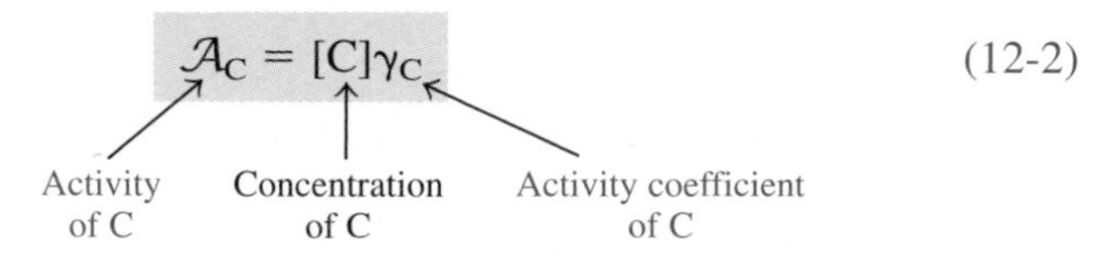

Do not confuse the terms *activity* and *activity coefficient.*

The activity of species C is its concentration multiplied by its **activity coefficient**. The activity coefficient depends on ionic strength. If there were no effect of ionic strength on the chemical reaction, the activity coefficient would be 1. The correct form of the equilibrium constant for the reaction $aA + bB \rightleftharpoons cC + dD$ is

This is the "real" equilibrium constant.

General form of equilibrium constant: $$K = \frac{\mathcal{A}_C^c \mathcal{A}_D^d}{\mathcal{A}_A^a \mathcal{A}_B^b} = \frac{[C]^c\gamma_C^c [D]^d\gamma_D^d}{[A]^a\gamma_A^a [B]^b\gamma_B^b} \qquad (12\text{-}3)$$

For the reaction $PbI_2(s) \rightleftharpoons Pb^{2+} + 2I^-$, the equilibrium constant is

$$K_{sp} = \mathcal{A}_{Pb^{2+}}\mathcal{A}_{I^-}^2 = [Pb^{2+}]\gamma_{Pb^{2+}}[I^-]^2\gamma_{I^-}^2 \qquad (12\text{-}4)$$

If the concentrations of Pb^{2+} and I^- are to *increase* when a second salt is added to increase ionic strength, the activity coefficients must *decrease* with increasing ionic strength. Conversely, at low ionic strength, activity coefficients approach unity.

Activity Coefficients of Ions

Detailed consideration of the ionic atmosphere model leads to the **extended Debye-Hückel equation**, relating activity coefficients to ionic strength:

Extended Debye-Hückel equation: $$\log \gamma = \frac{-0.51z^2\sqrt{\mu}}{1 + (\alpha\sqrt{\mu}/305)} \quad \text{(at 25°C)} \qquad (12\text{-}5)$$

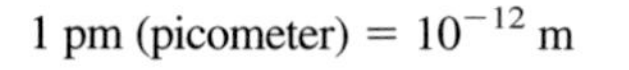
1 pm (picometer) = 10^{-12} m

In Equation 12-5, γ is the activity coefficient of an ion of charge $\pm z$ and size α (picometers, pm) in an aqueous solution of ionic strength μ. Table 12-1 gives sizes and activity coefficients for many ions.

The ion size α in Table 12-1 is an empirical parameter that provides agreement between measured activity coefficients and ionic strength up to $\mu \approx 0.1$ M. In theory, α is the diameter of the *hydrated ion*, which includes the ion and its tightly bound sheath of water molecules. Cations attract the negatively charged oxygen atom of H_2O and anions attract the positively charged H atoms.

H₂O ----- Li⁺ (200 pm)   ⁻Cl ----- H—O—H (220 pm; 330 pm)

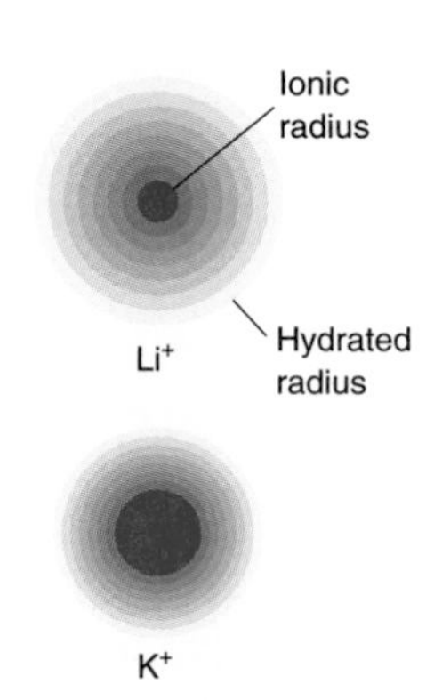

Figure 12-3 The *smaller* Li^+ ion binds water molecules more tightly than does the larger K^+ ion, so Li^+ has the *larger* hydrated diameter.

A small or more highly charged ion binds water more tightly and has a *larger* hydrated diameter than does a larger or less charged ion (Figure 12-3).

Sizes in Table 12-1 cannot be taken literally. For example, the diameter of Cs^+ ion in crystals is 340 pm. The hydrated Cs^+ ion in solution must be larger than the unhydrated ion in the crystal, but the size of Cs^+ given in Table 12-1 is only 250 pm. Even though ion sizes in Table 12-1 are empirical parameters, trends among sizes

TABLE 12-1 Activity coefficients for aqueous solutions at 25°C

Ion	Ion size (α, pm)	Ionic strength (μ, M)				
		0.001	0.005	0.01	0.05	0.1
Charge = ±1						
H^+	900	0.967	0.933	0.914	0.86	0.83
$(C_6H_5)_2CHCO_2^-$, $(C_3H_7)_4N^+$	800	0.966	0.931	0.912	0.85	0.82
$(O_2N)_3C_6H_2O^-$, $(C_3H_7)_3NH^+$, $CH_3OC_6H_4CO_2^-$	700	0.965	0.930	0.909	0.845	0.81
Li^+, $C_6H_5CO_2^-$, $HOC_6H_4CO_2^-$, $ClC_6H_4CO_2^-$, $C_6H_5CH_2CO_2^-$, $CH_2{=}CHCH_2CO_2^-$, $(CH_3)_2CHCH_2CO_2^-$, $(CH_3CH_2)_4N^+$, $(C_3H_7)_2NH_2^+$	600	0.965	0.929	0.907	0.835	0.80
$Cl_2CHCO_2^-$, $Cl_3CCO_2^-$, $(CH_3CH_2)_3NH^+$, $(C_3H_7)NH_3^+$	500	0.964	0.928	0.904	0.83	0.79
Na^+, $CdCl^+$, ClO_2^-, IO_3^-, HCO_3^-, $H_2PO_4^-$, HSO_3^-, $H_2AsO_4^-$, $Co(NH_3)_4(NO_2)_2^+$, $CH_3CO_2^-$, $ClCH_2CO_2^-$, $(CH_3)_4N^+$, $(CH_3CH_2)_2NH_2^+$, $H_2NCH_2CO_2^-$	450	0.964	0.928	0.902	0.82	0.775
$^+H_3NCH_2CO_2H$, $(CH_3)_3NH^+$, $CH_3CH_2NH_3^+$	400	0.964	0.927	0.901	0.815	0.77
OH^-, F^-, SCN^-, OCN^-, HS^-, ClO_3^-, ClO_4^-, BrO_3^-, IO_4^-, MnO_4^-, HCO_2^-, $H_2citrate^-$, $CH_3NH_3^+$, $(CH_3)_2NH_2^+$	350	0.964	0.926	0.900	0.81	0.76
K^+, Cl^-, Br^-, I^-, CN^-, NO_2^-, NO_3^-	300	0.964	0.925	0.899	0.805	0.755
Rb^+, Cs^+, NH_4^+, Tl^+, Ag^+	250	0.964	0.924	0.898	0.80	0.75
Charge = ±2						
Mg^{2+}, Be^{2+}	800	0.872	0.755	0.69	0.52	0.45
$CH_2(CH_2CH_2CO_2^-)_2$, $(CH_2CH_2CH_2CO_2^-)_2$	700	0.872	0.755	0.685	0.50	0.425
Ca^{2+}, Cu^{2+}, Zn^{2+}, Sn^{2+}, Mn^{2+}, Fe^{2+}, Ni^{2+}, Co^{2+}, $C_6H_4(CO_2^-)_2$, $H_2C(CH_2CO_2^-)_2$, $(CH_2CH_2CO_2^-)_2$	600	0.870	0.749	0.675	0.485	0.405
Sr^{2+}, Ba^{2+}, Cd^{2+}, Hg^{2+}, S^{2-}, $S_2O_4^{2-}$, WO_4^{2-}, $H_2C(CO_2^-)_2$, $(CH_2CO_2^-)_2$, $(CHOHCO_2^-)_2$	500	0.868	0.744	0.67	0.465	0.38
Pb^{2+}, CO_3^{2-}, SO_3^{2-}, MoO_4^{2-}, $Co(NH_3)_5Cl^{2+}$, $Fe(CN)_5NO^{2-}$, $C_2O_4^{2-}$, $Hcitrate^{2-}$	450	0.867	0.742	0.665	0.455	0.37
Hg_2^{2+}, SO_4^{2-}, $S_2O_3^{2-}$, $S_2O_6^{2-}$, $S_2O_8^{2-}$, SeO_4^{2-}, CrO_4^{2-}, HPO_4^{2-}	400	0.867	0.740	0.660	0.445	0.355
Charge = ±3						
Al^{3+}, Fe^{3+}, Cr^{3+}, Sc^{3+}, Y^{3-}, In^{3+}, lanthanides[a]	900	0.738	0.54	0.445	0.245	0.18
$citrate^{3-}$	500	0.728	0.51	0.405	0.18	0.115
PO_4^{3-}, $Fe(CN)_6^{3-}$, $Cr(NH)_6^{3+}$, $Co(NH_3)_6^{3+}$, $Co(NH_3)_5H_2O^{3+}$	400	0.725	0.505	0.395	0.16	0.095
Charge = ±4						
Th^{4+}, Zr^{4+}, Ce^{4+}, Sn^{4+}	1 100	0.588	0.35	0.255	0.10	0.065
$Fe(CN)_6^{4-}$	500	0.57	0.31	0.20	0.048	0.021

a. Lanthanides are elements 57–71 in the periodic table.

SOURCE: J. Kielland, *J. Am. Chem. Soc.* **1937**, *59*, 1675.

are sensible. Small, highly charged ions bind solvent more tightly and have larger effective sizes than do larger or less highly charged ions. For example, the order of sizes in Table 12-1 is $Li^+ > Na^+ > K^+ > Rb^+$, even though crystallographic radii are $Li^+ < Na^+ < K^+ < Rb^+$.

In Table 12-1, ions of the same size and the same charge appear in the same group and they have the same activity coefficients. For example, Ba^{2+} and succinate [$^-O_2CCH_2CH_2CO_2^-$, listed as $(CH_2CO_2^-)_2$] each have a size of 500 pm and are listed among the charge = ±2 ions. In a solution with an ionic strength of 0.001 M, both ions have an activity coefficient of 0.868.

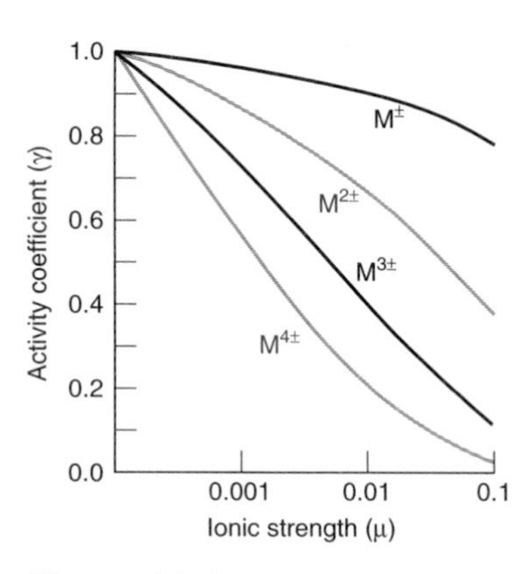

Figure 12-4 Activity coefficients for differently charged ions with a constant size α = 500 pm. At zero ionic strength, $\gamma = 1$. The greater the charge of the ion, the more rapidly γ decreases as ionic strength increases. Note that the abscissa is logarithmic.

Effect of Ionic Strength, Ion Charge, and Ion Size on the Activity Coefficient

Over the range of ionic strengths from 0 to 0.1 M, we find that

1. As ionic strength increases, the activity coefficient decreases (Figure 12-4). The activity coefficient (γ) approaches unity as the ionic strength (μ) approaches zero.
2. As the charge of the ion increases, the departure of its activity coefficient from unity increases. Activity corrections in Figure 12-4 are much more important for an ion with a charge of ±3 than for one with a charge of ±1. Note that activity coefficients in Table 12-1 depend on the magnitude of the charge but not on its sign.
3. The smaller the size (α) of the ion, the more important activity effects become.

Example **Using Table 12-1**

Find the activity coefficient of Mg^{2+} in a solution of 3.3 mM $Mg(NO_3)_2$.

SOLUTION The ionic strength is

$$\mu = \tfrac{1}{2}\{[Mg^{2+}] \cdot 2^2 + [NO_3^-] \cdot (-1)^2\}$$
$$= \tfrac{1}{2}\{(0.003\,3) \cdot 4 + (0.006\,6) \cdot 1\} = 0.010 \text{ M}$$

In Table 12-1, Mg^{2+} is listed under the charge ±2 and has a size of 800 pm. When $\mu = 0.010$ M, $\gamma = 0.69$.

Test Yourself Find the activity coefficient of SO_4^{2-} in 1.25 mM $MgSO_4$. (**Answer:** $\mu = 0.005$ M, $\gamma = 0.740$)

How to Interpolate

Interpolation is the estimation of a number that lies *between* two values in a table. Estimating a number that lies *beyond* values in a table is called *extrapolation*.

If you need to find an activity coefficient for an ionic strength that is between values in Table 12-1, you can use Equation 12-5. Alternatively, in the absence of a spreadsheet, it is usually easier to *interpolate* than to use Equation 12-5. In *linear interpolation*, we assume that values between two entries of a table lie on a straight line. For example, consider a table in which $y = 0.67$ when $x = 10$ and $y = 0.83$ when $x = 20$. What is the value of y when $x = 16$?

x value:	10	16	20
y value:	0.67	?	0.83

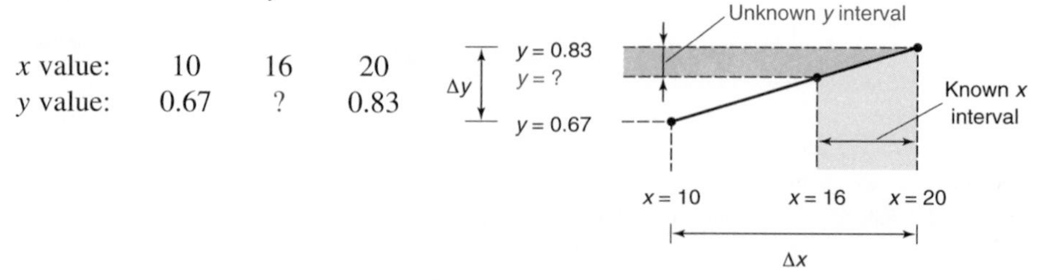

To interpolate a value of y, we can set up a proportion:

Interpolation:
$$\frac{\text{unknown } y \text{ interval}}{\Delta y} = \frac{\text{known } x \text{ interval}}{\Delta x} \tag{12-6}$$

This calculation is equivalent to saying

16 is 60% of the way from 10 to 20, so the y value will be 60% of the way from 0.67 to 0.83.

$$\frac{0.83 - y}{0.83 - 0.67} = \frac{20 - 16}{20 - 10}$$
$$\Rightarrow y = 0.76_6$$

For $x = 16$, our estimate of y is 0.76_6.

Example **Interpolating Activity Coefficients**

Calculate the activity coefficient of H^+ when $\mu = 0.025$ M.

SOLUTION H^+ is the first entry in Table 12-1.

	$\mu = 0.01$	0.025	0.05
H^+:	$\gamma = 0.914$	?	0.86

The linear interpolation is set up as follows:

$$\frac{\text{unknown } \gamma \text{ interval}}{\Delta \gamma} = \frac{\text{known } \mu \text{ interval}}{\Delta \mu}$$
$$\frac{0.86 - \gamma}{0.86 - 0.914} = \frac{0.05 - 0.025}{0.05 - 0.01}$$
$$\Rightarrow \gamma = 0.89_4$$

Another Solution A more accurate and slightly more tedious calculation uses Equation 12-5, with the size $\alpha = 900$ pm listed for H^+ in Table 12-1:

$$\log \gamma_{H^+} = \frac{(-0.51)(1^2)\sqrt{0.025}}{1 + (900\sqrt{0.025}/305)} = -0.054_{98}$$
$$\gamma_{H^+} = 10^{-0.054_{98}} = 0.88_1$$

The difference between this calculated value and the interpolated value is less than 2%. Equation 12-5 is easy to implement in a spreadsheet.

Test Yourself Find the activity coefficient of Hg^{2+} when $\mu = 0.06$ M by interpolation and with Equation 12-5. (**Answer:** 0.448, 0.440)

Activity Coefficients of Nonionic Compounds

Neutral molecules, such as benzene and acetic acid, have no ionic atmosphere because they have no charge. To a good approximation, their activity coefficients are unity when the ionic strength is less than 0.1 M. In this text, we set $\gamma = 1$ for neutral molecules. That is, *the activity of a neutral molecule will be assumed to be equal to its concentration.*

For neutral species, $\mathcal{A}_C \approx [C]$.

For gases such as H_2, the activity is written

$$\mathcal{A}_{H_2} = P_{H_2}\gamma_{H_2}$$

where P_{H_2} is pressure in bars. The activity of a gas is called its *fugacity*, and the activity coefficient is called the *fugacity coefficient*. Deviation of gas behavior from the ideal gas law results in deviation of the fugacity coefficient from unity. For most gases at or below 1 bar, $\gamma \approx 1$. Therefore, for all gases, *we will set* $\mathcal{A} = P$ *(bar)*.

For gases, $\mathcal{A} \approx P$ (bar).

High Ionic Strengths

At high ionic strength, γ increases with increasing μ.

The extended Debye-Hückel equation 12-5 predicts that the activity coefficient, γ, will decrease as ionic strength, μ, increases. In fact, above an ionic strength of approximately 1 M, activity coefficients of most ions *increase*, as shown for H^+ in $NaClO_4$ solutions in Figure 12-5. We should not be too surprised that activity coefficients in concentrated salt solutions are not the same as those in dilute aqueous solution. The "solvent" is no longer just H_2O but, rather, a mixture of H_2O and $NaClO_4$. Hereafter, we limit our attention to dilute aqueous solutions in which Equation 12-5 applies.

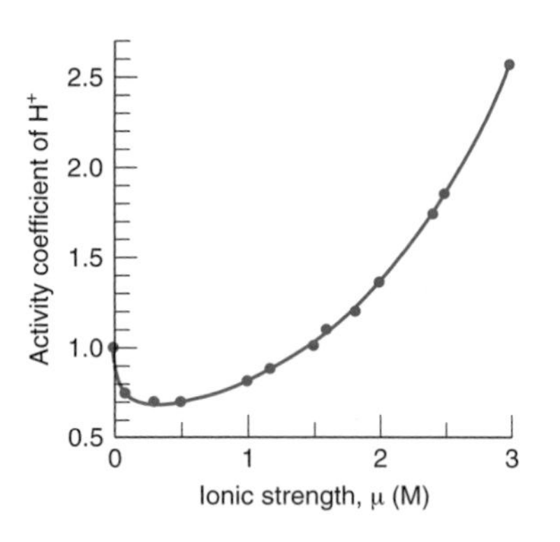

Figure 12-5 Activity coefficient of H^+ in solutions containing 0.010 0 M $HClO_4$ and various amounts of $NaClO_4$. [Data derived from L. Pezza, M. Molina, M. de Moraes, C. B. Melios, and J. O. Tognolli, *Talanta* **1996**, *43*, 1689.]

Example **A Better Estimate of the Solubility of PbI_2**

From the solubility product alone, you estimated in Ask Yourself 12-A that the concentration of dissolved iodine in a saturated solution of PbI_2 is 2.5 mM.

$$PbI_2(s) \overset{K_{sp}}{\rightleftharpoons} \underset{x}{Pb^{2+}} + \underset{2x}{2I^-}$$

$$x(2x)^2 = K_{sp} = 7.9 \times 10^{-9} \Rightarrow x = [Pb^{2+}] = 1.2_5 \times 10^{-3}\ M$$
$$2x = [I^-] = 2.5_0 \times 10^{-3}\ M$$

The observed concentration of dissolved iodine in the absence of KNO_3 in Figure 12-1 is 3.8 mM, which is 50% higher than the predicted concentration of I^- of 2.5 mM. The Pb^{2+} and I^- ions increase the ionic strength of the solution and therefore increase the solubility of PbI_2. Use activity coefficients to estimate the increased solubility.

SOLUTION The ionic strength of the solution is

$$\mu = \tfrac{1}{2}\{[Pb^{2+}] \cdot (+2)^2 + [I^-] \cdot (-1)^2\}$$
$$= \tfrac{1}{2}\{(0.001\ 2_5 \cdot 4) + (0.002\ 5_0 \cdot 1)\} = 0.003\ 7_5\ M$$

If $\mu = 0.003\ 7_5$ M, interpolation in Table 12-1 tells us that the activity coefficients are $\gamma_{Pb^{2+}} = 0.781$ and $\gamma_{I^-} = 0.937$. A better estimate of the solubility of PbI_2 is obtained by using these activity coefficients in the solubility product:

Successive approximations:
Use one approximation to find a better approximation. Repeat the process until successive approximations are in close agreement.

$$K_{sp} = [Pb^{2+}]\gamma_{Pb^{2+}}[I^-]^2\gamma_{I^-}^2 = (x_2)(0.781)(2x_2)^2(0.937)^2$$
$$\Rightarrow x_2 = [Pb^{2+}] = 1.4_2\ mM \quad \text{and} \quad [I^-] = 2x_2 = 2.8_4\ mM$$

We wrote a subscript 2 in x_2 to indicate that it is our second approximation. The new concentrations of Pb^{2+} and I^- give a new estimate of the ionic strength, $\mu = 0.004\ 2_6$ M, which gives new activity coefficients: $\gamma_{Pb^{2+}} = 0.765$ and $\gamma_{I^-} = 0.932$. Repeating the solubility computation gives

$$K_{sp} = [Pb^{2+}]\gamma_{Pb^{2+}}[I^-]^2\gamma_{I^-}^2 = (x_3)(0.765)(2x_3)^2(0.932)^2$$
$$\Rightarrow x_3 = [Pb^{2+}] = 1.4_4\ mM \quad \text{and} \quad [I^-] = 2x_3 = 2.8_8\ mM$$

This third estimate is only slightly different from the second estimate. With activity coefficients, we estimate $[I^-] = 2.9$ mM instead of 2.5 mM calculated without activity coefficients. The remaining difference between 2.9 mM and the observed solubility of 3.8 mM is that we have not accounted for other species (PbI^+ and $PbI_2(aq)$) in the solution.

Test Yourself This is not a short question. Use the procedure from this example to calculate the solubility of LiF (K_{sp} = 0.001 7) in water. (**Answer:** 1st iteration: 0.041 M; 2nd iteration: 0.049 M; 3rd iteration: 0.050 M)

The Real Definition of pH

The pH measured by a pH electrode is not the negative logarithm of the hydrogen ion *concentration*. The ideal quantity that we measure with the pH electrode is the negative logarithm of the hydrogen ion *activity*.

Real definition of pH:

$$\text{pH} = -\log \mathcal{A}_{H^+} = -\log([H^+]\gamma_{H^+}) \qquad (12\text{-}7)$$

A pH electrode measures $-\log \mathcal{A}_{H^+}$.

Example **Effect of Salt on Water Dissociation**

Consider the equilibrium $H_2O \rightleftharpoons H^+ + OH^-$, for which $K_w = \mathcal{A}_{H^+}\mathcal{A}_{OH^-} = [H^+]\gamma_{H^+}[OH^-]\gamma_{OH^-}$. What is the pH of pure water and of 0.1 M NaCl?

SOLUTION In pure water, the ionic strength is so low that the activity coefficients are close to 1. Putting $\gamma_{H^+} = \gamma_{OH^-} = 1$ into the equilibrium constant gives

$$\underset{}{H_2O} \rightleftharpoons \underset{x}{H^+} + \underset{x}{OH^-} \quad \Rightarrow \quad K_w = [H^+]\gamma_{H^+}[OH^-]\gamma_{OH^-}$$

$$1.0 \times 10^{-14} = (x)(1)(x)(1) \quad \Rightarrow \quad x = 1.0 \times 10^{-7}\text{ M}$$

$$\text{pH} = -\log([H^+]\gamma_{H^+}) = -\log([1.0 \times 10^{-7}](1)) = 7.00$$

Well, it is no surprise that the pH of pure water is 7.00.

In 0.1 M NaCl, the ionic strength is 0.1 M and the activity coefficients in Table 12-1 are $\gamma_{H^+} = 0.83$ and $\gamma_{OH^-} = 0.76$. Putting these values into the equilibrium constant gives

$$K_w = [H^+]\gamma_{H^+}[OH^-]\gamma_{OH^-}$$

$$1.0 \times 10^{-14} = (x)(0.83)(x)(0.76) \quad \Rightarrow \quad x = 1.2_6 \times 10^{-7}\text{ M}$$

The concentrations of H^+ and OH^- increase by 26% when 0.1 M NaCl is added to the water. This result is consistent with the notion that inert salts increase ion dissociation. However, the pH is not changed very much:

$$\text{pH} = -\log([H^+]\gamma_{H^+}) = -\log([1.2_6 \times 10^{-7}](0.83)) = 6.98$$

Test Yourself Find $[H^+]$ and the pH of 0.05 M $LiNO_3$. (**Answer:** $1.2_0 \times 10^{-7}$ M, 6.99)

Ask Yourself

12-B. **(a)** Using activity coefficients, calculate the solubility of $HgBr_2$ in water. By solubility, we mean the concentration of dissolved Hg^{2+}. The solubility is so small that the ionic strength is approximately 0 and the activity coefficients are approximately 1.

(**b**) Considering only the equilibrium $HgBr_2(s) \rightleftharpoons Hg^{2+} + 2Br^-$, and including activity coefficients, calculate the solubility of $HgBr_2$ in 0.050 M NaBr. The ionic strength is due almost entirely to 0.050 M NaBr.
(**c**) If the equilibrium $HgBr_2(s) + Br^- \rightleftharpoons HgBr_3^-$ also occurs, would the solubility of $HgBr_2$ be greater or less than that computed in (**b**)?

12-3 Charge and Mass Balances

Difficult equilibrium problems can be tackled by writing all the relevant chemical equilibria plus two more equations: the balances of charge and mass. We now examine these two conditions.

Charge Balance

Solutions must have zero total charge.

The **charge balance** is an algebraic statement of electroneutrality: *The sum of the positive charges in solution equals the sum of the negative charges in solution.*

Suppose that a solution contains the following ionic species: H^+, OH^-, K^+, $H_2PO_4^-$, HPO_4^{2-}, and PO_4^{3-}. The charge balance is

$$\underbrace{[H^+] + [K^+]}_{\text{Total positive charge}} = \underbrace{[OH^-] + [H_2PO_4^-] + 2[HPO_4^{2-}] + 3[PO_4^{3-}]}_{\text{Total negative charge}} \quad (12\text{-}8)$$

This statement says that the total charge contributed by H^+ and K^+ equals the magnitude of the charge contributed by all of the anions on the right side of the equation. *The coefficient in front of each species equals the magnitude of the charge on the ion.* A mole of, say, PO_4^{3-} contributes three moles of negative charge. If $[PO_4^{3-}] = 0.01$ M, the negative charge is $3[PO_4^{3-}] = 3(0.01) = 0.03$ M.

The coefficient of each term in the charge balance equals the magnitude of the charge on each ion. The coefficient of $[HPO_4^{2-}]$ in Equation 12-8 is 2 because the magnitude of its charge is 2. The coefficient of $[PO_4^{3-}]$ is 3 because the magnitude of its charge is 3.

Equation 12-8 appears unbalanced to many people. "The right side of the equation has much more charge than the left side!" you might think. But you would be wrong.

For example, consider a solution prepared by weighing out 0.025 0 mol of KH_2PO_4 plus 0.030 0 mol of KOH and diluting to 1.00 L. The concentrations of the species at equilibrium are

$$[H^+] = 5.1 \times 10^{-12}\text{ M} \qquad [H_2PO_4^-] = 1.3 \times 10^{-6}\text{ M}$$
$$[K^+] = 0.055\ 0\text{ M} \qquad [HPO_4^{2-}] = 0.022\ 0\text{ M}$$
$$[OH^-] = 0.002\ 0\text{ M} \qquad [PO_4^{3-}] = 0.003\ 0\text{ M}$$

Are the charges balanced? Yes, indeed. Plugging into Equation 12-8, we find

$$[H^+] + [K^+] = [OH^-] + [H_2PO_4^-] + 2[HPO_4^{2-}] + 3[PO_4^{3-}]$$
$$5.1 \times 10^{-12} + 0.055\ 0 = 0.002\ 0 + 1.3 \times 10^{-6} + 2(0.022\ 0) + 3(0.003\ 0)$$
$$0.055\ 0 = 0.055\ 0$$

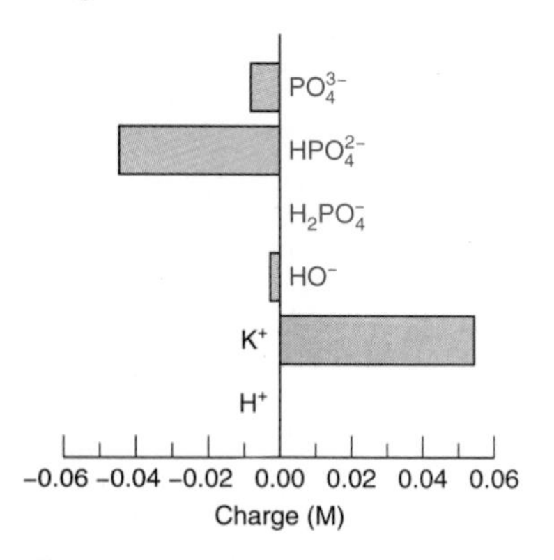

Figure 12-6 Charge contributed by each ion in 1.00 L of solution containing 0.025 0 mol KH_2PO_4 plus 0.030 0 mol KOH. The total positive charge equals the total negative charge.

The total positive charge is 0.055 0 M, and the total negative charge also is 0.055 0 M (Figure 12-6). Charges must balance in every solution. Otherwise a beaker with excess positive charge would glide across the lab bench and smash into another beaker with excess negative charge.

COLOR PLATE 1
Fajans Titration of Cl^- with $AgNO_3$, Using Dichlorofluorescein (Demonstration 6-1)
(*a*) Indicator before beginning titration. (*b*) AgCl precipitate before end point. (*c*) Indicator adsorbed on precipitate after end point.

COLOR PLATE 2 Colloids and Dialysis (Demonstration 7-1) (*a*) Ordinary aqueous Fe(III) (right) and colloidal Fe(III) (left). (*b*) Dialysis bags containing colloidal Fe(III) (left) and a solution of Cu(II) (right) immediately after placement in flasks of water. (*c*) After 24 h of dialysis, the Cu(II) has diffused out and is dispersed uniformly between the bag and the flask, but the colloidal Fe(III) remains inside the bag.

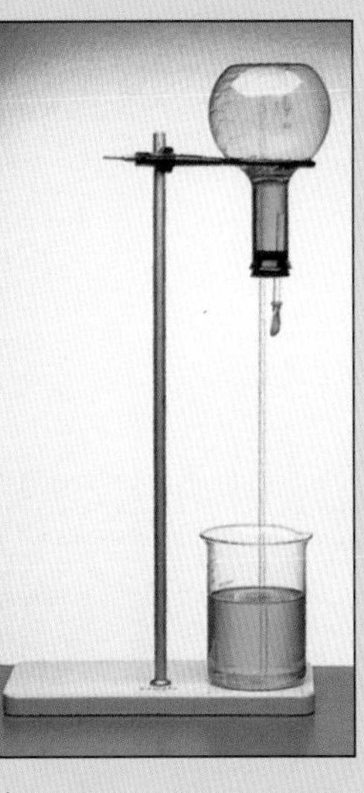

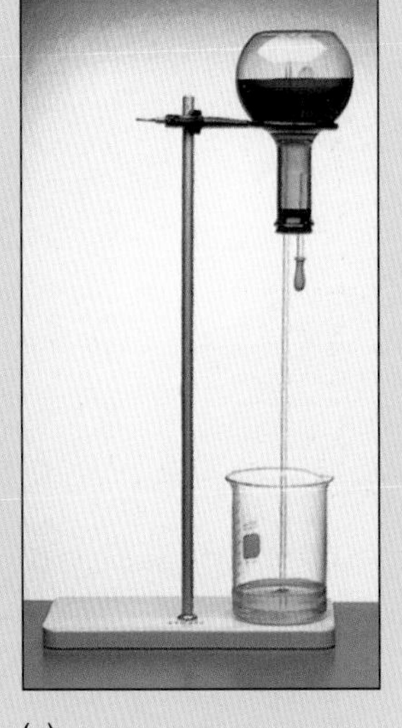

COLOR PLATE 3 HCl Fountain (Demonstration 8-1) (*a*) Basic indicator solution in beaker.
(*b*) Indicator is drawn into flask and changes to acidic color.
(*c*) Solution levels at end of experiment.

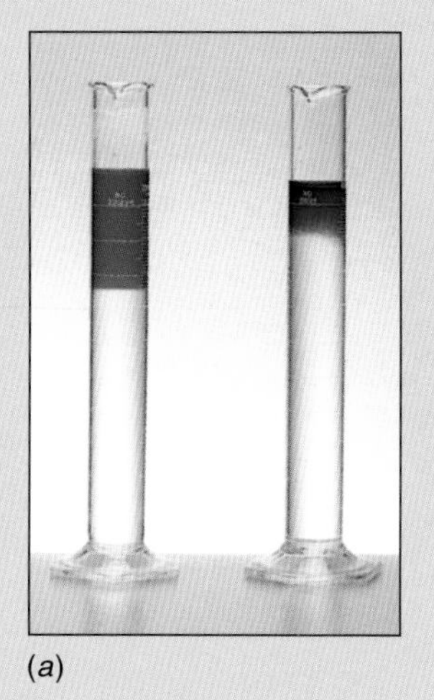

(*a*)

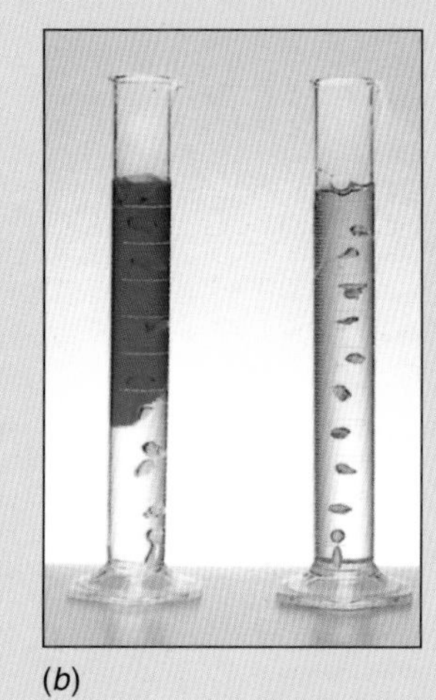

(*b*)

(*c*)

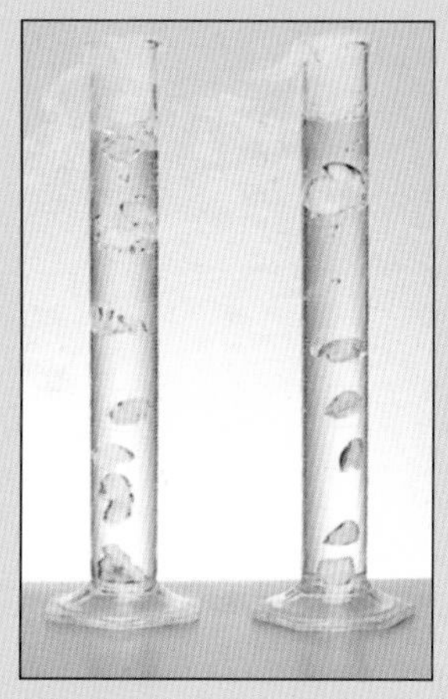

(*d*)

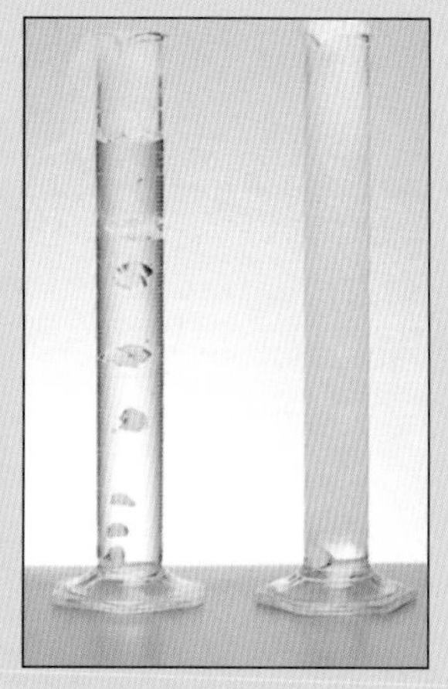

(*e*)

COLOR PLATE 4 Indicators and Acidity of CO_2 (Demonstration 9-2) (*a*) Cylinder before adding Dry Ice. Ethanol indicator solutions of phenolphthalein (left) and bromothymol blue (right) have not yet mixed with entire cylinder. (*b*) Adding Dry Ice causes bubbling and mixing. (*c*) Further mixing. (*d*) Phenolphthalein changes to its colorless acidic form. Color of bromothymol blue is due to mixture of acidic and basic forms. (*e*) After addition of HCl and stirring of right-hand cylinder, bubbles of CO_2 can be seen leaving solution, and indicator changes completely to its acidic color.

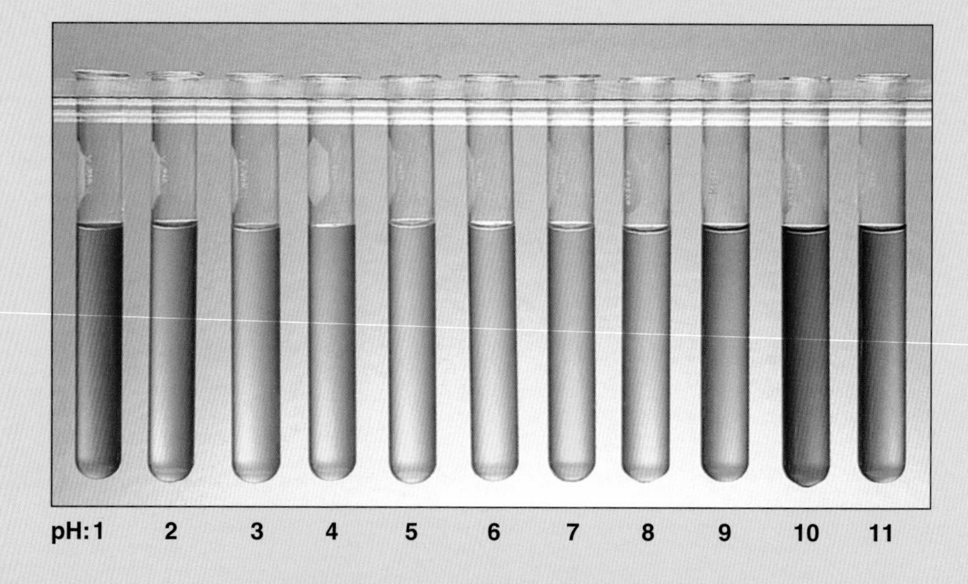

COLOR PLATE 5 Thymol Blue (Section 9-6) Acid-base indicator thymol blue between pH 1 (left) and 11 (right). The pK values are 1.7 and 8.9.

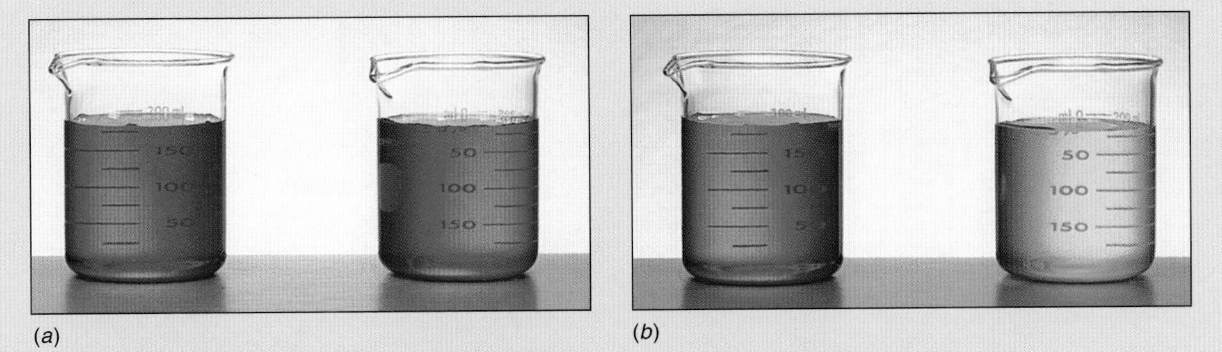

(*a*) (*b*)

COLOR PLATE 6 Effect of Ionic Strength on Ionic Dissociation (Demonstration 12-1) (*a*) Two beakers containing identical solutions with $FeSCN^{2+}$, Fe^{3+}, and SCN^-. (*b*) Color change when KNO_3 is added to the right-hand beaker.

COLOR PLATE 7 Titration of Cu(II) with EDTA, Using Auxiliary Complexing Agent (Section 13-2) *Left:* 0.02 M $CuSO_4$ before titration. *Center:* Color of Cu(II)-ammonia complex after addition of ammonia buffer, pH 10. *Right:* End-point color when all ammonia ligands have been displaced by EDTA. In this case, the auxiliary complexing agent accentuates the color change at the equivalence point.

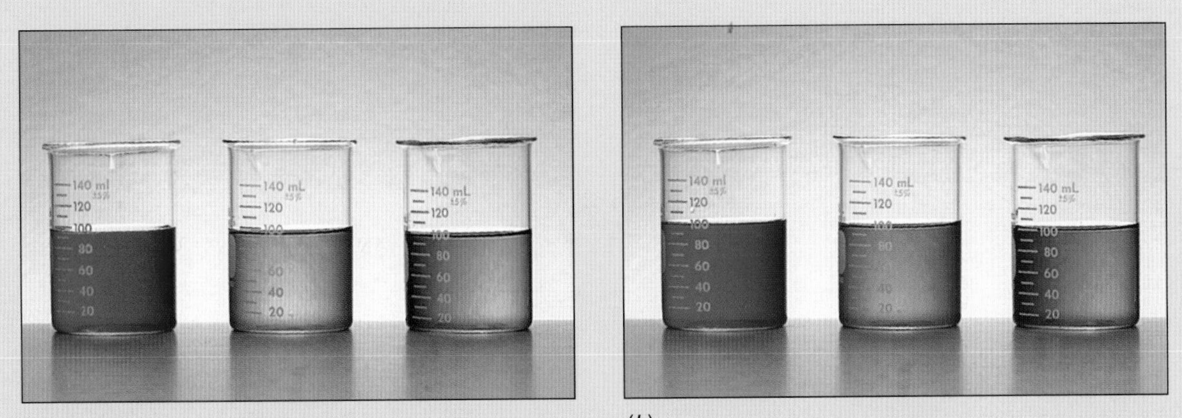

(*a*) (*b*)

COLOR PLATE 8 Titration of Mg^{2+} by EDTA, Using Eriochrome Black T Indicator (Demonstration 13-1) (*a*) Before (left), near (center), and after (right) equivalence point. (*b*) Same titration with methyl red added as inert dye to alter colors.

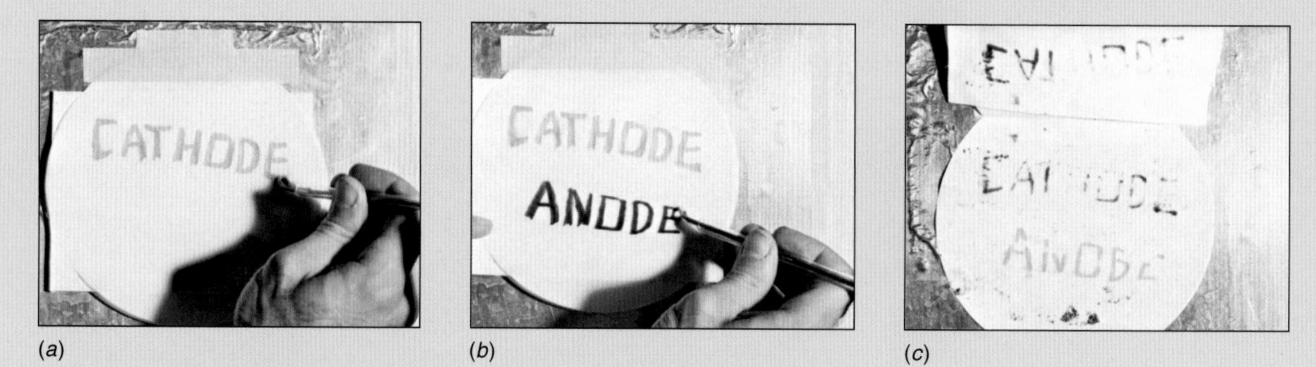

(*a*) (*b*) (*c*)

COLOR PLATE 9 Electrochemical Writing (Demonstration 14-1) (*a*) Stylus used as cathode. (*b*) Stylus used as anode. (*c*) Foil backing has a polarity opposite that of the stylus and produces reverse color on the bottom sheet of filter paper.

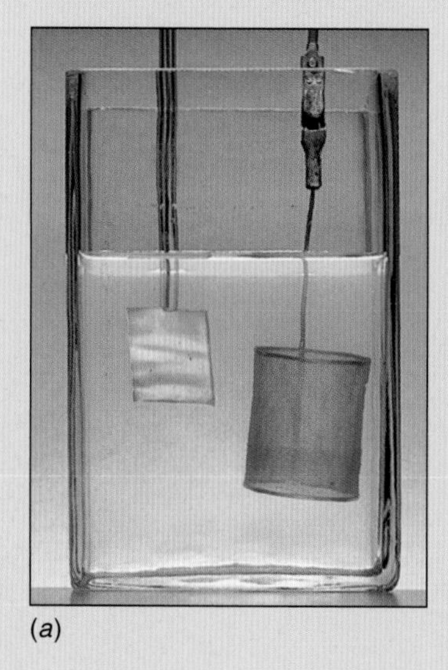

(*a*)

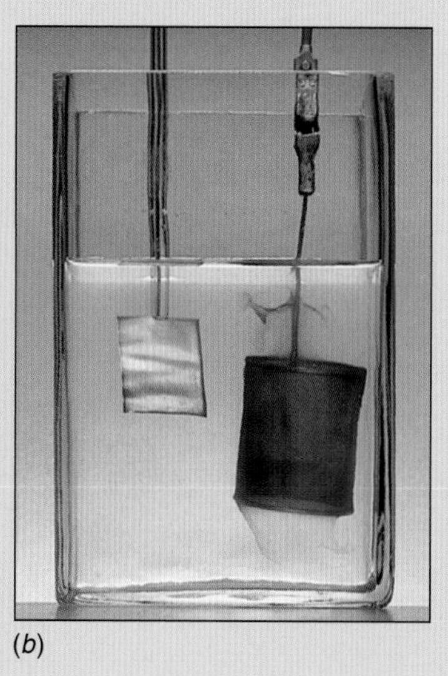

(*b*)

COLOR PLATE 10 Electrolysis of I^- in Solution to Make I_2 at the Anode (Demonstration 14-1) (*a*) Cu electrode (flat plate, left) and Pt electrode (mesh basket, right) immersed in solution containing KI and starch, with no electric current. (*b*) Starch-iodine complex forms at surface of Pt anode when current flows.

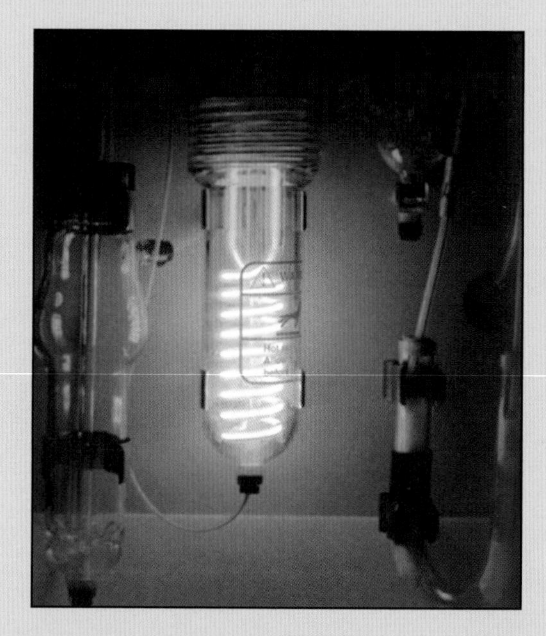

COLOR PLATE 11 Photolytic Environmental Carbon Analyzer (Box 16-1) A measured water sample is injected into the chamber at the left, where it is acidified with H_3PO_4 and sparged (bubbled with Ar or N_2) to remove CO_2 derived from HCO_3^- and CO_3^{2-}. The CO_2 is measured by its infrared absorbance. The sample is then forced into the digestion chamber, where $S_2O_8^{2-}$ is added and the sample is exposed to ultraviolet radiation from an immersion lamp (the coil at the center of the photo). Sulfate radicals (SO_4^-) formed by irradiation oxidize most organic compounds to CO_2, which is measured by infrared absorbance. The U-tube at the right contains Sn and Cu granules to scavenge volatile acids such as HCl and HBr liberated in the digestion. [Photo courtesy Ed Urbansky, U.S. Environmental Protection Agency, Cincinnati, OH.]

COLOR PLATE 12 Iodometric Titration (Section 16-3) *Left:* Initial I_3^- solution. *Left center:* I_3^- solution before end point in titration with $S_2O_3^{2-}$. *Right center:* I_3^- solution immediately before end point, with starch indicator present. *Right:* At the end point.

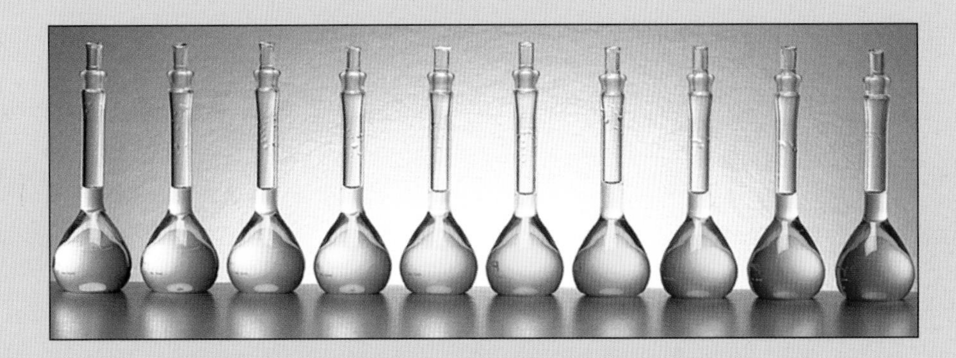

COLOR PLATE 13 Fe(phenanthroline)$_3^{2+}$ Standards for Spectrophotometric Analysis (Section 18-2) Volumetric flasks containing Fe(phenanthroline)$_3^{2+}$ solutions with iron concentrations ranging from 1 mg/L (left) to 10 mg/L (right).

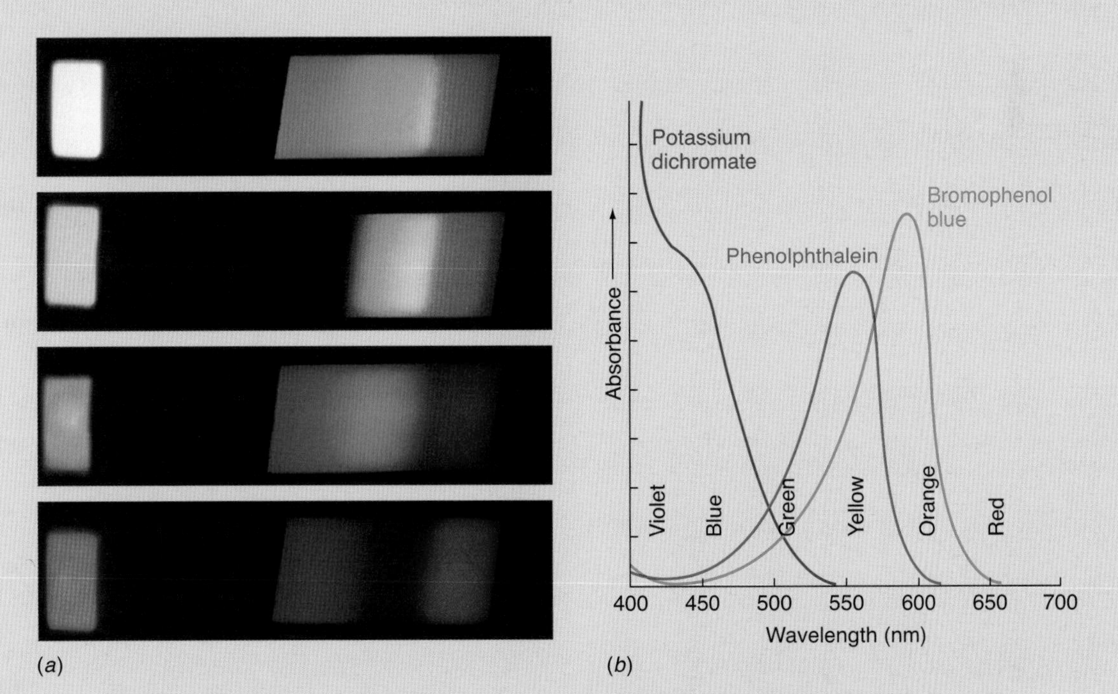

COLOR PLATE 14 Absorption Spectra (Demonstration 18-1) (*a*) Projected visible spectra of (from top to bottom) white light, potassium dichromate, bromophenol blue, and phenolphthalein. (*b*) Absorption spectra recorded with a spectrophotometer.

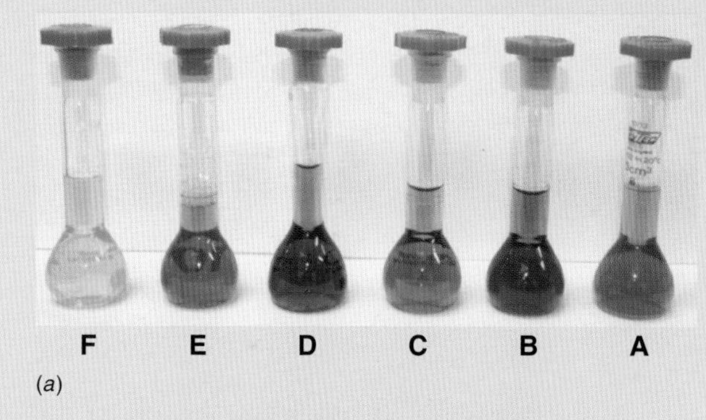

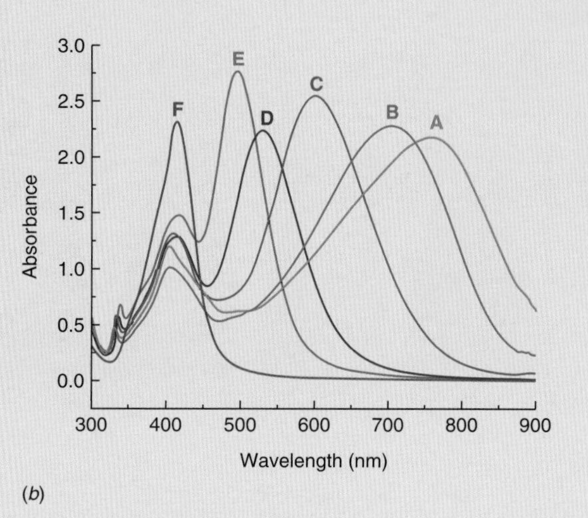

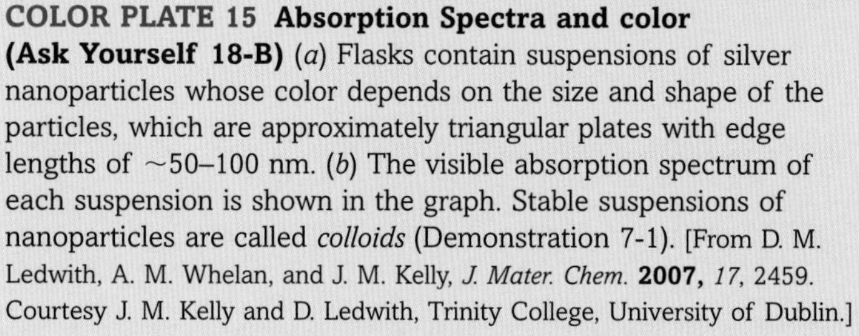

COLOR PLATE 15 Absorption Spectra and color (Ask Yourself 18-B) (*a*) Flasks contain suspensions of silver nanoparticles whose color depends on the size and shape of the particles, which are approximately triangular plates with edge lengths of ~50–100 nm. (*b*) The visible absorption spectrum of each suspension is shown in the graph. Stable suspensions of nanoparticles are called *colloids* (Demonstration 7-1). [From D. M. Ledwith, A. M. Whelan, and J. M. Kelly, *J. Mater. Chem.* **2007,** *17*, 2459. Courtesy J. M. Kelly and D. Ledwith, Trinity College, University of Dublin.]

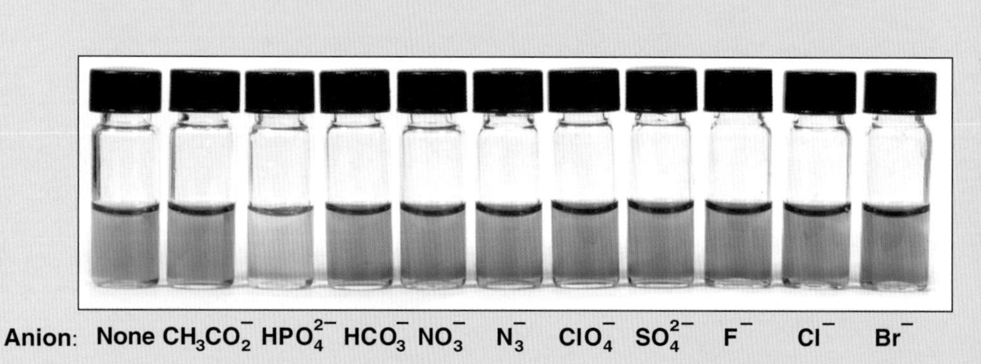

COLOR PLATE 16 Colorimetric Reagent for Phosphate (Box 18-2) The reagent in Box 18-2 was designed to turn yellow when phosphate is added, but not to respond to other common anions. Vials contain 50 μm colorimetric reagent plus 250 μM anion. [From M. S. Han and D. H. Kim, *Angew. Chem. Int. Ed.* **2002,** *41*, 3809. Courtesy D. H. Kim, Pohang University of Science and Technology, Korea.]

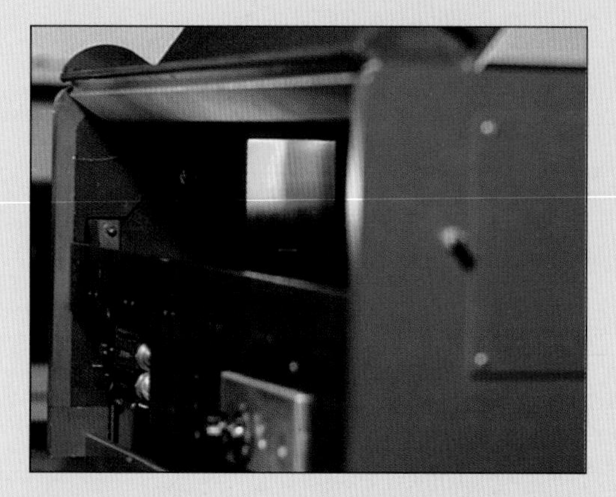

COLOR PLATE 17 Grating Dispersion (Section 19-1) Visible spectrum produced by grating inside spectrophotometer.

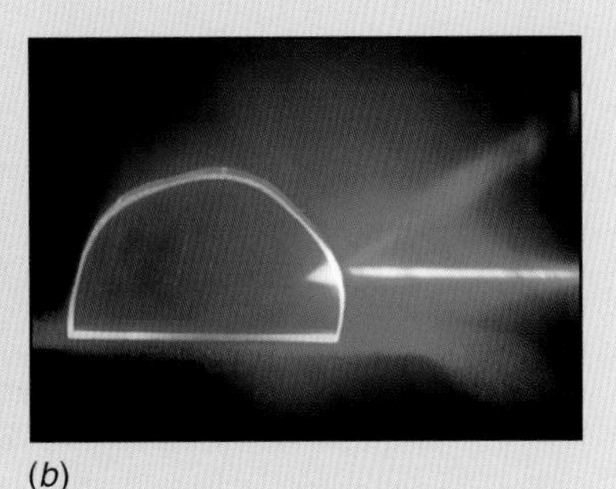

(*a*) (*b*)

COLOR PLATE 18 Transmission, Reflection, Refraction, and Absorption of Light (Section 19-1) (*a*) Blue-green laser is directed into a semicircular crystal of yttrium aluminum garnet containing a small amount of Er^{3+}, which emits yellow light when it absorbs the laser light. Light entering the crystal from the right is refracted (bent) and partly reflected at the right-hand surface of the crystal. The laser beam appears yellow inside the crystal because of luminescence from Er^{3+}. As it exits the crystal at the left side, the laser beam is refracted again, and partly reflected back into the crystal. (*b*) Same experiment, but with blue light instead of blue-green light. The blue light is absorbed by Er^{3+} and does not penetrate very far into the crystal. [Courtesy M. D. Seltzer, M. Johnson, and D. O'Connor, Michelson Laboratory, China Lake, CA.]

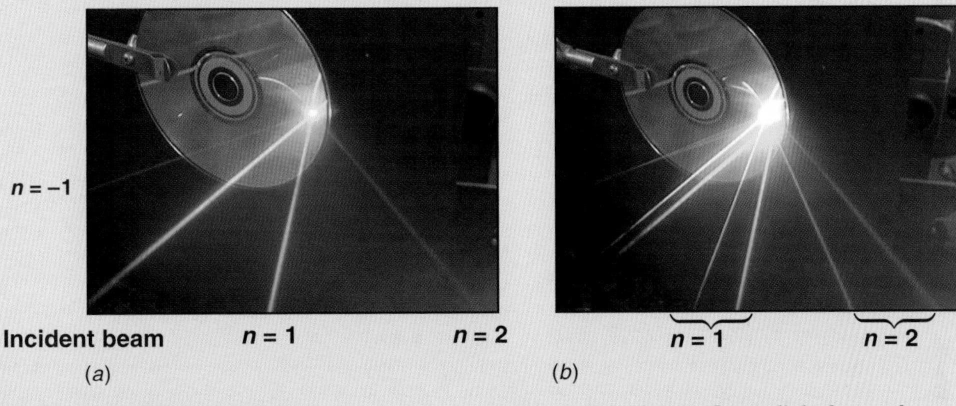

(*a*) (*b*)

COLOR PLATE 19 Laser Diffraction from a Compact Disk (Section 19-1) The grooves in an audio compact disk or a computer compact disk have a spacing of 1.6 μm. (*a*) When a red laser strikes disk at normal incidence ($\theta = 0$ in Figure 19-5 and Equation 19-2) three diffracted beams with orders $n = +1, +2$, and -1 are observed. (*b*) Red and green lasers strike the disk at normal incidence. Green light has a shorter wavelength than red light, and so, according to Equation 19-2, green light is diffracted at smaller angle (ϕ). Beams have been made visible by "fog" from liquid nitrogen. [Courtesy J. Tellinghuisen, Vanderbilt University. See J. Tellinghuisen, "Exploring the Diffraction Grating Using a He-Ne Laser and a CD-ROM," *J. Chem. Ed.* **2002,** *79,* 703.]

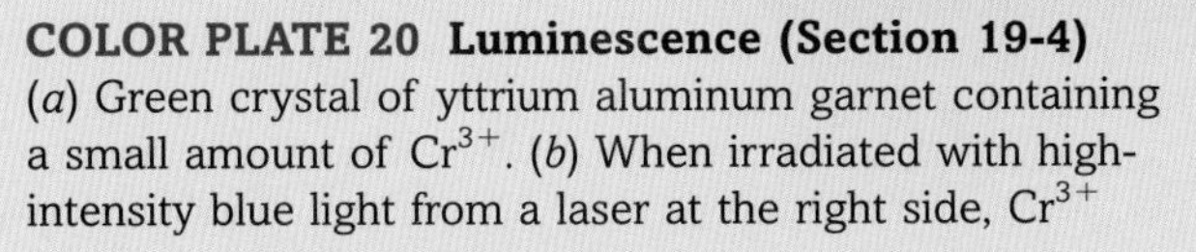

(*a*) (*b*)

COLOR PLATE 20 Luminescence (Section 19-4) (*a*) Green crystal of yttrium aluminum garnet containing a small amount of Cr^{3+}. (*b*) When irradiated with high-intensity blue light from a laser at the right side, Cr^{3+} absorbs blue light and emits lower-energy red light. When the laser is removed, the crystal appears green again. [Courtesy M. D. Seltzer, M. Johnson, and D. O'Connor, Michelson Laboratory, China Lake, CA.]

COLOR PLATE 21 Multielement Detection in Inductively Coupled Plasma Atomic Emission Spectrometry (Section 20-4) Light emitted from the plasma enters the polychromator at the upper right and is dispersed vertically by a prism and then horizontally by a grating. The resulting two-dimensional pattern of wavelengths from 165 to 1 000 nm is detected by a charge injection device (CID), which is a two-dimensional array with 262 000 pixels. Each emission wavelength lands at a different pixel. [Courtesy TJA Solutions, Franklin, MA.]

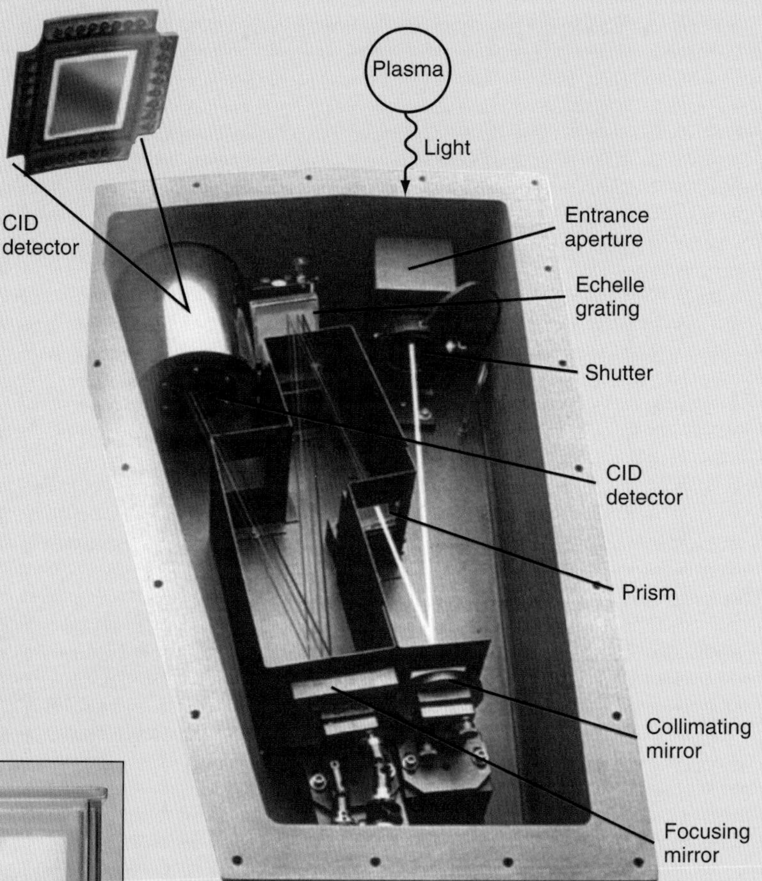

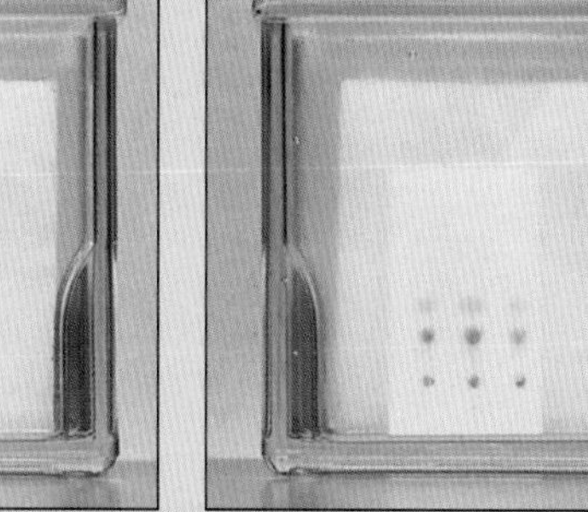

(*a*) (*b*)

COLOR PLATE 22 Thin-Layer Chromatography (Section 21-1) (*a*) Solvent ascends past mixture of dyes near bottom of flat plate coated with solid adsorbent. (*b*) Separation achieved after solvent has ascended most of the way up the plate.

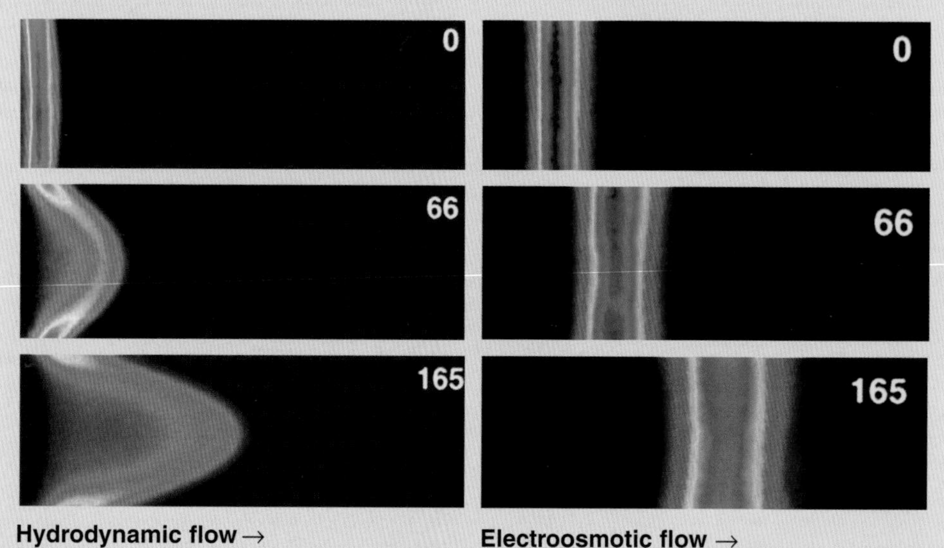

COLOR PLATE 23 Velocity Profiles for Hydrodynamic and Electroosmotic Flow (Section 23-6) A fluorescent dye was imaged inside a capillary tube at times of 0, 66, and 165 ms after initiating flow. The highest concentration of dye is represented by blue and the lowest concentration by red in these images in which different colors are assigned to different fluorescence intensities. [From P. H. Paul, M. G. Garguilo, and D. J. Rakestraw, *Anal. Chem.* **1998,** *70,* 2459.]

The general form of the charge balance for any solution is

Charge balance: $$n_1[C_1] + n_2[C_2] + \cdots = m_1[A_1] + m_2[A_2] + \cdots \quad (12\text{-}9)$$

Σ[positive charges] $= \Sigma$[negative charges]

Activity coefficients do not appear in the charge balance. The charge contributed by 0.1 M H^+ is *exactly* 0.1 M. Think about this.

where [C] is the concentration of a cation, n is the charge of the cation, [A] is the concentration of an anion, and m is the magnitude of the charge of the anion.

Example **Writing a Charge Balance**

Write the charge balance for a solution of lead(II) iodide containing the species Pb^{2+}, I^-, PbI^+, $PbI_2(aq)$, PbI_3^-, $PbOH^+$, H_2O, H^+, and OH^-.

SOLUTION The species $PbI_2(aq)$ and H_2O contribute no charge, so the charge balance is

$$2[Pb^{2+}] + [PbI^+] + [PbOH^+] + [H^+] = [I^-] + [PbI_3^-] + [OH^-]$$

Test Yourself Write a charge balance for H_2SO_4 in water. (**Answer:** $[H^+] = 2[SO_4^{2-}] + [HSO_4^-] + [OH^-]$)

Mass Balance

The **mass balance**, also called the *material balance*, is a statement of the conservation of matter. The mass balance states that *the quantity of all species in a solution containing a particular atom (or group of atoms) must equal the amount of that atom (or group) delivered to the solution.* Let's look at some examples.

The mass balance is a statement of the conservation of matter. It really refers to conservation of atoms, not to mass.

Suppose that a solution is prepared by dissolving 0.050 mol of acetic acid in water to give a total volume of 1.00 L. The acetic acid partially dissociates into acetate:

$$\underset{\text{Acetic acid}}{CH_3CO_2H} \rightleftharpoons \underset{\text{Acetate}}{CH_3CO_2^-} + H^+$$

The mass balance states that the quantity of dissociated and undissociated acetic acid in the solution must equal the amount of acetic acid put into the solution.

Mass balance for acetic acid in water: $$\underset{\text{What we put into the solution}}{0.050\ M} = \underset{\text{Undissociated product}}{[CH_3CO_2H]} + \underset{\text{Dissociated product}}{[CH_3CO_2^-]}$$

Activity coefficients do not appear in the mass balance. The concentration of each species counts *exactly* the number of atoms of that species.

When a compound dissociates in several ways, the mass balance must include all the products. Phosphoric acid (H_3PO_4), for example, dissociates to $H_2PO_4^-$, HPO_4^{2-}, and PO_4^{3-}. The mass balance for a solution prepared by dissolving 0.025 0 mol of H_3PO_4 in 1.00 L is

$$0.025\ 0\ M = [H_3PO_4] + [H_2PO_4^-] + [HPO_4^{2-}] + [PO_4^{3-}]$$

Now consider a saturated solution of K_2HPO_4 in water. We do not know the concentration, because we do not know how much K_2HPO_4 dissolves. However, we can say that, for every mole of phosphorus in solution, there are two moles of K^+. Phosphorus is in the forms H_3PO_4, $H_2PO_4^-$, HPO_4^{2-}, and PO_4^{3-}. Therefore, the mass balance is

$$[K^+] = 2\ \underbrace{\{[H_3PO_4] + [H_2PO_4^-] + [HPO_4^{2-}] + [PO_4^{3-}]\}}_{2\ \times\ \text{Total concentration of phosphorus atoms}}$$

The concentration of K is twice the total concentration of P. Think about which side of the equation should have the 2.

Now consider a solution prepared by dissolving PbI_2 in water to give Pb^{2+}, I^-, PbI^+, $PbI_2(aq)$, PbI_3^-, $PbOH^+$, H_2O, H^+, and OH^-. The source of all these species is PbI_2, so there must be two I atoms for every Pb atom in the solution. The mass balance is therefore

We do not know how much PbI_2 dissolved, but we do know that there must be two I atoms for every Pb atom in the solution.

$$\underbrace{2\,\{[Pb^{2+}] + [PbI^+] + [PbOH^+] + [PbI_2(aq)] + [PbI_3^-]\}}_{2\,\times\,\text{Total concentration of Pb atoms}}$$
$$= \underbrace{[I^-] + [PbI^+] + 2[PbI_2(aq)] + 3[PbI_3^-]}_{\text{Total concentration of I atoms}} \quad (12\text{-}10)$$

On the right side of Equation 12-10, there is a 2 in front of $[PbI_2(aq)]$ because each mole of $PbI_2(aq)$ contains two moles of I atoms. There is a 3 in front of $[PbI_3^-]$ because each mole of PbI_3^- contains three moles of I atoms.

Ask Yourself

12-C. Consider a buffer solution prepared by mixing 5.00 mmol of $Na_2C_2O_4$ (sodium oxalate) with 2.50 mmol of HCl in 0.100 L.
(a) List all the chemical species in the solution. Oxalate can accept one or two protons.
(b) What is the charge balance for the solution?
(c) Write separate mass balances for Na^+, oxalate, and Cl^-.
(d) Decide which species are negligible and simplify the expressions in **(b)** and **(c)**.

12-4 Systematic Treatment of Equilibrium

Now that we have learned about the charge and mass balances, we are ready for the systematic treatment of equilibrium. The general prescription follows these steps:

Step 1. Write the *pertinent reactions.*

Step 2. Write the *charge balance* equation. There is only one.

Step 3. Write *mass balance* equations. There may be more than one.

Step 4. Write the *equilibrium constant* for each chemical reaction. This step is the only one in which activity coefficients enter.

Activity coefficients enter *only* in step 4.

Step 5. *Count the equations and unknowns.* At this point, you should have as many equations as unknowns (chemical concentrations). If not, you must either find more equilibria or fix some concentrations at known values.

Step 6. By hook or by crook, *solve* for all the unknowns.

Steps 1 and 6 are the heart of the problem. Guessing what chemical equilibria exist in a given solution requires a fair degree of chemical intuition. In this text, you will usually be given help with step 1. Unless we know all the relevant equilibria, it is not possible to calculate the composition of a solution correctly. Because we do not know all the chemical reactions, we undoubtedly oversimplify many equilibrium problems.

Step 6 is likely to be your biggest challenge. With n equations involving n unknowns, the problem can always be solved, at least in principle. In the simplest cases, you can do this by hand; but, for most problems, approximations or a spreadsheet are employed.

A Simple Example: The pH of 10^{-8} M KOH

Here is a trick question: What is the pH of 1.0×10^{-8} M KOH? Your first response might be $[OH^-] = 1.0 \times 10^{-8}$ M, so $[H^+] = 1.0 \times 10^{-6}$ M, and so the pH is 6.00. However, adding base to a neutral solution could not possibly make it acidic. So let's see how the systematic treatment of equilibrium works in this case.

Step 1. *Pertinent reactions:* The only one is $H_2O \rightleftharpoons H^+ + OH^-$, which exists in every aqueous solution.

Step 2. *Charge balance:* The ions are K^+, H^+, and OH^-, so $[K^+] + [H^+] = [OH^-]$.

Step 3. *Mass balance:* You might be tempted to write $[K^+] = [OH^-]$, but this is false because OH^- comes from both KOH and H_2O. For every mole of K^+, one mole of OH^- is introduced into the solution. We also know that, for every mole of H^+ from H_2O, one mole of OH^- is introduced. Therefore one mass balance is $[OH^-] = [K^+] + [H^+]$, which is the same as the charge balance in this particularly simple example. A second mass balance is $[K^+] = 1.00 \times 10^{-8}$ M.

Step 4. *Equilibrium constants:* The only one is $K_w = [H^+]\gamma_{H^+}[OH^-]\gamma_{OH^-}$.

Step 5. *Count equations and unknowns:* At this point, you should have as many equations as unknowns (chemical species). There are three unknowns, $[K^+]$, $[H^+]$, and $[OH^-]$, and three equations:

If you have fewer equations than unknowns, look for another mass balance or a chemical equilibrium that you overlooked.

Charge balance:	$[K^+] + [H^+] = [OH^-]$
Mass balance:	$[K^+] = 1.0 \times 10^{-8}$ M
Equilibrium constant:	$K_w = [H^+]\gamma_{H^+}[OH^-]\gamma_{OH^-}$

Step 6. *Solve:* The ionic strength must be very low in this solution (~10^{-7} M), so it is safe to say that the activity coefficients are 1.00. Therefore the equilibrium constant simplifies to $K_w = [H^+][OH^-]$. Substituting 1.0×10^{-8} M for $[K^+]$ and $[OH^-] = K_w/[H^+]$ into the charge balance gives

$$[1.0 \times 10^{-8}] + [H^+] = K_w/[H^+]$$

Multiplying both sides by $[H^+]$ gives a quadratic equation

$$[1.0 \times 10^{-8}][H^+] + [H^+]^2 = K_w$$
$$[H^+]^2 + [1.0 \times 10^{-8}][H^+] - K_w = 0$$

whose two solutions are $[H^+] = 9.6 \times 10^{-8}$ and -1.1×10^{-7} M. Rejecting the negative solution (because the concentration cannot be negative), we find

$$pH = -\log([H^+]\gamma_{H^+}) = -\log([9.6 \times 10^{-8}](1.00)) = 7.02$$

It should not be too surprising that the pH is close to 7 and very slightly basic.

Coupled Equilibria: Solubility of CaF_2

The opening of this chapter gave an example of *coupled equilibria* in which calcium carbonate dissolves and the resulting bicarbonate reacts with H^+. The second reaction drives the first reaction forward.

Now we look at a similar case in which CaF_2 dissolves in water:

$$CaF_2(s) \xrightleftharpoons{K_{sp} = 3.9 \times 10^{-11}} Ca^{2+} + 2F^- \qquad (12\text{-}11)$$

The fluoride ion can then react with water to give HF(*aq*):

If we were very smart, we might also write the reaction

$$Ca^{2+} + OH^- \overset{K = 20}{\rightleftharpoons} CaOH^+$$

It turns out that this reaction is important only at high pH.

$$F^- + H_2O \overset{K_b = 1.5 \times 10^{-11}}{\rightleftharpoons} HF + OH^- \qquad (12\text{-}12)$$

Also, for every aqueous solution, we can write

$$H_2O \overset{K_w}{\rightleftharpoons} H^+ + OH^- \qquad (12\text{-}13)$$

If Reaction 12-12 takes place, then the solubility of CaF_2 is greater than that predicted by the solubility product because F^- produced in Reaction 12-11 is consumed in Reaction 12-12. According to Le Châtelier's principle, Reaction 12-11 will be driven to the right. The systematic treatment of equilibrium allows us to find the net effect of all three reactions.

Step 1. *Pertinent reactions:* The three reactions are 12-11 through 12-13.

Step 2. *Charge balance:* $[H^+] + 2[Ca^{2+}] = [OH^-] + [F^-]$ (12-14)

Step 3. *Mass balance:* If all fluoride remained in the form F^-, we could write $[F^-] = 2[Ca^{2+}]$ from the stoichiometry of Reaction 12-11. But some F^- reacts to give HF. The total number of moles of fluorine atoms is equal to the sum of F^- plus HF, and the mass balance is

$$\underbrace{[F^-] + [HF]}_{\text{Total concentration of fluorine atoms}} = 2[Ca^{2+}] \qquad (12\text{-}15)$$

Step 4. *Equilibrium constants:*

$$K_{sp} = [Ca^{2+}]\gamma_{Ca^{2+}}[F^-]^2\gamma_{F^-}^2 = 3.9 \times 10^{-11} \qquad (12\text{-}16)$$

$$K_b = \frac{[HF]\gamma_{HF}[OH^-]\gamma_{OH^-}}{[F^-]\gamma_{F^-}} = 1.5 \times 10^{-11} \qquad (12\text{-}17)$$

$$K_w = [H^+]\gamma_{H^+}[OH^-]\gamma_{OH^-} = 1.0 \times 10^{-14} \qquad (12\text{-}18)$$

For simplicity, we are generally going to ignore the activity coefficients.

Although we wrote activity coefficients in the equilibrium equations, we are not so masochistic as to use them in the rest of the problem. At this point, we are going to explicitly ignore the activity coefficients, which is equivalent to saying that they are unity. There will be some inaccuracy in the results, but you could go back after the calculation and compute the ionic strength and the activity coefficients and find a better approximation for the solution if you had to.

Step 5. *Count equations and unknowns:* There are five equations (12-14 through 12-18) and five unknowns: $[H^+]$, $[OH^-]$, $[Ca^{2+}]$, $[F^-]$, and [HF].

Step 6. *Solve:* This is no simple matter for these five equations. Instead, let us ask a simpler question: What will be the concentrations of $[Ca^{2+}]$, $[F^-]$, and [HF] if the pH is *fixed* at 3.00 by adding a buffer?

Once we know that $[H^+] = 1.0 \times 10^{-3}$ M, there is a straightforward procedure for solving the equations. From Equation 12-18, we know $[OH^-] = K_w/[H^+] = 1.0 \times 10^{-11}$ M. Putting this value of $[OH^-]$ into Equation 12-17 gives

$$\frac{[HF]}{[F^-]} = \frac{K_b}{[OH^-]} = \frac{1.5 \times 10^{-11}}{1.0 \times 10^{-11}} = 1.5$$

$$\Rightarrow [HF] = 1.5[F^-]$$

Substituting 1.5[F^-] for [HF] in the mass balance (Equation 12-15) gives

$$[F^-] + [HF] = 2[Ca^{2+}] \qquad (12\text{-}15)$$
$$[F^-] + 1.5[F^-] = 2[Ca^{2+}]$$
$$[F^-] = 0.80[Ca^{2+}]$$

Finally, we substitute 0.80[Ca^{2+}] for [F^-] in the solubility product (Equation 12-16):

$$[Ca^{2+}][F^-]^2 = K_{sp}$$
$$[Ca^{2+}](0.80[Ca^{2+}])^2 = K_{sp}$$
$$[Ca^{2+}] = \left(\frac{K_{sp}}{0.80^2}\right)^{1/3} = 3.9 \times 10^{-4}\ \text{M}$$

Challenge Use the concentration of Ca^{2+} that we just calculated to show that [F^-] = 3.1×10^{-4} M and [HF] = 4.7×10^{-4} M.

You should realize that the charge balance equation (12-14) is no longer valid if the pH is fixed by external means. To adjust the pH, an ionic compound must necessarily have been added to the solution. Equation 12-14 is therefore incomplete, because it omits those ions. However, we did not use Equation 12-14 to solve the problem, because we omitted [H^+] as a variable when we fixed the pH.

Fixing the pH invalidates the original charge balance because we added unspecified ions to the solution to fix the pH. There exists a new charge balance, but we do not know enough to write an equation for it.

If we had selected a pH other than 3.00, we would have found a different set of concentrations because of the coupling of Reactions 12-11 and 12-12. Figure 12-7 shows the pH dependence of the concentrations of Ca^{2+}, F^-, and HF. At high pH, there is very little HF, so [F^-] ≈ 2[Ca^{2+}]. At low pH, there is very little F^-, so [HF] ≈ 2[Ca^{2+}]. The concentration of Ca^{2+} increases at low pH, because Reaction 12-11 is drawn to the right by the reaction of F^- with H_2O to make HF in Reaction 12-12.

In general, many minerals are more soluble at low pH because the anions react with acid.[3] Box 12-1 describes an environmental consequence of solubility in acids.

Ask Yourself

12-D. (a) Ignoring activity coefficients, find the concentrations of Ag^+, CN^-, and HCN in a saturated solution of AgCN whose pH is fixed at 9.00. Consider the equilibria:

$$AgCN(s) \rightleftharpoons Ag^+ + CN^- \qquad K_{sp} = 2.2 \times 10^{-16}$$
$$CN^- + H_2O \rightleftharpoons HCN(aq) + OH^- \qquad K_b = 1.6 \times 10^{-5}$$

(b) What would be the mass balance if the following equilibria also occur?

$$Ag^+ + CN^- \rightleftharpoons AgCN(aq) \qquad AgCN(aq) + CN^- \rightleftharpoons Ag(CN)_2^-$$
$$Ag^+ + H_2O \rightleftharpoons AgOH(aq) + H^+$$

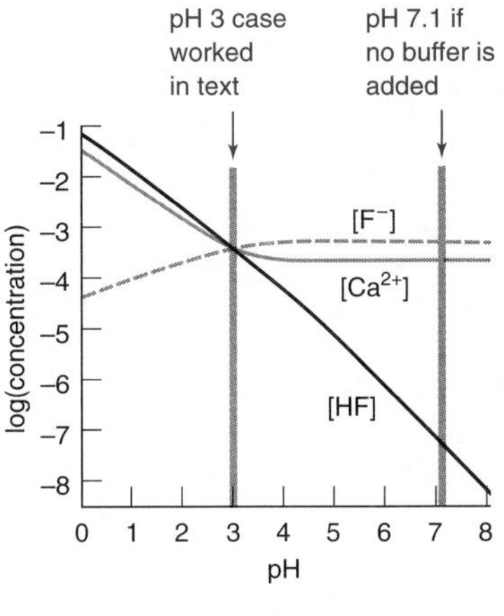

Figure 12-7 pH dependence of the concentrations of Ca^{2+}, F^-, and HF in a saturated solution of CaF_2. As the pH is lowered, H^+ reacts with F^- to make HF, and the concentration of Ca^{2+} increases. Note the logarithmic ordinate.

12-5 Fractional Composition Equations

As a final topic in our study of chemical equilibrium, we derive expressions for the fraction of a weak acid, HA, in each form (HA and A^-). These equations were

Box 12-1 Aluminum Mobilization from Minerals by Acid Rain

Aluminum is the third most abundant element on Earth (after oxygen and silicon), but it is tightly locked into insoluble minerals such as kaolinite ($Al_2(OH)_4Si_2O_5$) and bauxite (AlOOH). Acid rain from human activities is a recent change in the history of Earth, and it is introducing soluble forms of aluminum (and lead and mercury) into the environment.[4] The graph shows that at a pH below 5, aluminum is mobilized from minerals and its concentration in lake water rises rapidly. At a concentration of 130 μg/L, aluminum kills fish. In humans, high concentrations of aluminum cause dementia, softening of bones, and anemia.

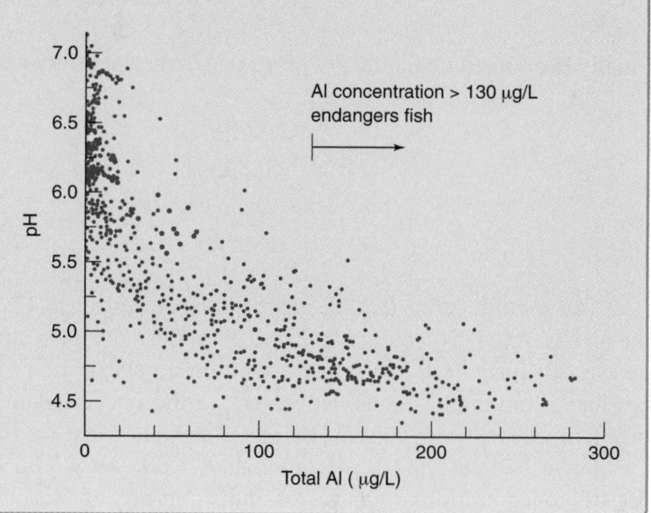

Relationship of total aluminum (including dissolved and suspended species) in 1 000 Norwegian lakes as a function of the pH of the lake water. The more acidic the water, the greater the aluminum concentration. [From G. Howells, *Acid Rain and Acid Waters,* 2nd ed. (Hertfordshire: Ellis Horwood, 1995).]

already used in Section 10-7 for an acid-base titration spreadsheet. In Equation 8-16, we defined the *fraction of dissociation* as

$$\text{fraction of HA in the form A}^- \equiv \alpha_{A^-} = \frac{[A^-]}{[A^-] + [HA]} \tag{12-19}$$

Similarly, we define the fraction in the form HA as

$$\text{fraction of HA in the form HA} \equiv \alpha_{HA} = \frac{[HA]}{[A^-] + [HA]} \tag{12-20}$$

Consider an acid with formal concentration F:

$$HA \overset{K_a}{\rightleftharpoons} H^+ + A^- \qquad K_a = \frac{[H^+][A^-]}{[HA]}$$

The mass balance is simply

$$F = [HA] + [A^-]$$

Rearranging the mass balance gives $[A^-] = F - [HA]$, which can be plugged into the K_a equilibrium to give

$$K_a = \frac{[H^+](F - [HA])}{[HA]}$$

or, with a little algebra,

$$[HA] = \frac{[H^+]F}{[H^+] + K_a} \tag{12-21}$$

Dividing both sides by F gives the fraction α_{HA}:

Fraction in the form HA:

$$\alpha_{HA} = \frac{[HA]}{F} = \frac{[H^+]}{[H^+] + K_a} \qquad (12\text{-}22)$$

If we substitute $[HA] = F - [A^-]$ into the K_a equation, we can rearrange and solve for the fraction α_{A^-}:

Fraction in the form A^-:

$$\alpha_{A^-} = \frac{[A^-]}{F} = \frac{K_a}{[H^+] + K_a} \qquad (12\text{-}23)$$

α_{HA} = fraction of species in the form HA

α_{A^-} = fraction of species in the form A^-

$\alpha_{HA} + \alpha_{A^-} = 1$

Figure 12-8 shows α_{HA} and α_{A^-} for a system with $pK_a = 5.00$. At low pH, almost all of the acid is in the form HA. At high pH, almost everything is in the form A^-. HA is the predominant species when pH < pK_a. A^- is the predominant species when pH > pK_a. Figure 11-1 is an analogous diagram of species in a diprotic system. At low pH, α_{H_2A} approaches 1 and, at high pH, $\alpha_{A^{2-}}$ approaches 1. At intermediate pH, α_{HA^-} is the largest fraction.

If you were dealing with the conjugate pair BH^+ and B instead of HA and A^-, Equation 12-22 gives the fraction in the form BH^+ and Equation 12-23 gives the fraction in the form B. In this case, K_a is the acid dissociation constant for BH^+ (which is K_w/K_b).

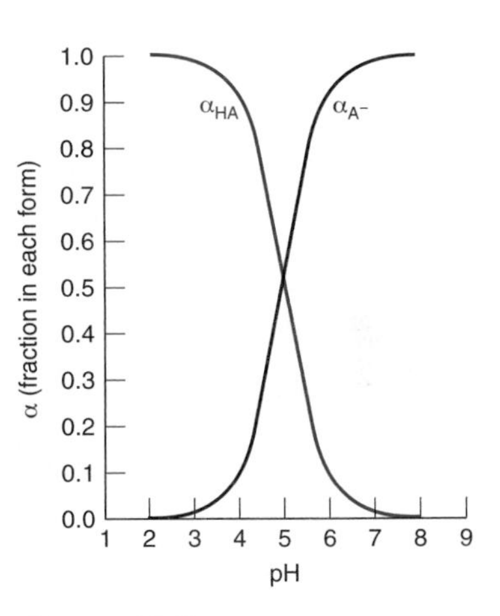

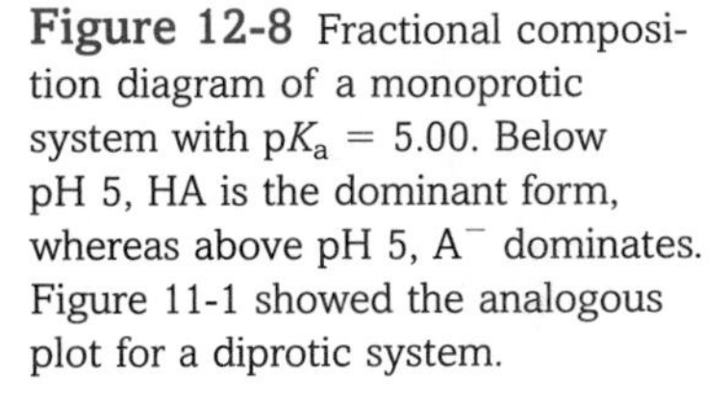

Figure 12-8 Fractional composition diagram of a monoprotic system with $pK_a = 5.00$. Below pH 5, HA is the dominant form, whereas above pH 5, A^- dominates. Figure 11-1 showed the analogous plot for a diprotic system.

Example **Fractional Composition for an Acid**

pK_a for benzoic acid (HA) is 4.20. Find the concentration of A^- at pH 5.31 if the formal concentration of HA is 0.021 3 M.

SOLUTION At pH = 5.31, $[H^+] = 10^{-5.31} = 4.9 \times 10^{-6}$ M.

$$\alpha_{A^-} = \frac{K_a}{[H^+] + K_a} = \frac{6.3 \times 10^{-5}}{(4.9 \times 10^{-6}) + (6.3 \times 10^{-5})} = 0.92_8$$

From Equation 12-23, we can say

$$[A^-] = \alpha_{A^-}F = (0.92_8)(0.021\ 3) = 0.020 \text{ M}$$

Test Yourself Find $[A^-]$ if pH = 4.31. (**Answer:** 0.012 M)

Example **Fractional Composition for a Base**

K_a for the ammonium ion, NH_4^+, is 5.69×10^{-10} ($pK_a = 9.245$). Find the fraction in the form BH^+ at pH 10.38.

SOLUTION At pH = 10.38, $[H^+] = 10^{-10.38} = 4.1_7 \times 10^{-11}$ M. Using Equation 12-22, with BH^+ in place of HA, we find

$$\alpha_{BH^+} = \frac{[H^+]}{[H^+] + K_a} = \frac{4.1_7 \times 10^{-11}}{(4.1_7 \times 10^{-11}) + (5.69 \times 10^{-10})} = 0.068$$

Problem 12-39 gives fractional composition equations for a diprotic acid, H_2A.

Test Yourself Find the fraction in the form NH_3 at pH 10.00. (**Answer:** 0.85)

Ask Yourself

12-E. The acid HA has $pK_a = 3.00$. Find the fraction in the form HA and the fraction in the form A^- at pH = 2.00, 3.00, and 4.00. Compute the quotient $[HA]/[A^-]$ at each pH.

Key Equations

Activity: $\mathcal{A}_C = [C]\gamma_C$ $\mathcal{A}$ = activity; γ = activity coefficient

Equilibrium constant: For the reaction $aA + bB \rightleftharpoons cC + dD$,

$$K = \frac{\mathcal{A}_C^c \mathcal{A}_D^d}{\mathcal{A}_A^a \mathcal{A}_B^b} = \frac{[C]^c\gamma_C^c [D]^d\gamma_D^d}{[A]^a\gamma_A^a [B]^b\gamma_B^b}$$

Ionic strength: $\mu = \frac{1}{2}(c_1z_1^2 + c_2z_2^2 + \cdots) = \frac{1}{2}\sum_i c_iz_i^2$

c = concentration; z = charge

Extended Debye-Hückel equation:

$$\log \gamma = \frac{-0.51z^2\sqrt{\mu}}{1 + (\alpha\sqrt{\mu}/305)}$$

z = charge; μ = ionic strength; α = size

Linear interpolation:

$$\frac{\text{unknown } y \text{ interval}}{\Delta y} = \frac{\text{known } x \text{ interval}}{\Delta x}$$

pH: $pH = -\log \mathcal{A}_{H^+} = -\log[H^+]\gamma_{H^+}$

Charge balance: positive charge in solution = negative charge in solution

Mass balance: Quantity of all species in a solution containing a particular atom (or group of atoms) must equal the amount of that atom (or group) delivered to the solution.

Systematic treatment of equilibrium:
1. Pertinent reactions
2. Charge balance
3. Mass balance
4. Equilibrium constants
5. Count equations/unknowns
6. Solve

Fraction of HA in acidic form:

$$\alpha_{HA} = \frac{[HA]}{F} = \frac{[H^+]}{[H^+] + K_a} = \alpha_{BH^+} = \frac{[BH^+]}{F}$$

Fraction of HA in basic form:

$$\alpha_{A^-} = \frac{[A^-]}{F} = \frac{K_a}{[H^+] + K_a} = \alpha_B = \frac{[B]}{F}$$

Important Terms

activity
activity coefficient
charge balance
extended Debye-Hückel equation
interpolation
ionic atmosphere
ionic strength
mass balance

Problems

12-1. What is an ionic atmosphere?

12-2. Explain why the solubility of an ionic compound increases as the ionic strength of the solution increases (at least up to ~0.5 M).

12-3. The graph on page 283 shows the quotient of concentrations $[CH_3CO_2^-][H^+]/[CH_3CO_2H]$ for the dissociation of acetic acid as a function of the concentration of KCl added to the solution. Explain the shape of the curve.

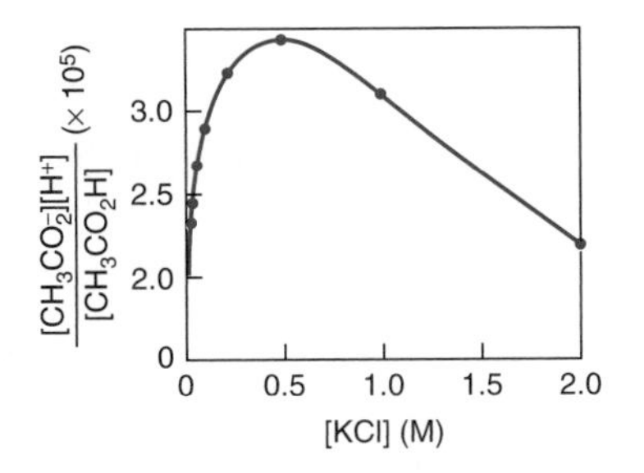

12-4. Which statements are true? In the ionic strength range 0–0.1 M, activity coefficients decrease with **(a)** increasing ionic strength; **(b)** increasing ionic charge; **(c)** decreasing hydrated size (α).

12-5. Explain the following observations:

(a) Mg^{2+} has a greater hydrated diameter than Ba^{2+}.

(b) Hydrated diameters decrease in the order $Sn^{4+} > In^{3+} > Cd^{2+} > Rb^{+}$.

(c) H^+ (which is really H_3O^+) has one of the largest hydrated sizes in Table 12-1. Consider the possibilities of hydrogen bonding to H_3O^+.

12-6. State in words the meaning of the charge balance equation.

12-7. State the meaning of the mass balance equation.

12-8. Why does the solubility of a salt of a basic anion increase with decreasing pH? Write chemical reactions for the minerals galena (PbS) and cerussite ($PbCO_3$) to explain how acid rain mobilizes trace quantities of toxic metallic elements from relatively inert forms into the environment, where the metals can be taken up by plants and animals. Why are the minerals kaolinite and bauxite in Box 12-1 more soluble in acidic solution than in neutral solution?

12-9. Assuming complete dissociation of the salts, calculate the ionic strength of **(a)** 0.2 mM KNO_3; **(b)** 0.2 mM Cs_2CrO_4; **(c)** 0.2 mM $MgCl_2$ plus 0.3 mM $AlCl_3$.

12-10. Find the activity coefficient of each ion at the indicated ionic strength:

(a) SO_4^{2-} (μ = 0.01 M)
(b) Sc^{3+} (μ = 0.005 M)
(c) Eu^{3+} (μ = 0.1 M)
(d) $(CH_3CH_2)_3NH^+$ (μ = 0.05 M)

12-11. Find the activity (not the activity coefficient) of $(C_3H_7)_4N^+$ (tetrapropylammonium ion) in a solution containing 0.0050 M $(C_3H_7)_4N^+Br^-$ plus 0.0050 M $(CH_3)_4N^+Cl^-$.

12-12. Interpolate in Table 12-1 to find the activity coefficient of H^+ when μ = **(a)** 0.030 M and **(b)** 0.042 M.

12-13. Calculate the activity coefficient of Zn^{2+} when μ = 0.083 M by using **(a)** Equation 12-5; **(b)** linear interpolation with Table 12-1.

12-14. Using activities, find the concentration of Ag^+ in a saturated solution of AgSCN in **(a)** 0.060 M KNO_3; **(b)** 0.060 M KSCN.

12-15. Find the activity coefficient of H^+ in a solution containing 0.010 M HCl plus 0.040 M $KClO_4$. What is the pH of the solution?

12-16. Using activities, calculate the pH and concentration of H^+ in pure water containing 0.050 M LiBr at 25°C.

12-17. Using activities, calculate the pH of a solution containing 0.010 M NaOH plus 0.0120 M $LiNO_3$. What would be the pH if you neglected activities?

12-18. Using activities, find the concentration of OH^- in a solution of 0.075 M $NaClO_4$ saturated with $Mn(OH)_2$. What is the pH of this solution?

12-19. Using activities, find the concentration of Ba^{2+} in a 0.100 M $(CH_3)_4N^+IO_3^-$ solution saturated with $Ba(IO_3)_2$. Assume that $Ba(IO_3)_2$ makes a negligible contribution to the ionic strength and verify your assumption when you are done.

12-20. Using activities, calculate $[Pb^{2+}]$ in a saturated solution of PbF_2. Consider only the equilibrium $PbF_2 \rightleftharpoons Pb^{2+} + 2F^-$. Follow the example "A Better Estimate of the Solubility of PbI_2" in Section 12-2 to find the ionic strength by successive approximations.

12-21. **(a)** Using activities and K_{sp} for $CaSO_4$, calculate the concentration of dissolved Ca^{2+} in a saturated aqueous solution of $CaSO_4$.

(b) The observed total concentration of dissolved calcium is 15–19 mM. Explain.

12-22. Write a charge balance for a solution containing H^+, OH^-, Ca^{2+}, HCO_3^-, CO_3^{2-}, $Ca(HCO_3)^+$, $Ca(OH)^+$, K^+, and ClO_4^-.

12-23. Write a charge balance for a solution of H_2SO_4 in water if the H_2SO_4 ionizes to HSO_4^- and SO_4^{2-}.

12-24. Write the charge balance for an aqueous solution of arsenic acid, H_3AsO_4, in which the acid can dissociate to $H_2AsO_4^-$, $HAsO_4^{2-}$, and AsO_4^{3-}. Look up the structure of arsenic acid in Appendix B and write the structure of $HAsO_4^{2-}$.

12-25. **(a)** Suppose that $MgBr_2$ dissolves to give Mg^{2+} and Br^-. Write a charge balance for this aqueous solution.

(b) What is the charge balance if, in addition to Mg^{2+} and Br^-, $MgBr^+$ is formed?

12-26. For a 0.1 M aqueous solution of sodium acetate, $Na^+CH_3CO_2^-$, one mass balance is simply $[Na^+]$ = 0.1 M. Write a mass balance involving acetate.

12-27. Suppose that $MgBr_2$ dissolves to give Mg^{2+} and Br^-.
(a) Write the mass balance for Mg^{2+} for 0.20 M $MgBr_2$.
(b) Write a mass balance for Br^- for 0.20 M $MgBr_2$.
Now suppose that $MgBr^+$ is formed in addition to Mg^{2+} and Br^-.
(c) Write a mass balance for Mg^{2+} for 0.20 M $MgBr_2$.
(d) Write a mass balance for Br^- for 0.20 M $MgBr_2$.

12-28. **(a)** Write the mass balance for CaF_2 in water if the reactions are $CaF_2(s) \rightleftharpoons Ca^{2+} + 2F^-$ and $F^- + H^+ \rightleftharpoons HF(aq)$.
(b) Write a mass balance for CaF_2 in water if, in addition to the preceding reactions, the following reaction takes place: $HF(aq) + F^- \rightleftharpoons HF_2^-$.

12-29. **(a)** Write a mass balance for an aqueous solution of $Ca_3(PO_4)_2$ if the aqueous species are Ca^{2+}, PO_4^{3-}, HPO_4^{2-}, $H_2PO_4^-$, and H_3PO_4.
(b) Write a mass balance for a solution of $Fe_2(SO_4)_3$ if the species are Fe^{3+}, $Fe(OH)^{2+}$, $Fe(OH)_2^+$, $FeSO_4^+$, SO_4^{2-}, and HSO_4^-.

12-30. Consider the dissolution of the compound X_2Y_3, which gives $X_2Y_2^{2+}$, X_2Y^{4+}, $X_2Y_3(aq)$, and Y^{2-}. Use the mass balance to find an expression for $[Y^{2-}]$ in terms of the other concentrations. Simplify your answer as much as possible.

12-31. Ignoring activity coefficients, calculate the concentration of each ion in a solution of 4.0×10^{-8} M $Mg(OH)_2$, which is completely dissociated to Mg^{2+} and OH^-.

12-32. Consider a saturated solution of $R_3NH^+Br^-$, where R is an organic group. Find the solubility (mol/L) of $R_3NH^+Br^-$ in a solution maintained at pH 9.50.

$$R_3NH^+Br^-(s) \rightleftharpoons R_3NH^+ + Br^- \qquad K_{sp} = 4.0 \times 10^{-8}$$
$$R_3NH^+ \rightleftharpoons R_3N + H^+ \qquad K_a = 2.3 \times 10^{-9}$$

12-33. **(a)** Ignoring activity coefficients, find the concentrations of Ag^+, CN^-, and HCN in a saturated solution of AgCN whose pH is somehow fixed at 9.00. Consider the following equilibria:

$$AgCN(s) \rightleftharpoons Ag^+ + CN^- \qquad K_{sp} = 2.2 \times 10^{-16}$$
$$CN^- + H_2O \rightleftharpoons HCN(aq) + OH^- \qquad K_b = 1.6 \times 10^{-5}$$

(b) *Activity problem.* Use activity coefficients to answer **(a)**. Assume that the ionic strength is fixed at 0.10 M by addition of an inert salt. When activities are used, the statement that the pH is 9.00 means that $-\log([H^+]\gamma_{H^+}) = 9.00$.

12-34. **(a)** Consider the equilibrium

$$PbO(s) + H_2O \rightleftharpoons Pb^{2+} + 2OH^- \qquad K = 5.0 \times 10^{-16}$$

How many moles of PbO will dissolve in a 1.00-L solution if the pH is fixed at 10.50?
(b) Answer the same question asked in **(a)**, but also consider the reaction

$$Pb^{2+} + H_2O \rightleftharpoons PbOH^+ + H^+ \qquad K_a = 2.5 \times 10^{-8}$$

(c) *Activity problem.* Answer **(a)** by using activity coefficients, assuming the ionic strength is fixed at 0.050 M.

12-35. Find the fraction of 1-naphthoic acid in the form HA and the fraction in the form A^- at pH = **(a)** 2.00; **(b)** 3.00; **(c)** 3.50.

12-36. Find the fraction of pyridine (B) in the form B and the fraction in the form BH^+ at pH = **(a)** 4.00; **(b)** 5.00; **(c)** 6.00.

12-37. The base B has $pK_b = 4.00$. Find the fraction in the form B and the fraction in the form BH^+ at pH = **(a)** 9.00; **(b)** 10.00; **(c)** 10.30.

12-38. Create a spreadsheet that uses Equations 12-22 and 12-23 to compute and plot the concentrations of HA and A^- in a 0.200 M solution of hydroxybenzene as a function of pH from pH 2 to pH 12.

12-39. From the mass balance and the equilibrium expressions, following the procedure in Section 12-5, we can derive fractional composition equations for a diprotic system, H_2A:

Fraction in the form H_2A:

$$\alpha_{H_2A} = \frac{[H_2A]}{F} = \frac{[H^+]^2}{[H^+]^2 + [H^+]K_1 + K_1K_2}$$

Fraction in the form HA^-:

$$\alpha_{HA^-} = \frac{[HA^-]}{F} = \frac{K_1[H^+]}{[H^+]^2 + [H^+]K_1 + K_1K_2}$$

Fraction in the form A^{2-}:

$$\alpha_{A^{2-}} = \frac{[A^{2-}]}{F} = \frac{K_1K_2}{[H^+]^2 + [H^+]K_1 + K_1K_2}$$

in which $F = [H_2A] + [HA^-] + [A^{2-}]$. Enter these equations into a spreadsheet to calculate the fraction of *trans*-butenedioic acid ($pK_1 = 3.02$, $pK_2 = 4.48$) as a function of pH from pH 0 to 8 in 0.2 pH units. Construct a graph of the results and compare it with Figure 11-1.

How Would You Do It?

12-40. Figure 12-1 shows student data for the total concentration of dissolved iodine in solutions saturated with $PbI_2(s)$ in the presence of added KNO_3. Dissolved iodine is present as iodide ion (I^-) or iodide attached to Pb^{2+}. Dissolved iodine was measured by adding nitrite to convert iodide into iodine (I_2):

$$\underset{\text{Iodide (colorless)}}{2I^-} + \underset{\text{Nitrite}}{2NO_2^-} + 4H^+ \longrightarrow \underset{\text{Iodine (orange-brown)}}{I_2(aq)} + \underset{\text{Nitric oxide}}{2NO} + 2H_2O$$

$$K = 5 \times 10^{15}$$

Only the product I_2 is colored, so it can be measured by its absorption of visible light. To collect the data in Figure 12-1, excess $PbI_2(s)$ was shaken with various concentrations of KNO_3. Solutions were then centrifuged and the clear supernatant liquid was removed for analysis by reaction with nitrite.

By assuming that the only dissolved species were Pb^{2+} and I^-, and taking activity coefficients into account, the people who measured the concentration of dissolved iodine calculated that the solubility product for PbI_2 is 1.64×10^{-8}. This number is higher than the value 7.9×10^{-9} in Appendix A because no account was made for species such as PbI^+, $PbI_2(aq)$, PbI_3^-, and $PbOH^+$.

(a) Write the chemical reaction that produces $PbOH^+$ and propose an experiment to measure the concentration of this species, using a pH meter.

(b) I can't think of a way to distinguish the species PbI^+ from I^- by measuring total dissolved iodine. Can you propose a different kind of experiment to measure the equilibrium constant for formation of PbI^+?

12-41. We use the approximation that the activity coefficient (γ) of neutral molecules is 1.00. A more accurate relation is $\log \gamma = k\mu$, where μ is ionic strength and $k \approx 0.11$ for NH_3 and CO_2 and $k \approx 0.2$ for organic molecules. Using activity coefficients for HA, A^-, and H^+, predict the quotient

$$\frac{[H^+][A^-]/[HA] \text{ (at } \mu = 0.1 \text{ M)}}{[H^+][A^-]/[HA] \text{ (at } \mu = 0 \text{ M)}}$$

for benzoic acid ($C_6H_5CO_2H$). The observed quotient is 0.63 ± 0.03.[5]

Notes and References

1. The primary source of equilibrium constants for serious work is the computer database compiled by A. E. Martell, R. M. Smith, and R. J. Motekaitis, *NIST Critically Selected Stability Constants of Metal Complexes*, NIST Standard Reference Database 46, Gaithersburg, MD, 2001.

2. D. R. Driscol, *J. Chem. Ed.* **1979**, *56*, 603. See also R. W. Ramette, *J. Chem. Ed.* **1963**, *40*, 252.

3. For a general spreadsheet approach to computing the solubility of salts in which the anion and cation can react with water, see J. L. Guiñón, J. Garcia-Antón, and V. Pérez-Herranz, *J. Chem. Ed.* **1999**, *76*, 1157.

4. R. B. Martin, *Acc. Chem. Res.* **1994**, *27*, 204.

5. E. Koort, P. Gans, K. Herodes, V. Pihl, and I. Leito, *Anal. Bional. Chem.* **2006**, *385*, 1124.

Further Reading

W. B. Guenther, *Unified Equilibrium Calculations* (New York: Wiley, 1991).

J. N. Butler, *Ionic Equilibrium: Solubility and pH Calculations* (New York: Wiley, 1998).

A. Martell and R. Motekaitis, *Determination and Use of Stability Constants* (New York: VCH Publishers, 1992).

M. Meloun, *Computation of Solution Equilibria* (New York: Wiley, 1988).

Nature's Ion Channels

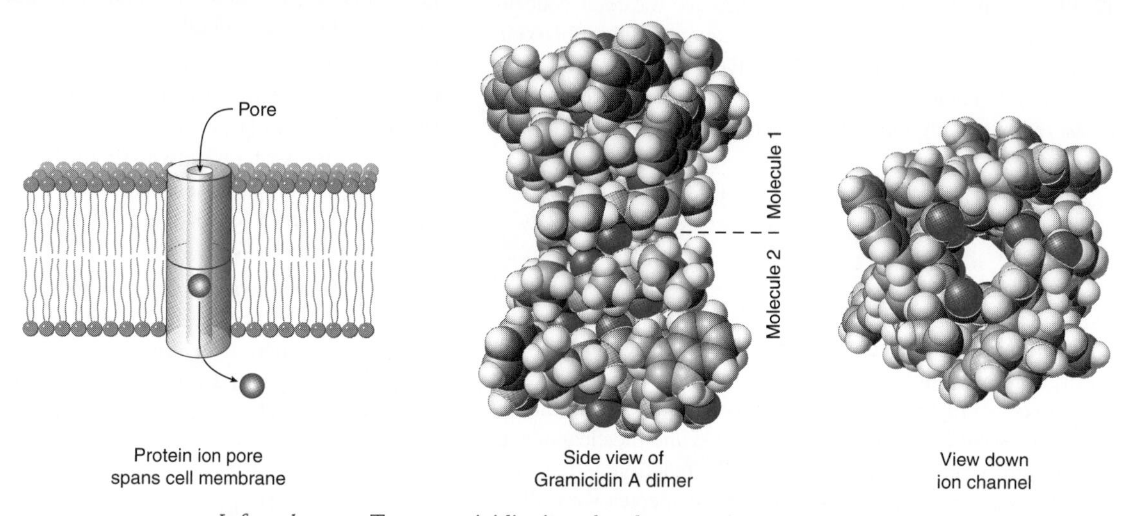

Left and center: Two gramicidin A molecules associate to span a cell membrane. *Right:* Axial view showing ion channel. [Structure from B. Roux, *Acc. Chem. Res.* **2002**, *35*, 366; based on solid-state nuclear magnetic resonance. Schematic at left from L. Stryer, *Biochemistry*, 4th ed. (New York: W. H. Freeman and Company, 1995).]

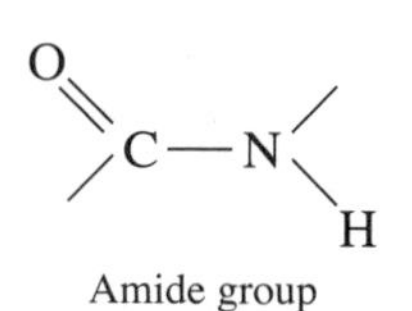

Amide group

Gramicidin A is an antibiotic that kills cells by making their membranes permeable to Na^+ and K^+. Gramicidin A is made of 15 amino acids wound into a helix with a 0.4-nm-diameter channel through the middle. The channel is lined by polar amide groups, and the outside of the peptide is covered by nonpolar hydrocarbon substituents. *Polar* groups have positive and negative regions that attract neighboring molecules by electrostatic forces. *Nonpolar* groups are not charged and are soluble inside the nonpolar cell membrane.

Metal cations are said to be *hydrophilic* ("water-loving") because they dissolve in water. Cell membranes are described as *hydrophobic* ("water-hating") because they are not soluble in water. Gramicidin A readily lodges in the cell membrane because the outside of the molecule is hydrophobic. Na^+ and K^+ pass through the hydrophilic pore at a rate of 10^7 ions/s through each ion channel. The pore excludes anions and more highly charged cations.

Part of the Nobel Prize in chemistry in 2003 was awarded to Roderick MacKinnon for elucidating the structure of channels that selectively permit K^+ to pass through membranes of cells such as nerves. Unlike gramicidin A channels, the potassium channels are selective for K^+ over Na^+. Amide oxygen atoms of the protein backbone in the channel are spaced just right to replace waters of hydration from $K(H_2O)_6^+$. There is little change in energy when hydrated K^+ sheds H_2O and binds inside the channel. The spacing of amide oxygens is too great by 0.04 nm to displace H_2O from $Na(H_2O)_6^+$. Hydrated Na^+ remains outside the channel, whereas hydrated K^+ sheds H_2O and binds inside the channel. K^+ passes at a rate of 10^8 ions/s per channel—100 times faster than Na^+.

CHAPTER 13

EDTA Titrations

EDTA is a merciful abbreviation for *ethylenediaminetetraacetic acid*, a compound that can be used to titrate most metal ions by forming strong 1:1 complexes. In addition to its place in chemical analysis, EDTA plays a larger role as a strong metal-binding agent in industrial processes and in household products such as soaps, cleaning agents, and food additives that prevent metal-catalyzed oxidation of food. EDTA participates in environmental chemistry. For example, the majority of nickel and a significant fraction of the iron, lead, copper, and zinc discharged into San Francisco Bay are EDTA complexes that pass unscathed through wastewater treatment plants.

13-1 Metal-Chelate Complexes

An atom or group of atoms bound to whatever atom you are interested in is called a **ligand**. A ligand with a pair of electrons to share can bind to a metal ion that can accept a pair of electrons. Electron pair acceptors are called **Lewis acids** and electron pair donors are called **Lewis bases**. Cyanide is said to be a **monodentate** ("one-toothed") **ligand** because it binds to a metal ion through only one atom (the carbon atom). A **multidentate ligand** binds to a metal ion through more than one ligand atom. EDTA in Figure 13-1 is *hexadentate*, binding to a metal through two N atoms and four O atoms.

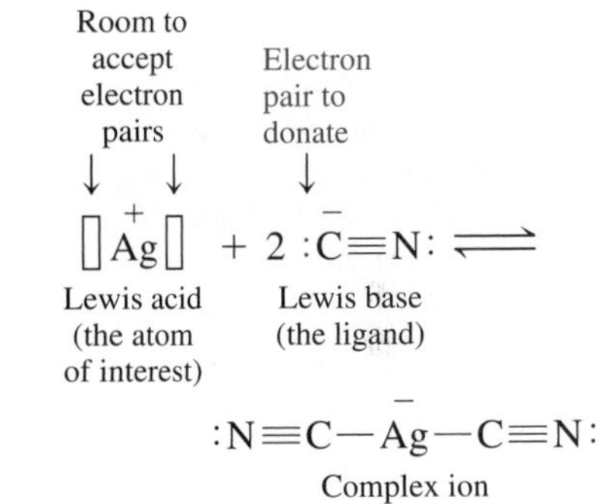

Lewis acid: electron pair acceptor
Lewis base: electron pair donor

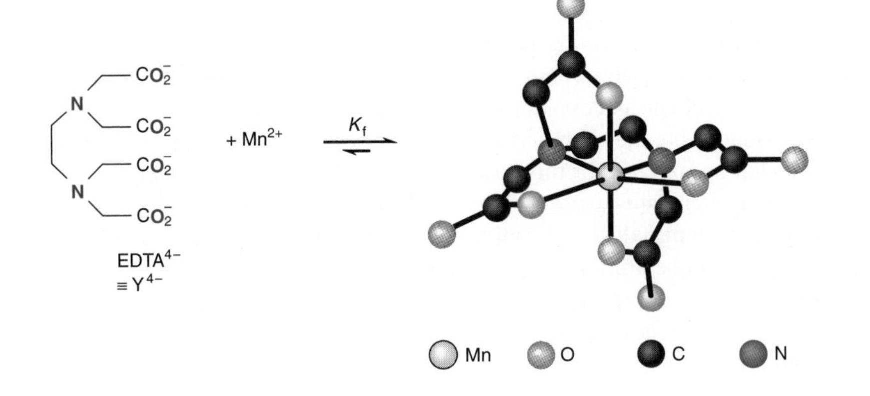

Figure 13-1 EDTA forms strong 1:1 complexes with most metal ions, binding through four oxygen and two nitrogen atoms. The six-coordinate structure of Mn^{2+}-EDTA is found in the compound $KMnEDTA \cdot 2H_2O$. [From J. Stein, J. P. Fackler, Jr., G. J. McClune, J. A. Fee, and L. T. Chan, *Inorg. Chem.* **1979**, *18*, 3511.]

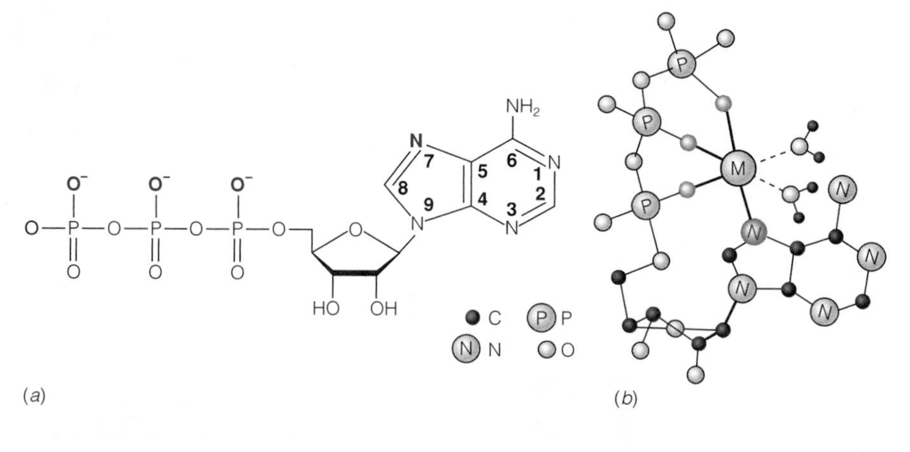

Figure 13-2 (*a*) Structure of adenosine triphosphate (ATP), with ligand atoms shown in color. (*b*) Possible structure of a metal-ATP complex, with four bonds to ATP and two bonds to H_2O ligands.

A multidentate ligand is also called a **chelating ligand**, or just a *chelate*, pronounced **KEE**-late. The term "chelate" is derived from the great claw, or *chela* (from the Greek *chēlē*) of the lobster. The chelating ligand engulfs a metal ion the way a lobster might grab an object with its claw. Proteins lining the ion channels described at the opening of this chapter behave as chelating ligands for ions passing through the channels.

Most transition metal ions bind six ligand atoms. An important *tetradentate* ligand is adenosine triphosphate (ATP), which binds to divalent metal ions (such as Mg^{2+}, Mn^{2+}, Co^{2+}, and Ni^{2+}) through four of their six coordination positions (Figure 13-2). The fifth and sixth positions are occupied by water molecules. The biologically active form of ATP is generally the Mg^{2+} complex.

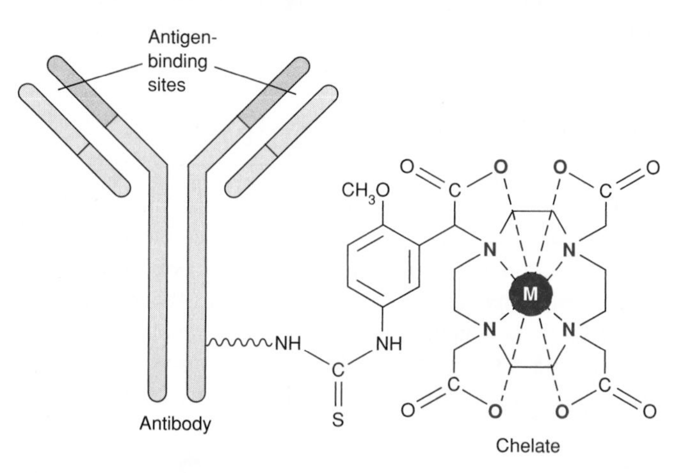

Figure 13-3 Synthetic chelate covalently attached to an antibody carries a metal isotope (M) to deliver lethal doses of radiation to tumor cells.

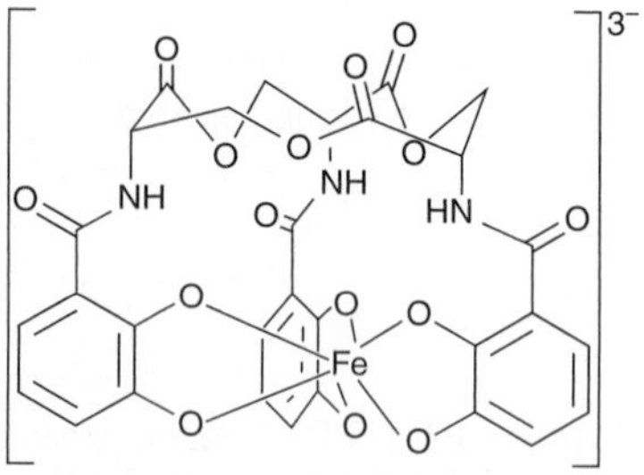

The synthetic *octadentate* ligand in Figure 13-3 is being evaluated as an anticancer agent. This chelate binds a metal tightly through four N atoms and four O atoms. The chelate is covalently attached to a *monoclonal antibody*, which is a protein produced by one specific type of cell in response to one specific foreign substance called an *antigen*. In this case, the antibody binds to a specific feature of a tumor cell. The chelate carries a short-lived radioisotope such as $^{90}Y^{3+}$ or $^{177}Lu^{3+}$, which delivers lethal doses of radiation to the tumor. Box 13-1 describes another important use of chelating agents in medicine.

Figure 13-4 Iron(III)-enterobactin complex. Certain bacteria secrete enterobactin to capture iron and bring it into the cell. Enterobactin is one of several known chelates—designated *siderophores*—released by microbes to capture iron to be used by the cell.

Metal-chelate complexes are ubiquitous in biology. Bacteria such as *Escherichia coli* and *Salmonella enterica* in your gut excrete a powerful iron chelate called enterobactin (Figure 13-4) to scavenge iron that is essential for bacterial

Box 13-1 Chelation Therapy and Thalassemia

Oxygen (O_2) in the human circulatory system is bound to iron in the protein hemoglobin, which consists of two pairs of subunits, designated α and β. β-Thalassemia major is a genetic disease in which β subunits are not synthesized in adequate quantities. A child afflicted with this disease survives only with frequent transfusions of normal red blood cells. However, the child accumulates 4–8 g of iron per year from hemoglobin in the transfused cells. Our bodies have no mechanism for excreting large quantities of iron, and most patients die by age 20 from toxic effects of iron overload. One reason why iron is toxic is that it catalyzes the formation of hydroxyl radical (HO·), a powerful and destructive oxidant.

To enhance iron excretion, intensive chelation therapy is used. The most successful drug is *desferrioxamine B*, a powerful Fe^{3+}-chelate produced by the microbe *Streptomyces pilosus.* The formation constant for binding Fe^{3+} to form the iron complex, ferrioxamine B, is $10^{30.6}$. The illustration shows structures of ferrioxamine complexes and a graph of treatment results. Used in conjunction with ascorbic acid (vitamin C, a reducing agent that reduces Fe^{3+} to the more soluble Fe^{2+}) desferrioxamine clears several grams of iron per year from an overloaded patient. The ferrioxamine complex is excreted in the urine.

Desferrioxamine reduces the incidence of heart and liver disease in thalassemia patients and maintains approximate iron balance. In patients for whom desferrioxamine effectively controls iron overload, there is a 91% rate of cardiac disease-free survival after 15 years of chelation therapy. Among the negative effects of desferrioxamine is that too high a dose stunts the growth of children.

Desferrioxamine is expensive and must be taken by continuous injection. It is not absorbed through the intestine. Many potent iron chelates have been tested to find an effective one that can be taken orally, but only deferiprone is currently used orally. In the long term, bone marrow transplants or gene therapy might cure the disease.

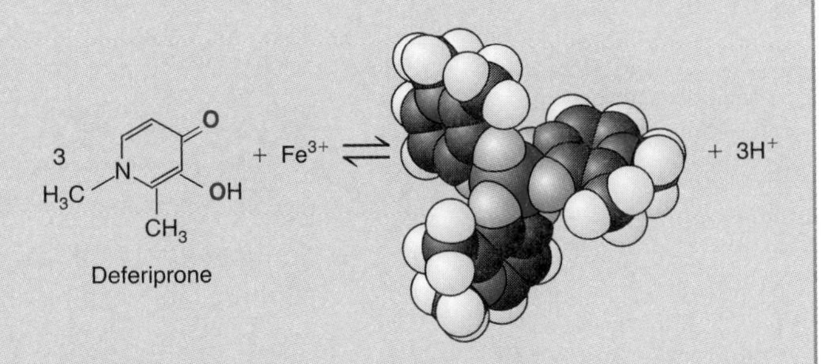

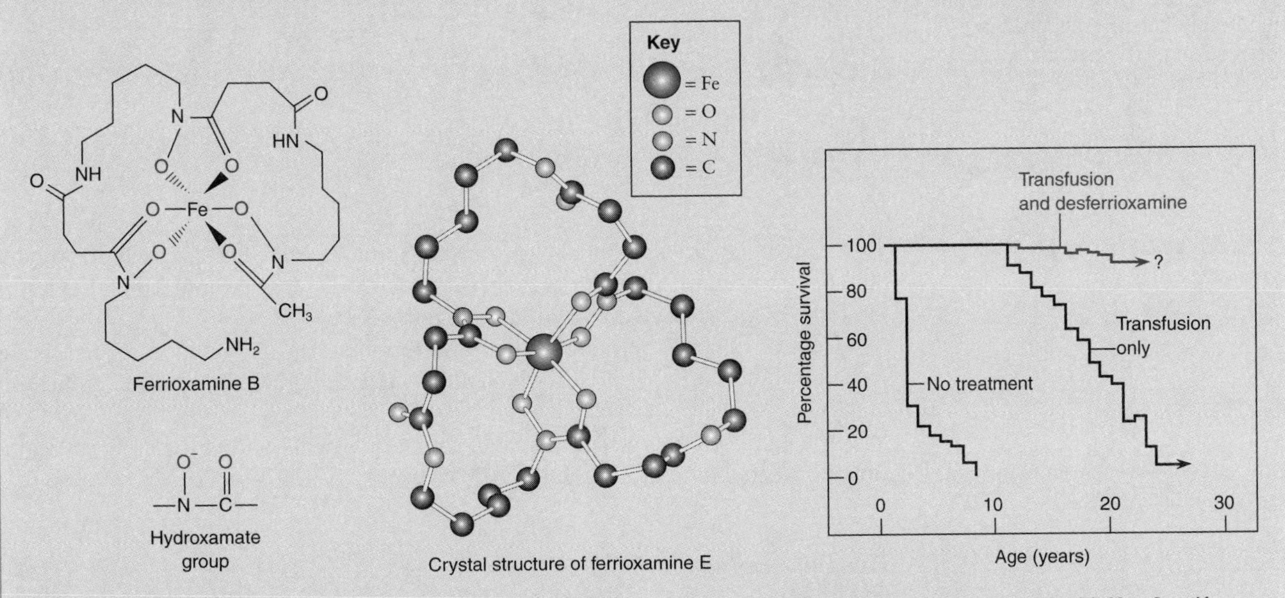

Structure of the iron complex ferrioxamine B and crystal structure of the related compound, ferrioxamine E, in which the chelate has a cyclic structure. Graph shows success of transfusions and transfusions plus chelation therapy. [Crystal structure kindly provided by M. Neu, Los Alamos National Laboratory, based on D. Van der Helm and M. Poling, *J. Am. Chem. Soc.* **1976**, *98*, 82. Graph from P. S. Dobbin and R. C. Hider, *Chem. Br.* **1990**, *26*, 565.]

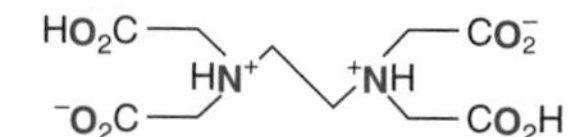

EDTA
Ethylenediaminetetraacetic acid
(also called ethylenedinitrilotetraacetic acid)

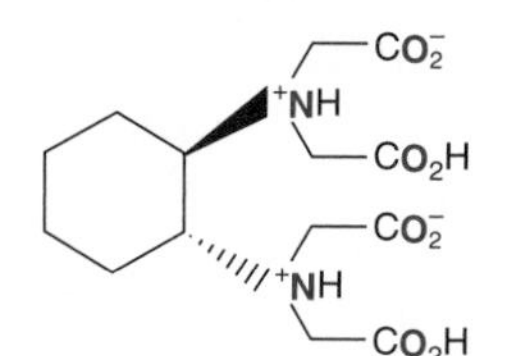

DCTA
trans-1,2-Diaminocyclohexanetetraacetic acid

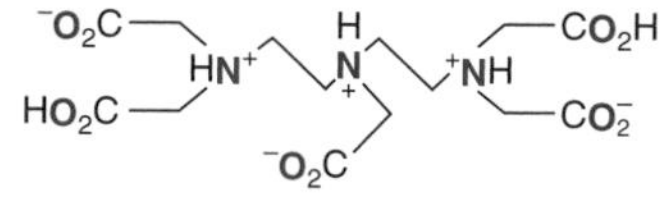

DTPA
Diethylenetriaminepentaacetic acid

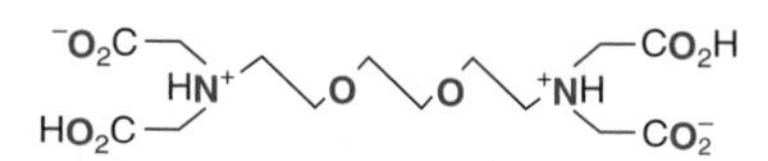

EGTA
Bis(aminoethyl)glycolether-*N*,*N*,*N*′,*N*′-tetraacetic acid

Figure 13-5 Analytically useful synthetic chelating agents that form strong 1:1 complexes with most metal ions.

growth.[1] Chelates excreted by microbes to gather iron are called siderophores. The iron-enterobactin complex is recognized at specific sites on the bacterial cell surface and taken into the cell, and iron is released by enzymatic disassembly of the chelate. To fight bacterial infection, your immune system produces a protein called siderocalin to sequester and inactivate enterobactin.[2]

The aminocarboxylic acids in Figure 13-5 are synthetic chelating agents whose nitrogen and carboxylate oxygen atoms can lose protons and bind to metal ions. Molecules in Figure 13-5 form strong 1:1 complexes with all metal ions, except univalent ions such as Li^+, Na^+, and K^+. *The stoichiometry is 1:1 regardless of the charge on the ion.* A titration based on complex formation is called a **complexometric titration**.

Ask Yourself

13-A. What is the difference between a monodentate and a multidentate ligand? Is a chelating ligand monodentate or multidentate?

13-2 EDTA

EDTA is, by far, the most widely used chelator in analytical chemistry. By direct titration or through an indirect sequence of reactions, virtually every element of the periodic table can be analyzed with EDTA.

EDTA is a hexaprotic system, designated H_6Y^{2+}. The highlighted acidic hydrogen atoms are the ones that are lost on metal-complex formation:

$$(\mathrm{HO_2CCH_2})_2\overset{+}{\mathrm{N}}\mathrm{HCH_2CH_2}\overset{+}{\mathrm{N}}\mathrm{H}(\mathrm{CH_2CO_2H})_2 \quad (\mathrm{H_6Y^{2+}})$$

$pK_1 = 0.0\ (CO_2\mathbf{H})$ $pK_4 = 2.69\ (CO_2\mathbf{H})$
$pK_2 = 1.5\ (CO_2\mathbf{H})$ $pK_5 = 6.13\ (\mathbf{N}H^+)$
$pK_3 = 2.00\ (CO_2\mathbf{H})$ $pK_6 = 10.37\ (\mathbf{N}H^+)$

One mole of EDTA reacts with *one* mole of metal ion.

The first four pK values apply to carboxyl protons, and the last two are for the ammonium protons. Below a pH of 10.24, most EDTA is protonated and is not in the form Y^{4-} that binds to metal ions (Figure 13-1).

Neutral EDTA is tetraprotic, with the formula H_4Y. A common reagent is the disodium salt, $Na_2H_2Y \cdot 2H_2O$, which attains the dihydrate composition on heating at 80°C.

Box 13-2 describes the notation used for formation constants.

The equilibrium constant for the reaction of a metal with a ligand is called the **formation constant**, K_f, or the *stability constant*:

$$\textit{Formation constant:} \quad M^{n+} + Y^{4-} \rightleftharpoons MY^{n-4} \qquad K_f = \frac{[MY^{n-4}]}{[M^{n+}][Y^{4-}]} \tag{13-1}$$

Only some of the EDTA is in the form Y^{4-}.

Table 13-1 shows that formation constants for EDTA complexes are large and tend to be larger for more positively charged metal ions. Note that K_f is defined for reaction of the species Y^{4-} with the metal ion. At low pH, most EDTA is in one of its protonated forms, not Y^{4-}.

Box 13-2 Notation for Formation Constants

Formation constants are equilibrium constants for complex formation. The **stepwise formation constants**, designated K_i, are defined as follows:

$$M + X \overset{K_1}{\rightleftharpoons} MX \qquad K_1 = [MX]/[M][X]$$

$$MX + X \overset{K_2}{\rightleftharpoons} MX_2 \qquad K_2 = [MX_2]/[MX][X]$$

$$MX_{n-1} + X \overset{K_n}{\rightleftharpoons} MX_n \qquad K_n = [MX_n]/[MX_{n-1}][X]$$

where M is a metal ion and X is a ligand. The **overall**, or **cumulative**, **formation constants** are denoted β_i:

$$M + 2X \overset{\beta_2}{\rightleftharpoons} MX_2 \qquad \beta_2 = [MX_2]/[M][X]^2$$

$$M + nX \overset{\beta_n}{\rightleftharpoons} MX_n \qquad \beta_n = [MX_n]/[M][X]^n$$

A useful relation is that $\beta_n = K_1K_2 \cdots K_n$. Page 265 shows cumulative formation constants for lead-iodide complexes used to calculate the composition of a saturated solution of PbI_2.

A metal-EDTA complex becomes unstable at low pH because H^+ competes with the metal ion for EDTA. At too high a pH, the EDTA complex is unstable because OH^- competes with EDTA for the metal ion and may precipitate the metal hydroxide or form unreactive hydroxide complexes. Figure 13-6 shows the pH ranges over which some common metal ions can be titrated. Pb^{2+}, for example, reacts "quantitatively" with EDTA between pH 3 and pH 12. Between pH 9 and 12, it is necessary to use an **auxiliary complexing agent**, which forms a weak complex with Pb^{2+} and keeps it in solution (Color Plate 7). The auxiliary complexing agent is displaced by EDTA during the titration. Auxiliary complexing agents, such as ammonia, tartrate, citrate, or triethanolamine, prevent metal ions from precipitating in the absence of EDTA. The titration of Pb^{2+} is carried out at pH 10 in the presence of tartrate, which complexes the metal ion and does not allow $Pb(OH)_2$ to precipitate.

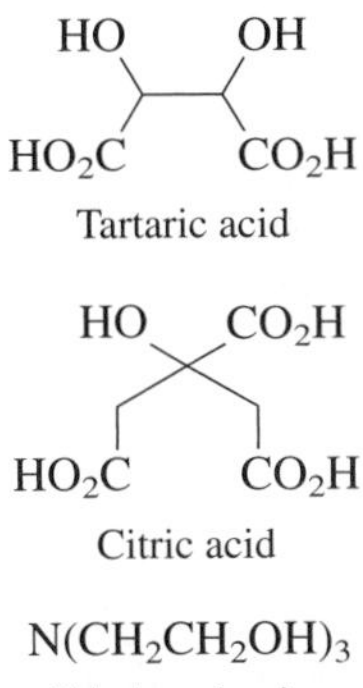

Table 13-1 Formation constants for metal-EDTA complexes

Ion	log K_f	Ion	log K_f	Ion	log K_f	Ion	log K_f
Li^+	2.95	V^{2+}	12.7[a]	Fe^{3+}	25.1	Sn^{2+}	18.3[b]
Na^+	1.86	Cr^{2+}	13.6[a]	Co^{3+}	41.4	Pb^{2+}	18.0
K^+	0.8	Mn^{2+}	13.89	Zr^{4+}	29.3	Al^{3+}	16.4
Be^{2+}	9.7	Fe^{2+}	14.30	VO^{2+}	18.7	Ga^{3+}	21.7
Mg^{2+}	8.79	Co^{2+}	16.45	VO_2^+	15.5	In^{3+}	24.9
Ca^{2+}	10.65	Ni^{2+}	18.4	Ag^+	7.20	Tl^{3+}	35.3
Sr^{2+}	8.72	Cu^{2+}	18.78	Tl^+	6.41	Bi^{3+}	27.8[a]
Ba^{2+}	7.88	Ti^{3+}	21.3	Pd^{2+}	25.6[a]	Ce^{3+}	15.93
Ra^{2+}	7.4	V^{3+}	25.9[a]	Zn^{2+}	16.5	Gd^{3+}	17.35
Sc^{3+}	23.1[a]	Cr^{3+}	23.4[a]	Cd^{2+}	16.5	Th^{4+}	23.2
Y^{3+}	18.08	Mn^{3+}	25.2	Hg^{2+}	21.5	U^{4+}	25.7
La^{3+}	15.36						

Note: The stability constant is the equilibrium constant for the reaction $M^{n+} + Y^{4-} \rightleftharpoons MY^{n-4}$. Values in table apply at 25°C and ionic strength 0.1 M unless otherwise indicated.

a. 20°C, ionic strength = 0.1 M.

b. 20°C, ionic strength = 1 M.

SOURCE: A. E. Martell, R. M. Smith, and R. J. Motekaitis, *NIST Critically Selected Stability Constants of Metal Complexes*, NIST Standard Reference Database 46, Gaithersburg, MD, 2001.

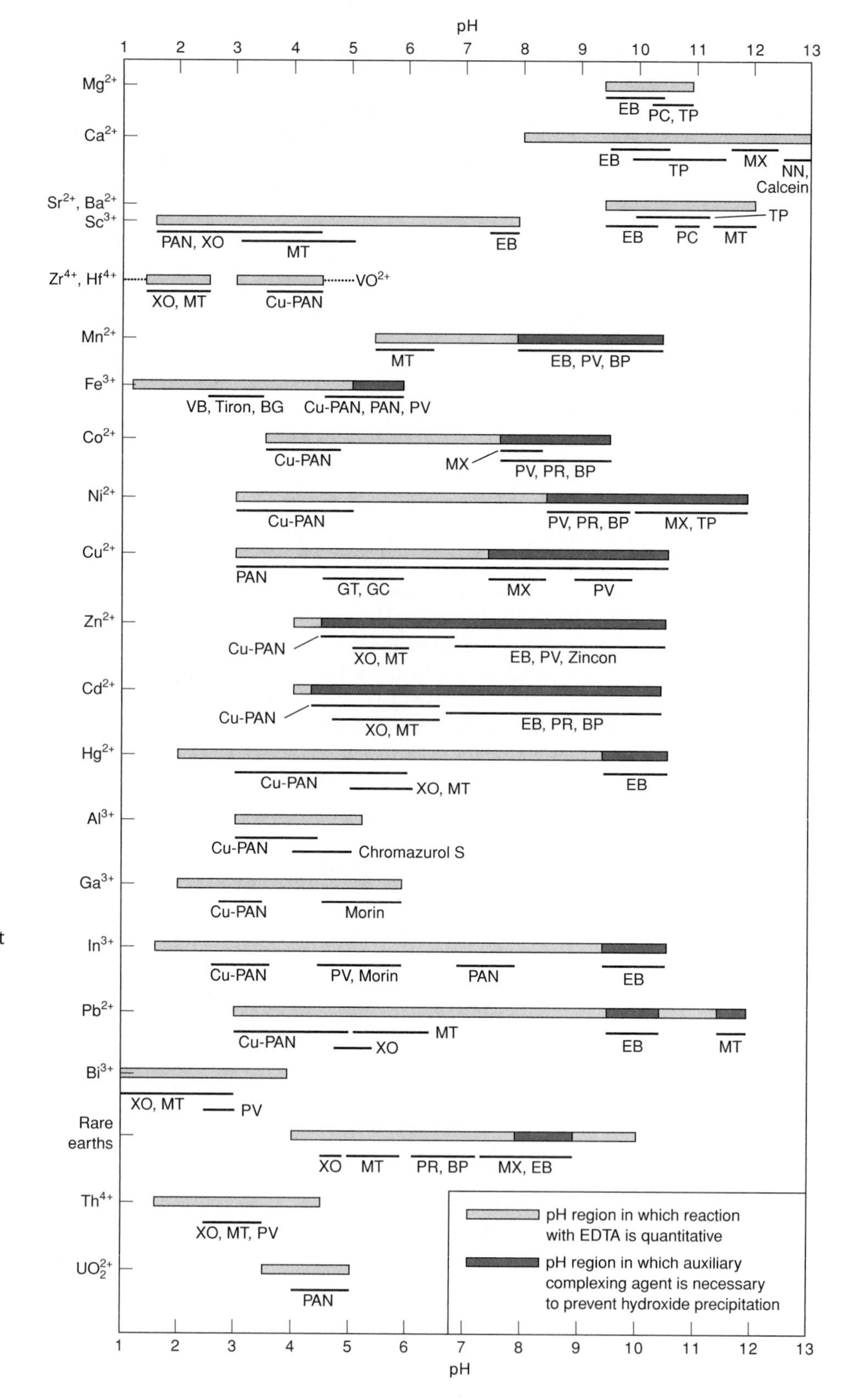

Figure 13-6 Guide to EDTA titrations of some common metals. Light color shows pH region in which reaction with EDTA is quantitative. Dark color shows pH region in which auxiliary complexing agent such as ammonia is required to prevent metal from precipitating. [Adapted from K. Ueno, *J. Chem. Ed.* **1965**, *42*, 432.]

Abbreviations for indicators:
BG, Bindschedler's green leuco base
BP, Bromopyrogallol red
Cu-PAN, PAN plus Cu-EDTA
EB, Eriochrome black T
GC, Glycinecresol red
GT, Glycinethymol blue
MT, Methylthymol blue
MX, Murexide
NN, Patton & Reeder's dye
PAN, Pyridylazonaphthol
PC, *o*-Cresolphthalein complexone
PR, Pyrogallol red
PV, Pyrocatechol violet
TP, Thymolphthalein complexone
VB, Variamine blue B base
XO, Xylenol orange

The lead-tartrate complex must be less stable than the lead-EDTA complex or the titration would not be feasible.

Figure 13-6 shows pH ranges in which various *metal ion indicators* (discussed in the next section) are useful for finding end points. The chart also provides a strategy for selective titration of one ion in the presence of another. For example, a solution containing both Fe^{3+} and Ca^{2+} could be titrated with EDTA at pH 4. At this pH, Fe^{3+} is titrated without interference from Ca^{2+}.

Ask Yourself

13-B. **(a)** Write the reaction whose equilibrium constant is the formation constant for EDTA complex formation and write the algebraic form of K_f.
(b) Why is EDTA complex formation less complete at lower pH?
(c) What is the purpose of an auxiliary complexing agent?
(d) The following diagram is analogous to those in Section 11-3 for polyprotic acids. It shows the pH at which each species of EDTA is predominant. Fill in the pH at each arrow. Explain the significance of the pH at the borders between regions and at the center of each region.

H_6Y^{2+}	H_5Y^+	H_4Y	H_3Y^-	H_2Y^{2-}	HY^{3-}	Y^{4-}

pH: 1.5 (at the border between H_5Y^+ and H_4Y); 8.25 (at the center of HY^{3-})

13-3 Metal Ion Indicators

A **metal ion indicator** is a compound whose color changes when it binds to a metal ion. Two common indicators are shown in Table 13-2. *For an indicator to be useful, it must bind metal less strongly than EDTA does.*

A typical analysis is illustrated by the titration of Mg^{2+} with EDTA, using Calmagite as the indicator:

$$\underset{\text{Red}}{\text{MgIn}} + \underset{\text{Colorless}}{\text{EDTA}} \longrightarrow \underset{\text{Colorless}}{\text{MgEDTA}} + \underset{\text{Blue}}{\text{In}} \tag{13-2}$$

The indicator must release its metal to EDTA.

At the start of the experiment, a small amount of indicator (In) is added to the colorless solution of Mg^{2+} to form a red complex. As EDTA is added, it reacts first with free, colorless Mg^{2+}. When free Mg^{2+} is used up, the last EDTA added before the equivalence point displaces indicator from the red MgIn complex. The change from the red of MgIn to the blue of unbound In signals the end point of the titration (Demonstration 13-1).

Most metal ion indicators are also acid-base indicators. Because the color of free indicator is pH dependent, most indicators can be used only in certain pH ranges. For example, xylenol orange (pronounced ZY-leen-ol) in Table 13-2 changes from yellow to red when it binds to a metal ion at pH 5.5. This is an easy color change to observe. At pH 7.5, the change from violet to red is rather difficult to see.

Table 13-2 Some metal ion indicators

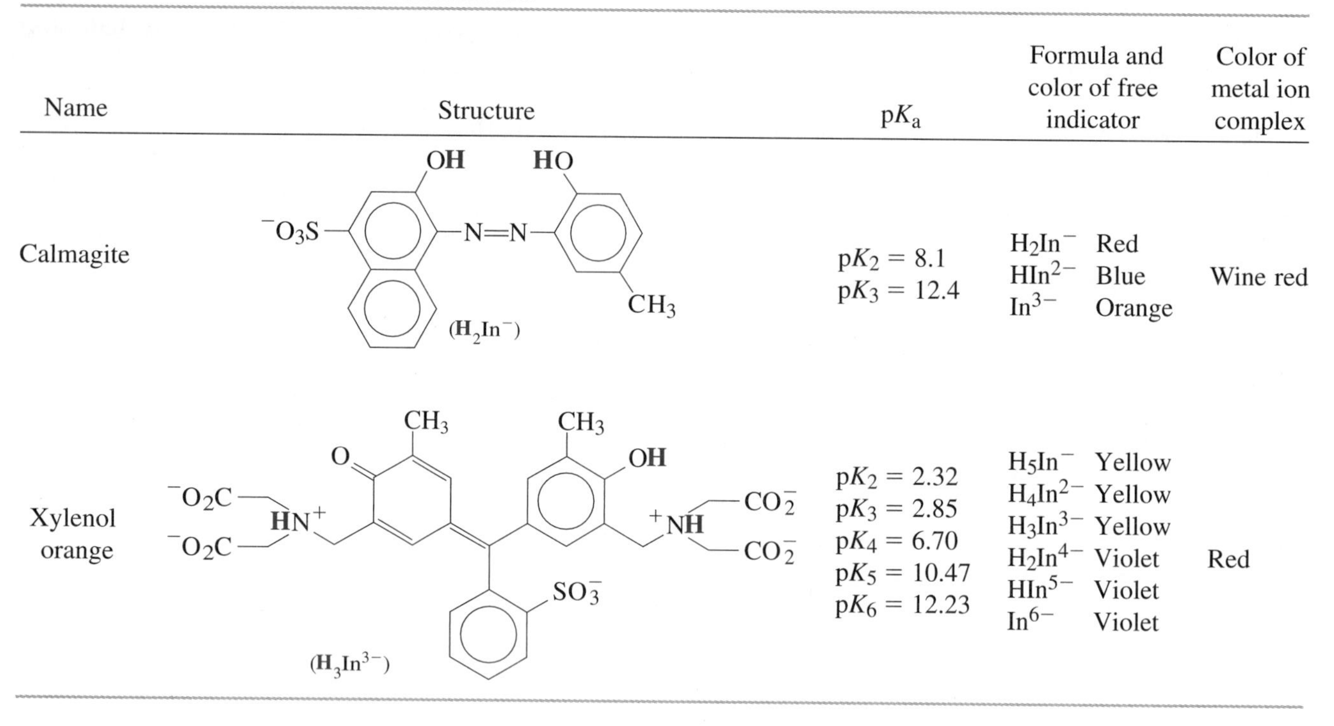

Name	Structure	pK_a	Formula and color of free indicator	Color of metal ion complex
Calmagite	(H_2In^-)	$pK_2 = 8.1$ $pK_3 = 12.4$	H_2In^- Red HIn^{2-} Blue In^{3-} Orange	Wine red
Xylenol orange	(H_3In^{3-})	$pK_2 = 2.32$ $pK_3 = 2.85$ $pK_4 = 6.70$ $pK_5 = 10.47$ $pK_6 = 12.23$	H_5In^- Yellow H_4In^{2-} Yellow H_3In^{3-} Yellow H_2In^{4-} Violet HIn^{5-} Violet In^{6-} Violet	Red

Demonstration 13-1 Metal Ion Indicator Color Changes

This demonstration illustrates the color change associated with Reaction 13-2 and shows how a second dye can be added to a solution to produce a more easily detected color change.

Stock solutions

Calmagite: Dissolve 0.05 g of indicator in 100 mL of water. Alternatively, dissolve 0.1 g Eriochrome black T indicator in 7.5 mL of triethanolamine plus 2.5 mL of absolute ethanol. Color changes are the same for both indicators.

Methyl red: Dissolve 0.02 g in 60 mL of ethanol; then add 40 mL of water.

Buffer: Add 142 mL of concentrated (14.5 M) aqueous ammonia to 17.5 g of ammonium chloride and dilute to 250 mL with water.

$MgCl_2$: 0.05 M

EDTA: 0.05 M $Na_2H_2EDTA \cdot 2H_2O$

Prepare a solution containing 25 mL of $MgCl_2$, 5 mL of buffer, and 300 mL of water. Add 6 drops of indicator and titrate with EDTA. Note the color change from wine red to pale blue at the end point (Color Plate 8*a*). The spectroscopic change accompanying the color change is shown in the figure.

For some people, the change of indicator color is not so easy to see. Adding 3 mL of methyl red (or other yellow dyes) produces an orange color prior to the end point and a green color after it. This sequence of colors is shown in Color Plate 8*b*.

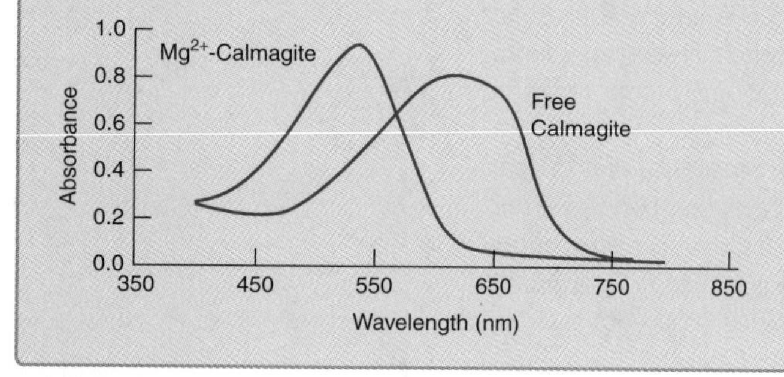

Visible spectra of Mg^{2+}-Calmagite and free Calmagite at pH 10 in ammonia buffer. [From C. E. Dahm, J. W. Hall, and B. E. Mattioni, *J. Chem. Ed.* **2004**, *81*, 1787.]

For an indicator to be useful in an EDTA titration, the indicator must give up its metal ion to EDTA. If metal does not freely dissociate from the indicator, the metal is said to **block** the indicator. Calmagite is blocked by Cu^{2+}, Ni^{2+}, Co^{2+}, Cr^{3+}, Fe^{3+}, and Al^{3+}. It cannot be used for the direct titration of any of these metals. It can be used for a back titration, however. For example, excess standard EDTA can be added to Cu^{2+}. Then indicator is added and excess EDTA is back-titrated with Mg^{2+}.

Question What will the color change be when the back titration is performed?

Ask Yourself

13-C. **(a)** Explain why the change from red to blue in Reaction 13-2 occurs suddenly at the equivalence point instead of gradually throughout the entire titration. **(b)** EDTA buffered to pH 5 was titrated with standard Pb^{2+}, using xylenol orange as indicator (Table 13-2).
- **(i)** Which is the principal species of the indicator at pH 5?
- **(ii)** What color was observed before the equivalence point?
- **(iii)** What color was observed after the equivalence point?
- **(iv)** What would the color change be if the titration were conducted at pH 8 instead of pH 5?

13-4 EDTA Titration Techniques

EDTA can be used directly or indirectly to analyze most elements of the periodic table. In this section, we discuss several important techniques.

Direct Titration

In a **direct titration**, analyte is titrated with standard EDTA. The analyte is buffered to an appropriate pH, at which the reaction with EDTA is essentially complete and the free indicator has a color distinctly different from that of the metal-indicator complex. An auxiliary complexing agent might be required to maintain the metal ion in solution in the absence of EDTA.

Back Titration

In a **back titration**, a known excess of EDTA is added to the analyte. Excess EDTA is then titrated with a standard solution of metal ion. A back titration is necessary if analyte precipitates in the absence of EDTA, if analyte reacts too slowly with EDTA, or if analyte blocks the indicator. The metal used in the back titration must not displace analyte from EDTA.

Example **A Back Titration**

Ni^{2+} can be analyzed by a back titration with standard Zn^{2+} at pH 5.5 and xylenol orange indicator. A solution containing 25.00 mL of Ni^{2+} in dilute HCl was treated with 25.00 mL of 0.052 83 M Na_2EDTA. The solution was neutralized with NaOH, and the pH was adjusted to 5.5 with acetate buffer. The solution turned yellow when a few drops of indicator were added. Titration with 0.022 99 M Zn^{2+} required 17.61 mL of Zn^{2+} to reach the red end point. What was the molarity of Ni^{2+} in the unknown?

SOLUTION The unknown was treated with 25.00 mL of 0.052 83 M EDTA, which contains (25.00 mL)(0.052 83 M) = 1.320 8 mmol of EDTA.

$$\underset{x \text{ mmol}}{Ni^{2+}} + \underset{1.320\,8 \text{ mmol}}{EDTA} \longrightarrow Ni(EDTA) + \underset{1.320\,8 - x \text{ mmol}}{EDTA}$$

Back titration required (17.61 mL)(0.022 99 M) = 0.404 9 mmol of Zn^{2+}.

$$\underset{0.404\,9 \text{ mmol}}{Zn^{2+}} + \underset{1.320\,8 - x \text{ mmol}}{EDTA} \longrightarrow \underset{0.404\,9 \text{ mmol}}{Zn(EDTA)}$$

The moles of Zn^{2+} required in the second reaction must equal the moles of excess EDTA from the first reaction:

$$0.404\,9 \text{ mmol } Zn^{2+} = 1.320\,8 \text{ mmol EDTA} - x \text{ mmol } Ni^{2+}$$
$$x = 0.915\,9 \text{ mmol } Ni^{2+}$$

The concentration of Ni^{2+} is 0.915 9 mmol/25.00 mL = 0.036 64 M.

Test Yourself Suppose that 0.040 00 M EDTA was used and back titration required 15.00 mL of Zn^{2+}. Find the concentration of Ni^{2+}. (**Answer:** 0.026 21 M)

An EDTA back titration can prevent precipitation of analyte. For example, $Al(OH)_3$ precipitates at pH 7 in the absence of EDTA. An acidic solution of Al^{3+} can be treated with excess EDTA, adjusted to pH 7 with sodium acetate, and boiled to ensure complete complexation. The Al^{3+}-EDTA complex is stable at pH 7. The solution is then cooled; Calmagite indicator is added; and back titration with standard Zn^{2+} is performed.

Displacement Titration

For some metal ions, there is no satisfactory indicator, but a **displacement titration** is feasible. In this procedure, the analyte is usually treated with excess $Mg(EDTA)^{2-}$ to displace Mg^{2+}, which is later titrated with standard EDTA.

Challenge Calculate the equilibrium constant for Reaction 13-3 if $M^{n+} = Hg^{2+}$. Why is $Mg(EDTA)^{2-}$ used for a displacement titration?

$$M^{n+} + MgY^{2-} \longrightarrow MY^{n-4} + Mg^{2+} \qquad (13\text{-}3)$$

Hg^{2+} is determined in this manner. The formation constant of $Hg(EDTA)^{2-}$ must be greater than the formation constant of $Mg(EDTA)^{2-}$ or else the displacement of Mg^{2+} from $Mg(EDTA)^{2-}$ would not occur.

There is no suitable indicator for Ag^+. However, Ag^+ will displace Ni^{2+} from the tetracyanonickelate(II) ion:

$$2Ag^+ + Ni(CN)_4^{2-} \longrightarrow 2Ag(CN)_2^- + Ni^{2+}$$

The liberated Ni^{2+} can then be titrated with EDTA to find out how much Ag^+ was added.

Indirect Titration

Anions that precipitate metal ions can be analyzed with EDTA by **indirect titration.** For example, sulfate can be analyzed by precipitation with excess Ba^{2+} at pH 1.

The $BaSO_4(s)$ is filtered, washed, and boiled with excess EDTA at pH 10 to bring Ba^{2+} back into solution as $Ba(EDTA)^{2-}$. The excess EDTA is back-titrated with Mg^{2+}.

Alternatively, an anion can be precipitated with excess metal ion. The precipitate is filtered and washed, and the excess metal ion in the filtrate is titrated with EDTA. CO_3^{2-}, CrO_4^{2-}, S^{2-}, and SO_4^{2-} can be determined in this manner.

Masking

Masking prevents one element from interfering in the analysis of another element. Box 13-3 describes an important application of masking.

A **masking agent** is a reagent that protects some component of the analyte from reaction with EDTA. For example, Mg^{2+} in a mixture of Mg^{2+} and Al^{3+} can be titrated by masking Al^{3+} with F^- to generate AlF_6^{3-}, which does not react with EDTA. Only the Mg^{2+} reacts with EDTA.

Cyanide is a masking agent that forms complexes with Cd^{2+}, Zn^{2+}, Hg^{2+}, Co^{2+}, Cu^+, Ag^+, Ni^{2+}, Pd^{2+}, Pt^{2+}, Fe^{2+}, and Fe^{3+}, but not with Mg^{2+}, Ca^{2+}, Mn^{2+}, or Pb^{2+}. When CN^- is added to a solution containing Cd^{2+} and Pb^{2+}, only the Pb^{2+} can react with EDTA. (**CAUTION:** Cyanide forms toxic gaseous HCN below pH 11. Cyanide solutions should be made strongly basic and handled in a hood.) Fluoride masks Al^{3+}, Fe^{3+}, Ti^{4+}, and Be^{2+}. (**CAUTION:** HF formed by F^- in acidic solution is extremely hazardous and should not contact skin or eyes. It may not be immediately painful, but the affected area should be flooded with water for 5 min and then treated with 2.5 wt% calcium gluconate gel that you have on hand *before* the accident. First aiders must wear rubber gloves to protect themselves. Damage

Box 13-3 What Is Hard Water?

Hardness refers to the total concentration of alkaline earth ions in water. The concentrations of Ca^{2+} and Mg^{2+} are usually much greater than those of other Group 2 ions, so hardness can be equated to $[Ca^{2+}] + [Mg^{2+}]$. Hardness is commonly expressed as the equivalent number of milligrams of $CaCO_3$ per liter. Thus, if $[Ca^{2+}] + [Mg^{2+}] = 1$ mM, we would say that the hardness is 100 mg $CaCO_3$ per liter, because 100 mg $CaCO_3$ = 1 mmol $CaCO_3$. Water whose hardness is less than 60 mg $CaCO_3$ per liter is considered to be "soft."

Hard water reacts with soap to form insoluble curds:

$$\underset{}{Ca^{2+}} + \underset{\text{Soap}}{2RSO_3^-} \longrightarrow \underset{\text{Precipitate}}{Ca(RSO_3)_2(s)} \quad \text{(A)}$$

Enough soap to consume the Ca^{2+} and Mg^{2+} must be used before the soap is useful for cleaning. Hard water is not thought to be unhealthy. Hardness is beneficial in irrigation water because the alkaline earth ions tend to *flocculate* (cause to aggregate) *colloidal* particles in soil and thereby increase the permeability of the soil to water. Colloids are soluble particles that are 1–500 nm in diameter (see Demonstration 7-1). Such small particles tend to plug the paths by which water can drain through soil.

To measure hardness, water is treated with ascorbic acid to reduce Fe^{3+} to Fe^{2+} and with cyanide to mask Fe^{2+}, Cu^+, and several other minor metal ions. Titration with EDTA at pH 10 in ammonia buffer gives $[Ca^{2+}] + [Mg^{2+}]$. $[Ca^{2+}]$ can be determined separately if the titration is carried out at pH 13 without ammonia. At this pH, $Mg(OH)_2$ precipitates and is inaccessible to the EDTA.

Insoluble carbonates are converted into soluble bicarbonates by excess carbon dioxide:

$$\underset{\text{Calcium carbonate}}{CaCO_3(s)} + CO_2 + H_2O \longrightarrow \underset{\text{Calcium bicarbonate}}{Ca(HCO_3)_2(aq)} \quad \text{(B)}$$

Heating reverses Reaction B to form a solid scale of $CaCO_3$ that clogs boiler pipes. The fraction of hardness due to $Ca(HCO_3)_2(aq)$ is called *temporary hardness*, because this calcium is lost (by precipitation of $CaCO_3$) on heating. Hardness arising from other salts (mainly dissolved $CaSO_4$) is called *permanent hardness*, because it is not removed by heating.

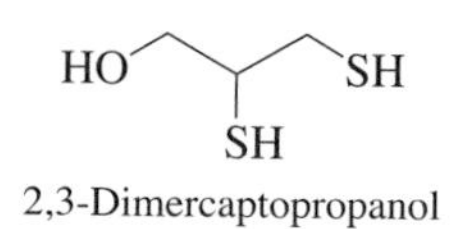

2,3-Dimercaptopropanol

from HF exposure can continue for several days after exposure. Exposure of 2% of your body to concentrated HF can kill you.[3]) Triethanolamine masks Al^{3+}, Fe^{3+}, and Mn^{2+}; and 2,3-dimercaptopropanol masks Bi^{3+}, Cd^{2+}, Cu^{2+}, Hg^{2+}, and Pb^{2+}. Selectivity afforded by masking and pH control allows individual components of complex mixtures to be analyzed by EDTA titration.

Ask Yourself

13-D. (a) A 50.0-mL sample containing Ni^{2+} was treated with 25.0 mL of 0.050 0 M EDTA to complex all the Ni^{2+} and leave excess EDTA in solution. How many millimoles of EDTA are contained in 25.0 mL of 0.050 0 M EDTA?
(b) The excess EDTA in **(a)** was then back-titrated, requiring 5.00 mL of 0.050 0 M Zn^{2+}. How many millimoles of Zn^{2+} are in 5.00 mL of 0.050 0 M Zn^{2+}?
(c) The millimoles of Ni^{2+} in the unknown is the difference between the EDTA added in **(a)** and the Zn^{2+} required in **(b)**. Find the number of millimoles of Ni^{2+} and the concentration of Ni^{2+} in the unknown.

13-5 The pH-Dependent Metal-EDTA Equilibrium

This is the point in the chapter where those who want to spend more time on instrumental methods of analysis might want to move to a new chapter. We now consider equilibrium calculations required to understand the shape of an EDTA titration curve.

Fractional Composition of EDTA Solutions

The fraction of EDTA in each of its protonated forms is plotted in Figure 13-7. The fraction, α, is defined here as it was in Section 12-5 for any weak acid. For example, $\alpha_{Y^{4-}}$ is

Fraction of EDTA in the form Y^{4-}:

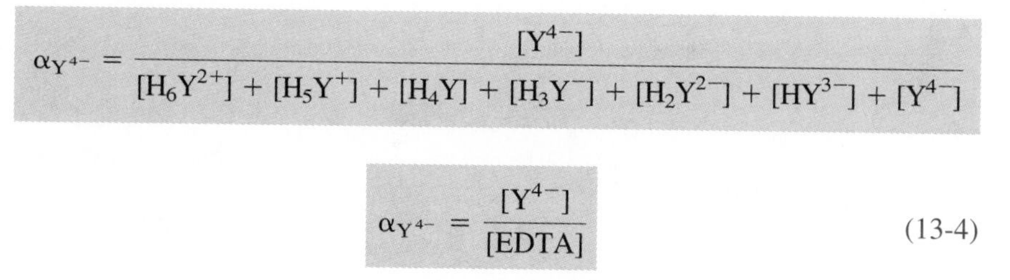

$$\alpha_{Y^{4-}} = \frac{[Y^{4-}]}{[H_6Y^{2+}] + [H_5Y^{+}] + [H_4Y] + [H_3Y^{-}] + [H_2Y^{2-}] + [HY^{3-}] + [Y^{4-}]}$$

$$\alpha_{Y^{4-}} = \frac{[Y^{4-}]}{[EDTA]} \qquad (13\text{-}4)$$

where [EDTA] is the total concentration of all *free* EDTA species in the solution. "Free" means EDTA that is not complexed to metal ions. Following a derivation similar to the one in Section 12-5, it can be shown that $\alpha_{Y^{4-}}$ is given by

$$\alpha_{Y^{4-}} = \frac{K_1K_2K_3K_4K_5K_6}{\{[H^+]^6 + [H^+]^5K_1 + [H^+]^4K_1K_2 + [H^+]^3K_1K_2K_3 + [H^+]^2K_1K_2K_3K_4 + [H^+]K_1K_2K_3K_4K_5 + K_1K_2K_3K_4K_5K_6\}} \qquad (13\text{-}5)$$

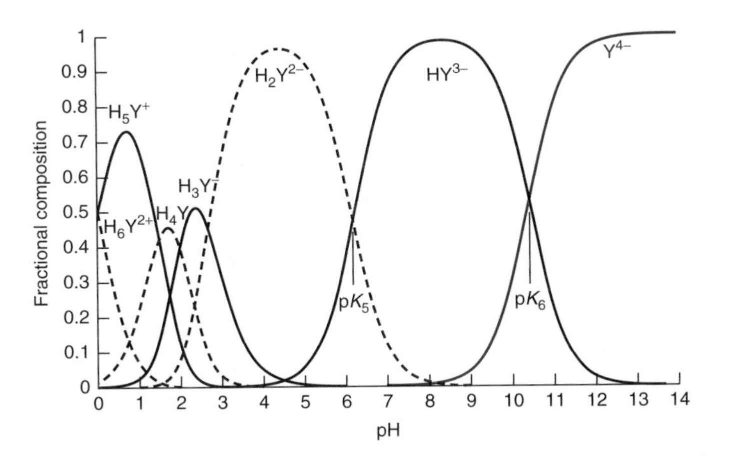

Figure 13-7 Fraction of EDTA in each of its protonated forms as a function of pH. This figure should remind you of Figure 11-1 for a diprotic acid. The fraction of Y^{4-}, shown by the colored curve at the right, is very small below pH 8.

Table 13-3 Values of $\alpha_{Y^{4-}}$ for EDTA at 20°C and μ = 0.10 M

pH	$\alpha_{Y^{4-}}$
0	1.3×10^{-23}
1	1.4×10^{-18}
2	2.6×10^{-14}
3	2.1×10^{-11}
4	3.0×10^{-9}
5	2.9×10^{-7}
6	1.8×10^{-5}
7	3.8×10^{-4}
8	4.2×10^{-3}
9	0.041
10	0.30
11	0.81
12	0.98
13	1.00
14	1.00

Table 13-3 gives values for $\alpha_{Y^{4-}}$ as a function of pH.

Example **What Does $\alpha_{Y^{4-}}$ Mean?**

The fraction of all free EDTA in the form Y^{4-}, called $\alpha_{Y^{4-}}$, is shown by the colored curve in Figure 13-7. At pH 6.00 and a formal concentration of 0.10 M, the composition of an EDTA solution is

$$[H_6Y^{2+}] = 8.9 \times 10^{-20}\text{ M} \quad [H_5Y^+] = 8.9 \times 10^{-14}\text{ M} \quad [H_4Y] = 2.8 \times 10^{-9}\text{ M}$$
$$[H_3Y^-] = 2.8 \times 10^{-5}\text{ M} \quad [H_2Y^{2-}] = 0.057\text{ M} \quad [HY^{3-}] = 0.043\text{ M}$$
$$[Y^{4-}] = 1.8 \times 10^{-6}\text{ M}$$

Find $\alpha_{Y^{4-}}$.

SOLUTION $\alpha_{Y^{4-}}$ is the fraction in the form Y^{4-}:

$$\alpha_{Y^{4-}} = \frac{[Y^{4-}]}{[H_6Y^{2+}] + [H_5Y^+] + [H_4Y] + [H_3Y^-] + [H_2Y^{2-}] + [HY^{3-}] + [Y^{4-}]}$$
$$= \frac{[1.8 \times 10^{-6}]}{[8.9 \times 10^{-20}] + [8.9 \times 10^{-14}] + [2.8 \times 10^{-9}] + [2.8 \times 10^{-5}] + [0.057] + [0.043] + [1.8 \times 10^{-6}]}$$
$$= 1.8 \times 10^{-5}$$

Test Yourself From Figure 13-7, estimate the fractions of HY^{3-} and H_2Y^{2-} at pH 7. (**Answer:** 0.9 and 0.1)

Conditional Formation Constant

Equation 13-1 should not be interpreted to mean that Y^{4-} is the only species that reacts with M^{n+}. It only says that the equilibrium constant is expressed in terms of the concentration of Y^{4-}.

The formation constant in Equation 13-1 describes the reaction between Y^{4-} and a metal ion. As you can see in Figure 13-7, most of the EDTA is not Y^{4-} below pH = pK_6 = 10.37. The species HY^{3-}, H_2Y^{2-}, and so on, predominate at lower pH. It is convenient to express the fraction of free EDTA in the form Y^{4-} by rearranging Equation 13-4 to give

$$[Y^{4-}] = \alpha_{Y^{4-}}[\text{EDTA}] \qquad (13\text{-}6)$$

where [EDTA] refers to the total concentration of all EDTA species not bound to metal ion.

The equilibrium constant for Reaction 13-1 can now be rewritten as

The product $\alpha_{Y^{4-}}$ [EDTA] accounts for the fact that only some of the free EDTA is in the form Y^{4-}.

$$K_f = \frac{[MY^{n-4}]}{[M^{n+}][Y^{4-}]} = \frac{[MY^{n-4}]}{[M^{n+}]\alpha_{Y^{4-}}[\text{EDTA}]}$$

If the pH is fixed by a buffer, then $\alpha_{Y^{4-}}$ is a constant that can be combined with K_f:

Conditional formation constant:

$$K'_f = \alpha_{Y^{4-}} K_f = \frac{[MY^{n-4}]}{[M^{n+}][\text{EDTA}]} \qquad (13\text{-}7)$$

The number $K'_f = \alpha_{Y^{4-}} K_f$ is called the **conditional formation constant** or the *effective formation constant.* It describes the formation of MY^{n-4} at any particular pH.

The conditional formation constant allows us to look at EDTA complex formation as if the uncomplexed EDTA were all in one form:

With the conditional formation constant, we can treat EDTA complex formation as if all free EDTA were in one form.

$$M^{n+} + \text{EDTA} \rightleftharpoons MY^{n-4} \qquad K'_f = \alpha_{Y^{4-}}K_f$$

At any given pH, we can find $\alpha_{Y^{4-}}$ and evaluate K'_f.

Example Using the Conditional Formation Constant

The formation constant in Table 13-1 for FeY^- is $10^{25.1} = 1.3 \times 10^{25}$. Calculate the concentration of free Fe^{3+} in a solution of 0.10 M FeY^- at pH 4.00 and at pH 1.00.

SOLUTION The complex formation reaction is

$$Fe^{3+} + \text{EDTA} \rightleftharpoons FeY^- \qquad K'_f = \alpha_{Y^{4-}}K_f$$

where EDTA on the left side of the equation refers to all forms of unbound EDTA ($=Y^{4-}$, HY^{3-}, H_2Y^{2-}, H_3Y^-, and so on). Using $\alpha_{Y^{4-}}$ from Table 13-3, we find

At pH 4.00: $K'_f = (3.0 \times 10^{-9})(1.3 \times 10^{25}) = 3.9 \times 10^{16}$

At pH 1.00: $K'_f = (1.4 \times 10^{-18})(1.3 \times 10^{25}) = 1.8 \times 10^{7}$

Because dissociation of FeY^- must produce equal quantities of Fe^{3+} and EDTA, we can write

	Fe^{3+}	+	EDTA	$\rightleftharpoons$	FeY^-
Initial concentration (M)	0		0		0.10
Final concentration (M)	x		x		$0.10 - x$

$$\frac{[FeY^-]}{[Fe^{3+}][EDTA]} = \frac{0.10 - x}{x^2} = K'_f = 3.9 \times 10^{16} \quad \text{at pH 4.00}$$
$$= 1.8 \times 10^7 \quad \text{at pH 1.00}$$

Solving for x, we find $[Fe^{3+}] = x = 1.6 \times 10^{-9}$ M at pH 4.00 and 7.4×10^{-5} M at pH 1.00. *Using the conditional formation constant at a fixed pH, we treat the dissociated EDTA as if it were a single species.*

Test Yourself Calculate $[Fe^{3+}]$ in 0.10 M FeY^- at pH 5.00 (**Answer:** 1.6×10^{-10} M)

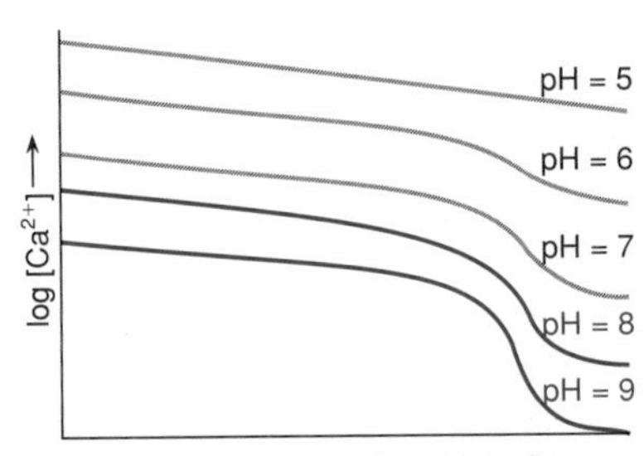

Figure 13-8 Titration of Ca^{2+} with EDTA as a function of pH. The experimental ordinate is the voltage difference between two electrodes (mercury and calomel) immersed in the titration solution. This voltage is a measure of $\log[Ca^{2+}]$. [From C. N. Reilley and R. W. Schmid, *Anal. Chem.* **1958**, *30*, 947.]

We see that a metal-EDTA complex becomes less stable at lower pH. For a titration reaction to be effective, it must go "to completion," which means that the equilibrium constant must be large—the analyte and titrant are essentially completely reacted (say, 99.9%) at the equivalence point. Figure 13-8 shows how pH affects the titration of Ca^{2+} with EDTA. Below pH ≈ 8, the break at the end point is not sharp enough to allow accurate determination. At lower pH, the inflection point disappears because the conditional formation constant for CaY^{2-} is too small.

Ask Yourself

13-E. The equivalence point in the titration of Ca^{2+} with EDTA is the same as a solution of pure CaY^{2-}. Suppose that the formal concentration $[CaY^{2-}] = 0.010$ M at the equivalence point in Figure 13-8. Let's see how complete the reaction is at low and high pH.
(a) Find the concentration of free Ca^{2+} at pH 5.00 at the equivalence point.
(b) What is the fraction of bound Ca^{2+} $(= [CaY^{2-}]/\{[CaY^{2-}] + [Ca^{2+}]\})$?
(c) Find the concentration of free Ca^{2+} at pH 9.00 and the fraction of bound Ca^{2+}.

13-6 EDTA Titration Curves

Now we calculate the concentration of free metal ion in the course of the titration of metal with EDTA. The titration reaction is

$$M^{n+} + EDTA \rightleftharpoons MY^{n-4} \qquad K'_f = \alpha_{Y^{4-}} K_f \qquad (13\text{-}8)$$

K'_f is the effective formation constant at the fixed pH of the solution.

If K'_f is large, we can consider the reaction to be complete at each point in the titration.

The titration curve is a graph of pM ($\equiv -\log[M]$) versus the volume of added EDTA. The curve is analogous to that obtained by plotting pH versus volume of titrant in an acid-base titration. There are three natural regions of the titration curve in Figure 13-9.

Region 1: Before the Equivalence Point

In this region, there is excess M^{n+} after EDTA has been consumed. The concentration of free metal ion is equal to the concentration of excess, unreacted M^{n+}. The dissociation of MY^{n-4} is negligible.

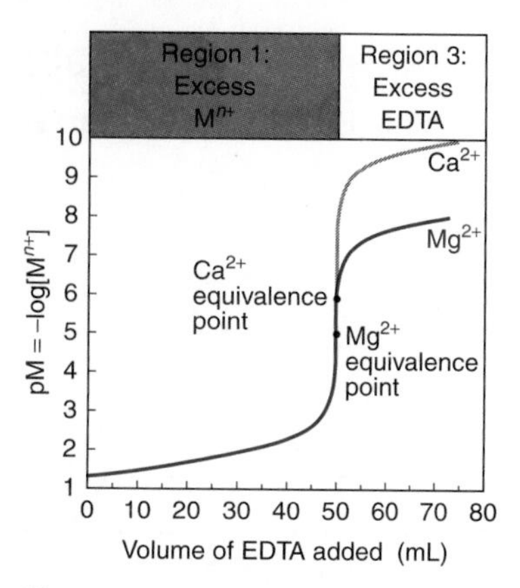

Figure 13-9 Three regions in an EDTA titration of 50.0 mL of 0.050 0 M Mg^{2+} or Ca^{2+} with 0.050 0 M EDTA at pH 10. Region 2 is the equivalence point. The concentration of free M^{n+} decreases as the titration proceeds.

Region 2: At the Equivalence Point

There is exactly as much EDTA as metal in the solution. We can treat the solution as if it were made by dissolving pure MY^{n-4}. Some free M^{n+} is generated by the slight dissociation of MY^{n-4}:

$$MY^{n-4} \rightleftharpoons M^{n+} + \text{EDTA}$$

In this reaction, EDTA refers to the total concentration of free EDTA in all of its forms. At the equivalence point, $[M^{n+}] = [\text{EDTA}]$.

Region 3: After the Equivalence Point

Now there is excess EDTA, and virtually all the metal ion is in the form MY^{n-4}. The concentration of free EDTA can be equated to the concentration of excess EDTA added after the equivalence point.

Titration Calculations

Let's calculate the titration curve for the reaction of 50.0 mL of 0.050 0 M Mg^{2+} (buffered to pH 10.00) with 0.050 0 M EDTA. The equivalence volume is 50.0 mL.

$\alpha_{Y^{4-}}$ comes from Table 13-3 and K_f comes from Table 13-1.

$$Mg^{2+} + \text{EDTA} \longrightarrow MgY^{2-}$$
$$K'_f = \alpha_{Y^{4-}} K_f = (0.30)(6.2 \times 10^8) = 1.9 \times 10^8$$

Because K'_f is large, it is reasonable to say that the reaction goes to completion with each addition of titrant. We want to make a graph in which pMg^{2+} ($=-\log[Mg^{2+}]$) is plotted versus milliliters of added EDTA.

Region 1: Before the Equivalence Point

Before the equivalence point, there is excess unreacted M^{n+}.

Consider the addition of 5.00 mL of EDTA. Because the equivalence point is 50.0 mL, one-tenth of the Mg^{2+} will be consumed and nine-tenths remains.

$$\text{initial mmol } Mg^{2+} = (0.050\,0 \text{ M } Mg^{2+})(50.0 \text{ mL}) = 2.50 \text{ mmol}$$
$$\text{mmol remaining} = (0.900)(2.50 \text{ mmol}) = 2.25 \text{ mmol}$$
$$[Mg^{2+}] = \frac{2.25 \text{ mmol}}{55.0 \text{ mL}} = 0.040\,9 \text{ M} \Rightarrow pMg^{2+} = -\log[Mg^{2+}] = 1.39$$

In a similar manner, we could calculate pMg^{2+} for any volume of EDTA less than 50.0 mL.

Region 2: At the Equivalence Point

At the equivalence point, the major species is MY^{n-4}, in equilibrium with small, equal amounts of free M^{n+} and EDTA.

Virtually all the metal is in the form MgY^{2-}. We began with 2.50 mmol Mg^{2+}, which is now close to 2.50 mmol MgY^{2-} in a volume of 50.0 + 50.0 = 100.0 mL.

$$[MgY^{2-}] = \frac{2.50 \text{ mmol}}{100.0 \text{ mL}} = 0.025\,0 \text{ M}$$

The concentration of free Mg^{2+} is small and unknown. We can write

	Mg^{2+}	+ EDTA	$\rightleftharpoons$ MgY^{2-}
Initial concentration (M)	—	—	0.025 0
Final concentration (M)	x	x	0.025 0 − x

$$\frac{[MgY^{2-}]}{[Mg^{2+}][EDTA]} = K'_f = 1.9 \times 10^8$$

$$\frac{0.025\,0 - x}{x^2} = 1.9 \times 10^8 \Rightarrow x = 1.1_5 \times 10^{-5}\ M$$

$$pMg^{2+} = -\log x = 4.94$$

[EDTA] refers to the total concentration of all forms of EDTA not bound to metal.

Region 3: After the Equivalence Point

In this region, virtually all the metal is in the form MgY^{2-}, and there is excess, unreacted EDTA. The concentrations of MgY^{2-} and excess EDTA are easily calculated. For example, when we have added 51.00 mL of EDTA, there is 1.00 mL of excess EDTA = (0.050 0 M)(1.00 mL) = 0.050 0 mmol.

After the equivalence point, virtually all metal is in the form MY^{n-4}. There is a known excess of EDTA. A small amount of free M^{n+} exists in equilibrium with MY^{n-4} and EDTA.

$$[EDTA] = \frac{0.050\,0\ mmol}{101.0\ mL} = 0.000\,495\ M$$

$$[MgY^{2-}] = \frac{2.50\ mmol}{101.0\ mL} = 0.024\,8\ M$$

The concentration of Mg^{2+} is governed by

$$\frac{[MgY^{2-}]}{[Mg^{2+}][EDTA]} = K'_f = 1.9 \times 10^8$$

$$\frac{[0.024\,8]}{[Mg^{2+}](0.000\,495)} = 1.9 \times 10^8$$

$$[Mg^{2+}] = 2.6 \times 10^{-7}\ M \Rightarrow pMg^{2+} = 6.58$$

The same sort of calculation can be used for any volume past the equivalence point.

The Titration Curve

The calculated titration curves for Mg^{2+} and for Ca^{2+} in Figure 13-9 show a distinct break at the equivalence point, where the slope is greatest. The break is greater for Ca^{2+} than for Mg^{2+} because the formation constant for CaY^{2-} is greater than K_f for MgY^{2-}. Notice the analogy between Figure 13-9 and acid-base titration curves. The greater the metal-EDTA formation constant, the more pronounced the break at the equivalence point. The stronger the acid HA, the greater the break at the equivalence point in the titration with OH^-.

The completeness of reaction (and hence the sharpness of the equivalence point) is determined by the conditional formation constant, $\alpha_{Y^{4-}}K_f$, which is pH dependent. Because $\alpha_{Y^{4-}}$ decreases as pH is lowered, pH is an important variable determining whether a titration is feasible. The end point is more distinct at high pH. However, the pH must not be so high that metal hydroxide precipitates. The effect of pH on the titration of Ca^{2+} was shown in Figure 13-8.

The lower the pH, the less distinct the end point.

Beware

The calculation that we just did was oversimplified because we neglected any other chemistry of M^{n+}, such as formation of MOH^+, $M(OH)_2(aq)$, $M(OH)_2(s)$, and $M(OH)_3^-$. These species decrease the concentration of available M^{n+} and decrease the sharpness of the titration curve. Mg^{2+} is normally titrated in ammonia buffer at pH 10 in which $Mg(NH_3)^{2+}$ also is present. The accurate calculation of metal-EDTA titration curves requires full knowledge of the chemistry of the metal with water and any other ligands present in the solution.

Ask Yourself

13-F. Find pCa^{2+} ($=-\log[Ca^{2+}]$) in the titration in Figure 13-9 at V_{EDTA} = 5.00, 50.00, and 51.00 mL. See that your answers agree with Figure 13-9.

Key Equations

Formation constant

$$M^{n+} + Y^{4-} \rightleftharpoons MY^{n-4} \qquad K_f = \frac{[MY^{n-4}]}{[M^{n+}][Y^{4-}]}$$

Fraction of EDTA as Y^{4-}

$$\alpha_{Y^{4-}} = \frac{[Y^{4-}]}{[EDTA]}$$

[EDTA] = total concentration of EDTA not bound to metal

Conditional formation constant

$$K'_f = \alpha_{Y^{4-}} K_f = \frac{[MY^{n-4}]}{[M^{n+}][EDTA]}$$

Titration calculations

Before V_e, there is a known excess of M^{n+}
$pM = -\log[M]$

At V_e: Some M^{n+} is generated by the dissociation of MY^{n-4}

$$\underset{x}{M^{n+}} + \underset{x}{EDTA} \overset{K'_f}{\rightleftharpoons} \underset{F-x}{MY^{n-4}}$$

After V_e, [EDTA] and $[MY^{n-4}]$ are known

$$\underset{x}{M^{n+}} + \underset{\text{Known}}{EDTA} \overset{K'_f}{\rightleftharpoons} \underset{\text{Known}}{MY^{n-4}}$$

Important Terms

auxiliary complexing agent
back titration
blocking
chelating ligand
complexometric titration
conditional formation constant
cumulative formation constant
direct titration
displacement titration
formation constant
indirect titration
Lewis acid
Lewis base
ligand
masking agent
metal ion indicator
monodentate ligand
multidentate ligand
overall formation constant
stepwise formation constant

Problems

13-1. How many milliliters of 0.050 0 M EDTA are required to react with 50.0 mL of **(a)** 0.010 0 M Ca^{2+} or **(b)** 0.010 0 M Al^{3+}?

13-2. Give three circumstances in which an EDTA back titration might be necessary.

13-3. Describe what is done in a displacement titration and give an example.

13-4. Give an example of the use of a masking agent.

13-5. What is meant by water hardness? Explain the difference between temporary and permanent hardness.

13-6. State the purpose of an auxiliary complexing agent and give an example of its use.

13-7. Draw a reasonable structure for a complex between Fe^{3+} and nitrilotriacetic acid, $N(CH_2CO_2H)_3$.

13-8. A 25.00-mL sample containing Fe^{3+} was treated with 10.00 mL of 0.036 7 M EDTA to complex all the Fe^{3+} and leave excess EDTA in solution. The excess EDTA was then back-titrated, requiring 2.37 mL of 0.046 1 M Mg^{2+}. What was the concentration of Fe^{3+} in the original solution?

13-9. A 50.0-mL solution containing Ni^{2+} and Zn^{2+} was treated with 25.0 mL of 0.045 2 M EDTA to bind all the metal. The excess unreacted EDTA required 12.4 mL of 0.012 3 M Mg^{2+} for complete reaction. An excess of the reagent 2,3-dimercapto-1-propanol was then added to displace the EDTA from zinc. Another 29.2 mL of Mg^{2+} were required for reaction with the liberated EDTA. Calculate the molarities of Ni^{2+} and Zn^{2+} in the original solution.

13-10. Sulfide ion was determined by indirect titration with EDTA. To a solution containing 25.00 mL of 0.043 32 M $Cu(ClO_4)_2$ plus 15 mL of 1 M acetate buffer (pH 4.5) was added 25.00 mL of unknown sulfide solution with vigorous stirring. The CuS precipitate was filtered and washed with hot water. Ammonia was added to the filtrate (which contains excess Cu^{2+}) until the blue color of $Cu(NH_3)_4^{2+}$ was observed. Titration of the filtrate with 0.039 27 M EDTA required 12.11 mL of EDTA to reach the end point with the indicator murexide. Find the molarity of sulfide in the unknown.

13-11. *Propagation of uncertainty.* The potassium ion in a 250.0 (±0.1)-mL water sample was precipitated with sodium tetraphenylborate:

$$K^+ + (C_6H_5)_4B^- \longrightarrow KB(C_6H_5)_4(s)$$

The precipitate was filtered, washed, and dissolved in an organic solvent. Treatment of the organic solution with an excess of Hg^{2+}-EDTA then gave the following reaction:

$$4HgY^{2-} + (C_6H_5)_4B^- + 4H_2O \longrightarrow H_3BO_3 + 4C_6H_5Hg^+ + 4HY^{3-} + OH^-$$

The liberated EDTA was titrated with 28.73 (±0.03) mL of 0.043 7 (±0.000 1) M Zn^{2+}. Find the concentration (and uncertainty) of K^+ in the original sample.

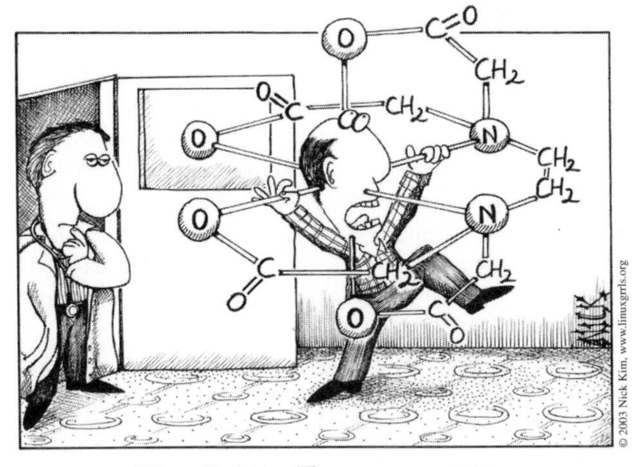

When Chelation Therapy goes wrong.

13-12. A 25.00-mL sample of unknown containing Fe^{3+} and Cu^{2+} required 16.06 mL of 0.050 83 M EDTA for complete titration. A 50.00-mL sample of the unknown was treated with NH_4F to protect the Fe^{3+}. Then the Cu^{2+} was reduced and masked by addition of thiourea. On addition of 25.00 mL of 0.050 83 M EDTA, the Fe^{3+} was liberated from its fluoride complex and formed an EDTA complex. The excess EDTA required 19.77 mL of 0.018 83 M Pb^{2+} to reach a xylenol orange end point. Find $[Cu^{2+}]$ in the unknown.

13-13. Cyanide recovered from the refining of gold ore can be determined indirectly by EDTA titration. A known excess of Ni^{2+} is added to the cyanide to form tetracyanonickelate(II):

$$4CN^- + Ni^{2+} \longrightarrow Ni(CN)_4^{2-}$$

When the excess Ni^{2+} is titrated with standard EDTA, $Ni(CN)_4^{2-}$ does not react. In a cyanide analysis, 12.7 mL of cyanide solution were treated with 25.0 mL of standard solution containing excess Ni^{2+} to form tetracyanonickelate. The excess Ni^{2+} required 10.1 mL of 0.013 0 M EDTA for complete reaction. In a separate experiment, 39.3 mL of 0.013 0 M EDTA were required to react with 30.0 mL of the standard Ni^{2+} solution. Calculate the molarity of CN^- in the 12.7-mL sample of unknown.

13-14. A mixture of Mn^{2+}, Mg^{2+}, and Zn^{2+} was analyzed as follows: The 25.00-mL sample was treated with 0.25 g of $NH_3OH^+Cl^-$ (hydroxylammonium chloride, a reducing agent that maintains manganese in the +2 state), 10 mL of ammonia buffer (pH 10), and a few drops of Calmagite indicator and then diluted to 100 mL. It was warmed to 40°C and titrated with 39.98 mL of 0.045 00 M EDTA to the blue end point. Then 2.5 g of NaF were added to displace Mg^{2+} from its EDTA complex. The liberated EDTA required 10.26 mL of standard 0.020 65 M Mn^{2+} for complete titration. After this second end point was reached, 5 mL of 15 wt% aqueous KCN were added to displace Zn^{2+} from its EDTA complex. This time the liberated EDTA required 15.47 mL of standard 0.020 65 M Mn^{2+}. Calculate the number of milligrams of each metal (Mn^{2+}, Zn^{2+}, and Mg^{2+}) in the 25.00-mL sample of unknown.

13-15. The sulfur content of insoluble sulfides that do not readily dissolve in acid can be measured by oxidation with Br_2 to SO_4^{2-}.[4] Metal ions are then replaced with H^+ by an ion-exchange column (Chapter 23), and sulfate is precipitated as $BaSO_4$ with a known excess of $BaCl_2$. The excess Ba^{2+} is then titrated with EDTA to determine how much was present. (To make the indicator end point clearer, a small, known quantity of Zn^{2+} is also added. The EDTA titrates both Ba^{2+} and Zn^{2+}.) Knowing the excess Ba^{2+}, we can calculate how much sulfur was in the original material. To analyze the mineral sphalerite (ZnS, FM 97.474), 5.89 mg of powdered solid were suspended in a mixture of carbon tetrachloride and water containing 1.5 mmol Br_2. After 1 h at 20° and 2 h at 50°C, the powder dissolved and the solvent and excess Br_2 were removed by heating. The residue was dissolved in 3 mL of water and passed through an ion-exchange column to replace Zn^{2+} with H^+. Then 5.000 mL of 0.014 63 M $BaCl_2$ were added to precipitate all sulfate as $BaSO_4$. After the addition of 1.000 mL of 0.010 00 M $ZnCl_2$ and 3 mL of ammonia buffer, pH 10, the excess Ba^{2+} and Zn^{2+} required 2.39 mL of 0.009 63 M EDTA to reach the Calmagite end point. Find the weight percent of sulfur in the sphalerite. What is the theoretical value?

13-16. State (in words) what $\alpha_{Y^{4-}}$ means. Calculate $\alpha_{Y^{4-}}$ for EDTA at **(a)** pH 3.50 and **(b)** pH 10.50.

13-17. The cumulative formation constants for the reaction of Co^{2+} with ammonia are $\log \beta_1 = 1.99$, $\log \beta_2 = 3.50$, $\log \beta_3 = 4.43$, $\log \beta_4 = 5.07$, $\log \beta_5 = 5.13$, and $\log \beta_6 = 4.39$.

(a) Write the chemical reaction whose equilibrium constant is β_4.

(b) Write the reaction whose stepwise formation constant is K_4 and find its numerical value.

13-18. (a) Find the conditional formation constant for $Mg(EDTA)^{2-}$ at pH 9.00.

(b) Find the concentration of free Mg^{2+} at pH 9.00 in 0.050 M $Na_2[Mg(EDTA)]$.

13-19. A 100.0-mL solution of the ion M^{n+} at a concentration of 0.050 0 M buffered to pH 9.00 was titrated with 0.050 0 M EDTA.

(a) What is the equivalence volume, V_e, in milliliters?

(b) Calculate the concentration of M^{n+} at $V = \frac{1}{2}V_e$.

(c) What fraction ($\alpha_{Y^{4-}}$) of free EDTA is in the form Y^{4-} at pH 9.00?

(d) The formation constant (K_f) is $10^{12.00}$. Calculate the value of the conditional formation constant K'_f $(= \alpha_{Y^{4-}}K_f)$.

(e) Calculate the concentration of M^{n+} at $V = V_e$.

(f) What is the concentration of M^{n+} at $V = 1.100V_e$?

13-20. Consider the titration of 25.0 mL of 0.020 0 M $MnSO_4$ with 0.010 0 M EDTA in a solution buffered to pH 6.00. Calculate pMn^{2+} at the following volumes of added EDTA and sketch the titration curve: 0, 20.0, 40.0, 49.0, 49.9, 50.0, 50.1, 55.0, and 60.0 mL.

13-21. Using the volumes from Problem 13-20, calculate pCa^{2+} for the titration of 25.00 mL of 0.020 00 M EDTA with 0.010 00 M $CaSO_4$ at pH 10.00 and sketch the titration curve.

13-22. Explain the analogies between the titration of a metal with EDTA and the titration of a weak acid (HA) with OH^-. Make comparisons in all three regions of the titration curve.

13-23. Calculate pCu^{2+} at each of the following points in the titration of 25.0 mL of 0.080 0 M $Cu(NO_3)_2$ with 0.040 0 M EDTA at pH 5.00: 0, 20.0, 40.0, 49.0, 50.0, 51.0, and 55.0 mL. Sketch a graph of pCu^{2+} versus volume of titrant.

13-24. *Effect of pH on titration curve.* Repeat the calculations of Problem 13-23 for pH = 7.00. Sketch the two titration curves on one graph and give a chemical explanation for the difference between the two curves.

13-25. *Metal ion buffers.* By analogy to a hydrogen ion buffer, a metal ion buffer tends to maintain a particular metal ion concentration in solution. A mixture of the acid HA and its conjugate base A^- is a hydrogen ion buffer that maintains a pH defined by the equation $K_a = [A^-][H^+]/[HA]$. A mixture of CaY^{2-} and Y^{4-} serves as a Ca^{2+} buffer governed by the equation $1/K'_f = [EDTA][Ca^{2+}]/[CaY^{2-}]$. How many grams of $Na_2H_2EDTA \cdot 2H_2O$ (FM 372.23) should be mixed with 1.95 g of $Ca(NO_3)_2 \cdot 2H_2O$ (FM 200.12) in a 500-mL volumetric flask to give a buffer with $pCa^{2+} = 9.00$ at pH 9.00?

How Would You Do It?

13-26. Consider the titration curve for Mg^{2+} in Figure 13-9. The curve was calculated under the simplifying assumption that all the magnesium not bound to EDTA is free Mg^{2+}. In fact, at pH 10 in a buffer containing 1 M NH_3, approximately 63% of the magnesium not bound to EDTA is $Mg(NH_3)^{2+}$ and 4% is $MgOH^+$. Both NH_3 and OH^- are readily displaced from Mg^{2+} by EDTA in the course of the titration.

(a) *Sketch* how the titration curve in Figure 13-9 would be different prior to the equivalence point if two-thirds of the magnesium not bound to EDTA were bound to the ligands NH_3 and OH^-.

(b) Consider a point that is 10% past the equivalence point. We know that almost all magnesium is bound to EDTA at this point. We also know that there is 10% excess unbound EDTA at this point. Is the concentration of free Mg^{2+} the same or different from what we calculated in Figure 13-9? Will the titration curve be the same or different from Figure 13-9 past the equivalence point?

13-27. A reprecipitation was employed to remove occluded nitrate from $BaSO_4$ precipitate prior to isotopic analysis of oxygen for geologic studies.[5] Approximately 30 mg of $BaSO_4$ crystals were mixed with 15 mL of 0.05 M DTPA (Figure 13-5) in 1 M NaOH. After dissolving the solid with vigorous shaking at 70°C, it was reprecipitated by adding 10 M HCl dropwise to obtain pH 3–4 and allowing the mixture to stand for 1 h. The solid was isolated by centrifugation, removal of the mother liquor, and resuspension in deionized water. Centrifugation and washing was repeated a second time to reduce the molar ratio NO_3^-/SO_4^{2-} from 0.25 in the original precipitate to 0.001 in the purified material. What will be the predominant species of sulfate and DTPA at pH 14 and pH 3? Explain why $BaSO_4$ dissolves in DTPA in 1 M NaOH and then reprecipitates when the pH is lowered to 3–4.

Notes and References

1. R. J. Abergel, J. A. Warner, D. K. Shuh, and K. N. Raymond, *J. Am. Chem. Soc.* **2006**, *128*, 8920.

2. R. J. Abergel, E. G. Moore, R. K. Strong, and K. N. Raymond, *J. Am. Chem. Soc.* **2006**, *128*, 10998.

3. *Chem. Eng. News,* 13 September 1999, p. 40.

4. T. Darjaa, K. Yamada, N. Sato, T. Fujino, and Y. Waseda, *Fresenius J. Anal. Chem.* **1998,** *361,* 442.

5. H. Bao, *Anal. Chem.* **2006**, *78*, 304.

Further Reading

A. E. Martell and R. D. Hancock, *Metal Complexes in Aqueous Solution* (New York: Plenum Press, 1996).

G. Schwarzenbach and H. Flaschka, *Complexometric Titrations,* H. M. N. H. Irving, trans. (London: Methuen, 1969).

A. Ringbom, *Complexation in Analytical Chemistry* (New York: Wiley, 1963).

Remediation of Underground Pollution with Emulsified Iron Nanoparticles

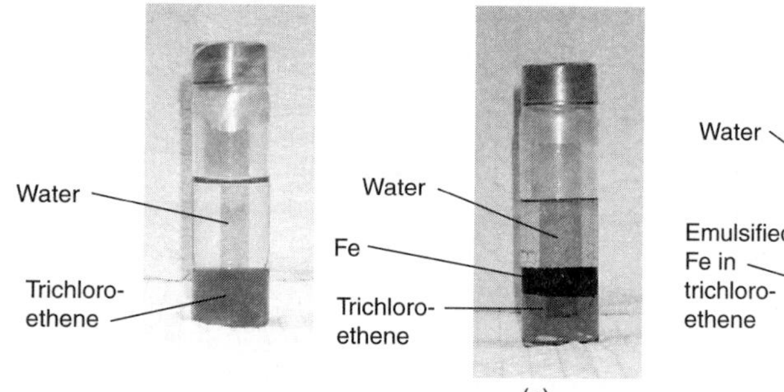

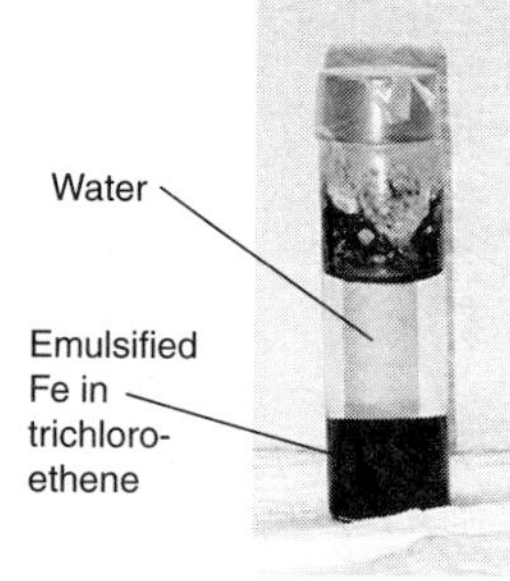

Left: Two-phase mixture of trichloroethene and water. *Center:* Nanometer-size iron particles form a separate layer that does not mix with trichloroethene. *Right:* Emulsified nano-iron mixes with organic layer, permitting Fe to react with trichloroethene. [J. Quinn, C. Geiger, C. Clausen, K. Brooks, C. Coon, S. O'Hara, T. Krug, D. Major, W.-S. Yoon, A. Gavaskar, and T. Holdsworth, *Environ. Sci. Technol.* **2005**, *39*, 1309. Courtesy J. W. Quinn, Kennedy Space Center.]

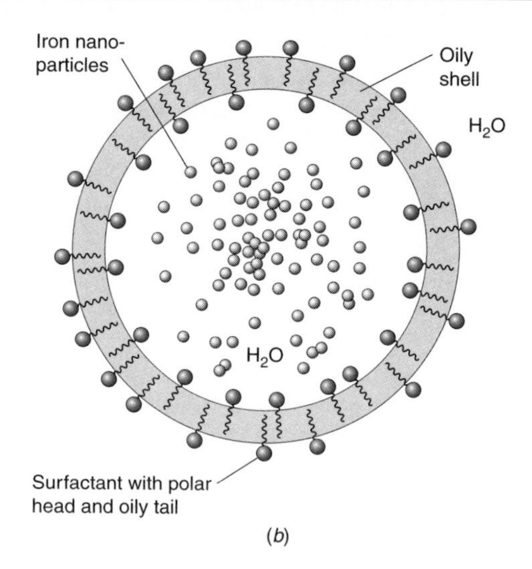

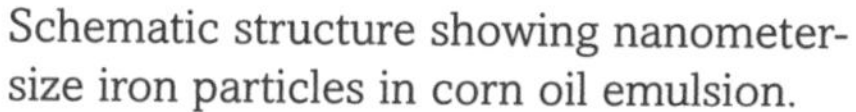

Schematic structure showing nanometer-size iron particles in corn oil emulsion.

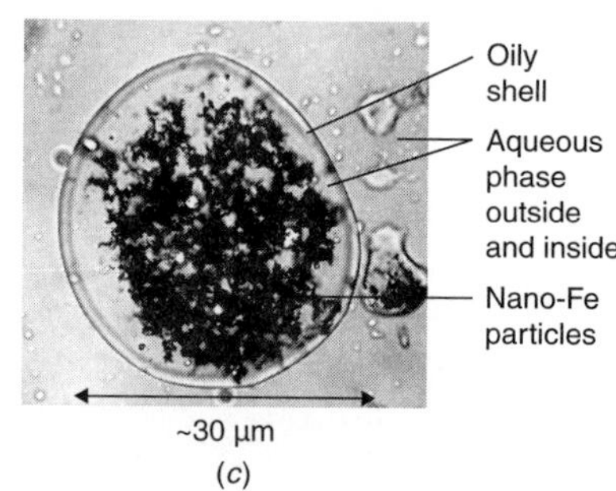

Micrograph of emulsion filled with nano-iron particles.

For decades, the cleaning solvents trichloroethene and tetrachloroethene were dumped or injected into wells for disposal. Only later did we realize that these carcinogenic liquids persist in the ground and pollute groundwater at approximately 20 000 sites in the United States. At Launch Complex 34 at Cape Canaveral, where rockets were launched from 1960 to 1968, chlorinated solvents used to wash rocket engines ran into the sand to form dense nonaqueous-phase liquids that persist underground today.

One way to remove chlorinated solvents is by reduction with Fe(0):[1]

$$\underset{\text{Trichloroethene}}{\mathrm{ClHC{=}CCl_2}} \xrightarrow{\text{Fe(0) particles}} \underset{\text{Ethene}}{\mathrm{H_2C{=}CH_2}} + \mathrm{Cl^-} + \text{other chlorinated hydrocarbons}$$

The smaller the iron particles, the more surface they have and the more reactive they are. One approach to mixing iron particles with trichloroethene is to encapsulate iron with a shell of biodegradable oil that is soluble in trichloroethene. An emulsion made of water, corn oil, biodegradable surfactant, and 17 wt% nano-iron contains iron particles surrounded by an oily shell. (An *emulsion* is a suspension of immiscible phases, such as oil and water, stabilized at the interface by surfactant that has affinity for both phases.)

In a test of the technology, 2 400 L of emulsion were injected into 8 wells in a 4.6 × 2.9 × 3.0-m deep test area at Launch Complex 34. After 5 months, three-fourths of the trichloroethene in the test plot was consumed.

CHAPTER 14

Electrode Potentials

Measurements of the acidity of rainwater, the fuel-air mixture in an automobile engine, and the concentrations of gases and electrolytes in your bloodstream are all made with electrochemical sensors. This chapter provides a foundation needed to discuss some of the most common electrochemical sensors in Chapter 15.

14-1 Redox Chemistry and Electricity

In a **redox reaction**, electrons are transferred from one species to another. A molecule is said to be **oxidized** when it *loses electrons*. It is **reduced** when it *gains electrons*. An **oxidizing agent**, also called an **oxidant**, takes electrons from another substance and becomes reduced. A **reducing agent**, also called a **reductant**, gives electrons to another substance and is oxidized in the process. In the reaction

$$\underset{\text{Oxidizing agent}}{Fe^{3+}} + \underset{\text{Reducing agent}}{V^{2+}} \longrightarrow Fe^{2+} + V^{3+}$$

Fe^{3+} is the oxidizing agent because it takes an electron from V^{2+}. V^{2+} is the reducing agent because it gives an electron to Fe^{3+}. As the reaction proceeds from left to right, Fe^{3+} is reduced (its oxidation state goes down from +3 to +2), and V^{2+} is oxidized (its oxidation state goes up from +2 to +3). Appendix D discusses oxidation numbers and balancing redox equations. You should be able to write a balanced redox reaction as the sum of two *half-reactions*, one an oxidation and the other a reduction.

Oxidation: loss of electrons
Reduction: gain of electrons

Oxidizing agent: takes electrons
Reducing agent: gives electrons

Michael Faraday (1791–1867) was a self-educated English "natural philosopher" (the old term for "scientist") who discovered that the extent of an electrochemical reaction is proportional to the electric charge that passes through the cell. More important, he discovered many fundamental laws of electromagnetism. He gave us the electric motor, electric generator, and electric transformer, as well as the terms *ion, cation, anion, electrode, cathode, anode,* and *electrolyte.* His gift for lecturing is best remembered from his Christmas lecture demonstrations for children at the Royal Institution. Faraday "took great delight in talking to [children], and easily won their confidence . . . . They felt as if he belonged to them; and indeed he sometimes, in his joyous enthusiasm, appeared like an inspired child."[2]

Chemistry and Electricity

Electric charge (q) is measured in **coulombs** (C). The magnitude of the charge of a single electron (or proton) is 1.602×10^{-19} C. A mole of electrons therefore has a charge of $(1.602 \times 10^{-19}\ \text{C})(6.022 \times 10^{23}/\text{mol}) = 9.649 \times 10^{4}$ C/mol, which is called the **Faraday constant**, F.

Relation between charge and moles:

$$\underset{\text{Coulombs}}{q} = \underset{\text{Moles}}{n} \cdot \underset{\frac{\text{Coulombs}}{\text{Mole}}}{\text{F}} \qquad (14\text{-}1)$$

The *quantity* of electrons flowing from a reaction is proportional to the quantity of analyte that reacts.

In the reaction $Fe^{3+} + V^{2+} \rightarrow Fe^{2+} + V^{3+}$, one electron is transferred to oxidize one atom of V^{2+} and to reduce one atom of Fe^{3+}. If we know how many moles of electrons are transferred from V^{2+} to Fe^{3+}, then we know how many moles of product have been formed.

Example **Relating Coulombs to Quantity of Reaction**

If 5.585 g of Fe^{3+} were reduced in the reaction $Fe^{3+} + V^{2+} \rightarrow Fe^{2+} + V^{3+}$, how many coulombs of charge must have been transferred from V^{2+} to Fe^{3+}?

SOLUTION First, we find that 5.585 g of Fe^{3+} equal 0.100 0 mol of Fe^{3+}. Because each Fe^{3+} ion requires one electron, 0.100 0 mol of electrons must have been transferred. Equation 14-1 relates coulombs of charge to moles of electrons:

Faraday constant:
$F \approx 9.649 \times 10^4$ C/mol

$$q = n\text{F} = (0.100\,0 \text{ mol e}^-)\left(9.649 \times 10^4 \frac{\text{C}}{\text{mol e}^-}\right) = 9.649 \times 10^3 \text{ C}$$

Test Yourself How many coulombs are released when H_2O is oxidized to liberate 1.00 mol O_2 in the reaction $2H_2O \rightarrow O_2 + 4H^+ + 4e^-$? (**Answer:** 3.86×10^5 C)

Electric Current Is Proportional to the Rate of a Redox Reaction

1 A = 1 C/s

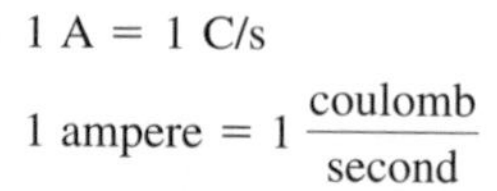

$$1 \text{ ampere} = 1 \frac{\text{coulomb}}{\text{second}}$$

Electric **current** (I) is the quantity of charge flowing each second past a point in an electric circuit. The unit of current is the **ampere** (A), which is a flow of 1 coulomb per second.

Consider Figure 14-1 in which electrons flow into a Pt wire dipped into a solution in which Sn^{4+} is reduced to Sn^{2+}:

$$Sn^{4+} + 2e^- \rightarrow Sn^{2+}$$

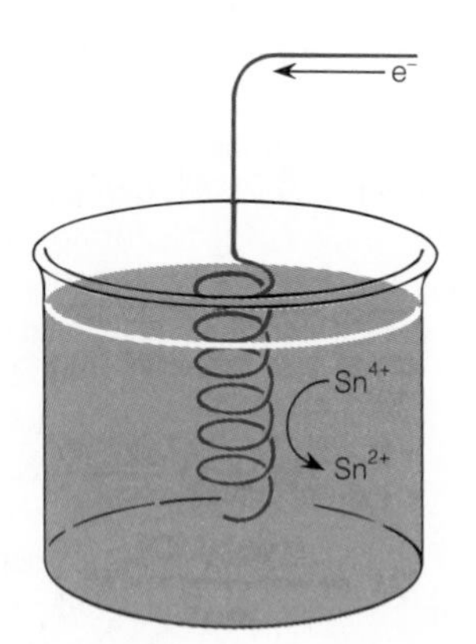

Figure 14-1 Electrons flowing into a coil of Pt wire at which Sn^{4+} ions in solution are reduced to Sn^{2+} ions. This process could not happen by itself, because there is no complete circuit. If Sn^{4+} is to be reduced at this Pt electrode, some other species must be oxidized at some other place.

The Pt wire is an **electrode**—a device to conduct electrons into or out of the chemicals involved in the redox reaction. Platinum is an *inert* electrode; it does not participate in the reaction except as a conductor of electrons. We call a molecule that can donate or accept electrons at an electrode an **electroactive species.** The rate at which electrons flow into the electrode is a measure of the rate of reduction of Sn^{4+}.

Example **Relating Current to the Rate of Reaction**

Suppose that Sn^{4+} is reduced to Sn^{2+} at a constant rate of 4.24 mmol/h in Figure 14-1. How much current flows into the solution?

SOLUTION *Two* electrons are required to reduce *one* Sn^{4+} ion to Sn^{2+}. If Sn^{4+} is reacting at a rate of 4.24 mmol/h, electrons flow at a rate of 2(4.24) = 8.48 mmol/h, which corresponds to

$$\frac{8.48 \text{ mmol/h}}{3\,600 \text{ s/h}} = 2.356 \times 10^{-3} \text{ mmol/s} = 2.356 \times 10^{-6} \text{ mol/s}$$

To find the current, we use the Faraday constant to convert moles of electrons per second into coulombs per second:

$$\text{current} = \frac{\text{coulombs}}{\text{second}} = \frac{\text{moles}}{\text{second}} \cdot \frac{\text{coulombs}}{\text{mole}}$$

$$= \left(2.356 \times 10^{-6}\,\frac{\text{mol}}{\text{s}}\right)\left(9.649 \times 10^{4}\,\frac{\text{C}}{\text{mol}}\right) = 0.227\ \text{C/s} = 0.227\ \text{A}$$

Test Yourself What current is required to oxidize water at an electrode to liberate 1.00 mol O_2/day in the reaction $2H_2O \rightarrow O_2 + 4H^+ + 4e^-$? (**Answer:** 4.47 A)

Voltage and Electrical Work

Because of their negative charge, electrons are attracted to positively charged regions and repelled from negatively charged regions. If electrons are attracted from one point to another, they can do useful work along the way. If we want to force electrons into a region from which they are repelled, we must do work on the electrons to push them along. *Work* has the dimensions of energy, whose units are *joules* (J).

It costs energy to move like charges toward each other. Energy is released when opposite charges move toward each other.

The difference in **electric potential** between two points measures the work that is needed (or can be done) when electrons move from one point to another. The greater the potential difference between points A and B, the more work can be done (or must be done) when electrons travel from point A to point B. Potential difference is measured in **volts** (V).

1 volt = 1 J/C

A good analogy for understanding current and potential is to think of water flowing through a garden hose (Figure 14-2). Electric current is the quantity of electric charge flowing per second through a wire. Electric current is analogous to the volume of water flowing per second through the hose. Electric potential is a measure of the force pushing on the electrons. The greater the force, the more current flows. Electric potential is analogous to the pressure on the water in the hose. The greater the pressure, the faster the water flows.

When a charge, q, moves through a potential difference, E, the work done is

Relation between work and voltage:

$$\underset{\text{Joules}}{\text{work}} = \underset{\text{Volts}}{E} \cdot \underset{\text{Coulombs}}{q} \tag{14-2}$$

One joule of energy is gained or lost when one coulomb of charge moves through a potential difference of one volt. Equation 14-2 tells us that the dimensions of volts are J/C.

Here is the garden hose analogy for work: Suppose that one end of a hose is raised 1 m above the other end and 1 L of water flows through the hose. The water could go through a mechanical device to do a certain amount of work. If one end of the hose is raised 2 m above the other, the amount of work that can be done by the same volume of water is twice as great. The elevation difference between the ends of the hose is analogous to electric potential difference and the volume of water is analogous to electric charge. The greater the electric potential difference between two points in a circuit, the more work can be done by the charge flowing between those two points.

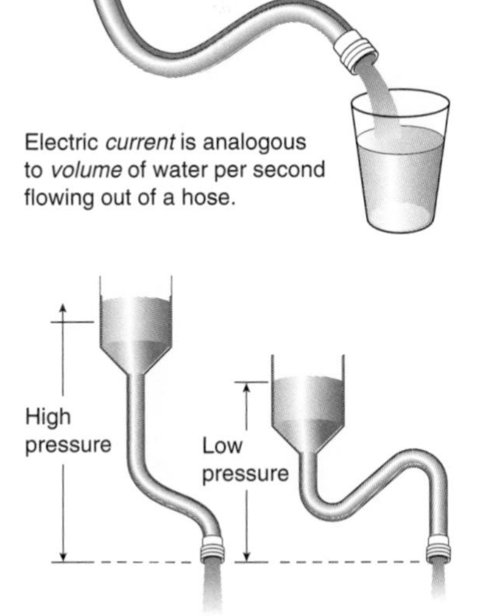

Electric *current* is analogous to *volume* of water per second flowing out of a hose.
High pressure
Low pressure
Electric *potential* is analogous to the hydrostatic *pressure* pushing water through a hose. High pressure gives high flow.

Figure 14-2 Analogy between the flow of water through a hose and the flow of electricity through a wire.

Example **Electrical Work**

How much work can be done when 2.36 mmol of electrons move "downhill" through a potential difference of 1.05 V? "Downhill" means that the flow is energetically favorable.

SOLUTION To use Equation 14-2, first use Equation 14-1 to convert moles of electrons into coulombs of charge:

$$q = n\mathrm{F} = (2.36 \times 10^{-3}\ \text{mol})\,(9.649 \times 10^{4}\ \text{C/mol}) = 2.277 \times 10^{2}\ \text{C}$$

The work that can be done by the moving electrons is

$$\text{work} = E \cdot q = (1.05\ \text{V})\,(2.277 \times 10^{2}\ \text{C}) = 239\ \text{J}$$

Test Yourself An electric current of 1.5 mA does 0.25 J of work in 1.00 min. How much voltage is driving the electrons? (**Answer:** 2.78 V)

Electrolytic production of aluminum by the Hall-Héroult process consumes 4.5% of the electrical output of the United States! Al^{3+} in a molten solution of Al_2O_3 and cryolite (Na_3AlF_6) is reduced to aluminum metal at the cathode of a cell that typically draws 250 000 A. This process was invented by **Charles Hall** in 1886 when he was 22 years old, just after graduating from Oberlin College. [Photo courtesy of Alcoa Co., Pittsburgh, PA.]

Electrolysis is a chemical reaction in which we apply a voltage to drive a redox reaction that would not otherwise occur. For example, electrolysis is used to make aluminum metal from Al^{3+} and to make chlorine gas (Cl_2) from Cl^- in seawater. Demonstration 14-1 is a great classroom illustration of electrolysis.

Ask Yourself

14-A. Consider the redox reaction

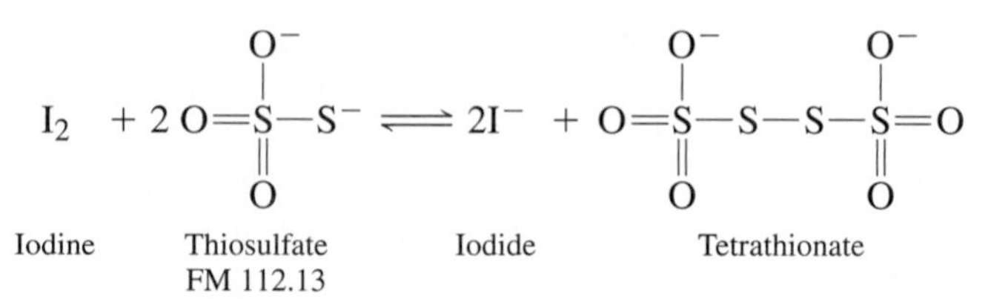

(a) Identify the oxidizing agent on the left side of the reaction and write a balanced oxidation half-reaction.
(b) Identify the reducing agent on the left side of the reaction and write a balanced reduction half-reaction.
(c) How many coulombs of charge are passed from reductant to oxidant when 1.00 g of thiosulfate reacts?
(d) If the rate of reaction is 1.00 g of thiosulfate consumed per minute, what current (in amperes) flows from reductant to oxidant?
(e) If the charge in **(c)** flows "downhill" through a potential difference of 0.200 V, how much work (in joules) can be done by the electric current?

14-2 Galvanic Cells

A *spontaneous reaction* is one in which it is energetically favorable for reactants to be converted into products. Energy released from the chemicals is available as electrical energy.

In a **galvanic cell**, a *spontaneous* chemical reaction generates electricity. To accomplish this, one reagent must be oxidized and another must be reduced. The two cannot be in contact, or electrons would flow directly from the reducing agent to the oxidizing agent without going through the external circuit. Therefore, the oxidizing and reducing agents are physically separated, and electrons flow through a wire to get from one reactant to the other.

Demonstration 14-1 Electrochemical Writing

Approximately 7% of electric power in the United States goes into electrolytic chemical production. The electrolysis apparatus shown here consists of a sheet of aluminum foil taped to a wooden surface. An area ~15 cm on a side is convenient for a classroom demonstration. On the metal foil is taped (at one edge only) a sandwich consisting of filter paper, printer paper, and another sheet of filter paper. A stylus is prepared from copper wire (18 gauge or thicker) looped at the end and passed through a length of glass tubing.

Prepare a fresh solution from 1.6 g of KI, 20 mL of water, 5 mL of 1 wt% starch solution, and 5 mL of phenolphthalein indicator solution. (If the solution darkens after standing for several days, decolorize it by adding a few drops of dilute $Na_2S_2O_3$.) Soak the three layers of paper with the KI-starch-phenolphthalein solution. Connect the stylus and foil to a 12-V DC power source and write on the paper with the stylus.

When the stylus is the cathode, pink color appears from the reaction of OH^- with phenolphthalein:

$$\text{Cathode:}\quad H_2O + e^- \rightarrow \tfrac{1}{2}H_2(g) + OH^-$$
$$OH^- + \text{phenolphthalein} \rightarrow \text{pink color}$$

When the polarity is reversed and the stylus is the anode, a black (very dark blue) color appears from the reaction of the newly generated I_2 with starch:

$$\text{Anode:}\quad I^- \rightarrow \tfrac{1}{2}I_2 + e^-$$
$$I_2 + \text{starch} \rightarrow \text{dark blue complex}$$

Pick up the top sheet of filter paper and the printer paper, and you will discover that the writing appears in the opposite color on the bottom sheet of filter paper. This sequence is shown in Color Plate 9. Color Plate 10 shows the same oxidation reaction in a solution where you can clearly see the process in action.

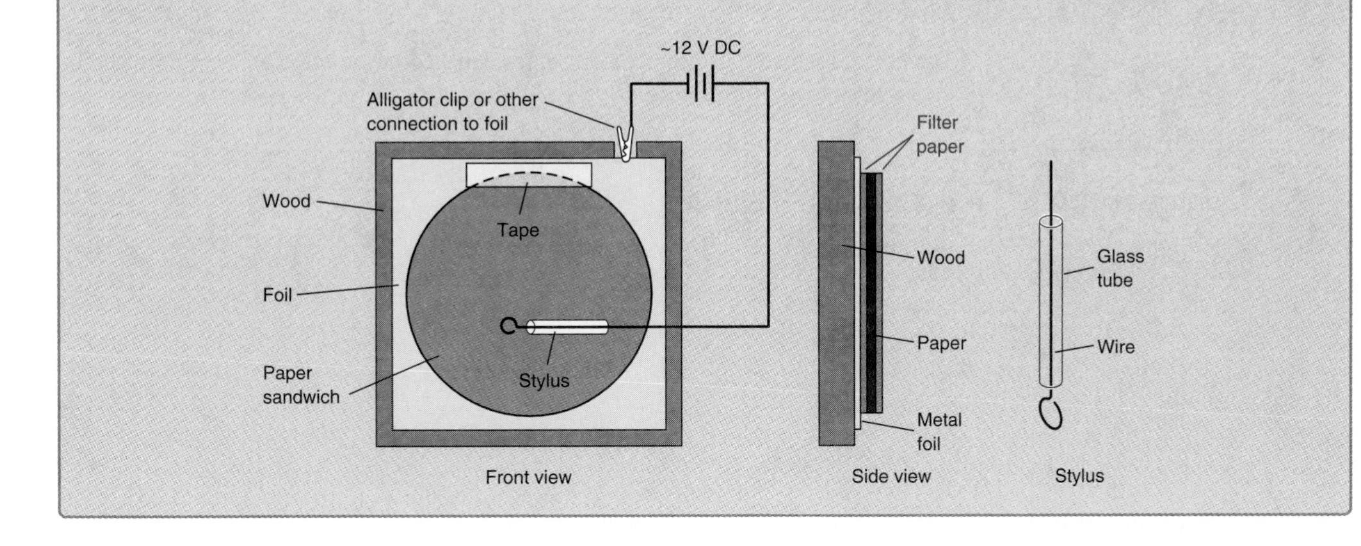

A Cell in Action

Figure 14-3 shows a galvanic cell consisting of two *half-cells* connected by a **salt bridge** through which ions migrate to maintain electroneutrality in each vessel.[3] The left half-cell has a zinc electrode dipped into aqueous $ZnCl_2$. The right half-cell has a copper electrode immersed in aqueous $CuSO_4$. The salt bridge is filled with a gel containing saturated aqueous KCl. The electrodes are connected by a *potentiometer* (a voltmeter) to measure the voltage difference between the two half-cells.

For more on salt bridges, see Demonstration 14-2.

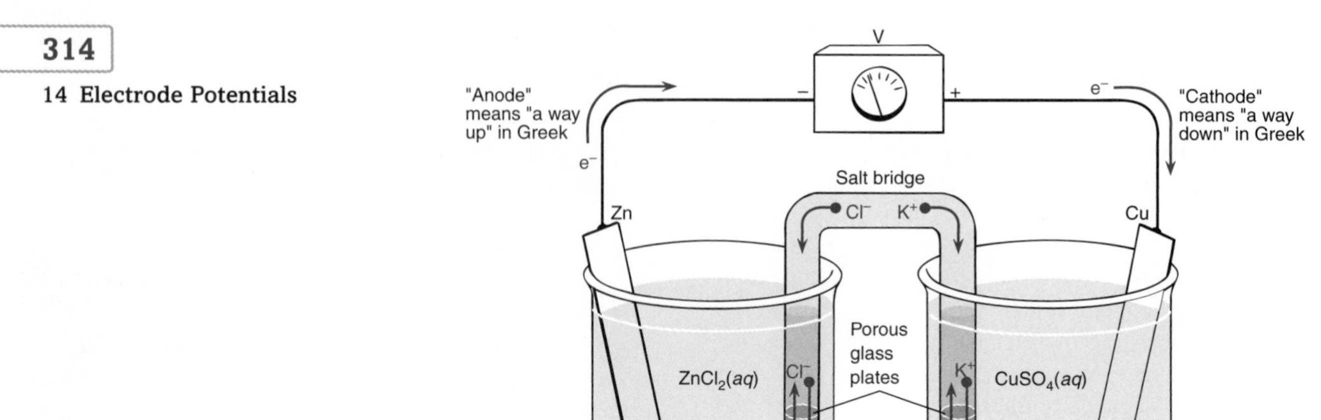

Figure 14-3 A galvanic cell consisting of two half-cells and a salt bridge through which ions can diffuse to maintain electroneutrality on each side.

The chemical reactions taking place in this cell are

Reduction half-reaction: $Cu^{2+}(aq) + \cancel{2e^-} \rightleftharpoons Cu(s)$

Oxidation half-reaction: $Zn(s) \rightleftharpoons Zn^{2+}(aq) + \cancel{2e^-}$

Net reaction: $Cu^{2+}(aq) + Zn(s) \rightleftharpoons Cu(s) + Zn^{2+}(aq)$

The two half-reactions are always written with equal numbers of electrons so that their sum includes no free electrons.

Demonstration 14-2 The Human Salt Bridge

A salt bridge is an ionic medium through which ions diffuse to maintain electroneutrality in each compartment of an electrochemical cell. One way to prepare a salt bridge is to heat 3 g of agar (the stuff used to grow bacteria in a Petri dish) with 30 g of KCl in 100 mL of water until a clear solution is obtained. Pour the solution into a U-tube and allow it to gel. Store the bridge in saturated aqueous KCl.

To conduct this demonstration, set up the galvanic cell in Figure 14-3 with 0.1 M $ZnCl_2$ on the left side and 0.1 M $CuSO_4$ on the right side. You can use a voltmeter or a pH meter to measure voltage. If you use a pH meter, the positive terminal is the connection for the glass electrode and the negative terminal is the reference electrode connection.

Write the two half-reactions for this cell and use the Nernst equations (14-6 and 14-7) to calculate the theoretical voltage. Measure the voltage with a conventional salt bridge. Then replace the salt bridge with one made of filter paper freshly soaked in NaCl solution and measure the voltage again. Finally, replace the filter paper with two fingers of the same hand and measure the voltage again. Your body is really just a bag of salt housed in a *semipermeable membrane* (skin) through which ions can diffuse. Small differences in voltage observed when the salt bridge is replaced can be attributed to the junction potential discussed in Section 15-2.

Challenge One hundred eighty students at Virginia Polytechnic Institute and State University made a salt bridge by holding hands.[4] (Their electrical resistance was lowered by a factor of 100 by wetting everyone's hands.) Can your school beat this record?

Oxidation of Zn(s) at the left side of Figure 14-3 produces $Zn^{2+}(aq)$. Electrons from the Zn metal flow through the potentiometer into the Cu electrode, where $Cu^{2+}(aq)$ is reduced to Cu(s). If there were no salt bridge, the left half-cell would soon build up positive charge (from excess Zn^{2+}) and the right half-cell would build up negative charge (by depletion of Cu^{2+}). In an instant, the charge buildup would oppose the driving force for the chemical reaction and the process would cease.

To maintain electroneutrality, Zn^{2+} on the left side diffuses into the salt bridge and Cl^- from the salt bridge diffuses into the left half-cell. In the right half-cell, SO_4^{2-} diffuses into the salt bridge and K^+ diffuses out of the salt bridge. The result is that positive and negative charges in each half-cell remain exactly balanced.

We call the electrode at which *reduction* occurs the **cathode**. The **anode** is the electrode at which *oxidation* occurs. In Figure 14-3, Cu is the cathode because reduction takes place at its surface ($Cu^{2+} + 2e^- \rightarrow Cu$) and Zn is the anode because it is oxidized ($Zn \rightarrow Zn^{2+} + 2e^-$).

cathode ↔ reduction
anode ↔ oxidation

Michael Faraday wanted to describe his discoveries with terms that would "advance the general cause of science" and not "retard its progress." He sought the aid of William Whewell in Cambridge, who coined words such as "anode" and "cathode."[5]

Line Notation

We often use a notation employing two symbols to describe electrochemical cells:

| phase boundary || salt bridge

The cell in Figure 14-3 is represented by the *line diagram*

$$Zn(s) \mid ZnCl_2(aq) \parallel CuSO_4(aq) \mid Cu(s)$$

Each phase boundary is indicated by a vertical line. The electrodes are shown at the extreme left- and right-hand sides of the diagram. The contents of the salt bridge are not specified.

The symbol || for a salf bridge represents two phase boundaries. In Figure 14-3, the $ZnCl_2$ aqueous phase in the left half-cell is separated from the aqueous phase of the salt bridge by a porous glass plate that allows species to slowly pass through. The porous glass plate on the right side separates the aqueous phase of the salt bridge from the $CuSO_4$ aqueous phase in the right half-cell.

Example Interpreting Line Diagrams of Cells

Write a line diagram for the cell in Figure 14-4. Write an anode reaction (an oxidation) for the left half-cell, a cathode reaction (a reduction) for the right half-cell, and the net cell reaction.

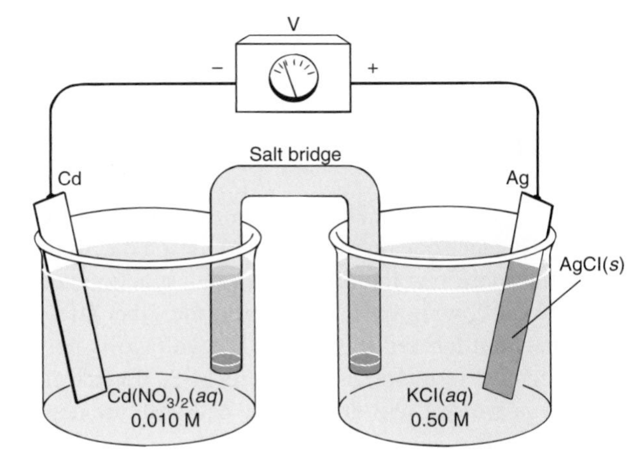

Figure 14-4 Another galvanic cell.

SOLUTION The right half-cell contains two solid phases (Ag and AgCl) and an aqueous phase. The left half-cell contains one solid phase and one aqueous phase. The line diagram is

$$Cd(s) \mid Cd(NO_3)_2(aq) \parallel KCl(aq) \mid AgCl(s) \mid Ag(s)$$

$Cd(NO_3)_2$ is a strong electrolyte, so it is completely dissociated to Cd^{2+} and NO_3^-; and KCl is dissociated into K^+ and Cl^-. In the left half-cell, we find cadmium in the oxidation states 0 and +2. In the right half-cell, we find silver in the 0 and +1 states. The Ag(I) is solid AgCl adhering to the Ag electrode. The electrode reactions are

Anode (oxidation): $Cd(s) \rightleftharpoons Cd^{2+}(aq) + 2e^-$

Cathode (reduction): $AgCl(s) + e^- \rightleftharpoons Ag(s) + Cl^-(aq)$

Net reaction: $Cd(s) + AgCl(s) \rightleftharpoons Cd^{2+}(aq) + Ag(s) + Cl^-(aq) + e^-$

Oops! We didn't write both half-reactions with the same number of electrons, so the net reaction still has electrons in it. Electrons are not allowed in a net reaction, so we double the coefficients of the second half-reaction:

Anode (oxidation): $Cd(s) \rightleftharpoons Cd^{2+}(aq) + 2e^-$

Cathode (reduction): $2AgCl(s) + 2e^- \rightleftharpoons 2Ag(s) + 2Cl^-(aq)$

Net reaction: $Cd(s) + 2AgCl(s) \rightleftharpoons Cd^{2+}(aq) + 2Ag(s) + 2Cl^-(aq)$

Test Yourself Write the diagram for the cell in Figure 14-4 if the left half-cell is replaced by a Pt electrode dipped into a solution containing $SnCl_4(aq)$ and $SnCl_2(aq)$. (**Answer:** $Pt(s) \mid SnCl_4(aq), SnCl_2(aq) \parallel KCl(aq) \mid AgCl(s) \mid Ag(s)$)

Ask Yourself

14-B. (a) What is the line notation for the cell in Figure 14-5?
(b) Draw a picture of the following cell. Write an oxidation half-reaction for the left half-cell, a reduction half-reaction for the right half-cell, and a balanced net reaction.

$$Au(s) \mid Fe(CN)_6^{4-}(aq), Fe(CN)_6^{3-}(aq) \parallel Ag(S_2O_3)_2^{3-}(aq), S_2O_3^{2-}(aq) \mid Ag(s)$$

14-3 Standard Potentials

The voltage measured in the experiment in Figure 14-3 is the difference in electric potential between the Cu electrode on the right and the Zn electrode on the left. The greater the voltage, the more favorable is the net cell reaction and the more work can be done by electrons flowing from one side to the other (Equation 14-2). The potentiometer terminals are labeled + and −. The voltage is positive when electrons flow into the negative terminal, as in Figure 14-3. If electrons flow the other way, the voltage is negative. We will draw every cell with the negative terminal of the potentiometer at the left.

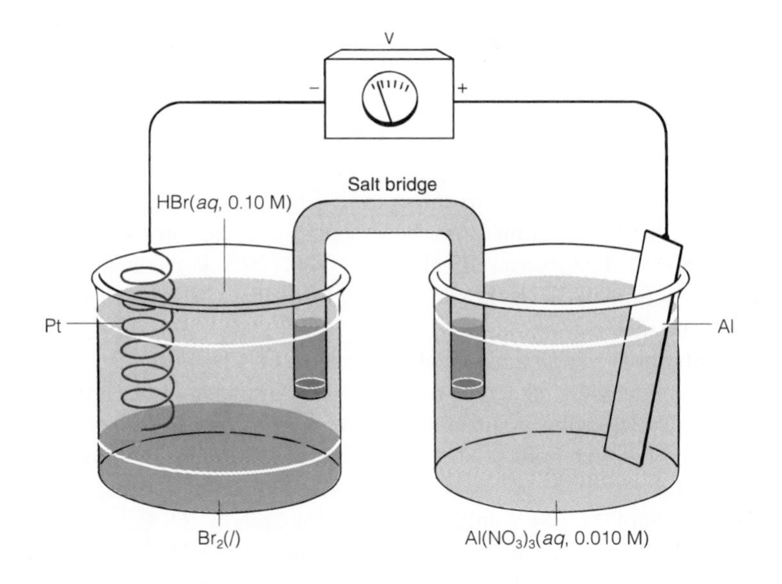

Figure 14-5 Cell for Ask Yourself 14-B.

To predict the voltage when different half-cells are connected to each other, the **standard reduction potential** (E°) for each half-cell is measured by an experiment shown in idealized form in Figure 14-6. The half-reaction of interest in this diagram is

$$Ag^+ + e^- \rightleftharpoons Ag(s) \tag{14-3}$$

taking place in the right-hand half-cell, which is connected to the *positive* terminal of the potentiometer. The term *standard* means that species are solids or liquids or their concentrations are 1 M or their pressures are 1 bar. We call these conditions the *standard states* of the reactants and products.

We oversimplify the standard half-cell in this text. The word *standard* really means that all *activities* are 1. Although activity (Section 12-2) is not the same as concentration, we will consider *standard* to mean that concentrations are 1 M or 1 bar for simplicity.

The left half-cell, connected to the *negative* terminal of the potentiometer, is called the **standard hydrogen electrode (S.H.E.)**. It consists of a catalytic Pt surface

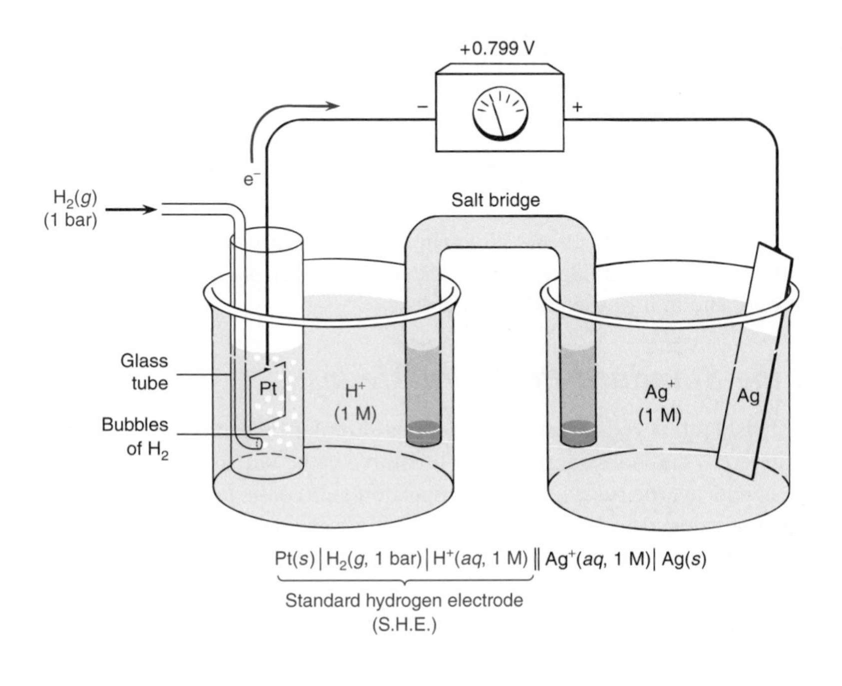

Figure 14-6 Setup used to measure the standard reduction potential (E°) for the half-reaction $Ag^+ + e^- \rightleftharpoons Ag(s)$. The left half-cell is called the standard hydrogen electrode (S.H.E.).

Question What is the pH of the standard hydrogen electrode?

in contact with an acidic solution in which $[H^+] = 1$ M. A stream of $H_2(g$, 1 bar) is bubbled past the electrode. The reaction at the surface of the Pt electrode is

We will write all half-reactions as *reductions*. By convention, $E° = 0$ for S.H.E.

$$H^+(aq, 1\text{ M}) + e^- \rightleftharpoons \tfrac{1}{2}H_2(g, 1\text{ bar}) \tag{14-4}$$

We *assign* a potential of 0 to the standard hydrogen electrode. The voltage measured in Figure 14-6 can therefore be *assigned* to Reaction 14-3, which occurs in the right half-cell. The measured value $E° = +0.799$ V is the standard reduction potential for Reaction 14-3. The positive sign tells us that electrons flow from left to right through the meter.

We can arbitrarily *assign* a potential to Reaction 14-4 because it serves as a point from which we can measure other half-cell potentials. An analogy is the arbitrary assignment of 0°C to the freezing point of water. Relative to this freezing point, hexane boils at 69°C and benzene boils at 80°C. The difference between the boiling points of benzene and hexane is $80° - 69° = 11°$. If we had assigned the freezing point of water to be 200°C instead of 0°C, we would say that hexane boils at 269°C and benzene boils at 280°C. The difference between their boiling points is still 11°. When we measure half-cell potentials relative to that of the standard hydrogen electrode, we are simply putting the potentials on a scale that allows us to measure differences. Regardless of where we set zero on the scale, differences between points remain constant.

The line notation for the cell in Figure 14-6 is

$$Pt(s) \mid H_2(g, 1\text{ bar}) \mid H^+(aq, 1\text{ M}) \parallel Ag^+(aq, 1\text{ M}) \mid Ag(s)$$

which is abbreviated

$$\text{S.H.E.} \parallel Ag^+(aq, 1\text{ M}) \mid Ag(s)$$

Measured voltage =
right-hand electrode potential
– left-hand electrode potential

The standard reduction potential is really the *difference* between the standard potential of the reaction of interest and the potential of the S.H.E., which we have arbitrarily set to 0.

If we wanted to measure the standard potential of the half-reaction

$$Cd^{2+} + 2e^- \rightleftharpoons Cd(s) \tag{14-5}$$

we would construct the cell

$$\text{S.H.E.} \parallel Cd^{2+}(aq, 1\text{ M}) \mid Cd(s)$$

Challenge Draw a picture of the cell S.H.E. || $Cd^{2+}(aq$, 1 M) | Cd(s) and show the direction of electron flow.

with the cadmium half-cell *at the right*. In this case, we would observe a *negative* voltage of −0.402 V. The negative sign means that electrons flow from Cd to Pt, a direction opposite that of the cell in Figure 14-6.

What the Standard Potential Means

Table 14-1 lists a few reduction half-reactions in order of decreasing $E°$ value. The more positive $E°$, the more energetically favorable is the half-reaction. The strongest oxidizing agents are the reactants at the upper-left side of the table because they have the strongest tendency to accept electrons. $F_2(g)$ is the strongest oxidizing agent in the table. Conversely, F^- is the weakest reducing agent because it has the least tendency to give up electrons to make F_2. The strongest reducing agents in Table 14-1 are at the lower-right side. Li(s) and K(s) are very strong reducing agents.

Table 14-1 Ordered redox potentials

	Oxidizing agent		Reducing agent		$E°$ (V)
Oxidizing power increases ⇑	$F_2(g) + 2e^-$	$\rightleftharpoons$	$2F^-$	Reducing power increases ⇓	2.890
	$O_3(g) + 2H^+ + 2e^-$	$\rightleftharpoons$	$O_2(g) + H_2O$		2.075
	$MnO_4^- + 8H^+ + 5e^-$	$\rightleftharpoons$	$Mn^{2+} + 4H_2O$		1.507
	$Ag^+ + e^-$	$\rightleftharpoons$	$Ag(s)$		0.799
	$Cu^{2+} + 2e^-$	$\rightleftharpoons$	$Cu(s)$		0.339
	$2H^+ + 2e^-$	$\rightleftharpoons$	$H_2(g)$		0.000
	$Cd^{2+} + 2e^-$	$\rightleftharpoons$	$Cd(s)$		−0.402
	$K^+ + e^-$	$\rightleftharpoons$	$K(s)$		−2.936
	$Li^+ + e^-$	$\rightleftharpoons$	$Li(s)$		−3.040

Formal Potential

Appendix C lists many standard reduction potentials. Sometimes, multiple potentials are listed for one reaction, as for the $AgCl(s) \mid Ag(s)$ half-reaction:

$$AgCl(s) + e^- \rightleftharpoons Ag(s) + Cl^- \qquad \begin{cases} 0.222 \text{ V} \\ 0.197 \text{ V saturated KCl} \end{cases}$$

Standard potential: 0.222 V

Formal potential for saturated KCl: 0.197 V

The value 0.222 V is the standard potential that would be measured in the cell

$$\text{S.H.E.} \parallel Cl^- (aq, 1 \text{ M}) \mid AgCl(s) \mid Ag(s)$$

The value 0.197 V is measured in a cell containing saturated KCl solution instead of 1 M Cl^-:

$$\text{S.H.E.} \parallel KCl(aq, \text{saturated}) \mid AgCl(s) \mid Ag(s)$$

The potential for a cell containing a specified concentration of reagent other than 1 M is called the **formal potential**.

Ask Yourself

14-C. Draw a line diagram and a picture of the cell used to measure the standard potential of the reaction $Fe^{3+} + e^- \rightleftharpoons Fe^{2+}$, which comes to equilibrium at the surface of a Pt electrode. Use Appendix C to find the cell voltage. From the sign of the voltage, show the direction of electron flow in your picture.

14-4 The Nernst Equation

Le Châtelier's principle tells us that, if we increase reactant concentrations, we drive a reaction to the right. Increasing the product concentrations drives a reaction to the left. The net driving force for a reaction is expressed by the **Nernst equation**, whose two terms include the driving force under standard conditions ($E°$, which applies when concentrations are 1 M or 1 bar) and a term that shows the dependence on concentrations.

Nernst Equation for a Half-Reaction

For the half-reaction

$$aA + ne^- \rightleftharpoons bB$$

the Nernst equation giving the half-cell potential, E, is

In the Nernst equation:

- E and E° are measured in volts
- solute concentration = mol/L
- gas concentration = bar
- solid, liquid, solvent omitted

Nernst equation:

$$E = E^\circ - \frac{0.059\,16}{n} \log\left(\frac{[B]^b}{[A]^a}\right) \quad \text{(at 25°C)} \qquad (14\text{-}6)$$

where E° is the standard reduction potential that applies when [A] = [B] = 1 M.

The logarithmic term in the Nernst equation is the *reaction quotient*, Q (= $[B]^b/[A]^a$). Q has the same form as the equilibrium constant, but the concentrations need not be at their equilibrium values. Concentrations of solutes are expressed as moles per liter and concentrations of gases are expressed as pressures in bars. Pure solids, pure liquids, and solvents are omitted from Q. When all concentrations are 1 M and all pressures are 1 bar, $Q = 1$ and $\log Q = 0$, thus giving $E = E^\circ$.

Example Writing the Nernst Equation for a Half-Reaction

Let's write the Nernst equation for the reduction of white phosphorus to phosphine gas:

$$\tfrac{1}{4}P_4(s,\text{ white}) + 3H^+ + 3e^- \rightleftharpoons \underset{\text{Phosphine}}{PH_3(g)} \qquad E^\circ = -0.046\text{ V}$$

SOLUTION We omit solids from the reaction quotient, and the concentration of phosphine gas is expressed as its pressure in bars (P_{PH_3}):

$$E = -0.046 - \frac{0.059\,16}{3} \log\left(\frac{P_{PH_3}}{[H^+]^3}\right)$$

Test Yourself Write the Nernst equation for a hydrogen electrode (Reaction 14-4) in which the concentrations of H_2 and H^+ can vary. (**Answer:** $E = 0 - (0.059\,16/1) \log P_{H_2}^{1/2}/[H^+]$)

Example Multiplication of a Half-Reaction

If you multiply a half-reaction by any factor, the value of E° does not change. However, the factor n before the log term and the exponents in the reaction quotient do change. Write the Nernst equation for the reaction in the preceding example, multiplied by 2:

$$\tfrac{1}{2}P_4(s,\text{ white}) + 6H^+ + 6e^- \rightleftharpoons 2PH_3(g) \qquad E^\circ = -0.046\text{ V}$$

SOLUTION

$$E = -0.046 - \frac{0.059\,16}{6} \log\left(\frac{P_{PH_3}^2}{[H^+]^6}\right)$$

$E°$ remains at -0.046 V, as in the preceding example. However, the factor in front of the log term and the exponents in the log term have changed.

Test Yourself Write the Nernst equation for Reaction 14-4 multiplied by 2. (**Answer:** $E = 0 - (0.059\ 16/2) \log P_{H_2}/[H^+]^2$)

Nernst Equation for a Complete Reaction

Consider the cell in Figure 14-4 in which the negative terminal of the potentiometer is connected to the Cd electrode and the positive terminal is connected to the Ag electrode. The voltage, E, is the difference between the potentials of the two electrodes:

Nernst equation for a complete cell:

$$E = E_+ - E_- \tag{14-7}$$

where E_+ is the potential of the half-cell attached to the positive terminal of the potentiometer and E_- is the potential of the half-cell attached to the negative terminal. The potential of each half-reaction (*written as a reduction*) is governed by the Nernst equation 14-6.

Here is a procedure for writing a net cell reaction and finding its voltage:

Both half-reactions are written as *reductions* when we use Equation 14-7.

Step 1. Write *reduction* half-reactions for both half-cells and find $E°$ for each in Appendix C. Multiply the half-reactions as necessary so that they each contain the same number of electrons. When you multiply a reaction by any number, *do not* multiply $E°$.

Step 2. Write a Nernst equation for the half-reaction in the right half-cell, which is attached to the positive terminal of the potentiometer. This is E_+.

Step 3. Write a Nernst equation for the half-reaction in the left half-cell, which is attached to the negative terminal of the potentiometer. This is E_-.

Step 4. Find the net cell voltage by subtraction: $E = E_+ - E_-$.

Step 5. To write a balanced net cell reaction, subtract the left half-reaction from the right half-reaction. (*This operation is equivalent to reversing the left half-reaction and adding.*)

If the net cell voltage, E $(= E_+ - E_-)$, is positive, then the net cell reaction is spontaneous in the forward direction. If the net cell voltage is negative, then the reaction is spontaneous in the reverse direction.

Example Nernst Equation for a Complete Reaction

Find the voltage of the cell in Figure 14-4 if the right half-cell contains 0.50 M $KCl(aq)$ and the left half-cell contains 0.010 M $Cd(NO_3)_2(aq)$. Write the net cell reaction and state whether it is spontaneous in the forward or reverse direction.

SOLUTION

Step 1. Write reduction half-reactions:

Right half-cell: $2AgCl(s) + 2e^- \rightleftharpoons 2Ag(s) + 2Cl^-$ $\quad E_+^\circ = 0.222$ V

Left half-cell: $Cd^{2+} + 2e^- \rightleftharpoons Cd(s)$ $\quad E_-^\circ = -0.402$ V

Pure solids, pure liquids, and solvents are omitted from Q.

Step 2. Nernst equation for right half-cell:

$$E_+ = E_+^\circ - \frac{0.059\ 16}{2}\log([Cl^-]^2) \quad (14\text{-}8)$$

$$= 0.222 - \frac{0.059\ 16}{2}\log([0.50]^2) = 0.240\ V$$

Step 3. Nernst equation for left half-cell:

$$E_- = E_-^\circ - \frac{0.059\ 16}{2}\log\left(\frac{1}{[Cd^{2+}]}\right) = -0.402 - \frac{0.059\ 16}{2}\log\left(\frac{1}{[0.010]}\right)$$

$$= -0.461\ V$$

Step 4. Cell voltage: $E = E_+ - E_- = 0.240 - (-0.461) = 0.701$ V

Step 5. Net cell reaction:

Subtracting a reaction is the same as *reversing the reaction and adding.*

$$2AgCl(s) + 2e^- \rightleftharpoons 2Ag(s) + 2Cl^-$$

$$-\quad Cd^{2+} + 2e^- \rightleftharpoons Cd(s)$$

$$Cd(s) + 2AgCl(s) \rightleftharpoons Cd^{2+} + 2Ag(s) + 2Cl^-$$

The *positive* voltage tells us that the net reaction is spontaneous in the *forward* direction. Cd(s) is oxidized to Cd^{2+} and AgCl(s) is reduced to Ag(s). Electrons flow from the left electrode to the right electrode. If the voltage were *negative*, the reaction would be spontaneous in the *reverse* direction.

Test Yourself Write the Nernst equation for the cell Pt(s) | H_2(g, 1.0×10^{-6} bar) | H^+(aq, 0.50 M) || Ag^+ (aq, 1.0×10^{-10} M) | Ag(s) and find the cell voltage. In what direction will electrons flow through the circuit? (**Answer:** 0.048 V, electrons flow from Pt to Ag)

What if you had written the Nernst equation for the right half-cell with just one electron instead of two: $AgCl(s) + e^- \rightleftharpoons Ag(s) + Cl^-$? Try this and you will discover that the half-cell potential is unchanged. *Neither $E°$ nor E changes when you multiply a reaction.*

Electrons flow toward more positive potential.

Electrons Flow Toward More Positive Potential

In the example above, we found that E for the silver half-cell was 0.240 V and E for the cadmium half-cell was −0.461 V. Place these values on the number line in Figure 14-7 and note that *negatively charged electrons flow toward more positive potential.* Therefore, electrons in the circuit flow from cadmium (−0.461 V) to silver (0.240 V). The diagram works the same way even if both half-cell potentials are positive or both are negative. Electrons always flow from left to right in a diagram like Figure 14-7.

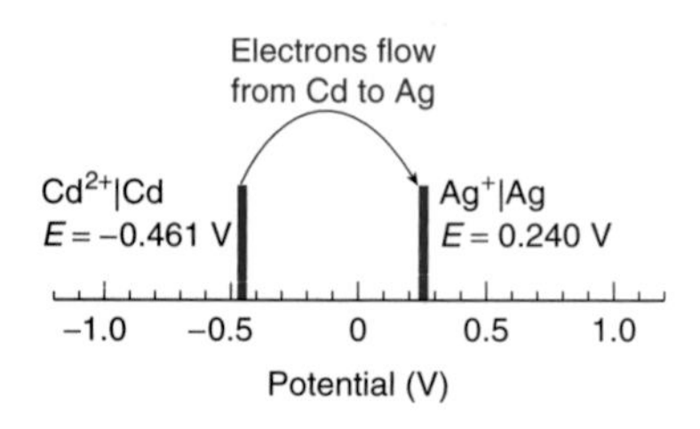

Figure 14-7 An intuitive view of cell potentials.[6] Electrons always flow to the right in this diagram.

Different Descriptions of the Same Reaction

We know that the right half-cell in Figure 14-4 must contain some $Ag^+(aq)$ in equilibrium with AgCl(s). Suppose that, instead of writing the reaction

$2AgCl(s) + 2e^- \rightleftharpoons 2Ag(s) + 2Cl^-$, a different, less handsome, author wrote the reaction

$$2Ag^+(aq) + 2e^- \rightleftharpoons 2Ag(s) \qquad E_+^\circ = 0.799 \text{ V}$$

$$E_+ = 0.799 - \frac{0.059\ 16}{2}\log\left(\frac{1}{[Ag^+]^2}\right) \tag{14-9}$$

Both descriptions of the right half-cell are valid. In both cases, Ag(I) is reduced to Ag(0).

If the two descriptions are equal, then they should predict the same voltage. To use Equation 14-9, you must know the concentration of Ag^+ in the right half-cell, which is not obvious. But you are clever and realize that you can find $[Ag^+]$ from the solubility product for AgCl and the concentration of Cl^-, which is 0.50 M.

$$K_{sp} = [Ag^+][Cl^-] \Rightarrow [Ag^+] = \frac{K_{sp}}{[Cl^-]} = \frac{1.8 \times 10^{-10}}{0.50} = 3.6 \times 10^{-10} \text{ M}$$

Putting this concentration into Equation 14-9 gives

$$E_+ = 0.799 - \frac{0.059\ 16}{2}\log\left(\frac{1}{(3.6 \times 10^{-10})^2}\right) = 0.240 \text{ V}$$

which is the same voltage computed in Equation 14-8! The two choices of half-reaction give the same voltage because they describe the same cell.

The cell voltage is a measured, experimental quantity that cannot depend on how we write the reaction!

Advice for Finding Relevant Half-Reactions

When faced with a cell drawing or line diagram, the first step is to write reduction reactions for each half-cell. To do this, *look for elements in two oxidation states.* For the cell

$$Pb(s) \mid PbF_2(s) \mid F^-(aq) \parallel Cu^{2+}(aq) \mid Cu(s)$$

we see lead in the oxidation states 0 in $Pb(s)$ and +2 in $PbF_2(s)$, and we see copper in the oxidation states 0 in $Cu(s)$ and +2 in Cu^{2+}. Thus, the half-reactions are

Right half-cell: $Cu^{2+} + 2e^- \rightleftharpoons Cu(s)$

Left half-cell: $PbF_2(s) + 2e^- \rightleftharpoons Pb(s) + 2F^-$ (14-10)

Don't write a reaction such as $F_2(g) + 2e^- \rightleftharpoons 2F^-$, because $F_2(g)$ is not shown in the line diagram of the cell. $F_2(g)$ is neither a reactant nor a product.

You might have chosen to write the lead half-reaction as

Left half-cell: $Pb^{2+} + 2e^- \rightleftharpoons Pb(s)$ (14-11)

because you know that, if $PbF_2(s)$ is present, there must be some Pb^{2+} in the solution. Reactions 14-10 and 14-11 are both valid descriptions of the cell, and each should predict the same cell voltage. Your choice of reaction depends on whether the F^- or Pb^{2+} concentration is more easily known to you.

Ask Yourself

14-D. (a) Arsine (AsH_3) is a poisonous gas used to make gallium arsenide for diode lasers. Find E for the half-reaction $As(s) + 3H^+ + 3e^- \rightleftharpoons AsH_3(g)$ if pH = 3.00 and $P_{AsH_3} = 0.010\,0$ bar.
(b) The cell in Demonstration 14-2 (and Figure 14-3) can be written

$$Zn(s) \mid Zn^{2+}(0.1\text{ M}) \parallel Cu^{2+}(0.1\text{ M}) \mid Cu(s)$$

Write a reduction half-reaction for each half-cell and use the Nernst equation to predict the cell voltage. Draw a diagram like Figure 14-7 and show the direction of electron flow.

14-5 $E°$ and the Equilibrium Constant

At equilibrium, E (not $E°$) = 0. $E°$ is the potential when all reactants and products are present in their standard state (1 M, 1 bar, pure solid, pure liquid).

A galvanic cell produces electricity because the cell reaction is not at equilibrium. If the cell runs long enough, reactants are consumed and products are created until the reaction comes to equilibrium, and the cell voltage, E, reaches 0. This is what happens to a battery when it runs down.[7]

If $E°_+$ is the standard reduction potential for the right half-cell and $E°_-$ is the standard reduction potential for the left half-cell, $E°$ for the net cell reaction is

$E°$ for net cell reaction:
$$E° = E°_+ - E°_- \qquad (14\text{-}12)$$

Now let's relate $E°$ to the equilibrium constant for the net cell reaction.

As an example, consider the cell in Figure 14-8. The left half-cell contains a silver electrode coated with solid AgCl and dipped into saturated aqueous KCl. The right half-cell has a platinum wire dipped into a solution containing Fe^{2+} and Fe^{3+}.

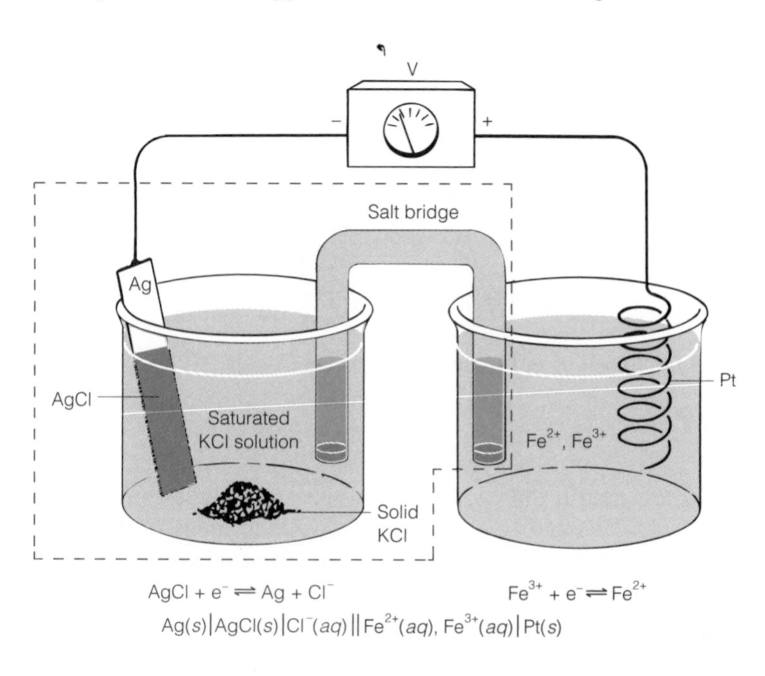

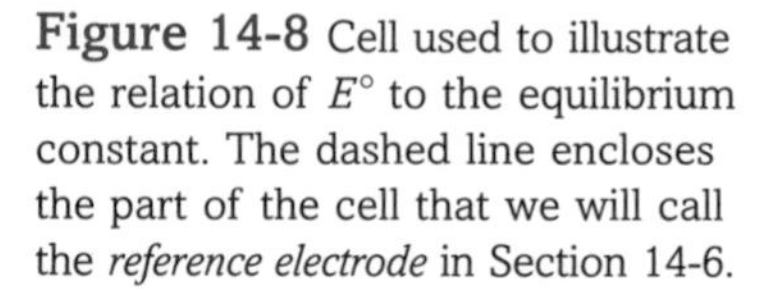

Figure 14-8 Cell used to illustrate the relation of $E°$ to the equilibrium constant. The dashed line encloses the part of the cell that we will call the *reference electrode* in Section 14-6.

The two half-reactions are

Right: $Fe^{3+} + e^- \rightleftharpoons Fe^{2+}$ $\qquad E^\circ_+ = 0.771$ V $\qquad$ (14-13)

Left: $AgCl(s) + e^- \rightleftharpoons Ag(s) + Cl^-$ $\qquad E^\circ_- = 0.222$ V $\qquad$ (14-14)

E°_+ and E_+ are for half-cell connected to positive terminal of potentiometer.

E°_- and E_- are for half-cell connected to negative terminal.

The net cell reaction is found by subtracting the left half-reaction from the right half-reaction:

Net reaction: $Fe^{3+} + Ag(s) + Cl^- \rightleftharpoons Fe^{2+} + AgCl(s)$

$$E^\circ = E^\circ_+ - E^\circ_- = 0.771 - 0.222 = 0.549 \text{ V} \qquad (14\text{-}15)$$

The two electrode potentials are given by the Nernst equation 14-6:

$$E_+ = E^\circ_+ - \frac{0.059\,16}{n}\log\left(\frac{[Fe^{2+}]}{[Fe^{3+}]}\right) \qquad (14\text{-}16)$$

$$E_- = E^\circ_- - \frac{0.059\,16}{n}\log([Cl^-]) \qquad (14\text{-}17)$$

where $n = 1$ for Reactions 14-13 and 14-14. The cell voltage is the difference $E_+ - E_-$:

$$E = E_+ - E_- = \left\{E^\circ_+ - \frac{0.059\,16}{n}\log\left(\frac{[Fe^{2+}]}{[Fe^{3+}]}\right)\right\} - \left\{E^\circ_- - \frac{0.059\,16}{n}\log([Cl^-])\right\}$$

$$= (E^\circ_+ - E^\circ_-) - \left\{\frac{0.059\,16}{n}\log\left(\frac{[Fe^{2+}]}{[Fe^{3+}]}\right) - \frac{0.059\,16}{n}\log([Cl^-])\right\} \qquad (14\text{-}18)$$

To combine logarithms, we use the equality $\log x - \log y = \log(x/y)$:

Algebra of logarithms:

$\log x + \log y = \log(xy)$

$\log x - \log y = \log(x/y)$

$$E = \underbrace{(E^\circ_+ - E^\circ_-)}_{E^\circ \text{ for net reaction}} - \frac{0.059\,16}{n}\log\underbrace{\left(\frac{[Fe^{2+}]}{[Fe^{3+}][Cl^-]}\right)}_{\text{Reaction quotient } (Q) \text{ for net reaction}} = E^\circ - \frac{0.059\,16}{n}\log Q \qquad (14\text{-}19)$$

Equation 14-19 is true at any time. *In the special case when the cell is at equilibrium, $E = 0$ and $Q = K$, the equilibrium constant.* At equilibrium,

$$0 = E^\circ - \frac{0.059\,16}{n}\log K \Rightarrow E^\circ = \frac{0.059\,16}{n}\log K \qquad (14\text{-}20)$$

To go from Equation 14-20 to 14-21:

$$\frac{0.059\,16}{n}\log K = E^\circ$$

$$\log K = \frac{nE^\circ}{0.059\,16}$$

$$10^{\log K} = 10^{nE^\circ/0.059\,16}$$

$$K = 10^{nE^\circ/0.059\,16}$$

Finding K from E°:

$$K = 10^{nE^\circ/0.059\,16} \quad \text{(at 25°C)} \qquad (14\text{-}21)$$

Equation 14-21 gives the equilibrium constant for a net cell reaction from E°.

A positive value of E° means that $K > 1$ in Equation 14-21 and a negative value means that $K < 1$. A reaction is spontaneous under standard conditions (that is, when all concentrations of reactants and products are 1 M or 1 bar) if E° is positive. Biochemists use a different potential, called $E^{\circ\prime}$, which is described in Box 14-1.

Box 14-1 Why Biochemists Use $E^{\circ\prime}$

Redox reactions are essential for life. For example, the enzyme-catalyzed reduction of pyruvate to lactate is a step in the anaerobic fermentation of sugar by bacteria:

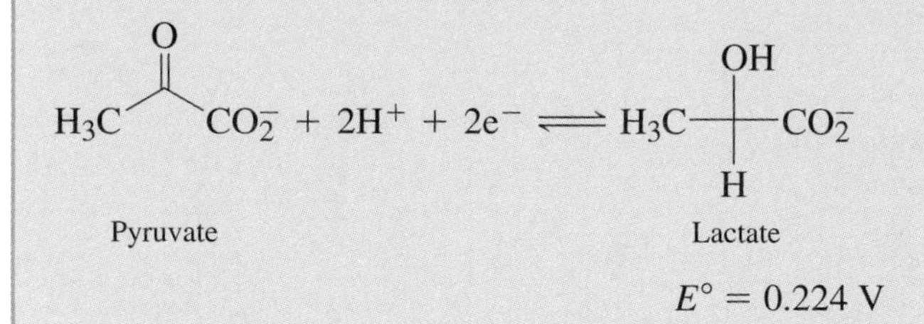

$$E^{\circ} = 0.224 \text{ V}$$

The same reaction causes lactate to build up in your muscles and make you feel fatigued during intense exercise, when the flow of oxygen cannot keep pace with your requirement for energy.

E° applies when the concentrations of reactants and products are 1 M. For a reaction involving H^+, E° applies when the pH is 0 (because $[H^+] = 1$ M and $\log 1 = 0$). Biochemists studying the energetics of fermentation or respiration are more interested in reduction potentials that apply near physiologic pH, not at pH 0. Therefore, biochemists use a formal potential designated $E^{\circ\prime}$, which applies at pH 7.

E° applies at pH 0
$E^{\circ\prime}$ applies at pH 7

For the conversion of pyruvate into lactate, the reactant and product are carboxylic acids at pH 0 and carboxylate anions at pH 7. The value of $E^{\circ\prime}$ at pH 7 is -0.190 V, which is quite different from $E^{\circ} = +0.224$ V at pH 0.

Example **Using E° to Find the Equilibrium Constant**

Find the equilibrium constant for the reaction $Fe^{3+} + Ag(s) + Cl^- \rightleftharpoons Fe^{2+} + AgCl(s)$, which is the net cell reaction in Figure 14-8.

SOLUTION Equation 14-15 states that $E^{\circ} = 0.549$ V. The equilibrium constant is computed with Equation 14-21, using $n = 1$, because one electron is transferred in each half-reaction:

$$K = 10^{nE^{\circ}/0.059\,16} = 10^{(1)(0.549)/(0.059\,16)} = 1.9 \times 10^9$$

Significant figures in logarithms and exponents were discussed in Section 3.2.

K has two significant figures because E° has three digits. One digit of E° is used for the exponent (9), and the other two are left for the multiplier (1.9).

Test Yourself Write the net cell reaction for Figure 14-6 with two electrons in each half-reaction. Find E° and K. If you had written half-reactions with one electron, what would be the equilibrium constant? (**Answer:** $2Ag^+ + H_2(g) \rightleftharpoons 2Ag(s) + 2H^+$, $E^{\circ} = 0.799$ V, $K = 1.0 \times 10^{27}$, 3×10^{13})

Ask Yourself

14-E. (a) Write the half-reactions for Figure 14-3. Calculate E° and the equilibrium constant for the net cell reaction.
(b) The solubility product reaction of AgBr is $AgBr(s) \rightleftharpoons Ag^+ + Br^-$. Use the reactions $Ag^+ + e^- \rightleftharpoons Ag(s)$ and $AgBr(s) + e^- \rightleftharpoons Ag(s) + Br^-$ to compute the solubility product of AgBr. Compare your answer with that in Appendix A.

14-6 Reference Electrodes

Imagine a solution containing an electroactive species whose concentration we wish to measure. We construct a half-cell by inserting an electrode (such as a Pt wire) into the solution to transfer electrons to or from the species of interest. Because this electrode responds directly to the analyte, it is called the **indicator electrode**. The potential of the indicator electrode is E_+. We then connect this half-cell to a second half-cell by a salt bridge. The second half-cell has a fixed composition that provides a known, constant potential, E_-. Because the second half-cell has a constant potential, it is called a **reference electrode**. The cell voltage ($E = E_+ - E_-$) is the difference between the variable potential that reflects changes in the analyte concentration and the constant reference potential.

Indicator electrode: responds to analyte concentration
Reference electrode: maintains a fixed (reference) potential

Suppose you have a solution containing Fe^{2+} and Fe^{3+}. If you are clever, you can make this solution part of a cell whose voltage tells you the relative concentrations of these species—that is, the value of $[Fe^{2+}]/[Fe^{3+}]$. Figure 14-8 shows one way to do this. A Pt wire acts as an indicator electrode through which Fe^{3+} can receive electrons or Fe^{2+} can lose electrons. The left half-cell completes the galvanic cell and has a known, constant potential.

The cell voltage in Figure 14-8 responds only to changes in the quotient $[Fe^{2+}]/[Fe^{3+}]$. Everything else is constant.

The two half-reactions were given in Reactions 14-13 and 14-14, and the two electrode potentials were given in Equations 14-16 and 14-17. The cell voltage is the difference $E_+ - E_-$:

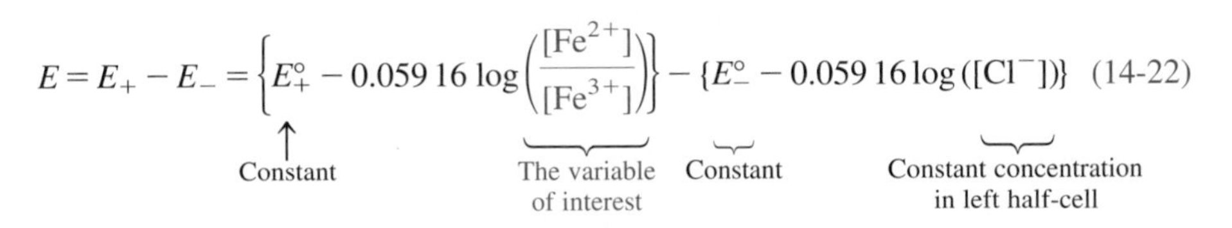

$$E = E_+ - E_- = \underbrace{\Big\{E_+^\circ}_{\text{Constant}} - 0.059\,16 \log\underbrace{\left(\frac{[Fe^{2+}]}{[Fe^{3+}]}\right)}_{\text{The variable of interest}}\Big\} - \{\underbrace{E_-^\circ}_{\text{Constant}} - 0.059\,16 \log \underbrace{([Cl^-])}_{\text{Constant concentration in left half-cell}}\} \qquad (14\text{-}22)$$

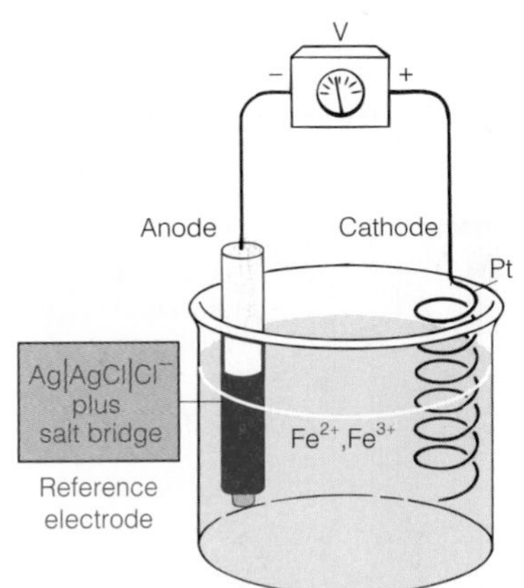

Figure 14-9 Another view of Figure 14-8. The contents of the dashed box in Figure 14-8 are now considered to be a reference electrode dipped into the analyte solution.

Because $[Cl^-]$ in the left half-cell is constant (fixed by the solubility of KCl, with which the solution is saturated), the cell voltage changes only when the quotient $[Fe^{2+}]/[Fe^{3+}]$ changes.

The half-cell on the left in Figure 14-8 can be thought of as a *reference electrode.* We can picture the cell and salt bridge enclosed by the dashed line as a single unit dipped into the analyte solution, as in Figure 14-9. The Pt wire is the indicator electrode, whose potential responds to changes in the quotient $[Fe^{2+}]/[Fe^{3+}]$. The reference electrode completes the redox reaction and provides a *constant potential* to the left side of the potentiometer. Changes in the cell voltage can be assigned to changes in the quotient $[Fe^{2+}]/[Fe^{3+}]$.

Silver-Silver Chloride Reference Electrode

The half-cell enclosed by the dashed line in Figure 14-8 is called a **silver-silver chloride electrode**. Figure 14-10 shows how the half-cell is reconstructed as a thin, glass-enclosed electrode that can be dipped into the analyte solution in Figure 14-9. The porous plug at the base of the electrode functions as a salt bridge. It allows ions to diffuse between solutions inside and outside the electrode with minimal physical mixing. We use silver-silver chloride or other reference electrodes because they are more convenient than a hydrogen electrode, which requires bubbling gas over a freshly prepared catalytic Pt surface.

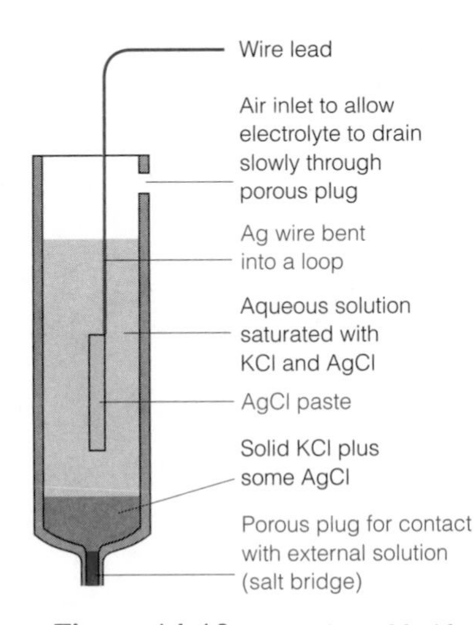

Figure 14-10 Silver-silver chloride reference electrode.

The standard reduction potential for AgCl | Ag is +0.222 V at 25°C. If the cell is saturated with KCl, the potential is +0.197 V. This is the value that we will use for all problems involving the AgCl | Ag reference electrode. The advantage of a saturated KCl solution is that the concentration of chloride does not change if some of the liquid evaporates.

Ag | AgCl electrode: $\quad AgCl(s) + e^- \rightleftharpoons Ag(s) + Cl^- \qquad E° = +0.222\ V$

$$E(\text{saturated KCl}) = +0.197\ V$$

Example Using a Reference Electrode

Calculate the cell voltage in Figure 14-9 if the reference electrode is a saturated silver-silver chloride electrode and $[Fe^{2+}]/[Fe^{3+}] = 10$.

SOLUTION We use Equation 14-22, noting that $E_- = 0.197$ V for a saturated silver-silver chloride electrode:

$$E = E_+ - E_- = \Big\{\underbrace{E_+^\circ}_{0.771\ \text{V}} - 0.059\ 16 \log \underbrace{\left(\frac{[Fe^{2+}]}{[Fe^{3+}]}\right)}_{10}\Big\} - \underbrace{0.197}_{\text{Reference electrode voltage}}$$

$$E = \{0.712\} - 0.197 = 0.515\ \text{V}$$

Test Yourself Find the voltage if $[Fe^{2+}]/[Fe^{3+}]$ increases to 100. (**Answer:** 0.456 V)

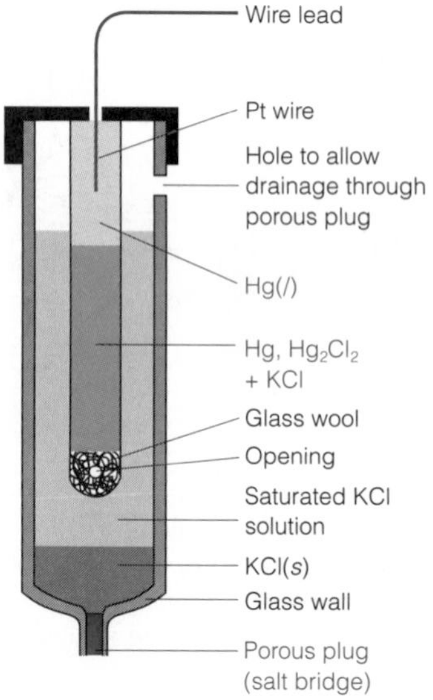

Calomel Reference Electrode

The *calomel electrode* in Figure 14-11 is based on the reaction

Calomel electrode: $\quad \underset{\text{Mercury(I) chloride (calomel)}}{\tfrac{1}{2}Hg_2Cl_2(s)} + e^- \rightleftharpoons Hg(l) + Cl^- \qquad E° = +0.268\ V$

$$E(\text{saturated KCl}) = +0.241\ V$$

If the cell is saturated with KCl, it is called a **saturated calomel electrode** and the cell potential is +0.241 V at 25°C. This electrode is encountered so frequently that it is abbreviated **S.C.E.**

Wire lead
Pt wire
Hole to allow drainage through porous plug
Hg(l)
Hg, Hg_2Cl_2 + KCl
Glass wool
Opening
Saturated KCl solution
KCl(s)
Glass wall
Porous plug (salt bridge)

Figure 14-11 Saturated calomel electrode (S.C.E.).

Voltage Conversions Between Different Reference Scales

It is sometimes necessary to convert potentials between different reference scales. If an electrode has a potential of −0.461 V with respect to a calomel electrode, what is the potential with respect to a silver-silver chloride electrode? What would be the potential with respect to the standard hydrogen electrode?

To answer these questions, Figure 14-12 shows the positions of the calomel and silver-silver chloride electrodes with respect to the standard hydrogen electrode. Point A, which is −0.461 V from S.C.E., is −0.417 V from the silver-silver chloride electrode and −0.220 V from S.H.E. What about point B, whose potential is +0.033 V with respect to silver-silver chloride? Its position is −0.011 V from S.C.E. and +0.230 V with respect to S.H.E. By keeping this diagram in mind, you can convert potentials from one scale to another.

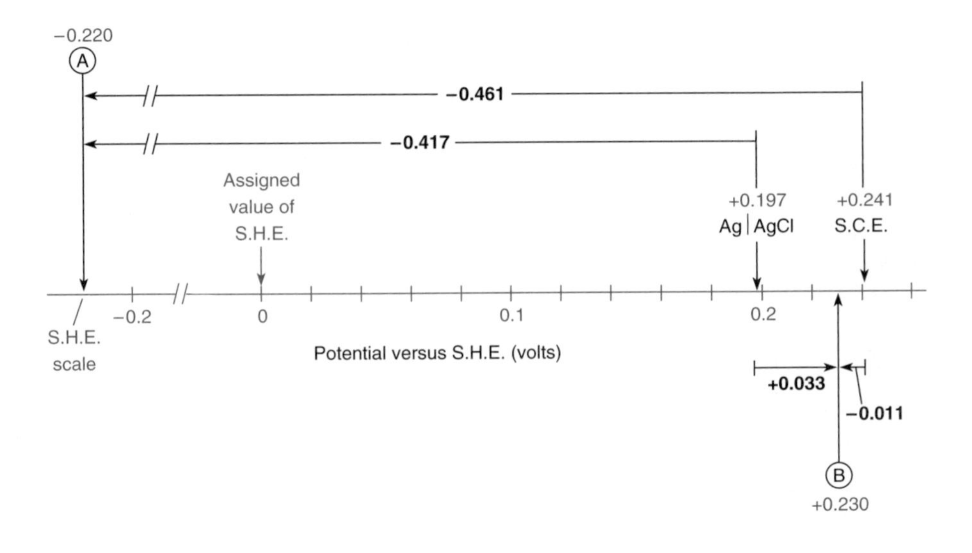

Figure 14-12 Converting potentials from one reference scale into another.

Ask Yourself

14-F. (a) Find the concentration ratio $[Fe^{2+}]/[Fe^{3+}]$ for the cell in Figure 14-9 if the measured voltage is 0.703 V.
(b) Convert the potentials listed below. The Ag | AgCl and calomel reference electrodes are saturated with KCl.

(i) 0.523 V versus S.H.E. = ? versus Ag | AgCl
(ii) 0.222 V versus S.C.E. = ? versus S.H.E.

Key Equations

Definitions	Oxidizing agent—takes electrons
	Reducing agent—gives electrons
	Anode—where oxidation occurs
	Cathode—where reduction occurs
Relation between charge and moles	$q = n\text{F}$
	q = electric charge (coulombs)
	n = moles of electrons
	F = Faraday constant
Relation between work and voltage	Work = Eq
	E = voltage difference through which charge q is moved
Standard potential	Measured by the cell S.H.E. ‖ half-reaction of interest, where S.H.E. is the standard hydrogen electrode and all reagents in the right half-cell are in their standard state (=1 M, 1 bar, pure solid, or pure liquid)
Nernst equation	For the half-reaction $a\text{A} + ne^- \rightleftharpoons b\text{B}$

$$E = E^\circ - \frac{0.059\ 16}{n}\log\left(\frac{[\text{B}]^b}{[\text{A}]^a}\right)\ \text{(at 25°C)}$$

Voltage of complete cell	$E = E_+ - E_-$
	E_+ = voltage of electrode connected to + terminal of meter
	E_- = voltage of electrode connected to − terminal of meter
E° for net cell reaction	$E^\circ = E^\circ_+ - E^\circ_-$
	E°_+ = standard potential for half-reaction in cell connected to + terminal of meter
	E°_- = standard potential for half-reaction in cell connected to − terminal of meter
Finding K from E°	$K = 10^{nE^\circ/0.059\ 16}$
	n = number of e^- in half-reaction

Important Terms

ampere	formal potential	reductant
anode	galvanic cell	reduction
cathode	indicator electrode	reference electrode
coulomb	Nernst equation	salt bridge
current	oxidant	saturated calomel electrode (S.C.E.)
electric potential	oxidation	silver-silver chloride electrode
electroactive species	oxidizing agent	standard hydrogen electrode (S.H.E.)
electrode	redox reaction	standard reduction potential
electrolysis	reducing agent	volt
Faraday constant		

Problems

14-1. **(a)** Explain the difference between electric charge (q, coulombs), electric current (I, amperes), and electric potential (E, volts).

(b) How many electrons are in 1 coulomb?

(c) How many coulombs are in 1 mole of charge?

14-2. **(a)** Identify the oxidizing and reducing agents among the following reactants and write a balanced half-reaction for each.

$$\underset{\text{Dithionite}}{2S_2O_4^{2-}} + \underset{\text{Tellurite}}{TeO_3^{2-}} + 2OH^- \rightleftharpoons \underset{\text{Sulfite}}{4SO_3^{2-}} + Te(s) + H_2O$$

(b) How many coulombs of charge are passed from reductant to oxidant when 1.00 g of Te is deposited?

(c) If Te is created at a rate of 1.00 g/h, how much current is flowing?

14-3. In the remediation experiment described at the chapter opening, 2 400 L of emulsion containing ~480 kg of Fe(0) consumed 17 kg of trichloroethene in 5 months.

(a) Identify oxidizing and reducing agents for the (unbalanced) reaction $Fe + C_2HCl_3 \longrightarrow Fe^{2+} + C_2H_4 + Cl^-$. Write a balanced half-reaction for each, using H_2O and H^+ to complete the balancing. Write a balanced net reaction.

(b) What percentage of injected Fe was used by this reaction in 5 months?

(c) How much current is flowing between the reactants if trichloroethene reacts at a constant rate of 17 kg/150 days?

14-4. Hydrogen ions can carry out useful work in a living cell (such as the synthesis of the molecule ATP that provides energy for chemical synthesis) when they pass from a region of high potential to a region of lower potential. How many joules of work can be done when 1.00 μmol of H^+ crosses a membrane and goes from a potential of +0.075 V to a potential of −0.090 V (that is, through a potential difference of 0.165 V)?

14-5. The basal rate of consumption of O_2 by a 70-kg human is 16 mol O_2/day. This O_2 oxidizes food and is reduced to H_2O, thereby providing energy for the organism:

$$O_2 + 4H^+ + 4e^- \rightleftharpoons 2H_2O$$

(a) To what current (in amperes = C/s) does this respiration rate correspond? (Current is defined by the flow of electrons from food to O_2.)

(b) If the electrons flow from reduced nicotinamide adenine dinucleotide (NADH) to O_2, they experience a potential drop of 1.1 V. How many joules of work can be done by 16 mol O_2?

14-6. Draw a picture of the following cell.

$$Pt(s) \mid Hg(l) \mid Hg_2Cl_2(s) \mid KCl(aq) \parallel ZnCl_2(aq) \mid Zn(s)$$

Write an oxidation for the left half-cell and a reduction for the right half-cell.

14-7. Redraw the cell in Figure 14-6, showing KNO_3 in the salt bridge. Noting the direction of electron flow through the circuit, show what happens to each reactant and product in each half-cell and show the direction of motion of each ion in each half-cell and in the salt bridge. Show the reaction at each electrode. When you finish this, you should understand what is happening in the cell.

14-8. Suppose that the concentrations of NaF and KCl were each 0.10 M in the cell

$$Pb(s) \mid PbF_2(s) \mid F^-(aq) \parallel Cl^-(aq) \mid AgCl(s) \mid Ag(s)$$

(a) Using the half-reactions $2AgCl(s) + 2e^- \rightleftharpoons 2Ag(s) + 2Cl^-$ and $PbF_2(s) + 2e^- \rightleftharpoons Pb(s) + 2F^-$, calculate the cell voltage.

(b) Now calculate the cell voltage by using the reactions $2Ag^+ + 2e^- \rightleftharpoons 2Ag(s)$ and $Pb^{2+} + 2e^- \rightleftharpoons Pb(s)$ and K_{sp} for AgCl and PbF_2 (see Appendix A).

14-9. Consider a circuit in which the left half-cell was prepared by dipping a Pt wire in a beaker containing an equimolar mixture of Cr^{2+} and Cr^{3+}. The right half-cell contained a Tl rod immersed in 1.00 M $TlClO_4$.

(a) Use line notation to describe this cell.

(b) Find the potential of each half-cell and calculate the cell voltage

(c) Draw a diagram like Figure 14-7 for this cell. When the electrodes are connected by a salt bridge and a wire, which terminal (Pt or Tl) is the anode?

(d) Write the spontaneous net cell reaction.

14-10. Consider the cell

$$Pt(s) \mid H_2(g, 0.100 \text{ bar}) \mid H^+(aq, \text{pH} = 2.54) \parallel Cl^-(aq, 0.200 \text{ M}) \mid Hg_2Cl_2(s) \mid Hg(l) \mid Pt(s)$$

(a) Write a reduction reaction and Nernst equation for each half-cell and find each half-cell potential. For the Hg_2Cl_2 half-reaction, $E° = 0.268$ V.

(b) Draw a diagram like Figure 14-7 for this cell. Which half-cell is the anode?

(c) Find E for the net cell reaction. From the sign of E, state whether reduction will occur at the left- or right-hand electrode.

14-11. **(a)** Calculate the cell voltage (E, not $E°$), and state the direction in which electrons will flow through the potentiometer in Figure 14-5. Write the spontaneous net cell reaction.

(b) Draw a diagram like Figure 14-7 for this cell. Which way do electrons flow?

(c) The left half-cell was loaded with 14.3 mL of $Br_2(l)$ (density = 3.12 g/mL). The aluminum electrode contains 12.0 g of Al. Which element, Br_2 or Al, is the limiting reagent in this cell? (That is, which reagent will be used up first?)

(d) If the cell is somehow operated under conditions in which it produces a constant voltage of 1.50 V, how much electrical work will have been done when 0.231 mL of $Br_2(l)$ has been consumed?

(e) If the current is 2.89×10^{-4} A, at what rate (grams per second) is Al(s) dissolving?

14-12. Calculate $E°$ and K for each of the following reactions:

(a) $Cu(s) + Cu^{2+} \rightleftharpoons 2Cu^+$

(b) $2F_2(g) + H_2O \rightleftharpoons F_2O(g) + 2H^+ + 2F^-$

14-13. In fuel cells[8] used in Apollo flights to the moon, $H_2(g)$ is oxidized to $H_2O(l)$ at a catalytic cathode and $O_2(g)$ is reduced to $H_2O(g)$ at a catalytic anode.

(a) Write the half-reactions and the net reaction. Find the cell voltage if H_2 and O_2 are each present at 1 bar, the cathode compartment pH is 0, and the anode compartment pH is 14.

(b) Find the equilibrium constant for the net cell reaction and write the equilibrium constant in terms of concentrations of reactants and products.

(c) If the cell produces a constant current of 10.0 A, how many days will it take to consume 1.00 kg of H_2? How many kilograms of O_2 will be consumed in the same time?

14-14. From the following half-reactions, calculate the solubility product of $Mg(OH)_2$.

$$Mg^{2+} + 2e^- \rightleftharpoons Mg(s) \qquad E° = -2.360 \text{ V}$$

$$Mg(OH)_2(s) + 2e^- \rightleftharpoons Mg(s) + 2OH^- \qquad E° = -2.690 \text{ V}$$

14-15. Select half-reactions from Appendix C to compute the formation constant for the reaction $Ca^{2+} + \text{acetate}^- \rightleftharpoons \text{Ca(acetate)}^+$. Find the value of K_f.

14-16. The nickel-metal hydride rechargeable battery used in early laptop computers is based on the following chemistry:

Cathode:

$$NiOOH(s) + H_2O + e^- \underset{\text{charge}}{\overset{\text{discharge}}{\rightleftharpoons}} Ni(OH)_2(s) + OH^-$$

Anode:

$$MH(s) + OH^- \underset{\text{charge}}{\overset{\text{discharge}}{\rightleftharpoons}} M(s) + H_2O + e^-$$

The anode material, MH, is a metal hydride in which the metal is one of several transition metal or rare earth alloys. Explain why the voltage of this cell remains nearly constant during its entire discharge cycle.

14-17. The cell in Ask Yourself 14-B(**b**) contains 1.3 mM $Fe(CN)_6^{4-}$, 4.9 mM $Fe(CN)_6^{3-}$, 1.8 mM $Ag(S_2O_3)_2^{3-}$, and 55 mM $S_2O_3^{2-}$.

(**a**) Find K for the net cell reaction.

(**b**) Find the cell voltage and state whether $Ag(s)$ is oxidized or reduced by the spontaneous cell reaction.

14-18. From the standard potentials for reduction of $Br_2(aq)$ and $Br_2(l)$ in Appendix C, calculate the solubility of Br_2 in water at 25°C. Express your answer as g/L.

14-19. A solution contains 0.010 0 M IO_3^-, 0.010 0 M I^-, 1.00×10^{-4} M I_3^-, and pH 6.00 buffer. Consider the reactions

$$2IO_3^- + I^- + 12H^+ + 10e^- \rightleftharpoons I_3^- + 6H_2O \quad E° = 1.210 \text{ V}$$

$$I_3^- + 2e^- \rightleftharpoons 3I^- \quad E° = 0.535 \text{ V}$$

(**a**) Write a balanced net reaction that can take place in this solution.

(**b**) Calculate $E°$ and K for the reaction.

(**c**) Calculate E for the conditions given.

(**d**) At what pH would the given concentrations of IO_3^-, I^-, and I_3^- listed above be in equilibrium?

14-20. (**a**) Write the half-reactions for the silver-silver chloride and calomel reference electrodes.

(**b**) Predict the voltage for the following cell:

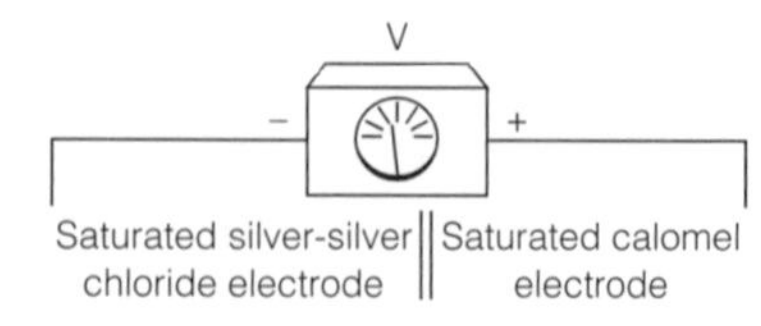

14-21. (**a**) Find the potential (versus S.H.E.) of the half-cell Pt | VO^{2+} (0.050 M), VO_2^+ (0.025 M), pH 2.00.

(**b**) What would be the voltage of the cell S.C.E. || VO^{2+} (0.050 M), VO_2^+ (0.025 M), pH 2.00 | Pt?

14-22. Convert the following potentials. The Ag | AgCl and calomel reference electrodes are saturated with KCl.

(**a**) −0.111 V versus Ag | AgCl = ? versus S.H.E.

(**b**) 0.023 V versus Ag | AgCl = ? versus S.C.E.

(**c**) −0.023 V versus S.C.E. = ? versus Ag | AgCl

14-23. Suppose that the silver-silver chloride electrode in Figure 14-8 is replaced by a saturated calomel electrode. Calculate the cell voltage if $[Fe^{2+}]/[Fe^{3+}] = 2.5 \times 10^{-3}$.

14-24. The formal reduction potential for $Fe^{3+} + e^- \rightleftharpoons Fe^{2+}$ in 1 M $HClO_4$ is 0.73 V. The formal reduction potential for the complex LFe(III) + e^- $\rightleftharpoons$ LFe(II) (where L is the chelate desferrioxamine B in Box 13-1) is −0.48 V.[9] What do these potentials tell you about the relative stability of the complexes LFe(III) and LFe(II)?

How Would You Do It?

14-25. A *concentration cell* has the same half-reaction in both half-cells, but the concentration of a reactant or product is different in the half-cells. Consider the cell

$$Ag(s) \mid Ag^+(aq, c_l) \parallel Ag^+(aq, c_r) \mid Ag(s)$$

where c_l is the concentration of Ag^+ in the left half-cell and c_r is the concentration of Ag^+ in the right half-cell.

(**a**) Write the Nernst equation for the cell, showing how the cell voltage depends on the concentrations c_l and c_r. What is the expected voltage when $c_l = c_r$?

(**b**) The *formation constant* is the equilibrium constant for the reaction of a metal with a ligand: M + L $\rightleftharpoons$ ML. Propose a method for measuring formation constants of silver by using a concentration cell.[10] Make up a hypothetical example to predict what voltage would be observed.

14-26. One measure of the capability of a battery is how much electricity it can produce per kilogram of reactants. The quantity of electricity could be measured in coulombs, but it is customarily measured in ampere · hours, where 1 A · h provides 1 A for 1 h. Thus if 0.5 kg of reactants can produce 3 A · h, the storage capacity would be 3 A · h/0.5 kg = 6 A · h/kg. Compare the capabilities of a conventional lead-acid car battery with that of a hydrogen-oxygen fuel cell in terms of A · h/kg.

Lead-acid battery: $Pb + PbO_2 + 2H_2SO_4 \longrightarrow 2PbSO_4 + 2H_2O$

FM of reactants = 207.2 + 239.2 + 2 × 98.079 = 642.6

Hydrogen-oxygen fuel cell: $2H_2 + O_2 \longrightarrow 2H_2O$

FM of reactants = 36.031

Notes and References

1. For a student experiment, see B. A. Balko and P. G. Tratnyek, "A Discovery-Based Experiment Illustrating How Iron Metal Is Used to Remediate Contaminated Groundwater," *J. Chem. Ed.* **2001**, *78*, 1661.

2. The quotation about Faraday was made by Lady Pollock and cited in J. Kendall, *Great Discoveries by Young Chemists* (New York: Thomas Y. Crowell Co., 1953, p. 63).

3. For demonstration of half-cells, see J. D. Ciparick, *J. Chem. Ed.* **1991**, *68*, 247 and P.-O. Eggen, T. Grønneberg, and L. Kvittengen, *J. Chem. Ed.* **2006**, *83*, 1201.

4. L. P. Silverman and B. B. Bunn, *J. Chem. Ed.* **1992**, *69*, 309.

5. J. Hamilton, *A Life of Discovery: Michael Faraday, Giant of the Scientific Revolution* (New York: Random House, 2004, pp. 258–260.

6. K. Rajeshwar and J. G. Ibanez, *Environmental Electrochemistry* (San Diego: Academic Press, 1997).

7. For an experiment on measuring battery lifetime, see M. J. Smith and C. A. Vincent, *J. Chem. Ed.* **2002**, *79,* 851.

8. To construct a demonstration methanol/oxygen fuel cell, see O. Zerbinati, *J. Chem. Ed.* **2002**, *79*, 829.

9. I. Spasojević, S. K. Armstrong, T. J. Brickman, and A. L. Crumbliss, *Inorg. Chem.* **1999**, *38*, 449.

10. For a student experiment using concentration cells to measure formation constants, see M. L. Thompson and L. J. Kateley, *J. Chem. Ed.* **1999**, *76*, 95.

Further Reading

A. Hamnett, C. H. Hamann, and W. Vielstich, *Electrochemistry* (New York: Wiley, 1998).

K. Rajeshwar and J. G. Ibanez, *Environmental Electrochemistry* (San Diego: Academic Press, 1997).

Z. Galus, *Fundamentals of Electrochemical Analysis* (New York: Ellis Horwood, 1994).

M. A. Brett and A. M. O. Brett, *Electrochemistry* (Oxford: Oxford University Press, 1993).

H. B. Oldham and J. C. Myland, *Fundamentals of Electrochemical Science* (San Diego: Academic Press, 1993).

Measuring Carbonate in Seawater with an Ion-Selective Electrode

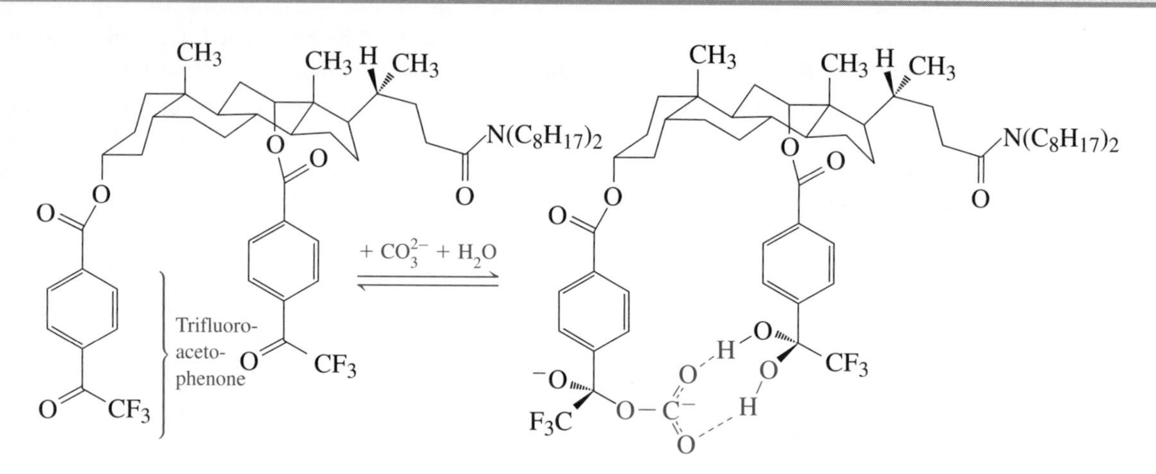

A "molecular-tweezer" that binds carbonate is the key component of an ion-selective electrode used to measure carbonate in seawater. There is strong evidence that carbonate is bound by both arms of the tweezer, but the exact mode of binding shown here is conjecture.

It is important to monitor the distribution of dissolved CO_2 in the ocean to understand the role of the ocean as a sink for CO_2 produced by the burning of fossil fuel. Ion-selective electrodes described in this chapter are simple and convenient for this purpose, but they must discriminate between CO_3^{2-} and high concentrations of Cl^- in the ocean. The key to an ion-selective electrode is a membrane that binds analyte specifically, reversibly, and rapidly.

The bidentate ligand, above, developed by years of painstaking synthesis and evaluation, has two trifluoroacetophenone groups that selectively bind a CO_3^{2-} ion. The response of an electrode made with this ligand is 10^7 times greater for $[CO_3^{2-}]$ than for $[Cl^-]$. The table shows that the ion-selective electrode gives the same results as two other methods. The advantage of the ion-selective electrode is that it gives a direct reading when placed in ocean water. By contrast, a CO_2 gas sensor requires adjustment of the pH of the water sample, and a titration of CO_3^{2-} is tedious.

Total carbonate in water from the Yellow Sea near Korea

Method	Total carbonate (mmol/kg seawater)*
Ion-selective electrode	1.94 ± 0.03 (15 measurements)
CO_2 gas sensor	1.93 ± 0.01 (3 measurements)
Potentiometric (pH) titration	1.95 ± 0.04 (8 measurements)

*Total carbonate = $[CO_3^{2-}] + [HCO_3^-] + [H_2CO_3] + [CO_2(aq)]$

SOURCE: Y. S. Choi, L. Lvova, J. H. Shin, S. H. Oh, C. S. Lee, B. H. Kim, G. S. Cha, and H. Nam, *Anal. Chem.* **2002**, *74*, 2435; H. J. Lee, I. J. Yoon, H.-J. Pyun, G. S. Cha, and H. Nam, *Anal. Chem.* **2000**, *72*, 4694.

CHAPTER 15

Electrode Measurements

A critically ill patient is wheeled into the emergency room and the doctor needs blood chemistry information quickly to help her make a diagnosis and begin treatment. Every analyte in Table 15-1, which is part of the critical care profile of blood chemistry, can be measured by electrochemical means. Ion-selective electrodes, introduced in this chapter, are the method of choice for Na^+, K^+, Cl^-, pH, and P_{CO_2}. The "Chem 7" test constitutes up to 70% of tests performed in the hospital lab. It measures Na^+, K^+, Cl^-, total CO_2, glucose, urea, and creatinine, the first four of which are analyzed with ion-selective electrodes. The use of voltage measurements to extract chemical information is called **potentiometry.**

Table 15-1 Critical care profile

Function	Analyte
Conduction	K^+, Ca^{2+}
Contraction	Ca^{2+}, Mg^{2+}
Energy level	Glucose, P_{O_2}, lactate, hematocrit
Ventilation	P_{O_2}, P_{CO_2},
Perfusion	Lactate, SO_2%, hematocrit
Acid-base	pH, P_{CO_2}, HCO_3^-
Osmolality	Na^+, glucose
Electrolyte balance	Na^+, K^+, Ca^{2+}, Mg^{2+}
Renal function	Blood urea nitrogen, creatinine

SOURCE: C. C. Young, *J. Chem. Ed.* **1997,** *74,* 177.

15-1 The Silver Indicator Electrode

In Chapter 14, we learned that the voltage of an electrochemical cell is related to the concentrations of species in the cell. We saw that some cells could be divided into a *reference electrode* that provides a constant electric potential and an *indicator electrode* whose potential varies in response to analyte concentration.

Chemically inert platinum, gold, and carbon indicator electrodes are frequently used to conduct electrons to or from species in solution. In contrast with chemically inert elements, silver participates in the reaction $Ag^+ + e^- \rightleftharpoons Ag(s)$.

Figure 15-1 shows how a silver electrode can be used in conjunction with a saturated calomel reference electrode to measure $[Ag^+]$ during the titration of halide ions by Ag^+ (as shown in Figures 6-4 and 6-5). The reaction at the silver indicator electrode is

$$Ag^+ + e^- \rightleftharpoons Ag(s) \qquad E^\circ_+ = 0.799 \text{ V}$$

and the reference half-cell reaction is

$$Hg_2Cl_2(s) + 2e^- \rightleftharpoons 2Hg(l) + 2Cl^- \qquad E_- = 0.241 \text{ V}$$

E_- = reference electrode potential with actual concentrations in the reference cell

E°_- = standard potential of reference half-reaction when all species are in their standard states (pure solid, pure liquid, 1 M, or 1 bar)

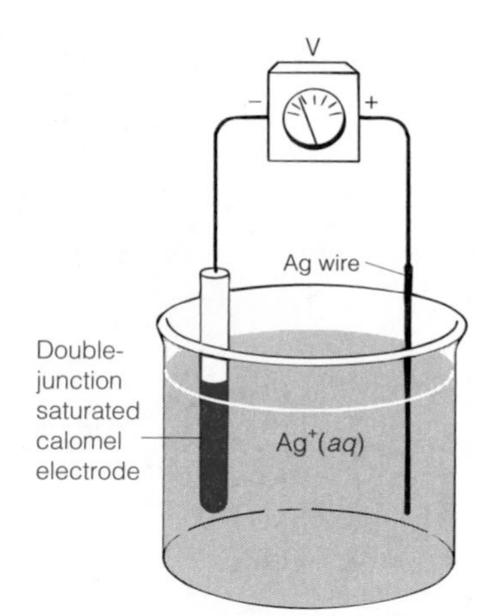

Figure 15-1 Use of silver and calomel electrodes to measure the concentration of Ag^+ in a solution. The calomel electrode has a double junction, like that in Figure 15-3. The outer compartment of the electrode is filled with KNO_3, so there is no direct contact between the KCl solution in the inner compartment and Ag^+ in the beaker.

The reference half-cell potential (E_-, not E°_-) is constant at 0.241 V because $[Cl^-]$ is fixed by the concentration of saturated KCl. The Nernst equation for the entire cell is therefore

$$E = E_+ - E_- = \underbrace{\left\{0.799 - 0.059\,16 \log \frac{1}{[Ag^+]}\right\}}_{\text{Potential of Ag | Ag}^+\text{ indicator electrode}} - \underset{\substack{\uparrow \\ \text{Constant potential of} \\ \text{S.C.E. reference electrode}}}{\{0.241\}}$$

Noting that $\log(1/[Ag^+]) = -\log[Ag^+]$, we rewrite the preceding expression as

$$E = 0.558 + 0.059\,16 \log[Ag^+] \tag{15-1}$$

The voltage changes by 0.059 16 V (at 25°C) for each factor-of-10 change in $[Ag^+]$.

The experiment in Figure 6-4 used a silver indicator electrode and a *glass* reference electrode. The glass electrode responds to the pH of the solution, which is held constant by a buffer. Therefore, the glass electrode remains at a constant potential.

Titration of a Halide Ion with Ag^+

Let's consider how the concentration of Ag^+ varies during the titration of I^- with Ag^+, as shown in Figure 6-5*b*. The titration reaction is

$$Ag^+ + I^- \longrightarrow AgI(s) \qquad K = \frac{1}{K_{sp}} = \frac{1}{8.3 \times 10^{-17}}$$

If you were monitoring the reaction with silver and calomel electrodes, you could use Equation 15-1 to compute the expected voltage at each point in the titration.

When Ag^+ is added to I^-:

- before V_e, there is a known excess of I^-: $[Ag^+] = K_{sp}/[I^-]$
- at V_e, $[Ag^+] = [I^-] = \sqrt{K_{sp}}$
- after V_e, there is a known excess of Ag^+

At any point prior to the equivalence point (V_e), there is a known excess of I^-, from which we can calculate $[Ag^+]$:

Before V_e:
$$K_{sp} = [Ag^+][I^-] \Rightarrow [Ag^+] = K_{sp}/[I^-] \tag{15-2}$$

At the equivalence point, the quantity of Ag^+ added is exactly equal to the I^- that was originally present. We can imagine that AgI(*s*) is made stoichiometrically and a little bit redissolves:

At V_e:
$$K_{sp} = \underset{x}{[Ag^+]}\underset{x}{[I^-]} \Rightarrow [Ag^+] = [I^-] = \sqrt{K_{sp}} \tag{15-3}$$

Beyond the equivalence point, the quantity of excess Ag^+ added from the buret is known, and the concentration is just

After V_e:
$$[Ag^+] = \frac{\text{moles of excess Ag}^+}{\text{total volume of solution}} \tag{15-4}$$

Example **Potentiometric Precipitation Titration**

A 20.00-mL solution containing 0.100 4 M KI was titrated with 0.084 5 M $AgNO_3$, using the cell in Figure 15-1. Calculate the voltage at volumes $V_{Ag^+} = 15.00$, V_e, and 25.00 mL.

SOLUTION The titration reaction is $Ag^+ + I^- \rightarrow AgI(s)$, and the equivalence volume is

$$\underbrace{(V_e\,(\text{mL}))(0.084\,5\text{ M})}_{\text{mmol Ag}^+} = \underbrace{(20.00\text{ mL})(0.100\,4\text{ M})}_{\text{mmol I}^-} \Rightarrow V_e = 23.76\text{ mL}$$

15.00 mL: We began with (20.00 mL)(0.100 4 M) = 2.008 mmol I^- and added (15.00 mL)(0.084 5 M) = 1.268 mmol Ag^+. The concentration of unreacted I^- is

$$[I^-] = \frac{(2.008 - 1.268)\text{ mmol}}{(20.00 + 15.00)\text{ mL}} = 0.021\,1\text{ M}$$

The concentration of Ag^+ in equilibrium with the solid AgI is therefore

$$[Ag^+] = \frac{K_{sp}}{[I^-]} = \frac{8.3 \times 10^{-17}}{0.021\,1\text{ M}} = 3.9 \times 10^{-15}\text{ M}$$

The cell voltage is computed with Equation 15-1:

$$E = 0.558 + 0.059\,16\log(3.9 \times 10^{-15}) = -0.294\text{ V}$$

At V_e: Equation 15-3 tells us that $[Ag^+] = \sqrt{K_{sp}} = 9.1 \times 10^{-9}$ M, so

$$E = 0.558 + 0.059\,16\log(9.1 \times 10^{-9}) = 0.082\text{ V}$$

25.00 mL: Now there is an excess of 25.00 − 23.76 = 1.24 mL of 0.084 5 M $AgNO_3$ in a total volume of 45.00 mL.

$$[Ag^+] = \frac{(1.24\text{ mL})(0.084\,5\text{ M})}{45.00\text{ mL}} = 2.33 \times 10^{-3}\text{ M}$$

and the cell voltage is

$$E = 0.558 + 0.059\,16\log(2.33 \times 10^{-3}) = 0.402\text{ V}$$

Test Yourself Find the voltage at V_{Ag^+} = 20.00 and 30.00 mL. (**Answer:** −0.294, 0.441 V)

The voltage in Figure 15-2 barely changes prior to the equivalence point because the concentration of Ag^+ is very low and relatively constant until I^- is used up. When I^- has been consumed, $[Ag^+]$ suddenly increases and so does the voltage. Figure 15-2 is upside down relative to curve *b* in Figure 6-5. The reason is that, in Figure 15-1, the indicator electrode is connected to the *positive* terminal of the potentiometer. In Figure 6-4, the indicator electrode is connected to the *negative* terminal because the glass pH electrode only fits into the positive terminal of the meter. In addition to their opposite polarities, the voltages in Figures 15-2 and 6-5 are different because each experiment uses a different reference electrode.

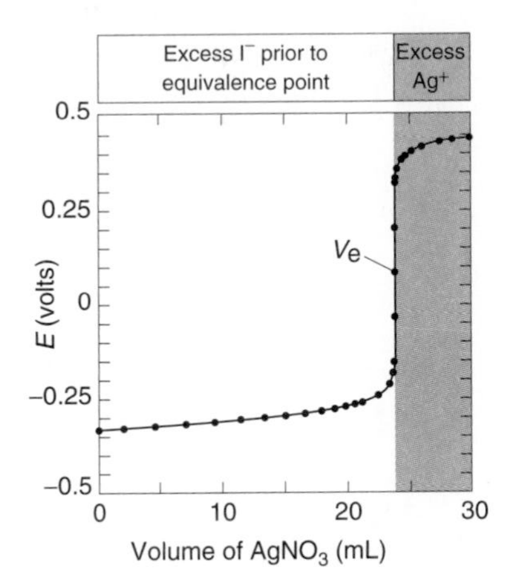

Figure 15-2 Calculated titration curve for the addition of 0.084 5 M Ag^+ to 20.00 mL of 0.100 4 M I^- in the cell in Figure 15-1.

Double-Junction Reference Electrode

Demonstration 15-1 uses a pair of electrodes to monitor a pretty amazing chemical reaction.

If you tried to titrate I^- with Ag^+ by using the cell in Figure 15-1, KCl solution would slowly leak into the titration beaker from the porous plug at the base of the

Demonstration 15-1 Potentiometry with an Oscillating Reaction

Principles of potentiometry are illustrated in a fascinating manner by *oscillating reactions* in which chemical concentrations oscillate between high and low values. An example is the Belousov-Zhabotinskii reaction:

$$\underset{\text{Malonic acid}}{3CH_2(CO_2H)_2} + \underset{\text{Bromate}}{2BrO_3^-} + 2H^+ \xrightarrow{Ce^{3+/4+}\text{ catalyst}} \underset{\text{Bromomalonic acid}}{2BrCH(CO_2H)_2} + 3CO_2 + 4H_2O$$

During this reaction, the quotient $[Ce^{3+}]/[Ce^{4+}]$ oscillates by a factor of 10 to 100. When the Ce^{4+} concentration is high, the solution is yellow. When Ce^{3+} predominates, the solution is colorless.

To start the show, combine the following solutions in a 300-mL beaker:

160 mL of 1.5 M H_2SO_4
40 mL of 2 M malonic acid
30 mL of 0.5 M $NaBrO_3$ (or saturated $KBrO_3$)
4 mL of saturated ceric ammonium sulfate, $Ce(SO_4)_2 \cdot 2(NH_4)_2SO_4 \cdot 2H_2O$

After an induction period of 5 to 10 min with magnetic stirring, you can initiate oscillations by adding 1 mL of ceric ammonium sulfate solution. The reaction is somewhat temperamental and may need more Ce^{4+} over a 5-min period to initiate oscillations.

Monitor the $[Ce^{3+}]/[Ce^{4+}]$ ratio with Pt and calomel electrodes. You should be able to write the cell reactions and a Nernst equation for this experiment.

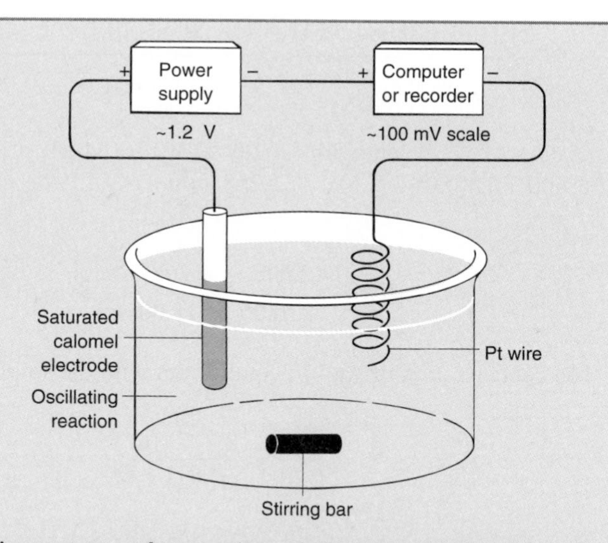

Apparatus used to monitor relative concentrations of Ce^{3+} and Ce^{4+} in an oscillating reaction. [George Rossman, California Institute of Technology.]

In place of a potentiometer (a pH meter), we use a computer to obtain a permanent record of the oscillations. The potential oscillates over a range of ~100 mV centered near ~1.2 V, so we offset the cell voltage by ~1.2 V with any available power supply. Trace *a* shows what is usually observed. The potential changes rapidly during the abrupt colorless-to-yellow transition and more gradually during the gentle yellow-to-colorless transition. Trace *b* shows two different cycles superimposed in the same solution.

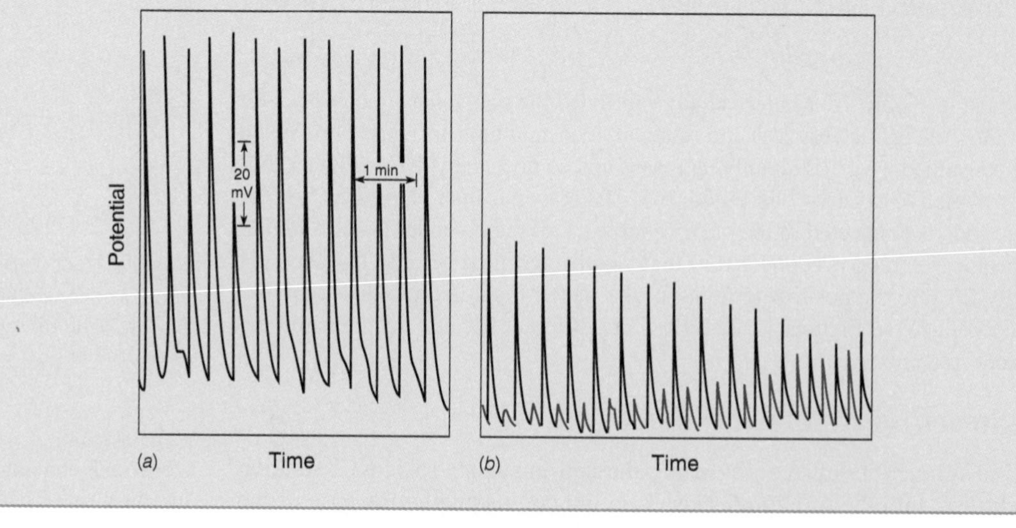

reference electrode (Figure 14-11). Cl^- introduces a titration error because it consumes Ag^+. The *double-junction reference electrode* in Figure 15-3 prevents the inner electrolyte solution from leaking directly into the titration vessel.

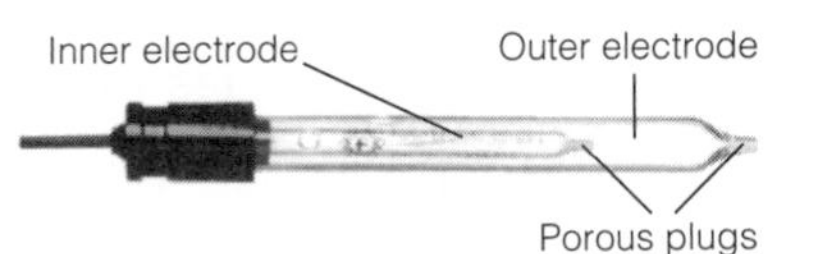

Figure 15-3 Double-junction reference electrode has an inner electrode identical to those in Figures 14-10 or 14-11. The outer compartment is filled with an electrolyte such as KNO_3 that is compatible with the titration solution. KCl electrolyte from the inner electrode slowly leaks into the outer electrode, so the outer electrolyte should be changed periodically. [Fisher Scientific, Pittsburgh, PA.]

Ask Yourself

15-A. Consider the titration of 40.0 mL of 0.050 0 M NaCl with 0.200 M $AgNO_3$, using the cell in Figure 15-1. The equivalence volume is V_e = 10.0 mL.
(a) Prior to V_e, there is a known excess of Cl^-. Find $[Cl^-]$ at the following volumes of added silver: V_{Ag^+} = 0.10, 2.50, 5.00, 7.50, and 9.90 mL. From $[Cl^-]$, use K_{sp} for AgCl to find $[Ag^+]$ at each volume.
(b) Find $[Cl^-]$ and $[Ag^+]$ at $V_{Ag^+} = V_e$ = 10.00 mL.
(c) After V_e, there is a known excess of Ag^+. Find $[Ag^+]$ at V_{Ag^+} = 10.10 and 12.00 mL.
(d) Find the cell voltage at each volume in **(a)**–**(c)** and make a graph of the titration curve.

15-2 What Is a Junction Potential?

When two dissimilar electrolyte solutions are placed in contact, a voltage difference called the **junction potential** develops at the interface. This small, unknown voltage (usually a few millivolts) exists at each end of a salt bridge connecting two half-cells. *The junction potential puts a fundamental limitation on the accuracy of direct potentiometric measurements,* because we usually do not know the contribution of the junction to the measured voltage.

$E_{observed} = E_{cell} + E_{junction}$
Because the junction potential is usually unknown, E_{cell} is uncertain.

To see why a junction potential occurs, consider a solution containing NaCl in contact with pure water (Figure 15-4). The Na^+ and Cl^- ions diffuse from the NaCl solution into the water phase. However, Cl^- ion has a greater *mobility* than Na^+. That is, Cl^- diffuses faster than Na^+. As a result, a region rich in Cl^-, with excess negative charge, develops at the front. Behind it is a positively charged region depleted of Cl^-. The result is an electric potential difference at the junction of the NaCl and H_2O phases.

Mobilities of ions are shown in Table 15-2, and several junction potentials are listed in Table 15-3. Because K^+ and Cl^- have similar mobilities, junction potentials of a KCl salt bridge are slight. This is why saturated KCl is used in salt bridges.

Direct Versus Relative Potentiometric Measurements

In a *direct potentiometric measurement,* we use an electrode such as a silver wire to measure $[Ag^+]$ or a pH electrode to measure $[H^+]$ or a calcium ion-selective electrode to measure $[Ca^{2+}]$. There is inherent inaccuracy in most direct potentiometric measurements because there is usually a liquid-liquid junction with an unknown voltage difference making the intended indicator electrode potential uncertain. For example, Figure 15-5 shows a 4% standard deviation among 14 measurements by direct potentiometry. Part of the variation could be attributed to differences in the indicator (ion-selective) electrodes and part could be from varying liquid junction potentials.

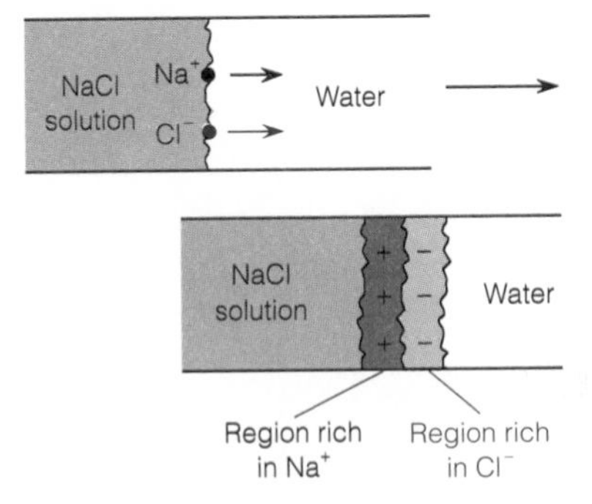

Figure 15-4 Development of the junction potential caused by unequal mobilities of Na^+ and Cl^-.

By contrast, in *relative potentiometric measurements,* changes in the potential observed during a titration, such as that in Figure 6-5, are relatively precise and

Table 15-2 Mobilities of ions in water at 25°C

Ion	Mobility $[m^2/(s \cdot V)]^a$	Ion	Mobility $[m^2/(s \cdot V)]^a$
H^+	36.30×10^{-8}	OH^-	20.50×10^{-8}
K^+	7.62×10^{-8}	SO_4^{2-}	8.27×10^{-8}
NH_4^+	7.61×10^{-8}	Br^-	8.13×10^{-8}
La^{3+}	7.21×10^{-8}	I^-	7.96×10^{-8}
Ba^{2+}	6.59×10^{-8}	Cl^-	7.91×10^{-8}
Ag^+	6.42×10^{-8}	NO_3^-	7.40×10^{-8}
Ca^{2+}	6.12×10^{-8}	ClO_4^-	7.05×10^{-8}
Cu^{2+}	5.56×10^{-8}	F^-	5.70×10^{-8}
Na^+	5.19×10^{-8}	$CH_3CO_2^-$	4.24×10^{-8}
Li^+	4.01×10^{-8}		

a. The mobility of an ion is the velocity achieved in an electric field of 1 V/m. Mobility = velocity/field. The units of mobility are therefore (m/s)/(V/m) = $m^2/(s \cdot V)$.

Table 15-3 Liquid junction potentials at 25°C

Junction	Potential (mV)
0.1 M NaCl | 0.1 M KCl	−6.4
0.1 M NaCl | 3.5 M KCl	−0.2
1 M NaCl | 3.5 M KCl	−1.9
0.1 M HCl | 0.1 M KCl	+27
0.1 M HCl | 3.5 M KCl	+3.1

Note: A positive sign means that the right side of the junction becomes positive with respect to the left side.

permit an end point to be identified with little uncertainty. Measuring the *absolute* Ag^+ concentration by direct potentiometry is inherently inaccurate, but measuring *changes* in Ag^+ can be done accurately and precisely.

Ask Yourself

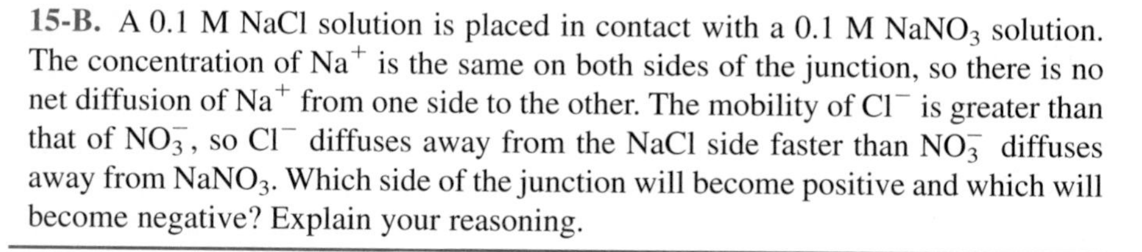

15-B. A 0.1 M NaCl solution is placed in contact with a 0.1 M $NaNO_3$ solution. The concentration of Na^+ is the same on both sides of the junction, so there is no net diffusion of Na^+ from one side to the other. The mobility of Cl^- is greater than that of NO_3^-, so Cl^- diffuses away from the NaCl side faster than NO_3^- diffuses away from $NaNO_3$. Which side of the junction will become positive and which will become negative? Explain your reasoning.

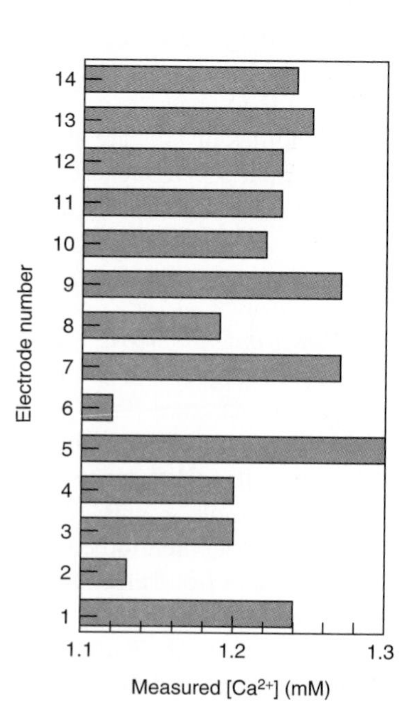

Figure 15-5 Response of 14 different Ca^{2+} ion-selective electrodes to identical human blood serum samples. The mean value is 1.22 ± 0.05 mM. [From M. Umemoto, W. Tani, K. Kuwa, and Y. Ujihira, *Anal. Chem.* **1994**, *66*, 352A.]

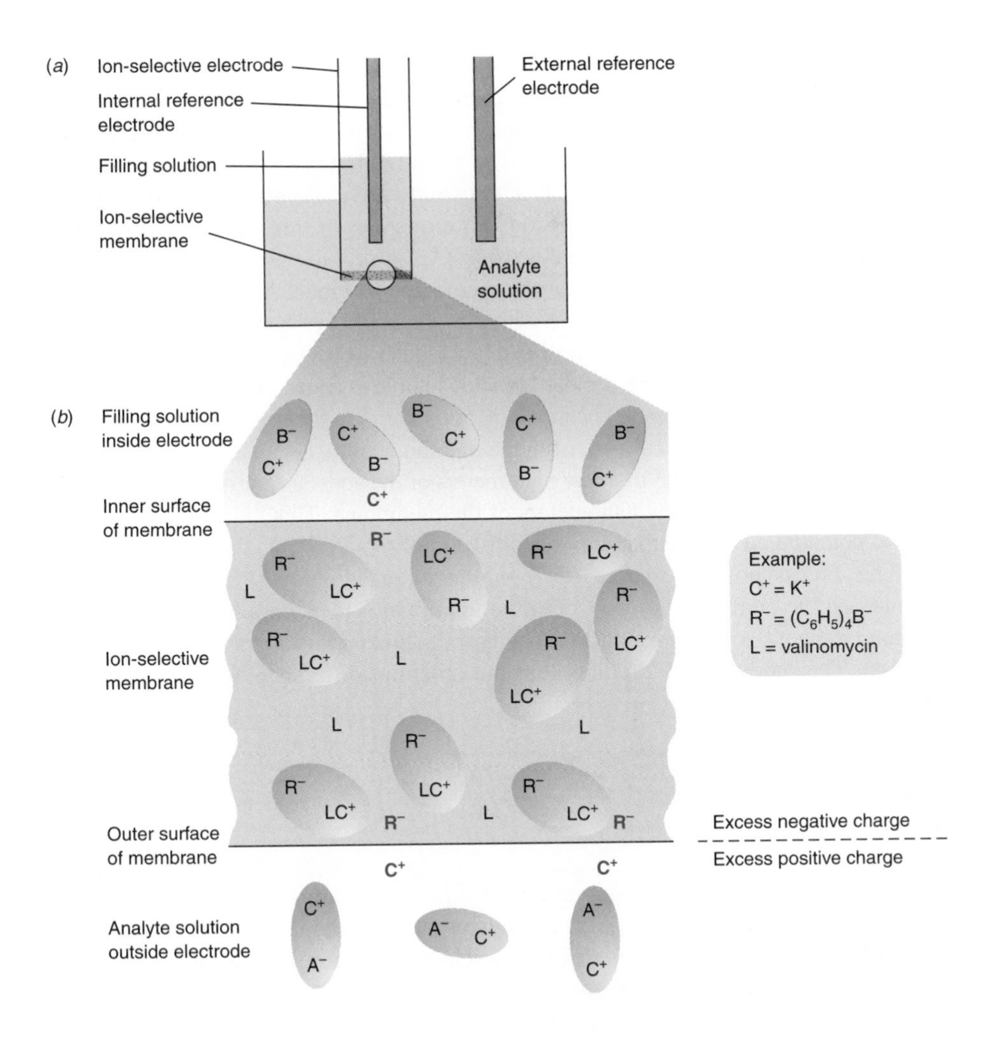

Figure 15-6 (*a*) Ion-selective electrode immersed in aqueous solution containing analyte cation C^+. Typically, the membrane is made of poly(vinyl chloride) impregnated with a nonpolar liquid containing the ion-selective ionophore L, the complex LC^+, and a hydrophobic anion R^-. (*b*) Close-up of membrane. Ellipses encircling pairs of ions are a guide for the eye to count the charge in each phase. Colored ions represent excess charge in each phase.

15-3 How Ion-Selective Electrodes Work

An **ion-selective electrode** responds preferentially to one species in a solution. Differences in concentration of the selected ion inside and outside the electrode produce a voltage difference across the membrane.[1]

Consider the *liquid-based ion-selective electrode* shown schematically in Figure 15-6*a*. This electrode develops a voltage that is related to the concentration of analyte cation C^+ in an unknown solution. The electrode is "liquid based" because the ion-selective membrane is a hydrophobic organic polymer impregnated with an organic liquid containing the hydrophobic anion, R^-, and a ligand, L, that selectively binds the analyte cation. R^- is an "ion exchanger" that reversibly associates with cations by electrostatic attraction. R^- is soluble in the organic phase, but not in water, so it is confined to the membrane.

Hydrophobic: "water hating" (does not mix with water)

Example of hydrophobic anion, R^-:

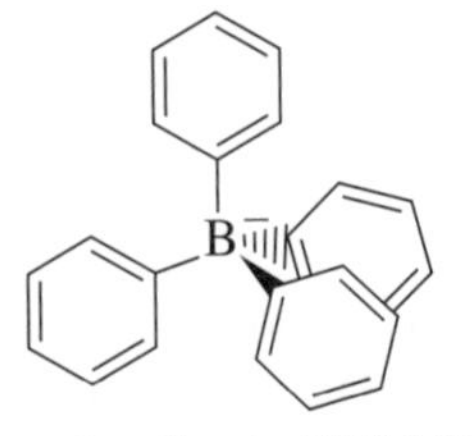

Tetraphenylborate, $(C_6H_5)_4B^-$

The aqueous filling solution inside the electrode contains the ions $C^+(aq)$ and $B^-(aq)$. The outside of the electrode is immersed in unknown aqueous solution containing analyte $C^+(aq)$ and anion $A^-(aq)$. Ideally, it does not matter what A^- and B^- are. The electric potential difference (the voltage) across the ion-selective membrane is measured by two reference electrodes, which might be Ag | AgCl. *If the concentration of C^+ in the unknown solution changes, the voltage changes.* With the use of a calibration curve, the voltage tells us the concentration of C^+ in the analyte solution.

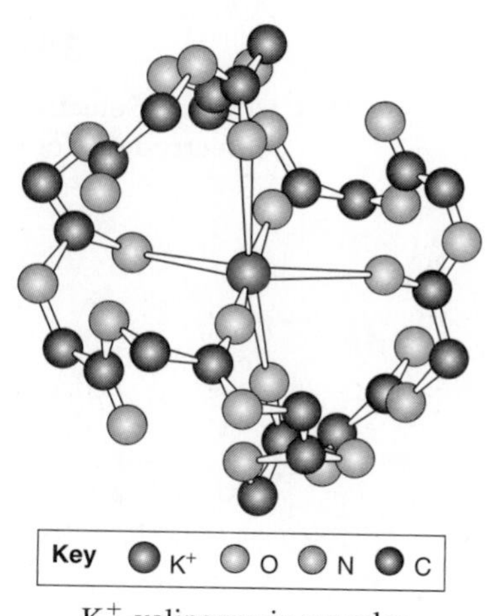

K^+-valinomycin complex
Valinomycin has a cyclic structure containing six amino acids and six carboxylic acids. Isopropyl and methyl substituents are not shown in this diagram [From L. Stryer, *Biochemistry,* 4th ed. (New York: W. H. Freeman and Company, 1995).]

Figure 15-6*b* shows how the electrode works. The key in this example is the ligand, L (called an *ionophore*), which is soluble inside the membrane and selectively binds analyte ion. In a potassium ion-selective electrode, for example, L could be valinomycin, a natural antibiotic secreted by certain microorganisms to carry K^+ ion across cell membranes. The ligand L is chosen to have a high affinity for analyte cation C^+ and low affinity for other ions. The opening of this chapter shows a ligand L for a carbonate ion-selective electrode.

Almost all the analyte ion inside the membrane in Figure 15-6*b* is bound in the complex LC^+, which is in equilibrium with a small amount of free C^+. The membrane also contains excess free L. C^+ can diffuse across the interface. In an ideal electrode, R^- cannot leave the membrane because it is not soluble in water, and the aqueous anion A^- cannot enter the membrane because it is not soluble in the organic phase. As soon as a tiny number of C^+ ions diffuse from the membrane into the aqueous phase, there is excess positive charge in the first few nanometers of the aqueous phase and excess negative charge in the outer few nanometers of the membrane. This imbalance creates an electric potential difference that opposes diffusion of more C^+ into the aqueous phase.

The excess positive charge (C^+) in the outer (unknown) aqueous solution depends on the concentration of C^+ in the unknown. The excess positive charge in the inner aqueous solution is constant because the inner solution has a constant composition. Thermodynamics predicts that the potential difference between the outer and the inner solutions is

The electrode really responds to the *activity* of analyte (Section 12-2), not the concentration. In this book, we will write concentrations instead of activities.

Electric potential difference for ion-selective electrode:

$$E = \frac{0.05916}{n} \log\left(\frac{[C^+]_{outer}}{[C^+]_{inner}}\right) \quad \text{(volts at 25°C)} \qquad (15\text{-}5)$$

where n is the charge of the analyte ion, $[C^+]_{outer}$ is its concentration in the outer (unknown) solution, and $[C^+]_{inner}$ is its concentration in the inner solution (which is constant). Equation 15-5 applies to any ion-selective electrode, including a glass pH electrode. If analyte is an anion, the sign of n is negative. Later, we will modify the equation to account for interfering ions.

If C^+ were K^+, then $n = +1$ and there would be a potential increase of +0.059 16 V for every factor-of-10 increase in $[K^+]$ in the analyte (outer) solution. If C^+ were Ca^{2+}, then $n = +2$ and there would be a potential increase of $+0.059\ 16/2 = +0.029\ 58$ V for every factor-of-10 increase in $[Ca^{2+}]$ in the unknown. For a carbonate electrode, $n = -2$ and there is a potential decrease of $-0.059\ 16/2$ V for every factor-of-10 increase in $[CO_3^{2-}]$.

Chemists use molecular modeling to design synthetic ligands with high selectivity for a specific ion.

The key feature of an ion-selective electrode is a membrane that selectively binds the analyte of interest. No membrane is perfectly selective, so there is always some interference from unintended species.

Two Classes of Indicator Electrodes

Metal electrodes such as silver or platinum function as surfaces on which redox reactions can take place:

Equilibrium on a silver electrode: $\quad Ag^+ + e^- \rightleftharpoons Ag(s)$

Equilibrium on a platinum electrode: $\quad Fe(CN)_6^{3+} + e^- \rightleftharpoons Fe(CN)_6^{2+}$

Metal electrode: surface on which redox reaction takes place
Ion-selective electrode: selectively binds one ion—no redox chemistry

Ion-selective electrodes such as a calcium electrode or a glass pH electrode selectively bind analyte ion. *There is no redox chemistry in the ion-selective electrode.* The potential across the electrode membrane depends on the concentration of analyte ion in the unknown.

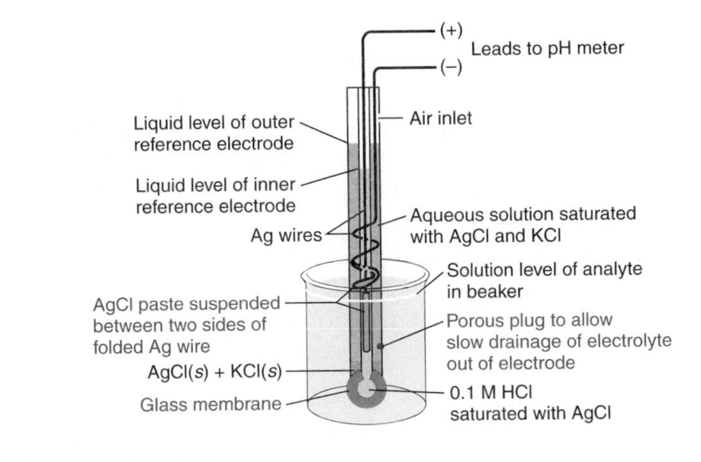

Figure 15-7 Glass combination electrode with a silver-silver chloride reference electrode. The glass electrode is immersed in a solution of unknown pH so that the porous plug on the lower right is below the surface of the liquid. The two Ag | AgCl electrodes measure the voltage across the glass membrane.

Ask Yourself

15-C. (a) Predict the change in voltage across the membranes of ion-selective electrodes for NH_4^+, F^-, and S^{2-} for a 10-fold increase in analyte concentration.
(b) The ion-selective membrane in Figure 15-6 contains the hydrophobic anion R^- = tetraphenylborate and the neutral ligand L = valinomycin to bind K^+. A CO_3^{2-} ion-selective electrode contains the neutral ligand L shown at the opening of this chapter that forms $L(CO_3^{2-})(H_2O)$. Which hydrophobic ion, $R^- = (C_6H_5)_4B^-$ or $R^+ = (C_{12}H_{25})NCH_3^+$ (tridodecylmethylammonium) is required for the ion-selective membrane? Why?

15-4 pH Measurement with a Glass Electrode

The most widely employed ion-selective electrode is the **glass electrode** for measuring pH. A pH electrode responds selectively to H^+, with a potential difference of 0.059 16 V for every factor-of-10 change in $[H^+]$. A factor-of-10 difference in $[H^+]$ is one pH unit, so a change of, say, 4.00 pH units leads to a change in electrode potential of $4.00 \times 0.059\,16 = 0.237$ V.

A **combination electrode,** incorporating both glass and reference electrodes in one body, is shown in Figure 15-7. The line diagram of this cell is

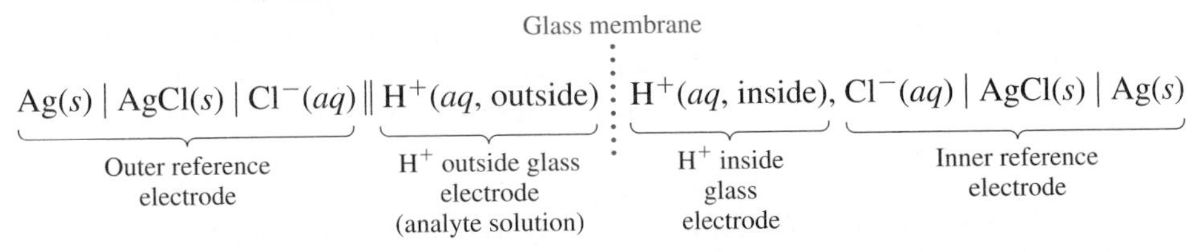

The pH-sensitive part of the electrode is the thin glass membrane in the shape of a bulb at the bottom of the electrodes in Figures 15-7 and 15-8.[2]

The glass membrane at the bottom of the pH electrode consists of an irregular network of SiO_4 tetrahedra through which Na^+ ions move sluggishly. Studies with tritium (the radioactive isotope ^3H) show that H^+ does *not* diffuse through the membrane. The glass surface contains exposed —O^- that can bind H^+ from the solutions on either side of the membrane (Figure 15-9). H^+ equilibrates with the glass surface, giving the side of the membrane exposed to the higher concentration of H^+ a more positive charge. To measure a potential difference, at least some tiny amount of electric current must flow through the complete circuit. Na^+ ions in the glass carry

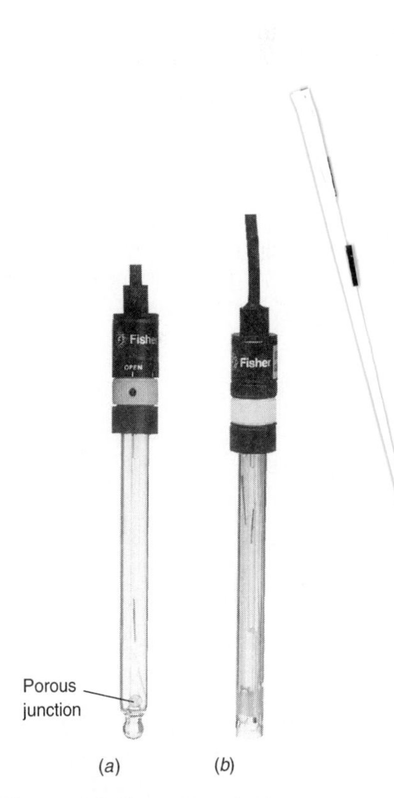

Figure 15-8 (*a*) Glass-body combination electrode with pH-sensitive glass bulb at the bottom. The porous junction is the salt bridge to the reference electrode compartment. (*b*) Polymer body surrounds glass electrode to protect the delicate bulb. [Fisher Scientific, Pittsburgh, PA.]

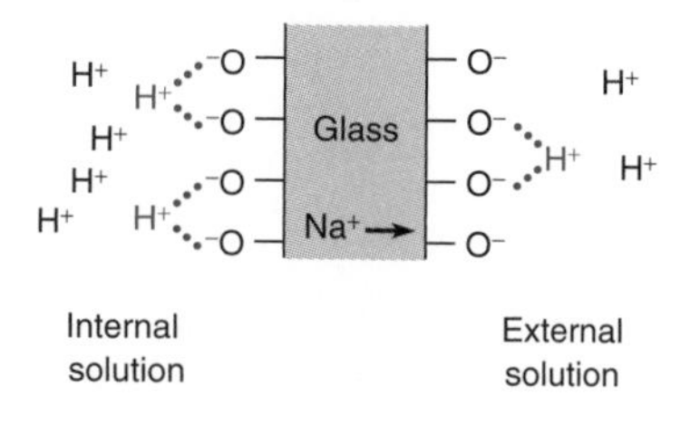

Figure 15-9 Ion-exchange equilibria on the inner and outer surfaces of the glass membrane. The pH of the internal solution is fixed. As the pH of the external solution (the sample) changes, the electric potential difference across the glass membrane changes.

electric current by migrating across the membrane. The resistance of the glass membrane is high, so little current flows across it.

The potential difference between the inner and outer silver-silver chloride electrodes in Figure 15-7 depends on $[Cl^-]$ in each electrode compartment and on the potential difference across the glass membrane. Because $[Cl^-]$ is fixed and because $[H^+]$ is constant inside the glass electrode, the only variable is the pH of the analyte solution outside the glass membrane.

Real glass electrodes are described by the equation

Response of glass electrode:

$$E = \text{constant} + \beta(0.059\ 16)\Delta\text{pH} \quad \text{(at 25°C)} \tag{15-6}$$

where ΔpH is the difference in pH between analyte and the solution inside the glass bulb. The factor β, which is ideally 1, is typically 0.98–1.00. The constant term, called the *asymmetry potential,* arises because no two sides of a real object are identical, so a small voltage exists even if the pH is the same on both sides of the membrane. We correct for asymmetry and we measure β by calibrating the electrode in solutions of known pH.

Calibrating a Glass Electrode

A pH electrode *must* be calibrated before use. It should be calibrated every 2 h in sustained use. Ideally, calibration standards should bracket the pH of the unknown.

Before using a pH electrode, be sure that the air inlet near the upper end of the electrode in Figure 15-7 is not capped. (This hole is capped during storage to prevent evaporation of the reference electrode filling solution.) Wash the electrode with distilled water and gently *blot* it dry with a tissue. Do not *wipe* it because this action might produce a static charge on the glass. Dip the electrode in a standard buffer whose pH is near 7 and allow the electrode to equilibrate with stirring for at least a minute. Following the manufacturer's instructions, press a key that might say "calibrate" or "read" on a microprocessor-controlled meter or adjust the reading of an analog meter to indicate the pH of the standard buffer. Wash the electrode with water, blot it dry, and immerse it in a second standard whose pH is further from 7 than the pH of the first standard. Enter the second buffer on the meter. If the electrode were ideal, the voltage would change by 0.059 16 V per pH unit at 25°C; the actual change may be slightly less. These two measurements establish the values of β and the constant in Equation 15-6. Finally, dip the electrode in unknown, stir the liquid, allow the reading to stabilize, and read the pH.

Don't leave a glass electrode out of water (or in a nonaqueous solvent) longer than necessary.

Store the glass electrode in aqueous solution to prevent dehydration of the glass. Ideally, the solution should be similar to that inside the reference compartment of the electrode. Distilled water is *not* a good storage medium. If the electrode is dry, recondition it in aqueous solution for several hours. If the electrode is to be used above pH 9, soak it in a high-pH buffer.

If electrode response becomes sluggish or if the electrode cannot be calibrated properly, try soaking it in 6 M HCl, followed by water. As a last resort, dip the electrode in 20 wt% aqueous ammonium bifluoride, NH_4HF_2 in a plastic beaker for 1 min. This reagent dissolves glass, exposing fresh surface. Wash the electrode with water and try calibrating it again. *Ammonium bifluoride must not contact your skin, because it produces HF burns.* (See page 297 for precautions with HF.)

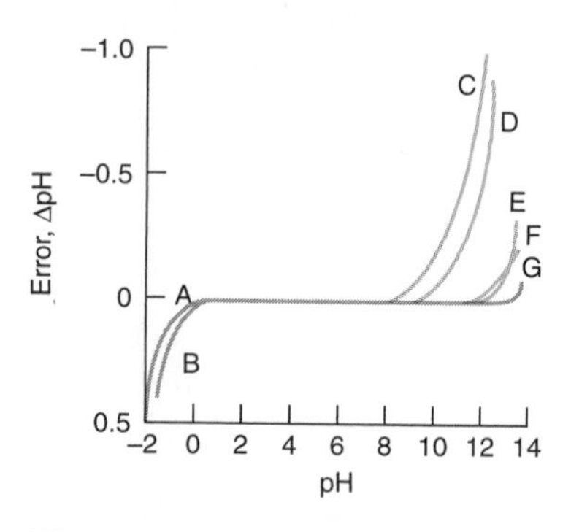

Figure 15-10 Acid and alkaline errors of some glass electrodes. A: Corning 015, H_2SO_4. B: Corning 015, HCl. C: Corning 015, 1 M Na^+. D: Beckman-GP, 1 M Na^+. E: L & N Black Dot, 1 M Na^+. F: Beckman Type E, 1 M Na^+. G: Ross electrode.[3] [From R. G. Bates, *Determination of pH: Theory and Practice,* 2nd ed. (New York: Wiley, 1973). Ross electrode data are from Orion *Ross pH Electrode Instruction Manual.*]

Errors in pH Measurement

To use a glass electrode intelligently, you should understand its limitations:

1. *Standards.* A pH measurement cannot be more accurate than our standards, which are typically ±0.01–0.02 pH units.

2. *Junction potential.* A junction potential exists at the porous plug near the bottom of the electrode in Figure 15-7. If the ionic composition of analyte is different from that of the standard buffer, the junction potential will change *even if the pH of the two solutions is the same.* This factor gives an uncertainty of at least ~0.01 pH unit. Box 15-1 describes how junction potentials affect the measurement of the pH of rainwater.

3. *Junction potential drift.* Most combination electrodes have a silver-silver chloride reference electrode containing saturated KCl solution. More than 350 mg of silver per liter dissolve in the KCl (mainly as $AgCl_4^{3-}$ and $AgCl_3^{2-}$). In the porous plug salt bridge in Figure 15-7, KCl is diluted and AgCl precipitates in the plug. If analyte contains a reducing agent, Ag(*s*) also can precipitate in the plug. Both effects change the junction potential, causing slow drift of the pH reading. Compensate for this error by recalibrating the electrode every 2 h.

4. *Sodium error.* When $[H^+]$ is very low and $[Na^+]$ is high, the electrode responds to Na^+ as if Na^+ were H^+, and the apparent pH is lower than the true pH. This response is called the *alkaline error* or *sodium error* (Figure 15-10).

Box 15-1 Systematic Error in Rainwater pH Measurement: The Effect of Junction Potential

The opening of Chapter 8 shows the pH of rainfall over the U.S. and Europe. Acidity in rainfall is partly a result of human activities and is slowly changing the nature of many ecosystems. Monitoring the pH of rainwater is a critical component of programs to reduce the production of acid rain.

To identify and correct systematic errors in the measurement of pH of rainwater, 8 samples were provided to each of 17 laboratories, along with explicit instructions for how to conduct the measurements. Each lab used two standard buffers to calibrate its pH meter.

The figure below shows typical results for the pH of rainwater. The average of the 17 measurements is given by the horizontal line at pH 4.14 and the letters s, t, u, v, w, x, y, and z identify the types of pH electrodes used for the measurements. Laboratories using electrode types s and w had relatively large systematic errors. The type s electrode was a combination electrode (Figure 15-7) containing a reference electrode liquid junction with an exceptionally large area. Electrode type w had a reference electrode filled with a gel.

Variability in the liquid junction potential (Section 15-2) was hypothesized to lead to the variability in the pH measurements. Standard buffers used for pH meter calibration typically have ion concentrations of ~0.05 M, whereas rainwater samples have ion concentrations two or more orders of magnitude lower. To test the hypothesis that junction potential caused systematic errors, a pure HCl solution with a concentration near 2×10^{-4} M was used as a pH calibration standard in place of high ionic strength buffers. The following data were obtained, with good results from all but the first lab. The standard deviation of all 17 measurements was reduced from 0.077 pH unit with the standard buffer to 0.029 pH unit with the HCl standard. It was concluded that junction potential is the cause of most of the variability between labs and that a low ionic strength standard is appropriate for rainwater pH measurements.

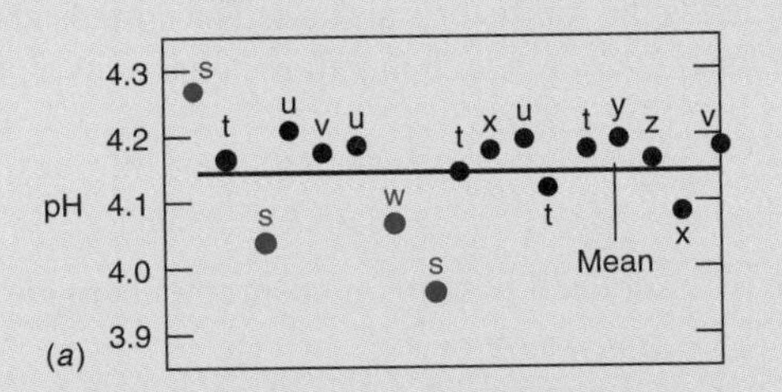

pH of rainwater from identical samples measured at 17 different labs using standard calibration buffers. Letters designate different types of pH electrodes.

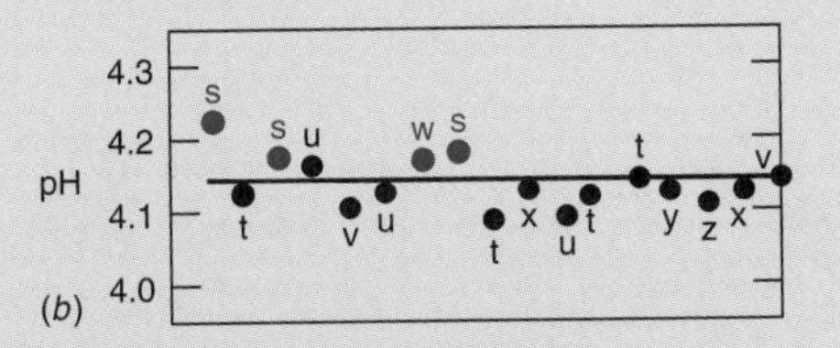

Rainwater pH measured after using low-ionic-strength HCl for calibration. [W. F. Koch, G. Marinenko, and R. C. Paule, *J. Res. National Bureau of Standards* **1986**, *91*, 23.]

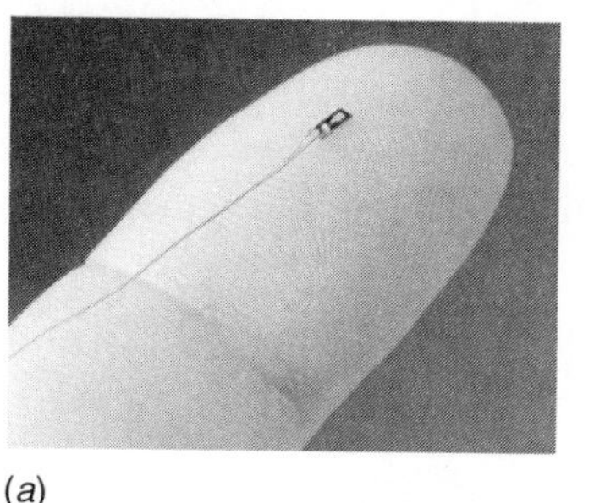

(a)

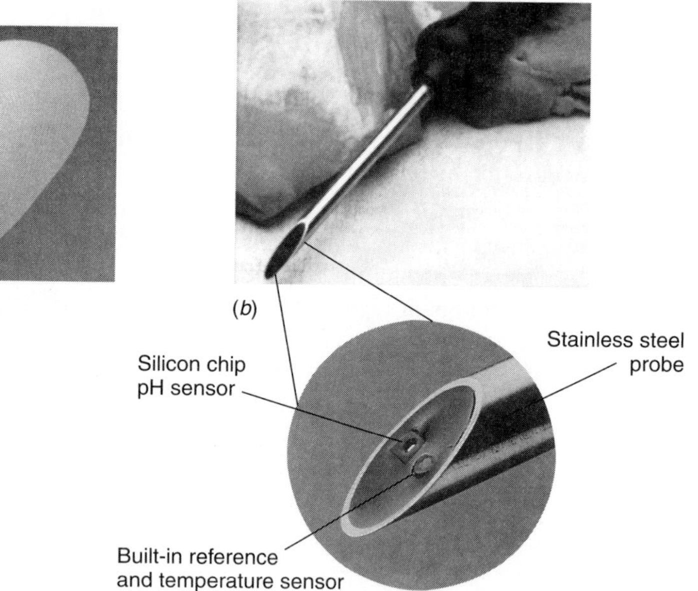

Figure 15-11 (*a*) pH-sensitive field effect transistor, (*b*) Rugged field effect transistor mounted in a steel shaft can be inserted into meat, poultry, or other damp solids to measure pH. [From Sentron, Gig Harbor, WA and IQ Scientific Instruments, San Diego, CA.]

5. *Acid error.* In strong acid, the measured pH is higher than the actual pH, for reasons that are not well understood (Figure 15-10).

6. *Equilibration time.* In a well-buffered solution with adequate stirring, equilibration of the glass with analyte solution takes seconds. In a poorly buffered solution near the equivalence point of a titration, it could take minutes.

7. *Hydration.* A dry electrode takes several hours in aqueous solution before it responds to H^+ correctly.

8. *Temperature.* A pH meter should be calibrated at the same temperature at which the measurement will be made. You cannot calibrate your equipment at one temperature and make accurate measurements at a second temperature.

Challenge Show that the potential of the glass electrode changes by 1.3 mV when the analyte H^+ concentration changes by 5.0%. Because 59 mV $\approx$ 1 pH unit, 1.3 mV = 0.02 pH unit.

Moral: A small uncertainty in voltage (1.3 mV) or pH (0.02 units) corresponds to a large uncertainty (5%) in H^+ concentration. Similar uncertainties arise in other potentiometric measurements.

Errors 1 and 2 limit the accuracy of pH measurements with the glass electrode to $\pm$0.02 pH unit, at best. Measurement of pH *differences* can be accurate to $\pm$0.002 pH unit, but knowledge of the true pH will still be at least an order of magnitude more uncertain. An uncertainty of $\pm$0.02 pH unit corresponds to an uncertainty of $\pm$5% in $[H^+]$.

Solid-State pH Sensors

Some pH sensors do not depend on a fragile glass membrane. The *field effect transistor* in Figure 15-11 is a tiny semiconductor device whose surface binds H^+ from the medium in which the transistor is immersed. The higher the concentration of H^+ in the external medium, the more positively charged is the transistor's surface. The surface charge regulates the flow of current through the transistor, which therefore behaves as a pH sensor.

Ask Yourself

15-D. **(a)** List the sources of error associated with pH measurements made with the glass electrode.
(b) When the difference in pH across the membrane of a glass electrode at 25°C is 4.63 pH units, how much voltage is generated by the pH gradient? Assume that the constant β in Equation 15-6 is 1.00.
(c) Why do glass electrodes indicate a pH lower than the actual pH in 0.1 M NaOH?

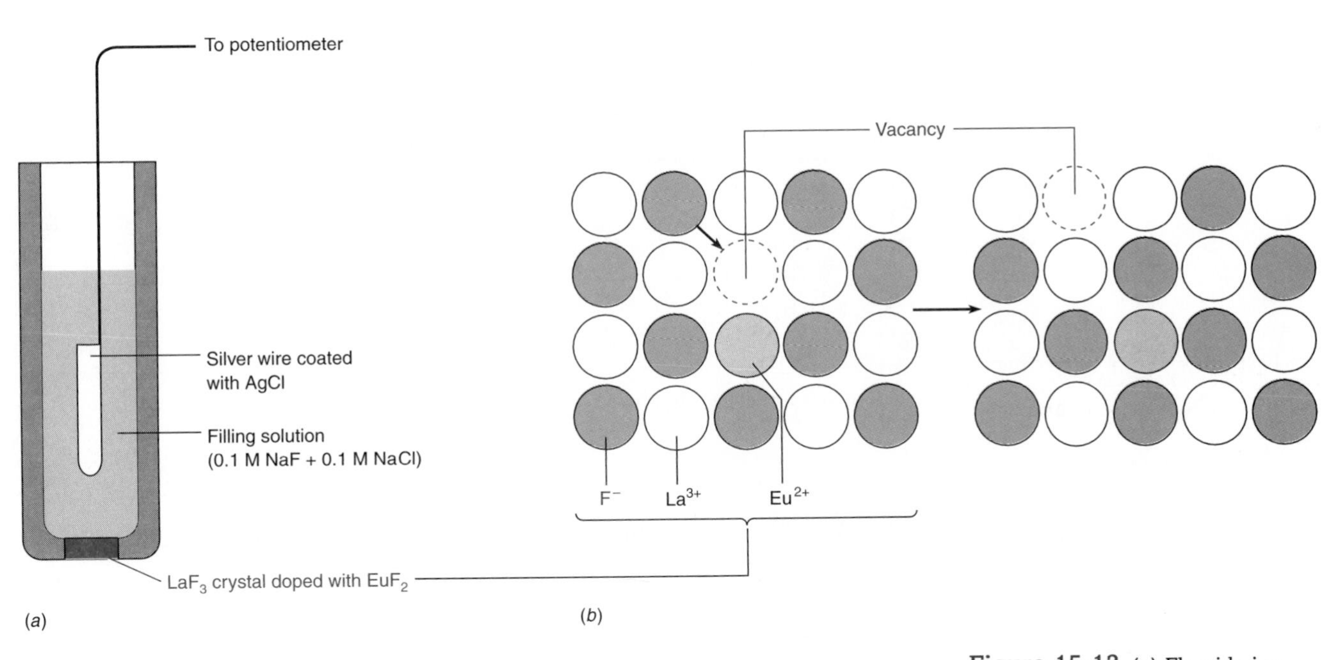

Figure 15-12 (*a*) Fluoride ion-selective electrode employing a crystal of LaF_3 doped with EuF_2 as the ion-selective membrane. (*b*) Migration of F^- through the doped crystal: For charge conservation, every Eu^{2+} is accompanied by an anion vacancy in the crystal. When a neighboring F^- jumps into the vacancy, another site becomes vacant. Repetition of this process moves F^- through the lattice.

15-5 Ion-Selective Electrodes

The glass pH electrode is an example of a *solid-state ion-selective electrode* whose operation depends on (1) an ion-exchange reaction of H^+ between the glass surface and analyte solution and (2) transport of Na^+ across the glass membrane. We now examine several ion-selective electrodes.

Solid-State Electrodes

The ion-sensitive component of a fluoride **solid-state ion-selective electrode** is a crystal of LaF_3 doped with EuF_2 (Figure 15-12*a*). *Doping* is the addition of an impurity (EuF_2 in this case) into the solid crystal (LaF_3). The inner surface of the crystal is exposed to filling solution with a constant concentration of F^-. The outer surface is exposed to a variable concentration of F^- in the unknown. Fluoride on each surface of the crystal equilibrates with F^- in the solution contacting that surface. Anion vacancies in doped LaF_3 permit F^- to jump from one site to the next, thereby transporting electric current across the crystal (Figure 15-12*b*).

The response of the electrode to F^- is

Response of F^- electrode:
$$E = \text{constant} - \beta(0.059\,16)\log[F^-]_{outside} \qquad (15\text{-}7)$$

Electrode response depends on $\log\left(\frac{[F^-]_{outside}}{[F^-]_{inside}}\right)$. The constant value of $[F^-]_{inside}$ is incorporated into the constant term in Equation 15-7.

where $[F^-]_{outside}$ is the concentration of F^- in analyte solution and β is close to 1.00. The electrode response is close to 59 mV per decade over a F^- concentration range from about 10^{-6} M to 1 M. The electrode is $>10^3$ times more responsive to F^- than to most other ions. However response to OH^- is one-tenth as great as the response to F^-, so OH^- provides serious interference. At low pH, F^- is converted into HF ($pK_a = 3.17$), to which the electrode is insensitive. Fluoride is added to drinking water to help prevent tooth decay. The F^- electrode is used to monitor and control the fluoridation of municipal water supplies. Several other solid-state ion-selective electrodes are listed in Table 15-4.

Table 15-4 Solid-state ion-selective electrodes

Ion	Concentration range (M)	Membrane crystal[a]	pH range	Interfering species
F^-	10^{-6}–1	LaF_3	5–8	OH^-
Cl^-	10^{-4}–1	AgCl	2–11	CN^-, S^{2-}, I^-, $S_2O_3^{2-}$, Br^-
Br^-	10^{-5}–1	AgBr	2–12	CN^-, S^{2-}, I^-
I^-	10^{-6}–1	AgI	3–12	S^{2-}
CN^-	10^{-6}–10^{-2}	AgI	11–13	S^{2-}, I^-
S^{2-}	10^{-5}–1	Ag_2S	13–14	

a. Electrodes containing silver-based crystals such as Ag_2S should be stored in the dark and protected from light during use to prevent light-induced chemical degradation.

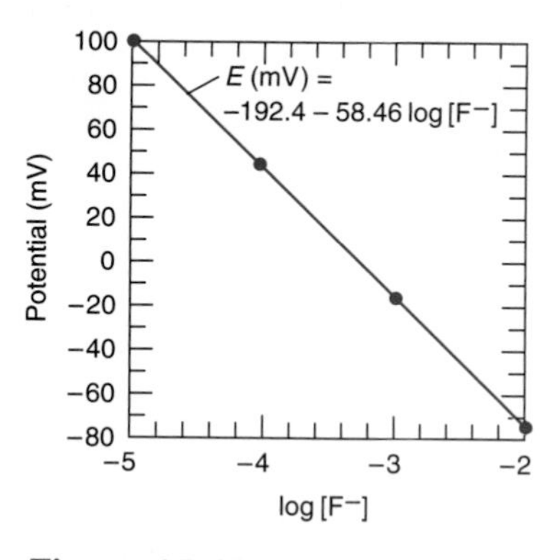

Figure 15-13 Calibration curve for fluoride ion-selective electrode.

Example Calibration Curve for an Ion-Selective Electrode

A fluoride electrode immersed in standard solutions gave the following potentials:

$[F^-]$ (M)	$\log[F^-]$	E (mV vs S.C.E.)
1.00×10^{-5}	5.00	100.0
1.00×10^{-4}	4.00	41.4
1.00×10^{-3}	3.00	−17.0
1.00×10^{-2}	2.00	−75.4

(a) What potential is expected if $[F^-] = 5.00 \times 10^{-5}$ M? **(b)** What concentration of F^- will give a potential of 0.0 V?

SOLUTION **(a)** Our strategy is to fit the calibration data with Equation 15-7 and then to substitute the concentration of F^- into the equation to find the potential:

$$\underbrace{E}_{y} = \underbrace{\text{constant}}_{\text{Intercept}} - \underbrace{m}_{\text{Slope}} \cdot \underbrace{\log[F^-]}_{x}$$

Using the method of least squares from Chapter 4, we plot E versus $\log[F^-]$ to find a straight line with a slope of −58.46 mV and an intercept of −192.4 mV (Figure 15-13). Setting $[F^-] = 5.00 \times 10^{-5}$ M gives

$$E = -192.4 - 58.46 \log[5.00 \times 10^{-5}] = 59.0 \text{ mV}$$

(b) If E = 0.0 mV, we can solve for the concentration of $[F^-]$:

$$0.0 = -192.4 - 58.46 \log[F^-] \Rightarrow [F^-] = 5.1 \times 10^{-4} \text{ M}$$

Test Yourself Find $[F^-]$ if E = −22.3 mV. (**Answer**: 1.23 mM)

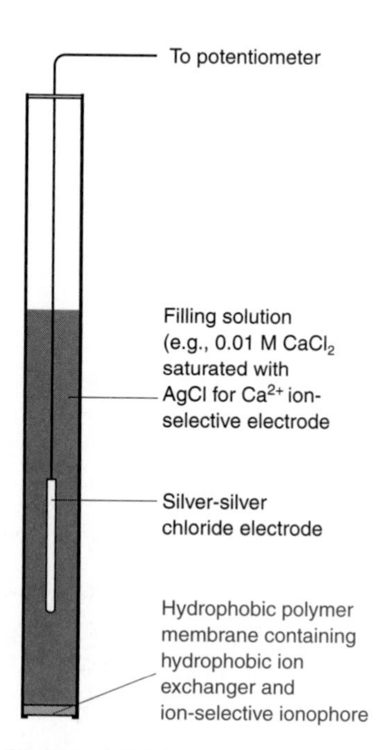

Figure 15-14 Calcium ion-selective electrode with a liquid ion exchanger. Figure 15-5 shows the variability in response from different electrodes.

Liquid-Based Ion-Selective Electrodes

The principle of a **liquid-based ion-selective electrode** was described in Figure 15-6. Figure 15-14 shows a Ca^{2+} ion-selective electrode, featuring a hydrophobic poly(vinyl chloride) membrane saturated with a neutral Ca^{2+}-binding ligand (L) and a salt

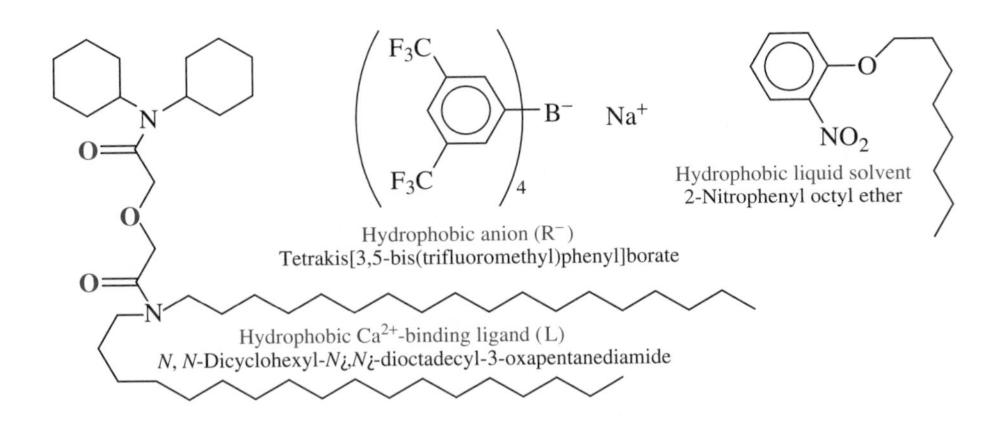

Figure 15-15 Components of the liquid phase in the membrane at the base of the Ca^{2+}-ion selective electrode in Figure 15-14.

of a hydrophobic anion (Na^+R^-) dissolved in a hydrophobic liquid (Figure 15-15). The response is

Response of Ca^{2+} electrode:

$$E = \text{constant} + \beta\left(\frac{0.059\ 16}{2}\right)\log[Ca^{2+}]_{\text{outside}} \tag{15-8}$$

where β is close to 1.00. Equations 15-8 and 15-7 have different signs before the log term because one involves an anion and the other a cation. The charge of the calcium ion requires a factor of 2 in the denominator before the logarithm. The liquid-based NH_4^+ ion-selective electrode used to measure ammonia in marine sediments in Box 6-1 is described in Box 15-2.

Selectivity Coefficient

No electrode responds exclusively to one kind of ion, but the glass electrode is among the most selective. A high-pH glass electrode responds to Na^+ only when $[H^+] \leq 10^{-12}$ M and $[Na^+] \geq 10^{-2}$ M (Figure 15-10).

If an electrode that measures ion A also responds to ion X, the **selectivity coefficient** is defined as

Selectivity coefficient:

$$k_{A,X} = \frac{\text{response to X}}{\text{response to A}} \tag{15-9}$$

The smaller the selectivity coefficient, the less the interference by X. A K^+ ion-selective electrode that uses valinomycin as the liquid ion exchanger has selectivity coefficients $k_{K^+,Na^+} = 1 \times 10^{-5}$, $k_{K^+,Cs^+} = 0.44$, and $k_{K^+,Rb^+} = 2.8$. These coefficients tell us that Na^+ hardly interferes with the measurement of K^+, but Cs^+ and Rb^+ strongly interfere.

For interfering ions X with the same charge as the primary ion A, the response of ion-selective electrodes is described by the equation

Response of ion-selective electrode:

$$E = \text{constant} + \beta\left(\frac{0.059\ 16}{n}\right)\log\left[[A] + \sum_X (k_{A,X}[X])\right] \tag{15-10}$$

Equation 15-10 describes the response of an electrode to its primary ion, A, and to interfering ions, X, of the same charge.

where n is the charge of A. β is near 1 for most electrodes.

The most serious interference for the Ca^{2+} electrode based on the liquids in Figure 15-15 comes from Sr^{2+}, with a selectivity coefficient $k_{Ca^{2+},Sr^{2+}} = 0.13$. That is, the response to Sr^{2+} is 13% as great as the response to an equal concentration of Ca^{2+}. For most cations, $k < 10^{-3}$. It is a good idea to hold pH and ionic strength of standards and unknowns constant when using ion-selective electrodes.

Box 15-2 Ammonium Ion-Selective Microelectrode

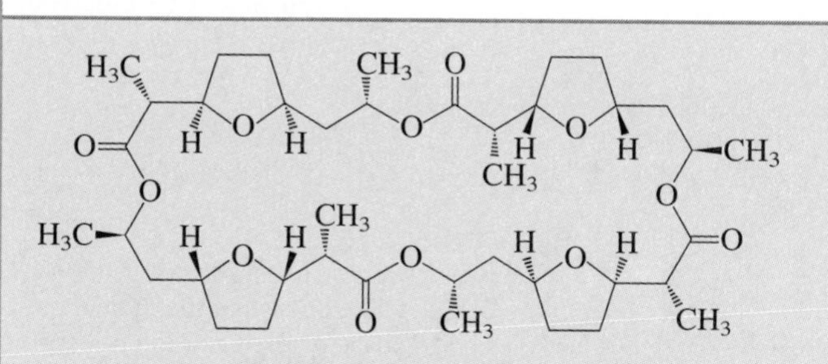

Nonactin—a natural antibiotic isolated from fermentation

Synthetic crown ether with high selectivity for NH_4^+ over K^+

Box 6-1 describes a marine ecosystem in which NH_3 is oxidized to NO_2^- (nitrite) and then to NO_3^- (nitrate). Ammonium ion in the top millimeter of sediment was measured with a microelectrode made by drawing a glass capillary tube to a fine point with an opening diameter of 1 μm. Liquid ion exchanger drawn into the tip of the capillary serves as the ion-selective membrane in Figure 15-6. The natural antibiotic nonactin is the ligand L in Figure 15-6. It selectively binds ammonia in a cage of ligand oxygen atoms. Other components of the ion exchanger are sodium tetraphenylborate to provide the hydrophobic anion R^- and *o*-nitrophenyl octyl ether as the hydrophobic solvent.

Current research is aimed at finding ligands that discriminate better between NH_4^+ and K^+. The synthetic crown ether shown in this box has a selectivity coefficient $k_{NH_4^+,K^+} = 0.03$, whereas the selectivity of nonactin is only $k_{NH_4^+,K^+} = 0.1$. Recall that the smaller the selectivity coefficient in Equation 15-9, the more selective is the ligand. The diagram compares selectivity coefficients for the two ligands with various interfering ions.

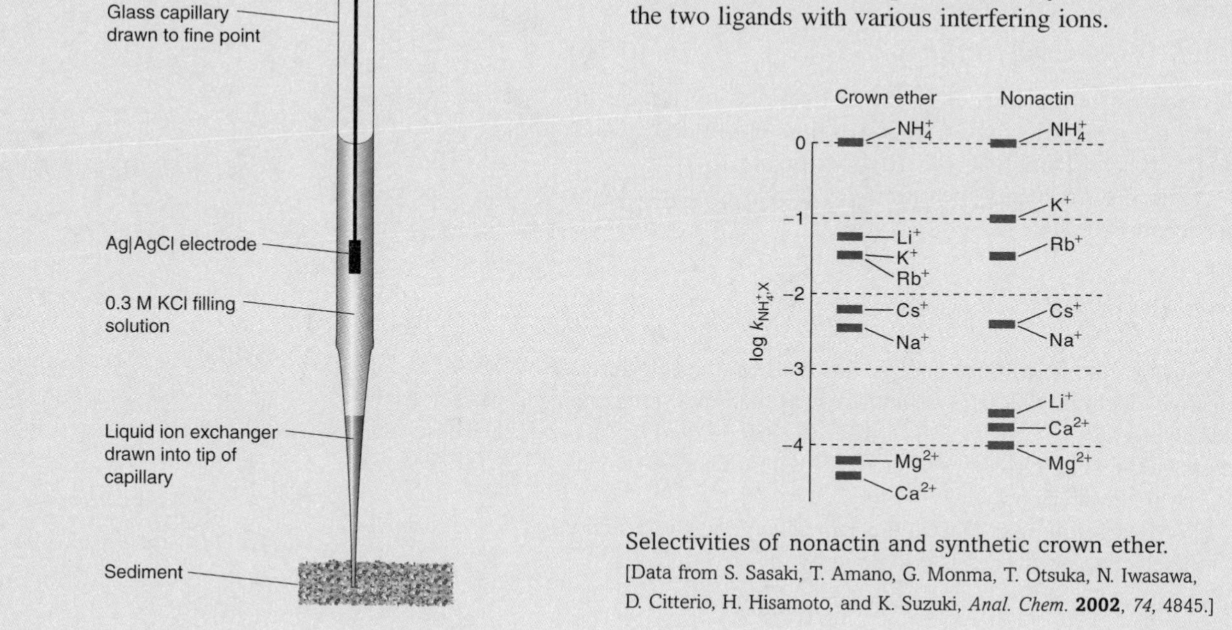

Selectivities of nonactin and synthetic crown ether. [Data from S. Sasaki, T. Amano, G. Monma, T. Otsuka, N. Iwasawa, D. Citterio, H. Hisamoto, and K. Suzuki, *Anal. Chem.* **2002**, *74*, 4845.]

Example Using the Selectivity Coefficient

A fluoride ion-selective electrode has a selectivity coefficient $k_{F^-,OH^-} = 0.1$. What will be the change in electrode potential when 1.0×10^{-4} M F^- at pH 5.5 is raised to pH 10.5?

SOLUTION If $n = -1$ and $\beta = 1$ in Equation 15-10, the potential with negligible OH^- at pH 5.5 is

$$E = \text{constant} - 0.059\ 16 \log[1.0 \times 10^{-4}] = \text{constant} + 236.6 \text{ mV}$$

At pH 10.50, $[OH^-] = 3.2 \times 10^{-4}$ M, so the electrode potential is

$$E = \text{constant} - 0.059\ 16 \log[1.0 \times 10^{-4} + (0.1)(3.2 \times 10^{-4})]$$
$$= \text{constant} + 229.5 \text{ mV}$$

The change is $229.5 - 236.6 = -7.1$ mV, which is quite significant. If you didn't know about the pH change, you would think that the concentration of F^- had increased by 32%.

Test Yourself Show that a change of -7.1 mV is equivalent to an increase in $[F^-]$ of 32%.

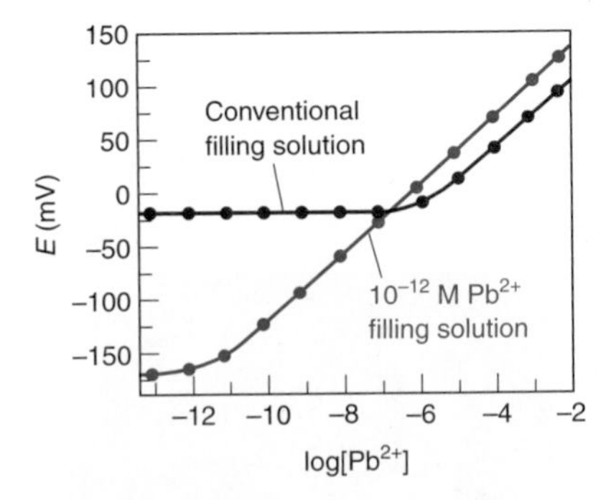

Figure 15-16 Response of Pb^{2+} liquid-based ion-selective electrode with (*black curve*) conventional filling solution containing 0.5 mM Pb^{2+} or (*colored curve*) metal ion buffer filling solution with $[Pb^{2+}] = 10^{-12}$ M. [T. Sokalski, A. Ceresa, T. Zwickl, and E. Pretsch, *J. Am. Chem. Soc.* **1997**, *119*, 11347.]

Ion-Selective Electrode Detection Limits[4]

The black curve in Figure 15-16 was typical of liquid-based ion-selective electrodes until recently. The response of this Pb^{2+} electrode levels off at an analyte concentration around 10^{-6} M. The electrode detects changes in concentration above 10^{-6} M but not below 10^{-6} M. The filling solution inside the electrode contains 0.5 mM $PbCl_2$.

The colored curve in Figure 15-16 was obtained with the same electrode components, but the internal filling solution was replaced by a *metal ion buffer* that fixes $[Pb^{2+}]$ at 10^{-12} M. Now the electrode responds to changes in analyte Pb^{2+} concentration down to $\sim 10^{-11}$ M.

Problem 15-26 describes the metal ion buffer.

The sensitivity of liquid-based ion-selective electrodes has been limited by leakage of the primary ion (Pb^{2+} in this case) from the internal filling solution through the ion-exchange membrane. By lowering the concentration of primary ion inside the electrode, the concentration of leaking ion outside the membrane is reduced by orders of magnitude and the detection limit is correspondingly reduced. Not only is the detection limit for Pb^{2+} improved by 10^5, but the observed selectivity for Pb^{2+} over other cations increases by several orders of magnitude. The sensitivity of a solid-state electrode cannot be lowered by changing the filling solution because analyte concentration is governed by the solubility of the inorganic salt crystal forming the ion-sensitive membrane.

Compound Electrodes

A **compound electrode** is a conventional electrode surrounded by a membrane that isolates (or generates) the analyte to which the electrode responds. The CO_2 gas-sensing electrode in Figure 15-17 is an ordinary glass pH electrode surrounded by electrolyte solution enclosed in a semipermeable membrane made of rubber, Teflon, or polyethylene.[5] A silver-silver chloride reference electrode is immersed in the electrolyte solution. When CO_2 diffuses through the semipermeable membrane, it lowers the pH in the electrolyte compartment. The response of the glass electrode to the change in pH is measured.

Other acidic or basic gases, including NH_3, SO_2, H_2S, NO_x (nitrogen oxides), and HN_3 (hydrazoic acid), can be detected in the same manner. These electrodes can be used to measure gases *in the gas phase* or dissolved in solution. Some ingenious compound electrodes contain a conventional electrode coated with an enzyme that catalyzes a reaction of the analyte. The product of the reaction is detected by the electrode. Compound electrodes based on enzymes are among the most selective because enzymes tend to be extremely specific in their reactivity with just the species of interest.

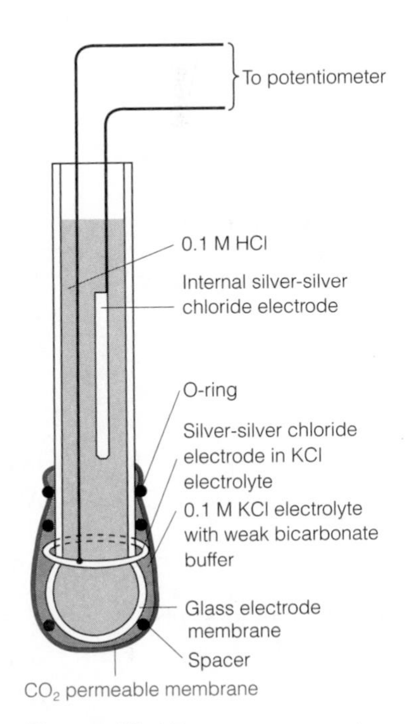

Figure 15-17 A CO_2 gas-sensing electrode.

Linda A. Hughes

Ask Yourself

15-E. Box 6-1 discussed nitrogen species found in a saltwater aquarium. Now we consider the measurement of ammonia in the fishtank, using an ammonia-selective compound electrode. The procedure is to mix 100.0 mL of unknown or standard with 1.0 mL of 10 M NaOH and then to measure NH_3 with an electrode. The purpose of NaOH is to raise the pH above 11 so that ammonia is in the form NH_3, not NH_4^+. (In a more rigorous procedure, EDTA is added prior to NaOH to bind metal ions and displace NH_3 from the metals.)

(a) A series of standards gave the following readings. Prepare a calibration curve of potential (mV) versus log (nitrogen concentration in parts per million) and determine the equation of the straight line by the method of least squares. (Calibration is done in terms of nitrogen in the original standards. There is no dilution factor to consider from the NaOH.)

NH_3 nitrogen concentration (ppm)	log[N]	Electrode potential (mV vs S.C.E.)
0.100	−1.000	72
0.500	−0.301	42
1.000	0.000	25

(b) Two students measured NH_3 in the aquarium and observed values of 106 and 115 mV. What NH_3 nitrogen concentration (in ppm) should be reported by each student?

(c) Artificial seawater for the aquarium is prepared by adding a commercial seawater salt mix to the correct volume of distilled water. There is an unhealthy level of NH_4Cl impurity in the salt mix, so the instructions call for several hours of aerating freshly prepared seawater to remove $NH_3(g)$ before adding the water to a tank containing live fish. A student measured the concentration of NH_3 in freshly prepared seawater prior to aeration and observed a potential of 56 mV. What is the concentration of NH_3 in the freshly prepared seawater?

Key Equations

Voltage of complete cell

$E = E_+ - E_-$ (repeated from Chapter 14)

E_+ = voltage of electrode connected to + terminal of meter

E_- = voltage of electrode connected to − terminal of meter

Titration of X^- with M^+

Before V_e: $[M^+] = K_{sp}/[X^-]$

At V_e: $[M^+] = [X^-] = \sqrt{K_{sp}}$

After V_e: $[M^+] = \dfrac{\text{mol excess } M^+}{\text{total volume}}$

Response of glass pH electrode

$E = \text{constant} + \beta(0.059\,16)\Delta\text{pH}$

ΔpH = (analyte pH) − (pH of internal solution)

β ($\approx$1.00) is measured with standard buffers

constant = asymmetry potential (measured by calibration)

Ion-selective electrode response

$$E = \text{constant} + \beta\left(\frac{0.059\ 16}{n}\right)\log\left[[A] + \sum_X (k_{A,X}[X])\right]$$

A = analyte ion with charge n

X = interfering ion with charge n

$k_{A,X}$ = selectivity coefficient

Important Terms

combination electrode	junction potential	selectivity coefficient
compound electrode	liquid-based ion-selective electrode	solid-state ion-selective electrode
glass electrode	potentiometry	
ion-selective electrode		

Problems

15-1. A cell was prepared by dipping a Cu wire and a saturated Ag | AgCl electrode into 0.10 M $CuSO_4$ solution. The Cu wire was attached to the positive terminal of a potentiometer and the reference electrode was attached to the negative terminal.

(a) Write a half-reaction for the Cu electrode.

(b) Write the Nernst equation for the Cu electrode.

(c) Calculate the cell voltage.

15-2. Pt and saturated calomel electrodes are dipped into a solution containing 0.002 17 M $Br_2(aq)$ and 0.234 M Br^-.

(a) Write the reaction that takes place at Pt and find the half-cell potential E_+.

(b) Find the net cell voltage, E.

15-3. A 50.0-mL solution of 0.100 M NaSCN was titrated with 0.200 M $AgNO_3$ in the cell in Figure 15-1. Find $[Ag^+]$ and E at V_{Ag^+} = 0.1, 10.0, 25.0, and 30.0 mL and sketch the titration curve.

15-4. A 10.0-mL solution of 0.050 0 M $AgNO_3$ was titrated with 0.025 0 M NaBr in the cell in Figure 15-1. Find the cell voltage at V_{Br^-} = 0.1, 10.0, 20.0, and 30.0 mL and sketch the titration curve.

15-5. A 25.0-mL solution of 0.050 0 M NaCl was titrated with 0.025 0 M $AgNO_3$ in the cell in Figure 15-1. Find $[Ag^+]$ and E at V_{Ag^+} = 1.0, 10.0, 50.0, and 60.0 mL and sketch the titration curve.

15-6. *A more advanced problem.* A 50.0-mL solution of 0.100 M NaCl was titrated with 0.100 M $Hg_2(NO_3)_2$ in a cell analogous to that in Figure 15-1, but with a mercury electrode instead of a silver electrode. The cell is S.C.E. || titration reaction | Hg(l).

(a) Write the titration reaction and find the equivalence volume.

(b) The electrochemical equilibrium at the mercury electrode is $Hg_2^{2+} + 2e^- \rightleftharpoons 2Hg(l)$. Derive an equation for the cell voltage analogous to Equation 15-1.

(c) Compute the cell voltage at the following volumes of added $Hg_2(NO_3)_2$: 0.1, 10.0, 25.0, 30.0 mL. Sketch the titration curve.

15-7. Which side of the liquid junction 0.1 M KNO_3 | 0.1 M NaCl will be negative? Explain your answer.

15-8. In Table 15-3, the liquid junction 0.1 M HCl | 0.1 M KCl has a voltage of +27 mV and the junction 0.1 M HCl | 3.5 M KCl has a voltage of +3.1 mV. Which side of each junction becomes positive? Why is the voltage so much less with 3.5 M KCl than with 0.1 M KCl?

15-9. If electrode C in Figure 15-10 is placed in a solution of pH 11.0, what will the pH reading be?

15-10. Suppose that the Ag | AgCl outer electrode in Figure 15-7 is filled with 0.1 M NaCl instead of saturated KCl. Suppose that the electrode is calibrated in a dilute buffer containing 0.1 M KCl at pH 6.54 at 25°C. The electrode is then dipped in a second buffer *at the same pH* and same temperature, but containing 3.5 M KCl.

(a) Use Table 15-3 to estimate the change in junction potential and how much the indicated pH will change.

(b) Suppose that a change in junction potential causes the apparent pH to change from 6.54 to 6.60. By what percentage does $[H^+]$ appear to change?

15-11. Why is measuring $[H^+]$ with a pH electrode somewhat inaccurate, whereas locating the end point in an acid-base titration with a pH electrode can be very accurate?

15-12. Explain the principle of operation of a liquid-based ion-selective electrode.

15-13. How does a compound electrode differ from a simple ion-selective electrode?

15-14. What does the selectivity coefficient tell us? Is it better to have a large or a small selectivity coefficient?

15-15. A micropipet H^+ ion-selective electrode similar to the NH_4^+ electrode in Box 15-2 was constructed to measure the pH inside large, live cells by impaling the cell with the electrode (and with a similarly small reference electrode).[6] The ion exchanger at the tip of the H^+ ion-selective electrode was made from 10 wt% tri(dodecyl)amine $[(C_{12}H_{25})_3N]$ and 0.7 wt% sodium tetraphenylborate, dissolved in o-nitrophenyl octyl either. The selectivity for H^+ relative to Na^+, K^+, Mg^{2+}, and Ca^{2+} was sufficient for intra- and extracelluar measurements without significant interference from these metal ions. Explain how this electrode works.

15-16. By how many volts will the potential of an ideal Mg^{2+} ion-selective electrode change if the electrode is transferred from 1.00×10^{-4} M $MgCl_2$ to 1.00×10^{-3} M $MgCl_2$?

15-17. When measured with a F^- ion-selective electrode that has a Nernstian response at 25°C, the potential due to F^- in unfluoridated groundwater in Foxboro, Massachusetts, was 40.0 mV more positive than the potential of tap water in Providence, Rhode Island. Providence maintains its fluoridated water at the recommended level of 1.00 ± 0.05 mg F^-/L. What is the concentration of F^- in milligrams per liter in groundwater in Foxboro? (Disregard the uncertainty.)

15-18. A cyanide ion-selective electrode obeys the equation E = constant − (0.059 16) log$[CN^-]$. The potential was −0.230 V when the electrode was immersed in 1.00×10^{-3} M NaCN.

(a) Evaluate the constant in the equation for the electrode.

(b) Find the concentration of CN^- if $E = -0.300$ V.

15-19. The selectivity coefficient, k_{Li^+,Na^+}, for a lithium electrode is 5×10^{-3}. When this electrode is placed in 3.44×10^{-4} M Li^+ solution, the potential is −0.333 V versus S.C.E. What would the potential be if Na^+ were added to give 0.100 M Na^+? If you did not know that Na^+ was interfering, what would be the apparent concentration of Li^+ that gives the same potential as the solution containing Na^+?

15-20. A Ca^{2+} ion-selective electrode has a selectivity coefficient $k_{Ca^{2+},Mg^{2+}} = 0.010$. What will be the electrode potential if 1.0 mM Mg^{2+} is added to 0.100 mM Ca^{2+}? By what percentage would $[Ca^{2+}]$ have to change to give the same voltage change?

15-21. An ammonia gas-sensing electrode gave the following calibration points when all solutions contained 1 M NaOH:

NH_3 (M)	E (mV)	NH_3 (M)	E (mV)
1.00×10^{-5}	268.0	5.00×10^{-4}	368.0
5.00×10^{-5}	310.0	1.00×10^{-3}	386.4
1.00×10^{-4}	326.8	5.00×10^{-3}	427.6

A dry food sample weighing 312.4 mg was digested by the Kjeldahl procedure (Section 10-6) to convert all the nitrogen into NH_4^+. The digestion solution was diluted to 1.00 L, and 20.0 mL were transferred to a 100-mL volumetric flask. The 20.0-mL aliquot was treated with 10.0 mL of 10.0 M NaOH plus enough NaI to complex the Hg catalyst from the digestion and diluted to 100.0 mL. When measured with the ammonia electrode, this solution gave a reading of 339.3 mV.

(a) From the calibration data, find $[NH_3]$ in the 100-mL solution.

(b) Calculate the wt% of nitrogen in the food sample.

15-22. The selectivities of a lithium ion-selective electrode are indicated in the following diagram. Which alkali metal (Group 1) ion causes the most interference?

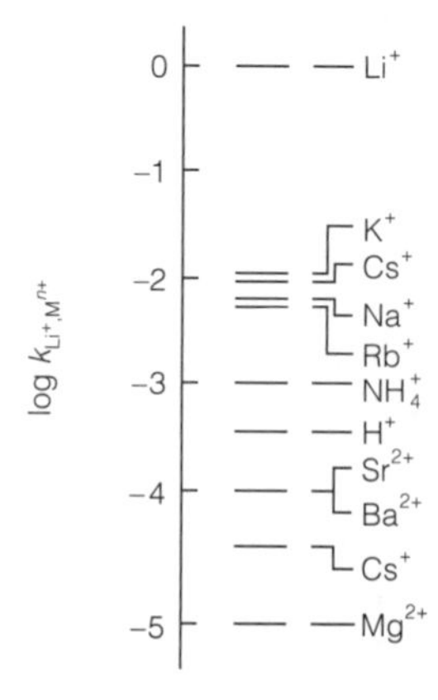

15-23. (a) Write an expression analogous to Equation 15-8 for the response of a La^{3+} ion-selective electrode to La^{3+} ion.

(b) If $\beta \approx 1.00$, by how many millivolts will the potential change when the electrode is removed from 1.00×10^{-4} M $LaClO_4$ and placed in 1.00×10^{-3} M $LaClO_4$?

(c) By how many millivolts will the potential of the electrode change if the electrode is removed from 2.36×10^{-4} M $LaClO_4$ and placed in 4.44×10^{-3} M $LaClO_4$?

(d) The electrode potential is +100 mV in 1.00×10^{-4} M $LaClO_4$ and the selectivity coefficient $k_{La^{3+},Fe^{3+}}$ is $\frac{1}{1\,200}$. What will the potential be when 0.010 M Fe^{3+} is added?

15-24. The following data were obtained when a Ca^{2+} ion-selective electrode was immersed in a series of standard solutions.

Ca^{2+} (M)	E (mV)
3.38×10^{-5}	−74.8
3.38×10^{-4}	−46.4
3.38×10^{-3}	−18.7
3.38×10^{-2}	+10.0
3.38×10^{-1}	+37.7

(a) Prepare a graph of E versus log$[Ca^{2+}]$. Calculate the slope and the y-intercept (and their standard deviations) of the best straight line through the points, using your least-squares spreadsheet from Chapter 4.

(b) Calculate the concentration of a sample that gave a reading of −22.5 mV.

(c) Your spreadsheet gives the uncertainty in $\log[Ca^{2+}]$. Using the upper and lower limits for $\log[Ca^{2+}]$, express the Ca^{2+} concentration as $[Ca^{2+}] = x \pm y$.

15-25. Fourteen ion-selective electrodes were used to measure Ca^{2+} in the same solution with the following results: $[Ca^{2+}]$ = 1.24, 1.13, 1.20, 1.20, 1.30, 1.12, 1.27, 1.19, 1.27, 1.22, 1.23, 1.23, 1.25, 1.24 mM. Find the 95% confidence interval for the mean. If the actual concentration is known to be 1.19 mM, are the results of the ion-selective electrode within experimental error of the known value at the 95% confidence level?

15-26. *Metal ion buffer.* Consider the reaction of Pb^{2+} with EDTA to form a metal complex: $Pb^{2+} + EDTA \overset{K_f'}{\rightleftharpoons} PbY^{2-}$, where EDTA represents all forms of EDTA not bound to metal (Equations 13-1 and 13-7). The effective formation constant, K_f', is related to the formation constant, K_f, by $K_f' = \alpha_{Y^{4-}} K_f$, where $\alpha_{Y^{4-}}$ is the fraction of unbound EDTA in the form Y^{4-}. We can make a lead ion buffer by fixing the concentrations of PbY^{2-} and EDTA. Knowing these two concentrations and the formation constant, we can compute $[Pb^{2+}]$. The lead ion buffer used in the electrode for the colored curve in Figure 15-16 was prepared by mixing 0.74 mL of 0.10 M $Pb(NO_3)_2$ with 100.0 mL of 0.050 M Na_2EDTA. At the measured pH of 4.34, $\alpha_{Y^{4-}} = 1.5 \times 10^{-8}$ (Equation 13-5). Show that $[Pb^{2+}] = 1.0 \times 10^{-12}$ M.

How Would You Do It?

15-27. The graph shows the effect of pH on the response of a liquid-based nitrite (NO_2^-) ion-selective electrode. Ideally, the response would be flat—independent of pH.

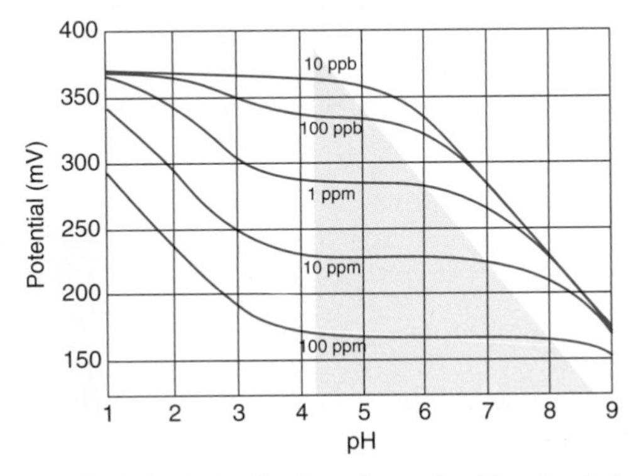

Response of nitrite ion-selective electrode. The shaded region is where response is nearly independent of pH. [From S. J. West and X. Wen, *Am. Environ. Lab.* September 1997, 15.]

(a) Nitrite is the conjugate base of nitrous acid. Why do the curves rise at low pH?

(b) Why do the curves fall at high pH?

(c) What is the optimum pH for using this electrode?

(d) Measure points on the graph at the optimum pH and construct a curve of millivolts versus $\log[NO_2^-]$. What is the lower concentration limit for linear response?

15-28. Consider the cell: $Ag(s) \mid Ag^+(aq, c_l) \parallel Ag^+(aq, c_r) \mid Ag(s)$, where c_l is the concentration of Ag^+ in the left half-cell and c_r is the concentration of Ag^+ in the right half-cell. When both cells contain 0.010 0 M $AgNO_3$, the measured voltage is very close to 0. When the solution in the right-hand cell is replaced by 15.0 mL of 0.020 0 M $AgNO_3$ plus 15.0 mL of 0.200 M NH_3, the voltage changes to −0.289 V. Under these conditions, nearly all the silver in the right half-cell is $Ag(NH_3)_2^+$. From the measured voltage, find the formation constant (called β_2) for the reaction $Ag^+ + 2NH_3 \rightleftharpoons Ag(NH_3)_2^+$.

Notes and References

1. E. Bakker, P. Bühlmann, and E. Pretsch, *Chem. Rev.* **1997**, *97*, 3083.
2. Make a glass electrode from a glass Christmas tree ornament in an instructive experiment: R. T. da Rocha, I. G. R. Gutz, and C. L. do Lago, *J. Chem. Ed.* **1995**, *72*, 1135.
3. The reference electrode in the Ross combination electrode is Pt | I_2, I^-. This electrode gives improved precision and accuracy over conventional pH electrodes [R. C. Metcalf, *Analyst* **1987**, *112*, 1573].
4. For a review, see E. Bakker and E. Pretsch, *Angew. Chem Int. Ed.* **2007**, *46*, 5660.
5. Build a CO_2 compound electrode: S Kocmur, E. Cortón, L. Haim, G. Locascio, and L. Galagosky, *J. Chem. Ed.* **1999**, *76*, 1253.
6. D. Ammann, F. Lanter, R. A. Steiner, P. Schulthess, Y. Shijo, and W. Simon, *Anal. Chem.* **1981**, *53*, 2267.

Further Reading

History of ion-selective electrodes: M. S. Frant, *J. Chem. Ed.* **1997**, *74*, 159; J. Ruzicka, *J. Chem. Ed.* **1997**, *74*, 167; T. S. Light, *J. Chem. Ed.* **1997**, *74*, 171; and C. C. Young, *J. Chem. Ed.* **1997**, *74*, 177.

High-Temperature Superconductors

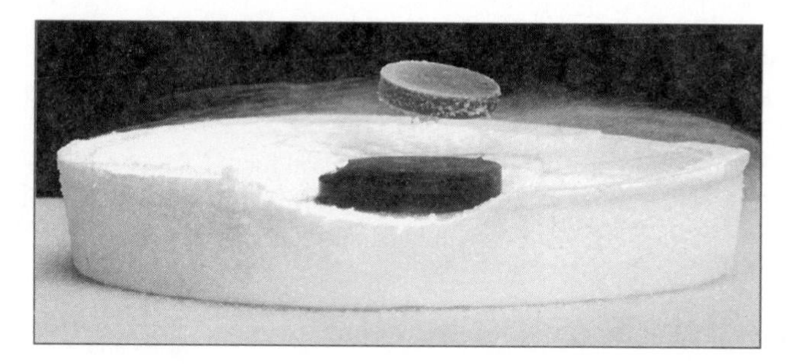

Permanent magnet levitates above superconducting disk cooled in a pool of liquid nitrogen. Redox titrations are crucial in measuring the chemical composition of a superconductor. [Photo courtesy D. Cornelius and T. Vanderah, Michelson Laboratory.]

Superconductors **are materials that lose all electrical resistance when cooled below a critical temperature. Prior to 1987, all known superconductors required cooling to temperatures near that of liquid helium (4 K), a process that is costly and impractical for all but a few applications. In 1987, a giant step was taken when "high-temperature" superconductors that retain superconductivity above the boiling point of liquid nitrogen (77 K) were discovered.**

The most startling characteristic of a superconductor is magnetic levitation, which is shown here. When a magnetic field is applied to a superconductor, current flows in the outer skin of the material such that the applied magnetic field is exactly canceled by the induced magnetic field, and the net field inside the specimen is 0. Expulsion of a magnetic field from a superconductor is called the ***Meissner effect*****.**

A prototypical high-temperature superconductor is yttrium barium copper oxide, $YBa_2Cu_3O_7$, in which two-thirds of the copper is in the +2 oxidation state and one-third is in the unusual +3 state. Another example is $Bi_2Sr_2(Ca_{0.8}Y_{0.2})Cu_2O_{8.295}$, in which the average oxidation state of copper is +2.105 and the average oxidation state of bismuth is +3.090 (which is formally a mixture of Bi^{3+} and Bi^{5+}). The most reliable means to unravel these complex formulas is through redox titrations described in the last problem of this chapter.

CHAPTER 16

Redox Titrations

Oxidation-reduction reactions are ubiquitous. Much of the energy transfer that occurs in living organisms takes place through redox reactions in processes such as photosynthesis and the metabolism of food. In environmental engineering, we saw at the opening of Chapter 14 how the *reducing agent*, Fe(0), can remediate underground pollutants. Conversely, the powerful *oxidizing agent* Fe(VI) in FeO_4^{2-} destroys other pollutants (Figure 17-2). For example, thiocyanate (SCN^-) from photofinishing, metal separation, electroplating, and coke production is oxidized by Fe(VI) to SO_4^{2-}, an environmentally benign product.

In analytical chemistry, a **redox titration** is based on an oxidation-reduction reaction between analyte and titrant. Common analytical oxidants include iodine (I_2), permanganate (MnO_4^-), cerium(IV), and dichromate ($Cr_2O_7^{2-}$). Titrations with reducing agents such as Fe^{2+} (ferrous ion) and Sn^{2+} (stannous ion) are less common because solutions of most reducing agents need protection from air to prevent reaction with O_2.

Box 16-1 explains how dichromate is used to measure *chemical oxygen demand* in environmental analysis.

16-1 Theory of Redox Titrations

Consider the titration of iron(II) with standard cerium(IV), the course of which could be monitored potentiometrically as shown in Figure 16-1. The titration reaction is

Titration reaction:
$$\underset{\text{Ceric titrant}}{Ce^{4+}} + \underset{\text{Ferrous analyte}}{Fe^{2+}} \longrightarrow \underset{\text{Cerous}}{Ce^{3+}} + \underset{\text{Ferric}}{Fe^{3+}} \qquad (16\text{-}1)$$

The *titration reaction* goes to completion after each addition of titrant. The equilibrium constant is given by Equation 14-21: $K = 10^{nE°/0.059\,16}$ at 25°C.

for which $K \approx 10^{16}$ in 1 M $HClO_4$. Each mole of ceric ion oxidizes 1 mol of ferrous ion rapidly and quantitatively. The titration reaction creates a mixture of Ce^{4+}, Ce^{3+}, Fe^{2+}, and Fe^{3+} in the beaker in Figure 16-1.

To follow the course of the reaction, we use a Pt indicator electrode and a saturated calomel (or other) reference electrode. At the *Pt indicator electrode*, *two* reactions each come to equilibrium:

Indicator half-reaction: $$Fe^{3+} + e^- \rightleftharpoons Fe^{2+} \qquad E° = 0.767\text{ V} \qquad (16\text{-}2)$$

Indicator half-reaction: $$Ce^{4+} + e^- \rightleftharpoons Ce^{3+} \qquad E° = 1.70\text{ V} \qquad (16\text{-}3)$$

Equilibria 16-2 and 16-3 are both established at the Pt electrode.

Box 16-1 Environmental Carbon Analysis and Oxygen Demand

Industrial waste streams are partly characterized and regulated on the basis of their carbon content or oxygen demand. *Total carbon* (TC) is defined by the amount of CO_2 evolved when a sample is completely oxidized by combustion:

$$\textit{Total carbon analysis:} \quad \text{all carbon} \xrightarrow[\text{catalyst}]{O_2/900^\circ C} CO_2$$

Color Plate 11 shows an alternative photochemical method to oxidize all carbon to CO_2 without the high temperature required for combustion with O_2.

Total carbon includes dissolved organic material (called *total organic carbon*, TOC) and dissolved CO_3^{2-} and HCO_3^- (called *inorganic carbon*, IC). By definition, TC = TOC + IC. To distinguish TOC from IC, the pH of a fresh sample is lowered below 2 to convert CO_3^{2-} and HCO_3^- into CO_2, which is purged from (bubbled out of) the solution with N_2. After IC has been removed, combustion analysis of the remaining material measures TOC. IC is the difference between the two experiments.

TOC is widely used to determine compliance with discharge laws. Minicipal wastewater might contain ~1 g of TOC per liter. At the other extreme, high-purity water required for microelectronic processes might contain ~1 μg of TOC per liter.

Total oxygen demand (TOD) tells us how much O_2 is required for complete combustion of pollutants in a waste stream. A volume of N_2 containing a known quantity of O_2 is mixed with the sample and complete combustion is carried out. The remaining O_2 is measured by a Clark electrode (Figure 17-3). This measurement is sensitive to the oxidation states of species in the waste stream. For example, urea, $(NH_2)_2C{=}O$, consumes five times as much O_2 as formic acid, HCO_2H, does. Species such as NH_3 and H_2S also contribute to TOD.

Pollutants can be oxidized by refluxing with dichromate, $Cr_2O_7^{2-}$. *Chemical oxygen demand* (COD) is defined as the O_2 that is chemically equivalent to the $Cr_2O_7^{2-}$ consumed in this process. Each $Cr_2O_7^{2-}$ consumes $6e^-$ (to make $2Cr^{3+}$) and each O_2 consumes $4e^-$ (to make $2H_2O$). Therefore 1 mol of $Cr_2O_7^{2-}$ is chemically equivalent to 1.5 mol of O_2 for this computation. COD analysis is carried out by refluxing polluted water for 2 h with excess standard $Cr_2O_7^{2-}$ in H_2SO_4 solution containing Ag^+ catalyst. Unreacted $Cr_2O_7^{2-}$ is then measured by titration with standard Fe^{2+} or by spectrophotometry. Many permits for industrial operations are specified in terms of COD analysis of the waste streams.

Biological oxygen demand (BOD) is defined as the O_2 required for biological degradation of organic materials by microorganisms. The procedure calls for incu-

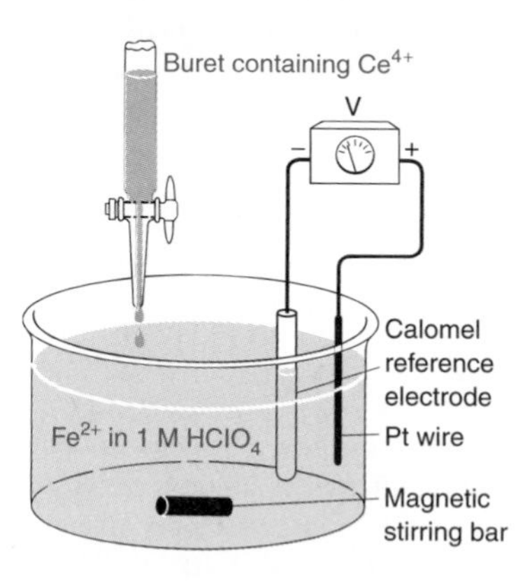

Figure 16-1 Apparatus for potentiometric titration of Fe^{2+} with Ce^{4+}.

E_+ is the potential of the Pt electrode connected to the positive terminal of the potentiometer in Figure 16-1. E_- is the potential of the calomel reference electrode connected to the negative terminal.

The potentials cited here are the formal potentials that apply in 1 M $HClO_4$.

We now calculate how the cell voltage changes as Fe^{2+} is titrated with Ce^{4+}. The titration curve has three regions.

Region 1: Before the Equivalence Point

As each aliquot of Ce^{4+} is added, titration reaction 16-1 consumes the Ce^{4+} and creates an equal number of moles of Ce^{3+} and Fe^{3+}. Prior to the equivalence point, excess unreacted Fe^{2+} remains in the solution. Therefore, we can find the concentrations of Fe^{2+} and Fe^{3+} without difficulty. On the other hand, we cannot find the concentration of Ce^{4+} without solving a fancy little equilibrium problem. Because the amounts of Fe^{2+} and Fe^{3+} are both known, it is *convenient* to calculate the cell voltage with Reaction 16-2 instead of 16-3.

$$E = E_+ - E_- = \underbrace{\left[0.767 - 0.059\ 16 \log\left(\frac{[Fe^{2+}]}{[Fe^{3+}]}\right)\right]}_{\substack{\text{Formal potential for}\\ Fe^{3+}\text{ reduction in}\\ 1\text{ M HClO}_4}} - \underbrace{0.241}_{\substack{\text{Potential of}\\ \text{saturated calomel}\\ \text{electrode}}} \tag{16-4}$$

$$E = 0.526 - 0.059\ 16 \log\left(\frac{[Fe^{2+}]}{[Fe^{3+}]}\right) \tag{16-5}$$

bating a sealed container of wastewater with no extra air space for 5 days at 20°C in the dark while microbes metabolize organic compounds in the waste. Dissolved O_2 is measured before and after incubation. The difference is BOD,[1] which also measures species such as HS^- and Fe^{2+} that may be in the water. Inhibitors are added to prevent oxidation of nitrogen species such as NH_3.

The San Joaquin River is an ecologically sensitive aquatic system that drains into San Francisco Bay. Dissolved O_2 in the river often decreases to <5 mg/L in summer and fall, inhibiting upstream migration of salmon and stressing or killing aquatic organisms. The charts show correlation between high BOD and high levels of algae. Microorganisms consume the algae and, in the process, consume O_2 from the river. Each gram of algal carbon is associated with 0.177 g of algal nitrogen. Oxidation of algae at a concentration of 1 mg/L to CO_2 and NO_3^- consumes 3.4 mg O_2/L. Algae thrive on nitrogen and phosphorus nutrients from fertilizer in agricultural runoff. Possible strategies to increase dissolved O_2 in the river are to decrease nutrient flow into the river and decrease algae seed sources upstream.

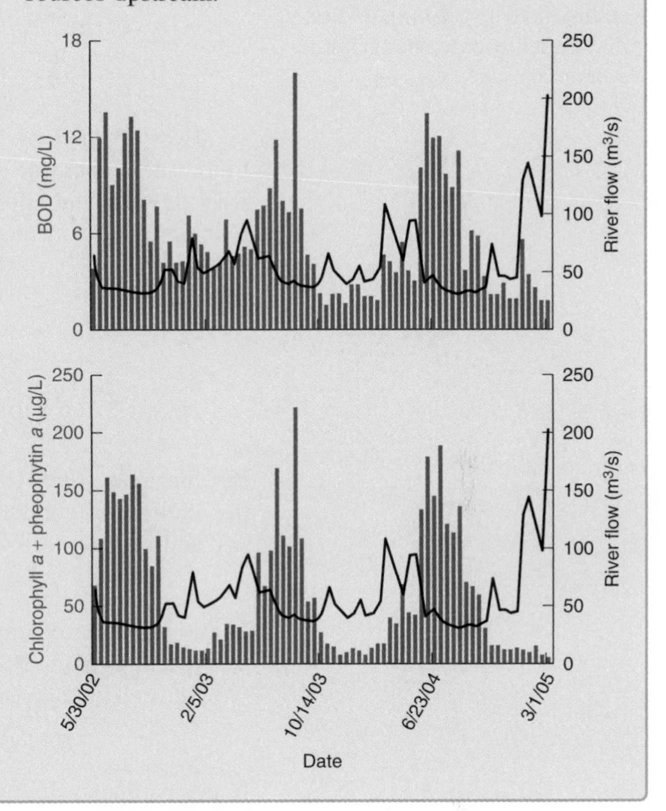

BOD and algae in the lower San Joaquin River at Mossdale, CA. Chlorophyll *a* and pheophytin *a* (a chlorophyll *a* degradation product) extracted from the river are surrogate measurements for algae. [From E. C. Volkmar and R. A. Dahlgren, *Environ. Sci. Technol.* **2006**, *40*, 5653.]

One special point is reached before the equivalence point. When the volume of titrant is one-half of the amount required to reach the equivalence point ($V = \frac{1}{2}V_e$), the concentrations of Fe^{3+} and Fe^{2+} are equal. In this case, the log term is 0, and $E_+ = E°$ for the $Fe^{3+} | Fe^{2+}$ couple. *The point at which $V = \frac{1}{2}V_e$ is analogous to the point at which pH = pK_a when $V = \frac{1}{2}V_e$ in an acid-base titration.*

For Reaction 16-2, $E_+ = E°$ for the $Fe^{3+} | Fe^{2+}$ couple when $V = \frac{1}{2}V_e$.

Region 2: At the Equivalence Point

Exactly enough Ce^{4+} has been added to react with all the Fe^{2+}. Virtually all cerium is in the form Ce^{3+}, and virtually all iron is in the form Fe^{3+}. Tiny amounts of Ce^{4+} and Fe^{2+} are present at equilibrium. From the stoichiometry of Reaction 16-1, we can say that

$$[Ce^{3+}] = [Fe^{3+}] \tag{16-6}$$

$$[Ce^{4+}] = [Fe^{2+}] \tag{16-7}$$

To understand why Equations 16-6 and 16-7 are true, imagine that *all* the cerium and the iron have been converted into Ce^{3+} and Fe^{3+}. Because we are at the equivalence point, $[Ce^{3+}] = [Fe^{3+}]$. Now let Reaction 16-1 come to equilibrium:

$$Fe^{3+} + Ce^{3+} \rightleftharpoons Fe^{2+} + Ce^{4+} \quad \text{(16-8, reverse of Reaction 16-1)}$$

If a little bit of Fe^{3+} goes back to Fe^{2+}, an equal number of moles of Ce^{4+} must be made. So $[Ce^{4+}] = [Fe^{2+}]$.

At the equivalence point, we use both Reactions 16-2 and 16-3 to calculate the cell voltage. This is just an algebraic convenience.

At any time, Reactions 16-2 and 16-3 are *both* in equilibrium at the Pt electrode. At the equivalence point, it is *convenient* to use both reactions to find the cell voltage. The Nernst equations are

$$E_+ = 0.767 - 0.059\,16 \log\left(\frac{[Fe^{2+}]}{[Fe^{3+}]}\right) \tag{16-9}$$

$$E_+ = 1.70 - 0.059\,16 \log\left(\frac{[Ce^{3+}]}{[Ce^{4+}]}\right) \tag{16-10}$$

Here is where we stand: Equation 16-9 and 16-10 are statements of algebraic truth. But neither one alone allows us to find E_+, because we do not know exactly what tiny concentrations of Fe^{2+} and Ce^{4+} are present. It is possible to solve the four simultaneous equations 16-6, 16-7, 16-9, and 16-10, by first *adding* Equations 16-9 and 16-10:

$$2E_+ = 0.767 + 1.70 - 0.059\,16 \log\left(\frac{[Fe^{2+}]}{[Fe^{3+}]}\right) - 0.059\,16 \log\left(\frac{[Ce^{3+}]}{[Ce^{4+}]}\right)$$

$\log a + \log b = \log ab$

$$2E_+ = 2.46_7 - 0.059\,16 \log\left(\frac{[Fe^{2+}][Ce^{3+}]}{[Fe^{3+}][Ce^{4+}]}\right)$$

But, because $[Ce^{3+}] = [Fe^{3+}]$ and $[Ce^{4+}] = [Fe^{2+}]$ at the equivalence point, the quotient of concentrations in the log term is unity. Therefore, the logarithm is 0 and

$$2E_+ = 2.46_7 \text{ V} \Rightarrow E_+ = 1.23 \text{ V}$$

In this particular example, E_+ is the average of the standard potentials for the two half-reactions at the Pt electrode.

The cell voltage is

$$E = E_+ - E(\text{calomel}) = 1.23 - 0.241 = 0.99 \text{ V} \tag{16-11}$$

In this particular titration, the equivalence-point voltage is independent of the concentrations and volumes of the reactants.

Region 3: After the Equivalence Point

After V_e, we use Reaction 16-3 because we know $[Ce^{3+}]$ and $[Ce^{4+}]$. It is not convenient to use Reaction 16-2, because we do not know $[Fe^{2+}]$, which has been "used up."

Now virtually all iron atoms are Fe^{3+}. The moles of Ce^{3+} equal the moles of Fe^{3+}, and there is a known excess of unreacted Ce^{4+}. Because we know both $[Ce^{3+}]$ and $[Ce^{4+}]$, it is *convenient* to use Reaction 16-3 to describe the chemistry at the Pt electrode:

$$E = E_+ - E(\text{calomel}) = \left[1.70 - 0.059\,16 \log\left(\frac{[Ce^{3+}]}{[Ce^{4+}]}\right)\right] - 0.241 \tag{16-12}$$

At the special point when $V = 2V_e$, $[Ce^{3+}] = [Ce^{4+}]$ and $E_+ = E°(Ce^{4+} \mid Ce^{3+}) =$ 1.70 V.

Before the equivalence point, the voltage is fairly steady near the value $E = E_+ - E(\text{calomel}) \approx E°(Fe^{3+} \mid Fe^{2+}) - 0.241 \text{ V} = 0.53$ V. After the equivalence point, the voltage levels off near $E \approx E°(Ce^{4+} \mid Ce^{3+}) - 0.241 \text{ V} = 1.46$ V. At the equivalence point, there is a rapid rise in voltage.

Example **Potentiometric Redox Titration**

Suppose that we titrate 100.0 mL of 0.050 0 M Fe^{2+} with 0.100 M Ce^{4+}, by using the cell in Figure 16-1. The equivalence point occurs when $V_{Ce^{4+}} = 50.0$ mL, because the Ce^{4+} is twice as concentrated as the Fe^{2+}. Calculate the cell voltage at 36.0, 50.0, and 63.0 mL.

SOLUTION

At 36.0 mL: This is 36.0/50.0 of the way to the equivalence point. Therefore, 36.0/50.0 of the iron is in the form Fe^{3+} and 14.0/50.0 is in the form Fe^{2+}. Putting $[Fe^{2+}]/[Fe^{3+}] = 14.0/36.0$ into Equation 16-5 gives $E = 0.550$ V.

At 50.0 mL: Equation 16-11 tells us that the cell voltage at the equivalence point is 0.99 V, regardless of the concentrations of reagents for this particular titration.

At 63.0 mL: The first 50.0 mL of cerium have been converted into Ce^{3+}. Because 13.0 mL of excess Ce^{4+} have been added, $[Ce^{3+}]/[Ce^{4+}] = 50.0/13.0$ in Equation 16-12, and $E = 1.424$ V.

Test Yourself Compute E at 37.0 and 64.0 mL. Do your answers make sense in comparison to the values at 36.0 and 63.0 mL? (**Answer**: 0.553 V, 1.426 V)

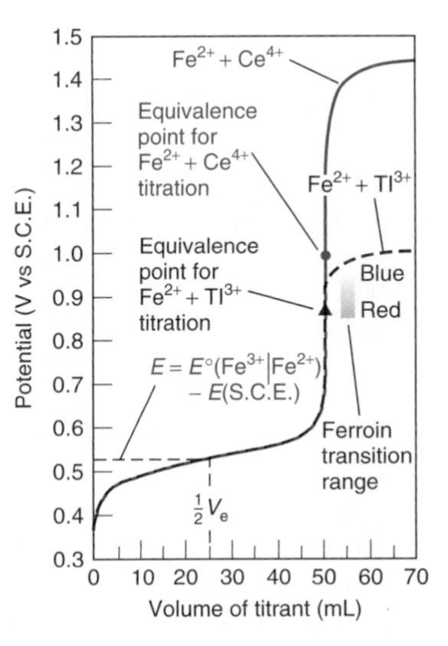

Figure 16-2 *Solid line:* Theoretical curve for titration of 100.0 mL of 0.050 0 M Fe^{2+} with 0.100 M Ce^{4+} in 1 M $HClO_4$. You cannot calculate the potential for zero titrant, but you can start at a small volume such as 0.1 mL. *Dashed line:* Theoretical curve for titration of 100.0 mL of 0.050 0 M Fe^{2+} with 0.050 0 M Tl^{3+} in 1 M $HClO_4$.

Shapes of Redox Titration Curves

The preceding calculations allow us to plot the solid titration curve for Reaction 16-1 in Figure 16-2, which shows the potential as a function of the volume of added titrant. The equivalence point is marked by a steep rise in the voltage. The calculated value of E_+ at $\frac{1}{2}V_e$ is the formal potential of the $Fe^{3+} \mid Fe^{2+}$ couple, because the quotient $[Fe^{2+}]/[Fe^{3+}]$ is unity at this point. The calculated voltage at any point in this titration depends only on the *concentration ratio* of reactants; their *absolute concentrations* make no difference in this example. We expect, therefore, that the curve in Figure 16-2 would not change if both reactants were diluted by a factor of 10.

The voltage at zero titrant volume cannot be calculated because we do not know how much Fe^{3+} is present. If $[Fe^{3+}] = 0$, the voltage calculated with Equation 16-9 would be $-\infty$. In fact, there must be some Fe^{3+} in each reagent, either as an impurity or from oxidation of Fe^{2+} by atmospheric oxygen. In any case, the voltage could not be lower than that needed to reduce the solvent ($H_2O + e^- \rightarrow \frac{1}{2}H_2 + OH^-$).

For Reaction 16-1, the titration curve in Figure 16-2 is symmetric near the equivalence point because the reaction stoichiometry is 1:1. For oxidation of Fe(II) by Tl(III)

$$2Fe^{2+} + Tl^{3+} \longrightarrow 2Fe^{3+} + Tl^{+} \qquad (16\text{-}13)$$

the dashed curve in Figure 16-2 is not symmetric about the equivalence point because the stoichiometry of reactants is 2:1, not 1:1. Still, the curve is so steep near the equivalence point that negligible error is introduced if the center of the steepest portion is taken as the end point. Demonstration 16-1 provides an example of an asymmetric titration curve whose shape also depends on the pH of the reaction medium.

The shape of the curve in Figure 16-2 is essentially independent of the concentrations of analyte and titrant. The solid curve is symmetric near V_e because the stoichiometry is 1:1.

The change in voltage near the equivalence point for the dashed curve in Figure 16-2 is smaller than the voltage change for the solid curve because Tl^{3+} is a weaker oxidizing agent than Ce^{4+}. Clearest results are achieved with the strongest oxidizing and reducing agents. The same rule applies to acid-base titrations where strong-acid or strong-base titrants give the sharpest break at the equivalence point.

You would not choose a weak acid to titrate a weak base because the break at V_e would not be very large.

Ask Yourself

16-A. A 20.0-mL solution of 0.005 00 M Sn^{2+} in 1 M HCl was titrated with 0.020 0 M Ce^{4+} to give Sn^{4+} and Ce^{3+}. What is the potential (versus S.C.E.) at the following volumes of Ce^{4+}: 0.100, 1.00, 5.00, 9.50, 10.00, 10.10, and 12.00 mL? Sketch the titration curve.

Demonstration 16-1 Potentiometric Titration of Fe^{2+} with MnO_4^-

The titration of Fe^{2+} with $KMnO_4$ nicely illustrates principles of potentiometric titrations.

$$\underset{\text{Titrant}}{MnO_4^-} + \underset{\text{Analyte}}{5Fe^{2+}} + 8H^+ \longrightarrow Mn^{2+} + 5Fe^{3+} + 4H_2O \quad \text{(A)}$$

Dissolve 0.60 g of $Fe(NH_4)_2(SO_4)_2 \cdot 6H_2O$ (FM 392.14; 1.5 mmol) in 400 mL of 1 M H_2SO_4. Titrate the well-stirred solution with 0.02 M $KMnO_4$ ($V_e \approx 15$ mL), using Pt and saturated calomel electrodes with a pH meter as a potentiometer. The reference socket of the pH meter is the negative input terminal. Before starting the titration, calibrate the meter by connecting the two input sockets directly to each other with a wire and setting the millivolt scale of the meter at 0.

The demonstration is more meaningful if you calculate points on the theoretical titration curve before performing the experiment. Then compare the theoretical and experimental results. Also note the coincidence of the potentiometric and visual end points.

Question Potassium permanganate is purple, and all the other species in this titration are colorless (or very faintly colored). What color change is expected at the equivalence point?

To calculate points on the theoretical titration curve, we use the following half-reactions:

$$Fe^{3+} + e^- \rightleftharpoons Fe^{2+} \qquad E^\circ = 0.68 \text{ V in 1 M } H_2SO_4 \quad \text{(B)}$$

$$MnO_4^- + 8H^+ + 5e^- \longrightarrow Mn^{2+} + 4H_2O \qquad E^\circ = 1.507 \text{ V} \quad \text{(C)}$$

Prior to the equivalence point, calculations are similar to those in Section 16-1 for the titration of Fe^{2+} by Ce^{4+}, but $E^\circ = 0.68$ V. After the equivalence point, you can find the potential by using Reaction C. For example, suppose that you titrate 0.400 L of 3.75 mM Fe^{2+} with 0.020 0 M $KMnO_4$. From the stoichiometry of Reaction A, the equivalence point is $V_e = 15.0$ mL. When you have added 17.0 mL of $KMnO_4$, the concentrations of species in Reaction C are $[Mn^{2+}] = 0.719$ mM, $[MnO_4^-] = 0.095\,9$ mM, and $[H^+] = 0.959$ M (neglecting the small quantity of H^+ consumed in the titration).

The cell voltage is

$$\begin{aligned} E &= E_+ - E(\text{calomel}) \\ &= \left[1.507 - \frac{0.059\,16}{5}\log\left(\frac{[Mn^{2+}]}{[MnO_4^-][H^+]^8}\right)\right] - 0.241 \\ &= \left[1.507 - \frac{0.059\,16}{5}\log\left(\frac{7.19 \times 10^{-4}}{(9.59 \times 10^{-5})(0.959)^8}\right)\right] \\ &\quad - 0.241 = 1.254 \text{ V} \end{aligned}$$

To calculate the voltage at the equivalence point, we add the Nernst equations for Reactions B and C, as we did for the cerium and iron reactions in Section 16-1. Before doing so, however, we multiply the permanganate equation by 5 so that we can add the log terms:

$$E_+ = 0.68 - 0.059\,16 \log\left(\frac{[Fe^{2+}]}{[Fe^{3+}]}\right)$$

$$5E_+ = 5\left[1.507 - \frac{0.059\,16}{5}\log\left(\frac{[Mn^{2+}]}{[MnO_4^-][H^+]^8}\right)\right]$$

Now we can add the two equations to get

$$6E_+ = 8.215 - 0.059\,16 \log\left(\frac{[Mn^{2+}][Fe^{2+}]}{[MnO_4^-][Fe^{3+}][H^+]^8}\right) \quad \text{(D)}$$

But the stoichiometry of titration reaction A tells us that at the equivalence point $[Fe^{3+}] = 5[Mn^{2+}]$ and $[Fe^{2+}] = 5[MnO_4^-]$. Substituting these values into Equation D gives

$$\begin{aligned} 6E_+ &= 8.215 - 0.059\,16 \log\left(\frac{\cancel{[Mn^{2+}]}\cancel{(5[MnO_4^-])}}{\cancel{[MnO_4^-]}\cancel{(5[Mn^{2+}])}[H^+]^8}\right) \\ &= 8.215 - 0.059\,16 \log\left(\frac{1}{[H^+]^8}\right) \end{aligned} \quad \text{(E)}$$

Inserting the concentration of $[H^+]$, which is (400/415)(1.00 M) = 0.964 M, we find

$$6E_+ = 8.215 - 0.059\,16 \log\left(\frac{1}{(0.964)^8}\right) \Rightarrow E_+ = 1.368 \text{ V}$$

The predicted cell voltage at V_e is $E = E_+ - E(\text{calomel}) = 1.368 - 0.241 = 1.127$ V.

16-2 Redox Indicators

An indicator (In) may be used to detect the end point of a redox titration, just as an indicator may be used in an acid-base titration. A **redox indicator** changes color when it goes from its oxidized to its reduced state. One common indicator is ferroin, whose color change is from pale blue (almost colorless) to red.

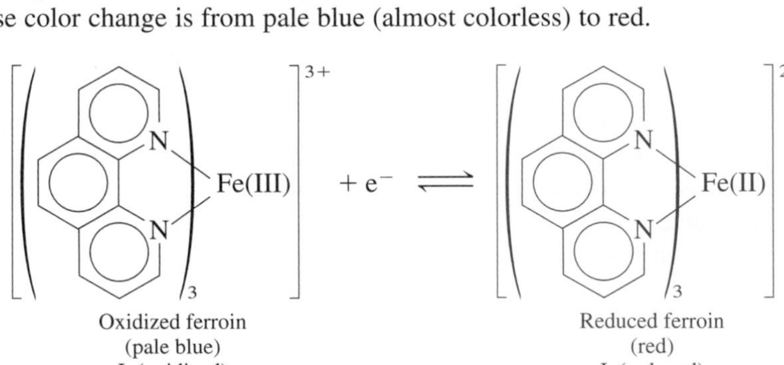

Oxidized ferroin (pale blue) In(oxidized)

Reduced ferroin (red) In(reduced)

To predict the potential range over which the indicator color will change, we first write a Nernst equation for the indicator.

$$\text{In(oxidized)} + ne^- \rightleftharpoons \text{In(reduced)}$$

$$E = E^\circ - \frac{0.059\ 16}{n} \log \frac{[\text{In(reduced)}]}{[\text{In(oxidized)}]}$$

As with acid-base indicators, the color of In(reduced) will be observed when

$$\frac{[\text{In(reduced)}]}{[\text{In(oxidized)}]} \gtrsim \frac{10}{1}$$

and the color of In(oxidized) will be observed when

$$\frac{[\text{In(reduced)}]}{[\text{In(oxidized)}]} \lesssim \frac{1}{10}$$

Putting these quotients into the Nernst equation for the indicator tells us that the color change will occur over the range

Redox indicator color change range:

$$E = \left(E^\circ \pm \frac{0.059\ 16}{n}\right) \text{volts} \tag{16-14}$$

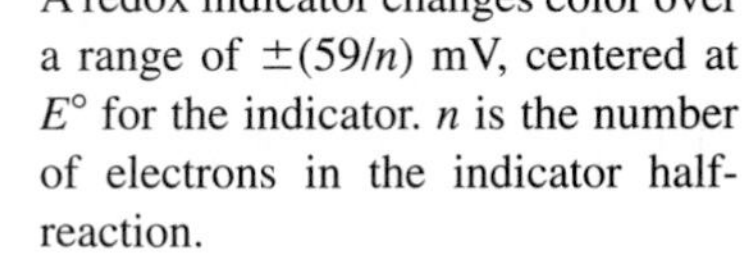

A redox indicator changes color over a range of $\pm(59/n)$ mV, centered at E° for the indicator. n is the number of electrons in the indicator half-reaction.

For ferroin, with $E^\circ = 1.147$ V (Table 16-1), we expect the color change to occur in the approximate range 1.088 V to 1.206 V with respect to the standard hydrogen electrode. If a saturated calomel electrode is used as the reference instead, the indicator transition range is

$$\begin{pmatrix}\text{indicator transition}\\ \text{range versus calomel}\\ \text{electrode (S.C.E)}\end{pmatrix} = \begin{pmatrix}\text{(transition range}\\ \text{versus standard hydrogen}\\ \text{electrode (S.H.E.)}\end{pmatrix} - E(\text{calomel}) \tag{16-15}$$

$$= (1.088 \text{ to } 1.206) - (0.241)$$

$$= 0.847 \text{ to } 0.965 \text{ V (versus S.C.E.)}$$

Figure 14-12 will help you understand Equation 16-15.

Ferroin would therefore be a useful indicator for the solid curve in Figure 16-2.

The indicator transition range should overlap the steep part of the titration curve.

The larger the difference in standard potential between titrant and analyte, the sharper the break in the titration curve at the equivalence point. A redox titration is

Table 16-1 Redox indicators

Indicator	Color		E°
	Reduced	Oxidized	
Phenosafranine	Colorless	Red	0.28
Indigo tetrasulfonate	Colorless	Blue	0.36
Methylene blue	Colorless	Blue	0.53
Diphenylamine	Colorless	Violet	0.75
4′-Ethoxy-2,4-diaminoazobenzene	Red	Yellow	0.76
Diphenylamine sulfonic acid	Colorless	Red-violet	0.85
Diphenylbenzidine sulfonic acid	Colorless	Violet	0.87
Tris(2,2′-bipyridine)iron	Red	Pale blue	1.120
Tris(1,10-phenanthroline)iron (ferroin)	Red	Pale blue	1.147
Tris(5-nitro-1,10-phenanthroline)iron	Red-violet	Pale blue	1.25
Tris(2,2′-bipyridine)ruthenium	Yellow	Pale blue	1.29

usually feasible if the difference between analyte and titrant is $\gtrsim$0.2 V. However, the end point of such a titration is not very sharp and is best detected potentiometrically. If the difference in formal potentials is $\gtrsim$0.4 V, then a redox indicator usually gives a satisfactory end point.

Ask Yourself

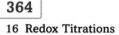

16-B. What would be the best redox indicator in Table 16-1 for the titration of $Fe(CN)_6^{4-}$ with Tl^{3+} in 1 M HCl? (*Hint:* The potential at the equivalence point must be between the potentials for each redox couple.) What color change would you look for?

16-3 Titrations Involving Iodine

Redox titrations (Tables 16-2 and 16-3) are available for many analytes with iodine (I_2, a mild oxidizing agent) or iodide (I^-, a mild reducing agent).

Iodine as oxidizing agent: $$I_2(aq) + 2e^- \longrightarrow 2I^- \qquad (16\text{-}16)$$

Iodide as reducing agent: $$2I^- \longrightarrow I_2(aq) + 2e^- \qquad (16\text{-}17)$$

For example, vitamin C in foods and the compositions of superconductors (shown at the opening of the chapter) can be measured with iodine. When a reducing analyte is titrated with iodine (I_2), the method is called *iodimetry. Iodometry* is the titration of iodine produced when an oxidizing analyte is added to excess I^-. The iodine is usually titrated with standard thiosulfate solution.

I_2 is only slightly soluble in water (1.3 mM at 20°C), but its solubility is enhanced by complexation with iodide:

$$\underset{\text{Iodine}}{I_2(aq)} + \underset{\text{Iodide}}{I^-} \rightleftharpoons \underset{\text{Triiodide}}{I_3^-} \qquad K = 7 \times 10^2 \qquad (16\text{-}18)$$

A typical 0.05 M solution of I_3^- for titrations is prepared by dissolving 0.12 mol of KI plus 0.05 mol of I_2 in 1 L of water. When we speak of using "iodine," we usually

Table 16-2 Iodimetric titrations: Titrations with standard iodine (actually I_3^-)

Species analyzed	Oxidation reaction	Notes
SO_2	$SO_2 + H_2O \rightleftharpoons H_2SO_3$ $H_2SO_3 + H_2O \rightleftharpoons SO_4^{2-} + 4H^+ + 2e^-$	Add SO_2 (or H_2SO_3 or HSO_3^- or SO_3^{2-}) to excess standard I_3^- in dilute acid and back-titrate unreacted I_3^- with standard thiosulfate.
H_2S	$H_2S \rightleftharpoons S(s) + 2H^+ + 2e^-$	Add H_2S to excess I_3^- in 1 M HCl and back-titrate with thiosulfate.
Zn^{2+}, Cd^{2+}, Hg^{2+}, Pb^{2+}	$M^{2+} + H_2S \longrightarrow MS(s) + 2H^+$ $MS(s) \rightleftharpoons M^{2+} + S + 2e^-$	Precipitate and wash metal sulfide. Dissolve in 3 M HCl with excess standard I_3^- and back-titrate with thiosulfate.
Cysteine, glutathione, mercaptoethanol	$2RSH \rightleftharpoons RSSR + 2H^+ + 2e^-$	Titrate the sulfhydryl compound at pH 4–5 with I_3^-.
$H_2C{=}O$	$H_2CO + 3OH^- \rightleftharpoons HCO_2^- + 2H_2O + 2e^-$	Add excess I_3^- plus NaOH to the unknown. After 5 min, add HCl and back-titrate with thiosulfate.
Glucose (and other reducing sugars)	$RC(=O)H + 3OH^- \rightleftharpoons RCO_2^- + 2H_2O + 2e^-$	Add excess I_3^- plus NaOH to the sample. After 5 min, add HCl and back-titrate with thiosulfate.

Table 16-3 Iodometric titrations: Titrations of iodine (actually I_3^-) produced by analyte

Species analyzed	Reaction	Notes
HOCl	$HOCl + H^+ + 3I^- \rightleftharpoons Cl^- + I_3^- + H_2O$	Reaction in 0.5 M H_2SO_4.
Br_2	$Br_2 + 3I^- \rightleftharpoons 2Br^- + I_3^-$	Reaction in dilute acid.
IO_3^-	$2IO_3^- + 16I^- + 12H^+ \rightleftharpoons 6I_3^- + 6H_2O$	Reaction in 0.5 M HCl.
IO_4^-	$2IO_4^- + 22I^- + 16H^+ \rightleftharpoons 8I_3^- + 8H_2O$	Reaction in 0.5 M HCl.
O_2	$O_2 + 4Mn(OH)_2 + 2H_2O \rightleftharpoons 4Mn(OH)_3$ $2Mn(OH)_3 + 6H^+ + 6I^- \rightleftharpoons 2Mn^{2+} + 2I_3^- + 6H_2O$	The sample is treated with Mn^{2+}, NaOH, and KI. After 1 min, it is acidified with H_2SO_4, and the I_3^- is titrated.
H_2O_2	$H_2O_2 + 3I^- + 2H^+ \rightleftharpoons I_3^- + 2H_2O$	Reaction in 1 M H_2SO_4 with NH_4MoO_3 catalyst.
O_3[a]	$O_3 + 3I^- + 2H^+ \rightleftharpoons O_2 + I_3^- + H_2O$	O_3 is passed through neutral 2 wt% KI solution. Add H_2SO_4 and titrate.
NO_2^-	$2HNO_2 + 2H^+ + 3I^- \rightleftharpoons 2NO + I_3^- + 2H_2O$	The nitric oxide is removed (by bubbling CO_2 generated in situ) prior to titration of I_3^-.
$S_2O_8^{2-}$	$S_2O_8^{2-} + 3I^- \rightleftharpoons 2SO_4^{2-} + I_3^-$	Reaction in neutral solution. Then acidify and titrate.
Cu^{2+}	$2Cu^{2+} + 5I^- \rightleftharpoons 2CuI(s) + I_3^-$	NH_4HF_2 is used as a buffer.
MnO_4^-	$2MnO_4^- + 16H^+ + 15I^- \rightleftharpoons 2Mn^{2+} + 5I_3^- + 8H_2O$	Reaction in 0.1 M HCl.
MnO_2	$MnO_2(s) + 4H^+ + 3I^- \rightleftharpoons Mn^{2+} + I_3^- + 2H_2O$	Reaction in 0.5 M H_3PO_4 or HCl.

a. The pH must be ≥ 7 when O_3 is added to I^-. In acidic solution, each O_3 produces 1.25 I_3^-, not 1 I_3^-. [N. V. Klassen, D. Marchington, and H. C. E. McGowan, *Anal. Chem.* **1994**, *66*, 2921.]

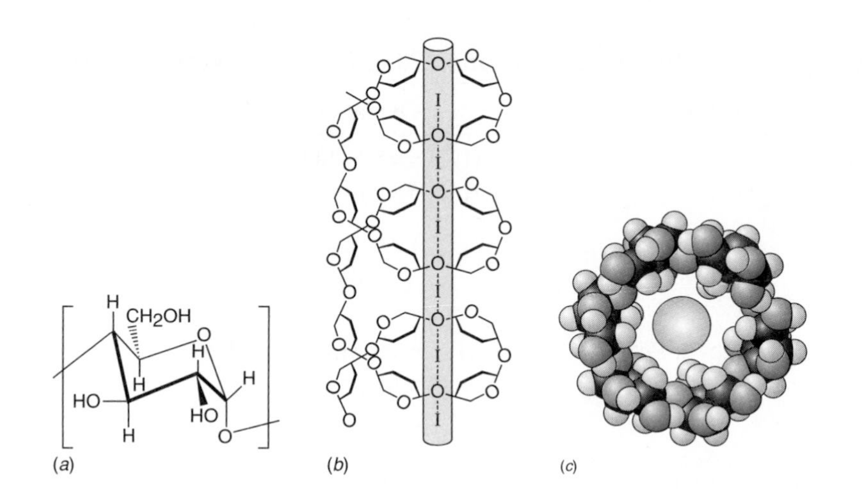

Figure 16-3 (*a*) Structure of the repeating unit of the sugar amylose found in starch. (*b*) In the starch-iodine complex, the sugar chain forms a helix around nearly linear I_6 units. [V. T. Calabrese and A. Khan, *J. Polymer Sci.* **1999**, *A37*, 2711.] (*c*) View down the starch helix. [Drawing from R. D. Hancock, Power Engineering, Salt Lake City.]

mean I_2 plus excess I^-. A mole of I_2 is equivalent to a mole of I_3^- through Reaction 16-18.

Starch Indicator

Starch is the indicator of choice for iodine because it forms an intense blue complex with iodine. The active fraction of starch is amylose, a polymer of the sugar α-D-glucose (Figure 16-3). The polymer coils into a helix, inside of which chains of I_6 (made from $3I_2$) form an intense blue color. In a solution with no other colored species, it is possible to see the color of $\sim 5 \times 10^{-6}$ M I_3^-. With starch, the limit of detection is extended by a factor of 10.

Starch is biodegradable, so either it should be freshly dissolved or the solution should contain a preservative, such as HgI_2 or thymol. A hydrolysis product of starch is glucose, which is a reducing agent. Partly hydrolyzed starch could be a source of error in a redox titration.

An alternative to using starch is to add a few milliliters of *p*-xylene to the vigorously stirred titration vessel. After each addition of reagent near the end point, stop stirring long enough to examine the color of the xylene. I_2 is 400 times more soluble in xylene than it is in water, and its color is readily detected in the xylene.

In iodi**m**etry (titration *with* I_3^-), starch can be added at the beginning of the titration. The first drop of excess I_3^- after the equivalence point causes the solution to turn dark blue. In iodo**m**etry (titration *of* I_3^-), I_3^- is present throughout the reaction up to the equivalence point. *Starch should not be added until immediately before the equivalence point*, as detected visually, by fading of the I_3^- (Color Plate 12). Otherwise some iodine tends to remain bound to starch particles after the equivalence point has been reached.

Preparation and Standardization of I_3^- Solutions

There is a significant vapor pressure of toxic I_2 above solid I_2 and aqueous I_3^-. Vessels containing I_2 or I_3^- should be sealed and kept in a fume hood. Waste solutions of I_3^- should not be dumped into a sink in the open lab.

Triiodide (I_3^-) is prepared by dissolving solid I_2 in excess KI. I_2 is seldom used as a primary standard because some sublimes (evaporates) during weighing. Instead, an approximate amount is rapidly weighed, and the solution of I_3^- is standardized with a pure sample of the intended analyte or with As_4O_6 or $Na_2S_2O_3$.

Acidic solutions of I_3^- are unstable because the excess I^- is slowly oxidized by air:

$$6I^- + O_2 + 4H^+ \longrightarrow 2I_3^- + 2H_2O \qquad (16\text{-}19)$$

Disproportionation means that an element in one oxidation state changes to the same element in both higher and lower oxidation states.

At neutral pH, oxidation is insignificant in the absence of heat, light, and metal ions. Above pH 11, iodine disproportionates to hypoiodous acid (HOI), iodate (IO_3^-), and iodide.

Box 16-2 Disinfecting Drinking Water with Iodine

Many hikers use iodine to disinfect water from streams and lakes to make it safe to drink. Iodine is more effective than filter pumps, which remove bacteria but not viruses, because viruses are small enough to pass through the filter. Iodine kills everything in the water.

When I hike, I carry a 60-mL glass bottle of water containing a few large crystals of solid iodine and a Teflon-lined cap. The crystals keep the solution saturated with I_2. I keep the bottle inside two layers of plastic bags to prevent I_2 vapor from attacking everything in my backpack.

I use the cap to measure out liquid from this bottle and add it to a 1-L bottle of water from a stream or lake. The required volume of saturated aqueous I_2 is shown in the table. For example, I use 4 caps of iodine solution when the air temperature is near 20°C to deliver approximately 13 mL of disinfectant to my 1-L water bottle. It is important to use just the supernatant liquid, not the crystals of solid iodine, because too much iodine is harmful to humans. After allowing 30 min for the iodine to kill any critters, the water is safe to drink. Each time I use I_2 solution, I refill the small bottle with water so that saturated aqueous I_2 is available at the next water stop.

Recipe for disinfecting drinking water

Temperature of saturated $I_2(aq)$	Volume to add to 1 L
3°C (37°F)	20 mL
20°C (68°F)	13 mL
25°C (77°F)	12.5 mL
40°C (104°F)	10 mL

Vitamin C, a reducing agent present in many foods, reacts rapidly with I_2:

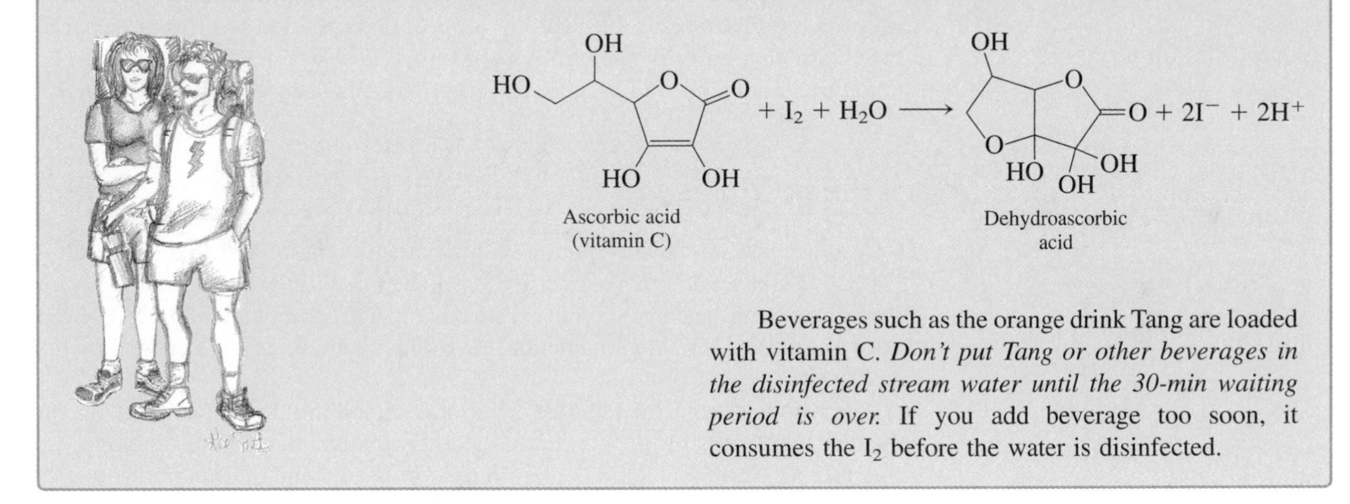

Beverages such as the orange drink Tang are loaded with vitamin C. *Don't put Tang or other beverages in the disinfected stream water until the 30-min waiting period is over.* If you add beverage too soon, it consumes the I_2 before the water is disinfected.

Standard I_3^- is made by adding a weighed quantity of pure potassium iodate to a small excess of KI. Addition of excess strong acid (to give pH ≈ 1) produces I_3^-:

Preparing standard I_3^-:

$$\underset{\text{Iodate}}{IO_3^-} + 8I^- + 6H^+ \rightleftharpoons 3I_3^- + 3H_2O \tag{16-20}$$

KIO_3 is a primary standard for the generation of I_3^-.

Freshly acidified iodate plus iodide can be used to standardize thiosulfate. The I_3^- reagent must be used immediately, because it is soon oxidized by air. The only disadvantage of KIO_3 is its low formula mass relative to the number of electrons it accepts. The small quantity of KIO_3 leads to a larger-than-desirable relative weighing error in preparing solutions.

Use of Sodium Thiosulfate

Sodium thiosulfate is the almost universal titrant for iodine. At pH < 9, iodine oxidizes thiosulfate cleanly to tetrathionate:

One mole of I_3^- in Reaction 16-21 is equivalent to one mole of I_2. I_2 and I_3^- are interchangeable through the equilibrium $I_2 + I^- \rightleftharpoons I_3^-$.

$$\underset{\text{Thiosulfate}}{I_3^- + 2S_2O_3^{2-}} \rightleftharpoons 3I^- + \underset{\text{Tetrathionate}}{O{=}\overset{\overset{O}{\|}}{\underset{\underset{O^-}{|}}{S}}{-}S{-}S{-}\overset{\overset{O}{\|}}{\underset{\underset{O^-}{|}}{S}}{=}O} \qquad (16\text{-}21)$$

The common form of thiosulfate, $Na_2S_2O_3 \cdot 5H_2O$, is not pure enough to be a primary standard. Instead, thiosulfate is standardized by reaction with a fresh solution of I_3^- prepared from KIO_3 plus KI.

A stable solution of $Na_2S_2O_3$ is prepared by dissolving the reagent in high-quality, freshly boiled distilled water. Dissolved CO_2 promotes disproportionation of $S_2O_3^{2-}$:

$$S_2O_3^{2-} + H^+ \rightleftharpoons \underset{\text{Bisulfite}}{HSO_3^-} + \underset{\text{Sulfur}}{S(s)} \qquad (16\text{-}22)$$

and metal ions catalyze atmospheric oxidation of thiosulfate. Thiosulfate solutions are stored in the dark with 0.1 g of Na_2CO_3 per liter to maintain optimum pH. Three drops of chloroform should be added to a thiosulfate solution to prevent bacterial growth. Although an acidic solution of thiosulfate is unstable, the reagent can be used to titrate iodine in acid because Reaction 16-21 is faster than Reaction 16-22.

Ask Yourself

16-C. **(a)** Potassium iodate solution was prepared by dissolving 1.022 g of KIO_3 (FM 214.00) in a 500-mL volumetric flask. Then 50.00 mL of the solution were pipetted into a flask and treated with excess KI (2 g) and acid (10 mL of 0.5 M H_2SO_4) to drive Reaction 16-20 to completion. How many moles of I_3^- are created by the reaction?
(b) The triiodide from **(a)** required 37.66 mL of sodium thiosulfate solution for Reaction 16-21. What is the concentration of the sodium thiosulfate solution?
(c) A 1.223-g sample of solid containing ascorbic acid and inert ingredients was dissolved in dilute H_2SO_4 and treated with 2 g of KI and 50.00 mL of potassium iodate solution from **(a)**. After Reaction 16-20 and the reaction with ascorbic acid in Box 16-2 went to completion, the excess, unreacted triiodide required 14.22 mL of sodium thiosulfate solution from **(b)** for complete titration. Find the moles of ascorbic acid and weight percent of ascorbic acid (FM 176.13) in the unknown.

Key Equations

Redox titration potential calculations	Cell voltage $= E_+ - E$(reference electrode) where E_+ is the indicator electrode potential
	Prior to equivalence point: Analyte is in excess; use analyte Nernst equation to find indicator electrode potential

At equivalence point: Add analyte and titrant Nernst equations (with equal numbers of electrons) and use stoichiometry to cancel many terms; if necessary, use known concentrations to evaluate log term

Past equivalence point: Titrant is in excess; use titrant Nernst equation to find indicator electrode potential

Redox indicator color change range

$$E = \left(E^\circ \pm \frac{0.059\ 16}{n}\right) \text{ volts}$$

n = number of electrons in indicator half-reaction

Important Terms

redox indicator | redox titration

Problems

16-1. Find E° and K for the titration reaction 16-1 in 1 F $HClO_4$ at 25°C.

16-2. Consider the titration of Fe^{2+} with Ce^{4+} in Figure 16-2.

(a) Write a balanced titration reaction.

(b) Write two half-reactions for the indicator electrode.

(c) Write two Nernst equations for the cell voltage.

(d) Calculate E at the following volumes of Ce^{4+}: 10.0, 25.0, 49.0, 50.0, 51.0, 60.0, and 100.0 mL. Compare your results with Figure 16-2.

16-3. Consider the titration of 100.0 mL of 0.010 0 M Ce^{4+} in 1 M $HClO_4$ by 0.040 0 M Cu^+ to give Ce^{3+} and Cu^{2+}, using Pt and saturated Ag | AgCl electrodes.

(a) Write a balanced titration reaction.

(b) Write two half-reactions for the indicator electrode.

(c) Write two Nernst equations for the cell voltage.

(d) Calculate E at the following volumes of Cu^+: 1.00, 12.5, 24.5, 25.0, 25.5, 30.0, and 50.0 mL. Sketch the titration curve.

(e) Select a suitable indicator for this titration from Table 16-1.

16-4. Consider the titration of 25.0 mL of 0.010 0 M Sn^{2+} by 0.050 0 M Tl^{3+} in 1 M HCl, using Pt and saturated calomel electrodes.

(a) Write a balanced titration reaction.

(b) Write two half-reactions for the indicator electrode.

(c) Write two Nernst equations for the cell voltage.

(d Calculate E at the following volumes of Tl^{3+}: 1.00, 2.50, 4.90, 5.00, 5.10, and 10.0 mL. Sketch the titration curve.

(e) Select a suitable indicator for this titration from Table 16-1.

16-5. Compute the titration curve for Demonstration 16-1, in which 400.0 mL of 3.75 mM Fe^{2+} are titrated with 20.0 mM MnO_4^- at a *fixed pH* of 0.00 in 1 M H_2SO_4. Calculate the cell voltage at titrant volumes of 1.0, 7.5, 14.0, 15.0, 16.0, and 30.0 mL and sketch the titration curve.

16-6. Consider the titration of 25.0 mL of 0.050 0 M Sn^{2+} with 0.100 M Fe^{3+} in 1 M HCl to give Fe^{2+} and Sn^{4+}, using Pt and saturated calomel electrodes.

(a) Write a balanced titration reaction.

(b) Write two half-reactions for the indicator electrode.

(c) Write two Nernst equations for the cell voltage.

(d) Calculate E at the following volumes of Fe^{3+}: 1.0, 12.5, 24.0, 25.0, 26.0, and 30.0 mL. Sketch the titration curve.

16-7. Ascorbic acid (0.010 0 M) (structure in Box 16-2) was added to 10.0 mL of 0.020 0 M Fe^{3+} in a solution buffered to pH 0.30, and the potential was monitored with Pt and saturated Ag | AgCl electrodes.

$$\text{dehydroascorbic acid} + 2H^+ + 2e^- \rightleftharpoons \text{ascorbic acid} + H_2O \quad E^\circ = 0.390 \text{ V}$$

(a) Write a balanced equation for the titration reaction.

(b) Using $E^\circ = 0.767$ V for the $Fe^{3+} \mid Fe^{2+}$ couple, calculate the cell voltage when 5.0, 10.0, and 15.0 mL of ascorbic acid have been added. (*Hint:* Whenever $[H^+]$ appears in a Nernst equation, use the numerical value $10^{-pH} = 10^{-0.30}$.)

16-8. Select indicators from Table 16-1 that would be suitable for finding the two end points in Figure 16-2. What color changes would be observed?

16-9. Would tris(2,2′-bipyridine)iron be a useful indicator for the titration of Sn^{2+} in 1 M HCl with $Mn(EDTA)^-$? (*Hint:* The potential at the equivalence point must be between the potentials for each redox couple.)

16-10. Why is iodine almost always used in a solution containing excess I^-?

16-11. Ozone (O_3) is a colorless gas with a pungent odor. It can be generated by passing a high-voltage electric spark through air. O_3 can be analyzed by its stoichiometric reaction with I^- in neutral solution:

$$O_3 + 3I^- + H_2O \longrightarrow O_2 + I_3^- + 2OH^-$$

(The reaction must be carried out in neutral solution. In acidic solution, more I_3^- is made than the preceding reaction indicates.) A 1.00-L bulb of air containing O_3 produced by an electric spark was treated with 25 mL of 2 M KI, shaken well, and left closed for 30 min so that all O_3 would react. The aqueous solution was then drained from the bulb, acidified with 2 mL of 1 M H_2SO_4, and required 29.33 mL of 0.050 44 M $S_2O_3^{2-}$ for titration of the I_3^-.

(a) What color would you expect the KI solution to be before and after reaction with O_3?

(b) Calculate the mass of O_3 in the 1.00-L bulb.

(c) Does it matter whether starch indicator is added at the beginning or near the end point in this titration? Why?

16-12. The Kjeldahl analysis in Section 10-6 is used to measure the nitrogen content of organic compounds, which are digested in boiling sulfuric acid to decompose to ammonia, which, in turn, is distilled into standard acid. The remaining acid is then back-titrated with base. Kjeldahl himself had difficulty discerning by lamplight in 1883 the methyl red indicator end point in the back titration. He could have refrained from working at night, but instead he chose to complete the analysis differently. After distilling the ammonia into standard sulfuric acid, he added a mixture of KIO_3 and KI to the acid. The liberated iodine was then titrated with thiosulfate, with starch for easy end point detection—even by lamplight. Explain how the thiosulfate titration is related to the nitrogen content of the unknown. Derive a relation between moles of NH_3 liberated in the digestion and moles of thiosulfate required for titration of iodine.

16-13. Sulfite (SO_3^{2-}) is added to many foods as a preservative. Some people have an allergic reaction to sulfite, so it is important to control the level of sulfite. Sulfite in wine was measured by the following procedure: To 50.0 mL of wine were added 5.00 mL of solution containing (0.804 3 g KIO_3 + 5 g KI)/100 mL. Acidification with 1.0 mL of 6.0 M H_2SO_4 quantitatively converted IO_3^- into I_3^- by Reaction 16-20. The I_3^- reacted with sulfite to generate sulfate, leaving excess I_3^- in solution. The excess I_3^- required 12.86 mL of 0.048 18 M $Na_2S_2O_3$ to reach a starch end point.

(a) Write the reaction that takes place when H_2SO_4 is added to KIO_3 + KI and explain why 5 g of KI were added to the stock solution. Is it necessary to measure out 5 g very accurately? Is it necessary to measure 1.0 mL of H_2SO_4 very accurately?

(b) Write a balanced reaction between I_3^- and sulfite.

(c) Find the concentration of sulfite in the wine. Express your answer in moles per liter and in milligrams of SO_3^{2-} per liter.

(d) *t test.* Another wine was found to contain 277.7 mg of SO_3^{2-}/L with a standard deviation of ±2.2 mg/L for three determinations by the iodimetric method. A spectrophotometric method gave 273.2 ± 2.1 mg/L in three determinations. Are these results significantly different at the 95% confidence level?

16-14. From the following reduction potentials,

$$I_2(s) + 2e^- \rightleftharpoons 2I^- \qquad E° = 0.535 \text{ V}$$

$$I_2(aq) + 2e^- \rightleftharpoons 2I^- \qquad E° = 0.620\text{V}$$

$$I_3^- + 2e^- \rightleftharpoons 3I^- \qquad E° = 0.535 \text{ V}$$

(a) Calculate the equilibrium constant for the reaction $I_2(aq) + I^- \rightleftharpoons I_3^-$.

(b) Calculate the equilibrium constant for the reaction $I_2(s) + I^- \rightleftharpoons I_3^-$.

(c) Calculate the solubility (g/L) of I_2 in water.

How Would You Do It?

16-15. Ozone (O_3) in smog is formed by the action of solar ultraviolet light on organic vapors plus nitric oxide (NO) in the air. An O_3 level of 100 to 200 ppb (nL per liter of air) for 1 h creates a "1st stage smog alert" and is considered unhealthful. A level above 200 ppb, which defines a "2nd stage smog alert," is very unhealthful.

(a) The ideal gas law tells us that $PV = nRT$, where P is pressure (bar), V is volume (L), n is moles, R is the gas constant (0.083 14 L · bar/(mol · K)), and T is temperature (K). If the pressure of air in a flask is 1 bar, the partial pressure of a 1-ppb component is 10^{-9} bar. Find the number of moles of O_3 in a liter of air if the concentration of O_3 is 200 ppb and the temperature is 300 K.

(b) Is it feasible to use the iodometric procedure of Problem 16-11 to measure O_3 at a level of 200 ppb in smog? State your reason.

16-16. *Idiometric analysis of a superconductor.* An analysis was carried out to find the effective copper oxidation state, and therefore the number of oxygen atoms, in the superconductor $YBa_2Cu_3O_{7-z}$, where z ranges from 0 to 0.5. Common oxidation states of yttrium and barium are Y^{3+} and Ba^{2+}, and common states of copper are Cu^{2+} and Cu^{+}. If copper were Cu^{2+}, the formula of the superconductor would be $(Y^{3+})(Ba^{2+})_2(Cu^{2+})_3(O^{2-})_{6.5}$, with a cation charge of +13 and an anion charge of −13. The composition $YBa_2Cu_3O_7$ formally requires Cu^{3+}, which is rather rare. $YBa_2Cu_3O_7$ can be thought of as $(Y^{3+})(Ba^{2+})_2(Cu^{2+})_2(Cu^{3+})(O^{2-})_7$, with a cation charge of +14 and an anion charge of −14.

An iodometric analysis of $YBa_2Cu_3O_x$ entails two experiments. In *Experiment 1*, $YBa_2Cu_3O_x$ is dissolved in dilute acid, in which Cu^{3+} is converted into Cu^{2+}. For simplicity, we write the equations for the formula $YBa_2Cu_3O_7$, but you could balance these equations for $x \neq 7$:

$$YBa_2Cu_3O_7 + 13H^+ \longrightarrow Y^{3+} + 2Ba^{2+} + 3Cu^{2+} + \tfrac{13}{2}H_2O + \tfrac{1}{4}O_2 \qquad \text{(A)}$$

The total copper content is measured by treatment with iodide

$$3Cu^{2+} + \tfrac{15}{2}I^- \longrightarrow 3CuI(s) + \tfrac{3}{2}I_3^- \qquad \text{(B)}$$

followed by titration of the liberated I_3^- with standard thiosulfate (Reaction 16-21). Each mole of Cu in $YBa_2Cu_3O_7$ is equivalent to 1 mol of $S_2O_3^{2-}$ in Experiment 1.

In *Experiment 2*, $YBa_2Cu_3O_x$ is dissolved in dilute acid containing I^-. Each mole of Cu^{3+} produces 1 mol of I_3^-, and each mole of Cu^{2+} produces 0.5 mol of I_3^-:

$$Cu^{3+} + 4I^- \longrightarrow CuI(s) + I_3^- \qquad \text{(C)}$$

$$Cu^{2+} + \tfrac{5}{2}I^- \longrightarrow CuI(s) + \tfrac{1}{2}I_3^- \qquad \text{(D)}$$

The moles of thiosulfate required in Experiment 1 equal the total moles of Cu in the superconductor. The difference in thiosulfate required between Experiments 2 and 1 gives the Cu^{3+} content.

(a) In Experiment 1, 1.00 g of superconductor required 4.55 mmol of $S_2O_3^{2-}$. In Experiment 2, 1.00 g of superconductor required 5.68 mmol of $S_2O_3^{2-}$. What is the value of z in the formula $YBa_2Cu_3O_{7-z}$ (FM 666.246 − 15.999 4z)?

(b) *Propagation of uncertainty.* In several replications of Experiment 1, the thiosulfate required was 4.55 (±0.10) mmol of $S_2O_3^{2-}$ per gram of $YBa_2Cu_3O_{7-z}$. In Experiment 2, the thiosulfate required was 5.68 (±0.05) mmol of $S_2O_3^{2-}$ per gram. Find the uncertainty of x in the formula $YBa_2Cu_3O_x$.

Notes and Reference

1. Biochemical oxygen demand (BOD) and chemical oxygen demand (COD) procedures are described in *Standard Methods for the Examination of Wastewater*, 21st ed. (Washington, DC: American Public Health Association, 2005), which is the standard reference for water analysis.

A Biosensor for Personal Glucose Monitoring

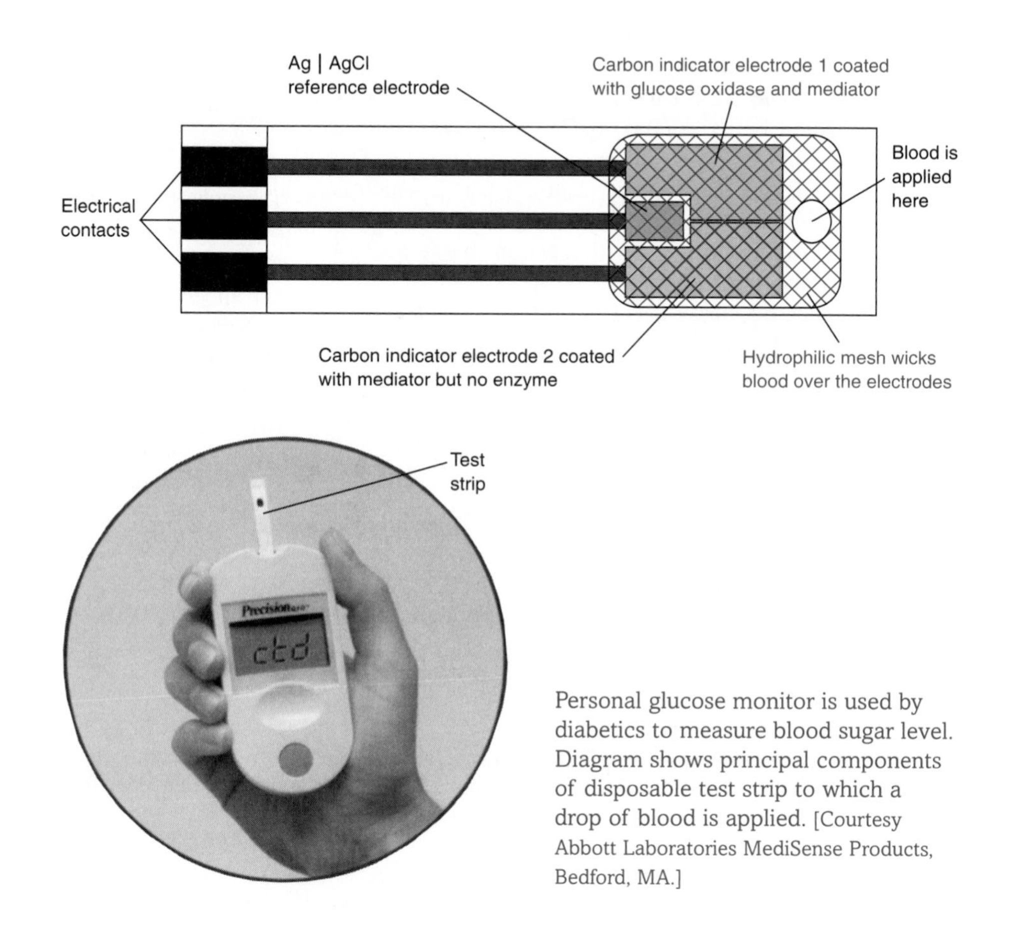

Personal glucose monitor is used by diabetics to measure blood sugar level. Diagram shows principal components of disposable test strip to which a drop of blood is applied. [Courtesy Abbott Laboratories MediSense Products, Bedford, MA.]

A *biosensor* is an analytical device that uses a biological component such as an enzyme, an antibody, or even whole cells for specific sensing of one substance. Many people with diabetes must monitor their blood sugar (glucose) levels several times a day to control the disease through diet and insulin injections. The photograph shows a home glucose monitor featuring a disposable test strip to which as little as 4 μL of blood is applied for each measurement. This biosensor uses the enzyme glucose oxidase to catalyze the oxidation of glucose. The electrodes measure an oxidation product. Section 17-2 explains how the sensor works. The market for glucose sensors is more then $3 billion per year.

CHAPTER 17

Instrumental Methods in Electrochemistry

We now introduce a variety of electrochemical methods used in chemical analysis. These techniques are used in applications such as home glucose monitors, quality control in food processing, and chromatography detectors.

17-1 Electrogravimetric and Coulometric Analysis

Electrolysis is a chemical reaction in which we apply a voltage to drive a redox reaction that would not otherwise occur. An **electroactive species** is one that can be oxidized or reduced at an electrode.

Electrogravimetric Analysis

One of the oldest electrolytic methods in quantitative analysis is **electrogravimetric analysis**, in which the analyte is plated out on an electrode and weighed. For example, an excellent procedure for the measurement of copper is to pass a current through a solution of a copper salt to deposit all of the copper on the cathode:

$$Cu^{2+}(aq) + 2e^- \longrightarrow Cu(s, \text{deposited on the cathode}) \qquad (17\text{-}1)$$

The increase in mass of the cathode tells us how much copper was present in the solution.

Figure 17-1 shows how this experiment might be done. Analyte is typically deposited on a carefully cleaned, chemically inert Pt gauze cathode with a large surface area.

How do you find out when electrolysis is complete? One way is to observe the disappearance of color in a solution from which a colored species such as Cu^{2+} is removed. Another way is to expose most, but not all, of the surface of the cathode to the solution during electrolysis. To test whether the reaction is complete, raise the beaker or add water so that fresh surface of the cathode is exposed to the solution. After an additional period of electrolysis (15 min, say), see whether the newly

Tests for completion of the deposition:

1. Disappearance of color
2. Deposition on freshly exposed electrode surface
3. Qualitative test for analyte in solution

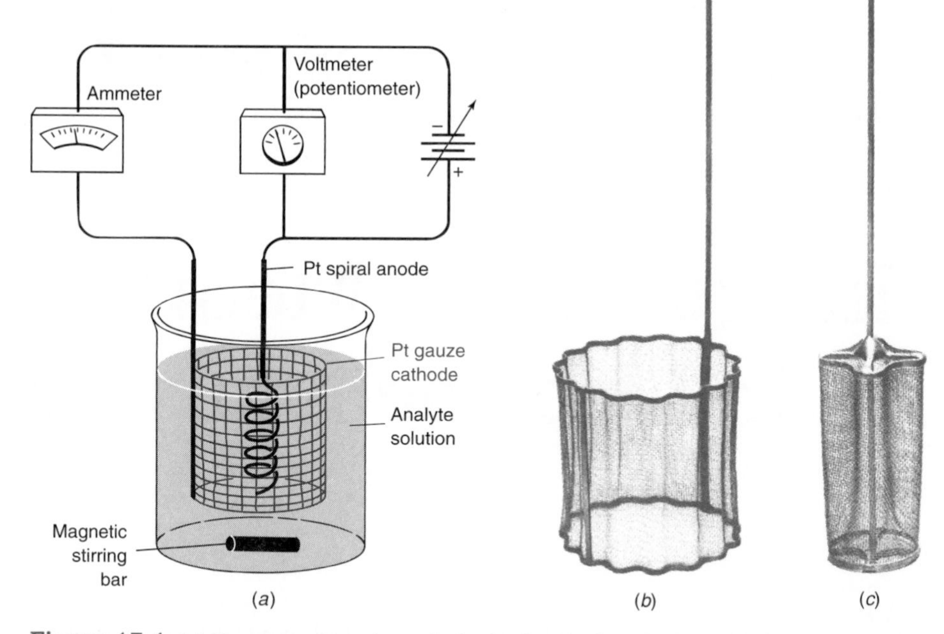

Figure 17-1 (*a*) Electrogravimetric analysis. Analyte is deposited on the large Pt gauze electrode. If analyte is to be oxidized, rather than reduced, the polarity of the power supply is reversed so that deposition is always on the large electrode. (*b*) Outer Pt gauze electrode. (*c*) Optional inner Pt gauze electrode designed to be spun by a motor in place of magnetic stirring.

exposed electrode surface has a deposit. If it does, repeat the procedure. If not, the electrolysis is finished. A third method is to remove a small sample of solution and perform a qualitative test for analyte.

Electrogravimetric analysis would be simple if there were only a single analyte in an otherwise inert solution. In practice, there may be other electroactive species that interfere. Water decomposes to H_2 at the cathode and to O_2 at the anode at sufficiently high voltage. Gas bubbles at an electrode interfere with deposition of solids. Because of these complications, control of electrode potential is important for successful analysis.

Coulometric Analysis

Review of Section 14-1:

Electric charge is measured in *coulombs* (C).

Electric current (charge per unit time) is measured in *amperes* (A).

$$1\ \text{A} = 1\ \text{C/s}$$

Faraday constant relates coulombs to moles:

$$F \approx 96\,485\ \text{C/mol}$$

$$\underset{\text{Coulombs}}{q} = \underset{\text{Moles of electrons}}{n} \cdot \underset{\text{C/mol}}{F}$$

In **coulometry,** electrons participating in a chemical reaction are counted to learn how much analyte reacted. For example, hydrogen sulfide (H_2S) can be measured by its reaction with I_2 generated at an anode:

Anode generates I_2: $$2I^- \longrightarrow I_2 + 2e^- \qquad (17\text{-}2)$$

Reaction in solution: $$I_2 + H_2S \longrightarrow S(s) + 2H^+ + 2I^- \qquad (17\text{-}3)$$

We measure the electric current and the time required to generate enough I_2 in Reaction 17-2 to reach the equivalence point of Reaction 17-3. From the current and time, we calculate how many electrons participated in Reaction 17-2 and therefore how many moles of H_2S took part in Reaction 17-3. A way to find the end point in this example would be to have some starch in the solution. As long as I_2 is consumed rapidly by H_2S, the solution remains colorless. After the equivalence point, the solution turns blue because excess I_2 accumulates.

Example Coulometry

Find the moles of H_2S in an unknown if the end point in Reaction 17-3 came after a current of 0.058 2 A flowed for 184 s in Reaction 17-2.

SOLUTION The quantity of charge in Reaction 17-2 was (0.058 2 C/s)(184 s) = 10.7_1 C. We use the Faraday constant to convert coulombs into moles of electrons:

$$n = \text{mol e}^- = \frac{q}{\text{F}} = \frac{10.7_1\ \text{C}}{96\,485\ \text{C/mol}} = 1.11_0 \times 10^{-4}\ \text{mol}$$

Because 2 electrons in Reaction 17-2 correspond to 1 mol of H_2S in Reaction 17-3, there must have been $\frac{1}{2}(1.11_0 \times 10^{-4}\ \text{mol}) = 5.55 \times 10^{-5}$ mol H_2S in the unknown.

Test Yourself How long would it take to titrate 1.00 mmol H_2S at a constant current of 100.0 mA? (**Answer:** 1.94×10^3 s)

Ask Yourself

17-A. The apparatus in Figure 17-2 generates the powerful oxidant Fe(VI) as FeO_4^{2-}, which can oxidize hazardous species in wastewater. For example, sulfide (S^{2-}) is converted to thiosulfate ($S_2O_3^{2-}$), cyanide (CN^-) is converted to cyanate (CNO^-), and arsenite (AsO_2^-) is converted to arsenate (AsO_4^{3-}).
(a) Write a balanced half-reaction for the Fe anode in basic solution.
(b) Write a balanced reaction for $FeO_4^{2-} + S^{2-} \longrightarrow Fe(OH)_3(s) + S_2O_3^{2-}$.
(c) How many moles of S^{2-} can be removed from the wastewater if a current of 16.0 A is applied for 1.00 h?
(d) What volume of wastewater containing 10.0 mM S^{2-} can be purified in 1.00 h?

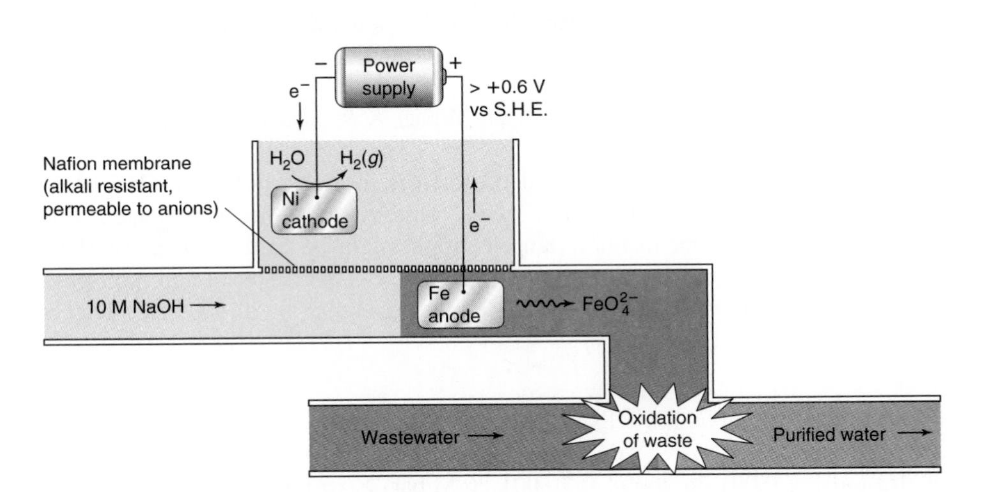

Figure 17-2 Oxidative purification of wastewater with electrochemically generated Fe(VI). [From S. Licht and X. Yu, *Environ. Sci. Technol.* **2005**, *39*, 8071.]

In *amperometry*, we measure an electric current that is proportional to the concentration of a species in solution.

In *coulometry*, we measure the total number of electrons (= current × time) that flow during a chemical reaction.

17-2 Amperometry

In **amperometry**, we measure the electric current between a pair of electrodes that are driving an electrolysis reaction. One of the reactants is the intended analyte, and the measured current is proportional to the concentration of analyte.

An important amperometric method is the measurement of dissolved O_2 with the **Clark electrode**,[1] such as that in Figure 17-3 used to measure O_2 in marine sediment in Box 6-1. The glass body is drawn to a fine point with a 5-μm opening at the base. Inside the opening is a 10- to 40-μm-long plug of silicone rubber, which is permeable to O_2. Oxygen diffuses into the electrode through the rubber and is reduced at the Au tip on the Pt wire, which is held at −0.75 V with respect to the Ag|AgCl reference electrode:

$$\text{Pt | Au cathode:} \quad O_2 + 4H^+ + 4e^- \longrightarrow 2H_2O \qquad (17\text{-}4)$$
$$\text{Ag | AgCl anode:} \quad 4Ag + 4Cl^- \longrightarrow 4AgCl + 4e^-$$

A Clark electrode is calibrated by placing it in solutions of known O_2 concentration, and a graph of current versus $[O_2]$ is constructed. The electrode in Figure 17-3 also contains a silver *guard electrode* extending most of the way to the bottom. The guard electrode is kept at a negative potential such that any O_2 diffusing in from the top of the electrode is reduced, and does not interfere with measurement of O_2 diffusing in through the silicone membrane at the bottom.

A Clark electrode can fit into the tip of a surgical catheter to measure O_2 in the umbilical artery of a newborn child to detect respiratory distress. The sensor responds within 20–50 s to administration of O_2 for breathing or to mechanical ventilation of the lungs.

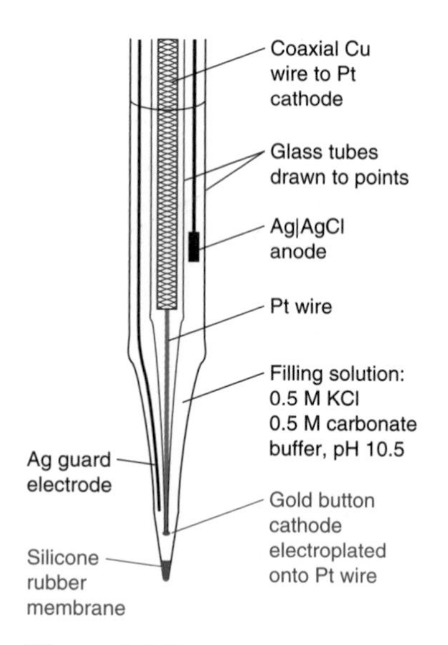

Figure 17-3 Clark oxygen microelectrode used to measure dissolved O_2 in marine sediment in Box 6-1. The tip of the cathode is plated with Au, which is less prone than Pt to fouling by adsorption of species from the test solution. [Adapted From N. P. Revsbech, *Limnol. Oceanogr.* **1989**, *34*, 474.]

Example A Digression on Henry's Law

Henry's law is the observation that, in dilute solutions, the concentration of a gaseous species dissolved in the liquid is proportional to the pressure of that species in the gas phase. For oxygen dissolved in water at 25°C, Henry's law takes the form

$$[O_2(aq)] = (0.001\ 26\ \text{M/bar}) \times P_{O_2}\ (\text{bar})$$

where P_{O_2} is expressed in bars and $[O_2(aq)]$ is in moles per liter. Clark electrodes are often calibrated in terms of P_{O_2} rather than $[O_2(aq)]$, because P_{O_2} is easier to measure. For example, an electrode might be calibrated in solutions bubbled with pure nitrogen ($P_{O_2} = 0$), dry air ($P_{O_2} \approx 0.21$ bar), and pure oxygen ($P_{O_2} \approx 1.0$ bar). If a Clark electrode gives a reading of "0.100 bar," what is the molarity of $O_2(aq)$?

SOLUTION Henry's law tells us that

$$[O_2(aq)] = 0.001\ 26 \times P_{O_2} = (0.001\ 26\ \text{M/bar}) \times (0.100\ \text{bar}) = 0.126\ \text{mM}$$

Test Yourself What is the molarity of O_2 in water saturated with air? (**Answer:** 0.26 mM)

Glucose Monitors

The blood glucose sensor at the opening of this chapter is probably the most widely used **biosensor**—a device that uses a biological component such as an *enzyme* or *antibody* for highly selective response to one analyte. Glucose monitors account for over 95% of all amperometic instruments sold each year. The disposable test strip shown at the opening of the chapter has two carbon indicator electrodes and a Ag | AgCl reference electrode. As little as 4 μL of blood applied in the circular opening at the right of the figure is wicked over all three electrodes by a thin *hydrophilic* ("water-loving") mesh. A 20-s measurement begins when liquid reaches the reference electrode.

Enzyme: A protein that catalyzes a biochemical reaction. The enzyme increases the rate of reaction by many orders of magnitude.

Antibody: A protein that binds to a specific target molecule called an *antigen*. Foreign cells that infect your body are marked by antibodies and destroyed by *lysis* (bursting them open with fluid) or gobbled up by macrophage cells.

Indicator electrode 1 is coated with the enzyme glucose oxidase and a *mediator*, described below. The enzyme is a protein that catalyzes the reaction of glucose with oxygen:

Reaction in coating above indicator electrode 1:

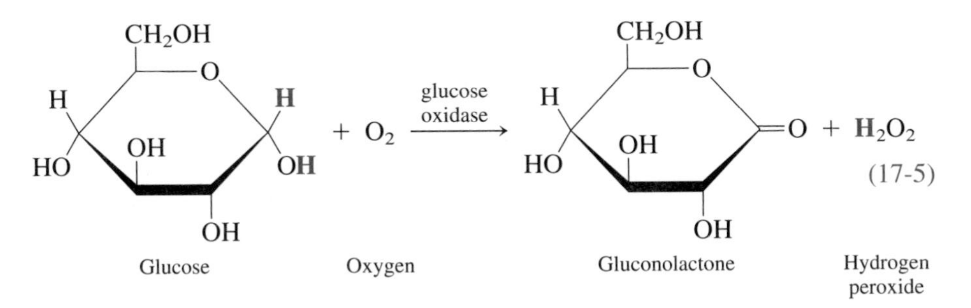

In the absence of enzyme, the rate of Reaction 17-5 is negligible.

Early glucose monitors measured H_2O_2 from Reaction 17-5 by oxidation at a single indicator electrode, which was held at +0.6 V versus Ag | AgCl:

Reaction at indicator electrode 1: $$H_2O_2 \longrightarrow O_2 + 2H^+ + 2e^- \quad (17\text{-}6)$$

The current is proportional to the concentration of H_2O_2, which, in turn, is proportional to the glucose concentration in blood (Figure 17-4).

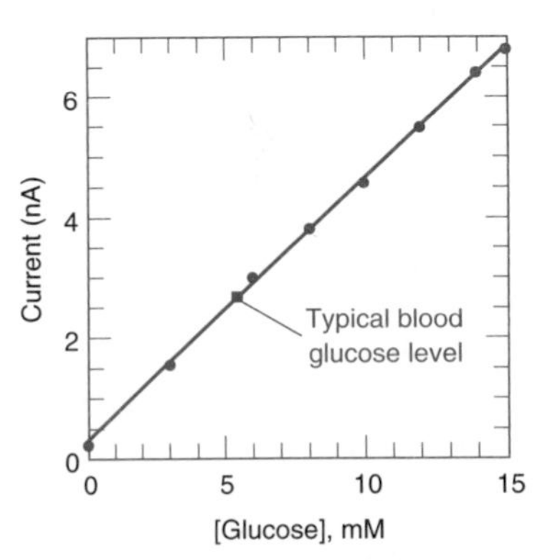

Figure 17-4 Response of an amperometric glucose electrode when the dissolved O_2 concentration corresponds to P_{O_2} = 0.027 bar, which is 20% lower than the typical concentration in subcutaneous tissue. [Data from S.-K. Jung and G. W. Wilson, *Anal. Chem.* **1996**, *68*, 591.]

A problem with early glucose monitors is that their response depended on the concentration of O_2 in the enzyme layer, because O_2 participates in Reaction 17-5. If the O_2 concentration was low, the monitor responded as though the glucose concentration were low.

A good way to reduce O_2 dependence is to incorporate into the enzyme layer a species that substitutes for O_2 in Reaction 17-5. A substance that transports electrons between the analyte (glucose, in this case) and the electrode is called a **mediator**. Ferricinium salts serve this purpose nicely:

A *mediator* transports electrons between analyte and the working electrode. The mediator undergoes no net reaction itself.

Reaction in coating above indicator electrode 1:

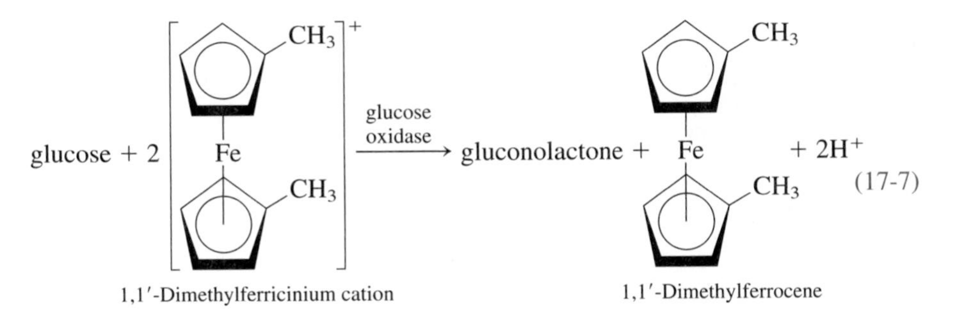

Ferrocene contains flat five-membered aromatic carbon rings, similar to benzene. Each ring formally carries one negative charge, so the oxidation state of iron is +2. The iron atom sits between the two flat rings. Because of its shape, this type of molecule is called a *sandwich complex*.

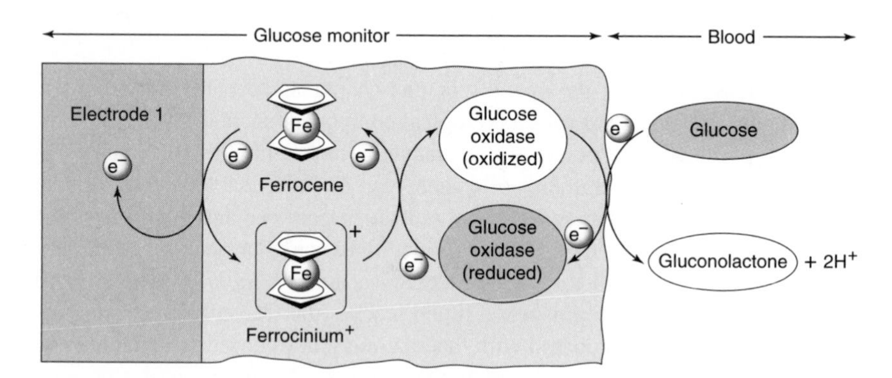

Figure 17-5 Electrons flow from glucose (in blood) to glucose oxidase to the ferricinium ion (coated on the electrode), and finally to electrode 1 of the glucose monitor. Colored species are in their reduced state.

Ferricinium mediator lowers the required working electrode potential from 0.6 V to 0.2 V versus Ag | AgCl, thereby improving the stability of the glucose sensor and eliminating some interference by other species in the blood.

You can build your own glucose biosensor for student experiments.[2]

The mediator consumed in Reaction 17-7 is then regenerated at the indicator electrode:

Reaction at indicator electrode 1:

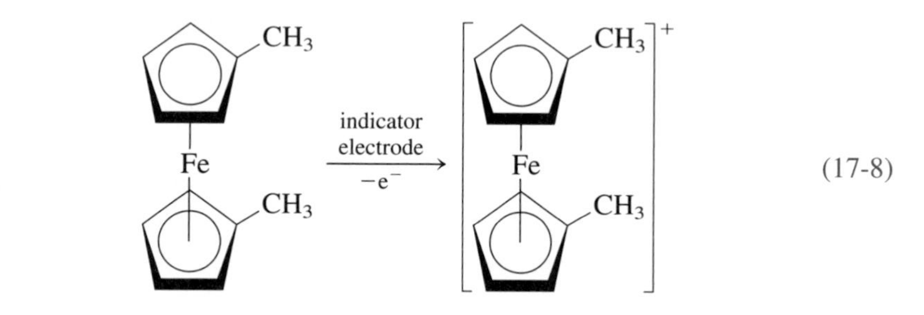

(17-8)

The sequence by which glucose is oxidized and electrons flow to the indicator electrode is shown in Figure 17-5. The current at the electrode is proportional to the concentration of ferrocene, which, in turn, is proportional to the concentration of glucose in the blood.

Another problem with glucose monitors is that other species found in blood can be oxidized at the same potential required to oxidize the mediator in Reaction 17-8. Interfering species include ascorbic acid (vitamin C), uric acid, and acetaminophen (Tylenol). To correct for this interference, the test strip at the opening of this chapter has a second indicator electrode coated with mediator *but not with glucose oxidase.* Interfering species that are reduced at electrode 1 are also reduced at electrode 2. The current due to glucose is the current at electrode 1 minus the current at electrode 2 (both measured with respect to the reference electrode). Now you see why the test strip has three electrodes.

A major challenge is to manufacture glucose monitors in such a reproducible manner that they do not require calibration. A user expects to add a drop of blood to the test strip and get a reliable reading without constructing a calibration curve from known concentrations of glucose in blood. Each lot of test strips must be highly reproducible and calibrated at the factory.

Cells with Three Electrodes

Cells discussed so far are based on two electrodes: an indicator electrode and a reference electrode. Current is measured between the two electrodes. The apparent

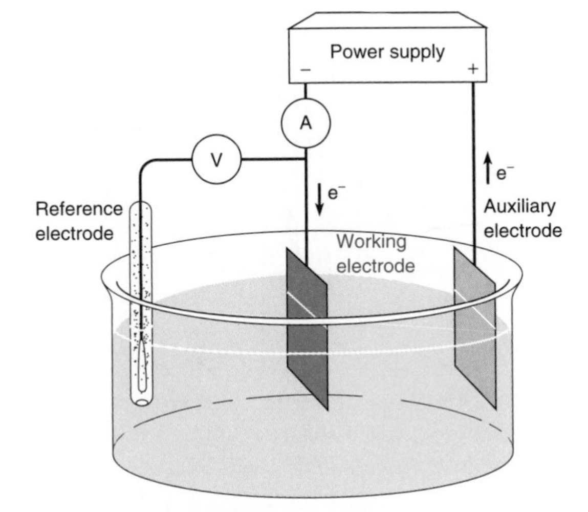

Figure 17-6 Controlled-potential electrolysis with a three-electrode cell. Voltage is measured between the working and reference electrodes. Current is measured between the working and auxiliary electrodes. Negligible current flows through the reference electrode. Ⓥ is a voltmeter (potentiometer) and Ⓐ is an ammeter.

exception—the glucose monitor at the opening of the chapter—has two indicator electrodes.

For many techniques, a cell with three electrodes is required for fine control of the electrochemistry. The cell in Figure 17-6 features a conventional **reference electrode** (such as calomel or silver-silver chloride), a **working electrode** at which the reaction of interest takes place, and an **auxiliary electrode** that is the current-carrying partner of the working electrode. The working electrode is equivalent to the indicator electrode of two-electrode cells. The auxiliary electrode is something new that we have not encountered before. *Current flows between the working and auxiliary electrodes. Voltage is measured between the working and reference electrodes.*

Reference electrode: Provides fixed reference potential with negligible current flow.

Working electrode: Analyte reacts here. Voltage is measured between working and reference electrodes.

Auxiliary electrode: Other half of the electrochemistry occurs here. Current flows between working and auxiliary electrodes.

Potentiostat: Controls potential difference between working and reference electrodes.

The voltage between the working and reference electrodes is controlled by a device called a **potentiostat**. Virtually no current flows through the reference electrode; it simply establishes a fixed reference potential with which to measure the working electrode potential. Current flows between the working and auxiliary electrodes. The potential of the auxiliary electrode varies with time in an uncontrolled manner in response to changing concentrations and current in an electrolysis cell. It is beyond the scope of this text to explain why the electrode potential varies. Suffice it to say, however, that in a two-electrode cell the potential of the working electrode can drift as the reaction proceeds. As it drifts, reactions other than the intended analytical reaction can take place. In a three-electrode cell, the potentiostat maintains the working electrode at the desired potential, while the auxiliary electrode potential drifts out of our control.

Figure 17-6 shows reduction of analyte at the working electrode, which is therefore the cathode in this figure. In other cases, the working electrode could be the anode. The working electrode is an indicator electrode at which analyte reacts.

Amperometric Detector for Chromatography

Figure 0-4 provided an example of chromatography used to separate caffeine from theobromine in a chemical analysis. Absorption of light and electrochemical reactions

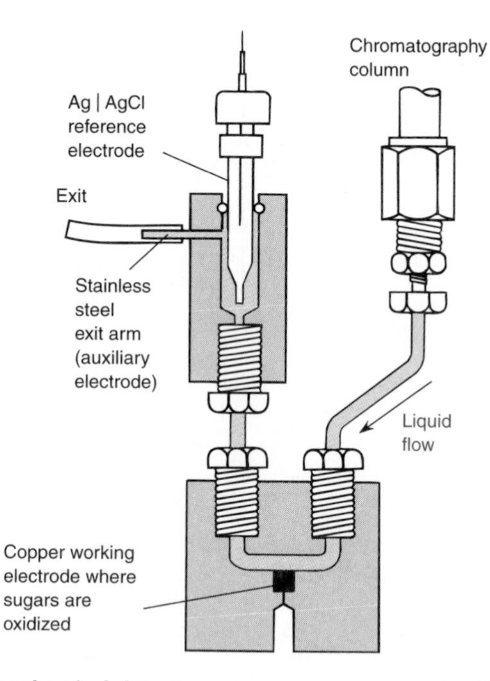

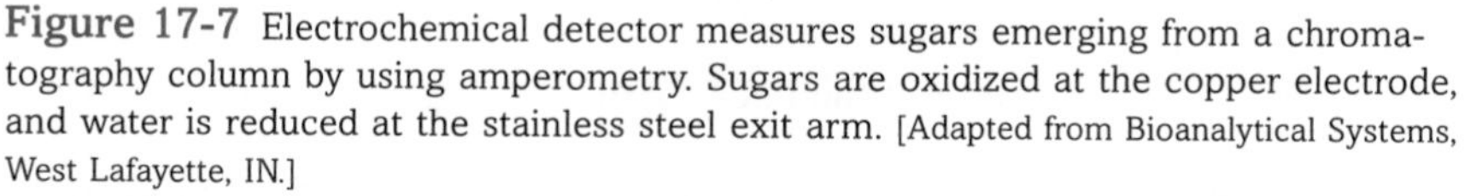

Figure 17-7 Electrochemical detector measures sugars emerging from a chromatography column by using amperometry. Sugars are oxidized at the copper electrode, and water is reduced at the stainless steel exit arm. [Adapted from Bioanalytical Systems, West Lafayette, IN.]

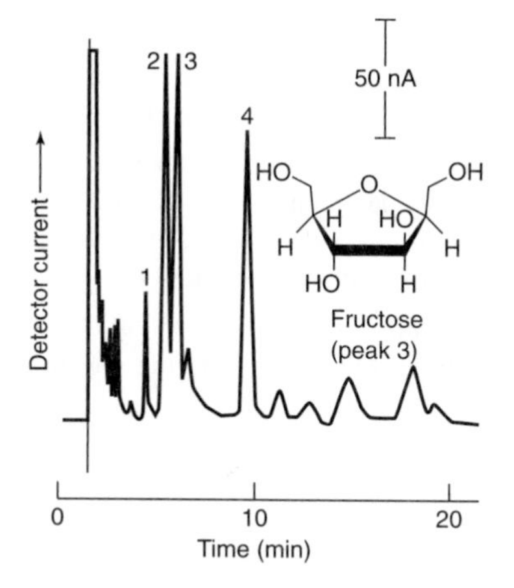

Figure 17-8 Anion-exchange chromatogram of Bud Dry beer diluted by a factor of 100 with water and filtered through a 0.45-μm membrane to remove particles. Column stationary phase is CarboPac PA1 and the mobile phase is 0.1 M NaOH. Labeled peaks are the sugars (1) arabinose, (2) glucose, (3) fructose, and (4) lactose. [From P. Luo, M. Z. Luo, and R. P. Baldwin, *J. Chem. Ed.* **1993**, *70*, 679.]

are common means to detect analytes as they emerge from the column. Sugars in beverages can be measured by separating them by anion-exchange chromatography (described in Chapter 23) and detecting them with an electrode as they emerge. The —OH groups of sugars such as glucose partly dissociate to $—O^-$ anions in 0.1 M NaOH. Anions are separated from one another as they pass through a column packed with particles having fixed positive charges.

The amperometric detector in Figure 17-7 features a Cu working electrode over which the liquid from the column flows. The Ag|AgCl reference electrode and a stainless steel auxiliary electrode are farther downstream at the upper left of the diagram. The working electrode is poised by a potentiostat at a potential of +0.55 V versus Ag|AgCl. As sugars emerge from the column, they are oxidized at the Cu surface. Reduction of water ($H_2O + e^- \longrightarrow \frac{1}{2}H_2 + OH^-$) takes place at the auxiliary electrode. Electric current flowing between the working and auxiliary electrodes is proportional to the concentration of each sugar exiting the column. Figure 17-8 shows the chromatogram, which is a trace of detector current versus time as different sugars emerge from the chromatography column. Table 17-1 shows the sugar contents in various beverages measured by this method.

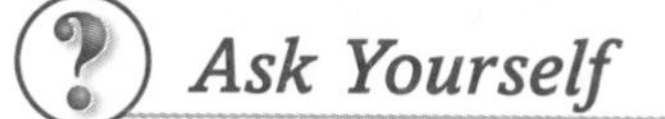

Ask Yourself

17-B. **(a)** How does the glucose monitor work?
(b) Why is a mediator advantageous in the glucose monitor?

Table 17-1 Partial list of sugars in beverages

Brand	Sugar concentration (g/L)			
	Glucose	Fructose	Lactose	Maltose
Budweiser	0.54	0.26	0.84	2.05
Bud Dry	0.14	0.29	0.46	—
Coca Cola	45.1	68.4	—	1.04
Pepsi	44.0	42.9	—	1.06
Diet Pepsi	0.03	0.01	—	—

SOURCE: P. Luo, M. Z. Luo, and R. P. Baldwin, *J Chem. Ed.* **1993**, *70*, 679.

17-3 Voltammetry

In **voltammetry**, current is measured while voltage between two electrodes is varied. (In amperometry, we held voltage fixed during the measurement of current.) Consider the apparatus in Figure 17-9 used to measure vitamin C (ascorbic acid) in fruit drinks. Oxidation of analyte takes place at the exposed tip of the graphite working electrode:

Graphite electrodes were chosen because they are inexpensive. The working electrode has a small exposed tip to decrease distortion of the electrochemical signal from electrical resistance of the solution and capacitance of the electrode.

Working electrode:

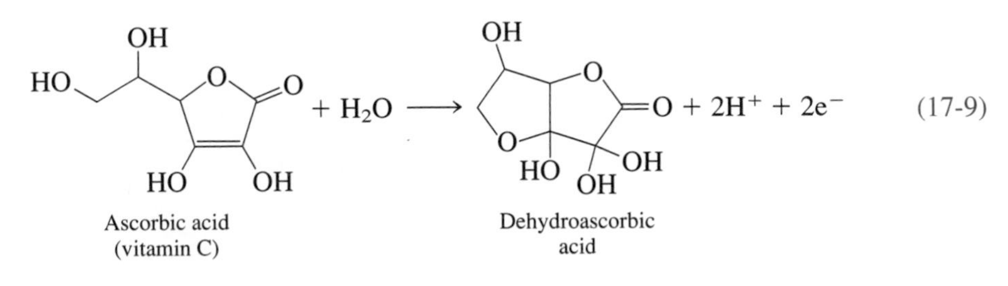

and reduction of H^+ occurs at the auxiliary electrode:

Auxiliary electrode: $2H^+ + 2e^- \longrightarrow H_2(g)$

We measure current between the working and auxiliary electrodes as the potential of the working electrode is varied with respect to the reference electrode.

To record the **voltammogram** (the graph of current versus potential) of orange juice in Figure 17-10, the working electrode was first held at a potential of −1.5 V (versus Ag | AgCl) for 2 min while the solution was stirred. This *conditioning* reduces and removes organic material from the tip of the electrode. The potential was then changed to −0.4 V and stirring continued for 30 s while bubbles of gas were dislodged from the electrode by gentle tapping. Stirring was then discontinued for 30 s so that the solution would be calm for the measurement. Finally, the voltage was scanned from −0.4 V to +1.2 V at a rate of +33 mV/s to record the lowest trace in Figure 17-10.

Conditioning is repeated before each measurement (including each standard addition) to obtain a clean, fairly reproducible electrode surface.

What happens as the voltage is scanned? At −0.4 V, there is no significant reaction and little current flows. At a potential near +0.2 V in Figure 17-10, ascorbic acid begins to be oxidized at the tip of the working electrode and current rises. Beyond ~+0.8 V, ascorbic acid in the vicinity of the electrode tip is depleted by the electrochemical reaction. The current falls slightly because analyte cannot diffuse fast enough to the electrode to maintain the peak reaction rate.

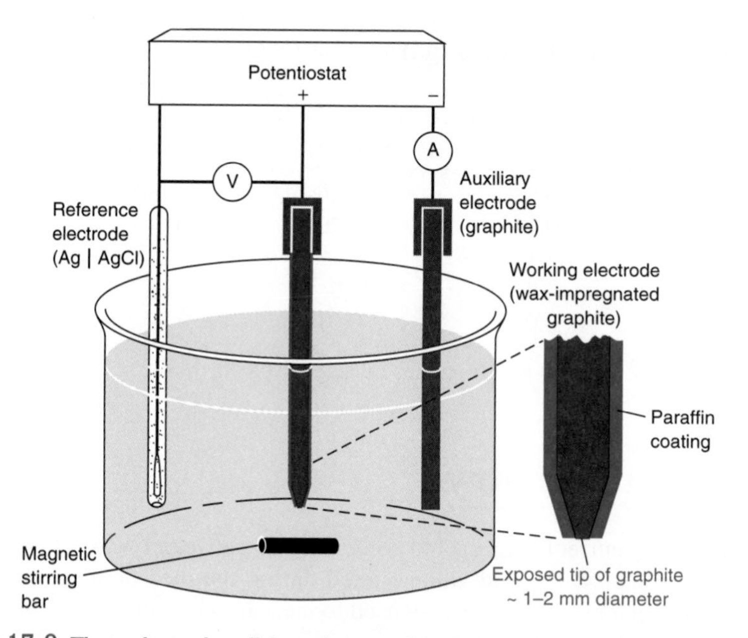

Figure 17-9 Three-electrode cell for voltammetric measurement of vitamin C in fruit drinks. The voltage between the working and reference electrodes is measured by the voltmeter Ⓥ, and the current between the working and auxiliary electrodes is measured by the ammeter Ⓐ. The potentiostat varies the voltage in a chosen manner.

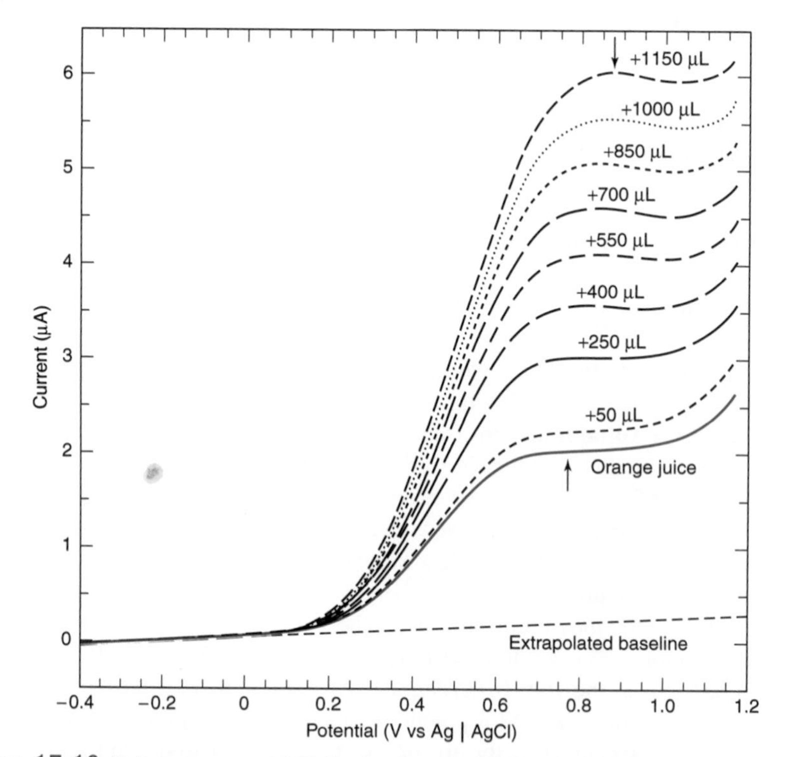

Figure 17-10 Voltammogram of 50.0 mL of orange juice and standard additions of 0.279 M ascorbic acid in 0.029 M HNO_3. Voltage was scanned at +33 mV/s with apparatus in Figure 17-9. Peak position marked by arrows in the lowest and highest curves changes slightly as standard is added because the solution becomes more acidic.

The peak current is proportional to the concentration of ascorbic acid in the orange juice. We measure peak current at the arrow in Figure 17-10 from the baseline extrapolated from the region between -0.4 and 0 V, where little reaction occurs. Any species in juice that is oxidized near $+0.8$ V will interfere with the analysis. We do not yet know the proportionality constant between current and ascorbic acid concentration. To complete the measurement, we make several *standard additions* of known quantities of ascorbic acid, shown by the dashed curves in Figure 17-10.

The method of standard addition was described in Section 5-3. Problem 5-19 gives an equation for the uncertainty in a standard addition graph.

Ask Yourself

17-C. If you have not worked Ask Yourself 5-C, now is the time to do it to get practice in the method of standard addition.

17-4 Polarography

Polarography is voltammetry conducted with a *dropping-mercury electrode.* The cell in Figure 17-11 has a dropping-mercury working electrode, a Pt auxiliary electrode, and a calomel reference electrode. An electronically controlled dispenser suspends one drop of mercury from the tip of a glass capillary tube immersed in analyte solution. A measurement is made in ~1 s, the drop is released, and a fresh drop is suspended for the next measurement. There is always fresh, reproducible metal surface for each measurement.

Polarography was invented in 1922 by Jaroslav Heyrovský, who received the Nobel Prize in 1959.

Mercury is particularly useful for reduction processes. At other working electrodes, such as Pt, Au, or carbon, H^+ is reduced to H_2 at modest negative potentials. High current from this reaction obscures the signal from reduction of analyte. Reduction of H^+ is difficult at a Hg surface and requires much more negative potentials. Conversely, Hg has little useful range for oxidations, because Hg itself is oxidized to Hg^{2+} at modest positive potentials. Therefore, a dropping-mercury electrode is usually used to reduce analytes. Platinum, gold, or carbon are used to oxidize analytes such as vitamin C in Reaction 17-9.

Potential limit (versus S.C.E.) for electrodes in 1 M H_2SO_4:

Pt	-0.2 to $+0.9$ V
Au	-0.3 to $+1.4$ V
Glassy carbon	-0.8 to $+1.1$ V
B-doped diamond	-1.5 to $+1.7$ V
Hg	-1.3 to $+0.1$ V

In the presence of 1 M Cl^-, Hg is oxidized near 0 V by the reaction $Hg(l) + 4Cl^- \rightarrow HgCl_4^{2-} + 2e^-$.

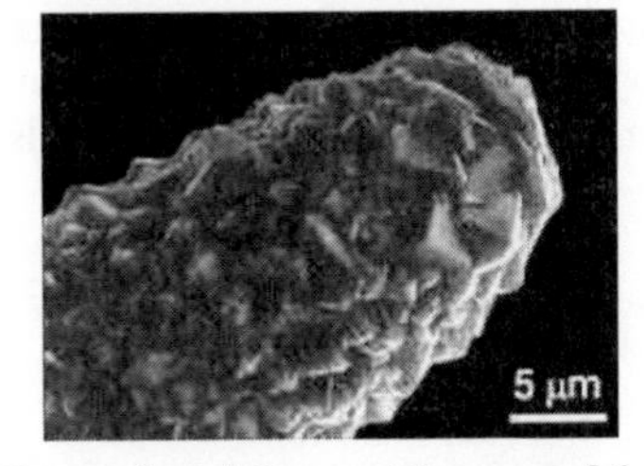

Boron-doped diamond has one of the widest available potential ranges and is chemically inert. [From J. Cvačka et al., *Anal. Chem.* **2003**, *75*, 2678. Courtesy G. M. Swain, Michigan State University.]

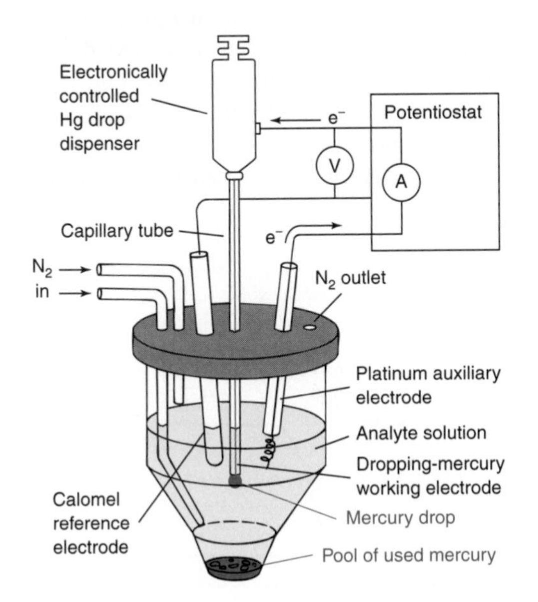

Figure 17-11 A cell for polarography.

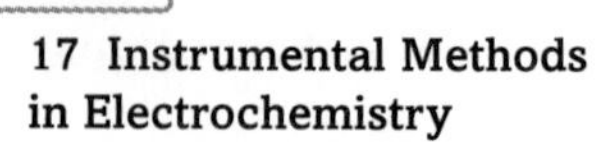

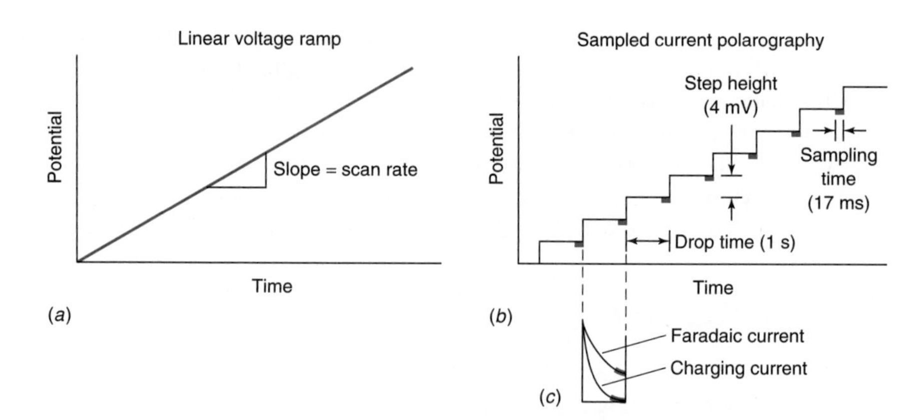

Figure 17-12 Voltage profiles for voltammetry: (*a*) linear voltage ramp used in vitamin C experiment; (*b*) staircase profile for *sampled current* polarography. Inset (*c*) shows how faradaic and charging currents decay after each potential step.

The Polarogram

To record the voltammogram of vitamin C in Figure 17-10, the potential applied to the working electrode was varied at a constant rate from −0.4 V to +1.2 V. We call this voltage profile a *linear voltage ramp* (Figure 17-12*a*).

One of many ways to conduct a polarography experiment is with a *staircase voltage ramp* (Figure 17-12*b*). When each drop of Hg is dispensed, the potential is made more negative by 4 mV. After almost 1 s, current is measured during the last 17 ms of the life of each Hg drop. The **polarogram** in Figure 17-13*a* is a graph of current versus voltage when Cd^{2+} is the analyte. The chemistry at the working electrode is

$$\textit{Reaction at working electrode:} \quad Cd^{2+} + 2e^- \longrightarrow Cd(\text{dissolved in Hg}) \qquad (17\text{-}10)$$

The product Cd(0) is dissolved in the liquid Hg drop. A solution of anything in Hg is called an **amalgam**. We call Figure 17-13*a* a *sampled current polarogram* because the current is measured only at the end of each drop life.

$E_{1/2}$ is characteristic of a particular analyte in a particular medium. Analytes can be distinguished from one another by their half-wave potentials.

The curve in Figure 17-13*a* is called a **polarographic wave**. The potential at which half the maximum current is reached is called the **half-wave potential** ($E_{1/2}$) in Figure 17-13*a*. The constant current in the plateau region is called the **diffusion current** because it is limited by the rate of diffusion of analyte to the electrode. *For quantitative analysis, diffusion current is proportional to the concentration of analyte.* Diffusion current is measured from the baseline recorded

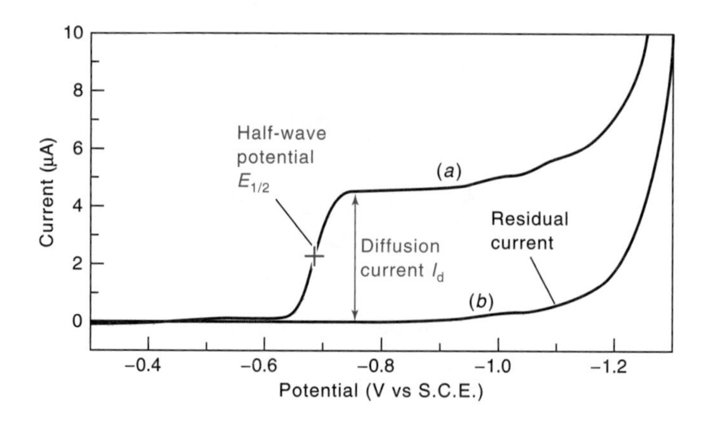

Figure 17-13 Sampled current polarogram of (*a*) 5 mM Cd^{2+} in 1 M HCl and (*b*) 1 M HCl alone.

without analyte in Figure 17-13*b*. The small **residual current** in the absence of analyte is due mainly to reduction of impurities in the solution and on the surface of the electrodes. At sufficiently negative potential (−1.2 V in Figure 17-13), current increases rapidly as reduction of H^+ to H_2 in the aqueous solution commences.

Quantitative analysis requires that peak current (the diffusion current) be governed by the rate at which analyte diffuses to the electrode. Analyte can also reach the electrode by convection and electrostatic attraction. We minimize convection by using an unstirred solution. Electrostatic attraction is decreased by high concentration of inert ions (called *supporting electrolyte*), such as 1 M HCl in Figure 17-13.

Oxygen must be absent because O_2 gives two polarographic waves when it is reduced first to H_2O_2 and then to H_2O. In Figure 17-11, N_2 is bubbled through analyte solution for 10 min to remove O_2. Then bubbling is suspended, but the liquid is maintained under a blanket of flowing N_2 to keep O_2 out. The liquid must be calm during a measurement to minimize convection of analyte to the electrode.

For the first 50 years of polarography, current was measured continuously as Hg flowed from an open capillary tube. Each drop grew until it fell off and was replaced by a new drop. The current oscillated from a low value when the drop was small to a high value when the drop was big. Polarograms in the older literature have large oscillations superimposed on the curve in Figure 17-13*a*.

Faradaic and Charging Currents

The current that we seek to measure in voltammetry is **faradaic current** due to reduction (or oxidation) of analyte at the working electrode. In Figure 17-13*a*, faradaic current is from reduction of Cd^{2+} at the Hg electrode. Another current, called **charging current** (or *capacitor current*) interferes with every measurement. To step the working electrode to a more negative potential, electrons are forced into the electrode from the potentiostat. In response, cations in solution flow toward the electrode, and anions flow away from the electrode. This flow of ions and electrons, called the *charging current*, is not from redox reactions. We try to minimize charging current because it obscures the faradaic current. The charging current usually controls the detection limit in polarography or voltammetry.

Faradaic current: Due to redox reaction at the electrode.

Charging current: Due to migration of ions toward or away from an electrode because of electrostatic attraction or repulsion. Redox reactions have no role in charging current.

Figure 17-12*c* shows the behavior of faradaic and charging currents after each potential step in Figure 17-12*b*. Faradaic current decays because analyte cannot diffuse to the electrode fast enough to sustain the high reaction rate. Charging current decays even faster because ions near the electrode redistribute themselves rapidly. Waiting 1 s after each potential step ensures that faradaic current is still significant and charging current is small.

By waiting after each potential step before measuring current, we observe significant faradaic current from the redox reaction with little interference from the charging current.

Square Wave Voltammetry

The most efficient voltage profile for polarography or voltammetry, called **square wave voltammetry**, uses the waveform in Figure 17-14, which consists of a square wave superimposed on a staircase.[3] During each cathodic pulse in Figure 17-14, there is a rush of analyte to be reduced at the electrode surface. During the anodic pulse, analyte that was just reduced is reoxidized. The square wave polarogram in Figure 17-15 is the *difference* in current between intervals 1 and 2 in Figure 17-14. Electrons flow from the electrode to analyte at point 1 and in the reverse direction at point 2. Because the two currents have opposite signs, their difference is larger than either current alone. Because the difference is plotted, the shape of the square wave polarogram in Figure 17-15 is essentially the derivative of the sampled current polarogram.

The optimum height of the square wave, E_p in Figure 17-14, is $50/n$ mV, where n is the number of electrons in the half-reaction. For Reaction 17-10, $n = 2$, so $E_p = 25$ mV.

The signal in square wave voltammetry is increased relative to a sampled current voltammogram, and the wave becomes peak shaped. The detection limit is reduced from $\sim 10^{-5}$ M for sampled current polarography to $\sim 10^{-7}$ M in square wave polarography. It is easier to resolve neighboring peaks than neighboring waves, so square wave polarography can resolve species whose half-wave potentials differ by ~0.05 V, whereas the potentials must differ by ~0.2 V to be resolved in

Advantages of square wave voltammetry:

- increased signal
- derivative (peak) shape provides better resolution of neighboring signals
- faster measurement

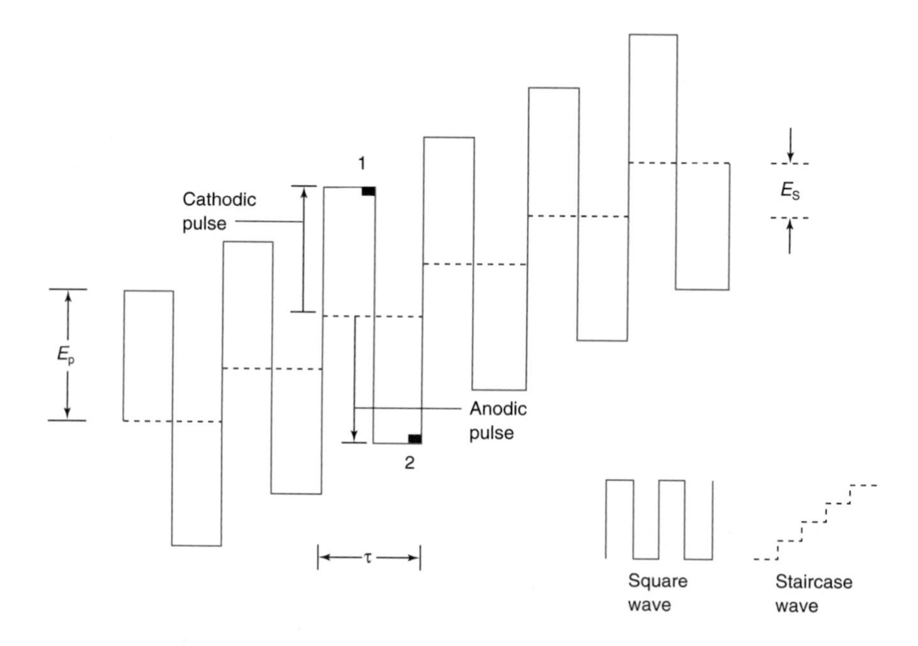

Figure 17-14 Waveform for square wave voltammetry. Typical parameters are pulse height (E_p) = 25 mV, step height (E_s) = 10 mV, and pulse period (τ) = 5 ms. Current is measured in regions 1 and 2.

sampled current polarography. Square wave voltammetry is faster than other voltammetric techniques. The square wave polarogram in Figure 17-15 was recorded in one-fifth of the time required for the sampled current polarogram. In principle, the shorter the pulse period, τ, in Figure 17-14, the greater the current that will be observed. In practice, a pulse period of 5 ms is a practical lower limit for common equipment.

Stripping Analysis

Stripping analysis:
1. Concentrate analyte into a drop of Hg by reduction.
2. Reoxidize analyte by making the potential more positive.
3. Measure polarographic signal during oxidation.

In **stripping analysis**, analyte from a dilute solution is first concentrated into a single drop of Hg (or a thin film of Hg or onto a solid electrode) by electroreduction. Analyte is then *stripped* from the electrode by making the potential more positive, thereby oxidizing it back into solution. Current measured during oxidation is proportional to the quantity of analyte that was initially deposited. Figure 17-16 shows an anodic stripping voltammogram of traces of Cd, Pb, and Cu from honey. Anodic stripping is used to measure Pb in blood and is a valuable tool in screening children for exposure to lead.

Stripping is the most sensitive voltammetric technique because analyte is concentrated from a dilute solution. The longer the period of concentration, the more sensitive is the analysis. Only a fraction of analyte from solution is deposited, so deposition must be done for a reproducible time (such as 5 min) with reproducible stirring. Detection limits are $\sim 10^{-10}$ M.

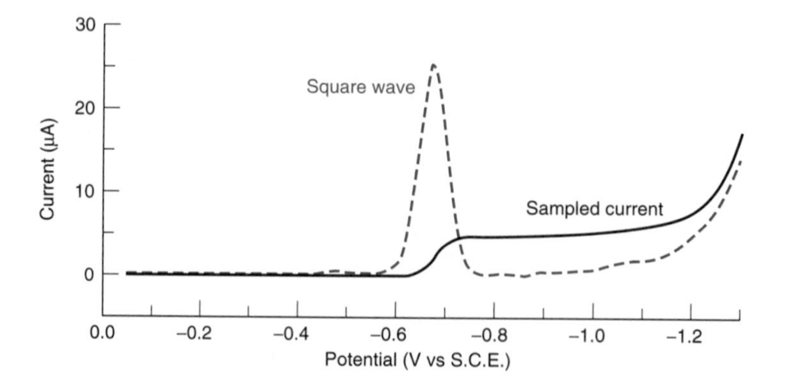

Figure 17-15 Comparison of polarograms of 5 mM Cd^{2+} in 1 M HCl. Operating parameters are defined in Figures 17-12*b* and 17-14. Sampled current: drop time = 1 s, step height = 4 mV, sampling time = 17 ms. Square wave: drop time = 1 s, step height (E_s) = 4 mV, pulse period (τ) = 67 ms, pulse height (E_p) = 25 mV, sampling time = 17 ms.

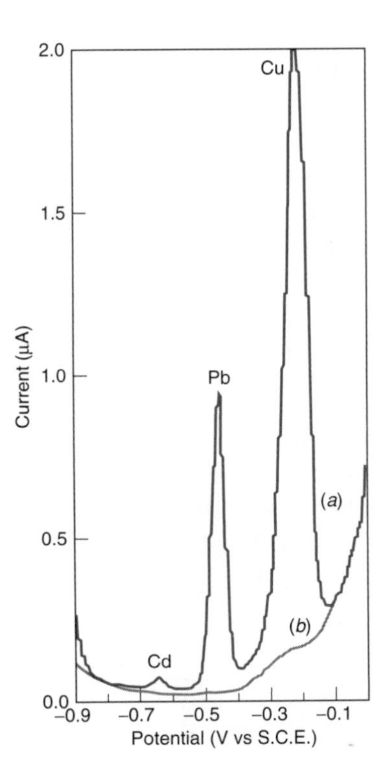

Figure 17-16 (*a*) Anodic stripping voltammogram of honey dissolved in water and acidified to pH 1.2 with HCl. Cd, Pb, and Cu were reduced from solution into a thin film of Hg for 5 min at −1.4 V (versus S.C.E.) prior to recording the voltammogram. (*b*) Voltammogram obtained without 5-min reduction step. The concentrations of Cd and Pb in the honey were 7 and 27 ng/g (ppb), respectively. The precision of the analysis was 2–4%. [From Y. Li, F. Wahdat, and R. Neeb, *Fresenius J. Anal. Chem.* **1995**, *351*, 678.]

Ask Yourself

17-D. (a) What is the difference between faradaic and charging current?
(b) Why is it desirable to wait 1 s after a potential pulse before recording the current in voltammetry?
(c) What are the advantages of square wave polarography over sampled current polarography?
(d) Explain what is done in anodic stripping voltammetry. Why is stripping the most sensitive polarographic technique?

Important Terms

amalgam	electrogravimetric analysis	potentiostat
amperometry	electrolysis	reference electrode
auxiliary electrode	faradaic current	residual current
biosensor	half-wave potential	square wave voltammetry
charging current	mediator	stripping analysis
Clark electrode	polarogram	voltammetry
coulometry	polarographic wave	voltammogram
diffusion current	polarography	working electrode
electroactive species		

Problems

17-1. (a) State the general idea behind electrogravimetric analysis.

(b) How can you know when an electrogravimetric deposition is complete?

17-2. How do the measurements of current and time in Reaction 17-2 allow us to measure the quantity of H_2S in Reaction 17-3?

17-3. In the following diagram, ——o is the symbol for the working electrode, ——| is the auxiliary electrode, and ——→ is the reference electrode. Which voltage, V_1 or V_2, is held constant in an electrolysis with three electrodes?

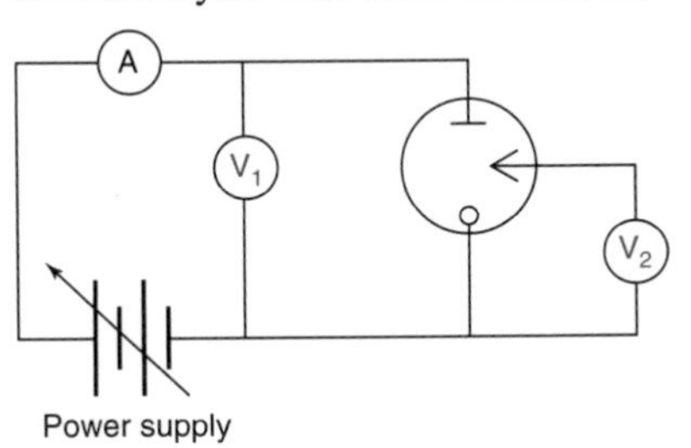

17-4. Explain the function of each electrode in the polarography cell in Figure 17-11.

17-5. What is the difference between faradaic and charging current and why do we wait 1 s after each voltage step in Figure 17-12*b* before measuring current?

17-6. A 50.0-mL aliquot of unknown Cu(II) solution was exhaustively electrolyzed to deposit all copper on the cathode. The mass of the cathode was 15.327 g prior to electrolysis and 16.414 g after electrolysis. Find the molarity of Cu(II) in the unknown.

17-7. A solution containing 0.402 49 g of $CoCl_2 \cdot xH_2O$ (a solid with an unknown number of waters of hydration) was exhaustively electrolyzed to deposit 0.099 37 g of metallic cobalt on a platinum cathode by the reaction $Co^{2+} + 2e^- \longrightarrow Co(s)$. Calculate the number of moles of water per mole of cobalt in the reagent.

17-8. Ions that react with Ag^+ can be determined electrogravimetrically by deposition on a silver anode: $Ag(s) + X^- \longrightarrow AgX(s) + e^-$. What will be the final mass of a silver anode used to electrolyze 75.00 mL of 0.023 80 M KSCN if the initial mass of the anode is 12.463 8 g?

17-9. A 0.326 8-g unknown containing lead lactate, $Pb(CH_3CHOHCO_2)_2$ (FM 385.3), plus inert material was electrolyzed to produce 0.111 1 g of PbO_2 (FM 239.2). Was the PbO_2 deposited at the anode or at the cathode? Find the weight percent of lead lactate in the unknown.

17-10. $H_2S(aq)$ is analyzed by titration with coulometrically generated I_2 in Reactions 17-2 and 17-3. To 50.00 mL of unknown H_2S sample were added 4 g of KI. Electrolysis required 812 s at 52.6 mA. Find the concentration of H_2S (μg/mL) in the sample.

17-11. OH^- generated at the right side of the apparatus in the diagram was used to titrate an unknown acid.

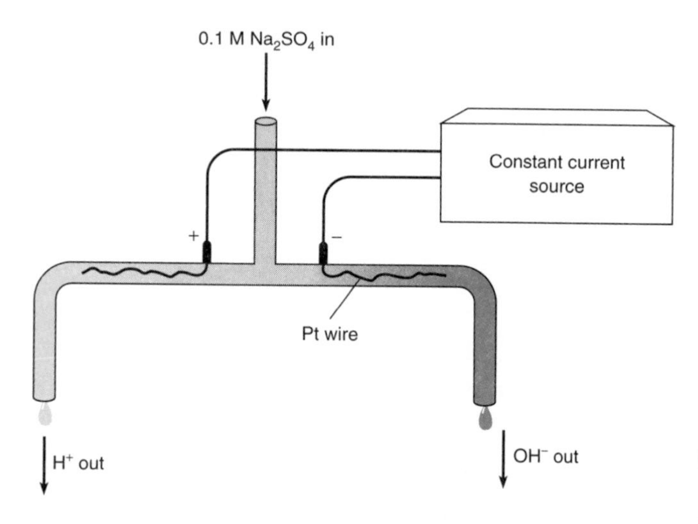

(a) What chemical reactions produce OH^- and H^+?

(b) If a current of 89.2 mA for 666 s was required to reach the end point in the titration of 5.00 mL of an unknown acid, HA, what was the molarity of HA?

17-12. A 1.00-L electrolysis cell initially containing 0.025 0 M Mn^{2+} and another metal ion, M^{3+}, is fitted with Mn and Pt electrodes. The reactions are

$$Mn(s) \longrightarrow Mn^{2+} + 2e^-$$

$$M^{3+} + 3e^- \longrightarrow M(s)$$

(a) Is the Mn electrode the anode or the cathode?

(b) A constant current of 2.60 A was passed through the cell for 18.0 min, causing 0.504 g of the metal M to plate out on the Pt electrode. What is the atomic mass of M?

(c) What will be the concentration of Mn^{2+} in the cell at the end of the experiment?

17-13. The sensitivity of a coulometer is governed by the delivery of its minimum current for its minimum time. Suppose that 5 mA can be delivered for 0.1 s.

(a) How many moles of electrons are delivered at 5 mA for 0.1 s?

(b) How many milliliters of a 0.01 M solution of a two-electron reducing agent are required to deliver the same number of electrons?

17-14. The electrolysis cell shown here was run at a constant current of 0.021 96 A. On one side, 49.22 mL of H_2 were produced (at 303 K and 0.996 bar); on the other side, Cu metal was oxidized to Cu^{2+}.

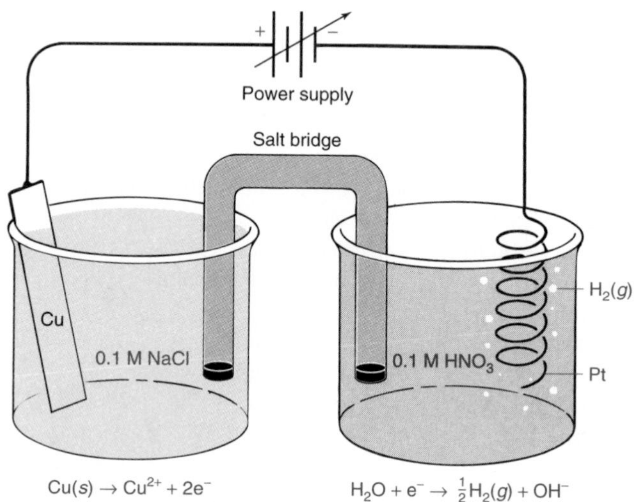

(a) How many moles of H_2 were produced? (See Problem 16-15 for the ideal gas law.)

(b) If 47.36 mL of EDTA were required to titrate the Cu^{2+} produced by the electrolysis, what was the molarity of the EDTA?

(c) For how many hours was the electrolysis run?

17-15. A mixture of trichloroacetate and dichloroacetate can be analyzed by selective reduction in a solution containing 2 M KCl, 2.5 M NH_3, and 1 M NH_4Cl. At a mercury cathode potential of −0.90 V (versus S.C.E.), only trichloroacetate is reduced:

$$Cl_3CCO_2^- + H_2O + 2e^- \longrightarrow Cl_2CHCO_2^- + OH^- + Cl^-$$

At a potential of −1.65 V, dichloroacetate reacts:

$$Cl_2CHCO_2^- + H_2O + 2e^- \longrightarrow ClCH_2CO_2^- + OH^- + Cl^-$$

A hygroscopic mixture of trichloroacetic acid (FM 163.39) and dichloroacetic acid (FM 128.94) containing an unknown quantity of water weighed 0.721 g. On controlled potential electrolysis, 224 C passed at −0.90 V, and 758 C were required to complete the electrolysis at -1.65 V. Calculate the weight percent of each acid in the mixture.

17-16. Chlorine has been used for decades to disinfect drinking water. An undesirable side effect of this treatment is the reaction of chlorine with organic impurities to create organochlorine compounds, some of which could be toxic. Monitoring total organic halide (designated TOX) is now required for many water providers. A standard procedure for TOX is to pass water through activated charcoal that adsorbs organic compounds. Then the charcoal is combusted to liberate hydrogen halides:

$$\text{organic halide (RX)} \xrightarrow{O_2/800^\circ C} CO_2 + H_2O + HX$$

The HX is absorbed into aqueous solution and measured by automatic coulometric titration with a silver anode:

$$X^-(aq) + Ag(s) \longrightarrow AgX(s) + e^-$$

When 1.00 L of drinking water was analyzed, a current of 4.23 mA was required for 387 s. A blank prepared by oxidizing charcoal required 6 s at 4.23 mA. Express the TOX of the drinking water as micromoles of halogen per liter. If all halogen is chlorine, express the TOX as micrograms of Cl per liter.

17-17. *Propagation of uncertainty.* In an extremely accurate measurement of the Faraday constant, a pure silver anode was oxidized to Ag^+ with a constant current of 0.203 639 0 ($\pm$0.000 000 4) A for 18 000.075 ($\pm$0.010) s to give a mass loss of 4.097 900 ($\pm$0.000 003) g from the anode. Given that the atomic mass of Ag is 107.868 2 ($\pm$0.000 2), find the value of the Faraday constant and its uncertainty.

17-18. **(a)** How does a Clark electrode measure the concentration of dissolved O_2?

(b) What does it mean when we say that the concentration of dissolved O_2 is "0.20 bar"? What is the actual molarity of O_2?

17-19. What are the advantages of a dropping-Hg electrode in polarography? Why is polarography used mainly to study reductions rather than oxidations?

17-20. Suppose that a peak current of 3.9 μA was observed in the oxidation of 50 mL of 2.4 mM ascorbic acid in the experiment in Figures 17-9 and 17-10. Suppose that this much current flowed for 10 min in the course of several measurements. From the current and time, calculate what fraction of the ascorbic acid is oxidized at the electrode. Is it fair to say that the ascorbic acid concentration is nearly constant during the measurements?

17-21. *Calibration curve and error estimate.* The following polarographic diffusion currents were measured at -0.6 V for $CuSO_4$ in 2 M NH_4Cl/2M NH_3. Use the method of least squares to estimate the molarity (and its uncertainty) of an unknown solution giving I_d = 15.6 μA.

$[Cu^{2+}]$ (mM)	I_d (μA)	$[Cu^{2+}]$ (mM)	I_d (μA)
0.039 3	0.256	0.990	6.37
0.078 0	0.520	1.97	13.00
0.158 5	1.058	3.83	25.0
0.489	3.06	8.43	55.8

17-22. The drug Librium gives a polarographic wave with $E_{1/2} = -0.265$ V (versus S.C.E.) in 0.05 M H_2SO_4. A 50.0-mL sample containing Librium gave a wave height of 0.37 μA. When 2.00 mL of 3.00 mM Librium in 0.05 M H_2SO_4 were added to the sample, the wave height increased to 0.80 μA. Find the molarity of Librium in the unknown.

17-23. A polarogram of reagent-grade methanol is shown in trace *a*. Trace *b* shows reagent methanol with added 0.001 00 wt% acetone, 0.001 00 wt% acetaldehyde, and 0.001 00 wt% formaldehyde. Estimate the weight percent of acetone in reagent-grade methanol.

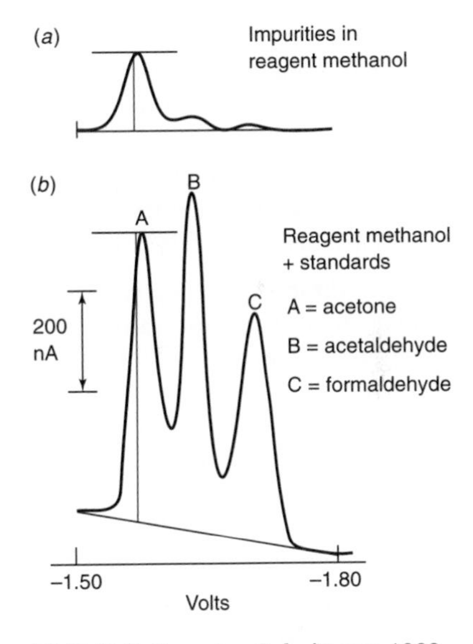

[D. B. Palladino, *Am. Lab.* August 1992, p. 56.]

Scales are the same in both panels. Solutions were prepared by diluting 25 mL of methanol up to 100 mL with water containing buffer and hydrazine sulfate, which reacts with carbonyl compounds to form electroactive hydrazones. An example is shown below:

$$\underset{\text{Acetone}}{(CH_3)_2C{=}O} + \underset{\text{Hydrazine}}{H_2N{-}NH_2} \longrightarrow$$

$$\underset{\text{Acetone hydrazone}}{(CH_3)_2C{=}N{-}NH_2} \xrightarrow{2e^- + 2H^+} (CH_3)_2CH{-}NH{-}NH_2$$

17-24. Problem 5-20 shows standard additions of Cu^{2+} to acidified tap water measured by anodic stripping voltammetry at a solid Ir electrode.

(a) What reaction occurs during the concentration stage of the analysis?

(b) What reaction takes place during the stripping stage of the analysis?

17-25. *Standard addition.* Chromium is an essential trace element present to the extent of ~3–10 ppb in blood. It can be measured by cathodic stripping voltammetry after digesting blood with a powerful oxidant to destroy organic matter.[4] Standard additions of Cr(VI) to blood prior to digestion gave the following (fictitious) stripping currents:

Standard addition (ppb)	Peak current (μA)
0	9
0.26	13
0.78	18

In each experiment, 0.50 mL of blood was treated with Cr(VI) standard and oxidant. After digestion, the solution was brought to a final volume of 20.0 mL prior to stripping analysis. Prepare a standard addition graph for the case of constant final volume. Find the concentration (and uncertainty) of Cr in the 20.0-mL volume and in the original 0.50 mL of blood.

17-26. *Coulometric titration of sulfite in wine.*[5] Sulfur dioxide is added to many foods as a preservative. In aqueous solution, the following species are in equilibrium:

$$\underset{\text{Sulfur dioxide}}{SO_2} \rightleftharpoons \underset{\text{Sulfurous acid}}{H_2SO_3} \rightleftharpoons \underset{\text{Bisulfite}}{HSO_3^-} \rightleftharpoons \underset{\text{Sulfite}}{SO_3^{2-}} \qquad \text{(A)}$$

Bisulfite reacts with aldehyde groups in food near neutral pH:

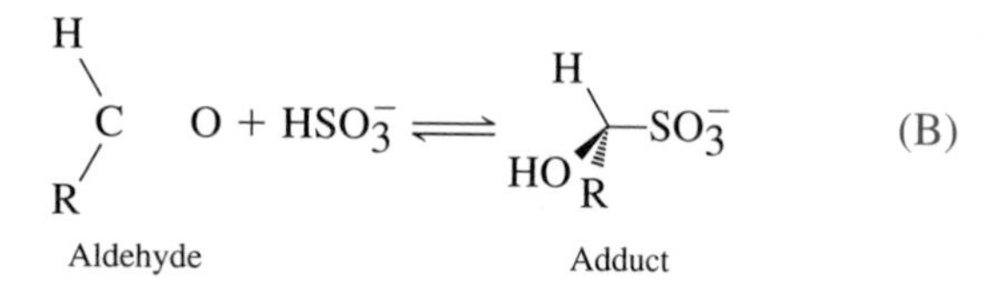

Sulfite is released from the adduct in 2 M NaOH and can be analyzed by its reaction with I_3^- to give I^- and sulfate. Excess I_3^- must be present for quantitative reaction.

Here is a coulometric procedure for analysis of total sulfite in white wine. Total sulfite means all species in Reaction A and the adduct in Reaction B. We use white wine so that we can see the color of a starch-iodine end point.

1. Mix 9.00 mL of wine plus 0.8 g of NaOH and dilute to 10.00 mL. The NaOH releases sulfite from its organic adducts.

2. Generate a known quantity of I_3^- at the working electrode (the anode) by passing a known current for a known time through the following cell.

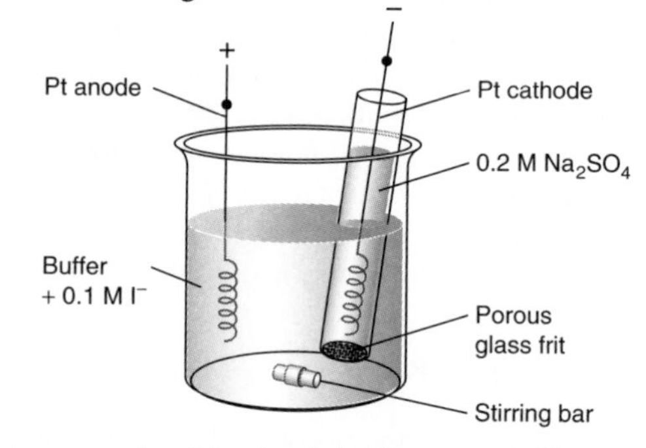

The beaker contains 30 mL of 1 M acetate buffer (pH 3.7) plus 0.1 M KI. The reaction at the cathode is reduction of H_2O to $H_2 + OH^-$. The cathode is contained in a glass tube with a porous glass frit through which ions slowly diffuse. The frit retards diffusion of OH^- into the main compartment, where it would react with I_3^- to give IO^-.

3. Generate I_3^- at the anode with a current of 10.0 mA for 4.00 min.

4. Inject 2.000 mL of the wine/NaOH solution into the cell, where the sulfite reacts with I_3^-, leaving excess I_3^-.

5. Add 0.500 mL of 0.050 7 M thiosulfate to consume I_3^- by Reaction 16-21 and leave excess thiosulfate.

6. Add starch indicator to the cell and generate fresh I_3^- with a constant current of 10.0 mA. A time of 131 s is required to consume excess thiosulfate and reach the starch end point.

(a) In what pH range is each form of sulfurous acid predominant?

(b) Write balanced half-reactions for the anode and cathode.

(c) At pH 3.7, the dominant form of sulfurous acid is HSO_3^- and the dominant form of sulfuric acid is SO_4^{2-}. Write balanced reactions between I_3^- and HSO_3^- and between I_3^- and thiosulfate.

(d) Find the concentration of total sulfite in undiluted wine.

How Would You Do It?

17-27. Nitrite (NO_2^-) is a potential carcinogen for humans, but it is also an important preservative in foods such as bacon and hot dogs. Nitrite and nitrate (NO_3^-) in foods can be measured by the following procedure:

1. A 10-g food sample is blended with 82 mL of 0.07 M NaOH, transferred to a 200-mL volumetric flask, and heated on a steam bath for 1 h. Then 10 mL of 0.42 M $ZnSO_4$ are added; after an additional 10 min of heating, the mixture is cooled and diluted to 200 mL.

2. After the solids settle, an aliquot of supernatant solution is mixed with 200 mg of charcoal to remove some organic solutes and filtered.

3. A 5.00-mL aliquot of the filtered solution is placed in a test tube containing 5 mL of 9 M H_2SO_4 and 150 μL of 85 wt% NaBr, which reacts with nitrite, but not nitrate:

$$HNO_2 + Br^- \longrightarrow NO(g) + Br_2$$

4. The NO is transported by bubbling with purified N_2 that is subsequently passed through a trapping solution to convert NO into diphenylnitrosamine:

$$\underset{\text{Diphenylamine}}{NO + (C_6H_5)_2NH} \xrightarrow{H^+ \text{ and catalyst}} \underset{\text{Diphenylnitrosamine}}{(C_6H_5)_2N{-}N{=}O}$$

5. The acidic diphenylnitrosamine is analyzed by amperometry at −0.66 V (versus Ag|AgCl) to measure the NO liberated from the food sample:

$$(C_6H_5)_2N{-}N{=}OH^+ + 4H^+ + 4e^- \longrightarrow (C_6H_5)_2N{-}NH_3^+ + H_2O$$

6. To measure nitrate, an additional 6 mL of 18 M H_2SO_4 are added to the sample tube in step 3. The acid promotes reduction of HNO_3 to NO, which is then purged with N_2 and trapped and analyzed as in steps 4 and 5.

$$HNO_3 + Br^- \xrightarrow{\text{strong acid}} NO(g) + Br_2$$

A 10.0-g bacon sample gave a current of 8.9 μA in step 5, which increased to 23.2 μA in step 6. (Step 6 measures the sum of signals from nitrite and nitrate, not just the signal from nitrate.) In a second experiment, the 5.00-mL sample in step 3 was spiked with a standard addition of 5.00 μg of NO_2^- ion. The analysis was repeated to give currents of 14.6 μA in step 5 and 28.9 μA in step 6.

(a) Find the nitrite content of the bacon, expressed as micrograms per gram of bacon.

(b) From the ratio of signals due to nitrate and nitrite in the first experiment, find the micrograms of nitrate per gram of bacon.

Notes and References

1. L. C. Clark, R. Wolf, D. Granger, and A. Taylor, "Continuous Recording of Blood Oxygen Tension by Polarography," *J. Appl. Physiol.* **1953**, *6*, 189. To construct an oxygen electrode, see J. E. Brunet, J. I. Gardiazabal, and R. Schrebler, *J. Chem. Ed.* **1983**, *60*, 677.

2. Make an enzymatic amperometric glucose electrode: M. C. Blanco-López, M. J. Lobo-Castañón, and A. J. Miranda-Ordieres, *J. Chem. Ed.* **2007**, *84*, 677.

3. For an excellent account of square wave voltammetry, see J. G. Osteryoung and R. A. Osteryoung, *Anal. Chem.* **1985**, *57*, 101A.

4. L. Yong, K. C. Armstrong, R. N. Dansby-Sparks, N. A. Carrington, J. Q. Chambers, and Z.-L. Xue, *Anal. Chem.* **2006**, *78*, 7582.

5. D. Lowinsohn and M. Bertotti, *J. Chem. Ed.* **2002,** *79,* 103. Some other species in wine, in addition to sulfite, react with I_3^-. A blank titration to correct for such reactions is described in this article.

Further Reading

J. Wang, *Analytical Electrochemistry*, 3rd ed. (New York: Wiley-VCH, 2006).

P. Zanello, *Inorganic Electrochemistry: Theory, Practice and Application* (Cambridge: Royal Society of Chemistry, 2003.)

A. M. Bond, *Broadening Electrochemical Horizons* (Oxford: Oxford University Press, 2002).

A. J. Bard and L. R. Faulkner, *Electrochemical Methods and Applications*, 2nd ed. (New York: Wiley, 2001).

A. J. Cunningham, *Introduction to Bioanalytical Sensors* (New York: Wiley, 1998).

D. Diamond, *Principles of Chemical and Biological Sensors* (New York: Wiley, 1998).

G. Ramsay, *Commercial Biosensors: Applications to Clinical, Bioprocess, and Environmental Samples* (New York: Wiley, 1998).

B. Eggins, *Biosensors* (Chichester, UK: Wiley Teubner, 1996).

P. Vanysek, *Modern Techniques in Electroanalysis* (New York: Wiley, 1996).

J. Wang, *Stripping Analysis: Principles, Instrumentation and Applications* (Deerfield Beach, FL: VCH Publishers, 1984).

M. Alvarez-Icaza and U. Bilitewski, "Mass Production of Biosensors," *Anal. Chem.* **1993**, *65*, 525A.

The Ozone Hole

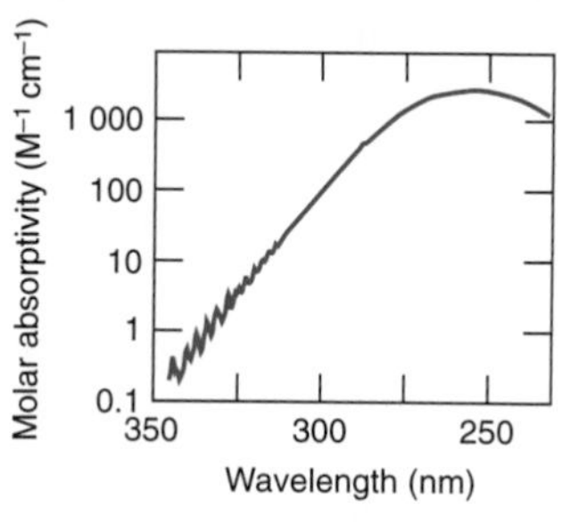

Spectrum of ozone, showing maximum absorption of ultraviolet radiation at a wavelength near 260 nm. At this wavelength, a layer of ozone is more opaque than a layer of gold of the same mass. [From R. P. Wayne, *Chemistry of Atmospheres* (Oxford: Clarendon Press, 1991).]

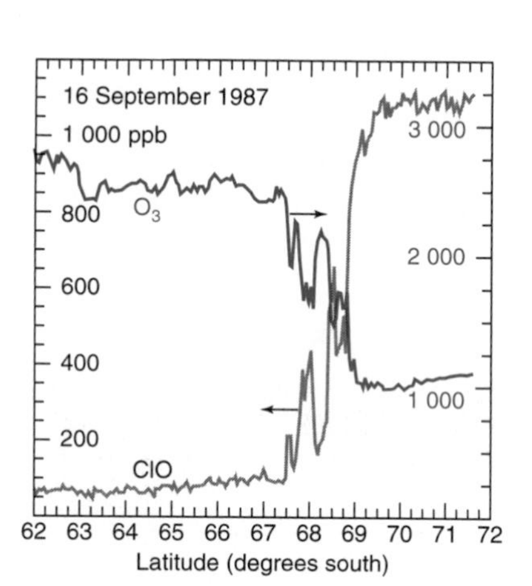

Spectroscopically measured concentrations of O_3 and ClO (measured in ppb = nL/L) in the stratosphere near the South Pole in 1987. Where ClO increases, O_3 decreases, in agreement with the reaction sequence on this page. [From J. G. Anderson, W. H. Brune, and M. H. Proffitt, *J. Geophys. Res.* **1989**, *94D*, 11465.]

Ozone, formed at altitudes of 20 to 40 km by the action of solar ultraviolet radiation ($h\nu$) on O_2, absorbs the ultraviolet radiation that causes sunburns and skin cancer:

$$O_2 \xrightarrow{h\nu} 2O \qquad O + O_2 \longrightarrow \underset{\text{Ozone}}{O_3}$$

In 1985, the British Antarctic Survey reported that ozone over Antarctica had decreased by 50% in early spring, relative to levels observed in the preceding 20 years. This "ozone hole" appeared only in early spring and deepened for four decades.[1]

Ozone destruction begins with chlorofluorocarbons such as Freon-12 (CCl_2F_2) from refrigerants. These long-lived compounds diffuse to the stratosphere, where they catalyze ozone decomposition:

$$(1)\quad CCl_2F_2 \xrightarrow{h\nu} CClF_2 + Cl \qquad \text{Photochemical Cl formation}$$

$$(2)\quad Cl + O_3 \longrightarrow ClO + O_2$$

$$(3)\quad O_3 \xrightarrow{h\nu} O + O_2$$

$$(4)\quad O + ClO \longrightarrow Cl + O_2$$

Net reaction of (2)–(4): Catalytic O_3 destruction

$$2O_3 \longrightarrow 3O_2$$

Cl produced in step 4 goes back to destroy another ozone molecule in step 2. A single Cl atom in this chain reaction can destroy $>10^5$ molecules of O_3. The chain is terminated when Cl or ClO reacts with hydrocarbons or NO_2 to form HCl or $ClONO_2$.

Stratospheric clouds catalyze the reaction of HCl with $ClONO_2$ to form Cl_2, which is split by sunlight into Cl atoms to initiate O_3 destruction:

$$HCl + ClONO_2 \xrightarrow[\text{polar clouds}]{\text{surface of}} Cl_2 + HNO_3 \qquad Cl_2 \xrightarrow{h\nu} 2Cl$$

The clouds require winter cold to form. When the sun rises at the South Pole in September and October and clouds are still present, conditions are right for O_3 destruction.

To protect life from ultraviolet radiation, international treaties now ban or phase out chlorofluorocarbons. However, so much has already been released that ozone depletion is not expected to return to historic values until late in the twenty-first century.

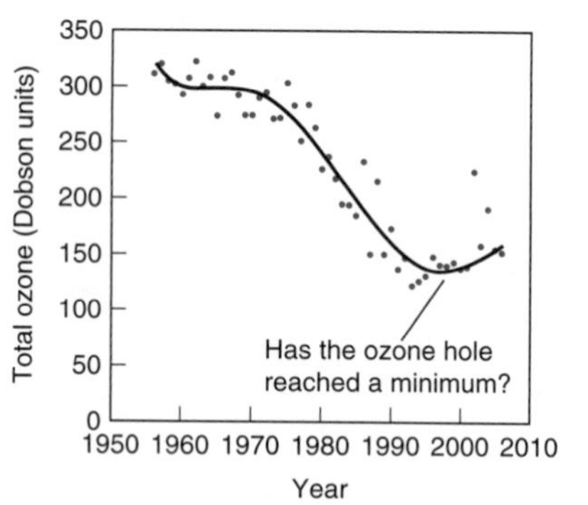

October average atmospheric ozone at Halley in Antarctica. *Dobson units* are a measure of total ozone. [From J. D. Shanklin, British Antarctic Survey, www.antarctica.ac.uk/met/jds/ozone.]

CHAPTER 18

Let There Be Light

Absorption and emission of *electromagnetic radiation* (a fancy term for light) are molecular characteristics used in quantitative and qualitative analysis. This chapter discusses basic aspects of **spectrophotometry**—the use of electromagnetic radiation to measure chemical concentrations—and Chapter 19 provides further detail on instrumentation and applications.

Following the discovery of the Antarctic ozone "hole" in 1985, atmospheric chemist Susan Solomon led the first expedition in 1986 specifically intended to make chemical measurements of the Antarctic atmosphere by using high-altitude balloons and ground-based spectroscopy. The expedition discovered that ozone depletion occurred after polar sunrise and that the concentration of chemically active chlorine in the stratosphere was 100 times greater than that predicted from gas-phase chemistry. Solomon's group identified chlorine as the culprit in ozone destruction and polar stratospheric clouds as the catalytic surface for the release of so much chlorine.

18-1 Properties of Light

Light can be described both as waves and as particles. Light waves consist of perpendicular, oscillating electric and magnetic fields (Figure 18-1). **Wavelength**, λ, is the crest-to-crest distance between waves. **Frequency**, ν, is the number of oscillations that the wave makes each second. The unit of frequency is *reciprocal seconds*, s^{-1}. One oscillation per second is also called 1 **hertz** (Hz). A frequency of $10^9\ s^{-1}$ is therefore said to be 10^9 Hz, or one *gigahertz* (GHz). The product of frequency times wavelength is c, the speed of light (2.998×10^8 m/s in vacuum):

Relation between frequency and wavelength:

$$\nu \lambda = c \qquad (18\text{-}1)$$

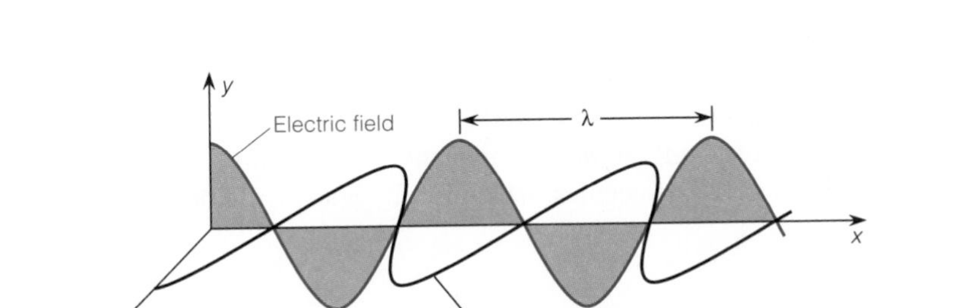

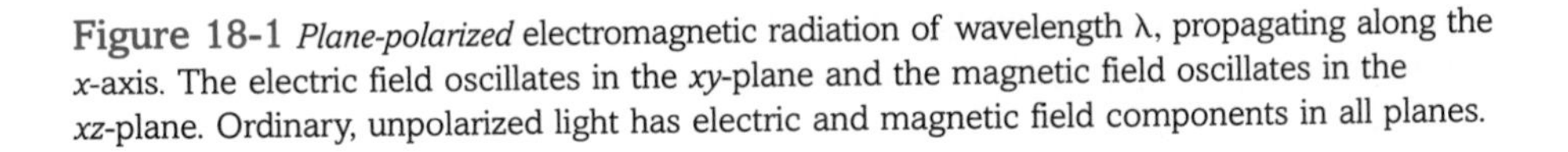
Figure 18-1 *Plane-polarized* electromagnetic radiation of wavelength λ, propagating along the x-axis. The electric field oscillates in the xy-plane and the magnetic field oscillates in the xz-plane. Ordinary, unpolarized light has electric and magnetic field components in all planes.

Example **Relating Wavelength and Frequency**

What is the wavelength of radiation in your microwave oven, whose frequency is 2.45 GHz?

SOLUTION First recognize that 2.45 GHz means 2.45×10^9 Hz = $2.45 \times 10^9\ s^{-1}$. From Equation 18-1, we write

$$\lambda = \frac{c}{\nu} = \frac{2.998 \times 10^8 \text{ m/s}}{2.45 \times 10^9 \text{ s}^{-1}} = 0.122 \text{ m}$$

Test Yourself What is the frequency of green light with λ = 500 nm? (**Answer:** $6.00 \times 10^{14}\ s^{-1}$ = 600 THz)

Light can also be thought of as particles called **photons**. The energy, E (measured in joules, J), of a photon is proportional to its frequency:

Relation between energy and frequency: $$E = h\nu \tag{18-2}$$

Physical constants are listed inside the book cover.

where h is *Planck's constant* ($= 6.626 \times 10^{-34}$ J · s).

Combining Equations 18-1 and 18-2, we can write

$$E = h\frac{c}{\lambda} = hc\frac{1}{\lambda} = hc\tilde{\nu} \tag{18-3}$$

Energy increases as
- frequency (ν) increases
- wavelength (λ) decreases
- wavenumber ($\tilde{\nu}$) increases

where $\tilde{\nu}$ ($= 1/\lambda$) is called the **wavenumber**. Energy is inversely proportional to wavelength and directly proportional to wavenumber. Red light, with a wavelength longer than that of blue light, is less energetic than blue light. The SI unit for wavenumber is m^{-1}. However, the most common unit of wavenumber is cm^{-1}, read "reciprocal centimeters" or "wavenumbers." Wavenumber units are most common in infrared spectroscopy.

Regions of the **electromagnetic spectrum** are shown in Figure 18-2. Visible light—the kind that our eyes detect—represents only a small fraction of the electromagnetic spectrum.

The lowest energy state of a molecule is called the **ground state**. When a molecule absorbs a photon, its energy increases and we say that the molecule is promoted to an **excited state** (Figure 18-3). If the molecule emits a photon, its energy decreases. Figure 18-2 indicates that microwave radiation stimulates molecules to rotate faster. A microwave oven heats food by increasing the rotational energy of water in the food. Infrared radiation excites vibrations of molecules. Visible and ultraviolet radiation promote electrons to higher energy states. (Molecules that absorb visible light are colored.) X-rays and short-wavelength ultraviolet radiation are harmful because they break chemical bonds and ionize molecules, which is why you should minimize your exposure to medical X-rays.

Example **Photon Energies**

By how many joules is the energy of a molecule increased when it absorbs **(a)** visible light with a wavelength of 500 nm or **(b)** infrared radiation with a wavenumber of 1 251 cm^{-1}?

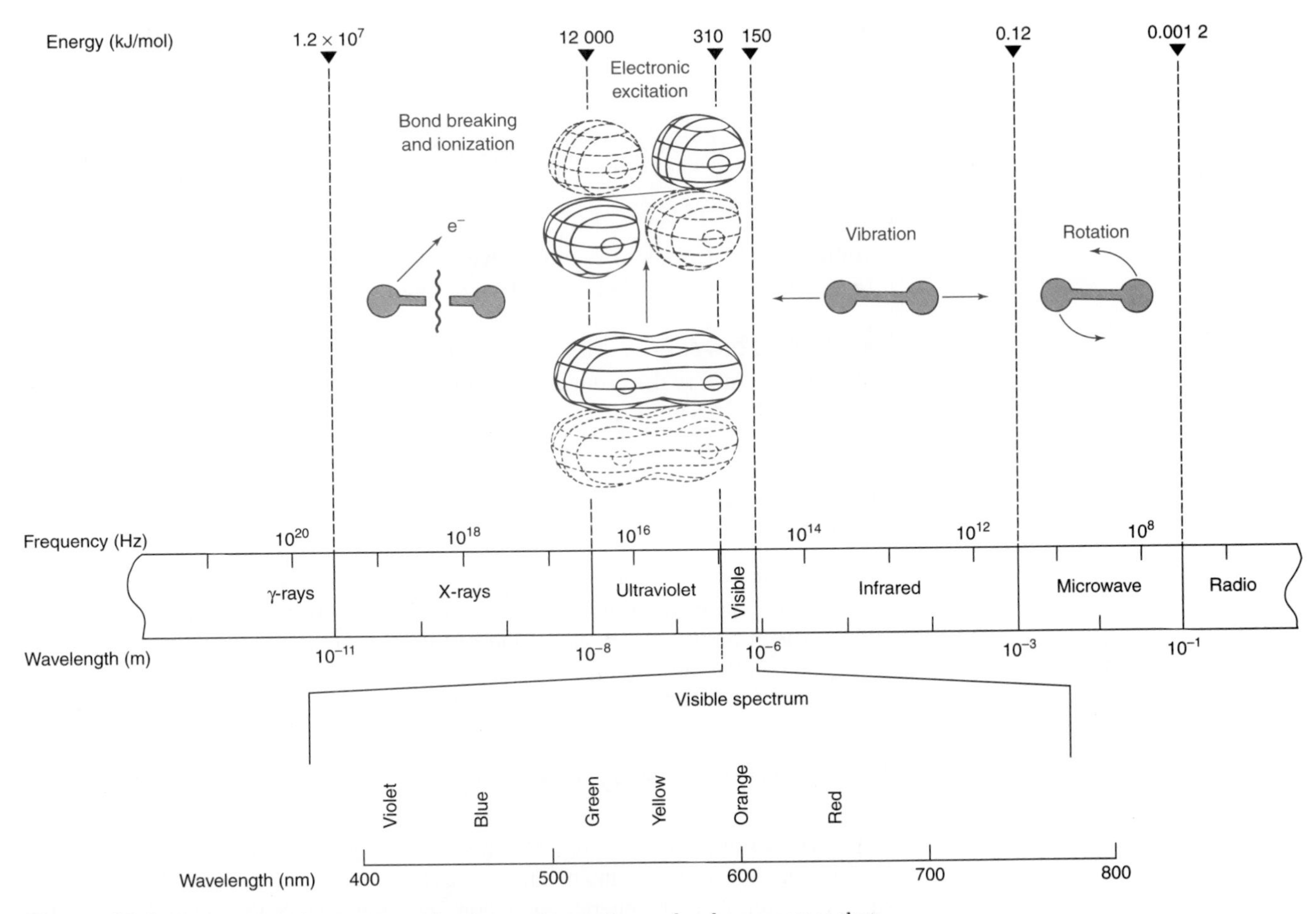

Figure 18-2 Electromagnetic spectrum, showing representative molecular processes that occur when radiation in each region is absorbed. The visible spectrum spans the wavelength range 380 to 780 nanometers (1 nm = 10^{-9} m).

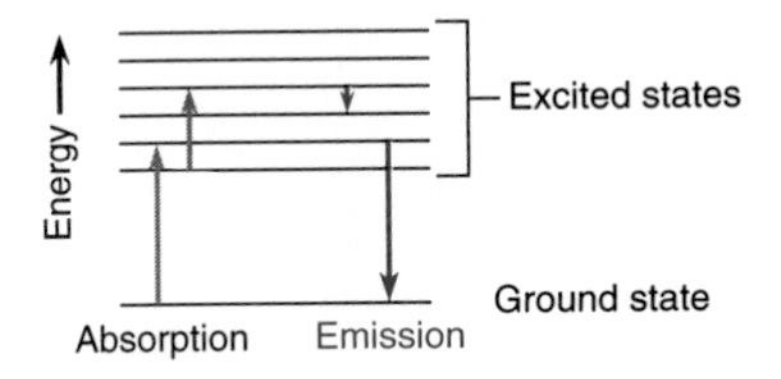

Figure 18-3 Absorption of light increases the energy of a molecule. Emission of light decreases its energy.

SOLUTION (a) The visible wavelength is 500 nm = 500×10^{-9} m.

$$E = h\nu = h\frac{c}{\lambda}$$

$$= (6.626 \times 10^{-34}\ \text{J} \cdot \text{s})\left(\frac{2.998 \times 10^{8}\ \text{m/s}}{500 \times 10^{-9}\ \text{m}}\right) = 3.97 \times 10^{-19}\ \text{J}$$

This is the energy of one photon absorbed by one molecule. If a mole of molecules absorbed a mole of photons, the energy increase is

$$E = \left(3.97 \times 10^{-19}\ \frac{\text{J}}{\text{molecule}}\right)\left(6.022 \times 10^{23}\ \frac{\text{molecules}}{\text{mol}}\right)$$

$$= 2.39 \times 10^{5}\ \frac{\text{J}}{\text{mol}}$$

$$\left(2.39 \times 10^{5}\ \frac{\text{J}}{\text{mol}}\right)\left(\frac{1\ \text{kJ}}{1\,000\ \text{J}}\right) = 239\ \frac{\text{kJ}}{\text{mol}}$$

(b) When given the wavenumber, we use Equation 18-3. First convert the wavenumber unit cm^{-1} to m^{-1} with the conversion factor 100 cm/m. The energy of one photon is

$$E = hc\tilde{\nu} = (6.626 \times 10^{-34}\ J \cdot s)\left(2.998 \times 10^{8}\ \frac{m}{s}\right)\underbrace{(1\,251\ cm^{-1})\left(100\ \frac{cm}{m}\right)}_{\text{Conversion of } cm^{-1} \text{ to } m^{-1}}$$

$$= 2.485 \times 10^{-20}\ J$$

Multiplying by Avogadro's number, we find that this photon energy corresponds to 14.97 kJ/mol, which falls in the infrared region and excites molecular vibrations.

Test Yourself We say that the first excited vibrational state of H_2 lies "4 160 cm^{-1} above the ground state." Find the energy (kJ/mol) of H_2 in this state. (**Answer:** 49.76 kJ/mol)

Ask Yourself

18-A. What is the frequency (Hz), wavenumber (cm^{-1}), and energy (kJ/mol) of light with a wavelength of **(a)** 100 nm; **(b)** 500 nm; **(c)** 10 μm; and **(d)** 1 cm? In which spectral region does each kind of radiation lie and what molecular process occurs when the radiation is absorbed?

18-2 Absorption of Light

A **spectrophotometer** measures transmission of light. If a substance absorbs light, the *radiant power* of a light beam decreases as it passes through the substance. Radiant power, P, is the energy per second per unit area of the beam. Light with a very narrow range of wavelength is said to be **monochromatic** ("one color"). In Figure 18-4, light passes through a *monochromator*, a device that selects a narrow band of wavelengths. This light with radiant power P_0 strikes a sample of length b. The radiant power of the beam emerging from the other side of the sample is P. Some of the light may be absorbed by the sample, so $P \leq P_0$.

Transmittance, Absorbance, and Beer's Law

Transmittance, T, is the fraction of incident light that passes through a sample.

Transmittance:
$$T = \frac{P}{P_0} \qquad (18\text{-}4)$$

Transmittance lies in the range 0 to 1. If no light is absorbed, the transmittance is 1. If all light is absorbed, the transmittance is 0. *Percent transmittance* ($100T$) ranges

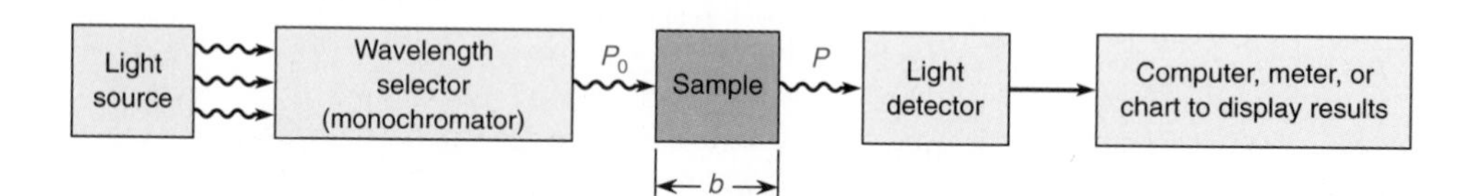

Figure 18-4 Schematic representation of a single-beam spectrophotometric experiment.

from 0% to 100%. A transmittance of 30% means that 70% of the light does not pass through the sample.

The most useful quantity for chemical analysis is **absorbance**, A, defined as

Absorbance:
$$A = \log\left(\frac{P_0}{P}\right) = -\log\left(\frac{P}{P_0}\right) = -\log T \qquad (18\text{-}5)$$

Of course, you remember that $\log\left(\frac{1}{x}\right) = -\log x$.

When no light is absorbed, $P = P_0$ and $A = 0$. If 90% of the light is absorbed, 10% is transmitted and $P = P_0/10$. This ratio gives $A = 1$. If 1% of the light is transmitted, $A = 2$.

P/P_0	% T	A
1	100	0
0.1	10	1
0.01	1	2

Example **Absorbance and Transmittance**

What absorbance corresponds to 99% transmittance? To 0.10% transmittance?

SOLUTION Use the definition of absorbance in Equation 18-5:

99% T: $\quad A = -\log T = -\log 0.99 = 0.004\,4$

0.10% T: $\quad A = -\log T = -\log 0.001\,0 = 3.0$

The higher the absorbance, the less light is transmitted through a sample.

Test Yourself What absorbance corresponds to 1% transmittance? To 50% transmittance? (**Answer:** 2.0, 0.30)

Absorbance is proportional to the concentration of light-absorbing molecules in the sample. Figure 18-5 shows that the absorbance of $KMnO_4$ is proportional to concentration over four orders of magnitude (from 0.6 μM to 3 mM).

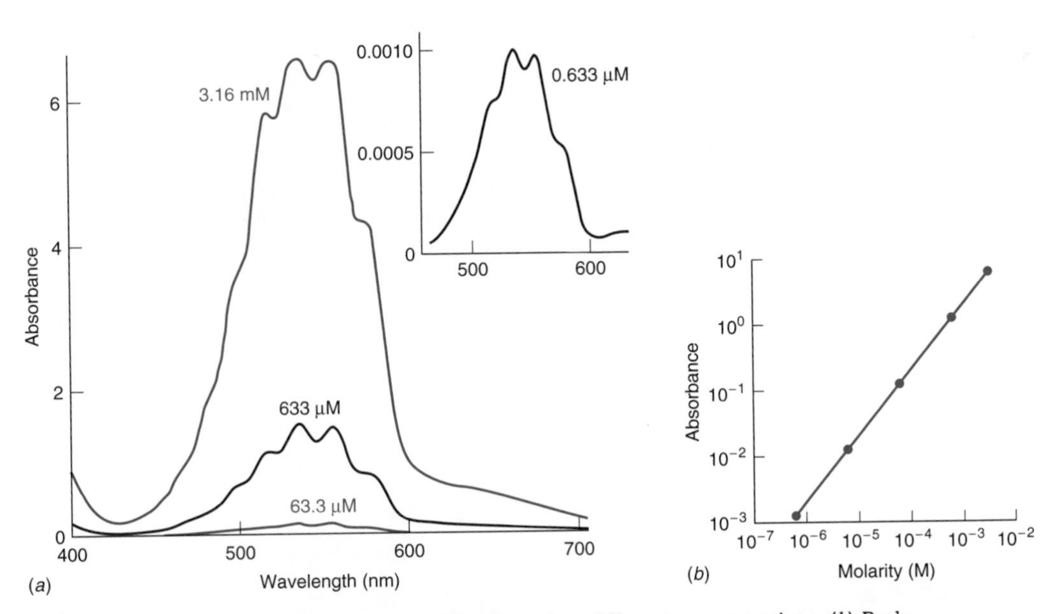

Figure 18-5 (*a*) Absorption spectrum of $KMnO_4$ at four different concentrations. (*b*) Peak absorbance at 555 nm is proportional to concentration from 0.6 μM to 3 mM. The Cary 5000 ultraviolet-visible-near infrared spectrophotometer used for this work has a wider operating range than many instruments. It is usually difficult to measure absorbance accurately above 2 or below 0.01. [From A. R. Hind, *Am. Lab.* December 2002, p. 32. Courtesy Varian, Inc., Palo Alto, CA.]

Absorbance is also proportional to the pathlength of substance through which light travels. The dependence on concentration and pathlength is expressed in **Beer's law**:

Beer's law: $$A = \varepsilon bc \quad (18\text{-}6)$$

Box 18-1 gives a physical picture of Beer's law that could be the basis for a classroom exercise.

Absorbance (A) is dimensionless. Concentration (c) has units of moles per liter (M), and pathlength (b, Figure 18-4) is commonly expressed in centimeters. The quantity ε (epsilon) is called the **molar absorptivity**. It has the units $M^{-1}\ cm^{-1}$ because the product εbc must be dimensionless. Molar absorptivity tells how much light is absorbed at a particular wavelength.

Color Plate 13 shows that color intensity increases as the concentration of the absorbing molecule increases. Absorbance is a measure of the color. The more intense the color, the greater the absorbance.

Example **Using Beer's Law**

The peak absorbance of 3.16×10^{-3} M $KMnO_4$ at 555 nm in a 1.000-cm-pathlength cell in Figure 18-5 is 6.54. **(a)** Find the molar absorptivity and percent

Box 18-1 Discovering Beer's Law[2]

Each photon passing through a solution has a certain probability of striking a light-absorbing molecule and being absorbed. Let's model this process by thinking of an inclined plane with holes representing absorbing molecules. The number of molecules is equal to the number of holes and the pathlength is equal to the length of the plane. Suppose that 1 000 small balls, representing 1 000 photons, are rolled down the incline. Whenever a ball drops through a hole, we consider it to have been "absorbed" by a molecule.

Let the plane be divided into 10 equal intervals and let the probability that a ball will fall through a hole in the first interval be 1/10. Of the 1 000 balls entering the first interval, one-tenth—100 balls—are absorbed (dropping through the holes) and 900 pass into the second interval. Of the 900 balls entering the second interval, one-tenth—90 balls—are absorbed and 810 proceed to the third interval. Of these 810 balls, 81 are absorbed and 729 proceed to the fourth interval. The table summarizes the action.

Transmittance is defined as

$$T = \frac{\text{number of surviving balls}}{\text{initial number of balls } (= 1\,000)}$$

Graph *a* shows that a plot of transmittance versus interval number (which is analogous to plotting trans-

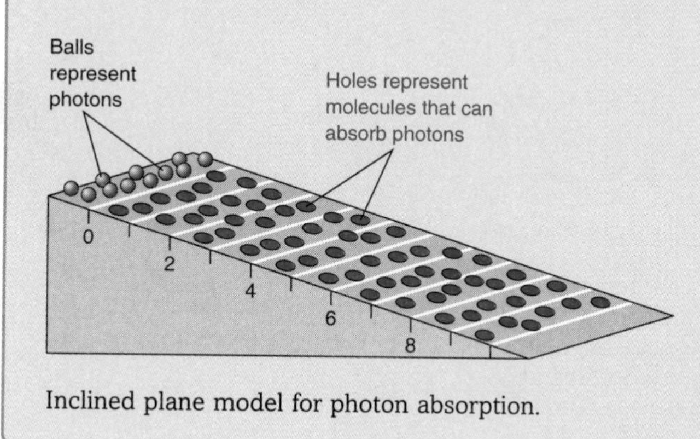

Inclined plane model for photon absorption.

Interval	Photons absorbed	Photons transmitted	Transmittance (P/P_0)
0		1 000	1.000
1	100	900	0.900
2	90	810	0.810
3	81	729	0.729
4	73	656	0.656
5	66	590	0.590
6	59	531	0.531
7	53	478	0.478
8	48	430	0.430
9	43	387	0.387
10	39	348	0.348

transmittance of this solution. **(b)** What would be the absorbance if the pathlength were 0.100 cm? **(c)** What would be the absorbance in a 1.000-cm cell if the concentration were decreased by a factor of 4?

SOLUTION The molar absorptivity in Beer's law is the constant of proportionality between absorbance and the product pathlength × concentration:

(a) $A = \varepsilon bc$

$$6.54 = \varepsilon(1.000\ \text{cm})(3.16 \times 10^{-3}\ \text{M}) \Rightarrow \varepsilon = 2.07 \times 10^{3}\ \text{M}^{-1}\ \text{cm}^{-1}.$$

Equation 18-5 tells us that $A = -\log T$ or $\log T = -A$. We solve for T by raising 10 to the power on both sides of the equation:

$$\log T = -A$$
$$\underbrace{10^{\log T}}_{10^{\log T}\ \text{is the same as}\ T} = 10^{-A}$$
$$T = 10^{-A} = 10^{-6.54} = 2.88 \times 10^{-7}$$

To evaluate $10^{-6.54}$ on your calculator, use y^x or *antilog*. If you use y^x, $y = 10$ and $x = -6.54$. If you use *antilog*, find the antilog of -6.54. Be sure that you can show that $10^{-6.54} = 2.88 \times 10^{-7}$.

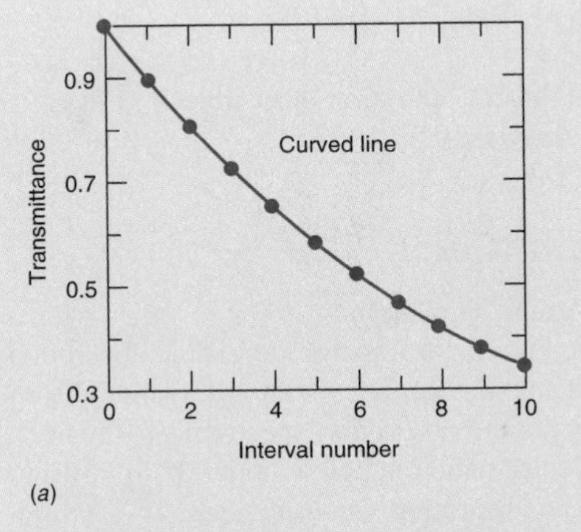

(a)

Transmittance versus interval is analogous to T versus pathlength.

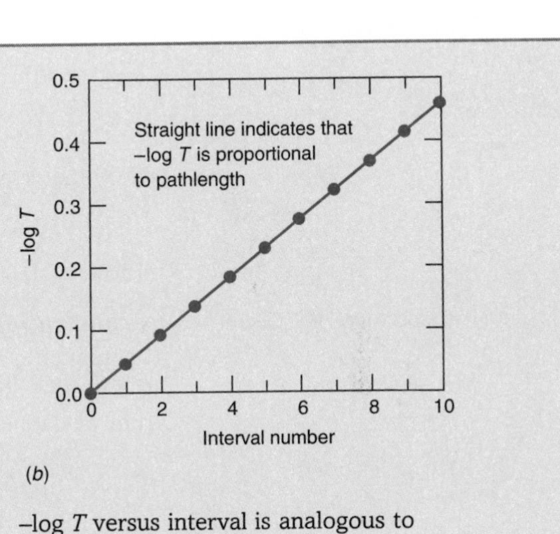

(b)

–log T versus interval is analogous to –log T versus pathlength.

mittance versus pathlength in a spectrophotometric experiment) is not linear. However, the plot of $-\log(\text{transmittance})$ versus interval number in graph *b* is linear and passes through the origin. Graph *b* shows that $-\log T$ is proportional to pathlength.

To investigate how transmittance depends on the concentration of absorbing molecules, you could do the same mental experiment with a different number of holes in the inclined plane. For example, try setting up a table to show what happens if the probability of absorption in each interval is 1/20 instead of 1/10. This change corresponds to decreasing the concentration of absorbing molecules to half of its initial value. You will discover that a graph of $-\log T$ versus interval number has a slope equal to one-half that in graph *b*. That is, $-\log T$ is proportional to concentration as well as to pathlength. So, we have just shown that $-\log T$ is proportional to both concentration and pathlength. Defining absorbance as $-\log T$ gives us the essential terms in Beer's law:

$$A \equiv -\log T \propto \text{concentration} \times \text{pathlength}$$

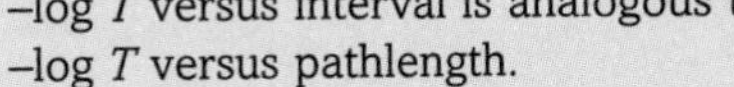

This symbol means "is proportional to."

Percent transmittance is $100T = 2.88 \times 10^{-5}\%$. When the absorbance is 6.54, transmittance is very tiny.
(b) If we decrease pathlength by a factor of 10, we decrease absorbance by a factor of 10 to $6.54/10 = 0.654$.
(c) If we decrease concentration by a factor of 4, we decrease absorbance by a factor of 4 to 6.54/4 to 1.64.

Test Yourself Find the absorbance of a 13.0-μM solution of a compound whose molar absorptivity is $4.64 \times 10^4\ M^{-1}\ cm^{-1}$ in a 1.00-cm cell and in a 2.00-cm cell. (**Answer:** 0.603, 1.206)

Example **Finding Concentration from Absorbance**

Gaseous ozone has a molar absorptivity of $2\,700\ M^{-1}\ cm^{-1}$ at the absorption peak near 260 nm in the spectrum at the beginning of this chapter. Find the concentration of ozone (mol/L) in air if a sample has an absorbance of 0.23 in a 10.0-cm cell. Air has negligible absorbance at 260 nm.

SOLUTION Rearrange Beer's law to solve for concentration:

$$c = \frac{A}{\varepsilon b} = \frac{0.23}{(2\,700\ M^{-1}\ cm^{-1})(10.0\ cm)} = 8.5 \times 10^{-6}\ M$$

Test Yourself What is the concentration of O_3 if $A = 0.18$ at 260 nm in a 2.00-meter-pathlength cell? (**Answer:** 0.33 μM)

Absorption Spectra and Color

The plural of "spectrum" is "spectra."

An **absorption spectrum** is a graph showing how A (or ε) varies with wavelength (or frequency or wavenumber). Figure 18-5 shows the visible absorption spectrum of $KMnO_4$, and the opening of this chapter shows the ultraviolet absorption spectrum of ozone. Figure 18-6 shows the absorption spectrum of a typical sunscreen lotion, which absorbs harmful solar radiation below about 350 nm. Demonstration 18-1 illustrates the meaning of an absorption spectrum.

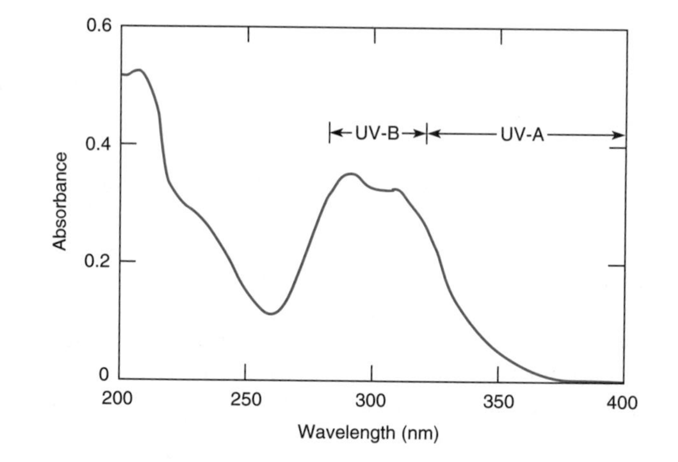

Figure 18-6 Absorption spectrum of typical sunscreen lotion shows absorbance versus wavelength in the ultraviolet region. Sunscreen was thinly coated onto a transparent window to make this measurement. [From D. W. Daniel, *J. Chem. Ed.* **1994**, *71*, 83.] Sunscreen makers refer to the region 400–320 nm as UV-A and 320–280 nm as UV-B.

Demonstration 18-1 Absorption Spectra[3,4]

The spectrum of visible light can be projected on a screen in a darkened room in the following manner: Four layers of plastic diffraction grating[†] are mounted on a cardboard frame that has a square hole large enough to cover the lens of an overhead projector. This assembly is taped over the projector lens facing the screen. An opaque cardboard surface with two 1 × 3 cm slits is placed on the working surface of the projector.

When the lamp is turned on, the white image of each slit is projected on the center of the screen. A visible spectrum appears on either side of each image. When a beaker of colored solution is placed over one slit, you can see color projected on the screen where the white image previously appeared. The spectrum beside the colored image loses its intensity in regions where the colored solution absorbs light.

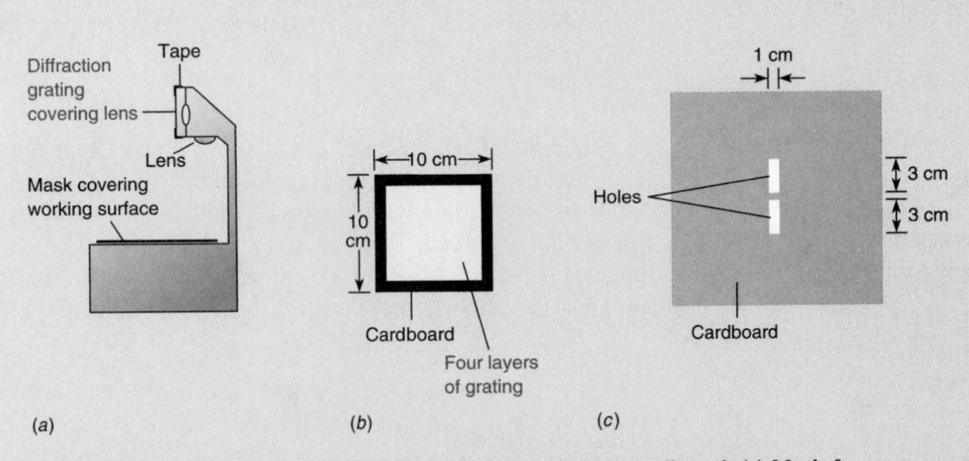

(*a*) Overhead projector. (*b*) Diffraction grating mounted on cardboard. (*c*) Mask for working surface.

Color Plate 14*a* shows the spectrum of white light and the absorption spectra of three different colored solutions. We see that potassium dichromate, which appears orange or yellow, absorbs blue wavelengths. Bromophenol blue absorbs yellow and orange wavelengths and appears blue to our eyes. The absorption of phenolphthalein is located near the center of the visible spectrum. For comparison, the spectra of these three solutions recorded with a spectrophotometer are shown in Color Plate 14*b*.

[†]Edmund Scientific Co., edmundoptics.com, catalog no. NT40-267.

Example How Effective Is Sunscreen?

What fraction of ultraviolet radiation is transmitted through the sunscreen in Figure 18-6 at the peak absorbance near 300 nm?

Problem 18-12 defines the "SPF" number of sunscreens.

SOLUTION From the spectrum in Figure 18-6, the absorbance at 300 nm is approximately 0.35. Therefore the transmittance is $T = 10^{-A} = 10^{-0.35} = 0.45 =$ 45%. Just over half the ultraviolet radiation (55%) is absorbed by sunscreen and does not reach your skin.

Test Yourself If you goop on more sunscreen to double its thickness, absorbance will be doubled. What will be the transmittance near 300 nm and what percentage of ultraviolet radiation is blocked? (**Answer:** $T = 0.20$; 80% is blocked)

Table 18-1 Colors of visible light

Wavelength of maximum absorption (nm)	Color absorbed	Color observed
380–420	Violet	Green-yellow
420–440	Violet-blue	Yellow
440–470	Blue	Orange
470–500	Blue-green	Red
500–520	Green	Purple-red
520–550	Yellow-green	Violet
550–580	Yellow	Violet-blue
580–620	Orange	Blue
620–680	Red	Blue-green
680–780	Red	Green

The color of a substance is the complement of the color that it absorbs.

White light contains all colors of the rainbow. A substance that absorbs visible light appears colored when white light is transmitted through it or reflected from it. The substance absorbs certain wavelengths of white light, and our eyes detect wavelengths that are not absorbed. Table 18-1 is a rough guide to colors. The observed color is the *complement* of the absorbed color. As an example, bromophenol blue in Color Plate 14 has a visible absorbance maximum at 591 nm, and its observed color is blue.

Ask Yourself

18-B. **(a)** What is the absorbance of a 2.33×10^{-4} M solution of a compound with a molar absorptivity of 1.05×10^3 M^{-1} cm^{-1} in a 1.00-cm cell?
(b) What is the transmittance of the solution in **(a)**?
(c) Find A and $\%T$ when the pathlength is doubled to 2.00 cm.
(d) Find A and $\%T$ when the pathlength is 1.00 cm but the concentration is doubled.
(e) What would be the absorbance in **(a)** for a different compound with twice as great a molar absorptivity ($\varepsilon = 2.10 \times 10^3$ M^{-1} cm^{-1})? The concentration and pathlength are unchanged from **(a)**.
(f) Color Plate 15 shows suspensions of silver nanoparticles whose color depends on the size and shape of the particles. From Table 18-1, predict the color of each solution from the wavelength of maximum absorption. Do observed colors agree with predicted colors?

18-3 Practical Matters

For chemical analysis, transmittance is converted into absorbance: $A = -\log T$.

Minimum requirements for a spectrophotometer were shown in Figure 18-4. The instrument measures the fraction of incident light (the transmittance) that passes through a sample to the detector. Sample is usually contained in a cell called a **cuvet** (Figure 18-7), which has flat, fused-silica faces. Fused silica (a glass made of SiO_2) transmits visible and ultraviolet radiation. Plastics and ordinary glass absorb ultraviolet radiation, so plastic or glass cuvets can be used only for measurements at visible wavelengths. Infrared cells are typically made of sodium chloride or potassium bromide crystals. Gases are more dilute than liquids and require cells with longer

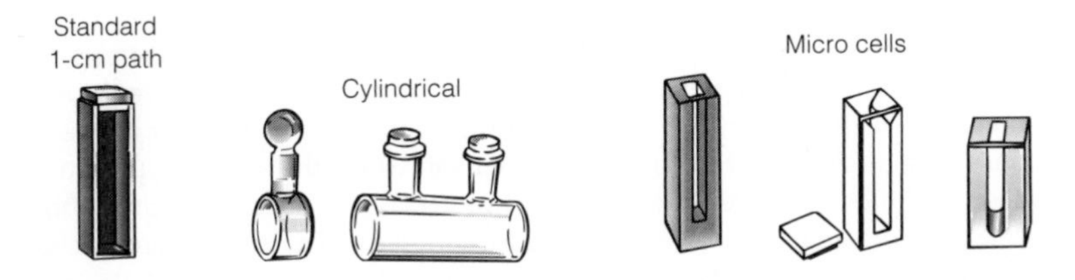

Figure 18-7 Common cuvets for ultraviolet and visible measurements. [From A. H. Thomas Co., Philadelphia, PA.]

pathlengths, typically ranging from 10 cm to many meters. A pathlength of many meters is obtained by reflecting light so that it traverses the sample many times before reaching the detector.

The instrument represented in Figure 18-4 is called a *single-beam spectrophotometer* because it has only one beam of light. We do not measure the incident radiant power, P_0, directly. Rather, the radiant power passing through a reference cuvet containing pure solvent is *defined* as P_0 in Equation 18-4. This cuvet is then removed and replaced by an identical one containing sample. Radiant power striking the detector is then taken as P in Equation 18-4, thus allowing T or A to be determined. The reference cuvet containing pure solvent compensates for reflection, scattering, or absorption of light by the cuvet and solvent. The radiant power reaching the detector would not be the same if the reference cuvet were removed from the beam. A double-beam spectrophotometer housing both a sample cuvet and a reference cuvet is described in Section 19-1.

Radiant power passing through cuvet filled with solvent $\equiv P_0$

Radiant power passing through cuvet filled with sample $\equiv P$

Transmittance $= P/P_0$

Good Operating Techniques

Cuvets should be handled with a tissue to avoid putting fingerprints on the cuvet faces and must be kept scrupulously clean. Fingerprints or contamination from previous samples can scatter or absorb light. Wash the cuvet and rinse it with distilled water as soon as you are finished using it. Let it dry upside down so that water drains out and no water marks are left on the walls. All vessels should be covered to protect them from dust, which scatters light and therefore makes it look as if the absorbance of the sample has increased. Another reason to cover a cuvet is to prevent evaporation of the sample.

Do not touch the clear faces of a cuvet with your fingers. Keep the cuvet scrupulously clean.

Use *matched* cuvets manufactured to have identical pathlength. Any mismatch leads to systematic error. Place each cuvet in the spectrophotometer as reproducibly as possible. One side of the cuvet should be marked so that the cuvet is always oriented the same way. Slight misplacement of the cuvet in its holder, or turning a flat cuvet around by 180°, or rotation of a circular cuvet, can lead to random errors in absorbance. If matched cuvets are not available, use the same cuvet to read the absorbance of both the sample and reference.

Modern spectrophotometers are most precise (reproducible) at intermediate levels of absorbance ($A \approx 0.3$ to 2). If too little light gets through the sample (high absorbance), intensity is hard to measure. If too much light gets through (low absorbance), it is hard to distinguish transmittance of the sample from that of the reference.

Figure 18-8 shows the relative standard deviation of replicate measurements made at 350 nm with a diode array spectrometer. The two curves show measurements made with a new cuvet holder (solid line) or a 10-year-old cuvet holder (dashed, colored line). For the new cuvet holder, removing the sample and replacing it in the holder between measurements had no significant effect. Relative standard deviation is below 0.1% in both cases for absorbance in the range 0.3 to 2. Results

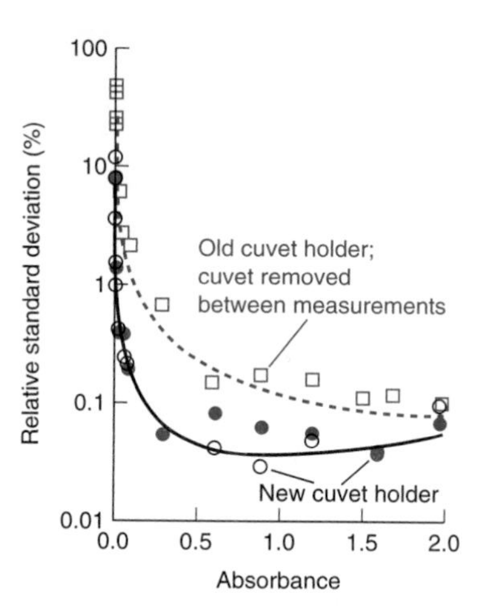

Figure 18-8 Precision of replicate absorbance measurements with a diode array spectrometer at 350 nm with a dichromate solution. Filled circles are from replicate measurements in which the sample was not removed from the cuvet holder between measurements. Open circles are from measurements in which the sample was removed and then replaced in the cuvet holder between measurements. Best reproducibility is observed at intermediate absorbance ($A \approx 0.3$ to 2). Note logarithmic ordinate. Lines are least-squares fit of data to theoretical equations. [Data from J. Galbán, S. de Marcos, I. Sanz, C. Ubide, and J. Zuriarrain, *Anal. Chem.* **2007**, *79*, 4763.]

shown by squares were obtained when a 10-year-old cuvet holder was used and the sample was removed and replaced in the holder between measurements. Variability in the position of the cuvet more than doubles the relative standard deviation. The conclusion is that modern spectrometers have excellent precision and modern cell holders provide excellent reproducibility. Precision was degraded when an old cell holder was used and the sample was removed and inserted between measurements.

For spectrophotometric analysis, measurements are made at a wavelength (λ_{max}) corresponding to a peak in the absorbance spectrum. This wavelength gives the greatest sensitivity—maximum response for a given concentration of analyte. Errors due to wavelength drift and the bandwidth selected by the monochromator are minimized because the spectrum varies least with wavelength at the absorbance maximum.

In measuring a spectrum, it is routine to first record a baseline with pure solvent or a reagent blank in *both* cuvets. In principle, the baseline absorbance should be 0. However, small mismatches between the two cuvets and instrumental imperfections lead to small positive or negative baseline absorbance. The absorbance of the sample is then recorded and the absorbance of the baseline is subtracted from that of the sample to obtain true absorbance.

Ask Yourself

18-C. **(a)** What precautions should you take in handling a cuvet and placing it in the spectrophotometer?
(b) Why is it most accurate to measure absorbances in the range $A = 0.4\text{–}0.9$?

18-4 Using Beer's Law

Spectrophotometric analysis with visible radiation is called *colorimetric* analysis. Box 18-2 gives an example of the rational design of a colorimetric analysis.

For a compound to be analyzed by spectrophotometry, it must absorb electromagnetic radiation, and this absorption should be distinguishable from that of other species in the sample. Biochemists assay proteins in the ultraviolet region at 280 nm because the aromatic amino acids tyrosine, phenylalanine, and tryptophan (Table 11-1) have maximum absorbance near 280 nm. Other common solutes such as salts, buffers, and carbohydrates have little absorbance at this wavelength. In this section, we use Beer's law for a simple analysis and then discuss the measurement of nitrite in an aquarium.

Benzene
C_6H_6

Example **Measuring Benzene in Hexane**

(a) A solution prepared by dissolving 25.8 mg of benzene (C_6H_6, FM 78.11) in hexane and diluting to 250.0 mL has an absorption peak at 256 nm, with an absorbance of 0.266 in a 1.000-cm cell. Hexane does not absorb at 256 nm. Find the molar absorptivity of benzene at this wavelength.

Box 18-2 Designing a Colorimetric Reagent to Detect Phosphate[5]

Chemists in Korea demonstrated a clever, rational approach to designing a reagent for the spectrophotometric analysis of phosphate. A ligand containing 6 N atoms and 1 O atom that could bind 2 Zn^{2+} ions was selected. The distance between Zn^{2+} ions is just right for the metal ion indicator pyrocatechol violet to bind, as shown at the left below. Recall from Chapter 13 that a metal ion indicator has one color when it binds to metal and a different color when free. Pyrocatechol violet is blue when bound to metal and yellow when free.

Near neutral pH, phosphate binds tightly to the two Zn^{2+} ions. When phosphate is added, indicator is displaced and the color changes from blue to yellow. The change in the absorption spectrum provides a quantitative measure of the amount of phosphate added. Color Plate 16 shows that common anions do not displace indicator from Zn^{2+} and therefore do not interfere in the analysis.

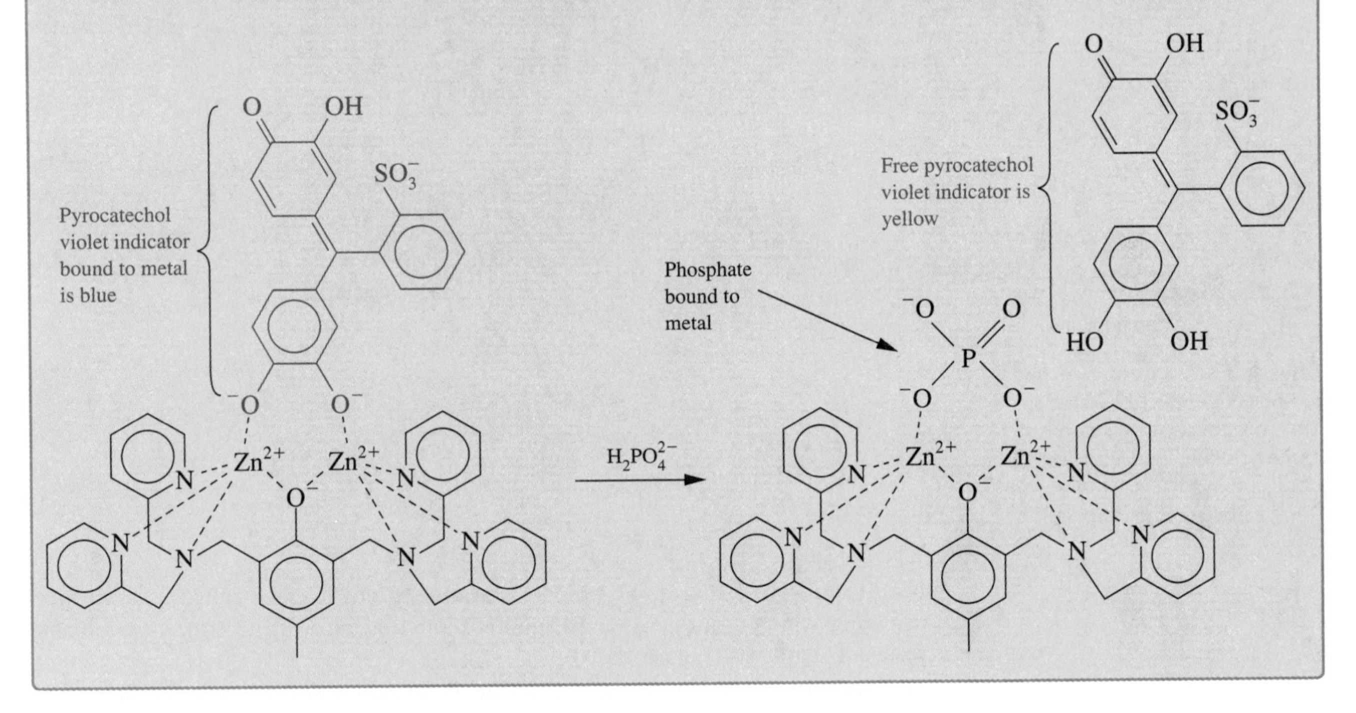

SOLUTION The concentration of benzene is

$$[C_6H_6] = \frac{(0.025\,8\ g)/(78.11\ g/mol)}{0.250\,0\ L} = 1.32_1 \times 10^{-3}\ M$$

We find the molar absorptivity from Beer's law:

$$\text{molar absorptivity} = \varepsilon = \frac{A}{bc} = \frac{0.266}{(1.000\ cm)(1.32_1 \times 10^{-3}\ M)} = 201._3\ M^{-1}\ cm^{-1}$$

(b) A sample of hexane contaminated with benzene has an absorbance of 0.070 at 256 nm in a cell with a 5.000-cm pathlength. Find the concentration of benzene.

SOLUTION Use the molar absorptivity from **(a)** in Beer's law:

$$[C_6H_6] = \frac{A}{\varepsilon b} = \frac{0.070}{(201._3\ M^{-1}\ cm^{-1})(5.000\ cm)} = 7.0 \times 10^{-5}\ M$$

Test Yourself Find the concentration of benzene in hexane if the absorbance is 0.188 in a 1.000-cm cell. (**Answer:** 0.934 mM)

Beer's law:

$$A = \varepsilon bc$$

A = absorbance (dimensionless)
ε = molar absorptivity ($M^{-1}\ cm^{-1}$)
b = pathlength (cm)
c = concentration (M)
ε has funny units so that the product εbc will be dimensionless.

Using a Standard Curve to Measure Nitrite

Box 6-1 stated that nitrogen compounds derived from animals and plants are broken down to ammonia by heterotrophic bacteria. Ammonia is oxidized first to nitrite (NO_2^-) and then to nitrate (NO_3^-) by nitrifying bacteria. In Section 6-3 we saw how a permanganate titration was used to standardize a nitrite stock solution. The nitrite solution is used here to prepare standards for a spectrophotometric analysis of nitrite in aquarium water.

The aquarium nitrite analysis is based on a reaction whose colored product has an absorbance maximum at 543 nm (Figure 18-9):

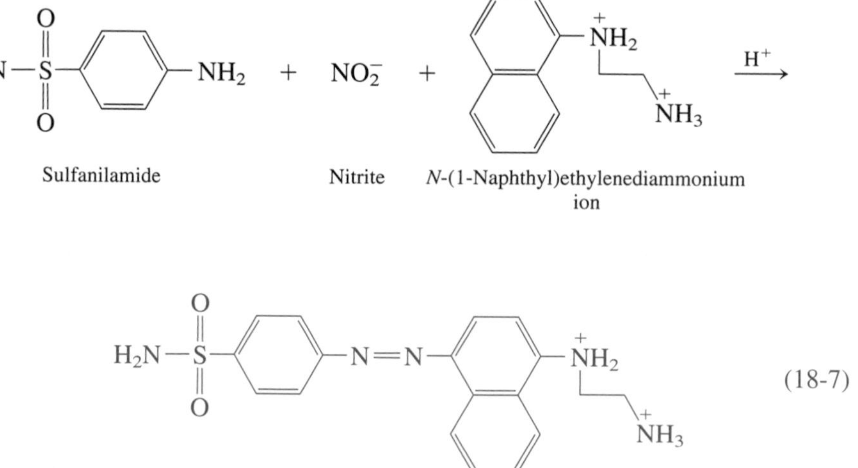

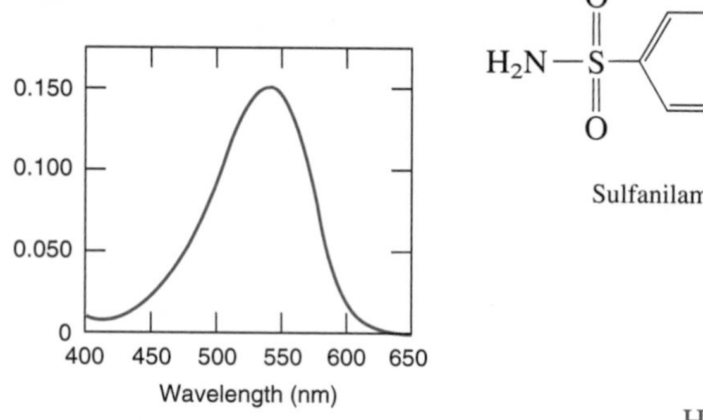

Figure 18-9 Spectrum of the red-purple product of Reaction 18-7, beginning with a standard nitrite solution containing 0.915 ppm nitrogen. [From Kenneth Hughes, Georgia Institute of Technology.]

For quantitative analysis, we prepare a **standard curve** (also called a *calibration curve*) in which absorbance at 543 nm is plotted against nitrite concentration in a series of standards (Figure 18-10).

The procedure for measuring nitrite is to add color-forming reagent to an unknown or standard, wait 10 min for the reaction to be completed, and measure the absorbance. A *reagent blank* is prepared with nitrite-free, artificial seawater in place of unknown or standards. *The absorbance of the blank is subtracted from the absorbance of all other samples prior to any calculations.* The purpose of the blank is to subtract absorbance at 543 nm arising from starting materials or impurities. Here are the details.

The saltwater aquarium is filled with artificial seawater made by adding water to a mixture of salts.

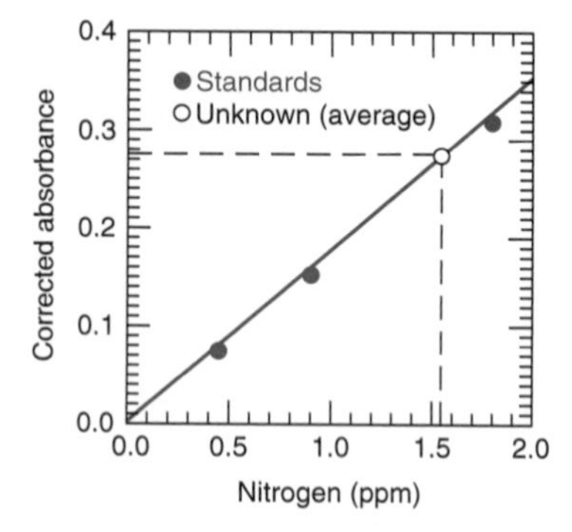

Figure 18-10 Calibration curve for nitrite analysis with corrected absorbance values from Table 18-2.

Reagents

1. *Color-forming reagent* is prepared by mixing 1.0 g of sulfanilamide, 0.10 g of *N*-(1-naphthyl)ethylenediamine dihydrochloride, and 10 mL of 85 wt% phosphoric acid and diluting to 100 mL. Store the solution in a dark bottle in the refrigerator to prevent thermal and photochemical degradation.
2. *Standard nitrite* (~0.02 M) is prepared by dissolving $NaNO_2$ (FM 68.995) in water and standardizing (measuring the concentration) by the titration in Section 6-3. Dilute the concentrated standard with artificial seawater (containing no nitrite) to prepare standards containing 0.5–3 ppm nitrite nitrogen.

Table 18-2 Aquarium nitrite analysis

Sample	Absorbance at 543 nm in 1.000-cm cuvet	Corrected absorbance (blank subtracted)
Blank	0.003	—
Standards		
0.457 5 ppm	0.085	0.082
0.915 0 ppm	0.167	0.164
1.830 ppm	0.328	0.325
Unknown	0.281	0.278
Unknown	0.277	0.274

Procedure

For each analysis, dilute 10.00 mL of standard (containing 0.5–3 ppm nitrite nitrogen) or unknown up to 100.0 mL with water. Diluted solutions will contain 0.05–0.3 ppm nitrite nitrogen. Place 25.00 mL of diluted solution in a flask and add 1.00 mL of color-forming reagent. After 10 min, measure the absorbance in a 1.000-cm cuvet.

1. Construct a *standard curve* from known nitrite solutions. Prepare a *reagent blank* by carrying artificial seawater through the same steps as a standard.
2. Analyze duplicate samples of *unknown* aquarium water that has been filtered prior to dilution to remove suspended solids. Several trial dilutions may be required before the aquarium water is dilute enough to have a nitrite concentration that falls within the calibration range.

Unknown should be adjusted to fall within the calibration range, because you have not verified a linear response outside the calibration range.

Table 18-2 and Figure 18-10 show typical results. The equation of the calibration line in Figure 18-10, determined by the method of least squares, is

$$\text{absorbance} = 0.176\,9\,[\text{ppm}] + 0.001\,5 \tag{18-8}$$

where [ppm] represents micrograms of nitrite nitrogen per milliliter. In principle, the intercept should be 0, but we will use the observed intercept (0.001 5) in our calculations. By plugging the average absorbance of unknown into Equation 18-8, we can solve for the concentration of nitrite (ppm) in the unknown.

Example Preparing Nitrite Standards

How would you prepare a nitrite standard containing approximately 2 ppm nitrite nitrogen from a concentrated standard containing 0.018 74 M $NaNO_2$?

SOLUTION First, let's find out how many ppm of nitrogen are in 0.018 74 M $NaNO_2$. Because 1 mol of nitrite contains 1 mol of nitrogen, the concentration of nitrogen in the concentrated standard is 0.018 74 M. The mass of nitrogen in 1 mL is

$$\frac{\text{g of N}}{\text{mL}} = \left(0.018\,74\,\frac{\cancel{\text{mol}}}{\cancel{\text{L}}}\right)\left(14.007\,\frac{\text{g N}}{\cancel{\text{mol}}}\right)\left(0.001\,\frac{\cancel{\text{L}}}{\text{mL}}\right) = 2.625 \times 10^{-4}\,\frac{\text{g N}}{\text{mL}}$$

Assuming that 1.00 mL of solution has a mass of 1.00 g, we use the definition of parts per million to convert the mass of nitrogen into ppm:

$$\text{ppm} = \frac{\text{g N}}{\text{g solution}} \times 10^6 = \frac{2.625 \times 10^{-4}\ \text{g N}}{1.00\ \text{g solution}} \times 10^6 = 262.5\ \text{ppm}$$

Dilution formula 1-5:

$$M_{conc} \cdot V_{conc} = M_{dil} \cdot V_{dil}$$

Use any units you like for M and V, but use the same units on both sides of the equation. M could be ppm and V could be mL.

To prepare a standard containing ~2 ppm N, you could dilute the concentrated standard by a factor of 100 to give 2.625 ppm N. This dilution could be done by pipeting 10.00 mL of concentrated standard into a 1-L volumetric flask and diluting to the mark.

Test Yourself How might you prepare a standard containing ~5 ppm N from 0.013 37 M $NaNO_2$? (**Answer:** stock concentration is 187.3 ppm N; dilute 25.00 mL to 1.000 L ⇒ 4.682 ppm N)

Example **Using the Standard Curve**

From the data in Table 18-2, find the molarity of nitrite in the aquarium.

SOLUTION The average corrected absorbance of unknowns in Table 18-2 is 0.276. Substituting this value into Equation 18-8 gives ppm of nitrite nitrogen in the aquarium:

$$0.276 = 0.176\,9\,[\text{ppm}] + 0.001\,5$$

$$[\text{ppm}] = \frac{0.276 - 0.001\,5}{0.176\,9} = 1.55\ \text{ppm} = 1.55\ \frac{\mu\text{g N}}{\text{mL}}$$

To find the molarity of nitrite nitrogen, we first find the mass of nitrogen in a liter, which is

$$1.55 \times 10^{-6}\ \frac{\text{g N}}{\cancel{\text{mL}}} \times 1\,000\ \frac{\cancel{\text{mL}}}{\text{L}} = 1.55 \times 10^{-3}\ \frac{\text{g N}}{\text{L}}$$

Then we convert mass of nitrogen into moles of nitrogen:

$$[\text{nitrite nitrogen}] = \frac{1.55 \times 10^{-3}\ \cancel{\text{g N}}/\text{L}}{14.007\ \cancel{\text{g N}}/\text{mol}} = 1.11 \times 10^{-4}\ \text{M}$$

Because 1 mol of nitrite (NO_2^-) contains 1 mol of nitrogen, the concentration of nitrite is also 1.11×10^{-4} M.

Test Yourself What is the molarity of nitrite if the observed (uncorrected) absorbance is 0.400? (**Answer:** 1.60×10^{-4} M)

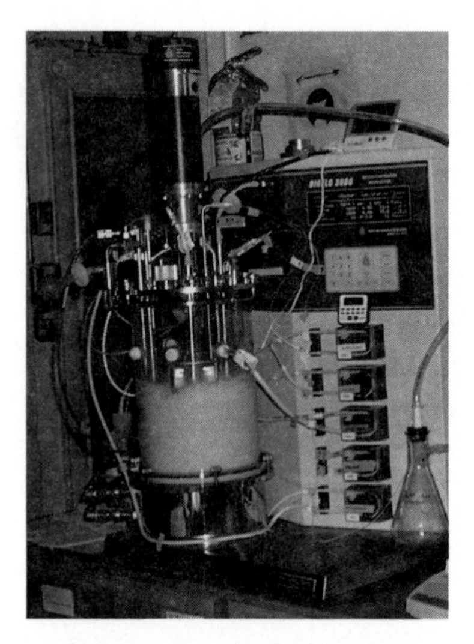

Figure 18-11 Nitrate reductase derived from corn leaves was not easy to produce for NO_3^- analysis. A stable, *recombinant* form of the enzyme is made on a commercial scale from yeast cells in the 14-L fermentor in this photo. Recombinant DNA technology permits nitrate reductase genes from *Arabidopsis thaliana* (a flowering plant also known as thale cress) to be expressed by *Pichia pastoris* (yeast). [W. H. Campbell, P. Song, and G. G. Barbier, *Environ. Chem. Lett.* **2006**, *4*, 69. Photo countesy W. H. Campbell, The Nitrate Elimination Co., Lake Linden, MI.]

Enzyme-Based Nitrate Analysis—A Green Idea

Nitrate (NO_3^-) in natural waters is derived from sources such as fertilizers and undertreated animal and human waste. U.S. environmental regulations set a maximum of 10 ppm NO_3^- nitrogen in drinking water. Nitrate is commonly analyzed by reduction to nitrite (NO_2^-), followed by a colorimetric assay of NO_2^-. Metallic Cd has been the most common reducing agent for NO_3^-. However, the use of toxic Cd should be curtailed to protect the environment.

Therefore, a field test for NO_3^- was developed with the biological reducing agent β-nicotinamide adenine dinucleotide (NADH, derived from the vitamin niacin) instead of Cd. The enzyme nitrate reductase (Figure 18-11) catalyzes the reduction:

$$NO_3^- + NADH + H^+ \xrightarrow[\text{pH 7}]{\text{nitrate reductase}} NO_2^- + NAD^+ + H_2O \qquad (18\text{-}9)$$

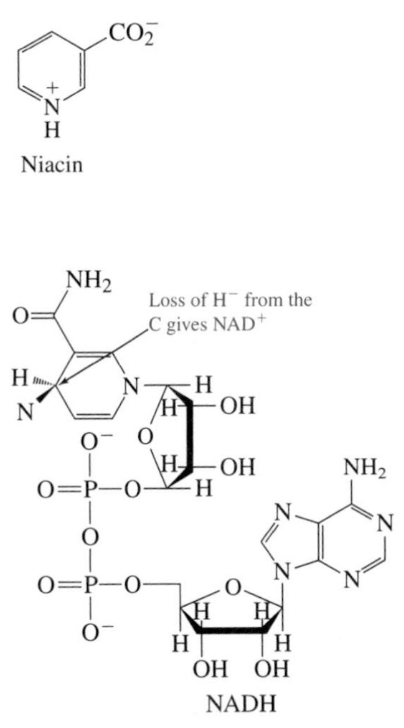

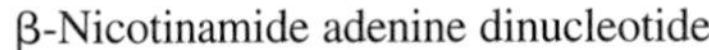
β-Nicotinamide adenine dinucleotide

Excess NADH is then oxidized to NAD^+ to eliminate interference with color development when NO_2^- from Reaction 18-9 is measured colorimetrically by reactions such as 18-7. For quantitative analysis in the field, a small, battery-operated spectrophotometer can be used with a set of nitrate standards. Alternatively, color can be compared visually with a chart showing colors from several standards. Commercial field kits allow analysis in the range 0.05–10 ppm nitrate nitrogen. Laboratory apparatus provides a precision of 2% when measuring 0.2 ppm NO_3^- nitrogen, with a detection limit of 3 ppb. The nitrate reductase procedure has been applied to the measurement of nitrate in a classroom aquarium.[6]

Ask Yourself

18-D. You have been sent to India to investigate the occurrence of goiter disease attributed to iodine deficiency. As part of your investigation, you make field measurements of traces of iodide (I^-) in groundwater. The procedure is to oxidize I^- to I_2 and convert I_2 into a colored complex with the dye brilliant green in the solvent toluene.

(a) A 3.15×10^{-6} M solution of the colored complex exhibited an absorbance of 0.267 at 635 nm in a 1.000-cm cuvet. A blank solution made from distilled water in place of groundwater had an absorbance of 0.019. Find the molar absorptivity of the colored complex.

(b) The absorbance of an unknown solution prepared from groundwater was 0.175. Subtract the blank absorbance from the unknown absorbance and use Beer's law to find the concentration of the unknown.

Key Equations

Frequency-wavelength relation	$\nu\lambda = c$
	ν = frequency $\quad \lambda$ = wavelength $\quad c$ = speed of light
Wavenumber	$\tilde{\nu} = 1/\lambda$
Photon energy	$E = h\nu = hc/\lambda = hc\tilde{\nu}$
	h = Planck's constant
Transmittance	$T = P/P_0$
	P_0 = radiant intensity of light incident on sample
	P = radiant intensity of light emerging from sample
Absorbance	$A = -\log T$
Beer's law	$A = \varepsilon bc$
	ε = molar absorptivity of absorbing species ($M^{-1}\ cm^{-1}$)
	b = pathlength (cm)
	c = concentration of absorbing species (M)

Important Terms

absorbance
absorption spectrum
Beer's law
cuvet
electromagnetic spectrum
excited state
frequency
ground state
hertz
molar absorptivity
monochromatic light
photon
spectrophotometer
spectrophotometry
standard curve
transmittance
wavelength
wavenumber

Problems

18-1. **(a)** When you double the frequency of electromagnetic radiation, you __________ the energy.

(b) When you double the wavelength, you __________ the energy.

(c) When you double the wavenumber, you __________ the energy.

18-2. How much energy (J) is carried by one photon of **(a)** red light with $\lambda = 650$ nm? **(b)** violet light with $\lambda = 400$ nm? After finding the energy of one photon of each wavelength, express the energy of a mole of each type of photons in kJ/mol.

18-3. What color would you expect for light transmitted through a solution with an absorption maximum at **(a)** 450; **(b)** 550; **(c)** 650 nm?

18-4. State the difference between transmittance, absorbance, and molar absorptivity. Which one is proportional to concentration?

18-5. An absorption spectrum is a graph of ________ or ________ versus ________.

18-6. Why does a compound that has a visible absorption maximum at 480 nm (blue-green) appear to be red?

18-7. What color would you expect to observe for a solution with a visible absorbance maximum at 562 nm?

18-8. Calculate the frequency (Hz), wavenumber (cm^{-1}), and energy (J/photon and kJ/mole of photons) of **(a)** ultraviolet light with a wavelength of 250 nm and **(b)** infrared light with a wavelength of 2.50 μm.

18-9. Industrial CO_2 lasers with $\lambda = 10.6$ μm are used in cutting and welding. How many photons/s are produced by a laser whose output is 5.0 kW? Recall that 1 watt = 1 joule/s.

18-10. Convert transmittance (*T*) into absorbance (*A*):

T:	0.99	0.90	0.50	0.10	0.010	0.001 0	0.000 10
A:				1.0			

18-11. Find the absorbance and percent transmittance of a 0.002 40 M solution of a substance with a molar absorptivity of 1.00×10^2 or 2.00×10^2 M^{-1} cm^{-1} in a cell with a 2.00-cm pathlength.

18-12. The "SPF number" of a sunscreen states how long you can be in the sun before your skin turns red relative to how long it would take without sunscreen:[7] SPF = $1/T$, where T is the transmission of UV-B radiation (Figure 18-6) through a uniformly applied layer of sunscreen at a concentration of 2 mg/cm^2.

(a) What are the transmittance and absorbance when SPF = 2? What fraction of UV-B radiation is absorbed by this sunscreen? Explain why your answer makes sense in terms of how long the sunscreen should protect you.

(b) Find the transmittance and absorbance when SPF = 10 and SPF = 20. What fraction of UB-B radiation is absorbed by each?

18-13. The absorbance of a 2.31×10^{-5} M solution is 0.822 at a wavelength of 266 nm in a 1.00-cm cell. Calculate the molar absorptivity at 266 nm.

18-14. The iron-transport protein in your blood is called transferrin. When its two iron-binding sites do not contain metal ions, the protein is called apotransferrin.

(a) Apotransferrin has a molar absorptivity of 8.83×10^4 M^{-1} cm^{-1} at 280 nm. Find the concentration of apotransferrin in water if the absorbance is 0.244 in a 0.100-cm cell.

(b) The formula mass of apotransferrin is 81 000. Express the concentration from **(a)** in g/L.

18-15. A 15.0-mg sample of a compound with a formula mass of 384.63 was dissolved in a 5-mL volumetric flask. A 1.00-mL aliquot was withdrawn, placed in a 10-mL volumetric flask, and diluted to the mark.

(a) Find the concentration of sample in the 5-mL flask.

(b) Find the concentration in the 10-mL flask.

(c) The 10-mL sample was placed in a 0.500-cm cuvet and gave an absorbance of 0.634 at 495 nm. Find the molar absorptivity at 495 nm.

18-16. **(a)** In Figure 18-6, measure the peak absorbance of sunscreen near 215 nm.

(b) What fraction of ultraviolet radiation is transmitted through the sunscreen near 215 nm?

18-17. **(a)** What value of absorbance corresponds to 45.0% T?

(b) When the concentration of a solution is doubled, the *absorbance* is doubled. If a 0.010 0 M solution exhibits 45.0% T at some wavelength, what will be the percent transmittance for a 0.020 0 M solution of the same substance?

18-18. A 0.267-g quantity of a compound with a formula mass of 337.69 was dissolved in 100.0 mL of ethanol. Then 2.000 mL were withdrawn and diluted to 100.0 mL. The spectrum of this solution exhibited a maximum absorbance of 0.728 at 438 nm in a 2.000-cm cell. Find the molar absorptivity of the compound.

18-19. Transmittance of vapor from the solid compound pyrazine was measured at a wavelength of 266 nm in a 3.00-cm cell at 298 K.

Pressure (μbar)	Transmittance (%)
4.3	83.8
11.4	61.6
20.0	39.6
30.3	24.4
60.7	5.76
99.7	0.857
134.5	0.147

Data from M. A. Muyskens and E. T. Sevy, *J. Chem. Ed.* **1997**, *74*, 1138.

(a) Use the ideal gas law (Problem 16-15) to convert pressure into concentration in mol/L. Convert transmittance into absorbance.

(b) Prepare a graph of absorbance versus concentration to see whether the data conform to Beer's law. Find the molar absorptivity from the slope of the graph.

18-20. **(a)** A 3.96×10^{-4} M solution of compound A exhibited an absorbance of 0.624 at 238 nm in a 1.000-cm cuvet. A blank solution containing only solvent had an absorbance of 0.029 at the same wavelength. Find the molar absorptivity of compound A.

(b) The absorbance of an unknown solution of compound A in the same solvent and cuvet was 0.375 at 238 nm. Subtract the blank absorbance from the unknown absorbance and use Beer's law to find the concentration of the unknown.

(c) A concentrated solution of compound A in the same solvent was diluted from an initial volume of 2.00 mL to a final volume of 25.00 mL and then had an absorbance of 0.733. What is the concentration of A in the 25.00-mL solution?

(d) Considering the dilution from 2.00 mL to 25.00 mL, what was the concentration of A in the 2.00-mL solution in **(c)**?

18-21. A compound with a formula mass of 292.16 was dissolved in solvent in a 5-mL volumetric flask and diluted to the mark. A 1.00-mL aliquot was withdrawn, placed in a 10-mL volumetric flask, and diluted to the mark. The absorbance measured at 340 nm was 0.427 in a 1.000-cm cuvet. The molar absorptivity for this compound at 340 nm is 6 130 M^{-1} cm^{-1}.

(a) Calculate the concentration of compound in the cuvet.

(b) What was the concentration of compound in the 5-mL flask?

(c) How many milligrams of compound were used to make the 5-mL solution?

18-22. When I was a boy, I watched Uncle Wilbur measure the iron content of runoff from his banana ranch. He acidified a 25.0-mL sample with HNO_3 and treated it with excess KSCN to form a red complex. (KSCN itself is colorless.) He then diluted the solution to 100.0 mL and put it in a variable-pathlength cell. For comparison, he treated a 10.0-mL reference sample of 6.80×10^{-4} M Fe^{3+} with HNO_3 and KSCN and diluted it to 50.0 mL. The reference was placed in a cell with a 1.00-cm pathlength. Runoff had the same absorbance as the reference when the pathlength of the runoff cell was 2.48 cm. What was the concentration of iron in Uncle Wilbur's runoff?

18-23. A nitrite analysis conducted by the procedure in Section 18-4 gave data in the following table. Fill in corrected absorbance, which is measured absorbance minus the average blank absorbance (0.023). Construct a calibration line to find **(a)** ppm nitrite nitrogen ± uncertainty and **(b)** molar concentration of nitrite in the aquarium. Use the average blank and the average unknown absorbances.

Sample	Absorbance	Corrected absorbance
Blank	0.022	—
Blank	0.024	—
Standards:		
0.538 ppm	0.121	0.098
1.076 ppm	0.219	
2.152 ppm	0.413	
3.228 ppm	0.600	
4.034 ppm	0.755	
Unknown	0.333	
Unknown	0.339	
Unknown	0.338	

18-24. Starting with 0.015 83 M $NaNO_2$ solution, explain how you would prepare standards containing *approximately* 0.5, 1, 2, and 3 ppm nitrogen (1 ppm = 1 μg/mL). Use any volumetric flasks and transfer pipets in Tables 2-2 and 2-3. What would be the exact concentrations of the standards prepared by your method?

18-25. Nitrate in a classroom aquarium was determined by the nitrate reductase enzyme procedure described at the end of Section 18-4.

(a) Write the sequence of chemical reactions beginning with nitrate and ending with the colored product. Which step is catalyzed by the enzyme?

(b) Each solution whose absorbance was measured contained 50.0 μL of standard or aquarium water mixed with reagents to give a total volume of 2.02 mL. Prepare a calibration curve showing absorbance versus nitrate nitrogen content in the 50-μL standards. No data are given for the blank, so do not subtract anything from the observed absorbance. Find the average concentration and uncertainty (Equation 4-16) of nitrate nitrogen (ppm) in the aquarium water.

Nitrate nitrogen (ppm) in 50-μL sample	Absorbance at 540 nm
0.250	0.062
0.500	0.069
1.00	0.108
1.50	0.126
2.50	0.209
5.00	0.423
7.50	0.592
10.00	0.761
aquarium	0.192
aquarium	0.201

Data from H. Van Ryswyk, E. W. Hall, S. J. Petesch, and A. E. Wiedeman, *J. Chem. Ed.* **2007**, *84*, 306.

(c) Nitrate nitrogen (ppm) means micrograms of nitrate N per gram of solution. The slope of the calibration line is (absorbance)/(ppm nitrate N in standard). Find the molarity of colored product if the standard contains 1.00 ppm nitrate N. Estimate the molar absorptivity of colored product, assuming that the cuvet pathlength is 1.00 cm and the density of standard solutions is near 1.00 g/mL.

18-26. Starting with 28.6 wt% NH_3, explain how to prepare NH_3 standards containing exactly 1.00, 2.00, 4.00, and 8.00 ppm nitrogen (1 ppm = 1 μg/mL) for a spectrophotometric calibration curve. Use a known *mass* of concentrated reagent and any flasks and pipets from Tables 2-2 and 2-3.

18-27. Ammonia (NH_3) is determined spectrophotometrically by reaction with phenol in the presence of hypochlorite (OCl^-):

$$\underset{\text{Colorless}}{\text{phenol}} + \underset{\text{Colorless}}{\text{ammonia}} \xrightarrow{OCl^-} \underset{\lambda_{max} = 625 \text{ nm}}{\text{blue product}}$$

1. A 4.37-mg sample of protein was chemically digested to convert its nitrogen into NH_3 and then diluted to 100.0 mL.

2. Then 10.0 mL of the solution were placed in a 50-mL volumetric flask and treated with 5 mL of phenol solution plus 2 mL of NaOCl solution. The sample was diluted to 50.0 mL, and the absorbance at 625 nm was measured in a 1.00-cm cuvet after 30 min.

3. A standard solution was prepared from 0.010 0 g of NH_4Cl (FM 53.49) dissolved in 1.00 L of water. A 10.0-mL aliquot of this standard was placed in a 50-mL volumetric flask and analyzed in the same manner as the unknown.

4. A reagent blank was prepared by using distilled water in place of unknown.

Sample	Absorbance at 625 nm
Blank	0.140
Standard	0.308
Unknown	0.592

(a) From step 3, calculate the molar absorptivity of the blue product.

(b) Using molar absorptivity, find [NH_3] in step 2.

(c) From **(b)**, find [NH_3] in the 100-mL solution in step 1.

(d) Find the weight percent of nitrogen in the protein.

18-28. Cu^+ reacts with neocuproine to form (neocuproine)$_2$-Cu^+, with an absorption maximum at 454 nm. Neocuproine reacts with few other metals. The copper complex is soluble in isoamyl alcohol, an organic solvent that does not dissolve appreciably in water. When isoamyl alcohol is added to water, a two-layered mixture results, with the denser water layer at the bottom. If (neocuproine)$_2Cu^+$ is present, virtually all of it goes into the organic phase. For the purpose of this problem, assume that no isoamyl alcohol dissolves in water and that the colored complex is only in the organic phase.

1. A rock containing copper is pulverized, and metals are extracted with strong acid. The acid is neutralized with base and made up to 250.0 mL in flask A.

2. 10.00 mL of solution are transferred to flask B and treated with 10.00 mL of reducing agent to reduce Cu^{2+} to Cu^+. Then 10.00 mL of buffer are added to bring the pH to a value suitable for complex formation with neocuproine.

3. 15.00 mL of solution are withdrawn and placed in flask C. To the flask are added 10.00 mL of aqueous neocuproine and 20.00 mL of isoamyl alcohol. After the flask has been shaken well and the phases allowed to separate, all (neocuproine)$_2Cu^+$ is in the organic phase.

For Problem 18-28.

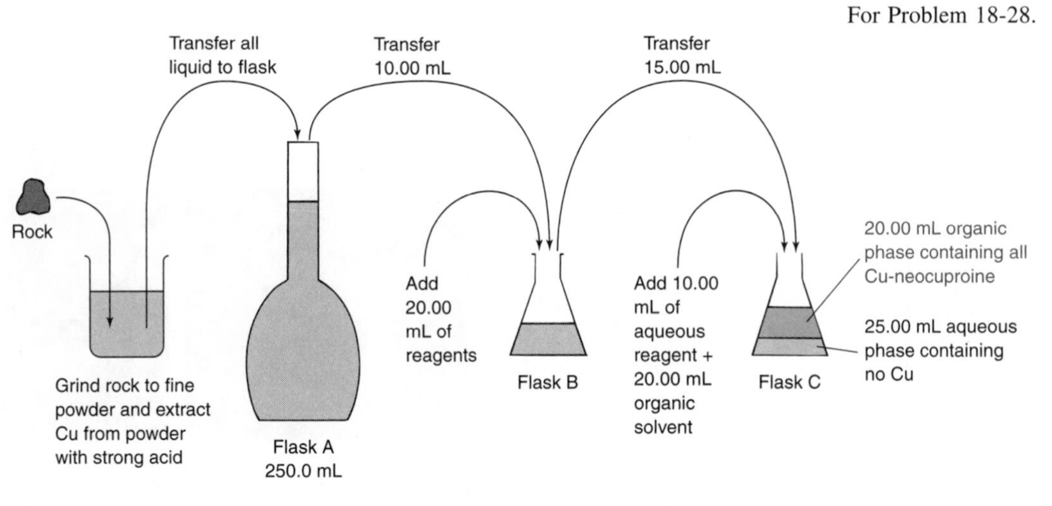

4. A few milliliters of the upper layer are withdrawn, and absorbance is measured at 454 nm in a 1.00-cm cell. A blank carried through the same procedure gave an absorbance of 0.056.

(a) Suppose that the rock contained 1.00 mg of Cu. What will be the concentration of copper (moles per liter) in the isoamyl alcohol phase?

(b) If the molar absorptivity of $(neocuproine)_2Cu^+$ is $7.90 \times 10^3\ M^{-1}\ cm^{-1}$, what will the observed absorbance be? Remember that a blank carried through the same procedure gave an absorbance of 0.056.

(c) A rock is analyzed and found to give a final absorbance of 0.874 (uncorrected for the blank). How many milligrams of copper are in the rock?

18-29. Spectrophotometric analysis of phosphate:

Standard solutions

A. KH_2PO_4 (potassium dihydrogen phosphate, FM 136.09): 81.37 mg dissolved in 500.0 mL H_2O

B. $Na_2MoO_4 \cdot 2H_2O$ (sodium molybdate): 1.25 g in 50 mL of 5 M H_2SO_4

C. $H_3NNH_3^{2+}SO_4^{2-}$ (hydrazine sulfate): 0.15 g in 100 mL H_2O

Procedure

Place sample (unknown or standard phosphate solution, A) in a 5-mL volumetric flask and add 0.500 mL of B and 0.200 mL of C. Dilute to almost 5 mL with water and heat at 100°C for 10 min to form a blue product ($H_3PO_4(MoO_3)_{12}$, 12-molybdophosphoric acid). Cool the flask to room temperature, dilute to the mark with water, mix well, and measure absorbance at 830 nm in a 1.00-cm cell.

(a) When 0.140 mL of solution A was analyzed, an absorbance of 0.829 was recorded. A blank carried through the same procedure gave an absorbance of 0.017. Find the molar absorptivity of blue product.

(b) A solution of the phosphate-containing iron-storage protein ferritin was analyzed. Unknown containing 1.35 mg ferritin was digested in a total volume of 1.00 mL to release PO_4^{3-} from the protein. Then 0.300 mL of this solution was analyzed and gave an absorbance of 0.836. A blank carried through the procedure gave an absorbance of 0.038. Find wt% phosphorus in the ferritin.

Notes and References

1. Web sites on ozone depletion: http://www.nas.nasa.gov/about/education/ozone/ and http://www.chemheritage.org/educationservices/faces/facehome.htm.

2. R. W. Ricci, M. A. Ditzler, and L. P. Nestor, *J. Chem. Ed.* **1994**, *71*, 983.

3. D. H. Alman and F. W. Billmeyer, Jr., *J. Chem. Ed.* **1976**, *53*, 166.

4. Classroom demonstrations of absorption and emission spectra with a Web camera and fiber-optic spectrophotometer are described by B. K. Niece, *J. Chem Ed.* **2006**, *83*, 761.

5. M. S. Han and D. H. Kim, *Angew. Chem. Int. Ed.* **2002**, *41*, 3809.

6. H. Van Ryswyk, E. W. Hall, S. J. Petesch, and A. E. Wiedeman, *J. Chem. Ed.* **2007**, *84*, 306.

7. C. Walters, A. Keeney, C. T. Wigal, C. R. Johnston, and R. D. Cornelius, *J. Chem. Ed.* **1997**, *74*, 99.

Flu Virus Identification with an RNA Array and Fluorescent Markers

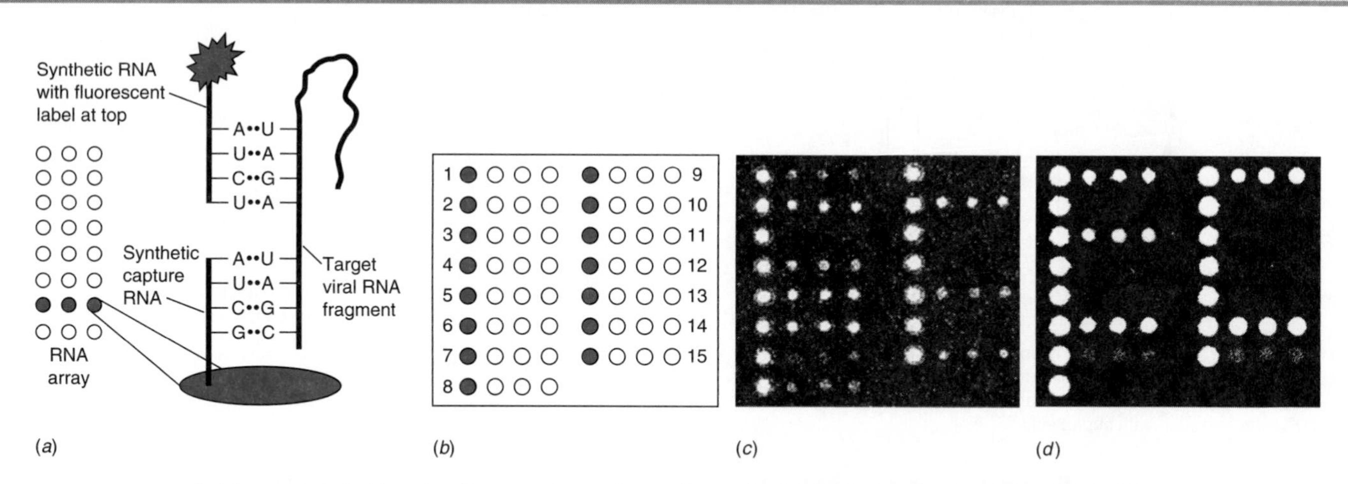

(*a*) Schematic hybridization fluorescence scheme. Bases in the RNA are adenine (A), uracil (U), guanine (G), and cytosine (C), which are hydrogen bonded A to U and G to C. (*b*) Layout of spots on a microarray. (*c*) Fluorescence image from H3N2 strain of influenza A. (*d*) Fluorescence image from H5N1 strain (avian flu). [Parts *b–d*, from E. Dawson, C. L. Moore, J. A. Smagala, D. M. Dankbar, M. Mehlmann, M. B. Townsend, C. B. Smith, N. J. Cox, R. D. Kuchta, and K. L. Rowlen, *Anal. Chem.* **2006**, *78*, 7610; *Anal. Chem.* **2007**, *79*, 378.]

Influenza virus is responsible for 36 000 deaths/yr in the U.S. The virus is classified into types A, B, and C, and subtypes ("strains") according to differences in viral proteins. One particular strain of avian virus ("bird flu") is of particular concern because it has the potential to cause widespread human disease. The World Health Organization identifies flu strains so that vaccines can be prepared. Conventional typing methods are expensive and require days or weeks. An RNA array has the potential to reduce cost and time.

The array above has 15 rows of spots containing synthetic "capture" RNA covalently attached to a glass slide. Three spots in each row contain identical capture RNA designed to bind to a short section of one strain of viral RNA. The spot at the left in each row is a control that will become fluorescent in every test and serves as an internal standard. Viral RNA extracted from patients is *amplified* (reproduced into many copies), and *digested* (cleaved into fragments). Capture RNA on the slide binds selected viral RNA fragments. Another synthetic RNA with a fluorescent tag is designed to bind to a different section of viral RNA. After allowing digested viral RNA to bind to capture RNA and to fluorescent RNA, excess fluorescent RNA is washed away. Fluorescence intensity in each spot is related to the amount of viral RNA bound at that spot.

Pattern recognition methods assign the relative brightness in different spots to a particular strain of flu. In its first trial with patients, 50 of 53 samples were correctly identified. There were also one *false positive* and two *false negatives*. A false positive states that the strain being sought is present when it is not. A false negative fails to find the strain being sought when it is there. The success rate is higher than that of existing rapid diagnostic tests, which should give you pause to appreciate errors possible in medical tests.

CHAPTER 19

Spectrophotometry: Instruments and Applications

In this chapter, we describe the components of a spectrophotometer, some of the physical processes that take place when light is absorbed by molecules, and a few important applications of spectrophotometry in analytical chemistry. New analytical instruments and procedures for medicine and biology, such as the RNA array, are being developed by combining sensitive optical methods with biologically specific recognition elements.

19-1 The Spectrophotometer

The minimum requirements for a *single-beam spectrophotometer* were shown in Figure 18-4. *Polychromatic light* from a lamp passes through a *monochromator* that separates different wavelengths from one another and selects one narrow band of wavelengths to pass through the sample. Transmittance is P/P_0, where P_0 is the radiant power reaching the detector when the sample cell contains a blank solution with no analyte and P is the power reaching the detector when analyte is present in the sample cell. In a single-beam spectrophotometer, two different samples must be placed alternately in the beam. Error occurs if the source intensity or detector response drifts between the two measurements.

Polychromatic light contains many wavelengths (literally, "many colors").

A refresher:

$$\text{transmittance} = T = \frac{P}{P_0}$$

$$\text{absorbance} = -\log T$$

The *double-beam spectrophotometer* in Figure 19-1 features a rotating mirror (the *beam chopper*), which alternately directs light through the sample or reference

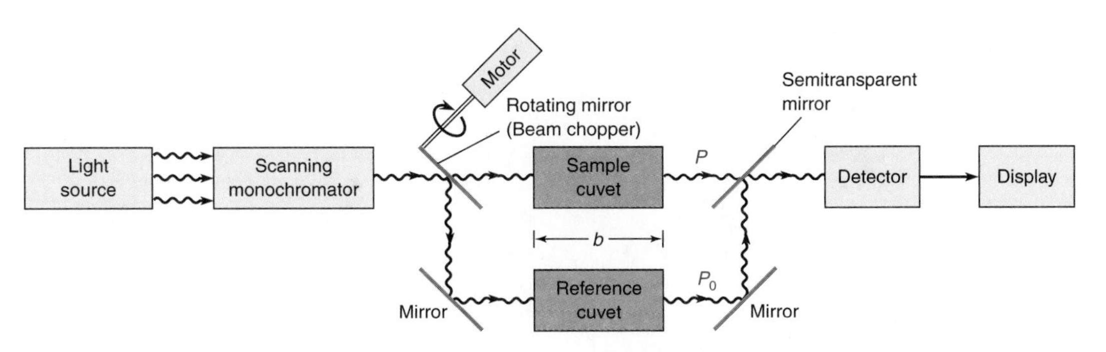

Figure 19-1 Double-beam scanning spectrophotometer. The incident beam is passed alternately through sample and reference cuvets by the rotating beam chopper.

cell several times per second. Radiant power emerging from the reference cell, which is filled with a blank solution or pure solvent, is P_0. Light emerging from the sample cell has power P. By measuring P and P_0 many times per second, the instrument compensates for drift in the source intensity or detector response. Figure 19-2 shows a double-beam instrument and the layout of its components. Let's examine the principal components.

Light Source

The two lamps in the upper part of Figure 19-2*b* provide visible or ultraviolet radiation. An ordinary *tungsten lamp*, whose filament glows at a temperature near 3 000 K, produces radiation in the visible and near-infrared regions at wavelengths of 320 to 2 500 nm (Figure 19-3). For ultraviolet spectroscopy, we normally employ a *deuterium arc lamp* in which a controlled electric discharge dissociates D_2 molecules, which then emit ultraviolet radiation from 200 to 400 nm (Figure 19-3). Typically, a switch is made between the deuterium and tungsten lamps when passing through 360 nm so that the source giving the most radiation is always employed. Other sources of visible and ultraviolet radiation are electric discharge (arc) lamps filled with mercury vapor or xenon. Infrared radiation (5 000 to 200 cm^{-1}) is commonly obtained from a silicon carbide rod called a *globar*, electrically heated to near 1 500 K. Lasers are extremely bright sources of light at just one or a few wavelengths.

Ultraviolet radiation is harmful to the naked eye. Do not view an ultraviolet source without protection.

Question What wavelengths in μm and nm correspond to 5 000 and 200 cm^{-1}?

Answer:
5 000 cm^{-1} = 2 μm = 2 000 nm
200 cm^{-1} = 50 μm = 50 000 nm

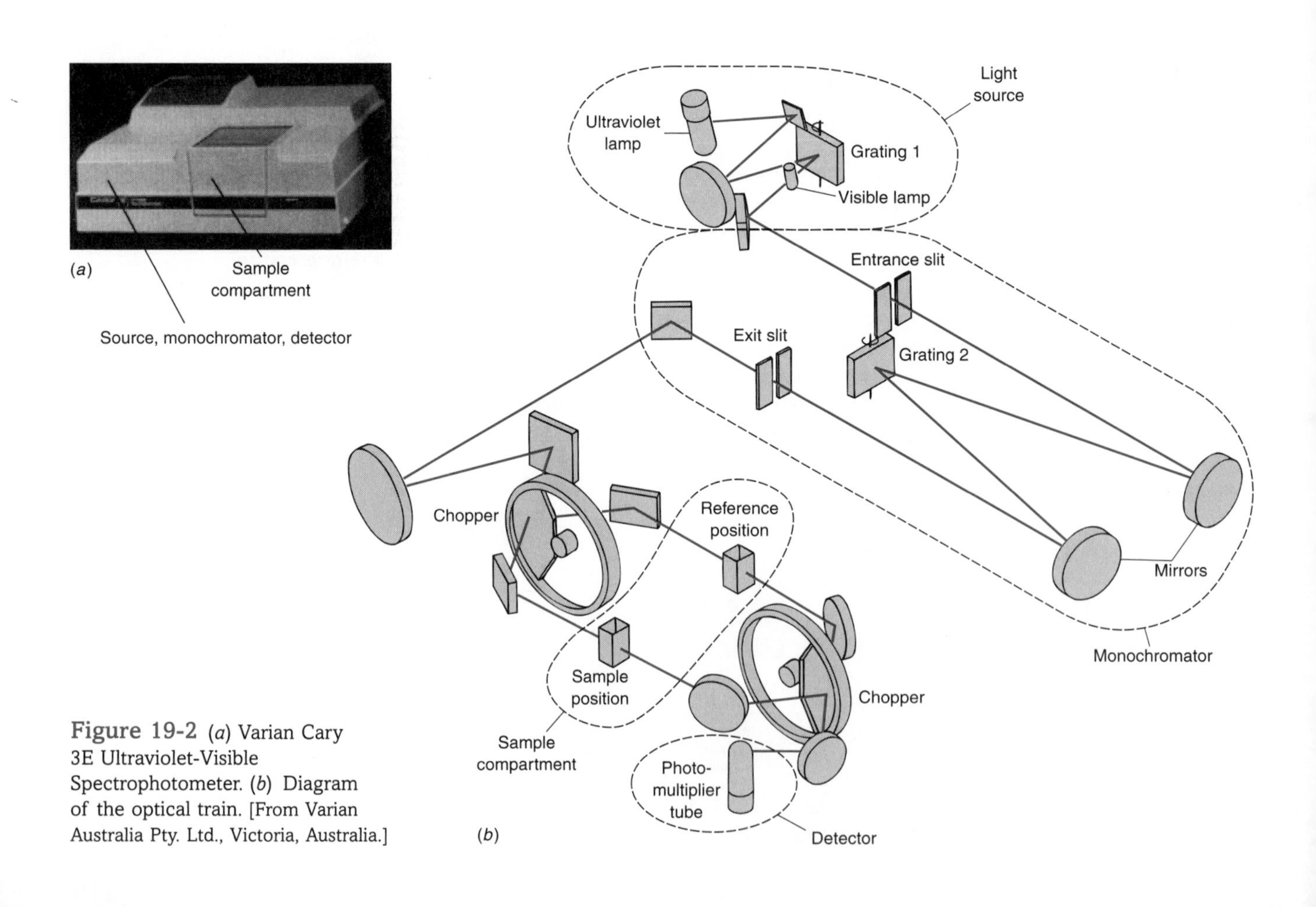

Figure 19-2 (*a*) Varian Cary 3E Ultraviolet-Visible Spectrophotometer. (*b*) Diagram of the optical train. [From Varian Australia Pty. Ltd., Victoria, Australia.]

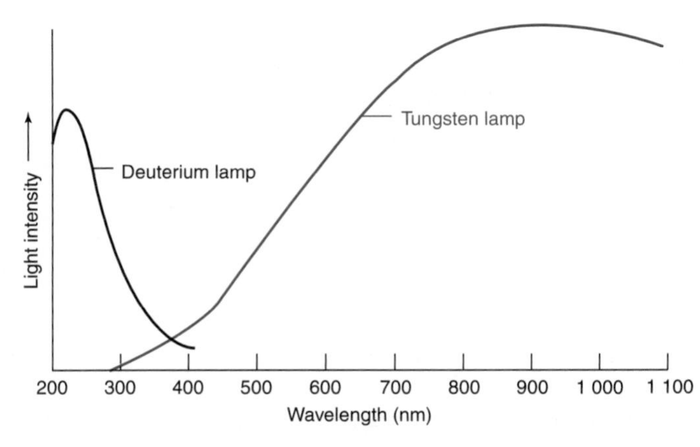

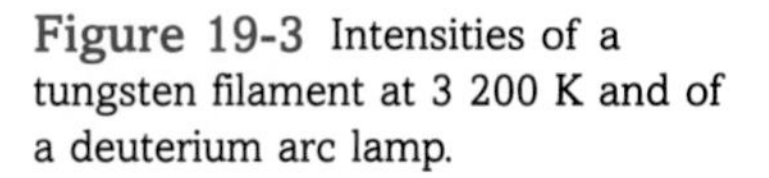
Figure 19-3 Intensities of a tungsten filament at 3 200 K and of a deuterium arc lamp.

Monochromator

A **monochromator** disperses light into its component wavelengths and selects a narrow band of wavelengths to pass through the sample. The monochromator in Figure 19-2*b* consists of entrance and exit slits, mirrors, and a *grating* to disperse the light. *Prisms* were used in older instruments to disperse light.

A **grating** has a series of closely ruled lines. When light is reflected from or transmitted through the grating, each line behaves as a separate source of radiation. Different wavelengths are reflected or transmitted at different angles from the grating (Color Plate 17). The bending of light rays by a grating is called **diffraction**. (In contrast, the bending of light rays by a prism or lens, which is called *refraction*, is shown in Color Plate 18.)

Grating: optical element with closely spaced lines
Diffraction: bending of light by a grating
Refraction: bending of light by a lens or prism

In the grating monochromator in Figure 19-4, *polychromatic* radiation from the entrance slit is *collimated* (made into a beam of parallel rays) by a concave mirror. These rays fall on a reflection grating, whereupon different wavelengths are diffracted at different angles. The light strikes a second concave mirror, which focuses each wavelength at a different point. The grating directs a narrow band of wavelengths to the exit slit. Rotation of the grating allows different wavelengths to pass through the exit slit.

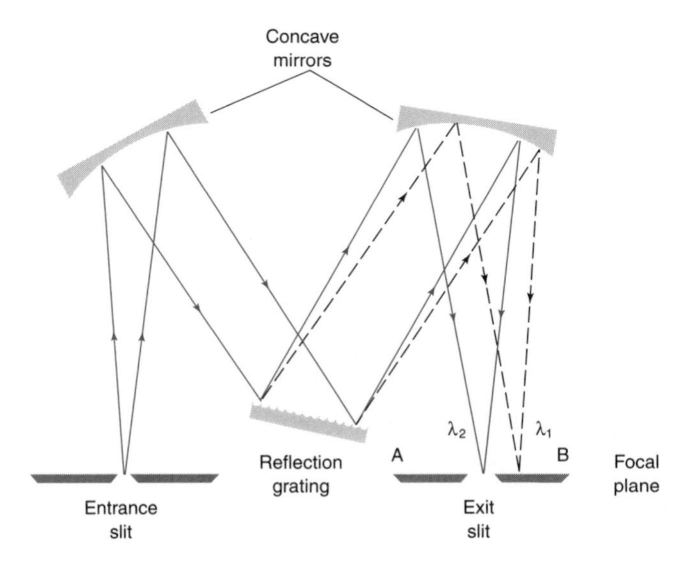

Figure 19-4 Czerny-Turner grating monochromator.

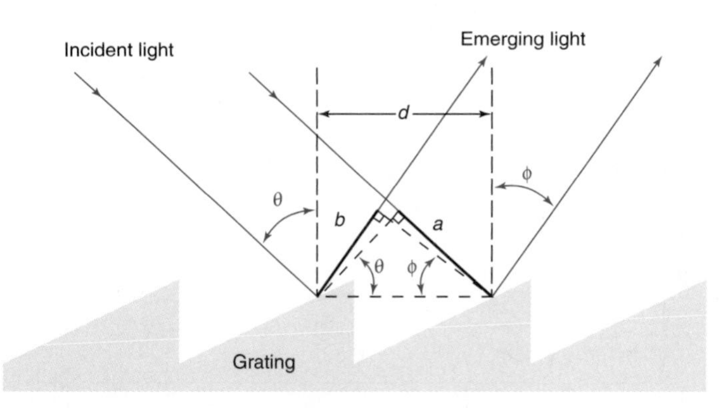

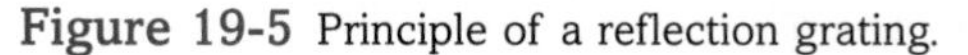

Figure 19-5 Principle of a reflection grating.

Diffraction at a reflection grating is shown in Figure 19-5. The closely spaced parallel grooves of the grating have a repeat distance d. When light is reflected from the grating, each groove behaves as a source of radiation. If adjacent light rays are in phase, they reinforce one another. If they are out of phase, they cancel one another (Figure 19-6).

Constructive interference occurs if the difference in pathlength $(a - b)$ traveled by the two rays in Figure 19-5 is an integer multiple of the wavelength:

$$n\lambda = a - b \tag{19-1}$$

where the diffraction order, n, is ±1, ±2, ±3, ±4, ... The maximum for which $n = \pm1$ is called *first-order diffraction.* When $n = \pm2$, we have *second-order diffraction*, and so on.

In Figure 19-5, the incident angle θ is defined to be positive. The diffraction angle ϕ in Figure 19-5 goes in the opposite direction from θ, so, by convention, ϕ is negative. It is possible for ϕ to be on the same side of the normal as θ, in which case ϕ would be positive. In Figure 19-5, $a = d \sin \theta$ and $b = -d \sin \phi$ (because ϕ

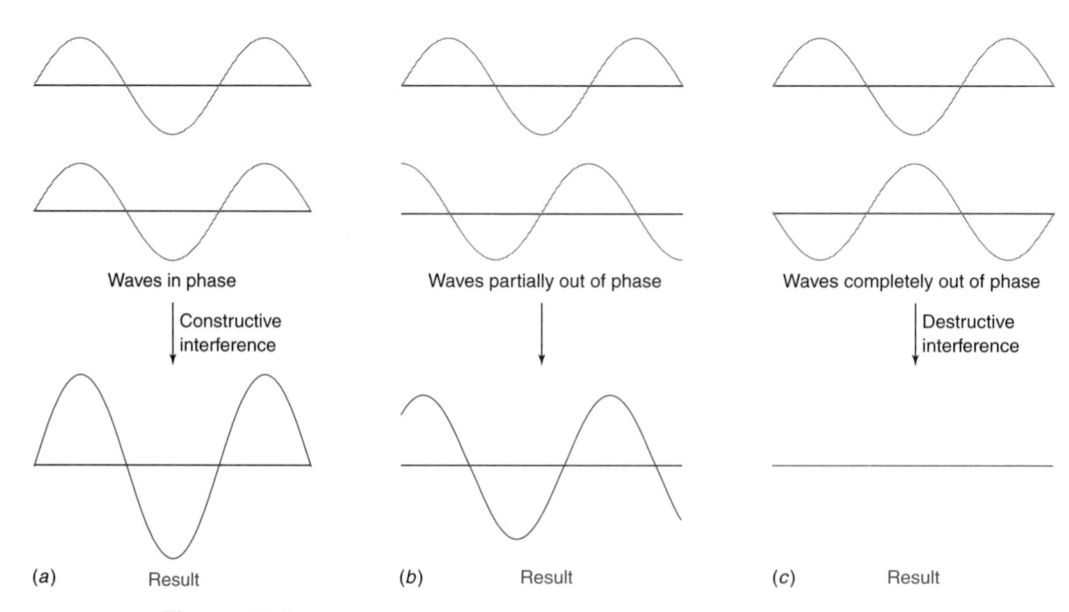

Figure 19-6 Interference of adjacent waves that are (*a*) 0°, (*b*) 90°, and (*c*) 180° out of phase.

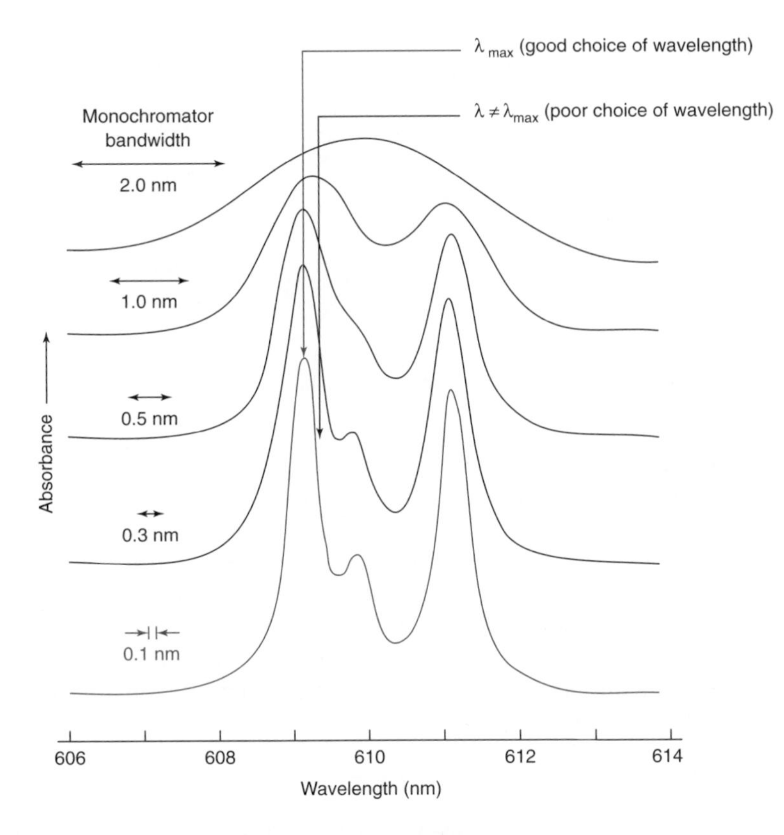

Figure 19-7 Choosing wavelength and monochromator bandwidth. For quantitative analysis, use a wavelength of maximum absorbance so that small errors in the wavelength do not change the absorbance very much. Choose a monochromator bandwidth (controlled by the exit slit width in Figure 19-4) small enough that it does not distort the band shape, but not so small that the spectrum is too noisy. Widening the slit width distorts the spectrum. In the lowest trace, a monochromator bandwidth that is one-fifth of the width of the sharp absorption bands (measured at half the peak height) prevents distortion. [Courtesy M. D. Seltzer, Michelson Laboratory, China Lake, CA.]

is negative and sin ϕ is negative). Substituting into Equation 19-1 gives the condition for constructive interference:

Grating equation:
$$n\lambda = d(\sin \theta + \sin \phi) \tag{19-2}$$

For each incident angle θ, there are diffraction angles ϕ at which a given wavelength will produce maximum constructive interference, as shown in Color Plate 19.

In general, first-order diffraction of one wavelength overlaps higher-order diffraction of another wavelength. Therefore *filters* that reject many wavelengths are used to select a desired wavelength while rejecting other wavelengths at the same diffraction angle. High-quality spectrophotometers use several gratings with different line spacings optimized for different wavelengths. Spectrophotometers with two monochromators in series (a *double monochromator*) reduce unwanted radiation by orders of magnitude.

Decreasing the exit slit width in Figure 19-4 decreases the selected bandwidth and decreases the energy reaching the detector. Thus, *resolution of closely spaced bands, which requires a narrow slit width, can be achieved at the expense of decreased signal-to-noise ratio.* For quantitative analysis, a monochromator bandwidth that is $\lesssim \frac{1}{5}$ of the width of the absorption band is reasonable (Figure 19-7).

Trade-off between resolution and signal: The narrower the exit slit, the greater the ability to resolve closely spaced peaks and the noisier the spectrum.

Detector

A detector produces an electric signal when it is struck by photons. Figure 19-8 shows that detector response depends on the wavelength of the incident photons. In a single-beam spectrophotometer, the 100% transmittance control must be readjusted each time the wavelength is changed because the maximum possible detector signal depends on the wavelength. Subsequent readings are scaled to the 100% reading.

Detector response is a function of wavelength of incident light.

A **photomultiplier tube** (Figure 19-9) is a very sensitive detector. When light of sufficient energy strikes a photosensitive cathode, electrons are emitted into the

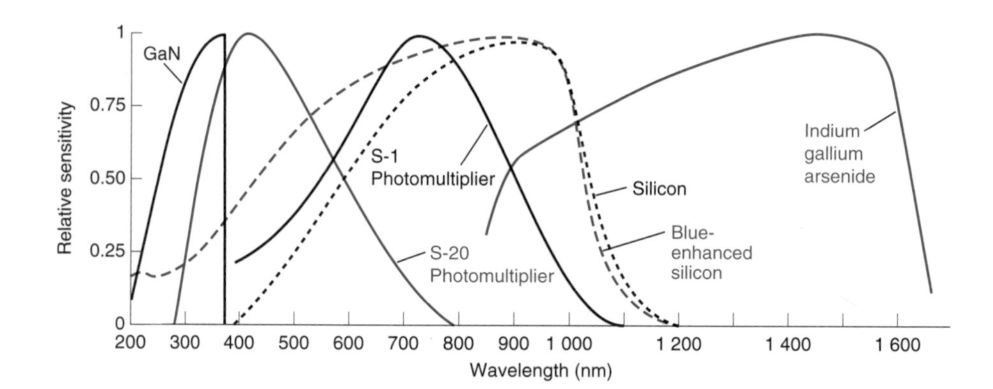

Figure 19-8 Detector response. Each curve is normalized to a maximum value of 1. [Courtesy Barr Associates, Inc., Westford, MA. GaN data from APA Optics, Blaine, MN. InGaAs data from Shimadzu Corp., Tokyo.]

vacuum inside the tube. Emitted electrons strike a second surface, called a *dynode*, which is positive with respect to the cathode. Electrons strike the dynode with more than their original kinetic energy. Each energetic electron knocks more than one electron from the dynode. These new electrons are accelerated toward a second dynode, which is more positive than the first dynode. On striking the second dynode, even more electrons are knocked off and accelerated toward a third dynode. This process is repeated so that more than 10^6 electrons are finally collected for each photon striking the cathode. Extremely low light intensities are translated into measurable electric signals.

Photodiode Array Spectrophotometer

A *dispersive* spectrophotometer spreads light from the source into its component wavelengths and then measures the absorption of one narrow band of wavelengths at a time.

A typical photodiode array responds to visible and ultraviolet radiation, with a response curve similar to that of the blue-enhanced silicon in Figure 19-8.

Dispersive spectrophotometers described so far scan through a spectrum one wavelength at a time. A *diode array spectrophotometer* records the entire spectrum at once. The entire spectrum of a compound emerging from a chromatography column can be recorded in a fraction of a second by a photodiode array spectrophotometer. At the heart of rapid spectroscopy is a **photodiode array** such as the one in Figure 19-10, which contains 1 024 individual semiconductor detector elements (diodes) in a row.

In the diode array spectrophotometer in Figure 19-11, *white light* (with all wavelengths) passes through the sample. The beam then enters a **polychromator**, which disperses light into its component wavelengths and directs the light to the diode array. *A different wavelength band strikes each diode.* The resolution, which is typically 1 to 3 nm, depends on how closely spaced the diodes are and how much

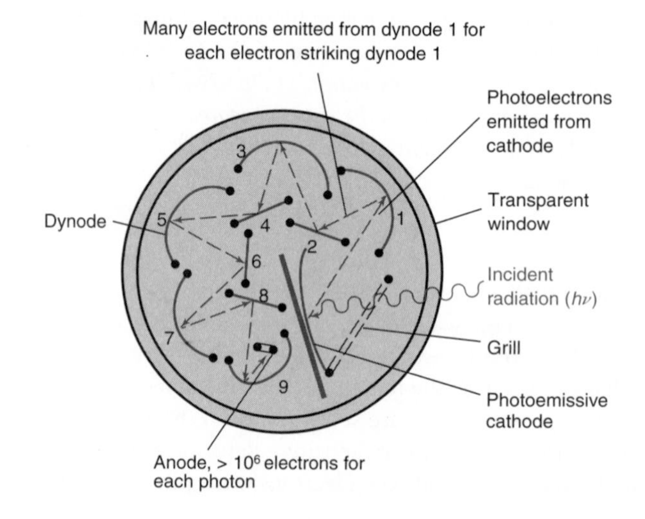

Figure 19-9 *Left:* Diagram of photomultiplier tube with nine dynodes. Amplification occurs at each dynode, which is approximately 90 volts more positive than the preceding dynode. *Right:* Photomultiplier tube. [Photograph by David J. Green/Alamy.]

Figure 19-10 Photodiode array with 1 024 elements, each 25 μm wide and 2.5 mm high. The entire chip is 5 cm long. [Courtesy Oriel Corporation, Stratford, CT.]

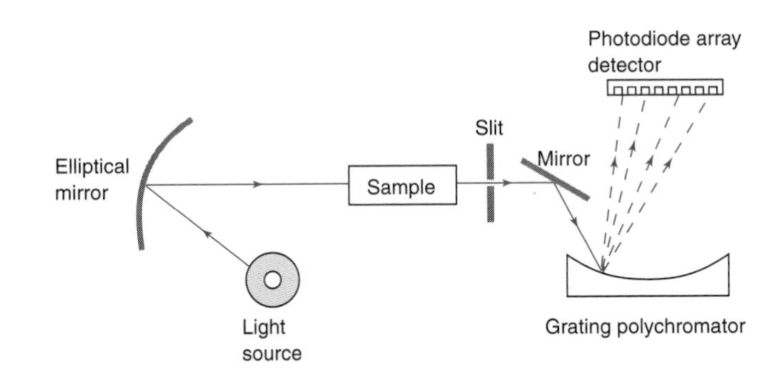

Figure 19-11 Diode array spectrophotometer. The spectrum in Figure 20-16 (in the next chapter) was produced with a photodiode array detector.

dispersion is produced by the polychromator. By comparison, high-quality dispersive spectrophotometers can resolve features that are 0.1 nm apart. Diode array spectrophotometers are faster than dispersive spectrophotometers because the array measures all wavelengths at once, instead of one at a time. A diode array spectrophotometer is usually a single-beam instrument, subject to absorbance errors from drift in the source intensity and detector response between calibrations.

Attributes of diode array spectrophotometer:

- speed (~1 s per spectrum)
- excellent wavelength repeatability (because the grating does not rotate)
- simultaneous measurements at multiple wavelengths
- relatively insensitive to errors from stray light
- relatively poor resolution (1 to 3 nm)

Ask Yourself

19-A. Explain what each component in the optical train in Figure 19-2*b* does, beginning with the lamp and ending with the detector.

19-2 Analysis of a Mixture

When there is more than one absorbing species in a solution, *the absorbance at a particular wavelength is the sum of absorbances from all species at that wavelength:*

Absorbance of a mixture:

$$A = \varepsilon_X b[X] + \varepsilon_Y b[Y] + \varepsilon_Z b[Z] + \cdots \qquad (19\text{-}3)$$

Absorbance is additive.

where ε is the molar absorptivity of each species (X, Y, Z, and so on) and b is the pathlength. If we measure the spectra of the pure components in a separate experiment, we can mathematically disassemble the spectrum of the mixture into those of its components.

Figure 19-12 shows spectra of titanium and vanadium complexes and an unknown mixture of the two. Let's denote the titanium complex by X and the vanadium complex by Y. To analyze a mixture, we usually choose wavelengths of maximum absorption for the individual components. Accuracy is improved if compound Y absorbs weakly at the maximum for compound X and if compound X absorbs weakly at the maximum for compound Y. In Figure 19-12, the two spectra overlap badly, so there will be some loss of accuracy.

We normally choose wavelengths at absorbance maxima. The absorbance of the mixture should not be too small or too great, so that uncertainty in absorbance is small.

Choosing wavelengths λ′ and λ″ as the absorbance maxima in Figure 19-12, we can write a Beer's law expression for each wavelength:

$$A' = \varepsilon'_X b[X] + \varepsilon'_Y b[Y] \qquad A'' = \varepsilon''_X b[X] + \varepsilon''_Y b[Y] \qquad (19\text{-}4)$$

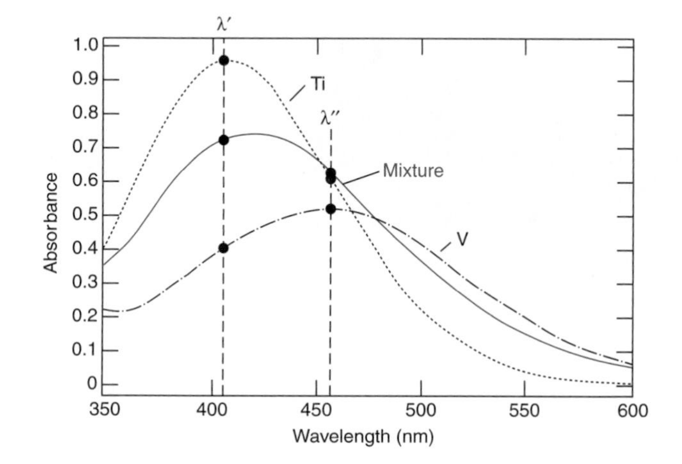

Figure 19-12 Visible spectra of hydrogen peroxide complexes of Ti(IV) (1.32 mM), V(V) (1.89 mM), and an unknown mixture containing both. All solutions contain 0.5 wt% H_2O_2 and ~0.01 M H_2SO_4 in a 1.00-cm-pathlength cell. [From M. Blanco, H. Iturriaga, S. Maspoch, and P. Tarín, *J. Chem. Ed.* **1989**, *66*, 178. Consult this article for a more accurate method for finding the composition of the mixture by using more than two wavelengths.]

Solving Equations 19-4 for [X] and [Y], we find

Analysis of a mixture when spectra are resolved:

$$[X] = \frac{1}{D}(A'\varepsilon''_Y - A''\varepsilon'_Y)$$
$$[Y] = \frac{1}{D}(A''\varepsilon'_X - A'\varepsilon''_X) \qquad (19\text{-}5)$$

where $D = b(\varepsilon'_X\varepsilon''_Y - \varepsilon'_Y\varepsilon''_X)$. To analyze the mixture, we measure absorbances at two wavelengths and must know ε at each wavelength for each compound.

Example Analysis of a Mixture with Equations 19-5

The molar absorptivities of X (the Ti complex) and Y (the V complex) in Figure 19-12 were measured with pure samples of each:

	ε (M^{-1} cm^{-1})	
λ (nm)	X	Y
$\lambda' \equiv 406$	$\varepsilon'_X = 720$	$\varepsilon'_Y = 212$
$\lambda'' \equiv 457$	$\varepsilon''_X = 479$	$\varepsilon''_Y = 274$

A mixture of X and Y in a 1.00-cm cell had an absorbance of $A' = 0.722$ at 406 nm and $A'' = 0.641$ at 457 nm. Find the concentrations of X and Y in the mixture.

SOLUTION Using Equations 19-5 and setting $b = 1.00$ cm, we find

$$D = b(\varepsilon'_X\varepsilon''_Y - \varepsilon'_Y\varepsilon''_X) = (1.00)[(720)(274) - (212)(479)] = 9.57_3 \times 10^4$$

$$[X] = \frac{1}{D}(A'\varepsilon''_Y - A''\varepsilon'_Y) = \frac{(0.722)(274) - (0.641)(212)}{9.57_3 \times 10^4} = 6.47 \times 10^{-4}\text{ M}$$

$$[Y] = \frac{1}{D}(A''\varepsilon'_X - A'\varepsilon''_X) = \frac{(0.641)(720) - (0.722)(479)}{9.57_3 \times 10^4} = 1.21 \times 10^{-3}\text{ M}$$

Test Yourself The absorbance of the mixture is 0.600 at 406 nm and 0.500 at 457 nm. Find [X] and [Y]. (**Answer:** 0.610 mM, 0.758 mM)

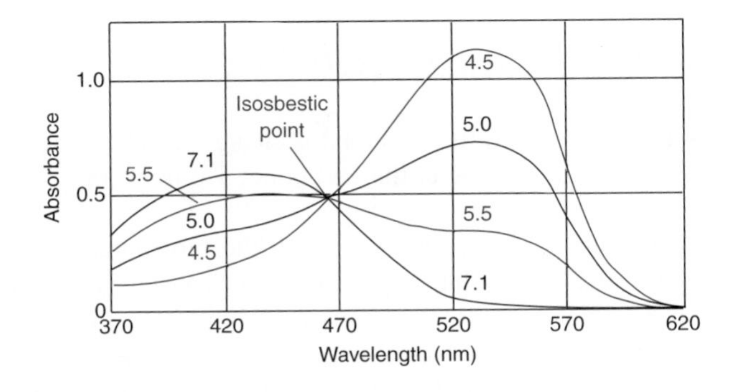

Figure 19-13 Absorption spectrum of 3.7×10^{-4} M methyl red as a function of pH between pH 4.5 and 7.1. [From E. J. King, *Acid-Base Equilibria* (Oxford: Pergamon Press, 1965).]

Isosbestic Points

If species X is converted into species Y in the course of a chemical reaction, a spectrum of the mixture has an obvious, characteristic behavior, shown in Figure 19-13. If the spectra of pure X and pure Y cross each other at some wavelength, then every spectrum recorded during this chemical reaction crosses at the same point, called an **isosbestic point**. *An isosbestic point observed during a chemical reaction is good evidence that only two principal species are present.*

The acid-base indicator methyl red, shown in the margin, changes between red (HIn) and yellow (In^-) near pH 5.1. The spectra of HIn and In^- at equal concentration happen to cross at 465 nm in Figure 19-13, so *all* spectra cross at this point. (If the spectra of HIn and In^- crossed at several points, each would be an isosbestic point.)

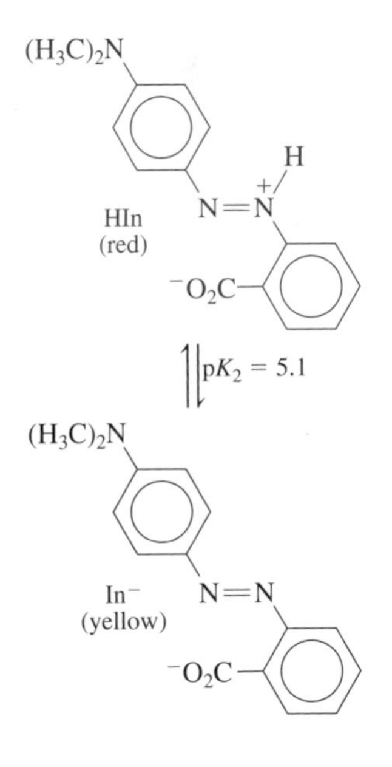

To see why there is an isosbestic point, we write an equation for the absorbance of the solution at 465 nm:

$$A^{465} = \varepsilon_{HIn}^{465}b[HIn] + \varepsilon_{In^-}^{465}b[In^-] \qquad (19\text{-}6)$$

But the spectra of pure HIn and pure In^- at equal concentration cross at 465 nm, so ε_{HIn}^{465} must be equal to $\varepsilon_{In^-}^{465}$. Setting $\varepsilon_{HIn}^{465} = \varepsilon_{In^-}^{465} = \varepsilon^{465}$, we rewrite Equation 19-6 in the form

$$A^{465} = \varepsilon^{465}b([HIn] + [In^-]) \qquad (19\text{-}7)$$

In Figure 19-13, all solutions contain the same total concentration of methyl red ($= [HIn] + [In^-]$). Only the pH varies. Therefore the sum of concentrations in Equation 19-7 is constant, and there is an isosbestic point because A^{465} is constant.

An isosbestic point occurs when $\varepsilon_X = \varepsilon_Y$ and $[X] + [Y]$ is constant.

Ask Yourself

19-B. (a) In the example beneath Equations 19-5, a mixture of X and Y in a *0.100-cm cell* had an absorbance of 0.233 at 406 nm and 0.200 at 457 nm. Find [X] and [Y].

(b) If the total concentration of methyl red were increased by 37% from whatever concentration gave Figure 19-13, would there still be an isosbestic point at 465 nm? Why?

19-3 Spectrophotometric Titrations

In a **spectrophotometric titration**, we monitor changes in absorption or emission of electromagnetic radiation to detect the end point. We now consider an example from biochemistry.

Iron for biosynthesis is transported through the bloodstream by the protein *transferrin* (Figure 19-14). A solution of transferrin can be titrated with iron to measure its iron-binding capacity. Transferrin without iron, called *apotransferrin*, is colorless. Each protein molecule with a formula mass of 81 000 has two Fe^{3+}-binding sites. When Fe^{3+} binds to the protein, a red color with an absorbance maximum at 465 nm develops. The color intensity allows us to follow the course of the titration of an unknown amount of apotransferrin with standard Fe^{3+}.

$$\underset{\text{Colorless}}{\text{apotransferrin}} + 2Fe^{3+} \longrightarrow \underset{\text{Red}}{(Fe^{3+})_2\text{transferrin}} \qquad (19\text{-}8)$$

Ferric nitrilotriacetate is used because Fe^{3+} precipitates as $Fe(OH)_3$ in neutral solution. Nitrilotriacetate binds Fe^{3+} through four **bold** atoms:

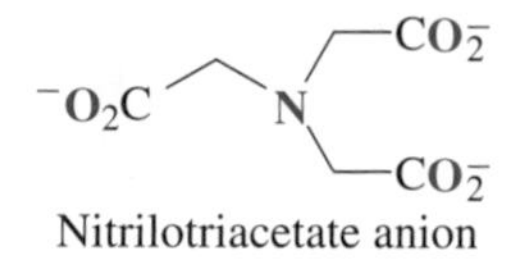

Nitrilotriacetate anion

Figure 19-15 shows the titration of 2.000 mL of apotransferrin with 1.79×10^{-3} M ferric nitrilotriacetate. As iron is added to protein, red color develops and absorbance increases. When protein is saturated with iron, no more iron can bind and the curve levels off. The end point is the extrapolated intersection of the two straight lines at 203 μL. Absorbance rises slowly after the equivalence point because ferric nitrilotriacetate has some absorbance at 465 nm.

To construct the graph in Figure 19-15, we must account for the volume change as titrant is added. Each point plotted on the graph represents the absorbance that would be observed *if the solution had not been diluted from its original volume of 2.000 mL.*

$$\text{corrected absorbance} = \left(\frac{\text{total volume}}{\text{initial volume}}\right)(\text{observed absorbance}) \qquad (19\text{-}9)$$

Example **Correcting Absorbance for the Effect of Dilution**

The absorbance measured after adding 125 μL (= 0.125 mL) of ferric nitrilotriacetate to 2.000 mL of apotransferrin was 0.260. Calculate the corrected absorbance that should be plotted in Figure 19-15.

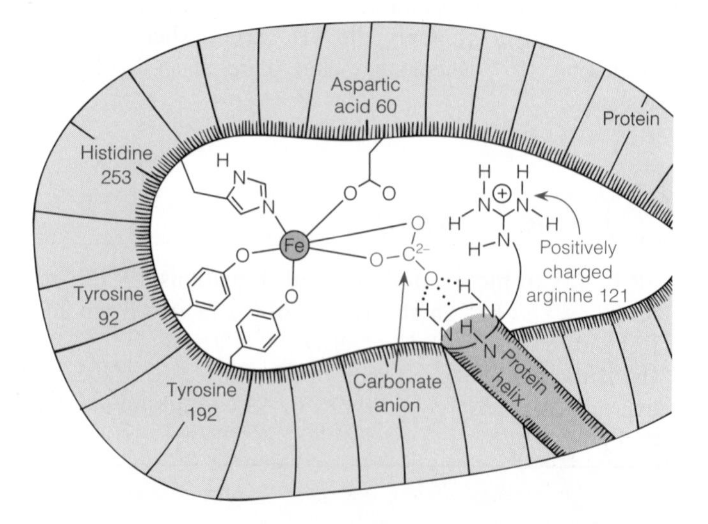

Figure 19-14 Each of the two Fe-binding sites of transferrin is located in a cleft in the protein. Each site has one nitrogen ligand from the amino acid histidine and three oxygen ligands from tyrosine and aspartic acid. Two oxygen ligands come from a carbonate anion (CO_3^{2-}) anchored by electrostatic attraction to positively charged arginine and by hydrogen bonding to the protein helix. When transferrin is taken up by a cell, it is brought into a vesicle (Box 1-1) whose pH is lowered to 5.5. H^+ reacts with carbonate to make HCO_3^- and H_2CO_3, thereby releasing Fe^{3+} from the protein. [Adapted from E. N. Baker, B. F. Anderson, H. M. Baker, M. Haridas, G. E. Norris, S. V. Rumball, and C. A. Smith, *Pure Appl. Chem.* **1990**, *62*, 1067.]

SOLUTION The total volume was 2.000 + 0.125 = 2.125 mL. If the volume had been 2.000 mL, the absorbance would have been greater than 0.260 by a factor of 2.125/2.000.

$$\text{corrected absorbance} = \left(\frac{2.125 \text{ mL}}{2.000 \text{ mL}}\right)(0.260) = 0.276$$

The absorbance plotted in the graph is 0.276.

Test Yourself The absorbance after adding 100 μL of ferric nitrilotriacetate was 0.210. Calculate the corrected absorbance to be plotted. (**Answer:** 0.221)

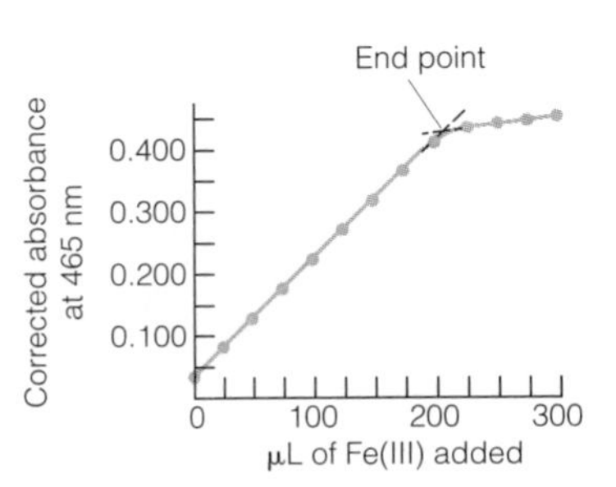

Figure 19-15 Spectrophotometric titration of apotransferrin with ferric nitrilotriacetate. Absorbance is corrected for dilution. The initial absorbance of the solution, before iron is added, is due to a colored impurity.

Ask Yourself

19-C. A 2.00-mL solution of apotransferrin, titrated as in Figure 19-15, required 163 μL of 1.43 mM ferric nitrilotriacetate to reach the end point.
(a) How many moles of Fe^{3+} were required to reach the end point?
(b) Each apotransferrin molecule binds two Fe^{3+} ions. Find the concentration of apotransferrin in the 2.00-mL solution.
(c) Why does the slope in Figure 19-15 change abruptly at the equivalence point?

19-4 What Happens When a Molecule Absorbs Light?

When a molecule absorbs a photon, the molecule is promoted to a more energetic *excited state* (Figure 18-3). Conversely, when a molecule emits a photon, the energy of the molecule falls by an amount equal to the energy of the photon that is given off. Figure 18-2 showed that molecules are promoted to excited electronic, vibrational, and rotational states by radiation in different regions of the electromagnetic spectrum.

As an example, we will discuss formaldehyde, whose ground state and one excited state are shown in Figure 19-16. The ground state is planar, with a double bond between carbon and oxygen. The double bond consists of a sigma bond between carbon and oxygen and a pi bond made from the $2p_y$ (out-of-plane) atomic orbitals of carbon and oxygen.

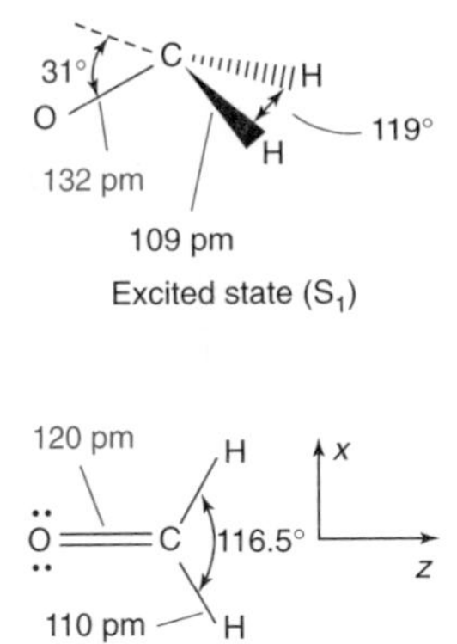

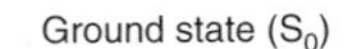

Figure 19-16 Geometry of formaldehyde in its ground state (S_0) and lowest excited singlet state (S_1).

Electronic States of Formaldehyde

Molecular orbitals describe the distribution of electrons in a molecule, just as *atomic orbitals* describe the distribution of electrons in an atom. In Figure 19-17, four low-lying orbitals of formaldehyde, labeled σ_1 through σ_4, are each occupied by a pair of electrons with opposite spin (spin quantum numbers = $+\frac{1}{2}$ and $-\frac{1}{2}$ represented by $\uparrow$ and $\downarrow$). At higher energy is a pi bonding orbital (π), made of the p_y atomic orbitals of carbon and oxygen. The highest-energy occupied orbital is a nonbonding orbital (n), composed principally of the oxygen $2p_x$ atomic orbital. The lowest-energy unoccupied orbital is a pi antibonding orbital (π^*). An electron in this orbital produces repulsion, rather than attraction, between the carbon and oxygen atoms.

In *sigma* orbitals, electrons are localized between atoms. In *pi* orbitals, electrons are concentrated on either side of the plane of the formaldehyde molecule.

Figure 19-17 Molecular orbital diagram of formaldehyde, showing energy levels and orbital shapes. The coordinate system was shown in Figure 19-16. [From W. L. Jorgensen and L. Salem, *The Organic Chemist's Book of Orbitals* (New York: Academic Press, 1973).]

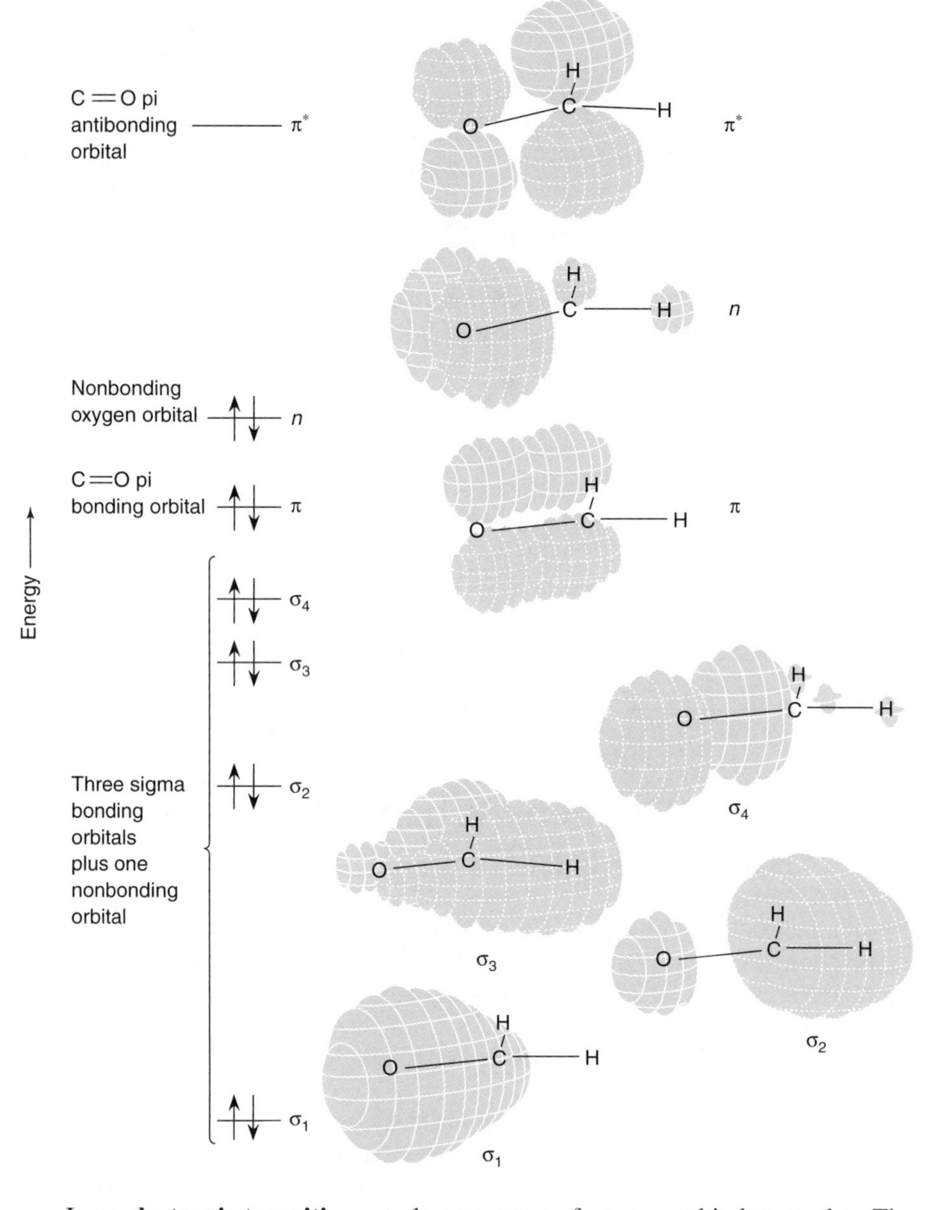

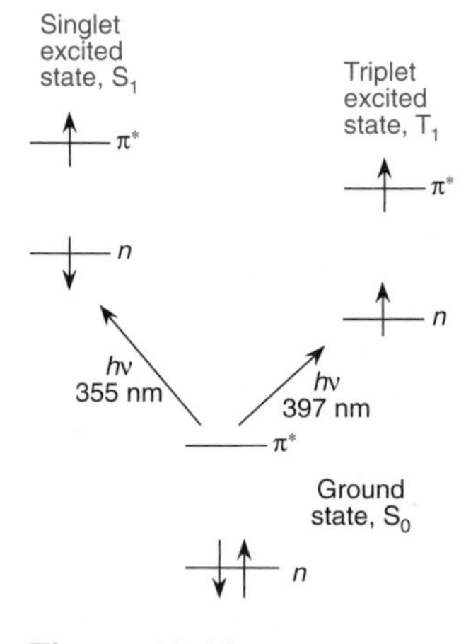

Figure 19-18 Two possible electronic states arising from an $n \rightarrow \pi^*$ transition. The terms *singlet* and *triplet* are used because a triplet state splits into three slightly different energy levels in a magnetic field, but a singlet state is not split.

In an **electronic transition**, an electron moves from one orbital to another. The lowest-energy electronic transition of formaldehyde promotes a nonbonding (n) electron to the antibonding pi orbital (π^*). There are actually two possible transitions, depending on the spin quantum numbers in the excited state. The state in which the spins are opposed in Figure 19-18 is called a **singlet state**. If spins are parallel, the excited state is a **triplet state**.

The lowest-energy excited singlet and triplet states are called S_1 and T_1. In general, T_1 has lower energy than S_1. In formaldehyde, the weakly absorbing transition $n \rightarrow \pi^*(T_1)$ requires visible light with a wavelength of 397 nm. The more intense $n \rightarrow \pi^*(S_1)$ transition takes 355-nm ultraviolet radiation.

Although formaldehyde is planar in its ground state (S_0), it is pyramidal in both the S_1 (Figure 19-16) and T_1 excited states. Promotion of a nonbonding electron to an antibonding C—O orbital weakens and lengthens the C—O bond and changes the molecular geometry.

Vibrational and Rotational States of Formaldehyde

Infrared and microwave radiation are not energetic enough to induce electronic transitions, but they can change the vibrational or rotational motion of the molecule. The six modes of vibration of formaldehyde are shown in Figure 19-19. When formaldehyde absorbs an infrared photon with a wavenumber of 1 746 cm^{-1}, for example, C—O stretching is stimulated: Oscillations of the atoms increase in amplitude and the energy of the molecule increases.

Rotational energies of a molecule are even smaller than vibrational energies. Absorption of microwave radiation increases the rotational speed of a molecule.

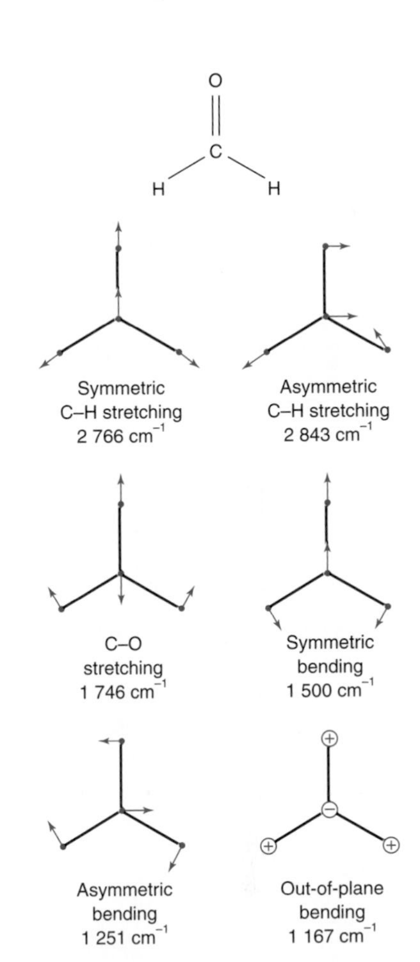

Figure 19-19 The six modes of vibration of formaldehyde. The wavenumber of infrared radiation needed to stimulate each kind of motion is given in units of reciprocal centimeters, cm^{-1}.

Combined Electronic, Vibrational, and Rotational Transitions

In general, when a molecule absorbs light of sufficient energy to cause an electronic transition, **vibrational** and **rotational transitions**—changes in the vibrational and rotational states—occur as well. Formaldehyde can absorb one photon with just the right energy to (1) promote the molecule from S_0 to the S_1 electronic state; (2) increase the vibrational energy from the ground vibrational state of S_0 to an excited vibrational state of S_1; and (3) change from one rotational state of S_0 to a different rotational state of S_1. Electronic absorption bands are usually very broad (~100 nm in Figures 18-5 and 18-9) because many different vibrational and rotational levels are excited at slightly different energies.

What Happens to Absorbed Energy?

Suppose that absorption of a photon promotes a molecule from the ground electronic state, S_0, to a vibrationally and rotationally excited level of the excited electronic state S_1 (Figure 19-20). Usually, the first event after absorption is *vibrational relaxation* to the lowest vibrational level of S_1. In this process, labeled R_1 in Figure 19-20, energy is lost to other molecules (solvent, for example) through collisions. The net effect is to convert part of the energy of the absorbed photon into heat spread through the entire medium.

From S_1, the molecule could enter a highly excited vibrational level of S_0 having the same energy as S_1. This process is called *internal conversion*. Then the molecule can relax back to the ground vibrational state, transferring energy to

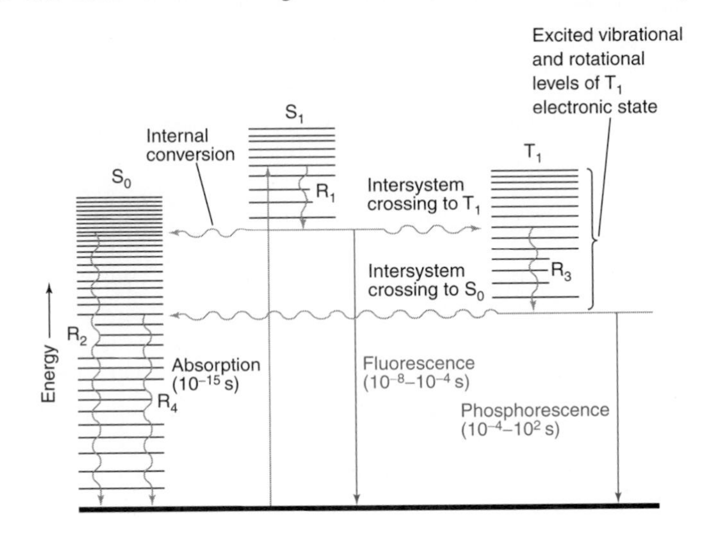

Figure 19-20 Physical processes that can occur after a molecule absorbs an ultraviolet or visible photon. S_0 is the ground electronic state. S_1 and T_1 are the lowest excited singlet and triplet states, respectively. Straight arrows represent processes involving photons, and wavy arrows are radiationless transitions. R is vibrational relaxation.

Internal conversion: radiationless transition between states with the same spin (e.g., $S_1 \rightarrow S_0$).

Intersystem crossing: radiationless transition between states with different spins (e.g., $T_1 \rightarrow S_0$).

Fluorescence: emission of a photon from a transition between states with the same spin (e.g., $S_1 \rightarrow S_0$).

Phosphorescence: emission of a photon from a transition between states with different spins (e.g., $T_1 \rightarrow S_0$).

neighboring molecules through collisions. If a molecule follows the path absorption $\rightarrow R_1 \rightarrow$ internal conversion $\rightarrow R_2$ in Figure 19-20, the entire energy of the photon will have been converted into heat.

Alternatively, the molecule could cross from S_1 into an excited vibrational level of T_1. Such an event is known as *intersystem crossing*. Following relaxation R_3, the molecule finds itself at the lowest vibrational level of T_1. From here, the molecule might undergo a second intersystem crossing to S_0, followed by relaxation R_4, which liberates heat.

Alternatively a molecule could relax from S_1 or T_1 to S_0 by emitting a photon. The transition $S_1 \rightarrow S_0$ is called **fluorescence** (Demonstration 19-1), and $T_1 \rightarrow S_0$ is called **phosphorescence**. (Fluorescence and phosphorescence can terminate in any of the vibrational levels of S_0, not just the ground state shown in Figure 19-20.) The rates of internal conversion, intersystem crossing, fluorescence, and phosphores-

Demonstration 19-1 In Which Your Class Really Shines[1,2,3]

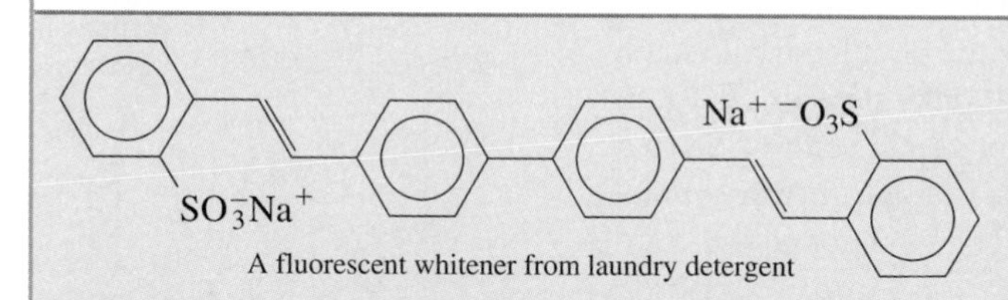

A fluorescent whitener from laundry detergent

White fabrics are sometimes made "whiter" with a fluorescent dye. Turn on an ultraviolet lamp in a darkened classroom and illuminate people standing at the front of the room. (*The victims should not look directly at the lamp*, because ultraviolet light is harmful to eyes.) You will discover emission from white fabrics, including shirts, pants, shoelaces, and unmentionables. You may also be surprised to see fluorescence from teeth and from recently bruised areas of skin that show no surface damage.

A fluorescent lamp is a glass tube filled with Hg vapor; the inner walls are coated with a *phosphor* (luminescent substance) consisting of a calcium halophosphate ($Ca_5(PO_4)_3F_{1-x}Cl_x$) doped with Mn^{2+} and Sb^{3+}. (*Doping* means adding an intentional impurity, called a *dopant*.) Hg atoms, promoted to an excited state by electricity passing through the lamp, return to the ground state and emit mostly ultraviolet radiation at 254 and 185 nm. This radiation is absorbed by Sb^{3+}, and some energy is passed on to Mn^{2+}. Sb^{3+} emits blue light and Mn^{2+} emits yellow light, with the combined emission shown in the spectrum appearing white.

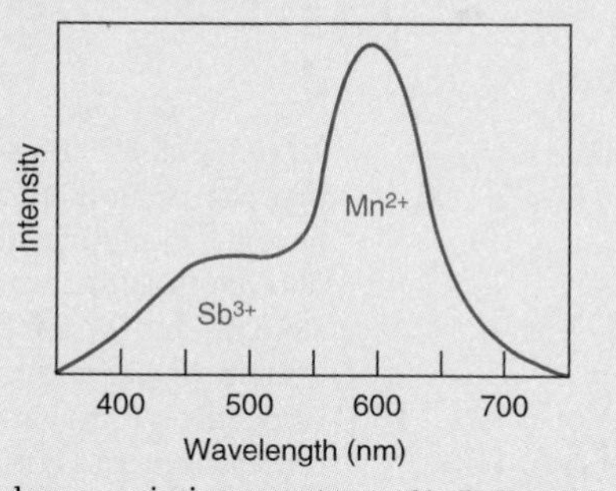

Fluorescent lamp emission spectrum. [A. DeLuca, *J. Chem. Ed.* **1980**, *57*, 541.]

Fluorescent lamps are more efficient than incandescent lamps in converting electricity into light. In the near future, LED (light-emitting diode) lamps could become more efficient than fluorescent lamps. An easy way for you to reduce greenhouse gas emission is to replace incandescent light bulbs with fluorescent bulbs. Replacing a 75-W incandescent bulb with an 18-W compact fluorescent bulb saves 57 W. Over the 10 000-h lifetime of the fluorescent bulb, you will reduce CO_2 emission by ~600 kg and will put 10 kg less SO_2 into the atmosphere (see Problem 19-21). Alas, fluorescent bulbs contain Hg and should be recycled at a collection center where Hg will be captured from the bulbs. They should not be discarded as ordinary waste.

Lamp	Efficiency (lumens per watt)
32-W fluorescent	85–95
Compact fluorescent	48–60
T3 tubular halogen	20
100-W incandescent	17
Flashlight (incandescent)	<6

Lumen (lm) is a measure of luminous flux. 1 lm = radiant energy emitted in a solid angle of 1 steradian (sr) from a source that radiates 1/683 W/sr uniformly in all directions at a frequency of 540 THz (near the middle of the visible spectrum).

SOURCE: http://www.otherpower.com/otherpower_lighting.html.

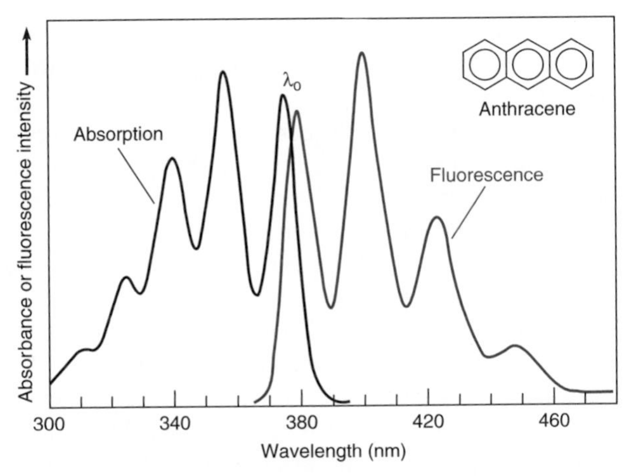

Figure 19-21 Spectra of anthracene show typical approximate mirror image relationship between absorption and fluorescence. Fluorescence comes at lower energy (longer wavelength) than absorption. [From C. M. Byron and T. C. Werner, *J. Chem. Ed.* **1991**, *68*, 433.]

cence depend on the solvent and conditions such as temperature and pressure. We see in Figure 19-20 that phosphorescence occurs at lower energy (longer wavelength) than does fluorescence.

Molecules generally decay from the excited state through collisions—not by emitting light. The *lifetime* of fluorescence is always very short (10^{-8} to 10^{-4} s). The lifetime of phosphorescence is much longer (10^{-4} to 10^{2} s). Phosphorescence is rarer than fluorescence, because a molecule in the T_1 state has a good chance of collisional deactivation before phosphorescence can occur.

Figure 19-21 compares absorption and fluorescence spectra of anthracene. Fluorescence comes at lower energy and is roughly the mirror image of absorption. To understand the mirror image relation, consider the energy levels in Figure 19-22.

An example of emission at lower energy (longer wavelength) than absorption is seen in Color Plate 20, in which blue light absorbed by a crystal gives rise to *red* emission.

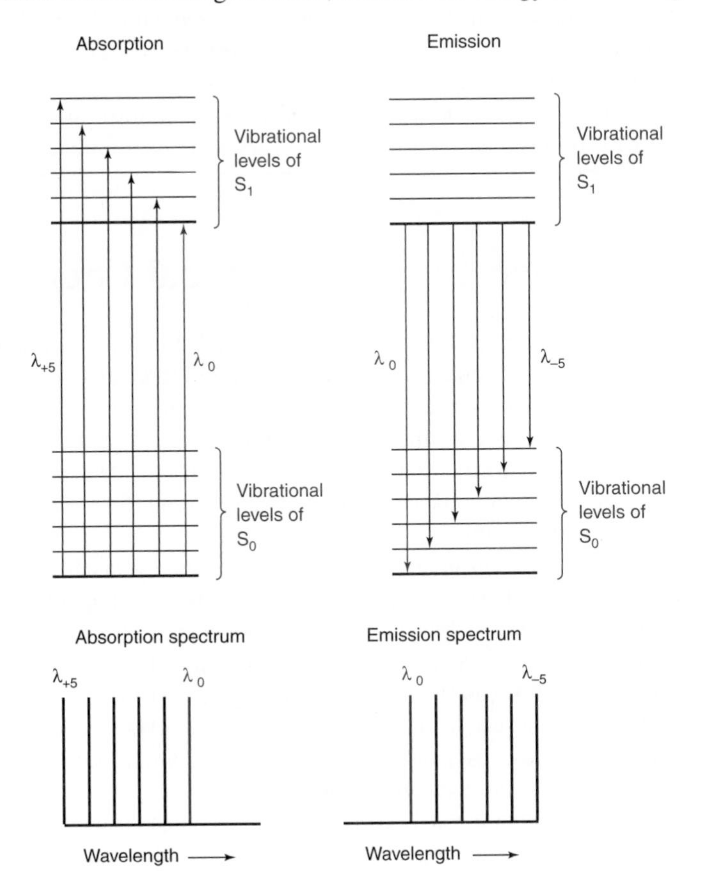

Figure 19-22 Energy-level diagram showing why structure is seen in the absorption and emission spectra and why the spectra are roughly mirror images. In absorption, wavelength λ_0 comes at lowest energy, and λ_{+5} is at highest energy. In emission, wavelength λ_0 comes at highest energy, and λ_{-5} is at lowest energy.

In the absorption spectrum, wavelength λ_0 corresponds to a transition from the ground vibrational level of S_0 to the lowest vibrational level of S_1. Absorption maxima at higher energy (shorter wavelength) correspond to the $S_0 \rightarrow S_1$ transition accompanied by absorption of one or more quanta of vibrational energy. In polar solvents, vibrational structure is often broadened beyond recognition, and only a broad envelope of absorption is observed. In Figure 19-21, the solvent is cyclohexane, which is nonpolar, and the vibrational structure is easily seen.

Following absorption, the vibrationally excited S_1 molecule relaxes back to the lowest vibrational level of S_1 prior to emitting any radiation. Emission from S_1 can go to any of the vibrational levels of S_0 in Figure 19-22. The highest-energy transition comes at wavelength λ_0, with a series of peaks following at longer wavelength. The absorption and emission spectra will have an approximate mirror image relation if spacings between vibrational levels are roughly equal and if the transition probabilities are similar.

Another possible consequence of the absorption of light is the breaking of chemical bonds. **Photochemistry** is a chemical reaction initiated by absorption of light (as in the reaction $O_2 \xrightarrow{h\nu} 2O$ in the upper atmosphere mentioned at the opening of Chapter 18). Some chemical reactions (not initiated by light) release energy in the form of light, which is called **chemiluminescence**. The light from a firefly or a light stick[4] is chemiluminescence.

? Ask Yourself

19-D. **(a)** What is the difference between electronic, vibrational, and rotational transitions?
(b) How can the energy of an absorbed photon be released without emission of light?
(c) What processes lead to fluorescence and phosphorescence? Which comes at higher energy? Which is faster?
(d) Why does fluorescence tend to be the mirror image of absorption?
(e) What is the difference between photochemistry and chemiluminescence?

19-5 Luminescence in Analytical Chemistry

Luminescence is any emission of electromagnetic radiation and includes fluorescence, phosphorescence, and other possible processes. In Figure 19-23, luminescence is measured by exciting a sample at a wavelength that it absorbs ($\lambda_{excitation}$) and observing at the wavelength of maximum emission ($\lambda_{emission}$). Luminescence is observed perpendicular to the incident direction to minimize the detection of scattered radiation. Scattered radiation is incident light that is scattered to the side by particles or large molecules in the sample.

Luminescence is more sensitive than absorbance for detecting very low concentrations of analyte.

Luminescence is more sensitive than absorption. Imagine yourself in a stadium at night with the lights off, but each of the 50 000 raving fans is holding a lighted candle. If 500 people blow out their candles, you will hardly notice the difference. Now imagine that the stadium is completely dark, and then 500 people light their candles. The change would be dramatic. The first case is analogous to changing transmittance from 100% to 99%. It is hard to measure such a small change because the 50 000-candle background is so bright. The second case is analogous to observing luminescence from 1% of the molecules in a sample. Against the dark background, luminescence is easy to detect.

Luminescence is such a sensitive technique that scientists can observe emission from a *single molecule*. For example, the mechanism by which the protein myosin

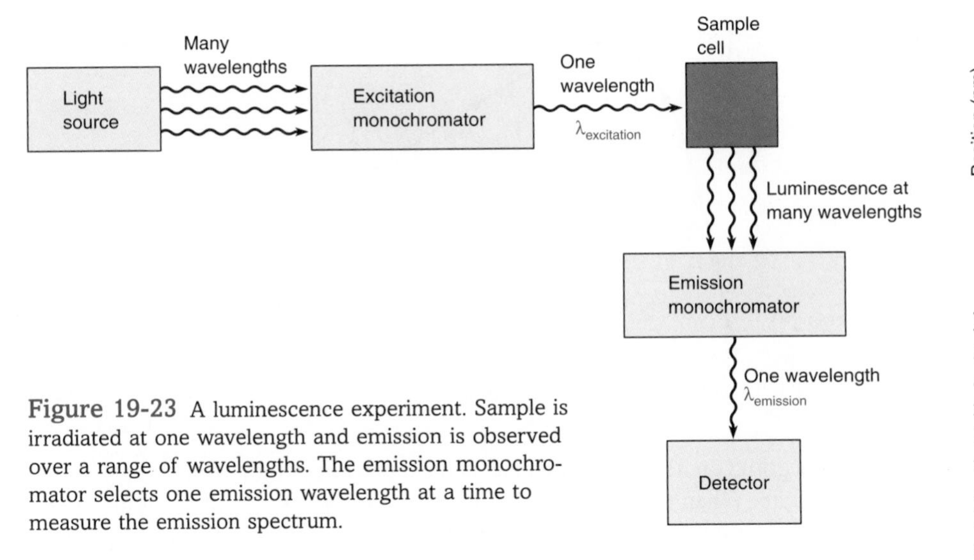

Figure 19-23 A luminescence experiment. Sample is irradiated at one wavelength and emission is observed over a range of wavelengths. The emission monochromator selects one emission wavelength at a time to measure the emission spectrum.

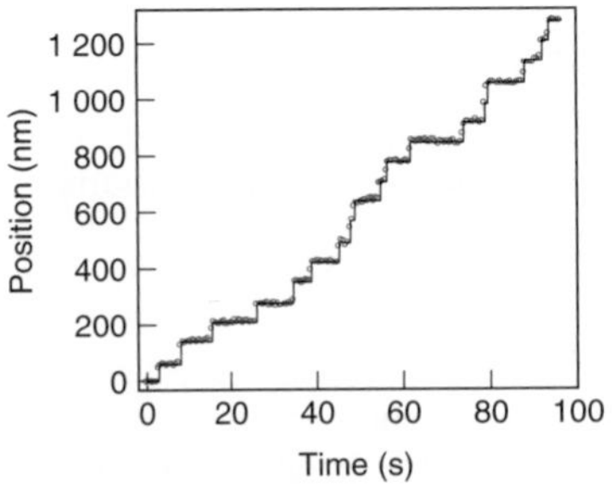

Figure 19-24 Steps taken by *one molecule* of myosin moving along an actin fiber observed with fluorescent dye attached to myosin. Measurement of 231 steps by 32 molecules gave a mean step length of 74 ± 5 nm. Knowledge of step length allows one of two postulated mechanisms of motion to be rejected. [From A. Yildiz and P. R. Selvin, *Acc. Chem. Res.* **2005**, *38*, 574.]

"walks" along fibers of the protein actin to make muscles contract has been studied by labeling myosin with a highly fluorescent dye. As myosin steps along actin—powered by hydrolysis of adenosine triphosphate (ATP)—the position of the dye can be located to within 1 nm by statistical analysis of its emission observed with a light microscope (Figure 19-24).

For quantitative analysis, the intensity of luminescence (I) is proportional to the concentration of the emitting species (c) over some limited concentration range:

Relation of emission intensity to concentration:

$$I = kP_0c \tag{19-10}$$

Luminescence can be increased by increasing the incident radiant power.

where P_0 is the incident radiant power and k is a constant.

Fluorimetric Assay of Selenium in Brazil Nuts

Selenium is a trace element essential to life. For example, the selenium-containing enzyme glutathione peroxidase catalyzes the destruction of peroxides (ROOH) that are harmful to cells. Conversely, at high concentration, selenium can be toxic.

To measure selenium in Brazil nuts, 0.1 g of nut is digested with 2.5 mL of 70 wt% HNO_3 in a Teflon bomb in a microwave oven (Figure 2-18). Hydrogen selenate (H_2SeO_4) in the digest is reduced to hydrogen selenite (H_2SeO_3) with hydroxylamine (NH_2OH). Selenite is then **derivatized** to form a fluorescent product that is extracted into cyclohexane.

Derivatization is the chemical alteration of an analyte so that it can be detected conveniently or separated from other species easily.

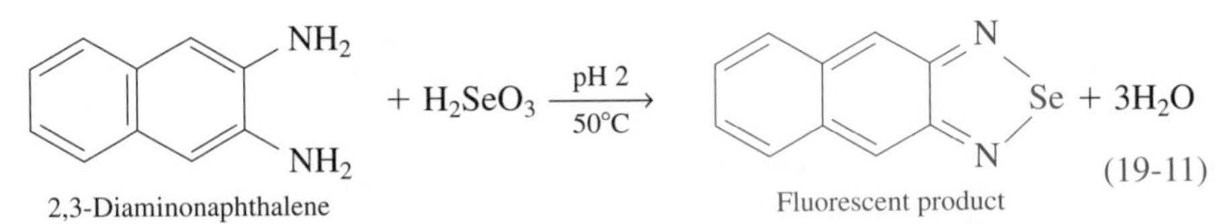

(19-11)

Maximum response of the fluorescent product was observed with an excitation wavelength of 378 nm and an emission wavelength of 518 nm. The fluorescence calibration curve in Figure 19-25 is linear, obeying Equation 19-10, only up to ~0.1 μg Se/mL. Beyond 0.1 μg Se/mL, the response becomes curved, eventually reaches a maximum, and finally *decreases* with increasing selenium concentration.

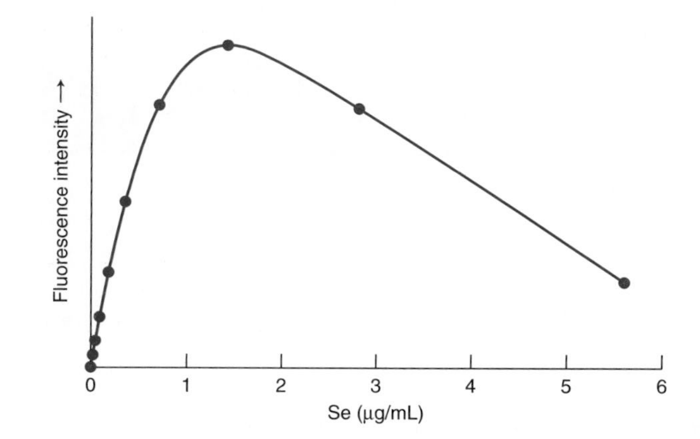

Figure 19-25 Fluorescence calibration curve for the selenium-containing product in Reaction 19-11. The curvature and maximum are due to self-absorption. [From M.-C. Sheffield and T. M. Nahir, *J. Chem. Ed.* **2002**, *79*, 1345.]

What is going on here? Fluorescence is the emission of light from a molecule in an excited state. When the concentration becomes too great, neighboring unexcited molecules absorb light from the excited molecule before the light can escape from the cuvet. Absorption of excitation energy by neighboring molecules of the same substance is called **self-absorption**. Some fraction of excitation energy absorbed by the neighbors is converted to heat. The higher the concentration, the more the analyte absorbs its own fluorescence and the less emission we observe. We say that the molecule *quenches* its own emission.

The behavior in Figure 19-25 is general. At low concentration, luminescence intensity is proportional to analyte concentration. At higher concentration, self-absorption becomes important and, eventually, luminescence reaches a maximum. Equation 19-10 applies only at low concentration.

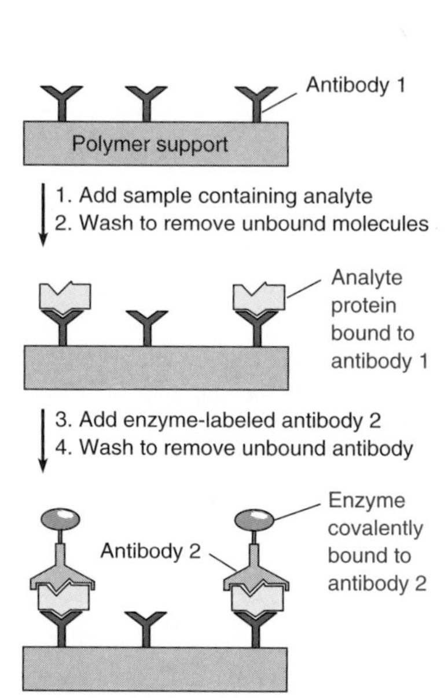

Figure 19-26 Enzyme-linked immunosorbent assay. Antibody 1, which is specific for the analyte of interest, is bound to a polymer support and treated with unknown. After excess, unbound molecules have been washed away, analyte remains bound to antibody 1. Bound analyte is then treated with antibody 2, which recognizes a different site on the analyte. An enzyme is covalently attached to antibody 2. After unbound material has been washed away, each molecule of analyte is coupled to an enzyme, which will be used in Figure 19-27.

Immunoassays

An important application of luminescence is in **immunoassays**, which employ antibodies to detect analyte. An **antibody** is a protein produced by the immune system of an animal in response to a foreign molecule, which is called an **antigen**. An antibody specifically recognizes and binds to the antigen that stimulated its synthesis.

Figure 19-26 illustrates the principle of an *enzyme-linked immunosorbent assay*, abbreviated ELISA in biochemical literature. Antibody 1, which is specific for the analyte of interest (the antigen), is bound to a polymer support. In steps 1 and 2, analyte is incubated with the polymer-bound antibody to form the antibody-antigen complex. The fraction of antibody sites that bind analyte is proportional to the concentration of analyte

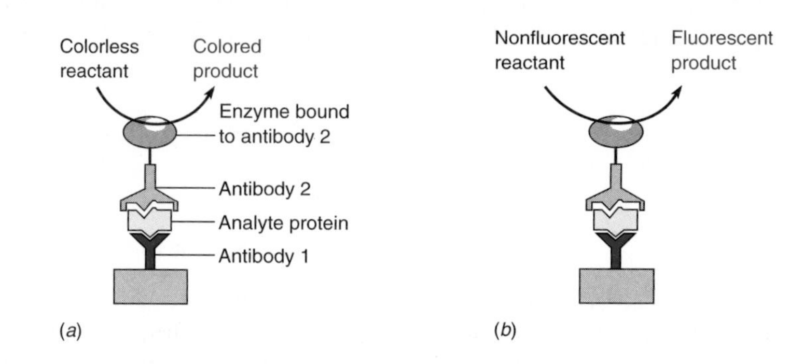

Figure 19-27 Enzyme bound to antibody 2 can catalyze reactions that produce (*a*) colored or (*b*) fluorescent products. Each molecule of analyte bound in the immunoassay leads to many molecules of colored or fluorescent product that are easily measured.

in the unknown. The surface is then washed to remove unbound substances. In steps 3 and 4, the antibody-antigen complex is treated with antibody 2, which recognizes a different region of the analyte. An enzyme that will be used later was covalently attached to antibody 2 (prior to step 3). Again, excess unbound substances are washed away.

Figure 19-27 shows two ways in which the enzyme attached to antibody 2 is used for quantitative analysis. In Figure 19-27*a*, the enzyme transforms a colorless reactant into a colored product. Because one enzyme molecule catalyzes the same reaction many times, many molecules of colored product are created for each molecule of antigen. The enzyme thereby *amplifies* the signal in chemical analysis. The higher the concentration of analyte in the unknown, the more enzyme is bound and the greater the extent of the enzyme-catalyzed reaction. In Figure 19-27*b*, the enzyme converts a nonfluorescent reactant into a fluorescent product. Enzyme-linked immunosorbent assays are sensitive to <1 ng of analyte. Pregnancy tests are based on the immunoassay of a placental protein in urine. Box 19-1 shows how immunoassays are used in field-portable environmental analyses.

Ask Yourself

19-E. **(a)** Why is Figure 19-25 curved and why does it reach a maximum?
(b) How is signal amplification achieved in enzyme-linked immunosorbent assays?

Key Equations

Absorption of a mixture of species X and Y

$$A = \varepsilon_X b[X] + \varepsilon_Y b[Y]$$

A = absorbance at wavelength λ
ε_i = molar absorptivity of species i at wavelength λ
b = pathlength

You should be able to use Equations 19-5 to analyze the spectrum of a mixture

Fluorescence intensity (at low concentration)

$$I = kP_0c$$

I = fluorescence intensity
k = constant
P_0 = radiant power of incident radiation
c = concentration of fluorescing species

Important Terms

antibody
antigen
chemiluminescence
derivatization
diffraction
electronic transition
fluorescence
grating
immunoassay
isosbestic point
luminescence
molecular orbital
monochromator
phosphorescence
photochemistry
photodiode array
photomultiplier tube
polychromator
rotational transition
self-absorption
singlet state
spectrophotometric titration
triplet state
vibrational transition

Problems

19-1. State the differences between single- and double-beam spectrophotometers and explain how each measures the transmittance of a sample. What source of error in a single-beam instrument is absent in the double-beam instrument?

19-2. Would you use a tungsten or a deuterium lamp as a source of 300-nm radiation?

19-3. What are the advantages and disadvantages of decreasing monochromator slit width?

Box 19-1 Immunoassays in Environmental Analysis

Immunoassays are available for use in the field to monitor pesticides, industrial chemicals, and microbial toxins at the parts per trillion to parts per million levels in groundwater, soil, and food. An advantage of screening in the field is that uncontaminated regions that require no further attention are readily identified. In some cases, the immunoassay field test is 20–40 times less expensive than a chromatographic analysis in the laboratory. Immunoassays require less than a milliliter of sample and can be completed in 2–3 h. Chromatographic analyses might require as long as 2 days, because analyte must first be extracted or concentrated from liter-quantity samples to obtain sufficient concentration.

The diagram shows how an assay works. In step 1, antibody for the intended analyte is adsorbed to the bottom of a microtiter well, which is a depression in a plate having as many as 96 wells for simultaneous analyses. In step 2, known volumes of sample and standard solution containing enzyme-labeled analyte are added to the well. In step 3, analyte in the sample competes with enzyme-labeled analyte for binding sites on the antibody. This is the key step: *The greater the concentration of analyte in the unknown sample, the more it will bind to antibody and the less enzyme-labeled analyte will bind.* After an incubation period, unbound sample and standard are washed away. In step 4, a *chromogenic* substance is added to the well. This is a colorless substance that reacts in the presence of enzyme to make a colored product. In step 5, colored product is measured visually by comparison with standards or quantitatively with a spectrophotometer. The greater the concentration of analyte in the unknown sample, the lighter the color in step 5.

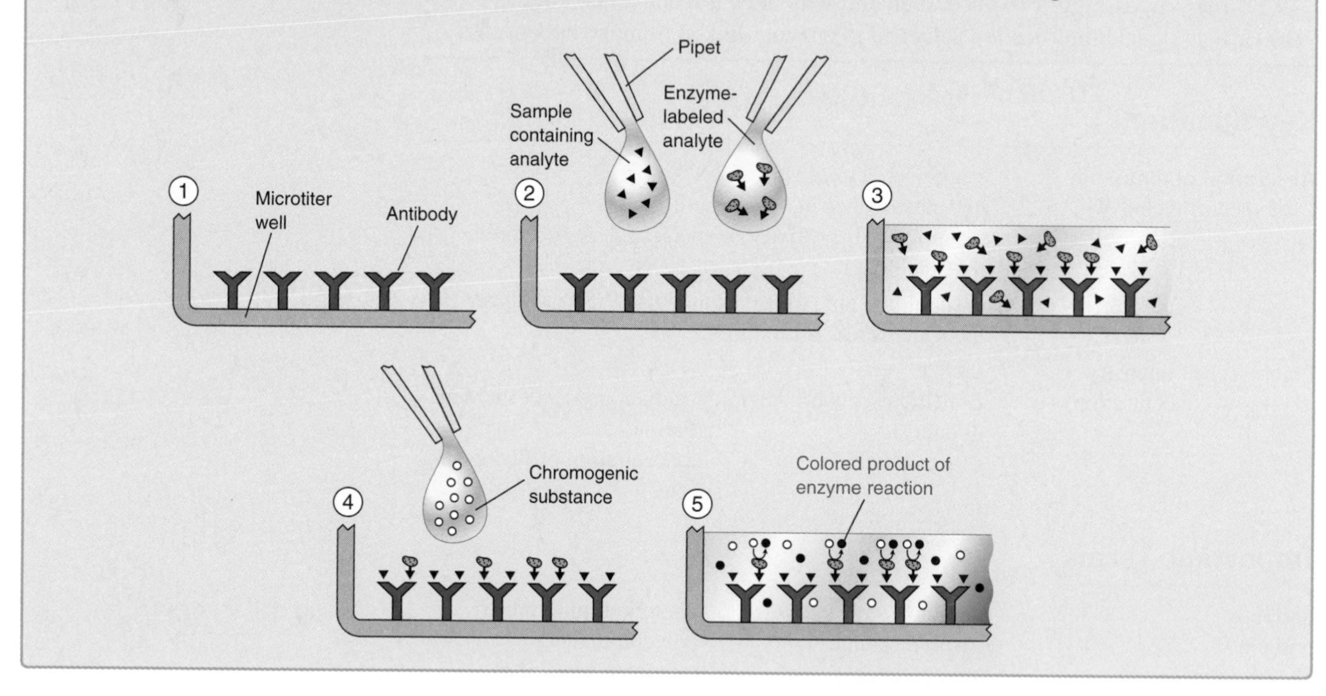

19-4. Consider a reflection grating operating with an incident angle of 40° in Figure 19-5.

(a) How many lines per centimeter should be etched in the grating if the first-order diffraction angle for 600 nm (visible) light is to be $-30°$?

(b) Answer the same question for 1 000 cm^{-1} (infrared) light.

19-5. **(a)** In Color Plate 19*a*, red light with a wavelength of 633 nm strikes a grating at normal incidence ($\theta = 0$). The grating spacing is $d = 1.6$ μm. At what angles are the $n = -1$, $n = +1$, and $n = +2$ diffracted beams expected?

(b) Explain why the $n = 3$ diffracted beam is not observed.

19-6. Why is a photomultiplier such a sensitive photodetector?

19-7. What characteristic makes a photodiode array spectrophotometer suitable for measuring the spectrum of a compound as it emerges from a chromatography column and a dispersive spectrophotometer not suitable? What is the disadvantage of the photodiode array spectrophotometer?

19-8. When are isosbestic points observed and why?

19-9. Molar absorptivities of compounds X and Y were measured with pure samples of each:

λ (nm)	ε (M^{-1} cm^{-1}) X	Y
$\lambda' \equiv 272$	$\varepsilon'_X = 16\,440$	$\varepsilon'_Y = 3\,870$
$\lambda'' \equiv 327$	$\varepsilon''_X = 3\,990$	$\varepsilon''_Y = 6\,420$

A mixture of X and Y in a 1.000-cm cell had an absorbance of $A' = 0.957$ at 272 nm and $A'' = 0.559$ at 327 nm. Find the concentrations of X and Y in the mixture.

19-10. *Spreadsheet for simultaneous equations.* Write a spreadsheet for the analysis of a mixture by using Equations 19-5. The input will be the sample pathlength, the observed absorbances at two wavelengths, and the molar absorptivities of the two pure compounds at two wavelengths. The output will be the concentration of each component of the mixture. Test your spreadsheet with numbers from Problem 19-9.

19-11. Ultraviolet absorbance for 1.00×10^{-4} M MnO_4^-, 1.00×10^{-4} M $Cr_2O_7^{2-}$, and an unknown mixture of both (all in a 1.000-cm cell) are given in the following table. Find the concentration of each species in the mixture.

Wavelength (nm)	MnO_4^- standard	$Cr_2O_7^{2-}$ standard	Mixture
266	0.042	0.410	0.766
320	0.168	0.158	0.422

19-12. Transferrin is the iron-transport protein found in blood. It has a molecular mass of 81 000 and carries two Fe^{3+} ions. Desferrioxamine B (Box 13-1) is a potent iron chelator used to treat patients with iron overload. It has a molecular mass of about 650 and can bind one Fe^{3+}. Desferrioxamine can take iron from many sites within the body and is excreted (with its iron) through the kidneys. The molar absorptivities of these compounds (saturated with iron) at two wavelengths are given in the following table. Both compounds are colorless (no visible absorption) in the absence of iron.

λ (nm)	ε (M^{-1} cm^{-1}) Transferrin	Desferrioxamine
428	3 540	2 730
470	4 170	2 290

(a) A solution of transferrin has an absorbance of 0.463 at 470 nm in a 1.000-cm cell. Find the concentration of transferrin in mg/mL and the concentration of iron in μg/mL.

(b) After addition of desferrioxamine (which dilutes the sample), the absorbance at 470 nm was 0.424 and the absorbance at 428 nm was 0.401. Calculate the fraction of iron in transferrin. Remember that transferrin binds two Fe^{3+} ions and desferrioxamine binds only one.

19-13. *Finding pK_a by spectrophotometry.* An indicator has a molar absorptivity of 2 080 M^{-1} cm^{-1} for HIn and 14 200 M^{-1} cm^{-1} for In^-, at a wavelength of 440 nm.

$$\text{HIn} \underset{}{\overset{K_{\text{HIn}}}{\rightleftharpoons}} \text{H}^+ + \text{In}^-$$

(a) Write a Beer's law expression for the absorbance at 440 nm of a solution in a 1.00-cm cuvet containing the concentrations [HIn] and [In^-].

(b) A solution of HIn is adjusted to pH 6.23 contains a mixture of HIn and In^- with a total concentration of 1.84×10^{-4} M. The absorbance at 440 nm is 0.868. From your expression from **(a)** and the mass balance [HIn] + [In^-] = 1.84×10^{-4} M, calculate pK_{HIn}.

19-14. *Graphical method to find pK_a by spectrophotometry.* This method requires a series of solutions containing a compound of unknown but constant concentration at different pH values. The figure shows that we choose a wavelength at which one of the species, say In^-, has maximum absorbance (A_{In^-}) and HIn has a different absorbance (A_{HIn}).

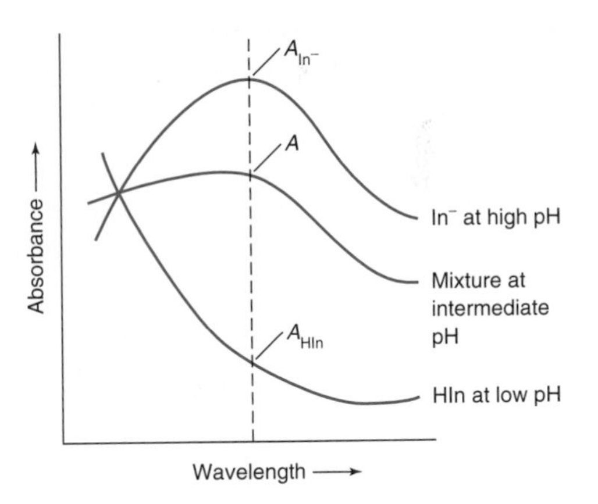

At intermediate pH, the absorbance (A) is between the two extremes. Let the total concentration be c_0 = [HIn] + [In^-]. At high pH, the absorbance is $A_{In^-} = \varepsilon_{In^-}bc_0$ and at low pH the absorbance is $A_{HIn} = \varepsilon_{HIn}bc_0$, where ε is molar absorptivity and b is pathlength. At intermediate pH, both species are present and the absorbance is $A = \varepsilon_{HIn}b[\text{HIn}] + \varepsilon_{In^-}b[\text{In}^-]$. You can combine these expressions to show that

$$\frac{[\text{In}^-]}{[\text{HIn}]} = \frac{A - A_{\text{HIn}}}{A_{\text{In}^-} - A}$$

Placing this expression into the Henderson-Hasselbalch equation gives

$$\mathrm{pH} = \mathrm{p}K_{\mathrm{HIn}} + \log\left(\frac{[\mathrm{In}^-]}{[\mathrm{HIn}]}\right) \Rightarrow \log\left(\frac{A - A_{\mathrm{HIn}}}{A_{\mathrm{In}^-} - A}\right) = \mathrm{pH} - \mathrm{p}K_{\mathrm{HIn}}$$

For several solutions of intermediate pH, a graph of log $[(A - A_{\mathrm{HIn}})/(A_{\mathrm{In}^-} - A)]$ versus pH should be a straight line with a slope of 1 that crosses the x-axis at $\mathrm{p}K_{\mathrm{HIn}}$.

pH	Absorbance
~2	$0.006 \equiv A_{\mathrm{HIn}}$
3.35	0.170
3.65	0.287
3.94	0.411
4.30	0.562
4.64	0.670
~12	$0.818 \equiv A_{\mathrm{In}^-}$

Data at 590 nm for HIn = bromophenol blue from G. S. Patterson, *J. Chem. Ed.* **1999**, *76*, 395.

Prepare a graph of log $[(A - A_{\mathrm{HIn}})/(A_{\mathrm{In}^-} - A)]$ versus pH. Find the slope and intercept and $\mathrm{p}K_{\mathrm{HIn}}$.

19-15. Infrared spectra are customarily recorded on a transmittance scale so that weak and strong bands can be displayed on the same scale. The region near 2 000 cm^{-1} in the infrared spectra of compounds A and B is shown in the figure below. Absorption corresponds to a *downward* peak on this scale. Spectra of a 0.010 0 M solution of each compound were obtained with a 0.005 00-cm-pathlength cell. A mixture of A and B in a 0.005 00-cm cell had a transmittance of 34.0% at 2 022 cm^{-1} and 38.3% at 1 993 cm^{-1}. Find [A] and [B] with the spreadsheet from Problem 19-10.

Wavenumber	Pure A	Pure B
2 022 cm^{-1}	31.0% *T*	97.4% *T*
1 993 cm^{-1}	79.7% *T*	20.0% *T*

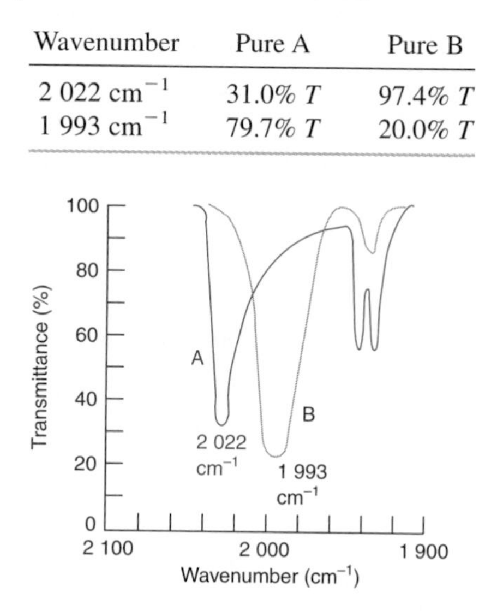

19-16. The iron-binding site of transferrin in Figure 19-14 accommodates certain other metal ions besides Fe^{3+} and certain other anions besides CO_3^{2-}. Data are given in the table for the titration of transferrin (3.57 mg in 2.00 mL) with 6.64 mM Ga^{3+} in the presence of the anion oxalate, $C_2O_4^{2-}$, and in the absence of a suitable anion. Prepare a graph similar to Figure 19-15 showing both sets of data. Mark the theoretical equivalence point for binding two Ga^{3+} ions per molecule of protein. How many Ga^{3+} ions are bound to transferrin in the presence and absence of oxalate?

Titration in presence of $C_2O_4^{2-}$		Titration in absence of anion	
Total μL Ga^{3+} added	Absorbance at 241 nm	Total μL Ga^{3+} added	Absorbance at 241 nm
0.0	0.044	0.0	0.000
2.0	0.143	2.0	0.007
4.0	0.222	6.0	0.012
6.0	0.306	10.0	0.019
8.0	0.381	14.0	0.024
10.0	0.452	18.0	0.030
12.0	0.508	22.0	0.035
14.0	0.541	26.0	0.037
16.0	0.558		
18.0	0.562		
21.0	0.569		
24.0	0.576		

19-17. The metal-binding compound semi-xylenol orange is yellow at pH 5.9 but turns red (λ_{max} = 490 nm) when it reacts with Pb^{2+}. A 2.025-mL sample of semi-xylenol orange was titrated with 7.515×10^{-4} M $Pb(NO_3)_2$, with the following results:

Total μL Pb^{2+} added	Absorbance at 490 nm in 1-cm cell	Total μL Pb^{2+} added	Absorbance at 490 nm in 1-cm cell
0.0	0.227	42.0	0.425
6.0	0.256	48.0	0.445
12.0	0.286	54.0	0.448
18.0	0.316	60.0	0.449
24.0	0.345	70.0	0.450
30.0	0.370	80.0	0.447
36.0	0.399		

Make a graph of corrected absorbance versus microliters of Pb^{2+} added. Corrected absorbance is what would be observed if the volume were not changed from its initial value of 2.025 mL. Assuming that the reaction of semi-xylenol orange with Pb^{2+}

has a 1:1 stoichiometry, find the molarity of semi-xylenol orange in the original solution.

19-18. An *immunoassay* to measure explosives such as trinitrotoluene (TNT) in organic solvent extracts of soil employs a *flow cytometer*, which counts small particles (such as living cells) flowing through a narrow tube past a detector. The cytometer in this experiment irradiates the particles with a green laser and measures fluorescence from each particle as it flows past the detector.

1. Antibodies that bind TNT are chemically attached to 5-μm-diameter latex beads.

2. The beads are incubated with a fluorescent derivative of TNT to saturate the antibodies and excess TNT derivative is removed.

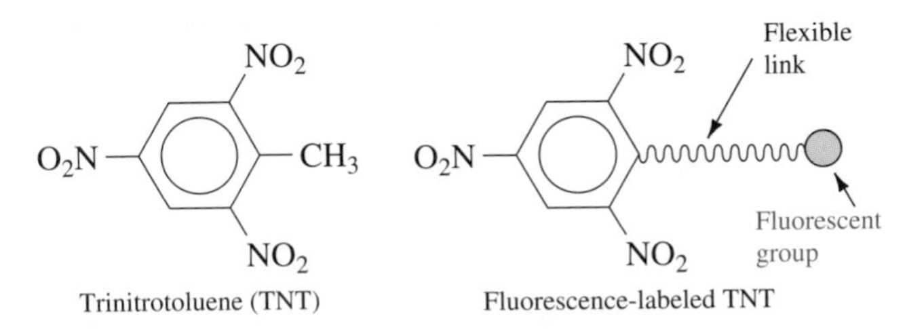

3. 5 μL of a suspension of beads are added to 100 μL of sample. TNT in the sample displaces some derivatized TNT from the antibodies. The higher the concentration of TNT, the more derivatized TNT is displaced.

4. Sample/bead suspension is injected into the flow cytometer, which measures fluorescence of individual beads as they pass the detector. The figure below shows median fluorescence intensity ± standard deviation. TNT can be quantified in the ppb to ppm range.

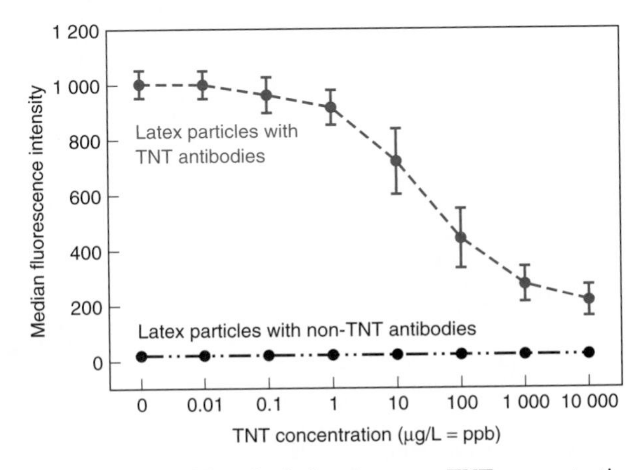

Fluorescence of TNT-antibody-beads versus TNT concentration. [From G. P. Anderson, S. C. Moreira, P. T. Charles, I. L. Medintz, E. R. Goldman, M. Zeinali, and C. R. Taitt, *Anal. Chem.* **2006**, *78*, 2279.]

Draw pictures showing the state of the beads in steps 1, 2 and 3 and explain how this method works.

19-19. *Standard addition.* Selenium from 0.108 g of Brazil nuts was converted into the fluorescent product in Reaction 19-11, which was extracted into 10.0 mL of cyclohexane. Then 2.00 mL of the cyclohexane solution were placed in a cuvet for fluorescence measurements.

(a) Standard additions of fluorescent product containing 1.40 μg Se/mL are given in the table. Construct a standard addition graph like Figure 5-5 to find the concentration of Se in the 2.00-mL unknown solution. Find the wt% Se in the nuts.

Volume of standard added (μL)	Fluorescence intensity (arbitrary units)
0	41.4
10.0	49.2
20.0	56.4
30.0	63.8
40.0	70.3

(b) With the formula in Problem 5-19, find the uncertainty in the x-intercept and the uncertainty in wt% Se.

19-20. *pH measurement in living cells.* A fluorescent indicator called C-SNARF-1 has acidic (HIn) and basic (In^-) forms with different emission spectra.

$$\text{HIn} \xrightleftharpoons{pK_a = 7.50} \text{In}^- + \text{H}^+$$

The following figure shows emission from the indicator dissolved inside human lymphoblastoid cells and emission from cells in the absence of indicator.

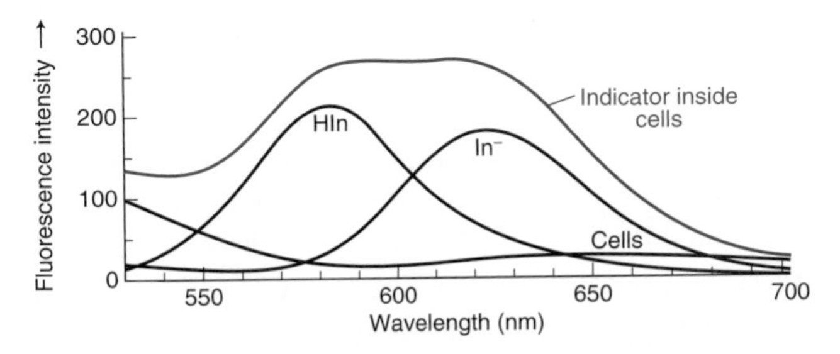

Fluorescence of C-SNARF-1 dissolved in human lymphoblastoid cells and from cells that lack indicator. Emission spectra of HIn and In^- are superimposed with the correct intensities so that intensities from HIn, In^-, and pure cells add up to emission from cells + indicator. [From A.-C. Ribou, J. Vigo, and J.-M. Salmon, *J. Chem. Ed.* **2002**, *79*, 1471.]

In a separate experiment, emission from pure HIn (in acidic solution) and pure In^- (in basic solution) at equal concentrations was measured. Results are shown in the following table.

	Fluorescence intensity (arbitrary units)		
	590 nm	625 nm	Intensity ratio (I_{590}/I_{625})
Cells + indicator	269	258	
Cells only	13	21	
Difference	256	237	$1.08_0 \equiv R$
HIn	14 780	4 700	$3.14_5 \equiv R_{HIn}$
In^-	3 130	9 440	$0.332 \equiv R_{In^-}$

Emission intensities are additive. At wavelengths $\lambda' = 590$ nm and $\lambda'' = 625$ nm, we can write

$$I' = a'_{HIn}\,[HIn] + a'_{In^-}\,[In^-] \quad \text{(A)}$$

$$I'' = a''_{HIn}\,[HIn] + a''_{In^-}\,[In^-] \quad \text{(B)}$$

where the coefficients a relate emission intensity to concentration. We designate the ratio of emission intensities from the unknown mixture of HIn + In^- dissolved in the cells as $R = I'/I''$. Similarly, the ratios for HIn and In^- are designated R_{HIn} and R_{In^-}. Values are given in the table. We can rearrange the simultaneous equations A and B to find

$$\frac{[In^-]}{[HIn]} = \left(\frac{R - R_{HIn}}{R_{In^-} - R}\right)\frac{a''_{HIn}}{a''_{In^-}} \quad \text{(C)}$$

The quotient a''_{HIn}/a''_{In^-} is the ratio of emission from HIn and In^- at wavelength $\lambda'' = 625$ nm when they are at equal concentration. From the table, this quotient is $a''_{HIn}/a''_{In^-} = 4\,700/9\,440 = 0.497_9$. The wavelengths 590 and 625 were chosen so that ratios in Equation C can be measured with some accuracy. Find the quotient $[In^-]/[HIn]$ inside the cells and from this quotient find the pH inside the cells.

19-21. *Greenhouse gas reduction.* An 18-W compact fluorescent bulb produces approximately the same amount of light as a 75-W incandescent bulb that screws into the same socket. The fluorescent bulb lasts ~10 000 h and the incandescent bulb lasts ~750 h. Over the life of the fluorescent bulb, the electricity savings is $(75 - 18\text{ W})(10^4\text{ h}) = 570$ kW · h. One kilogram of coal produces ~2 kW · h. of electricity. If coal contains 60 wt% carbon, how many more kg of CO_2 are produced by running the incandescent bulb than the fluorescent bulb? If the coal contains 2 wt% sulfur, how many more kg of SO_2 are produced?

How Would You Do It?

19-22. The metal ion indicator xylenol orange (see Table 13-2) is yellow at pH 6 (λ_{max} = 439 nm). The spectral changes that occur as VO^{2+} (vanadyl ion) is added to the indicator at pH 6 are shown in the figure below. The mole ratio VO^{2+}/xylenol orange at each point is shown in the following table. Suggest a sequence of chemical reactions to explain the spectral changes, especially the isosbestic points at 457 and 528 nm.

Trace	Mole ratio	Trace	Mole ratio	Trace	Mole ratio
0	0	6	0.60	12	1.3
1	0.10	7	0.70	13	1.5
2	0.20	8	0.80	14	2.0
3	0.30	9	0.90	15	3.1
4	0.40	10	1.0	16	4.1
5	0.50	11	1.1		

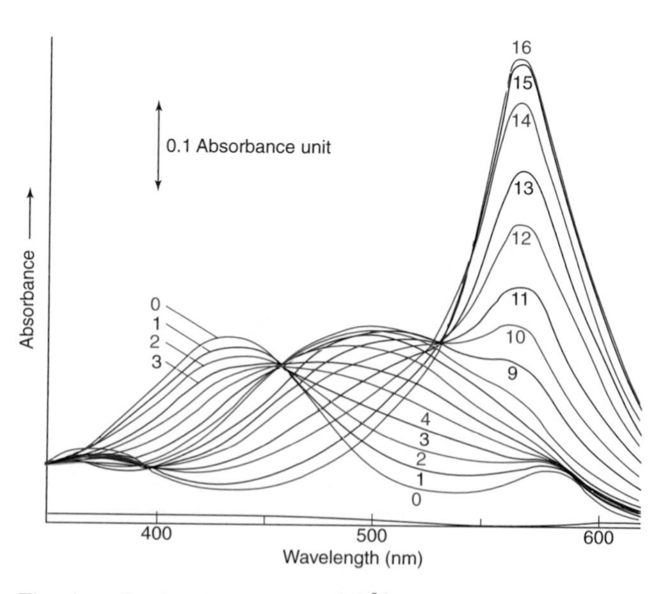

Titration of xylenol orange with VO^{2+} at pH 6.0. [From D. C. Harris and M. H. Gelb, *Biochim. Biophys. Acta* **1980**, *623*, 1.]

Notes and References

1. J. A. DeLuca, *J. Chem. Ed.* **1980**, *57,* 541.

2. Demonstrations with a spectrophotometer and fiber-optic probe: J. P. Blitz, D. J. Sheeran, and T. L. Becker, "Classroom Demonstrations of Concepts in Molecular Fluorescence," *J. Chem. Ed.* **2006**, *83*, 758.

3. R. B. Weinberg, *J. Chem. Ed.* **2007**, *84*, 797. Demonstration of fluorescence quenching clock reaction with laundry detergent and household chemicals.

4. C. Salter, K. Range, and G. Salter, *J Chem. Ed.* **1999**, *76*, 84.

Further Reading

A. Manz, N. Pamme, and D. Iossifidis, *Bioanalytical Chemistry* (London: Imperial College Press, 2004).

Historical Record of Mercury in the Snow Pack

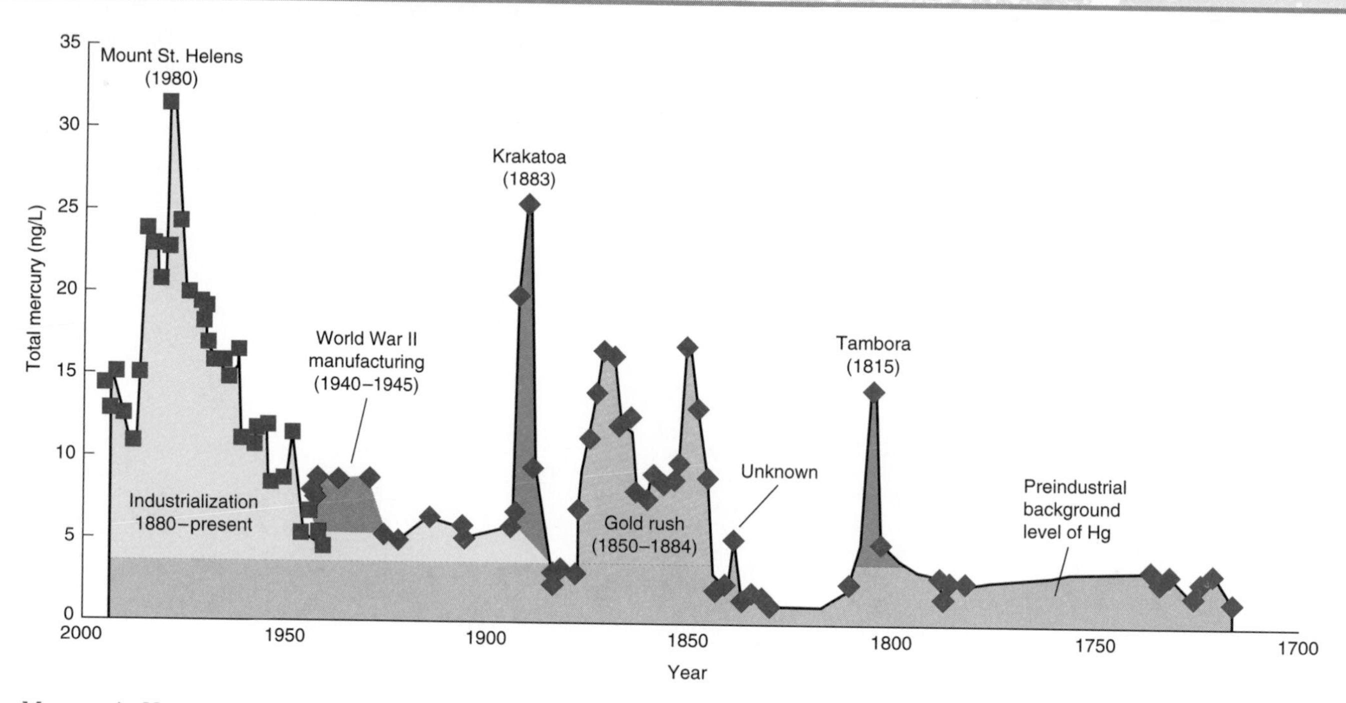

Mercury in Upper Fremont Glacier. [From P. F. Schuster, D. P. Krabbenhoft, D. L. Naftz, L. D. Cecil, M. L. Olson, J. F. Dewild, D. D. Susong, J. R. Green, and M. L. Abbott, *Environ. Sci. Technol.* **2002**, *36*, 2303. For a record of Hg going back 3 000 years, see F. Roos-Barraclough and W. Shotyk, *Environ. Sci. Technol.* **2003**, *37*, 235.]

Mercury at parts per trillion (ng/L) levels in glacial ice drilled to a depth of 160 m in Wyoming provides a record of events since the year 1720. Calibration of depths for the years 1958 and 1963 was determined by radioactive debris from nuclear bomb tests. Depths for 1815 and 1883 were found from peaks in electrical conductivity from acid produced by the volcanoes Krakatoa and Tambora.

Anthropogenic Hg is observed in 1850–1884 from the California gold rush when tons of Hg were used to extract Au from ore. Use of Hg for recovering Au was limited in 1884. High concentrations of Hg in the twentieth century are attributed to burning of coal, waste incineration, and use of Hg to manufacture Cl_2 in the chlor-alkali process. Atmospheric Hg has declined from its peak since Hg use was limited by international agreements late in the 20th century. Coal burning remains a major source of Hg. Studies of "pristine" locations identify atmospheric deposition as the source of methylmercury (CH_3Hg^+) in fish.[1] High levels of mercury led to warnings to limit the quantity of fish in our diets.

Hg from melted ice was measured by reduction to Hg(0), which was purged from solution by bubbling Ar gas. Hg(*g*) was trapped by metallic Au coated on sand. (Hg is soluble in gold.) For analysis, the trap was heated to liberate Hg, which passed into a cuvet. The cuvet was irradiated with a mercury lamp and fluorescence from Hg vapor was observed. The detection limit was 0.04 ng/L. Blanks prepared by performing all steps with pure water in place of melted glacier had 0.66 ± 0.25 ng Hg/L, which was subtracted from glacier readings. All steps in trace analysis are carried out in a scrupulously clean environment.

CHAPTER 20

Atomic Spectroscopy

Atomic spectroscopy is a principal tool for measuring metallic elements at major and trace levels in industrial and environmental laboratories. Automated with mechanical sample-changing devices, each instrument can turn out hundreds of analyses per day.

20-1 What Is Atomic Spectroscopy?

In the atomic spectroscopy experiment in Figure 20-1, a liquid sample is *aspirated* (sucked) through a plastic tube into a flame that is hot enough to break molecules apart into atoms. The concentration of an element in the flame is measured by absorption or emission of radiation. For **atomic absorption spectroscopy,** radiation of the correct frequency is passed through the flame (Figure 20-2) and the intensity of transmitted radiation is measured. For **atomic emission spectroscopy,** no lamp is required. Radiation is emitted by hot atoms whose electrons have been promoted to excited states in the flame. For both experiments in Figure 20-2, a monochromator selects the wavelength that will reach the detector. Analyte concentrations at the parts per million level are measured with a precision of 2%. To analyze major constituents, a sample must be diluted to reduce concentrations to the ppm level.

Atomic spectroscopy:
- absorption (requires lamp with light that is absorbed by atoms)
- emission (luminescence from excited atoms—no lamp required)

Parts per million (ppm) means micrograms of solute per gram of solution. The density of dilute aqueous solutions is close to 1.00 g/mL, so ppm usually refers to μg/mL. 1 ppm Fe = 1 μg Fe/mL ≈ 2×10^{-5} M.

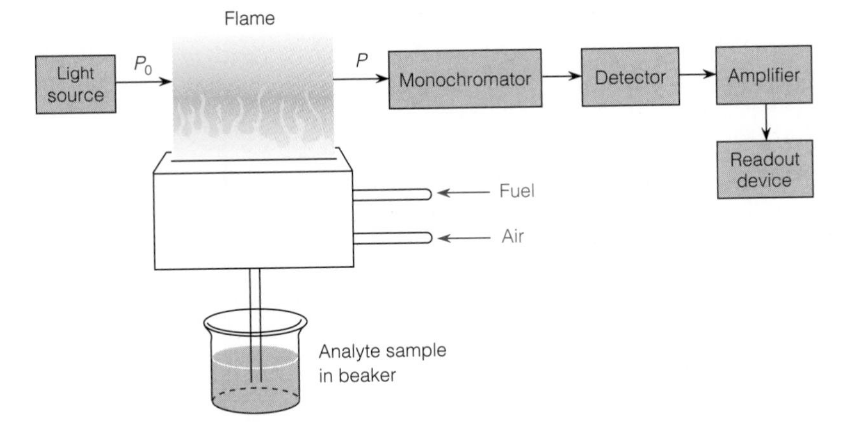

Figure 20-1 Atomic absorption experiment.

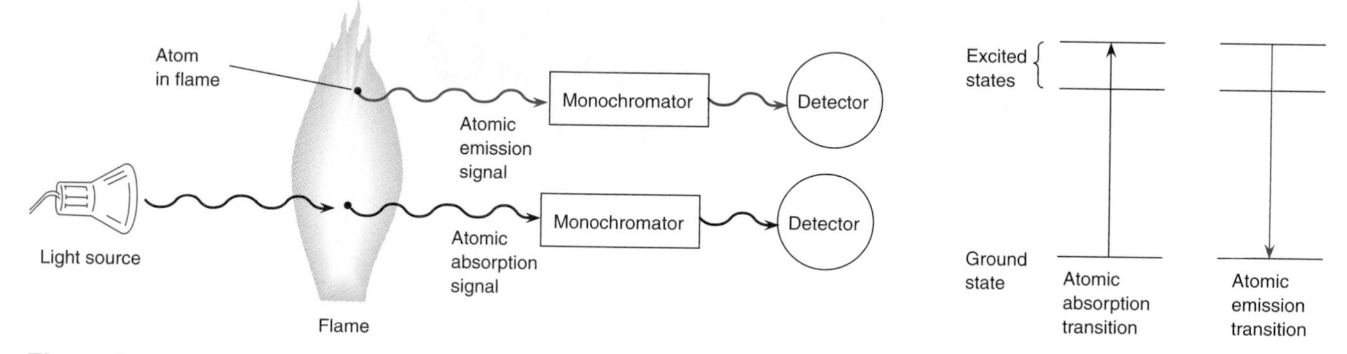

Figure 20-2 Absorption and emission of light by atoms in a flame. In atomic absorption, atoms absorb light from the lamp and unabsorbed light reaches the detector. In atomic emission, light is emitted by excited atoms in the flame.

Molecules in solution typically have absorption and emission bands that are ~10 to 100 nm wide (Figure 19-21). In contrast, gaseous atoms in a flame have extremely sharp lines with widths of 10^{-3} to 10^{-2} nm (Figure 20-3). Because the lines are so sharp, there is usually little overlap between spectra of different elements in the same sample. The lack of overlap allows some instruments to measure more than 70 elements in a sample simultaneously.

Ask Yourself

20-A. What is the difference between atomic absorption and atomic emission spectroscopy?

20-2 Atomization: Flames, Furnaces, and Plasmas

Atomization method	Quantitation method
flame	absorption or emission
graphite furnace	absorption or emission
plasma	emission or mass spectrometry

Atomization is the process of breaking analyte into gaseous atoms, which then are measured by their absorption or emission of radiation. Older atomic absorption spectrometers—and the apparatus generally found in teaching laboratories—use a combustion flame to decompose analyte into atoms, as in Figure 20-1. Modern instruments employ an inductively coupled argon plasma or an electrically heated graphite tube (also called a graphite furnace) for atomization. The combustion flame and the graphite furnace are used for both atomic absorption and atomic emission measurements. The plasma is so hot that many atoms are in excited states and their

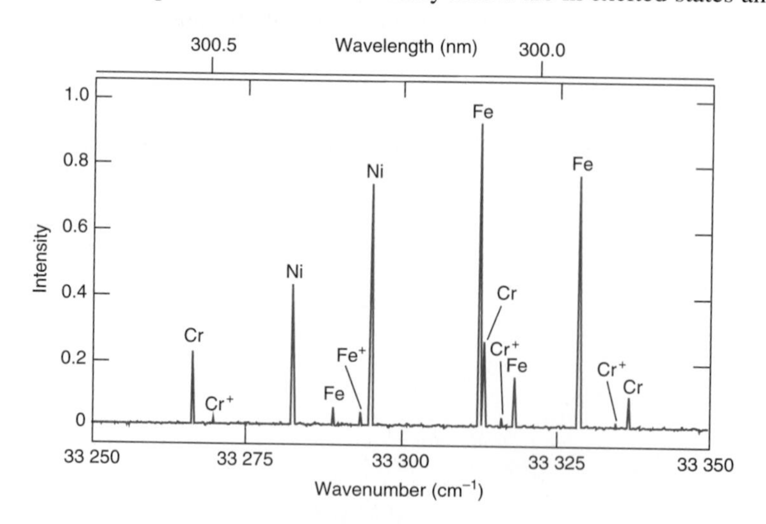

Figure 20-3 Part of the emission spectrum of a steel hollow-cathode lamp, showing sharp lines characteristic of gaseous Fe, Ni, and Cr atoms and weak lines from Cr^+ and Fe^+. Resolution is 0.001 nm, which is about half of the true linewidths of the signals. [From A. P. Thorne, *Anal. Chem.* **1991,** *63,* 57A.]

emission is readily observed. Plasmas are used almost exclusively for atomic emission measurements. The newest technique found in industrial, environmental, and research laboratories, is to atomize the sample with a plasma and measure the concentration of ions in the plasma with a *mass spectrometer.* This method does not involve absorption or emission of light and is not a form of spectroscopy.

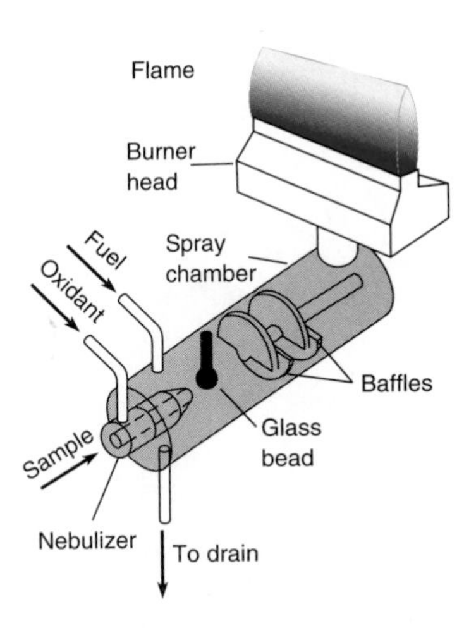

Figure 20-4 Premix burner with a pneumatic nebulizer. The slot in the burner head is typically 10 cm long and 0.5 mm wide.

Flame

Most flame spectrometers use a *premix burner,* such as that in Figure 20-4, in which the sample, oxidant, and fuel are mixed before being introduced into the flame. Sample solution is drawn in by rapid flow of oxidant and breaks into a fine mist when it leaves the tip of the *nebulizer* and strikes a glass bead. The formation of small droplets is termed *nebulization.* The mist flows past a series of baffles, which promote further mixing and block large droplets of liquid (which flow out to the drain). A fine mist containing about 5% of the initial sample reaches the flame. The remainder flows out the drain at the bottom of the burner.

After solvent evaporates in the flame, the remaining sample vaporizes and decomposes to atoms. Many metal atoms (M) form oxides (MO) and hydroxides (MOH) as they rise through the flame. Molecules do not have the same spectra as atoms, so the atomic signal is lowered. If the flame is relatively rich in fuel (a "rich" flame), excess carbon species tend to reduce MO and MOH back to M and thereby increase sensitivity. The opposite of a rich flame is a "lean" flame, which has excess oxidant and is hotter. We choose lean or rich flames to provide optimum conditions for different elements.

The most common fuel-oxidant combination is acetylene and air, which produces a flame temperature of 2 400–2 700 K (Table 20-1). When a hotter flame is required to vaporize *refractory* elements (those with high boiling points), acetylene and nitrous oxide is usually the mixture of choice. The height above the burner head at which maximum atomic absorption or emission is observed depends on the element being measured, as well as flow rates of sample, fuel, and oxidant. These parameters can be optimized for a given analysis.

Furnace

The electrically heated **graphite furnace** in Figure 20-5*a* provides greater sensitivity than a flame and requires less sample. A 1- to 100-μL sample is injected into the furnace through the hole at the center. The light beam travels through windows at each end of the tube. The maximum recommended temperature for a graphite furnace is 2 550°C for not more than 7 s. Surrounding the graphite with an atmosphere of Ar helps prevent oxidation.

Furnaces offer increased sensitivity and require less sample than a flame.

Graphite is a form of carbon. At high temperature in air, $C(s) + O_2 \longrightarrow CO_2$.

A graphite furnace has high sensitivity because it confines atoms in the optical path for several seconds. In flame spectroscopy, the sample is diluted during nebulization, and its residence time in the optical path is only a fraction of a second. Flames require a sample volume of at least 10 mL, because sample is constantly

Table 20-1 Maximum flame temperatures

Fuel	Oxidant	Temperature (K)
Acetylene	Air	2 400–2 700
Acetylene	Nitrous oxide	2 900–3 100
Acetylene	Oxygen	3 300–3 400
Hydrogen	Air	2 300–2 400
Hydrogen	Oxygen	2 800–3 000
Cyanogen	Oxygen	4 800

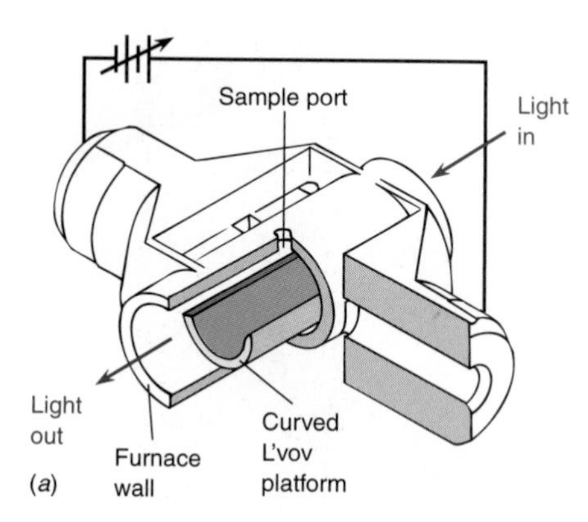

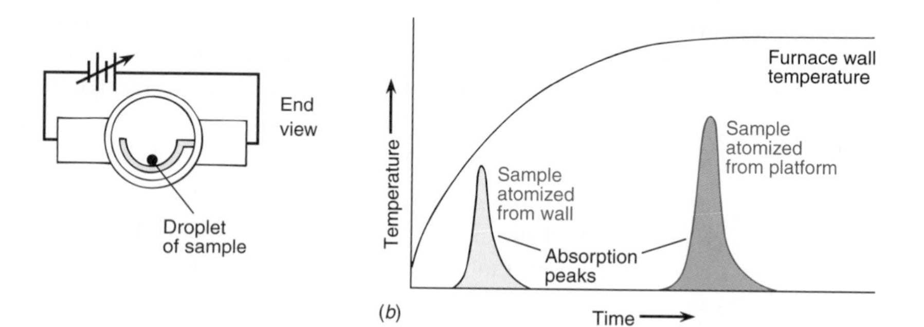

Figure 20-5 (*a*) Electrically heated graphite furnace for atomic spectroscopy. Sample is injected through the port at the top. L'vov platform inside the furnace is heated by radiation from the outer wall. Platform is attached to the wall by one small connection hidden from view. [Courtesy Perkin-Elmer Corp., Norwalk, CT.] (*b*) Heating profile comparing analyte evaporation from wall and from platform.

flowing into the flame. The graphite furnace requires only tens of microliters. In an extreme case, when only nanoliters of kidney tubular fluid were available, a method was devised to reproducibly deliver 0.1 nL to a furnace for analysis of Na and K. Precision with a furnace is rarely better than 5–10% with manual sample injection, but automated injection improves reproducibility.

Sample is injected onto the *L'vov platform* inside the furnace in Figure 20-5*a*. Analyte does not vaporize until the furnace wall has reached constant temperature (Figure 20-5*b*). If sample were injected directly onto the inside wall of the furnace, atomization would occur while the wall is heating up at a rate of 2 000 K/s (Figure 20-5*b*) and the signal would be much less reproducible than with the platform.

The operator must determine reasonable time and temperature for each stage of the analysis. Once a program is established, it can be applied to similar samples.

A skilled operator must determine heating conditions for three or more steps to properly atomize a sample. To analyze Fe in the iron-storage protein ferritin, 10 μL of sample containing ~0.1 ppm Fe are injected into the cold graphite furnace. The furnace is programmed to *dry* the sample at 125°C for 20 s to remove solvent. Drying is followed by 60 s of *charring* (also called *pyrolysis*) at 1 400°C to destroy organic matter, which creates smoke that would interfere with the optical measurement. *Atomization* is then carried out at 2 100°C for 10 s, during which absorbance reaches a maximum and then decreases as Fe evaporates from the furnace. The time-integrated absorbance (the peak area) is taken as the analytical signal. Finally, the furnace is heated to 2 500°C for 3 s to vaporize any residue.

The temperature needed to char the sample **matrix** (the medium containing the analyte) might also vaporize analyte. *Matrix modifiers* can retard evaporation of analyte until the matrix has charred away. Alternatively, a matrix modifier might increase the evaporation of matrix and thereby reduce interference by the matrix during atomization. For example, the matrix modifier ammonium nitrate added to seawater reduces interference by NaCl. Figure 20-6*a* shows a graphite furnace heating profile used to analyze Mn in seawater. When 0.5 M NaCl is subjected to this profile, signals are observed in Figure 20-6*b* at the analytical wavelength of Mn. Much of the apparent absorbance is probably due to scattering of light by smoke created by heating NaCl. The NaCl signal at the start of atomization interferes with the measurement of Mn. Adding NH_4NO_3 to the sample in Figure 20-6*c* reduces the matrix signal by forming NH_4Cl and $NaNO_3$, which cleanly evaporate instead of making smoke.

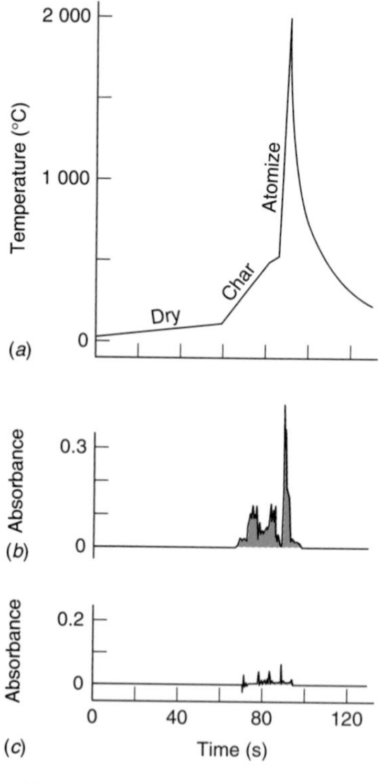

Figure 20-6 Reduction of interference by matrix modifier. (*a*) Graphite furnace temperature profile for analysis of Mn in seawater. (*b*) Absorbance profile when 10 μL of 0.5 M reagent-grade NaCl are subjected to the temperature profile. Absorbance is monitored at the Mn wavelength of 279.5 nm with a bandwidth of 0.5 nm. (*c*) Greatly reduced absorbance from 10 μL of 0.5 M NaCl plus 10 μL of 50 wt% NH_4NO_3 matrix modifier. [From M. N. Quigley and F. Vernon, *J. Chem. Ed.* **1996,** *73,* 980.]

When using a graphite furnace, it is important to monitor the absorption signal as a function of time, as in Figure 20-6. Peak shapes help you to adjust time and temperature for each step to obtain a clean signal from analyte. Also, a graphite furnace has a finite lifetime. Degradation of peak shape, loss of precision, or a large change in the slope of the calibration curve tells you that it is time to change the furnace.

It is possible to *preconcentrate* a sample by injecting and evaporating multiple aliquots in the graphite furnace prior to charring and atomization. For example, to measure trace levels of arsenic in drinking water, a 30-μL aliquot of water plus matrix modifier was injected and evaporated. The procedure was repeated five more times so that the total sample was 180 μL. The detection limit for As was 0.3 μg/L (parts per billion). Without preconcentration, the detection limit would have been approximately six times higher (1.8 μg/L). The increased capability with preconcentration is critical because As is a health hazard at concentrations of just a few parts per billion.

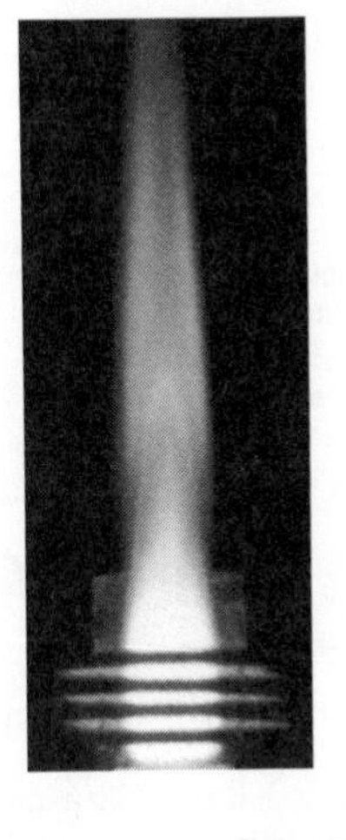

Inductively Coupled Plasma

The **inductively coupled plasma** in Figure 20-7 reaches a much higher temperature than that of combustion flames. Its high temperature and stability eliminate many problems encountered with conventional flames. The plasma's disadvantage is its expense to purchase and operate.

The plasma is energized by a radio-frequency induction coil wrapped around the quartz torch. High-purity argon gas plus analyte aerosol are fed into the torch. After a spark from a Tesla coil ionizes the Ar gas, free electrons accelerated by the radio-frequency field heat the gas to 6 000 to 10 000 K by colliding with atoms.

In the pneumatic nebulizer in Figure 20-4, liquid sample is sucked in by the flow of gas and breaks into small drops when it strikes the glass bead. The concentration of analyte needed for adequate signal (Table 20-2) is reduced by an order of magnitude with the *ultrasonic nebulizer* in Figure 20-8, in which liquid sample is directed onto a quartz crystal oscillating at 1 MHz. The vibrating crystal creates a fine aerosol that is carried by a stream of Ar through a heated tube, where solvent evaporates. The stream then passes through a cool zone in which solvent condenses and is removed. Analyte reaches the plasma as an aerosol of dry, solid particles. Plasma energy is not wasted in evaporating solvent, so more energy is available for atomization. Also, a larger fraction of the original sample reaches the flame than with a conventional nebulizer. Detection limits for the inductively coupled plasma can be further improved by a factor of 3 to 10 by viewing emission along the length of the plasma, rather than at right angles to the plasma.

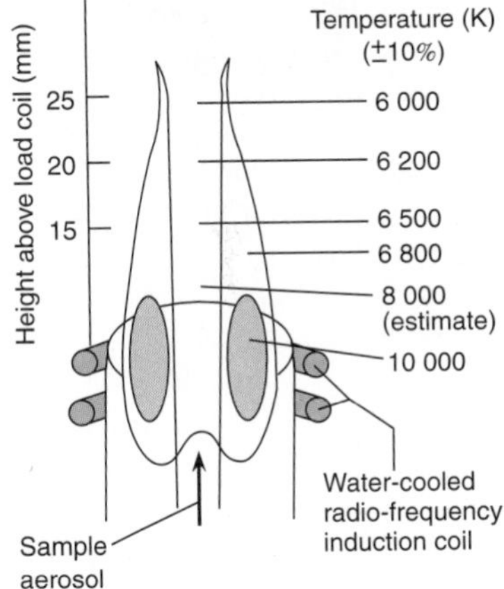

Figure 20-7 Temperature profile of a typical inductively coupled plasma used in analytical spectroscopy. [From V. A. Fassel, *Anal. Chem.* **1979,** *51,* 1290A.]

Detection Limits

The *detection limit* in Equation 5-4 is found by measuring the standard deviation of replicate samples whose analyte concentration is ~1 to 5 times the detection limit. Detection limits for a furnace are typically 100 times lower than those of a flame

Table 20-2 Comparison of detection limits for Ni^+ ion at 231 nm

Technique	Detection limits for different instruments (ng/mL)
Inductively coupled plasma–atomic emission (pneumatic nebulizer)	3–50
Inductively coupled plasma–atomic emission (ultrasonic nebulizer)	0.3–4
Graphite furnace–atomic absorption	0.02–0.06
Inductively coupled plasma–mass spectrometry	0.001–0.2

SOURCE: J.-M. Mermet and E. Poussel, *Appl. Spectros.* **1995,** *49,* 12A.

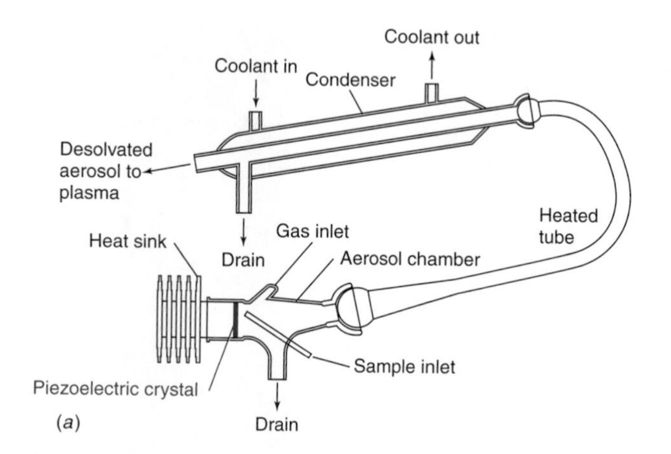

(b)

Figure 20-8 (*a*) Ultrasonic nebulizer lowers the detection limit for atomic spectroscopy for most elements by an order of magnitude. (*b*) Mist created when sample is sprayed against vibrating crystal. [Courtesy Cetac Technologies, Omaha, NE.]

(Figure 20-9), because the sample is confined in a small volume for a relatively long time in the furnace.

Ask Yourself

20-B. (a) Why can we detect smaller samples with lower concentrations with a furnace than with a flame or a plasma?
(b) Tin can be leached (dissolved) from the tin-plated steel of a food can. A provisional tolerable weekly intake of tin in the diet is 14 mg Sn/kg body mass. How many kilograms of canned tomato juice will exceed the tolerable weekly intake for a 55-kg woman?

Tin in canned foods (mg Sn/kg food) measured by inductively coupled plasma

Tomato juice	241	Fruit cocktail	73	Chili con carne	2
Grapefruit juice	182	Peach halves	58	Hungarian soup	2
Pineapple	114	Chocolate drink	2		

From L. Perring and M. Basic-Dvorzak, *Anal. Bioanal. Chem.* **2002,** *374,* 235.

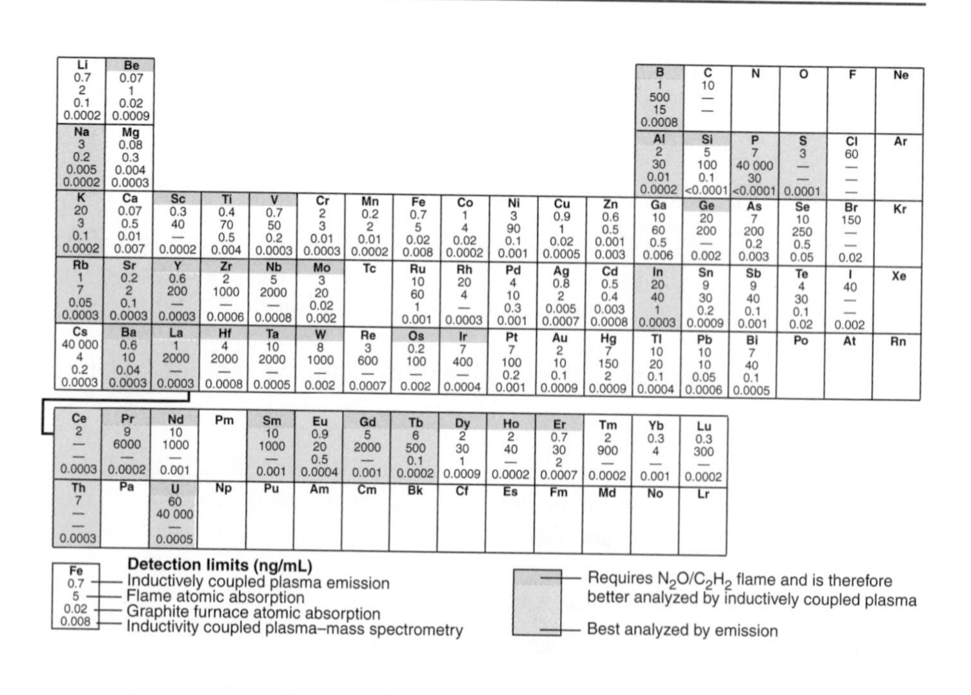

Figure 20-9 Flame, furnace, and inductively coupled plasma emission, and inductively coupled plasma–mass spectrometry detection limits (ng/g = ppb). [From R. J. Gill, *Am. Lab.*, November 1993, p. 24F; T. T. Nham, *Am. Lab.*, August 1998, p. 17A; and V. B. E. Thomsen, G. J. Roberts, and D. A. Tsourides, *Am. Lab.*, August 1997, p. 18H.] Accurate quantitative analysis requires concentrations 10–100 times the detection limit.

20-3 How Temperature Affects Atomic Spectroscopy

Temperature determines the degree to which a sample breaks down to atoms and the extent to which an atom is found in its ground, excited, or ionized state. Each effect influences the strength of the observed signal.

The Boltzmann Distribution

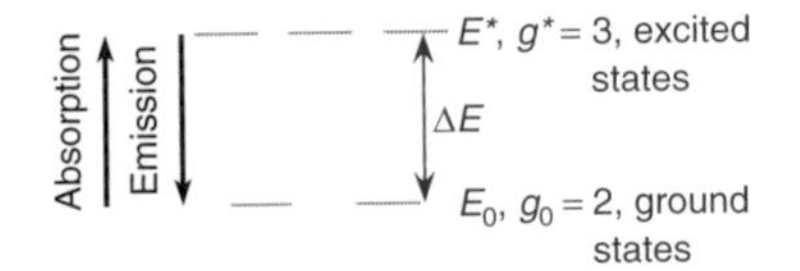

Figure 20-10 Two energy levels with degeneracies g_0 and g^*. Ground-state atoms can absorb light to be promoted to the excited state. Excited-state atoms can emit light to return to the ground state.

Consider a molecule with two energy levels (Figure 20-10) separated by energy ΔE. Call the lower level E_0 and the upper level E^*. An atom (or molecule) may have more than one state available at a given energy. In Figure 20-10, we show three states at E^* and two at E_0. The number of states at each energy is called the *degeneracy*. Let the degeneracies be g_0 and g^*.

The **Boltzmann distribution** describes the relative populations of different states at thermal equilibrium. If equilibrium exists (which is not true in all parts of a flame), the relative population (N^*/N_0) of any two states is

Boltzmann distribution:
$$\frac{N^*}{N_0} = \left(\frac{g^*}{g_0}\right) e^{-\Delta E/kT} \tag{20-1}$$

where T is temperature (K) and k is Boltzmann's constant (1.381×10^{-23} J/K).

The Effect of Temperature on the Excited-State Population

The lowest excited state of a sodium atom lies 3.371×10^{-19} J/atom above the ground state. The degeneracy of the excited state is 2, whereas that of the ground state is 1. Let's calculate the fraction of Na atoms in the excited state in an acetylene-air flame at 2 600 K:

$$\frac{N^*}{N_0} = \left(\frac{2}{1}\right) e^{-(3.371 \times 10^{-19}\ \text{J})/[(1.381 \times 10^{-23}\ \text{J/K})(2\ 600\ \text{K})]} = 0.000\ 167$$

Fewer than 0.02% of the atoms are in the excited state.

How would the fraction of atoms in the excited state change if the temperature were 2 610 K instead?

$$\frac{N^*}{N_0} = \left(\frac{2}{1}\right) e^{-(3.371 \times 10^{-19}\text{J})/[(1.381 \times 10^{-23}\ \text{J/K})(2\ 610\ \text{K})]} = 0.000\ 174$$

A 10-K temperature rise changes the excited-state population by 4% in this example.

The fraction of atoms in the excited state is still less than 0.02%, but that fraction has increased by $[(1.74 - 1.67)/1.67] \times 100 = 4\%$.

The Effect of Temperature on Absorption and Emission

We see that 99.98% of the sodium atoms are in their ground state at 2 600 K. *Varying the temperature by 10 K hardly affects the ground-state population and would not noticeably affect the signal in atomic absorption.*

How would emission intensity be affected by a 10-K rise in temperature? In Figure 20-10, we see that absorption arises from ground-state atoms, but emission arises from excited-state atoms. Emission intensity is proportional to the population of the excited state. *Because the excited-state population changes by 4% when the temperature rises 10 K, the emission intensity rises by 4%.* It is critical in atomic *emission* spectroscopy that the flame be very stable, or the emission intensity will vary. In atomic *absorption* spectroscopy, flame temperature variation is not as critical.

Atomic absorption is not as sensitive as atomic emission is to temperature variation.

Table 20-3 **Effect of energy separation and temperature on population of excited states**

Wavelength separation of states (nm)	Energy separation of states (J)	Excited-state fraction $(N^*/N_0)^a$ 2 500 K	6 000 K
250	7.95×10^{-19}	1.0×10^{-10}	6.8×10^{-5}
500	3.97×10^{-19}	1.0×10^{-5}	8.3×10^{-3}
750	2.65×10^{-19}	4.6×10^{-4}	4.1×10^{-2}

a. Based on the equation $N^*/N_0 = (g^*/g_0)e^{-\Delta E/kT}$ in which $g^* = g_0 = 1$.

The inductively coupled plasma is almost always used for emission, not absorption, because it is so hot that a substantial fraction of atoms and ions are excited and because no lamp is required for emission measurements. Table 20-3 compares excited-state populations for a flame at 2 500 K and a plasma at 6 000 K. Although the fraction of excited atoms is small, each atom emits many photons per second because it is rapidly promoted back to the excited state by collisions.

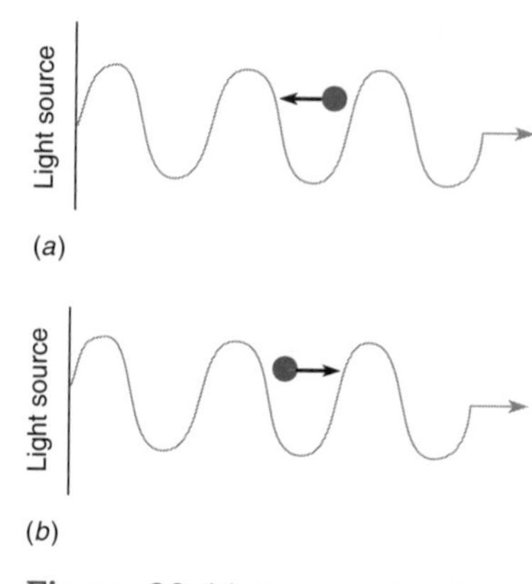

Figure 20-11 The Doppler effect. A molecule moving (*a*) toward the radiation source "feels" the electromagnetic field oscillate more often than one moving (*b*) away from the source.

Ask Yourself

20-C. **(a)** A ground-state atom absorbs light of wavelength 400 nm to be promoted to an excited state. Find the energy difference (in joules) between the two states. **(b)** If both states have a degeneracy of $g_0 = g^* = 1$, find the fraction of excited-state atoms (N^*/N_0) at thermal equilibrium at 2 500 K.

20-4 Instrumentation

Requirements for atomic absorption were shown in Figure 20-1. Principal differences between atomic and solution spectroscopy lie in the light source, the sample container (the flame, furnace, or plasma), and the need to subtract background emission from the observed signal.

The Linewidth Problem

The linewidth of the source must be narrower than the linewidth of the atomic vapor for Beer's law (Section 18-2) to be obeyed. The terms "linewidth" and "bandwidth" are used interchangeably, but "lines" are narrower than "bands."

For absorbance to be proportional to analyte concentration, the linewidth of radiation being measured must be substantially narrower than the linewidth of the absorbing atoms. Atomic absorption lines are very sharp, with an inherent width of $\sim 10^{-4}$ nm.

Two mechanisms broaden atomic spectra. One is the *Doppler effect,* in which an atom moving toward the lamp samples the oscillating electromagnetic wave more frequently than one moving away from the lamp (Figure 20-11). That is, an atom moving toward the source "sees" higher-frequency light than that encountered by one moving away. Linewidth is also affected by *pressure broadening* from collisions between atoms. Colliding atoms absorb a broader range of frequencies than do isolated atoms. Broadening is proportional to pressure. The Doppler effect and pressure broadening are similar in magnitude and yield linewidths of 10^{-3} to 10^{-2} nm in atomic spectroscopy.

Doppler and pressure effects broaden atomic lines by 1–2 orders of magnitude relative to their inherent linewidths.

Hollow-Cathode Lamps

To produce narrow lines of the correct frequency for atomic absorption, we use a **hollow-cathode lamp** whose cathode is made from the element we want to observe.

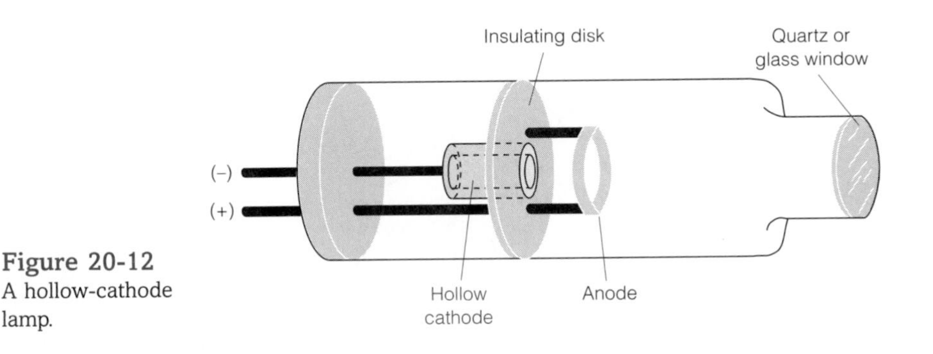

Figure 20-12 A hollow-cathode lamp.

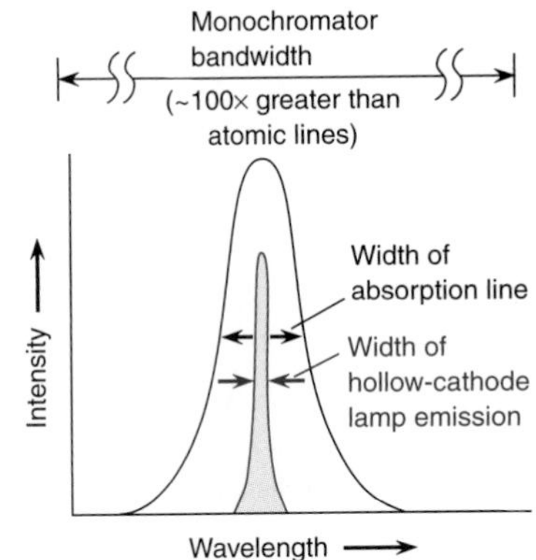

Figure 20-13 Relative linewidths of hollow-cathode emission, atomic absorption, and a monochromator. The linewidth from the hollow cathode is narrowest because gas temperature in the lamp is lower than flame temperature (so there is less Doppler broadening) and the pressure in the lamp is lower than flame pressure (so there is less pressure broadening).

The lamp in Figure 20-12 is filled with Ne or Ar at a pressure of 130–700 Pa. A high voltage between the anode and cathode ionizes the gas and accelerates cations toward the cathode. Ions striking the cathode "sputter" atoms from the metallic cathode into the gas phase. Gaseous metal atoms excited by collisions with high-energy electrons emit photons to return to the ground state. Atomic radiation shown in Figure 20-3 has the same frequency as that absorbed by atoms in the flame or furnace. The linewidth in Figure 20-13 is sufficiently narrow for Beer's law to hold. *A lamp with different cathode material is required for each element.*

Background Correction

Atomic spectroscopy requires **background correction** to distinguish analyte signal from absorption, emission, and optical scattering by the sample matrix, the flame, plasma, or white-hot graphite furnace. For example, Figure 20-14 shows the absorption spectrum of Fe, Cu, and Pb in a graphite furnace. Sharp atomic signals with maximum absorbance near 1.0 are superimposed on a flat background absorbance of 0.3. If we did not subtract background absorbance, significant errors would result. Background correction is most critical for graphite furnaces, which tend to be filled with smoke from the charring step. Optical scatter from smoke must somehow be distinguished from optical absorption by analyte.

Background signal arises from absorption, emission, or scatter by the flame, plasma, or furnace, and from everything in the sample besides analyte (the *matrix*).

For atomic absorption, *Zeeman background correction* (pronounced ZAY-man) is most commonly found in modern instruments for research and industrial laboratories. When a strong magnetic field is applied parallel to the light path through a flame or furnace, the absorption (or emission) lines of the atoms are split into three components (Figure 20-15). Two are shifted to slightly lower and higher wavelengths. The third component is unshifted, but has the wrong electromagnetic polarization to absorb light from the lamp. Therefore, analyte absorption decreases markedly in the presence of the magnetic field. For Zeeman background correction, the magnetic field is pulsed on and

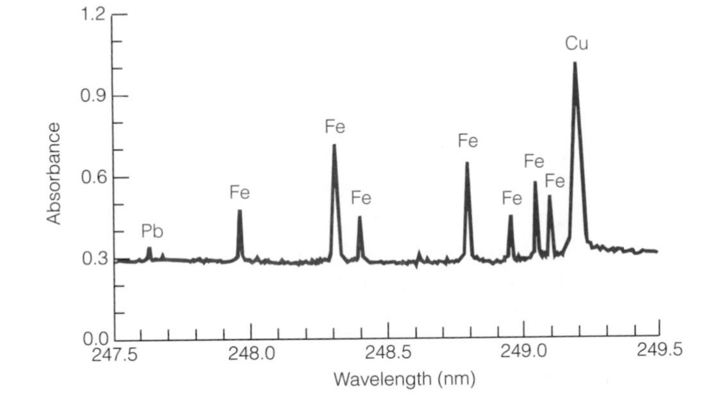

Figure 20-14 Graphite furnace absorption spectrum of bronze dissolved in HNO_3. Notice the high, constant background absorbance of 0.3 in this narrow region of the spectrum. [From B. T. Jones, B. W. Smith, and J. D. Winefordner, *Anal. Chem.* **1989**, *61*, 1670.]

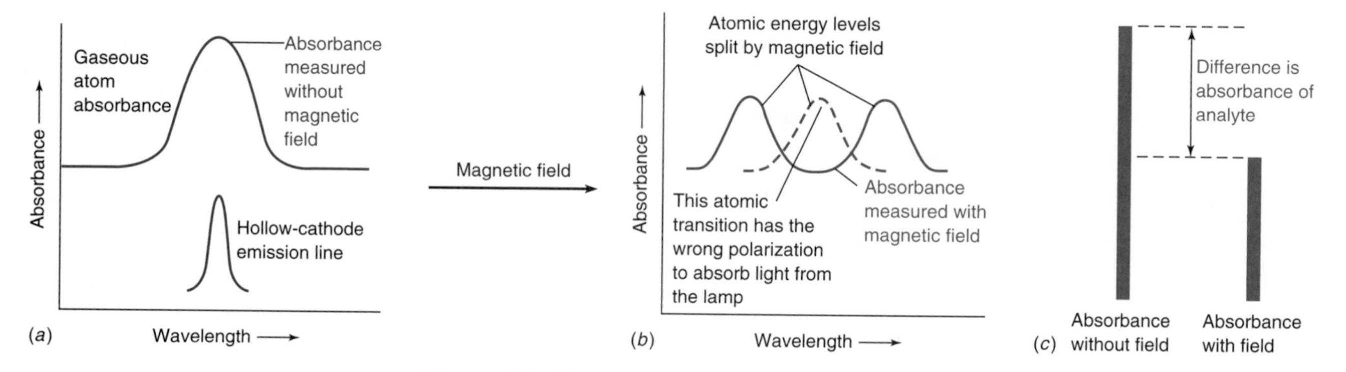

Figure 20-15 Principle of Zeeman background correction for atomic absorption spectroscopy. (*a*) In the absence of a magnetic field, we observe the sum of the absorbances of analyte and background. (*b*) In the presence of a magnetic field, the analyte absorbance is split away from the hollow cathode wavelength and the absorbance is due to background only. (*c*) The desired signal is the difference between those observed without and with the magnetic field.

off. Absorption from the sample plus background is observed when the field is off and absorption from the background alone is observed when the field is on. The difference between the two absorbances is due to analyte.

Atomic emission spectrometers provide background correction by measuring the signal at the peak emission wavelength and then at wavelengths slightly above and slightly below the peak, where the signal has returned to background level. The mean signal on either side of the peak is subtracted from the peak signal.

Multielement Analysis with the Inductively Coupled Plasma

Comparison of methods:

Flame atomic absorption
- lowest cost equipment
- different lamp required for each element
- poor sensitivity

Furnace atomic absorption
- expensive equipment
- different lamp required for each element
- high background signals
- very high sensitivity

Plasma emission
- expensive equipment
- no lamps
- low background and low interference
- moderate sensitivity

Inductively coupled plasma–mass spectrometry
- very expensive equipment
- no lamps
- least background and interference
- highest sensitivity

Inductively coupled plasma–atomic emission is more versatile than atomic absorption because emission does not require a lamp for each element. An element emits light at many characteristic frequencies. As many as 70 elements can be measured simultaneously. One way to measure many elements in a single sample is to scan with the monochromator, directing one wavelength at a time to the detector. This process requires time to integrate signal at each wavelength (typically 1 to 10 s at each wavelength) and to rotate the grating in the monochromator from one wavelength to another. Tens of milliliters of sample may be required in the time needed to measure several elements.

Alternatively, in Color Plate 21, atomic emission entering from the top right is dispersed in the vertical plane by a prism and in the horizontal plane by a grating. The radiation forms a two-dimensional pattern that lands on a semiconductor detector similar to that in a digital camera. Each pixel receives a different wavelength and therefore responds to a different element.

The spectrometer can be purged with N_2 or Ar to exclude O_2, thereby allowing ultraviolet wavelengths in the 100- to 200-nm range to be observed. This spectral region permits more sensitive detection of some elements that are normally detected at longer wavelengths and allows halogens, P, S, and N to be measured (with poor detection limits of tens of parts per million). These elements cannot be observed at wavelengths above 200 nm. In one application, N in fertilizers is measured along with other major elements. The plasma torch is specially designed to be purged with Ar to exclude N_2 from air. Unknowns are purged with He to remove dissolved air. Emission from N is observed near 174 nm.

Ask Yourself

20-D. How is background correction accomplished in atomic absorption and atomic emission spectroscopy?

20-5 Interference

Interference is any effect that changes the signal when analyte concentration remains unchanged. Interference can be corrected by counteracting the source of interference or by preparing standards that exhibit the same interference.

Types of interference:

- *spectral:* unwanted signals overlap analyte signal
- *chemical:* chemical reactions decrease the concentration of analyte atoms
- *ionization:* ionization of analyte atoms decreases the concentration of neutral atoms

Types of Interference

Spectral interference occurs when analyte signal overlaps signals from other species in the sample or signals due to the flame or furnace. Interference from the flame is subtracted by background correction. The best means of dealing with overlap between lines of different elements in the sample is to choose another wavelength for analysis. The spectrum of a molecule is much broader than that of an atom, so spectral interference can occur at many wavelengths. Figure 20-16 shows an example of a plasma containing Y and Ba atoms in addition to YO molecules. Elements that form stable oxides in the flame commonly give spectral interference.

Chemical interference is caused by any substance that decreases the extent of atomization of analyte. For example, SO_4^{2-} and PO_4^{3-} hinder the atomization of Ca^{2+}, perhaps by forming nonvolatile salts. *Releasing agents* can be added to a sample to decrease chemical interference. EDTA and 8-hydroxyquinoline protect Ca^{2+} from the interfering effects of SO_4^{2-} and PO_4^{3-}. La^{3+} is also a releasing agent, apparently because it preferentially reacts with PO_4^{3-} and frees the Ca^{2+}. A fuel-rich flame reduces oxidized analyte species that would otherwise hinder atomization. Higher flame temperatures eliminate many kinds of chemical interference.

Ionization interference is a problem in the analysis of alkali metals, which have the lowest ionization potentials. For any element, we can write a gas-phase ionization reaction:

$$M(g) \rightleftharpoons M^+(g) + e^-(g) \qquad K = \frac{[M^+][e^-]}{[M]} \tag{20-2}$$

At 2 450 K and a pressure of 0.1 Pa, Na is 5% ionized. With its lower ionization potential, K is 33% ionized under the same conditions. Ionized atoms have energy levels different from those of neutral atoms, so the desired signal is decreased.

An *ionization suppressor* is an element added to a sample to decrease the ionization of analyte. For example, 1 mg/mL of CsCl is added to the sample for the analysis of potassium, because cesium is more easily ionized than potassium. By producing a high concentration of electrons in the flame, ionization of Cs reverses Reaction 20-2 for K. This reversal is an example of Le Châtelier's principle.

The *method of standard addition,* described in Section 5-3, compensates for many types of interference by adding known quantities of analyte to the unknown in its complex matrix. For example, Figure 20-17 shows the analysis of strontium in aquarium water by standard addition. The slope of the standard addition curve is 0.018 8 absorbance units/ppm. If, instead, Sr is added to distilled water, the slope is 0.030 8 absorbance units/ppm. That is, in distilled water, the absorbance increases 0.030 8/0.018 8 = 1.64 times more than it does in aquarium water for each addition of standard Sr. We attribute the lower response in aquarium water to interference by

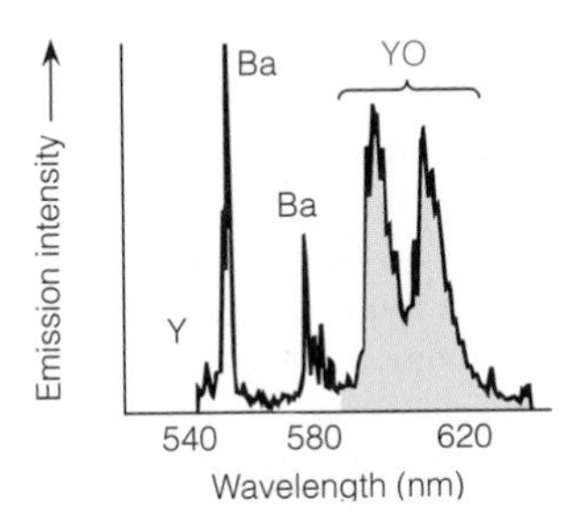

Figure 20-16 Emission from a plasma produced by laser irradiation of the high-temperature superconductor $YBa_2Cu_3O_7$. Solid is vaporized by the laser, and excited atoms and molecules in the gas phase emit light at characteristic wavelengths. [From W. A. Weimer, *Appl. Phys. Lett.* **1988**, *52*, 2171.]

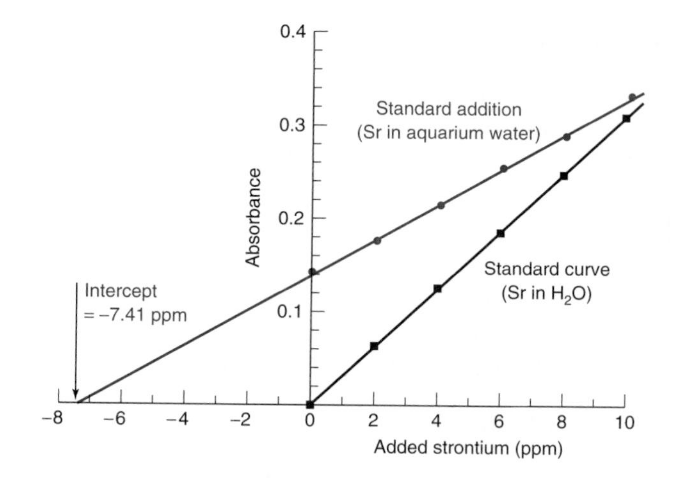

Figure 20-17 Atomic absorption calibration curve for Sr added to distilled water and standard addition of Sr to aquarium water. All solutions are made up to a constant volume, so the ordinate is the concentration of added Sr. [Data from L. D. Gilles de Pelichy, C. Adams, and E. T. Smith, *J. Chem. Ed.* **1997**, *74*, 1192.]

other species that are present. The negative x-intercept of the standard addition curve, 7.41 ppm, is a reliable measure of Sr in the aquarium. If we had just measured the atomic absorption of Sr in aquarium water and used the calibration curve for distilled water, we would have overestimated the Sr concentration by 64%.

Virtues of the Inductively Coupled Plasma

An Ar plasma eliminates common interferences. The plasma is twice as hot as a conventional flame, and the residence time of analyte in the plasma is about twice as long. Atomization is more complete, and the signal is correspondingly enhanced. Formation of analyte oxides and hydroxides is negligible. The plasma is relatively free of background radiation.

Self-absorption: Ground-state atoms in cooler, outer part of flame absorb emission from excited atoms at center of flame, thereby decreasing overall emission. Higher analyte concentration gives higher self-absorption and nonlinear calibration curve.

In flame emission spectroscopy, the concentration of electronically excited atoms in the cooler, outer part of the flame is lower than in the warmer, central part of the flame. Emission from the central region is absorbed in the outer region. This *self-absorption* increases with increasing concentration of analyte and gives nonlinear calibration curves. In a plasma, the temperature is more uniform, and self-absorption is not nearly so important. Table 20-4 states that plasma emission calibration curves are linear over five orders of magnitude compared with just two orders of magnitude for flames and furnaces.

Table 20-4 Comparison of atomic analysis methods

	Flame absorption	Furnace absorption	Plasma emission	Plasma–mass spectrometry
Detection limits (ng/g)	10–1 000	0.01–1	0.1–10	0.000 01–0.000 1
Linear range	10^2	10^2	10^5	10^8
Precision				
Short term (5–10 min)	0.1–1%	0.5–5%	0.1–2%	0.5–2%
Long term (hours)	1–10%	1–10%	1–5%	<5%
Interferences				
Spectral	Very few	Very few	Many	Few
Chemical	Many	Very many	Very few	Some
Mass	—	—	—	Many
Sample throughput	10–15 s/element	3–4 min/element	6–60 elements/min	All elements in 2–5 min
Dissolved solid	0.5–5%	>20% slurries and solids	1–20%	0.1–0.4%
Sample volume	Large	Very small	Medium	Medium
Purchase cost	1	2	4–9	10–15

SOURCE: Adapted from TJA Solutions, Franklin, MA.

20-E. What is meant by **(a)** spectral, **(b)** chemical, and **(c)** ionization interference? **(d)** Bone consists of the protein collagen and the mineral hydroxyapatite, $Ca_{10}(PO_4)_6(OH)_2$. The Pb content of archeological human skeletons measured by graphite furnace atomic absorption sheds light on customs and economic status of individuals in historic times.[2] Explain why La^{3+} is added to bone samples to suppress matrix interference in Pb analysis.

20-6 Inductively Coupled Plasma–Mass Spectrometry

Inductively coupled plasma–mass spectrometry is one of the most sensitive techniques available for trace analysis. Analyte ions produced in the plasma are directed into the inlet of a mass spectrometer, which separates ions on the basis of their mass-to-charge ratio. Ions are measured with an extremely sensitive detector that is similar to a photomultiplier tube in its operation. The linear range listed in Table 20-4 extends over eight orders of magnitude and the detection limit is 100–1 000 times lower than that of furnace atomic absorption.

Figure 20-18 shows an example in which coffee beans were extracted with trace-metal-grade nitric acid and the aqueous extract was analyzed by inductively coupled plasma–mass spectrometry. Coffee brewed from either Cuban or Hawaiian beans contained ~15 ng Pb/mL. However, the Cuban beans also contained Hg at a concentration similar to that of Pb.

A problem that is unique to mass spectrometry is **isobaric interference** in which ions of similar mass-to-charge ratio cannot be distinguished from each other. For example, $^{40}Ar^{16}O^+$ found in an Ar plasma has nearly the same mass as $^{56}Fe^+$. Doubly ionized $^{138}Ba^{2+}$ interferes with $^{69}Ga^+$ because each has nearly the same mass-to-charge ratio (138/2 = 69/1). If an element has multiple isotopes, you can check for

Ar is an "inert" gas with virtually no chemistry. However, Ar^+ has the same electronic configuration of Cl, and its chemistry is similar to that of halogens. $^{40}Ar^{16}O^+$ and $^{56}Fe^+$ differ by 0.02 atomic mass units.

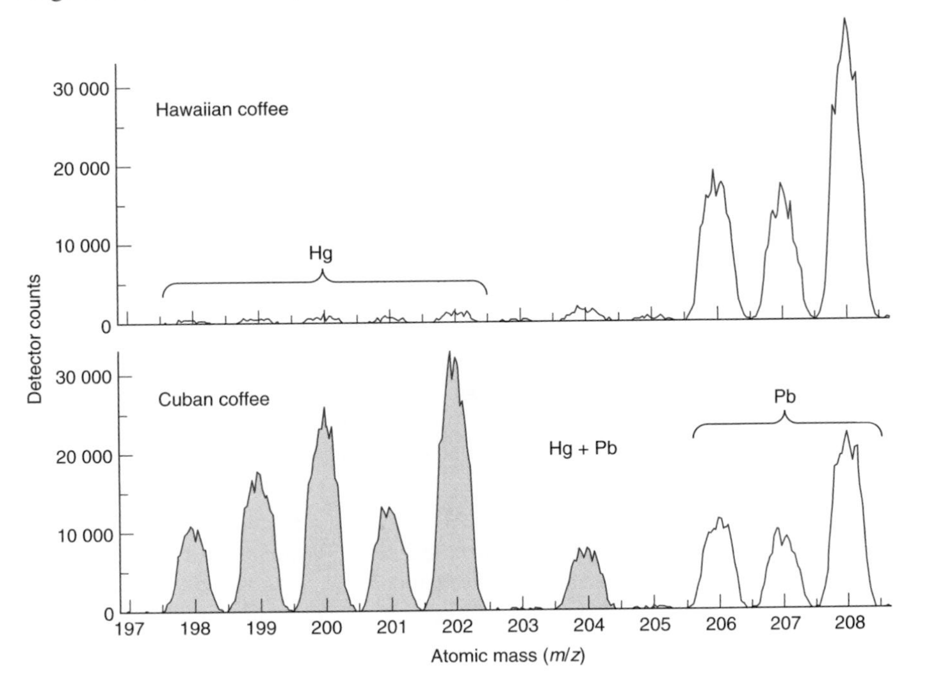

Figure 20-18 Partial elemental profile of coffee beans by inductively coupled plasma–mass spectrometry. Both beans have similar Pb content, but the Cuban beans had a much higher Hg content than the Hawaiian beans. A blank has not been subtracted from either spectrum, so the small amount of Hg in the upper spectrum could be in the blank. [Courtesy G. S. Ostrom and M. D. Seltzer, Michelson Laboratory, China Lake, CA.]

isobaric interference by measuring isotope ratios. For example, if the ratio of Se isotopes agrees with those found in nature ($^{74}Se : ^{76}Se : ^{77}Se : ^{78}Se : ^{80}Se : ^{82}Se = 0.008\,7 : 0.090 : 0.078 : 0.235 : 0.498 : 0.092$), then it is unlikely that there is interference at any of these masses.

Detection limits are so low that solutions must be made from extremely pure water and trace-metal-grade HNO_3 in Teflon or polyethylene vessels protected from dust. HCl and H_2SO_4 are avoided because they create isobaric interferences. The plasma–mass spectrometer interface generally cannot tolerate high concentrations of dissolved solids, which clog a small orifice between the plasma and the mass spectrometer. The plasma reduces organic matter to carbon, which can also clog the orifice. Organic material can be analyzed if some O_2 is fed into the plasma to oxidize the carbon.

Matrix effects on the yield of ions in the plasma are important, so calibration standards should be in the same matrix as the unknown. Internal standards can be used if they have nearly the same ionization energy as that of analyte. For example, Tm can be used as an internal standard for U. The ionization energies of these two elements are 5.81 and 6.08 eV, respectively; so they ionize to nearly the same extent in different matrices. If possible, internal standards with just one major isotope should be selected for maximum response.

Ask Yourself

20-F. Mercury has six isotopes with significant natural abundance: ^{198}Hg (10.0%), ^{199}Hg (16.9%), ^{200}Hg (23.1%), ^{201}Hg (13.2%), ^{202}Hg (29.9%), ^{204}Hg (6.9%). Lead has four isotopes: ^{204}Pb (1.4%), ^{206}Pb (24.1%), ^{207}Pb (22.1%), ^{208}Pb (52.4%). In Figure 20-18, why do we see six peaks for Hg and three for Pb? Why are there gaps at masses of 203 and 205?

Key Equation

Boltzmann distribution

$$\frac{N^*}{N_0} = \left(\frac{g^*}{g_0}\right) e^{-\Delta E/kT}$$

N^* = population of excited state
N_0 = population of ground state
ΔE = energy difference between excited and ground states
g^* = number of states with energy E^*
g_0 = number of states with energy E_0

Important Terms

atomic absorption spectroscopy	chemical interference	isobaric interference
atomic emission spectroscopy	graphite furnace	matrix
atomization	hollow-cathode lamp	spectral interference
background correction	inductively coupled plasma	
Boltzmann distribution	ionization interference	

Problems

20-1. Compare the advantages and disadvantages of furnaces and flames in atomic absorption spectroscopy.

20-2. Compare the advantages and disadvantages of the inductively coupled plasma and flames in atomic spectroscopy.

20-3. In which technique, atomic absorption or atomic emission, is flame temperature stability more critical? Why?

20-4. Atomic *emission* spectrometers provide background correction by measuring the signal at the peak emission wavelength and then at wavelengths slightly above and slightly below the peak, where the signal has returned to background. The mean background is subtracted from the peak signal. Why can't we use the same technique for background correction in atomic *absorption*? (*Hint:* Think about the lamp.)

20-5. What is the purpose of a matrix modifier in atomic spectroscopy?

20-6. The first excited state of Ca is reached by absorption of 422.7-nm light.

(a) What is the energy difference (J) between the ground and excited states? (*Hint:* See Section 18-1.)

(b) The degeneracies are $g^*/g_0 = 3$ for Ca. Find N^*/N_0 at 2 500 K.

(c) By what percentage will the fraction in **(b)** be changed by a 15-K rise in temperature?

(d) Find N^*/N_0 at 6 000 K.

20-7. The first excited state of Cu is reached by absorption of 327-nm radiation.

(a) What is the energy difference (J) between the ground and excited states?

(b) The ratio of degeneracies is $g^*/g_0 = 3$ for Cu. Find N^*/N_0 at 2 400 K.

(c) By what percentage will the fraction in **(b)** be changed by a 15-K rise in temperature?

(d) What will the ratio N^*/N_0 be at 6 000 K?

20-8. *Detection limit.* (Refer to Section 5-2.) To estimate the detection limit for arsenic in tap water by furnace atomic absorption, eight tap water samples with a low concentration of As were measured. Then eight tap water samples spiked with 0.50 ppb (ng/mL) As were measured.

Tap water: 0.014, 0.005, 0.011, 0.001, −0.002, 0.002, 0.010, 0.008

Tap water + 0.50 ppb As: 0.046, 0.043, 0.036, 0.037, 0.041, 0.031, 0.039, 0.034

(a) A calibration curve is a graph of signal versus analyte concentration: signal = m[As] + blank signal, where [As] is the concentration in ppb and m is the slope. Spiked tap water has 0.50 ppb more As than the unspiked tap water. Find the slope of the calibration line by solving for m in the equation

(mean spike signal − mean unspiked signal) = m[0.50 ppb]

(b) From Equation 5-4, find the detection limit.

(c) From Equation 5-5, find the lower limit of quantitation.

20-9. *Standard curve.* Potassium standards gave the following data at 404.3 nm.

(a) Subtract the blank from each emission intensity and construct a calibration curve (Section 4-6) of corrected emission intensity versus sample concentration (μg/mL). Find m, b, s_m, s_b, and s_y.

(b) Find [K^+] (μg/mL) and its uncertainty in the unknown.

Sample (μg K/mL)	Emission intensity
Blank	6
5.00	130
10.0	249
20.0	492
30.0	718
Unknown	423

20-10. *Standard addition.* Suppose that 5.00 mL of blood serum containing an unknown concentration of potassium, $[K^+]_i$, gave an atomic emission signal of 3.00 mV. After the addition of 1.00 mL of 30.0 mM K^+ standard and dilution of the mixture to 10.00 mL, the emission signal increased to 4.00 mV.

(a) Find the concentration of added standard, $[S]_f$, in the mixture.

(b) The initial 5.00-mL sample was diluted to 10.00 mL. Therefore the serum potassium concentration decreased from $[K^+]_i$ to $[K^+]_f = \frac{1}{2}[K^+]_i$. Use Equation 5-6 to find the original K^+ content of the serum, $[K^+]_i$.

20-11. *Standard addition.* A blood serum sample containing Na^+ gave an emission signal of 4.27 mV. A small volume of concentrated standard Na^+ was then added to increase the Na^+ concentration by 0.104 M, without significantly diluting the sample. This "spiked" serum sample gave a signal of 7.98 mV in atomic emission. Find the original concentration of Na^+ in the serum.

20-12. *Standard addition.* An unknown sample of Cu^{2+} gave an absorbance of 0.262 in an atomic absorption analysis. Then 1.00 mL of solution containing 100.0 ppm (= 100.0 μg/mL) Cu^{2+} was mixed with 95.0 mL of unknown, and the mixture was diluted to 100.0 mL in a volumetric flask. The absorbance of the new solution was 0.500. Find the original concentration of Cu^{2+} in the unknown.

20-13. *Standard addition.* An unknown containing element X was mixed with aliquots of a standard solution of element X for atomic absorption spectroscopy. The standard solution contained 1.000×10^3 μg of X per milliliter.

Volume of unknown (mL)	Volume of standard (mL)	Total volume (mL)	Absorbance
10.00	0	100.0	0.163
10.00	1.00	100.0	0.240
10.00	2.00	100.0	0.319
10.00	3.00	100.0	0.402
10.00	4.00	100.0	0.478

(a) Calculate the concentration (μg X/mL) of added standard in each solution.

(b) Prepare a graph as in Section 5-3 to determine the concentration of X in the unknown. All solutions have the same volume, so the ordinate is observed signal and the abscissa is the diluted concentration of unknown.

(c) Use the formula in Problem 5-19 to find the uncertainty in **(b)**.

20-14. *Standard addition.* Li^+ was determined by atomic emission, using standard additions of solution containing 1.62 μg Li/mL. From the following data, prepare a standard addition graph to find the concentration of Li in pure unknown. Use the formula in Problem 5-19 to find the uncertainty in concentration.

Unknown (mL)	Standard (mL)	Final volume (mL)	Emission intensity (arbitrary units)
10.00	0.00	100.0	309
10.00	5.00	100.0	452
10.00	10.00	100.0	600
10.00	15.00	100.0	765
10.00	20.00	100.0	906

20-15. *Internal standard.* A solution was prepared by mixing 10.00 mL of unknown (X) with 5.00 mL of standard (S) containing 8.24 μg S/mL and diluting to 50.0 mL. The measured signal quotient was (signal due to X/signal due to S) = 1.69. In a separate experiment, it was found that, for the concentration of X equal to 3.42 times the concentration of S, the signal due to X was 0.93 times as intense as the signal due to S. Find the concentration of X in the unknown.

20-16. *Internal standard.* Mn was used as an internal standard for measuring Fe by atomic absorption. A standard mixture containing 2.00 μg Mn/mL and 2.50 μg Fe/mL gave a quotient (Fe signal/Mn signal) = 1.05. A mixture with a volume of 6.00 mL was prepared by mixing 5.00 mL of unknown Fe solution with 1.00 mL containing 13.5 μg Mn/mL. The absorbance of this mixture at the Mn wavelength was 0.128, and the absorbance at the Fe wavelength was 0.185. Find the molarity of the unknown Fe solution.

20-17. *Quality assurance.* Tin can be leached (dissolved) into canned foods from the tin-plated steel can.

(a) For analysis by inductively coupled plasma–atomic emission, food is digested by microwave heating in a Teflon bomb (Figure 2-18) in three steps with HNO_3, H_2O_2, and HCl. CsCl is added to the final solution at a concentration of 1 g/L. What is the purpose of the CsCl?

(b) Calibration data for the 189.927-nm emission line of Sn are shown in the following table. Use the Excel LINEST function (Section 4-7) to find the slope and intercept and their standard deviations and R^2, which is a measure of the goodness of fit of the data to a line. Draw the calibration curve.

Sn (μg/L)	Emission intensity (arbitrary units)
0	4.0
10.0	8.5
20.0	19.6
30.0	23.6
40.0	31.1
60.0	41.7
100.0	78.8
200.0	159.1

From L. Perring and M. Basic-Dvorzak, *Anal. Bioanal. Chem.* **2002,** *374,* 235.

(c) Interference by high concentrations of other elements was assessed at different emission lines of Sn. Foods containing little tin were digested and spiked with Sn at 100.0 μg/L. Then other elements were deliberately added. The following table shows selected results. Which elements interfere at each of the two wavelengths? Which wavelength is preferred for the analysis?

Element added at 50 mg/L	Sn found (μg/L) with 189.927-nm emission line	Sn found (μg/L) with 235.485-nm emission line
None	100.0	100.0
Ca	96.4	104.2
Mg	98.9	92.6
P	106.7	104.6
Si	105.7	102.9
Cu	100.9	116.2
Fe	103.3	intense emission
Mn	99.5	126.3
Zn	105.3	112.8
Cr	102.8	76.4

From L. Perring and M. Basic-Dvorzak, *Anal. Bioanal. Chem.* **2002,** *374,* 235.

(d) *Limits of detection and quantitation.* The slope of the calibration curve in **(b)** is 0.782 units per (μg/L) of Sn. Food containing little Sn gave a mean signal of 5.1 units for seven replicates. Food spiked with 30.0 μg Sn/L gave a mean signal of 29.3 units with a standard deviation of 2.4 units for seven replicates. Use Equations 5-4 and 5-5 to estimate the limits of detection and quantitation.

(e) During sample preparation, 2.0 g of food are digested and eventually diluted to 50 mL for analysis. Express the limit of quantitation from **(d)** in terms of milligrams of Sn per kilogram of food.

20-18. Titanocene dichloride $(\pi\text{-}C_5H_5)_2TiCl_2$, is a potential antitumor drug thought to be carried to cancer cells by the protein transferrin (Figure 19-14). ($\pi\text{-}C_5H_5$ is the cyclopentadienyl group seen in ferrocene in Figure 17-5.) To measure the Ti(IV) binding capacity of transferrin, the protein was treated with excess titanocene dichloride. After allowing time for Ti(IV) to bind to the protein, excess small molecules were removed by dialysis (Demonstration 7-1). The protein was then digested with 2 M NH_3 and used to prepare a series of solutions with standard additions for chemical analysis. All solutions were made to the same total volume. Titanium and sulfur in each solution were measured by inductively coupled plasma–atomic emission spectrometry, with results in the table. Each transferrin molecule contains 39 sulfur atoms. Find the molar ratio Ti/transferrin in the protein.

Added Ti (mg/L)	Signal	Added S (mg/L)	Signal
0	0.86	0	0.0174
3.00	1.10	37.0	0.0221
6.00	1.34	74.0	0.0268
12.0	1.82	148.0	0.0362

Data derived from A. Cardona and E. Meléndez, *Anal. Bioanal. Chem.* **2006**, *386*, 1689.

How Would You Do It?

20-19. The measurement of Li in brine (salt water) is used by geochemists to help determine the origin of this fluid in oil fields. Flame atomic emission and absorption of Li are subject to interference by scattering, ionization, and overlapping spectral emission from other elements. Atomic absorption analysis of replicate samples of a marine sediment gave the results in the following table.

Sample and treatment	Li found (μg/g)	Analytical method	Flame type
1. None	25.1	standard curve	air/C_2H_2
2. Dilute to 1/10 with H_2O	64.8	standard curve	air/C_2H_2
3. Dilute to 1/10 with H_2O	82.5	standard addition	air/C_2H_2
4. None	77.3	standard curve	N_2O/C_2H_2
5. Dilute to 1/10 with H_2O	79.6	standard curve	N_2O/C_2H_2
6. Dilute to 1/10 with H_2O	80.4	standard addition	N_2O/C_2H_2

From B. Baraj, L. F. H. Niencheski, R. D. Trapaga, R. G. França, V. Cocoli, and D. Robinson, *Fresenius J. Anal. Chem.* **1999**, *364*, 678.

(a) Suggest a reason for the increasing apparent concentration of Li in samples 1 to 3.

(b) Why do samples 4 to 6 give an almost constant result?

(c) What value would you recommend for the real concentration of Li in the sample?

Notes and References

1. J. G. Weiner, B. C. Knights, M. B. Sandheinrich, J. D. Jeremiason, M. E. Brigham, D. R. Engstrom, L. G. Woodruff, W. F. Cannon, and S. J. Balogh, *Environ. Sci. Technol.* **2006**, *40*, 6261; D. M. Orihel, M. J. Paterson, C. C. Gilmour, R. A. Bodaly, P. J. Blanchfield, H. Hintelmann, R. C. Harris, and J. W. M. Rudd, *Environ. Sci. Technol.* **2006**, *40*, 5992.

2. L. Wittmers, Jr., A. Aufderheide, G. Rapp, and A. Alich, *Acc. Chem. Res.* **2002**, *35*, 669.

Further Reading

L. H. J. Lajunen and P. Perämäki, *Spectrochemical Analysis by Atomic Absorption* (Cambridge: Royal Society of Chemistry, 2004).

M. Cullen, *Atomic Spectroscopy in Elemental Analysis* (Oxford: Blackwell, 2003).

J. R. Dean and D. J. Ando, *Atomic Absorption and Plasma Spectroscopy* (New York: Wiley, 2002).

J. A. C. Broekaert, *Analytical and Atomic Spectrometry with Flames and Plasmas* (Weinheim: Wiley-VCH, 2002).

L. Ebdon, E. H. Evans, A. S. Fisher, and S. J. Hill, *An Introduction to Analytical Atomic Spectrometry* (Chichester: Wiley, 1998).

D. J. Butcher and J. Sneddon, *A Practical Guide to Graphite Furnace Atomic Absorption Spectrometry* (New York: Wiley, 1998).

R. Thomas, *Practical Guide to ICP-MS* (New York: Marcel Dekker, 2004).

H. E. Taylor, *Inductively Coupled Plasma–Mass Spectrometry* (San Diego: Academic Press, 2001).

S. J. Hill, ed., *Inductively Coupled Plasma Spectrometry and Its Applications* (Sheffield, England: Sheffield Academic Press, 1999).

A. Montaser, ed., *Inductively Coupled Plasma Mass Spectrometry* (New York: Wiley, 1998).

Katia and Dante

Courageous Katia Krafft samples vapors from a volcano. [Photo by Maurice Krafft from F. Press and R. Siever, *Understanding Earth* (New York: W. H. Freeman and Company, 1994), p. 107.] Katia and Maurice Krafft died in the eruption of Mt. Unzen in Japan in 1991.

Robot Dante was equipped with a gas chromatograph to sample emissions from volcanic Mt. Erebus in Antarctica. [Philip R. Kyle, New Mexico Tech.]

Over billions of years, volcanoes released gases that created today's atmosphere and condensed into the Earth's oceans. Generally 70–95% of volcanic gas is H_2O, followed by smaller volumes of CO_2 and SO_2, and traces of N_2, H_2, CO, sulfur, Cl_2, HF, HCl, and H_2SO_4. To reduce wear and tear on brave scientists who have ventured near a volcano to collect gases, the eight-legged robot Dante was designed to rappel 250 m into the sheer crater of Mt. Erebus in Antarctica in 1992 to sample volcanic emissions with a gas chromatograph and return information through a fiber-optic cable. Dante also had a quartz crystal microbalance (page 40) to measure aerosols, an infrared thermometer, and a γ-ray spectrometer to measure K, Th, and U in the crater's soil. Dante failed when its optical fiber broke after just 6 m of descent into the crater. In 1994, an improved Dante II successfully probed the volcano Mt. Spurr in Alaska.

CHAPTER 21

Principles of Chromatography and Mass Spectrometry

Chromatography is the most powerful tool in an analytical chemist's arsenal for separating and measuring components of a complex mixture. With mass spectrometric detection, we can identify the components as well. This chapter discusses principles of chromatography and mass spectrometry, and the following chapters take up specific methods of separation.

Chromatography is widely used for

Quantitative analysis: How much of a component is present?

Qualitative analysis: What is the identity of the component?

21-1 What Is Chromatography?

Chromatography is a process in which we separate compounds from one another by passing a mixture through a column that retains some compounds longer than others. In Figure 21-1, a solution containing compounds A and B is placed on top of a column previously packed with solid particles and filled with solvent. When the outlet is opened, A and B flow down into the column and are washed through with fresh solvent applied to the top of the column. If solute A is more strongly *adsorbed* than solute B on the solid

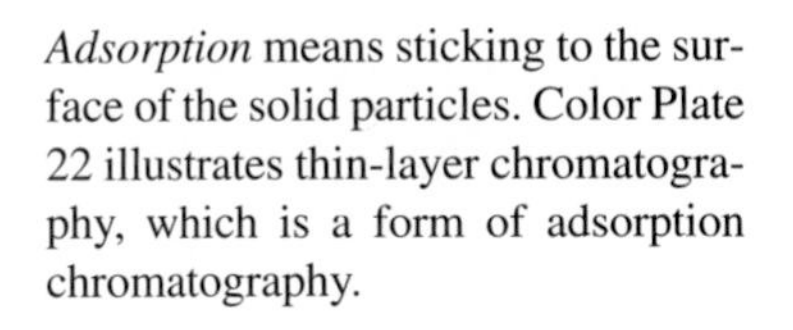

Adsorption means sticking to the surface of the solid particles. Color Plate 22 illustrates thin-layer chromatography, which is a form of adsorption chromatography.

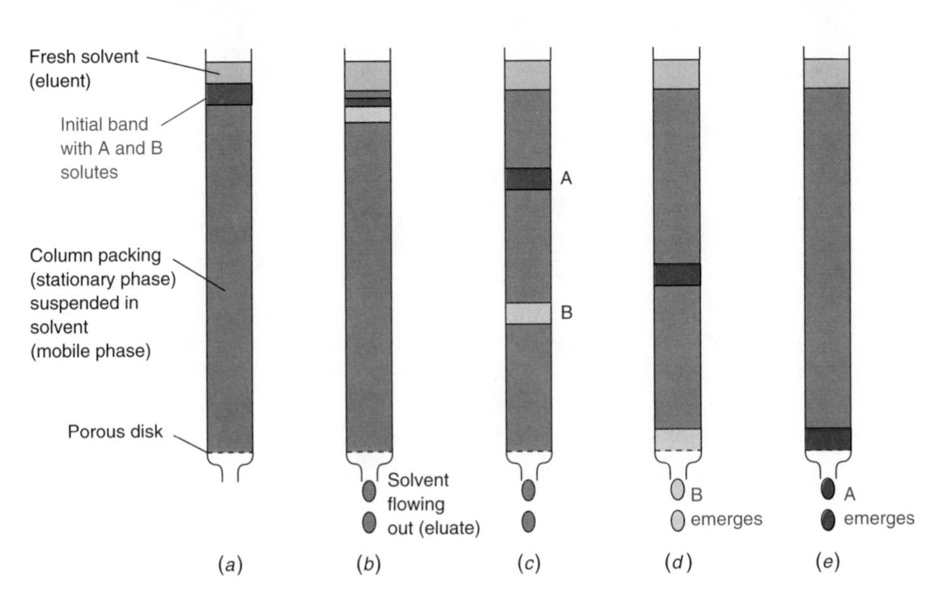

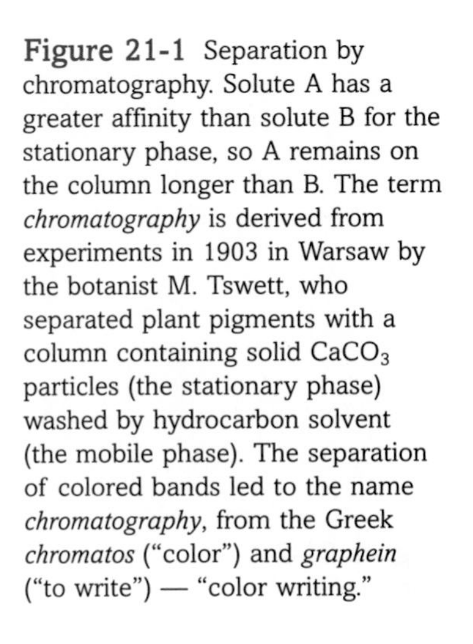

Figure 21-1 Separation by chromatography. Solute A has a greater affinity than solute B for the stationary phase, so A remains on the column longer than B. The term *chromatography* is derived from experiments in 1903 in Warsaw by the botanist M. Tswett, who separated plant pigments with a column containing solid $CaCO_3$ particles (the stationary phase) washed by hydrocarbon solvent (the mobile phase). The separation of colored bands led to the name *chromatography*, from the Greek *chromatos* ("color") and *graphein* ("to write") — "color writing."

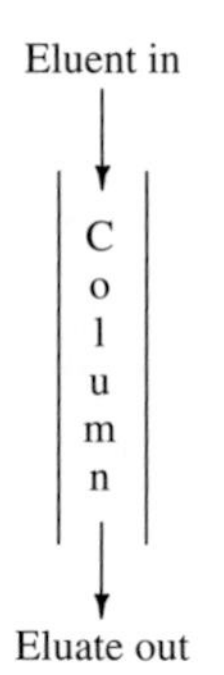

particles, then A spends a smaller fraction of the time free in solution. Solute A moves down the column more slowly than B and emerges at the bottom after B.

The **mobile phase** (solvent moving through the column) in chromatography is either a liquid or a gas. The **stationary phase** (the substance that stays fixed inside the column) is either a solid or a liquid that is usually covalently bonded to solid particles or to the inside wall of a hollow capillary column. Partitioning of solutes between the mobile and stationary phases gives rise to separation. In **gas chromatography**, the mobile phase is a gas; and, in **liquid chromatography**, the mobile phase is a liquid.

Fluid entering the column is called **eluent**. Fluid exiting the column is called **eluate**. The process of passing liquid or gas through a chromatography column is called **elution**.

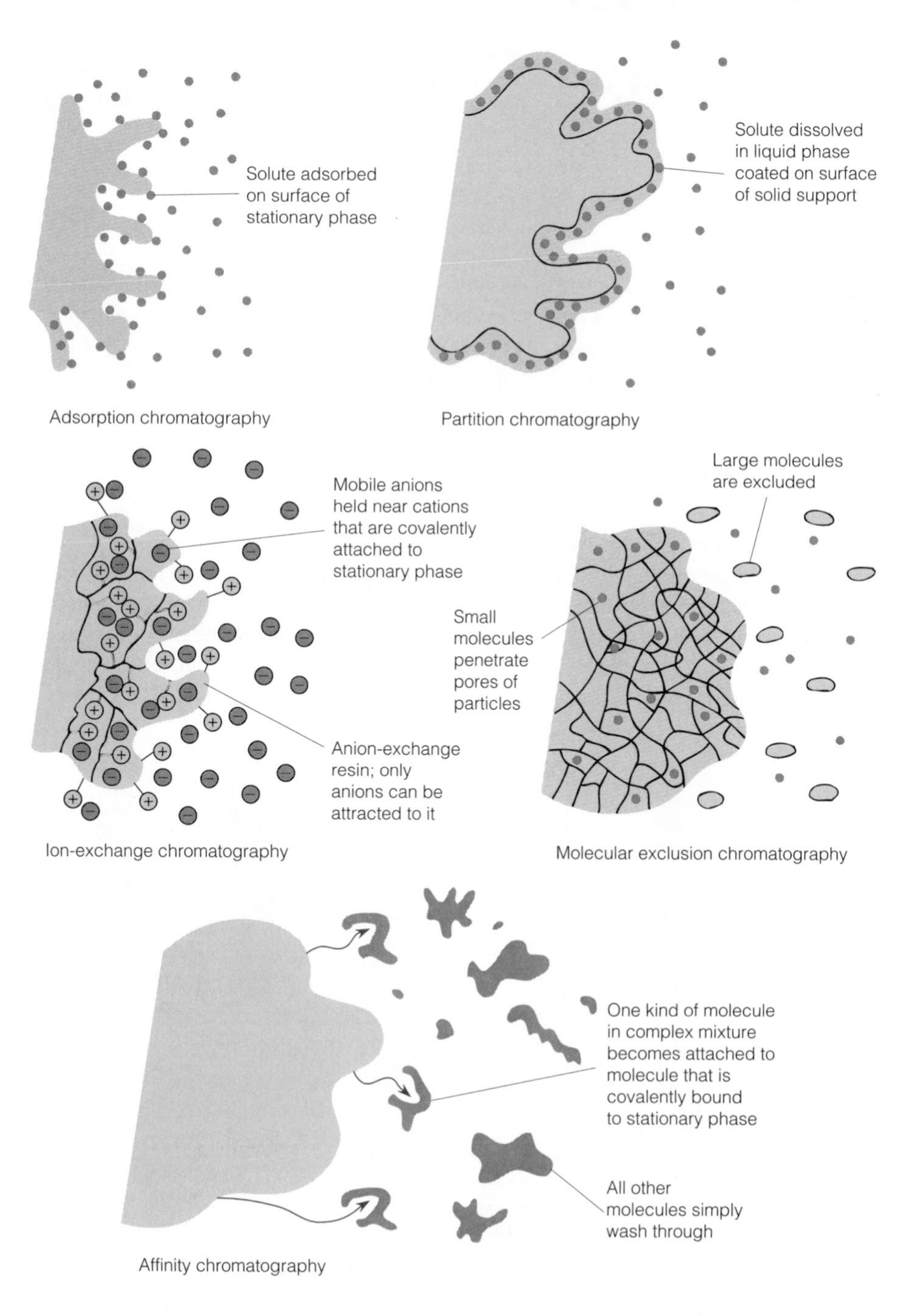

Figure 21-2 Types of chromatography.

Chromatography can be classified by the type of interaction of the solute with the stationary phase, as shown in Figure 21-2.

Adsorption chromatography uses a solid stationary phase and a liquid or gaseous mobile phase. Solute is adsorbed on the surface of the solid particles.

Partition chromatography involves a thin liquid stationary phase coated on the surface of a solid support. Solute equilibrates between the stationary liquid and the mobile phase.

Ion-exchange chromatography features ionic groups such as $—SO_3^-$ or $—N(CH_3)_3^+$ covalently attached to the stationary solid phase, which is usually a *resin*. Solute ions are attracted to the stationary phase by electrostatic forces. The mobile phase is a liquid.

Molecular exclusion chromatography (also called *size exclusion, gel filtration*, or *gel permeation* chromatography) separates molecules by size, with larger molecules passing through fastest. There is no attractive interaction between the stationary phase and the solute. The stationary phase has pores small enough to exclude large molecules, but not small ones. Large molecules stream past without entering the pores. Small molecules take longer to pass through the column because they enter the pores and therefore must flow through a larger volume before leaving the column.

Affinity chromatography, the most selective kind of chromatography, employs specific interactions between one kind of solute molecule and a second molecule that is covalently attached (immobilized) to the stationary phase. For example, the immobilized molecule might be an antibody to a particular protein. When a mixture containing a thousand different proteins is passed through the column, only that particular protein binds to the antibody on the column. After other proteins have been washed from the column, the desired protein is dislodged by changing the pH or ionic strength.

For pioneering work on liquid-liquid partition chromatography in 1941, A. J. P. Martin and R. L. M. Synge received the Nobel Prize in 1952.

B. A. Adams and E. L. Holmes developed synthetic ion-exchange resins in 1935. *Resins* are relatively hard, amorphous (noncrystalline) organic solids. *Gels* are relatively soft.

In molecular exclusion, large molecules pass through the column *faster* than small molecules.

Ask Yourself

21-A. Match the terms (1–5) with their definitions (A–E):

1. Adsorption chromatography
2. Partition chromatography
3. Ion-exchange chromatography
4. Molecular exclusion chromatography
5. Affinity chromatography

A. Mobile-phase ions attracted to stationary-phase ions.
B. Solute attracted to specific groups attached to stationary phase.
C. Solute equilibrates between mobile phase and surface of stationary phase.
D. Solute equilibrates between mobile phase and stationary liquid film.
E. Solute penetrates voids in stationary phase. Largest solutes eluted first.

21-2 How We Describe a Chromatogram

Detectors discussed in Chapter 22 respond to solutes as they exit the chromatography column. A **chromatogram** shows detector response as a function of time (or elution volume) in a chromatography separation (Figure 21-3). Each peak corresponds to a

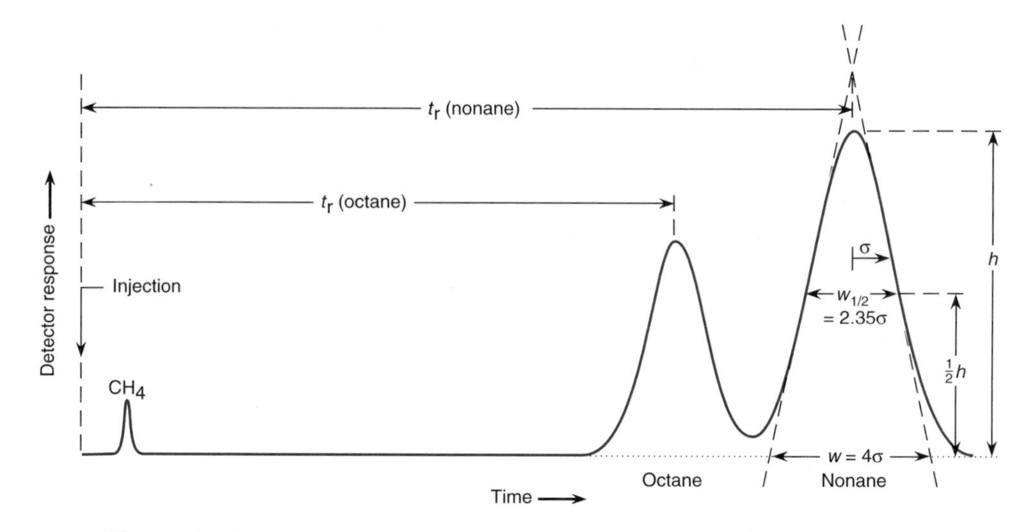

Figure 21-3 Schematic gas chromatogram showing measurement of retention time (t_r) and width at half-height ($w_{1/2}$). The width at the base (w) is found by drawing tangents to the steepest parts of the Gaussian curve and extrapolating down to the baseline. The standard deviation of the Gaussian curve is σ. In gas chromatography, a small volume of CH_4 injected with the 0.1- to 2-μL sample is usually the first component to be eluted.

different substance eluted from the column. **Retention time**, t_r, is the time needed after injection for an individual solute to reach the detector.

Theoretical Plates

An ideal chromatographic peak has a Gaussian shape, like that in Figure 21-3. If the height of the peak is h, the *width at half-height*, $w_{1/2}$, is measured at $\frac{1}{2}h$. For a Gaussian peak, $w_{1/2}$ is equal to 2.35σ, where σ is the standard deviation of the peak. The width of the peak at the baseline, as shown in Figure 21-3, is 4σ.

In days of old, distillation was the most powerful means for separating volatile compounds. A distillation column was divided into sections (*plates*) in which liquid and vapor equilibrated with each other. The more plates on a column, the more equilibration steps and the better the separation between compounds with different boiling points.

Nomenclature from distillation carried over to chromatography. We speak of a chromatography column as if it were divided into discrete sections (called **theoretical plates**) in which a solute molecule equilibrates between the mobile and stationary phases. Retention of a compound on a column can be described by the number of theoretical equilibration steps between injection and elution. The more equilibration steps (the more theoretical plates), the narrower the bandwidth when the compound emerges.

Although we speak of discrete "theoretical plates," chromatography is a continuous process. Theoretical plates are an imaginary way to picture the process.

For any peak in Figure 21-3, the number of theoretical plates is computed by measuring the retention time and the width at half-height:

Number of plates on column:

$$N = \frac{5.55\, t_r^2}{w_{1/2}^2} \qquad (21\text{-}1)$$

Test a column for degradation by injecting a standard periodically and looking for peak asymmetry and changes in the number of plates.

Retention and width can be measured in units of time or volume (such as milliliters of eluate). Both t_r and $w_{1/2}$ must be expressed in the same units in Equation 21-1.

If a column is divided into N theoretical plates (only in our minds), then the **plate height**, H, is the length of one plate. H is the length of the column (L) divided by the number of theoretical plates:

Plate height: $$H = L/N \qquad (21\text{-}2)$$

small plate height ⇒ narrow peaks ⇒ better separations

The smaller the plate height, the narrower the peaks. The ability of a column to separate components of a mixture is improved by decreasing plate height. An efficient column has more theoretical plates than an inefficient column. Different solutes behave as if the column has somewhat different plate heights, because different compounds equilibrate between the mobile and stationary phases at different rates. Plate heights are ~100 to 1 000 μm in gas chromatography, ~10 μm in high-performance liquid chromatography, and <1 μm in capillary electrophoresis.

Example Measuring Plates

A solute with a retention time of 400.0 s has a width at half-height of 8.0 s on a column 12.2 m long. Find the number of plates and the plate height.

SOLUTION

$$\text{number of plates} = N = \frac{5.55\, t_r^2}{w_{1/2}^2} = \frac{5.55 \cdot 400.0^2}{8.0^2} = 1.39 \times 10^4$$

$$\text{plate height} = H = \frac{L}{N} = \frac{12.2\text{ m}}{1.39 \times 10^4} = 0.88\text{ mm}$$

Test Yourself If the number of plates is constant at 1.39×10^4, what is the width of a compound eluted at 600 s? (**Answer:** 12.0 s)

Resolution

The **resolution** of neighboring peaks is the peak separation (Δt_r) divided by the average peak width (w_{av}) measured at the base, as in Figure 21-3:

Resolution: $$\text{resolution} = \frac{\Delta t_r}{w_{av}} = \frac{0.589\Delta t_r}{w_{1/2av}} \qquad (21\text{-}3)$$

In the second equality, $w_{1/2av}$ is the average width at half-height, which is used more often than w_{av} because $w_{1/2av}$ is easier to measure. The better the resolution, the more complete the separation between neighboring peaks. Figure 21-4 shows peaks with a

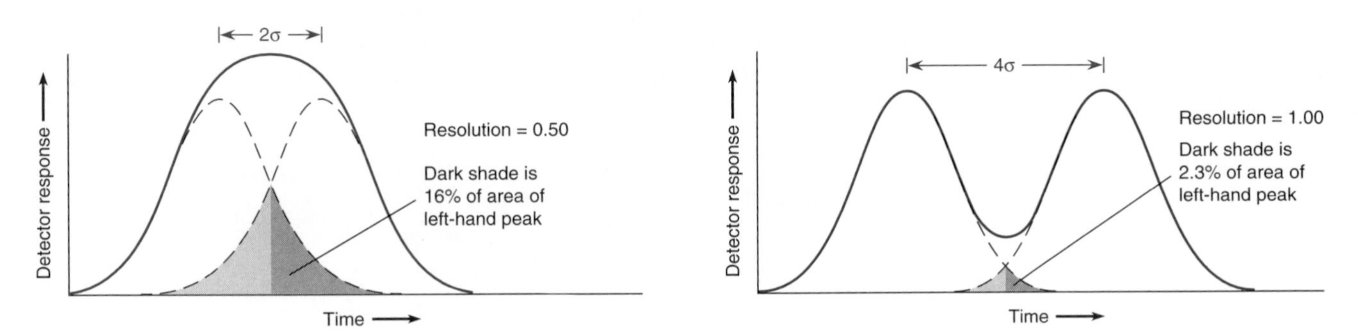

Figure 21-4 Resolution of Gaussian peaks of equal area and amplitude. Dashed lines show individual peaks; solid lines are the sum of two peaks. Overlapping area is shaded.

resolution $\propto \sqrt{\text{column length}}$

(The symbol $\propto$ means "is proportional to.")

resolution of 0.50 and 1.00. For quantitative analysis, resolution ≥ 2 is desirable for negligible overlap. If you double the length of an ideal chromatography column, you will improve resolution by $\sqrt{2}$.

Qualitative and Quantitative Analysis

For *qualitative analysis*, the simplest way to identify a chromatographic peak is to compare its retention time with that of an authentic sample of the suspected compound. The most reliable way to do this is by **spiking**, also called *co-chromatography*, in which authentic sample is added to the unknown. If the added compound is identical to one component of unknown, the relative size of that one peak will increase. Two different compounds might have the same retention time on a particular column. However, it is less likely that they will have the same retention time on different stationary phases.

For qualitative analysis, each chromatographic peak can be directed into a mass spectrometer or infrared spectrophotometer to record a spectrum as the substance is eluted from the column. The compound can be identified by comparing its spectrum with a library of spectra stored in a computer.

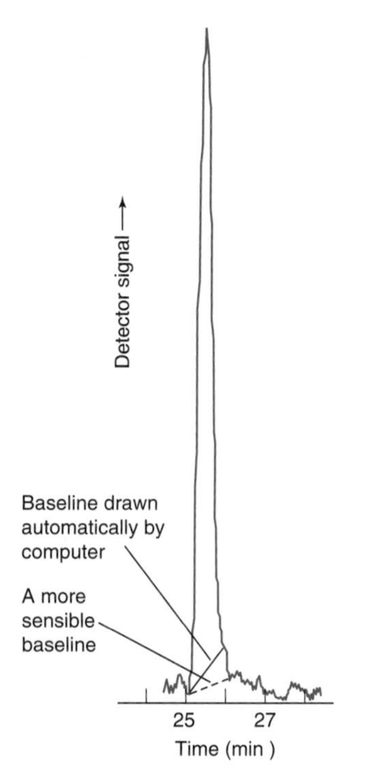

Figure 21-5 Baselines for integrating the area of a chromatographic peak. Computer algorithms do not necessarily produce the same baseline that a human being would choose.

For *quantitative analysis,* the area of a chromatographic peak is proportional to the quantity of analyte. Internal standards (Section 5-4) are frequently used in chromatography because the injected volume and exact chromatographic conditions vary somewhat from run to run. However, effects of variable conditions are usually the same for internal standard and analyte. By comparing the area of the analyte peak with that of the internal standard, we obtain a good measure of analyte concentration.

Computer-controlled chromatographs will find peak areas automatically. However, Figure 21-5 illustrates the point that computers and humans may not choose the same baseline for measuring area. Inspect the baseline proposed by the computer to see that it is sensible and modify the baseline when you disagree.

Scaling Up a Separation

Analytical chromatography is conducted on a small scale to separate, identify, and measure components of a mixture. *Preparative* chromatography is carried out on a large scale to isolate a significant quantity of one or more components of a mixture. Analytical chromatography typically uses a long, thin column to obtain good resolution. Preparative chromatography usually employs a short, fat column that handles larger quantities of material but gives inferior resolution. (Long, fat columns may be prohibitively expensive to buy or operate.)

Analytical chromatography: small-scale analysis

Preparative chromatography: large-scale separation

If you have developed a procedure to separate 2 mg of a mixture on a column with a diameter of 1.0 cm, what size column should you use to separate 20 mg of the mixture? The most straightforward way to scale up is to maintain column length and increase cross-sectional area to maintain a constant ratio of unknown to column volume. The cross-sectional area is proportional to the square of the column radius (r), so

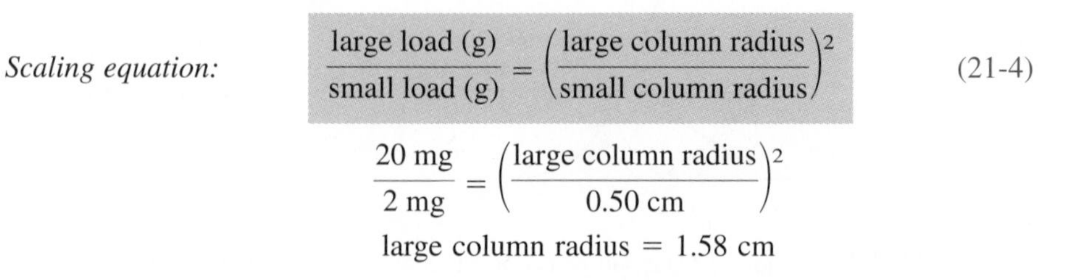

Scaling equation:

$$\frac{\text{large load (g)}}{\text{small load (g)}} = \left(\frac{\text{large column radius}}{\text{small column radius}}\right)^2 \qquad (21\text{-}4)$$

$$\frac{20 \text{ mg}}{2 \text{ mg}} = \left(\frac{\text{large column radius}}{0.50 \text{ cm}}\right)^2$$

$$\text{large column radius} = 1.58 \text{ cm}$$

A column with a diameter near 3 cm would be appropriate.

To reproduce the conditions of the smaller column in the larger column, the *volume flow rate* (mL/min) should be increased in proportion to the cross-sectional area of the column. If the area of the large column is 10 times greater than that of the small column, the volume flow rate should also be 10 times greater. If the volume of the small sample is V, the volume of the large sample can be $10V$.

Rules for scaling up without losing resolution:

- cross-sectional area of column $\propto$ mass of analyte
- keep column length constant
- volume flow rate $\propto$ cross-sectional area of column
- sample volume $\propto$ mass of analyte

Ask Yourself

21-B. **(a)** Use a ruler to measure retention times and widths at the base (w) of octane and nonane in Figure 21-3 to the nearest 0.1 mm.
(b) Calculate the number of theoretical plates for octane and nonane.
(c) If the column is 1.00 m long, find the plate height for octane and nonane.
(d) Use your measurements from **(a)** to compute the resolution between octane and nonane.
(e) Suppose that the sample size for Figure 21-3 was 3.0 mg, the column dimensions were 4.0 mm diameter × 1.00 m length, and the flow rate was 7.0 mL/min. What size column and what flow rate should be used to obtain the same quality of separation of 27.0 mg of sample?

21-3 Why Do Bands Spread?

If a solute is applied to a column in an ideal manner as an infinitesimally thin band, the band will broaden as it moves through a chromatography column (Figure 21-6). Broadening occurs because of diffusion, slow equilibration of solute between the mobile and stationary phases, and irregular flow paths on the column.

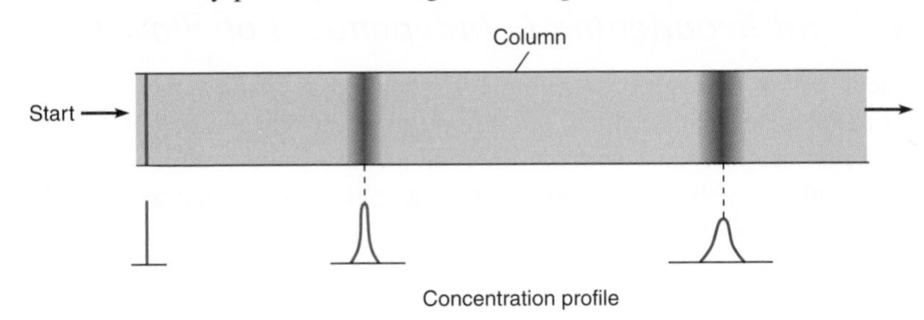

Figure 21-6 Broadening of an initially sharp band of solute as it moves through a chromatography column.

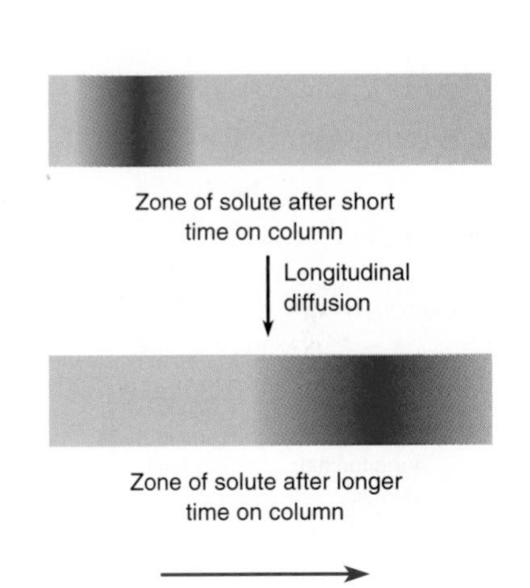

Figure 21-7 In longitudinal diffusion, solute continuously diffuses away from the concentrated zone at the center of its band. The greater the flow rate, the less time is spent on the column and the less longitudinal diffusion occurs.

Bands Diffuse

An infinitely narrow band of solute that is stationary inside the column slowly broadens because solute molecules diffuse away from the center of the band in both directions. This inescapable process, called *longitudinal diffusion*, begins at the moment solute is injected into the column. In chromatography, the farther a band has traveled, the more time it has had to diffuse and the broader it becomes (Figure 21-7).

The faster the flow rate, the less time a band spends in the column and the less time there is for diffusion to occur. The faster the flow, the sharper the peaks. Broadening by longitudinal diffusion is inversely proportional to flow rate:

Broadening by longitudinal diffusion:
$$\text{broadening} \propto \frac{1}{u} \qquad (21\text{-}5)$$

where u is the flow rate, usually measured in milliliters per minute.

Strictly speaking, flow rate in Equations 21-5 through 21-7 is *linear* flow rate (cm/min). This is the rate at which solvent goes past stationary phase. For a given column diameter, *volume* flow rate (mL/min) is proportional to linear flow rate.

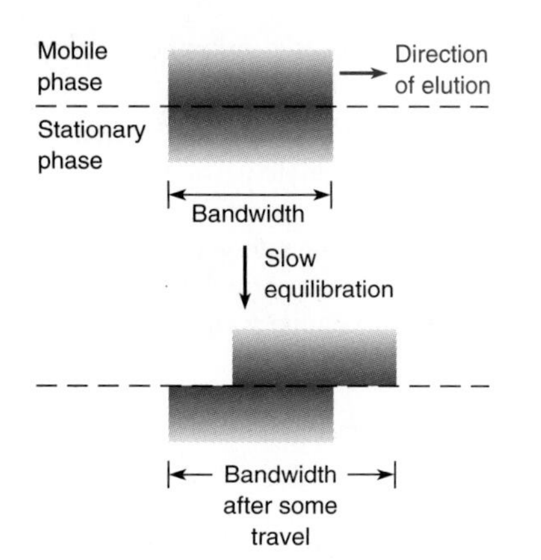

Figure 21-8 Solute requires a finite time to equilibrate between the mobile and stationary phases. If equilibration is slow, solute in the stationary phase tends to lag behind that in the mobile phase, causing the band to spread. The slower the flow rate, the less the zone broadening by this mechanism.

Solute Requires Time to Equilibrate Between Phases

Imagine a solute that is distributed between the mobile and stationary phases at some moment in time at some position in a column with zero flow rate. Now set the flow into motion. If the solute cannot equilibrate rapidly enough between the phases, then solute in the stationary phase tends to lag behind solute in the mobile phase (Figure 21-8). This broadening due to the *finite rate of mass transfer* between phases becomes worse as the flow rate increases:

Broadening by finite rate of mass transfer:

$$\text{broadening} \propto u \tag{21-6}$$

A Separation Has an Optimum Flow Rate

When trying to separate closely spaced bands, we want to minimize band broadening. If the bands broaden too much, they will not be resolved from one another. Because broadening by longitudinal diffusion decreases with increasing flow rate (Equation 21-5) but broadening by the finite rate of mass transfer increases with increasing flow rate (Equation 21-6), there is an intermediate flow rate that gives minimum broadening and optimum resolution (Figure 21-9). Part of the science and art of chromatography is to find conditions such as flow rate and solvent composition to obtain adequate separation between components of a mixture.

The rate of mass transfer between phases increases with temperature. Raising the column temperature might improve the resolution or allow faster separations without reducing resolution.

Some Band Broadening Is Independent of Flow Rate

Some mechanisms of band broadening are independent of flow rate. Figure 21-10 shows a mechanism that is called *multiple paths* and occurs in any column packed with solid particles. Because some of the random flow paths are longer than others, solute molecules entering the column at the same time on the left are eluted at different times on the right.

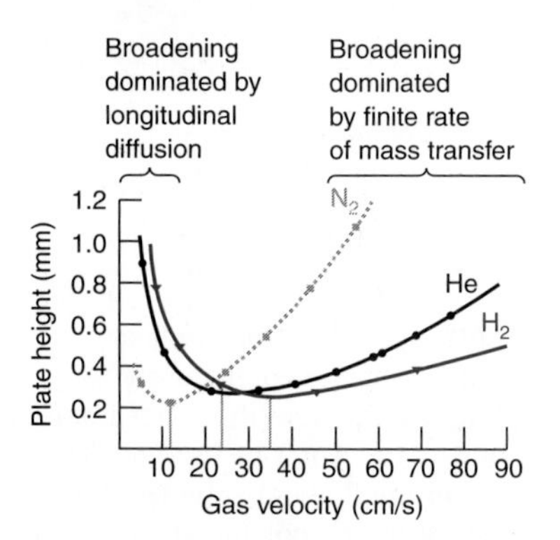

Figure 21-9 Optimum resolution (minimum plate height) occurs at an intermediate flow rate. Curves show measured plate height in gas chromatography of n-$C_{17}H_{36}$ at 175°C, using N_2, He, or H_2 mobile phase. [From R. R. Freeman, ed., *High Resolution Gas Chromatography* (Palo Alto, CA: Hewlett Packard Co., 1981).]

Plate Height Equation

The **van Deemter equation** for plate height (H) as a function of flow rate (u) is the net result of the three band-broadening mechanisms just discussed:

van Deemter equation for plate height:

$$H \approx A + \frac{B}{u} + Cu \tag{21-7}$$

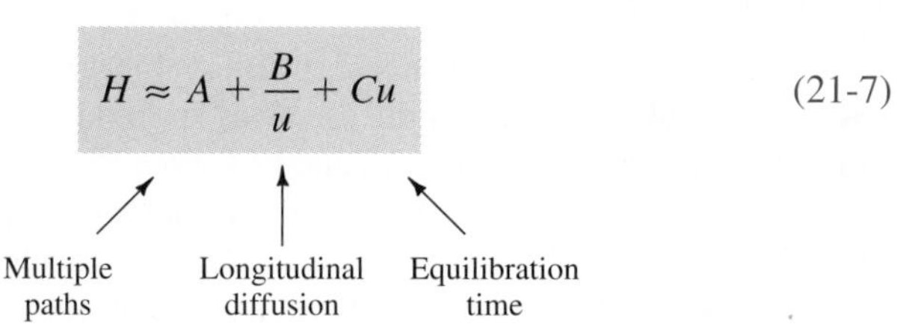

where A, B, and C are constants determined by the column, stationary phase, mobile phase, and temperature. Each of the curves in Figure 21-9 is described by Equation 21-7 with different values of A, B, and C.

Equation 21-7 describes the broadening of a band of solute as it passes through a chromatography column. If the band has some finite width when it is applied to the

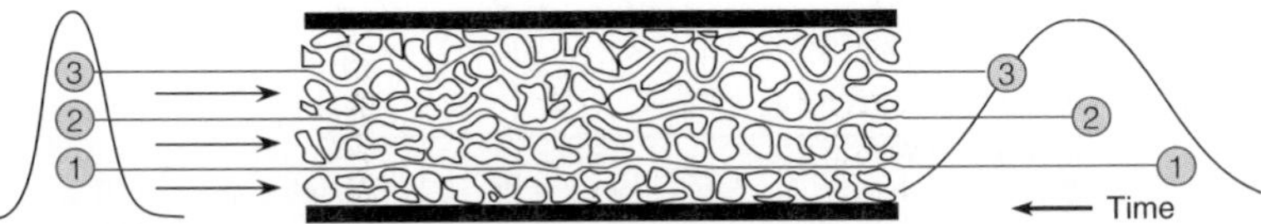

Figure 21-10 Band spreading from multiple flow paths. The smaller the stationary-phase particles, the less serious is this problem. This process is absent in an open tubular column. [From H. M. McNair and E. J. Bonelli, *Basic Gas Chromatography* (Palo Alto, CA: Varian Instruments Division, 1968).]

column, the eluted band emerging from the other end will be broader than we predict with the van Deemter equation. Bands can broaden outside the column if there is too much tubing to flow through or if the detector has too large a volume. Best results are obtained if the lengths and diameters of all tubing outside the column are kept to a minimum. Also, there should be no void spaces inside the column where mixing can occur.

Open Tubular Columns

In contrast with a **packed column** that is filled with solid particles coated with stationary phase, an **open tubular column** is a hollow capillary whose inner wall is coated with a thin layer of stationary phase (Figure 21-11). In an open tubular column, there is no broadening from multiple paths, because there are no particles of stationary phase in the flow path. Therefore a given length of open tubular column generally gives better resolution than the same length of packed column. We say that the open tubular column has more theoretical plates (or a smaller plate height) than the packed column.

A packed gas chromatography column has greater resistance to gas flow than does an open tubular column. Therefore an open tubular column can be made much longer than a packed column with the same operating pressure. Because of its resistance to gas flow, a packed gas chromatography column is usually just 2–3 m in length, whereas an open tubular column can be 100 m in length. The greater length and smaller plate height of the open tubular column provide much better resolution than a packed column does.

Open tubular columns give better separations than packed columns because

- there is no multiple path band broadening ($A = 0$ in van Deemter equation)
- the open tubular column can be much longer

An open tubular column cannot handle as much solute as a packed column, because there is less stationary phase in the open tubular column. Therefore open tubular columns are useful for analytical separations but not for preparative separations.

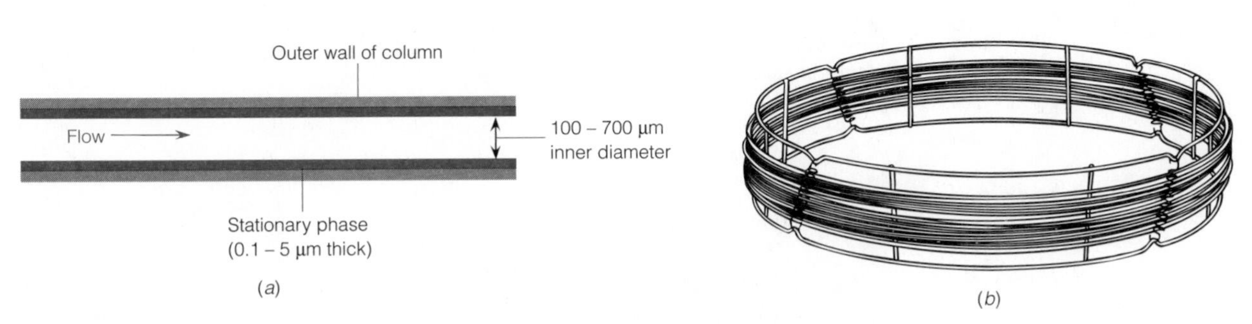

Figure 21-11 (*a*) Typical dimensions of open tubular column for gas chromatography. (*b*) A fused silica column with a length of 15–100 m is wound in a small coil to fit inside the chromatograph.

Box 21-1 Polarity

Polar compounds have positive and negative regions that attract neighboring molecules by electrostatic forces.

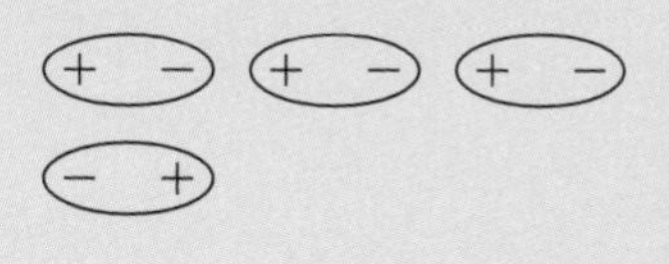

Polar molecules attract each other by electrostatic forces. Positive attracts negative.

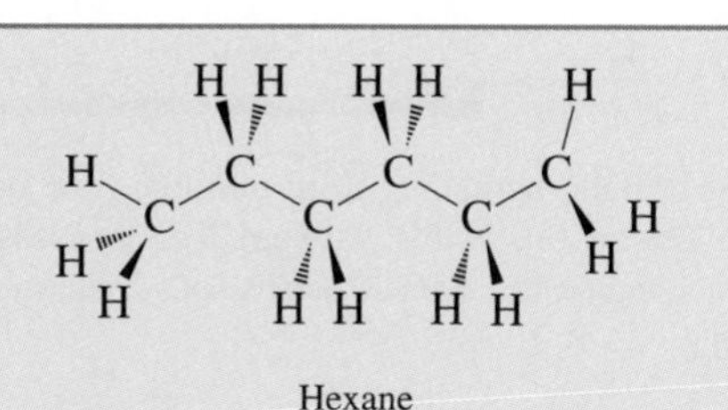

Hexane

Polarity arises because different atoms have different electronegativity—different abilities to attract electrons from the chemical bonds. For example, oxygen is more electronegative than carbon. In acetone, the oxygen attracts electrons from the C═O double bond and takes on a partial negative charge, designated δ−. This shift leaves a partial positive charge (δ+) on the neighboring carbon.

$O\,\delta-$
$\|$
$H_3C{-}\overset{\delta+}{C}{-}CH_3$

Acetone

In contrast with acetone, hexane is considered to be a **nonpolar compound** because there is little charge separation within a hexane molecule.

Water is a very polar molecule that forms hydrogen bonds between the electronegative oxygen atom in one molecule and the electropositive hydrogen atom of another molecule:

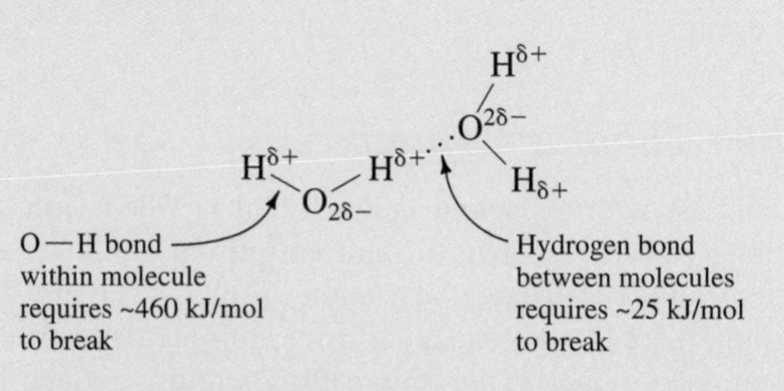

Ionic compounds usually dissolve in water. In general, polar organic compounds tend to be most soluble in polar solvents and least soluble in nonpolar solvents. Nonpolar compounds tend to be most soluble in nonpolar solvents. "Like dissolves like."

Typical nonpolar and weakly polar compounds		Typical polar compounds	
	octane (C_8H_{18})	CH_3OH	methanol
	benzene	CH_3CH_2OH	ethanol
$-CH_3$	toluene	$CHCl_3$	chloroform
CCl_4	carbon tetrachloride	$CH_3-C(=O)OH$	acetic acid
O	diethyl ether ($C_4H_{10}O$)	$CH_3C{\equiv}N$	acetonitrile

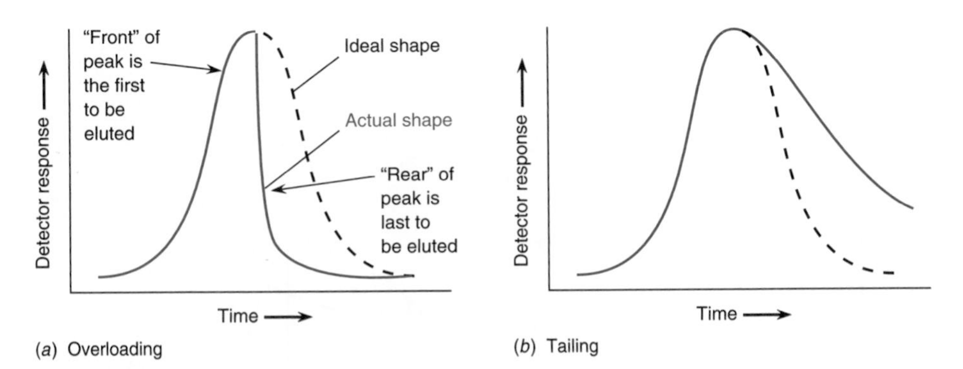

Figure 21-12 (*a*) Overloading gives rise to a chromatographic band with an ordinary front and an abruptly cut off rear. (*b*) Tailing is a peak shape with a normal front and an elongated rear.

Sometimes Bands Have Funny Shapes

When a column is overloaded by too much solute in one band, the band emerges from the column with a gradually rising front and an abruptly cut off back (Figure 21-12*a*). The reason for this behavior is that a compound is most soluble in itself. As the concentration rises from 0 at the front of the band to some high value inside the band, *overloading* can occur. When overloaded, the solute is so soluble in the concentrated part of the band that little trails behind the concentrated region.

Tailing is an asymmetric peak shape in which the trailing part of the band is elongated (Figure 21-12*b*). Such a peak occurs when there are strongly polar, highly adsorptive sites (such as exposed —OH groups) in the stationary phase that retain solute more strongly than other sites. To reduce tailing, we use a chemical treatment called *silanization* to convert polar —OH groups into nonpolar $—OSi(CH_3)_3$ groups. Increased tailing during the life of a column is a signal that the column needs to be replaced.

Box 21-1 discusses *polarity.*

Ask Yourself

21-C. **(a)** Why is longitudinal diffusion a more serious problem in gas chromatography than in liquid chromatography? (Think—the answer is not in the text.)
(b) **(i)** In Figure 21-9, what is the optimal flow rate for best separation of solutes with He mobile phase? **(ii)** Why does plate height increase at high flow rate? **(iii)** Why does plate height increase at low flow rate?
(c) Why does an open tubular gas chromatography column give better resolution than that of the same length of packed column?
(d) Why is it desirable to use a very long open tubular column? What difference between packed and open tubular columns allows much longer open tubular columns to be used?

21-4 Mass Spectrometry

Mass spectrometry measures the masses and abundances of ions in the gas phase. A mass spectrometer is the most powerful detector for chromatography because the spectrometer is sensitive to low concentrations of analyte, provides both qualitative and quantitative information, and can distinguish different substances with the same retention time.

Francis W. Aston (1877–1945) developed a "mass spectrograph" in 1919 that could separate ions and focus them onto a photographic plate. Aston immediately found that neon consists of two isotopes (^{20}Ne and ^{22}Ne) and went on to discover 212 of the 281 naturally occurring isotopes. He received the Nobel Prize in 1922.

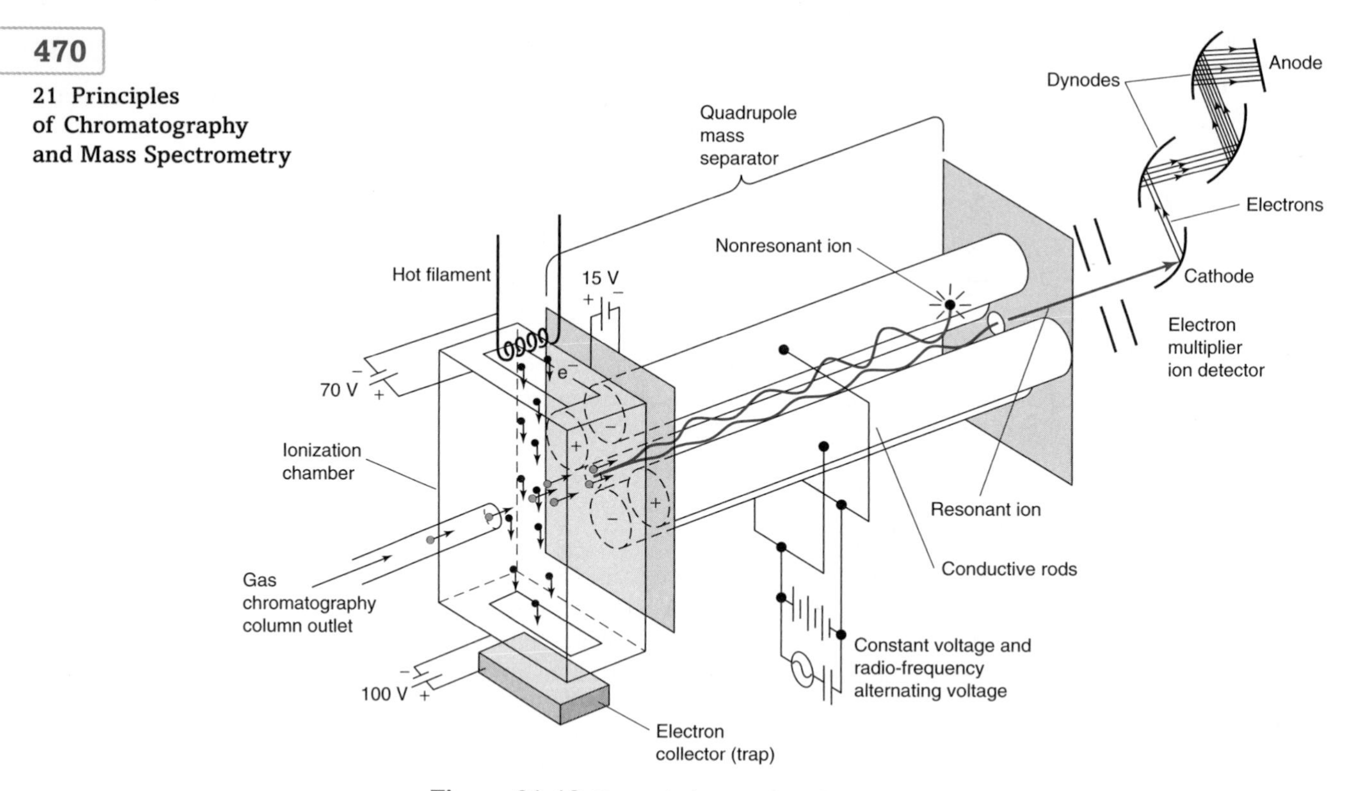

Figure 21-13 Transmission quadrupole mass spectrometer.

Atomic mass units are called *daltons*, Da. The mass of $^{37}Cl^-$ is 36.966 Da.

Prior to separation in the mass spectrometer, molecules must be converted into ions. These ions are then separated according to their mass-to-charge ratio, m/z. For an ion with a charge of $z = \pm 1$, such as $^{37}Cl^-$, m/z is numerically equal to m, which is close to 37. For the protein cytochrome *c* with 16 excess protons, $m = 12\ 230$, $z = +16$, and $m/z = 12\ 230/16 = 764.4$.

A Mass Spectrometer

Figure 21-13 shows a **transmission quadrupole mass spectrometer**, which is the most common mass separator in use today. It is connected to a gas chromatography column at the left to record the spectrum of each component as it is eluted. Compounds exiting the column pass through a heated connector into the ionization chamber where they are converted into ions and accelerated by 15 V before entering the quadrupole mass separator.

The mass separator consists of four parallel metal rods to which a constant voltage and a radio-frequency oscillating voltage are applied. The electric field deflects ions in complex trajectories as they migrate from the ionization chamber toward the detector, allowing only ions with one particular mass-to-charge ratio to reach the detector. Other (nonresonant) ions collide with the rods and are lost before they reach the detector. The mass spectrometer is evacuated to $\sim 10^{-9}$ bar to minimize collisions of ions with background gas as they pass through the quadrupole. Ions of different masses are selected to reach the detector by changing the voltages applied to the rods. Transmission quadrupoles can record from two to eight complete spectra per second, reaching as high as 4 000 m/z units. They can resolve peaks separated by m/z 0.3.

The *electron multiplier* ion detector at the right of Figure 21-13 is similar to the photomultiplier tube in Figure 19-9. Ions strike a cathode, dislodging electrons that are then accelerated into a more positive dynode. Electrons knocked off the first dynode are accelerated into a second dynode, where even more electrons are dislodged. Approximately 10^5 to 10^6 electrons reach the anode for each ion striking the cathode.

Ionization

Two common methods to convert molecules into ions are *electron ionization* and *chemical ionization.* Molecules entering the ionization chamber in Figure 21-13 are converted into ions by **electron ionization**. Electrons emitted from a hot filament (like the one in a light bulb) are accelerated by 70 V before encountering incoming molecules. Some (~0.01%) analyte molecules (M) absorb enough energy (9–15 electron volts, eV) to ionize:

Although we give examples of mass spectrometry of cations in this section, anions also can be produced and separated by mass spectrometry.

$$\mathrm{M} + \underset{70\ \mathrm{eV}}{\mathrm{e}^-} \longrightarrow \underset{\text{Molecular ion}}{\mathrm{M}^+} + \underset{\sim 55\ \mathrm{eV}}{\mathrm{e}^-} + \underset{0.1\ \mathrm{eV}}{\mathrm{e}^-}$$

The resulting cation, M^+, is called the **molecular ion**. After ionization, M^+ usually has enough residual internal energy (~1 eV) to break into fragments.

You can find electron ionization mass spectra of many compounds at http://webbook.nist.gov/chemistry.

There might be so little M^+ that its peak is small or absent in the mass spectrum. The electron ionization mass spectrum at the left side of Figure 21-14 does not exhibit an M^+ peak, which would be at *m/z* 226. Instead, fragments appear at *m/z* 197, 156, 141, 112, 98, 69, and 55. These peaks provide clues about the structure of the molecule. A computer search is commonly used to match the spectrum of an unknown to similar spectra in a library.

The most intense peak in a mass spectrum is called the **base peak**. Intensities of other peaks are expressed as a percentage of the base peak intensity. In the electron ionization spectrum in Figure 21-14, the base peak is at *m/z* 141.

Chemical ionization usually produces less fragmentation than electron ionization. For chemical ionization, the ionization chamber contains a *reagent gas* such as methane at a pressure of ~1 mbar. Energetic electrons (100–200 eV) convert CH_4 into a variety of products:

$$\mathrm{CH_4} + \mathrm{e}^- \longrightarrow \mathrm{CH_4^+} + 2\mathrm{e}^-$$
$$\mathrm{CH_4^+} + \mathrm{CH_4} \longrightarrow \mathrm{CH_5^+} + \mathrm{CH_3}$$

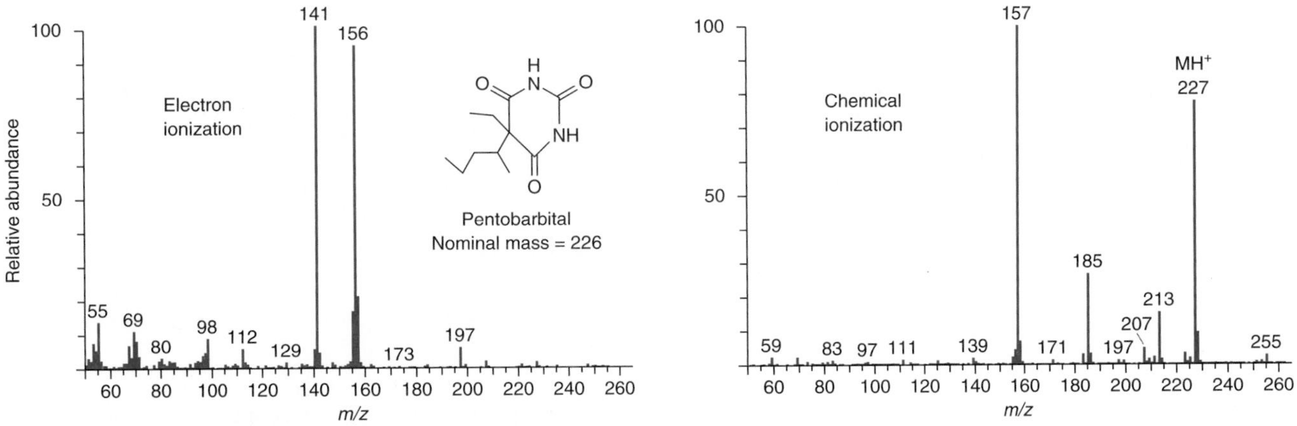

Figure 21-14 Mass spectra of the sedative pentobarbital with electron ionization (*left*) or chemical ionization (*right*). The molecular ion (*m/z* 226) is not evident with electron ionization. The dominant ion from chemical ionization is MH^+. The peak at *m/z* 255 in the chemical ionization spectrum is $M(C_2H_5)^+$. $C_2H_5^+$ can be formed through chemical ionization by the reactions (1) $CH_4^+ \rightarrow CH_3^+ + H$ and (2) $CH_3^+ + CH_4 \rightarrow C_2H_5^+ + H_2$. [Courtesy Varian Associates, Sunnyvale, CA.]

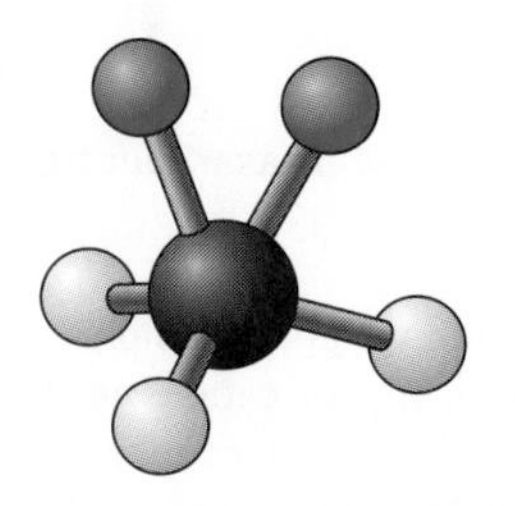

Figure 21-15 CH_5^+ is described as a CH_3 tripod with an added $\mathbf{H_2}$ unit.[1] [**H**--C--**H**] is held together by two electrons distributed over three atoms. Atoms of the $\mathbf{H_2}$ unit rapidly exchange with atoms of the CH_3 unit.

CH_5^+ (Figure 21-15) is a potent proton donor that reacts with analyte to give the *protonated molecule*, MH^+, which is usually the most abundant ion.

$$CH_5^+ + M \longrightarrow CH_4 + MH^+$$

In the chemical ionization mass spectrum in Figure 21-14, MH^+ at m/z 227 is a strong peak and there are fewer fragments than in the electron ionization spectrum.

Total Ion and Selected Ion Chromatograms

One way to use a mass spectrometer as a detector for chromatography is to record the total current from all ions produced by eluate. Box 21-2 shows a **reconstructed total ion chromatogram** of vapors from candy. This chromatogram is "reconstructed" by a computer from individual mass spectra recorded during chromatography. In this case, the spectrometer measures all ions above m/z 34. Therefore, it responds to all compounds shown in the figure, but not to carrier gas, H_2O, N_2, or O_2.

Figure 21-16*a* is a reconstructed total ion chromatogram showing all ions from seven opium alkaloids found in street heroin. Traces *b–h* are **selected ion chromatograms** in which the mass spectrometer is set to respond to just one mass

Box 21-2 Volatile Flavor Components of Candy

Students at Indiana State University identify components of candy and gum by using gas chromatography with a mass spectrometer as a detector. The extremely simple sampling technique for this qualitative analysis is called *headspace analysis.* A crushed piece of hard candy or a piece of gum is placed in a vial and allowed to stand for a few minutes so that *volatile* compounds (compounds with a high vapor pressure) can evaporate and fill the gas phase (the *headspace*) with vapor. A 5-μL sample of the vapor phase is then drawn into a syringe and injected into a gas chromatograph. Peaks are identified by comparison of their retention times and mass spectra with those of known compounds.

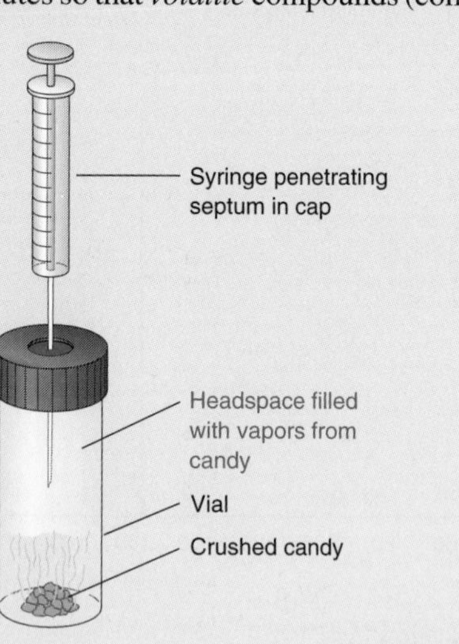

Headspace sampling. A syringe is inserted through a rubber *septum* (a rubber disk) in the cap of the vial to withdraw gas for analysis.

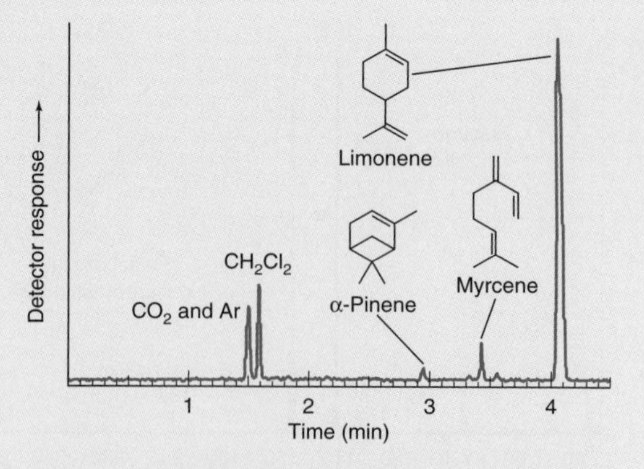

Reconstructed total ion gas chromatogram of headspace vapors from Orange Life Savers®. The mass spectral detector measures ions above 34 atomic mass units. CO_2 and Ar are from air and CH_2Cl_2 is the solvent used to clean the syringe. [From R. A. Kjonaas, J. L. Soller, and L. A. McCoy, *J. Chem. Ed.* **1997**, *74*, 1104.]

in each trace. In trace *f*, the spectrometer responds only to *m*/*z* 370, corresponding to the protonated ion (MH^+) of heroin. The peak in trace *f* arises only from heroin because other components of this mixture do not have significant intensity at *m*/*z* 370. Even if heroin were eluted at the same time as another component, only heroin would be observed in the selected ion chromatogram.

The selected ion chromatogram simplifies chromatographic analysis and improves the signal-to-noise ratio for the desired analyte. Signal-to-noise is increased because more time is spent collecting data at the selected value of *m*/*z*.

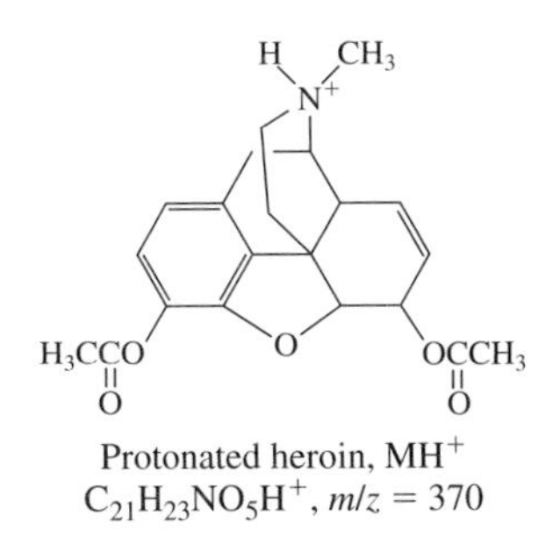

Protonated heroin, MH^+
$C_{21}H_{23}NO_5H^+$, $m/z = 370$

Ask Yourself

21-D. **(a)** What is the difference between a reconstructed total ion chromatogram and a selected ion chromatogram? Why does the selected ion chromatogram have a higher signal-to-noise ratio than that of the total ion chromatogram?
(b) Why does trace *h* in Figure 21-16 have just one peak, even though a mixture of seven compounds was injected into the chromatograph?

21-5 Information in a Mass Spectrum

The mass spectrum of a molecule provides information about its structure. At the frontier of mass spectrometry today, scientists are elucidating the sequences of amino acids in proteins and the structures of complex carbohydrates by their fragmentation patterns. In this section we touch on some of the simplest information available from mass spectrometry.

Nominal Mass

The unit of atomic mass is the dalton, Da, defined as 1/12 of the mass of ^{12}C. **Atomic mass** is the weighted average of the masses of the isotopes of an element. Table 21-1 tells us that bromine consists of 50.69% ^{79}Br with a mass of 78.918 34 Da and 49.31% ^{81}Br with a mass of 80.916 29 Da. In the weighted average, each mass is multiplied by its abundance. Therefore the atomic mass of Br is (0.506 9)(78.918 34) + (0.493 1)(80.916 29) = 79.904 Da.

The **molecular mass** of a molecule or ion is the sum of atomic masses listed in the periodic table. For 1-bromobutane, C_4H_9Br, the molecular mass is $(4 \times 12.010\,7) + (9 \times 1.007\,94) + (1 \times 79.904) = 137.018$.

The **nominal mass** of a molecule or ion is the *integer* mass of the species with the most abundant isotope of each of the constituent atoms. For carbon, hydrogen, and bromine, the most abundant isotopes are ^{12}C, ^{1}H, and ^{79}Br. Therefore the nominal mass of C_4H_9Br is $(4 \times 12) + (9 \times 1) + (1 \times 79) = 136$.

Fragmentation Patterns

The electron ionization mass spectrum of 1-bromobutane in Figure 21-17 has two peaks of almost equal intensity at *m*/*z* 136 and 138. The peak at *m*/*z* 136 is the molecular ion $C_4H_9{}^{79}Br^+$. Because bromine has almost equal abundance of the isotopes ^{79}Br and ^{81}Br, the second peak of almost equal intensity is $C_4H_9{}^{81}Br^+$. Any molecule or fragment containing just one Br will have pairs of peaks of nearly equal

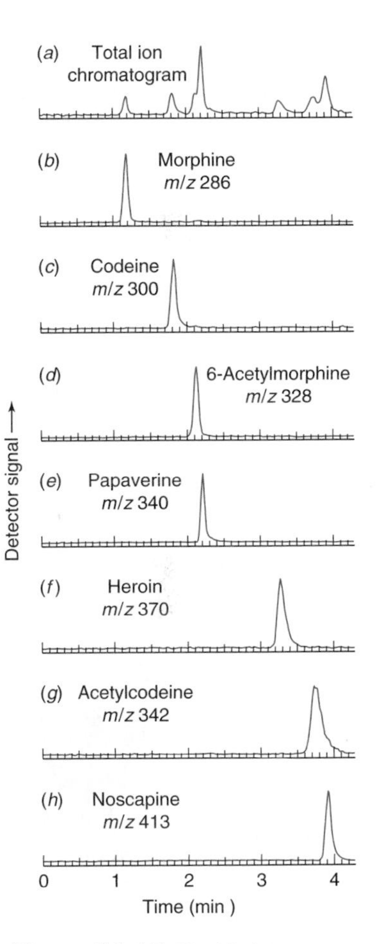

Figure 21-16 Liquid chromatography of opium alkaloids found in street heroin. Trace *a* is the reconstructed total ion chromatogram showing all masses in the range *m*/*z* 100–450. Traces *b*–*h* are selected ion chromatograms monitoring just a single value of *m*/*z* in each case. [From R. Dams, T. Benjits, W. Günther, W. Lambert, and A. De Leenheer, *Anal. Chem.* **2002**, *74*, 3206.]

Table 21-1 Isotopes of selected elements

Element	Mass number	Mass (Da)[a]	Abundance (atom %)[b]	Element	Mass number	Mass (Da)[a]	Abundance (atom %)[b]
Electron	—	0.000 548 580	—	F	19	18.998 40	100
H	1	1.007 825	99.988	P	31	30.973 76	100
	2	2.014 10	0.012	S	32	31.972 07	94.93
C	12	12 (exact)	98.93		33	32.971 46	0.76
	13	13.003 35	1.07		34	33.967 87	4.29
N	14	14.003 07	99.632		36	35.967 08	0.02
	15	15.000 11	0.368	Cl	35	34.968 85	75.78
O	16	15.994 91	99.757		37	36.965 90	24.22
	17	16.999 13	0.038	Br	79	78.918 34	50.69
	18	17.999 16	0.205		81	80.916 29	49.31

a. 1 dalton (Da) ≡ 1/12 of the mass of ^{12}C = $1.660\ 538\ 86\ (28) \times 10^{-27}$ kg (http://physics.nist.gov/constants). Nuclide masses from G. Audi, A. H. Wapsta, and C. Thibault, *Nucl. Phys.* **2003**, *A729*, 337 (found at www. nndc.bnl.gov/masses/).

b. Abundance is representative of what is found in nature. See J. K. Böhlke et al., *J. Phys. Chem. Ref. Data* **2005**, *34*, 57.

intensity in its mass spectrum. Other major peaks at *m/z* 107, 57, and 41 are explained by rupture of the bonds of 1-bromobutane in Figure 21-18. The peak at 107 has an equal-intensity partner at 109, so it must contain Br. Peaks at 57 and 41 do not have equal-intensity partners, so they cannot contain Br.

Isotope Patterns and the Nitrogen Rule

Peaks for ^{79}Br and ^{81}Br in Figure 21-17 are a characteristic *isotope pattern.* Information on the composition of organic compounds is obtained from the relative intensities at M + 1 and M^+, where M + 1 is one mass unit above the molecular ion. Table 21-1 tells us that ^{12}C is the common isotope of carbon and ^{13}C has a natural abundance of 1.1%. Other common elements in organic compounds, H, O, and N, each

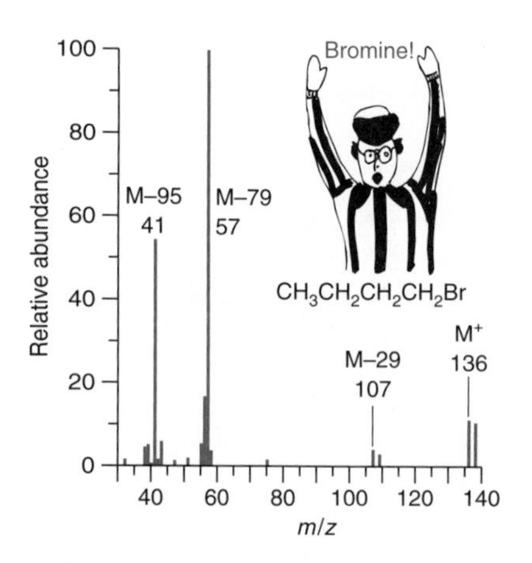

Figure 21-17 Electron ionization mass spectrum of 1-bromobutane. [From A. Illies, P. B. Shevlin, G. Childers, M. Peschke, and J. Tsai, *J. Chem. Ed.* **1995**, *72*, 717. Referee from Maddy Harris.]

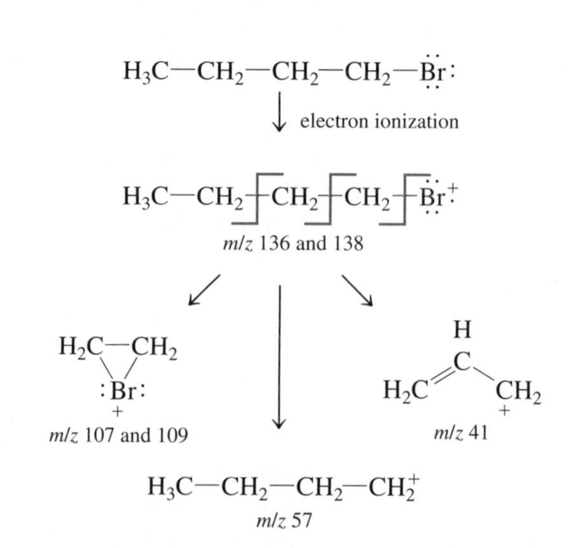

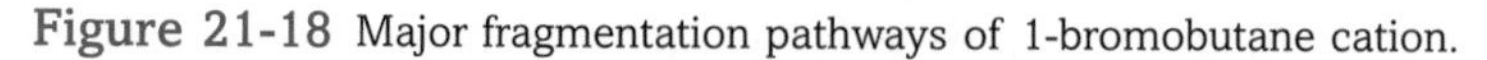

Figure 21-18 Major fragmentation pathways of 1-bromobutane cation.

have just one major isotope with little of the next-higher-mass isotope. Therefore, the compound $C_nH_xO_yN_z$ has a ratio of intensities of the molecular ion given by

Intensity of M + 1 relative to M^+ for $C_nH_xO_yN_z$:

$$\text{intensity} = n \times 1.1\% \tag{21-8}$$

Figure 21-19*a* shows the molecular ion region of the mass spectrum of benzene. For C_6H_6, Equation 21-8 predicts relative intensities $(M + 1)/M^+ = 6 \times 1.1\% = 6.6\%$. The observed ratio is 6.5%. Ordinary mass spectral intensity ratios are not more accurate than ±10%, so a value in the range 5.9–7.3% is within expected uncertainty from 6.6%.

The **nitrogen rule** helps us propose compositions for molecular ions: If a compound has an odd number of nitrogen atoms—in addition to any number of C, H, halogens, O, S, Si, and P—then M^+ has an odd nominal mass. For a compound with an even number of nitrogen atoms (0, 2, 4, and so on), M^+ has an even nominal mass. A molecular ion at *m/z* 128 can have 0 or 2 N atoms, but it cannot have 1 N atom.

Example Elemental Information from the Mass Spectrum

Figure 21-19*b* shows the molecular ion region of the spectrum of biphenyl. M^+ is observed at *m/z* 154 and the intensity of M + 1 is 12.9% of M^+. What formulas of the type $C_nH_xO_yN_z$ are consistent with the spectrum?

SOLUTION From the even nominal mass of the molecular ion (154), there must be an even number of N atoms (0, 2, 4, and so on). From the intensity ratio $(M + 1)/M^+ = 12.9\%$, we use Equation 21-8 to estimate that the number of carbon atoms is $12.9\%/1.1\% = 11.7 \approx 12$. A possible formula is $C_{12}H_{10}$ because $12 \times 12 + 10 \times 1 = 154$. Another plausible formula is $C_{11}H_6O$, whose nominal mass also is 154. The predicted intensity ratio for $C_{11}H_6O$ is $(M + 1)/M^+ = 11 \times 1.1\% = 12.1\%$, which is consistent with the observed value of 12.9%. The formula $C_{10}H_6N_2$ would have $(M + 1)/M^+ = 10 \times 1.1\% = 10.8\%$, which is somewhat low to match the observed ratio.

Test Yourself What formula $C_nH_xO_yN_z$ is consistent with a molecular ion at *m/z* 94 and the intensity ratio $(M + 1)/M^+ = 6.8\%$? (**Answer:** $n = 6$ and z is even. Possible formulas are C_6H_6O and C_6H_{22}, but the most H atoms that can bond to 6 C atoms is 14. $C_5H_6N_2$ would have $(M + 1)/M = 5.5\%$, which is too low to match the observed value of 6.8%. C_7H_{12} would have $(M + 1)/M = 7.7\%$, which is too high. The best answer is C_6H_6O.)

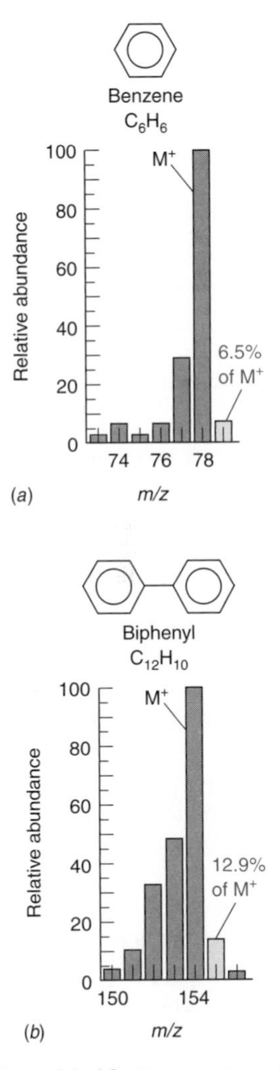

Figure 21-19 Electron ionization mass spectra of molecular ion region of (*a*) benzene (C_6H_6) and (*b*) biphenyl ($C_{12}H_{10}$). [From NIST/EPA/NIH Mass Spectral Database, SRData@enh.nist.gov.]

Chromatographic and isotopic analyses are used in testing athletes for illegal use of synthetic testosterone to build muscle mass. In men, the natural ratio of testosterone to its stereoisomer, epitestosterone, is typically near 1:1 and rarely exceeds 4:1. A ratio exceeding 4:1 in urine measured by chromatography suggests that the athlete is taking testosterone. If a second sample obtained at the same time as the first sample replicates the result, there is reason to suspect that the athlete is taking synthetic testosterone. Isotope measurements distinguish natural from synthetic testosterone. Testosterone from urine is isolated by gas chromatography and combusted to CO_2. The $^{13}C/^{12}C$ ratio in the CO_2 is measured accurately by *isotope ratio mass spectrometry*, which uses two calibrated detectors to collect *m/z* 44 ($^{12}CO_2^+$) and *m/z* 45 ($^{13}CO_2^+$) ions. Synthetic testosterone is made from plant oils whose $^{13}C/^{12}C$ ratio is ~0.5% lower than the ratio in the human body. A high testosterone : epitestosterone ratio coupled with a low $^{13}C/^{12}C$ ratio strongly suggests illegal use of testosterone. Athletes have been stripped of hard-won titles on the basis of gas chromatography–isotope ratio mass spectrometry tests.

Ask Yourself

21-E. $C_nH_xO_yN_z$ has a nominal mass of 194 with $(M + 1)/M^+ = 8.8\%$. How many C atoms are in the formula and what number of N atoms are allowed? Write the possible formulas for the compound.

Key Equations

Number of theoretical plates

$$N = \frac{5.55\, t_r^2}{w_{1/2}^2}$$

t_r = retention time of analyte
$w_{1/2}$ = peak width at half-height
} Both must be measured in the same units

Plate height

$H = L/N$ (L = length of column)

Resolution

$$\text{resolution} = \frac{\Delta t_r}{w_{av}} = \frac{0.589\,\Delta t_r}{w_{1/2av}}$$

Δt_r = difference in retention times between two peaks
w_{av} = average width of two peaks at baseline
$w_{1/2av}$ = average width of two peaks at half-height
(Δt_r and w must be measured in the same units.)

Scaling equation

$$\frac{\text{large load}}{\text{small load}} = \left(\frac{\text{large column radius}}{\text{small column radius}}\right)^2$$

van Deemter equation

$$H \approx A + \frac{B}{u} + Cu$$

H = plate height; u = flow rate
A = constant due to multiple flow paths
B = constant due to longitudinal diffusion of solute
C = constant due to equilibration time of solute between phases

Important Terms

adsorption chromatography
affinity chromatography
atomic mass
base peak
chemical ionization
chromatogram
chromatography
electron ionization
eluate
eluent
elution
gas chromatography
ion-exchange chromatography
liquid chromatography
mass spectrometry
mobile phase
molecular exclusion chromatography
molecular ion
molecular mass
nitrogen rule
nominal mass
nonpolar compound
open tubular column
packed column
partition chromatography
plate height
polar compound
reconstructed total ion chromatogram
resolution
retention time
selected ion chromatogram
spiking
stationary phase
transmission quadrupole mass spectrometer
theoretical plate
van Deemter equation

Problems

21-1. What is the difference between eluent and eluate?

21-2. Which column gives narrower bands in a chromatographic separation: plate height = 0.1 mm or 1 mm?

21-3. Explain why a chromatographic separation normally has an optimum flow rate that gives the best separation.

21-4. (a) Why is longitudinal diffusion a more serious problem in gas chromatography than in liquid chromatography?
(b) Suggest a reason why the optimum linear flow rate is much higher in gas chromatography than in liquid chromatography.

21-5. Why does silanization reduce tailing of chromatographic peaks?

21-6. What kind of information does a mass spectrometer detector give in gas chromatography that is useful for qualitative analysis? For quantitative analysis?

21-7. How is spiking used in qualitative analysis? Why are several different types of columns necessary to make a convincing case for the identity of a compound?

21-8. (a) How many theoretical plates produce a chromatography peak eluting at 12.83 min with a width at half-height of 8.7 s?
(b) The length of the column is 15.8 cm. Find the plate height.

21-9. A gas chromatogram of a mixture of toluene and ethyl acetate is shown below.

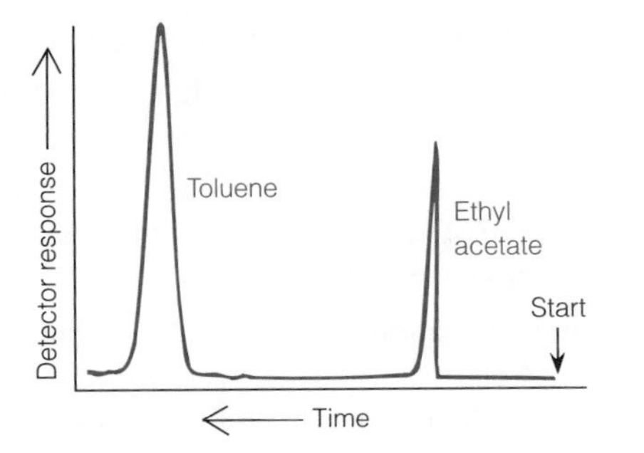

(a) Measure $w_{1/2}$ for each peak to the nearest 0.1 mm. When the thickness of the pen trace is significant relative to the length being measured, it is important to take the pen width into account. It is best to measure from the edge of one trace to the corresponding edge of the other trace, as shown at the bottom of the left column.

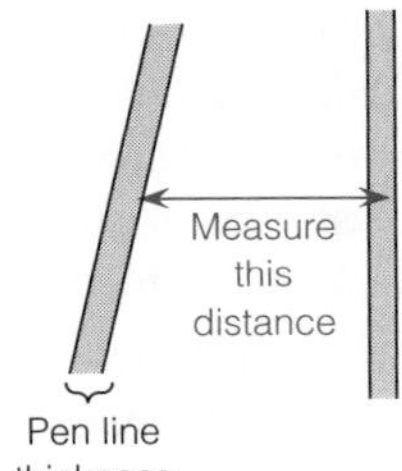

(b) Find the number of theoretical plates and the plate height for each peak.

21-10. Two components of a 12-mg sample are adequately separated by chromatography through a column 1.5 cm in diameter and 25 cm long at a flow rate of 0.8 mL/min. What size column and what flow rate should be used to obtain similar separation with a 250-mg sample?

21-11. The chromatogram in the figure below has a peak for isooctane at 13.81 min. The column is 30.0 m long.

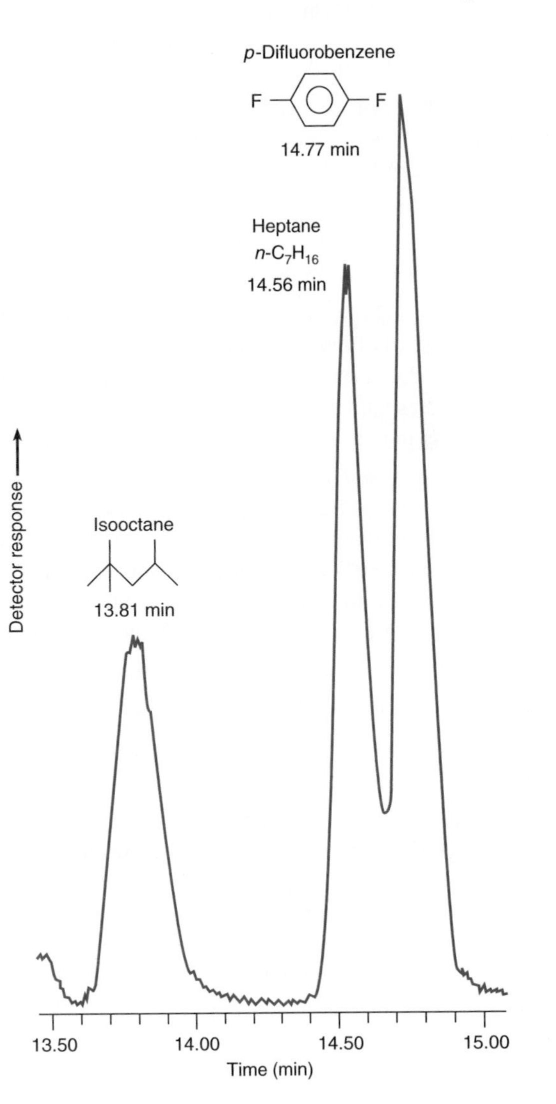

(a) Measure $w_{1/2}$ and find the number of theoretical plates for this peak.

(b) Find the plate height.

(c) Figure 21-3 tells us that the ratio $w/w_{1/2}$ for a Gaussian peak is $4\sigma/2.35\sigma = 1.70$. Measure the width at the base for isooctane in the chromatogram for this problem. Compare the measured ratio $w/w_{1/2}$ to the theoretical ratio.

21-12. Consider the peaks for heptane (14.56 min) and *p*-difluorobenzene (14.77 min) in the chromatogram for Problem 21-11. The column is 30.0 m long.

(a) Measure $w_{1/2}$ for each peak and calculate the number of plates and the plate height for each peak.

(b) From $w_{1/2}$, compute w for each peak and find the resolution between the two peaks.

21-13. A column 3.00 cm in diameter and 32.6 cm long gives adequate resolution of a 72.4-mg mixture of unknowns, initially dissolved in 0.500 mL.

(a) If you wished to scale down to 10.0 mg of the same mixture with minimum use of chromatographic stationary phase and solvent, what length and diameter column would you use?

(b) In what volume would you dissolve the sample?

(c) If the flow rate in the large column is 1.85 mL/min, what should be the flow rate in the small column?

21-14. In Figure 21-9, flow rate is expressed as gas velocity in cm/s. The gas is flowing through an open tubular column with an inner diameter of 0.25 mm. What volume flow rate (mL/min) corresponds to a gas velocity of 50 cm/s? (The volume of a cylinder is $\pi r^2 \times$ length, where r is the radius.)

21-15. *Internal standard.* A solution containing 3.47 mM X (analyte) and 1.72 mM S (standard) gave peak areas of 3 473 and 10 222, respectively, in a chromatographic analysis. Then 1.00 mL of 8.47 mM S was added to 5.00 mL of unknown X, and the mixed solution was diluted to 10.0 mL. This solution gave peak areas of 5 428 and 4 431 for X and S, respectively.

(a) Find the response factor for X relative to S in Equation 5-9.

(b) Find [S] (mM) in the 10.0 mL of mixed solution.

(c) Find [X] (mM) in the 10.0 mL of mixed solution.

(d) Find [X] in the original unknown.

21-16. *Internal standard.* A known mixture of compounds C and D gave the following chromatography results:

Compound	Concentration (μg/mL) in mixture	Peak area (cm^2)
C	236	4.42
D	337	5.52

A solution was prepared by mixing 1.23 mg of D in 5.00 mL with 10.00 mL of unknown containing just C and diluting to 25.00 mL. Peak areas of 3.33 and 2.22 cm^2 were observed for C and D, respectively. Find the concentration of C (μg/mL) in the unknown.

21-17. **(a)** The plate height in a particular packed gas chromatography column is characterized by the van Deemter equation $H(\text{mm}) = A + B/u + Cu$, where $A = 1.50$ mm, $B = 25.0$ mm · mL/min, $C = 0.025\,0$ mm · min/mL, and u is flow rate in mL/min. Construct a graph of plate height versus flow rate and find the optimum flow rate for minimum plate height.

(b) In the van Deemter equation, B is proportional to the rate of longitudinal diffusion. Predict whether the optimum flow rate would increase or decrease if the rate of longitudinal diffusion were doubled. To confirm your prediction, increase B to 50.0 mm · mL/min, construct a new graph, and find the optimum flow rate. Does the optimum plate height increase or decrease?

(c) The parameter C is inversely proportional to the rate of equilibration of solute between the mobile and stationary phase ($C \propto 1/\text{rate of equilibration}$). Predict whether the optimum flow rate would increase or decrease if the rate of equilibration between phases were doubled (that is, $C = 0.012\,5$ mm · min/mL). Does the optimum plate height increase or decrease?

21-18. **(a)** Using the van Deemter parameters in Problem 21-17**(a)**, find the plate height for a flow rate of 20.0 mL/min.

(b) How many plates are on the column if the length is 2.00 m?

(c) What will be the width at half-height of a peak eluted in a time of 8.00 min?

21-19. Explain the difference between molecular mass and nominal mass. Give the value of the molecular mass and nominal mass of benzene, C_6H_6.

21-20. From information in Table 21-1, compute the atomic mass of Cl and compare your answer with the value in the periodic table on the inside cover of this book.

21-21. *Elemental analysis by mass spectrometry.* High-resolution mass spectrometers can measure m/z to an accuracy of ~1 part in 10^5. This means that m/z 100 can be measured to an accuracy of 0.001 (and precision into the next decimal place). A molecular ion thought to be $C_4H_{11}N_3S^+$ or $C_4H_{11}N_3O_2^{\dot{+}}$ was observed at m/z 133.068 6. Compute the expected mass of each ion by adding the masses of the correct isotope of each atom and subtracting the mass of an electron, because each molecule has lost one electron. Which formula is correct?

21-22. Suggest molecular formulas for the major peaks at m/z 31, 41, 43, and 56 in the mass spectrum of 1-butanol, $CH_3CH_2CH_2CH_2OH$.

21-23. For each case below, suggest a plausible formula of the type $C_nH_xO_yN_z$:

(a) Nominal mass of $M^+ = 79$; $(M + 1)/M^+ = 5.9\%$

(b) Nominal mass of $M^+ = 123$; $(M + 1)/M^+ = 6.1\%$

(c) Nominal mass of $M^+ = 148$; $(M + 1)/M^+ = 7.4\%$

(d) Nominal mass of $M^+ = 168$; $(M + 1)/M^+ = 12.5\%$

21-24. Use Table 21-1 to predict relative intensities at *m/z* 36, 37, and 38 for HCl. Let the intensity of the molecular ion be 100 and disregard contributions <0.1%.

21-25. Use Table 21-1 to predict relative intensities at *m/z* 34, 35, and 36 for H_2S. Let the intensity of the molecular ion be 100 and disregard contributions <0.1%.

21-26. A student experiment in headspace gas chromatography–mass spectrometry of glue vapors identified tetrachloroethene in some household adhesives.[2] In the region of the molecular ion M^+, the following peaks were found (with relative intensities in parentheses: *m/z* 164 (779), 166 (999), 168 (479), 170 (101), 172 (10). Draw a stick diagram showing the peaks and their intensities. Label which species accounts for each peak and give a qualitative explanation for the pattern of intensities. Refer to isotopic abundance in Table 21-1.

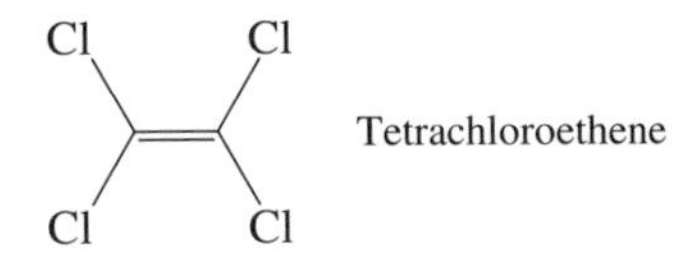

How Would You Do It?

21-27. Caffeine can be measured by chromatography with $^{13}C_3$-caffeine as an internal standard.[3] $^{13}C_3$-caffeine has the same retention time as ordinary ^{12}C-caffeine.

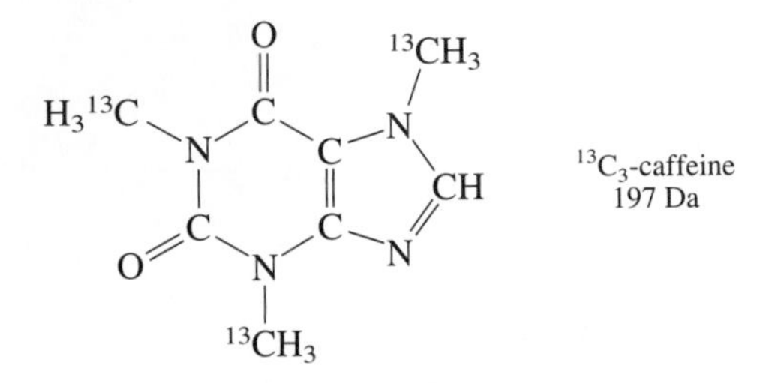

Caffeine can be extracted from aqueous solution by *solid-phase microextraction* (Section 22-4). In this procedure, a fused-silica (SiO_2) fiber coated with a polymer is dipped into the liquid and solutes from the liquid distribute themselves between the polymer phase and the liquid. Then the fiber is withdrawn from the liquid and heated in the port of a gas chromatograph. Solutes evaporate from the polymer and are carried into the column. Suggest a procedure for the quantitative analysis of caffeine in coffee by using solid-phase microextraction with a $^{13}C_3$-caffeine internal standard.

Notes and References

1. O. Asvany, P. Kumar P, B. Redlich, I. Hegemann, S. Schlemmer, and D. Marx, *Science* **2005**, *309*, 1219.

2. For headspace analysis of glue vapors, see J. Richer, J. Spencer, and M. Baird, *J. Chem. Ed.* **2006**, *83*, 1196.

3. M. J. Yang, M. L. Orton, and J. Pawliszyn, *J. Chem. Ed.* **1997**, *74*, 1130.

Further Reading

C. F. Poole, *The Essence of Chromatography* (Amsterdam: Elsevier, 2003).

J. M. Miller, *Chromatography: Concepts and Contrasts*, 2nd ed. (Hoboken, NJ: Wiley, 2005).

J. C. Giddings, *Unified Separation Science* (New York: Wiley, 1991).

J. H. Gross, *Mass Spectrometry: A Textbook* (Berlin: Springer-Verlag, 2004).

C. G. Herbert and R. A. W. Johnstone, *Mass Spectrometry Basics* (Boca Raton, FL: CRC Press, 2002).

J. Barker, *Mass Spectrometry*, 2nd ed. (Chichester, UK: Wiley, 1999).

J. T. Watson, *Introduction to Mass Spectrometry*, 3rd ed. (Philadelphia: Lippincott-Raven, 1997).

R. A. W. Johnstone and M. E. Rose, *Mass Spectrometry for Chemists and Biochemists* (Cambridge: Cambridge University Press, 1996).

E. de Hoffmann and V. Stroobant, *Mass Spectrometry: Principles and Applications*, 2nd ed. (Chichester: Wiley, 2002).

C. Dass, *Fundamentals of Contemporary Mass Spectrometry* (Weinheim: Wiley-VCH, 2007).

C. Dass, *Principles and Practice of Biological Mass Spectrometry* (New York: Wiley, 2001).

Protein Electrospray

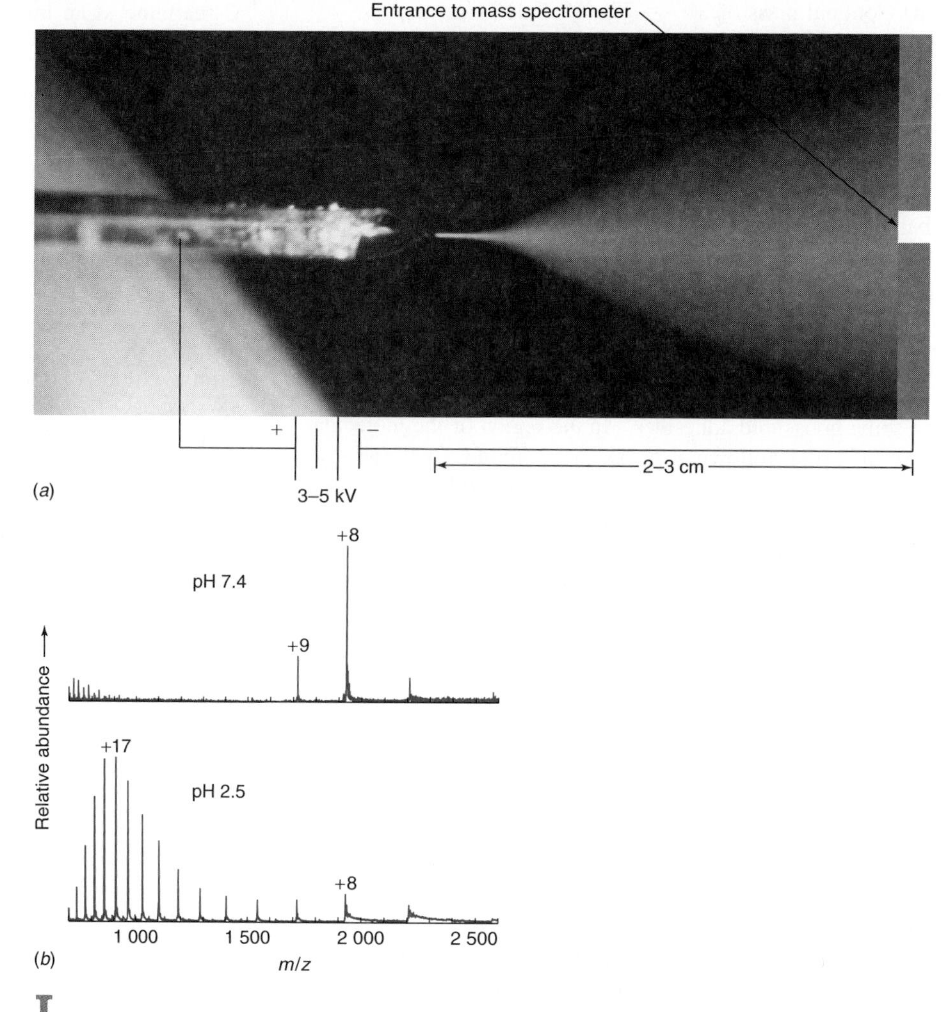

(a) Electrospray of liquid from a capillary held at a potential of ~5 kV with respect to the entrance of a nearby mass spectrometer. [Courtesy R. D. Smith, Pacific Northwest Laboratory, Richland, WA.]

(b) Electrospray mass spectrum of cellular retinoic acid binding protein I. [A. Dobo and I. Kaltashov, *Anal. Chem.* **2001**, *73*, 4763.]

Liquid exiting a chromatography column can be converted into the fine mist in panel *a* by electrospray, in which a high voltage is applied between the column and the entrance to a mass spectrometer. Micrometer-size droplets rapidly evaporate, leaving their solutes—including ions—free in the gas phase.

Electrospray has had a major impact in biological chemistry because it is one of the few means to introduce macromolecules into a mass spectrometer. Panel *b* shows the mass spectrum of a protein with a molecular mass near 16 000 Da. At pH 7.4, most of the molecules have a net positive charge of +8 from protonation of arginine, lysine, and histidine (Table 11-1). The +8 protein is observed at $m/z \approx 16\,000/8 = 2\,000$. At pH 2.5, many more amino acid substituent groups are protonated. The most frequent charge states are +17 and +18, and molecules with a charge of +22 are observed.

Electrospray was developed in the 1980s by John B. Fenn, who shared the Nobel Prize in 2002. More than 50 years earlier, Fenn's poor showing on an algebra exam came back with his teacher's comment, "Don't ever try to be a scientist or engineer."[1]

CHAPTER 22

Gas and Liquid Chromatography

Gas and liquid chromatography are workhorses of the analytical and environmental chemistry laboratories. This chapter describes equipment and basic techniques.[2]

22-1 Gas Chromatography

In **gas chromatography**, a gaseous mobile phase transports gaseous solutes through a long, thin column containing stationary phase. We begin the process in Figure 22-1 by injecting a volatile liquid through a rubber *septum* (a thin disk) into a heated port, which vaporizes the sample. The sample is swept through the column by He, N_2, or H_2 *carrier gas*, and the separated solutes flow through a detector, whose response is displayed on a computer. The column must be hot enough to produce sufficient vapor pressure for each solute to be eluted in a reasonable time. The detector is maintained at a higher temperature than that of the column so that all solutes are gaseous. The injected sample size for a liquid is typically 0.1–2 μL for analytical chromatography, whereas preparative columns can handle 20–1 000 μL. Gases can be introduced in volumes of 0.5–10 mL by a gas-tight syringe or a gas-sampling valve.

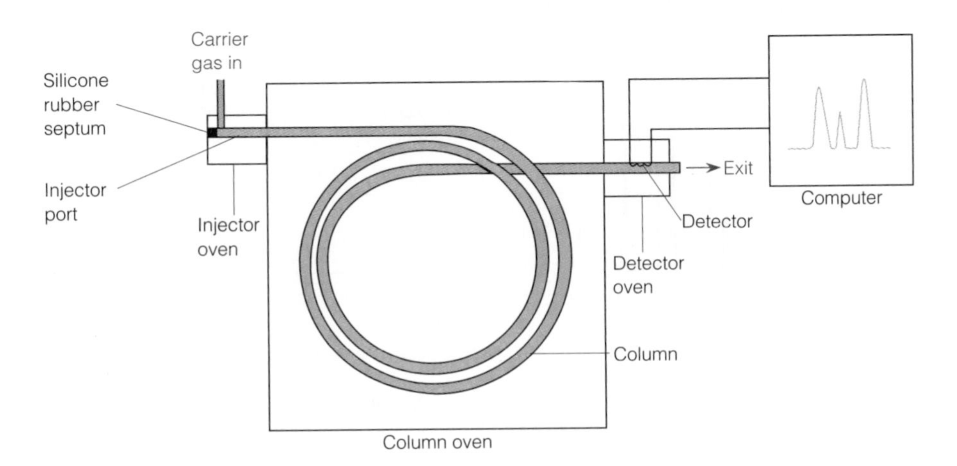

Figure 22-1 Schematic representation of a gas chromatograph.

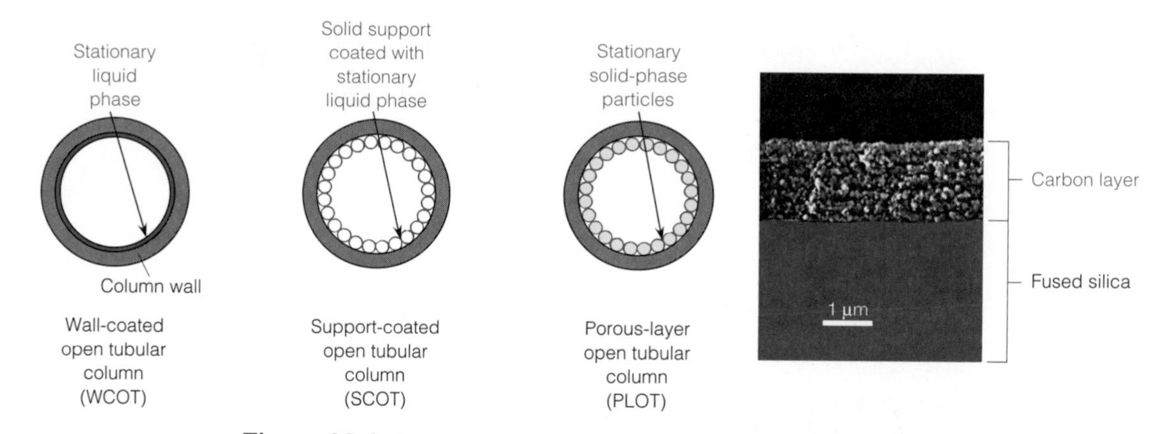

Figure 22-2 Cross-sectional view of wall-coated, support-coated, and porous-layer columns. Micrograph shows porous carbon stationary phase on inside wall of a fused-silica open tubular column.

Columns

Open tubular columns (Figure 21-11) have a liquid or solid stationary phase coated on the inside wall (Figure 22-2). Open tubular columns are usually made of fused silica (SiO_2). As the column ages, stationary phase bakes off and exposes silanol groups (Si—O—H) on the silica surface. The silanol groups strongly retain some polar compounds by hydrogen bonding, thereby causing *tailing* (Figure 21-12*b*) of chromatographic peaks. To reduce the tendency of a stationary phase to *bleed* from the column at high temperature, the stationary phase is normally *bonded* (chemically attached) to the silica surface and *cross-linked* to itself by covalent chemical bridges.

Polarity was discussed in Box 21-1.

Liquid stationary phases in Table 22-1 have a range of polarities. The choice of liquid phase for a given problem is based on the rule "like dissolves like." Nonpolar columns are usually best for nonpolar solutes, and polar columns are usually best for polar solutes.

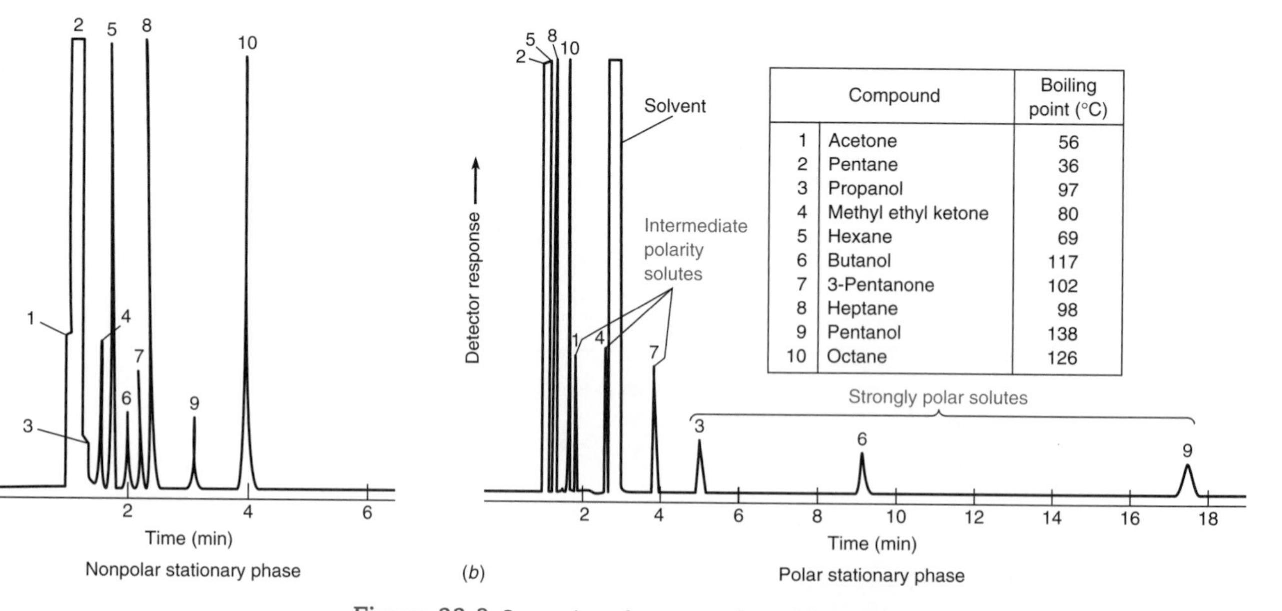

	Compound	Boiling point (°C)
1	Acetone	56
2	Pentane	36
3	Propanol	97
4	Methyl ethyl ketone	80
5	Hexane	69
6	Butanol	117
7	3-Pentanone	102
8	Heptane	98
9	Pentanol	138
10	Octane	126

Figure 22-3 Separation of compounds on (*a*) nonpolar poly(dimethylsiloxane) and (*b*) strongly polar polyethylene glycol stationary phases (1 μm thick) in open tubular columns (0.32 mm diameter × 30 m long) at 70°C. [Courtesy Restek Co., Bellefonte, PA.]

Table 22-1 Common stationary phases in capillary gas chromatography

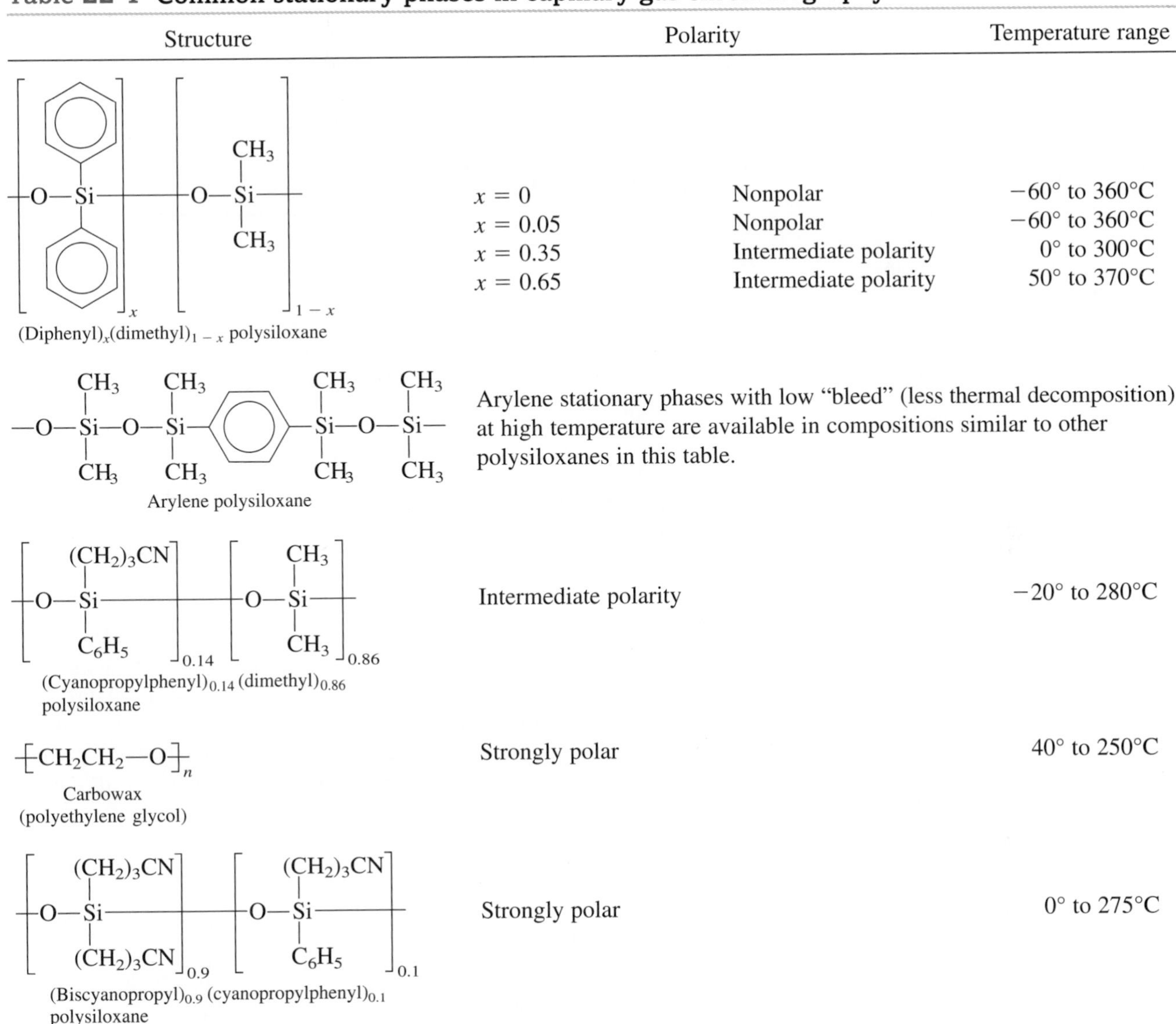

Structure		Polarity	Temperature range
(Diphenyl)$_x$(dimethyl)$_{1-x}$ polysiloxane	$x = 0$	Nonpolar	$-60°$ to 360°C
	$x = 0.05$	Nonpolar	$-60°$ to 360°C
	$x = 0.35$	Intermediate polarity	0° to 300°C
	$x = 0.65$	Intermediate polarity	50° to 370°C
Arylene polysiloxane	Arylene stationary phases with low "bleed" (less thermal decomposition) at high temperature are available in compositions similar to other polysiloxanes in this table.		
(Cyanopropylphenyl)$_{0.14}$ (dimethyl)$_{0.86}$ polysiloxane		Intermediate polarity	$-20°$ to 280°C
$\text{-}[CH_2CH_2\text{—}O]_n\text{-}$ Carbowax (polyethylene glycol)		Strongly polar	40° to 250°C
(Biscyanopropyl)$_{0.9}$ (cyanopropylphenyl)$_{0.1}$ polysiloxane		Strongly polar	0° to 275°C

We see the effects of column polarity on a separation in Figure 22-3. In Figure 22-3*a*, 10 compounds are eluted nearly in order of increasing boiling point from a nonpolar stationary phase: The higher the vapor pressure, the faster the compound is eluted. In Figure 22-3*b*, the strongly polar stationary phase strongly retains the polar solutes. The three alcohols (with —OH groups) are the last to be eluted, following the three ketones (with $\rangle$C=O groups), which follow four alkanes (having only C—H bonds). Hydrogen bonding between solute and the stationary phase is probably the strongest force causing retention. Figure 22-4 illustrates how changing the stationary phase can affect a separation. In this case, *trans* fatty acid methyl esters are resolved better on the stationary phase HP-88 than on DB-23, thus providing a more accurate measure of *trans* fat for food labels (Figure 5-3). Both stationary phases have intermediate polarity, but HP-88 has aryl groups (benzene rings) absent in DB-23.

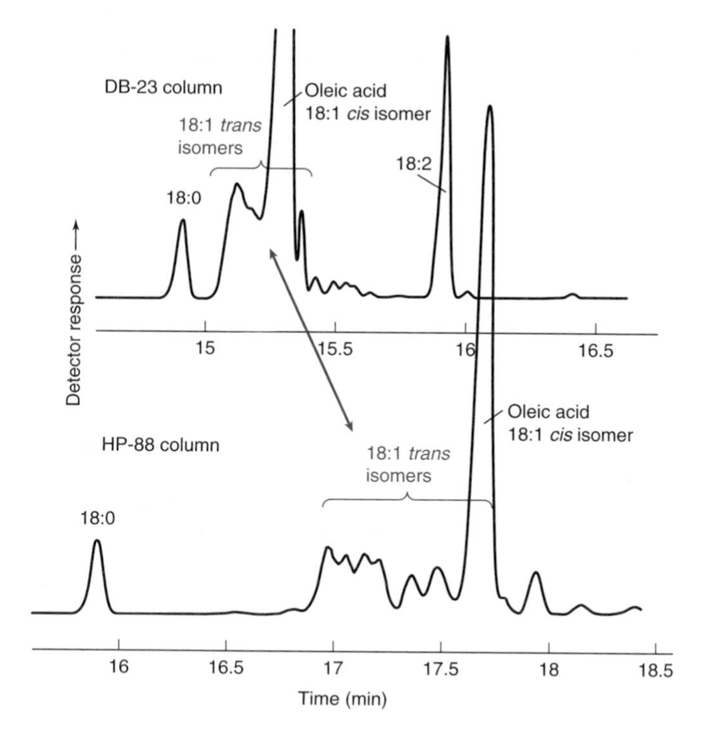

Figure 22-4 Resolution of *trans* fatty acids in hydrogenated food oil improves when the stationary phase is changed from DB-23 to HP-88. Both stationary phases have intermediate polarity with cyanopropyl and methyl groups, but HP-88 also has aryl groups. Fatty acids were converted (*derivatized*) to methyl esters to make them volatile enough for gas chromatography. The notation 18:*n* refers to an 18-carbon fatty acid with *n* double bonds. Oleic acid was shown in Figure 5-3. Wall-coated open tubular column (0.25 mm diameter × 60 m long with 0.20-μm-thick film) was run at 180°C with a flame ionization detector. [From A. K. Vickers, *Am. Lab. News Ed.*, January 2007, p. 18. Chromatograms courtesy P. Sandra and F. David.]

Common solid stationary phases include porous carbon (Figure 22-2 micrograph) and *molecular sieves*, which are inorganic materials with nanometer-size cavities that retain and separate small molecules such as H_2, O_2, N_2, CO_2, and CH_4. Figure 22-5 compares the separation of gases by molecular sieves in a wall-coated open tubular column and a *packed column* filled with particles of the solid stationary phase. Open tubular columns typically give better separations (narrower peaks), but packed columns

Molecular sieves are also used to dry gases because sieves strongly retain water. Sieves are regenerated (freed of water) by heating to 300°C in vacuum.

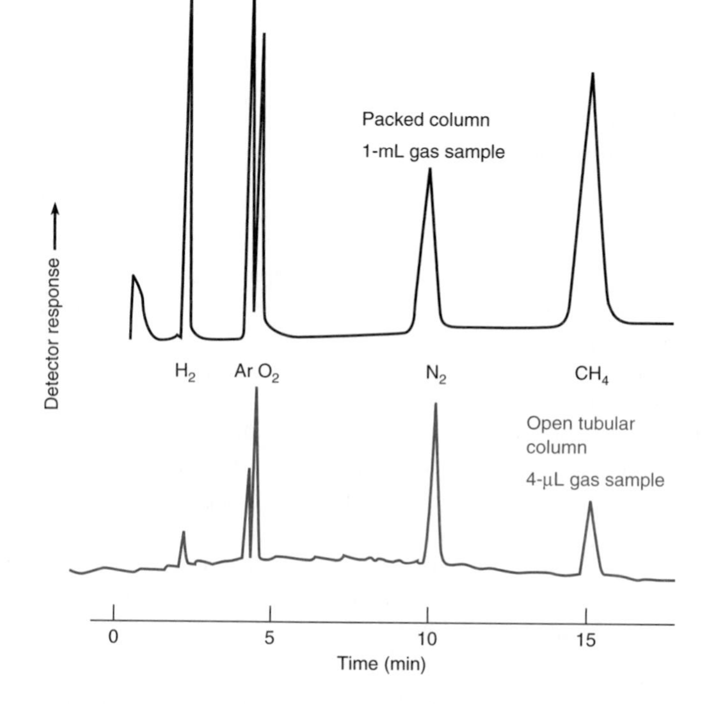

Figure 22-5 Gas chromatography with 5A molecular sieves. Upper chromatogram was obtained with a packed column (3.2 mm diameter × 4.6 m long) at 40°C, by using 1 mL of sample containing 2 ppm (by volume) of each analyte in He. Lower chromatogram was obtained with an open tubular column (0.32 mm diameter × 30 m long) at 30°C, by using 4 μL of the same sample. [From J. Madabushi, H. Cai, S. Steams, and W. Wentworth, *Am. Lab.* October 1995, p. 21.]

can handle larger samples. In Figure 22-5, the sample injected into the packed column was 250 times larger than the sample injected into the open tubular column.

Chromatographers often use a 5- to 10-m-long **guard column** attached to the front of a chromatography column. The guard column contains no stationary phase and its inner walls are *silanized* (page 469) to minimize retention of solutes. The purpose of the guard column is to collect nonvolatile components that would otherwise be injected into the chromatography column and never be eluted. Nonvolatile "junk" eventually ruins a chromatography column. A buildup of nonvolatiles in the guard column is manifested by distortion of chromatographic peaks. When this happens, we cut off and discard the beginning of the guard column.

Temperature Programming

Raising column temperature
- decreases retention time
- sharpens peaks

If you increase the column temperature in Figure 22-1, solute vapor pressure increases and retention times decrease. To separate compounds with a wide range of boiling points or polarities, we raise the column temperature *during* the separation, a technique called **temperature programming**. Figure 22-6 shows the effect of temperature programming on the separation of nonpolar compounds with a range of boiling points from 69°C for C_6H_{14} to 356°C for $C_{21}H_{44}$. At a constant column temperature of 150°C, low-boiling compounds emerge close together, and the higher-boiling compounds may not be eluted. If temperature is programmed to increase from 50° to 250°C, all compounds are eluted and the separation

We refer to constant-temperature conditions as *isothermal* conditions.

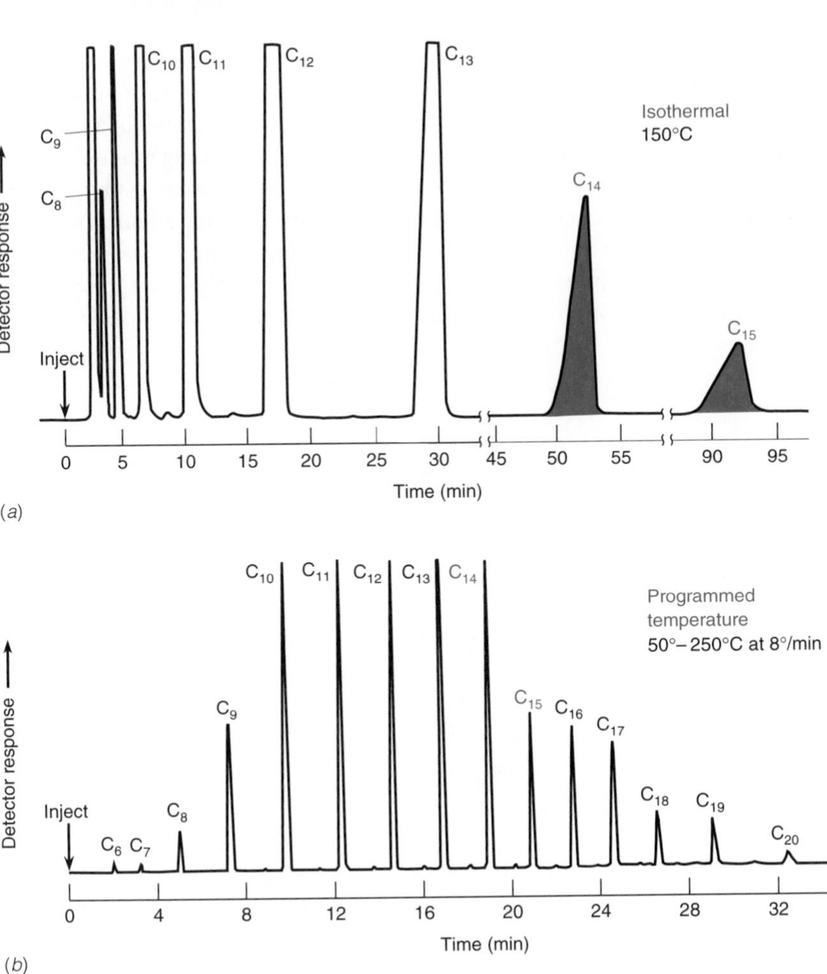

Figure 22-6 (*a*) Isothermal and (*b*) programmed temperature chromatography of linear alkanes through a packed column with a nonpolar stationary phase. Detector sensitivity is 16 times greater in (*a*) than in (*b*). [From H. M. McNair and E. J. Bonelli, *Basic Gas Chromatography* (Palo Alto, CA: Varian Instrument Division, 1968).]

of peaks is fairly uniform. Even though 250°C is below the boiling point of some compounds in the mixture, these compounds have sufficient vapor pressure to be eluted.

Carrier Gas

Figure 21-9 showed that H_2 and He give better resolution (smaller plate height) than N_2 at high flow rate. The reason is that solutes diffuse more rapidly through H_2 and He and therefore equilibrate between the mobile and stationary phases more rapidly than they can in N_2. To help protect the stationary phase, carrier gas is passed through traps to remove traces of O_2, H_2O, and hydrocarbons prior to entering the chromatograph. Metal tubing is strongly recommended over any kind of plastic tubing to maintain gas purity.

Sample Injection

Liquids are injected through a rubber septum into a heated glass port. Carrier gas sweeps the vaporized sample into the chromatography column. A complete injection usually contains too much material for an open tubular capillary column. In **split injection** (Figure 22-7*a*), only 0.1–10% of the injected sample reaches the column. The remainder is blown out to waste. If the entire sample is not vaporized during injection, however, higher boiling components are not completely injected and there will be errors in quantitative analysis.

Injection into open tubular columns:
- *split:* routine method
- *splitless:* best for quantitative analysis
- *on-column:* best for thermally unstable solutes

For quantitative analysis and for analysis of trace components of a mixture, **splitless injection** (Figure 22-7*b*) is appropriate. (*Trace components* are those present at extremely low concentrations.) For this purpose, a dilute sample in a low-boiling solvent is injected at a column temperature 40° below the boiling point of the solvent. Solvent condenses at the beginning of the column and traps a thin band of solute. (Hence, this technique is called **solvent trapping**.) After additional vapors have been purged from the injection port, the column temperature is raised and chromatography is begun. In splitless injection, ~80% of the sample is applied to the column, and little fractionation (selective evaporation of components) occurs during injection.

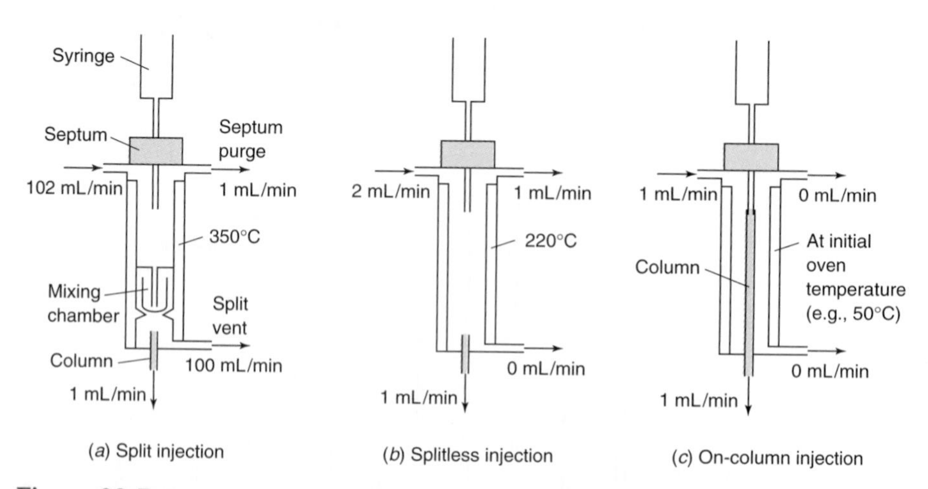

Figure 22-7 Injection port operation for (*a*) split, (*b*) splitless, and (*c*) on-column injection into an open tubular column. A slow flow of gas past the inside surface of the septum out to waste cools the rubber and prevents volatile emissions from the rubber from entering the chromatography column.

A technique called **cold trapping** is used to focus high-boiling solutes at the beginning of a column. In this case, the column is initially 150° lower than the boiling points of solutes of interest. Solvent and low-boiling solutes are eluted rapidly, but high-boiling solutes condense in a narrow band at the start of the column. The column is later warmed to initiate chromatography for the components of interest.

For sensitive compounds that decompose above their boiling temperature, we use **on-column injection** of solution directly into the column (Figure 22-7*c*), without going through a hot port. Analytes are focused in a narrow band by solvent trapping or cold trapping. Warming the column initiates chromatography.

"Wide-bore" columns (diameter ≥0.53 mm) are wide enough to accept the common syringe needle for on-column injection. Narrower columns (typically 0.10–0.32 mm diameter) provide higher resolution (sharper peaks) but have less capacity and require higher operating pressure. Diameters ≥0.32 mm are too large to be used with mass spectral detection because the mass flow rate is too high for most vacuum pumps to handle.

Flame Ionization Detector

In the **flame ionization detector** in Figure 22-8, eluate is burned in a mixture of H_2 and air. Carbon atoms (except carbonyl and carboxyl carbon atoms) produce CH radicals, which go on to produce CHO^+ ions in the flame:

$$CH + O \longrightarrow CHO^+ + e^-$$

The flow of ions and electrons to the electrodes produces the detector signal. Only about 1 in 10^5 carbon atoms produces an ion, but ion production is proportional to the number of susceptible carbon atoms entering the flame. Response to organic compounds is proportional to solute mass over seven orders of magnitude. The detector is sensitive enough for narrow-bore columns. It responds to most hydrocarbons and is insensitive to nonhydrocarbons such as H_2, He, N_2, O_2, CO, CO_2, H_2O, NH_3, NO, H_2S, and SiF_4. The detection limit is 100 times smaller than that of the thermal conductivity detector and is best with N_2 carrier gas. For open tubular columns eluted with H_2 or He, N_2 *makeup gas* is added to the stream before the stream reaches the detector. Makeup gas provides the higher flow rate needed by the detector and improves sensitivity.

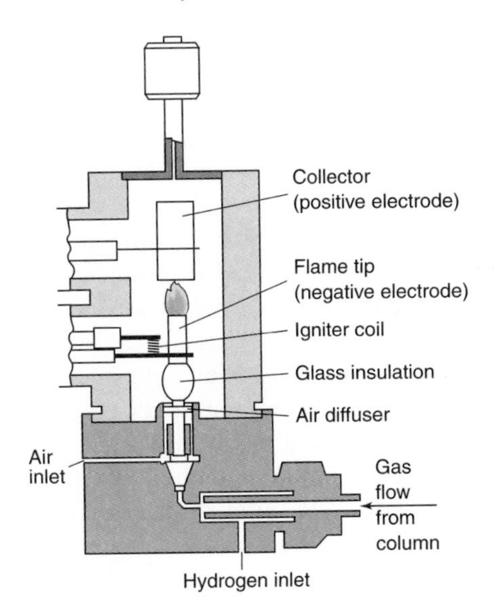

Figure 22-8 Flame ionization detector. [Courtesy Varian Associates, Palo Alto, CA.]

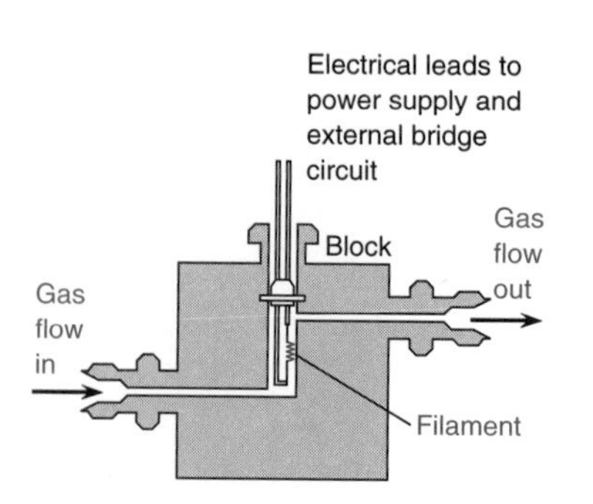

Figure 22-9 Thermal conductivity detector. [Courtesy Varian Associates, Palo Alto, CA.]

Thermal Conductivity Detector

Thermal conductivity measures the ability of a substance to transport heat. In the **thermal conductivity detector** in Figure 22-9, gas emerging from the chromatography column flows over a hot tungsten-rhenium filament. When solute emerges from the column, the thermal conductivity of the gas stream decreases, the filament gets hotter, its electrical resistance increases, and the voltage across the filament increases. The voltage change is the detector signal. Thermal conductivity detection is more sensitive at lower flow rates. To prevent overheating and oxidation of the filament, the detector should never be left on unless carrier gas is flowing.

The detector responds to *changes* in thermal conductivity, so the conductivities of solute and carrier gas should be as different as possible. Because H_2 and He have the highest thermal conductivity, these are the carriers of choice for thermal conductivity detection. A thermal conductivity detector responds to every substance except the carrier gas.

Thermal conductivity detectors are generally not sensitive enough to detect the small quantity of analyte eluted from open tubular columns smaller than 0.53 mm in diameter. For narrower columns, other detectors must be used.

Electron Capture Detector

Electron capture detector:

$$^{63}\text{Ni} \longrightarrow \underset{\text{High-energy electron}}{\beta^-}$$

$$\beta^- + N_2 \longrightarrow \underset{\text{Collected at anode}}{N_2^+ + 2e^-}$$

$$\text{analyte} + e^- \longrightarrow \underset{\text{Too slow to reach anode}}{\text{analyte}^-}$$

The **electron capture detector** is extremely sensitive to halogen-containing molecules, such as chlorinated pesticides, but relatively insensitive to hydrocarbons, alcohols, and ketones. Carrier gas entering the detector is ionized by high-energy electrons ("β-rays") emitted from a foil containing radioactive ^{63}Ni. Electrons liberated from the gas are attracted to an anode, producing a small, steady current. When analyte molecules with a high electron affinity enter the detector, they capture some of the electrons and reduce the current. The detector responds by varying the frequency of voltage pulses between anode and cathode to maintain constant current. Electron capture is extremely sensitive, with a detection limit of ~5 fg (femtogram, 10^{-15} g), comparable to that of selected ion monitoring by mass spectrometry. The carrier gas is usually N_2 or 5 vol% CH_4 in Ar. For open tubular columns, chromatography is conducted with H_2 or He at low flow rate and N_2 makeup gas is added to the stream prior to the detector.

Other Detectors

Gas chromatography detectors:
- *flame ionization:* responds to compounds with C—H
- *thermal conductivity:* responds to everything, but not sensitive enough for columns <0.53 mm in diameter
- *electron capture:* halogens, conjugated C=O, —C≡N, $—NO_2$
- *flame photometer:* P and S
- *alkali flame:* P and N
- *sulfur chemiluminescence:* S
- *mass spectrometer:* responds to everything

A *flame photometric detector* measures optical emission from phosphorus and sulfur compounds. When eluate passes through a H_2-air flame, excited sulfur- and phosphorus-containing species emit characteristic radiation, which is detected with a photomultiplier tube. Radiant emission is proportional to analyte concentration.

The *alkali flame detector*, also called a *nitrogen-phosphorus* detector, is a modified flame ionization detector that is selectively sensitive to phosphorus and nitrogen. It is especially important for drug analysis. Ions such as NO_2^-, CN^-, and PO_2^-, produced by these elements when they contact a Rb_2SO_4-containing glass bead at the burner tip create the current that is measured. N_2 from air is inert to this detector and does not interfere. The bead must be replaced periodically because Rb_2SO_4 is consumed.

A *sulfur chemiluminescence detector* mixes the exhaust from a flame ionization detector with O_3 to form an excited state of SO_2 that emits light, which is detected.

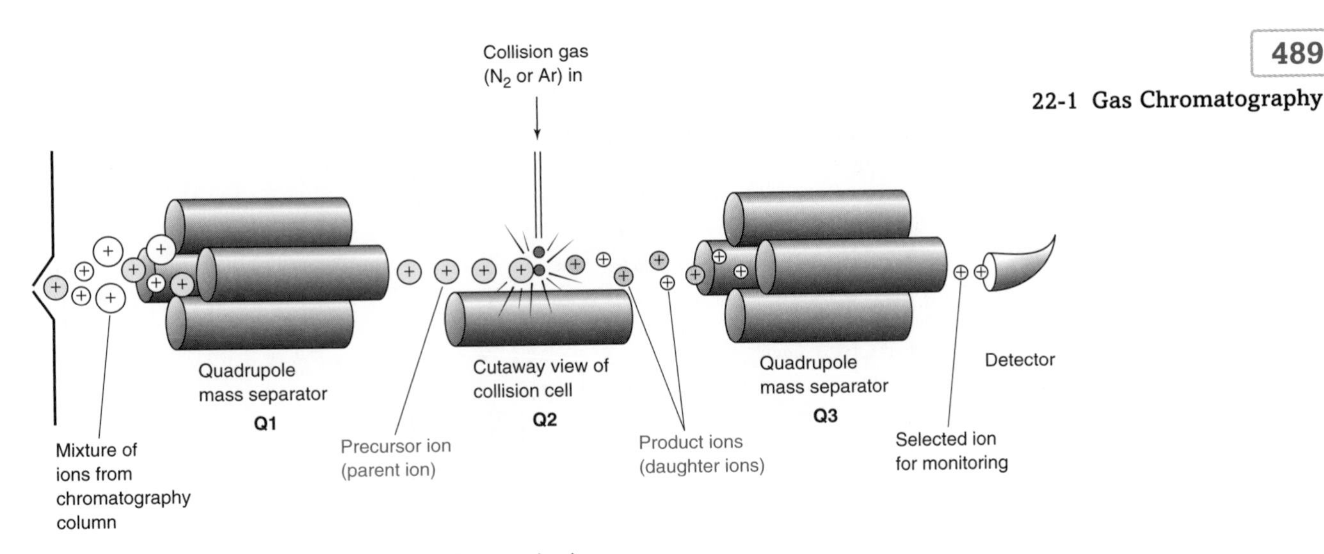

Figure 22-10 Principle of selected reaction monitoring.

Mass Spectrometric Detection and Selected Reaction Monitoring

A *mass spectrometer* such as that described in Section 21-4 is the single most versatile detector. The *reconstructed total ion chromatogram* (Figure 21-16*a*) shows all components in a mixture. The mass spectrum of each component recorded as it is eluted provides qualitative identification. Alternatively, *selected ion monitoring* at one value of *m/z* (Figures 21-16*b–h*) responds to one or a few components of a mixture. Because it does not respond to everything in the sample, selected ion monitoring reduces interference by overlapping chromatographic peaks.

Interference is further reduced and the *signal-to-noise ratio* is further increased by a powerful mass spectral technique called **selected reaction monitoring**. Figure 22-10 shows a *triple quadrupole mass spectrometer* in which a mixture of ions enters quadrupole Q1, which passes just one selected *precursor ion* to the second stage, Q2. The second stage passes all ions of all masses straight on to the third stage, Q3. However, while inside Q2, which is called a *collision cell*, the precursor ion collides with N_2 or Ar at a pressure of $\sim 10^{-8}$ to 10^{-6} bar and breaks into fragments called *product ions*. Quadrupole Q3 allows only specific product ions to reach the detector.

Signal arises from what we seek to measure, and *noise* is random variation in the instrument response. At the *signal detection limit* (Section 5-2), signal is 3 times greater than noise. At the *lower limit of quantitation*, the signal-to-noise ratio is 10, which we can measure with moderate precision.

Selected reaction monitoring is extremely selective for the analyte of interest. For example, we can monitor traces of caffeine in natural waters to detect contamination by domestic wastewater. The global daily average consumption of caffeine in Europe and North America is ~200–400 mg per person, much of which ends up in sewage. If caffeine shows up in a municipal water supply, it probably migrated from wastewater. To measure parts per trillion levels of caffeine, 1 L of water was passed through a column containing 10 mL of adsorbent polystyrene beads that retain caffeine and a host of other organic compounds. Caffeine and many other substances were then eluted with organic solvent, dried, and evaporated to 0.1–1 mL. By this *solid-phase extraction* procedure, caffeine was *preconcentrated* by a factor of 10^3–10^4 prior to chromatography.

Because it uses two consecutive mass spectrometers, selected reaction monitoring is also called *tandem mass spectrometry* or *mass spectrometry–mass spectrometry*, or just *ms-ms*.

Figure 22-11*a* shows the electron ionization mass spectrum of caffeine. For selected reaction monitoring, the *m/z* 194 precursor ion was selected by quadrupole

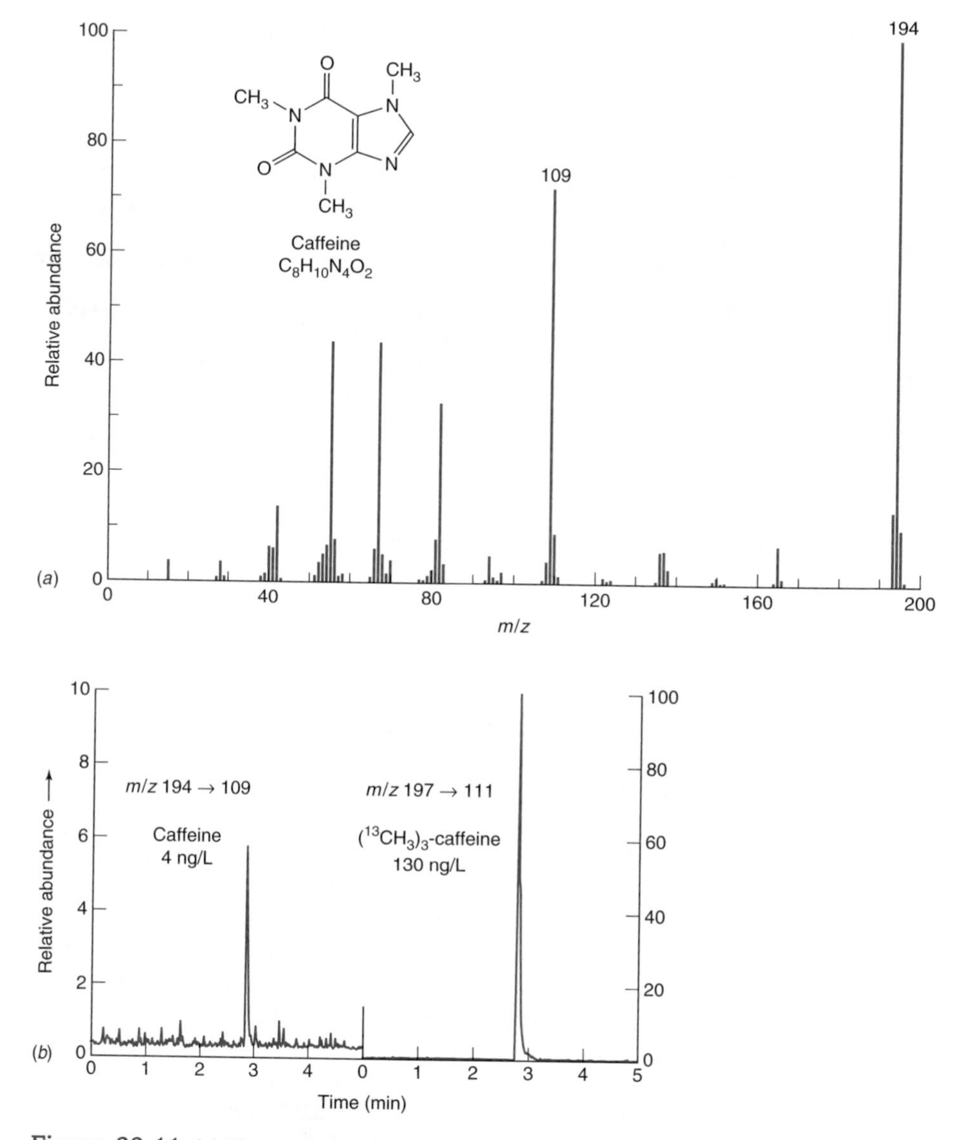

Figure 22-11 (*a*) Electron ionization mass spectrum of caffeine. [From NIST/EPA/NIH Mass Spectral Database.] (*b*) Selected reaction monitoring gas chromatogram of water from 5 m below the surface of the Mediterranean Sea containing 4 ng of caffeine per liter. [From I. J. Buerge, T. Poiger, M. D. Müller, and H.-R. Buser, *Environ. Sci. Technol.* **2003**, *37*, 691.]

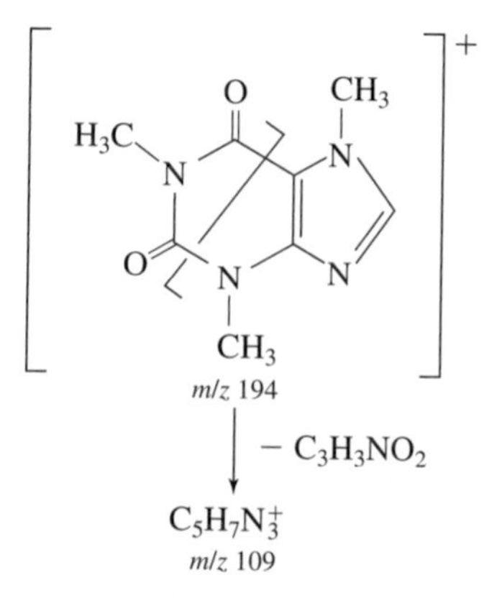

Q1 in Figure 22-10 and the *m/z* 109 product ion was selected by Q3 for detection. Few compounds other than caffeine produce the *m/z* 194 precursor ion, and it would be rare for any of them to give the *m/z* 109 product ion because they are unlikely to break into the same fragments as caffeine. The chromatogram in Figure 22-11*b* shows only one significant peak, even though the caffeine content was only 4 ng/L (parts per trillion) in the water sample and, undoubtedly, many other compounds were present at higher concentrations.

For quantitative analysis, $^{13}C_3$-caffeine was injected into the initial water sample as an *internal standard.* This isotopic molecule is eluted at the same time as ordinary caffeine, but the two are detected separately by selected reaction monitoring. The internal standard is detected by its precursor ion at *m/z* 197 and product ion at *m/z* 111. The paper cited in Figure 22-11 found that 1–4% of caffeine discharged by the population in the catchment area of one Swiss lake entered the lake. Most of the caffeine apparently entered the lake during rain events when the capacity of wastewater treatment plants was exceeded and sewage flowed directly into the lake.

Ask Yourself

22-A. **(a)** What is the advantage of temperature programming in gas chromatography?
(b) What is the advantage of an open tubular column over a packed column? Does a narrower or wider open tubular column provide higher resolution? What is the advantage of a packed column over an open tubular column?
(c) Why do H_2 and He allow more rapid linear flow rates in gas chromatography than does N_2, without loss of column efficiency (see Figure 21-9)?
(d) When would you use split, splitless, or on-column injection?
(e) To which kinds of analytes do the following detectors respond: **(i)** flame ionization; **(ii)** thermal conductivity; **(iii)** electron capture; **(iv)** flame photometric; **(v)** alkali flame; **(vi)** sulfur chemiluminescence; **(vii)** mass spectrometer?
(f) What information is displayed in a reconstructed total ion chromatogram, selected ion monitoring, and selected reaction monitoring? Which technique is most selective and which is least selective and why?

22-2 Classical Liquid Chromatography

Modern chromatography evolved from the experiment in Figure 21-1 in which sample is applied to the top of an open, gravity-fed column containing stationary phase. The next section describes high-performance liquid chromatography, which uses closed columns under high pressure and is the most common form of liquid chromatography today. However, open columns are used for preparative separations in biochemistry and chemical synthesis.

There is an art to pouring uniform columns, applying samples evenly, and obtaining symmetric elution bands. Stationary solid phase is normally poured into a column by first making a *slurry* (a mixture of solid and liquid) and pouring the slurry gently down the wall of the column. Try to avoid creating distinct layers, which form when some of the slurry is allowed to settle before more is poured in. Do not drain solvent below the top of the stationary phase, because air spaces and irregular flow patterns will be created. Solvent should be directed gently down the wall of the column. In *no* case should the solvent be allowed to dig a channel into the stationary phase. Maximum resolution demands slow flow.

Stationary Phase

For adsorption chromatography, *silica* ($SiO_2 \cdot xH_2O$, also called silicic acid) is a common stationary phase. Its active adsorption sites are Si—O—H (silanol) groups, which are slowly deactivated by adsorption of water from the air. Silica is activated by heating to 200°C to drive off the water. *Alumina* ($Al_2O_3 \cdot xH_2O$) is the other common adsorbent. Preparative chromatography in the biochemistry lab is most often based on molecular exclusion and ion exchange, described in Chapter 23.

Solvents

In *adsorption chromatography*, solvent competes with solute for adsorption sites on the stationary phase. *The relative abilities of different solvents to elute a given solute*

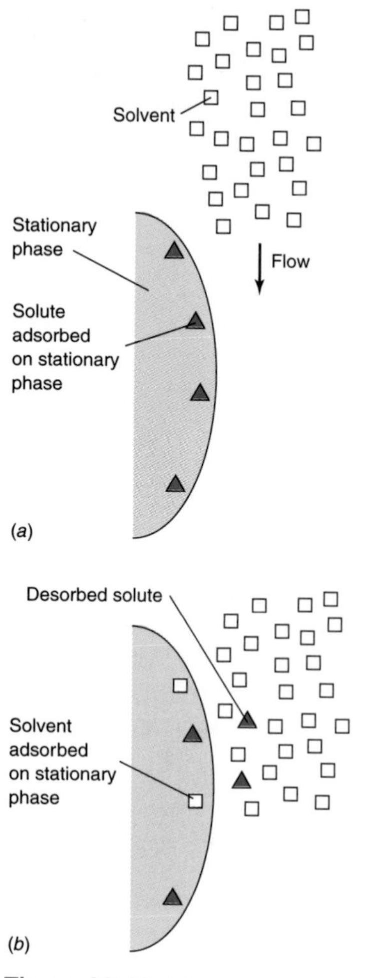

Figure 22-12 Solvent molecules compete with solute molecules for binding sites on the stationary phase. The more strongly the solvent binds to the stationary phase, the greater the *eluent strength* of the solvent.

Gradient elution in liquid chromatography is analogous to temperature programming in gas chromatography. Increased eluent strength is required to elute more strongly retained solutes.

Table 22-2 Eluotropic series and ultraviolet cutoff wavelengths of solvents for adsorption chromatography on silica

Solvent	Eluent strength ($\varepsilon°$)	Ultraviolet cutoff (nm)
Pentane	0.00	190
Hexane	0.01	195
Heptane	0.01	200
Trichlorotrifluoroethane	0.02	231
Toluene	0.22	284
Chloroform	0.26	245
Dichloromethane	0.30	233
Diethyl ether	0.43	215
Ethyl acetate	0.48	256
Methyl *t*-butyl ether	0.48	210
Dioxane	0.51	215
Acetonitrile	0.52	190
Acetone	0.53	330
Tetrahydrofuran	0.53	212
2-Propanol	0.60	205
Methanol	0.70	205

Ultraviolet cutoff is the approximate minimum wavelength at which solutes can be detected above the strong ultraviolet absorbance of solvent. The ultraviolet cutoff for water is 190 nm.

SOURCES: L. R. Snyder in *High-Performance Liquid Chromatography* (C. Horváth, ed.), vol. 3 (New York: Academic Press, 1983); *Burdick & Jackson Solvent Guide*, 3rd ed. (Muskegon, MI: Burdick & Jackson Laboratories, 1990).

from the column are nearly independent of the nature of the solute. Elution can be described as a displacement of solute from the adsorbent by solvent (Figure 22-12).

An *eluotropic series* ranks solvents by their relative abilities to displace solutes from a given adsorbent. **Eluent strength** in Table 22-2 is a measure of solvent adsorption energy, with the value for pentane defined as 0. The more polar the solvent, the greater its eluent strength. The greater the eluent strength, the more rapidly solutes will be eluted from the column.

A *gradient* (steady change) of eluent strength is used for many separations. First, weakly retained solutes are eluted with a solvent of low eluent strength. Then a second solvent is mixed with the first, either in discrete steps or continuously, to increase eluent strength and elute more strongly adsorbed solutes. A small amount of polar solvent markedly increases the eluent strength of a nonpolar solvent.

Ask Yourself

22-B. Why are the relative eluent strengths of solvents in adsorption chromatography fairly independent of solute?

22-3 High-Performance Liquid Chromatography

High-performance liquid chromatography (HPLC) uses high pressure to force eluent through a closed column packed with micrometer-size particles that provide exquisite separations. The analytical HPLC equipment in Figure 22-13 uses columns

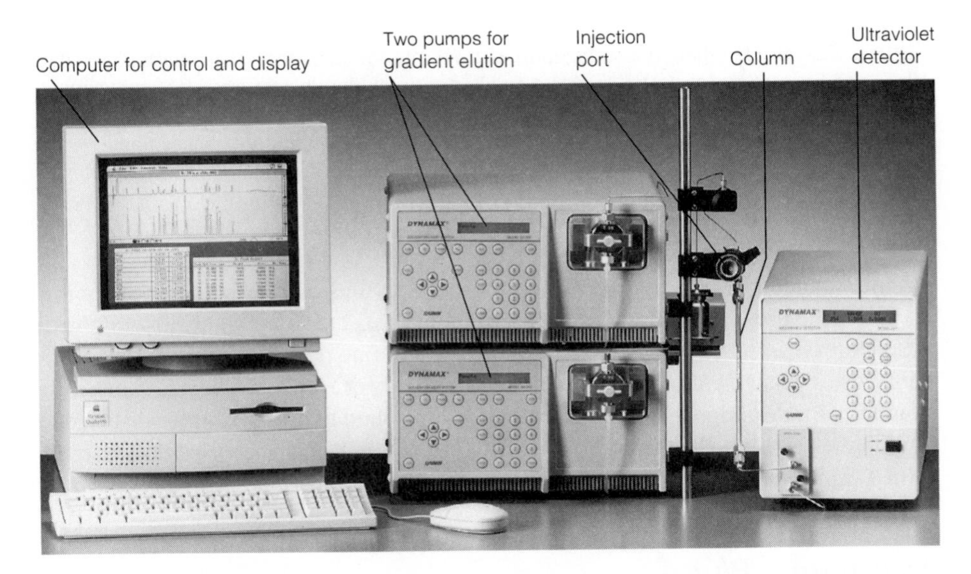

Figure 22-13 Typical laboratory equipment for high-performance liquid chromatography (HPLC). [Courtesy Rainin Instrument Co., Emeryville, CA.]

Figure 22-14 A 300-liter preparative chromatography column can purify a kilogram of material. [Courtesy Prochrom, Inc., Indianapolis, IN.]

Question According to the scaling rules in Section 21-2, if column diameter is increased from 4 mm to 40 mm, how much larger can the sample be to achieve the same resolution?

with diameters of 1–5 mm and lengths of 5–30 cm, yielding 50 000 to 100 000 plates per meter. Essential components include a solvent delivery system, a sample injection valve, a detector, and a computer to control the system and display results. The industrial preparative column in Figure 22-14 can handle as much as 1 kg of sample.

Under optimum conditions, resolution increases as stationary-phase particle size decreases. Notice how much sharper the peaks become in Figure 22-15 when particle

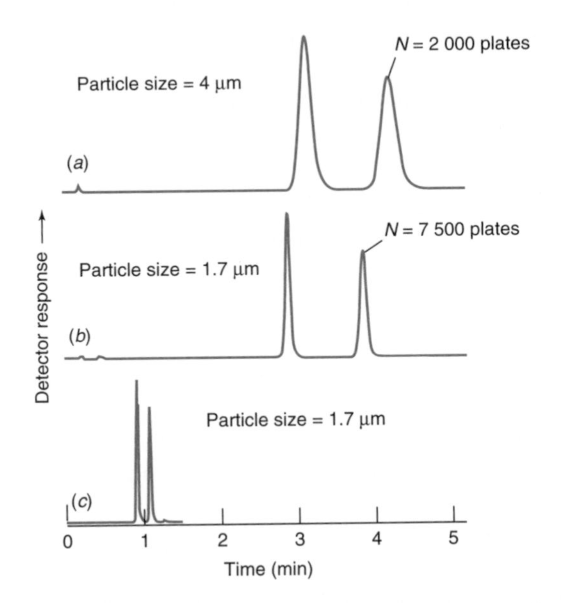

Figure 22-15 (*a* and *b*) Chromatograms of the same sample run at the same linear velocity on 5.0-cm-long columns packed with C_{18}-silica of different particle size. (*c*) A stronger solvent was used to elute solutes more rapidly from the column in panel *b*. [Y. Yang and C. C. Hodges, "Assay Transfer from HPLC to UPLC for Higher Analysis Throughput," *LCGC Supplement*, May 2005, p. 31.]

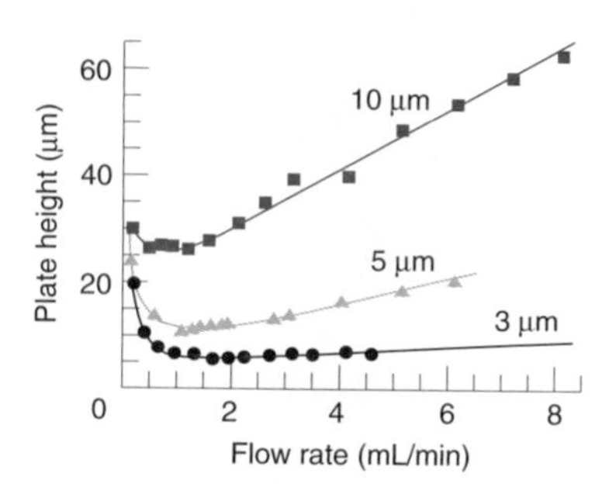

Figure 22-16 Plate height as a function of flow rate for stationary-phase particle sizes of 10, 5, and 3 μm. Smaller particles give a smaller optimal plate height, thereby producing sharper chromatographic peaks. Particle sizes of 5 or 3 μm are used routinely. Strictly speaking, comparison should be made for *linear flow rate* (cm/min), not *volume flow rate* (mL/min). However, all columns in this example have the same diameter, so volume flow rate is proportional to linear flow rate. [Courtesy Perkin-Elmer Corp., Norwalk, CT.]

Decreased particle size increases resolution but requires high pressure to obtain a reasonable flow rate.

size is changed from 4 μm to 1.7 μm. Figure 22-16 illustrates how decreasing particle size decreases optimum plate height and allows us to run a higher flow rate with little increase in place height. The penalty for using fine particles is resistance to solvent flow. Pressures of ~70–1 000 bar are required for flow rates of 0.5–5 mL/min. Smaller particles improve resolution by allowing solute to diffuse shorter distance to equilibrate between the stationary and mobile phases. Small particles decrease the term C in the van Deemter equation 21-7. Also, smaller particles decrease irregular flow paths (the term A).

Stationary Phase

Normal-phase chromatography: polar stationary phase and less polar solvent

Reversed-phase chromatography: low-polarity stationary phase and polar solvent.

Normal-phase chromatography uses a polar stationary phase and a less polar solvent. *Eluent strength is increased by adding a more polar solvent.* **Reversed-phase chromatography** is the more common scheme in which the stationary phase is nonpolar or weakly polar and the solvent is more polar. *Eluent strength is increased by adding a less polar solvent.* In general, eluent strength is increased by making the mobile phase more like the stationary phase. Reversed-phase chromatography eliminates tailing (Figure 21-12*b*) arising from adsorption of polar compounds by polar packings. Reversed-phase chromatography is also relatively insensitive to polar impurities (such as water) in the eluent.

Microporous particles of silica with diameters of 1.5–10 μm are the most common solid stationary-phase support. These particles are permeable to solvent and have a surface area as large as 500 m^2 per gram of silica. Solute adsorption occurs directly on the silica surface.

Most commonly, liquid-liquid partition chromatography is conducted with a **bonded stationary phase** covalently attached to silanol groups on the silica surface.

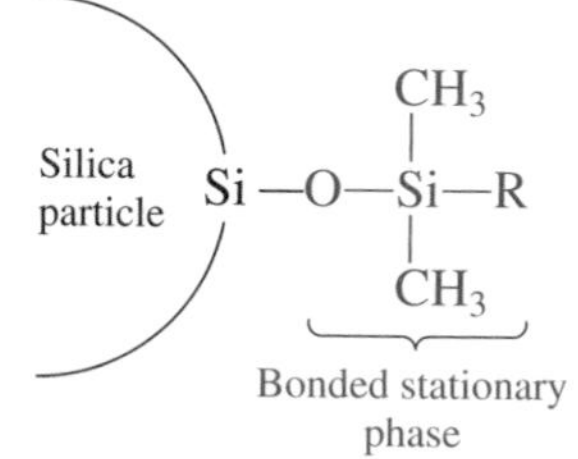

Bonded polar phases		Bonded nonpolar phases	
$R = (CH_2)_3NH_2$	amino	$R = (CH_2)_{17}CH_3$	octadecyl
$R = (CH_2)_3C{\equiv}N$	cyano	$R = (CH_2)_7CH_3$	octyl
$R = (CH_2)_2OCH_2CH(OH)CH_2OH$	diol	$R = (CH_2)_3C_6H_5$	phenyl

The octadecyl (C_{18}) stationary phase is, by far, the most common in HPLC. The Si—O—Si bond that attaches the stationary phase to the silica is stable only over the pH range 2–8. Strongly acidic or basic eluents generally cannot be used with silica.

Optical isomers are mirror image compounds such as D- and L-amino acids that cannot be superimposed on one another. Most compounds with four different groups attached to one tetrahedral carbon atom exist in two mirror image (optical) isomers. Optical isomers can be separated from each other by chromatography on a stationary phase containing just one optical isomer of the bonded phase (called a *chiral* column). The pharmaceutical industry is motivated to separate optical isomers of some drugs because one isomer could be biologically active while the other is inactive or toxic. Figure 22-17 shows separation of the two optical isomers of the anti-inflammatory drug Naproxen.

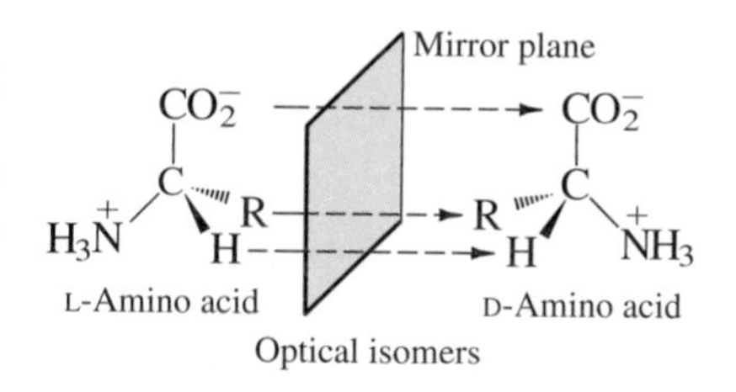

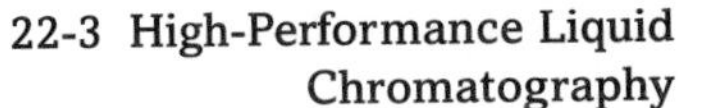

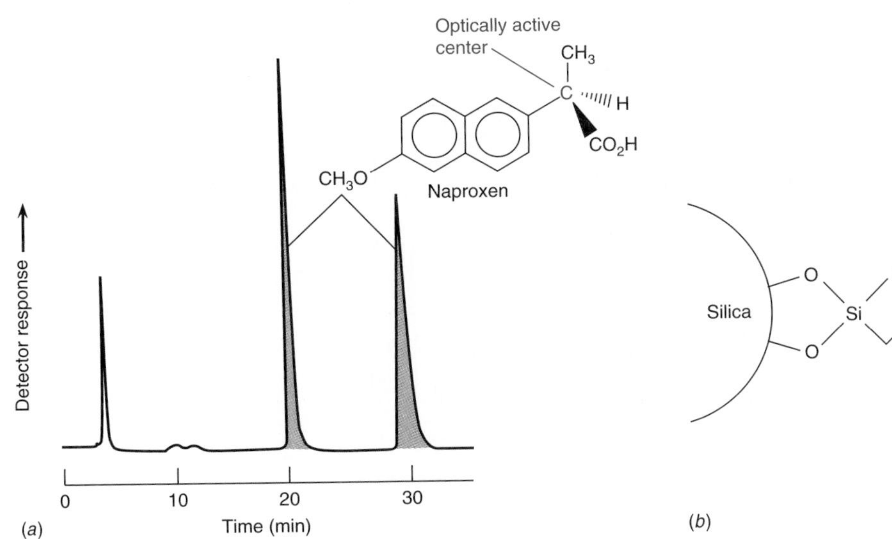

Figure 22-17 (*a*) HPLC separation of the two optical isomers (mirror image isomers) of the drug Naproxen eluted with 0.05 M ammonium acetate in methanol. Naproxen is the active ingredient of the anti-inflammatory drug Aleve®. (*b*) Structure of the bonded stationary phase. [Courtesy Phenomenex, Torrance, CA.]

The Column

High-performance liquid chromatography columns are expensive and easily degraded by irreversible adsorption of impurities from samples and solvents. Therefore we place a short, disposable **guard column** containing the same stationary phase as that of the main column at the entrance to the main column (Figure 22-18). Substances that would bind irreversibly to the main column are instead bound in the guard column, which is periodically discarded.

Because the column is under high pressure, a special technique is required to inject sample. The *injection valve* in Figure 22-19 has interchangeable steel sample

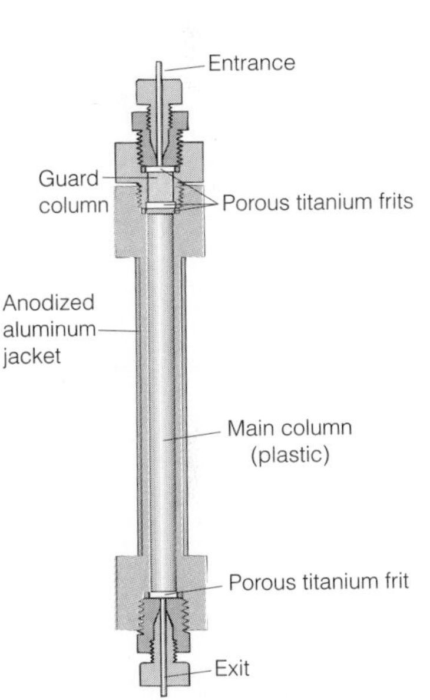

Figure 22-18 HPLC column with replaceable guard column to collect irreversibly adsorbed impurities. Titanium frits distribute the liquid evenly over the diameter of the column. [Courtesy Upchurch Scientific, Oak Harbor, WA.]

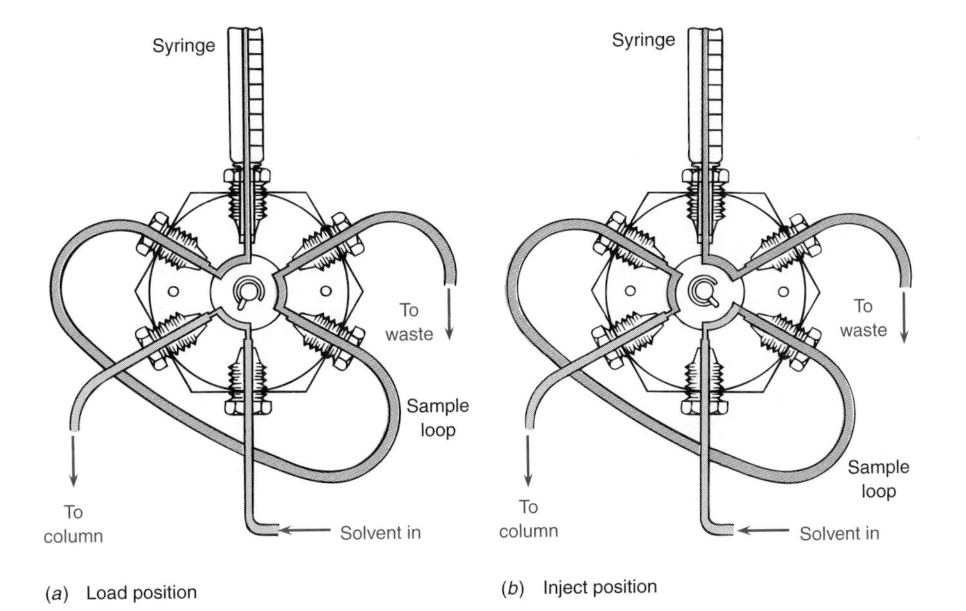

Figure 22-19 Injection valve for HPLC. Replaceable sample loop comes in various fixed-volume sizes.

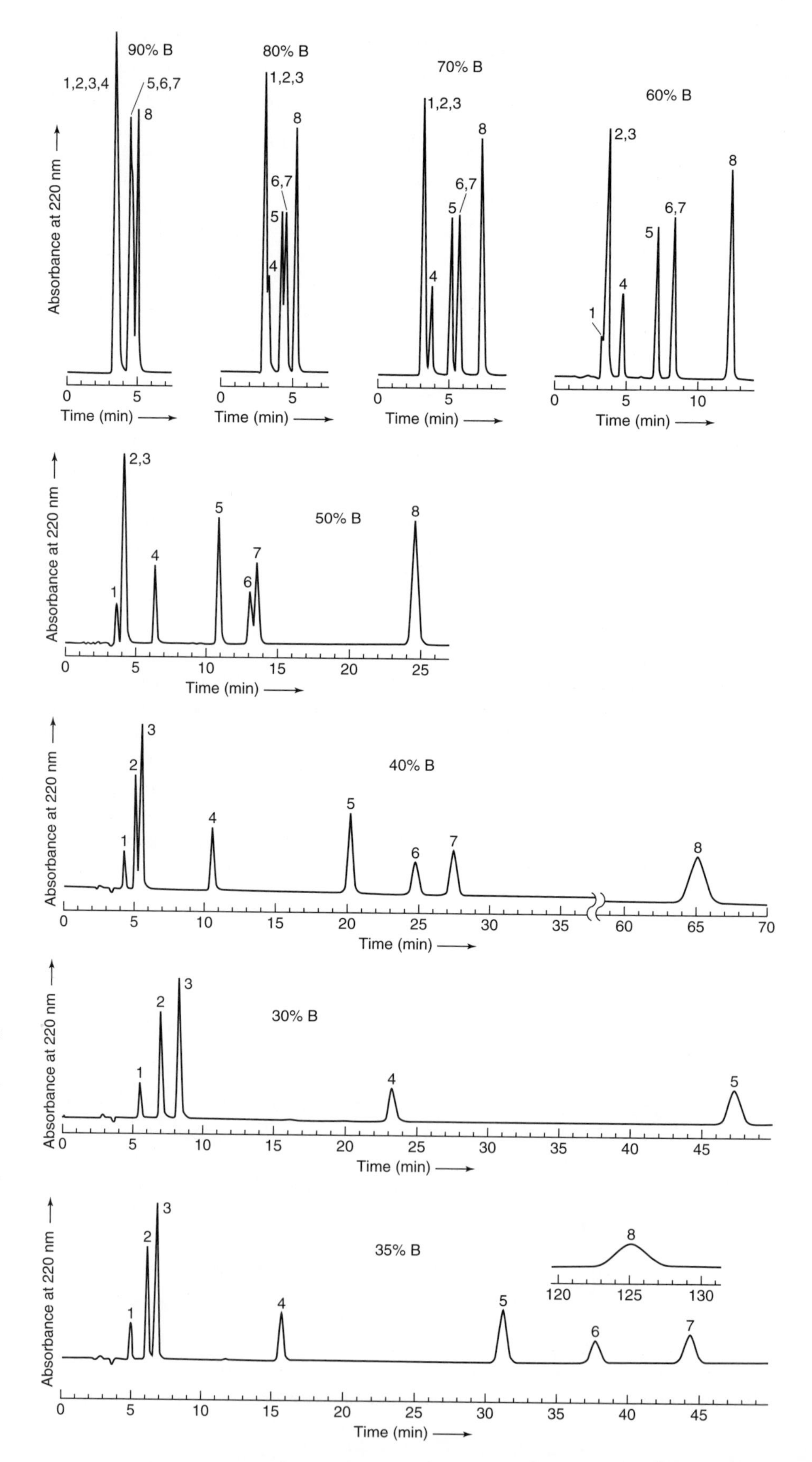

Figure 22-20 Isocratic HPLC separation of a mixture of aromatic compounds at 1.0 mL/min on a 0.46 × 25 cm Hypersil ODS column (C_{18} on 5-μm silica) at ambient temperature (~22°C): (1) benzyl alcohol; (2) phenol; (3) 3′,4′-dimethoxyacetophenone; (4) benzoin; (5) ethyl benzoate; (6) toluene; (7) 2,6-dimethoxytoluene; (8) *o*-methoxybiphenyl. Eluent consisted of aqueous buffer (designated A) and acetonitrile (designated B). The notation "90% B" in the first chromatogram means 10 vol% A and 90 vol% B. The buffer contained 25 mM KH_2PO_4 plus 0.1 g/L sodium azide adjusted to pH 3.5 with HCl.

loops that hold fixed volumes from 2 μL to 1 000 μL. In the load position, a syringe is used to wash and load the loop with fresh sample at atmospheric pressure. When the valve is rotated 60° counterclockwise, the content of the sample loop is injected into the column at high pressure.

Samples should be passed through a 0.5- to 2-μm filter prior to injection to prevent contaminating the column with particles, plugging the tubing, and damaging the pump. For liquid chromatography, use a *blunt nose syringe*, not the pointy needle from gas chromatography. There should also be a 0.5-μm frit (a filter) between the injection valve or autosampler and the chromatography column to provide additional protection against small particles that can ruin an expensive column. In-line filters should be used between solvent reservoirs and the pump.

Solvents

Elution with a single solvent or a constant solvent mixture is called **isocratic elution**. If one solvent does not discriminate adequately between the components of a mixture or if the solvent does not provide sufficiently rapid elution of all components, then **gradient elution** can be used. In gradient elution, solvent is changed continuously from weak to stronger eluent strength by mixing more strong solvent into the weak solvent during chromatography.

Figure 22-20 shows the effect of eluent strength in the *isocratic* elution of eight compounds from a reversed-phase column. In a reversed-phase separation, eluent strength increases as the solvent becomes *less* polar. The first chromatogram (upper left) was obtained with a solvent consisting of 90 vol% acetonitrile and 10 vol% aqueous buffer. Acetonitrile is an organic solvent that is less polar than water. Acetonitrile has a high eluent strength, and all compounds are eluted rapidly. In fact, only three peaks are observed because of overlap. It is customary to call the aqueous solvent A and the organic solvent B. The first chromatogram was obtained with 90% B. When eluent strength is *reduced* by changing the solvent to 80% B, there is slightly more separation and five peaks are observed. At 60% B, we begin to see a sixth peak. At 40% B, there are eight clear peaks, but compounds 2 and 3 are not fully resolved. At 30% B, all peaks are resolved, but the separation takes too long. Backing up to 35% B (the bottom trace) separates all peaks in ~2 h (which is still too long for many purposes).

To prepare a mixture of aqueous buffer with an organic solvent, adjust the pH of the buffer to the desired value *prior* to mixing it with organic solvent. Once you mix in the organic solvent, the meaning of "pH" is not well defined.

From the isocratic elutions in Figure 22-20, the *gradient* in Figure 22-21 was selected to resolve all peaks while reducing the time from 2 h to 38 min. First, 30% B was run for 8 min to separate components 1, 2, and 3. The eluent strength was then increased steadily over 5 min to 45% B and held there for 15 min to elute peaks 4 and 5. Finally, the solvent was changed to 80% B over 2 min and held there to elute the last peaks.

Pure HPLC solvents are expensive, and most organic solvents are expensive to dispose of in an environmentally sound manner. To reduce waste in isocratic separations, we can use a device that discards solvent from a column when it is contaminated with solutes but recycles pure solvent emerging between or after the chromatographic peaks. Box 22-1 describes the use of superheated water as a green solvent for chromatography.

Detectors

An **ultraviolet detector** is most common, with a flow cell such as that in Figure 22-22. Simple systems employ the intense 254-nm emission of a mercury lamp. More versatile instruments use a deuterium or xenon lamp and monochromator, so you can choose the optimum wavelength for your analytes. In some systems, the ultraviolet-visible absorption spectrum of each solute can be recorded with a photodiode array

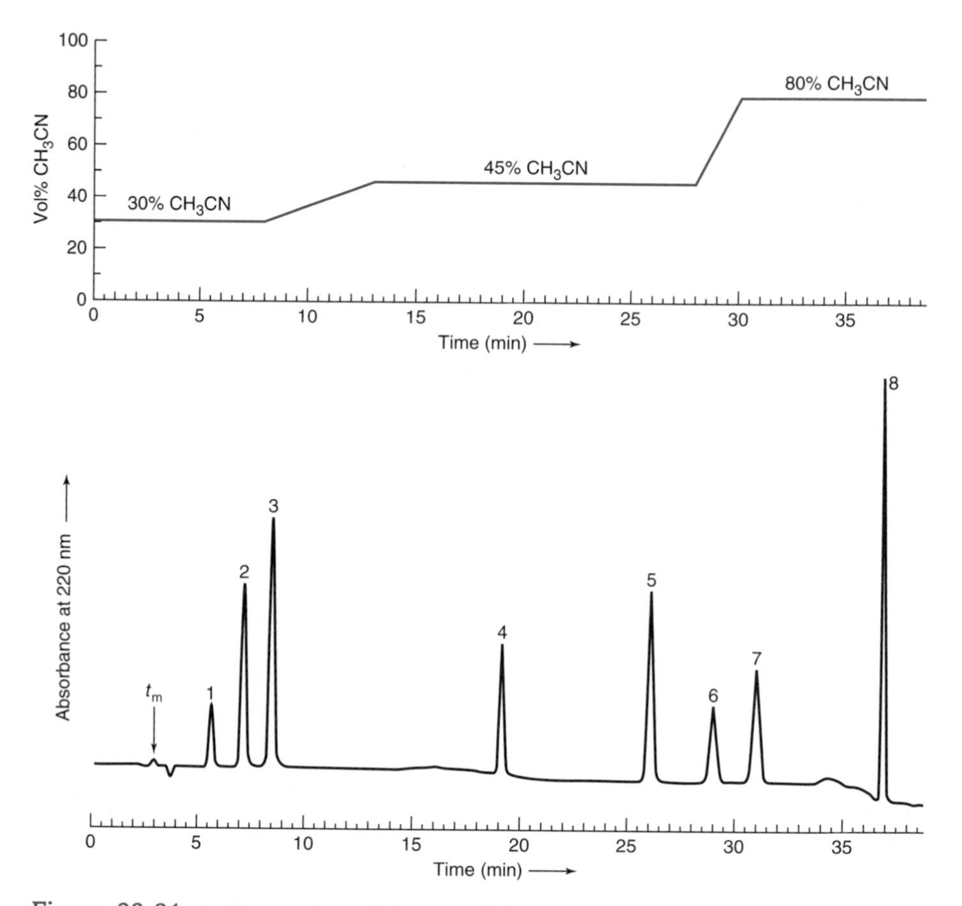

Figure 22-21 Gradient elution of the same mixture of aromatic compounds shown in Figure 22-20 with the same column, flow rate, and solvents. The upper trace is the *segmented gradient* profile, so named because it is divided into several different segments.

spectrophotometer (Figure 19-11) as it is eluted. Clearly, an ultraviolet detector can respond only to analytes with sufficient ultraviolet absorbance.

A more universal sensor is the *refractive index detector*, but its detection limit is about 1 000 times poorer than that of an ultraviolet detector, and it is not useful for gradient elution, because the baseline changes as the solvent changes. In general, a solute has a different refractive index (ability to deflect a ray of light) from that of the solvent. The detector measures the deflection of a light ray by the eluate. An *evaporative light-scattering detector* is another nearly universal detector in which eluate is evaporated to make an aerosol of fine particles of nonvolatile solutes that can be detected by light scattering.

The *charged aerosol detector* is the newest and most sensitive universal detector, and it is compatible with gradient elution. As in evaporative light scattering, eluate evaporates to leave an aerosol of nonvolatile solute. The fine particles mix with a stream of N_2^+ ions formed in a high-voltage discharge. N_2^+ becomes adsorbed to aerosol particles, giving them positive charge. Unadsorbed N_2^+ is separated from the aerosol by an electric field, and the charged aerosol flows to a collector. The chromatogram displays charge reaching the collector as a function of time. The *dynamic*

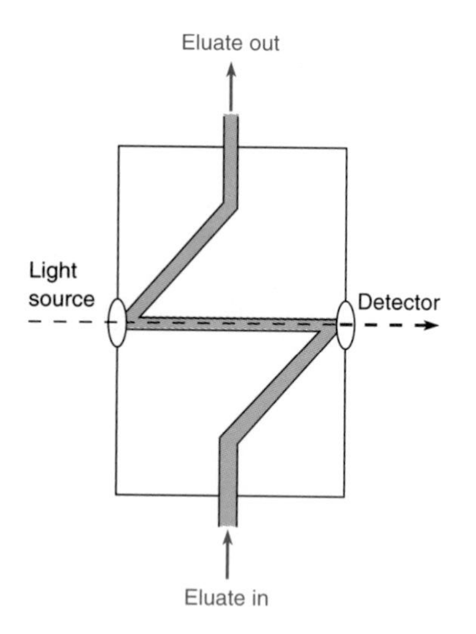

Figure 22-22 Light path in a micro flow cell of a spectrophotometric detector. Cells are available with a 0.5-cm pathlength containing only 10 μL of liquid.

Box 22-1 A "Green" Idea: Superheated Water As Solvent for HPLC

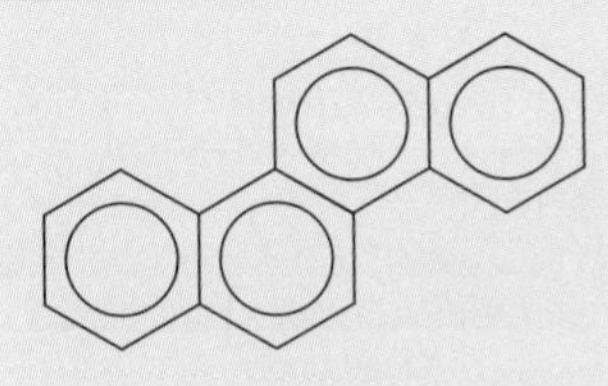

Chrysene
Solubility in H_2O:
8.0 μg/kg H_2O at 25°C/32 bar
0.96 g/kg H_2O at 225°C/62 bar

Water is the ultimate "green" solvent, but it is ordinarily not useful for nonpolar organic solutes. However, at elevated temperature, the polarity of water decreases and it behaves like a polar organic solvent. At 225°C, water resembles methanol in its polarity. The solubility of chrysene in water increases by a factor of 10^5 when the water is heated from 25° to 225°C.

The chromatogram in panel *a* shows the separation of a mixture of alcohols using superheated water as solvent. A gradient of increasing eluent strength was created by raising the temperature from 140° to 180°C at 7°/min. Ordinary HPLC pressure maintains water in its liquid state (panel *b*). For example, at 300°C the vapor pressure of H_2O is 86 bar. Most silica-based stationary phases are not stable in water above ~80°–90°C, so a poly(styrene-divinylbenzene) stationary phase was employed. Other stationary phases suitable for use in superheated water include porous graphitic carbon and zirconia.

Superheated water is compatible with flame ionization detection. This most useful gas chromatography detector cannot be used in liquid chromatography with organic solvents, because the detector responds to organic solvents.

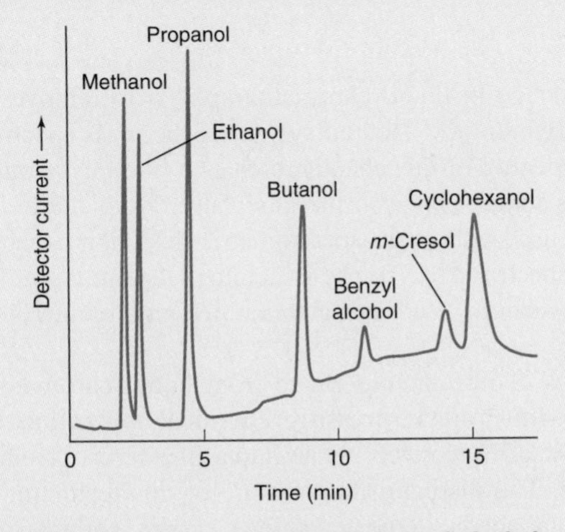

(*a*) Temperature-gradient HPLC separation of alcohols with superheated water and flame ionization detection. [From R. M. Smith, *Anal. Bioanal. Chem.* **2006**, *385*, 419.]

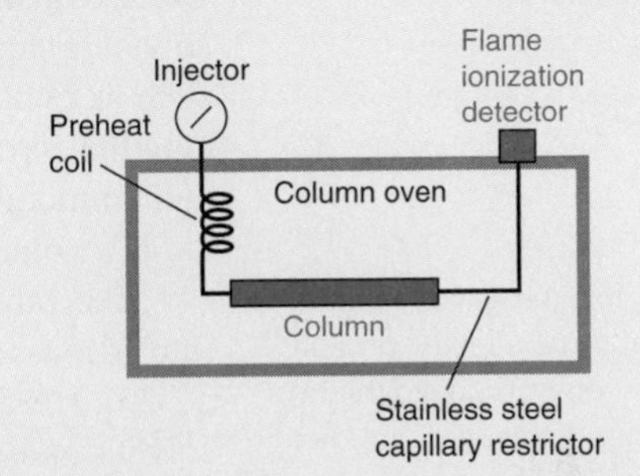

(*b*) HPLC setup. Stainless steel capillary restrictor with inside diameter of 57 μm maintains pressure necessary to keep superheated water in liquid state. [Adapted from D. J. Miller and S. B. Hawthorne, *Anal. Chem.* **1997**, *69*, 623.]

range (Figure 5-1) spans analyte mass of ~5 ng to 10^5 ng. Equal masses of different analytes give equal response within ~15%.

The *electrochemical detector* in Figure 17-7 responds to analytes that can be oxidized or reduced at an electrode over which eluate passes. Electric current is proportional to solute concentration. *Fluorescence detectors* are especially sensitive but respond only to analytes that fluoresce (Figure 19-20) or can be made fluorescent by *derivatization*. A fluorescence detector works by irradiating eluate at one wavelength and monitoring emission at a longer wavelength. Emission intensity is proportional to solute concentration.

Liquid Chromatography–Mass Spectrometry

As in gas chromatography, a *mass spectrometer* is generally the most powerful detector for liquid chromatography because of its capability for both quantitative and

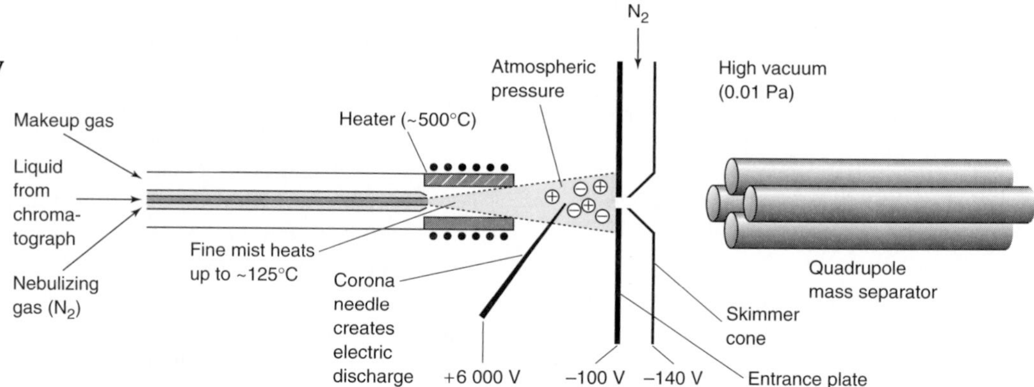

Figure 22-23 Atmospheric pressure chemical ionization interface between liquid chromatography column and mass spectrometer. Aerosol is produced by the nebulizing gas flow and the heater. Electric discharge from the corona needle creates gaseous ions from analyte. [Adapted from E. C. Huang, T. Wachs, J. J. Conboy, and J. D. Henion, *Anal. Chem.* **1990**, *62*, 713A.]

qualitative analysis. The challenge in liquid chromatography is to remove solvent from analyte so as not to overwhelm the vacuum system of the mass spectrometer. **Electrospray**, shown at the opening of this chapter, creates a fine mist from which solvent evaporates and leaves ionic solutes in the gas phase. Nonvolatile buffers, such as phosphate, cannot be used with mass spectrometric detection because they clog the entrance to the mass spectrometer. To obtain acidic pH, ammonium formate and ammonium acetate buffers can be used. For alkaline pH, ammonium bicarbonate is a volatile buffer.

Electrospray: Ions observed in mass spectrometer were already present in solution on chromatography column. Neutral analytes are not converted into ions by electrospray.

Atmospheric pressure chemical ionization: New ions such as protonated analyte are created by chemical ionization.

The other common means of introducing eluate from liquid chromatography into a mass spectrometer is **atmospheric pressure chemical ionization** (Figure 22-23). Heat and a coaxial flow of N_2 convert eluate into a fine aerosol from which solvent and analyte evaporate. The distinguishing feature of this technique is that high voltage is applied to a metal needle in the aerosol. An electric corona (a plasma containing charged particles) forms around the needle, injecting electrons into the aerosol, where a sequence of reactions can create both positive and negative ions. Common ionic products include protonated analyte (MH^+) and M^-. Voltages in a mass spectrometer can be reversed to measure either cations or anions.

In electrospray and atmospheric pressure chemical ionization, there is usually little fragmentation to provide structural information. However, in both techniques, the voltage between the entrance plate and the skimmer cone in Figure 22-23 can be varied to accelerate ions entering the mass separator. Fast-moving ions collide with background N_2 gas and break into smaller fragments that aid in qualitative identification. This widely used process is called *collisionally activated dissociation.*

Ask Yourself

22-C. **(a)** What is the difference between normal-phase and reversed-phase chromatography?

(b) What is the difference between isocratic and gradient elution?

(c) Why does eluent strength increase in normal-phase chromatography when a more polar solvent is added?

(d) Why does eluent strength increase in reversed-phase chromatography when a less polar solvent is added?

(e) What is the purpose of a guard column?

(f) Which chromatography–mass spectrometry interface, electrospray or atmospheric pressure chemical ionization, creates new ions in the gas phase and which just introduces existing solution-phase ions into the gas phase?

22-4 Sample Preparation for Chromatography

Example of derivatization: The alcohol RCH_2OH is converted to $RCH_2OSi(CH_3)_3$ to make it more volatile for gas chromatography and to give characteristic mass spectral peaks that aid in identification.

Sample preparation is the process of transforming a sample into a form that is suitable for analysis. This process might entail extracting analyte from a complex matrix, *preconcentrating* very dilute analytes to get a concentration high enough to measure, removing or masking interfering species, or chemically transforming (*derivatizing*) the analyte into a form that is easier to separate or to detect. *Solid-phase microextraction* and *purge and trap* are sample preparation techniques that are especially important for gas chromatography. *Solid-phase extraction* is useful for both liquid or gas chromatography.

Solid-phase microextraction is a simple way to extract compounds for gas chromatography from liquids, air, or even sludge without using any solvent. The key component is a fused-silica fiber coated with a 10- to 100-μm-thick film of nonvolatile liquid stationary phase similar to those used in gas chromatography. Figure 22-24 shows the fiber attached to the base of a syringe with a fixed metal needle. The fiber can be extended from the needle or retracted inside the needle. Figure 22-25 demonstrates the procedure of exposing the fiber to a sample solution (or the gaseous headspace above the liquid) for a fixed length of time while stirring and, perhaps, heating. Only a fraction of the analyte in the sample is extracted into the fiber.

After sampling, the fiber is retracted and the needle is introduced into a narrow injection port (0.7 mm inner diameter) of a gas chromatograph. The port maintains desorbed analyte in as narrow a band as possible. The fiber is extended inside the hot injection liner, where analyte is thermally desorbed from the fiber in the splitless

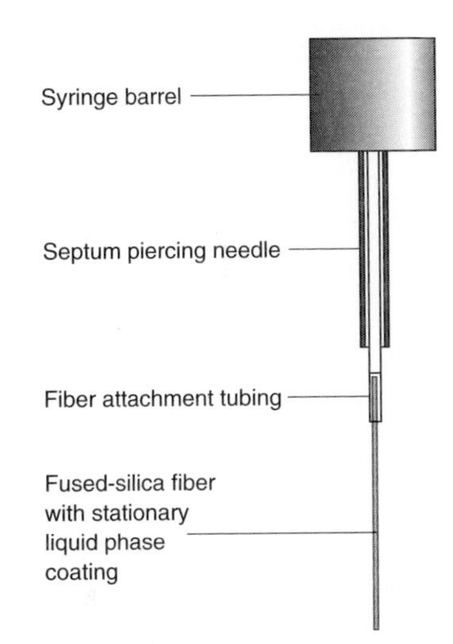

Figure 22-24 Syringe for solid-phase microextraction. The fused-silica fiber is withdrawn inside the steel needle after sample collection and whenever the syringe is used to pierce a septum.

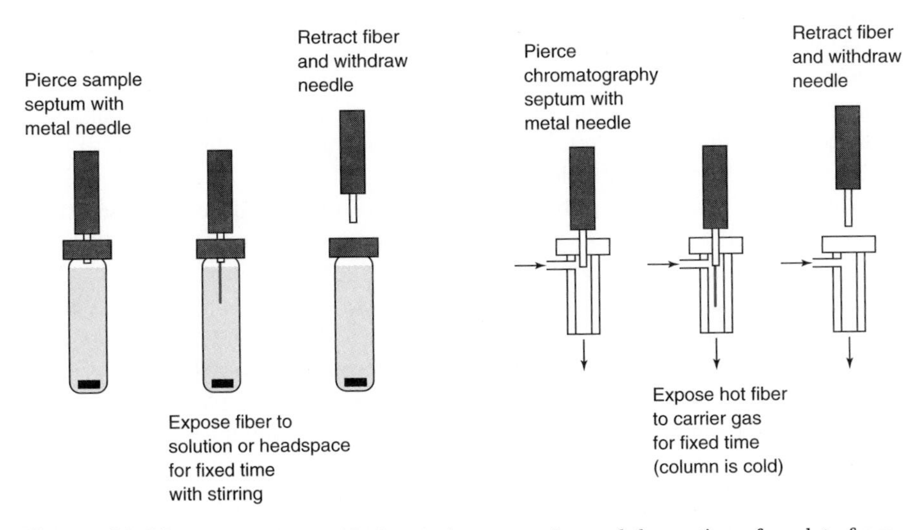

Figure 22-25 Sampling by solid-phase microextraction and desorption of analyte from the coated fiber into a gas chromatograph. [Adapted from Supelco Chromatography Products catalog, Bellefonte, PA.]

mode for a fixed time. *Cold trapping* (page 487) focuses the desorbed analyte at the head of the column prior to chromatography. Solid-phase microextraction can be adapted to liquid chromatography by using a special injection port in which the fiber is washed with a strong solvent.

Purge and trap is a method for removing volatile analytes from liquids or solids (such as groundwater or soil), concentrating the analytes, and introducing them into a gas chromatograph. In contrast with solid-phase microextraction, which removes only a portion of analyte from the sample, the goal in purge and trap is to remove 100% of the analyte from the sample. Quantitative removal of polar analytes from polar matrices can be difficult.

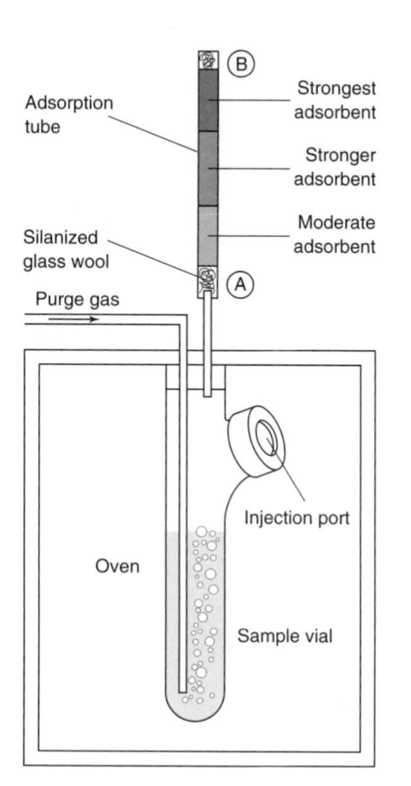

Figure 22-26 Purge and trap apparatus for extracting volatile substances from a liquid or solid by flowing gas. You need to establish the time and temperature required to purge 100% of the analyte from the sample in separate control experiments.

Figure 22-26 shows apparatus for extracting volatile flavor components from beverages. Helium purge gas from a stainless steel needle is bubbled through the beverage in the sample vial, which is heated to 50°C to aid evaporation of analytes. Gas exiting the sample vial passes through an adsorption tube containing three layers of adsorbent compounds with increasing adsorbent strength. For example, the moderate adsorbent could be a nonpolar phenylmethylpolysiloxane, the stronger adsorbent could be the polymer Tenax, and the strongest adsorbent could be carbon molecular sieves.

In the purge and trap process, gas flows through the adsorbent tube from end A to end B in Figure 22-26. After analyte has been purged from the sample, gas flow is reversed to go from B to A and the trap is purged at 25°C to remove as much water or other solvent as possible from the adsorbents. Outlet A of the adsorption tube is then directed to the injection port of a gas chromatograph operating in splitless mode and the trap is heated to ~200°C. Desorbed analytes flow into the chromatography column, where they are concentrated by cold trapping. After complete desorption from the trap, the chromatography column is warmed up to initiate the separation.

Solid-phase extraction uses a liquid chromatography solid phase in a short, open column to partially purify and *preconcentrate* analyte. For example, in drug tests of athletes, urine is passed through a short C_{18}-silica column that retains steroids and many other organic compounds. More polar and ionic substances in the urine pass through the column and are removed. Retained compounds are then washed from the column with a small volume of hexane or ethyl acetate. Solvent is evaporated and the residue dissolved in a minimum volume of solvent for application to a chromatography column. By similar means, trace quantities of benzoylecgonine were isolated from large volumes of Po River water in Italy to estimate cocaine use by the population upsteam of the sampling site (Chapter 0 opener.)

Sample cleanup refers to the removal of undesirable components of an unknown that interfere with measurement of analyte. In the urine-steroid analysis, polar molecules that are not retained by C_{18}-silica wash right through the column and are separated from the steroids.

Molecularly imprinted polymers are the newest media for solid-phase extraction. For example, a commercial molecularly imprinted polymer is available for extraction and preconcentration of the antibiotic chloramphenicol from milk. This antibiotic causes aplastic anemia and is a suspected carcinogen. Therefore it was banned from use in farm animals in North America and Europe. It is still used elsewhere, so certain foods are screened for chloramphenicol. The molecularly imprinted polymer is made by polymerizing monomeric building blocks in the presence of chloramphenicol (Figure 22-27). For extracting chloramphenicol from milk, the milk is centrifuged to separate fat from the aqueous phase, which is passed through a small cartridge containing molecularly imprinted polymer that had been washed with methanol and water. The polymer is then eluted with a sequence of solvents that remove most absorbed molecules, but not chloramphenicol. The antibiotic is finally removed by a small volume of methanol : acetic acid : water (89 : 1 : 10 vol/vol/vol).

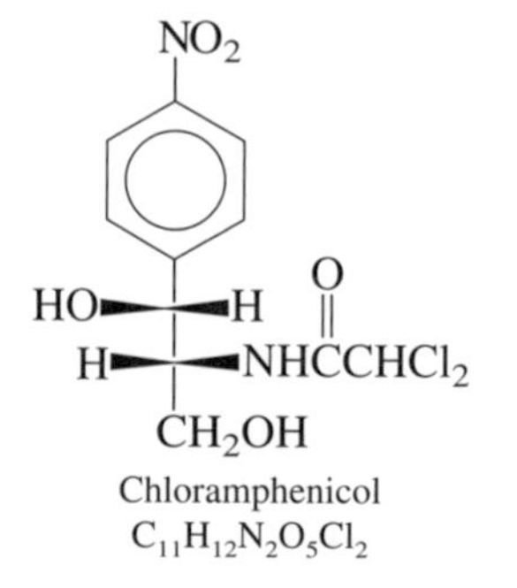

Chloramphenicol
$C_{11}H_{12}N_2O_5Cl_2$

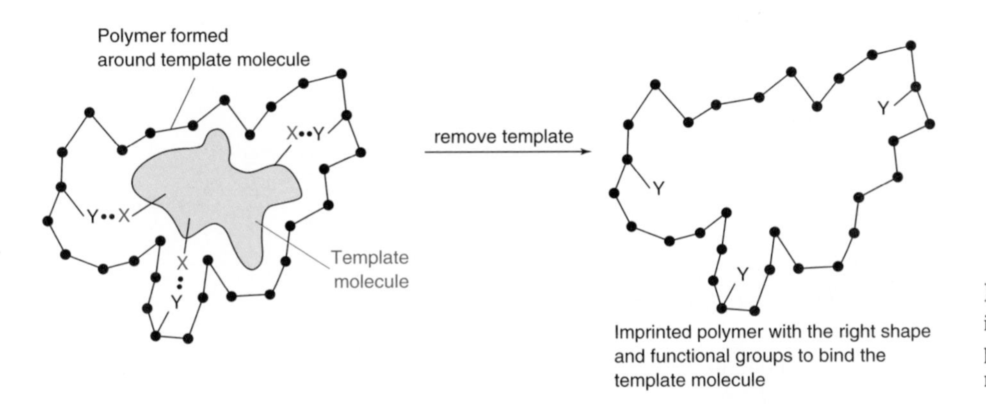

Figure 22-27 A molecularly imprinted polymer has a binding pocket for a specific template molecule.

Eluate is evaporated to dryness and dissolved in 150 μL of HPLC mobile phase for chromatography with mass spectrometric selected ion monitoring.

Ask Yourself

22-D. **(a)** Why is it necessary to use cold trapping on the gas chromatography column if sample is introduced from a solid-phase microextraction syringe or from a purge and trap absorption tube?
(b) What is the purpose of solid-phase extraction? Why is it advantageous to use large particles (50 μm) for solid-phase extraction, but small particles (5 μm) for chromatography?
(c) What kind of sample cleanup does solid-phase extraction accomplish in the analysis of steroids in urine?
(d) Chloramphenicol, shown on page 502, is measured by HPLC–electrospray mass spectrometry, observing negative ions. The base peak is at m/z 321. The second strongest peak, 60% as intense as the base peak, is m/z 323. Assign a molecular formula to each peak.

Important Terms

atmospheric pressure chemical ionization
bonded stationary phase
cold trapping
electron capture detector
electrospray
eluent strength
flame ionization detector
gas chromatography
gradient elution
guard column
high-performance liquid chromatography
isocratic elution
molecularly imprinted polymer
normal-phase chromatography
on-column injection
purge and trap
reversed-phase chromatography
sample cleanup
sample preparation
selected reaction monitoring
solid-phase extraction
solid-phase microextraction
solvent trapping
split injection
splitless injection
temperature programming
thermal conductivity detector
ultraviolet detector

Problems

22-1. **(a)** Explain the difference between wall-coated, support-coated, and porous-layer open tubular columns for gas chromatography.
(b) What is the advantage of a bonded stationary phase in gas chromatography?

(c) Why do we use a makeup gas for some gas chromatography detectors?

(d) Explain how solvent trapping and cold trapping work in splitless injection.

22-2. Explain why plate height increases at **(a)** very low and **(b)** very high flow rates in Figure 22-16. (*Hint:* Refer to Section 21-3.)

22-3. Why does a thermal conductivity detector respond to all analytes except the carrier gas? Why isn't the flame ionization detector universal?

22-4. Consider a narrow-bore (0.25 mm diameter) open tubular gas chromatography column coated with a thin film (0.10 μm thick) of stationary phase that provides 5 000 plates per meter. Consider also a wide-bore (0.53 mm diameter) column with a thick film (5.0 μm thick) of stationary phase that provides 1 500 plates per meter.

(a) Why does the thin film provide more theoretical plates per meter than the thick film?

(b) The density of stationary phase is 1.0 g/mL. What mass of stationary phase in each column is in a length equivalent to one theoretical plate?

(c) How many nanograms of analyte can be injected into each column if the mass of analyte is not to exceed 1.0% of the mass of stationary phase in one theoretical plate?

22-5. The adjusted retention time (t'_r) for a chromatographic peak is the observed retention time (t_r) minus the retention time of an unretained substance. Many gas chromatography columns do not retain methane, so a little methane is added to a sample to produce an unretained peak. Then $t'_r = t_r - t_r(\text{methane})$. For isothermal elution of a homologous series of compounds (those with similar structures, but differing by the number of CH_2 groups in a chain), log t'_r is usually a linear function of the number of carbon atoms. A compound was known to be a member of the family $(CH_3)_2CH(CH_2)_nCH_2OSi(CH_3)_3$. For the following retention times, use a spreadsheet to prepare a graph of log t'_r versus n. Use the Excel function LINEST (Section 4-7) to find the slope and intercept of the least-squares straight line through the three known points. From the equation of the line, calculate the value of n for the unknown.

$n = 7$	4.0 min	CH_4	1.1 min
$n = 8$	6.5 min	Unknown	42.5 min
$n = 14$	86.9 min		

22-6. The gasoline additive methyl *t*-butyl ether (MTBE) has been leaking into groundwater ever since its introduction in the 1990s. MTBE can be measured at parts per billion levels by solid-phase microextraction from groundwater to which 250 g/L NaCl has been added to lower the solubility of MTBE. After microextraction, analytes are thermally desorbed from the fiber in the port of a gas chromatograph. The figure above shows a reconstructed total ion chromatogram and selected ion monitoring of substances desorbed from the extraction fiber.

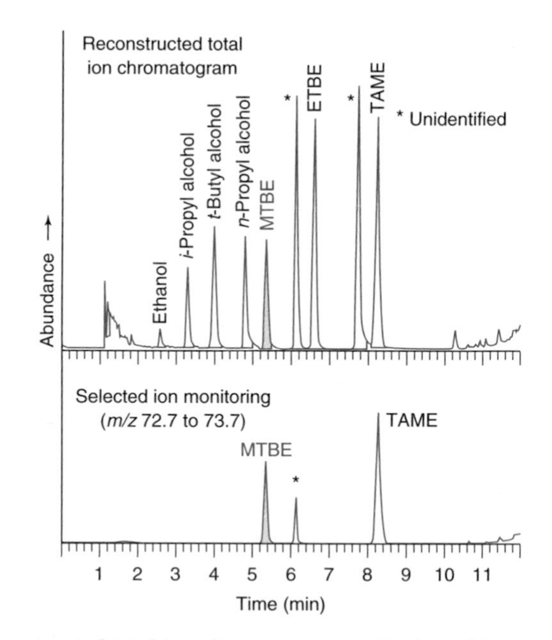

Reconstructed total ion chromatogram and selected ion monitoring of solid-phase microextract of groundwater. [From D. A. Cassada, Y. Zhang, D. D. Snow, and R. F. Spalding, *Anal. Chem.* **2000**, *72*, 4654.]

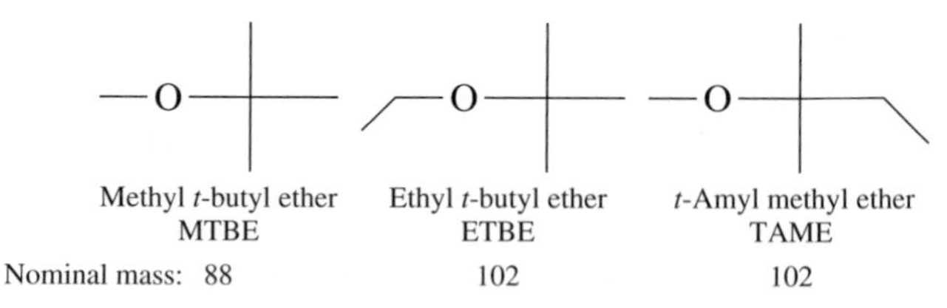

(a) What nominal mass is being observed in selected ion monitoring? Why are only three peaks observed?

(b) Here is a list of major ions above m/z 50 in the positive ion electron ionization mass spectra. Given that MTBE and TAME have an intense peak at m/z 73, and there is no significant peak at m/z 73 for ETBE, suggest a structure for m/z 73. Suggest a structure for m/z 87 for ETBE and TAME.

MTBE	ETBE	TAME
73	87	87
57	59	73
	57	71
		55

22-7. (a) In the selected reaction monitoring of caffeine in Figure 22-11, the transition m/z 194 → 109 was used. What is the nominal mass of the molecular ion of caffeine? Why was the transition m/z 197 → 111 used for the internal standard $(^{13}CH_3)_3$-caffeine.

(b) Why is an isotopic variant of the analyte an excellent internal standard?

(c) The relative intensities for caffeine in Figure 22-11 are m/z 195/194 = 10.3%. What is the isotopic composition of the ion at m/z 195? What is the expected intensity ratio m/z 195/194? (*Hint:* See Equation 21-8.) (The observed ratio is higher than the expected ratio because some of the intensity at m/z 194 is not the molecular ion but the isotopic partner of m/z 193.)

22-8. Nonpolar aromatic compounds were separated by HPLC on a bonded phase containing octadecyl groups [$—(CH_2)_{17}CH_3$] covalently attached to silica particles. The eluent was 65 vol% methanol in water. How would the retention times be affected if 90% methanol were used instead?

22-9. Polar solutes were separated by HPLC with a bonded phase containing polar diol substituents [$—CH(OH)CH_2OH$]. How would the retention times be affected if the eluent were changed from 40 vol% to 60 vol% acetonitrile in water? Acetonitrile ($CH_3C{\equiv}N$) is less polar than water.

22-10. Normal-phase chromatography is commonly conducted with a polar stationary phase and a mixture of organic solvents. Eluent strength is increased by increasing the content of more polar organic solvent. *Aqueous normal-phase chromatography*, also called *hydrophilic interaction chromatography*, is a variant of normal-phase chromatography in which mobile phase contains water (typically <30%) in a polar organic solvent. Retention is described as a partitioning of analyte between mobile phase and an aqueous phase adsorbed on the stationary phase. For gradient elution, would you increase or decrease the water content of the mobile phase?

22-11. Draw the chemical structures of two bonded nonpolar phases and two bonded polar phases in HPLC. Begin with a silicon atom at the surface of a silica particle.

22-12. (a) Why is high pressure needed in HPLC?

(b) Why does the efficiency (decreased plate height) of liquid chromatography increase as the stationary phase particle size is reduced?

22-13. A 15-cm-long HPLC column packed with 5-μm particles has an optimum plate height of 10.0 μm in Figure 22-16. What will be the half-width of a peak eluting at 10.0 min? If the particle size were 3 μm and the plate height were 5.0 μm, what would be the half-width?

22-14. Octanoic acid and 1-aminooctane were separated by HPLC on a bonded phase containing octadecyl groups [$—(CH_2)_{17}CH_3$]. The eluent was 20 vol% methanol in water adjusted to pH 3.0 with HCl.

$CH_3(CH_2)_6CO_2H$ — Octanoic acid

$CH_3(CH_2)_7NH_2$ — 1-Aminooctane

(a) Draw the predominant form (neutral or ionic) of a carboxylic acid and an amine at pH 3.0.

(b) State which compound is expected to be eluted first and why.

22-15. (a) When you try separating an unknown mixture by reversed-phase chromatography with 50% acetonitrile-50% water, the peaks are eluted between 1 and 3 min, and they are too close together to be well resolved for quantitative analysis. Should you use a higher or lower percentage of acetonitrile in the next run?

(b) Suppose that you try separating an unknown mixture by normal-phase chromatography with the solvent mixture 50% hexane-50% methyl *t*-butyl ether (which is more polar than hexane). The peaks are too close together and are eluted rapidly. Should you use a higher or lower percentage of hexane in the next run?

22-16. After an isocratic elution has been optimized with several solvents, the chromatogram has a resolution of 1.2 between the two closest peaks. How might you increase the resolution without changing solvents or the type of stationary phase?

22-17. (a) Sketch a graph of the van Deemter equation (plate height versus flow rate). What would the curve look like if the multiple path term were 0? If the longitudinal diffusion term were 0? If the finite equilibration time term were 0?

(b) Explain why the van Deemter curve for 3-μm particles in Figure 22-16 is nearly flat at high flow rate. What can you say about each of the terms in the van Deemter equation for 3-μm particles?

22-18. *Internal standard.* Compounds C and D gave the following HPLC results:

Compound	Concentration (mg/mL) in mixture	Peak area (cm^2)
C	1.03	10.86
D	1.16	4.37

A solution was prepared by mixing 12.49 mg of D plus 10.00 mL of unknown containing just C, and diluting to 25.00 mL. Peak areas of 5.97 and 6.38 cm^2 were observed for

C and D, respectively. Find the concentration (mg/mL) of C in the unknown.

22-19. Spherical, microporous silica particles used in chromatography have a density of 2.20 g/mL, a diameter of 10.0 μm, and a measured surface area of 300 m^2/g.

(a) The volume of a spherical particle is $\frac{4}{3}\pi r^3$, where r is the radius. The mass of the sphere is volume × density (= mL × g/mL). How many particles are in 1.00 g of silica?

(b) The surface area of a sphere is $4\pi r^2$. Calculate the surface area of 1.00 g of solid, spherical silica particles.

(c) By comparing the calculated and measured surface areas, what can you say about the porosity of the particles?

22-20. Here are guidelines for flushing a C_{18} column and changing mobile phase. Explain the rationale for each.

1. To remove strongly retained solutes, flush the column initially with 5–10 column volumes of the most recent mobile phase (such as 60 : 40 H_2O-acetonitrile) *without* buffer. That is, replace buffer with pure water.

2. Then flush with 10–20 volumes of strong solvent (such as 10 : 90 H_2O-acetonitrile).

3. Store the column in the strong solvent.

4. Equilibrate with 10–20 volumes of new, desired, buffered mobile phase.

5. Use a standard mixture to check for correct retention times and plate numbers.

22-21. *Chromatography–mass spectrometry.* Cocaine metabolism in rats can be studied by injecting the drug and periodically withdrawing blood to measure levels of metabolites by HPLC–mass spectrometry. For quantitative analysis, isotopically labeled internal standards are mixed with the blood sample. Blood was analyzed by reversed-phase chromatography with an acidic eluent and atmospheric pressure chemical ionization mass spectrometry for detection. The mass spectrum of the collisionally activated dissociation products from the m/z 304 positive ion is shown in panel *a* below. Selected reaction monitoring (m/z 304 from mass filter Q1 and m/z 182 from Q3 in Figure 22-10) gave a single chromatographic peak at 9.22 min for cocaine (panel *b*). The internal standard $^2H_5$-cocaine gave a single peak at 9.19 min for m/z 309 (Q1) → 182 (Q3).

(a) Draw the structure of the ion at m/z 304 and suggest a structure for m/z 182.

(b) Intense peaks at m/z 182 and 304 do not have isotopic partners at m/z 183 and 305. Explain why.

(c) Rat plasma is exceedingly complex. Why does the chromatogram show just one clean peak?

(d) Given that $^2H_5$-cocaine has only two major mass spectral peaks at m/z 309 and 182, which atoms are labeled with deuterium?

(e) Explain how you would use $^2H_5$-cocaine for measuring cocaine in blood.

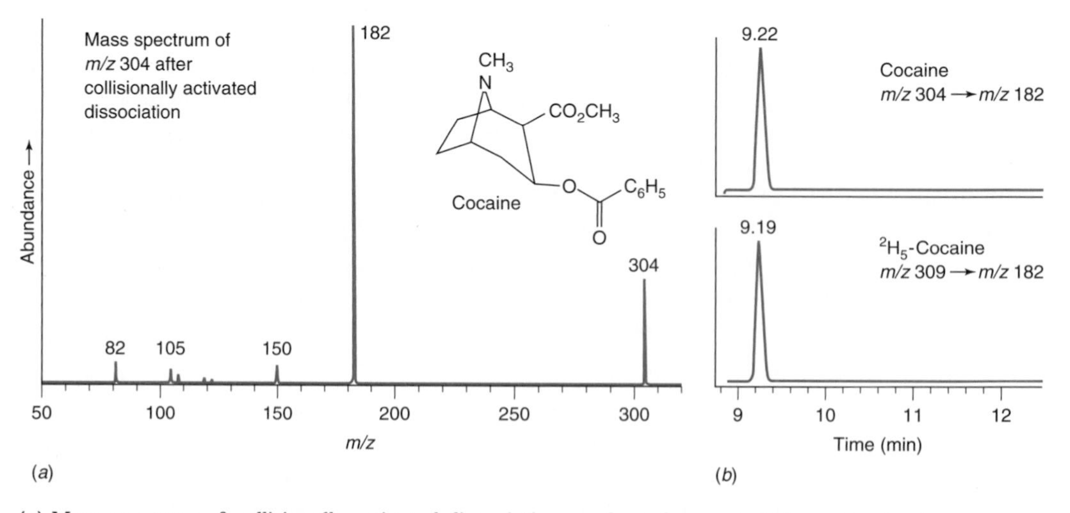

(*a*) Mass spectrum of collisionally activated dissociation products from m/z 304 positive ion from atmospheric pressure chemical ionization mass spectrum of cocaine. (*b*) Chromatograms obtained by selected reaction monitoring. [From G. Singh, V. Arora, P. T. Fenn, B. Mets, and I. A. Blair, *Anal. Chem.* **1999**, *71*, 2021.]

22-22. Why is splitless injection used with purge and trap sample preparation?

22-23. What is the purpose of derivatization in chromatography? Give an example.

22-24. Explain how solid-phase microextraction works. Why is cold trapping necessary during injection with this technique? Is all the analyte in an unknown extracted into the fiber in solid-phase microextraction?

22-25. Why does a molecularly imprinted polymer selectively bind a desired analyte?

22-26. Here is a student procedure to measure nicotine in urine. A 1.00-mL sample of biological fluid was placed in a 12-mL vial containing 0.7 g Na_2CO_3 powder. After 5.00 μg of the internal standard 5-aminoquinoline were injected, the vial was capped with a Teflon-coated silicone rubber septum. The vial was heated to 80°C for 20 min and then a solid-phase microextraction needle was passed through the septum and left in the headspace for 5.00 min. The fiber was retracted and then inserted into a gas chromatography injection port. Volatile substances were desorbed from the fiber at 250°C for 9.5 min in the injection port of a chromatograph whose column was at 60°C. The column temperature was then raised to 260°C at 25°C/min and eluate was monitored by electron ionization mass spectrometry with selected ion monitoring at m/z 84 for nicotine and m/z 144 for internal standard. Calibration data from replicate standard mixtures taken through the entire procedure are given in the table.

Nicotine in urine (μg/L)	Area ratio m/z 84/144
12	0.05_6, 0.05_9
51	0.40_2, 0.39_1
102	0.68_4, 0.66_9
157	1.01_1, 1.06_3
205	1.27_8, 1.35_5

Based on A. E. Wittner, D. M. Klinger, X. Fan, M. Lam, D. T. Mathers, and S. A. Mabury, *J. Chem. Ed.* **2002**, *79*, 1257.

(a) Why was the vial heated to 80°C before and during extraction?

(b) Why was the chromatography column kept at 60°C during thermal desorption of the extraction fiber?

(c) Suggest a structure for m/z 84 from nicotine. What is the m/z 144 ion from the internal standard, 5-aminoquinoline?

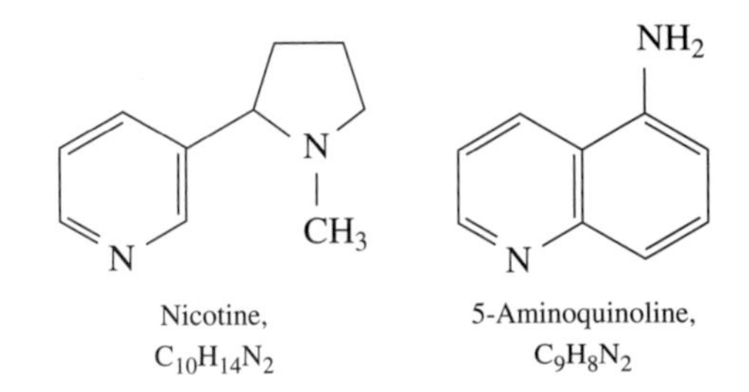

Nicotine, $C_{10}H_{14}N_2$ 5-Aminoquinoline, $C_9H_8N_2$

(d) Urine from an adult female nonsmoker had an area ratio m/z 84/144 = 0.51 and 0.53 in replicate determinations. Urine from a nonsmoking girl whose parents are heavy smokers had an area ratio 1.18 and 1.32. Find the nicotine concentration (μg/L) and its uncertainty in the urine of each person.

How Would You Do It?

22-27. The compound reserpine has the formula $C_{33}H_{40}N_2O_9$ with a nominal mass of 608 Da. The following positive ion mass spectra were recorded:

(i) Electron ionization of reserpine from gas chromatography column.

(ii) Electrospray of reserpine from liquid chromatography column.

(iii) Electrospray followed by tandem mass spectroscopy. The base peak from **(ii)** went through a collision cell and a full scan mass spectrum of the resulting fragments was obtained.

Here are descriptions of the three spectra in random order. Which description belongs to which spectrum—**(i)**, **(ii)**, and **(iii)**? Explain your reasoning.

a. Base peak at m/z 609 with 40% intensity at m/z 610. No other major peaks.

b. Base peak at m/z 195 with many other significant peaks. The two highest mass peaks are m/z 608 = 19% and m/z 609 = 6%. There is no peak at m/z 610.

c. Base peak at m/z 609 with nothing significant at m/z 610. There are several significant peaks at lower m/z.

22-28. Nitric oxide (NO) is a cell signaling agent in physiologic processes such as vasodilation, inhibition of clotting, and inflammation. A sensitive chromatography–mass spectrometry method was developed to measure two of its metabolites, nitrite (NO_2^-) and nitrate (NO_3^-), in biological fluids. Internal standards $^{15}NO_2^-$ and $^{15}NO_3^-$ were added to the fluid at concentrations of 80.0 and 800.0 μM, respectively. The naturally occurring $^{14}NO_2^-$ and $^{14}NO_3^-$ plus the internal standards were then converted into volatile derivatives in aqueous acetone:

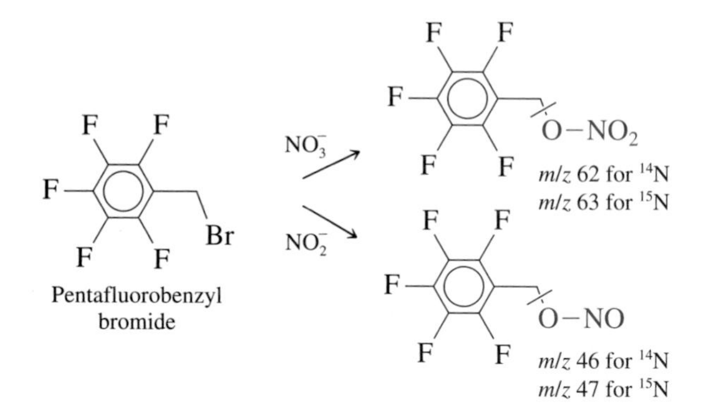

Because biological fluids are so complex, the derivatives were first isolated by high-performance liquid chromatography. For quantitative analysis, liquid chromatography peaks corresponding to the two products were injected into a gas chromatograph, ionized by *negative ion* chemical ionization (giving major peaks for NO_2^- and NO_3^-), and the products were measured by selected ion monitoring. Results are shown in the figure. If the ^{15}N internal standards undergo the same reactions and same separations at the same rate as the ^{14}N analytes, then the concentrations of analytes are simply

$$[^{14}NO_x^-] = [^{15}NO_x^-](R - R_{blank})$$

where R is the measured peak area ratio (m/z 46/47 for nitrite and m/z 62/63 for nitrate) and R_{blank} is the measured ratio of peak areas in a blank prepared from the same buffers and reagents with no added nitrate or nitrite. In the figure below, the ratios of peak areas are m/z 46/47 = 0.062 and m/z 62/63 = 0.538. The ratios for the blank were m/z 46/47 = 0.040 and m/z 62/63 = 0.058. Find the concentrations of nitrite and nitrate in the urine.

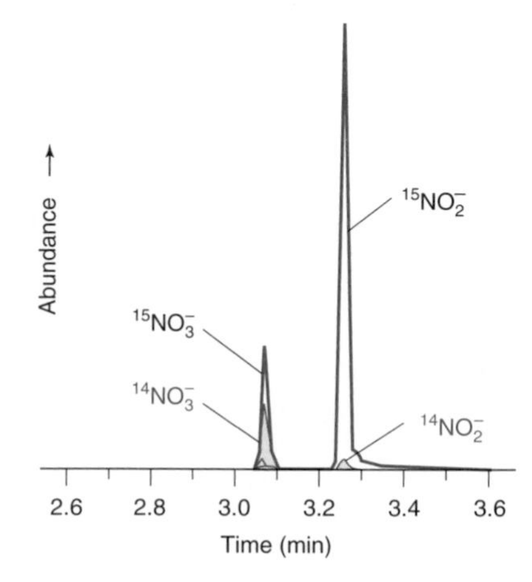

Selected ion chromatogram showing *negative ions* at m/z 46, 47, 62, and 63 obtained by derivatizing nitrite and nitrate plus internal standards ($^{15}NO_2^-$ and $^{15}NO_3^-$) in urine. [From D. Tsikas, *Anal. Chem.* **2000**, *72*, 4064.]

Notes and Reference

1. D. J. Frederick, "John Fenn: Father of Electrospray Ionization," *Chemistry*, Winter 2003, p. 13.

2. For instructive exercises to complement laboratory experience in chromatography, see D. C. Stone, "Teaching Chromatography using Using Virtual Laboratory Exercises," *J. Chem. Ed.* **2007,** *84,* 1488.

Further Reading

L. R. Snyder, J. J. Kirkland, and J. L. Glajch, *Practical HPLC Method Development* (New York: Wiley, 1997).

L. R. Snyder and J. W. Dolan, *High-Performance Gradient Elution* (Hoboken, NJ: Wiley, 2007).

K. Grob, *Split and Splitless Injection for Quantitative Gas Chromatography* (New York: Wiley, 2001).

S. Kromidas, *Practical Problem Solving in HPLC* (New York: Wiley, 2000).

V. R. Meyer, *Practical High-Performance Liquid Chromatography*, 4th ed. (Chichester: Wiley, 2004).

T. Hanai and R. M. Smith, *HPLC: A Practical Guide* (New York: Springer Verlag, 1999).

V. R. Meyer, *Pitfalls and Errors of HPLC in Pictures* (New York: Wiley, 1998).

H. M. McNair and J. M. Miller, *Basic Gas Chromatography* (New York: Wiley, 1998).

R. Eksteen, P. Schoenmakers, and N. Miller, eds., *Handbook of HPLC* (New York: Marcel Dekker, 1998).

R. P. W. Scott, *Introduction to Analytical Gas Chromatography* (New York: Marcel Dekker, 1998).

U. D. Neue, *HPLC Columns: Theory, Technology, and Practice* (New York: Wiley, 1997).

W. Jennings, E. Mittlefehldt, and P. Stremple, *Analytical Gas Chromatography*, 2nd ed. (San Diego, CA: Academic Press, 1997).

R. L. Grob, *Modern Practice of Gas Chromatography* (New York: Wiley, 1995).

H.-J. Hübschmann, *Handbook of GC/MS: Fundamentals and Applications* (Weinheim: Wiley-VCH, 2001).

M. McMaster and C. McMaster, *GC/MS: A Practical User's Guide* (New York: Wiley, 1998).

W. M. A. Niessen, *Liquid Chromatography–Mass Spectrometry*, 3rd ed. (Boca Raton, FL: Taylor & Francis, 2006).

J. V. Hinshaw and L. S. Ettre, *Introduction to Open Tubular Gas Chromatography* (Cleveland, OH: Advanstar Communications, 1994).

Capillary Electrophoresis in Medicine

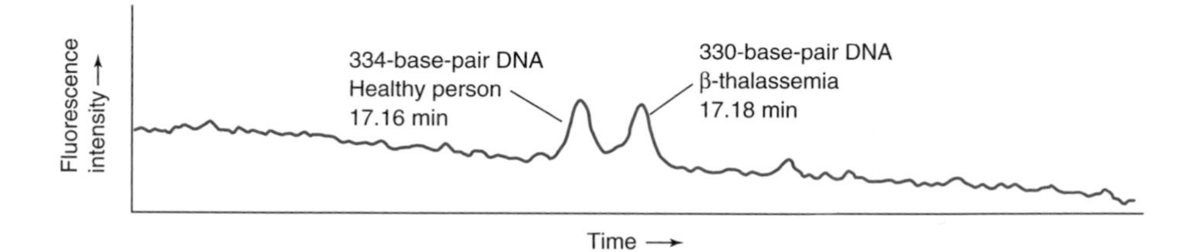

Separation of DNA fragments from a healthy person and a person with β-thalassemia. DNA is separated according to size by sieving through 1.5 wt% poly(ethylene oxide). Longer DNA emerges first, as explained at the end of Section 23-7. The dye ethidium bromide added to the sample becomes strongly fluorescent when it binds to DNA. Laser-induced fluorescence provides sensitive detection for the DNA. [From P.-L. Chang, I.-T. Kuo, T.-C. Chiu, and H.-T. Chang, *Anal. Bioanal. Chem.* **2004**, *379*, 404.]

***Capillary electrophoresis* separates and measures ions on the basis of their different migration rates in an electric field. Methods are available to separate small molecules or macromolecules. A tiny needle attached to a capillary can be inserted into a single living cell to analyze picoliter volumes. Capillary electrophoresis is the workhorse analytical tool that enabled the human genome to be sequenced.**

Capillary electrophoresis can be used to diagnose the genetic disease thalassemia (Box 13-1). A specific segment of a patient's DNA is amplified by the polymerase chain reaction to produce 2^{15} copies. The segment is selected by the nucleotide sequence of synthetic primer DNA that is added to begin the amplification process. After amplification, capillary electrophoresis can distinguish a healthy person's 334-base-pair DNA from a thalassemia patient's 330-base-pair DNA.

CHAPTER 23

Chromatographic Methods and Capillary Electrophoresis

The liquid chromatography discussed in Chapter 22 separates solutes by *adsorption* or *partition* mechanisms. In this chapter, we consider separations by *ion-exchange, molecular exclusion*, and *affinity* chromatography, which were illustrated in Figure 21-2. We also consider *capillary electrophoresis*, which separates species on the basis of their different rates of migration in an electric field.

23-1 Ion-Exchange Chromatography

Ion-exchange chromatography is based on the attraction between solute ions and charged sites in the stationary phase (Figure 21-2). **Anion exchangers** have positively charged groups on the stationary phase that attract solute anions. **Cation exchangers** contain negatively charged groups that attract solute cations.

Anion exchangers contain bound *positive* groups.
Cation exchangers contain bound *negative* groups.

The stationary phase for ion-exchange chromatography is usually a *resin*, such as polystyrene, which consists of amorphous (noncrystalline) particles. Polystyrene is made into a cation exchanger when negative sulfonate ($—SO_3^-$) or carboxylate ($—CO_2^-$) groups are attached to the benzene rings (Figure 23-1). Polystyrene is an

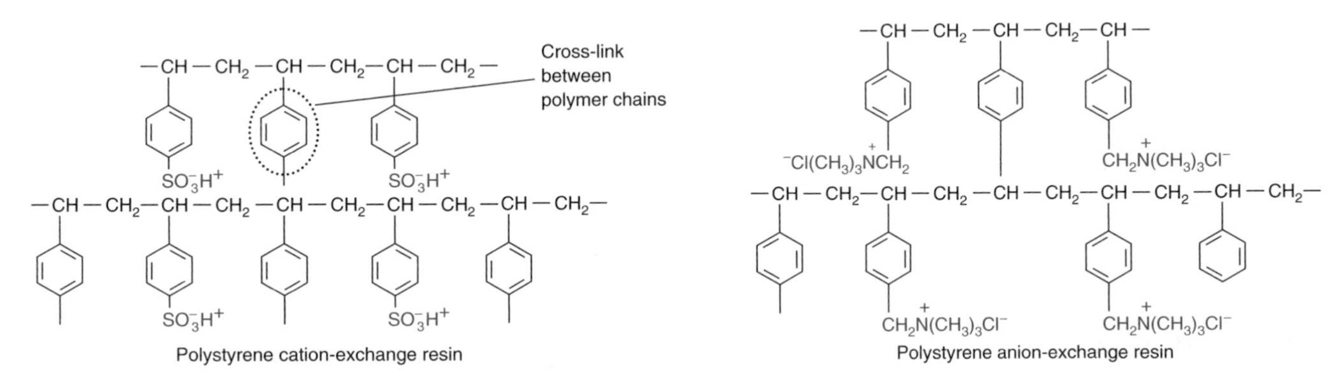

Figure 23-1 Structures of polystyrene ion-exchange resins. *Cross-links* are covalent bridges between polymer chains.

anion exchanger if ammonium groups ($—NR_3^+$) are attached. Cross-links (covalent bonds) between polystyrene chains in Figure 23-1 control the pore sizes into which solutes can diffuse.

Gel particles are softer than resin particles. Cellulose and dextran ion-exchange gels, which are polymers of the sugar glucose, possess larger pore sizes and lower charge densities. Gels are better suited than resins for ion exchange of macromolecules, such as proteins.

Ion-Exchange Selectivity

Consider the competition of K^+ and Li^+ for sites on the cation-exchange resin, R^-:

$$R^-K^+ + Li^+ \rightleftharpoons R^-Li^+ + K^+ \qquad K = \frac{[R^-Li^+][K^+]}{[R^-K^+][Li^+]} \tag{23-1}$$

The equilibrium constant is called the *selectivity coefficient*, because it describes the relative affinities of the resin for Li^+ and K^+. Discrimination between different ions tends to increase with the extent of cross-linking, because the resin pore size shrinks as cross-linking increases.

$H^{\delta+}$
$2\delta-$
$Li^+ \cdots O$
$H^{\delta+}$
Electrostatic attraction

The **hydrated radius** of an ion is the effective size of the ion plus its tightly bound sheath of water molecules, which are attracted by the positive or negative charge of the ion. The large species $Li(H_2O)_x^+$ does not have as much access to the resin as the smaller species $K(H_2O)_y^+$.

More highly charged ions bind more tightly to ion-exchange resins. For ions of the same charge, the larger the hydrated radius, the less tightly the ion is bound. An approximate order of selectivity for some cations is

$$Pu^{4+} \gg La^{3+} > Y^{3+} > Sc^{3+} > Al^{3+} \gg Ba^{2+} > Pb^{2+} > Sr^{2+} >$$
$$Ca^{2+} > Ni^{2+} > Cd^{2+} > Cu^{2+} > Co^{2+} > Zn^{2+} > Mg^{2+} \gg Tl^+ >$$
$$Ag^+ > Cs^+ > Rb^+ > K^+ > NH_4^+ > Na^+ > H^+ > Li^+$$

A large excess of one ion will displace another ion from the resin.

Reaction 23-1 can be driven in either direction. Washing a column containing K^+ with a substantial excess of Li^+ will replace K^+ with Li^+. Washing a column in the Li^+ form with excess K^+ will convert the resin into the K^+ form.

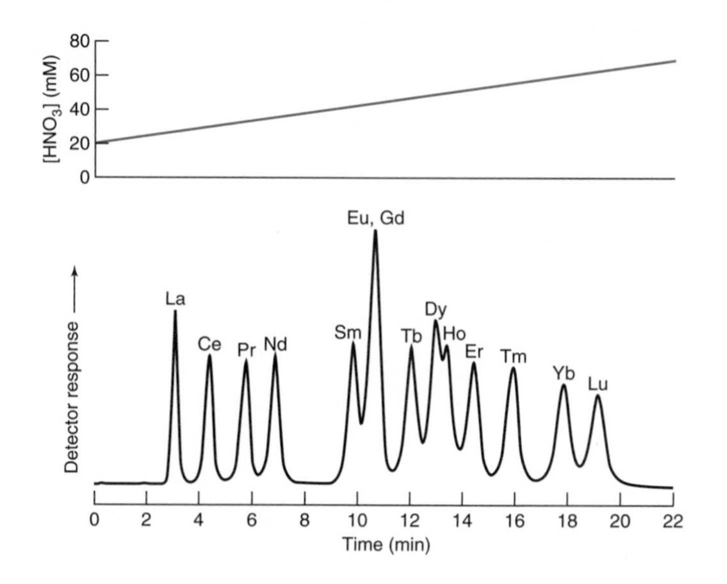

Figure 23-2 Elution of lanthanide(III) ions from a cation exchanger, by using a gradient of H^+ (20–80 mM HNO_3 over 25 min) to drive off more strongly retained cations. The higher the atomic number of the lanthanide, the smaller its ionic radius and the more strongly it binds to chelating groups on the resin. Lanthanides were detected spectrophotometrically by reaction with a color-forming reagent after elution. [From Y. Inoue, H. Kumagai, Y. Shimomura, T. Yokoyama, and T. M. Suzuki, *Anal. Chem.* **1996**, *68*, 1517.]

An ion exchanger loaded with one ion will bind a small amount of a different ion nearly quantitatively (completely). A resin loaded with K^+ binds small amounts of Li^+ quantitatively, even though the selectivity is greater for K^+. The same resin binds large quantities of Ni^{2+} because the selectivity for Ni^{2+} is greater than that for K^+. Even though Fe^{3+} is bound more tightly than H^+, Fe^{3+} can be quantitatively removed from the resin by washing with excess acid.

"Quantitative" is chemists' jargon for "complete."

To separate one ion from another by ion-exchange chromatography, *gradient elution* with increasing ionic strength (ionic concentration) in the eluent is extremely valuable. In Figure 23-2, a gradient of $[H^+]$ was used to separate lanthanide cations (M^{3+}). The more strongly bound metal ions require a higher concentration of H^+ for elution. An ionic strength gradient is analogous to a solvent gradient in HPLC or a temperature gradient in gas chromatography. Box 23-1 shows some applications of ion-exchange chromatography.

Box 23-1 Applications of Ion Exchange

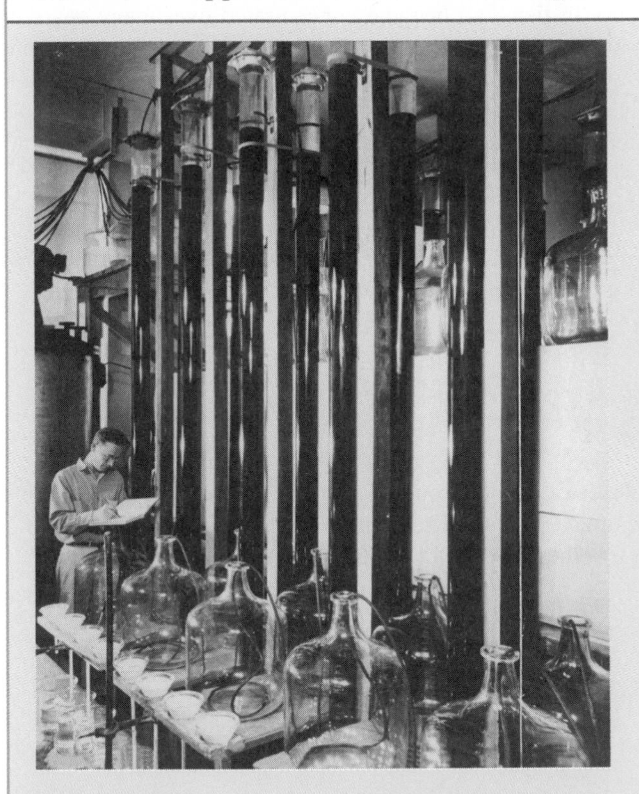

Preparative-scale ion-exchange columns used to separate rare earths for the Manhattan Project during World War II. [Iowa State University Library Special Collections Department.]

In the rush to create an atomic bomb during World War II, it was necessary to isolate significant quantities of pure lanthanide elements (rare earth elements 57 to 71).[1] The photograph shows some of 12 ion-exchange columns (10 cm diameter × 3.0 m long) in a pilot plant at Iowa State College. Each column took several *weeks* to separate a 50- to 100-g mixture of rare earth chlorides.

In the 1950s, W. H. Stein and S. Moore conducted pioneering work at Rockefeller Institute to understand the structure of the enzyme ribonuclease, for which they were awarded the Noble Prize in 1972. Continual need to measure amino acids led them to explore amino acid separations by ion exchange.[2] By 1958, their design led to the first commercial automated amino acid analyzer—a revolutionary advance for biochemistry.

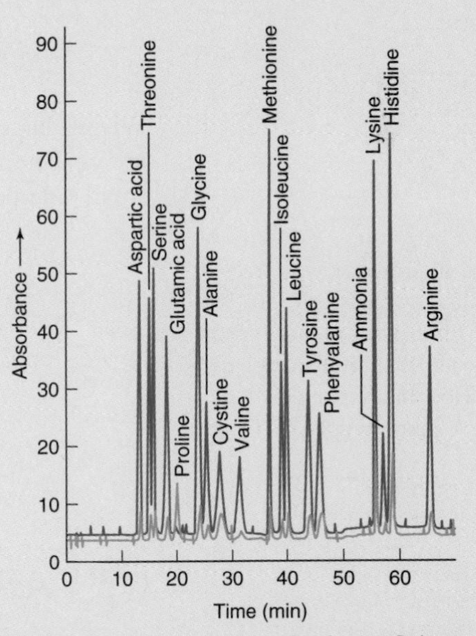

Ion-exchange chromatogram from Beckman-Spinco model 121MB amino acid analyzer introduced in 1969. After separation, amino acids were derivatized with ninhydrin to form colored products detected by visible absorption. [Courtesy Beckman-Coulter, Fullerton, CA.]

What Is Deionized Water?

Home water softeners use ion exchange to replace Ca^{2+} and Mg^{2+} from "hard" water with Na^+ (Box 13-3).

Deionized water is prepared by passing water through an anion-exchange resin loaded with OH^- and a cation-exchange resin loaded with H^+. Suppose, for example, that $Cu(NO_3)_2$ is present in the water. The cation-exchange resin binds Cu^{2+} and replaces it with $2H^+$. The anion-exchange resin binds NO_3^- and replaces it with OH^-. The H^+ and OH^- combine, so the eluate is pure water:

Charge is conserved during ion exchange. One Cu^{2+} displaces $2H^+$ from a cation-exchange column. It takes $3H^+$ to displace one Fe^{3+}. One SO_4^{2-} displaces $2OH^-$ from an anion-exchange column.

$$\left.\begin{array}{l} Cu^{2+} \xrightarrow{H^+ \text{ ion exchange}} 2H^+ \\ 2NO_3^- \xrightarrow{OH^- \text{ ion exchange}} 2OH^- \end{array}\right\} 2H^+ + 2OH^- \longrightarrow \text{pure } H_2O$$

Preconcentration

Measuring extremely low levels of analyte is called **trace analysis**. Trace analysis is especially important for environmental problems in which low concentrations of substances, such as mercury in fish, can become concentrated over many years in people who eat large quantities of fish. For trace analysis, analyte concentration may be so low that it cannot be measured without **preconcentration**, a process in which analyte is brought to a higher concentration prior to analysis.

Metals in natural waters can be preconcentrated with a cation-exchange column:

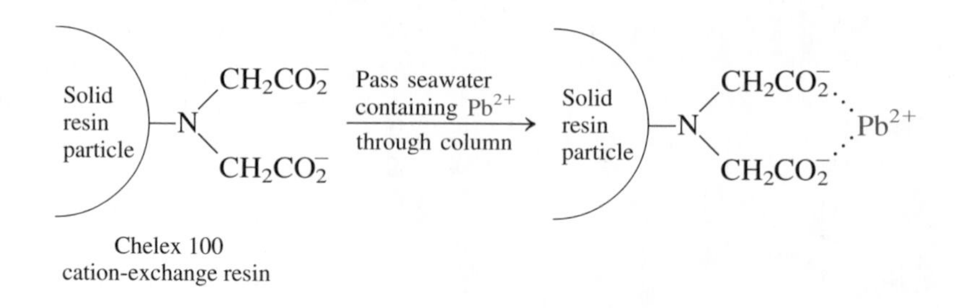

When a large volume of water is passed through a small volume of resin, the cations are concentrated into the small column. The cations can then be displaced into a small volume of solution by eluting the column with concentrated acid:

When lead in seawater was preconcentrated by a factor of 50, the detection limit was reduced from 15 ng of lead per liter of seawater to 0.3 ng/L. How many grams is 1 ng? Is 1 ng/L 1 ppm, 1 ppb, or 1 ppt?

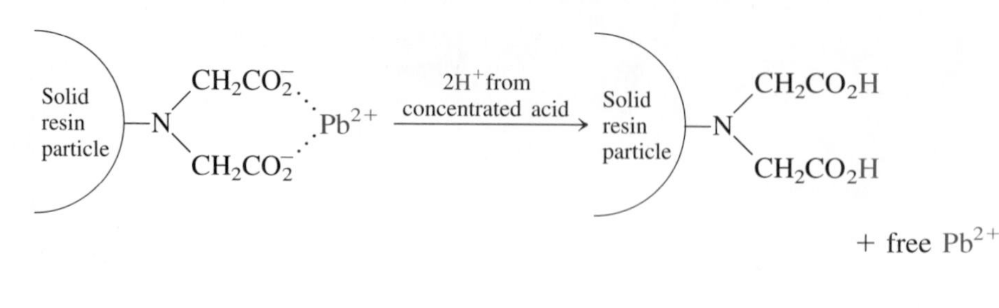

Ask Yourself

23-A. (a) What is deionized water? What kind of impurities are not removed by deionization?

(b) Why is gradient elution used in Figure 23-2?

(c) Explain how preconcentration of cations with an ion exchanger works. Why must the concentrated acid eluent be very pure?

23-2 Ion Chromatography

Ion chromatography is a high-performance version of ion-exchange chromatography, with a key modification that removes eluent ions before detecting analyte ions. Ion chromatography is the method of choice for anion analysis. It is used in the semiconductor industry to monitor anions and cations at 0.1-ppb levels in deionized water. Figure 23-3 shows an example of anion chromatography in environmental analysis.

In ion chromatography, anions are separated by ion exchange and detected by their electrical conductivity. The conductivity of the electrolyte in the eluent is ordinarily high enough to make it difficult or impossible to detect the conductivity change when analyte ions are eluted. Therefore, the key feature of *suppressed-ion* chromatography is removal of unwanted electrolyte prior to conductivity measurement.

In Figure 23-4, a sample containing $NaNO_3$ and $CaSO_4$ is injected into the *separator column*, which is an anion-exchange column with CO_3^{2-} at the anion-exchange sites. Upon elution with KOH, NO_3^- and SO_4^{2-} equilibrate with the resin and are slowly displaced by OH^-. Na^+ and Ca^{2+} are not retained and simply wash through. Eventually, KNO_3 and K_2SO_4 are eluted from the separator column. These species cannot be easily detected, however, because eluate contains KOH, whose high conductivity obscures that of the analytes.

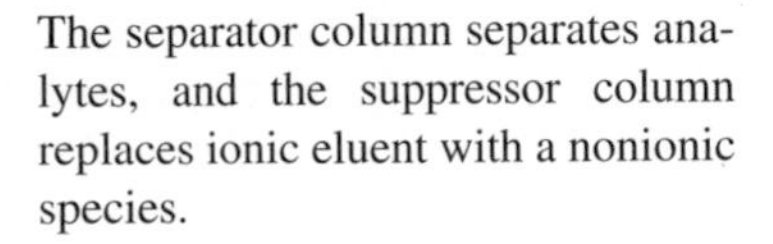
The separator column separates analytes, and the suppressor column replaces ionic eluent with a nonionic species.

To remedy this problem, the solution next passes through a *suppressor*, in which cations are replaced by H^+. H^+ exchanges with K^+, in this example, through a cation-exchange membrane in the suppressor. H^+ diffuses from high concentration outside the membrane to low concentration inside the membrane. K^+ diffuses from high concentration inside to low concentration outside. K^+ is carried away outside the membrane, so its concentration is always low on the outside. The net result is that KOH eluent, which has high conductivity, is converted into H_2O, which has low conductivity. When analyte is present, HNO_3 or H_2SO_4 with high conductivity is produced and detected. Automated systems generate eluent and

Figure 23-3 Converting one human hazard into another. Chlorination of drinking water converts some organic compounds into potential carcinogens, such as $CHCl_3$. To reduce this risk, ozone (O_3) has replaced Cl_2 in some municipal purification systems. Unfortunately, O_3 converts bromide (Br^-) into bromate (BrO_3^-), another carcinogen that must be monitored. The figure shows an anion chromatographic separation of ions found in drinking water. With preconcentration of the water, the detection limit for bromate is ~2 ppb. [From R. J. Joyce, *Am. Environ. Lab.*, May 1994, p. 1.]

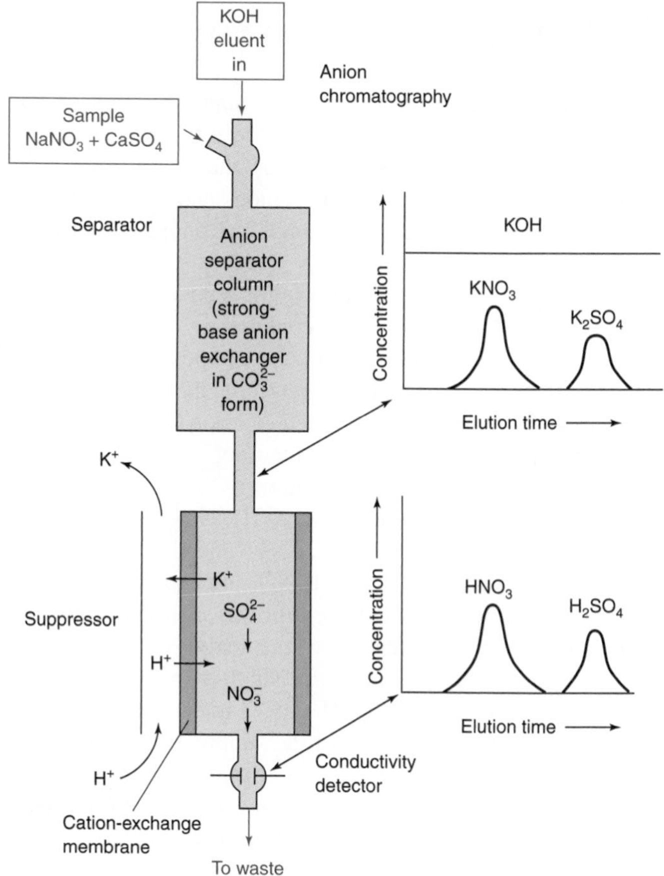

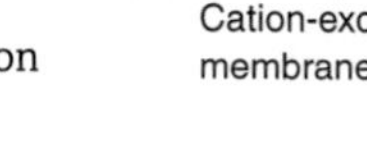

Figure 23-4
Suppressed-ion anion chromatography.

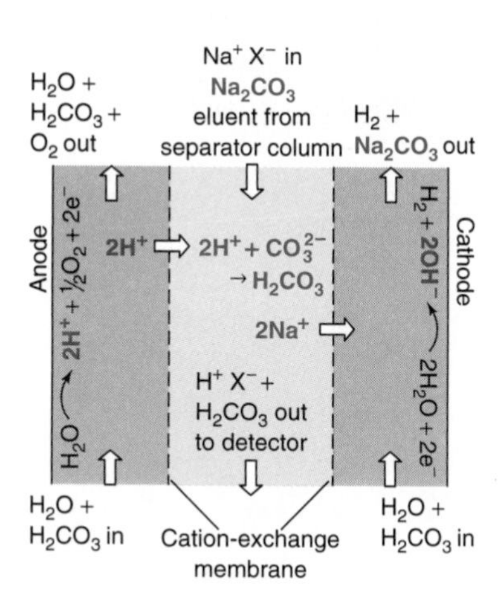

Electrolytic suppressor for anion chromatography. [Adapted from Y. Liu, Z. Lu, C. Pohl, J. Madden, and N. Shirakawa, *Am. Lab.* February 2007, p. 17.]

suppressor electrolytically. It is only necessary to add deionized water to compensate for evaporation to allow an automated system to run for a month.

Ask Yourself

23-B. **(a)** What do the separator and suppressor do in ion chromatography?
(b) Carbonate eluent for ion chromatography is suppressed by conversion to H_2CO_3, which has low conductivity. The diagram in the margin shows the operation of an electrolytic suppressor. Explain how it works, paying attention to the stoichiometry of the electrode reactions. Why does this suppressor use cation-exchange membranes, not anion-exchange membranes?

23-3 Molecular Exclusion Chromatography

Large molecules pass through the column *faster* than small molecules do.

In **molecular exclusion chromatography** (also called *gel filtration, gel permeation,* or *size exclusion chromatography*), molecules are separated according to their size. Small molecules enter the small pores in the stationary phase, but large molecules do not (Figure 21-2). Because small molecules must pass through an effectively larger volume in the column, large molecules are eluted first (Figure 23-5). This

Figure 23-5 (*a*) A mixture of large and small molecules is applied to the top of a molecular exclusion chromatography column. (*b*) Large molecules cannot penetrate the pores of the stationary phase, but small molecules can. Therefore less of the volume is available to large molecules and they move down the column faster.

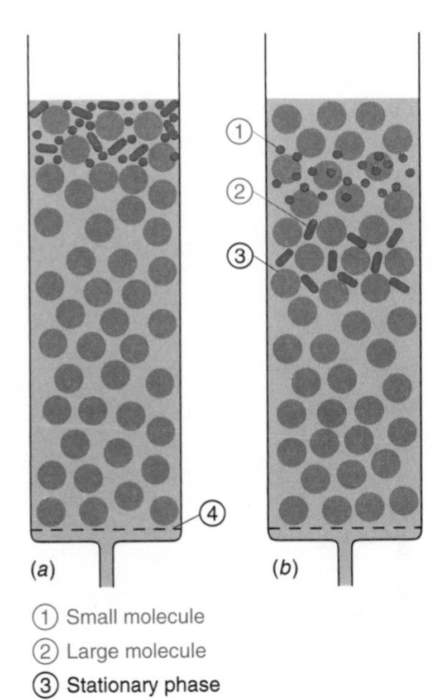

technique is widely used in biochemistry and polymer chemistry to purify macromolecules and to measure molecular mass (Figure 23-6). Figure 23-7 shows an application in materials chemistry.

In molecular exclusion chromatography, the volume of mobile phase (the solvent) in the column outside the stationary phase is called the *void volume*, V_0. Large molecules that are excluded from the stationary phase are eluted in the void volume. Void volume is measured by passing through the column a molecule that is too large to enter the pores. The dye Blue Dextran (2×10^6 Da) is commonly used.

Molecular Mass Determination

Retention volume is the volume of mobile phase required to elute a particular solute from the column. Each stationary phase has a range over which there is a logarithmic relation between molecular mass and retention volume. We can estimate the molecular mass of an unknown by comparing its retention volume with those of

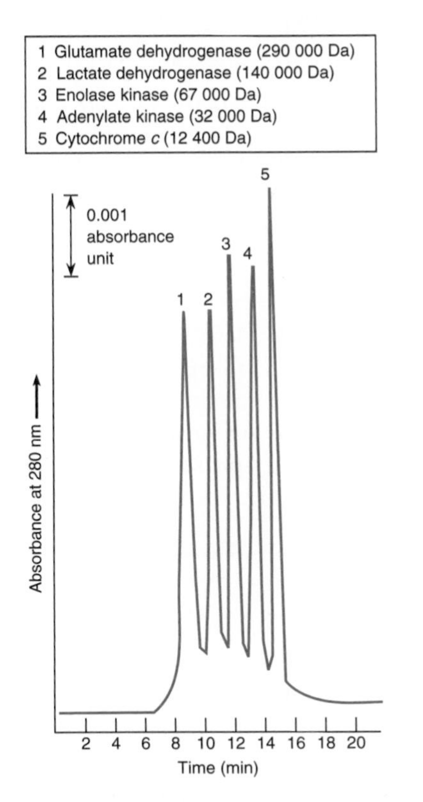

Figure 23-6 Separation of proteins by molecular exclusion chromatography, using a TSK 3000SW HPLC column. The highest molecular masses are eluted first. [Courtesy Varian Associates, Palo Alto, CA.]

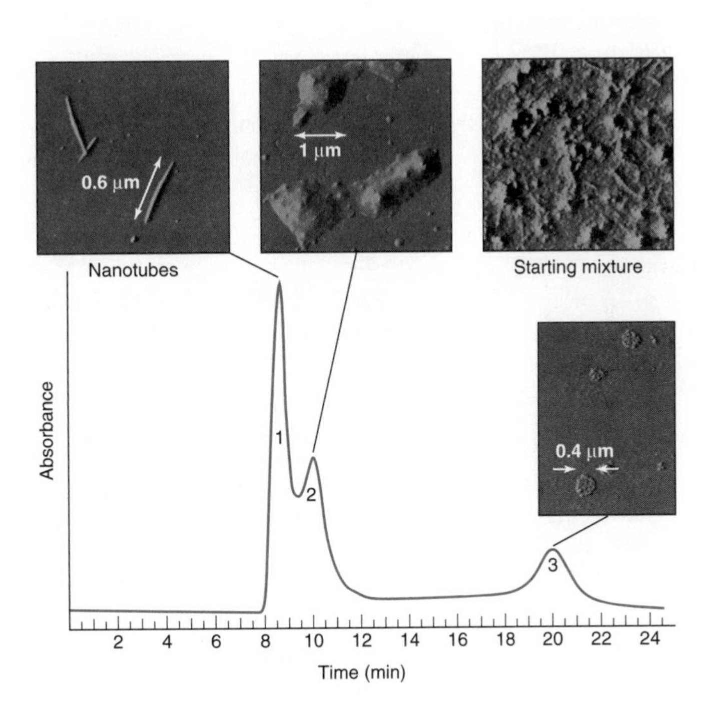

Figure 23-7 Purification of carbon nanotubes by molecular exclusion chromatography. An electric arc struck between graphite rods creates nanometer-size carbon products, including tubes with extraordinary strength and possible use in electronic devices. Molecular exclusion chromatography separates nanotubes (fraction 1) from other forms of carbon in fractions 2 and 3. The stationary phase is PLgel MIXED-A, a polystyrene-divinylbenzene resin with pore sizes corresponding to a molecular mass range of 2 000–40 000 000 Da. Images of carbon in each fraction were made by atomic force microscopy. [From B. Zao, H. Hu, S. Niyogi, M. E. Itkis, M. A. Hamon, P. Bhowmik, M. S. Meier, and R. C. Haddon, *J. Am. Chem. Soc.* **2001**, *123*, 11673.]

standards. For proteins, it is important to use eluent with an ionic strength (such as 0.05 M NaCl) high enough to eliminate electrostatic adsorption of solute by occasional charged sites on the gel.

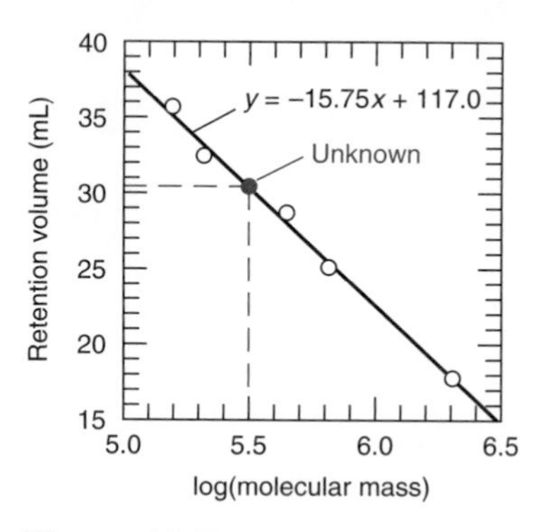

Figure 23-8 Calibration curve used to estimate the molecular mass of an unknown protein by molecular exclusion chromatography.

Example **Molecular Mass Determination by Gel Filtration**

Proteins were passed through a gel filtration column and retention volumes (V_r) were measured. Estimate the molecular mass (MM) of the unknown.

Compound	V_r (mL)	Molecular mass	Log(molecular mass)
Blue Dextran 2000	17.7	2×10^6	6.301
Aldolase	35.6	158 000	5.199
Catalase	32.3	210 000	5.322
Ferritin	28.6	440 000	5.643
Thyroglobulin	25.1	669 000	5.825
Unknown	30.3	?	

SOLUTION Figure 23-8 plots V_r versus log(MM). Putting the retention volume of unknown into the least-squares fit to the five calibration standards allows us to solve for the molecular mass of the unknown:

$$V_r\ (\text{mL}) = -15.75[\log(\text{MM})] + 117.0$$
$$30.3 = -15.75[\log(\text{MM})] + 117.0$$
$$\Rightarrow \log(\text{MM}) = 5.505 \Rightarrow \text{MM} = 10^{5.505} = 320\ 000$$

Test Yourself What is the expected retention volume of a protein with a molecular mass of 888 000? (**Answer:** 23.3 mL)

Ask Yourself

23-C. A gel filtration column has a radius (r) of 0.80 cm and a length (l) of 20.0 cm.
(a) Calculate the total volume of the column, which is equal to $\pi r^2 l$.
(b) Blue Dextran was eluted in a volume of 18.2 mL. What volume is occupied by the stationary phase plus the solvent inside the pores of the stationary phase?
(c) Suppose that the pores occupy 60.0% of the stationary-phase volume. Over what volume range (from x mL for the largest molecules to y mL for the smallest molecules) are all solutes expected to be eluted?

23-4 Affinity Chromatography

Affinity chromatography is used to isolate a single compound from a complex mixture. The technique is based on specific binding of that one compound to the stationary phase (Figure 21-2). When sample is passed through the column, only one solute is bound. After everything else has washed through, the one adhering solute is eluted by changing conditions such as pH or ionic strength to weaken its binding. Affinity chromatography is especially applicable in biochemistry and is based on specific interactions between enzymes and substrates, antibodies and antigens, or receptors and hormones.

Figure 23-9 shows the isolation of the protein immunoglobulin G (IgG) by affinity chromatography on a column containing covalently bound *protein A*. Protein

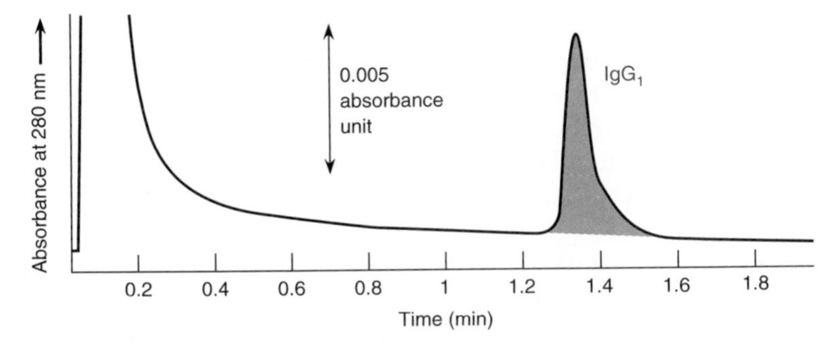

Figure 23-9 Purification of monoclonal antibody IgG by affinity chromatography on a column (4.6 mm diameter × 5 cm long) containing protein A covalently attached to a polymer support. Other proteins in the sample are eluted from 0 min to 0.3 min at pH 7.6. When the eluent pH is lowered to 2.6, IgG is freed from protein A and emerges from the column. [From B. J. Compton and L. Kreilgaard, *Anal. Chem.* **1994**, *66*, 1175A.]

A binds to one specific region of IgG at pH $\gtrsim$ 7.2. When a crude mixture containing IgG and other proteins was passed through the column at pH 7.6, everything except IgG was eluted within 0.3 min. At 1 min, the eluent pH was lowered to 2.6 and IgG was cleanly eluted at 1.3 min.

23-5 What Is Capillary Electrophoresis?

Electrophoresis is the migration of ions in an electric field. Anions are attracted to the anode, and cations are attracted to the cathode. Different ions migrate at different speeds. **Capillary electrophoresis** is a high-resolution separation technique conducted with solutions of ions in a narrow capillary tube. A clever modification of the technique allows us to separate neutral analytes also. Capillary electrophoresis applies with equal ease to the separation of macromolecules, such as proteins and DNA, and small species, such as Na^+ and benzene. Capillary electrophoresis can analyze the contents of a single cell.

Cations are attracted to the negative terminal (the cathode). Anions are attracted to the positive terminal (the anode).

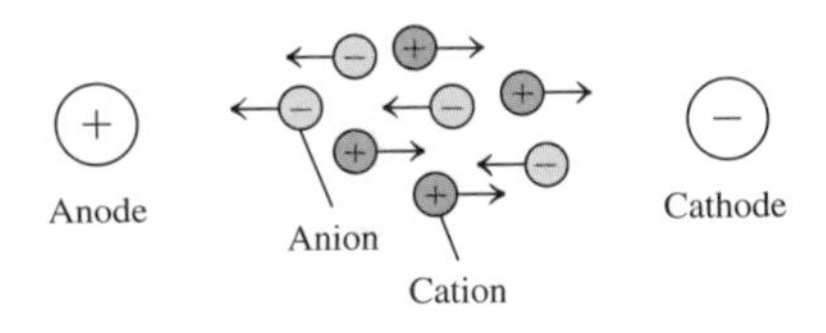

The typical experiment in Figure 23-10 features a fused-silica (SiO_2) capillary that is 50 cm long and has an inner diameter of 25–75 μm. The capillary is

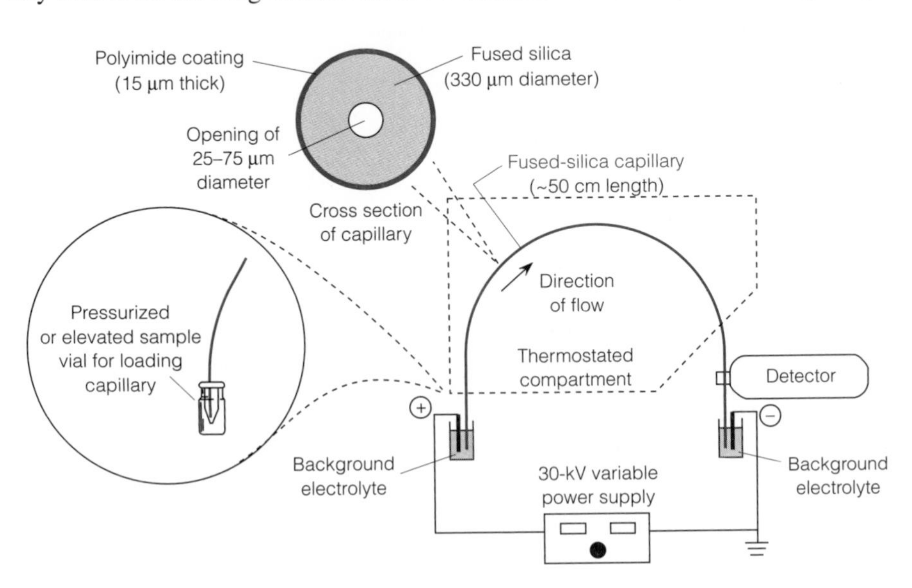

Figure 23-10 Capillary electrophoresis. Sample is injected by elevating or applying pressure to the sample vial or by applying suction at the outlet of the capillary.

Background electrolyte, also called *run buffer*, is the solution in the electrode reservoirs. It controls pH and ionic composition in the capillary.

The greater the charge on the ion, the faster it migrates in the electric field. The greater the size of the ion, the slower it migrates.

Working ranges for chloride analysis:

Ion-selective electrode

Capillary electrophoresis

Ion chromatography

1 ppb 1 ppm 1 000 ppm

1 ng/mL 1 μg/mL 1 mg/mL

immersed in *background electrolyte* solution at each end. At the start of the experiment, one end is dipped into a sample vial and ~10 nL (nanoliters, 10^{-9} L) of liquid are introduced by applying pressure to the sample vial or by applying an electric field between the sample vial and the column. After the capillary is placed back into the electrolyte solution, 20–30 kV is applied to the electrodes to cause ions in the capillary to migrate. Different ions migrate at different speeds, so they separate from one another as they travel through the capillary. Ions are detected inside the capillary near the far end with an ultraviolet absorbance monitor or other detector. The graph of detector response versus time in Figure 23-11 is called an *electropherogram.* (In chromatography, we call the same graph a *chromatogram.*) Capillary electrophoresis is not as sensitive as ion chromatography, but it is more sensitive than many currently available ion-selective electrodes.

Capillary electrophoresis can provide extremely narrow bands. Three mechanisms of band broadening in chromatography are longitudinal diffusion (B in the van Deemter equation 21-7), the finite rate of mass transfer between the stationary and mobile phases (C in the van Deemter equation), and multiple flow paths around particles (A in the van Deemter equation). An open tubular column in chromatography or electrophoresis reduces band broadening (relative to that of a packed column) by eliminating multiple flow paths (the A term). Capillary electrophoresis further reduces broadening by eliminating the mass transfer problem (the C term) because *there is no stationary phase.* The only source of broadening under ideal conditions is longitudinal diffusion of solute as it migrates through the capillary. The routine separation efficiency of 50 000 to 500 000 theoretical plates in capillary electrophoresis is an order of magnitude greater than that of chromatography.

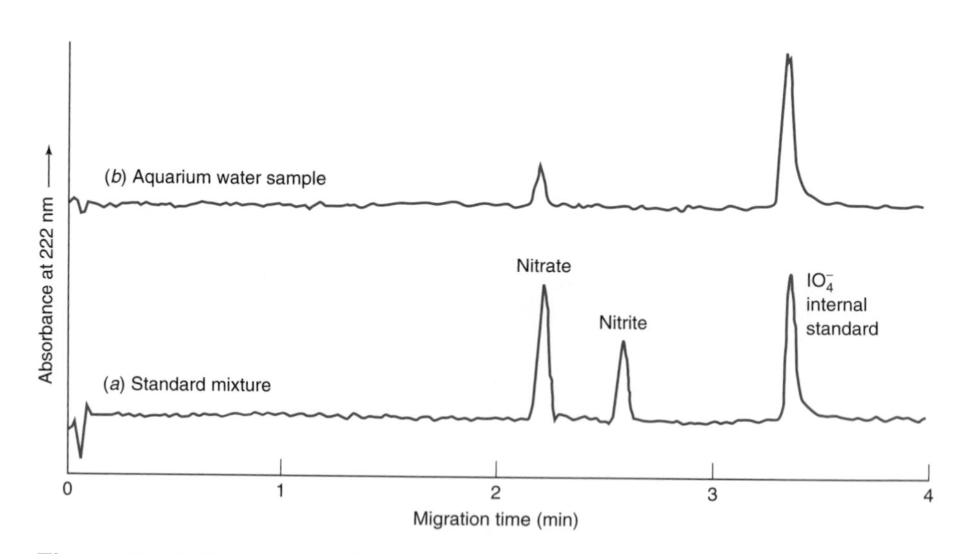

Figure 23-11 Measurement of nitrate in aquarium water (Box 6-1) by capillary electrophoresis. At the detection wavelength of 222 nm, many species in the water have too little absorbance to be observed. (*a*) Standard mixture containing 15 μg/mL nitrate (NO_3^-), 5 μg/mL nitrite (NO_2^-), and 10 μg/mL of the internal standard, periodate (IO_4^-). (*b*) 1:100 dilution of aquarium water with distilled water containing internal standard. This aquarium has NO_3^- but no detectable NO_2^-. [From D. S. Hage, A. Chattopadhyay, C. A. C. Wolfe, J. Grundman, and P. B. Kelter, *J. Chem. Ed.* **1998**, *75*, 1588.] Periodate is a questionable standard because it is an oxidizing agent that might react with organic matter in the aquarium water.

Ask Yourself

23-D. Capillary electrophoresis is noteworthy for analyzing small sample volumes and for producing high-resolution separations.
(a) A typical injected sample occupies a 5-mm length of the capillary. What volume of sample is this if the inside diameter of the capillary is 25 μm? 50 μm? (The volume of a cylinder of radius r is $\pi r^2 \times$ length.)
(b) Which mechanisms of band broadening that operate in chromatography are absent in capillary electrophoresis?

23-6 How Capillary Electrophoresis Works

Capillary electrophoresis involves two simultaneous processes called *electrophoresis* and *electroosmosis.* Electrophoresis is the migration of ions in an electric field. Electroosmosis pumps the entire solution through the capillary from the anode toward the cathode. Superimposed on this one-way flow are the flow of cations, which are attracted to the cathode, and the flow of anions, which are attracted to the anode. In Figure 23-10, cations migrate from the injection end at the left toward the detector at the right. Anions migrate toward the left in Figure 23-10. Both cations and anions are swept from left to right by electroosmosis. Cations arrive at the detector before anions. Neutral molecules swept along by electroosmosis arrive at the detector after the cations and before the anions.

Two processes operate in capillary electrophoresis:

- *electrophoresis:* migration of cations to the cathode and anions to the anode
- *electroosmosis:* migration of bulk fluid toward the cathode

Electroosmosis

Electroosmosis is the propulsion of fluid inside a fused-silica capillary from the anode toward the cathode caused by the applied electric field. To understand electroosmosis, consider what happens at the inside wall of the capillary. The wall is covered with silanol (Si—OH) groups that are negatively charged (Si—O^-) above pH 2. Figure 23-12*a* shows that the capillary wall and the solution immediately adjacent to the wall form an *electric double layer.* The double layer is composed of (1) a negative charge fixed to the wall and (2) an equal positive charge in solution adjacent to the wall. The thickness of the positive layer, called the *diffuse part of the double layer,* is approximately 1 nm. When an electric field is applied, cations are attracted to the cathode and anions are attracted to the anode. Excess cations in the diffuse part of the double layer drive the entire solution in the capillary toward the cathode (Figure 23-12*b*). The greater the applied electric field, the faster the flow.

Ions in the diffuse part of the double layer adjacent to the capillary wall are the "pump" that drives electroosmotic flow.

Color Plate 23 shows a critical distinction between electroosmotic flow induced by an electric field and ordinary hydrodynamic flow induced by a pressure difference. Because it is driven by ions on the walls of the capillary, electroosmotic flow is uniform across the diameter of the liquid, as shown schematically in Figure 23-12*c*. The only mechanism that broadens the moving band is diffusion. In contrast, hydrodynamic flow has a parabolic velocity profile, with fastest motion at the center of the capillary and little velocity at the wall. A parabolic profile creates broad bands.

Electroosmosis decreases at low pH because the wall loses its negative charge when Si—O^- is converted into Si—OH and the number of cations in the double layer diminishes. Electroosmotic velocity is measured by adding an ultraviolet-absorbing neutral solute, such as methanol, to the sample and measuring the time it takes (called the *migration time*) to reach the detector. In one experiment with 30 kV

Migration time in electrophoresis is analogous to retention time in chromatography.

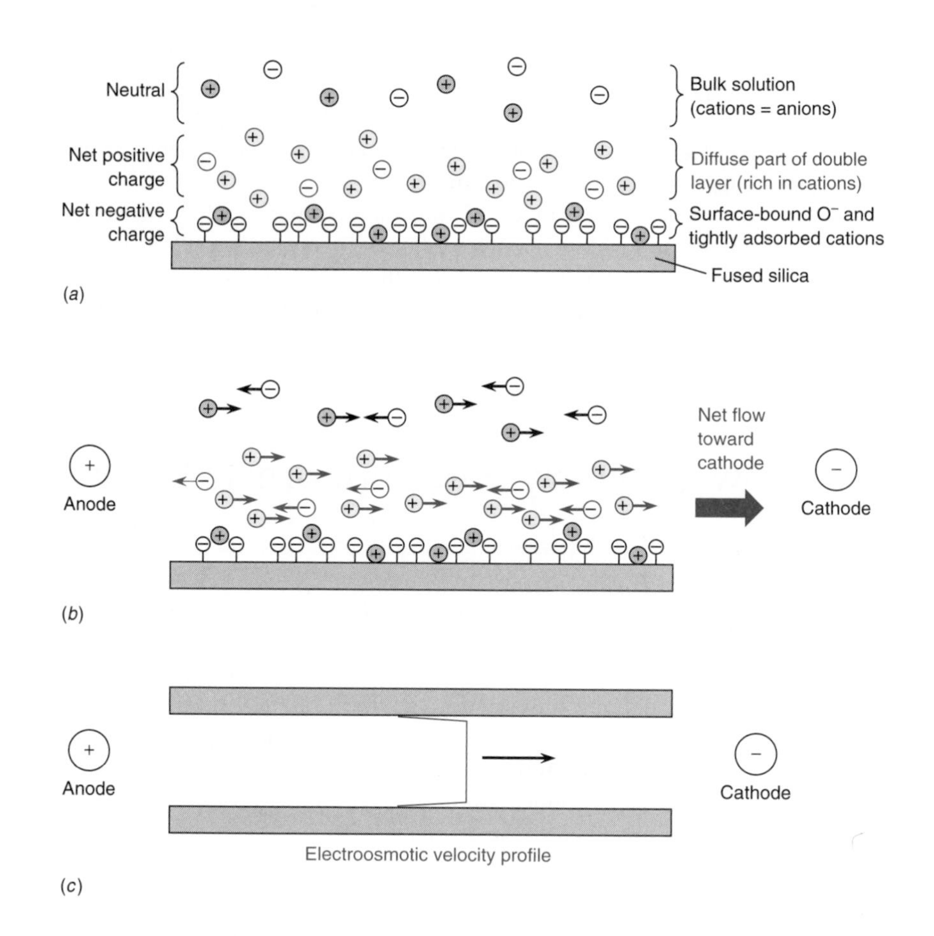

Figure 23-12 (*a*) Electric double layer is created by negative silica surface and excess cations in the diffuse part of the double layer in the solution near the wall. The wall is negative and the diffuse part of the double layer is positive.
(*b*) Predominance of cations in diffuse part of the double layer produces net electroosmotic flow toward the cathode when an external field is applied.
(*c*) Electroosmotic velocity profile is uniform over more than 99.9% of the cross section of the capillary. A capillary is required to maintain constant temperature in the liquid. Temperature variation in larger-diameter tubes causes bands to broaden.

across a 50-cm capillary, the electroosmotic velocity was 4.8 mm/s at pH 9 and 0.8 mm/s at pH 3.

In Figures 23-10 and 23-12, electroosmosis is from left to right because cations in the double layer are attracted to the cathode. Superimposed on electroosmosis of the bulk fluid, electrophoresis transports cations to the right and anions to the left. At neutral or high pH, electroosmosis is faster than electrophoresis and the net flow of anions is to the right. At low pH, electroosmosis is weak and anions may flow to the left and never reach the detector. To separate anions at low pH, you can reverse the polarity to make the sample side negative and the detector side positive.

Detectors

The most common detector is an *ultraviolet absorbance monitor* set to a wavelength near 200 nm, where many solutes absorb. It is not possible to use such short wavelengths with larger-diameter columns, because the solvent absorbs too much radiation. A *fluorescence detector* works for fluorescent analytes or fluorescent derivatives. To measure either absorbance or fluorescence, the protective polyimide coating on the capillary in Figure 23-10 is removed at the location of the detector. *Electrochemical detection* is sensitive to analytes that can gain or lose electrons at an electrode. Eluate can be directed into a *mass spectrometer* with an electrospray interface to provide information on the quantity and molecular structure of analyte. *Conductivity detection* with ion-exchange suppression of the background electrolyte (as in ion chromatography, Figure 23-4) gives a sensitivity of 1–10 ppb for small ions.

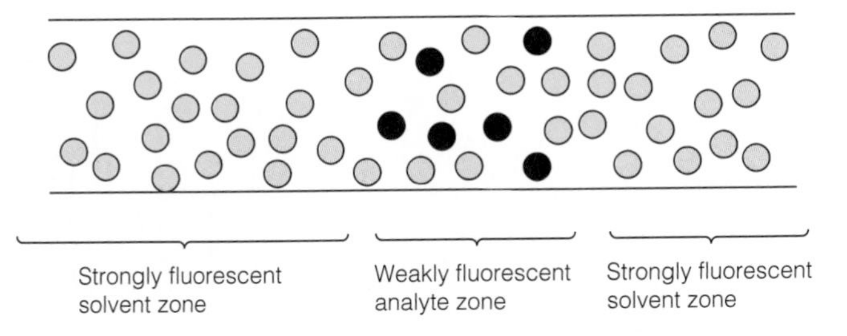

Figure 23-13 *Principle of indirect detection:* When analyte emerges from the capillary, the strong background signal *decreases.*

In contrast with direct detection of analyte discussed so far, **indirect detection** relies on measuring a strong signal from background electrolyte and a weak signal from analyte as it passes the detector. Figure 23-13 illustrates indirect fluorescence detection, but the same principle applies to any type of detection. A fluorescent ion with the same sign of charge as the analyte is added to background electrolyte to provide a steady background signal. In the analyte zone, there is less background ion because electroneutrality must be preserved. If the analyte ion is not fluorescent, the fluorescence level decreases when analyte emerges. We observe a *negative* signal.

Figure 23-14 shows indirect ultraviolet detection of Cl^- in the presence of ultraviolet-absorbing chromate, CrO_4^{2-}. In the absence of analyte, CrO_4^{2-} gives a steady absorbance at 254 nm. When Cl^- reaches the detector, there is less CrO_4^{2-} present and Cl^- does not absorb; therefore the detector signal *decreases.*

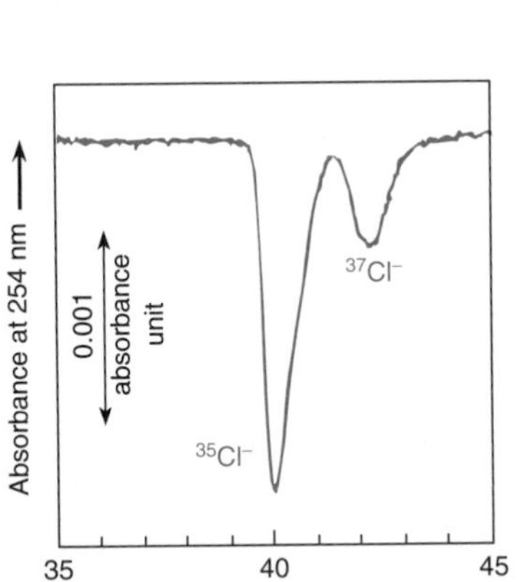

Figure 23-14 Separation of natural isotopes of 0.56 mM Cl^- by capillary electrophoresis with indirect spectrophotometric detection at 254 nm. The background electrolyte contains 5 mM CrO_4^{2-} to provide absorbance at 254 nm. There are few ways to separate isotopes so cleanly. This impressive separation was done by an undergraduate student at the University of Calgary. [From C. A. Lucy and T. L. McDonald, *Anal. Chem.* **1995**, *67*, 1074.]

Ask Yourself

23-E. **(a)** Capillary electrophoresis was conducted at a pH of 9, at which electroosmotic velocity is greater than electrophoretic velocity. Draw a picture of the capillary, showing the anode, cathode, injector, and detector. Show the directions of electroosmotic and electrophoretic flow of a cation and an anion. Show the direction of net flow for each ion.
(b) If the pH is reduced to 3, electroosmotic velocity is less than electrophoretic velocity. In what directions will cations and anions migrate?
(c) Explain why the detector signal is negative in Figure 23-14.

23-7 Types of Capillary Electrophoresis

The type of electrophoresis discussed so far is called **capillary zone electrophoresis**, in which separation is based on different electrophoretic velocities of different ions. Electroosmotic flow of the bulk fluid is toward the cathode (Figure 23-12*b*). Cations migrate faster than the bulk fluid and anions migrate slower than the bulk fluid. Therefore the order of elution is cations before neutrals before anions. If electrode polarity is reversed, the order of elution is anions before neutrals before cations. Neither scheme separates neutral molecules from one another.

Order of elution in capillary zone electrophoresis:
1. cations (highest mobility first)
2. all neutrals (unseparated)
3. anions (highest mobility last)

Capillary zone electrophoresis can separate optical isomers if a suitable *optically active* complexing agent is added to the background electrolyte. An optically active substance, also called a *chiral* substance, is one that is not superimposable on its mirror image. The chiral crown ether in Figure 23-15 can bind to the ammonium group of an amino acid through NH···O hydrogen bonds. The crown ether has

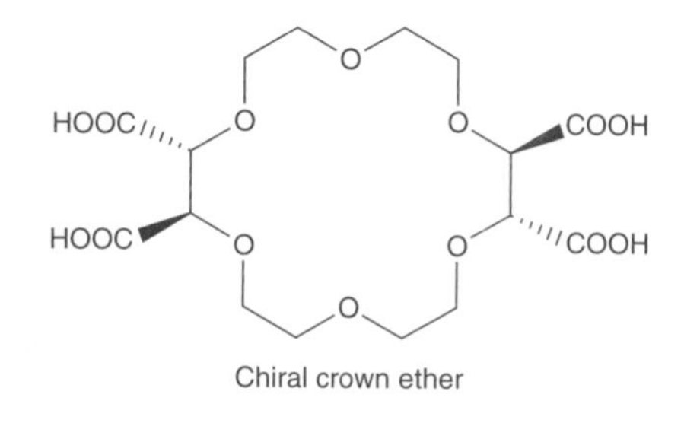

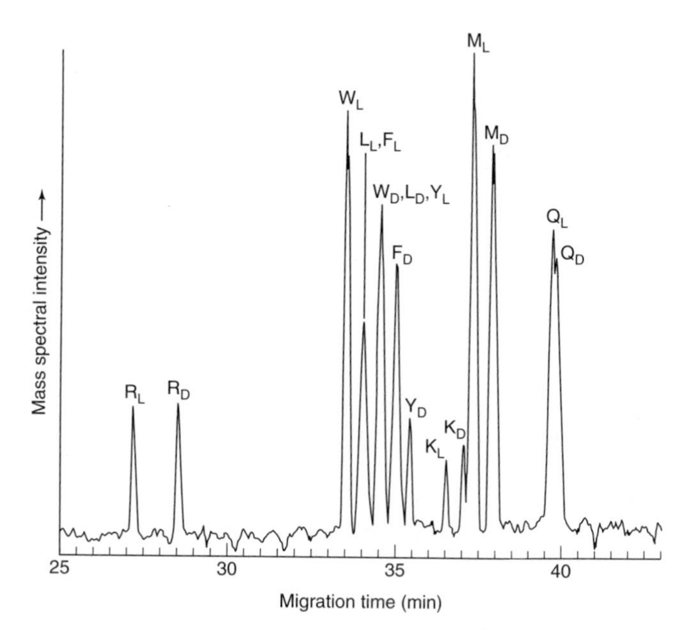

Figure 23-15 Capillary zone electrophoretic separation of D- and L-amino acids by adding a *chiral* crown ether to background electrolyte. Amino acids are designated by one-letter abbreviations listed in Table 11-1. The crown ether binds D-amino acids more strongly than it binds L-amino acids and therefore changes the migration time of the D-amino acid more than that of the L-amino acid. Wide-scan (m/z 74.5–250) mass spectra of liquid exiting the column were recorded continuously. The electropherogram was reconstructed by displaying the sum of ion intensities at m/z 132, 147, 150, 166, 175, 182, and 205. For example, m/z 150 responds only to protonated methionine, designated M in the electropherogram. [From C. L. Schultz and M. Moini, *Anal. Chem.* **2003**, *75*, 1508.]

greater affinity for D-amino acids than for L-amino acids (see structures on page 494). Amino acids migrating through an electrophoresis column spend part of the time complexed to the crown ether, during which time they migrate at a rate different from that of the free amino acid. Migration times of D- and L-amino acids differ if one spends more time bound to the crown ether. Figure 23-15 shows the separation of D- and L-amino acids in the presence of the chiral crown ether.

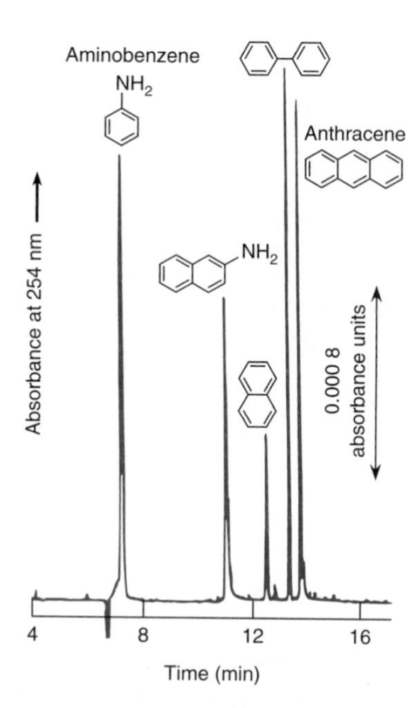

Figure 23-16 Separation of neutral molecules by micellar electrokinetic capillary electrophoresis. The average plate count in this experiment is 250 000 in 50 cm of capillary length. [From J. T. Smith, W. Nashabeh, and Z. E. Rassi, *Anal. Chem.* **1994**, *66*, 1119.]

Micellar Electrokinetic Capillary Chromatography

This mouthful of words describes a form of capillary electrophoresis that separates neutral molecules as well as ions (Figure 23-16). The key modification in **micellar electrokinetic capillary chromatography** is that the capillary solution contains *micelles*, described in Box 23-2, which you should read now.

To understand how neutral molecules are separated, suppose that the background electrolyte contains negatively charged micelles. In Figure 23-17, electroosmotic flow is to the right. Electrophoretic migration of the negatively charged micelles is to the left, but net motion is to the right because electroosmotic flow is faster than electrophoretic flow.

In the absence of micelles, all neutral molecules reach the detector together at a time we designate t_0. Micelles injected with the sample reach the detector at time t_{mc}, which is longer than t_0 because micelles are negative and migrate upstream. If a neutral molecule equilibrates between free solution and the inside of the micelles, its migration time is increased, because it migrates at the slower rate of the micelle part of the time. In this case, the neutral molecule reaches the detector at a time between t_0 and t_{mc}.

Box 23-2 What Is a Micelle?

A **micelle** is an aggregate of molecules with ionic headgroups and long, nonpolar tails. Such molecules are called *surfactants*, an example of which is sodium dodecyl sulfate:

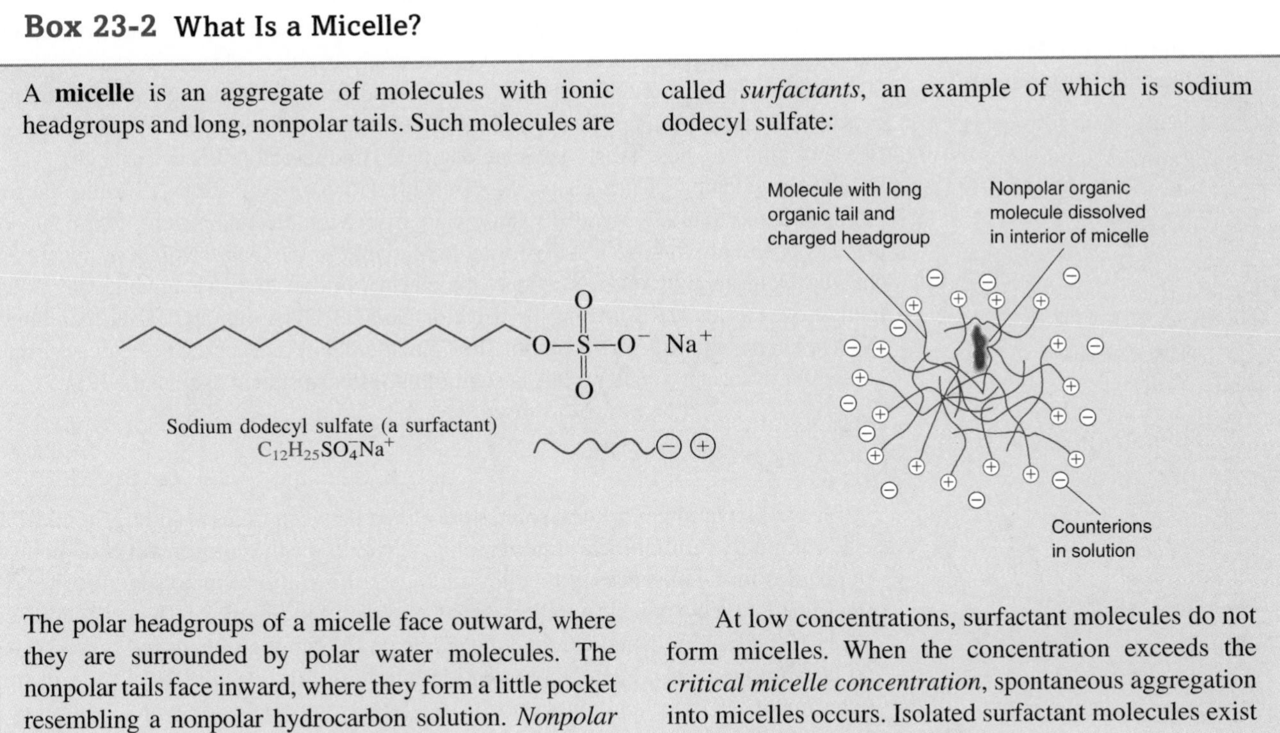

The polar headgroups of a micelle face outward, where they are surrounded by polar water molecules. The nonpolar tails face inward, where they form a little pocket resembling a nonpolar hydrocarbon solution. *Nonpolar solutes are soluble inside the micelle.*

At low concentrations, surfactant molecules do not form micelles. When the concentration exceeds the *critical micelle concentration*, spontaneous aggregation into micelles occurs. Isolated surfactant molecules exist in equilibrium with micelles.

The more soluble the neutral molecule is in the micelle, the more time it spends inside the micelle and the longer is its migration time. The nonpolar interior of a sodium dodecyl sulfate micelle dissolves nonpolar solutes best. Polar solutes are not as soluble in the micelles and have a shorter retention time than nonpolar solutes do. Migration times of cations and anions also can be affected by micelles because ions might associate with micelles. Micellar electrokinetic capillary chromatography is truly a form of chromatography because micelles behave like a pseudostationary phase. Solutes partition between the mobile phase (the aqueous solution) and the pseudostationary micelles.

Micellar electrokinetic capillary electrophoresis: The more time a solute spends inside the micelle, the longer its migration time.

Capillary Gel Electrophoresis

In **capillary gel electrophoresis**, macromolecules are separated by *sieving* as they migrate through a gel inside a capillary tube. Large molecules become entangled in the gel and their motion is slowed. Small molecules travel faster than large molecules

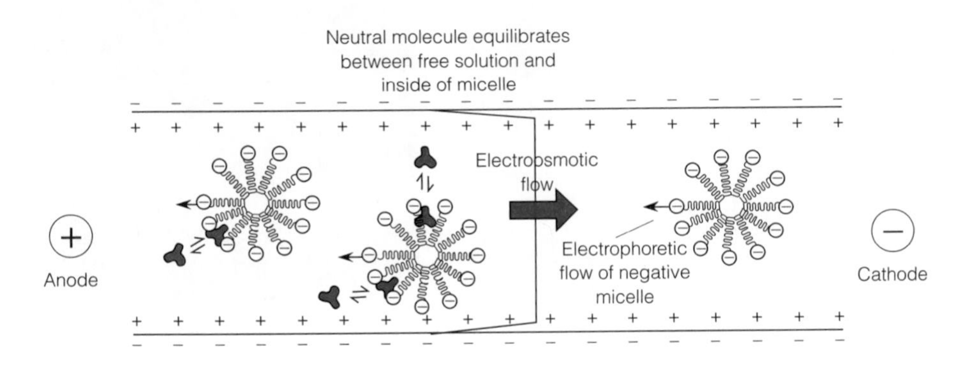

Figure 23-17 Negatively charged sodium dodecyl sulfate micelles migrate upstream against electroosmotic flow. Neutral molecules are in dynamic equilibrium between free solution and the inside of the micelle. The more time spent in the micelle, the more the neutral molecule lags behind the electroosmotic flow.

through the gel. This behavior is the opposite of that in molecular exclusion chromatography, in which large molecules are excluded from stationary-phase particles and move faster than small ones. For DNA sequencing, capillary electrophoresis separates 500 different lengths of DNA in <20 min. In capillary gel electrophoresis, gels are not cross-linked polymers, but just solutions of polymers such as poly(ethylene oxide) or polyacrylamide, whose long chains are entangled and therefore behave as a gel.

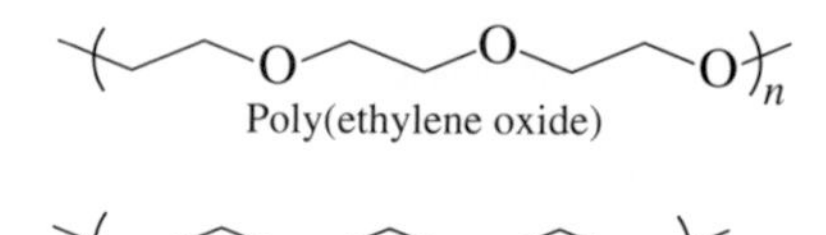

Poly(ethylene oxide)

CONH₂ CONH₂ CONH₂ CONH₂

Polyacrylamide

At the opening of this chapter, DNA with 334 base pairs from a healthy person migrated *faster* than DNA with 330 base pairs from a thalassemia patient. This behavior is the opposite of what we just described for sieving. In the thalassemia test, negatively charged DNA migrates slowly against the electroosmotic flow. Meanwhile, electroosmotic flow carries DNA toward the detector. Smaller DNA migrates faster than larger DNA *upstream* against electroosmotic flow. Smaller DNA reaches the detector *later* than larger DNA because smaller DNA is swimming faster upstream against the flow.

Ask Yourself

23-F. (a) Explain why neutral solutes are eluted between times t_0 and t_{mc} in micellar electrokinetic capillary chromatography, where t_0 is the elution time of neutral molecules in the absence of micelles and t_{mc} is the elution time of the micelles. **(b)** Micellar electrokinetic capillary chromatography in Figure 23-16 was conducted at pH 10 with anionic micelles and the anode on the sample side, as in Figure 23-10. **(i)** Is aminobenzene ($C_6H_5NH_2$) a cation, an anion, or a neutral molecule in this experiment? **(ii)** Explain how Figure 23-16 allows you to decide which compound, aminobenzene or anthracene, is more soluble in the micelles.

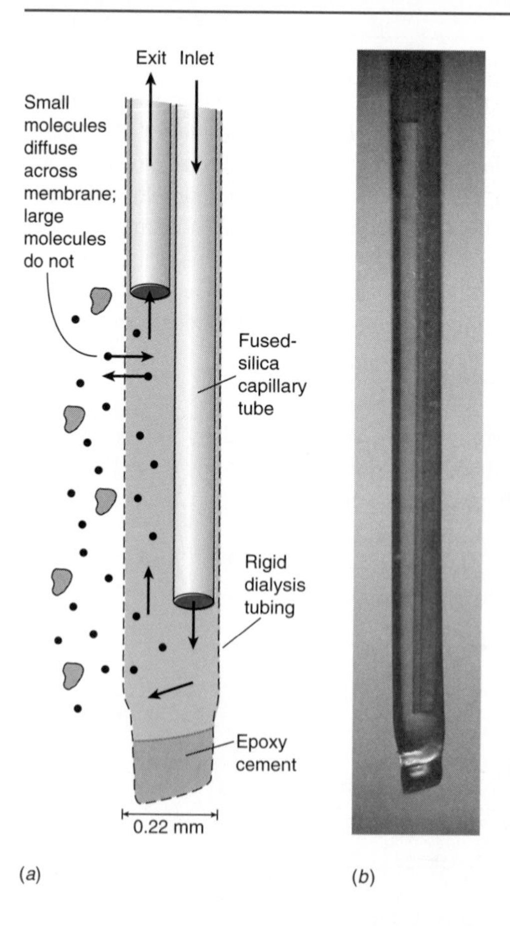

Figure 23-18 (*a*) Diagram and (*b*) photo of microdialysis probe. Small molecules pass through the semipermeable membrane, but large molecules cannot. (*c*) Probe inserted into anesthetized rat to sample chemicals in brain. [Courtesy R. T. Kennedy and Z. D. Sandlin, University of Michigan.]

23-8 Lab on a Chip: Probing Brain Chemistry

One of the most exciting and rapidly developing areas of analytical chemistry is the "lab on a chip." These glass or plastic chips, typically the size of a microscope slide, employ electroosmosis (Figure 23-12) to move liquid with precise control through micrometer-size channels. Chemical reaction are conducted by moving picoliters of fluid from different reservoirs, mixing them, and analyzing products on the chip with a variety of detectors. Chips that manipulate small volumes of liquid are called *microfluidic chips*. We describe one that is coupled to a microdialysis probe to monitor chemicals in the brain.

Dialysis is the process in which small molecules diffuse across a *semipermeable membrane* with pores large enough to pass small molecules but not large molecules. The *microdialysis probe* in Figure 23-18 has a rigid semipermeable tube that can be inserted into the brain of an anesthetized rat to collect neurotransmitter molecules. Fluid pumped through the probe at a rate of 3 μl/min transports small molecules that diffused into the probe. Fluid exiting the probe (*dialysate*) is routed to the sample introduction channel at the lower left of the microfluidic chip in Figure 23-19*a*.

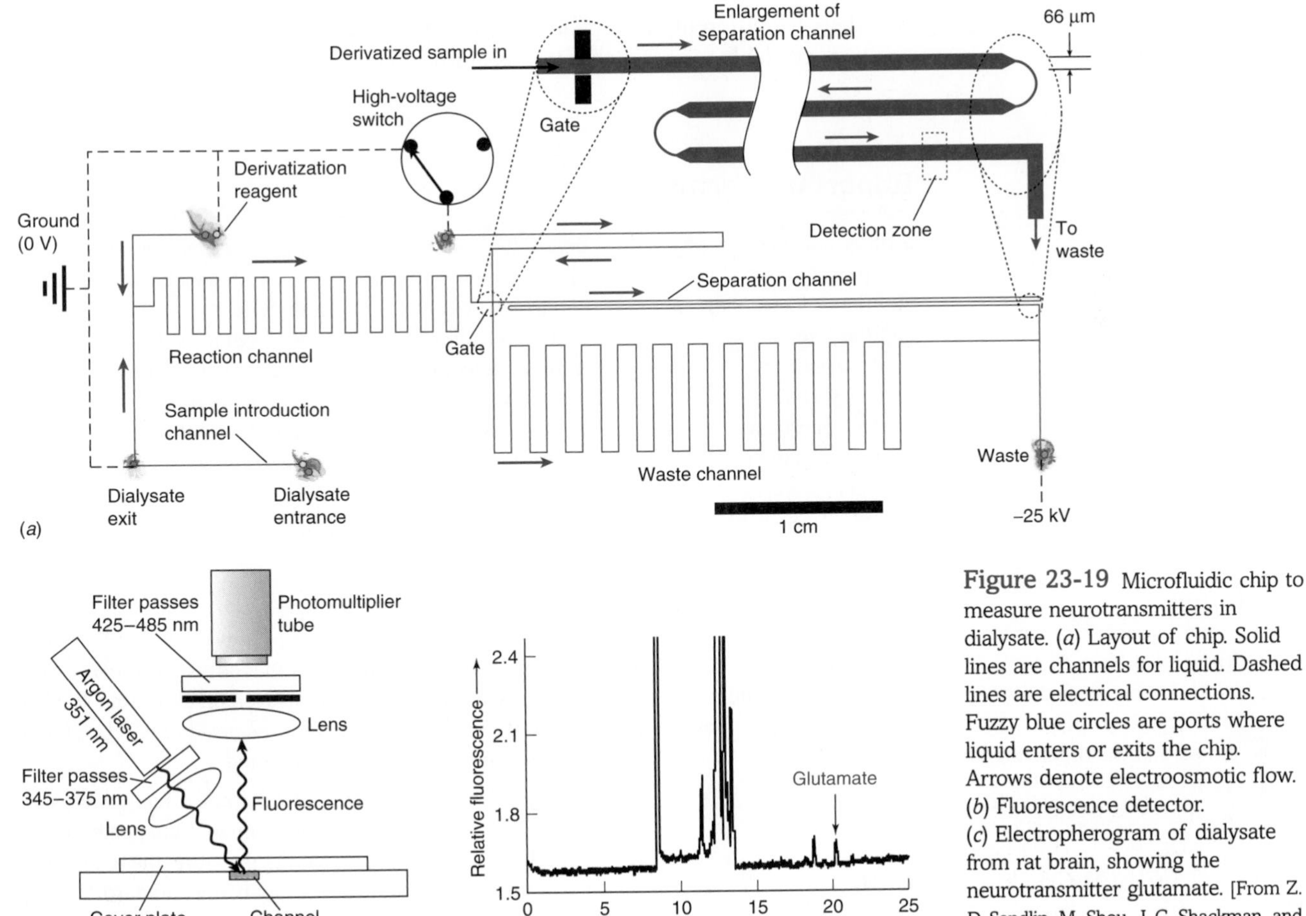

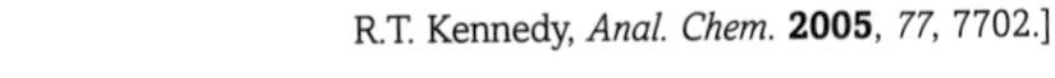

Figure 23-19 Microfluidic chip to measure neurotransmitters in dialysate. (*a*) Layout of chip. Solid lines are channels for liquid. Dashed lines are electrical connections. Fuzzy blue circles are ports where liquid enters or exits the chip. Arrows denote electroosmotic flow. (*b*) Fluorescence detector. (*c*) Electropherogram of dialysate from rat brain, showing the neurotransmitter glutamate. [From Z. D. Sandlin, M. Shou, J. G. Shackman, and R.T. Kennedy, *Anal. Chem.* **2005**, *77*, 7702.]

Derivatization:

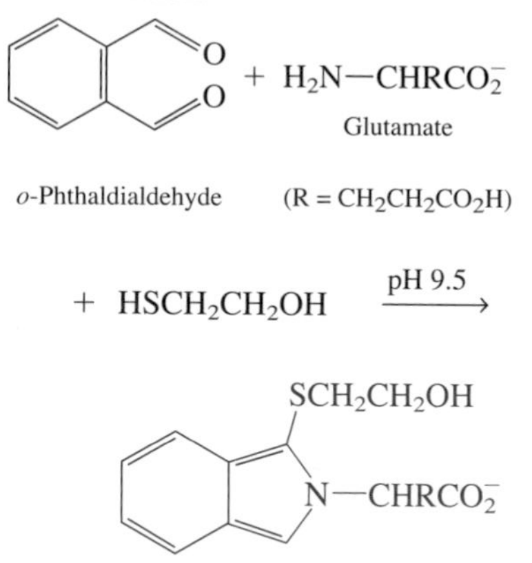

Liquid in the chip in Figure 23-19*a* is connected to electrical ground (0 V) at the left and to −25 kV through the waste outlet at the right. Electroosmotic flow proceeds from left to right through the channels. Some liquid from the sample introduction channel is drawn toward the reaction channel by electroosmotic flow. Derivatization reagent that reacts with the sample to make it fluorescent is also drawn in and mixes with the sample. The reaction channel provides ~1 min for derivatization before liquid enters the separation channel. Dimensions of the channels determine the electric potential at different points and direct fluid along the desired path. When the high-voltage switch is in the closed position shown in Figure 23-19*a*, liquid from the reaction channel flows to the waste channel. When the high-voltage switch is momentarily opened, a small plug of solution from the reaction channel is diverted to the separation channel for electrophoretic analysis. Fluorescent products are observed by the detector in Figure 23-19*b* when they reach the detection zone. The 9-cm-long separation channel provides $\sim 10^5$ theoretical plates with a detection limit of 0.2 μM for the neurotransmitter glutamate. Dialysate analyzed in Figure 23-19*c* contained 3.3 μM glutamate. The system is capable of measuring changes in glutamate in response to physiologic stimulation with a time resolution of 2 to 4 min. Changes in glutamate concentration occurring in less than 2 to 4 min would not be resolved from each other.

The microfluidic chip simplifies the handling of dialysate, making microdialysis sampling accessible to more neuroscience laboratories. The chip could be modified to conduct multiple analyses by directing aliquots of dialysate to different chambers for different derivatizations. Microfluidic chips promise to make complex analyses available to more laboratories at lower cost.

Important Terms

affinity chromatography
anion exchanger
capillary electrophoresis
capillary gel electrophoresis
capillary zone electrophoresis
cation exchanger
dialysis
deionized water
electroosmosis
electrophoresis
hydrated radius
indirect detection
ion chromatography
ion-exchange chromatography
micellar electrokinetic capillary chromatography
micelle
molecular exclusion chromatography
preconcentration
retention volume
trace analysis

Problems

23-1. **(a)** Hexanoic acid and 1-aminohexane, adjusted to pH 12 with NaOH, were passed through a cation-exchange column loaded with NaOH at pH 12. State the principal species that will be eluted and the order in which they are expected.

$CH_3CH_2CH_2CH_2CH_2CO_2H$

Hexanoic acid

$CH_3CH_2CH_2CH_2CH_2CH_2NH_2$

1-Aminohexane

(b) Hexanoic acid and 1-aminohexane, adjusted to pH 3 with HCl, were passed through a cation-exchange column loaded with HCl at pH 3. State the principal species that will be eluted and the order in which they are expected.

23-2. The exchange capacity of an ion-exchange resin is defined as the number of moles of charged sites per gram of dry resin. Describe how you would measure the exchange capacity of an anion-exchange resin by using standard NaOH, standard HCl, or any other reagent you wish.

23-3. Commercial vanadyl sulfate ($VOSO_4$, FM 163.00) is contaminated with H_2SO_4 and H_2O. A solution was prepared by dissolving 0.244 7 g of impure $VOSO_4$ in 50.0 mL of water. Spectrophotometric analysis indicated that the concentration of the blue VO^{2+} ion was 0.024 3 M. A 5.00-mL sample was passed through a cation-exchange column loaded with H^+. VO^{2+} is exchanged for $2H^+$ by this process. H_2SO_4 is unchanged by the cation-exchange column.

$VOSO_4 \longrightarrow$ [Cation-exchange column loaded with H^+] $\longrightarrow H_2SO_4$

$H_2SO_4 \longrightarrow$ [Cation-exchange column loaded with H^+] $\longrightarrow H_2SO_4$

H^+ eluted from the column required 13.03 mL of 0.022 74 M NaOH for titration. Find the weight percents of $VOSO_4$, H_2SO_4, and H_2O in the vanadyl sulfate.

23-4. Consider a protein with a net negative charge tightly adsorbed on an anion-exchange gel at pH 8.

(a) How will a gradient from pH 8 to some lower pH be useful for eluting the protein?

(b) How would a gradient of increasing ion concentration (at constant pH) be useful for eluting the protein?

23-5. Look up the pK_a values for trimethylamine, dimethylamine, methylamine, and ammonia. Predict the order of elution of these compounds from a cation-exchange column eluted with a gradient of increasing pH, beginning at pH 7.

23-6. One mole of 1,2-ethanediol consumes one mole of periodate in the reaction

$$\underset{\text{1,2-Ethanediol (FM 62.068)}}{CH_2OH{-}CH_2OH} + \underset{\text{Periodate}}{IO_4^-} \longrightarrow \underset{\text{Formaldehyde}}{2CH_2O} + H_2O + \underset{\text{Iodate}}{IO_3^-}$$

To analyze 1,2-ethanediol, oxidation with excess IO_4^- is followed by passage of the reaction solution through an anion-exchange resin that binds both IO_4^- and IO_3^-. IO_3^- is then quantitatively removed from the resin by elution with NH_4Cl. The absorbance of eluate is measured at 232 nm to find the quantity of IO_3^- (molar absorptivity (ε) = 900 M^{-1} cm^{-1}) produced by the reaction. In one experiment, 0.213 9 g of aqueous 1,2-ethanediol was dissolved in 10.00 mL. Then 1.000 mL of the solution was treated with 3 mL of 0.15 M KIO_4 and subjected to ion-exchange separation of IO_3^- from unreacted IO_4^-. The eluate (diluted to 250.0 mL) had an absorbance of A_{232} = 0.521 in a 1.000-cm cell, and a blank had A_{232} = 0.049. Find the weight percent of 1,2-ethanediol in the original sample.

23-7. The table gives the mean ionic composition of clouds at a mountaintop in Germany, measured by ion chromatography.

Ion	Concentration (μM)	Ion	Concentration (μM)
Cl^-	101	H^+	131
NO_3^-	360	Na^+	100
SO_4^{2-}	156	NH_4^+	472
		K^+	1.3
		Ca^{2+}	26
		Mg^{2+}	12

Data from K. Acker, D. Möller, W. Wieprecht, D. Kalaß, and R. Auel, *Fresenius J. Anal. Chem.* **1998**, *361*, 59.

(a) What is the pH of the cloud water?

(b) Do the anion charges equal the magnitude of the cation charges? What does your answer suggest about the quality of the analysis?

(c) What is the total mass of dissolved ions in each milliliter of water?

23-8. In *ion-exclusion chromatography*, ions are separated from nonelectrolytes (uncharged molecules) by an ion-exchange column. Nonelectrolytes penetrate the stationary phase, whereas ions with charge of the same sign as that of the stationary phase are repelled by the stationary phase. Because electrolytes have access to less of the column volume, they are eluted before nonelectrolytes. A mixture of trichloroacetic acid (TCA, pK_a = 0.66), dichloroacetic acid (DCA, pK_a = 1.30), and monochloroacetic acid (MCA, pK_a = 2.86) was separated by passage through a cation-exchange resin eluted with 0.01 M HCl. The order of elution was TCA $<$ DCA $<$ MCA. Explain why the three acids are separated and the order of elution.

23-9. Polystyrene standards of known molecular mass gave the following calibration data in a molecular exclusion column. Prepare a plot of log(molecular mass) versus retention time (t_r) and find the equation of the line. Find the molecular mass of an unknown with a retention time of 13.00 min.

Molecular mass	Retention time, t_r (min)
8.50×10^6	9.28
3.04×10^6	10.07
1.03×10^6	10.88
3.30×10^5	11.67
1.56×10^5	12.14
6.60×10^4	12.74
2.85×10^4	13.38
9.20×10^3	14.20
3.25×10^3	14.96
5.80×10^2	16.04

23-10. A molecular exclusion column has a diameter of 7.8 mm and a length of 30 cm. The solid portion of the particles occupies 20% of the volume, the pores occupy 40%, and the volume between particles occupies 40%.

(a) At what volume would totally excluded molecules be expected to emerge?

(b) At what volume would the smallest molecules be expected?

(c) A mixture of polymers of various molecular masses is eluted between 23 and 27 mL. What does this imply about the retention mechanism for these solutes on the column?

23-11. *Immunoaffinity measurements of drug concentration.* At therapeutic levels in blood, ~90% of the antiepileptic drug phenytoin is bound to the protein serum albumin. The unbound 10% is thought to be the active form of the drug. Free phenytoin can be measured with a thin-layer (0.94 mm tall × 2.1 mm diameter) affinity column at 37°C containing antiphenytoin antibodies covalently bound to silica.

Step 1. A fluorescent phenytoin derivative was applied to the column at time $t = 0$ to saturate the antibodies. Excess derivative was washed off with pH 7.4 buffer.

Step 2. At $t = 6$ min, 5 μL of serum was injected. As it flowed through the column, free phenytoin displaced some fluorescently labeled phenytoin from the silica. Fluorescence was measured at 820 nm with laser excitation at 785 nm.

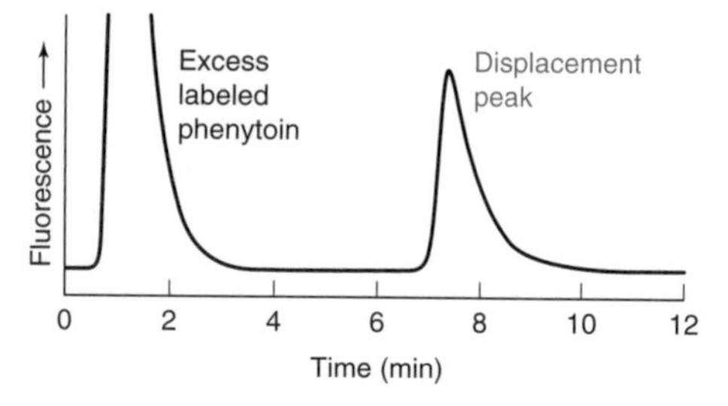

Fluorescence signal observed during phenytoin analysis. [From C. M. Ohnmacht, J. E. Schiel, and D. S. Hage, *Anal. Chem.* **2006**, *78*, 7547.]

(a) With the following symbols, draw what happens on the column before and after step 1 and during analysis in step 2.

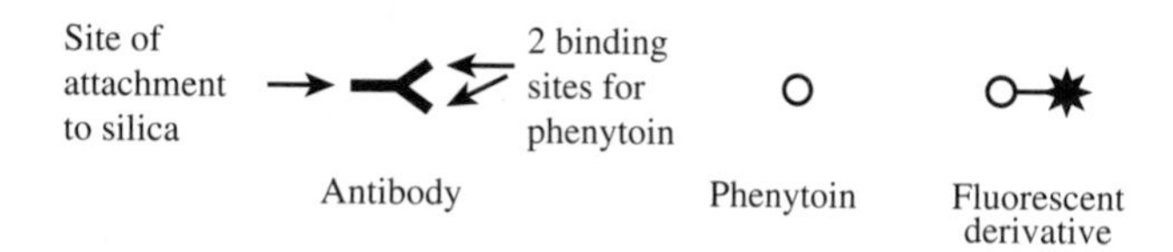

(b) Why are there two peaks in the chromatogram?

(c) The analysis was designed so that the residence time of serum on the column is short enough that insignificant dissociation of phenytoin from albumin occurs. If liquid occupies ~50% of the column volume, and the elution rate is 1.2 mL/min, what is the residence time of sample in the column?

(d) *Confidence level.* A calibration curve was constructed by plotting the area of the peak near 8 min versus concentration of free phenytoin applied to the column. Serum with 40.0 μM total phenytoin had 5.99 ± 0.14 μM free phenytoin according to the calibration curve. Uncertainty is the standard deviation for three replicates. A different assay (ultrafiltration) gave 6.11 ± 0.44 μM free phenytoin for three replicates. Do the two methods differ at the 95% confidence level?

23-12. **(a)** If a capillary is set up as in Figure 23-10 with the injector end positive and the detector end negative, in what order will cations, anions, and neutral molecules be eluted?

(b) The electropherograms in Figure 23-11 were obtained with the injector end *negative* and a run buffer of pH 4.0. Will there be high or low electroosmotic flow? Are the anions moving with or against the electroosmotic flow? Will cations injected with the sample reach the detector before or after the anions?

23-13. **(a)** What is electroosmosis?

(b) Why is the electroosmotic flow in a silica capillary five times faster at pH 9 than at pH 3?

(c) When the Si—OH groups on a silica capillary wall are converted into $Si—O(CH_2)_{17}CH_3$ groups, the electroosmotic flow is small and nearly independent of pH. Explain why.

23-14. Why is the detector response *negative* in indirect spectrophotometric detection?

23-15. Explain how neutral molecules can be separated by micellar electrokinetic capillary chromatography. Why is this a form of chromatography?

23-16. A van Deemter plot for capillary electrophoresis is a graph of plate height versus migration velocity, where migration velocity is governed by the net sum of electroosmotic flow and electrophoretic flow.

(a) What is the principal source of band broadening in ideal capillary zone electrophoresis? Sketch what you expect the van Deemter curve to look like.

(b) What are the sources of band broadening in micellar electrokinetic capillary chromatography? Sketch what you expect the van Deemter curve to look like.

23-17. **(a)** Measure the migration time and peak width of $^{35}Cl^-$ in Figure 23-14 and calculate the number of theoretical plates.

(b) The distance from injection to the detector is 40 cm in Figure 23-14. From your answer to **(a)**, find the plate height.

(c) Why are the peaks negative in Figure 23-14?

23-18. The water-soluble vitamins niacinamide (a neutral compound), riboflavin (a neutral compound), niacin (an anion), and thiamine (a cation) were separated by micellar electrokinetic capillary chromatography in 15 mM borate buffer (pH 8.0) with 50 mM sodium dodecyl sulfate. The migration times were niacinamide, 8.1 min; riboflavin, 13.0 min; niacin, 14.3 min; and thiamine, 21.9 min. What would the order have been in the absence of sodium dodecyl sulfate? Which compound is most soluble in the micelles?

23-19. *Molecular mass by capillary gel electrophoresis.* Protein molecular mass can be estimated by sodium dodecyl sulfate (SDS)-gel electrophoresis. Proteins are first *denatured* (unfolded) by sodium dodecyl sulfate (Box 23-2), which binds to hydrophobic regions and gives the protein a negative charge that is approximately proportional to the length of the protein. Also, disulfide bonds (—S—S—) are reduced to sulfhydryl (—SH) by excess 2-mercaptoethanol ($HSCH_2CH_2OH$). Denatured proteins are separated by electrophoresis through a gel that behaves as a sieve. Large molecules are retarded more than small molecules—behavior opposite that of size exclusion chromatography. The logarithm of molecular mass of the SDS-coated protein is proportional to 1/(migration time) of the protein through the gel. Absolute migration times are somewhat variable from run to run, so relative migration times are measured. The relative migration time is the migration time of a protein divided by the migration time of a fast-moving small dye molecule. Migration times for protein standards and unknowns are given in the table below.

Unknowns are the light and heavy chains of ferritin, the iron-storage protein found in animals, plants, and microbes. Ferritin is a hollow shell containing 24 subunits that are a mixture of heavy (H) and light (L) chains, arranged in octahedral symmetry. The hollow core has a diameter of 8 nm and can hold as many as 4 500 iron atoms in the approximate form of the mineral ferrihydrite ($5Fe_2O_3 \cdot 9H_2O$). Iron(II) enters the protein through any of six pores located on the threefold symmetry axes of the octahedron. Oxidation to Fe(III) takes place at catalytic sites on the H chains. Sites on the inside of L chains appear to nucleate the crystallization of ferrihydrite.

Protein	Molecular mass (Da)	Migration time (min)
Orange G marker dye	small molecule	13.17
α-Lactalbumin	14 200	16.46
Carbonic anhydrase	29 000	18.66
Ovalbumin	45 000	20.16
Bovine serum albumin	66 000	22.36
Phosphorylase B	97 000	23.56
β-Galactosidase	116 000	24.97
Myosin	205 000	28.25
Ferritin light chain		17.07
Ferritin heavy chain		17.97

Data from J. K. Grady, J. Zang, T. M. Laue, P. Arosio, and N. D. Chasteen, *Anal. Biochem.* **2002**, *302*, 263.

Prepare a graph of log(molecular mass) versus 1/(relative migration time), where relative migration time = (migration time)/(migration time of marker dye). Compute the molecular mass of the ferritin light and heavy chains. True masses computed from amino acid sequences are 19 766 and 21 099 Da.

23-20. *Limits of detection and quantitation.* An ion-chromatographic method was developed to measure sub-parts-per-billion levels of the disinfectant by-products iodate (IO_3^-), chlorite (ClO_2^-), and bromate (BrO_3^-) in drinking water. As the oxyhalides are eluted, they react with Br^- to make Br_3^-, which is measured by its strong absorption at 267 nm (absorptivity = 40 900 M^{-1} cm^{-1}). For example, each mole of BrO_3^- makes three moles of Br_3^-: $BrO_3^- + 8Br^- + 6H^+ \longrightarrow 3Br_3^- + 3H_2O$.

(a) Bromate near its detection limit gave the following chromatographic peak heights and standard deviations. The blank is 0 because chromatographic peak height is measured from the baseline adjacent to the peak. For each concentration, estimate the limit of detection and the limit of quantitation. Find the mean of the four values of detection and quantitation limit.

Bromate concentration (μg/L)	Peak height (arbitrary units)	Relative standard deviation (%)	Number of measurements
0.2	17	14.4	8
0.5	31	6.8	7
1.0	56	3.2	7
2.0	111	1.9	7

Data from H. S. Weinberg and H. Yamada, *Anal. Chem.* **1998**, *70*, 1.

(b) What is the absorbance of Br_3^- in the 6.00-mm-pathlength detection cell of the chromatograph if its concentration is at the mean bromate detection limit?

How Would You Do It?

23-21. (a) To obtain the best separation of two weak acids by electrophoresis, it makes sense to use the pH at which their charge difference is greatest. Explain why.

(b) Prepare a spreadsheet to examine the charges of *ortho*- and *para*-hydroxybenzoic acids as a function of pH. At what pH is the difference greatest?

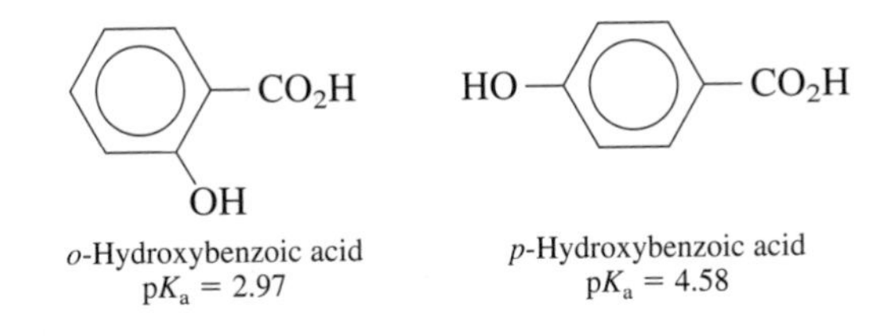

o-Hydroxybenzoic acid $pK_a = 2.97$ *p*-Hydroxybenzoic acid $pK_a = 4.58$

23-22. An aqueous solution of NaCl, $NaNO_3$, and Na_2SO_4 was passed through a C_{18}-silica reversed-phase liquid chromatography column eluted with water. None of the cations or anions is retained by the C_{18} stationary phase, so all three salts were eluted in a single, sharp band with a retention time of 0.9 min. Then, the column was equilibrated with aqueous 10 mM pentylammonium formate, whose hydrophobic tail is soluble in the C_{18} stationary phase.

Hydrocarbon tail soluble in C_{18} stationary phase $\overset{+}{N}H_3$ HCO_2^-

When the mixture of NaCl, $NaNO_3$, and Na_2SO_4 was passed through the column and eluted with 10 mM pentylammonium formate, the cations all came out in a single peak with a retention time of 0.9 min. However, anions were separated, with retention times of 1.9 min (Cl^-), 2.1 min (NO_3^-), and 4.1 min (SO_4^{2-}). Explain why this column behaves as an anion exchanger. Why is sulfate eluted last?

23-23. *Cosmogenic* ^{35}S.[3] Radioactive ^{35}S is produced in the atmosphere by the action of cosmic rays on Ar atoms. ^{35}S atoms are oxidized to SO_4^{2-} and fall to the ground in rain or as dry solids. Counting ^{35}S radioactive disintegrations allows us to measure the removal time for sulfur in the atmosphere and the residence time for sulfur species in different parts of the environment. To analyze minute quantities of ^{35}S in rain or lake water, 30-L volumes were passed through a 0.45-μm filter, acidified to pH 3-4 with HCl, and then 20 mg of Na_2SO_4 (containing no ^{35}S) were added. The whole volume was passed through 50 g of anion-exchange resin. SO_4^{2-} was quantitatively eluted with 300 mL of 3 M NaCl. After adjusting eluate to pH 3-4 with HCl, 5 mL of 10 wt% $BaCl_2 \cdot 3H_2O$ were added. After 5 h, precipitate was quantitatively recovered on a filter and ^{35}S was measured by scintillation counting. Explain the purpose of (**a**) initial filtration through a 0.45-μm filter; (**b**) passage through anion-exchange resin; (**c**) addition of $BaCl_2$. What is the final chemical form of ^{35}S that is measured by scintillation counting? (**d**) Why was Na_2SO_4 added prior to anion exchange?

23-24. A buffer containing 1 mM $MgSO_4$ and 1 mM $CaCl_2$ strongly reduces electroosmotic flow in capillary electrophoresis.[4] Electroosmosis is restored by adding 3 mM EDTA to the buffer. Suggest an explanation.

Notes and References

1. L. S. Ettre, *LCGC* **1999**, *17*, 1104; F. H. Spedding, *Disc. Faraday Soc.* **1949**, *7*, 214. See also F. A. Settle, "Analytical Chemistry and the Manhattan Project," *Anal. Chem.* **2002**, *74*, 36A.

2. L. S. Ettre, *LCGC* **2006**, *24*, 390.

3. Y.-L. Hong and G. Kim, *Anal. Chem.* **2005**, *77*, 3390.

4. Z. D. Sandlin, M. Shou, J. G. Shackman, and R.T. Kennedy, *Anal. Chem.* **2005**, *77*, 7702.

Further Reading

A. Manz, N. Pamme, and D. Iossifidis, *Bioanalytical Chemistry* (London: Imperial College Press, 2004).

J. S. Fritz and D. T. Gjerde, *Ion Chromatography*, 3rd ed. (New York: Wiley-VCH, 2000).

J. Weiss, *Handbook of Ion Chromatography* (Weinheim: Wiley-VCH, 2004).

www.virtualcolumn.com–An Ion Chromatography Simulator. (See P. R. Haddad *et al.*, *J. Chem. Ed.* **2004**, *81*, 1293).

C.-S. Wu, ed., *Handbook of Size Exclusion Chromatography*, 2nd ed. (New York: Marcel Dekker, 2004).

G. A. Marson and B. Baptista, *Principles of Gel Permeation Chromatography: Interactive Software*, J. Chem. Ed. WebWare (http:/www.jce.divched.org/JCEDLib/WebWare/).

R. Weinberger, *Practical Capillary Electrophoresis*, 2nd ed. (San Diego: Academic Press, 2000).

J. R. Petersen and A. A. Mohammad, eds., *Clinical and Forensic Applications of Capillary Electrophoresis* (Totowa, NJ: Humana Press, 2001).

Appendix A
Solubility Products*

Formula	K_{sp}
Azides: L = N_3^-	
CuL	4.9×10^{-9}
AgL	2.8×10^{-9}
Hg_2L_2	7.1×10^{-10}
TlL	2.2×10^{-4}
$PdL_2(\alpha)$	2.7×10^{-9}
Bromates: L = BrO_3^-	
$BaL \cdot H_2O$	7.8×10^{-6}
AgL	5.5×10^{-5}
TlL	1.7×10^{-4}
PbL_2	7.9×10^{-6}
Bromides: L = Br^-	
CuL	5×10^{-9}
AgL	5.0×10^{-13}
Hg_2L_2	5.6×10^{-23}
TlL	3.6×10^{-6}
HgL_2	1.3×10^{-19}
PbL_2	2.1×10^{-6}
Carbonates: L = CO_3^{2-}	
MgL	3.5×10^{-8}
CaL (calcite)	4.5×10^{-9}
CaL (aragonite)	6.0×10^{-9}
SrL	9.3×10^{-10}
BaL	5.0×10^{-9}
Y_2L_3	2.5×10^{-31}
La_2L_3	4.0×10^{-34}
MnL	5.0×10^{-10}
FeL	2.1×10^{-11}
CoL	1.0×10^{-10}
NiL	1.3×10^{-7}
CuL	2.3×10^{-10}
Ag_2L	8.1×10^{-12}
Hg_2L	8.9×10^{-17}
ZnL	1.0×10^{-10}
CdL	1.8×10^{-14}
PbL	7.4×10^{-14}
Chlorides: L = Cl^-	
CuL	1.9×10^{-7}
AgL	1.8×10^{-10}
Hg_2L_2	1.2×10^{-18}
TlL	1.8×10^{-4}
PbL_2	1.7×10^{-5}
Chromates: L = CrO_4^{2-}	
BaL	2.1×10^{-10}
CuL	3.6×10^{-6}
Ag_2L	1.2×10^{-12}
Hg_2L	2.0×10^{-9}
Tl_2L	9.8×10^{-13}
Cobalticyanides: L = $Co(CN)_6^{3-}$	
Ag_3L	3.9×10^{-26}
$(Hg_2)_3L_2$	1.9×10^{-37}
Cyanides: L = CN^-	
AgL	2.2×10^{-16}
Hg_2L_2	5×10^{-40}
ZnL_2	3×10^{-16}
Ferrocyanides: L = $Fe(CN)_6^{4-}$	
Ag_4L	8.5×10^{-45}
Zn_2L	2.1×10^{-16}
Cd_2L	4.2×10^{-18}
Pb_2L	9.5×10^{-19}
Fluorides: L = F^-	
LiL	1.7×10^{-3}
MgL_2	6.6×10^{-9}
CaL_2	3.9×10^{-11}
SrL_2	2.9×10^{-9}
BaL_2	1.7×10^{-6}
LaL_3	2×10^{-19}
ThL_4	5×10^{-29}
PbL_2	3.6×10^{-8}
Hydroxides: L = OH^-	
MgL_2 (amorphous)	6×10^{-10}
MgL_2 (brucite crystal)	7.1×10^{-12}
CaL_2	6.5×10^{-6}
$BaL_2 \cdot 8H_2O$	3×10^{-4}
YL_3	6×10^{-24}
LaL_3	2×10^{-21}
CeL_3	6×10^{-22}
UO_2 ($\rightleftharpoons U^{4+} + 4OH^-$)	6×10^{-57}

* Solubility products generally apply at 25°C and zero ionic strength. The designations α, β, or γ after some formulas refer to particular crystalline forms.

(continued)

Formula	K_{sp}
UO_2L_2 ($\rightleftharpoons UO_2^{2+} + 2OH^-$)	4×10^{-23}
MnL_2	1.6×10^{-13}
FeL_2	7.9×10^{-16}
CoL_2	1.3×10^{-15}
NiL_2	6×10^{-16}
CuL_2	4.8×10^{-20}
VL_3	4.0×10^{-35}
CrL_3	1.6×10^{-30}
FeL_3	1.6×10^{-39}
CoL_3	3×10^{-45}
VOL_2 ($\rightleftharpoons VO^{2+} + 2OH^-$)	3×10^{-24}
PdL_2	3×10^{-29}
ZnL_2 (amorphous)	3.0×10^{-16}
CdL_2 (β)	4.5×10^{-15}
HgO (red) ($\rightleftharpoons Hg^{2+} + 2OH^-$)	3.6×10^{-26}
Cu_2O ($\rightleftharpoons 2Cu^+ + 2OH^-$)	4×10^{-30}
Ag_2O ($\rightleftharpoons 2Ag^+ + 2OH^-$)	3.8×10^{-16}
AuL_3	3×10^{-6}
AlL_3 (α)	3×10^{-34}
GaL_3 (amorphous)	10^{-37}
InL_3	1.3×10^{-37}
SnO ($\rightleftharpoons Sn^{2+} + 2OH^-$)	6×10^{-27}
PbO (yellow) ($\rightleftharpoons Pb^{2+} + 2OH^-$)	8×10^{-16}
PbO (red) ($\rightleftharpoons Pb^{2+} + 2OH^-$)	5×10^{-16}
Iodates: L = IO_3^-	
CaL_2	7.1×10^{-7}
SrL_2	3.3×10^{-7}
BaL_2	1.5×10^{-9}
YL_3	7.1×10^{-11}
LaL_3	1.0×10^{-11}
CeL_3	1.4×10^{-11}
ThL_4	2.4×10^{-15}
UO_2L_2 ($\rightleftharpoons UO_2^{2+} + 2IO_3^-$)	9.8×10^{-8}
CrL_3	5×10^{-6}
AgL	3.1×10^{-8}
Hg_2L_2	1.3×10^{-18}
TlL	3.1×10^{-6}
ZnL_2	3.9×10^{-6}
CdL_2	2.3×10^{-8}
PbL_2	2.5×10^{-13}
Iodides: L = I^-	
CuL	1×10^{-12}
AgL	8.3×10^{-17}
CH_3HgL ($\rightleftharpoons CH_3Hg^+ + I^-$)	3.5×10^{-12}
CH_3CH_2HgL ($\rightleftharpoons CH_3CH_2Hg^+ + I^-$)	7.8×10^{-5}
TlL	5.9×10^{-8}
Hg_2L_2	1.1×10^{-28}
SnL_2	8.3×10^{-6}
PbL_2	7.9×10^{-9}
Oxalates: L = $C_2O_4^{2-}$	
CaL	1.3×10^{-8}
SrL	4×10^{-7}
BaL	1×10^{-6}

Formula	K_{sp}
La_2L_3	1×10^{-25}
ThL_2	4.2×10^{-22}
UO_2L ($\rightleftharpoons UO_2^{2+} + C_2O_4^{2-}$)	2.2×10^{-9}
Phosphates: L = PO_4^{3-}	
$MgHL \cdot 3H_2O$ ($\rightleftharpoons Mg^{2+} + HL^{2-}$)	1.7×10^{-6}
$CaHL \cdot 2H_2O$ ($\rightleftharpoons Ca^{2+} + HL^{2-}$)	2.6×10^{-7}
SrHL ($\rightleftharpoons Sr^{2+} + HL^{2-}$)	1.2×10^{-7}
BaHL ($\rightleftharpoons Ba^{2+} + HL^{2-}$)	4.0×10^{-8}
LaL	3.7×10^{-23}
$Fe_3L_2 \cdot 8H_2O$	1×10^{-36}
$FeL \cdot 2H_2O$	4×10^{-27}
$(VO)_3L_2$ ($\rightleftharpoons 3VO^{2+} + 2L^{3-}$)	8×10^{-26}
Ag_3L	2.8×10^{-18}
Hg_2HL ($\rightleftharpoons Hg_2^{2+} + HL^{2-}$)	4.0×10^{-13}
$Zn_3L_2 \cdot 4H_2O$	5×10^{-36}
Pb_3L_2	3.0×10^{-44}
GaL	1×10^{-21}
InL	2.3×10^{-22}
Sulfates: L = SO_4^{2-}	
CaL	2.4×10^{-5}
SrL	3.2×10^{-7}
BaL	1.1×10^{-10}
RaL	4.3×10^{-11}
Ag_2L	1.5×10^{-5}
Hg_2L	7.4×10^{-7}
PbL	6.3×10^{-7}
Sulfides: L = S^{2-}	
MnL (pink)	3×10^{-11}
MnL (green)	3×10^{-14}
FeL	8×10^{-19}
CoL (α)	5×10^{-22}
CoL (β)	3×10^{-26}
NiL (α)	4×10^{-20}
NiL (β)	1.3×10^{-25}
NiL (γ)	3×10^{-27}
CuL	8×10^{-37}
Cu_2L	3×10^{-49}
Ag_2L	8×10^{-51}
Tl_2L	6×10^{-22}
ZnL (α)	2×10^{-25}
ZnL (β)	3×10^{-23}
CdL	1×10^{-27}
HgL (black)	2×10^{-53}
HgL (red)	5×10^{-54}
SnL	1.3×10^{-26}
PbL	3×10^{-28}
In_2L_3	4×10^{-70}
Thiocyanates: L = SCN^-	
CuL	4.0×10^{-14}
AgL	1.1×10^{-12}
Hg_2L_2	3.0×10^{-20}
TlL	1.6×10^{-4}
HgL_2	2.8×10^{-20}

Appendix B
Acid Dissociation Constants

Name	Structure*	pK_a[†]	K_a
Acetic acid (ethanoic acid)	$CH_3CO_2\mathbf{H}$	4.756	1.75×10^{-5}
Alanine	NH_3^+–$CHCH_3$–$CO_2\mathbf{H}$	2.344 (CO_2H) 9.868 (NH_3)	4.53×10^{-3} 1.36×10^{-10}
Aminobenzene (aniline)	C_6H_5–NH_3^+	4.601	2.51×10^{-5}
2-Aminobenzoic acid (anthranilic acid)	$C_6H_4(NH_3^+)(CO_2\mathbf{H})$	2.08 (CO_2H) 4.96 (NH_3)	8.3×10^{-3} 1.10×10^{-5}
2-Aminoethanol (ethanolamine)	$HOCH_2CH_2NH_3^+$	9.498	3.18×10^{-10}
2-Aminophenol	$C_6H_4(O\mathbf{H})(NH_3^+)$	4.70 (NH_3) (20°) 9.97 (OH) (20°)	2.0×10^{-5} 1.05×10^{-10}
Ammonia	NH_4^+	9.245	5.69×10^{-10}
Arginine	NH_3^+–$CHCH_2CH_2CH_2NHC(=NH_2^+)NH_2$ with $CO_2\mathbf{H}$	1.823 (CO_2H) 8.991 (NH_3) (12.1) (NH_2)	1.50×10^{-2} 1.02×10^{-9} 8×10^{-13}
Arsenic acid (hydrogen arsenate)	$\mathbf{H}O$–$As(=O)(O\mathbf{H})$–$O\mathbf{H}$	2.31 7.05 11.9	4.9×10^{-3} 8.9×10^{-8} 1.3×10^{-12}
Arsenious acid (hydrogen arsenite)	$As(O\mathbf{H})_3$	9.29	5.1×10^{-10}
Asparagine	NH_3^+–$CHCH_2C(=O)NH_2$ with $CO_2\mathbf{H}$	2.16 (CO_2H) 8.73 (NH_3)	6.9×10^{-3} 1.86×10^{-9}

(continued)

* Each acid is written in its protonated form. The acidic protons are indicated in **bold** type.
[†] pK_a values refer to 25°C unless otherwise indicated. Values in parentheses are considered to be less reliable.

SOURCE: A. E. Martell, R. M. Smith, and R. J. Motekaitis, *NIST Critically Selected Stability Constants of Metal Complexes,* NIST Standard Reference Database 46, Gaithersburg, MD, 2001.

Name	Structure*	pK_a[†]	K_a
Aspartic acid	$\alpha \rightarrow$ CO_2H–$CH(NH_3^+)CH_2CO_2H$ (β)	1.990 (α-CO_2H) 3.900 (β-CO_2H) 10.002 (NH_3)	1.02×10^{-2} 1.26×10^{-4} 9.95×10^{-11}
Benzene-1,2,3-tricarboxylic acid (hemimellitic acid)	$C_6H_3(CO_2H)_3$	2.88 4.75 7.13	1.32×10^{-3} 1.78×10^{-5} 7.4×10^{-8}
Benzoic acid	C_6H_5–CO_2H	4.202	6.28×10^{-5}
Benzylamine	C_6H_5–$CH_2NH_3^+$	9.35	4.5×10^{-10}
2,2′-Bipyridine	(pyridyl)–(pyridyl), N–H^+	4.34	4.6×10^{-5}
Boric acid (hydrogen borate)	$B(OH)_3$	9.237 (12.74) (20°) (13.80) (20°)	5.79×10^{-10} 1.82×10^{-13} 1.58×10^{-14}
Bromoacetic acid	$BrCH_2CO_2H$	2.902	1.25×10^{-3}
Butane-2,3-dione dioxime (dimethylglyoxime)	$HON{=}C(CH_3)C(CH_3){=}NOH$	10.66 12.0	2.2×10^{-11} 1×10^{-12}
Butanoic acid	$CH_3CH_2CH_2CO_2H$	4.818	1.52×10^{-5}
cis-Butenedioic acid (maleic acid)	*cis*-$HO_2CCH{=}CHCO_2H$	1.92 6.27	1.20×10^{-2} 5.37×10^{-7}
trans-Butenedioic acid (fumaric acid)	*trans*-$HO_2CCH{=}CHCO_2H$	3.02 4.48	9.5×10^{-4} 3.3×10^{-5}
Butylamine	$CH_3CH_2CH_2CH_2NH_3^+$	10.640	2.29×10^{-11}
Carbonic acid (hydrogen carbonate)	$HO{-}C({=}O){-}OH$	6.351 10.329	4.46×10^{-7} 4.69×10^{-11}
Chloroacetic acid	$ClCH_2CO_2H$	2.865	1.36×10^{-3}
Chlorous acid (hydrogen chlorite)	$HOCl{=}O$	1.96	1.10×10^{-2}
Chromic acid (hydrogen chromate)	$HO{-}Cr({=}O)_2{-}OH$	−0.2 (20°) 6.51	1.6 3.1×10^{-7}
Citric acid (2-hydroxypropane-1,2,3-tricarboxylic acid)	$HO_2CCH_2C(OH)(CO_2H)CH_2CO_2H$	3.128 4.761 6.396	7.44×10^{-4} 1.73×10^{-5} 4.02×10^{-7}

Name	Structure*	pK_a[†]	K_a
Cyanoacetic acid	$NCCH_2CO_2H$	2.472	3.37×10^{-3}
Cyclohexylamine	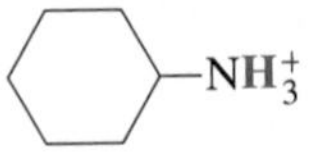 $C_6H_{11}NH_3^+$ (cyclohexyl–NH_3^+)	10.567	2.71×10^{-11}
Cysteine	NH_3^+–$CHCH_2SH$–CO_2H	(1.7) (CO_2H) 8.36 (SH) 10.74 (NH_3)	2×10^{-2} 4.4×10^{-9} 1.82×10^{-11}
Dichloroacetic acid	Cl_2CHCO_2H	(1.1)	8×10^{-2}
Diethylamine	$(CH_3CH_2)_2NH_2^+$	11.00	1.10×10^{-11}
1,2-Dihydroxybenzene (catechol)	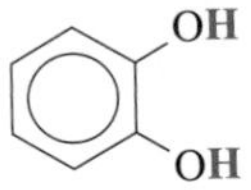 benzene ring with OH, OH (ortho)	9.41 (13.3)	3.9×10^{-10} 5.0×10^{-14}
1,3-Dihydroxybenzene (resorcinol)	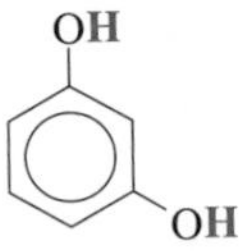 benzene ring with OH, OH (meta)	9.30 11.06	5.0×10^{-10} 8.7×10^{-12}
D-2,3-Dihydroxybutanedioic acid (D-tartaric acid)	$HO_2CCH(OH)CH(OH)CO_2H$	3.036 4.366	9.20×10^{-4} 4.31×10^{-5}
Dimethylamine	$(CH_3)_2NH_2^+$	10.774	1.68×10^{-11}
2,4-Dinitrophenol	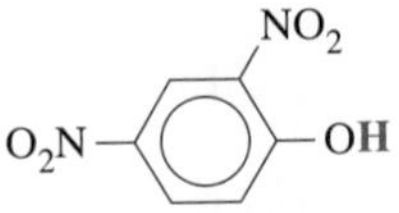 O_2N–$C_6H_3(NO_2)$–OH	4.114	7.69×10^{-5}
Ethane-1,2-dithiol	$HSCH_2CH_2SH$	8.85 (30°) 10.43 (30°)	1.4×10^{-9} 3.7×10^{-11}
Ethylamine	$CH_3CH_2NH_3^+$	10.673	2.12×10^{-11}
Ethylenediamine (1,2-diaminoethane)	$H_3\overset{+}{N}CH_2CH_2\overset{+}{N}H_3$	6.848 9.928	1.42×10^{-7} 1.18×10^{-10}
Ethylenedinitrilotetraacetic acid (EDTA)	$(HO_2CCH_2)_2\overset{+}{N}HCH_2CH_2\overset{+}{N}H(CH_2CO_2H)_2$	(0.0)(CO_2H) (1.5)(CO_2H) 2.00 (CO_2H) 2.69 (CO_2H) 6.13 (NH) 10.37 (NH)	1.0 0.032 0.010 0.002 0 7.4×10^{-7} 4.3×10^{-11}
Formic acid (methanoic acid)	HCO_2H	3.744	1.80×10^{-4}
Glutamic acid	NH_3^+–$CHCH_2CH_2CO_2H$ (γ)–CO_2H (α)	2.16 (α-CO_2H) 4.30 (γ-CO_2H) 9.96 (NH_3)	6.9×10^{-3} 5.0×10^{-5} 1.10×10^{-10}
Glutamine	NH_3^+–$CHCH_2CH_2C(=O)NH_2$–CO_2H	2.19 (CO_2H) 9.00 (NH_3)	6.5×10^{-3} 1.00×10^{-9}

(continued)

Name	Structure*	pK_a[†]	K_a
Glycine (aminoacetic acid)	$NH_3^+CH_2CO_2H$	2.350 (CO_2H) 9.778 (NH_3)	4.47×10^{-3} 1.67×10^{-10}
Guanidine	$H_2N—C(=NH_2^+)—NH_2$	(13.5) (27°)	3×10^{-14}
1,6-Hexanedioic acid (adipic acid)	$HO_2CCH_2CH_2CH_2CH_2CO_2H$	4.424 5.420	3.77×10^{-5} 3.80×10^{-6}
Histidine	$NH_3^+CH(CO_2H)CH_2$–imidazolium (NH, N⁺H)	(1.6) (CO_2H) 5.97 (NH) 9.28 (NH_3)	2.5×10^{-2} 1.07×10^{-6} 5.2×10^{-10}
Hydrazoic acid (hydrogen azide)	$HN=\overset{+}{N}=\overset{-}{N}$	4.65	2.2×10^{-5}
Hydrogen cyanate	$HOC\equiv N$	3.48	3.3×10^{-4}
Hydrogen cyanide	$HC\equiv N$	9.21	6.2×10^{-10}
Hydrogen fluoride	HF	3.17	6.8×10^{-4}
Hydrogen peroxide	HOOH	11.65	2.2×10^{-12}
Hydrogen sulfide	H_2S	7.02 14.0	9.5×10^{-8} 1×10^{-14}
Hydrogen thiocyanate	$HSC\equiv N$	(1.1)	0.08
Hydroxyacetic acid (glycolic acid)	$HOCH_2CO_2H$	3.832	1.48×10^{-4}
Hydroxybenzene (phenol)	C_6H_5–OH	9.997	1.01×10^{-10}
2-Hydroxybenzoic acid (salicylic acid)	$C_6H_4(CO_2H)(OH)$	2.972 (CO_2H) (13.7) (OH)	1.07×10^{-3} 2×10^{-14}
Hydroxylamine	$HO\overset{+}{N}H_3$	5.96	1.10×10^{-6}
8-Hydroxyquinoline (oxine)	quinolinium (NH^+) with HO at position 8	4.94 (NH) 9.82 (OH)	1.15×10^{-5} 1.51×10^{-10}
Hypochlorous acid (hydrogen hypochlorite)	HOCl	7.53	3.0×10^{-8}
Hypophosphorous acid (hydrogen hypophosphite)	$H_2P(=O)OH$	(1.3)	5×10^{-2}
Imidazole (1,3-diazole)	imidazolium (NH^+, N–H)	6.993	1.02×10^{-7}
Iminodiacetic acid	$H_2\overset{+}{N}(CH_2CO_2H)_2$	1.85 (CO_2H) 2.84 (CO_2H) 9.79 (NH_2)	1.41×10^{-2} 1.45×10^{-3} 1.62×10^{-10}
Iodic acid (hydrogen iodate)	$HO—I(=O)=O$	0.77	0.17

Name	Structure*	pK_a[†]	K_a
Iodoacetic acid	ICH_2CO_2H	3.175	6.68×10^{-4}
Isoleucine	NH_3^+–$CHCH(CH_3)CH_2CH_3$–CO_2H	2.318 (CO_2H) 9.758 (NH_3)	4.81×10^{-3} 1.75×10^{-10}
Leucine	NH_3^+–$CHCH_2CH(CH_3)_2$–CO_2H	2.328 (CO_2H) 9.744 (NH_3)	4.70×10^{-3} 1.80×10^{-10}
Lysine	$\alpha \rightarrow NH_3^+$–$CHCH_2CH_2CH_2CH_2NH_3^+$ (ε)–CO_2H	(1.77) (CO_2H) 9.07 (α-NH_3) 10.82 (ε-NH_3)	1.70×10^{-2} 8.5×10^{-10} 1.51×10^{-11}
Malonic acid (propanedioic acid)	$HO_2CCH_2CO_2H$	2.847 5.696	1.42×10^{-3} 2.01×10^{-6}
Mercaptoacetic acid (thioglycolic acid)	$HSCH_2CO_2H$	3.64 (CO_2H) 10.61 (SH)	2.3×10^{-4} 2.5×10^{-11}
2-Mercaptoethanol	$HSCH_2CH_2OH$	9.7_5	1.8×10^{-10}
Methionine	NH_3^+–$CHCH_2CH_2SCH_3$–CO_2H	2.18 (CO_2H) 9.08 (NH_3)	6.6×10^{-3} 8.3×10^{-10}
Methylamine	$CH_3\overset{+}{N}H_3$	10.645	2.26×10^{-11}
4-Methylaniline (*p*-toluidine)	 CH_3–C_6H_4–$\overset{+}{N}H_3$	5.080	8.32×10^{-6}
2-Methylphenol (*o*-cresol)	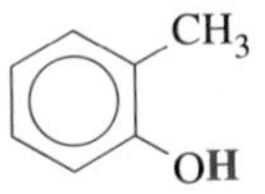 C_6H_4 with CH_3 and OH (ortho)	10.31	4.9×10^{-11}
4-Methylphenol (*p*-cresol)	 CH_3–C_6H_4–OH	10.26	5.5×10^{-11}
Morpholine (perhydro-1,4-oxazine)	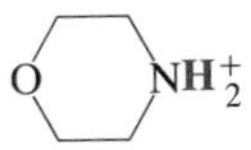 O(CH_2CH_2)$_2NH_2^+$ (ring)	8.492	3.22×10^{-9}
1-Naphthoic acid	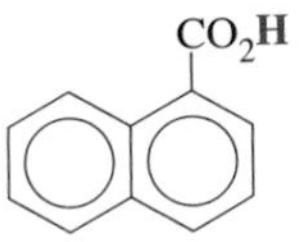 naphthalene–CO_2H (1-position)	3.67	2.1×10^{-4}
2-Naphthoic acid	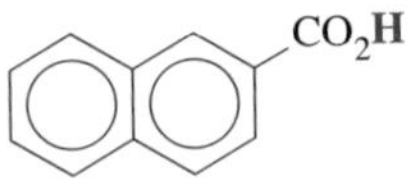 naphthalene–CO_2H (2-position)	4.16	6.9×10^{-5}
1-Naphthol	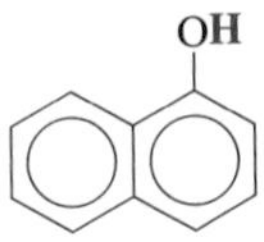 naphthalene–OH (1-position)	9.416	3.84×10^{-10}

(continued)

Name	Structure*	pK_a[†]	K_a
2-Naphthol	(2-naphthyl)–OH	9.573	2.67×10^{-10}
Nitrilotriacetic acid	$H\overset{+}{N}(CH_2CO_2H)_3$	(1.0) (CO_2H) (25°) 2.0 (CO_2H) (25°) 2.940 (CO_2H) (20°) 10.334 (NH) (20°)	0.10 0.010 1.15×10^{-3} 4.63×10^{-11}
4-Nitrobenzoic acid	O_2N–C_6H_4–CO_2H	3.442	3.61×10^{-4}
Nitroethane	$CH_3CH_2NO_2$	8.57	2.7×10^{-9}
4-Nitrophenol	O_2N–C_6H_4–OH	7.149	7.10×10^{-8}
N-Nitrosophenylhydroxylamine (cupferron)	C_6H_5–N(NO)(OH)	4.16	6.9×10^{-5}
Nitrous acid	HON=O	3.15	7.1×10^{-4}
Oxalic acid (ethanedioic acid)	HO_2CCO_2H	1.27 4.266	5.37×10^{-2} 5.42×10^{-5}
Oxoacetic acid (glyoxylic acid)	$HC(=O)CO_2H$	3.46	3.5×10^{-4}
Oxobutanedioic acid (oxaloacetic acid)	$HO_2CCH_2C(=O)CO_2H$	2.56 4.37	2.8×10^{-3} 4.3×10^{-5}
2-Oxopentanedioic (α-ketoglutaric acid)	$HO_2CCH_2CH_2C(=O)CO_2H$	1.90 4.44	1.26×10^{-2} 3.6×10^{-5}
2-Oxopropanoic acid (pyruvic acid)	$CH_3C(=O)CO_2H$	2.48	3.3×10^{-3}
1,5-Pentanedioic acid (glutaric acid)	$HO_2CCH_2CH_2CH_2CO_2H$	4.345 5.422	4.52×10^{-5} 3.78×10^{-6}
1,10-Phenanthroline	(diprotonated 1,10-phenanthroline, $\overset{+}{N}H$ $H\overset{+}{N}$)	1.8 4.91	0.016 1.38×10^{-5}
Phenylacetic acid	C_6H_5–CH_2CO_2H	4.310	4.90×10^{-5}
Phenylalanine	NH_3^+–CH(CO_2H)–CH_2–C_6H_5	2.20 (CO_2H) 9.31 (NH_3)	6.3×10^{-3} 4.9×10^{-10}
Phosphoric acid (hydrogen phosphate)	$HO-P(=O)(OH)-OH$	2.148 7.198 12.375	7.11×10^{-3} 6.34×10^{-8} 4.22×10^{-13}

Name	Structure*	pK_a[†]	K_a
Phosphorous acid (hydrogen phosphite)	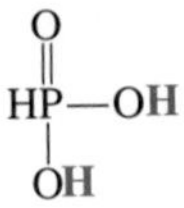 $HP(=O)(OH)OH$	(1.5) 6.78	3×10^{-2} 1.66×10^{-7}
Phthalic acid (benzene-1,2-dicarboxylic acid)	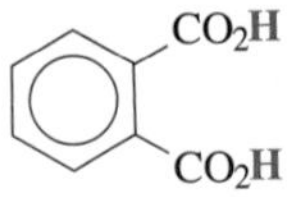 CO_2H, CO_2H	2.950 5.408	1.12×10^{-3} 3.90×10^{-6}
Piperazine (perhydro-1,4-diazine)	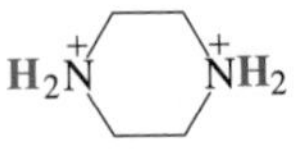 $H_2\overset{+}{N}$, $\overset{+}{N}H_2$	5.333 9.731	4.65×10^{-6} 1.86×10^{-10}
Piperidine	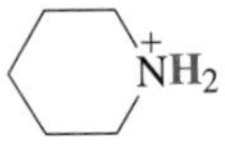 $\overset{+}{N}H_2$	11.125	7.50×10^{-12}
Proline	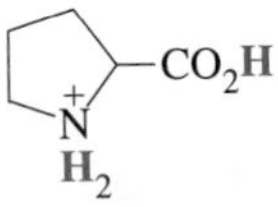 CO_2H, $\overset{+}{N}H_2$	1.952 (CO_2H) 10.640 (NH_2)	1.12×10^{-2} 2.29×10^{-11}
Propanoic acid	$CH_3CH_2CO_2H$	4.874	1.34×10^{-5}
Propenoic acid (acrylic acid)	$H_2C{=}CHCO_2H$	4.258	5.52×10^{-5}
Propylamine	$CH_3CH_2CH_2NH_3^+$	10.566	2.72×10^{-11}
Pyridine (azine)	 NH^+	5.20	6.3×10^{-6}
Pyridine-2-carboxylic acid (picolinic acid)	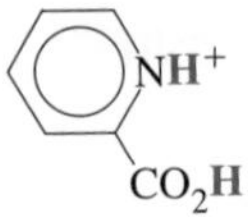 NH^+, CO_2H	(1.01) (CO_2H) 5.39 (NH)	9.8×10^{-2} 4.1×10^{-6}
Pyridine-3-carboxylic acid (nicotinic acid)	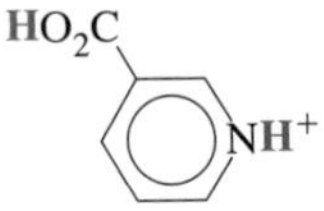 HO_2C, NH^+	2.05 (CO_2H) 4.81 (NH)	8.9×10^{-3} 1.55×10^{-5}
Pyridoxal-5-phosphate	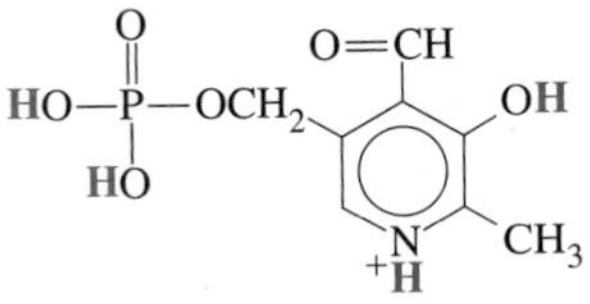 $HO{-}P(=O)(HO){-}OCH_2$, $O{=}CH$, OH, CH_3, $\overset{+}{N}H$	1.4 (POH) 3.44 (OH) 6.01 (POH) 8.45 (NH)	0.04 3.6×10^{-4} 9.8×10^{-7} 3.5×10^{-9}
Pyrophosphoric acid (hydrogen diphosphate)	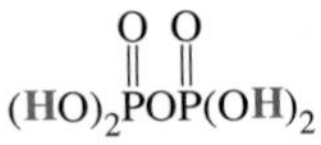 $(HO)_2P(=O)OP(=O)(OH)_2$	0.83 2.26 6.72 9.46	0.15 5.5×10^{-3} 1.9×10^{-7} 3.5×10^{-10}
Serine	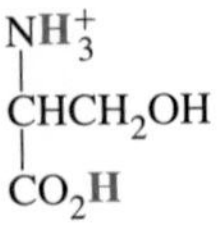 $NH_3^+CH(CH_2OH)CO_2H$	2.187 (CO_2H) 9.209 (NH_3)	6.50×10^{-3} 6.18×10^{-10}
Succinic acid (butanedioic acid)	$HO_2CCH_2CH_2CO_2H$	4.207 5.636	6.21×10^{-5} 2.31×10^{-6}

(continued)

Name	Structure*	pK_a[†]	K_a
Sulfuric acid (hydrogen sulfate)	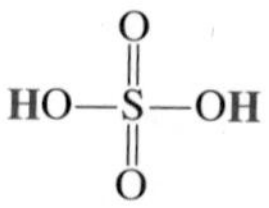 $HO-S(=O)_2-OH$	1.987 (pK_2)	1.03×10^{-2}
Sulfurous acid (hydrogen sulfite)	 HOS(=O)OH	1.857 7.172	1.39×10^{-2} 6.73×10^{-8}
Thiosulfuric acid (hydrogen thiosulfate)	 HOS(=O)(=O)SH	(0.6) (1.6)	0.3 0.03
Threonine	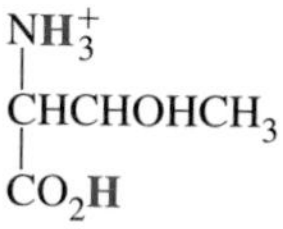 NH_3^+–CHCHOHCH$_3$–CO_2H	2.088 (CO_2H) 9.100 (NH_3)	8.17×10^{-3} 7.94×10^{-10}
Trichloroacetic acid	Cl_3CCO_2H	(0.5)	0.3
Triethanolamine	$(HOCH_2CH_2)_3NH^+$	7.762	1.73×10^{-8}
Triethylamine	$(CH_3CH_2)_3NH^+$	10.72	1.9×10^{-11}
1,2,3-Trihydroxybenzene (pyrogallol)	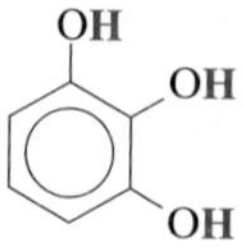 benzene ring with OH, OH, OH	8.96 11.00 (14.0) (20°)	1.10×10^{-9} 1.00×10^{-11} 10^{-14}
Trimethylamine	$(CH_3)_3NH^+$	9.799	1.59×10^{-10}
Tris(hydroxymethyl)aminomethane (tris or tham)	$(HOCH_2)_3CNH_3^+$	8.072	8.47×10^{-9}
Tryptophan	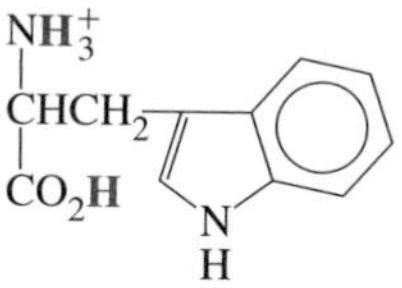 NH_3^+–CHCH$_2$–(indol-3-yl, N–H)–CO_2H	2.37 (CO_2H) 9.33 (NH_3)	4.3×10^{-3} 4.7×10^{-10}
Tyrosine	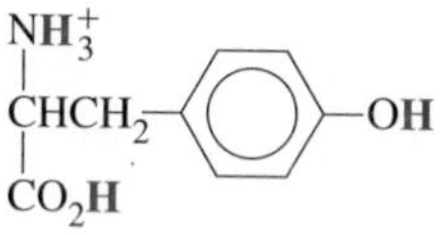 NH_3^+–CHCH$_2$–(C$_6$H$_4$)–OH, CO_2H	2.41 (CO_2H) 8.67 (NH_3) 11.01 (OH)	3.9×10^{-3} 2.1×10^{-9} 9.8×10^{-12}
Valine	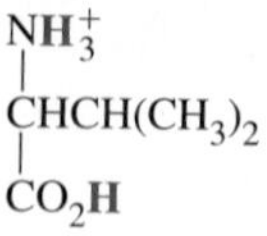 NH_3^+–CHCH$(CH_3)_2$–CO_2H	2.286 (CO_2H) 9.719 (NH_3)	5.18×10^{-3} 1.91×10^{-10}

Appendix C
Standard Reduction Potentials

Reaction*	E° (volts)
Aluminum	
$Al^{3+} + 3e^- \rightleftharpoons Al(s)$	−1.677
$Al(OH)_4^- + 3e^- \rightleftharpoons Al(s) + 4OH^-$	−2.328
Arsenic	
$H_3AsO_4 + 2H^+ + 2e^- \rightleftharpoons H_3AsO_3 + H_2O$	0.575
$H_3AsO_3 + 3H^+ + 3e^- \rightleftharpoons As(s) + 3H_2O$	0.247 5
$As(s) + 3H^+ + 3e^- \rightleftharpoons AsH_3(g)$	−0.238
Barium	
$Ba^{2+} + 2e^- \rightleftharpoons Ba(s)$	−2.906
Beryllium	
$Be^{2+} + 2e^- \rightleftharpoons Be(s)$	−1.968
Boron	
$2B(s) + 6H^+ + 6e^- \rightleftharpoons B_2H_6(g)$	−0.150
$B_4O_7^{2-} + 14H^+ + 12e^- \rightleftharpoons 4B(s) + 7H_2O$	−0.792
$B(OH)_3 + 3H^+ + 3e^- \rightleftharpoons B(s) + 3H_2O$	−0.889
Bromine	
$BrO_4^- + 2H^+ + 2e^- \rightleftharpoons BrO_3^- + H_2O$	1.745
$HOBr + H^+ + e^- \rightleftharpoons \frac{1}{2}Br_2(l) + H_2O$	1.584
$BrO_3^- + 6H^+ + 5e^- \rightleftharpoons \frac{1}{2}Br_2(l) + 3H_2O$	1.513
$Br_2(aq) + 2e^- \rightleftharpoons 2Br^-$	1.098
$Br_2(l) + 2e^- \rightleftharpoons 2Br^-$	1.078
$Br_3^- + 2e^- \rightleftharpoons 3Br^-$	1.062
$BrO^- + H_2O + 2e^- \rightleftharpoons Br^- + 2OH^-$	0.766
$BrO_3^- + 3H_2O + 6e^- \rightleftharpoons Br^- + 6OH^-$	0.613
Cadmium	
$Cd^{2+} + 2e^- \rightleftharpoons Cd(s)$	−0.402
$Cd(NH_3)_4^{2+} + 2e^- \rightleftharpoons Cd(s) + 4NH_3$	−0.613
Calcium	
$Ca(s) + 2H^+ + 2e^- \rightleftharpoons CaH_2(s)$	0.776
$Ca^{2+} + 2e^- \rightleftharpoons Ca(s)$	−2.868
$Ca(acetate)^+ + 2e^- \rightleftharpoons Ca(s) + acetate^-$	−2.891
$CaSO_4(s) + 2e^- \rightleftharpoons Ca(s) + SO_4^{2-}$	−2.936
Carbon	
$C_2H_2(g) + 2H^+ + 2e^- \rightleftharpoons C_2H_4(g)$	0.731
$O{=}C_6H_4{=}O + 2H^+ + 2e^- \rightleftharpoons HO{-}C_6H_4{-}OH$	0.700
$CH_3OH + 2H^+ + 2e^- \rightleftharpoons CH_4(g) + H_2O$	0.583
dehydroascorbic acid $+ 2H^+ + 2e^- \rightleftharpoons$ ascorbic acid $+ H_2O$	0.390
$(CN)_2(g) + 2H^+ + 2e^- \rightleftharpoons 2HCN(aq)$	0.373
$H_2CO + 2H^+ + 2e^- \rightleftharpoons CH_3OH$	0.237
$C(s) + 4H^+ + 4e^- \rightleftharpoons CH_4(g)$	0.131 5
$HCO_2H + 2H^+ + 2e^- \rightleftharpoons H_2CO + H_2O$	−0.029
$CO_2(g) + 2H^+ + 2e^- \rightleftharpoons CO(g) + H_2O$	−0.103 8
$CO_2(g) + 2H^+ + 2e^- \rightleftharpoons HCO_2H$	−0.114
$2CO_2(g) + 2H^+ + 2e^- \rightleftharpoons H_2C_2O_4$	−0.432
Cerium	
$Ce^{4+} + e^- \rightleftharpoons Ce^{3+}$	1.72 1.70 1 F $HClO_4$ 1.44 1 F H_2SO_4 1.61 1 F HNO_3 1.47 1 F HCl
$Ce^{3+} + 3e^- \rightleftharpoons Ce(s)$	−2.336
Cesium	
$Cs^+ + e^- \rightleftharpoons Cs(s)$	−3.026
Chlorine	
$HClO_2 + 2H^+ + 2e^- \rightleftharpoons HOCl + H_2O$	1.674
$HClO + H^+ + e^- \rightleftharpoons \frac{1}{2}Cl_2(g) + H_2O$	1.630
$ClO_3^- + 6H^+ + 5e^- \rightleftharpoons \frac{1}{2}Cl_2(g) + 3H_2O$	1.458
$Cl_2(aq) + 2e^- \rightleftharpoons 2Cl^-$	1.396
$Cl_2(g) + 2e^- \rightleftharpoons 2Cl^-$	1.360 4
$ClO_4^- + 2H^+ + 2e^- \rightleftharpoons ClO_3^- + H_2O$	1.226
$ClO_3^- + 3H^+ + 2e^- \rightleftharpoons HClO_2 + H_2O$	1.157
$ClO_3^- + 2H^+ + e^- \rightleftharpoons ClO_2 + H_2O$	1.130
$ClO_2 + e^- \rightleftharpoons ClO_2^-$	1.068
Chromium	
$Cr_2O_7^{2-} + 14H^+ + 6e^- \rightleftharpoons 2Cr^{3+} + 7H_2O$	1.36
$CrO_4^{2-} + 4H_2O + 3e^- \rightleftharpoons Cr(OH)_3$ (*s*, hydrated) $+ 5OH^-$	−0.12
$Cr^{3+} + e^- \rightleftharpoons Cr^{2+}$	−0.42
$Cr^{3+} + 3e^- \rightleftharpoons Cr(s)$	−0.74
$Cr^{2+} + 2e^- \rightleftharpoons Cr(s)$	−0.89
Cobalt	
$Co^{3+} + e^- \rightleftharpoons Co^{2+}$	1.92 1.817 8 F H_2SO_4 1.850 4 F HNO_3
$Co(NH_3)_6^{3+} + e^- \rightleftharpoons Co(NH_3)_6^{2+}$	0.1
$CoOH^+ + H^+ + 2e^- \rightleftharpoons Co(s) + H_2O$	0.003

* All species are aqueous unless otherwise indicated.

(continued)

Reaction*	E° (volts)
$Co^{2+} + 2e^- \rightleftharpoons Co(s)$	−0.282
$Co(OH)_2(s) + 2e^- \rightleftharpoons Co(s) + 2OH^-$	−0.746
Copper	
$Cu^+ + e^- \rightleftharpoons Cu(s)$	0.518
$Cu^{2+} + 2e^- \rightleftharpoons Cu(s)$	0.339
$Cu^{2+} + e^- \rightleftharpoons Cu^+$	0.161
$CuCl(s) + e^- \rightleftharpoons Cu(s) + Cl^-$	0.137
$Cu(IO_3)_2(s) + 2e^- \rightleftharpoons Cu(s) + 2IO_3^-$	−0.079
$Cu(ethylenediamine)_2^+ + e^- \rightleftharpoons Cu(s) + 2\ ethylenediamine$	−0.119
$CuI(s) + e^- \rightleftharpoons Cu(s) + I^-$	−0.185
$Cu(EDTA)^{2-} + 2e^- \rightleftharpoons Cu(s) + EDTA^{4-}$	−0.216
$Cu(OH)_2(s) + 2e^- \rightleftharpoons Cu(s) + 2OH^-$	−0.222
$Cu(CN)_2^- + e^- \rightleftharpoons Cu(s) + 2CN^-$	−0.429
$CuCN(s) + e^- \rightleftharpoons Cu(s) + CN^-$	−0.639
Fluorine	
$F_2(g) + 2e^- \rightleftharpoons 2F^-$	2.890
$F_2O(g) + 2H^+ + 4e^- \rightleftharpoons 2F^- + H_2O$	2.168
Gallium	
$Ga^{3+} + 3e^- \rightleftharpoons Ga(s)$	−0.549
$GaOOH(s) + H_2O + 3e^- \rightleftharpoons Ga(s) + 3OH^-$	−1.320
Germanium	
$Ge^{2+} + 2e^- \rightleftharpoons Ge(s)$	0.1
$H_4GeO_4 + 4H^+ + 4e^- \rightleftharpoons Ge(s) + 4H_2O$	−0.039
Gold	
$Au^+ + e^- \rightleftharpoons Au(s)$	1.69
$Au^{3+} + 2e^- \rightleftharpoons Au^+$	1.41
$AuCl_2^- + e^- \rightleftharpoons Au(s) + 2Cl^-$	1.154
$AuCl_4^- + 2e^- \rightleftharpoons AuCl_2^- + 2Cl^-$	0.926
Hydrogen	
$2H^+ + 2e^- \rightleftharpoons H_2(g)$	0.000 0
$H_2O + e^- \rightleftharpoons \frac{1}{2}H_2(g) + OH^-$	−0.828 0
Indium	
$In^{3+} + 3e^- \rightleftharpoons In(s)$	−0.338
$In^{3+} + 2e^- \rightleftharpoons In^+$	−0.444
$In(OH)_3(s) + 3e^- \rightleftharpoons In(s) + 3OH^-$	−0.99
Iodine	
$IO_4^- + 2H^+ + 2e^- \rightleftharpoons IO_3^- + H_2O$	1.589
$H_5IO_6 + 2H^+ + 2e^- \rightleftharpoons HIO_3 + 3H_2O$	1.567
$HOI + H^+ + e^- \rightleftharpoons \frac{1}{2}I_2(s) + H_2O$	1.430
$ICl_3(s) + 3e^- \rightleftharpoons \frac{1}{2}I_2(s) + 3Cl^-$	1.28
$ICl(s) + e^- \rightleftharpoons \frac{1}{2}I_2(s) + Cl^-$	1.22
$IO_3^- + 6H^+ + 5e^- \rightleftharpoons \frac{1}{2}I_2(s) + 3H_2O$	1.210
$IO_3^- + 5H^+ + 4e^- \rightleftharpoons HOI + 2H_2O$	1.154
$I_2(aq) + 2e^- \rightleftharpoons 2I^-$	0.620
$I_2(s) + 2e^- \rightleftharpoons 2I^-$	0.535
$I_3^- + 2e^- \rightleftharpoons 3I^-$	0.535
$IO_3^- + 3H_2O + 6e^- \rightleftharpoons I^- + 6OH^-$	0.269
Iron	
$Fe(phenanthroline)_3^{3+} + e^- \rightleftharpoons Fe(phenanthroline)_3^{2+}$	1.147
$Fe(bipyridyl)_3^{3+} + e^- \rightleftharpoons Fe(bipyridyl)_3^{2+}$	1.120
$FeOH^{2+} + H^+ + e^- \rightleftharpoons Fe^{2+} + H_2O$	0.900
$FeO_4^{2-} + 3H_2O + 3e^- \rightleftharpoons FeOOH(s) + 5OH^-$	0.80
$Fe^{3+} + e^- \rightleftharpoons Fe^{2+}$	0.771
	0.732 1 F HCl
	0.767 1 F $HClO_4$
	0.746 1 F HNO_3
	0.68 1 F H_2SO_4
$FeOOH(s) + 3H^+ + e^- \rightleftharpoons Fe^{2+} + 2H_2O$	0.74
$ferricinium^+ + e^- \rightleftharpoons ferrocene$	0.400
$Fe(CN)_6^{3-} + e^- \rightleftharpoons Fe(CN)_6^{4-}$	0.356
$FeOH^+ + H^+ + 2e^- \rightleftharpoons Fe(s) + H_2O$	−0.16
$Fe^{2+} + 2e^- \rightleftharpoons Fe(s)$	−0.44
$FeCO_3(s) + 2e^- \rightleftharpoons Fe(s) + CO_3^{2-}$	−0.756
Lanthanum	
$La^{3+} + 3e^- \rightleftharpoons La(s)$	−2.379
Lead	
$Pb^{4+} + 2e^- \rightleftharpoons Pb^{2+}$	1.69 1 F HNO_3
$PbO_2(s) + 4H^+ + SO_4^{2-} + 2e^- \rightleftharpoons PbSO_4(s) + 2H_2O$	1.685
$PbO_2(s) + 4H^+ + 2e^- \rightleftharpoons Pb^{2+} + 2H_2O$	1.458
$3PbO_2(s) + 2H_2O + 4e^- \rightleftharpoons Pb_3O_4(s) + 4OH^-$	0.269
$Pb_3O_4(s) + H_2O + 2e^- \rightleftharpoons 3PbO(s, red) + 2OH^-$	0.224
$Pb_3O_4(s) + H_2O + 2e^- \rightleftharpoons 3PbO(s, yellow) + 2OH^-$	0.207
$Pb^{2+} + 2e^- \rightleftharpoons Pb(s)$	−0.126
$PbF_2(s) + 2e^- \rightleftharpoons Pb(s) + 2F^-$	−0.350
$PbSO_4(s) + 2e^- \rightleftharpoons Pb(s) + SO_4^{2-}$	−0.355
Lithium	
$Li^+ + e^- \rightleftharpoons Li(s)$	−3.040
Magnesium	
$Mg(OH)^+ + H^+ + 2e^- \rightleftharpoons Mg(s) + H_2O$	−2.022
$Mg^{2+} + 2e^- \rightleftharpoons Mg(s)$	−2.360
$Mg(C_2O_4)(s) + 2e^- \rightleftharpoons Mg(s) + C_2O_4^{2-}$	−2.493
$Mg(OH)_2(s) + 2e^- \rightleftharpoons Mg(s) + 2OH^-$	−2.690
Manganese	
$MnO_4^- + 4H^+ + 3e^- \rightleftharpoons MnO_2(s) + 2H_2O$	1.692
$Mn^{3+} + e^- \rightleftharpoons Mn^{2+}$	1.56
$MnO_4^- + 8H^+ + 5e^- \rightleftharpoons Mn^{2+} + 4H_2O$	1.507
$Mn_2O_3(s) + 6H^+ + 2e^- \rightleftharpoons 2Mn^{2+} + 3H_2O$	1.485
$MnO_2(s) + 4H^+ + 2e^- \rightleftharpoons Mn^{2+} + 2H_2O$	1.230
$Mn(EDTA)^- + e^- \rightleftharpoons Mn(EDTA)^{2-}$	0.825
$MnO_4^- + e^- \rightleftharpoons MnO_4^{2-}$	0.56
$3Mn_2O_3(s) + H_2O + 2e^- \rightleftharpoons 2Mn_3O_4(s) + 2OH^-$	0.002
$Mn_3O_4(s) + 4H_2O + 2e^- \rightleftharpoons 3Mn(OH)_2(s) + 2OH^-$	−0.352
$Mn^{2+} + 2e^- \rightleftharpoons Mn(s)$	−1.182
$Mn(OH)_2(s) + 2e^- \rightleftharpoons Mn(s) + 2OH^-$	−1.565
Mercury	
$2Hg^{2+} + 2e^- \rightleftharpoons Hg_2^{2+}$	0.908
$Hg^{2+} + 2e^- \rightleftharpoons Hg(l)$	0.852
$Hg_2^{2+} + 2e^- \rightleftharpoons 2Hg(l)$	0.796
$Hg_2SO_4(s) + 2e^- \rightleftharpoons 2Hg(l) + SO_4^{2-}$	0.614
$Hg_2Cl_2(s) + 2e^- \rightleftharpoons 2Hg(l) + 2Cl^-$	0.268
	0.241 (saturated calomel electrode)
$Hg(OH)_3^- + 2e^- \rightleftharpoons Hg(l) + 3OH^-$	0.231

Reaction*	$E°$ (volts)
$Hg(OH)_2 + 2e^- \rightleftharpoons Hg(l) + 2OH^-$	0.206
$Hg_2Br_2(s) + 2e^- \rightleftharpoons 2Hg(l) + 2Br^-$	0.140
$HgO(s, \text{yellow}) + H_2O + 2e^- \rightleftharpoons Hg(l) + 2OH^-$	0.098 3
$HgO(s, \text{red}) + H_2O + 2e^- \rightleftharpoons Hg(l) + 2OH^-$	0.097 7
Molybdenum	
$MoO_4^{2-} + 2H_2O + 2e^- \rightleftharpoons MoO_2(s) + 4OH^-$	−0.818
$MoO_4^{2-} + 4H_2O + 6e^- \rightleftharpoons Mo(s) + 8OH^-$	−0.926
$MoO_2(s) + 2H_2O + 4e^- \rightleftharpoons Mo(s) + 4OH^-$	−0.980
Nickel	
$NiOOH(s) + 3H^+ + e^- \rightleftharpoons Ni^{2+} + 2H_2O$	2.05
$Ni^{2+} + 2e^- \rightleftharpoons Ni(s)$	−0.236
$Ni(CN)_4^{2-} + e^- \rightleftharpoons Ni(CN)_3^{2-} + CN^-$	−0.401
$Ni(OH)_2(s) + 2e^- \rightleftharpoons Ni(s) + 2OH^-$	−0.714
Nitrogen	
$HN_3 + 3H^+ + 2e^- \rightleftharpoons N_2(g) + NH_4^+$	2.079
$N_2O(g) + 2H^+ + 2e^- \rightleftharpoons N_2(g) + H_2O$	1.769
$2NO(g) + 2H^+ + 2e^- \rightleftharpoons N_2O(g) + H_2O$	1.587
$NO^+ + e^- \rightleftharpoons NO(g)$	1.46
$2NH_3OH^+ + H^+ + 2e^- \rightleftharpoons N_2H_5^+ + 2H_2O$	1.40
$NH_3OH^+ + 2H^+ + 2e^- \rightleftharpoons NH_4^+ + H_2O$	1.33
$N_2H_5^+ + 3H^+ + 2e^- \rightleftharpoons 2NH_4^+$	1.250
$HNO_2 + H^+ + e^- \rightleftharpoons NO(g) + H_2O$	0.984
$NO_3^- + 4H^+ + 3e^- \rightleftharpoons NO(g) + 2H_2O$	0.955
$NO_3^- + 3H^+ + 2e^- \rightleftharpoons HNO_2 + H_2O$	0.940
$NO_3^- + 2H^+ + e^- \rightleftharpoons \frac{1}{2}N_2O_4(g) + H_2O$	0.798
$N_2(g) + 8H^+ + 6e^- \rightleftharpoons 2NH_4^+$	0.274
$N_2(g) + 5H^+ + 4e^- \rightleftharpoons N_2H_5^+$	−0.214
$N_2(g) + 2H_2O + 4H^+ + 2e^- \rightleftharpoons 2NH_3OH^+$	−1.83
$\frac{3}{2}N_2(g) + H^+ + e^- \rightleftharpoons HN_3$	−3.334
Oxygen	
$OH + H^+ + e^- \rightleftharpoons H_2O$	2.56
$O(g) + 2H^+ + 2e^- \rightleftharpoons H_2O$	2.430 1
$O_3(g) + 2H^+ + 2e^- \rightleftharpoons O_2(g) + H_2O$	2.075
$H_2O_2 + 2H^+ + 2e^- \rightleftharpoons 2H_2O$	1.763
$HO_2 + H^+ + e^- \rightleftharpoons H_2O_2$	1.44
$\frac{1}{2}O_2(g) + 2H^+ + 2e^- \rightleftharpoons H_2O$	1.229 1
$O_2(g) + 2H^+ + 2e^- \rightleftharpoons H_2O_2$	0.695
$O_2(g) + H^+ + e^- \rightleftharpoons HO_2$	−0.05
Palladium	
$Pd^{2+} + 2e^- \rightleftharpoons Pd(s)$	0.915
$PdO(s) + 2H^+ + 2e^- \rightleftharpoons Pd(s) + H_2O$	0.79
$PdCl_6^{4-} + 2e^- \rightleftharpoons Pd(s) + 6Cl^-$	0.615
$PdO_2(s) + H_2O + 2e^- \rightleftharpoons PdO(s) + 2OH^-$	0.64
Phosphorus	
$\frac{1}{4}P_4(s, \text{white}) + 3H^+ + 3e^- \rightleftharpoons PH_3(g)$	−0.046
$\frac{1}{4}P_4(s, \text{red}) + 3H^+ + 3e^- \rightleftharpoons PH_3(g)$	−0.088
$H_3PO_4 + 2H^+ + 2e^- \rightleftharpoons H_3PO_3 + H_2O$	−0.30
$H_3PO_4 + 5H^+ + 5e^- \rightleftharpoons \frac{1}{4}P_4(s, \text{white}) + 4H_2O$	−0.402
$H_3PO_3 + 2H^+ + 2e^- \rightleftharpoons H_3PO_2 + H_2O$	−0.48
$H_3PO_2 + H^+ + e^- \rightleftharpoons \frac{1}{4}P_4(s) + 2H_2O$	−0.51
Platinum	
$Pt^{2+} + 2e^- \rightleftharpoons Pt(s)$	1.18
$PtO_2(s) + 4H^+ + 4e^- \rightleftharpoons Pt(s) + 2H_2O$	0.92
$PtCl_4^{2-} + 2e^- \rightleftharpoons Pt(s) + 4Cl^-$	0.755
$PtCl_6^{2-} + 2e^- \rightleftharpoons PtCl_4^{2-} + 2Cl^-$	0.68
Potassium	
$K^+ + e^- \rightleftharpoons K(s)$	−2.936
Rubidium	
$Rb^+ + e^- \rightleftharpoons Rb(s)$	−2.943
Scandium	
$Sc^{3+} + 3e^- \rightleftharpoons Sc(s)$	−2.09
Selenium	
$SeO_4^{2-} + 4H^+ + 2e^- \rightleftharpoons H_2SeO_3 + H_2O$	1.150
$H_2SeO_3 + 4H^+ + 4e^- \rightleftharpoons Se(s) + 3H_2O$	0.739
$Se(s) + 2H^+ + 2e^- \rightleftharpoons H_2Se(g)$	−0.082
$Se(s) + 2e^- \rightleftharpoons Se^{2-}$	−0.67
Silicon	
$Si(s) + 4H^+ + 4e^- \rightleftharpoons SiH_4(g)$	−0.147
$SiO_2(s, \text{quartz}) + 4H^+ + 4e^- \rightleftharpoons Si(s) + 2H_2O$	−0.990
$SiF_6^{2-} + 4e^- \rightleftharpoons Si(s) + 6F^-$	−1.24
Silver	
$Ag^{2+} + e^- \rightleftharpoons Ag^+$	1.989
$Ag^{3+} + 2e^- \rightleftharpoons Ag^+$	1.9
$AgO(s) + H^+ + e^- \rightleftharpoons \frac{1}{2}Ag_2O(s) + \frac{1}{2}H_2O$	1.40
$Ag^+ + e^- \rightleftharpoons Ag(s)$	0.799 3
$Ag_2C_2O_4(s) + 2e^- \rightleftharpoons 2Ag(s) + C_2O_4^{2-}$	0.465
$AgN_3(s) + e^- \rightleftharpoons Ag(s) + N_3^-$	0.293
$AgCl(s) + e^- \rightleftharpoons Ag(s) + Cl^-$	0.222 0.197 (saturated KCl)
$AgBr(s) + e^- \rightleftharpoons Ag(s) + Br^-$	0.071
$Ag(S_2O_3)_2^{3-} + e^- \rightleftharpoons Ag(s) + 2S_2O_3^{2-}$	0.017
$AgI(s) + e^- \rightleftharpoons Ag(s) + I^-$	−0.152
$Ag_2S(s) + H^+ + 2e^- \rightleftharpoons 2Ag(s) + SH^-$	−0.272
Sodium	
$Na^+ + \frac{1}{2}H_2(g) + e^- \rightleftharpoons NaH(s)$	−2.367
$Na^+ + e^- \rightleftharpoons Na(s)$	−2.714 3
Strontium	
$Sr^{2+} + 2e^- \rightleftharpoons Sr(s)$	−2.889
Sulfur	
$S_2O_8^{2-} + 2e^- \rightleftharpoons 2SO_4^{2-}$	2.01
$S_2O_6^{2-} + 4H^+ + 2e^- \rightleftharpoons 2H_2SO_3$	0.57
$4SO_2 + 4H^+ + 6e^- \rightleftharpoons S_4O_6^{2-} + 2H_2O$	0.539
$SO_2 + 4H^+ + 4e^- \rightleftharpoons S(s) + 2H_2O$	0.450
$2H_2SO_3 + 2H^+ + 4e^- \rightleftharpoons S_2O_3^{2-} + 3H_2O$	0.40
$S(s) + 2H^+ + 2e^- \rightleftharpoons H_2S(g)$	0.174
$S(s) + 2H^+ + 2e^- \rightleftharpoons H_2S(aq)$	0.144
$S_4O_6^{2-} + 2H^+ + 2e^- \rightleftharpoons 2HS_2O_3^-$	0.10
$5S(s) + 2e^- \rightleftharpoons S_5^{2-}$	−0.340
$2S(s) + 2e^- \rightleftharpoons S_2^{2-}$	−0.50
$2SO_3^{2-} + 3H_2O + 4e^- \rightleftharpoons S_2O_3^{2-} + 6OH^-$	−0.566
$SO_3^{2-} + 3H_2O + 4e^- \rightleftharpoons S(s) + 6OH^-$	−0.659
$SO_4^{2-} + 4H_2O + 6e^- \rightleftharpoons S(s) + 8OH^-$	−0.751
$SO_4^{2-} + H_2O + 2e^- \rightleftharpoons SO_3^{2-} + 2OH^-$	−0.936
$2SO_3^{2-} + 2H_2O + 2e^- \rightleftharpoons S_2O_4^{2-} + 4OH^-$	−1.130
$2SO_4^{2-} + 2H_2O + 2e^- \rightleftharpoons S_2O_6^{2-} + 4OH^-$	−1.71

(continued)

Reaction*	$E°$ (volts)
Thallium	
$Tl^{3+} + 2e^- \rightleftharpoons Tl^+$	1.280
	0.77 1 F HCl
	1.22 1 F H_2SO_4
	1.23 1 F HNO_3
	1.26 1 F $HClO_4$
$Tl^+ + e^- \rightleftharpoons Tl(s)$	−0.336
Tin	
$Sn(OH)_3^+ + 3H^+ + 2e^- \rightleftharpoons Sn^{2+} + 6H_2O$	0.142
$Sn^{4+} + 2e^- \rightleftharpoons Sn^{2+}$	0.139 1 F HCl
$SnO_2(s) + 4H^+ + 2e^- \rightleftharpoons Sn^{2+} + 2H_2O$	−0.094
$Sn^{2+} + 2e^- \rightleftharpoons Sn(s)$	−0.141
$SnF_6^{2-} + 4e^- \rightleftharpoons Sn(s) + 6F^-$	−0.25
$Sn(OH)_6^{2-} + 2e^- \rightleftharpoons Sn(OH)_3^- + 3OH^-$	−0.93
$Sn(s) + 4H_2O + 4e^- \rightleftharpoons SnH_4(g) + 4OH^-$	−1.316
$SnO_2(s) + H_2O + 2e^- \rightleftharpoons SnO(s) + 2OH^-$	−0.961
Titanium	
$TiO^{2+} + 2H^+ + e^- \rightleftharpoons Ti^{3+} + H_2O$	0.1
$Ti^{3+} + e^- \rightleftharpoons Ti^{2+}$	−0.9
$TiO_2(s) + 4H^+ + 4e^- \rightleftharpoons Ti(s) + 2H_2O$	−1.076
$TiF_6^{2-} + 4e^- \rightleftharpoons Ti(s) + 6F^-$	−1.191
$Ti^{2+} + 2e^- \rightleftharpoons Ti(s)$	−1.60
Tungsten	
$W(CN)_8^{3-} + e^- \rightleftharpoons W(CN)_8^{4-}$	0.457
$W^{6+} + e^- \rightleftharpoons W^{5+}$	0.26 12 F HCl
$WO_3(s) + 6H^+ + 6e^- \rightleftharpoons W(s) + 3H_2O$	−0.091
$W^{5+} + e^- \rightleftharpoons W^{4+}$	−0.3 12 F HCl
$WO_2(s) + 2H_2O + 4e^- \rightleftharpoons W(s) + 4OH^-$	−0.982
$WO_4^{2-} + 4H_2O + 6e^- \rightleftharpoons W(s) + 8OH^-$	−1.060

Reaction*	$E°$ (volts)
Uranium	
$UO_2^+ + 4H^+ + e^- \rightleftharpoons U^{4+} + 2H_2O$	0.39
$UO_2^{2+} + 4H^+ + 2e^- \rightleftharpoons U^{4+} + 2H_2O$	0.273
$UO_2^{2+} + e^- \rightleftharpoons UO_2^+$	0.16
$U^{4+} + e^- \rightleftharpoons U^{3+}$	−0.577
$U^{3+} + 3e^- \rightleftharpoons U(s)$	−1.642
Vanadium	
$VO_2^+ + 2H^+ + e^- \rightleftharpoons VO^{2+} + H_2O$	1.001
$VO^{2+} + 2H^+ + e^- \rightleftharpoons V^{3+} + H_2O$	0.337
$V^{3+} + e^- \rightleftharpoons V^{2+}$	−0.255
$V^{2+} + 2e^- \rightleftharpoons V(s)$	−1.125
Xenon	
$H_4XeO_6 + 2H^+ + 2e^- \rightleftharpoons XeO_3 + 3H_2O$	2.38
$XeF_2 + 2H^+ + 2e^- \rightleftharpoons Xe(g) + 2HF$	2.2
$XeO_3 + 6H^+ + 6e^- \rightleftharpoons Xe(g) + 3H_2O$	2.1
Yttrium	
$Y^{3+} + 3e^- \rightleftharpoons Y(s)$	−2.38
Zinc	
$ZnOH^+ + H^+ + 2e^- \rightleftharpoons Zn(s) + H_2O$	−0.497
$Zn^{2+} + 2e^- \rightleftharpoons Zn(s)$	−0.762
$Zn(NH_3)_4^{2+} + 2e^- \rightleftharpoons Zn(s) + 4NH_3$	−1.04
$ZnCO_3(s) + 2e^- \rightleftharpoons Zn(s) + CO_3^{2-}$	−1.06
$Zn(OH)_3^- + 2e^- \rightleftharpoons Zn(s) + 3OH^-$	−1.183
$Zn(OH)_4^{2-} + 2e^- \rightleftharpoons Zn(s) + 4OH^-$	−1.199
$Zn(OH)_2(s) + 2e^- \rightleftharpoons Zn(s) + 2OH^-$	−1.249
$ZnO(s) + H_2O + 2e^- \rightleftharpoons Zn(s) + 2OH^-$	−1.260
$ZnS(s) + 2e^- \rightleftharpoons Zn(s) + S^{2-}$	−1.405

Appendix D
Oxidation Numbers and Balancing Redox Equations

The *oxidation number, or oxidation state*, is a bookkeeping device used to keep track of the number of electrons formally associated with a particular element. The oxidation number is meant to tell how many electrons have been lost or gained by a neutral atom when it forms a compound. Because oxidation numbers have no real physical meaning, they are somewhat arbitrary, and not all chemists will assign the same oxidation number to a given element in an unusual compound. However, there are some ground rules that provide a useful start.

1. The oxidation number of an element by itself—e.g., Cu(*s*) or $Cl_2(g)$—is 0.
2. The oxidation number of H is almost always +1, except in metal hydrides—e.g., NaH, in which H is −1.
3. The oxidation number of oxygen is almost always −2. The only common exceptions are peroxides, in which two oxygen atoms are connected and each has an oxidation number of −1. Two examples are hydrogen peroxide (H—O—O—H) and its anion ($H—O—O^-$). The oxidation number of oxygen in gaseous O_2 is 0.
4. The alkali metals (Li, Na, K, Rb, Cs, Fr) almost always have an oxidation number of +1. The alkaline earth metals (Be, Mg, Ca, Sr, Ba, Ra) are almost always in the +2 oxidation state.
5. The halogens (F, Cl, Br, I) are usually in the −1 oxidation state. Exceptions are when two different halogens are bound to each other or when a halogen is bound to more than one atom. When different halogens are bound to each other, we assign the oxidation number −1 to the more electronegative halogen.

The sum of the oxidation numbers of each atom in a molecule must equal the charge of the molecule. In H_2O, for example, we have

$$\begin{array}{lrr} \text{2 hydrogen} = 2(+1) & = & +2 \\ \text{oxygen} & = & \underline{-2} \\ \text{net charge} & & 0 \end{array}$$

In SO_4^{2-}, sulfur must have an oxidation number of +6 so that the sum of the oxidation numbers will be −2:

$$\begin{array}{lrr} \text{oxygen} = 4(-2) & = & -8 \\ \text{sulfur} & = & \underline{+6} \\ \text{net charge} & & -2 \end{array}$$

In benzene (C_6H_6), the oxidation number of each carbon must be −1 if hydrogen is assigned the number +1. In cyclohexane (C_6H_{12}), the oxidation number of each carbon must be −2 for the same reason. The carbons in benzene are in a higher oxidation state than those in cyclohexane.

The oxidation number of iodine in ICl_2^- is +1. This is unusual, because halogens are usually −1. However, because chlorine is more electronegative than iodine, we assign Cl as −1, thereby forcing I to be +1.

The oxidation number of As in As_2S_3 is +3, and the value for S is −2. This is arbitrary but reasonable. Because S is more electronegative than As, we make S negative and As positive; and, because S is in the same family as oxygen, which is usually −2, we assign S as −2, thus leaving As as +3.

The oxidation number of S in $S_4O_6^{2-}$ (tetrathionate) is +2.5. The *fractional oxidation state* comes about because six O atoms contribute −12. Because the charge is −2, the four S atoms must contribute +10. The average oxidation number of S must be $+\frac{10}{4} = 2.5$.

The oxidation number of Fe in $K_3Fe(CN)_6$ is +3. To make this assignment, we first recognize cyanide (CN^-) as a common ion that carries a charge of −1. Six cyanide ions give −6, and three potassium ions (K^+) give +3. Therefore, Fe should have an oxidation number of +3 for the whole formula to be neutral. In this approach, it is not necessary to assign individual oxidation numbers to carbon and nitrogen, as long as we recognize that the charge of CN is −1.

Problems

Answers are given at the end of this appendix.

1. Write the oxidation state of the boldface atom in each of the following species.

(a) Ag Br
(b) $S_2O_3^{2-}$
(c) SeF_6
(d) $HS_2O_3^-$
(e) HO_2
(f) NO
(g) Cr^{3+}
(h) MnO_2
(i) $Pb(OH)_3^-$
(j) $Fe(OH)_3$
(k) ClO^-
(l) $K_4Fe(CN)_6$
(m) ClO_2
(n) ClO_2^-
(o) $Mn(CN)_6^{4-}$
(p) N_2
(q) NH_4^+
(r) $N_2H_5^+$
(s) $HAsO_3^{2-}$
(t) $(CH_3)_4Li_4$
(u) P_4O_{10}
(v) C_2H_6O
(w) $VO(SO_4)$
(x) Fe_3O_4

2. Identify the oxidizing agent and the reducing agent on the left side of each of the following reactions.

(a) $Cr_2O_7^{2-} + 3Sn^{2+} + 14H^+ \longrightarrow 2Cr^{3+} + 3Sn^{4+} + 7H_2O$

(b) $4I^- + O_2 + 4H^+ \rightarrow 2I_2 + 2H_2O$

(c) $5CH_3\overset{\overset{\large O}{\|}}{C}H + 2MnO_4^- + 6H^+ \longrightarrow 5CH_3\overset{\overset{\large O}{\|}}{C}OH + 2Mn^{2+} + 3H_2O$

Balancing Redox Reactions

To balance a reaction involving oxidation and reduction, we must first identify which element is oxidized and which is reduced. We then break the net reaction into two imaginary *half-reactions*, one of which involves only oxidation and the other only reduction. Although free electrons never appear in a balanced net reaction, they do appear in balanced half-reactions. If we are dealing with aqueous solutions, we proceed to balance each half-reaction by using H_2O and either H^+ or OH^-, as necessary. *A reaction is balanced when the number of atoms of each element is the same on both sides and the net charge is the same on both sides.*

Acidic Solutions

Here are the steps that we will follow:

1. Assign oxidation numbers to the elements that are oxidized or reduced.

2. Break the reaction into two half-reactions, one involving oxidation and the other reduction.

3. For each half-reaction, balance the number of atoms that are oxidized or reduced.

4. Balance the electrons to account for the change in oxidation number by adding electrons to one side of each half-reaction.

5. Balance oxygen atoms by adding H_2O to one side of each half-reaction.

6. Balance the hydrogen atoms by adding H^+ to one side of each half-reaction.

7. Multiply each half-reaction by the number of electrons in the other half-reaction so that the number of electrons on each side of the total reaction will cancel. Then add the two half-reactions and simplify to the smallest integral coefficients.

Example Balancing a Redox Equation

Balance the following equation by using H^+ but not OH^-:

$$\underset{+2}{Fe^{2+}} + \underset{\substack{+7\\ \text{Permanganate}}}{MnO_4^-} \rightleftharpoons \underset{+3}{Fe^{3+}} + \underset{+2}{Mn^{2+}}$$

Solution

1. *Assign oxidation numbers.* These are assigned for Fe and Mn in each species in the preceding reaction.

2. *Break the reaction into two half-reactions.*

$$\text{oxidation half-reaction:}\quad \underset{+2}{Fe^{2+}} \rightleftharpoons \underset{+3}{Fe^{3+}}$$

$$\text{reduction half-reaction:}\quad \underset{+7}{MnO_4^-} \rightleftharpoons \underset{+2}{Mn^{2+}}$$

3. *Balance the atoms that are oxidized or reduced.* Because there is only one Fe or Mn in each species on each side of the equation, the atoms of Fe and Mn are already balanced.

4. *Balance electrons.* Electrons are added to account for the change in each oxidation state.

$$Fe^{2+} \rightleftharpoons Fe^{3+} + e^-$$
$$MnO_4^- + 5e^- \rightleftharpoons Mn^{2+}$$

In the second case, we need $5e^-$ on the left side to take Mn from +7 to +2.

5. *Balance oxygen atoms.* There are no oxygen atoms in the Fe half-reactions. There are four oxygen atoms on the left side of the Mn reaction, so we add four molecules of H_2O to the right side:

$$MnO_4^- + 5e^- \rightleftharpoons Mn^{2+} + 4H_2O$$

6. *Balance hydrogen atoms.* The Fe equation is already balanced. The Mn equation needs $8H^+$ on the left.

$$MnO_4^- + 5e^- + 8H^+ \rightleftharpoons Mn^{2+} + 4H_2O$$

At this point, each half-reaction must be completely balanced (the same number of atoms and charge on each side) *or you have made a mistake.*

7. *Multiply and add the reactions.* We multiply the Fe equation by 5 and the Mn equation by 1 and add:

$$5Fe^{2+} \rightleftharpoons 5Fe^{3+} + \cancel{5e^-}$$
$$MnO_4^- + \cancel{5e^-} + 8H^+ \rightleftharpoons Mn^{2+} + 4H_2O$$
$$\overline{5Fe^{2+} + MnO_4^- + 8H^+ \rightleftharpoons 5Fe^{3+} + Mn^{2+} + 4H_2O}$$

The total charge on each side is +17, and we find the same number of atoms of each element on each side. The equation is balanced.

Basic Solutions

The method preferred by many people for basic solutions is to balance the equation first with H^+. The answer can then be converted into one in which OH^- is used instead. This is done by adding to each side of the equation a number of hydroxide ions equal to the number of H^+ ions appearing in the equation. For example, to balance Equation D-1 with OH^- instead of H^+, proceed as follows:

$$2I_2 + IO_3^- + 10Cl^- + 6H^+ \rightleftharpoons 5ICl_2^- + 3H_2O \qquad \text{(D-1)}$$
$$+ 6OH^- \qquad\qquad + 6OH^-$$
$$\overline{2I_2 + IO_3^- + 10Cl^- + \underbrace{6H^+ + 6OH^-}_{\cancel{6}H_2O \Downarrow 3H_2O} \rightleftharpoons 5ICl_2^- + \cancel{3H_2O} + 6OH^-}$$

Realizing that $6H^+ + 6OH^- = 6H_2O$, and canceling $3H_2O$ on each side, gives the final result:

$$2I_2 + IO_3^- + 10Cl^- + 3H_2O \rightleftharpoons 5ICl_2^- + 6OH^-$$

Problems

3. Balance the following reactions by using H^+ but not OH^-.
(a) $Fe^{3+} + Hg_2^{2+} \rightleftharpoons Fe^{2+} + Hg^{2+}$
(b) $Ag + NO_3^- \rightleftharpoons Ag^+ + NO$
(c) $VO^{2+} + Sn^{2+} \rightleftharpoons V^{3+} + Sn^{4+}$
(d) $SeO_4^{2-} + Hg + Cl^- \rightleftharpoons SeO_3^{2-} + Hg_2Cl_2$
(e) $CuS + NO_3^- \rightleftharpoons Cu^{2+} + SO_4^{2-} + NO$
(f) $S_2O_3^{2-} + I_2 \rightleftharpoons I^- + S_4O_6^{2-}$
(g) $Cr_2O_7^{2-} + CH_3\overset{O}{\overset{\|}{C}}H \rightleftharpoons CH_3\overset{O}{\overset{\|}{C}}OH + Cr^{3+}$
(h) $MnO_4^{2-} \rightleftharpoons MnO_2 + MnO_4^-$
(i) $ClO_3^- \rightleftharpoons Cl_2 + O_2$

4. Balance the following equations by using OH^- but not H^+.
(a) $PbO_2 + Cl^- \rightleftharpoons ClO^- + Pb(OH)_3^-$
(b) $HNO_2 + SbO^+ \rightleftharpoons NO + Sb_2O_5$
(c) $Ag_2S + CN^- + O_2 \rightleftharpoons S + Ag(CN)_2^- + OH^-$
(d) $HO_2^- + Cr(OH)_3^- \rightleftharpoons CrO_4^{2-} + OH^-$
(e) $ClO_2 + OH^- \rightleftharpoons ClO_2^- + ClO_3^-$
(f) $WO_3^- + O_2 \rightleftharpoons HW_6O_{21}^{5-} + OH^-$

Answers to Problems

1.

(a)	+1	(i)	+2	(q)	−3
(b)	+2	(j)	+3	(r)	−2
(c)	+6	(k)	+1	(s)	+3
(d)	+2	(l)	+2	(t)	−4
(e)	−1/2	(m)	+4	(u)	+5
(f)	+2	(n)	+3	(v)	−2
(g)	+3	(o)	+2	(w)	+4
(h)	+4	(p)	0	(x)	+8/3

2.

	Oxidizing agent	Reducing agent
(a)	$Cr_2O_7^{2-}$	Sn^{2+}
(b)	O_2	I^-
(c)	MnO_4^-	CH_3CHO

3. (a) $2Fe^{3+} + Hg_2^{2+} \rightleftharpoons 2Fe^{2+} + 2Hg^{2+}$
(b) $3Ag + NO_3^- + 4H^+ \rightleftharpoons 3Ag^+ + NO + 2H_2O$
(c) $4H^+ + 2VO^{2+} + Sn^{2+} \rightleftharpoons 2V^{3+} + Sn^{4+} + 2H_2O$
(d) $2Hg + 2Cl^- + SeO_4^{2-} + 2H^+ \rightleftharpoons Hg_2Cl_2 + SeO_3^{2-} + H_2O$
(e) $3CuS + 8NO_3^- + 8H^+ \rightleftharpoons 3Cu^{2+} + 3SO_4^{2-} + 8NO + 4H_2O$
(f) $2S_2O_3^{2-} + I_2 \rightleftharpoons S_4O_6^{2-} + 2I^-$
(g) $Cr_2O_7^{2-} + 3CH_3CHO + 8H^+ \rightleftharpoons 2Cr^{3+} + 3CH_3CO_2H + 4H_2O$
(h) $4H^+ + 3MnO_4^{2-} \rightleftharpoons MnO_2 + 2MnO_4^- + 2H_2O$
(i) $2H^+ + 2ClO_3^- \rightleftharpoons Cl_2 + \frac{5}{2}O_2 + H_2O$

4. (a) $H_2O + OH^- + PbO_2 + Cl^- \rightleftharpoons Pb(OH)_3^- + ClO^-$
(b) $4HNO_2 + 2SbO^+ + 2OH^- \rightleftharpoons 4NO + Sb_2O_5 + 3H_2O$
(c) $Ag_2S + 4CN^- + \frac{1}{2}O_2 + H_2O \rightleftharpoons S + 2Ag(CN)_2^- + 2OH^-$
(d) $2HO_2^- + Cr(OH)_3^- \rightleftharpoons CrO_4^{2-} + OH^- + 2H_2O$
(e) $2ClO_2 + 2OH^- \rightleftharpoons ClO_2^- + ClO_3^- + H_2O$
(f) $12WO_3^- + 3O_2 + 2H_2O \rightleftharpoons 2HW_6O_{21}^{5-} + 2OH^-$

Glossary

abscissa The horizontal (*x*) axis of a graph.
absolute uncertainty The margin of uncertainty associated with a measurement. Absolute error could also refer to the difference between a measured value and the "true" value.
absorbance, *A* Defined as $A = \log(P_0/P)$, where P_0 is the radiant power of light (power per unit area) striking the sample on one side and P is the radiant power emerging from the other side. Also called *optical density.*
absorption Occurs when a substance is taken up *inside* another. See also **adsorption.**
absorption spectrum A graph of absorbance or transmittance of light versus wavelength, frequency, or wavenumber.
accuracy A measure of how close a measured value is to the "true" value.
acid A substance that increases the concentration of H^+ when added to water.
acid-base titration One in which the reaction between analyte and titrant is an acid-base reaction.
acid dissociation constant, K_a The equilibrium constant for the reaction of an acid, HA, with H_2O:

$$HA + H_2O \overset{K_a}{\rightleftharpoons} A^- + H_3O^+ \qquad K_a = \frac{[A^-][H_3O^+]}{[HA]} = \frac{[A^-][H^+]}{[HA]}$$

acid error Occurs in strongly acidic solutions, where glass electrodes tend to indicate a value of pH that is too high.
acid wash Procedure in which glassware is soaked in 3–6 M HCl for >1 h (followed by rinsing well with distilled water and soaking in distilled water) to remove traces of cations adsorbed on the surface of the glass and to replace them with H^+.
acidic solution One in which the concentration of H^+ is greater than the concentration of OH^-.
activity, $\mathcal{A}$ The value that replaces concentration in a thermodynamically correct equilibrium expression. The activity of X is given by $\mathcal{A}_X = [X]\gamma_X$, where γ_X is the activity coefficient and [X] is the concentration.
activity coefficient, γ The number by which the concentration must be multiplied to give activity.
adduct The product formed when a Lewis base combines with a Lewis acid.
adjusted retention time, t'_r In chromatography, this parameter is $t'_r = t_r - t_m$, where t_r is the retention time of a solute and t_m is the time needed for mobile phase to travel the length of the column.
adsorption Occurs when a substance becomes attached to the *surface* of another substance. See also **absorption**.
adsorption chromatography A technique in which solute equilibrates between the mobile phase and adsorption sites on the stationary phase.
adsorption indicator Used for precipitation titrations, it becomes attached to a precipitate and changes color when the surface charge of the precipitate changes sign at the equivalence point.
aerosol A suspension of very small liquid or solid particles in air or gas. Examples include fog and smoke.
affinity chromatography A technique in which a particular solute is retained on a column by virtue of a specific interaction with a molecule covalently bound to the stationary phase.
aliquot Portion.
alkali flame detector Modified flame ionization detector that responds to N and P, which produce ions when they contact a Rb_2SO_4-containing glass bead in the flame. Also called *nitrogen-phosphorus detector.*
alkaline error See **sodium error**.
amalgam A solution of anything in mercury.
amine A compound with the general formula RNH_2, R_2NH, or R_3N, where R is any group of atoms.
amino acid One of 20 building blocks of proteins, having the general structure

$$H_3\overset{+}{N}\underset{}{C}(R)CO_2^-$$

where R is a different substituent for each acid.
ammonium ion *The* ammonium ion is NH_4^+. *An* ammonium ion is any ion of the type RNH_3^+, $R_2NH_2^+$, R_3NH^+, or R_4N^+, where R is an organic substituent.
ampere, A One ampere is the current that will produce a force of exactly 2×10^{-7} N/m when that current flows through two "infinitely" long, parallel conductors of negligible cross section, with a spacing of 1 m, in a vacuum.
amperometry The measurement of electric current for analytical purposes.
amphiprotic molecule One that can act as both a proton donor and a proton acceptor. The intermediate species of a polyprotic acid is amphiprotic.
analyte The substance being analyzed.
analytical chromatography Chromatography of small quantities of material conducted for the purpose of qualitative or quantitative analysis or both.
analytical concentration See **formal concentration**.
anhydrous Describes a substance from which all water has been removed.
anion A negatively charged ion.
anion exchanger An ion exchanger with positively charged groups covalently attached to the support. It can reversibly bind anions.
anode The electrode at which oxidation takes place. In electrophoresis, it is the positive electrode.
antibody A protein manufactured by an organism to sequester foreign molecules and mark them for destruction.
antigen A molecule that is foreign to an organism and causes the organism to make antibodies.
antilogarithm The antilogarithm of a is b if $10^a = b$.

aqua regia A 3:1 (vol/vol) mixture of concentrated (37 wt%) HCl and concentrated (70 wt%) HNO_3.
aqueous In water (as an *aqueous* solution).
argentometric titration One using Ag^+ ion.
ashless filter paper Specially treated paper that leaves a negligible residue after ignition. It is used for gravimetric analysis.
assessment In quality assurance, the process of (1) collecting data to show that analytical procedures are operating within specified limits and (2) verifying that final results meet use objectives.
asymmetry potential When the activity of analyte is the same on the inside and outside of an ion-selective electrode, there should be no voltage across the membrane. In fact, the two surfaces are never identical, and some voltage (called the asymmetry potential) is usually observed.
atmosphere, atm One atmosphere is defined as a pressure of 101 325 Pa. It is equal to the pressure exerted by a column of Hg 760 mm in height at the Earth's surface.
atmospheric pressure chemical ionization A method for interfacing liquid chromatography to mass spectrometry. Liquid is nebulized into a fine aerosol by a coaxial gas flow and the application of heat. Electrons from a high-voltage corona discharge create cations and anions from analyte exiting the chromatography column. The most common species observed with this interface is MH^+, the protonated analyte, with little fragmentation.
atomic absorption spectroscopy A technique in which the absorption of light by free gaseous atoms or ions in a flame, furnace, or plasma is used to measure concentration.
atomic emission spectroscopy A technique in which the emission of light by thermally excited atoms or ions in a flame, furnace, or plasma is used to measure concentration.
atomic mass The number of grams of an element containing Avogadro's number of atoms.
atomization The process in which a compound is decomposed into its atoms at high temperature.
autoprotolysis The reaction in which two molecules of the same species transfer a proton from one to the other; e.g., $CH_3OH + CH_3OH \rightleftharpoons CH_3OH_2^+ + CH_3O^-$.
autoprotolysis constant The equilibrium constant for an autoprotolysis reaction.
autotitrator A device that dispenses measured amounts of titrant into a solution and monitors a property such as pH or electrode potential after each addition. The instrument performs the titration automatically and can determine the end point automatically. Data from the titration can be transferred to a spreadsheet for further interpretation.
auxiliary complexing agent A species, such as ammonia, that is added to a solution to stabilize another species and keep that other species in solution. It binds loosely enough to be displaced by a titrant.
auxiliary electrode The current-carrying partner of the working electrode in an electrolysis. Also called *counterelectrode*.
average The sum of several values divided by the number of values. Also called *mean*.
Avogadro's number The number of atoms in exactly 0.012 kg of ^{12}C, approximately 6.022×10^{23}.

back titration One in which an excess of standard reagent is added to react with analyte. Then the excess reagent is titrated with a second reagent or with a standard solution of analyte.
background correction In atomic spectroscopy, a means of distinguishing signal due to analyte from signal due to absorption, emission, or scattering by the flame, furnace, plasma, or sample matrix.
background electrolyte In capillary electrophoresis, the buffer in which separation is carried out. Also called *run buffer*.
base A substance that decreases the concentration of H^+ when added to water.
base "dissociation" constant A misnomer for *base hydrolysis constant, K_b*.
base hydrolysis constant, K_b The equilibrium constant for the reaction of a base, B, with H_2O:

$$B + H_2O \overset{K_b}{\rightleftharpoons} BH^+ + OH^- \qquad K_b = \frac{[BH^+][OH^-]}{[B]}$$

base peak The most intense peak in a mass spectrum.
basic solution One in which the concentration of OH^- is greater than the concentration of H^+.
Beer's law Relates the absorbance (A) of a sample to its concentration (c), pathlength (b), and molar absorptivity (ε): $A = \varepsilon bc$.
biosensor Device that uses biological components such as enzymes, antibodies, or DNA, in combination with electrical, optical, or other signals, to achieve selective response to one analyte.
blank A sample not intended to contain analyte. It could be made from all reagents—except unknown—that would be used in an analytical procedure. Analyte signal measured with a blank solution could be due to impurities in the reagents or, possibly, interference.
blank titration One in which a solution containing all reagents except analyte is titrated. The volume of titrant needed in the blank titration is subtracted from the volume needed to titrate unknown.
blind sample See **performance test sample**.
blocking Occurs when a metal ion binds tightly to a metal ion indicator. A blocked indicator is unsuitable for a titration because no color change is observed at the end point.
Boltzmann distribution The relative population of two states at thermal equilibrium:

$$\frac{N_2}{N_1} = \frac{g_2}{g_1} e^{-(E_2 - E_1)/kT}$$

where N_i is the population of the state, g_i is the degeneracy of the state, E_i is the energy of the state, k is Boltzmann's constant, and T is temperature in kelvins. Degeneracy refers to the number of states with the same energy.
bomb Sealed vessel for conducting high-temperature, high-pressure reactions.
bonded stationary phase In HPLC, a stationary liquid phase covalently attached to the solid support.
Brønsted-Lowry acid A proton (hydrogen ion) donor.
Brønsted-Lowry base A proton (hydrogen ion) acceptor.
buffer A mixture of a weak acid and its conjugate base. A buffered solution is one that resists changes in pH when acids or bases are added.
buffer capacity, β A measure of the ability of a buffer to resist changes in pH. The larger the buffer capacity, the greater the resistance

to pH change. The definition of buffer capacity is $\beta = dC_b/d\text{pH} = -dC_a/d\text{pH}$, where C_a and C_b are the number of moles of strong acid or base per liter needed to produce a unit change in pH. Also called *buffer intensity.*
bulk sample Material taken from lot being analyzed—usually chosen to be representative of the entire lot. Also called *gross sample.*
buoyancy Upward force exerted on an object in a liquid or gaseous fluid. An object weighed in air appears lighter than its actual mass by an amount equal to the mass of air that it displaces.
buret A calibrated glass tube with a stopcock at the bottom. Used to deliver known volumes of liquid.

calibration Process of relating the actual physical quantity (such as mass, volume, force, or electric current) to the quantity indicated on the scale of an instrument.
calibration check In a series of analytical measurements, a calibration check is an analysis of a solution formulated by the analyst to contain a known concentration of analyte. It is the analyst's own check that procedures and instruments are functioning correctly.
calibration curve A graph showing the value of some property versus concentration of analyte. When the corresponding property of an unknown is measured, its concentration can be determined from the graph.
calomel electrode A common reference electrode based on the half-reaction $Hg_2Cl_2(s) + 2e^- \rightleftharpoons 2Hg(l) + 2Cl^-$.
capillary electrophoresis Separation of a mixture into its components by a strong electric field imposed between the two ends of a narrow capillary tube filled with electrolyte solution.
capillary gel electrophoresis A form of capillary electrophoresis in which the tube is filled with a polymer gel that serves as a sieve for macromolecules. The largest molecules migrate slowest through the gel.
capillary zone electrophoresis A form of capillary electrophoresis in which ionic solutes are separated because of differences in their electrophoretic mobility.
carboxylate anion The conjugate base (RCO_2^-) of a carboxylic acid.
carboxylic acid A molecule with the general structure RCO_2H, where R is any group of atoms.
carcinogen A cancer-causing agent.
carrier gas The mobile-phase gas in gas chromatography.
cathode The electrode at which reduction takes place. In electrophoresis, it is the negative electrode.
cation A positively charged ion.
cation exchanger An ion exchanger with negatively charged groups covalently attached to the support. It can reversibly bind cations.
characteristic The part of a logarithm to the left of the decimal point.
charge balance A statement that the sum of all positive charge in solution equals the magnitude of the sum of all negative charge in solution.
charged aerosol detector Sensitive, nearly universal detector for liquid chromatography in which solvent is evaporated from eluate to leave an aerosol of fine particles of nonvolatile solute. Aerosol particles are charged by adsorption of N_2^+ ions and flow to a collector that measures total charge reaching the detector versus time.
charging current Electric current arising from charging or discharging of the electric double layer at the electrode-solution interface. Also called *capacitor current* or *condenser current.*
charring In a gravimetric analysis, the precipitate and filter paper are first *dried* gently. Then the filter paper is *charred* at intermediate temperature to destroy the paper without letting it inflame. Finally, the precipitate is *ignited* at high temperature to convert it to its analytical form.
chelate effect The observation that a single multidentate ligand forms metal complexes that are more stable than those formed by several individual ligands with the same ligand atoms.
chelating ligand A ligand that binds to a metal through more than one atom.
chemical interference In atomic spectroscopy, any chemical reaction that decreases the efficiency of atomization.
chemical ionization A gentle method of producing ions for a mass spectrometer without extensive fragmentation of the analyte molecule, M. A reagent gas such as CH_4 is bombarded with electrons to make CH_5^+, which transfers H^+ to M, giving MH^+.
chemiluminescence Emission of light by an excited-state product of a chemical reaction.
chiral molecule One that is not superimposable on its mirror image in any accessible conformation. Also called an *optically active molecule,* a chiral molecule rotates the plane of polarization of light.
chromatogram A graph showing chromatography detector response as a function of elution time or volume.
chromatograph A machine used to perform chromatography.
chromatography A technique in which molecules in a mobile phase are separated because of their different affinities for a stationary phase. The greater the affinity for the stationary phase, the longer the molecule is retained.
Clark electrode One that measures dissolved oxygen by amperometry.
co-chromatography See **spiking**.
coefficient of variation The standard deviation (s) expressed as a percentage of the mean value $\bar{x}$: coefficient of variation $= 100 \times s/\bar{x}$. Also called *relative standard deviation.*
cold trapping Splitless gas chromatography injection technique in which solute is condensed far below its boiling point in a narrow band at the start of the column.
collimated light Light in which all rays travel in parallel paths.
collimation The process of making light rays travel parallel to one another.
collisionally activated dissociation Fragmentation of an ion in a mass spectrometer by high-energy collisions with gas molecules. In atmospheric pressure chemical ionization or electrospray interfaces, collisionally activated dissociation at the inlet to the mass filter can be promoted by varying the cone voltage. In tandem mass spectrometry, dissociation occurs in a collision cell between the two mass separators.
collision cell The middle stage of a tandem mass spectrometer in which the precursor ion selected by the first stage is fragmented by collisions with gas molecules.
colloid A dissolved particle with a diameter in the approximate range 1–500 nm. It is too large to be considered one molecule but too small to simply precipitate.
combination electrode Consists of a glass electrode with a concentric reference electrode built on the same body.
combustion analysis A technique in which a sample is heated in an atmosphere of O_2 to oxidize it to CO_2 and H_2O, which are collected

and weighed or measured by gas chromatography. Modifications permit the simultaneous analysis of N, S, and halogens.
common ion effect Occurs when a salt is dissolved in a solution already containing one of the ions of the salt. The salt is less soluble than it would be in a solution without that extra ion. An application of Le Châtelier's principle.
complex ion Historical name for any ion containing two or more ions or molecules that are each stable by themselves; e.g., $CuCl_3^-$ contains $Cu^+ + 3Cl^-$.
complexometric titration One in which the reaction between analyte and titrant involves complex formation.
composite sample A representative sample prepared from a heterogeneous material. If the material consists of distinct regions, the composite is made of portions of each region, with relative amounts proportional to the size of each region.
compound electrode An ion-selective electrode consisting of a conventional electrode surrounded by a barrier that is selectively permeable to the analyte of interest. Alternatively, the barrier region might contain an enzyme that converts external analyte into a species to which the inner electrode is sensitive.
concentration An expression of the quantity per unit volume or unit mass of a substance. Common measures of concentration are molarity (mol/L) and molality (mol/kg of solvent).
conditional formation constant The equilibrium constant for formation of a complex under a particular stated set of conditions, such as pH, ionic strength, and concentration of auxiliary complexing species. Also called *effective formation constant.*
confidence interval The range of values within which there is a specified probability that the true value will be found.
conjugate acid-base pair An acid and a base that differ only through the gain or loss of a single proton.
constant mass In gravimetric analysis, the product is heated and cooled to room temperature in a desiccator until successive weighings are "constant." There is no standard definition of "constant mass"; but, for ordinary work, it is usually taken to be about ± 0.3 mg. Constancy is usually limited by the irreproducible regain of moisture during cooling and weighing.
control chart A graph in which periodic observations of a process are recorded to determine whether the process is within specified control limits.
coprecipitation Occurs when a substance whose solubility is not exceeded precipitates along with one whose solubility is exceeded.
correlation coefficient The square of the correlation coefficient, R^2, is a measure of goodness of fit of data points to a straight line. The closer R^2 is to 1, the better the fit.
coulomb, C The amount of charge per second that flows past any point in a circuit when the current is 1 ampere. There are approximately 96 485 coulombs in a mole of electrons.
coulometry A technique in which the quantity of analyte is determined by measuring the number of coulombs needed for complete electrolysis.
counterelectrode The current-carrying partner of the working electrode. Same as *auxiliary electrode.*
counterion An ion with a charge opposite that of the ion of interest.
coupled equilibria Reversible chemical reactions in which the product of one reaction is a reactant in another reaction.
cross-linking Covalent linkage between different strands of a polymer.
cumulative formation constant, β_n The equilibrium constant for a reaction of the type $M + nX \rightleftharpoons MX_n$. Also called *overall formation constant.*
current, I The amount of charge flowing through a circuit per unit time (A/s).
cuvet A cell with transparent walls used to hold samples for spectrophotometric measurements.

Debye-Hückel equation Gives the activity coefficient (γ) as a function of ionic strength (μ). The *extended Debye-Hückel equation*, applicable to ionic strengths up to about 0.1 M, is $\log \gamma = [-0.51z^2 \sqrt{\mu}]/[1 + (\alpha\sqrt{\mu}/305)]$, where z is the ionic charge and α is the effective hydrated diameter in picometers.
decant To pour liquid off a solid or, perhaps, a denser liquid. The denser phase is left behind.
degrees of freedom In statistics, the number of independent observations upon which a result is based.
deionized water Water that has been passed through a cation exchanger (in the H^+ form) and an anion exchanger (in the OH^- form) to remove ions from the solution.
density Mass per unit volume.
derivatization Chemical alteration to attach a group to a molecule so that it can be detected conveniently. Alternatively, treatment can alter volatility or solubility.
desiccant A drying agent.
desiccator A sealed chamber in which samples can be dried in the presence of a desiccant or by vacuum pumping or both.
detection limit The smallest quantity of analyte that is "significantly different" from a blank. The detection limit is often taken as the mean signal for blanks plus 3 times the standard deviation of a low-concentration sample. Also called *lower limit of detection.*
determinate error See **systematic error**.
dialysis A technique in which solutions are placed on either side of a semipermeable membrane that allows small molecules, but not large molecules, to cross. Small molecules in the two solutions diffuse across and equilibrate between the two sides. Large molecules are retained on their original side.
diffraction Occurs when electromagnetic radiation passes through or is reflected from slits with a spacing comparable to the wavelength. Interference of waves from adjacent slits produces a spectrum of radiation, with each wavelength emerging at a different angle.
diffuse part of the double layer Region of solution near a charged surface in which excess counterions are attracted to the charge. The thickness of this layer is 0.3–10 nm.
diffusion current In polarography, the current observed when the rate of reaction is limited by the rate of diffusion of analyte to the electrode.
digestion (1) The process in which a precipitate is left (usually warm) in the presence of mother liquor to promote particle recrystallization and growth. Purer, more easily filterable crystals result. (2) Any chemical treatment in which a substance is decomposed to transform the analyte into a form suitable for analysis.

dilution factor The factor (initial volume of reagent)/(total volume of solution) used to multiply the initial concentration of reagent to find the diluted concentration.
dimer A molecule made from two identical units.
diprotic acid One that can donate two protons.
direct titration One in which the analyte is treated with titrant, and the volume of titrant required for complete reaction is measured.
displacement titration An EDTA titration procedure in which analyte is treated with excess $MgEDTA^{2-}$ to displace Mg^{2+}: $M^{n+} + MgEDTA^{2-} \rightleftharpoons MEDTA^{n-4} + Mg^{2+}$. The liberated Mg^{2+} is then titrated with EDTA. This procedure is useful if there is no suitable indicator for direct titration of M^{n+}.
disproportionation A reaction in which an element in one oxidation state gives products containing that element in both higher and lower oxidation states; e.g., $2Cu^{+} \rightleftharpoons Cu^{2+} + Cu(s)$.
dynamic range The range of analyte concentration over which a change in concentration gives a change in detector response.

$E°$ The standard reduction potential.
$E°'$ The effective standard reduction potential at pH 7 (or at some other specified conditions).
EDTA (ethylenediaminetetraacetic acid) $(HO_2CCH_2)_2NCH_2CH_2N(CH_2CO_2H)_2$, the most widely used reagent for complexometric titrations. It forms 1:1 complexes with virtually all cations with a charge of 2 or more.
effective formation constant See **conditional formation constant**.
electric potential The electric potential (in volts) at a point is the energy (in joules) needed to bring one coulomb of positive charge from infinity to that point. The potential difference between two points is the energy needed to transport one coulomb of positive charge from the negative point to the positive point.
electroactive species Any species that can be oxidized or reduced at an electrode.
electrochemical detector Liquid chromatography detector that measures current when an electroactive solute emerges from the column and passes over a working electrode held at a fixed potential with respect to a reference electrode. Also called *amperometric detector.*
electrochemistry Use of electrical measurements on a chemical system for analytical purposes. Also refers to use of electricity to drive a chemical reaction or use of a chemical reaction to produce electricity.
electrode A device through which electrons flow into or out of chemical species involved in a redox reaction.
electrogravimetric analysis A technique in which the mass of an electrolytic deposit is used to quantify the analyte.
electrolysis The process in which the passage of electric current causes a chemical reaction to occur.
electrolyte A substance that produces ions when dissolved.
electromagnetic spectrum The whole range of electromagnetic radiation, including visible light, radio waves, X-rays, etc.
electron capture detector Gas chromatography detector that is particularly sensitive to compounds with halogen atoms, nitro groups, and other groups with high electron affinity. Makeup gas (N_2 or 5% CH_4 in Ar) is ionized by β-rays from ^{63}Ni to liberate electrons that produce a small, steady current. High-electron-affinity analytes capture some of the electrons and reduce the detector current.
electronic balance A weighing device that uses an electromagnetic servomotor to balance the load on the pan. The mass of the load is proportional to the current needed to balance it.
electronic transition One in which an electron is promoted from one energy level to another.
electron ionization The interaction of analyte molecules (M) with high-energy electrons in the ion source of a mass spectrometer to give the cation, M^{+}, and fragments derived from M^{+}.
electron multiplier An ion detector that works like a photomultiplier tube. Cations striking a cathode liberate electrons. A series of *dynodes* multiplies the number of electrons by ~10^5 before they reach the anode.
electroosmosis Bulk flow of fluid in a capillary tube induced by an electric field. Mobile ions in the diffuse part of the double layer at the wall of the capillary serve as the "pump." Also called *electroendosmosis.*
electropherogram A graph of detector response versus time for electrophoresis.
electrophoresis Migration of ions in solution in an electric field. Cations move toward the cathode and anions move toward the anode.
electrospray A method for interfacing liquid chromatography to mass spectrometry. A high potential applied to the liquid at the column exit creates charged droplets in a fine aerosol. Gaseous ions are derived from ions that were already in the mobile phase on the column. It is common to observed protonated bases (BH^{+}), ionized acids (A^{-}), and complexes formed between analyte, M (which could be neutral or charged), and stable ions such as NH_4^{+}, Na^{+}, HCO_2^{-}, or $CH_3CO_2^{-}$ that were already in solution.
eluate or effluent What comes out of a chromatography column.
eluent The solvent applied to the beginning of a chromatography column.
eluent strength, ε° A measure of the ability of a solvent to elute solutes from a chromatography column. Eluent strength is a measure of the adsorption energy of a solvent on the stationary phase in chromatography. Also called *solvent strength.*
eluotropic series Ranks solvents according to their ability to displace solutes from the stationary phase in adsorption chromatography.
elution The process of passing a liquid or a gas through a chromatography column.
emission spectrum A graph of luminescence intensity versus luminescence wavelength (or frequency or wavenumber), obtained with a fixed excitation wavelength.
emulsion A stable dispersion of immiscible liquids, which might be made by vigorous shaking. Milk is an emulsion of cream in an aqueous phase. Emulsions usually require an emulsifying agent (a surfactant) for stability. The emulsifying agent stabilizes the interface between the two phases by its affinity for both phases.
end point The point in a titration at which there is a sudden change in a physical property, such as indicator color, pH, conductivity, or absorbance. Used as a measure of the equivalence point.
enthalpy change, ΔH The heat absorbed or released when a reaction occurs at constant pressure.
entropy A measure of the "disorder" of a substance.
enzyme A protein that catalyzes a chemical reaction.
equilibrium The state in which the forward and reverse rates of all reactions are equal; so the concentrations of all species remain constant.

equilibrium constant, K For the reaction $aA + bB \rightleftharpoons cC + dD$, $K = \mathcal{A}_C^c \mathcal{A}_D^d / \mathcal{A}_A^a \mathcal{A}_B^b$, where $\mathcal{A}_i$ is the activity of the ith species. If we ignore activity coefficients, which we usually do in this book, the equilibrium constant is written in terms of concentrations: $K = [C]^c[D]^d/[A]^a[B]^b$.

equivalence point The point in a titration at which the quantity of titrant is exactly sufficient for stoichiometric reaction with the analyte.

equivalent For a redox reaction, the amount of reagent that can donate or accept one mole of electrons. For an acid-base reaction, the amount of reagent that can donate or accept one mole of protons.

equivalent weight The mass of substance containing one equivalent.

evaporative light-scattering detector A liquid chromatography detector that makes a fine mist of eluate and evaporates solvent from the mist in a heated zone. The remaining particles of liquid or solid solute flow past a laser beam and are detected by their ability to scatter the light.

excited state Any state of an atom or a molecule having more than the minimum possible energy.

extended Debye-Hückel equation See **Debye-Hückel equation**.

extraction The process in which a solute is transferred from one phase to another. Analyte is sometimes removed from a sample by extraction into a solvent that dissolves the analyte.

Fajans titration A precipitation titration in which the end point is signaled by adsorption of a colored indicator on the precipitate.

false negative A conclusion that the concentration of analyte is below a certain limit when, in fact, the concentration is above the limit.

false positive A conclusion that the concentration of analyte exceeds a certain limit when, in fact, the concentration is below the limit.

faradaic current That component of current in an electrochemical cell due to oxidation and reduction reactions.

Faraday constant, F $9.648\,533\,99 \times 10^4$ C/mol of charge.

field blank A blank sample exposed to the environment at the sample collection site and transported in the same manner as other samples between the lab and the field.

filtrate Portion of a sample (usually the liquid) that passes through a filter.

first derivative The slope of a curve ($\Delta y/\Delta x$) measured at each point along the curve. The first derivative reaches a maximum value at the steepest point on the curve.

flame ionization detector Gas chromatography detector in which solute is burned in a H_2-air flame to produce CHO^+ ions. The current carried through the flame by these ions is proportional to the concentration of susceptible species in the eluate.

flame photometer A device that uses flame atomic emission and a filter photometer to quantify Li, Na, K, and Ca in liquid samples.

flame photometric detector Gas chromatography detector that measures optical emission from S and P in H_2-O_2 flame.

fluorescence The process in which a molecule emits a photon 10^{-8} to 10^{-4} s after absorbing a photon. It results from a transition between states of the same spin multiplicity (e.g., singlet → singlet).

fluorescence detector Liquid chromatography detector that uses a strong light or laser to irradiate eluate emerging from a column and detects radiant emission from fluorescent solutes.

flux In sample preparation, flux is a solid that is melted to dissolve a sample.

formal concentration The molarity of a substance if it did not change its chemical form on being dissolved. It represents the total number of moles of substance dissolved in a liter of solution, regardless of any reactions that take place when the solute is dissolved. Also called *analytical concentration*.

formality, F Same as *formal concentration*.

formal potential The potential of a half-reaction (relative to a standard hydrogen electrode) when the formal concentrations of reactants and products are unity. Any other conditions (such as pH, ionic strength, and concentrations of ligands) also must be specified.

formation constant, K_f The equilibrium constant for the reaction of a metal with its ligands to form a metal-ligand complex. Same as *stability constant*.

formula mass, FM The mass containing one mole of the indicated chemical formula of a substance. For example, the formula mass of $CuSO_4 \cdot 5H_2O$ is the sum of the masses of copper, sulfate, and five water molecules.

fortification See **spike**.

fraction of association, α For the reaction of a base (B) with H_2O, the fraction of base in the form BH^+. $\alpha = [BH^+]/([B] + [BH^+])$.

fraction of dissociation, α For the dissociation of an acid (HA), the fraction of acid in the form A^-. $\alpha = [A^-]/([HA] + [A^-])$.

frequency, ν The number of oscillations of a wave per second.

fusion The process in which an otherwise insoluble substance is dissolved in a molten salt such as Na_2CO_3, Na_2O_2, or KOH. Once the substance has dissolved, the melt is cooled, dissolved in aqueous solution, and analyzed.

galvanic cell One that produces electricity by means of a spontaneous chemical reaction. Also called a *voltaic cell*.

gas chromatography A form of chromatography in which the mobile phase is a gas.

gathering A process in which a trace constituent of a solution is intentionally coprecipitated with a major constituent.

Gaussian distribution Theoretical bell-shaped distribution of measurements when all error is random. The center of the curve is the mean (μ) and the width is characterized by the standard deviation (σ). A *normalized* Gaussian distribution, also called the *normal error curve*, has an area of unity and is given by

$$y = \frac{1}{\sigma\sqrt{2\pi}} e^{-(x-\mu)^2/2\sigma^2}$$

gel Chromatographic stationary-phase particles, such as Sephadex or polyacrylamide, which are soft and pliable.

gel filtration chromatography See **molecular exclusion chromatography**.

gel permeation chromatography See **molecular exclusion chromatography**.

glass electrode One that has a thin glass membrane across which a pH-dependent voltage develops. The voltage (and hence pH) is measured by a pair of reference electrodes on either side of the membrane.

gradient elution Chromatography in which the composition of the mobile phase is progressively changed to increase the eluent strength of the solvent.

graduated cylinder or **graduate** A tube with volume calibrations along its length.

gram-atom The amount of an element containing Avogadro's number of atoms; it is the same as a mole of the element.
graphite furnace A graphite tube that can be heated electrically to about 2 500 K to decompose and atomize a sample for atomic spectroscopy.
grating Either a reflective or a transmitting surface etched with closely spaced lines; used to disperse light into its component wavelengths.
gravimetric analysis Any analytical method that relies on measuring the mass of a substance (such as a precipitate) to complete the analysis.
gravimetric titration A titration in which the mass of titrant is measured, instead of the volume. Titrant concentration is conveniently expressed as mol reagent/kg titrant solution. Gravimetric titrations can be more accurate and precise than volumetric titrations.
green chemistry Principles intended to change our behavior in a manner that will help sustain the habitability of the Earth. Green chemistry seeks to design chemical products and processes to reduce the use of resources and energy and the generation of hazardous waste.
gross sample See **bulk sample**.
ground state The state of an atom or a molecule with the minimum possible energy.
Grubbs test Statistical test used to decide whether to discard a datum that appears discrepant.
guard column In HPLC, a short, disposable column packed with the same material as the main column and placed between the injector and the main column. The guard column removes impurities that might irreversibly bind to the main column and degrade it. Also called *precolumn*. In gas chromatography, the guard column is a length of empty tubing with chemically deactivated walls to minimize retention. The guard column collects nonvolatile components of the sample that would remain on and degrade the analytical column.

half-cell Part of an electrochemical cell in which half of an electrochemical reaction (either the oxidation or the reduction reaction) occurs.
half-reaction Any redox reaction can be conceptually broken into two half-reactions, one involving only oxidation and one involving only reduction.
half-wave potential Potential at the midpoint of the rise in the current of a polarographic wave.
hardness Total concentration of alkaline earth ions in natural water expressed as mg $CaCO_3$ per liter of water as if all of the alkaline earths present were $CaCO_3$.
Henderson-Hasselbalch equation A logarithmic rearranged form of the acid dissociation equilibrium equation:

$$pH = pK_a + \log \frac{[A^-]}{[HA]}$$

Henry's law The partial pressure of a gas in equilibrium with gas dissolved in a solution is proportional to the concentration of dissolved gas: $P = k$[dissolved gas]. The constant k is called the *Henry's law constant*. It is a function of the gas, the liquid, and the temperature.
hertz, Hz The unit of frequency, s^{-1}, also called *reciprocal seconds*.
heterogeneous Not uniform throughout.
HETP (height equivalent to a theoretical plate) The length of a chromatography column divided by the number of theoretical plates in the column.
hexadentate ligand One that binds to a metal atom through six ligand atoms.
high-performance liquid chromatography (HPLC) A chromatographic technique using small stationary-phase particles and high pressure to force solvent through the column.
hollow-cathode lamp One that emits sharp atomic lines characteristic of the element from which the cathode is made.
homogeneous Having the same composition everywhere.
homogeneous precipitation A technique in which a precipitating agent is generated slowly by a reaction in homogeneous solution, effecting a slow crystallization instead of a rapid precipitation of product.
hydrated radius The effective size of an ion or a molecule plus its associated water molecules in solution.
hydrolysis "Reaction with water." The reaction $B + H_2O \rightleftharpoons BH^+ + OH^-$ is called hydrolysis of a base.
hydronium ion, H_3O^+ What we really mean when we write $H^+(aq)$.
hydrophilic substance One that is soluble in water or attracts water to its surface.
hydrophobic substance One that is insoluble in water or repels water from its surface.
hygroscopic substance One that readily picks up water from the atmosphere.

ignition Heating a gravimetric precipitate to high temperature to convert it into a known, constant composition that can be weighed.
immiscible liquids Two liquids that do not form a single phase when mixed together.
immunoassay An analytical measurement using antibodies.
inclusion An impurity that occupies lattice sites in a crystal.
indeterminate error See **random error**.
indicator A compound having a physical property (usually color) that changes abruptly near the equivalence point of a chemical reaction.
indicator electrode One that develops a potential whose magnitude depends on the activity of one or more species in contact with the electrode.
indicator error The difference between the indicator end point of a titration and the true equivalence point.
indirect detection Chromatographic detection based on the *absence* of signal from a background species. For example, in ion chromatography, a light-absorbing ionic species can be added to the eluent. Nonabsorbing analyte replaces an equivalent amount of light-absorbing eluent when analyte emerges from the column, thereby decreasing the absorbance of eluate.
indirect titration One that is used when the analyte cannot be directly titrated. For example, analyte A may be precipitated with excess reagent R. The product is filtered, and the excess R washed away. Then AR is dissolved in a new solution, and R can be titrated.
inductively coupled plasma A high-temperature plasma that derives its energy from an oscillating radio-frequency field. It is used to atomize a sample for atomic emission spectroscopy.
inflection point One at which the derivative of the slope is 0: $d^2y/dx^2 = 0$. That is, the slope reaches a maximum or minimum value.

inorganic carbon In a natural water or industrial effluent sample, the quantity of dissolved carbonate and bicarbonate.

intensity Power per unit area of a beam of electromagnetic radiation (W/m^2). Also called *radiant power* or *irradiance*.

intercept For a straight line whose equation is $y = mx + b$, b is the intercept. It is the value of y when $x = 0$.

interference A phenomenon in which the presence of one substance changes the signal in the analysis of another substance.

internal conversion A radiationless, isoenergetic, electronic transition between states of the same electron-spin multiplicity.

internal standard A known quantity of a compound other than analyte added to a solution containing an unknown quantity of analyte. The concentration of analyte is then measured relative to that of the internal standard.

interpolation The estimation of the value of a quantity that lies between two known values.

iodimetry The use of triiodide (or iodine) as a titrant.

iodometry A technique in which an oxidant is treated with I^- to produce I_3^-, which is then titrated (usually with thiosulfate).

ion chromatography HPLC ion-exchange separation of ions.

ion-exchange chromatography A technique in which solute ions are retained by oppositely charged sites in the stationary phase.

ionic atmosphere The region of solution around an ion or a charged particle. It contains an excess of oppositely charged ions.

ionic radius The effective size of an ion in a crystal.

ionic strength, μ Given by $\mu = \frac{1}{2}\Sigma_i c_i z_i^2$, where c_i is the concentration of the ith ion in solution and z_i is the charge on that ion. The sum extends over all ions in solution, including the ions whose activity coefficients are being calculated.

ionization interference In atomic spectroscopy, a lowering of signal intensity as a result of ionization of analyte atoms.

ionization suppressor An element used in atomic spectroscopy to decrease the extent of ionization of the analyte.

ionophore A molecule with a hydrophobic outside and a polar inside that can engulf an ion and carry the ion through a hydrophobic phase (such as a cell membrane).

ion pair A closely associated anion and cation, held together by electrostatic attraction. In solvents less polar than water, ions are usually found as ion pairs.

ion-selective electrode One whose potential is selectively dependent on the concentration of one particular ion in solution.

irradiance Power per unit area of a beam of electromagnetic radiation (W/m^2). Also called *radiant power* or *intensity.*

isobaric interference In mass spectrometry, overlap of two peaks with nearly the same mass. For example, $^{41}K^+$ and $^{40}ArH^+$ differ by 0.01 atomic mass unit and appear as a single peak unless the spectrometer resolution is great enough to separate them.

isocratic elution Chromatography using a single solvent for the mobile phase.

isoelectric focusing A technique in which a sample containing polyprotic molecules is subjected to a strong electric field in a medium with a pH gradient. Each species migrates until it reaches the region of its isoelectric pH. In that region, the molecule has no net charge, ceases to migrate, and remains focused in a narrow band.

isosbestic point A wavelength at which the absorbance spectra of two species cross each other. The appearance of isosbestic points in a solution in which a chemical reaction is occurring is evidence that there are only two components present, with a constant total concentration.

joule, J SI unit of energy. One joule is expended when a force of 1 N acts over a distance of 1 m. This energy is equivalent to that required to raise 102 g (about $\frac{1}{4}$ pound) by 1 m at sea level.

junction potential An electric potential that exists at the junction between two different electrolyte solutions or substances. It arises in solutions as a result of unequal rates of diffusion of different ions.

kelvin, K The absolute unit of temperature defined such that the temperature of water at its triple point (where water, ice, and water vapor are at equilibrium) is 273.16 K and the absolute zero of temperature is 0 K.

kilogram, kg SI unit of mass equal to the mass of a particular Pt-Ir cylinder kept at the International Bureau of Weights and Measures, Sèvres, France.

Kjeldahl nitrogen analysis Procedure for the analysis of nitrogen in organic compounds. The compound is digested with boiling H_2SO_4 to convert nitrogen into NH_4^+, which is treated with base and distilled as NH_3 into a standard acid solution. The moles of acid consumed equal the moles of NH_3 liberated from the compound.

laboratory sample Portion of bulk sample taken to the lab for analysis. Must have the same composition as the bulk sample.

law of mass action States that for the chemical reaction $aA + bB \rightleftharpoons cC + dD$, the condition at equilibrium is $K = \mathcal{A}_C^c\mathcal{A}_D^d / \mathcal{A}_A^a\mathcal{A}_B^b$, where $\mathcal{A}_i$ is the activity of the ith species. The law is usually used in approximate form, in which activities are replaced by concentrations.

least squares Process of fitting a mathematical function to a set of measured points by minimizing the sum of the squares of the distances from the points to the curve.

Le Châtelier's principle If a system at equilibrium is disturbed, the direction in which it proceeds back to equilibrium is such that the disturbance is partly offset.

Lewis acid One that can form a chemical bond by sharing a pair of electrons donated by another species.

Lewis base One that can form a chemical bond by sharing a pair of its electrons with another species.

ligand An atom or a group attached to a central atom in a molecule. The term is often used to mean any group attached to anything else of interest.

linear flow rate In chromatography, the distance per unit time traveled by the mobile phase.

linear interpolation A form of interpolation in which it is assumed that the variation in some quantity is linear. For example, to find the value of b when $a = 32.4$ in the following table,

a:	32	32.4	33
b:	12.85	x	17.96

you can set up the proportion

$$\frac{32.4 - 32}{33 - 32} = \frac{x - 12.85}{17.96 - 12.85}$$

which gives $x = 14.89$.

linearity A measure of how well data in a graph follow a straight line.
linear range The concentration range over which the change in detector response is proportional to the change in analyte concentration.
linear response The case in which the analytical signal is directly proportional to the concentration of analyte.
liquid-based ion-selective electrode One that has a hydrophobic membrane separating an inner reference electrode from the analyte solution. The membrane contains an ion exchanger dissolved in nonpolar solvent. The ion-exchange equilibrium of analyte between the liquid ion exchanger and the aqueous solution gives rise to the electrode potential.
liquid chromatography A form of chromatography in which the mobile phase is a liquid.
liter, L Common unit of volume equal to exactly 1 000 cm^3.
logarithm The base 10 logarithm of n is a if $10^a = n$ (which means $\log n = a$). The natural logarithm of n is a if $e^a = n$ (which means $\ln n = a$). The number e (= 2.718 28 . . .) is called the base of the natural logarithm.
longitudinal diffusion Diffusion of solute molecules parallel to the direction of travel through a chromatography column.
lot Entire material that is to be analyzed. Examples are a bottle of reagent, a lake, or a truckload of gravel.
lower limit of detection See **detection limit**.
lower limit of quantitation Smallest amount of analyte that can be measured with reasonable accuracy. Usually taken as 10 times the standard deviation of a low-concentration sample.
luminescence Any emission of light by a molecule.
L'vov platform Platform on which sample is placed in a graphite-tube furnace for atomic spectroscopy to prevent sample vaporization before the walls reach constant temperature.

makeup gas Gas added to the exit stream from a gas chromatography column for the purpose of changing flow rate or gas composition to optimize detection of analyte.
mantissa The part of a logarithm to the right of the decimal point.
masking The process of adding a chemical substance (a *masking agent*) to a sample to prevent one or more components from interfering in a chemical analysis.
masking agent A reagent that selectively reacts with one (or more) component(s) of a solution to prevent the component(s) from interfering in a chemical analysis.
mass balance A statement that the sum of the moles of any element in all of its forms in a solution must equal the moles of that element delivered to the solution.
mass spectrometer An instrument that converts gaseous molecules into ions, accelerates them in an electric field, separates them according to their mass-to-charge ratio, and detects the amount of each species.
mass spectrometry A technique in which gaseous molecules are ionized, accelerated by an electric field, and then separated according to their mass.
mass spectrum In mass spectrometry, a graph showing the relative abundance of each ion as a function of its mass-to-charge ratio.
mass titration One in which the mass of titrant, instead of the volume, is measured.
matrix The medium containing analyte. For many analyses, it is important that standards be prepared in the same matrix as the unknown.
matrix effect A change in analytical signal caused by anything in the sample other than analyte.
matrix modifier Substance added to sample for atomic spectroscopy to make the matrix more volatile or the analyte less volatile so that the matrix evaporates before analyte does.
mean The average of a set of all results.
mechanical balance A balance having a beam that pivots on a fulcrum. Standard masses are used to measure the mass of an unknown.
median For a set of data, that value above and below which there are equal numbers of data.
mediator In electrolysis, a molecule that carries electrons between the electrode and the intended analyte. Used when the analyte cannot react directly at the electrode or when analyte concentration is so low that other reagents react instead. Mediator is recycled indefinitely by oxidation or reduction at the counterelectrode.
meniscus The curved surface of a liquid.
metal ion buffer Consists of a metal-ligand complex plus excess free ligand. The two serve to fix the concentration of free metal ion through the reaction $M + nL \rightleftharpoons ML_n$.
metal ion indicator A compound whose color changes when it binds to a metal ion.
meter, m SI unit of length defined as the distance that light travels in a vacuum during $\frac{1}{299\,792\,458}$ of a second.
method blank A sample without deliberately added analyte. The method blank is taken through all steps of a chemical analysis, including sample preparation. The blank measures the response of the analytical method to impurities in the reagents or any other effects caused by any component other than the analyte.
method of least squares Process of fitting a mathematical function to a set of measured points by minimizing the sum of the squares of the distances from the points to the curve.
method validation The process of proving that an analytical method is acceptable for its intended purpose.
micellar electrokinetic capillary chromatography A form of capillary electrophoresis in which a micelle-forming surfactant is present. Migration times of solutes depend on the fraction of time spent in the micelles.
micelle An aggregate of molecules with ionic headgroups and long, nonpolar tails. The inside of the micelle resembles hydrocarbon solvent, whereas the outside interacts strongly with aqueous solution.
microporous particles Chromatographic stationary phase consisting of porous particles 1.5–10 μm in diameter, with high efficiency and high capacity for solute.
migration Electrostatically induced motion of ions in a solution under the influence of an electric field.
miscible liquids Two liquids that form a single phase when mixed in any ratio.
mobile phase In chromatography, the phase that travels through the column.
mobility The terminal velocity that an ion reaches in a field of 1 V/m. Velocity = mobility × field.
molality A measure of concentration equal to the number of moles of solute per kilogram of solvent.
molar absorptivity, ε The constant of proportionality in Beer's law: $A = \varepsilon bc$, where A is absorbance, b is pathlength, and c is the molarity of the absorbing species. Also called *extinction coefficient.*

molarity, M A measure of concentration equal to the number of moles of solute per liter of solution.
mole, mol SI unit for the amount of substance that contains as many molecules as there are atoms in 12 g of ^{12}C. There are approximately 6.022×10^{23} molecules per mole.
molecular exclusion chromatography A technique in which the stationary phase has a porous structure into which small molecules can enter but large molecules cannot. Molecules are separated by size, with larger molecules moving faster than smaller ones. Also called *size exclusion, gel filtration,* or *gel permeation chromatography.*
molecularly imprinted polymer A polymer synthesized in the presence of a template molecule. After the template is removed, the polymer has a void with the right shape to hold the template, and polymer functional groups are positioned correctly to bind to template functional groups.
molecular ion In mass spectrometry, an ion that has not lost or gained any atoms during ionization.
molecular mass The number of grams of a substance that contains Avogadro's number of molecules.
molecular orbital Describes the distribution of an electron within a molecule.
mole fraction The number of moles of a substance in a mixture divided by the total number of moles of all components present.
monochromatic light Light of a single wavelength (color).
monochromator A device (usually a prism, grating, or filter) that disperses light into its component wavelengths and selects a narrow band of wavelengths to pass through the exit slit.
monodentate ligand One that binds to a metal ion through only one atom.
monoprotic acids and bases Compounds that can donate or accept one proton.
mortar and pestle A mortar is a hard ceramic or steel vessel in which a solid sample is ground with a hard tool called a pestle.
mother liquor The solution from which a substance has crystallized.
multidentate ligand One that binds to a metal ion through more than one atom.

natural logarithm The natural logarithm (ln) of a is b if $e^b = a$. See also **logarithm.**
nebulization The process of breaking the liquid sample into a mist of fine droplets.
nebulizer In atomic spectroscopy, this device breaks the liquid sample into a mist of fine droplets.
Nernst equation Relates the voltage of a cell (E) to the activities of reactants and products:

$$E = E^\circ - \frac{RT}{n\mathrm{F}} \ln Q$$

where R is the gas constant, T is temperature in kelvins, F is the Faraday constant, Q is the reaction quotient, and n is the number of electrons transferred in the balanced reaction. E° is the cell voltage when all activities are unity.
neutralization The process in which a stoichiometric equivalent of acid is added to a base (or vice versa).
newton, N SI unit of force. One Newton will accelerate a mass of 1 kg by 1 m/s^2.
nitrogen rule A compound with an odd number of nitrogen atoms—in addition to C, H, halogens, O, S, Si and P—will have an odd molecular mass. A compound with an even number of nitrogen atoms (0, 2, 4, etc.) will have an even molecular mass.
noise Signals originating from sources other than those intended to be measured.
nominal mass The *integer* mass of the species with the most abundant isotope of each of the constituent atoms.
nonpolar substance A substance, such as a hydrocarbon, with little charge separation within the molecule and no net ionic charge. Nonpolar substances interact with other substances by weak van der Waals forces and are generally not soluble in water.
normal error curve A Gaussian distribution whose area is unity.
normal hydrogen electrode (N.H.E.) Same as standard hydrogen electrode (S.H.E.)
normality n times the molarity of a redox reagent, where n is the number of electrons donated or accepted by that species in a particular chemical reaction. For acids and bases, it is also n times the molarity, but n is the number of protons donated or accepted by the species.
normal-phase chromatography A chromatographic separation utilizing a polar stationary phase and a less polar mobile phase.
nucleation The process whereby molecules in solution come together randomly to form small crystalline aggregates that can grow into larger crystals.
null hypothesis In statistics, the supposition that two quantities do not differ from each other or that two methods do not give different results.

occlusion An impurity that becomes trapped (sometimes with solvent) in a pocket within a growing crystal.
on-column injection Used in gas chromatography to place a thermally unstable sample directly on the column without excessive heating in an injection port. Solute is condensed at the start of the column by low temperature, and then the temperature is raised to initiate chromatography.
open tubular column In chromatography, a capillary column whose walls are coated with stationary phase.
optical isomers Molecules that are the mirror image of each other and cannot be superimposed on each other.
order of magnitude A power of 10.
ordinate The vertical (y) axis of a graph.
outlier A datum that is far from the other points in a data set.
overall formation constant, β_n same as **cumulative formation constant**.
oxidant See **oxidizing agent**.
oxidation A loss of electrons or a raising of the oxidation state.
oxidation number Same as *oxidation state.*
oxidation state A bookkeeping device used to tell how many electrons have been gained or lost by a neutral atom when it forms a compound. Also called *oxidation number.*
oxidizing agent or **oxidant** A substance that takes electrons in a chemical reaction.

p function The negative logarithm (base 10) of a quantity: $pX = -\log X$.
packed column A chromatography column filled with stationary-phase particles.

parallax error The apparent displacement of an object when the observer changes position. Occurs when the scale of an instrument is viewed from a position that is not perpendicular to the scale. The apparent reading is not the true reading.
particle growth The stage of crystal growth in which solute crystallizes onto the surface of preexisting crystals.
partition chromatography A technique in which separation is achieved by equilibration of solute between two phases.
parts per billion, ppb An expression of concentration denoting nanograms (10^{-9} g) of solute per gram of solution.
parts per million, ppm An expression of concentration denoting micrograms (10^{-6} g) of solute per gram of solution.
pascal, Pa SI unit of pressure equal to 1 N/m^2. There are 10^5 Pa in 1 bar and 101 325 Pa in 1 atm.
peptization Occurs when washing some ionic precipitates with distilled water causes the ions that neutralize the charges of individual particles to be washed away. The particles then repel one another, disintegrate, and pass through the filter with the wash liquid.
performance test sample In a series of analytical measurements, a performance test sample is inserted to see whether the procedure gives correct results when the analyst does not know the right answer. The performance test sample is formulated by someone other than the analyst to contain a known concentration of analyte. Also called a *quality control sample* or *blind sample.*
permanent hardness Component of water hardness not due to dissolved alkaline earth bicarbonates. This hardness remains in the water after boiling.
pH Defined as $\text{pH} = -\log \mathcal{A}_{\text{H}^+}$ where $\mathcal{A}_{\text{H}^+}$ is the activity of H^+. In most approximate applications, the pH is taken as $-\log[\text{H}^+]$.
phosphorescence Emission of light during a transition between states of different spin multiplicity (e.g., triplet $\rightarrow$ singlet). Phosphorescence is slower than fluorescence, with emission occurring $\sim 10^{-4}$ to 10^2 s after absorption of a photon.
photochemistry Chemical reaction initiated by absorption of a photon.
photodiode array An array of semiconductor diodes used to detect light. The array is normally used to detect light that has been spread into its component wavelengths. One small band of wavelengths falls on each detector.
photomultiplier tube One in which the cathode emits electrons when struck by light. The electrons then strike a series of dynodes (plates that are positive with respect to the cathode), and more electrons are released each time a dynode is struck. As a result, more than 10^6 electrons may reach the anode for every photon striking the cathode.
photon A "particle" of light with energy $h\nu$, where h is Planck's constant and ν is the frequency of the light.
pipet A glass tube calibrated to deliver a fixed or variable volume of liquid.
p*K* The negative logarithm (base 10) of an equilibrium constant: $\text{p}K = -\log K$.
plasma A gas that is hot enough to contain free ions and electrons, as well as neutral molecules.
plate height, *H* The length of a chromatography column divided by the number of theoretical plates in the column.
polar substance A substance, such as an alcohol, that has positive and negative regions that attract neighboring molecules by electrostatic forces. Polar substances tend to be soluble in water and insoluble in nonpolar substances, such as hydrocarbons.
polarogram A graph showing the relation between current and potential during a polarographic experiment.
polarograph An instrument used to obtain and record a polarogram.
polarographic wave The flattened S-shaped increase in current during a redox reaction in polarography.
polarography A voltammetry experiment using a dropping-mercury electrode.
polychromatic light Light of many wavelengths.
polychromator A device that spreads light into its component wavelengths and directs each small band of wavelengths to a different region where it is detected by a photodiode array.
polyprotic acids and bases Compounds that can donate or accept more than one proton.
postprecipitation The adsorption of otherwise soluble impurities on the surface of a precipitate after the precipitation is over.
potential See **electric potential.**
potentiometer A device that measures electric potential. A potentiometer measures the same quantity as that measured by a voltmeter, but the potentiometer is designed to draw much less current from the circuit being measured.
potentiometry An analytical method in which an electric potential difference (a voltage) of a cell is measured.
potentiostat An electronic device that maintains a chosen constant or time-varying voltage between a pair of electrodes.
power The amount of energy per unit time (J/s=watts, W) being expended.
ppb (parts per billion) An expression of concentration denoting nanograms (10^{-9} g) of solute per gram of solution.
ppm (parts per million) An expression of concentration denoting micrograms (10^{-6} g) of solute per gram of solution.
precipitant A substance that precipitates a species from solution.
precipitation Occurs when a substance leaves solution rapidly (to form either microcrystalline or amorphous solid).
precipitation titration One in which the analyte forms a precipitate with the titrant.
precision A measure of the reproducibility of a measurement.
precolumn See **guard column.**
preconcentration The process of concentrating trace components of a mixture prior to their analysis.
precursor ion In tandem mass spectrometry (selected reaction monitoring), the ion selected by the first mass separator for fragmentation in the collision cell.
premix burner In atomic spectroscopy, one in which the sample is nebulized and simultaneously mixed with fuel and oxidant before being fed into the flame.
preoxidation In some redox titrations, adjustment of the analyte oxidation state to a higher value so that it can be titrated with a reducing agent.
preparative chromatography Chromatography of large quantities of material conducted for the purpose of isolating pure material.
prereduction The process of reducing an analyte to a lower oxidation state prior to performing a titration with an oxidizing agent.
pressure Force per unit area, commonly measured in pascals (N/m) or bars.

pressure broadening In spectroscopy, line broadening due to collisions between molecules.
primary standard A reagent that is pure enough and stable enough to be used directly after weighing. The entire mass is considered to be pure reagent.
prism A transparent, triangular solid. Each wavelength of light passing through the prism is bent (refracted) at a different angle.
product The species created in a chemical reaction. Products appear on the right side of the chemical equation.
product ion In tandem mass spectrometry (selected reaction monitoring), the fragment ion from the collision cell selected by the final mass separator for passage through to the detector.
protic solvent One with an acidic hydrogen atom.
protocol In quality assurance, written directions stating what must be documented and how the documentation is to be done.
proton The ion H^+.
proton acceptor A Brønsted-Lowry base: a molecule that combines with H^+.
protonated molecule In mass spectrometry, the ion MH^+ resulting from addition of H^+ to the analyte.
proton donor A Brønsted-Lowry acid: a molecule that can provide H^+ to another molecule.
purge To force a fluid (usually gas) to flow through a substance or a chamber, usually to extract something from the substance being purged or to replace the fluid in the chamber with the purging fluid.
purge and trap A method for removing volatile analytes from liquids or solids, concentrating the analytes, and introducing them into a gas chromatograph. A carrier gas bubbled through a liquid or solid extracts volatile analytes, which are then trapped in a tube containing adsorbent. After analyte has been collected, the adsorbent tube is heated and purged to desorb the analytes, which are collected by cold trapping at the start of a gas chromatography column.
pyrolysis Thermal decomposition of a substance.

quadratic equation An equation that can be rearranged to the form $ax^2 + bx + c = 0$.
qualitative analysis The process of determining the identity of the constituents of a substance.
quality assurance Practices that demonstrate the reliability of analytical data.
quality control Active measures taken to ensure the required accuracy and precision of a chemical analysis.
quality control sample See **performance test sample.**
quantitative analysis The process of measuring how much of a constituent is present in a substance.
quantitative transfer To transfer the entire contents from one vessel to another. This process is usually accomplished by rinsing the first vessel several times with fresh liquid and pouring each rinse into the receiving vessel.

radiant power Power per unit area (W/m^2) of a beam of electromagnetic radiation. Also called *irradiance* or *intensity*.
random error A type of error, which can be either positive or negative and cannot be eliminated, based on the ultimate limitations on a physical measurement. Also called *indeterminate error.*
random heterogeneous material A material in which there are differences in composition with no pattern or predictability and on a fine scale. When you collect a portion of the material for analysis, you obtain some of each of the different compositions.
random sample Bulk sample constructed by taking portions of the entire lot at random.
range The difference between the highest and lowest values in a set of data; also called *spread.* With respect to an analytical method, range is the concentration interval over which linearity, accuracy, and precision are all acceptable.
raw data Individual values of a measured quantity, such as peak areas from a chromatogram or volumes from a buret.
reactant The species that is consumed in a chemical reaction. It appears on the left side of a chemical equation.
reaction quotient, Q Expression having the same form as the equilibrium constant for a reaction. However, the reaction quotient is evaluated for a particular set of existing activities (concentrations), which are generally not the equilibrium values. At equilibrium, $Q = K$.
reagent blank A solution prepared from all of the reagents, but no analyte. The blank measures the response of the analytical method to impurities in the reagents or any other effects caused by any component other than the analyte. A reagent blank is similar to a method blank, but it has not been subjected to all sample preparation procedures.
reagent-grade chemical A high-purity chemical generally suitable for use in quantitative analysis and meeting purity requirements set by organizations such as the American Chemical Society.
reconstructed total ion chromatogram In chromatography, a graph of the sum of intensities of all ions detected at all masses (above a selected cutoff) versus time.
redox couple A pair of reagents related by electron transfer; e.g., $Fe^{3+} \mid Fe^{2+}$ or $MnO_4^- \mid Mn^{2+}$.
redox indicator A compound used to find the end point of a redox titration because its various oxidation states have different colors. The standard potential of the indicator must be such that its color changes near the equivalence point of the titration.
redox reaction A chemical reaction in which electrons are transferred from one element to another.
redox titration One in which the reaction between analyte and titrant is an oxidation-reduction reaction.
reducing agent or **reductant** A substance that donates electrons in a chemical reaction.
reductant See **reducing agent**.
reduction A gain of electrons or a lowering of the oxidation state.
reference electrode One that maintains a constant potential against which the potential of another half-cell may be measured.
refraction Bending of light when it passes between media with different refractive indexes.
refractive index, n The speed of light in any medium is c/n, where c is the speed of light in vacuum and n is the refractive index of the medium. The refractive index also measures the angle at which a light ray is bent when it passes from one medium into another. Snell's law states that $n_1 \sin \theta_1 = n_2 \sin \theta_2$, where n_i is the refractive index for each medium and θ_i is the angle of the ray with respect to the normal between the two media.
refractive index detector Liquid chromatography detector that measures the change in refractive index of eluate as solutes emerge from the column.

relative uncertainty The uncertainty of a quantity divided by the value of the quantity. It is usually expressed as a percentage of the measured quantity.
reporting limit Concentration below which regulations dictate that an analyte is reported as "not detected." The reporting limit is typically set 5 to 10 times higher than the detection limit.
reprecipitation Sometimes a gravimetric precipitate can be freed of impurities only by redissolving it in fresh solvent and reprecipitating it. The impurities are present at lower concentration during the second precipitation and are less likely to coprecipitate.
residual current Small current observed in the absence of analyte reaction in an electrolysis.
resin An ion exchanger, such as polystyrene with ionic substituents, which exists as small, hard particles.
resolution How close two bands in a spectrum or a chromatogram can be to each other and still be seen as two peaks. In chromatography, it is defined as the difference in retention times of adjacent peaks divided by their width.
response factor, F The relative response of a detector to analyte (X) and internal standard (S): (signal from X)/[X] = F(signal from S)/[S]. Once you have measured F with a standard mixture, you can use it to find [X] in an unknown if you know [S] and the quotient (signal from X)/(signal from S).
results What we ultimately report after applying statistics to treated data.
retention time, t_r The time, measured from injection, needed for a solute to be eluted from a chromatography column.
retention volume, V_r The volume of solvent needed to elute a solute from a chromatography column.
reversed-phase chromatography A technique in which the stationary phase is less polar than the mobile phase.
robustness The ability of an analytical method to be unaffected by small changes in operating conditions.
rotational transition Occurs when a molecule changes its rotation energy.
rubber policeman A glass rod with a flattened piece of rubber on the tip. The rubber is used to scrape solid particles from glass surfaces in gravimetric analysis.
run buffer See **background electrolyte**.

salt An ionic solid.
salt bridge A conducting ionic medium in contact with two electrolyte solutions. It allows ions to flow without allowing immediate diffusion of one electrolyte solution into the other.
sample cleanup Removal of portions of the sample that do not contain analyte and may interfere with analysis.
sample preparation Transforming a sample into a state that is suitable for analysis. This process can include concentrating a dilute analyte and removing or masking interfering species.
sampling The process of collecting a representative sample for analysis.
saturated calomel electrode (S.C.E.) A calomel electrode saturated with KCl. The electrode half-reaction is $Hg_2Cl_2(s) + 2e^- \rightleftharpoons 2Hg(l) + 2Cl^-$. The compound Hg_2Cl_2 is called calomel.
saturated solution One that contains the maximum amount of a compound that can dissolve at equilibrium.
S.C.E. See **saturated calomel electrode**.
schlieren Streaks in a liquid mixture observed before the two phases have mixed. Streaks arise from regions that refract light differently.
second, s SI unit of time equal to the duration of 9 192 631 770 periods of the radiation corresponding to the transition between two hyperfine levels of the ground state of ^{133}Cs.
second derivative The slope of the slope, $\Delta(\text{slope})/\Delta x$, of a curve, measured at each point along the curve. When the slope reaches a maximum or minimum, the second derivative is 0.
segregated heterogeneous material A material in which differences in composition are on a large scale. Different regions have obviously different composition.
selected ion chromatogram A graph of detector response versus time when a mass spectrometer monitors just one or a few species of selected mass-to-charge ratio emerging from a chromatograph.
selected ion monitoring Use of a mass spectrometer to monitor species with just one or a few mass-to-charge ratios (m/z).
selected reaction monitoring A technique in which a precursor ion selected by one mass separator passes through a collision cell in which the precursor breaks into several product ions. A second mass separator then selects one (or a few) of these ions for detection. Selected reaction monitoring improves chromatographic signal-to-noise ratio because it is insensitive to almost everything other than the intended analyte. Also called *mass spectrometry–mass spectrometry (ms–ms)* or *tandem mass spectrometry*.
selectivity The capability of an analytical method to distinguish analyte from other species in the sample. Also called *specificity*.
selectivity coefficient With respect to an ion-selective electrode, a measure of the relative response of the electrode to two different ions. In ion-exchange chromatography, the selectivity coefficient is the equilibrium constant for displacement of one ion by another from the resin.
self-absorption In a luminescence measurement, a high concentration of analyte molecules can absorb excitation energy from excited analyte. If the absorbed energy is dissipated as heat instead of light, fluorescence does not increase in proportion to analyte concentration. Analyte concentration can be so high that fluorescence *decreases* with increasing concentration. In flame emission atomic spectroscopy, there is a lower concentration of excited-state atoms in the cool, outer part of the flame than in the hot, inner flame. The cool atoms can absorb emission from the hot ones and thereby decrease the observed signal.
semipermeable membrane A thin layer of material that allows some substances, but not others, to pass across the material. A dialysis membrane allows small molecules to pass, but not large molecules.
sensitivity The response of an instrument or method to a given amount of analyte.
separator column Ion-exchange column used to separate analyte species in ion chromatography.
septum A disk, usually made of silicone rubber, covering the injection port of a gas chromatograph. The sample is injected by syringe through the septum.
SI units The international system of units based on the meter, kilogram, second, ampere, kelvin, candela, mole, radian, and steradian.
sieving In electrophoresis, the separation of macromolecules by migration through a polymer gel. The smallest molecules move fastest and the largest move slowest.

signal averaging Improvement of a signal by averaging successive scans. The signal increases in proportion to the number of scans accumulated. The noise increases in proportion to the square root of the number of scans. Therefore, the signal-to-noise ratio improves in proportion to the square root of the number of scans collected.
significant figure The number of significant digits in a quantity is the minimum number of digits needed to express the quantity in scientific notation. In experimental data, the first uncertain figure is the last significant figure.
silanization Treatment of a chromatographic solid support or glass column with hydrophobic silicon compounds that bind to the most reactive Si—OH groups. It reduces irreversible adsorption and tailing of polar solutes.
silver-silver chloride electrode A common reference electrode containing a silver wire coated with AgCl paste and dipped in a solution saturated with AgCl and (usually) KCl. The half-reaction is $AgCl(s) + e^- \rightleftharpoons Ag(s) + Cl^-$.
singlet state One in which all electron spins are paired.
size exclusion chromatography See **molecular exclusion chromatography**.
slope For a straight line whose equation is $y = mx + b$, the value of m is the slope. It is the ratio $\Delta y/\Delta x$ for any segment of the line.
slurry A suspension of a solid in a solvent.
sodium error Occurs when a glass pH electrode is placed in a strongly basic solution containing very little H^+ and a high concentration of Na^+. The electrode begins to respond to Na^+ as if it were H^+, so the pH reading is lower than the actual pH. Also called *alkaline error.*
solid-phase extraction Preconcentration procedure in which a solution is passed through a short column of chromatographic stationary phase, such as C_{18}-silica. Trace solutes adsorbed on the column can be eluted with a small volume of solvent of high eluent strength.
solid-phase microextraction Extraction of compounds from liquids or gases into a coated fiber dispensed from a syringe needle. After extraction, the fiber is withdrawn into the needle and the needle is injected through the septum of a chromatograph. The fiber is extended inside the injection port and adsorbed solutes are desorbed by heating (for gas chromatography) or solvent (for liquid chromatography).
solid-state ion-selective electrode An ion-selective electrode that has a solid membrane made of an inorganic salt crystal. Ion-exchange equilibria between the solution and the surface of the crystal account for the electrode potential.
solubility product, K_{sp} The equilibrium constant for the dissociation of a solid salt to give its ions in solution. For the reaction $M_mN_n(s) \rightleftharpoons mM^{n+} + nN^{m-}$, $K_{sp} = [M^{n+}]^m[N^{m-}]^n$.
solute A minor component of a solution.
solvation The interaction of solvent molecules with solute. Solvent molecules orient themselves around solute to minimize the energy through dipole and van der Waals forces.
solvent The major constituent of a solution.
solvent strength See **eluent strength**.
solvent trapping Splitless gas chromatography injection technique in which solvent is condensed near its boiling point at the start of the column. Solutes dissolve in a narrow band in the condensed solvent.
species Chemists refer to any element, compound, or ion of interest as a *species*. The word *species* is both singular and plural.
specific adsorption Process in which molecules are held tightly to a surface by van der Waals or electrostatic forces.
specific gravity A dimensionless quantity equal to the mass of a substance divided by the mass of an equal volume of water at 4°C. Specific gravity is virtually identical with density in g/mL.
specifications In quality assurance, specifications are written statements describing how good analytical results need to be and what precautions are required in an analytical method.
specificity See **selectivity**
spectral interference In atomic spectroscopy, any physical process that affects the light intensity at the analytical wavelength. Created by substances that absorb, scatter, or emit light of the analytical wavelength.
spectrophotometer A device used to measure absorption of light. It includes a source of light, a wavelength selector (monochromator), and an electrical means of detecting light.
spectrophotometric analysis Any method in which light absorption, emission, reflection, or scattering is used to measure chemical concentrations.
spectrophotometric titration One in which absorption of light is used to monitor the progress of the chemical reaction.
spectrophotometry In a broad sense, any method using light to measure chemical concentrations.
specular reflection Reflection of light at an angle equal to the angle of incidence.
spike Addition of a known compound (usually at a known concentration) to an unknown. In isotope dilution mass spectrometry, the spike is the added, unusual isotope. *Spike* is a noun and a verb. Also called a *fortification.*
spike recovery The fraction of a spike eventually found by chemical analysis of the spiked sample.
spiking In chromatography, the simultaneous chromatography of a known compound with an unknown. If a known and an unknown have the same retention time on several different columns, they are probably identical. Also called *co-chromatography.*
split injection Used in capillary gas chromatography to inject a small fraction of sample onto the column; the rest of the sample is blown out to waste.
splitless injection Used in capillary gas chromatography for trace analysis and quantitative analysis. The entire sample in a low-boiling solvent is directed to the column, where the sample is concentrated by *solvent trapping* (condensing the solvent below its boiling point) or *cold trapping* (condensing solutes far below their boiling range). The column is then warmed to initiate separation.
spontaneous process One that is energetically favorable. It will eventually occur, but thermodynamics makes no prediction about how long it will take.
spread See **range**.
square wave voltammetry A form of *voltammetry* in which the potential waveform consists of a square wave superimposed on a voltage staircase. The technique is faster and more sensitive than voltammetry with other waveforms.
stability constant See **formation constant**.
stacking In electrophoresis, the process of concentrating a dilute electrolyte into a narrow band by an electric field. Stacking occurs because the electric field in the dilute electrolyte is stronger than the field in more concentrated surrounding electrolyte.

standard addition A technique in which an analytical signal due to an unknown is first measured. Then a known quantity of analyte is added, and the increase in signal is recorded. From the response, it is possible to calculate what quantity of analyte was in the unknown.
standard curve A graph showing the response of an analytical technique to known quantities of analyte. Also called *calibration curve*.
standard deviation A statistic measuring how closely data are clustered about the mean value. For a finite set of data, the standard deviation, s, is computed from the formula

$$s = \sqrt{\frac{\Sigma_i(x_i - \bar{x})^2}{n - 1}} = \sqrt{\frac{\Sigma_i(x_i^2)}{n - 1} - \frac{(\Sigma_i x_i)^2}{n(n - 1)}}$$

where n is the number of results, x_i is an individual result, and $\bar{x}$ is the mean result. For a large number of measurements, s approaches σ, the true standard deviation of the population, and $\bar{x}$ approaches μ, the true population mean.
standard hydrogen electrode (S.H.E.) or **normal hydrogen electrode (N.H.E.)** One that contains $H_2(g)$ bubbling over a catalytic Pt surface immersed in aqueous H^+. The activities of H_2 and H^+ are both unity in the hypothetical standard electrode. The reaction is $H^+ + e^- \rightleftharpoons \frac{1}{2}H_2(g)$.
standard reduction potential, $E°$ The voltage that would be measured when a hypothetical cell containing the desired half-reaction (with all species present at unit activity) is connected to a standard hydrogen electrode anode.
Standard Reference Materials Certified samples sold by the U.S. National Institute of Standards and Technology (or national measurement institutes of other countries) containing known concentrations or quantities of particular analytes. Used to standardize testing procedures in different laboratories.
standard solution A solution whose composition is known by virtue of the way that it was made from a reagent of known purity or by virtue of its reaction with a known quantity of a standard reagent.
standard state The standard state of a solute is 1 M and the standard state of a gas is 1 bar. Pure solids and liquids are considered to be in their standard states. In equilibrium constants, dimensionless concentrations are expressed as a ratio of the concentration of each species to its concentration in its standard state.
standardization The process of determining the concentration of a reagent by reaction with a known quantity of a second reagent.
stationary phase In chromatography, the phase that does not move through the column.
stepwise formation constant, K_n The equilibrium constant for a reaction of the type $ML_{n-1} + L \rightleftharpoons ML_n$.
steradian, sr Unit of solid angle. There are 4π steradians in a complete sphere.
stoichiometry Ratios of substances participating in a chemical reaction.
stripping analysis A sensitive polarographic technique in which analyte is concentrated from dilute solution by reduction into a drop (or a film) of Hg. It is then analyzed polarographically during an anodic redissolution process. Some analytes can be oxidatively concentrated onto an electrode other than Hg and stripped in a reductive process.
strong acids and bases Those that are completely dissociated (to H^+ or OH^-) in water.
strong electrolyte One that mostly dissociates into ions in solution.
Student's *t* A statistical tool used to express confidence intervals and to compare results from different experiments.
sulfur chemiluminescence detector Gas chromatography detector for the element sulfur. Exhaust from a flame ionization detector is mixed with O_3 to form an excited state of SO_2 that emits light, which is detected.
supernatant liquid Liquid remaining above the solid after a precipitation. Also called *supernate*.
supersaturated solution One that contains more dissolved solute than would be present at equilibrium.
supporting electrolyte An unreactive salt added in high concentration to solutions for voltammetric measurements. Supporting electrolyte carries most of the ion-migration current and therefore decreases the coulombic migration of electroactive species to a negligible level. The electrolyte also decreases the resistance of the solution.
suppressed-ion chromatography Separation of ions by using an ion-exchange column followed by a suppressor (membrane or column) to remove ionic eluent.
suppressor In ion chromatography, a device that transforms ionic eluent into a nonionic form.
surfactant A molecule with an ionic or polar headgroup and a long, nonpolar tail. Surfactants aggregate in aqueous solution to form micelles. Surfactants derive their name from the fact that they accumulate at boundaries between polar and nonpolar phases and modify the surface tension, which is the free energy of formation of the surface. Soaps are surfactants.
syringe A device having a calibrated barrel into which liquid is sucked by a plunger. The liquid is expelled through a needle by pushing on the plunger.
systematic error Error due to procedural or instrumental factors that cause a measurement to be consistently too large or too small. The error can, in principle, be discovered and corrected. Also called *determinate error*.
systematic treatment of equilibrium A method that uses the charge balance, mass balance(s), and equilibria to completely specify the system's composition.

***t* test** Statistical test used to decide whether the results of two experiments are within experimental uncertainty of each other. The uncertainty must be specified to within a certain probability.
tailing Asymmetric chromatographic band in which the later part elutes very slowly. It often results from adsorption of a solute onto a few active sites on the stationary phase.
tandem mass spectrometry See **selected reaction monitoring**.
tare As a noun, *tare* is the mass of an empty vessel used to receive a substance to be weighed. As a verb, *tare* means setting the balance reading to 0 when an empty vessel or weighing paper is placed on the pan.
temperature programming Raising the temperature of a gas chromatography column during a separation to reduce the retention time of late-eluting components.
temporary hardness Component of water hardness due to dissolved alkaline earth bicarbonates. It is temporary because boiling causes precipitation of the carbonates.
test portion Part of the laboratory sample used for one analysis. Also called *aliquot*.

theoretical plate An imaginary construct in chromatography denoting a segment of a column in which one equilibration of solute occurs between the stationary and mobile phases. The number of theoretical plates on a column with Gaussian bandshapes is defined as $N = t_r^2/\sigma^2$, where t_r is the retention time of a peak and σ is the standard deviation of the band.
thermal conductivity, κ Rate at which a substance transports heat (energy per unit time per unit area) through a temperature gradient (degrees per unit distance). Energy flow $[J/(s \cdot m^2)] = -\kappa(dT/dx)$, where κ is the thermal conductivity $[W/(m \cdot K)]$ and dT/dx is the temperature gradient (K/m).
thermal conductivity detector A device that detects substances eluted from a gas chromatography column by measuring changes in the thermal conductivity of the gas stream.
thermogravimetric analysis A technique in which the mass of a substance is measured as the substance is heated. Changes in mass indicate decomposition of the substance, often to well-defined products.
thin-layer chromatography Liquid chromatography in which the stationary phase is coated on a flat glass or plastic plate. Solute is spotted near the bottom of the plate. The bottom edge of the plate is placed in contact with solvent, which creeps up the plate by capillary action.
titer A measure of concentration, usually defined as how many milligrams of reagent B will react with 1 mL of reagent A. One milliliter of $AgNO_3$ solution with a titer of 1.28 mg NaCl/mL will be consumed by 1.28 mg NaCl in the reaction $Ag^+ + Cl^- \rightarrow AgCl(s)$. The same solution of $AgNO_3$ has a titer of 0.993 mg of KH_2PO_4/mL, because 1 mL of AgNO solution will be consumed by 0.993 mg KH_2PO_4 to precipitate Ag_3PO_4.
titrant The substance added to the analyte in a titration.
titration A procedure in which one substance (titrant) is carefully added to another (analyte) until complete reaction has occurred. The quantity of titrant required for complete reaction tells how much analyte is present.
titration curve A graph showing how the concentration of a reactant or a physical property of the solution varies as one reactant (the titrant) is added to another (the analyte).
titration error The difference between the observed end point and the true equivalence point in a titration.
tolerance Manufacturer's stated uncertainty in the accuracy of a device such as a buret or volumetric flask. A 100-mL flask with a tolerance of $\pm$0.08 mL may contain 99.92 to 100.08 mL and be within tolerance.
total carbon In a natural water or industrial effluent sample, the quantity of CO_2 produced when the sample is completely oxidized by oxygen at 900°C in the presence of a catalyst.
total ion chromatogram A graph of detector response versus time when a mass spectrometer monitors all ions above a selected m/z ratio emerging from a chromatograph.
total organic carbon In a natural water or industrial effluent sample, the quantity of CO_2 produced when the sample is first acidified and purged to remove carbonate and bicarbonate and then completely oxidized by oxygen at 900°C in the presence of a catalyst.
total oxygen demand In a natural water or industrial effluent sample, the quantity of O_2 required for complete oxidation of species in the water at 900°C in the presence of a catalyst.
trace analysis Chemical analysis of very low levels of analyte, typically ppm and lower.
transition range For an acid-base indicator, the pH range over which the color change occurs. For a redox indicator, the potential range over which the color change occurs.
transmission quadrupole mass spectrometer A mass spectrometer that separates ions by passing them between four metallic cylinders to which are applied direct current and oscillating electric fields. Resonant ions with the right mass-to-charge ratio pass through the chamber to the detector, while nonresonant ions are deflected into the cylinders and are lost.
transmittance, T Defined as $T = P/P_0$, where P_0 is the radiant power of light striking the sample on one side and P is the radiant power of light emerging from the other side of the sample.
treated data Concentrations or amounts of analyte found from raw data with a calibration curve or some other calibration method.
triplet state An electronic state in which there are two unpaired electrons.

ultraviolet detector Liquid chromatography detector that measures ultraviolet absorbance of solutes emerging from the column.
use objectives In quality assurance, use objectives are a written statement of how results will be used. Use objectives are required before specifications can be written for the method.

validation See **method validation.**
van Deemter equation Describes the dependence of chromatographic plate height (H) on linear flow rate (u, in units such as m/s): $H = A + B/u + Cu$. The constant A depends on band-broadening processes such as multiple flow paths that are independent of flow rate. B depends on the rate of diffusion of solute in the mobile phase. C depends on the rate of mass transfer between the stationary and mobile phases.
variance, σ^2 The square of the standard deviation.
vibrational transition Occurs when a molecule changes its vibrational energy.
viscosity Resistance to flow in a fluid.
void volume, V_0 The volume of the mobile phase in a molecular exclusion chromatography column.
volatile Easily vaporized.
Volhard titration Titration of Ag^+ with SCN^- in the presence of Fe^{3+}. Formation of red $Fe(SCN)^{2+}$ marks the end point.
volt, V Unit of electric potential difference. If the potential difference between two points is one volt, then one joule of energy is required to move one coulomb of charge between the two points.
voltammetry An analytical method in which the relationship between current and voltage is observed during an electrochemical reaction.
voltammogram A graph of current versus electrode potential in an electrochemical cell.
volume flow rate In chromatography, the volume of mobile phase per unit time eluted from the column.
volume percent Defined as (volume of solute/volume of solution) $\times$ 100.
volumetric analysis A technique in which the volume of material needed to react with the analyte is measured.

volumetric flask One having a tall, thin neck with a calibration mark. When the liquid level is at the calibration mark, the flask contains its specified volume of liquid.

watt, W SI unit of power equal to an energy flow of one joule per second. When an electric current of one ampere flows through a potential difference of one volt, the power is one watt.
wavelength, λ distance between consecutive crests of a wave.
wavenumber, $\tilde{\nu}$ The reciprocal of the wavelength, $1/\lambda$.
weak acids and bases Those whose dissociation constants are not large.
weak electrolyte One that only partly dissociates into ions when it dissolves.
weighing paper Paper on which to place a solid reagent on a balance. Weighing paper has a very smooth surface, from which solids fall easily for transfer to a vessel.
weight percent, wt% (mass of solute/mass of solution) $\times$ 100.
weight/volume percent (mass of solute/volume of solution) $\times$ 100.
Work Energy required or released when an object is moved from one point to another.
working electrode One at which the reaction of interest occurs.

***y*-intercept** The value of y at which a line crosses the y-axis.

Zeeman background correction Technique used in atomic spectroscopy in which analyte signals are shifted outside the detector monochromator range by applying a strong magnetic field to the sample. Signal that remains is the background.
zwitterion A molecule with a positive charge localized at one position and a negative charge localized at another position.

Solutions to "Ask Yourself" Questions

Chapter 0

0-A. **(a)** A heterogeneous material has different compositions in different regions. A homogeneous material is the same everywhere.

(b) The composition of a random heterogeneous material varies from place to place with no pattern or predictability to the variation. A segregated heterogeneous material has relatively large regions of distinctly different compositions.

(c) A random sample is selected by taking material at random from the lot. That is, the locations from which material is selected should follow no pattern. You could do this by dividing the lot into many imaginary regions and assigning a number to each. Then use your computer or calculator to generate random numbers and take a sample from each region whose number is selected by the computer. A composite sample is selected deliberately by taking predetermined portions of the lot from selected regions. Random sampling is appropriate for a random heterogeneous lot. Composite sampling is appropriate when the lot is segregated into regions of different composition.

Chapter 1

1-A. **(a)** See Table 1-3

(b) $10^{-19}\ \cancel{g} \times \left(\frac{1\ zg}{10^{-21}\ \cancel{g}}\right) = 100\ zg\ (\text{or } 0.1\ ag)$

$10^{7}\ \cancel{g} \times \left(\frac{1\ Mg}{10^{6}\ \cancel{g}}\right) = 10\ Mg$

$10^{4}\ \cancel{W} \times \left(\frac{1\ kW}{10^{3}\ \cancel{W}}\right) = 10\ kW$

1-B. **(a)** Office worker: $2.2 \times 10^{6}\ \frac{\cancel{cal}}{day} \times 4.184\ \frac{J}{\cancel{cal}}$

$= 9.2 \times 10^{6}\ J/day$

Mountain climber: $3.4 \times 10^{6}\ \frac{\cancel{cal}}{day} \times 4.184\ \frac{J}{\cancel{cal}}$

$= 14.2 \times 10^{6}\ J/day$

(b) $60\ \frac{s}{\cancel{min}} \times 60\ \frac{\cancel{min}}{\cancel{h}} \times 24\ \frac{\cancel{h}}{day} = 8.64 \times 10^{4}\ s/day$

(c) $\dfrac{9.2 \times 10^{6}\ \frac{J}{\cancel{day}}}{8.64 \times 10^{4}\ \frac{s}{\cancel{day}}} = 1.1 \times 10^{2}\ W;$

$\dfrac{14.2 \times 10^{6}\ J/\cancel{day}}{8.64 \times 10^{4}\ s/\cancel{day}} = 1.6 \times 10^{2}\ W$

(d) The office worker consumes more power than does the light bulb.

1-C. **(a)** $\left(1.67\ \frac{g\ solution}{\cancel{mL}}\right)\left(1\ 000\ \frac{\cancel{mL}}{L}\right) = 1\ 670\ \frac{g\ solution}{L}$

(b) $\left(0.705\ \frac{g\ HClO_4}{\cancel{g\ solution}}\right)\left(1\ 670\ \frac{\cancel{g\ solution}}{L}\right)$

$= 1.18 \times 10^{3}\ \frac{g\ HClO_4}{L}$

(c) $\left(1.18 \times 10^{3}\ \frac{g}{L}\right)\Big/\left(100.458\ \frac{g}{mol}\right) = 11.7\ \frac{mol}{L}$

1-D. **(a)** $\left(1.50\ \frac{g\ solution}{\cancel{mL\ solution}}\right)\left(1\ 000\ \frac{\cancel{mL\ solution}}{L\ solution}\right)$

$= 1.50 \times 10^{3}\ \frac{g\ solution}{L\ solution}$

(b) $\left(0.480\ \frac{g\ HBr}{\cancel{g\ solution}}\right)\left(1.50 \times 10^{3}\ \frac{\cancel{g\ solution}}{L\ solution}\right)$

$= 7.20 \times 10^{2}\ \frac{g\ HBr}{L\ solution}$

(c) $\left(7.20 \times 10^{2}\ \frac{\cancel{g\ HBr}}{L\ solution}\right)\Big/\left(\frac{80.912\ \cancel{g\ HBr}}{mol}\right) = 8.90\ M$

(d) $M_{con} \cdot V_{con} = M_{dil} \cdot V_{dil}$

$(8.90\ M)(x\ mL) = (0.160\ M)(250\ mL) \Rightarrow x = 4.49\ mL$

1-E. **(a)** and **(b)**

$HOBr + OCl^- \rightleftharpoons HOCl + OBr^- \qquad K_1 = 1/15$

$HOCl \rightleftharpoons H^+ + OCl^- \qquad K_2 = 3.0 \times 10^{-8}$

$HOBr \rightleftharpoons H^+ + OBr^- \qquad K = K_1K_2 = 2 \times 10^{-9}$

(c) Consumption of a product drives the reaction forward (to the right).

Chapter 2

2-A. The lab notebook must (1) state what was done; (2) state what was observed; and (3) be understandable to a stranger.

2-B. **(a)** $$m = \frac{(24.913\text{ g})\left(1 - \dfrac{0.001\,2\text{ g/mL}}{8.0\text{ g/mL}}\right)}{\left(1 - \dfrac{0.001\,2\text{ g/mL}}{1.00\text{ g/mL}}\right)} = 24.939\text{ g}$$

(b) $$\text{density} = \frac{\text{mass}}{\text{volume}} \Rightarrow \text{volume} = \frac{\text{mass}}{\text{density}}$$

$$\text{volume} = (24.939\text{ g})/(0.998\,00\text{ g/mL}) = 24.989\text{ mL}$$

2-C. Dissolve (0.250 0 L)(0.150 0 mol/L) = 0.037 50 mol of K_2SO_4 in less than 250 mL of water in a 250-mL volumetric flask. Add more water and mix. Dilute to the 250.0-mL mark and invert the flask many times for complete mixing.

2-D. Transfer pipet. The adjustable 100-μL micropipet has a tolerance of ±1.8% at 10 μL and ±0.6% at 100 μL. The uncertainty at 10 μL is ±1.8% of 10 μL = (0.018) × (10 μL) = ±0.18 μL. The uncertainty at 100 μL is ±0.6% of 100 μL = (0.006)(100 μL) = ±0.6 μL.

2-E. mass delivered × conversion factor
$$= (10.000\,0\text{ g})(1.002\,0\text{ mL/g}) = 10.020\text{ mL}$$

2-F. $S^{2-} + 2H^+ \rightarrow H_2S$. $H_2S(g)$ is lost when the solution is boiled to dryness.

Chapter 3

3-A. **(a)** 5 **(b)** 4 **(c)** 3

3-B. **(a)** 3.71 **(b)** 10.7 **(c)** 4.0×10^1 **(d)** 2.85×10^{-6} **(e)** 12.625 1 **(f)** 6.0×10^{-4} **(g)** 242

3-C. **(a)** Carmen **(b)** Cynthia **(c)** Chastity **(d)** Cheryl

3-D. **(a)** percent relative uncertainty in mass
$$= (0.002/4.635) \times 100 = 0.04_3\%$$
percent relative uncertainty in volume
$$= (0.05/1.13) \times 100 = 4._4\%$$

(b) $$\text{density} = \frac{4.635 \pm 0.002\text{ g}}{1.13 \pm 0.05\text{ mL}} = \frac{4.635\,(\pm 0.04_3\%)\text{ g}}{1.13(\pm 4._4\%)\text{ mL}} = 4.10 \pm ?\text{ g/mL}$$

$$\text{uncertainty} = \sqrt{(0.04_3)^2 + (4._4)^2} = 4._4\%$$

$4._4\%$ of 4.10 = 0.18

The answer can be written 4.1 ± 0.2 g/mL.

3-E. −196°C = 77 K = −321°F

Chapter 4

4-A. $$\bar{x} = \frac{821 + 783 + 834 + 855}{4} = 823._2$$

$$s = \sqrt{\frac{(821 - 823._2)^2 + (783 - 823._2)^2 + (834 - 823._2)^2 + (855 - 823._2)^2}{4 - 1}} = 30._3$$

relative standard deviation
$$= (30._3/823._2) \times 100 = 3.6_8\%$$

median = (821 + 834)/2 = $827._5$

range = 855 − 783 = 72

4-B. **(a)** $$\bar{x} = \frac{117 + 119 + 111 + 115 + 120}{5} = 116._4$$

$$s = \sqrt{\frac{(117 - 116._4)^2 + (119 - 116._4)^2 + \cdots + (120 - 116._4)^2}{5 - 1}} = 3._{58}$$

(b) $$s_{\text{pooled}} = \sqrt{\frac{2.8^2\,(4 - 1) + 3.58^2\,(5 - 1)}{4 + 5 - 2}} = 3.27$$

$$t_{\text{calculated}} = \frac{|111.0 - 116.4|}{3.27}\sqrt{\frac{4 \cdot 5}{4 + 5}} = 2.46$$

$t_{\text{table}} = 2.365$ for 95% confidence and 4 + 5 − 2 = 7 degrees of freedom.

$t_{\text{calculated}} > t_{\text{table}}$, so the difference is significant at the 95% confidence level.

4-D. $\bar{x} = 201.8$; $s = 9.34$

$G_{\text{calculated}} = |216 - 201.8|/9.34 = 1.52$

$G_{\text{table}} = 1.672$ for five measurements

Because $G_{\text{calculated}} < G_{\text{table}}$, we should not reject 216.

4-E.

x_i	y_i	x_iy_i	x_i^2	$d_i\,(= y_i - mx_i - b)$	d_i^2
1	3	3	1	−0.167	0.027 89
3	2	6	9	0.333	0.110 89
5	0	0	25	−0.167	0.027 89
$\Sigma x_i = 9$	$\Sigma y_i = 5$	$\Sigma x_iy_i = 9$	$\Sigma(x_i^2) = 35$		$\Sigma(d_i^2) = 0.166\,67$

$$D = n\,\Sigma(x_i^2) - (\Sigma x_i)^2 = 3 \cdot 35 - 9^2 = 24$$

$$m = [n\,\Sigma x_iy_i - \Sigma x_i\Sigma y_i]/D = [3 \cdot 9 - 9 \cdot 5]/24 = -0.750$$

$$b = [\Sigma(x_i^2)\Sigma y_i - \Sigma(x_iy_i)\Sigma x_i]/D = [35 \cdot 5 - 9 \cdot 9]/24 = 3.917$$

$$s_y = \sqrt{\frac{\Sigma(d_i^2)}{3 - 2}} = \sqrt{\frac{0.166\,67}{1}} = 0.408\,2$$

$$s_m = s_y\sqrt{\frac{n}{D}} = 0.408\,2\sqrt{\frac{3}{24}} = 0.144$$

$$s_b = s_y\sqrt{\frac{\Sigma(x_i^2)}{D}} = 0.408\,2\sqrt{\frac{35}{24}} = 0.493$$

$$y\,(\pm 0.4_{08}) = -0.75_0\,(\pm 0.14_4)x + 3.9_{17}\,(\pm 0.4_{93})$$

4-E. See the table at the bottom of the previous page.

4-F.
$$x = \frac{y - b}{m} = \frac{1.00 - 3.9_{17}}{-0.750} = 3.89$$

$$\bar{x} = (1 + 3 + 5)/3 = 3; \quad \bar{y} = (3 + 2 + 0)/3 = 1.667$$

$$\text{uncertainty in } x = \frac{s_y}{|m|}\sqrt{\frac{1}{k} + \frac{1}{n} + \frac{(y - \bar{y})^2}{m^2 \sum(x_i - \bar{x})^2}}$$

$$= \frac{0.408\,2}{0.750} \times$$

$$\sqrt{\frac{1}{5} + \frac{1}{3} + \frac{(1.00 - 1.667)^2}{(-0.750)^2[(1 - 3)^2 + (3 - 3)^2 + (5 - 3)^2]}}$$

$$= 0.43$$

Final answer: $x = 3.9 \pm 0.4$

Chapter 5

5-A. The three parts of quality assurance are defining use objectives, setting specifications, and assessing results.

Use objectives:
Question: Why do I want the data and results and how will I use them?
Actions: Write use objectives.

Specifications:
Question: How good do the numbers have to be?
Actions: Write specifications and pick an analytical method to meet the specifications. Consider requirements for sampling, precision, accuracy, selectivity, sensitivity, detection limit, robustness, and allowed rate of false results. Plan to employ blanks, fortification, calibration checks, quality control samples, and control charts.

Assessment:
Question: Did I meet the specifications?
Actions: Compare data and results with specifications. Document procedures and keep records suitable for meeting use objectives. Verify that the use objectives were met.

5-B. *Precision* is demonstrated by the repeatability of analyses of replicate samples and replicate portions of the same sample. *Accuracy* is demonstrated by fortification recovery, calibration checks, blanks, and quality control samples (blind samples).

5-C. The equation of the least-squares line through the data points is $y = 0.886x + 1.634$. To find the x-intercept, set $y = 0$.

$$0 = 0.886x + 1.634 \Rightarrow x = -1.84 \text{ mM}$$

The concentration of ascorbic acid in the orange juice is 1.84 mM.

	A	B	C	D	E
1	Vitamin C standard addition experiment				
2	Add 25.0 mM ascorbic acid to 50.0 mL of orange juice				
3					
4		Vs =			
5	Vo (mL) =	mL ascorbic	x-axis function	I(s+x) =	y-axis function
6	50	acid added	Si*Vs/Vo	signal (μA)	I(s+x)*V/Vo
7	[S]i (mM) =	0.000	0.000	1.66	1.660
8	25	1.000	0.500	2.03	2.071
9		2.000	1.000	2.39	2.486
10		3.000	1.500	2.79	2.957
11		4.000	2.000	3.16	3.413
12		5.000	2.500	3.51	3.861
13					
14	C7 = $A$8*B7/$A$6		E7 = D7*($A$6+B7)/$A$6		

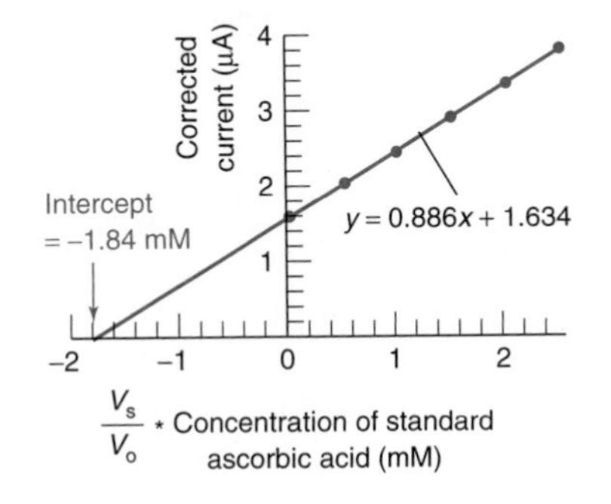

5-D. First evaluate the response factor from the known mixture:

$$\frac{A_X}{[X]} = F\left(\frac{A_S}{[S]}\right)$$

$$\frac{0.644}{52.4 \text{ nM}} = F\left(\frac{1.000}{38.9 \text{ nM}}\right) \Rightarrow F = 0.478_1$$

For the unknown mixture, we can write

$$\frac{A_X}{[X]} = F\left(\frac{A_S}{[S]}\right)$$

$$\frac{1.093}{[X]} = 0.478_1\left(\frac{1.000}{742 \text{ nM}}\right) \Rightarrow$$

$$[X] = 1.70 \times 10^3 \text{ nM} = 1.70 \text{ μM}$$

Chapter 6

6-A. **(a)** The concentrations of reagents used in an analysis are determined either by weighing out pure primary standards or by reaction with such standards. If the standards are not pure, none of the concentrations will be correct.

(b) In a blank titration, the quantity of titrant required to reach the end point in the absence of analyte is measured. This quantity is subtracted from the quantity of titrant needed in the presence of analyte.

(c) In a direct titration, titrant reacts directly with analyte. In a back titration, a known excess of reagent that reacts with analyte is used. The excess is then titrated.

(d) The uncertainty in the equivalence point is constant at ±0.04 mL. The relative uncertainty in delivering

20 mL is 0.04/20 = 0.2%. The relative uncertainty in delivering 40 mL is only half as great: 0.04/40 = 0.1%.

6-B. **(a)** 0.197 0 g of ascorbic acid = $1.118\,6 \times 10^{-3}$ mol, which requires $1.118\,6 \times 10^{-3}$ mol I_3^-. $[I_3^-] = 1.118\,6 \times 10^{-3}$ mol/0.029 41 L = 0.038 03 M

(b) mol I_3^- required to react with 0.424 2 g of powder from vitamin C tablet = (0.031 63 L I_3^-)(0.038 03 M) = 1.203×10^{-3} mol I_3^-
Because 1 mol I_3^- reacts with 1 mol ascorbic acid, there must have been 1.203×10^{-3} mol ascorbic acid in the sample that was titrated.

(c) mass of ascorbic acid titrated
$= (1.203 \times 10^{-3}$ mol ascorbic acid)(176.12 g/mol)
= 0.211 9 g ascorbic acid
wt% ascorbic acid in tablet

$$= \frac{0.211\,9 \text{ g ascorbic acid}}{0.424\,2 \text{ g tablet}} \times 100 = 49.95 \text{ wt\%}$$

6-C. **(a)** moles of $Na_2C_2O_4$ in 1 L = (3.514 g)/(134.00 g/mol) = 0.026 22_4 mol

$C_2O_4^{2-}$ in 25.00 mL = (0.026 22_4 M)(0.025 00 L) = $6.556_0 \times 10^{-4}$ mol

$$\text{mol } MnO_4^- = (\text{mol } C_2O_4^{2-})\left(\frac{2 \text{ mol } MnO_4^-}{5 \text{ mol } C_2O_4^{2-}}\right)$$
$$= (6.556_0 \times 10^{-4} \text{ mol})\left(\frac{2}{5}\right)$$
$$= 2.622_4 \times 10^{-4} \text{ mol}$$

equivalence volume of $KMnO_4$ = 24.44 − 0.03 = 24.41 mL

$$[MnO_4^-] = \frac{2.622_4 \times 10^{-4} \text{ mol}}{0.024\,41 \text{ L}} = 0.010\,74_3 \text{ M}$$

(b) moles of $KMnO_4$ consumed
= (0.025 00 L)(0.010 74_3 M) = $2.685_8 \times 10^{-4}$ mol
mol $NaNO_2$ reacting with $KMnO_4$

$$= (2.685_8 \times 10^{-4} \text{ mol } KMnO_4)\left(\frac{5 \text{ mol } NaNO_2}{2 \text{ mol } KMnO_4}\right)$$
$$= 6.714_4 \times 10^{-4} \text{ mol}$$
$$[NaNO_2] = \frac{6.714_4 \times 10^{-4} \text{ mol}}{0.038\,11 \text{ L}} = 0.017\,62 \text{ M}$$

6-D. **(a)**
$$PbBr_2(s) \overset{K_{sp}}{\rightleftharpoons} \underset{x}{Pb^{2+}} + \underset{2x}{2Br^-}$$
$$x(2x)^2 = 2.1 \times 10^{-6} \Rightarrow$$
$$x = [Pb^{2+}] = 8.0_7 \times 10^{-3} \text{ M}$$

(b)
$$PbBr_2(s) \overset{K_{sp}}{\rightleftharpoons} \underset{x}{Pb^{2+}} + \underset{0.10 \text{ M}}{2Br^-}$$
$$[Pb^{2+}](0.10)^2 = 2.1 \times 10^{-6} \Rightarrow$$
$$[Pb^{2+}] = 2.1 \times 10^{-4} \text{ M}$$

6-E. **(a)** $Ag^+ + Br^- \rightleftharpoons AgBr(s)$
$K = 1/K_{sp}(AgBr) = 2.0 \times 10^{12}$

$Ag^+ + Cl^- \rightleftharpoons AgCl(s)$
$K = 1/K_{sp}(AgCl) = 5.6 \times 10^9$

AgBr precipitates first.

(b) mol Br^- = mol Ag^+ at first end point
= (0.015 55 L)(0.033 33 M)
= 5.183×10^{-4} mol
The original concentration of Br^- was therefore
$[Br^-] = (5.183 \times 10^{-4}$ mol)/(0.025 00 L)
= 0.020 73 M

(c) mol Cl^- = mol Ag^+ at second end point
− mol Ag^+ at first end point
= (0.042 23 L − 0.015 55 L)(0.033 33 M)
= 8.892×10^{-4} mol
The original concentration of Cl^- was therefore
$[Cl^-] = (8.892 \times 10^{-4}$ mol)/(0.025 00 L)
= 0.035 57 M

6-F. **(a)** AgCl is more soluble than AgSCN and will slowly dissolve in the presence of excess SCN^-. This reaction consumes the SCN^- and causes the red end point color to fade. If the AgCl is removed by filtration, it is no longer present when SCN^- is added in the back titration.

(b) Consider the titration of C^+ (in a flask) by A^- (from a buret). Before the equivalence point, there is excess C^+ in solution. Selective adsorption of C^+ on the CA crystal surface gives the crystal a positive charge. After the equivalence point, there is excess A^- in solution. Selective adsorption of A^- on the CA crystal surface gives it a negative charge.

(c) Beyond the equivalence point, there is excess $Fe(CN)_6^{4-}$ in solution. Adsorption of this anion on the precipitate will make the particles *negative*.

Chapter 7

7-A. **(a)**
$$\frac{0.214\,6 \text{ g AgBr}}{187.772 \text{ g AgBr/mol}} = 1.142\,9 \times 10^{-3} \text{ mol AgBr}$$

(b)
$$[NaBr] = \frac{1.142\,9 \times 10^{-3} \text{ mol}}{50.00 \times 10^{-3} \text{ L}} = 0.022\,86 \text{ M}$$

7-B. **(a)** Absorbed impurities are taken into a substance. Adsorbed impurities are found on the surface of a substance.

(b) Included impurities occupy lattice sites of the host crystal. Occluded impurities are trapped in a pocket inside the host.

(c) An ideal gravimetric precipitate should be insoluble, be easily filtered, possess a known, constant composition, and be stable to heat.

(d) High supersaturation often leads to formation of colloidal product with a large amount of impurities.

(e) Supersaturation can be decreased by increasing temperature (for most solutions), mixing well during addition of precipitant, and using dilute reagents. Homogeneous precipitation also reduces supersaturation.

(f) Washing with electrolyte preserves the electric double layer and prevents peptization.

(g) The volatile HNO_3 evaporates during drying. $NaNO_3$ is nonvolatile and will lead to a high mass for the precipitate.

(h) During the first precipitation, the concentration of impurities in the solution is high, giving a relatively high concentration of impurities in the precipitate. In the reprecipitation, the level of solution impurities is reduced, thus giving a purer precipitate.

(i) In thermogravimetric analysis, the mass of a sample is measured as the sample is heated. The mass lost during decomposition provides information about the composition of the sample.

7-C. **(a)** $\dfrac{0.104 \text{ g } CeO_2}{172.115 \text{ g } CeO_2\text{/mol}} = 6.043 \times 10^{-4} \text{ mol } CeO_2$

$6.043 \times 10^{-4} \text{ mol } CeO_2 \times \dfrac{1 \text{ mol Ce}}{1 \text{ mol } CeO_2} = 6.043 \times 10^{-4} \text{ mol Ce}$

$(6.043 \times 10^{-4} \text{ mol Ce})(140.116 \text{ g Ce/mol Ce}) = 0.084\,66 \text{ g Ce}$

(b) $\text{wt\% Ce} = \dfrac{0.084\,66 \text{ g Ce}}{4.37 \text{ g unknown}} \times 100 = 1.94 \text{ wt\%}$

7-D. **(a)** In *combustion*, a substance is heated in the presence of excess O_2 to convert carbon into CO_2 and hydrogen into H_2O. In *pyrolysis*, the substance is decomposed by heating in the absence of added O_2. All oxygen in the sample is converted into CO by passage through a suitable catalyst.

(b) WO_3 catalyzes the complete combustion of C to CO_2 in the presence of excess O_2. Cu reduces SO_3 to SO_2 and removes excess O_2 from the gas stream.

(c) The tin capsule melts and is oxidized to SnO_2 to liberate heat and crack the sample. Tin uses the available oxygen immediately, ensures that sample oxidation occurs in the gas phase, and acts as an oxidation catalyst.

(d) By dropping the sample in before very much O_2 is present, pyrolysis of the sample to give gaseous products occurs prior to oxidation. This practice minimizes the formation of nitrogen oxides.

(e) $C_8H_7NO_2SBrCl + 9\frac{1}{4}O_2 \rightarrow 8CO_2 + \frac{5}{2}H_2O + \frac{1}{2}N_2 + SO_2 + HBr + HCl$

Chapter 8

8-A. neutralize . . . conjugate

8-B. **(a)** Our strategy is to write the solubility product for $Mg(OH)_2$ and substitute the known value of $[Mg^{2+}] = 0.050$ M. Then we can solve for $[OH^-]$.

$$\underset{0.050\text{ M}}{[Mg^{2+}]}\underset{?}{[OH^-]^2} = K_{sp} = 7.1 \times 10^{-12} = (0.050)[OH^-]^2$$

$$\Rightarrow [OH^-] = 1.1_9 \times 10^{-5} \text{ M}$$

(b) Knowing the concentration of OH^- we can compute $[H^+]$ and pH.

$[H^+] = K_w/[OH^-] = 8.3_9 \times 10^{-10} \text{ M} \Rightarrow$

$\text{pH} = -\log [H^+] = 9.08$

8-C. The stronger acid is **A**, which has the larger value of K_a.

A: $Cl_2CHCO_2H \xrightleftharpoons{K_a = 0.08} Cl_2CHCO_2^- + H^+$

B: $ClCH_2CO_2H \xrightleftharpoons{K_a = 0.001\,36} ClCH_2CO_2^- + H^+$

The stronger base is **C**, which has the larger value of K_b.

C: $H_2NNH_2 + H_2O \xrightleftharpoons{K_b = 3.0 \times 10^{-6}} H_2NNH_3^+ + OH^-$

D: $H_2NCONH_2 + H_2O \xrightleftharpoons{K_b = 1.5 \times 10^{-14}} H_2NCONH_3^+ + OH^-$

8-D. **(a)** **(i)** $\text{pH} = -\log[H^+] = -\log(1.0 \times 10^{-3}) = 3.00$

(ii) $\text{pH} = -\log[H^+] = -\log(K_w/[OH^-]) = -\log(1.0 \times 10^{-14}/1.0 \times 10^{-2}) = 12.00$

(b) **(i)** $\text{pH} = -\log(3.2 \times 10^{-5}) = 4.49$

(ii) $\text{pH} = -\log(1.0 \times 10^{-14}/0.007\,7) = 11.89$

(c) $[H^+] = 10^{-4.44} = 3.6 \times 10^{-5}$ M

(d) $[H^+] = K_w/[OH^-] = 1.0 \times 10^{-14}/0.007\,7 = 1.3 \times 10^{-12}$ M

H^+ is derived from $H_2O \rightleftharpoons H^+ + OH^-$

(e) The concentration of the strong base is too low to perturb the pH of pure water. The pH is very close to 7.00.

8-E. **(a)** $pK_a = 3$ **(b)** $pK_b = 3$

(c) $HCO_2H \rightleftharpoons HCO_2^- + H^+$ **(d)** HCO_2^-

(e) $K_a = \dfrac{[HCO_2^-][H^+]}{[HCO_2H]} = 1.80 \times 10^{-4}$

(f) $K_b = \dfrac{[HCO_2H][OH^-]}{[HCO_2^-]}$

(g) $K_b = \dfrac{K_w}{K_a} = 5.56 \times 10^{-11}$

8-F. **(a)** Let $x = [H^+] = [A^-]$ and $0.100 - x = [HA]$.

$\dfrac{x^2}{0.100 - x} = 1.00 \times 10^{-5} \Rightarrow$

$x = 9.95 \times 10^{-4} \text{ M} \Rightarrow \text{pH} = -\log x = 3.00$

fraction of dissociation

$= \dfrac{[A^-]}{[A^-] + [HA]} = \dfrac{9.95 \times 10^{-4}}{0.100} = 9.95 \times 10^{-3}$

(b) Write the acid dissociation equation and substitute $[H^+] = 10^{-\text{pH}} = 10^{-2.78}$. But, we also know that $[A^-] = [H^+]$ and $[HA] = F - [A^-]$.

$$\underset{0.045\,0\,-\,10^{-2.78}}{HA} \rightleftharpoons \underset{10^{-2.78}}{H^+} + \underset{10^{-2.78}}{A^-}$$

$$K_a = \frac{(10^{-2.78})^2}{0.045\,0 - 10^{-2.78}} = 6.4 \times 10^{-5} \Rightarrow pK_a = 4.19$$

(c) We are told that the fraction of dissociation is 0.60%, which means that $\dfrac{[A^-]}{[A^-] + [HA]} = 0.006\,0$. Writing the weak-acid equilibrium and setting $[H^+] = [A^-] = x$ and setting $[HA] = F - x$, we can solve for x:

$$\underset{F-x}{HA} \rightleftharpoons \underset{x}{H^+} + \underset{x}{A^-}$$

$$\frac{[A^-]}{[HA] + [A^-]} = \frac{x}{F - x + x} = 0.006\,0$$

With $F = 0.045\,0$ M, $x = 2.7 \times 10^{-4}$ M

$$\Rightarrow K_a = \frac{[A^-][H^+]}{[HA]} = \frac{x^2}{F - x} = 1.6 \times 10^{-6}$$

$$\Rightarrow pK_a = 5.79$$

8-G. **(a)** Let $x = [OH^-] = [BH^+]$ and $0.100 - x = [B]$.

$$\frac{x^2}{0.100 - x} = 1.00 \times 10^{-5}$$

$$\Rightarrow x = 9.95 \times 10^{-4}\,\text{M} \Rightarrow [H^+] = \frac{K_w}{x}$$

$$= 1.005 \times 10^{-11} \Rightarrow pH = 11.00$$

$$\frac{[BH^+]}{[B] + [BH^+]} = \frac{9.95 \times 10^{-4}}{0.100} = 9.95 \times 10^{-3}$$

(b) We know that $[H^+] = 10^{-pH} = 10^{-9.28}$. We also know that $[OH^-] = K_w/[H^+]$. In the weak-base equilibrium, $[BH^+] = [OH^-]$ and $[B] = F - [BH^+]$. Inserting all these values lets us solve for the equilibrium constant.

$$\underset{0.10 - (K_w/10^{-9.28})}{B} + H_2O \rightleftharpoons \underset{K_w/10^{-9.28}}{BH^+} + \underset{K_w/10^{-9.28}}{OH^-}$$

$$K_b = \frac{(K_w/10^{-9.28})^2}{0.10 - (K_w/10^{-9.28})} = 3.6 \times 10^{-9}$$

(c) We are told that the fraction of association is 2.0%, which means that $\frac{[BH^+]}{[B] + [BH^+]} = 0.020$. Writing the weak-base equilibrium and setting $[BH^+] = [OH^-] = x$ and setting $[B] = F - x$, we can solve for x:

$$\underset{0.10-x}{B} + H_2O \overset{K_b}{\rightleftharpoons} \underset{x}{BH^+} + \underset{x}{OH^-}$$

$$\frac{[BH^+]}{[BH^+] + [B]} = 0.020$$

$$\frac{x}{0.10 - x + x} = 0.020 \Rightarrow x = 0.002\,0$$

$$K_b = \frac{[BH^+][OH^-]}{[B]} = \frac{x^2}{0.10 - x}$$

$$= \frac{(0.002\,0)^2}{0.10 - 0.002\,0} = 4.1 \times 10^{-5}$$

Chapter 9

9-A. **(a)** $pH = pK_a + \log\left(\frac{[A^-]}{[HA]}\right)$

$$= 5.00 + \log\left(\frac{0.050}{0.100}\right) = 4.70$$

(b) $pH = 3.744 + \log\left(\frac{[HCO_2^-]}{[HCO_2H]}\right) = 3.744 + \log R$,

where we abbreviate the quotient $[HCO_2^-]/[HCO_2H]$ by R. At pH 3.00, we solve for R as follows:

$$3.00 = 3.744 + \log R$$

$$\log R = 3.00 - 3.744 = -0.744$$

$$10^{\log R} = 10^{-0.744}$$

$$R = 0.180$$

pH:	3.000	3.744	4.000
$[HCO_2^-]/[HCO_2H]$:	0.180	1.00	1.80

9-B. **(a)** $pH = pK_a + \log\left(\frac{[\text{tris}]}{[\text{trisH}^+]}\right)$

$$= 8.07 + \log\left(\frac{(10.0\text{ g})/(121.14\text{ g/mol})}{(10.0\text{ g})/(157.60\text{ g/mol})}\right) = 8.18$$

(b) The number of mmol of $HClO_4$ added is (10.5 mL)(0.500 M) = 5.25 mmol

	tris	+	H^+	→	trisH$^+$	+	H_2O
Initial mmol:	82.55		5.25		63.45		
Final mmol:	77.30		—		68.70		

$$pH = 8.07 + \log\left(\frac{77.30}{68.70}\right) = 8.12$$

(c) The number of mmol of NaOH added is (10.5 mL)(0.500 M) = 5.25 mmol.

	trisH$^+$	+	OH^-	→	tris	+	H_2O
Initial mmol:	63.45		5.25		82.55		
Final mmol:	58.20		—		87.80		

$$pH = 8.075 + \log\left(\frac{87.80}{58.20}\right) = 8.25$$

9-C. **(a)** mol tris = 10.0 g/(121.14 g/mol)
= 0.082 5_5 mol = 82.5_5 mmol

	tris	+	H^+	→	trisH$^+$ + H_2O
Initial mmol:	82.5_5		x		—
Final mmol:	82.5_5 − x		—		x

$$pH = pK_a + \log\left(\frac{[\text{tris}]}{[\text{trisH}^+]}\right)$$

$$7.60 = 8.07 + \log\left(\frac{82.5_5 - x}{x}\right)$$

$$-0.47 = \log\left(\frac{82.5_5 - x}{x}\right) \Rightarrow$$

$$10^{-0.47} = 10^{\log[(82.5_5 - x)/x]}$$

$$0.33_{88} = \frac{82.5_5 - x}{x} \Rightarrow x = 61._{66}\text{ mmol}$$

volume of HCl required
$= (0.061_{66}\text{ mol})/(1.20\text{ M}) = 51._4\text{ mL}$

(b) I would weigh out 0.020 0 mol of acetic acid (= 1.201 g) and place it in a beaker with ~75 mL of water. While monitoring the pH with a pH electrode, I would add 3 M NaOH (~4 mL are required) until the pH is exactly 5.00. I would then pour the solution into a 100-mL volumetric flask and wash the beaker several times with a few milliliters of distilled water. Each washing would be added to the volumetric flask, to ensure quantitative transfer from the beaker to the flask. After swirling the volumetric flask to mix the solution, I would carefully add water up to the 100-mL mark, insert the cap, and invert many times to ensure complete mixing.

9-D. (a)

Compound	pK_a
Hydroxybenzene	9.98
Propanoic acid	4.87
Cyanoacetic acid	2.47 ← Most suitable, because pK_a is closest to pH
Sulfuric acid	1.99

(b) Buffer capacity is based on the ability of buffer to react with added acid or base, without making a large change in the ratio of concentrations $[A^-]/[HA]$. The greater the concentration of each component, the less relative change is brought about by reaction with a small increment of added acid or base.

(c) From the given values of K_b, compute $K_a = K_w/K_b$ and $pK_a = -\log K_a$.

Compound	pK_a (for conjugate acid)
Ammonia	9.26 ← Most suitable, because pK_a is closest to pH
Aniline	4.60
Hydrazine	8.48
Pyridine	5.20

9-E. (a) The quotient $[HIn]/[In^-]$ changes from 10:1 when pH $= pK_{HIn} - 1$ to 1:10 when pH $= pK_{HIn} + 1$. This change is generally sufficient to cause a complete color change.

(b) red, orange, yellow

Chapter 10

10-A. The titration reaction is $H^+ + OH^- \rightarrow H_2O$ and V_e = 5.00 mL because

$$\underbrace{(V_e(\text{mL}))(0.100\text{ M})}_{\text{mmol HCl at }V_e} = \underbrace{(50.00\text{ mL})(0.010\,0\text{ M})}_{\text{Initial mmol NaOH}} \Rightarrow$$

$$V_e = 5.00\text{ mL}$$

Representative calculations:

$V_a = 1.00$ mL:

$$OH^- = \underbrace{(50.00\text{ mL})(0.010\,0\text{ M})}_{\text{Initial mmol NaOH}} - \underbrace{(1.00\text{ mL})(0.100\text{ M})}_{\text{Added mmol HCl}}$$

$$= 0.400\text{ mmol}$$

$$[OH^-] = (0.400\text{ mmol})/(51.00\text{ mL}) = 0.007\,84\text{ M}$$

$$[H^+] = K_w/(0.007\,84\text{ M}) = 1.28 \times 10^{-12}\text{ M}$$

$$pH = -\log(1.28 \times 10^{-12}) = 11.89$$

$V_a = V_e = 5.00$ mL:

$$\underset{}{H_2O} \rightleftharpoons \underset{x}{H^+} + \underset{x}{OH^-}$$

$$K_w = x^2 \Rightarrow x = 1.00 \times 10^{-7}\text{ M} \Rightarrow pH = 7.00$$

$V_a = 5.01$ mL:

$$\text{excess } H^+ = (0.01\text{ mL})(0.100\text{ M}) = 0.001\text{ mmol}$$

$$[H^+] = (0.001\text{ mmol})/(55.01\text{ mL}) = 1._8 \times 10^{-5}\text{ M}$$

$$pH = -\log(1._8 \times 10^{-5}) = 4.74$$

V_a (mL)	pH	V_a (mL)	pH	V_a (mL)	pH
0.00	12.00	4.50	10.96	5.10	3.74
1.00	11.89	4.90	10.26	5.50	3.05
2.00	11.76	4.99	9.26	6.00	2.75
3.00	11.58	5.00	7.00	8.00	2.29
4.00	11.27	5.01	4.74	10.00	2.08

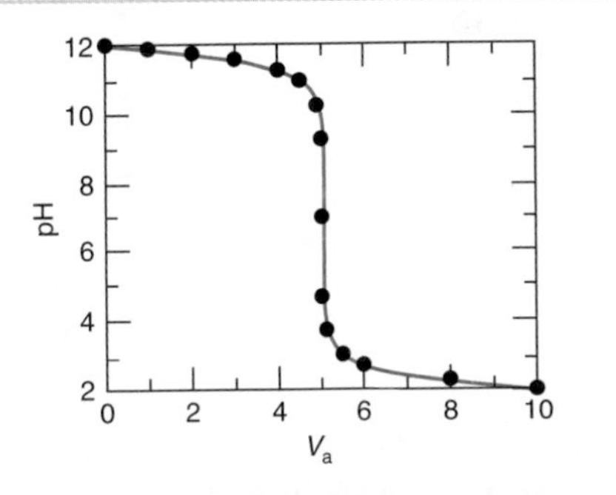

10-B. Titration reaction: $HCO_2H + OH^- \rightarrow HCO_2^- + H_2O$

$$\underbrace{(V_e(\text{mL}))(0.050\,0\text{ M})}_{\text{mmol KOH at }V_e} = \underbrace{(50.0\text{ mL})(0.050\,0\text{ M})}_{\text{Initial mmol }HCO_2H}$$

$$\Rightarrow V_e = 50.0\text{ mL}$$

Representative calculations:

$V_b = 0$ mL: $\underset{0.050\,0 - x}{HA} \rightleftharpoons \underset{x}{H^+} + \underset{x}{A^-}$

$K_a = 1.80 \times 10^{-4}$ $\quad pK_a = 3.744$

$$\frac{x^2}{0.050\,0 - x} = K_a \Rightarrow x = 2.91 \times 10^{-3} \Rightarrow pH = 2.54$$

$V_b = 48.0$ mL:

	HA	+	OH^-	→	A^-	+ H_2O
Initial mmol:	2.50		2.40		—	
Final mmol:	0.10		—		2.40	

$$pH = pK_a + \log\frac{[A^-]}{[HA]} = 3.744 + \log\frac{2.40}{0.10} = 5.12$$

$V_b = 50.0$ mL: formal conc. of A^-

$$= \frac{(50.0\text{ mL})(0.050\,0\text{ M})}{100.0\text{ mL}} = 0.025\,0\text{ M}$$

$$\underset{0.025\,0 - x}{A^-} + H_2O \rightleftharpoons \underset{x}{HA} + \underset{x}{OH^-}$$

$$K_b = K_w/K_a = 5.56 \times 10^{-11}$$

$$\frac{x^2}{0.025\,0 - x} = K_b \Rightarrow x = 1.18 \times 10^{-6} \Rightarrow$$

$$\text{pH} = -\log\left(\frac{K_w}{x}\right) = 8.07$$

$$V_b = 60.0 \text{ mL: excess } [\text{OH}^-] = \frac{(10.0 \text{ mL})(0.050\,0 \text{ M})}{110.0 \text{ mL}}$$

$$= 4.55 \times 10^{-3} \text{ M} \Rightarrow \text{pH} = 11.66$$

V_b (mL)	pH	V_b (mL)	pH	V_b (mL)	pH
0.0	2.54	45.0	4.70	50.5	10.40
10.0	3.14	48.0	5.13	51.0	10.69
20.0	3.57	49.0	5.44	52.0	10.99
25.0	3.74	49.5	5.74	55.0	11.38
30.0	3.92	50.0	8.07	60.0	11.66
40.0	4.35				

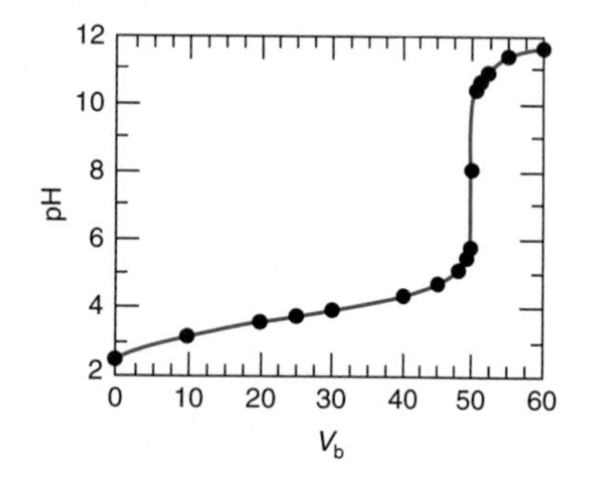

At $V_b = \frac{1}{2}V_e$, the pH should equal pK_a, which it does.

10-C. (a) Titration of a weak base, B, produces the conjugate acid, BH^+, which is necessarily acidic.

(b) The titration reaction is $B + H^+ \rightarrow BH^+$.

Find V_e: $(V_e)(0.200 \text{ M}) = (100.0 \text{ mL})(0.100 \text{ M})$

$$\Rightarrow V_e = 50.0 \text{ mL}$$

Representative calculations:

$V_a = 0.0$ mL: $\underset{0.100 - x}{B} + H_2O \rightleftharpoons \underset{x}{BH^+} + \underset{x}{OH^-}$

$$K_b = 2.6 \times 10^{-6}$$

$$\frac{x^2}{0.100 - x} = K_b = 2.6 \times 10^{-6} \Rightarrow x = 5.09 \times 10^{-4}$$

$$\text{pH} = -\log\left(\frac{K_w}{x}\right) = 10.71$$

$V_a = 20.0$ mL:

	B	+	H^+	$\rightarrow$	BH^+
Initial mmol:	10.00		4.00		—
Final mmol:	6.00		—		4.00

$$\text{pH} = \underset{K_a = K_w/K_b}{\underset{\uparrow}{pK_a}} \text{ (for BH}^+) + \log\left(\frac{[\text{B}]}{[\text{BH}^+]}\right)$$

$$= 8.41 + \log\left(\frac{6.00}{4.00}\right) = 8.59$$

$V_a = V_e = 50$ mL: All B has been converted into the conjugate acid, BH^+. The formal concentration of BH^+ is

$$F' = \frac{(100.0 \text{ mL})(0.100 \text{ M})}{150.0 \text{ mL}} = 0.066\,7 \text{ M}$$

The pH is determined by the reaction

$$\underset{0.066\,7 - x}{BH^+} \rightleftharpoons \underset{x}{B} + \underset{x}{H^+}$$

$$\frac{x^2}{0.066\,7 - x} = K_a = \frac{K_w}{K_b} \Rightarrow x = 1.60 \times 10^{-5}$$

$$\Rightarrow \text{pH} = 4.80$$

$V_a = 51.0$ mL: There is excess $[H^+]$.

$$\text{excess } [\text{H}^+] = \frac{(1.0 \text{ mL})(0.200 \text{ M})}{(151.0 \text{ mL})}$$

$$= 1.32 \times 10^{-3} \Rightarrow \text{pH} = 2.88$$

V_a (mL)	pH	V_a (mL)	pH	V_a (mL)	pH
0.0	10.71	30.0	8.23	50.0	4.80
10.0	9.01	40.0	7.81	50.1	3.88
20.0	8.59	49.0	6.72	51.0	2.88
25.0	8.41	49.9	5.71	60.0	1.90

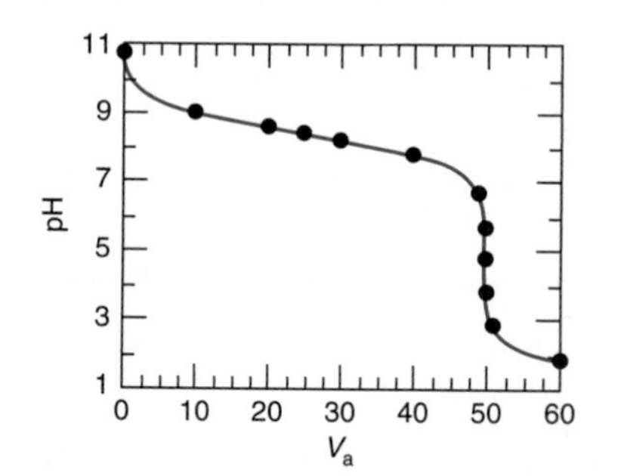

10-D. (a) Figure 10-1: bromothymol blue: blue $\rightarrow$ yellow
Figure 10-2: thymol blue: yellow $\rightarrow$ blue
Figure 10-11 ($pK_a = 8$): thymolphthalein: colorless $\rightarrow$ blue

(b) The derivatives are shown in the spreadsheet on page 575. In the first derivative graph, the maximum is near 119 mL. In Figure 10-6, the second derivative graph gives an end point of 118.9 μL.

10-E. (a) Tris(hydroxymethyl)aminomethane ($H_2NC(CH_2OH)_3$), sodium carbonate (Na_2CO_3), or borax ($Na_2B_4O_7 \cdot 10H_2O$) can be used to standardize HCl. Potassium hydrogen phthalate ($HO_2C—C_6H_4—CO_2^-K^+$) or potassium hydrogen iodate ($KH(IO_3)_2$) can be used to standardize NaOH.

(b) 30 mL of 0.05 M OH^- = 1.5 mmol OH^- = 1.5 mmol potassium hydrogen phthalate. $(1.5 \times 10^{-3} \text{ mol}) \times (204.22 \text{ g/mol}) = 0.30$ g of potassium hydrogen phthalate.

Spreadsheet for 10-D:

	A	B	C	D	E	F
1			First Derivative		Second Derivative	
2	μL NaOH	pH	μL	Derivative	μL	Derivative
3	107	6.921				
4	110	7.117	108.5	6.533E-02		
5	113	7.359	111.5	8.067E-02	110	5.11E-03
6	114	7.457	113.5	9.800E-02	112.5	8.67E-03
7	115	7.569	114.5	1.120E-01	114	1.40E-02
8	116	7.705	115.5	1.360E-01	115	2.40E-02
9	117	7.878	116.5	1.730E-01	116	3.70E-02
10	118	8.090	117.5	2.120E-01	117	3.90E-02
11	119	8.343	118.5	2.530E-01	118	4.10E-02
12	120	8.591	119.5	2.480E-01	119	–5.00E-03
13	121	8.794	120.5	2.030E-01	120	–4.50E-02
14	122	8.952	121.5	1.580E-01	121	–4.50E-02
15						
16	C4 = (A4+A3)/2			E5 = (C5+C4)/2		
17	D4 = (B4-B3)/(A4-A3)			F5 = (D5-D4)/(C5-C4)		

10-F. **(a)** 5.00 mL of 0.033 6 M HCl = 0.168_0 mmol
6.34 mL of 0.010 0 M NaOH = 0.063 4 mmol
HCl consumed by NH_3 = $0.168_0 - 0.063\ 4 = 0.104_6$ mmol
mol NH_3 = mol HCl = 0.104_6 mmol

(b) mol nitrogen = mol NH_3 = 0.104_6 mmol
$(0.104_6 \times 10^{-3}$ mol N)(14.006 7 g/mol) $= 1.46_5$ mg of nitrogen

(c) $(256\ \mu L)\left(\frac{1\ mL}{1\ 000\ \mu L}\right)(37.9\ mg\ protein/mL) = 9.70_2$ mg protein

(d) $$\text{wt\% N} = \frac{1.46_5\ \text{mg N}}{9.70_2\ \text{mg protein}} \times 100 = 15.1\ \text{wt\%}$$

10-G. **(a)** Your spreadsheet should reproduce the results in Figure 10-11.

(b)

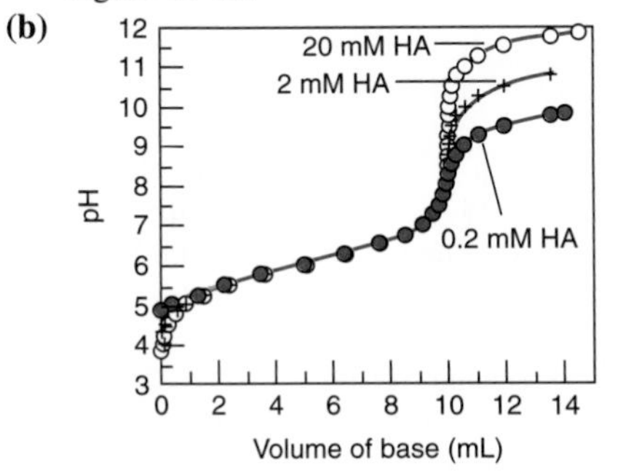

Chapter 11

11-A. **(a)** **(i)** $$H_3\overset{+}{N}CH_2CH_2\overset{+}{N}H_3 \overset{K_{a1} = 1.42 \times 10^{-7}}{\rightleftharpoons} H_2NCH_2CH_2\overset{+}{N}H_3 + H^+$$

$$H_2NCH_2CH_2\overset{+}{N}H_3 \overset{K_{a2} = 1.18 \times 10^{-10}}{\rightleftharpoons} H_2NCH_2CH_2NH_2 + H^+$$

(ii) $$^-O_2CCH_2CO_2^- + H_2O \overset{K_{b1} = K_w/K_{a2} = 4.98 \times 10^{-9}}{\rightleftharpoons} HO_2CCH_2CO_2^- + OH^-$$

$$HO_2CCH_2CO_2^- + H_2O \overset{K_{b2} = K_w/K_{a1} = 7.04 \times 10^{-12}}{\rightleftharpoons} HO_2CCH_2CO_2H + OH^-$$

(b) **(iii)** $\overset{+}{N}H_3$–$CHCH_2CO_2H$ (with CO_2H) $\overset{K_1}{\rightleftharpoons}$ $\overset{+}{N}H_3$–$CHCH_2CO_2H$ (with CO_2^-) $\overset{K_2}{\rightleftharpoons}$

Aspartic acid

$\overset{+}{N}H_3$–$CHCH_2CO_2^-$ (with CO_2^-) $\overset{K_3}{\rightleftharpoons}$ NH_2–$CHCH_2CO_2^-$ (with CO_2^-)

(iv) $H_3\overset{+}{N}$, HO_2C–$CHCH_2CH_2CH_2NHC(=\overset{+}{N}H_2)NH_2$ $\overset{K_1}{\rightleftharpoons}$

$H_3\overset{+}{N}$, ^-O_2C–$CHCH_2CH_2CH_2NHC(=\overset{+}{N}H_2)NH_2$ $\overset{K_2}{\rightleftharpoons}$

H_2N, ^-O_2C–$CHCH_2CH_2CH_2NHC(=\overset{+}{N}H_2)NH_2$ $\overset{K_3}{\rightleftharpoons}$

Arginine

H_2N, ^-O_2C–$CHCH_2CH_2CH_2NHC(=NH)NH_2$

11-B. **(a)** We treat H_2SO_3 as a weak monoprotic acid:

$$\underset{0.050 - x}{H_2SO_3} \rightleftharpoons \underset{x}{HSO_3^-} + \underset{x}{H^+}$$

$$\frac{x^2}{0.050 - x} = K_1 = 1.39 \times 10^{-2} \Rightarrow x = 2.03 \times 10^{-2}$$

$[HSO_3^-] = [H^+] = 2.03 \times 10^{-2}$ M $\Rightarrow$ pH = 1.69
$[H_2SO_3] = 0.050 - x = 0.030$ M

$$[SO_3^{2-}] = \frac{K_2[HSO_3^-]}{[H^+]} = K_2 = 6.73 \times 10^{-8}\ M$$

(b) HSO_3^- is the intermediate form of a diprotic acid.
pH $\approx \frac{1}{2}(pK_1 + pK_2) = \frac{1}{2}(1.857 + 7.172) = 4.51$
$[H^+] = 10^{-pH} = 3.1 \times 10^{-5}$ M; $[HSO_3^-] \approx 0.050$ M

$$[H_2SO_3] = \frac{[H^+][HSO_3^-]}{K_1} = \frac{(3.1 \times 10^{-5})(0.050)}{1.39 \times 10^{-2}} = 1.1 \times 10^{-4}\ M$$

$$[SO_3^{2-}] = \frac{K_2[HSO_3^-]}{[H^+]} = \frac{(6.73 \times 10^{-8})(0.050)}{3.1 \times 10^{-5}}$$
$$= 1.1 \times 10^{-4}\ M$$

(c) We treat SO_3^{2-} as if it were monobasic:

$$\underset{0.050 - x}{SO_3^{2-}} + H_2O \rightleftharpoons \underset{x}{HSO_3^-} + \underset{x}{OH^-}$$

$$\frac{x^2}{0.050 - x} = K_{b1} = \frac{K_w}{K_{a2}} = 1.49 \times 10^{-7} \Rightarrow$$
$$x = 8.62 \times 10^{-5}$$

$[HSO_3^-] = 8.62 \times 10^{-5}$ M;

$$[H^+] = \frac{K_w}{x} = 1.16 \times 10^{-10}\ M \Rightarrow pH = 9.94$$

$$[SO_3^{2-}] = 0.050 - x = 0.050\ M$$

$$[H_2SO_3] = \frac{[H^+][HSO_3^-]}{K_1} = 7.2 \times 10^{-13}\ M$$

11-C. **(a)** pH 9.00 | pH 11.00

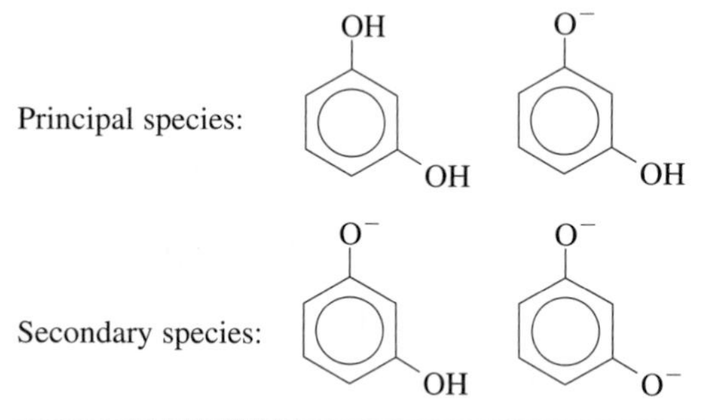

Reasoning: The acid dissociation constants are $pK_1 = 9.30$ and $pK_2 = 11.06$. pH 9.00 is below pK_1, so H_2A is the predominant form. If H_2A is predominant, HA^- must be the second most abundant species. pH 11.00 is just below $pK_2 = 11.06$. At pH = 11.06, HA^- and A^{2-} would be present in equal concentration. At pH 11.00, the more acidic HA^- will be slightly more abundant than the basic form, A^{2-}.

(b) HC^-: $pH \approx \frac{1}{2}(pK_2 + pK_3) = \frac{1}{2}(8.36 + 10.74) = 9.55$

11-D. **(a)** $\underbrace{(50.0\ mL)(0.050\ 0\ M)}_{mmol\ H_2A} = \underbrace{V_{e1}(0.100\ M)}_{mmol\ OH^-}$

$$\Rightarrow V_{e1} = 25.0\ mL$$

$V_{e2} = 2V_{e1} = 50.0$ mL

(b) 0 mL: Treat H_2A as a monoprotic weak acid

$$\underset{0.050\,0 - x}{H_2A} \overset{K_1}{\rightleftharpoons} \underset{x}{H^+} + \underset{x}{HA^-}$$

$$\frac{x^2}{0.050\,0 - x} = K_1 = 1.42 \times 10^{-3}$$

$\Rightarrow x = 7.75 \times 10^{-3} \Rightarrow pH = 2.11$

12.5 mL: We are halfway to the first equivalence point and there is a 1:1 mixture of H_2A and HA^-: $pH = pK_1 = 2.85$.

25.0 mL: This is the first equivalence point. H_2A has been converted into HA^-, the intermediate form of a diprotic acid.

$$pH \approx \tfrac{1}{2}(pK_1 + pK_2) = \tfrac{1}{2}(2.847 + 5.696) = 4.27$$

37.5 mL: Half of HA^- has been converted into A^{2-}, so $pH = pK_2 = 5.70$.

50.0 mL: H_2A has been converted into A^{2-} at a formal concentration of

$$F = \frac{(50.0\ mL)(0.050\ M)}{100.0\ mL} = 0.025\ 0\ M$$

$$\underset{0.025\,0 - x}{A^{2-}} + H_2O \rightleftharpoons \underset{x}{HA^-} + \underset{x}{OH^-}$$

$$\frac{x^2}{0.025\,0 - x} = K_{b1} = \frac{K_w}{K_{a2}}$$
$$\Rightarrow x = 1.12 \times 10^{-5}$$
$$\Rightarrow pH = -\log \frac{K_w}{x} = 9.05$$

55.0 mL: There is an excess of 5.0 mL of OH^-:

$$[OH^-] = \frac{(5.0\ mL)(0.100\ 0\ M)}{105.0\ mL}$$
$$= 0.004\ 76\ M \Rightarrow pH = 11.68$$

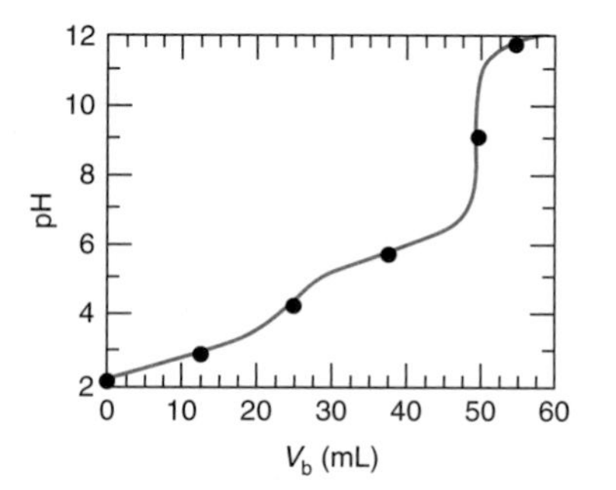

Chapter 12

12-A. **(a)** $$PbI_2(s) \overset{K_{sp}}{\rightleftharpoons} \underset{x}{Pb^{2+}} + \underset{2x}{2I^-}$$

$x(2x)^2 = K_{sp} = 7.9 \times 10^{-9} \Rightarrow 2x = [I^-] = 2.5$ mM

The observed concentration of dissolved iodine in pure water is approximately 3.8 mM, or 50% higher than predicted. One reason for this is that there are more species in solution than just Pb^{2+} and I^-, such as PbI^+. A second reason is that as PbI_2 dissolves, it increases its own solubility by adding ions to the solution to create ionic atmospheres around the dissolved ions and decreasing their attraction for each other. When KNO_3 is added, the number of ions in the ionic atmosphere increases further, thereby decreasing the attraction of Pb^{2+} and I^- for each other, and increasing the solubility of PbI_2.

(b) $\mu = \frac{1}{2}\{[Pb^{2+}] \cdot (+2)^2 + [I^-] \cdot (-1)^2\}$
$= \frac{1}{2}\{(0.001\ 0 \cdot 4) + (0.002\ 0 \cdot 1)\} = 0.003\ 0\ M$

12-B. **(a)** $HgBr_2(s) \rightleftharpoons \underset{x}{Hg^{2+}} + \underset{2x}{2Br^-} \Rightarrow$

$$K_{sp} = [Hg^{2+}]\gamma_{Hg^{2+}}[Br^-]^2\gamma^2_{Br^-}$$
$$1.3 \times 10^{-19} = (x)(1)(2x)^2(1)$$
$$1.3 \times 10^{-19} = 4x^3$$
$$x = [Hg^{2+}] = \sqrt[3]{\frac{1.3 \times 10^{-19}}{4}} = 3.2 \times 10^{-7}\ M$$

The value of K_{sp} comes from Appendix A. To find the cube root of a number, you can raise the number to the 1/3 power on your calculator.

(b) The concentration of Br^- is 0.050 M from NaBr. The ionic strength is 0.050 M. The activity coefficients in Table 12-1 are $\gamma_{Hg^{2+}} = 0.465$ and $\gamma_{Br^-} = 0.805$.

$HgBr_2(s) \rightleftharpoons Hg^{2+} + 2Br^- \Rightarrow$
$$K_{sp} = [Hg^{2+}]\gamma_{Hg^{2+}}[Br^-]^2\gamma^2_{Br^-}$$
$$1.3 \times 10^{-19} = [Hg^{2+}](0.465)(0.050)^2(0.805)^2$$
$$[Hg^{2+}] = \frac{1.3 \times 10^{-19}}{(0.465)(0.050)^2(0.805)^2} = 1.7 \times 10^{-16}\ M$$

(c) If the equilibrium $HgBr_2(s) + Br^- \rightleftharpoons HgBr_3^-$ also occurs, additional mercuric ion would be in solution in the form $HgBr_3^-$. The solubility of $HgBr_2$ would be greater than that calculated in **(b)**.

12-C. **(a)** $C_2O_4^{2-} \xrightarrow{H^+} HC_2O_4^- \xrightarrow{H^+} H_2C_2O_4$

The species are Na^+, $C_2O_4^{2-}$, $HC_2O_4^-$, $H_2C_2O_4$, Cl^-, H^+, OH^-, and H_2O.

(b) Charge balance: $[Na^+] + [H^+] = 2[C_2O_4^{2-}] + [HC_2O_4^-] + [Cl^-] + [OH^-]$

(c) One mass balance states that the total concentration of Na must be $2 \times \frac{5.00\ mmol}{0.100\ L} \Rightarrow [Na^+] = 0.100$ M. A second mass balance states that the total moles of oxalate are $\frac{5.00\ mmol}{0.100\ L} \Rightarrow 0.050\ 0\ M = [C_2O_4^{2-}] + [HC_2O_4^-] + [H_2C_2O_4]$. A third mass balance is that $[Cl^-] = 0.025\ 0$ M.

(d) We are adding 2.50 mmol H^+ to 5.00 mmol of the base, oxalate. The predominant species are $C_2O_4^{2-} + HC_2O_4^-$, with a negligible amount of $H_2C_2O_4$. The pH is going to be near pK_2 for oxalic acid, which is 4.27. Therefore $[H^+] \approx 10^{-4.27}$ M and $[OH^-] = K_w/[H^+] \approx 10^{-9.73}$ M in this solution. The charge and mass balances can be simplified by ignoring $[H_2C_2O_4]$, $[H^+]$, and $[OH^-]$ in comparison with the concentrations of major species:

Charge balance: $[Na^+] \approx 2[C_2O_4^{2-}] + [HC_2O_4^-] + [Cl^-]$

Mass balance: $0.050\ 0\ M \approx [C_2O_4^{2-}] + [HC_2O_4^-]$

12-D. **(a)** *Pertinent reactions:* Two given in the problem plus $H_2O \rightleftharpoons H^+ + OH^-$.

Charge balance: Invalid because pH is fixed

Mass balance: $[Ag^+] = [CN^-] + [HCN]$ (A)

Equilibrium constants:

$K_{sp} = [Ag^+][CN^-] = 2.2 \times 10^{-16}$ (B)

$K_b = \frac{[HCN][OH^-]}{[CN^-]} = 1.6 \times 10^{-5}$ (C)

$K_w = [H^+][OH^-] = 1.0 \times 10^{-14}$ (D)

Count equations and unknowns: There are four equations (A–D) and four unknowns: $[Ag^+]$, $[CN^-]$, $[HCN]$, and $[OH^-]$. $[H^+]$ is known.

Solve: Because $[H^+] = 10^{-9.00}$ M, $[OH^-] = 10^{-5.00}$ M. Putting this value of $[OH^-]$ into Equation C gives

$$[HCN] = \frac{K_b}{[OH^-]}[CN^-] = 1.6[CN^-]$$

Substituting into Equation A gives

$$[Ag^+] = [CN^-] + [HCN]$$
$$= [CN^-] + 1.6[CN^-] = 2.6[CN^-]$$

Substituting into Equation B gives

$$K_{sp} = 2.2 \times 10^{-16} = [Ag^+][CN^-]$$
$$= (2.6[CN^-])[CN^-] \Rightarrow [CN^-] = 9.2_0 \times 10^{-9}\ M$$
$$[Ag^+] = K_{sp}/[CN^-] = 2.2 \times 10^{-16}/9.2_0 \times 10^{-9}$$
$$= 2.3_9 \times 10^{-8}\ M$$
$$[HCN] = 1.6[CN^-] = 1.6(9.2_0 \times 10^{-9})$$
$$= 1.4_7 \times 10^{-8}\ M$$

(b) *Mass balance:* Because all species are derived from AgCN(*s*), the moles of silver must equal the moles of cyanide:

$$\underbrace{[Ag^+] + [AgCN(s)] + [Ag(CN)_2^-] + [AgOH(aq)]}_{\text{mol silver}}$$
$$= \underbrace{[CN^-] + [HCN] + [AgCN(s)] + 2[Ag(CN)_2^-]}_{\text{mol cyanide}}$$

which simplifies to $[Ag^+] + [AgOH(aq)] = [CN^-] + [HCN] + [Ag(CN)_2^-]$

12-E. At pH 2.00: $\alpha_{HA} = \frac{10^{-2.00}}{10^{-2.00} + 10^{-3.00}} = 0.90_9$

$$\alpha_{A^-} = \frac{10^{-3.00}}{10^{-2.00} + 10^{-3.00}} = 0.090_9$$
$$\frac{[HA]}{[A^-]} = \frac{0.90_9}{0.090_9} = 10._0$$

The results for all three pH values are

pH	α_{HA}	α_{A^-}	$[HA]/[A^-]$
2.00	0.90_9	0.090_9	$10._0$
3.00	0.50_0	0.50_0	1.0_0
4.00	0.090_9	0.90_9	0.10_0

Of course, you already knew these results from the Henderson-Hasselbalch equation.

Chapter 13

13-A. A monodentate ligand binds to a metal ion through one ligand atom. A multidentate ligand binds through more than one ligand atom. A chelating ligand is a multidentate ligand.

13-B. **(a)** $M^{n+} + Y^{4-} \rightleftharpoons MY^{n-4}$

$$K_f = [MY^{n-4}]/([M^{n+}][Y^{4-}])$$

(b) At low pH, H^+ competes with M^{n+} in binding to the ligand atoms of EDTA.

(c) The auxiliary complexing agent prevents the metal ion from precipitating with hydroxide at high pH. EDTA displaces the auxiliary complexing agent from the metal ion.

(d) The dividing line between each region is a pK value. The midpoint of each region is the average of the two surrounding pK values. The pH at the dividing line between two regions is the pH at which the concentrations of the species in the neighboring regions are equal. For example, at pH 2.69, $[H_3Y^-] = [H_2Y^{2-}]$. The pH at the center of a region is the pH containing the "pure" species in that region. For example, the pH of a solution made by dissolving a salt of H_2Y^{2-} (such as Na_2H_2Y) is 4.41.

H_6Y^{2+}	H_5Y^+	H_4Y	H_3Y^-	H_2Y^{2-}	HY^{3-}	Y^{4-}

Dividing lines (pK): 0.0, 1.5, 2.00, 2.69, 6.13, 10.37

pH: 0.75, 1.75, 2.34, 4.41, 8.25

13-C. (a) Only a small amount of indicator is employed. Most of the Mg^{2+} is not bound to indicator. The free Mg^{2+} reacts with EDTA before MgIn reacts. Therefore the concentration of MgIn is constant until all of the Mg^{2+} has been consumed. Only when MgIn begins to react does the color change.

(b) (i) Between pH 2.85 and pH 6.70, the predominant indicator species is H_3In^{3-}, which is **(ii)** yellow. The metal-indictor complex is **(iii)** red. At pH 8, the predominant indicator species is the violet H_2In^{4-}, so the titration color change is **(iv)** violet → red.

13-D. (a) (25.0 mL)(0.050 0 M) = 1.25 mmol EDTA

(b) (5.00 mL)(0.050 0 M) = 0.25 mmol Zn^{2+}

(c) mmol Ni^{2+} = mmol EDTA − mmol Zn^{2+} = 1.25 − 0.25 = 1.00 mmol Ni^{2+}

$[Ni^{2+}]$ = (1.00 mmol)/(50.0 mL) = 0.020 0 M

13-E. (a) At pH 5.00: $K_f' = \alpha_{Y^{4-}} K_f = (2.9 \times 10^{-7})(10^{10.65}) = 1.3 \times 10^4$

	Ca^{2+}	+ EDTA ⇌	CaY^{2-}
Initial concentration (M):	0	0	0.010
Final concentration (M):	x	x	$0.010 - x$

$$\frac{[CaY^{2-}]}{[Ca^{2+}][EDTA]} = \frac{0.010 - x}{x^2} = K_f' = 1.3 \times 10^4$$

$$\Rightarrow x = [Ca^{2+}] = 8.4 \times 10^{-4}\ M$$

(b) fraction of bound calcium

$$= \frac{[CaY^{2-}]}{[CaY^{2-}] = [Ca^{2+}]} = \frac{[0.010 - 0.000\ 84]}{0.010} = 0.92$$

92% of Ca is bound to EDTA at the equivalence point at pH 5.00.

(c) At pH 9.00: $K_f' = (0.041)(10^{10.65}) = 1.8 \times 10^9$

$$\frac{[CaY^{2-}]}{[Ca^{2+}][EDTA]} = \frac{0.010 - x}{x^2} = 1.8 \times 10^9$$

$$\Rightarrow x = [Ca^{2+}] = 2.4 \times 10^{-6}\ M$$

$$\text{fraction of bound calcium} = \frac{[CaY^{2-}]}{[CaY^{2-}] + [Ca^{2+}]} = \frac{[0.010 - 2.4 \times 10^{-6}]}{0.010} = 0.999\ 8$$

99.98% of Ca is bound to EDTA at the equivalence point at pH 9.00.

13-F. At pH 10.0, $K_f' = (0.30)(10^{10.65}) = 1.3_4 \times 10^{10}$

At V_{EDTA} = 5.00 mL, calculations are identical with those of Mg^{2+}:

initial mmol Ca^{2+} = (0.050 0 M Ca^{2+})(50.0 mL) = 2.50 mmol

mmol remaining = (0.900)(2.50 mmol) = 2.25 mmol

$$[Ca^{2+}] = \frac{2.25\ mmol}{55.0\ mL} = 0.040\ 9\ M$$

$$\Rightarrow pCa^{2+} = -\log[Ca^{2+}] = 1.39$$

V_{EDTA} = 50.00 mL is the equivalence point:

$$[Ca^{2+}] = \frac{2.25\ mmol}{100.0\ mL} = 0.025\ 0\ M$$

	Ca^{2+}	+ EDTA ⇌	CaY^{2-}
Initial concentration (M):	—	—	0.025 0
Final concentration (M):	x	x	$0.025\ 0 - x$

$$\frac{[CaY^{2-}]}{[Ca^{2+}][EDTA]} = K_f' = 1.3_4 \times 10^{10}$$

$$\frac{0.025\ 0 - x}{x^2} = 1.3_4 \times 10^{10} \Rightarrow$$

$$x = 1.3_7 \times 10^{-6}\ M \Rightarrow pCa^{2+} = -\log x = 5.86$$

At V_{EDTA} = 51.00 mL, there is 1.00 mL of excess EDTA:

$$[EDTA] = \frac{0.050\ 0\ mmol}{101.0\ mL} = 0.000\ 49_5\ M$$

$$[CaY^{2-}] = \frac{2.50\ mmol}{101.0\ mL} = 0.024\ 8\ M$$

$$\frac{[CaY^{2-}]}{[Ca^{2+}][EDTA]} = K_f' = 1.3_4 \times 10^{10}$$

$$\frac{[0.024\ 8]}{[Ca^{2+}](0.000\ 49_5)} = 1.3_4 \times 10^{10}$$

$$\Rightarrow [Ca^{2+}] = 3.7 \times 10^{-9}\ M \Rightarrow pCa^{2+} = 8.43$$

Chapter 14

14-A. (a) $I_2 + 2e^- \rightleftharpoons 2I^-$
Oxidant

(b) $2S_2O_3^{2-} \rightleftharpoons S_4O_6^{2-} + 2e^-$
Reductant

(c) $1.00 \text{ g } S_2O_3^{2-}/(112.13 \text{ g/mol}) = 8.92 \text{ mmol } S_2O_3^{2-} = 8.92 \text{ mmol e}^-$

$(8.92 \times 10^{-3} \text{ mol})(9.649 \times 10^4 \text{ C/mol}) = 861 \text{ C}$

(d) current (A) = coulombs/s = 861 C/60 s = 14.3 A

(e) work = $E \cdot q$ = (0.200 V)(861 C) = 172 J

14-B. (a) $Pt(s) \mid Br_2(l) \mid HBr(aq, 0.10 \text{ M}) \parallel Al(NO_3)_3(aq, 0.010 \text{ M}) \mid Al(s)$

(b)

$$\begin{array}{ll} \text{Right:} & Ag(S_2O_3)_2^{3-} + e^- \rightleftharpoons Ag(s) + 2S_2O_3^{2-} \\ \text{Left:} & Fe(CN)_6^{4-} \rightleftharpoons Fe(CN)_6^{3-} + e^- \\ \hline \text{Net:} & Fe(CN)_6^{4-} + Ag(S_2O_3)_2^{3-} \rightleftharpoons Fe(CN)_6^{3-} + Ag(s) + 2S_2O_3^{2-} \end{array}$$

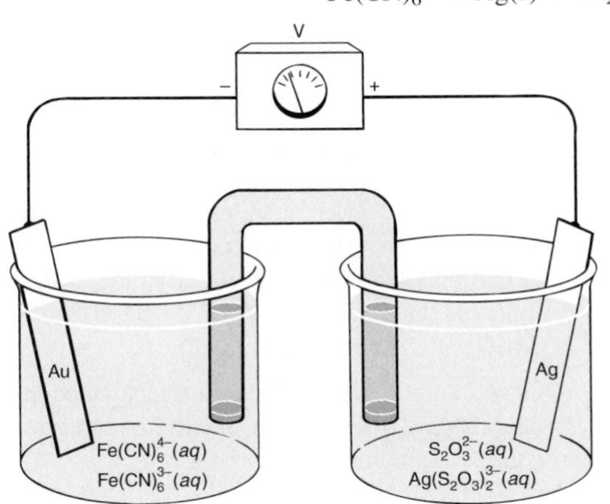

14-C. $Pt(s) \mid H_2(g, 1 \text{ bar}) \mid H^+(aq, 1 \text{ M}) \parallel Fe^{2+}(aq, 1 \text{ M}), Fe^{3+}(aq, 1 \text{ M}) \mid Pt(s)$

$E°$ for the reaction $Fe^{3+} + e^- \rightleftharpoons Fe^{2+}$ is 0.771 V. Therefore electrons flow from left to right through the meter.

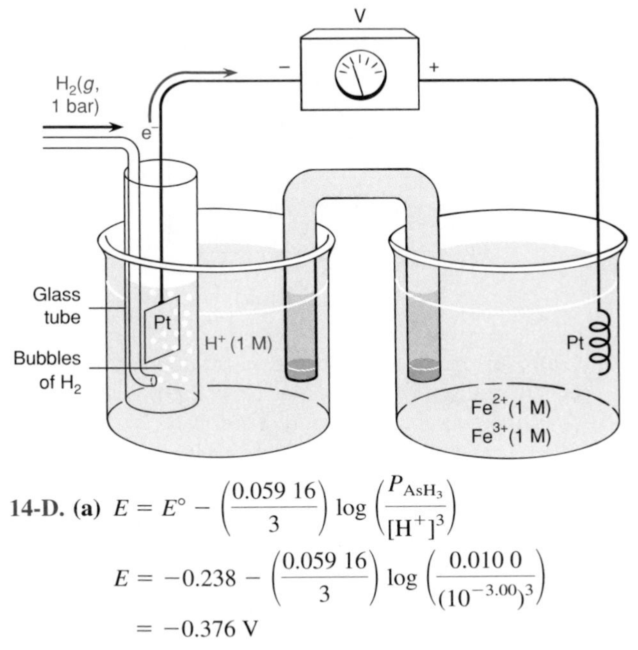

14-D. (a)

$$E = E° - \left(\frac{0.059\ 16}{3}\right) \log \left(\frac{P_{AsH_3}}{[H^+]^3}\right)$$

$$E = -0.238 - \left(\frac{0.059\ 16}{3}\right) \log \left(\frac{0.010\ 0}{(10^{-3.00})^3}\right) = -0.376 \text{ V}$$

(b) $Zn(s) \mid Zn^{2+}(0.1 \text{ M}) \parallel Cu^{2+}(0.1 \text{ M}) \mid Cu(s)$

Right half-cell: $Cu^{2+} + 2e^- \rightleftharpoons Cu(s)$ $\quad E°_+ = 0.339 \text{ V}$

Left half-cell: $Zn^{2+} + 2e^- \rightleftharpoons Zn(s)$ $\quad E°_- = -0.762 \text{ V}$

$$E = \left\{0.339 - \left(\frac{0.059\ 16}{2}\right) \log \left(\frac{1}{0.1}\right)\right\} - \left\{-0.762 - \left(\frac{0.059\ 16}{2}\right) \log \left(\frac{1}{0.1}\right)\right\} = 1.101 \text{ V}$$

The positive voltage tells us that electrons are transferred from Zn to Cu. The net reaction is $Cu^{2+} + Zn(s) \rightleftharpoons Cu(s) + Zn^{2+}$.

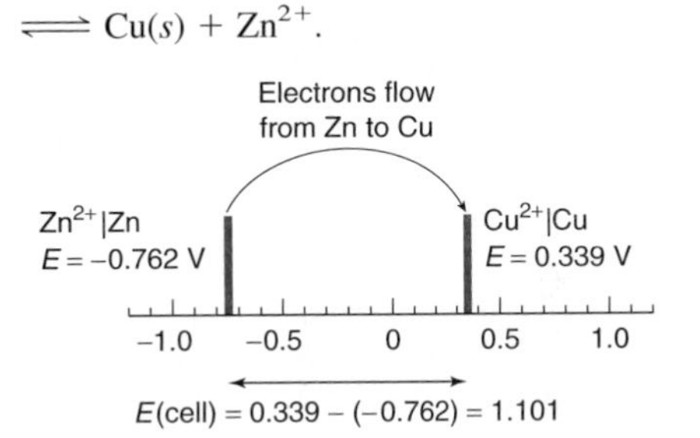

14-E. (a) Right half-cell: $Cu^{2+} + 2e^- \rightleftharpoons Cu(s)$ $\quad E°_+ = 0.339 \text{ V}$

Left half-cell: $Zn^{2+} + 2e^- \rightleftharpoons Zn(s)$ $\quad E°_- = -0.762 \text{ V}$

$E° = E°_+ - E°_- = 1.101 \text{ V}$

$K = 10^{nE°/0.059\ 16} = 10^{2(1.101)/0.059\ 16} = 1.7 \times 10^{37}$

(b)

$$\begin{array}{lll} & AgBr(s) + e^- \rightleftharpoons Ag(s) + Br^- & E°_+ = 0.071 \text{ V} \\ - & Ag^+ + e^- \rightleftharpoons Ag(s) & E°_- = 0.799 \text{ V} \\ \hline & AgBr(s) \rightleftharpoons Ag^+ + Br^- & E° = 0.071 - 0.799 = -0.728 \text{ V} \end{array}$$

$K_{sp} = 10^{1E°/0.059\ 16} = 5 \times 10^{-13}$

14-F. (a)

$$E = E_+ - E_- = \left\{\underbrace{E°_+}_{0.771 \text{ V}} - (0.059\ 16) \log \underbrace{\left(\frac{[Fe^{2+}]}{[Fe^{3+}]}\right)}_{?}\right\} - \underbrace{0.197}_{\text{Reference electrode voltage}}$$

$$0.703 = \left\{0.771 - 0.059\ 16 \log \left(\frac{[Fe^{2+}]}{[Fe^{3+}]}\right)\right\} - 0.197$$

$$0.059\ 16 \log \left(\frac{[Fe^{2+}]}{[Fe^{3+}]}\right) = -0.129$$

$$\Rightarrow \log \left(\frac{[Fe^{2+}]}{[Fe^{3+}]}\right) = -2.18$$

$$\Rightarrow \left(\frac{[Fe^{2+}]}{[Fe^{3+}]}\right) = 10^{-2.18} = 6.6 \times 10^{-3}$$

(b) **(i)** A potential of 0.523 V vs. S.H.E. lies to the right of the origin in Figure 14-12. The Ag|AgCl potential lies 0.197 V to the right of the origin. The difference between these two points is 0.523 − 0.197 = 0.326 V.

(ii) A potential of 0.222 V vs. S.C.E. lies 0.222 V to the right of S.C.E. in Figure 14-12. Because S.C.E. lies 0.241 V to the right of S.H.E., the distance from S.H.E. is 0.222 + 0.241 = 0.463 V.

Chapter 15

15-A. **(a)** At 0.10 mL:

initial Cl^- = 2.00 mmol; added Ag^+ = 0.020 mmol

$$[Cl^-] = \frac{(2.00 - 0.020)\text{ mmol}}{(40.0 + 0.10)\text{ mL}} = 0.049\ 4\text{ M}$$

$$[Ag^+] = K_{sp}/[Cl^-] = (1.8 \times 10^{-10})/(0.049\ 4) = 3.6 \times 10^{-9}\text{ M}$$

In a similar manner, we find

2.50 mL: $[Cl^-] = 0.035\ 3$ M $[Ag^+] = 5.1 \times 10^{-9}$ M
5.00 mL: $[Cl^-] = 0.022\ 2$ M $[Ag^+] = 8.1 \times 10^{-9}$ M
7.50 mL: $[Cl^-] = 0.010\ 5$ M $[Ag^+] = 1.7 \times 10^{-8}$ M
9.90 mL: $[Cl^-] = 0.000\ 401$ M $[Ag^+] = 4.5 \times 10^{-7}$ M

(b) At V_e: $[Ag^+][Cl^-] = x^2 = K_{sp} \Rightarrow$

$$[Cl^-] = [Ag^+] = \sqrt{K_{sp}} = 1.3 \times 10^{-5}\text{ M}$$

(c) At 10.10 mL: There is 0.10 mL of excess Ag^+. Therefore

$$[Ag^+] = \frac{(0.10\text{ mL})(0.200\text{ M})}{50.1\text{ mL}} = 4.0 \times 10^{-4}\text{ M}$$

At 12.00 mL: $$[Ag^+] = \frac{(2.00\text{ mL})(0.200\text{ M})}{52.0\text{ mL}} = 7.7 \times 10^{-3}\text{ M}$$

(d) At 0.10 mL:

$E = 0.558 + (0.059\ 16) \log [Ag^+]$

$E = 0.558 + (0.059\ 16) \log (3.6 \times 10^{-9}) = 0.059$ V

In a similar manner, we find the following results:

mL $AgNO_3$	E(V)	mL	E
0.10	0.059	9.90	0.182
2.50	0.067	10.00	0.270
5.00	0.079	10.10	0.357
7.50	0.099	12.00	0.433

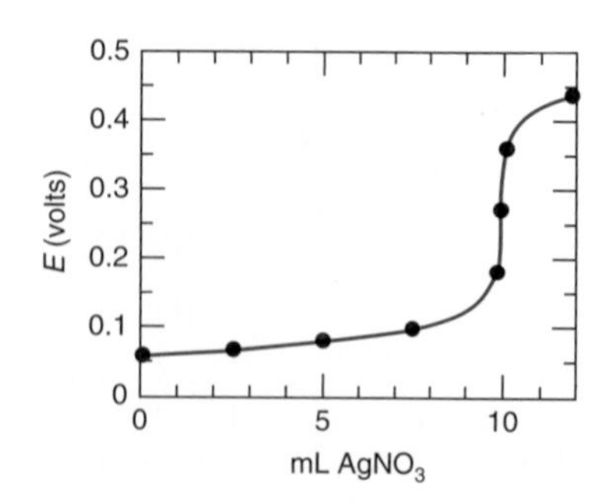

15-B. Cl^- diffuses into the $NaNO_3$, and NO_3^- diffuses into the NaCl. The mobility of Cl^- is greater than that of NO_3^-, so the NaCl region is depleted of Cl^- faster than the $NaNO_3$ region is depleted of NO_3^-. The $NaNO_3$ side becomes negative, and the NaCl side becomes positive.

15-C. **(a)** $E = \frac{0.059\ 16}{n} \log \left(\frac{[NH_4^+]_{outer}}{[NH_4^+]_{inner}}\right)$, where $n = +1$. The concentration $[NH_4^+]_{inner}$ is fixed. If $[NH_4^+]_{outer}$ increases by a factor of 10, then E increases by (0.059 16/1) log 10 = 0.059 16 V.

Fluoride electrode: $E = \frac{0.059\ 16}{n} \log \left(\frac{[F^-]_{outer}}{[F^-]_{inner}}\right)$, where $n = -1$. If $[F^-]_{outer}$ increases by a factor of 10, E changes by (−0.059 16) log 10 = −0.059 16 V.

Sulfide electrode, $E = \frac{0.059\ 16}{n} \log \left(\frac{[S^{2-}]_{outer}}{[S^{2-}]_{inner}}\right)$, where $n = -2$. If $[S^{2-}]_{outer}$ increases by a factor of 10, E changes by (−0.059 16/2) log 10 = −0.029 58 V.

(b) A hydrophobic *cation* is required in the membrane to balance the charge of $L(CO_3^{2-})(H_2O)$ and retain $L(CO_3^{2-})(H_2O)$ in the membrane.

15-D. **(a)** Uncertainty in pH of standard buffers, junction potential, alkaline or acid errors at extreme pH values, and equilibration time for electrode.

(b) (4.63)(0.059 16 V) = 0.274 V

(c) Na^+ competes with H^+ for cation-exchange sites on the glass surface. The glass responds as if some H^+ were present, and the apparent pH is lower than the actual pH.

15-E. **(a)**

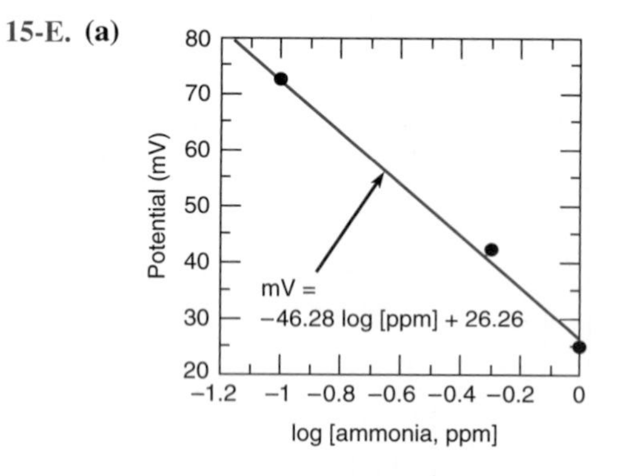

(b) Plugging the experimental values of potential into the equation of the calibration curve in **(a)** gives

106 = −46.28 log[ppm] + 26.26

$\Rightarrow$ [ppm] = 0.019 ppm

115 = −46.28 log[ppm] + 26.26

$\Rightarrow$ [ppm] = 0.012 ppm

(Note that the unknown values lie beyond the calibration points, which is not good practice. It would be better to obtain calibration points at lower concentration to include the full range of unknown points.)

(c) 56 = −46.28 log[ppm] + 26.26 $\Rightarrow$ [ppm] = 0.23 ppm

Chapter 16

16-A. Titration reaction:

$Sn^{2+} + 2Ce^{4+} \rightarrow Sn^{4+} + 2Ce^{3+} \quad V_e = 10.0$ mL

Indicator electrode half-reactions:

$Sn^{4+} + 2e^- \rightleftharpoons Sn^{2+} \quad E^\circ = 0.139$ V

$Ce^{4+} + e^- \rightleftharpoons Ce^{3+} \quad E^\circ = 1.47$ V (1 M HCl)

Indicator electrode Nernst equations:

$$E_+ = 0.139 - \frac{0.059\,16}{2}\log\frac{[Sn^{2+}]}{[Sn^{4+}]} \quad \text{(A)}$$

$$E_+ = 1.47 - 0.059\,16\log\frac{[Ce^{3+}]}{[Ce^{4+}]} \quad \text{(B)}$$

Representative calculations:

At 0.100 mL: The ratio $[Sn^{2+}]/[Sn^{4+}]$ is 9.90/0.100.

$$E_+ = 0.139 - \frac{0.059\,16}{2}\log\frac{[Sn^{2+}]}{[Sn^{4+}]}$$

$$= 0.139 - \frac{0.059\,16}{2}\log\frac{9.90}{0.100} = 0.080 \text{ V}$$

$$E = E_+ - E_- = 0.080 - 0.241 = -0.161 \text{ V}$$

At 10.00 mL: To add Nernst equations A and B, the factor in front of the log term needs to be the same in both. Therefore we multiply equation A by 2 before carrying out the addition:

$$2E_+ = 2(0.139) - 0.059\,16\log\frac{[Sn^{2+}]}{[Sn^{4+}]}$$

$$E_+ = 1.47 - 0.059\,16\log\frac{[Ce^{3+}]}{[Ce^{4+}]}$$

$$3E_+ = 1.748 - 0.059\,16\log\frac{[Sn^{2+}][Ce^{3+}]}{[Sn^{4+}][Ce^{4+}]} \quad \text{(C)}$$

At the equivalence point, $[Ce^{3+}] = 2[Sn^{4+}]$ and $[Ce^{4+}] = 2[Sn^{2+}]$. Putting these equalities into Equation C makes the log term 0.

Therefore $3E_+ = 1.748$ and $E_+ = 0.583$ V.

$E = E_+ - E_- = 0.583 - 0.241 = 0.342$ V

At 10.10 mL: The ratio $[Ce^{3+}]/[Ce^{4+}]$ is 10.00/0.10

$$E_+ = 1.47 - 0.059\,16\log\frac{[Ce^{3+}]}{[Ce^{4+}]}$$

$$E_+ = 1.47 - 0.059\,16\log\frac{10.00}{0.10} = 1.35_2 \text{ V}$$

$$E = E_+ - E_- = 1.35_2 - 0.241 = 1.11 \text{ V}$$

mL	E (V)	mL	E (V)	mL	E (V)
0.100	−0.161	9.50	−0.064	10.10	1.11
1.00	−0.130	10.00	0.342	12.00	1.19
5.00	−0.102				

16-B. $Fe(CN)_6^{3-} + e^- \rightleftharpoons Fe(CN)_6^{4-} \quad E^\circ = 0.356$ V

$Tl^{3+} + 2e^- \rightleftharpoons Tl^+ \quad E^\circ = 0.77$ V (in 1 M HCl)

The end point will be between 0.356 and 0.77 V. Methylene blue with $E^\circ = 0.53$ V is closest to the midpoint of the steep part of the titration curve. The color change would be from blue to colorless.

16-C. **(a)** 50.00 mL contains exactly 1/10 of the KIO_3 = 0.102 2 g = 0.477 5_7 mmol KIO_3. Each mol of iodate makes 3 mol of triiodide, so $I_3^- = 3(0.477\,5_7) = 1.432_7$ mmol.

(b) Two moles of thiosulfate react with one mole of I_3^-. Therefore there must have been $2(1.432_7) = 2.865_4$ mmol of thiosulfate in 37.66 mL, so the concentration is $(2.865_4$ mmol$)/(37.66$ mL$) = 0.076\,08_7$ M.

(c) 50.00 mL of KIO_3 make 1.432_7 mmol I_3^-. The unreacted I_3^- requires 14.22 mL of sodium thiosulfate = (14.22 mL)(0.076 08_7 M) = 1.082_0 mmol, which reacts with $\frac{1}{2}(1.082_0$ mmol$) = 0.541_0$ mmol I_3^-. The ascorbic acid must have consumed the difference = $1.432_7 - 0.541_0 = 0.891_7$ mmol I_3^-. Each mole of ascorbic acid consumes one mole of I_3^-, so mol ascorbic acid = 0.891_7 mmol, which has a mass of $(0.891_7 \times 10^{-3}$ mol)(176.13 g/mol) = 0.157_1 g. Ascorbic acid in the unknown = $100 \times (0.157_1$ g$)/(1.223$ g$)$ = 12.8 wt%.

Chapter 17

17-A. **(a)** Anode: $Fe + 8OH^- \rightarrow FeO_4^{2-} + 4H_2O + 6e^-$

(b) $4FeO_4^{2-} + 3S^{2-} + 11.5H_2O \rightarrow 4Fe(OH)_3(s) + 1.5S_2O_3^{2-} + 11OH^-$

(c) The anode reaction produces $6e^-$ for each FeO_4^{2-}. Reaction with sulfide requires $4FeO_4^{2-}$ to consume $3S^{2-}$.

$$\left(\frac{6e^-}{FeO_4^{2-}}\right)\left(\frac{4FeO_4^{2-}}{3S^{2-}}\right) = \frac{8e^-}{S^{2-}}$$

The current provides

$$\frac{(16.0 \text{ C/s})(3\,600 \text{ s})}{96\,485 \text{ C/mol}} = 0.597 \text{ mmol } e^-$$

which will react with

$$\frac{0.597 \text{ mol } e^-}{8 \text{ mol } e^-/\text{mol } S^{2-}} = 0.074\,6 \text{ mol } S^{2-}$$

(d) $$\frac{0.074\,6 \text{ mol } S^{2-}}{0.010\,0 \text{ mol } S^{2-}/\text{L}} = 7.46 \text{ L}$$

17-B. **(a)** The glucose monitor has a test strip with two carbon indicator electrodes and a silver-silver chloride reference electrode. Indicator electrode 1 is coated with glucose oxidase and a mediator. When a drop of blood is placed on the test strip, glucose from the blood is oxidized near indicator electrode 1 by mediator to gluconolactone and the mediator is reduced. The enzyme glucose oxidase catalyzes the oxidation. Reduced mediator is re-oxidized at the indicator electrode whose potential is kept at +0.2 V with respect to the Ag | AgCl reference electrode. Electric current between indicator electrode 1 and the reference electrode is proportional to the rate of oxidation of mediator, which is proportional to the concentration of glucose plus any interfering species in the blood. Indicator electrode 2 has mediator but no glucose oxidase. Current measured between indicator electrode 2 and the reference electrode is proportional to the concentration of interfering species in the blood. The difference between the two currents is proportional to the concentration of glucose in the blood.

(b) In the absence of a mediator, the rate of oxidation of glucose depends on the concentration of O_2 in the blood. If $[O_2]$ is low, the current will be low and the monitor will give an incorrect, low reading for the glucose concentration. A mediator such as 1,1′-dimethylferrocene replaces O_2 in the glucose oxidation and is subsequently reduced at the indicator electrode. The concentration of mediator is constant and high enough so that variations in electrode current are due mainly to variations in glucose concentration. Also, lowering the required electrode potential for oxidation of the mediator reduces the possible interference by other species in the blood.

17-C. See solution to 5-C.

17-D. (a) Faradaic current arises from redox reactions at the electrode. It is what we are trying to measure. Charging current comes from the flow of ions toward or away from the electrode as electrons flow into or out of the electrode when a potential step is applied. Charging current is unrelated to any electrochemical reactions.

(b) Charging current decays more rapidly than faradaic current after a potential step. By waiting 1 s after a step before measuring current, we find that charging current has largely decayed and there is still significant faradaic current.

(c) Square wave polarography is much faster than other forms of polarography. It gives increased signal because current is measured from both reduction and oxidation of analyte in each cycle. The derivative peak shape makes it easier to resolve closely spaced signals.

(d) In anodic stripping voltammetry, analyte is reduced and concentrated at the working electrode at a controlled potential for a constant time. The potential is then ramped in a positive direction to reoxidize analyte, during which current is measured. The height of the oxidation wave is proportional to the concentration of analyte. Stripping is the most sensitive polarographic technique because analyte is concentrated from a dilute solution. The longer the period of concentration, the more sensitive is the analysis.

Chapter 18

18-A. (a) $\nu = c/\lambda = (2.998 \times 10^8 \text{ m/s})/(100 \times 10^{-9} \text{ m})$
$= 2.998 \times 10^{15} \text{ s}^{-1} = 2.998 \times 10^{15} \text{ Hz}$

$\tilde{\nu} = 1/\lambda = 1/(100 \times 10^{-9} \text{ m}) = 10^7 \text{ m}^{-1}$

$$(10^7 \text{ m}^{-1})\left(\frac{1 \text{ m}}{100 \text{ cm}}\right) = 10^5 \text{ cm}^{-1}$$

$E = h\nu = (6.626\,2 \times 10^{-34} \text{ J}\cdot\text{s})(2.998 \times 10^{15} \text{ s}^{-1})$
$= 1.986 \times 10^{-18} \text{ J}$

$(1.986 \times 10^{-18} \text{ J/photon})(6.022 \times 10^{23} \text{ photons/mol})$
$= 1\,196 \text{ kJ/mol}$

(b), (c), (d) These are done in an analogous manner:

λ	ν (Hz)	$\tilde{\nu}$ (cm^{-1})	E (kJ/mol)	Spectral region	Molecular process
100 nm	2.998×10^{15}	10^5	1.196×10^3	ultraviolet	electronic excitation
500 nm	5.996×10^{14}	2×10^4	239.3	visible	electronic excitation
10 μm	2.998×10^{13}	1 000	11.96	infrared	molecular vibration
1 cm	2.998×10^{10}	1	0.011 96	microwave	molecular rotation

18-B. (a) $A = \varepsilon bc = (1.05 \times 10^3 \text{ M}^{-1} \text{ cm}^{-1})(1.00 \text{ cm})(2.33 \times 10^{-4} \text{ M}) = 0.245$

(b) $T = 10^{-A} = 0.569 = 56.9\%$

(c) Doubling b will double $A \Rightarrow A = 0.489 \Rightarrow T = 10^{-A} = 32.4\%$

(d) Doubling c also doubles $A \Rightarrow A = 0.489 \Rightarrow T = 10^{-A} = 32.4\%$

(e) $A = \varepsilon bc = (2.10 \times 10^3 \text{ M}^{-1} \text{ cm}^{-1})(1.00 \text{ cm})(2.33 \times 10^{-4} \text{ M}) = 0.489$

(f)

Curve	Absorption peak (nm)	Predicted color (Table 18-1)	Observed color
A	760	green	green
B	700	green	blue green
C	600	blue	blue
D	530	violet	violet
E	500	red or purple red	red
F	410	green-yellow	yellow

18-C. **(a)** Don't touch the cuvet with your fingers. Wash the cuvet as soon as you are finished with it and drain out the rinse water. Use matched cuvets. Place the cuvet in the instrument with the same orientation each time. Cover the cuvet to prevent evaporation and to keep dust out.

(b) If absorbance is too high, too little light reaches the detector for accurate measurement. If absorbance is too low, there is too little difference between sample and reference for accurate measurement.

18-D. **(a)** $\varepsilon = \frac{A}{cb} = \frac{0.267 - 0.019}{(3.15 \times 10^{-6}\ \text{M})(1.000\ \text{cm})}$
$= 7.87 \times 10^4\ \text{M}^{-1}\ \text{cm}^{-1}$

(b) $c = \frac{A}{\varepsilon b} = \frac{0.175 - 0.019}{(7.87 \times 10^4\ \text{M}^{-1}\ \text{cm}^{-1})(1.00\ \text{cm})}$
$= 1.98 \times 10^{-6}\ \text{M}$

Chapter 19

19-A. The source contains a visible lamp and an ultraviolet lamp, only one of which is used at a time. Grating 1 apparently selects some limited range of wavelengths to leave the source. In the monochromator, grating 2 is the principal element that selects the wavelength to reach the exit slit. The width of the exit slit determines how wide a range of wavelengths leaves the monochromator. The rotating chopper is a mirror that alternately directs the monochromatic light through the sample or reference cuvets in the sample compartment. The rotating chopper after the sample compartment directs light from each path to the detector, which is a photomultiplier tube.

19-B. **(a)** We use Equations 19-5 with $b = 0.100$ cm:

$D = b(\varepsilon'_X \varepsilon''_Y - \varepsilon'_Y \varepsilon''_X)$
$= (0.100)[(720)(274) - (212)(479)]$
$= 9.57_3 \times 10^3$ (the units are $\text{M}^{-2}\ \text{cm}^{-1}$)

$[X] = \frac{1}{D}(A'\varepsilon''_Y - A''\varepsilon'_Y)$
$= \frac{(0.233)(274) - (0.200)(212)}{9.57_3 \times 10^3} = 2.24\ \text{mM}$

$[Y] = \frac{1}{D}(A''\varepsilon'_X - A'\varepsilon''_X)$
$= \frac{(0.200)(720) - (0.233)(479)}{9.57_3 \times 10^3} = 3.38\ \text{mM}$

(b) There would still be an isosbestic point at 465 nm because $\varepsilon^{465}_{\text{HIn}} = \varepsilon^{465}_{\text{In}^-}$.

19-C. **(a)** $(163 \times 10^{-6}\ \text{L})(1.43 \times 10^{-3}\ \text{M Fe(III)})$
$= 2.33 \times 10^{-7}$ mol Fe(III)

(b) 1.17×10^{-7} mol apotransferrin/2.00×10^{-3} L
$\Rightarrow 5.83 \times 10^{-5}$ M apotransferrin

(c) Prior to the equivalence point, all added Fe(III) binds to the protein to form a red complex whose absorbance is shown in the figure. After the equivalence point, no more protein binding sites are available. The slight increase in absorbance arises from the color of the iron reagent in the titrant.

19-D. **(a)** In electronic transitions, energy is absorbed or liberated when the distribution of electrons in a molecule is changed. In vibrational transitions, the amplitude of vibrations increases or decreases. A rotational transition causes the molecule to rotate faster or slower.

(b) A molecule in an excited state can collide with other molecules and transfer kinetic energy without emitting a photon. The molecule that absorbs the energy in the collision ends up moving faster or vibrating with greater amplitude or rotating faster.

(c) Fluorescence is a transition from an excited singlet electronic state to the ground singlet state. Phosphorescence is a transition from an excited triplet state to the ground singlet state. Fluorescence is at higher energy than phosphorescence and fluorescence occurs faster than phosphorescence.

(d) Fluorescence involves a set of transitions opposite those for absorption. Instead of going from the ground vibrational state of S_0 to various states of S_1 in absorption, fluorescence takes a molecule from the ground vibrational state of S_1 to various states of S_0. Absorption has a series of peaks from λ_0 to higher energy. Fluorescence has a series of peaks from λ_0 to lower energy.

(e) Photochemistry is the breaking of chemical bonds initiated by absorption of a photon. Chemiluminescence is the emission of light as a result of a chemical reaction.

19-E. **(a)** At low concentration, fluorescence is proportional to analyte concentration. As concentration increases, neighboring unexcited molecules absorb some of the fluorescence before it can leave the cuvet. Some excited neighbors return to the ground state by emission of heat, rather than light. Therefore fluorescence becomes less efficient as concentration increases. Eventually, a point is reached at which self-absorption is so likely that further increase in analyte concentration decreases the fluorescence.

(b) Each molecule of analyte bound to antibody 1 also binds one molecule of antibody 2 that is linked to one enzyme molecule. Each enzyme molecule catalyzes many cycles of reaction in which a colored or fluorescent product is created. Therefore, each analyte molecule results in many product molecules.

Chapter 20

20-A. In atomic absorption spectroscopy, light of a specific frequency is passed through a flame containing free atoms. Absorbance of light is measured and is proportional to the concentration of atoms. In atomic emission spectroscopy, no lamp is used. The intensity of light emitted by excited atoms in the flame is measured and is proportional to the concentration of atoms.

20-B. **(a)** Atoms in a furnace are confined in a small volume for a relatively long time. The detection limit is small because the concentration of atoms in the gas phase is relatively high. A large sample volume is not required because gaseous atoms do not rapidly escape from the furnace. In a flame or plasma, the volume of gas phase is relatively large, so the concentration of atoms is relatively low. A great deal of sample is required because the atoms are rapidly moving through and escaping from the flame or plasma.

(b) For a body mass of 55 kg, the tolerable weekly intake of tin is (14 mg Sn/kg body mass)(55 kg) = 770 mg. One kilogram of tomato juice contains 241 mg Sn, so the number of kilograms of tomato juice containing 770 mg is (770 mg)/(241 mg/kg) = 3.2 kg.

20-C. **(a)** First find the frequency of light:

$$\nu = c/\lambda = (2.998 \times 10^8 \text{ m/s})/(400 \times 10^{-9} \text{ m}) = 7.495 \times 10^{14} \text{ s}^{-1}$$

$$\text{energy} = h\nu = (6.626 \times 10^{-34} \text{ J} \cdot \text{s}) \times (7.495 \times 10^{14} \text{ s}^{-1}) = 4.966 \times 10^{-19} \text{ J}$$

(b)

$$\frac{N^*}{N_0} = \left(\frac{g^*}{g_0}\right) e^{-\Delta E/kT} = \left(\frac{1}{1}\right) e^{-(4.966 \times 10^{-19} \text{ J})/[(1.381 \times 10^{-23} \text{ J/K})(2\,500 \text{ K})]} = 5.7 \times 10^{-7}$$

20-D. *Absorption* is measured in the presence and absence of a strong magnetic field. The Zeeman effect splits the analyte absorption peak when a magnetic field is applied and very little light is absorbed at the hollow-lamp wavelength. The difference between absorption with and without the magnetic field is due to analyte. *Emission* is measured at the peak wavelength and at the surrounding baseline at slightly lower and higher wavelengths. Emission due to analyte is the peak emission minus the average baseline intensity of the two off-wavelength measurements.

20-E. **(a)** *Spectral interference* arises from overlap of analyte absorption or emission lines with absorptions or emissions from other elements or molecules in the sample or the flame.

(b) *Chemical interference* is caused by any substance that decreases the extent of atomization of analyte.

(c) *Ionization interference* is a reduction of the concentration of free atoms by ionization of the atoms.

(d) La^{3+} acts as releasing agent by binding tightly to PO_4^{3-} and freeing Pb^{2+}.

20-F. The six peaks that we see for Hg correspond to the six natural isotopes at their relative abundances. For example, ^{202}Hg is most abundant and ^{200}Hg is the second most abundant. We see only three of the four isotopes of Pb because ^{204}Pb has only 1.4% abundance and it is hidden beneath ^{204}Hg. The coincidence of ^{204}Hg and ^{204}Pb is an example of isobaric interference. There are gaps at masses of 203 and 205 because neither Hg nor Pb has isotopes at these masses.

Chapter 21

21-A. 1-C, 2-D, 3-A, 4-E, 5-B

21-B. Your answer to **(a)** will be different from mine, which was measured on a larger figure than the one in your textbook. However, your answers to **(b)** through **(e)** should be the same as mine.

(a) t_r: octane, 200 units; nonane, 260 units
w: octane, 40 units; nonane, 52 units

(b) octane: $N = \dfrac{5.55\,(200)^2}{(40)^2} = 139$

nonane: $N = \dfrac{5.55\,(260)^2}{(52)^2} = 139$

(c) octane: $H = L/N = (1.00 \text{ m}/139) = 7.2 \text{ mm}$
nonane: $H = (1.00 \text{ m}/139) = 7.2 \text{ mm}$

(d) $\text{resolution} = \dfrac{\Delta t_r}{w_{av}} = \dfrac{260 - 200}{\frac{1}{2}(40 + 52)} = 1.30$

(e)

$$\frac{\text{large load}}{\text{small load}} = \left(\frac{\text{large column radius}}{\text{small column radius}}\right)^2$$

$$\frac{27.0 \text{ mg}}{3.0 \text{ mg}} = \left(\frac{\text{large column radius}}{2.0 \text{ mm}}\right)^2 \Rightarrow$$

large column radius = 6.0 mm
large column diameter = 12.0 mm

Flow rate should be proportional to the cross-sectional area of the column, which is proportional to the square of the radius.

$$\frac{\text{large column flow rate}}{\text{small column flow rate}} = \left(\frac{\text{large column radius}}{\text{small column radius}}\right)^2 = \left(\frac{6.0 \text{ mm}}{2.0 \text{ mm}}\right)^2 = 9.0$$

If the small column flow rate is 7.0 mL/min, the large column flow rate should be (9.0)(7.0 mL/min) = 63 mL/min.

21-C. **(a)** Molecules in the gas phase move faster than molecules in liquid. Therefore, longitudinal diffusion in the gas phase is much faster than longitudinal diffusion in the liquid phase, so band broadening by this mechanism is more rapid in gas chromatography than in liquid chromatography.

(b) **(i)** Optimum flow rate gives minimum plate height (23 cm/s for He). **(ii)** When flow is too fast, there is not adequate time for solute to equilibrate between the phases as the mobile phase streams past the stationary phase. This also broadens the band. **(iii)** When flow is too slow, plate height increases (that is, band broadening gets worse) because solute spends a long time on the column and is broadened by longitudinal diffusion.

(c) There is no broadening by multiple flow paths in an open tubular column.

(d) The longer the column, the better the resolution. We can use a longer open tubular column than a packed column because the particles in the packed column resist flow and require high pressures for high flow rate.

21-D. (a) In a total ion chromatogram, the mass spectrometer is set to respond to a wide range of *m/z* values. Any compound eluted from the column will give a mass spectral signal, so all compounds are observed in the chromatogram. In the selected ion chromatogram, the spectrometer is set to respond to just one value (or perhaps a few values) of *m/z*. Only those compounds that produce ions with the set value of *m/z* are observed. The selected ion chromatogram has a higher signal-to-noise ratio than that of the total ion chromatogram because more time is spent collecting data at the chosen value of *m/z*..

(b) The mass spectrometer is set to respond only to *m/z* 413 in trace h. Prior to the elution of noscapine, none of the components of the mixture produce ions with *m/z* 413. The chromatogram is flat until noscapine is eluted. Even though six other compounds are eluted, and acetylcodeine partly overlaps noscapine, only noscapine is observed.

21-E. The even mass of the molecular ion at *m/z* 194 tells us that there are an even number of N atoms. The intensity ratio (M+1)/M = 8.8% suggests that the number of C atoms is 8.8%/1.1% = 8. Formulas with 8 C atoms, an even number of N atoms, and a nominal mass of 194 are $C_8H_2O_6$, $C_8H_{18}O_5$, $C_8H_6O_4N_2$, $C_8H_{22}O_3N_2$, $C_8H_{10}O_2N_4$, $C_8H_{26}ON_4$ and $C_8H_{14}N_6$. However, there are no possible structures you could draw for $C_8H_{22}O_3N_2$ or $C_8H_{26}ON_4$ if C makes 4 bonds, H makes 1 bond, O makes 2 bonds, and N makes 3 bonds. There are too many H atoms in these two formulas.

Chapter 22

22-A. (a) Low-boiling solutes are separated well at low temperature, and the retention of high-boiling solutes is reduced to a reasonable time at high temperature.

(b) An open tubular column gives higher resolution than a packed column. The narrower the column, the higher the resolution that can be attained. A packed column can handle much more sample than an open tubular column, which is critical for preparative separations in which we are trying to isolate some quantity of the separated components.

(c) Diffusion of solute in H_2 and He is more rapid than in N_2. Therefore equilibration of solute between mobile phase and stationary phase is faster. The column can be run faster without excessive broadening from the finite rate of mass transfer between the mobile and stationary phases.

(d) Split injection is the ordinary mode for open tubular columns. Splitless injection is useful for trace and quantitative analysis. On-column injection is useful for thermally sensitive solutes that might decompose during a high-temperature injection.

(e) **(i)** carbon atoms bearing hydrogen atoms; **(ii)** all analytes; **(iii)** molecules with halogens, conjugated C=O, CN, NO_2; **(iv)** P and S; **(v)** P and N; **(vi)** S; **(vii)** all analytes

(f) A *reconstructed total ion chromatogram* is created by summing all ion intensities (above a selected value of *m/z*) in each mass spectrum at each time interval in a chromatography experiment. The technique responds to essentially everything eluted from the column and has no selectivity at all. In *selected ion monitoring*, intensities at just one or a few values of *m/z* are monitored. Only species with ions at those *m/z* values are detected, so the selectivity is much greater than that of the reconstructed total ion chromatogram. Signal is increased because ions are collected at each *m/z* for a longer time than would be allowed if the entire spectrum were being scanned. Noise is decreased because other eluates are less likely to contribute signal intensity at the selected value of *m/z*.

Selected reaction monitoring is most selective. One ion from the first mass separator is passed through a collision cell where it breaks into product ions that are separated by a second mass separator. The intensities of one or a few of these product ions are plotted as a function of elution time. The selectivity is high because few species from the column produce the first selected ion and even fewer break into the same fragments in the collision cell. This technique is so selective that it can transform a poor chromatographic separation into a highly specific determination of one component with virtually no interference.

22-B. Solvent is competing with solute for adsorption sites. The strength of the solvent-adsorbent interaction is independent of solute.

22-C. (a) In normal-phase chromatography, the stationary phase is more polar than the mobile phase. In reversed-phase chromatography, the stationary phase is less polar than the mobile phase.

(b) In isocratic elution, the composition of eluent is constant. In gradient elution, the composition of the eluent is changed—usually continuously—from low eluent strength to high eluent strength.

(c) In normal-phase chromatography, polar solvent must compete with analyte for polar sites on the stationary phase. The more polar the solvent, the better it binds to the stationary phase and the greater is its ability to displace analyte from the stationary phase.

(d) In reversed-phase chromatography, nonpolar analyte adheres to the nonpolar stationary phase. A polar solvent does not compete with analyte for nonpolar sites on the stationary phase. Making the solvent less polar gives it greater ability to displace analyte from the stationary phase.

(e) A guard column is a small, disposable column containing the same stationary phase as the main column. Sample and solvent pass through the guard column first. Any irreversibly bound impurities stick to the guard column, which is eventually thrown away. This guard column prevents junk from being irreversibly bound to the expensive main column and eventually ruining it.

(f) Electrospray introduces solution-phase ions into the gas phase. Atmospheric pressure chemical ionization produces new ions.

22-D. (a) Analytes are released into the chromatography column over a long period of time (possibly many minutes) from the heated fiber or the heated absorption tube. If analytes were not cold trapped on the column prior to chromatography, they would be eluted in extremely broad bands instead of sharp peaks.

(b) Solid-phase extraction uses a short column containing a chromatographic stationary phase. The column carries out gross separations of one type of analyte from other types of analyte (for example, separating nonpolar from polar analytes). Large particle size allows sample to drain through the solid-phase extraction column without applying high pressure. In chromatography, small particle size increases the efficiency of separation, but high pressure is necessary to force solvent through the column.

(c) In the urine steroid analysis, the solid-phase extraction column retains relatively nonpolar solutes from urine and lets everything else, such as polar molecules and salts, pass right through. When the solid-phase extraction column is washed with a nonpolar solvent, the steroids are eluted and can be concentrated into a small volume for application to the HPLC column. The polar molecules and salts from the urine were removed in the sample cleanup and not applied to the HPLC column.

(d) The nominal mass of chloramphenicol, $C_{11}H_{12}N_2O_5Cl_2$, is $11 \times 12 + 12 \times 1 + 2 \times 14 + 5 \times 16 + 2 \times 35 = 322$. The base peak at *m/e* 321 must be $[M - H]^-$, which is $[C_{11}H_{11}N_2O_5{}^{35}Cl_2]^-$. The 60% abundant ion at *m/e* 323 is $[C_{11}H_{11}N_2O_5{}^{35}Cl^{37}Cl]^-$.

Chapter 23

23-A. (a) Deionized water has been passed through ion-exchange columns to convert cations to H^+ and anions to OH^-, making H_2O. Nonionic impurities (such as neutral organic compounds) are not removed by this process.

(b) Gradient elution with increasing concentration of H^+ is required to displace more and more strongly bound cations from the column.

(c) Cations from a large volume of water are collected on a small ion-exchange column and then eluted in a small volume of concentrated acid. If the acid were not very pure, impurities in the acid could be greater than the concentration of trace species collected from the large volume of water.

23-B. (a) The separator is a column that separates ions by ion exchange. The suppressor exchanges the counterion to reduce the conductivity of eluent and allow analytes to be detected by their electrical conductivity.

(b) X^- (analyte anion), Na^+, and Na_2CO_3 from the separator column enter the suppressor. The anode generates $2H^+ + \frac{1}{2}O_2$ for every $2e^-$ flowing between anode and cathode. The cathode generates $H_2 + 2OH^-$ for every $2e^-$. $2H^+$ from the anode diffuse across the cation-exchange membrane and react with Na_2CO_3 to generate $H_2CO_3 + 2Na^+$. $2Na^+$ from the suppressor diffuse across the cation-exchange membrane to the cathode to compensate for the charge of $2OH^-$ generated at the cathode. Cation-exchange membranes permit cations to diffuse between the electrodes and the suppressor, but retain analyte anion, X^-. Anion-exchange membranes would not work for anion chromatography.

23-C. (a) total volume $= \pi(0.80 \text{ cm})^2(20.0 \text{ cm}) = 40.2$ mL

(b) If the void volume (= volume of mobile phase excluded from gel) is 18.2 mL, the stationary phase plus its included solvent must occupy $40.2 - 18.2 = 22.0$ mL.

(c) If pores occupy 60.0% of the stationary-phase volume, the pore volume is $(0.600)(22.0 \text{ mL}) = 13.2$ mL. Large molecules excluded from the pores are eluted in the void volume of $x = 18.2$ mL. The smallest molecules (which can enter all of the pores) are eluted in a volume of $y = 18.2 + 13.2 = 31.4$ mL. (In fact, there is usually some adsorption of solutes on the stationary phase, so retention volumes in real columns can be greater than 31.4 mL.)

23-D. (a) 25 μm diameter:

$$\text{volume} = \pi r^2 \times \text{length} = \pi(12.5 \times 10^{-6} \text{ m})^2(5 \times 10^{-3} \text{ m}) = 2.5 \times 10^{-12} \text{ m}^3$$

$$(2.5 \times 10^{-12} \text{ m}^3)\left(\frac{1 \text{ L}}{10^{-3} \text{ m}^3}\right) = 2.5 \times 10^{-9} \text{ L} = 2.5 \text{ nL}$$

50 μm diameter:
volume = $\pi(25 \times 10^{-6}\ \text{m})^2(5 \times 10^{-3}\ \text{m}) = 9.8\ \text{nL}$

(b) Capillary electrophoresis eliminates peak broadening from (1) the finite rate of mass transfer between the mobile and the stationary phases and (2) multiple flow paths around stationary-phase particles.

23-E. (a)

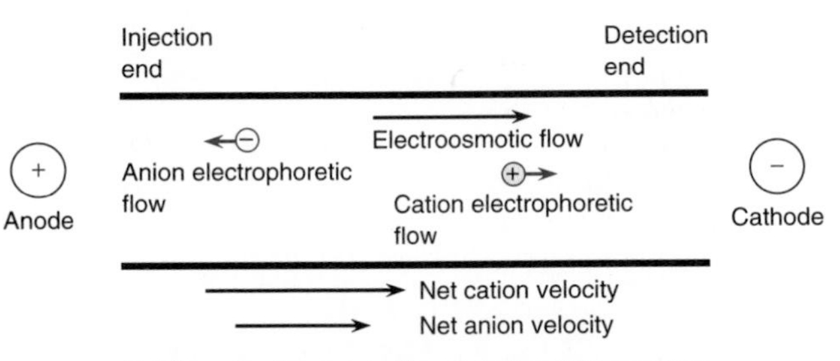

(b) At pH 3, net flow of cations is to the *right* and the net flow of anions is to the *left*.

(c) With no analyte present, the constant concentration of chromate in background buffer gives a steady ultraviolet absorbance at 254 nm. When Cl^- emerges, it displaces some chromate anion (to maintain electroneutrality). Because Cl^- does not absorb at 254 nm, the absorbance *decreases*.

23-F. (a) In the absence of micelles, all neutral molecules move with the electroosmotic speed of the bulk solvent and arrive at the detector at time t_0. Negative micelles migrate upstream and arrive at time $t_{mc} > t_0$. Neutral molecules partition between bulk solvent and micelles, so they arrive between t_0 and t_{mc}. The more time a neutral molecule spends in micelles, the closer is its migration time to t_{mc}.

(b) (i) pK_a for $C_6H_5NH_3^+$ is 4.60. At pH 10, the predominant form is neutral. **(ii)** Anthracene arrives at the detector last, so it spends more time inside micelles and therefore it is more soluble in the micelles.

Answers to Problems

Complete solutions to all problems can be found in the *Student Solutions Manual for Exploring Chemical Analysis*, Fourth Edition. Numerical answers and other short answers are given here.

Chapter 1

1-2. **(a)** milliwatt = 10^{-3} watt
(b) picometer = 10^{-12} meter
(c) kiloohm = 10^{3} ohm
(d) microcoulomb = 10^{-6} coulomb
(e) terajoule = 10^{12} joule
(f) nanosecond = 10^{-9} second
(g) femtogram = 10^{-15} gram
(h) decipascal = 10^{-1} pascal

1-3. **(a)** 100 fJ or 0.1 pJ **(b)** 43.172 8 nC **(c)** 299.79 THz **(d)** 0.1 nm or 100 pm **(e)** 21 TW **(f)** 0.483 amol or 483 zmol

1-4. **(a)** 7.457×10^4 W **(b)** 7.457×10^4 J/s **(c)** 1.782×10^4 cal/s **(d)** 6.416×10^7 cal/h

1-5. **(a)** 0.025 4 m, 39.37 inches
(b) 0.214 mile/s, 770 mile/h
(c) 1.04×10^3 m, 1.04 km, 0.643 mile

1-7. 1.10 M

1-8. 0.054 8 ppm, 54.8 ppb

1-9. 4.4×10^{-3} M, 6.7×10^{-3} M

1-10. **(a)** 70.5 g **(b)** 29.5 g **(c)** 0.702 mol

1-11. 6.18 g

1-12. **(a)** 1.7×10^3 L **(b)** 2.4×10^5 g

1-13. 8.0 g

1-14. **(a)** 55.6 mL **(b)** 1.80 g/mL

1-15. 5.48 g

1-16. 10^{-3} g/L, 10^3 μg/L, 1 μg/mL, 1 mg/L

1-17. 7×10^{-10} M

1-18. **(a)** 804 g solution, 764 g ethanol **(b)** 16.6 M

1-19. **(a)** 0.228 g Ni **(b)** 1.06 g/mL

1-20. 1.235 M

1-21. Dilute 8.26 mL of 12.1 M HCl to 100.0 mL.

1-22. **(a)** 3.40 M **(b)** 14.7 mL

1-23. Shredded Wheat: 3.6 Cal/g, 102 Cal/oz; doughnut: 3.9, 111; hamburger: 2.8, 79; apple: 0.48, 14

1-25. **(a)** $K = 1/[Ag^+]^3\,[PO_4^{3-}]$ **(b)** $K = P_{CO_2}^{6}/P_{O_2}^{15/2}$

1-26. **(a)** $P_A = 0.028$ bar, $P_E = 48._0$ bar **(b)** 1.2×10^{10}

1-27. unchanged

1-28. 4.5×10^3

1-29. **(a)** 3.6×10^{-7} **(b)** 3.6×10^{-7} M **(c)** 3.0×10^4

1-31. If mean concentration is assumed to be 2.3 mg nitrate nitrogen/L, flow ≈ 5000 tons/yr.

Chapter 2

2-6. 5.403 1 g

2-7. 14.85 g

2-8. 0.296 1 g

2-9. 9.980 mL

2-10. 5.022 mL

2-11. 15.631 mL

2-12. 0.70%

Chapter 3

3-1. **(a)** 1.237 **(b)** 1.238 **(c)** 0.135 **(d)** 2.1 **(e)** 2.00

3-2. **(a)** 0.217 **(b)** 0.216 **(c)** 0.217 **(d)** 0.216

3-3. **(a)** 4 **(b)** 4 **(c)** 4

3-4. **(a)** 12.3 **(b)** 75.5 **(c)** 5.520×10^3 **(d)** 3.04 **(e)** 3.04×10^{-10} **(f)** 11.9 **(g)** 4.600 **(h)** 4.9×10^{-7}

3-5. **(a)** 12.01 **(b)** 10.9 **(c)** 14 **(d)** 14.3 **(e)** −17.66 **(f)** 5.97×10^{-3} **(g)** 2.79×10^{-5}

3-6. **(a)** 208.233 **(b)** 560.594

3-7. 389.977

3-9. **(b)** 25.031, systematic; ±0.009, random
(c) 1.98 and 2.03, systematic; ±0.01 and ±0.02, random
(d) random **(e)** random
(f) Mass is systematically low because empty funnel was not dried; also, there is always random error, but we do not know how much in one experiment.

3-10. **(a)** 3.124 (±0.005) or 3.123_6 ($\pm 0.005_2$)
(b) 3.124 (±0.2%) or 3.123_6 ($\pm 0.1_7$%)

3-11. **(a)** 2.1 ± 0.2 (or 2.1 ± 11%)
(b) 0.151 ± 0.009 (or 0.151 ± 6%)
(c) $0.22_3 \pm 0.02_4$ (±11%)
(d) $0.097_1 \pm 0.002_2$ ($\pm 2._3$%)

3-12. **(a)** 21.0_9 ($\pm 0.1_6$), or 21.1 (±0.2); relative uncertainty = ±0.8%
(b) 27.4_3 ($\pm 0.8_6$); relative uncertainty = $\pm 3._1$%
(c) $(14._9 \pm 1._3) \times 10^4$, or $(15 \pm 1) \times 10^4$; relative uncertainty = ±9%

3-13. **(a)** 10.18 (±0.07) (±0.7%) **(b)** 174 (±3) (±2%)
(c) 0.147 (±0.003) (±2%)
(d) 7.86 (±0.01) (±0.1%)
(e) 2 185.8 (±0.8) (±0.04%)

3-14. **(a)** 6.0 ± 0.2 (±4%) **(b)** $1.30_8 \pm 0.09_2$ ($\pm 7._0$%)
(c) 1.30_8 ($\pm 0.09_2$) $\times 10^{-11}$ ($\pm 7._0$%)
(d) $2.7_2 \pm 0.7_8$ (±29%)

3-15. 78.112 ± 0.005

3-16. 95.978 ± 0.009

3-17. **(a)** 58.443 ± 0.002 g/mol
(b) 0.450 7 (±0.000 5) M

3-18. **(a)** 0.020 77 ± 0.000 03 M **(b)** yes

3-19. 1.235 ± 0.002 M

3-20. **(a)** 16.6_6 mL **(b)** 0.169 (±0.002) M

3-21. formula in cell F3:
=B3*$A$4+C3*$A$6+D3*$A$8+E3*$A$10

3-23. Method 2 is more accurate. The relative uncertainty in mass in Method 1 is far greater than any other uncertainty in either procedure. Method 1 gives $0.002\ 72_6 \pm 0.000\ 01_8$ M $AgNO_3$. Method 2 gives $0.002\ 726 \pm 0.000\ 006_5$ M $AgNO_3$.

Chapter 4

4-2. 0.683, 0.955, 0.997

4-3. **(a)** 1.527 67 **(b)** 0.001 26, 0.082 5%
(c) 1.527 93 ± 0.000 10

4-5. 108.6_4, 7.1_4, $108.6_4 \pm 6.8_1$

4-6. $s_{pooled} = 0.000\ 4_{49}$; $t = 2._{32} > t_{table}\ (95\%) = 2.306$, so the difference *is* significant

4-7. 2.299 47 ± 0.001 15, 2.299 47 ± 0.001 71

4-8. $s_{pooled} = 8._{90}$; $t = 1.67 < t_{table}\ (95\%) = 2.306$, so the difference is *not* significant

4-9. For 1 and 2: $s_{pooled} = 0.001\ 864\ 8$; $t = 15.60 > t_{table}$ (95%, 44 degrees of freedom) ≈ 2.02; difference is significant
For 2 and 3: $s_{pooled} = 0.001\ 075\ 8$; $t = 1.39 < t_{table}$; difference is not significant

4-10. $s_{pooled} = 9.20$; $t_{calculated} = 2.75 > t_{table}$ (= 2.365 for 95% confidence and 4 + 5 − 2 = 7 degrees of freedom), so difference is significant

4-11. $s_{pooled} = 0.000\ 021_3$; $t_{calculated} = 2.48 > t_{table}$ (= 2.262 for 95% confidence and 5 + 6 − 2 = 9 degrees of freedom), so difference is significant

4-12. **(a)** no **(b)** yes **(c)** yes **(d)** Buret and pipet are within manufacturer's tolerance. Flask is below manufacturer's tolerance.

4-13. $G_{calculated} = 1.98 > 1.822$; discard 0.195

4-14. outlier = 0.169; $G_{calculated} = 2.26 > 2.176$; discard 0.169

4-15. $y = -2x + 15$

4-16. $-1.299\ (\pm 0.001) \times 10^4$, $3\ (\pm 3) \times 10^2$

4-17. **(a)** $2.0_0 \pm 0.3_8$ **(b)** $2.0_0 \pm 0.2_6$

4-18. **(a)** $y\ (\pm 0.005_7) = 0.021\ 7_7\ (\pm 0.000\ 1_9)x + 0.004_6\ (\pm 0.004_4)$
(c) protein = $23.0_7 \pm 0.2_9$ μg

4-20. It does not appear to be either $CuCO_3$ or $CuCO_3 \cdot xH_2O$. The 99% confidence intervals for the class and instructor data exceed 51.43 wt% Cu expected in $CuCO_3$. If the material were a hydrate, the Cu content would be even less.

4-21. $^{87}Sr/^{86}Sr$ ratios agree with each other within their 95% confidence limits at both sites for the LGM time. Therefore LGM dust could have come from the same source at both sites. The isotope ratios are not the same at the two sites at the EH time. Therefore EH dust came from different sources for each site.
pg/g = 10^{-12} g Sr per g ice = parts per trillion.

Chapter 5

5-8. c

5-9. 50% of red wells should be green and 8% of green wells should be red. It would be worse if there were a 50% false negative rate.

5-10. yes ($t_{calculated} = 4.01 > t_{table} = 2.262$ for 95% confidence and 9 degrees of freedom)

5-11. **(a)** 0.003_{12} **(b)** 8.6×10^{-8} M (c) 2.9×10^{-7} M

5-12. One observation (day 101) is outside the action line. None of the other criteria are violated.

5-13. yes: 7 consecutive measurements all above or all below the center line

5-14. **(a)** 22.2 ng/mL: precision = 23.8%, accuracy = 6.6%
88.2 ng/mL: precision = 13.9%, accuracy = −6.5%
314 ng/mL: precision = 7.8%, accuracy = −3.6%
(b) signal detection limit = $129._6$; detection limit = 4.8×10^{-8} M; quantitation limit = 1.6×10^{-7} M

5-15. 96%, 0.064 μg/L (=3*s*)

5-16. 0.644 mM

5-17. 1.21 mM

5-18. **(a)** *x*-intercept of standard addition graph = −8.72 ppb
(b) Sr = 116 ppm

5-19. **(a)** 8.72 ± 0.43 ppb **(b)** 116 (±6) ppm

5-20. 313 ppb

5-21. 11.9 μM

5-22. 7.49 μg/mL

5-23. 0.47 mmol

Chapter 6

6-2. 43.2 mL, 270.0 mL

6-3. 4.300×10^{-2} M

6-4. 0.149 M

6-5. 32.0 mL

6-6. **(a)** 0.045 00 M **(b)** 36.42 mg/mL

6-7. 947 mg

6-8. 1.72 mg

6-9. **(a)** 0.020 34 M **(b)** 0.125 7 g **(c)** 0.019 83 M

6-10. **(a)** 0.105 3 mol/kg solution
(b) 0.286_9 mol/kg solution

6-11. **(a)** 0.001 492 8 (±0.06%) mol
(b) 0.001 434 (±0.14%) mol
(c) $\frac{\text{volumetric uncertainty}}{\text{gravimetric uncertainty}} = 2.2$; largest uncertainty in gravimetric delivery is mass of $AgNO_3$; largest uncertainty in volumetric delivery is pipet volume

6-12. 89.07 wt%

6-13. 9.066 mM

6-14. 3.555 mM
6-15. 30.5 wt%
6-16. 0.020 6 (±0.000 7) M
6-17. **(a)** $7._1 \times 10^{-5}$ M **(b)** $1._0 \times 10^{-3}$ g/100 mL
6-18. **(a)** $6.6_9 \times 10^{-5}$ M **(b)** 14.4 ppm
6-19. 1 400 ppb, 76 ppb, 0.98 ppb
6-20. **(a)** $[Hg_2^{2+}] = 6.8_8 \times 10^{-7}$ M; $[IO_3^-] = 1.3_8 \times 10^{-6}$ M
(b) $[Hg_2^{2+}] = 1.3 \times 10^{-14}$ M
6-21. I^- before Br^- before Cl^- before CrO_4^{2-}
6-22. 0.106 0 M, 11.55 M
6-23. **(a)** 17 L **(b)** 793 L **(c)** 1.05×10^3 L
6-24. **(a)** x = 0.000 945 4
6-24. **(a)** 0.500 mmol tris, 0.545 mmol pyridine
(b) liberates heat

Chapter 7

7-2. 2.03
7-3. 0.085 38 g
7-4. 50.79 wt% Ni
7-5. 7.22 mL
7-6. 8.665 wt%
7-7. 0.339 g
7-8. **(a)** 5.5 mg/100 mL **(b)** 5.834 mg, yes
7-9. Ba, 47.35 wt%; K, 8.279 wt%; Cl, 31.95 wt%
7-10. **(a)** 19.98 wt%
7-11. 11.69 mg CO_2, 2.051 mg H_2O
7-12. **(a)** 51.36 wt% C, 3.639 wt% H **(b)** C_6H_5
7-13. $C_4H_9NO_2$
7-14. 104.1 ppm
7-15. **(a)** 0.027 36 M **(b)** systematic
7-16. 75.40 wt%
7-17. $C_8H_{9.06\pm0.17}N_{0.997\pm0.010}$
7-18. 12.4 wt%
7-19. **(a)** 98.3, 104.0, 98.6, 97.6, <0.3, 36.5, 6.4, <0.3, 4.2
(b) Fe^{3+}, Pb^{2+}, Cd^{2+}, and In^{3+} are gathered quantitatively. **(c)** 10
7-20. $s_{pooled} = 0.036_{36}$ and $t = 5.1_3 \Rightarrow$ difference is significant above 99% confidence level

Chapter 8

8-1. **(a)** HCN/CN^-, HCO_2H/HCO_2^-
(b) H_2O/OH^-, HPO_4^{2-}/PO_4^{3-}
(c) H_2O/OH^-, HSO_3^-/SO_3^{2-}
8-2. $[H^+] > [OH^-]$, $[OH^-] > [H^+]$
8-3. **(a)** 4 **(b)** 9 **(c)** 3.24 **(d)** 9.76
8-4. **(a)** 7.46 **(b)** 2.9×10^{-7} M
8-5. 2.5×10^{-5} M
8-6. 7.8
8-7. See Table 8-1.
8-8. weak acids: carboxylic acids, ammonium salts, aqueous metal ion with charge ≥2
weak bases: carboxylate anions and amines
8-9. **(a)** 0.010 M, 2.00 **(b)** $2.8_6 \times 10^{-13}$ M, pH 12.54
(c) 0.030 M, 1.52 **(d)** 3.0 M, −0.48
(e) 1.0×10^{-12} M, 12.00
8-10. **(b)** trichloroacetic acid
8-11. **(b)** sodium 2-mercaptoethanol
8-12. $2H_2SO_4 \rightleftharpoons H_3SO_4^+ + HSO_4^-$
8-13. (piperidine) NH + $H_2O \rightleftharpoons$ (piperidinium) $\overset{+}{N}H_2 + OH^-$
(benzoate) $C_6H_5{-}CO_2^- + H_2O \rightleftharpoons C_6H_5{-}CO_2H + OH^-$
8-14. $OCl^- + H_2O \rightleftharpoons HOCl + OH^-$, 3.3×10^{-7}
8-15. 9.78
8-16. 2.2×10^{-12} M
8-17. 3.02, 9.51×10^{-2}
8-18. 5.41, 2.59×10^{-5}
8-19. 3.14, 7.29×10^{-4} M, 0.084 3 M, 0.085 0 M
8-20. 5.00
8-21. 3.70
8-22. 7.30
8-23. 5.51, 3.1×10^{-6} M, 0.060 M
8-24. **(a)** 3.03, 0.094 **(b)** 7.00, 0.999
8-25. 5.50
8-26. $K_a = 2.71 \times 10^{-11}$, $K_b = 3.69 \times 10^{-4}$
8-28. 11.28, 0.058 M, 1.9×10^{-3} M
8-29. pH = 8.88, 0.007 56%; pH = 8.38, 0.023 9%; pH = 7.00, 0.568%
8-30. 10.95
8-31. 9.97, 0.003 6
8-32. 3.4×10^{-6}
8-33. 2.2×10^{-7}
8-35. 2.93, 0.118

Chapter 9

9-3. 4.13
9-4. **(b)** 1/1000 **(c)** $pK_a - 4$
9-5. pH 10–red, pH 8–orange, pH 6–yellow
9-6. **(a)** 1.5×10^{-7} **(b)** 0.15
9-7. **(a)** 14 **(b)** 1.4×10^{-7}
9-8. **(a)** 2.26×10^{-7} **(b)** 1.00 **(c)** 22.6
9-9. 3.59
9-10. **(a)** 8.37 **(b)** 0.423 g **(c)** 8.33 **(d)** 8.41
9-11. **(b)** 7.18 **(c)** 7.00 **(d)** 6.86 mL
9-12. **(a)** 2.56 **(b)** 2.86
9-13. 3.38 mL
9-14. 13.7 mL
9-15. 4.68 mL
9-16. **(a)** citric acid or acetic acid
(b) imidazole hydrochloride
(c) CAPS **(d)** CHES, boric acid, or ammonia

9-17. ii

9-18. **(a)** NaOH

9-19. **(b)** HCl

9-20. **(a)** 90.8 mL HCl

9-21. **(a)** red **(b)** orange **(c)** yellow **(d)** red

9-22. **(a)** $p = 0.940\,6$ mol, $q = 0.059\,35$ mol **(b)** Δ(pH) $= -0.072$

9-23. 9.13

Chapter 10

10-6. $V_e = 10.0$ mL; pH = 13.00, 12.95, 12.68, 11.96, 10.96, 7.00, 3.04, 1.75

10-7. $V_e = 12.5$ mL; pH = 1.30, 1.35, 1.60, 2.15, 3.57, 7.00, 10.43, 11.11

10-8. $V_e = 5.00$ mL; pH = 2.66, 3.40, 4.00, 4.60, 5.69, 8.33, 10.96, 11.95

10-9. $V_e = 5.00$ mL; pH = 6.32, 10.64, 11.23, 11.85

10-10. 2.80, 3.65, 4.60, 5.56, 8.65, 11.96

10-11. 8.18

10-12. 3.72

10-13. 0.091 8 M

10-14. **(a)** 0.025 92 M **(b)** 0.020 31 M **(c)** 9.69 **(d)** 10.07

10-15. $V_e = 10.0$ mL; pH = 11.00, 9.95, 9.00, 8.05, 7.00, 5.02, 3.04, 1.75

10-16. $V_e = 47.79$ mL; pH = 8.74, 5.35, 4.87, 4.40, 3.22, 2.58.

10-17. **(a)** 2.2×10^9 **(b)** 10.92, 9.57, 9.35, 8.15, 5.53, 2.74.

10-18. **(a)** 9.44 **(b)** 2.55 **(c)** 5.15

10-19. $V_e = 10.0$ mL; pH = 9.85, 7.95, 6.99, 6.04, 5.00, 4.22, 3.45, 2.17.

10-20. no

10-21. yellow, green, blue

10-23. **(a)** colorless $\longrightarrow$ pink **(b)** systematically requires too much NaOH

10-24. cresol red (orange $\longrightarrow$ red) or phenophthalein (colorless $\longrightarrow$ red)

10-25. 10.727 mL

10-26. 0.063 56 M

10-27. **(a)** 0.087 99 **(b)** 25.74 mg **(c)** 2.860 wt%

10-28. **(a)** 5.62 **(b)** methyl red

10-31. tribasic, 0.015 3 M

Chapter 11

11-2. $K_{a2} = 1.03 \times 10^{-2}$, $K_{b2} = 1.86 \times 10^{-13}$

11-3. 7.09×10^{-3}, 6.33×10^{-8}, 4.2×10^{-13}

11-4. Two pK values apply to the carboxylic acid and ammonium groups of all amino acids. Some amino acids have a substituent that is an acid or base, which is responsible for a third pK value.

11-5. piperazine: $K_{b1} = 5.38 \times 10^{-5}$, $K_{b2} = 2.15 \times 10^{-9}$
phthalate: $K_{b1} = 2.56 \times 10^{-9}$, $K_{b2} = 8.93 \times 10^{-12}$

11-7. 2.49×10^{-8}, 5.78×10^{-10}, 1.34×10^{-11}

11-8. 1.62×10^{-5}, 1.54×10^{-12}

11-9. **(a)** 1.95, 0.089 M, 1.12×10^{-2} M, 2.01×10^{-6} M
(b) 4.27, 3.8×10^{-3} M, 0.100 M, 3.8×10^{-3} M
(c) 9.35, 7.04×10^{-12} M, 2.23×10^{-5} M, 0.100 M

11-10. **(a)** 11.00, 0.099 0 M, 9.95×10^{-4} M, 1.00×10^{-9} M
(b) 7.00, 1.0×10^{-3} M, 0.100 M, 1.0×10^{-3} M
(c) 3.00, 1.00×10^{-9} M, 9.95×10^{-4} M, 0.099 0 M

11-11. 11.60, [B] = 0.296 M, $[BH^+] = 3.99 \times 10^{-3}$ M, $[BH_2^{2+}] = 2.15 \times 10^{-9}$ M

11-12. 7.53, $[BH_2^{2+}] = 9.4_8 \times 10^{-4}$ M, $[BH^+] \approx 0.150$ M, $[B] = 9.4_9 \times 10^{-4}$ M

11-13. 5.60

11-14. **(a)** pK_1: $[H_2A] = [HA^-]$; $\frac{1}{2}(pK_1 + pK_2)$: $[H_2A] = [A^{2-}]$; pK_2: $[HA^-] = [A^{2-}]$
(b) monoprotic: pK_a: $[HA] = [A^-]$
triprotic: pK_1: $[H_3A] = [H_2A^-]$; $\frac{1}{2}(pK_1 + pK_2)$: $[H_3A] = [HA^{2-}]$;
pK_2: $[H_2A^-] = [HA^{2-}]$; $\frac{1}{2}(pK_2 + pK_3)$: $[H_2A^-] = [A^{3-}]$; pK_3: $[HA^{2-}] = [A^{3-}]$

11-15. **(a)** HA **(b)** A^- **(c)** **(i)** 1.0, **(ii)** 0.10

11-16. **(a)** 4.00 **(b)** 8.00 **(c)** H_2A **(d)** HA^- **(e)** A^{2-}

11-17.

pH	Dominant form	Second form
2	H_3PO_4	$H_2PO_4^-$
3, 4	$H_2PO_4^-$	H_3PO_4
5, 6, 7	$H_2PO_4^-$	HPO_4^{2-}
8, 9	HPO_4^{2-}	$H_2PO_4^-$
10, 11, 12	HPO_4^{2-}	PO_4^{3-}
13	PO_4^{3-}	HPO_4^{2-}

11-18. **(a)** 9.00 **(b)** 9.00 **(c)** BH^+ **(d)** 1.0×10^3

11-19. **(a)** 6.85, 9.93 **(b)** 9.93 **(c)** 6.85
(d)

pH	Dominant form	Second form
4, 5, 6	BH_2^{2+}	BH^+
7, 8	BH^+	BH_2^{2+}
9	BH^+	B
10	B	BH^+

(e) 1.2×10^2 **(f)** 7.1×10^4

11-21. **(a)** 11.00 **(b)** 10.5

11-22.

O, $^-$O—P—O, O$^-$, O, O$^-$, N, $^+$H, CH_3

11-23. 5.41, $[H_3Arg^{2+}] = 1.3 \times 10^{-5}$ M, $[H_2Arg^+] = 0.050$ M, $[HArg] = 1.3 \times 10^{-5}$ M, $[Arg^-] = 3 \times 10^{-12}$ M

11-24. -2

11-25. $V_{e1} = 10.0$ mL, $V_{e2} = 20.0$ mL; pH = 2.51, 4.00, 6.00, 8.00, 10.46, 12.21

11-26. $V_{e1} = 10.0$ mL, $V_{e2} = 20.0$ mL; pH = 11.49, 10.00, 8.00, 6.00, 3.54, 1.79

11-27. **(a)** phenolphthalein (colorless ⟶ red)
(b) *p*-nitrophenol (colorless ⟶ yellow)
(c) bromothymol blue (yellow ⟶ green)
or bromocresol purple (yellow ⟶ yellow + purple)
(d) thymolphthalein (colorless ⟶ blue)

11-28. $V_{e1} = 40.0$ mL, $V_{e2} = 80.0$ mL; pH = 11.36, 9.73, 7.53, 5.33, 3.41, 1.85

11-29. $V_{e1} = 20.0$ mL, $V_{e2} = 40.0$ mL, $V_{e3} = 60.0$ mL;
pH = 1.86, 2.15, 4.67, 7.20, 9.79, 11.43
(pH cannot rise as fast as we calculated at 42 mL because even adding 0.04 M OH^- to water cannot raise the pH as fast as we estimated. HPO_4^{2-} is too weak an acid to react appreciably with dilute OH^-.)

11-30. $V_{e1} = 25.0$ mL, $V_{e2} = 50.0$ mL; pH = 7.62, 5.97, 3.8, 1.9

11-31. **(a)** 2.56, 3.46, 4.37, 8.42, 11.45 **(b)** second
(c) thymolphthalein; first trace of blue

11-32. mean molecular mass = 327.0
Mean formula is

$$HOCH_2CH_2—[OCH_2CH_2]_5—OCH_2CH_2OH.$$

11-33. **(a)** $[CO_3^{2-}] = K_{a2}\,K_{a1}\,K_H\,P_{CO_2}/[H^+]^2$
(b) 0°C: 6.6×10^{-5} mol kg^{-1};
30°C: 1.8×10^{-4} mol kg^{-1}
(c) 0°C: $[Ca^{2+}][CO_3^{2-}] = 6.6 \times 10^{-7}$ mol^2 kg^{-2} (aragonite dissolves, calcite does not);
30°C: $[Ca^{2+}][CO_3^{2-}] = 1.8 \times 10^{-6}$ mol^2 kg^{-1} (neither dissolves)

Chapter 12

12-4. **(a)** true **(b)** true **(c)** true

12-9. **(a)** 0.2 mM **(b)** 0.6 mM **(c)** 2.4 mM

12-10. **(a)** 0.660 **(b)** 0.54 **(c)** 0.18 **(d)** 0.83

12-11. 0.004 6

12-12. **(a)** 0.88_7 **(b)** 0.87_1

12-13. **(a)** 0.42_2 **(b)** 0.43_2

12-14. **(a)** 1.3×10^{-6} M **(b)** 2.9×10^{-11} M

12-15. $\gamma_{H^+} = 0.86$, pH = 2.07

12-16. $[H^+] = 1.2 \times 10^{-7}$ M; pH = 6.99

12-17. 11.94, 12.00

12-18. $[OH^-] = 1.1 \times 10^{-4}$ M; pH = 9.92

12-19. 6.6×10^{-7} M

12-20. 2.5×10^{-3} M

12-21. **(a)** 9.2 mM **(b)** Remainder is $CaSO_4(aq)$.

12-22. $[H^+] + 2[Ca^{2+}] + [Ca(HCO_3)^+] + [Ca(OH)^+] + [K^+] = [OH^-] + [HCO_3^-] + 2[CO_3^{2-}] + [ClO_4^-]$

12-23. $[H^+] = [OH^-] + [HSO_4^-] + 2[SO_4^{2-}]$

12-24. $[H^+] = [OH^-] + [H_2AsO_4^-] + 2[HAsO_4^{2-}] + 3[AsO_4^{3-}]$

12-25. **(a)** $2[Mg^{2+}] + [H^+] = [Br^-] + [OH^-]$
(b) $2[Mg^{2+}] + [H^+] + [MgBr^+] = [Br^-] + [OH^-]$

12-26. $[CH_3CO_2^-] + [CH_3CO_2H] = 0.1$ M

12-27. **(a)** $0.20\text{ M} = [Mg^{2+}]$ **(b)** $0.40\text{ M} = [Br^-]$
(c) $0.20\text{ M} = [Mg^{2+}] + [MgBr^+]$
(d) $0.40\text{ M} = [Br^-] + [MgBr^+]$

12-28. **(a)** $[F^-] + [HF] = 2[Ca^{2+}]$
(b) $[F^-] + [HF] + 2[HF_2^-] = 2[Ca^{2+}]$

12-29. **(a)** $2[Ca^{2+}] = 3\{[PO_4^{3-}] + [HPO_4^{2-}] + [H_2PO_4^-] + [H_3PO_4]\}$
(b) $3\{[Fe^{3+}] + [Fe(OH)^{2+}] + [Fe(OH)_2^+] + [FeSO_4^+]\} = 2\{[SO_4^{2-}] + [HSO_4^-] + [FeSO_4^+]\}$

12-30. $[Y^{2-}] = [X_2Y_2^{2+}] + 2[X_2Y^{4+}]$

12-31. $[OH^-] = 1.4_8 \times 10^{-7}$ M; $[H^+] = 6.8 \times 10^{-8}$ M; $[Mg^{2+}] = 4.0 \times 10^{-8}$ M

12-32. 5.8×10^{-4} M

12-33. **(a)** $[Ag^+] = 2.4 \times 10^{-8}$ M; $[CN^-] = 9.2 \times 10^{-9}$ M; $[HCN] = 1.5 \times 10^{-8}$ M
(b) $[Ag^+] = 2.9 \times 10^{-8}$ M; $[CN^-] = 1.3 \times 10^{-8}$ M; $[HCN] = 1.6 \times 10^{-8}$ M

12-34. **(a)** 5.0×10^{-9} mol **(b)** 4.0×10^{-6} mol
(c) 1.1×10^{-8} mol

12-35. **(a)** 0.98, 0.02 **(b)** 0.82, 0.18 **(c)** 0.60, 0.40

12-36. **(a)** 0.06, 0.94 **(b)** 0.39, 0.61 **(c)** 0.86, 0.14

12-37. **(a)** 0.09, 0.91 **(b)** 0.50, 0.50 **(c)** 0.67, 0.33

12-41. 0.63

Chapter 13

13-1. **(a)** 10.0 mL **(b)** 10.0 mL

13-8. 0.010 3 M

13-9. $[Ni^{2+}] = 0.012\,4$ M, $[Zn^{2+}] = 0.007\,18$ M

13-10. 0.024 30 M

13-11. 1.256 (±0.003) mM

13-12. 0.014 68 M

13-13. 0.092 6 M

13-14. 5.150 mg Mg, 20.89 mg Zn, 69.64 mg Mn

13-15. 32.7 wt% (theoretical = 32.90 wt%)

13-16. **(a)** 2.7×10^{-10} **(b)** 0.57

13-17. **(a)** $Co^{2+} + 4NH_3 \rightleftharpoons Co(NH_3)_4^{2+}$
(b) $Co(NH_3)_3^{2+} + NH_3 \rightleftharpoons Co(NH_3)_4^{2+}$, $\log K_4 = 0.64$

13-18. **(a)** 2.5×10^7 **(b)** 4.5×10^{-5} M

13-19. **(a)** 100.0 mL **(b)** 0.016 7 M **(c)** 0.041
(d) 4.1×10^{10} **(e)** 7.8×10^{-7} M **(f)** 2.4×10^{-10} M

13-20. **(a)** 1.70 **(b)** 2.18 **(c)** 2.81
(d) 3.87 **(e)** 4.87 **(f)** 5.66
(g) 6.46 **(h)** 8.15 **(i)** 8.45

13-21. **(a)** ∞ **(b)** 10.30 **(c)** 9.52
(d) 8.44 **(e)** 7.43 **(f)** 6.15
(g) 4.88 **(h)** 3.20 **(i)** 2.93

13-22. EDTA behaves as a weak base (A^-) and the metal ion behaves as H^+.

13-23. pCu^{2+} = 1.10, 1.57, 2.21, 3.27, 6.91, 10.54, 11.24

13-24. At pH 7, pCu^{2+} = 1.10, 1.57, 2.21, 3.27, 8.47, 13.66, 14.36

13-25. 5.6 g

13-26. Reaction with NH_3 and OH^- raises the titration curve by 0.48 log units before V_e and leaves it unchanged after V_e.

Chapter 14

14-1. **(b)** 6.242×10^{18} e$^-$/C
(c) 9.649×10^4 C/mol

14-2. **(a)** oxidant: TeO_3^{2-}; reductant: $S_2O_4^{2-}$
(b) 3.02×10^3 C **(c)** 0.840 A

14-3. **(a)** oxidant: C_2HCl_3; reductant: Fe **(b)** 4.5%
(c) 5.78 A

14-4. 0.015 9 J

14-5. **(a)** $71._5$ A **(b)** 6.8×10^6 J

14-6. oxidation: $2Hg(l) + 2Cl^- \rightleftharpoons Hg_2Cl_2(s) + 2e^-$
reduction: $Zn^{2+} + 2e^- \rightleftharpoons Zn(s)$

14-8. **(a)** 0.572 V **(b)** 0.568 V
Disagreement between **(a)** and **(b)** is probably within the uncertainty of the standard potentials and solubility products.

14-9. **(a)** $Pt(s) \mid Cr^{2+}(aq), Cr^{3+}(aq) \parallel Tl^+(aq) \mid Tl(s)$
(b) 0.08_4 V **(c)** Pt
(d) $Tl^+ + Cr^{2+} \rightleftharpoons Tl(s) + Cr^{3+}$

14-10. **(a)** right half-cell: $E = 0.309_4$ V; left half-cell: $E = -0.120_7$ V **(b)** anode is hydrogen electrode
(c) 0.430 V; reduction at right-hand electrode

14-11. **(a)** electrons flow right to left; $\frac{3}{2}Br_2(l) + Al(s) \rightleftharpoons 3Br^- + Al^{3+}$ **(b)** electron flow: Al $\rightarrow$ Br_2
(c) Br_2 **(d)** 1.31 kJ **(e)** 2.69×10^{-8} g/s

14-12. **(a)** -0.357 V; 9.2×10^{-7} **(b)** 0.722 V; 7×10^{48}

14-13. **(a)** 2.057 V

$$\textbf{(b)}\ K = \frac{1}{P_{O_2}^{1/2} P_{H_2} [H^+]^2 [OH^-]^2} = 3.5 \times 10^{69}$$

(c) 111 days, 7.94 kg O_2

14-14. 7×10^{-12}

14-15. 6.0

14-17. **(a)** 1.9×10^{-6} **(b)** -0.386 V; oxidized

14-18. For $Br_2(l) \rightleftharpoons Br_2(aq)$, we find $E° = -0.020$ V and $[Br_2(aq)] = 34$ g/L.

14-19. **(b)** 0.675 V; 10^{114} **(c)** 0.178 V **(d)** 8.5

14-20. **(b)** 0.044 V

14-21. **(a)** 0.747 V **(b)** 0.506 V

14-22. **(a)** 0.086 V **(b)** -0.021 V **(c)** 0.021 V

14-23. 0.684 V

14-24. The reaction L + Fe(III) $\rightleftharpoons$ LFe(III) has a greater formation constant than does the reaction L + Fe(II) $\rightleftharpoons$ LFe(II).

14-25. **(a)** $E_{cell} = 0.059\,16 \log(c_r/c_l)$. If $c_r = c_l$, $E_{cell} = 0$.

14-26. lead-acid battery: 83.42 A · h/kg; hydrogen-oxygen fuel cell: 2 975 A · h/kg

Chapter 15

15-1. **(a)** $Cu^{2+} + 2e^- \rightleftharpoons Cu(s)$ **(c)** 0.112 V

15-2. **(a)** $Br_2(aq) + 2e^- \rightleftharpoons 2Br^-$, $E_+ = 1.057$ V
(b) 0.816 V

15-3. 0.1 mL: $[Ag^+] = 1.1 \times 10^{-11}$ M, $E = -0.090$ V
10.0 mL: $[Ag^+] = 2.2 \times 10^{-11}$ M; $E = -0.073$ V
25.0 mL: $[Ag^+] = 1.0_5 \times 10^{-6}$ M; $E = 0.204$ V
30.0 mL: $[Ag^+] = 0.012\,5$ M; $E = 0.445$ V

15-4. 0.1 mL, $E = 0.481$ V; 10.0 mL, 0.445 V; 20.0 mL, 0.194 V; 30.0 mL, -0.039 V

15-5. 1 mL, $E = 0.060$ V; 10.0 mL, 0.073 V; 50.0 mL, 0.270 V; 60.0 mL, 0.408 V

15-6. **(a)** $2Cl^- + Hg_2^{2+} \longrightarrow Hg_2Cl_2(s)$, $V_e = 25.0$ mL

$$\textbf{(b)}\ E = 0.555 + \frac{0.059\,16}{2} \log\,[Hg_2^{2+}]$$

(c) 0.1 mL, $E = 0.084$ V; 10.0 mL, 0.102 V; 25.0 mL, 0.372 V; 30.0 mL, 0.490 V

15-7. Left side is negative.

15-8. Right side is positive.

15-9. 10.67

15-10. **(a)** +0.10 **(b)** 13%

15-14. small

15-16. +0.029 58 V

15-17. 0.211 mg/L

15-18. **(a)** -0.407 V **(b)** $1.5_5 \times 10^{-2}$ M

15-19. constant = -0.128 V; with Na^+, $E = -0.310$ V; apparent concentration of $Li^+ = 8.4 \times 10^{-4}$ M

15-20. +1.2 mV, 10%

15-21. **(a)** 1.60×10^{-4} M **(b)** 3.59 wt%

15-22. K^+

$$\textbf{15-23. (a)}\ E = \text{constant} + \beta\left(\frac{0.059\,16}{3}\right)\log\,([La^{3+}]_{outside})$$

(b) 19.7 mV **(c)** +25.1 mV **(d)** +100.7 mV

15-24. **(a)** $E = 51.09\,(\pm 0.24) + 28.14\,(\pm 0.08_5) \log\,[Ca^{2+}]$ $(s_y = 0.2_7)$ **(b), (c)** $2.43\,(\pm 0.06) \times 10^{-3}$ M

15-25. $1.22_1 \pm 0.02_9 = 1.19_2$ to 1.25_0

15-27. **(c)** ~4.5 **(d)** ~3 μM

15-28. 1.2×10^7

Chapter 16

16-1. $E° = 0.93_3$; $K = 6 \times 10^{15}$ or just 10^{16} based on significant digits

16-2. **(d)** 0.490, 0.526, 0.626, 0.99, 1.36, 1.42, 1.46 V

16-3. **(d)** 1.58, 1.50, 1.40, 0.733, 0.065, 0.005, -0.036 V
(e) Diphenylbenzidine sulfonic acid (violet $\longrightarrow$ colorless) or diphenylamine sulfonic acid (red-violet $\longrightarrow$ colorless) would be suitable. Other indicators such as diphenylamine (violet $\longrightarrow$ colorless) or tris(2,2′-bipyridine)iron (pale blue $\longrightarrow$ red) also would work.

16-4. **(d)** -0.120, -0.102, -0.052, 0.21, 0.48, 0.53 V
(e) methylene blue (colorless $\longrightarrow$ blue)

16-5. 0.371, 0.439, 0.507, 1.128, 1.252, 1.266 V

16-6. **(d)** -0.143, -0.102, -0.061, 0.096, 0.408, 0.450 V

16-7. **(b)** 0.570, 0.307, 0.184 V

16-8. Solid curve: diphenylbenzidine sulfonic acid (colorless $\longrightarrow$ violet) or tris(2,2′-bipyridine)iron (red $\longrightarrow$ pale blue)

Dashed curve: diphenylamine sulfonic acid (colorless $\longrightarrow$ red violet) or diphenylbenzidine sulfonic acid (colorless $\longrightarrow$ violet)

16-9. no

16-10. I^- reacts with I_2, to give I_3^-, which increases the solubility of I_2 and decreases its volatility.

16-11. **(a)** colorless $\longrightarrow$ pale red **(b)** 35.50 mg **(c)** It does matter.

16-12. mol NH_3 = 2(initial mol $H_2SO_4 - \frac{1}{2} \times$ mol thiosulfate)

16-13. **(a)** We do not need to measure KI or H_2SO_4 accurately. **(b)** $I_3^- + SO_3^{2-} + H_2O \longrightarrow 3I^- + SO_4^{2-} + 2H^+$ **(c)** $[SO_3^{2-}] = 5.07\,9 \times 10^{-3}$ M = 406.6 mg/L **(d)** no

16-14. **(a)** 7×10^2 **(b)** 1.0 **(c)** 0.34 g I_2/L

16-15. **(a)** 8 nmol **(b)** maybe just barely

16-16. **(a)** 0.125 **(b)** $6.87_5 \pm 0.03_8$

Chapter 17

17-3. V_2

17-6. 0.342 M

17-7. 6.04

17-8. 12.567 5 g

17-9. 54.77 wt%

17-10. 151 μg/mL

17-11. **(b)** 0.123 M

17-12. **(a)** anode **(b)** 52.0 g/mol **(c)** 0.039 6 M

17-13. **(a)** $5._2 \times 10^{-9}$ mol e^- **(b)** $0.000\,2_6$ mL

17-14. **(a)** 1.946 mmol H_2 **(b)** 0.041 09 M **(c)** 4.750 h

17-15. trichloroacetic acid = 26.3%; dichloroacetic acid = 49.5%

17-16. organohalide = 16.7 μM; 592 μg Cl/L

17-17. $96\,486.6_7 \pm 0.2_8$ C/mol

17-18. **(b)** 0.25 mM

17-20. 0.010% ascorbic acid consumed in 10 min; yes

17-21. 2.37 (±0.02) mM

17-22. 0.096 mM

17-23. 0.000 35 wt%

17-24. **(a)** $Cu^{2+} + 2e^- \longrightarrow Cu(s)$ **(b)** $Cu(s) \longrightarrow Cu^{2+} + 2e^-$

17-25. 9.1 ± 1.6 ppb in blood

17-26. **(a)** $H_2SO_3 <$ pH 1.86; pH 1.86 $< HSO_3^- <$ pH 7.17; $SO_3^{2-} >$ pH 7.17
(b) cathode: $H_2O + e^- \longrightarrow \frac{1}{2}H_2(g) + OH^-$
anode: $3I^- \longrightarrow I_3^- + 2e^-$
(c) $I_3^- + HSO_3^- + H_2O \longrightarrow 3I^- + SO_4^{2-} + 3H^+$
$I_3^- + 2S_2O_3^{2-} \rightleftharpoons 3I^- + S_4O_6^{2-}$ **(d)** 3.64 mM

17-27. **(a)** 31.2 μg nitrite/g bacon **(b)** 67.6 μg nitrate/g bacon

Chapter 18

18-1. **(a)** double **(b)** halve **(c)** double

18-2. **(a)** 3.06×10^{-19} J/photon, 184 kJ/mol **(b)** 4.97×10^{-19} J/photon, 299 kJ/mol

18-3. **(a)** orange **(b)** violet-blue or violet **(c)** blue-green

18-5. absorbance or molar absorptivity versus wavelength

18-7. violet-blue

18-8. **(a)** 1.20×10^{15} Hz, 4.00×10^4 cm^{-1}, 7.95×10^{-19} J/photon, 479 kJ/mol **(b)** 1.20×10^{14} Hz, 4 000 cm^{-1}, 7.95×10^{-20} J/photon, 47.9 kJ/mol

18-9. 2.7×10^{23} photons/s

18-10. 0.004 4, 0.046, 0.30, 1.00, 2.00, 3.00, 4.00

18-11. 0.480, 33.1%; 0.960, 11.0%

18-12. **(a)** $T = 0.5$, $A = 0.30$; half of the radiation is absorbed **(b)** $T = 0.10$, $A = 1.00$, 90% absorbed; $T = 0.05$, $A = 1.30$, 95% absorbed

18-13. 3.56×10^4 M^{-1} cm^{-1}

18-14. **(a)** 2.76×10^{-5} M **(b)** 2.24 g/L

18-15. **(a)** 7.80×10^{-3} M **(b)** 7.80×10^{-4} M **(c)** 1.63×10^3 M^{-1} cm^{-1}

18-16. **(a)** 0.052 **(b)** 89%

18-17. **(a)** 0.347 **(b)** 20.2%

18-18. 2.30×10^3 M^{-1} cm^{-1}

18-19. **(b)** 1.74×10^5 M^{-1} cm^{-1}

18-20. **(a)** 1.50×10^3 M^{-1} cm^{-1} **(b)** 2.31×10^{-4} M **(c)** 4.69×10^{-4} M **(d)** 5.87×10^{-3} M

18-21. **(a)** 6.97×10^{-5} M **(b)** 6.97×10^{-4} M **(c)** 1.02 mg

18-22. 2.19×10^{-4} M

18-23. **(a)** 1.74 ± 0.02 ppm (±0.01_7 ppm) **(b)** 1.24×10^{-4} M

18-25. **(b)** $2.2_1 \pm 0.1_4$ ppm or 2.2 ± 0.1 ppm **(c)** 4.2×10^4 M^{-1} cm^{-1}

18-27. **(a)** $4.49_3 \times 10^3$ M^{-1} cm^{-1} **(b)** $1.00_6 \times 10^{-4}$ M **(c)** $5.03_0 \times 10^{-4}$ M **(d)** 16.1 wt%

18-28. **(a)** 1.57×10^{-5} M **(b)** 0.180 **(c)** 6.60 mg

18-29. **(a)** $2.42_5 \times 10^4$ M^{-1} cm^{-1} **(b)** 1.26 wt%

Chapter 19

19-2. deuterium

19-4. **(a)** 2.38×10^3 **(b)** 143

19-5. **(a)** −23°, +23°, +52°

19-9. **(a)** $[X] = 4.42 \times 10^{-5}$ M **(b)** $[Y] = 5.96 \times 10^{-5}$ M

19-11. $[MnO_4^-] = 8.35 \times 10^{-5}$ M; $[Cr_2O_7^{2-}] = 1.78 \times 10^{-4}$ M

19-12. **(a)** [transferrin] = 8.99 mg/mL, [Fe] = 12.4 μg/mL **(b)** fraction of Fe in transferrin = 73.7%

19-13. **(a)** $A = 2\,080[HIn] + 14\,200[In^-]$ **(b)** 6.79

19-14. slope = 0.962, intercept = −3.803, $pK_{HIn} = 3.95$

19-15. [A] = 0.009 11 M; [B] = 0.004 68 M

19-16. ~2 and 0

19-17. 1.73×10^{-5} M

19-19. **(a)** [Se] in unknown solution = 0.038 4 μg/mL; wt% Se in nuts = 3.56×10^{-4} wt% **(b)** 3.56 (±0.07) $\times 10^{-4}$ wt%

19-20. $[In^-]/[HIn] = 1.37$, pH = 7.64

19-21. 620 kg CO_2, 11 kg SO_2

Chapter 20

20-6. **(a)** 4.699×10^{-19} J **(b)** 3.67×10^{-6}
(c) +8.4% **(d)** 0.010 3
20-7. **(a)** 6.07×10^{-19} J **(b)** 3.3×10^{-8}
(c) +12% **(d)** 0.002 0
20-8. **(a)** 0.064_6 ppb^{-1} **(b)** 0.23 ppb **(c)** 0.76 ppb
20-9. **(a)** slope = $23.5_9 \pm 0.2_8$ (μg/mL)$^{-1}$; intercept = $7._9 \pm 5._2$; $s_y = 5._3$ **(b)** 17.3 ± 0.3 μg/mL
20-10. **(a)** 3.00 mM **(b)** 3.60 mM
20-11. 0.120 M
20-12. 1.04 ppm
20-13. **(a)** 0, 10.0, 20.0, 30.0, 40.0 μg/mL
(b) and **(c)** 204 ± 3 μg/mL
20-14. 1.64 ± 0.05 μg/mL
20-15. 25.6 μg/mL
20-16. 8.33×10^{-5} M
20-17. **(a)** CsCl inhibits ionization of Sn
(b) slope = 0.782 ± 0.019; intercept = 0.86 ± 1.56; $R^2 = 0.997$
(c) Little interference at 189.927 nm, which is the better choice of wavelengths. At 235.485 nm, there is interference from Fe, Cu, Mn, Zn, Cr, and, perhaps, Mg.
(d) limit of detection = 9 μg/L; limit of quantitation = 31 μg/L
(e) 0.8 mg/kg
20-18. Ti/transferrin = 2.05

Chapter 21

21-2. 0.1 mm
21-8. **(a)** 4.3×10^4 **(b)** 3.6 μm
21-9. **(a)** $w_{1/2}$ = 0.8 mm for ethyl acetate and 2.6 mm for toluene
(b) $N = 1.1 \times 10^3$ for ethyl acetate and 1.1×10^3 for toluene
21-10. 6.8 cm diameter × 25 cm long, 17 mL/min
21-11. **(a)** $w_{1/2}$ = 0.172 min, $N = 3.58 \times 10^4$ **(b)** 0.838 mm
(c) w (measured) = 0.311 min; $w/w_{1/2}$ (measured) = 1.81
21-12. **(a)** heptane: $w_{1/2}$ = 0.126 min, 7.4×10^4 plates, 0.40 mm plate height
$C_6H_4F_2$: $w_{1/2}$ = 0.119 min, 8.6×10^4 plates, 0.35 mm plate height
(b) w = 0.311 min, resolution = 1.01
21-13. **(a)** 1.11 cm diameter × 32.6 cm long **(b)** 0.069 mL
(c) 0.256 mL/min
21-14. 1.47 mL/min
21-15. **(a)** 0.168_4 **(b)** 0.847 mM **(c)** 6.16 mM
(d) 12.3 mM
21-16. 161 μg/mL
21-17. **(a)** u_{optimum} = 31.6 mL/min
(b) u_{optimum} = 44.7 mL/min, H_{optimum} increases
(c) u_{optimum} increases, H_{optimum} decreases
21-18. **(a)** 3.25 mm **(b)** 615 plates **(c)** 0.760 min
21-19. molecular mass = 78.111 8, nominal mass = 78
21-20. 35.453
21-21. $C_4H_{11}N_3S^+$ (predicted mass = 133.066 81)
21-22. 31 = CH_2OH^+; 41 = $C_3H_5^+$; 43 = $C_3H_7^+$; 56 = $C_4H_8^+$ (loss of H_2O)
21-23. **(a)** C_5H_5N **(b)** $C_5H_5ON_3$ or $C_6H_5O_2N$
(c) $C_7H_4O_2N_2$, C_7O_4, $C_6H_{12}O_4$, $C_6H_{16}O_2N_2$
(d) $C_{11}H_{20}O$, $C_{11}H_4O_2$, $C_{11}H_8N_2$, $C_{12}H_{24}$, $C_{12}H_8O$
21-24. intensity ratio 36 : 37 : 38 = 100 : 0 : 31.96
21-25. intensity ratio 34 : 35 : 36 = 100 : 0.80 : 4.52
21-26. tallest peak is $^{12}C_2{}^{35}Cl_3{}^{37}Cl$

Chapter 22

22-4. **(b)** narrow bore: 16 ng; wide bore: 5.6 μg
(c) narrow bore: 0.16 ng; wide bore: 56 ng
22-5. $n = 12.44 \approx 12$ or 13 CH_2 groups
22-6. **(a)** nominal mass in selected ion chromatogram = 73
(b) m/z 73 is M − 15 (loss of CH_3) for MTBE and M − 29 (loss of C_2H_5) for TAME. Loss of C_2H_5 bound to C in TAME suggests that CH_3 lost from MTBE was also bound to C, not O. If CH_3 bound to O were easily lost from MTBE and TAME, we would expect to see C_2H_5 bound to O lost from ETBE, but there is no M − 29 (m/z 73) for ETBE. By similar reasoning, m/z 87 for ETBE and TAME represents loss of CH_3 that was bound to C in both molecules.
22-7. **(a)** $(^{13}CH_3)_3$-caffeine = $^{13}C_3{}^{12}C_5H_{10}N_4O_2$ (197 Da). The transition m/z 197⟶111 represents loss of $^{13}C^{12}C_2H_3NO_2$.
(c) expected intensity (M+1)/M = 8 × 1.1% = 8.8%
22-8. Retention time decreases.
22-9. Retention time increases.
22-10. increase H_2O
22-13. 0.19_2 min, 0.13_6 min
22-14. **(a)** RCO_2H, RNH_3^+ **(b)** Amine will be eluted first.
22-15. **(a)** lower **(b)** higher
22-16. Use slower flow rate, longer column, or smaller particle size.
22-17. **(b)** C term is very small and A term decreases as particles become smaller.
22-18. 0.418 mg/mL
22-19. **(a)** 8.68×10^8 **(b)** 0.273 m^2
(c) Particles must be very porous.
22-21. **(a)** CocaineH$^+$ (protonated on nitrogen) is $C_{17}H_{22}NO_4$, m/z 304. m/z 182 is probably cocaineH$^+$ minus $C_6H_5CO_2H$.
(b) m/z 304 = $^{12}C_{17}H_{22}NO_4$ was selected by Q1. $^{13}C^{12}C_{16}H_{22}NO_4$ was blocked by Q1. m/z 304 is isotopically pure, so it gives no ^{13}C-containing fragments.
(c) Q1 selects m/z 304, eliminating substances without a signal at m/z 304. Q3 selects the fragment m/z 182

derived from m/z 304, eliminating any other m/z 304 that does not give a fragment at m/z 182.
(d) phenyl group

22-26. **(d)** 78 ± 5 μg/L; 192 ± 6 μg/L
22-27. **(i)** b **(ii)** a **(iii)** c
22-28. $[NO_2^-]$ = 1.8 μM; $[NO_3^-]$ = 384 μM

Chapter 23

23-1. **(a)** Mixture contains RCO_2^-, RNH_2, Na^+, and OH^-; all pass directly through the cation-exchange column.
(b) Mixture contains RCO_2H, RNH_3^+, H^+, and Cl^-; RNH_3^+ is retained and the others are eluted.
23-3. $VOSO_4$ = 80.9 wt%, H_2SO_4 = $10._7$ wt%, H_2O = $8._4$ wt%
23-5. $NH_3 < (CH_3)_3N < CH_3NH_2 < (CH_3)_2NH$
23-6. 38.0 wt%
23-7. **(a)** 3.88 **(b)** −773 μM and +780 μM
(c) 0.053 2 mg/mL
23-9. 4.9×10^4
23-10. **(a)** 5.7 mL **(b)** 11.5 mL **(c)** adsorption occurs
23-11. **(b)** Peak near 2 min is excess fluorescence-labeled phenytoin that exceeds capacity of antibody on the column. Peak near 8 min is fluorescence-labeled phenytoin displaced from column by free phenytoin in unknown serum.
(c) 0.07 s
(d) s_{pooled} = 0.33 μM; $t_{calculated} = 0.44 < t_{table}$ = 2.776, so difference is *not* significant
23-12. **(a)** cations < neutrals < anions
(b) low, against, never!
23-16. **(a)** longitudinal diffusion: $H \approx B/u$
(b) longitudinal diffusion and mass transfer: $H \approx B/u + Cu$
23-17. **(a)** t = 40.1 min, $w_{1/2}$ = 0.75 min, 1.6×10^4 plates
(b) plate height = 25 μm
23-18. thiamine < (niacinamide + riboflavin) < niacin; thiamine is most soluble
23-19. light chain: 17 300; heavy chain: 23 500
23-20. **(a)** mean detection limit = 0.1 μg/L; mean quantitation limit = 0.3 μg/L
(b) 0.000 058
23-21. **(b)** 3.78
23-22. Pentylammonium cation is anchored to the stationary phase because its hydrophobic tail is soluble in the C_{18} phase. The cationic headgroup behaves as an ion-exchange site for anions.
23-23. **(a)** Filtration eliminates solids from the analysis.
(b) Anion exchange concentrates SO_4^{2-} by a factor of 100.
(c) Final produce is solid $BaSO_4$.
(d) Na_2SO_4 is a *carrier* to ensure that there is enough product to handle at the end of the analysis.
23-24. Perhaps Mg^{2+} and Ca^{2+} neutralize negative charge on the capillary wall by binding to —Si—O$^-$ groups. EDTA binds Mg^{2+} and Ca^{2+} more tightly than the wall does.

Index

Abbreviations:

b = box; d = demonstration; i = illustration; m = marginal note; p = problem; r = reference; t = table; CP = color plate

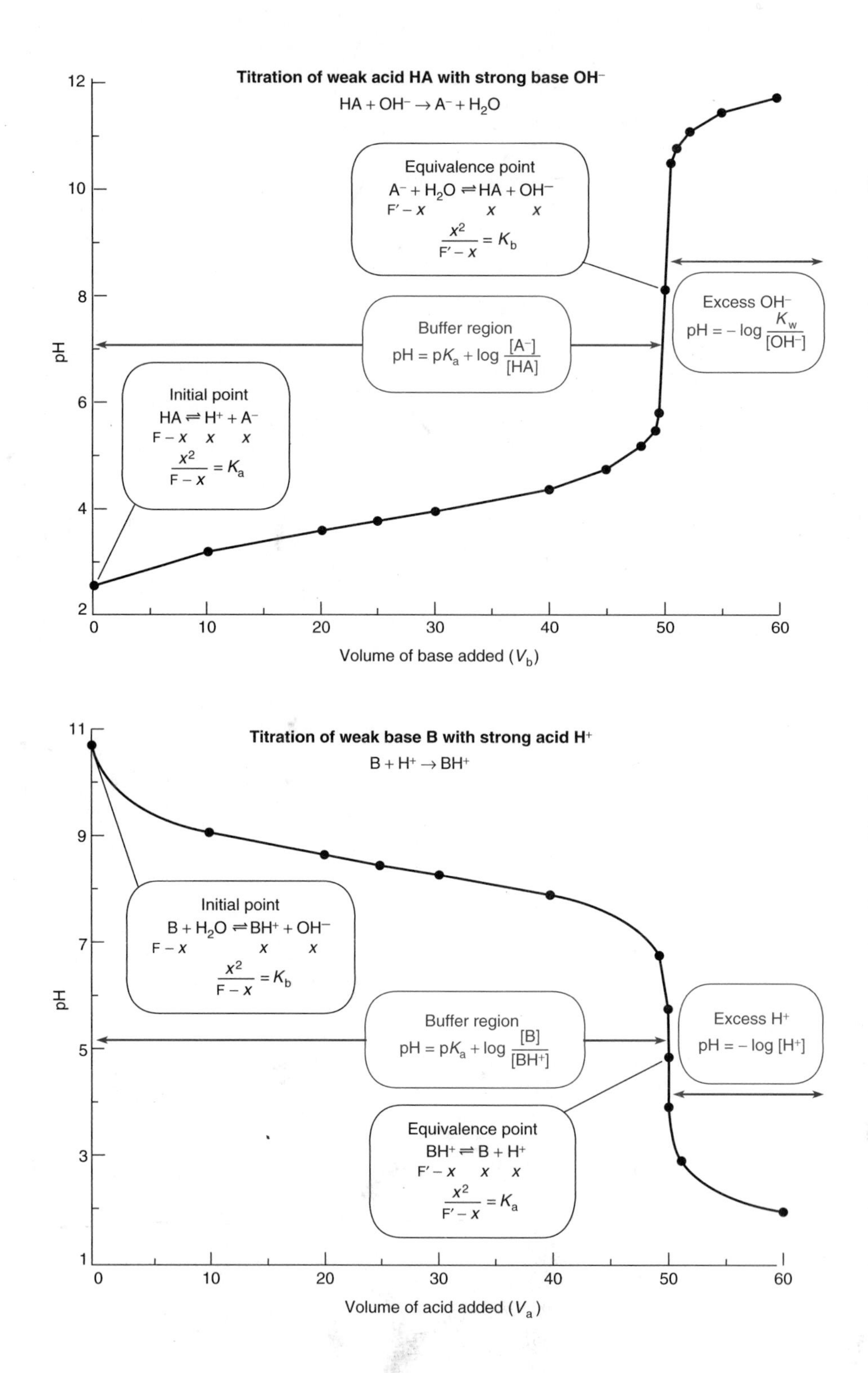
Titration of weak acid HA with strong base OH⁻
HA + OH⁻ → A⁻ + H₂O
Equivalence point
A⁻ + H₂O ⇌ HA + OH⁻
F′ − x x x
x²/(F′ − x) = K_b
Excess OH⁻
pH = −log K_w/[OH⁻]
Buffer region
pH = pK_a + log [A⁻]/[HA]
Initial point
HA ⇌ H⁺ + A⁻
F − x x x
x²/(F − x) = K_a
pH
Volume of base added (V_b)
Titration of weak base B with strong acid H⁺
B + H⁺ → BH⁺
Initial point
B + H₂O ⇌ BH⁺ + OH⁻
F − x x x
x²/(F − x) = K_b
Buffer region
pH = pK_a + log [B]/[BH⁺]
Excess H⁺
pH = −log [H⁺]
Equivalence point
BH⁺ ⇌ B + H⁺
F′ − x x x
x²/(F′ − x) = K_a
pH
Volume of acid added (V_a)